Metallic and Ceramic Coatings

Production, High Temperature Properties and Applications

Metallic and Ceramic Coatings

Production, High Temperature Properties and Applications

M G Hocking, V Vasantasree, and P S Sidky
Materials Department, Imperial College, London
Corrosion & Materials Consultants Ltd., London

Longman
Scientific &
Technical

Copublished in the United States with
John Wiley & Sons, Inc., New York

Longman Scientific & Technical
Longman Group UK Limited
Longman House, Burnt Mill, Harlow
Essex CM20 2JE, England
and Associated Companies throughout the world.

Copublished in the United States with
John Wiley & Sons Inc., 605 Third Avenue, New York, NY 10158

© Longman Group UK Limited 1989

First published in 1989

British Library Cataloguing in Publication Data
Hocking, M.G.
  Metallic and ceramic coatings: production, high
  temperature properties and applications.
  1. High temperature corrosion resistant coatings
  I. Title II. Vasantasree, V. III. Sidky, P.S.
  667'.9

  ISBN 0-582-03305-5

Library of Congress Cataloguing in Publication Data available

Printed and Bound in Great Britain
at the Bath Press, Avon

Contents

Foreword

Coatings enable the attributes of two or more materials [the substrate and the coatings(s)] to be combined to form a composite having characteristics not readily or economically available in a monolithic material. Examples are tribological properties and high strength coupled with corrosion resistance. In the high temperature field economic and technical pressures to achieve extended lives and greater reliability plus a need to conserve certain relatively scarce (and hence expensive) or strategic alloying elements, have dictated increasing recource to coatings. The evolution of stronger, more creep resistant alloys to enhance thermal efficiency resulted in loss of oxidation and corrosion resistance. The need for surface stabilization led to the growing need for coatings and rapid developments in the field of surface engineering. Operation of materials at progressively higher temperatures close to the melting points of conventional alloys, where adequate cooling leads to loss of efficiency, has led to the development of ceramic thermal barrier coatings. Coatings may also be essential in high temperature materials strengthed by fibres, to achieve compatibility between the matrix and reinforcement.

As in most fields of technology, the field of high temperatures has also witnessed the practical and sometimes serendipitous application of materials in advance of theoretical understanding of service problems. Between 1950 and 1970, very active scientific contributions were made to the development and use of high temperature materials and understanding of their response to their operating environment. The eighties saw a shift in emphasis to conservation of strategic materials and studies of ways and means of minimising loss and damage to expensive components with attendant loss of operating time and revenue. Although the literature on coatings can be traced back to the era of electro-deposition in general, and aluminising of iron in the early forties, much of the information on coatings relevant to achievement of optimum high temperature performance – production, control and replacement economics – has accumulated significantly in the seventies and to date. This book surveys the means of production of coatings, and their application over the broad range of high temperatures, above 400°C. The coating techniques discussed however, are actually used on substrates employed over entire groups of temperature ranges, from cryogenic to very high temperature. Thus the methods presented are selectively adapted in practice, for use as and on semi-conductor materials both in the regular and the rapidly growing areas of VLSI-components and surface mounting technology, as well as on materials used in power and energy generation and tribology.

The authors are to be congratulated on their diligence and scholarship in bringing together such a comprehensive survey in the important field of surface engineering in high temperature technology.

J.F.G. Condé
Senior Consultant, lately
Head, Metallurgy & Ceramics Division
Admiralty Research Establishment,
PEMoD, Holton Heath, Poole, Dorset, UK.

Preface

Definite limits are being approached, or have been reached, in single materials development. A solution is to produce a composite material which combines the best properties of different metals or ceramics. This may be a coating of one material on another. Hybrid materials bring their own problems, especially in property matching at the interface, and coatings technology has thus emerged as a challenging field both for the fundamental and applied research worker.

This book was written to fill a long-existing need for a comprehensive source of information on metallic and ceramic coatings with all basic information and data presented in one compendium. Coating production methods as well as their application in various environments are discussed, with emphasis directed towards their high temperature properties. The first chapter gives the background, and the range of high temperature application of coatings is indicated in Chapter 2. Chapters 3–6 survey the several techniques available for the production of coatings of various kinds and thicknesses on both metallic and ceramic substrates. Coatings properties are surveyed in the next two chapters – physical properties in Chapter 7 and chemical behaviour in Chapter 8. The influence of mechanical and thermal factors are interwoven in both these chapters. Coating testing and inspection methods: NDT, microscopy and spectroscopy and coating repair are discussed in Chapter 9 along with a tabulation of coatings function in the various high temperature areas. Chapter 10 surveys the success record of high temperature coating application, its role in saving strategic metals together with the scope of coating technology and further areas for investigation.

A survey book of this nature aims to be a comprehensive source of an enormous amount of information. The need for brevity restricts descriptive discussions. A general bibliography is given which provides material for further reading on much of the basic information given here. A separate list is provided of more than 2000 references referred to in the text. These and related papers of the same authors were used in reviewing the specific topics concerned. Further references to the work of individual authors can be accessed through the references listed. The literature output in the fields of fuel-cell and semi-conductor technology has been prolific, and their pace of expansion needs to be justified with a survey exclusively devoted to them. References in this area have been kept peripheral and a number of diagrams and tables have been edited and redrawn to suit the requirements of this book. The authors would like to extend a collective acknowledgement to all those whose contribution and expertise have made this review possible. Tables, figures, specific information and or data adapted for this book are acknowleged with a reference to the authors by the usual norm of scientific publication. Thanks are also due to Elsevier Sequoia and to Dr. G. Perugini for granting permission for the reproduction of figures 2.4 to 2.7 and 8.30.

If this book proves useful to industrialists and practicing engineers who seek state-of-the-art information, provides the junior and senior academic population with good information on the sources and background in both theoretical and practical fields in metallic and ceramic coatings and surface modification, and offers ideas and relevant lines of investigation to investigators who look for areas of further work, then the authors' efforts have been worthwhile.

Our special thanks are due to Mr J.F.G. Condé for writing a foreword to the book, and to all our colleagues in the Departments of Materials and of Mechanical Engineering at Imperial College, for their help and co-operation at various stages of its preparation.

The Authors
February 1989

CHAPTER 1

High temperature coatings – the background

1.1. INTRODUCTION

A major challenge in technological development is to continue to meet requirements for new materials for use in progressively more stringent conditions. Usually one or more of a material's properties are incompatible with the conditions prevailing in the operating environment. In the material-environment configuration, the surface of a component is a vital parameter in determining its optimum usefulness. This is the basis for the development of coating technology.

The outermost surface and the rest of the component can be considered together to form a system. The primary requirement of a protective surface is to have qualities superior to that of the substrate in order to shield the component from an aggressive environment. The system is invariably hybrid, whether it has been achieved by means of a surface modification of the component substrate itself or one or more other materials have been applied as a coating to the component surface. In either case, surface treatment is involved, the choice of which is vast and varied. The combination of coated surface and substrate is called a coating system and it includes surface modification although it may be more accurate to designate the latter as a surface modified system.

The need for coating systems in the field of high temperature arose when improved performance criteria could not be met adequately in spite of new materials with superior physical, mechanical and metallurgical properties. Operating efficiency and production economy had to be considered and an improvement sought.

In this chapter a very brief account is given of high temperature materials and systems where coatings have been necessary for enhancing component endurance and thereby the system function.

1.2. HIGH TEMPERATURE MATERIALS

1.2.1. METALLIC MATERIALS:

High temperature metallic materials have developed rapidly beyond the early series of conventional ferrous alloys consisting of steels and stainless steels of various compositions, physical, thermal and mechanical history and microstructure. Chromium additions, up to 25 wt.% have been used and contribute to an improvement in high temperature strength and oxidation resistance of steels. Rupture strength increases were realised up to 340 MN m^{-2} with 15 wt.% Cr, and a higher temperature capability at 25 wt.%Cr at the cost of 100 MN m^{-2} drop in rupture strength. Ferrous alloy development is well documented and Fe-base alloys are satisfactory up to 1000°C under most oxidising conditions. Several high strength nickel- and cobalt-based alloys were developed in order to meet more complex high temperature corrosion phenomena. Superalloys - viz. Nimonics, Inco-alloy series, MAR- and Rene- series of cobalt alloys etc., came in the course of development of gas turbine alloys. Chromium content has been varied over a wide range from 10-50 wt.%, in the Fe-, Ni-, Co- based alloys but it is the effect on properties by all other additives viz. Al, Ti, Si, Mo, Nb, Hf, Y, Ce, Zr, W etc., that resulted in the considerable expansion of superalloy uses.

Superalloy components were prepared as cast and wrought, first by air melting and later by special vacuum melting procedures. Directional solidification, single crystal material preparation, the powder metallurgy route of fabrication (PM alloys), mechanical alloying (MAP alloys), dispersion strengthening and oxide dispersion strengthening (DS and ODS) have been the various other process routes resorted to in improving superalloy performance. High temperature creep, strength and corrosion resistance have been improved but the melting points of these alloys are still in the order of 1250-1300°C. Material improvements between 1940-1985 have yielded a 300 degrees (C) increase in turbine entry temperature, while the benefits of cooling have given an added advantage of operation up to about 1320°C (Byworth 1986; Alexander & Driver 1986). The role of alloying additives particularly those with m.p. above 1400°C in increasing the temperature range of superalloys has been examined (Jena & Chaturvedi 1984; Fleischer 1988). But a further increase in operating temperatures solely on the basis of superalloy component fabrication seems unlikely. Coatings are necessary.

Progress in the development of creep resistant refractory alloys, viz. Mo-, Nb-, Ta- and W-based alloys has been explored for space

-nuclear applications (Wadsworth et al 1988). Elements with high melting points such as C, Cr, Mo, Nb, V, W, Zr etc., fail due to excessive oxidation. Volatile oxide products and molten oxide formation are two other undesirable features (Table 1:1). Alloys of these can operate at high temperatures but when not threatened by oxidation. They are also brittle and component welding has been a problem. Coating failure on such alloys can be catastrophic. Metals with bcc structure have, in general, a ductile-brittle transition temperature (DBTT) giving poor mechanical properties at lower temperatures, resulting in cracking during thermal cycling. Additives modify DBTT, eg. Re added to Mo, W and Cr. Alternatives to Re and extensions of this beneficial effect are desirable, as are mechanistic studies of the phenomenon.

TABLE 1:1

HIGH TEMPERATURE MATERIAL CONSTITUENTS: MELTING POINTS

Element	Melting Point, $^{\circ}$C	Element	Melting Point, $^{\circ}$C
Al	660	Cr	1840
Mn	1260		
Si	1420	Nb	1950
Ni	1455	Rh	1960
Y	1475		
Co	1495	Hf	2200
		Ir	2440
Fe	1539	Mo	2620
Pd	1552	Ta	2850
Zr	1600		
V	1710	Re	3167
Ti	1725	W	3390
Pt	1769	C	3500

Cr,Pt,Si can form volatile oxides under certain conditions; C, W & Mo form gaseous oxides; Mo, Nb, Ta, W, & Zr undergo excessive oxidation; V forms molten oxides at $>650^{\circ}$C; Ti dissolves oxygen. (Partington 1961)

Resistance heating elements function longer when coated, e.g. aluminized superalloy by surface oxidation self-protection. Other examples of oxidation protection are 18-8 stainless steels plasma sprayed with NiAl, siliconized refractory metals and plasma coated NiCr on steel used in annealing mills. High temperature metallic materials have to be assessed further but together with surface modification as a feature.

1.2.2. CERAMIC MATERIALS:

Ceramics appear at first an explorable alternative as high temperature materials because they have much higher melting points. Their brittleness is the principal drawback, and they are difficult to shape. Ceramics cannot function under conditions of mechanical and thermal shock (or cycling). Their high notch sensitivity renders them vulnerable to fracture under impact conditions. However, they have excellent corrosion resistance and low thermal conductivity. Ceramics are good candidates for thermal barrier materials. It is a two-fold problem to develop ceramics as coatings: to find the most suitable types of coating systems and to establish a satisfactory method of coating consolidation.

Ceramics are brittle (inductile) in that fracture occurs due to lack of significant plastic deformation (ductility) to absorb stresses by increasing the energy required for fracture propagation (Godfrey 1983/4). If a stress rate is high, e.g. explosive shock, it may exceed the dislocation propagation rate in a metal; the metal ductility then becomes irrelevant and it shows brittle behaviour; in this special case a ceramic may be stronger. But in general use, the non-ductile fracture behaviour of ceramics limits their use at high loads.

The low ductility of ceramics can be partly compensated for by a large modulus of elasticity, freedom of the microstructure from weak phases and a grain shape and size which maximises the work of fracture. Phase transformation toughening is also an area worth exploring. Big improvements in fracture toughness and microstructure free of holes and flaws has been achieved by hot pressing Si_3N_4 and SiC and adding a coating of a lower density reaction bonded surface layer containing Si (Kirchener & Seretsky 1974).

Large object impact on ceramics is destructive but in situations where only very small projectiles are possible, e.g. sand or dust erosion in an industrial compressor, hard ceramics perform much better than metals. Hot pressed silicon nitride eroded 5 times less than a superalloy in a dust erosion conditions test (Napier & Metcalfe 1977).

In general, above 1000°C ceramics have better strength, creep and oxidation resistance than superalloys. This temperature is considerably lower if comparison is made with ordinary alloys. Alumina has a high thermal expansion and thus undergoes thermal stress cracking; zirconia is stabilized by additives to reduce a destructive phase change at 1000°C, but thermal cycling above 1000°C may still cause problems. It also has a high thermal expansion but its low thermal conductivity and very high toughness and strength below 1000°C make it suitable for diesel engines, which operate in a maximum temperature zone of 800°C. A zirconia coating life depends on its tetragonal phase content, and methods of zirconia stabilization for diesel engine coatings have been discussed (Kvernes 1983).

SiC and silicon nitride have very low thermal expansion. At high temperature, oxidation evolves CO or N_2 which causes porosity in the protective layer on the ceramic. Above $1500^{\circ}C$ oxidation becomes severe. Production methods such as hot isostatic pressing (HIP), pyrolytic deposition and use of fugitive sinterers, e.g. Al in SiC, offer scope for further research. Sintering without pressing is feasible for Sialon, and materials comparable with hot pressed Si_3N_4 are obtainable (Godfrey 1983/4).

1.2.3. COMPOSITE MATERIALS:

The challenge for new materials has reached an exciting stage with the development of composites. Dispersion strengthened alloys, e.g. TD-NiCr, are the forerunners of the composite alloy group. Metal matrix composites (MMC) have advanced considerably from aerospace applications into light alloy technology, particularly the aluminium alloy series. Powder, particulates and fibres have been incorporated, mostly via the powder metallurgy route. Composites in matrices of Mg and Ti have also been tried but to a more limited extent. Ni- and Co- base alloys with incorporated oxides have been fabricated via the mechanical alloying route.

Although MMC materials have been available for the last two decades they are yet to be tested to their full potential for high temperature behaviour and as coating materials (Harris 1988; Stacey 1988). The development of carbon-carbon composites started in 1958 but intense research did not begin until the space shuttle project gathered momentum (Buckley 1988). Ceramic coatings for carbon composites was a logical consequent material modification in the process of optimising the application of the composite and combat its vulnerability to oxidation (Strife & Sheehan 1988). Composite coatings and composite/multilayer coatings have shown a considerable range and potential for development and application in the fields of high temperature and space technology (Wadsworth et al 1988; Das & Davis 1988; Lewis 1987).

1.2.4. RAPIDLY SOLIDIFIED MATERIALS:

Rapid solidification processing (RSP) of metallic materials has opened a totally new facet of alloy technology. The new materials were first called metallic glasses implying that their structure is amorphous. Materials prepared by RSP are mostly limited to narrow strips. The technique can be adapted to incorporate dispersed phases to consolidate the property advantages offered by MMC, and the conjoint product promises to provide a superior surface modification system yet to be fully explored (Hancock 1987; Metallurgia report,1987; Waterman 1987; Easterling, 1988). More details are given in section 1.6 of this chapter.

1.3. HIGH TEMPERATURE SYSTEMS.

1.3.1. GENERAL:

Power and energy systems where corrosion at high temperature is a severe problem are considered in this section. Component failure is often due to the synergistic action of several features in operation at high temperature conditions - the temperature and time scale of actual operation, local high temperature changes, the material used, the fuel employed, the prevailing environment and changes in it, mechanical and thermal stress cycles in the course of operation as well as shut down and rest periods. The insidious nature of high temperature corrosion lies in the fact that it causes a small section of the component to fail but the consequential costs for repair and re-installation stages are very high.

Lai (1985) gives a comprehensive survey of a number of other high temperature systems where molten salts, molten glass, hot metal and halogens comprise the corrosive media. Most of them use the high temperature alloys discussed here. A more detailed discussion on material degradation is given in a later chapter. Fig.1-1 indicates the prevailing conditions for a number of high temperature systems discussed below (Natesan, 1985).

1.3.2. COAL, OIL & FLUIDIZED-BED SYSTEMS:

Combined cycle electricity generation from coal gasification offers a potentially large increase in overall efficiency compared to the more conventional coal-fired and oil-fired boiler systems. Coal composition varies widely resulting in a wide range of chemical composition of both the gasifiable material as well as the ash and erodent products. Temperature regimes can extend from $350^{\circ}C$ for pressurised steam to fluid ash products at $1695^{\circ}C$ in combustion beds. Heat extractors operate at around $565^{\circ}C$ in general in the UK. A heat flux prevails; the evaporator tubes operate at a metal surface temperature of 450° while the super-heater tubes go up to $650^{\circ}C$. Corrosion rates of 100-500 nm/h and in extreme cases >1000 nm/h have been recorded (Meadowcroft, 1987). The corrosive medium is aggressive to both Fe and Ni but not Cr.

The gasification process operates at much higher temperatures than boiler units. The physical conditions can thus be severe, with hot, abrasive-erosive particles and ash impingement. The chemical environment is predominantly reducing with oxygen potentials (pO_2) as low as 10^{-14} to 10^{-18} at a bed temperature of 850-$870^{\circ}C$. Sulphur potentials of 10^{-5} to 10^{-9}, and carbon activity of 3×10^{-3} (sometimes approaching unit activity), have also been recorded (Weber,1986). Coal gasification atmospheres may contain CO and H_2O, each 10%, H_2S 0.5% and balance N_2. 90% of the

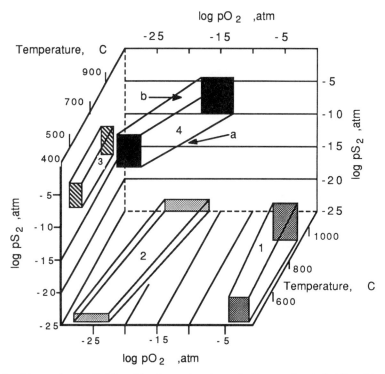

1: Atmospheric fluidized-bed combustors & Pressurized fluidized-bed combustors turbines;

2: Gas-cooled Reactor; 3: Oil refining processes; 4: Coal gasification

Fig. 1-1: Prevalent Sulphur & Oxygen Partial Pressures over
400 - 1100   C in Various High Temperature systems

(Natesan 1985)

chloride content in coal emerges as 80 vol. ppm HCl during combustion (Meadowcroft,1987).

Attack by alkali sulphate phases stabilised with high localized potentials of SO_3 have been noted in coal-fired boilers. A large amount of Na_2SO_4 is the predominant corrodent in the combustion zone of black liquor boilers. Sulphur, alkali metals and vanadium cause catastrophic attack in oil-fired boilers. Sulphidation and oxidation by the action of H_2S and CO is the most common mode of attack. Chlorides promote low melting eutectics with sulphates, which affect low pressure boilers (100 bar) operating at $370^{\circ}C$. Laboratory simulations have not been informative enough to coordinate the chlorine effect in a predominantly reducing medium with H_2S at higher operating pressures and temperatures.

Several drawbacks appear in the use of materials in boiler and gasification fields. In brief, austenitic steels cannot be used on the chloride-containing water side because they are prone to stress corrosion cracking (SCC). Ferritic steels with 12% or less Cr are not cost-justified since they exhibit only a marginal improvement in corrosion resistance over mild steel. A higher chromium content in this alloy class results in embrittlement at 475°C. Heat exchangers of plain carbon steel, low alloy ferritic steels, e.g. Fe-2.25Cr-1Mo, and type 304 steel (18Cr-10Ni) suffer corrosive attack. Erosion-hot corrosion is also a major problem.

Improvements in performance require surface modification. Cladding has been used as a control measure. Ni-50Cr is used for structural members, and also for co-extrusion in black liquor boilers. Controlling excess air over that required for stoichiometric combustion, and fuel additives based on MgO control the medium to reduce vanadic corrosion. High-Cr alloys have shown good resistance to vanadic attack. Co-extruded tubes made from Incoloy 800H with an outer layer of Ni-50Cr has been effective. AISI type 310 steel (Fe-25Cr-20Ni) co-extruded over carbon steel or mild steel has been used mostly in boilers operating at lower pressures of 100 bar (Meadowcroft,1987; Stringer,1987). Fe-Cr-Al- alloys with and without pre-oxidation, MAP alloys and coatings have been tested to improve component performance (Weber,1986; Hocking & Sidky,1987).

1.3.3. GAS TURBINE & DIESEL ENGINE SYSTEMS:

Three main types of gas turbine may be listed: (i) Aero-engine, (ii) Industrial and (iii) Marine. Aero-engines burn clean fuel, but ingest impurities in the air-intake. This is aggravated in the marine-gas turbines since the air intake includes sea salt. Industrial engines burn inferior fuel containing vanadium and sodium, but the air intake can be filtered. Turbine components are made of Ni- or Co- based alloys. The blades and vanes are subject to cyclic temperature corrosion in alkali sulphate-oxide +/- chloride , +/- vanadic environment. Advanced turbines operate at high inlet temperatures (1350°C) and the industrial turbines at 1150°C. The blades are cooled to 850-900°C. Catastrophic corrosion occurs at lower temperatures 610-750° for Ni-based alloys and 750-850°C for Co-based alloys. Coatings have been tested extensively in the field of gas turbines and these include multi-component alloy compositions and alloy-ceramic formulae as well as thermal barrier coatings. Superalloys form 50 wt% of gas turbine engines, the rest being about equal parts of Ti-alloys, steels and composites. Development trends in the superalloy systems are inclined towards directional solidification (DS), powder metallurgy (PM) and mechanical alloy processing (MAP).

Diesel engines are the most fuel-efficient production heat engines. Exhaust valves, turbochargers and exhaust systems are subject to high temperature corrosion. Increasing efficiency by reducing heat losses has led to the concept of adiabatic diesel engines. Reducing heat losses from the cooling system, and in the combustion chamber, and harnessing the exhaust gas energy into work all demand thermal barrier surface modified systems which can offer durable and cost effective adiabatic diesel energy. Increase in operating efficiency has been attained by using gas turbine alloys, and design advances in airfoils and turbo-chargers (multi-stage and sequential). The corrosion problems are similar to those of industrial gas turbines (Fairbanks, 1987).

1.3.4. NUCLEAR POWER SYSTEMS:

Low-alloy carbon steels, ferritic and austenitic steels are some of the materials used in nuclear reactors, the selection for component fabrication depending on the temperature, environment and service conditions. Nuclear industry requires a fail-safe feature and cannot afford a major failure with catastrophic consequences. A high degree of predictability and reliability of component performance is vital and the temperature range is wide, from 100 to $650^{\circ}C$.

Fast breeder reactor systems in the UK use liquid sodium coolant. Despite control of coolant purity and cold-trap devices, circuits contain typically a few ppm oxygen which initiates product formation on the various austenitic AISI 316 stainless steel, low alloy steels, Stellite hardfacings (Co-based alloys) and IN-series (Ni-based) superalloys. Advanced gas-cooled reactor systems (AGR) use CO_2 as coolant and oxidation reactions predominate. Both the systems are subject to a number of tribological problems and rely on metallurgical coatings for long-term, maintenance-free service (Lewis,1987).

CHAPTER 2

High temperature coating systems

2.1. GENERAL

The predominant aspect of coating technology is the life expectancy of a coated system. System optimisation considers the coating composition, structure, porosity and adhesion together with operating and coating temperatures, substrate/coating compatibility, material availability and costs, coating renewal and in-service repair and maintenance aspects. Tables 2:1 and 2:2 give the requirements of a coating system for an optimum service life.

Weak coatings undergoing thickness loss may affect the mechanical properties of the substrate by effectively reducing the component total cross section. Interdiffusion may add or deplete alloy constituents. Thermal processing conditions may cause substrate microstructural changes and introduce residual stress. The system properties affected would be strength (in its various aspects), ductility, toughness, fatigue and creep resistance. Rapid temperature cycling has the greatest effect on the mechanical properties; temperature, exposure time and the environment interaction govern the chemical effects.

Mechanically bonded coatings often do not provide the required adherence. Diffusion bonded coatings are superior. A good example of a successful diffusion coating is the aluminide series applied to ferrous and many non-ferrous substrates alike.

TABLE 2:1

DESIRABLE PROPERTIES OF COATINGS & SELECTION CRITERIA

GAS TURBINE COATINGS

1. Adherent, Crack- and Pore-free.

2. Structure - Non-columnar &
   Defect tolerant.

3. Microstructurally stable,
   Precipitation strengthened,
   no Brittle Intermetallics.

4. Thermal expansion compatible
   to substrate, Resistance to
   Thermal Fatigue

5. Predictable reaction & interdiffusion with substrate
   which can be stopped or manipulated to suit requirements.

6. Coating should not affect mechanical and structural properties
   of the substrate. Should improve
   bowing & cracking defects.

7. Fatigue & Creep resistant

8. Resistant to brittle fracture,
   DBTT at a low temperature.

9. Isothermal & Cyclic - Oxidation
   (aircraft) & Hot Corrosion (marine
   & industrial) resistant; Impact &
   Erosion resistant; Intergranular
   corrosion resistant. Self-healing
   scratches and damages.

10. Spall-resistant oxide scales;
    No molten reaction products;
    No vaporisation; have thermal
    barrier effect.

11. Locally repairable, applicable
    to partly corroded items & cost
    effective.

AEROSPACE COATINGS

1-8. All aspects indicated for
     gas turbines are valid.

9. Hot Corrosion not such a
   threat. All other aspects
   indicated are applicable
   here.

10. As in previous column.

11. In-situ repair difficult
    to implement. Otherwise
    repair and cost-effective-
    ness must be considered.

NUCLEAR REACTORS

1-8. All aspects indicated for
     gas turbines are valid.

9. Carburization and Hydrogen
   damage resistant. All other
   aspects indicated are valid
   Hot corrosion not considered

   Wear resistant and hardness
   coatings required.

10. Low neutron capture

11. Increase fuel life & prevent
    burn up

12. Drag-sliding protection in
    Magnox reactors.

TABLE 2:2

**GENERAL PROPERTY CRITERIA FOR COATING SYSTEMS FOR
HIGH TEMPERATURE SERVICE**

Component System: Criteria/Property	Coating: Criteria/Property
Aerodynamic Property	Smooth surface finish for the coating. Must conform to the appearance of a precision cast blade.
Mechanical Strength and Microstructural Stability	Coating has to be resistant to all types of stress – impact, fatigue, creep and thermal, the system will be exposed to.
System Adhesion, Bonding & Interface Stability	Coating/Substrate must be compatible without gross thermal or structural mismatch. Diffusion rates at interface must be at a minimum at operating temperatures, and also composition changes. Development of embrittling phases must be avoided.
Surface Resistance to Erosion & Oxidation/Hot Corrosion	Coating composition must have sufficient reserve of all reactant constituents to meet scale reformation without a marked deterioration in protecting ability. Coating must be ductile, must develop uniform, adherent and ductile scale at low rates.

===

2.2. METALLIC COATING SYSTEMS

2.2.1. DIFFUSION COATINGS:

Aluminiding, is one of the first high temperature metallic coating systems. Alumina scale-forming coatings have given the best protection in high velocity gas turbines. Uncoated alloys containing Al as a minor constituent get fast depleted of it as Al diffuses to the surface to maintain alumina formation due to recurring oxide spallation. Diffusion-coated alloys form a surface alloy layer eg. NiAl and CoAl, which develop the protective alumina layer, under more controlled kinetics, and reform on spalling without affecting the substrate alloy properties. A renewal is possible before excessive depletion occurs. Brittle intermetallics can be a danger in this system.

A diffusion barrier layer between the coating and the substrate becomes necessary to prevent interdiffusion of elements within the coating system. Multicomponent diffusion coatings (Al, B, Cr, Si, Ti, Zr) to protect machines from liquid Al and for brass corrosion have been used (Samsonov, 1973). More work is needed in this area.

On steel, Al coatings are good up to 500°C, above which brittle intermetallics are formed. Thermal stress-induced cracks then propagate into the substrate metal. Aluminide coatings protect steel from oxidation and corrosion in hydrocarbon and sulphur-containing atmospheres. Aluminised steel is better than stainless steel where oxidation carburization occurs (Sivakumar & Rao 1982). Chromised steel (diffused in) is air oxidation resistant up to 700°C. Above 800°C Cr diffuses into the steel reducing oxidation resistance. At higher temperatures brittle intermetallics form. Good chromised sheet can be bent 180° without damage and is suitable for most firebox and heat exchanger applications up to 600°C. Addition of Al or Si to the chromising pack process confers oxidation resistance up to 900°C on mild steel, although continuous use at 900°C causes brittle intermetallics and consequent cracking on thermal cycling.

On superalloys alumina-forming diffusion-bonded coatings provide an Al-rich surface to gas turbine environment. The superalloy aluminiding involves more than one phase formation in the Ni-Al (and Co-Al) systems. Heat treatment is given to stabilise the NiAl phase. NiAl with some Cr and Ti improve the hot corrosion resistance. Oxide particles, e.g. Y_2O_3, reduce spalling; defects like pinholes, blisters and cracks may be avoided by a combination of minor additives.

Aluminide coatings lack ductility below 750°C and on thermal cycling, undergo surface cracking resulting in spalling of the alumina scale. To overcome these two problems the coating composition was adjusted to embed the brittle beta-NiAl or beta-CoAl in a ductile gamma solid solution matrix. Addition of yttrium improved oxide adherence. Improvements in mechanical properties were achieved by HIP-densified, argon-atomised pre-alloyed powder ingots; tensile ductilities of over 20% were produced by the smaller precipitate and its better distribution (Lane & Geyer 1966). Much work was done on adding minor amounts of Si, Fe and Ti to improve the scale, but little improvement resulted (Ubank 1977). Pt electroplate followed by aluminiding gives improved oxidation resistance which offsets the higher cost (Wing & McGill 1981).

Co-base superalloys for higher temperature but less stressed gas turbine vanes, have no Al and this limits the aluminide coating thickness which can be applied without spalling. Superalloy compositions avoiding sigma and other embrittling phases can be destabilized by coating inter-diffusion (Boone 1981). Structural strengtheners like sub-micron oxides in ODS-alloys and carbides in DS-eutectic alloys can also limit coating selection (Jacobson

& Bunshah 1981). Further work is needed in these areas, increasingly, as higher performance alloys will narrow down the choice of acceptable coatings. In a recent 5 year period over 30 production coatings became necessary to replace an original selection of only one to two compositions (Boone 1981).

2.2.2. OVERLAY COATINGS:

Diffusion-type coatings, used successfully on early gas turbines, were tied to the substrate composition, microstructure and design. Later some changes were introduced - (i) in superalloy composition, such as reduction in Cr and increase in other refractory metals, (ii) in microstructure, by castings with more segregation, and (iii) in design, by air cooling and with thin walls (which introduced higher thermal stresses). These changes required coatings which were much more independent of the substrate. Overlay coatings met this necessity.

Overlay coatings also overcome the process restrictions encountered in diffusion coatings, especially the variants, viz. Cr/Al, Ta+Cr or the Pt-aluminides all of which give better stability and oxide-hot corrosion resistance than Al alone. MCrAlY compositions (M=Ni,Co,Fe alone or in combination) are the principals in the series of overlay coatings developed by electron beam evaporated physical vapour deposition (EBPVD) technique for multiple load use (Hill & Boone,1982; Hill,1976; Goward,1970). Fig.2-1 to 2-3 show the many composition variations produced. 3 million aerofoils have been successfully processed by EBPVD.

MCrAlY overlay used in gas turbines are usually Ni and/or Co with high Cr, 5-15% Al and Y addition around less than 1% for stability during cyclic oxidation. They are multi-phase alloys with ductile matrix, e.g. gamma Co-Cr, containing a high fraction of brittle phase, e.g. beta CoAl. The Cr provides oxidation hot corrosion resistance but too much Cr affects substrate phase stability. The success of most overlay coatings is the presence (and perhaps location) of oxygen-active elements like Y and Hf which promote alumina layer adherence during thermal cycling, giving increased coating protectivity at lower Al levels. Y mostly appears along grain boundaries if a MCrAlY is cast but is homogeneous if plasma sprayed. Thus MCrAlY with 12% Al are more protective than the more brittle diffusion aluminides with 30% Al.

Overlay claddings deposited by hot isostatic processing (HIP), electron beam evaporation or sputtering methods, are diffusion bonded at the substrate/coating interface, but the intention here is not to convert the whole coating thickness to NiAl or CoAl. There is thus more freedom in coating composition, whose properties can be maximised to the type required. Compositions based on NiCr, CoCr, NiCrAl, CoCrAl, NiCrAlY, CoCrAlY, FeCrAlY and NiCrSi have been successful in gas turbine engines. They are generally

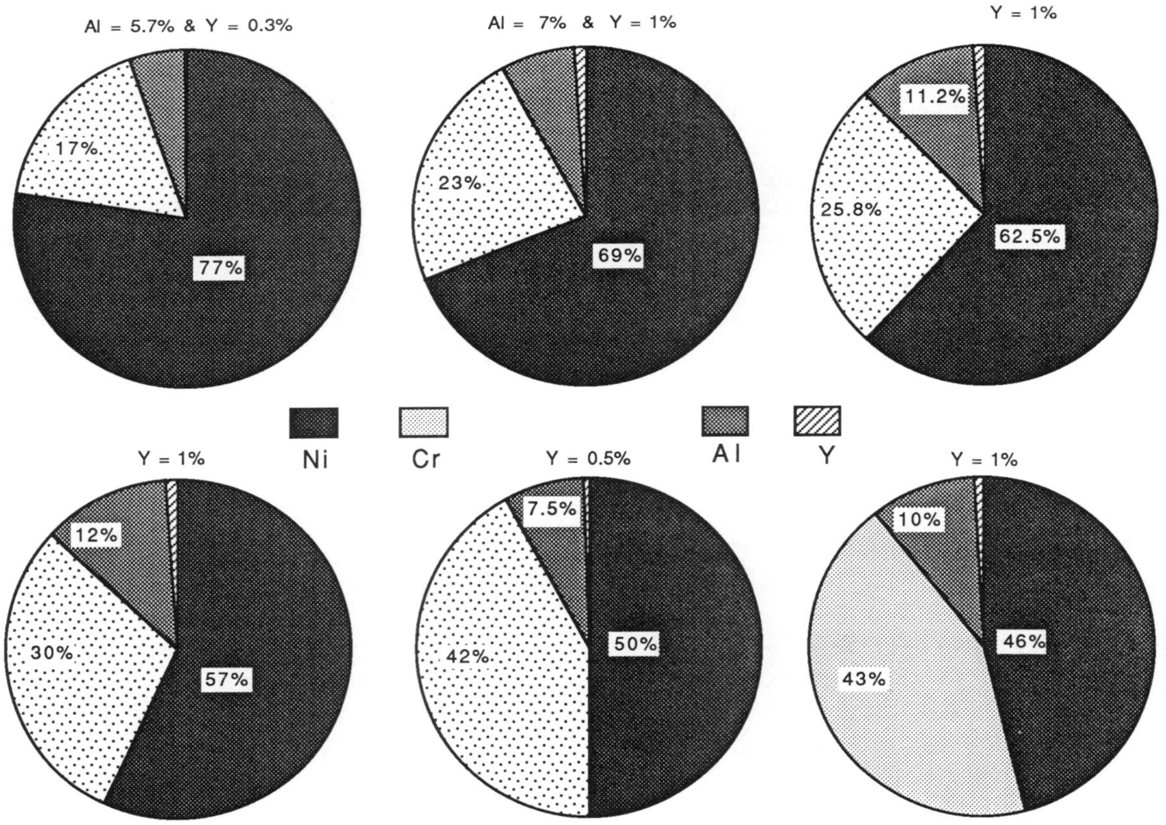

Fig.2-1: NiCrAlY COATING COMPOSITIONS , wt.% - A Selection of Coatings Tested on gas turbine blades

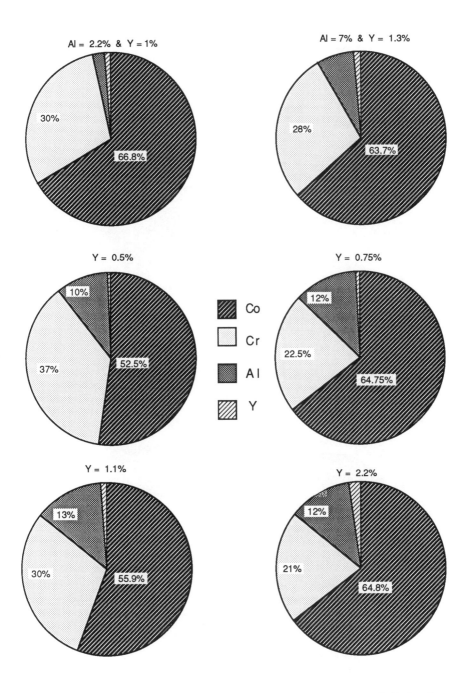

Fig.2-2: CoCrAlY Coating Compositions, wt.% - A Selection
Tested on Gas Turbine Blades

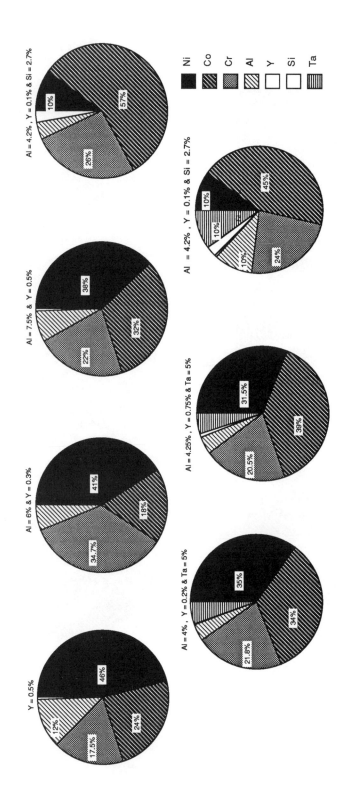

Fig.2-3: NiCoCrAlY Coating Compositions, wt.% - a Selection Tested on Gas Turbine Blades

18

alumina formers with only 10%Al unlike the 30% in nickel aluminide coatings. Cr increases the Al activity which allows this
feature. Higher Al levels cause brittleness and a higher DBTT,
and the Al levels are generally held below 12% (5-10% preferred).
The coatings are also more ductile than NiAl and CoAl and can be
rolled and bonded by HIP. In general, NiCrAlY give best results
against high temperature oxidation while CoCrAlY are best for hot
corrosion.

2.2.3. OTHER METALLIC COATINGS:

Refractory metals may be coated with oxidation resistant alloys
such as Hf-Ta and Nb-Ti-Al-Cr which tend to deform without cracking and can withstand impact damage. If a Ta-containing coating
is to be used on an alloy with a small carbon content, eg. IN
738, TaC will precipitate at the bondline and impair adhesion.
This is prevented by using a Ni electroplate at the interface. Al
foil placed at a superalloy/cladding interface acts as a source
of Al for that lost into the oxide. It melts and diffusion bonding occurs simultaneously with formation of beta NiAl at the
bondline (Lane & Gayer 1966). A refractory metal, e.g. Nb,Ta
interlayer between NiAl and superalloy substrate prevents interdiffusion of Al and Ni (Lakhtin et al 1977).

Plasma sprayed coatings have a large fraction of open and closed
porosity, which lowers strength and decreases resistance to corrosion. Heat treatment to reduce the porosity must avoid melting
of a thin substrate or degradation of the substrate/coating
interface. Even rapid CW-laser surface fusion may be unsatisfactory for air-sprayed coatings due to expansion of gas in the
pores giving a sponge structure. Better results are obtained with
a pulsed TEA laser which melts only the upper surface of the coating (Dallaire & Cielo 1982).

Ion implantation of Al in Fe reduces oxidation even when the
oxide formed is much thicker than the ion implanted layer (Pons
et al 1982). Y implantation in NiCr, FeNiCr and Fe41Ni25Cr10Al
reduces oxidation and improves scale adherence, of interest for
nuclear power materials (Pivin et al 1980). Ion implanted coatings produce results beyond what would be expected from just
their shallow effective depth, e.g. Ce penetrates only 0.1 micron
into stainless steel (Bennett 1981). Research is needed on the
possible effects of ion bombardment damage, ion bombardment induced sputtering of the substrate (perhaps increasing the oxide
nucleation sites giving smaller oxide grains with thus an increased and grain boundary diffusion).

Surface welded systems are used for wear and corrosion resistance. The thickness is usually at least 3 mm. Substrate pre-
heating prevents weld distortion. Multilayer deposits of hard
alloys can be interleaved with ductile layers to stop crack

propagation. Laser surface melting modifies the surface structure and can improve strength, wear and corrosion resistance. Phase change or stabilization, structural changes, additive and amorphous alloy forming are several other features of coating systems achievable using laser as a "coating" tool (Gregory 1980).

2.2.4. METAL COATINGS ON CERAMICS:

Cu, Ni and Co can be cementation coated onto carbon fibres for aeronautical applications (Kulkarni et al 1979). A Cu layer beneath the Ni layer prevents Ni/C interaction (Shiota & Watanabe 1979) as C does not diffuse through Cu. Graphite can be protected by coatings of Ir (expensive metal). Metallising thermal barrier coatings for obtaining temperature monitoring contacts by surface thermocouples is another feature of ceramic substrate/metal coating systems. Joining a metal to ceramic or glass is achieved by high temperature means in the region of $1550^{\circ}C$ (Jones 1985; Morrell & Nicholas 1984; Twentyman & Hancock 1981, Twentyman & Popper 1975; Twentyman 1975), and also at much lower temperatures, from 900 - $1100^{\circ}C$ (Tomlinson 1986; Tentarelli et al 1966).

2.3. CERAMIC COATING SYSTEMS

2.3.1. GENERAL:

Ceramic coatings on metals is the major topic considered here. Some brief information is provided on ceramic-on-ceramic coating systems before concluding this section. Fig.2-4 and 2-5 show the extent of applications of ceramic coatings over a wide range of operating temperatures (Perugini 1976).

Ideal coatings which satisfy all the requirements of Table 2:2 do not exist and the use of good substrates is necessary to assist the coating. The choice of a coating, as in other coating systems, depends on the environment, substrate material, coating availability and cost. Ceramic coatings are used severally as thermal and corrosion barrier coatings as well as wear and erosion resistant and tribological coatings. Wear resistant coatings are very important even just on economic terms, e.g. in the U.S.A. the cost of machining is about $70 billion/year (1978), and the cost of tools is $900 million/year (about half divided between cemented carbide and high speed steel tools).

The scope of ceramic coatings in high temperature media has been discussed by Perugini (1976) and Eriksson et al (1982). Coatings for gas turbines have been reviewed (Godfrey et al 1983; Godfrey 1983/4,1981, 1978, & 1974). A survey of the requirements for diesel engines (Timoney 1978) and a review on coatings for diesel engines are available (Kvernes 1983).

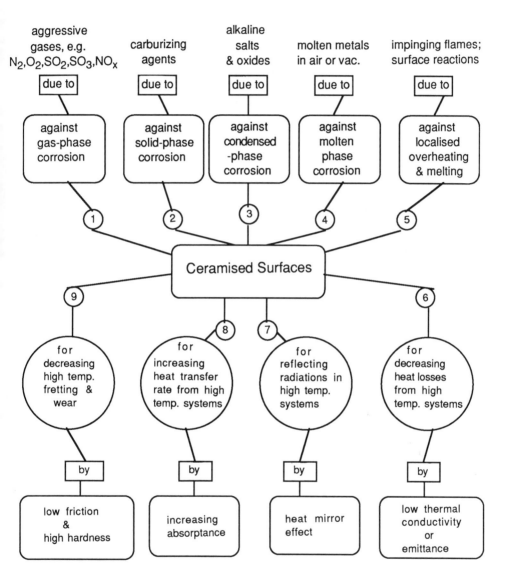

Fig.2-4: Ceramics & Their Role in High Temperature Environment

(Perugini 1976)

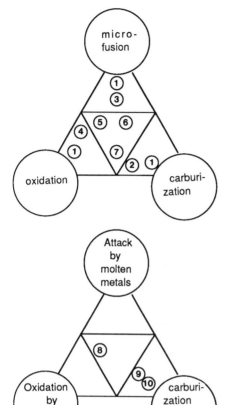

1. Burners & combustion chambers of reactors for the cracking of hydrocarbons in petrochemical industry.
2. Tube reactors of ethylene furnace in petrochemical industry.
3. Rocket nozzles.
4. Oxygen blow-pipe head in iron industry.
5. Piston heads; valves of big diesel engines.
6. Antipollution of thermal reactors in automobile industry.
7. Preheating or cooling manifolds.

8. Metallic parts under corrosive conditions. (caused by non-ferrous molten metals)
9. Metal moulds for casting iron and iron-alloys lubricated with graphite -additive oils.
10. Graphite crucibles for metallizing (eg. Al) under vacuum.

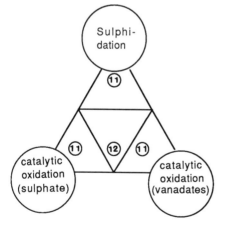

11. Industrial boiler tube configuration.
12. Sections operating under catastrophic corrosion conditions.

Fig.2-5: Applications of Ceramic Coatings in High Temperature Technology
(Perugini 1976)

Protection requirements for internal combustion engines include insulation of piston caps, shields for cylinder heads, top liner, exhaust ports and exhaust manifold. Improved high temperature lubrication and valve modifications are recommended, and development of ceramic cam followers offer advantages over metallic counterparts as the former are lighter than metals and thus may cause less wear. Plasma sprayed ZrO_2, liner inserts made of Al_2O_3 and ZrO_2, alumina titanate exhaust ports and ceramic bearings are being investigated. Si_3N_4 has excellent tribological properties (Godfrey & Taylor 1968; Burke et al 1978; Godfrey 1978; Hamburg et al 1981).

TiC has been coated on steel at $450^{\circ}C$ from Ti vapour and acetylene by activated reactive evaporation (ARE) (Raghuram & Bunshah 1982; Nimmagadda & Bunshah 1975) and at $300^{\circ}C$ by reactive sputtering (RS) (Nakamura et al 1974). TiN has been coated onto steel etc., using $Ti(NR_2)_4$, Ar, H_2, N_2 mixtures by chemical vapour deposition (CVD) (Sugiyama 1975) at $250-600^{\circ}C$, and using Ti vapour, N_2 (or NH_3) by ARE at $20^{\circ}C$ (Bunshah & Raghuram 1972). ZrO_2 has been coated onto Si using Zr acetylacetonate, O_2, He, at $450^{\circ}C$ by CVD (Balog et al 1972; Sladek & Gilbert 1972). W_2C + W_3C have been deposited on steel from WF_6, C_6H_6, H_2 at $300-700^{\circ}C$ (Archer 1975). The wear of this tungsten carbide composite is second to only TiC. Ti(C,N) coatings have been deposited on stainless steel compressor blades at $600^{\circ}c$ by CVD from $TiCl_4$, $N(CH_3)_3$, N_2, H_2. Cr_7C_3 coatings are produced by CVD of dicumene chromium at $450^{\circ}C$, with a thin interlayer of Ni to improve adhesion and mechanical properties of the system (Gates et al 1976). Other examples are summarised further in Chapter 4.

Future needs discussed by Mikoni & Green (1974) for oxidation resistance coatings for superalloys, refractory metals and graphite would be a useful source from which later developments may be checked. Many systems rely on Al_2O_3 or SiO_2 formation for protection, but multi-layer coatings offering resistance to corrosion and diffusion, coatings forming complex oxides, thermal barrier coatings and surface modifications to form dense glasses or incorporate dispersed oxides etc., have attracted much attention in recent years (Restall,1979; Nicholls & Hancock,1987; Stringer,1987; also see bibliography).

Ceramic coatings tend to be thick and brittle; they may also tend to devitrify. On thermal cycling the coatings tend to spall. But they offer excellent resistance to oxidation and corrosion and are suitable for static components like combustion liners, but further research is needed for use on rotating components. Ceramic coatings are often complex silicates with additions of ZrO_2, TiO_2, Al_2O_3, CeO_2, etc., and give protection up $1260^{\circ}C$. Ni-CrSi are used on industrial turbines (Tamarin & Dodonova,1977) and a liquid Sn-Al phase has been used to seal all cracks and fissures in porous silicide coatings on Mo- and Nb- alloys (Stecher & Lux,1968; Priceman & Sama,1968). Nitrides have been widely investigated for wear resistance and thermal barrier layers. An improved approach indicated is to add elements to coatings to resist cracking and to extend tolerance to cracks if formed.

Steel coated with vitreous enamel is probably one of the earliest of ceramic-on-metal systems and has one of the best resistances to the high temperature corrosion of gas appliances. The surface is non-porous, hard and wear-resistant and useable up to $600^{\circ}C$. However, tensile cracks can occur due to inadequate thermal shock resistance or to impact. A design guide from the Vitreous Enamel Development Council (U.K.) recommends using low carbon steel, preferably zero carbon and enamelling steels, with good hot strength and enough thickness and radius to eliminate stress points. Ferrous alloy supports in kilns are coated with thermally sprayed Al_2O_3 or NiAl. Plasma coated NiCr on steel is used in annealing mills. Thermocouples need protection from oxidation, e.g. chromel-alumel, is coated with vitreous enamel.

Chlorine is bubbled through melts in Al production to purify it and the steel tube used reaches $650^{\circ}C$; ceramic coatings prevent corrosion of the tube by chlorine and Al. Casting channels and moulds for Al are iron-based materials coated with alumina or NiAl; W has been used with silica coatings, fused silica has been coated with alumina and alumina-coated iron has also been used for Al casting.

The most important properties for vacuum vessel materials are the adsorption and desorption rates of gases on the surface. These properties are strongly affected by the surface composition of materials. This was investigated using Auger electron spectroscopy for SUS321 stainless steel with TiC precipitated onto the surface and for laboratory melted SUS304 stainless steel doped with boron and nitrogen. The latter steel proved superior because it formed a hexagonal lamellar boron nitride precipitate on the surface which was inert to the surface and self-healing occured.

Plasma deposited coatings of dielectrics, especially refractory oxides, are good high temperature electrical insulators (Dudko et al 1982). The adhesion of plasma sprayed ceramics to metals is generally poor but can be considerably improved if a sprayed bond-coat or interlayer is used, e.g. NiAl, Mo (Campbell 1970). Refractory metals for aerospace and other applications are coated with silicides (MSi_2, M_5Si_3), aluminides, borides (HfB_2, ZrB_2, TiB_2) and oxides (ZrO_2, HfO_2, Al_2O_3 and ThO_2). AlB_{12} coatings on Mo and W wires are in use up to 1800K as thermoelectric converter components for nuclear reactors (Motojima et al 1981). Protection of Nb and Ta against carburization can be achieved by NbB_2 and TaB_2 coatings (Samsonov 1973).

Energy saving refractory coatings have high emissivity for furnace walls and roofs radiating heat back into the furnace. Consequent time and fuel saving reduce production costs. Pyrolytic carbon coatings on nuclear reactor fuel particles form a low permeability encapsulation for containing fission product gases (Anon., Metallurgia 1980,1981). SiC is not stable above $1500^{\circ}C$ as a barrier to escape of fission products from nuclear fuel particles and ZrC-C mixtures have been proposed as an alternative (Ogawa et al 1979).

Extrusion dies plasma coated with Al_2O_3, ZrO_2, ZrB_2 and HfB_2 are used for W at $1800^{\circ}C$. But die life is short and improved coatings are needed. Ti is often glass coated for protection from contamination and from formation of the undesired alpha phase. But these coatings are hard to remove. Hot rolling requires thermally insulating coatings to avoid temperature drop of the billet and reduce furnace requirements for reheating between passes. Alumina and zirconia plasma sprayed coatings have been used for this purpose . Die casting of Al requires dies which last for over 10^5 injections and coatings are used of oxide layers generated either by oxidation during initial use or by wash coats. Steel die cores have been chromized to extend life about threefold mainly by improving the thermal resistance of the core materials. Coatings are needed for die casting steel; uncoated W or Mo alloy dies have been tried (Nat.Acd.Sci.Rep.1970).

2.3.2. OXIDATION RESISTANT CVD COATINGS:

Chromizing, aluminizing and siliconizing are widely used. The first two would be classed as metallics but silicides are more in the ceramic area. Siliconizing is good for refractory metals and alloys since $MoSi_2$, WSi_2 and VSi_2 have good oxidation resistance. Reactions between 20%Cr/25%Ni/Nb-stabilized stainless steel fuel cladding and the CO_2-based coolant in AGCR-s (Advanced Gas Cooled Reactor) causes oxidation of the steel and spalling of highly active oxide. C deposition from the CO_2 also affects heat transfer from the fuel pin. A coating can prevent these effects and factors include good heat transfer (thin coating, if ceramic), low neutron capture and no chemical interaction between coating and substrate. SiO_2 CVD coatings are used to protect 9% chrome steels from oxidation in high pressure CO_2 AGCR-s (Graham 1974; Brown et al 1978) but thermal shock cracks can occur.

CVD coatings of ZrC_xN_y are used for solar energy absorption, wear resistance for carbide cutting tools, and refractory protective coatings for nuclear fuel elements (Zirinsky & Irene 1978). Si_3N_4 CVD coatings are used on Mo (Galasso et al 1978). TiB_2 CVD coatings are very resistant to chemical and oxidation attack (Besmann & Spear 1977). Ti_5Si_3 CVD coating on Ti greatly reduces the oxidation rate but cracks occur after long times (Abba et al 1982).

A CVD SiO_2 coating on both clean and on corroded steel greatly reduced the oxidation rate (Brown et al 1978). A CVD SiO_2 coating on 20Cr25NiNb stainless steel reduced the oxidation rate by 5 times, in CO_2 at $825^{\circ}C$ (Bennett et al 1982). Its good adherence prevented oxide spallation. This is important for use in AGC reactor fuel cladding, as spalled oxide would be radioactive. Ceria and Y or Ce ion implantation is an alternative.

Wear resistant CVD coatings are widely applied, e.g. diffusion coatings by case-hardening (Krier & Gunderson 1965): carburizing, nitriding, carbonitriding, siliconizing and chromizing. Boriding (Matuschka 1979) and ion-nitriding are also used. CVD of TiC, TiN and alumina occurs at over 800°C , too high for normal steels and Cu alloys. Research is needed to overcome this. The number of wear resistant CVD coatings below 800°C is very limited at present. W_2C and W_3C coatings are suitable for normal steel, being deposited (CVD) at $300-700^{\circ}$C (Archer & Yee 1976), and their wear performance is somewhat less than TiC.

2.3.3. HARD OVERLAY COATINGS:

Wear resistance is an important feature of this class of coatings. A number of methods are available for surface modification for improved tribological applications (Sundgren & Hentzell 1986). Plasma-assisted vapour deposition processes have also been promoted to produce wear-resistant coatings (Bunshah & Deshpandey 1986). Coatings of TiN, TiC and Ti(C,N) are used on tools, e.g. TiC on punching dies, thread guides, cemented carbide tool inserts, on steel ball bearings for use in severe conditions and on pistons and valves (Gass & Hintermann 1973). Ti(C,N) is deposited on carbon steel nailheads in an ultrasonic field (19 kHz at 15 W/cm_2) to improve adherence and toughness (Takahashi & Itoh 1979). Similarly, TiB_2 is deposited on Fe, giving an adherent 160 micron thick film (Takahashi & Itoh 1977). About 30% of the tool inserts used in the automobile industry are TiC coated (Schintlmeister & Pacher 1974, 1975). The TiC (4 to 8 microns thick) protects the Co binder. Its low thermal conduction gives a cooler tool and hotter chip. Its success is also due to its compatibility with the substrate (Hintermann 1972). In high speed cutting, the tool tip can reach 1500°C, where TiC and TiN oxidise and alpha alumina coatings are superior (Pollard & Woodward 1950;Yee 1978). The alumina is put on a TiC layer to improve its adhesion. A number of coating methods are available for wear applications including aqueous and molten salt deposition, as well as PVD and CVD methods.

Carbide Tools: TiC coated cemented tungsten carbide disposable inserts first appeared about 18 years ago, coated by CVD. Graded coatings like TiC-Ti(C+N)-TiN appeared later, offering life improvements of up to 10 times longer for continuous turning jobs. PVD TiC coatings are better than CVD TiC and are equal to CVD graded coatings (Bunshah 1978). PVD TiN coatings are better than CVD due to increased crack propagation resistance due to inherent porosity and to absence of the substrate embrittlement which occurs with CVD. TiN and TiC coatings are based on cobalt cemented WC-TiC-TaC tool tips (Peterson 1974). Coated tools show much lower forces during cutting, than uncoated tools.

High Speed Steel Tools: CVD often needs 1000°C which would cause softening of a high speed steel (HSS) tool. PVD requires only

450°C (which is below the steel tempering temperature). TiC coatings by the ARE process improves the life of HSS tools up to 8 times in continuous cutting, and tool forces are reduced 50% (Bunshah 1986,1978). Sputter TiN coatings on HSS tools give similar results. Life tests are needed for interrupted cutting mode tools such as end-mills, drills, reamers, etc.

Other Wear Applications: PVD hard coatings offer improved wear resistance to abrasive and adhesive wear and impact erosion wear. Sputtered WC-6Co and ARE-deposited TiC gave the necessary wear resistance to rotary bearing of Be and BeO in vacuum. Wear rates of PVD chromium carbide and nitride coatings on stainless steel were 4 to 8 times less than for electroplated hard Cr and 300 times less than for the uncoated substrate. TiC coated stainless steel wears 34 times better than electrolytic hard Cr plate currently used in industry (Bunshah 1978). A TiC coating is smoother than the rough electroless Ni, advantageous, e.g. in the textile industry. Impact erosion of quartz dust on stainless steel caused a mass loss whereas on a TiC coating there was no mass loss in the same conditions (Yamanaka & Inomoto 1982). Such hard coatings are needed for turbine fan blades subject to dust erosion. Ion plated Al_2O_3 deposits increase the wear resistance of stainless steel for coal gasification plant (Bunshah 1978).

Ni-Cr-B-Si alloys are used for hardfacing on tools and in the glass industry (Knotek & Lugscheider 1974). CVD TaC, 30 microns thick is used as a subterrene drill coating, able to withstand melting rock temperatures of 1350 to 2450K, in conditions of wear and corrosion (Stark et al 1974).

2.3.4. RUBBING SEALS:

Rubbing seals are used in ceramic regenerators in gas turbine engines; low leakage rates, corrosion and thermal shock, wear and fatigue resistance are the property requirements. Graphite and refractory metal dichalcogenides oxidise below 500°C and are thus unsuitable; ceramic and metal seals are in demand. 85% NiO + 15% CaF_2 or SrF_2 have been suggested (Moore & Ritter 1974) for their wear and corrosion resistance. Rubbing surfaces are also required in liquid metal cooled fast breeder reactors, able to resist wear, high temperature liquid metal (Na,230-650°C) corrosion and nuclear irradiation. WC spark coating 250 microns thick is economically produced by CVD with excellent adhesion but Cr_3C_2 was best in liquid Na at 200 - 625°C, among coatings tested (Johnson et al 1974). Studies of adhesion and effects of friction are lacking. Coatings of solid lubricants or inert barrier materials to prevent adhesion between a moving ceramic and a metal, are desirable, e.g. yttria. Laboratory research, rather than the much more expensive in-engine testing, would be very cost effective at the present stage (Lewis 1987).

2.3.5. THERMAL BARRIER COATINGS:

Thermal barrier coatings were first tested on a research engine in 1976. The current state-of-the-art coating system for gas turbine applications is a plasma-sprayed $ZrO_2-(6\%-8\%)Y_2O_3$ ceramic layer over an MCrAlY (M=Ni,Co or NiCo) bondcoat layer plasma sprayed at low pressure. However, the current coating concepts and plasma coating technology are as yet, inadequate to meet long-range needs (Miller 1987). Erosion and corrosion resistant thermal barrier coatings are very necessary for preventing over-heating of metal parts in engines and some types of heat exchangers.

Poor adherence causing flaking has been a problem with some plasma sprayed coatings. The adhesion of plasma sprayed ceramic coatings to metals is generally poor but can be considerably improved if a sprayed bond-coat or interlayer is used e.g. NiAl or Mo. $MgO-Al_2O_3-SiO_2$ glass + powdered NiAl are applied on components exposed to vanadate containing fuels. $MgZrO_3$ thermal barrier coatings are used in turbine combustion chambers and yttria stabilised zirconia on turbine aerofoils (Freche & Ault,1976).

Thermal barrier coats reduce substrate air cooling requirements in gas turbine blades. An oxide acts as a thermal barrier, but oxides on MCrAl coatings are typically only 1 micron thick. To improve this, thermal barrier coatings are applied over existing coatings (Kear & Thompson 1980; Herman & Shankar 1987). This has been successful on gas turbine burner cans and exhaust liners. Yttria-stabilised zirconia plasma-sprayed layers 120-500 microns thick are used. To minimise thermal expansion stress the coatings are usually graded, changing gradually from substrate composition to ZrO_2, which is feasible by plasma spraying. Reductions of 100 degrees (C) in metal temperatures were achieved with 250 microns of yttria-stabilised zirconia over 100 microns of NiCrAl.

2.3.6. EROSION RESISTANT COATINGS:

Erosion resistant coatings are also needed for leading edges of gas turbine blades and pack cementation is used to give erosion and corrosion resistant aluminide coatings. Hard overlay coatings are needed for better protection, such as CVD Ti(C,N) on steel compressor blades (Wakefield et al 1974), using a thin Ni interlayer for better adhesion. Replacing the Ti(C,N) with CVD Cr_7C_3 gives much better thermal and mechanical shock resistance and also resistance to erosion, the latter ceramic being less brittle. Erosion of rocket nozzles is reduced by CVD coatings of Re, W and BC (Yee 1978) and pyrolytic carbon (Nickel 1974). NiCrAl/bentonite thermal spray powders for high temperature abradable seals have been tested (Clegg & Mehta 1988).

2.3.7. DIFFUSION BARRIER COATINGS:

TiC is used as a carbon diffusion barrier for Ta coatings on steel and TiN is a nitrogen diffusion barrier coating on SiC heating elements. Barrier coatings on metals prevent inward (and outward) C diffusion (carburization and decarburization), important for AGCR-s. TiC barrier coatings on BC particles dispersed in a W-Co matrix, prevent the BC reacting with the matrix during hot presssing at 1350°C. Diffusion barriers are used on fibre-reinforced materials; B, SiC, C, W fibres in metals interact (during hot pressing fabrication or high temperature use) causing embrittlement, loss of strength of the fibres and adverse effects on the metal structures. Development of diffusion barrier coatings is an important research area. Application is being delayed. The thermodynamics of interfaces between metals and ceramics is discussed in terms of surface energies and heats of formation of alloys (Wynblat & McCune 1980). Thus Pt alloying with Al explains why Pt reacts with alumina and the Pt surface absorbs 7% Al at 1527°C in H_2. This is known as 'solid state reaction bonding' and the bond remains durable even after long times at high temperatures. At magnifications of over 100000 an intermediate liquid phase is seen which wets the ceramics, and not recrystallising on cooling; the bond mechanism is not understood (Bailey & Borbidge 1980).

In older methods of metal-to-ceramic bonding, brazing is used in which a liquid metal or glass filler wets the ceramic to form the bond; this is limited to the few metals that will wet ceramics, e.g. Ti alloys. Metallizing ceramic surfaces prior to electroplating or brazing, involves Ag or Au alloys. Reducing or inert gases are needed throughout. The main limitation is the restricted temperature range for use of such bonds. Reaction bonding is a high pressure solid state process to join metals to ceramics (Allen & Borbidge 1983). Noble and transition metals can be bonded to Al_2O_3, MgO, ZrO_2, SiO_2 and BeO_2, for example. Such bonds retain their strength at high temperatures and breakage under load is often in the ceramic and not at the bond site. Solid state reaction bonding is performed at about 90% of the m.p. of the lowest component, usually in air, with a light clamping action for times between a few seconds and about 3 hours; the ceramic surface is polished to optical flatness for maximum bond strength. Pt has an expansion coefficient similar to that of most ceramics. Ni is suitable in inert atmospheres and bonding is due to a spinel, $NiO-Al_2O_3$.

2.3.8. CERAMIC COATINGS ON CERAMICS:

ZrN whiskers deposited on quartz tubes give a superhard surface (Motojima et al 1979). Si_3N_4 can be deposited on sintered vitreous silica by CVD at 800 to 1200°C (Mellottee et al 1976). TiB_2 can be deposited on uncoated and Ti-coated WC for tool tips by CVD (Zeman et al 1982). Si_3N_4 coatings on graphite, alumina,

BN, SiO_2 and hot-pressed Si_3N_4 are deposited by CVD (Galasso et al 1978). Si_3N_4 on graphite is used as rocket nozzles or high temperature blades. An inner SiC layer and an outer glaze based on B_2O_3, P_2O_5 or SiO_2 have been tried on carbon-carbon composites (Strife & Sheehan 1988). BP CVD coatings on Si,Mo,WC and C have been little researched. BP is very refractory and hard (Morojima 1979). Polycrystalline Si coatings on graphite susceptors give oxidation resistance (Van den Brekel 1977).

TiZrB coatings are used on graphite, giving good mechanical shock resistance, high hardness and melting point (Takahashi & Kamiya 1977). ZrC-C coatings on semicircular graphite/Mo are used as fission product barriers in gas cooled nuclear reactors (Ogawa et al 1979). Low cost refractory crucibles for metal melting are lined with a coating inert to the molten metal. A selected few are listed in Table 2:3.

Coatings protect resistance heating elements, e.g. SiC coated graphite, SiO_2 coated SiC. For molten metal probes, 2-layer ceramic coatings are used - a porous thermal shock resistant layer over a dense chemically resistant layer (Nat. Acad. Sci. Rep.1970).

TABLE 2:3

REFRACTORY LINING COATINGS TO GRAPHITE

Lining	Melt
AlN	Al
AlN + SiC	Al
alumina	Al, Ni, Co, Fe
clay	Fe, Al
$MgZrO_3$	U
mullite	Al
thoria	most metals
TiN, ZrN	Co, Fe
zirconia	Ru, Rh, Cr

==

2.3.9. DESIRABLE CERAMIC COATING PROPERTIES:

Elasticity in a ceramic coating is desirable to combat thermal cycling spalling. The elasticity of ceramics increases with porosity, but porosity is bad for corrosion. Fig.2-6 (Perugini 1976) shows the balance of these two effects. Fig.2-7 shows how self-sealing ceramic coatings offer a solution since their surface layer presents a closed porosity to the gas or other corrodent. The as-sprayed coating has an open porosity but this is closed by

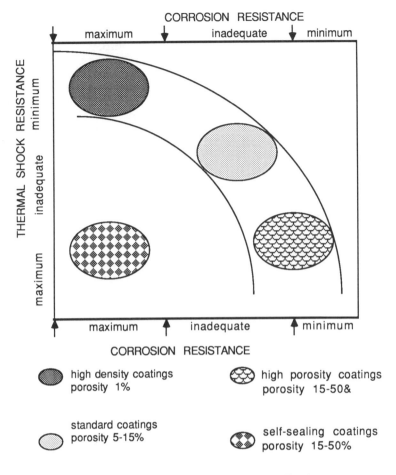

Fig.2-6: Effect of Coating Porosity on Resistance to
Thermal Shock and Corrosion

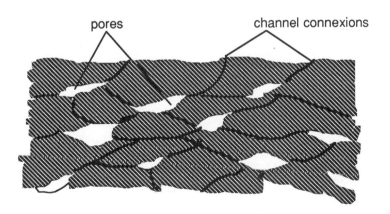

Fig.2-7: Porous Structure   of Self-Sealing Coatings

(Perugini 1976)

31

topochemical reactions (Perugini 1976). A graded bond to the metal is essential, e.g. Ni or Ni alloy (m.p.800 to 1350°C) forms a metal bond to the metallic substrate, Cr on top bonds to the Ni and by oxidising also seals the inter-pore channels in the ceramic above. Between the Cr and the ceramic a thin layer of stabilized zirconia (for example) is used. Thermal conductivity and heat capacity decrease from inner to outer layer while elastic modulus and melting point increase. The inner and second layers may include the third as a cermet. The self-sealing effect also confers the ability to heal cracks. Tensile and shear strengths >3kg/mm^2 and bending angles of 5° without detachment are reported (Perugini 1976).

2.4. COMPOSITE COATING SYSTEMS

Composite coatings may be made up of two or more individual and/or single coating materials, either in a layer-mode or mixed mode. A special category of the mixed mode is termed a cermet, meaning that it is made up of a ceramic and a metal. Cermet coatings combine the heat resistance and strength of ceramics with the ductility and thermal conductivity of metals. Particle and fibre-reinforced composites are currently being investigated to identify those which can incorporate ceramics with a negative coefficient of thermal expansion (Chu et al 1987). Composite production in powder form and mechanical alloying has progressed rapidly. Rapid solidification processing - RSP (Das & Davis 1988; Wadsworth et al 1988; Hays & Harris 1987; 6th Inter-Nat. Conf. Proc. 1987 in Mat. Sci. Eng. Vol. 97-99, 1987; Panitz 1986; Kear & Gissen 1984) and mechanical alloy processing (MAP) are two proven new methods for special alloy and cermet production.

Cermet coated enamelling steel is good up to 600°C under thermal cycling, but rigid quality control is needed during coating. Life is longer than for Al coatings and no brittle intermetallics form. Cermet coated steels are not yet recommended for gas flues and further research is needed. Cermet oxy-acetylene flame sprayed coatings of Ni+5% Al+30% alumina on steel give good oxidation protection with excellent thermal shock and cycling resistance, in gas fired applications. A rough surface gives better heat transfer and this feature may simplify heat exchanger design. For solar power units, coatings are used on stainless steels and superalloys and resist daily thermal cycling and thermal gradients. Absorbing coatings (black) are essential for this purpose, e.g. oxidised Cr, Ni and Co (Perkins 1981).

Multilayer films can give selective solar absorption by interference effects but are expensive and unstable due to high temperature interdiffusion (Mattox & Sowell 1974). For nuclear fusion reactors, coatings will be used for first wall armour and limiters. Suitable low neutron capture coatings are borides (TiB$_2$, VB$_{12}$), carbides (B$_4$C, TiC, SiC) and elements (B, C, Be). For fission reactors coatings are used for HTGR heat exchangers. Suitable contamination control coatings are Cr_3C_2, $Cr_{23}C_6$ + NiCr. These can be produced by CVD, PVD, sputtering, plasma spray and detonation gun (D-Gun) (Perkins 1981; Lewis 1987).

Electrodeposition of TaC is the first successful electrolytic coating of a refractory carbide from a high temperature electrolyte (Stern & Gadomski 1983) and offers promise for the coating of complex shapes. Further research to develop this is important. Studies of the mechanism of carbonate reduction in fluoride melts and of simultaneous reduction to Ta and C would give a better understanding of the coating composition. A dispersion of Al_2O_3 or TiO_2 in a metal confers a high resistance to recrystallization and grain growth, thus retarding loss of hot strength (Greco & Baldauf 1968; Sautter 1963).

SiO_2, Al_2O_3, TiO_2, ZrO_2, SiC, WC, BC, diamond and glass powder have all been codeposited from aqueous electrolytes, e.g. with Ni from a Watts electroplating bath. These coatings have wear resistant properties. ZrB in Cr gives oxidation resistance. Wear resistance is due to the presence of hard particles and also to the dispersion hardening of the matrix, which however still retains some ductility so that the coatings have both ductility and toughness (Pushparaman et al 1974). Electrodeposited cermet-coated steels have been successful in prolonging the service life of hot risers in sour gas wells (Hocking et al, 1984; Report, Ind. Corrosion 1987). Cermets have also been electrodeposited for well-proven application in gas turbine systems (Kedward et al 1987).

2.5. AMORPHOUS COATINGS

Certain groups of alloys with well characterized crystal structure and metallic properties can be rendered amorphous in structure but yet retain much of their metallic properties when they are subjected to rapid solidification from their molten state. They were first called metallic glasses in the early seventies, but in coating and powder production areas the terminology is less used and replaced by amorphous metals or rapidly solidified products (RSP). The materials are of interest in the field of coatings because of their remarkably high corrosion resistance and wear characteristics. Laser-oriented coating processes and powder metallurgy technology have enabled the exploration of this class of coating systems together with cermets.

Amorphous alloy films were first produced in the 1950s (Buckel & Hilsch 1954). Using quench techniques capable of achieving cooling rates of the order of 10^6 K/s the process of nucleation and crystalline phases could be kinetically by-passed to yield a frozen alloy melt - metallic glass (Duwez etal 1960, 1966; Turnbull 1961). The material has been deposited from the vapour phase by PVD and CVD techniques, while melt spinning, splat cooling and liquid atomization adopt the liquid quench approach. Atomization is a fast-developing powder metallurgy technique. Early production of amorphous alloys was mainly in the form of 2-3 mm wide ribbons. But since the development of powder metallurgy and laser techniques, amorphous alloys are a new class of high temperature material and coating system poised to move onto further development (Johnson 1986).

Table 2:4 gives a summary of coating systems studied for high temperature materials.

TABLE 2:4

HIGH TEMPERATURE SUBSTRATES – TYPES OF COATINGS USED

Substrates	Coatings
Fe-, Ni- & Co-base alloys	Al, Cr, Si; MCrAlX (M = Fe and or Ni and or Co; X = Y,Hf,Ce,Zr,Ti,Ta etc. with variations).
	Cermets: $Co-Cr_2C_3$; Ni- and or Co- with ceramics – Al_2O_3 etc.
	Ceramics – a number of oxides, carbides, borides and nitrides
	Platinides; other refractory metal depositions.
	Surface modification by various PVD, CVD and Laser methods.
Refractory metals – Nb, Ta, W, Mo & Cr	Silicide coatings: Si –Mo, –Nb, –Cr-Ti, –Al-Cr with Al, B, Cr, Ti and V additions.
	Aluminide coatings: Al –Nb, Ta, Ni with Cr, Si, Sn and Ti additions.
	Beryllide coatings: Be –Ta, –Ti, –Nb, –Cr, –Mo and Zr.
	Others: Nb-Zn
Graphites	Ir, SiC, oxides – HfO_2, ThO_2, Zr_2, borides – HfB_2, ZrB_2

2.6. COATING PROCESSES

This section gives a general introduction to a number of physical and chemical coating processes available and the several categorization of the coatings and coating methods. Chapters 3-6 cover the processes in greater detail.

2.6.1. GENERAL:

Almost all the processes given can be adapted for producing thick or thin films. In distinguishing coating parameters the demarkation of thin or thick films is often ambiguous. 10000 Angstrom (10^{-3}mm) is more or less accepted as the boundary between them. Another viewpoint is that a film can be considered thick or thin depending on whether it exhibits surface-like or bulk-like properties (Perkins 1981). Various units appear in coatings literature, viz. 1 mil = 0.001 in. = 25.4 micron; micron 1 = 1 micrometre = 10^{-3} mm. Coating performance is not necessarily pro-rata to its thickness nor entirely ascribable to the coating composition. It is difficult to specify a critical threshold on the basis of the coating material itself because it is not a stand-alone parameter. Each coating has to be assessed in conjunction with the substrate and the environment.

A thickness criterion for a coating is subjective. The same process can be adapted to produce opaque or transparent coatings and films. Most of the coatings considered for power and energy applications are opaque. Their hardness, corrosion resistance together with their compatibility to substrate service-conditions for physico- and chemico- mechanical and -thermal responses are the deciding factors. Hard, insulating and optically transparent films, on the other hand, have various other applications in metallurgy, surface finishing, semi-conductor, and optical industries. Their physical properties are very high hardness (up to 6000 Kg/mm^2 VHN), low coefficient of friction (0.04 - 0.08 steel/steel), corrosion resistance, chemical inertness, high resistivity (10^{12} ohm.cm), and the infrared refractive index (2 for carbon).

2.6.2. COATINGS & COATING PROCESS CLASSIFICATION:

Several ways of classification are reported in literature. Table 2:5 shows a categorization on the basis of a solid, liquid, gas and plasma media of matter from which the coating material is obtained. Table 2:6 gives a further breakdown of the material state and medium - that the coating is obtained from various particle states - atomic, ionic, bulk - micro-, macro-, and structure modified states. Some of the processes are carried out in a variety of routes; these are given in table 2:7. Some of the well known general features of coating processes are indicated in table 2:8 (Biswas 1986; Chatterji et al 1976; Nat. Acad.

Sci. Rep. 1970).

Pack cementation was one of the first and cheapest high temperature coating processes to be developed for aluminiding and chromizing individually or together. Table 2:9 gives a comparative rating for it regarding process costs and Table 2:10 gives the current status of a number of diffusion aluminiding coating processes (Nicholls & Hancock 1987; EUR 1982). Superalloy coating processes are surveyed in Table 2.11.

TABLE 2:5

COATING METHODS SHOWN BELOW COATING MEDIA AVAILABLE

Solid state	Liquid State	Semi-liquid or Paste	Gaseous State (atomic/ionic/electron inter-action)	Dissolved Solutions	Plasma
Bonding	Hot Dipping	Sol-Gel	PVD	Chemical	
Cladding	Spraying	Slurry	CVD	Galvanic	
Sintering	Welding	Brazing		Electro-galvanic	
	Surfacing				

TABLE 2:6

COATING PRODUCTION PROCESSES: CATEGORISATION BY MATERIAL STATE & FABRICATION ROUTES

Group 1 Atomic/Ionic State	Group 2 Macro Particle State	Group 3 Bulk Material State	Group 4 Structure Modified State

Vacuum Medium
  Vacuum Evaporation
  Ion Beam Deposition
  Molecular Beam Epitaxy
	Impact Coating	Overlaying
		  Weld coating
			Laser Surface Modification (LSM)

Plasma Medium
  Sputter Deposition
  (Ion; Magnetron)
  Ion Plating
  Plasma Polymerization
  Activated Reactive
  Evaporation
  Cathodic Arc Deposition
	Fusion Coating
	Thick Film Inking
	Enamelling
	Electrophoretic
	
	Thermal Coating
	Flame Spraying
	Plasma Spraying
	Detonation Gun
	(D-Gun) Spraying
		Cladding
		Explosive-
		Roll Bonding
		
		Laser Melting
		Wetting Method
		Painting
		Dip Coating
			Thermal Treatment
			Ion Implantation
			
			Surface Enrichment
			Diffusion from Bulk
			
			Sputtering

Chemical Vapour Medium
  Chemical-
  Vapour Deposition
  Reduction
  Decomposition
  Plasma Enhanced
  Spray Pyrolysis
	Sol-Gel Medium	Electrostatic Method
		Spin Coating
		Spray Printing
			Leaching
			
			Chemical-
			Conversion
			Liquid
			Vapour (Thermal, Plasma)
			Diffusion

Electrolyte Medium
  Electroplating
  Electroless Plating
  Molten Salt Electrolysis
  Chemical Displacement
			Electrolytic-
			Anodization
			Molten Salt
			Heat Treatment
			
			Mechanical Method
			Shot Peening

TABLE 2:7

TYPE CATEGORIZATION OF COATING PROCESSES

Mechanical	Physical	Chemical	Electro-chemical	Spraying	Welding
Cladding	Vapour deposition (PVD)	Vapour deposition (CVD)	Aqueous	Detonation gun	Laser
Bonding	Vacuum coating;	Electroless	Molten salt	Electric arc	Manual Metal Arc (MMA)
					Metal Inert Gas (MIG)
	Thermal evaporation			Metallising	Oxy-Acetylene
	Sputtering			Plasma	Plasma tranfered arc
	Ion Plating			Flame- Powder Wire	Plasma welding
					Spray fusion
					Submerged arc
					Tungsten inert arc (TIG)

Two Other Methods:- (no alternate techniques which can be categorised individually):

Ion Implantation & Hot Dipping

===

39

TABLE 2:8

GENERAL SURVEY OF COATING PROCESS CAPABILITY & DRAWBACKS

Method	Capability	Drawbacks
Physical Vapour Deposition (PVD)	Versatile; all solid elements, and materials can be deposited. Thin films and reasonably thick coatings are possible. Several variations in technique. * 5-260 microns.	Operations are mostly line-of-sight. Poor throwing power. Equipment expensive.
Chemical Vapour Deposition (CVD)	Competitive to PVD; elements & compounds chemically reactive & in vaporized state are coated. Good throwing power. * 5-260 μm	Heat sources play an important role. Generally operates at higher temperatures than PVD. Substrate overheating possible. Undesirable directional depositions possible.
Cementation	Good uniformity & close dimensional tolerance. A very cost-effective process. Most coated materials — Al & Cr. Hard deposits. * 5-80 microns	Substrate size a limitation. Not for high temperature-sensitive substrates.Thinner films than other diffusion processes. Coating embrittlement possible.
Spraying	Technique & starting material controlled process. Thick & uniform coverage possible. * 75-400 microns	Skilled operator dependent. Substrates have to be heat- and impact tolerant. Porous & rough coatings with possible inclusions unless controlled.
Cladding	Thick coatings possible. Large substrates can be handled. * 5-10% of substrate thickness.	Substrate distortion possible. Suitable for robust substrates.
Electro-deposition (also electroless; electrophoresis)	Cost-effective aqueous process. Fused salt electrolytes yield precious metal and refractory deposits. Cermet production in industrial use. Electroless & electrophoresis are limited to a few elements and substrate sizes. *0.25 - 250 microns	Good plant design to ensure good throwing power. Molten salt electrolytes require rigorous control to prevent moisture ingress & oxidation. Melt evaporation a problem. Coatings can be porous & under stress. Limited to specialised areas in high temperature.
Hot dipping	Relatively thick coatings. Fast coating method.* 25-130 microns.	Limited to Al for high temperture applications. Coating can porous and discontinuous.

==
*Coating Thickness

TABLE 2:9

PACK COATING : PROCESS STATUS & COSTS

Process	Availability	Relative Cost
Pack aluminising	Widely used coating with lowest unit cost. Euipment can be installed at moderate capital cost, enabling facilities to be available throughout the world.	1
Chromium- or Cr-Ta-modified aluminising	Equipment similar to that for pack aluminising. Process available in Europe and America.	2-3
Platinum- modified aluminising	Equipment similar to that for pack aluminising, plus special facilities for Pt-electroplating from fused salts. Process available in Europe and America.	3-5
Vapour- deposited overlay coatings	Sophisticated vacuum equipment and analytical control facilities required, therefore very high plant cost. Production capability available mostly in the U.S.A. Branches of US firms in UK. Limited facilities in France, Germany, Sweden and Switzerland, and independently in USSR.	7-10

TABLE 2:10

DIFFUSION ALUMINIDING COATING PROCESSES & THEIR CURRENT STATUS

Process Route	Coating Designation & Manufacturers	Process Status
PackAluminizing - Low & High Activity	Proprietary names - Gas turbine manufacturers and individual firms.	Active industrial
Pt- electroplated coating + Pack Aluminizing	LDC2/manuf. TEW; RT22/manuf. Chromalloy	Active industrial
Pt- metalliding from fused salts + Pack Aluminizing	JML1 & JML2/ manuf. Johnson Matthey	Engine Trials
Pack Chromizing	PWA70 (also HC12;MDC3V; RT5.A$_1$ etc.)/several manufacturers	Active industrial
Pack- co-deposited Aluminizing & Chromizing	HI 15/manuf. Alloy Surfaces	Laboratory & Burner rig tests
2-step Pack Chromizing + Aluminizing (some with siliconizing - SiO$_2$ safer than Si)	PWA62 (also HI32;RT17; SylCrAl & others)	Active industrial
Ta coating + Pack Aluminizing & Chromizing	Tan CrAl	-- as above --
Si/Ti Slurry Diffusion	Elcoat 360/Elbar	-- as above --
Al/Si Slurry diffusion	Sermaloy J/Sermetal	Burner rig tests

(Nicholls & Hancock 1987)

TABLE 2:11

KEY COATING PROCESSES FOR SUPERALLOYS: A BRIEF SURVEY

Process	Number of Steps	Environment	Comments
Pack Cementation (also halogen Streaming and slip pack methods) Al, Cr, Al-Cr, Al-Cr-X (also Be)	Normally single pack; could be two or more as required.	Inert to start with; vacuum in some systems. Reacting medium, eg.chloride is supplied when pack reaches required temperature.	Slip-pack process requires components to be supported. Special holding fixtures when used, cause blind spots in coating. Regular pack process eliminates this. Disadvantage: long heating and cooling times.
Slurry Al, Cr, Al-Cr, Al-Cr-X	Multiple	Heating cycle uses both inert and vacuum atmosphere. Slurry application in vacuum yields good coverage.	No heating and cooling cycles, but special holding fixtures are needed. (blind spots in coatings).
Physical Vapour Deposition (PVD): Al, Cr, Si, and a large range of alloys.	Normally two.	Pumped down to vacuum.	Sophisticated and expensive. Poor throwing power; not readily repairable. Allows versatile combination of elements.
Chemical Vapour Deposition (CVD): Al, Cr; technical limitation - none.	As for PVD	Vacuum as for PVD; inert gas also used.	Compares with pack cementation. Requires holding fixtures. Can yield undesirable directional deposition.
Fused Salt Electrolysis Electrolytic; electroless: a number of refractory metals, and rare earths.	One or two. Multi-component systems need additional steps.	Inert gas cover; moisture avoidance very important.	Limited to small components. Difficult and expensive to maintain; can be toxic and and hazardous. But enables 'hi-tech' coating.
Vitreous Enamelling (glassy-bonded refractory)	Generally multiple	Selective: air or inert.	Yield brittle coatings; relatively thick gauge. Limited to about $1000^\circ C$ in service.
Hot-dipping Al, Al-X	Generally multiple	Presence of flux or inert.	Short process times. Scope limited.
Electrophoresis Al-rich	Multiple	Dielectric organic solvent.	Limited to relatively small components.

43

2.7. LIST OF ACRONYMS

Acronyms found in the literature of deposition processes, techniques, analyses and applications are cited here. New words and letters get introduced at a fast rate. The list, therefore, is not exhaustive. Acronyms are used for brevity, and should be 'coined' with care to avoid ambiguity, absurdity or sometimes serious errors. C2S (di-calcium silicate), C3S(tri-calcium silicate), C3A (tri-calcium aluminate), C4AF (tetra-calcium alumino ferrite) etc., by civil engineers to express chemical formulae are a few examples of such use. A more serious error occurs if a chemical formula is allowed as abbreviation; for instance, CO for Cr_2O_3; the authors report on mechanical testing of plasma sprayed coating of chromic oxide on cast iron (Colin et al 1988).

Coating, Deposition, & Plating

CVD	– Chemical Vapour Deposition
CCVD	– Conventional CVD
LCVD	– Laser CVD
LACVD	– Laser Assisted CVD
LECVD	– Laser Enhanced CVD
LPCVD	– Low Pressure CVD
MFCVD	– Molecular Flow CVD
MOCVD	– Metal Organic CVD
MPCVD	– Magnetron Plasma (enhanced) CVD
IHPACVD	– Induction Heated PACVD
PACVD	– Plasma Assisted CVD
PECVD	– Plasma Enhanced CVD
RPECVD	– Remote PECVD
CVI/CVD	– Chemical Vapour Impregnated CVD &/or
	– Chemical Vapour Infiltration CVD
CCRS	– Controlled Composition Reaction Sintering
HTLA	– High Temperature Low Activity (Pack process)
LTHA	– Low Temperature High Activity (Pack process)
PPP	– Pressure Pulse Pack
AVID	– Arc Vapour Ion Deposition
CAPD	– Cathode Arc Plasma Deposition
EBPVD	– Electron Beam PVD
ESD	– Electro Spark Deposition
PVD	– Physical Vapour Deposition
PAPVD	– Plasma Assisted PVD
RPAPVD	– Reactive Plasma Assisted PVD
IAC	– Ion-assisted Coatings
IAD	– Ion-Assisted Deposition
IBAD	– Ion Beam Assisted Deposition
IBED	– Ion Beam Enhanced Deposition
IVD	– Ion Vapour Deposition

```
RE          - Reactive Evaporation
ARE         - Activated Reactive Evaporation
BARE        - Biased ARE
PARE        - Plasma ARE
RSAE        - Random & Steered Arc Evaporation

RS          - Reactive Sputtering
              (also used for Raman Spectroscopy)

CIP         - Cathode Ion Plating
IAC         - Ion Assisted Coating
IP          - Ion Plating
SIP         - Sputter Ion Plating
RIP         - Reactive Ion Plating
RTIP        - Reactive Triode Ion Plating

HHC         - Hot Hollow Cathode
HCD         - Hollow Cathode Discharge
ICB         - Ionized Cluster Beam (Process Technique)

LPAS        - Low Pressure Arc Spray (-ing)
LPPS        - Low Pressure Plasma Spray
RPS         - Reduced Pressure Plasma Spray
SPS         - Shrouded Plasma Spray
UPS         - Underwater Plasma Spray
VPS         - Vacuum Plasma Spray
```

Laser Process

```
LASER       - Light Amplified Stimulated Emission Radiation
cw or CW     - Continuous wave
Nd YAG      - Neodymium Yttrium Aluminium Garnet
TEA         - Transversely Excited Atmospheric (laser)
LW          - Laser Welding
LAPP        - Laser Physical Properties
LSA         - Laser Surface Alloying
LASAP       - Laser Surface Alloying Parameter
LST         - Laser Surface Treatment
LSM         - Laser Surface Melting
LAZ         - Laser Affected Zone
LACVD       - Laser Asssisted Chemical Vapour Deposition
LECVD       - Laser Enhanced Chemical Vapour Deposition
LAPVD       - Laser Assisted Physical Vapour Deposition
LAPC        - Laser  Assisted Plasma Coating

RSP         - Rapid Solidification Processing
```

Analyses: Microscopy, Spectroscopy

```
AEM         - Analytical Electron Microscopy
AES         - Auger Electron Spectroscopy
ATIS        - Automatic Thermal Impedance Scanning
```

```
EASE    - Electron(ic) Absorption Spectroscopy Experiments
EELS    - Electron Energy Loss Spectroscopy
ETM     - Electron Tunnelling Microscopy
EDAX    - Electron Dispersive X-ray Analysis
EDX     - as above
EPMA    - Electron Probe Microanalysis
ESCA    - Electron Scattered Chemical Analysis
GDMS    - Glow Discharge Mass Spectroscopy
MS      - Mass Spectroscopy
OES     - Optical Emission Spectroscopy
ISS     - Ion Scattering Spectroscopy
NMR     - Nuclear Magnetic Resonance
PMS     - Plasma MS
RS      - Raman Spectroscopy (to note context in text)
RBS     - Rutherford Backscattering Spectroscopy
SIMS    - Secondary Ion MS
SAM     - Scanning Auger Microscopy
SEE     - Secondary Electron Emission
SEM     - Scanning Electron Microscopy
SXS     - Soft X-ray Spectroscopy
STEM    - Scanning Transmission Electron Microscopy
STM     - Scanning Tunnelling Microscopy
TEM     - Transmission Electron Microscopy
UPS     - Ultraviolet Photoemission Spectroscopy
WDX     - Wave Dispersive X-ray analysis
XPS     - X-ray Photo-Spectroscopy
XRD     - X-Ray Diffraction
```

Miscellaneous

```
ac/AC   - Alternating Current
AFBC    - Advanced fluidized-bed Combustion
AGR     - Advanced Gas (cooled) Reactor
AGCR    - Advanced Gas Cooled Reactor
DBTT    - Ductile Brittle Tranformation Temperature
dc/DC   - Direct Current
DRM     - Dynamic Recoil Mixing
DS      - Directionally Solidified (alloys, superalloys etc.)
EDCC    - Electro-Deposited Composite Coating
EXAFS   - Extended X-ray Absorption Fine Structure
FET     - Field Effect Transistor
HCD     - Hollow Cathode Discharge
HSS     - High Speed Steel
HTGR    - High Temperature Gas (cooled) Reactor
IH      - Induction Heating
ODS     - Oxide Dispersion Strengthened (alloys, products etc.)
MAP     - Mechanically Alloyed Products
MMC     - Metal Matrix Composites
MNS     - Metal-Nitride Silicon (devices)
MIOS    - Metal Insulator Oxide Silicon (devices)
NDE     - Non-Destructive Evaluation
NDT     - Non-Destructive Testing
PFBC    - Pressurized Fluidized-bed Combustion
```

46

PM   – Powder Metallurgy (components, products etc.)
rf/RF   – Radio Frequency
RSP   – Rapid Solidification Processing
SCC   – Stress Corrosion Cracking
TESS   – Tribologically Engineered Surface Selection
UHV   – Ultra High Vacuum
VHN   – Vickers Hardness Number
VLSI   – Very Large Scale Integrated Circuits

CHAPTER 3

Physical vapour deposition (PVD)

3.1. INTRODUCTION

The term Physical Vapour Deposition (PVD) covers three major techniques; evaporation, sputtering and ion plating. Originally PVD was used to deposit single metals by transport of vapour in a vacuum, without involving a chemical reaction. PVD technology has now developed to be extremely versatile enabling deposition of a wide range of inorganic materials – metals, alloys, compounds or their mixtures as well as some organic materials. In general, a vapour source and the substrate on which deposition occurs are contained in a vacuum chamber. Variations in atmosphere, e.g. presence of a neutral gas such as argon or a reactive gas, heating method of vapour source, e.g. by induction or electron beam, as well as electrical voltage of the substrate give rise to the different techniques of deposition which in turn determine the structure, properties and deposition rate of the coating.

Deposition takes place, broadly according to the steps outlined below (Bunshah 1981):

1. Synthesis of the material to be deposited.

   a) Transition from a condensed phase (solid or liquid) to the vapour phase,

   b) For deposition of compounds, a reaction between the components of the compound some of which may be introduced into the chamber as a gas or vapour.

2. Transport of the vapours between the source and substrate.

3. Condensation of vapours (and gases) followed by film nucleation and growth.

These steps can be independently controlled in a PVD process giving it an advantage over CVD.

Conventional PVD is generally an overlay coating where the coating is just an add-on as in other overlay coatings achieved by other means. The coating-substrate interface of a PVD deposit is a distinct region, whereas for ion implantation or diffusion coatings this is not the case. PVD methods are versatile as mentioned earlier, and any metal, alloy or compound can be deposited. Purity, structure and adhesion can all be controlled. Vacuum evaporation is much faster than vacuum sputtering; alloy composition and deposition rate are easier to control by sputtering (Teer 1983). Ion plating is preferred for very good adhesion.

Both chemical and physical vapour deposition techniques have been reviewed for tool coating (Doi & Doi 1982). PVD wear resistant TiC, TiN and Ti(C,N) coatings onto low temperature substrates normally gives poor film morphology and adherence. A low N_2 pressure improves adherence, as does a thin Ti film , and vaporization in C_2H_4 + N_2 (Inomoto et al 1982). PVD successfully coats steel tools with TiN at less than 500°C whereas CVD may require 1000°C. 500°C is lower than the tempering level for high speed steels but for CVD coatings heat treatment is needed after coating and the necessary quenching may cause distortion and adhesion loss. TiN is one of the easiest and best PVD coatings. TiC is harder but more brittle and more difficult to apply as are WC, Cr_2C_3 and ZrC (Boston 1983). For carbides, a hydrocarbon reactant gas is needed and H_2 is produced.

Sputtering is slower than evaporation but is justified if a substrate will not tolerate much heating. Electron beam evaporation is much faster and creates a large plasma volume. Continuous PVD processes are feasible, with components, e.g. tools for coating, passing in and out via load locks, but the vacuum system interior would also be coated and require cleaning and the coating source would require renewing. The major effort needed to develop the equipment may not be justified (Boston 1983).

Plasmas can be used to provide heat to substrates to improve adhesion and structure and to increase reactivity (for reactive deposition processes). An ion passing through 100 eV reaches a kinetic energy equivalent to 10^6 °C, forming coatings with unusual properties, e.g. C coatings are produced which are hard, transparent and diamond-like (Teer 1983).

There are several advantages of PVD processes over competitive processes such as electrodeposition, CVD and plasma spray. These have beeen outlined by Bunshah (1981) and are as follows:

(i)   Extreme versatility in composition of the deposit. Virtually any metal, alloy, refractory or intermetallic compound, some polymeric type materials, and their mixtures can be easily deposited. In this regard the PVD processes are superior to any other deposition processes.

(ii) Possibility to vary the substrate temperature within very wide limits, from sub-zero to high temperatures.

(iii)Ability to produce coatings of self-supported shapes at high deposition rates.

(iv)   Very high purity of the deposits.

(v)   Excellent bonding to the substrate.

(vi) Excellent surface finish which can be equal to that of the substrate, thus minimizing or eliminating post-deposition machining or grinding.

A review of the difference between electron beam (EB) evaporation and high-rate sputtering suggests that they supplement each other and will therefore co-exist in the future. A description of the magnetic-field-enhanced gas discharge process (magnetron or plasmatron discharge) is given. The coating properties as well as applicability of the method are discussed (Schiller et al 1982).

Vacuum vapour-coating processes are reviewed in a general way (Hayashi 1982; Nicholls & Lawson 1984; Biswas 1986). Table 3:1 gives a brief group survey of vacuum deposition techniques.

TABLE 3:1

Techniques in Vacuum Deposition: Group Survey

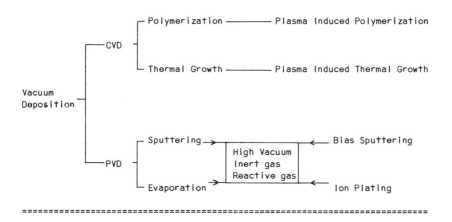

Improvements have been implemented in some of the PVD techniques in recent years. Several of these appear in the various conference proceedings published during 1985-1988. Broad-beam ion sources (Kaufman 1986), ionized cluster beam deposition technique (Younger 1985), the hot hollow cathode heating device (Kuo, Bunshah & Okrent 1986), steered arc evaporation technique (Boelens & Veltrop 1987), and, cathodic-spot arc coating method (Boxman & Goldsmith, 1987) may be mentioned here. Several hybrid PVD techniques are reported. For instance, sputter ion plating – SIP (Jacobs 1986) and reactive arc vapour ion deposition (Johansen et al 1987; Freller & Haessler 1987; Martin et al 1987). Amongst these, broad-beam ion sources are a product of NASA-commissioned research work of the 1960s but very little was published about them until 1974 as Kaufman's review indicates.

3.2. EVAPORATION PROCESS

3.2.1. THE BASIC PROCESS

Producing coatings by evaporation involves two basic processes, viz. evaporation of the coating metal followed by its deposition onto the substrate material. At first this was done in vaccum for the deposition of single metallic coats. Later variations in the atmosphere by introducing plasmas and reactive gases, gave rise to the different evaporation processes and more complex coatings as shown in Tables 3:2 and 3:3.

In all cases the chamber housing the components is evacuated down to about 10^{-5} torr. This is then followed by introducing any gases required, usually at low pressures between 10^{-3} and 10^{-1} torr.

The widescale application of vacuum evaporation in industrial processes dates from 1946 with the development of vacuum techniques. Evaporated thin films are covered in a detailed review by Glang (1970) and in other chapters of the Handbook of Thin film Technology (Maissel & Glang 1970). Thick films and bulk deposits have been reviewed by Paton, Movchan & Demichishin (1973) and by Bunshah (1976). A comprehensive review on evaporation was compiled by Bunshah (1981, 1986, 1987) as well as Teer (1983).

The evaporation process is governed by thermodynamic considerations, i.e. phase transitions from which the equilibrium vapour phase pressure of materials can be derived, as well as the kinetic aspects of nucleation and growth which determine the evolution and microstructure of the deposit. These have been treated in detail by Glang (1970).

TABLE 3:2

E v a p o r a t i o n   P r o c e s s   C h a r a c t e r i s t i c s

Method	Direct Evaporation		Gas Scattering Evaporation	
Type	Vacuum Evaporation	Activated Evaporation	Reactive Evaporation	Activated Reactive Evaporation
Coatings produced	Single elements; alloys	Alloys; compounds	Compounds	Refractory Compounds
Evaporation source	Any suitable source Single source for elements or alloys; wire or rod feed. Multiple source for alloys – but with reduced deposition rate.	Any source; usually electron beam.	Usually electron beam	Usually electron beam
Atmosphere	Vacuum	Plasma	Reaction gas	Reactive gas & plasma
Substrate	Floating, Grounded or electrically biased	As in Col.2	As in Col.2	Biased ARE BARE — Substrate is -ve biased; also called Reactive Ion Plating (RIP). Enhanced ARE EARE — Uses a thermionic electron emitter together with electron beam heating

TABLE 3:3

Features of PVD Evaporation Processes

Type	Arc	Electron Beam			
Code	E1	E2a	E2b	E3a	E3b
Source of Power	Low pressure arc	Hollow Cathode Low Voltage, High Current		Hot Filament High Voltage, Low Current	
Gas Medium	Nitrogen	Argon & Nitrogen	Nitrogen	Argon & Nitrogen	Nitrogen
Operating Pressure	0.1Pa		0.1Pa	1Pa	0.1Pa
Activating Species	[–Low Energy Primary Electrons––]			Primary & Secondary electrons	Secondary electrons
Target Material	Arc-eroded solid	[–– Molten Pool in a Heated Crucible ––]			
Equipment Features	Target can be inverted			High Deposition Rates	
Type of Plasma		[–––––– High Intensity Plasma –––––––––]			
Drawbacks	Needs ignition system			Require High voltage Source	
	Target tends to spit.	Difficult to control rate.		Low reaction rate.	
Nomenclature	APD Accelerated Plasma Deposition;	HCD Hollow Cathode Deposition	BARE Biased Activated Reaction Evaporation	RIP Reactive Ion Plating	ARE Activated Reactive Evaporation
	PUSK Plasmen 11 Uskoritel				

===

3.2.2. APPARATUS

(a) The Vacuum Chamber: This is a simple glass bell jar or steel rectangular box for small scale laboratory production. The chamber must have adequate access for loading substrates and coating materials and for general servicing and cleaning. Viton or other good quality 'O' rings can be used for sealing flanges but for particulary high vacuum ($<10^{-8}$ torr), soft metal seals are preferred. Provision should exist to heat the chamber to 50°C when it is open to the atmosphere, to prevent condensation of water vapour leading to excessively long pump down cycles. Chamber lines simplify cleaning procedures. For industrial production application, fast cycle coaters have the deposition chamber attached to loading and unloading chambers by manifolds with isolation high vacuum valves. A semi-continuous in-line system is another method where a strip substrate stored in the vacuum chamber can be fed continuously over the source or a continuous system where the strip or sheet substrate is inserted and removed from the deposition chamber through air-to-air seals (Soddy 1967; Smith & Hunt 1965) as shown in Fig.3-1.

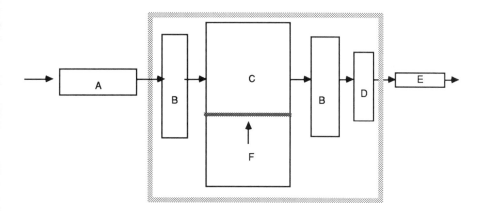

A : Degreasing & Rinsing; B : Vacuum Pumping; C : Heating Zone at high vacuum
D : Cooling; E : Rewinding; F : Electron beam input

Fig.3-1: Continuous High Vacuum Strip Processing Flow Chart

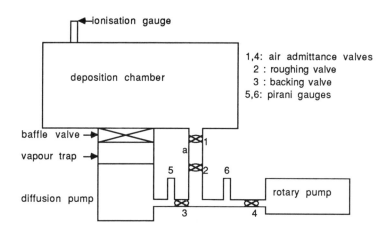

ionisation gauge

deposition chamber

1,4: air admittance valves
2 : roughing valve
3 : backing valve
5,6: pirani gauges

baffle valve
vapour trap
diffusion pump
rotary pump

Fig.3-2: Schematic Diagram of a Pumping System
for Vacuum Deposition

(b) The Vacuum Pumping System: The main problem is the out-gassing from chamber walls which is increased by the radiant heat from the vapour source and substrate heaters. The pumping system is usually based on an oil diffusion pump backed by a mechanical pump (Fig.3-2). The diffusion pump usually incorporates a baffle which can be water or liquid nitrogen cooled to prevent oil vapour entering the chamber. For high purity coatings, ultra high vacuum U.H.V. ($<10^{-9}$ torr) is needed and an ion pump can be used. Recently turbo-molecular and cryo-pumped systems have been used.

(c) Pressure Measurement: The pressure in the backing and roughing line is measured by a Pirani or a thermocouple gauge. The pressure in the vacuum chamber is measured by an ionization gauge and or a capacitance manometer and the partial pressures of contaminating gases or leaks is measured using a mass spectro-meter.

(d) Substrate Holders: Because vacuum evaporation is a line-of-sight process, intricate shapes require complicated movements to ensure a uniform coating. Hence substrate jigs can vary from simple holders to complex rotating carousels. Adhesion and struc-ture is sometimes improved by substrate heating. This is accom-plished by radiant heating from quartz lamps, or by controlled electrical heating or direct heating by a scanning or diffuse electron beam.

3.2.3. EVAPORATION SOURCES

The mode of heating used to convert the solid or liquid evaporant to the vapour phase classifies the evaporation source.

Not all materials can be evaporated from the different types of sources. The reason could be chemical interaction between the source material and the evaporant which would lead to impurities in the deposit or insufficient power density to attain a tempera-ture sufficient to cause evaporation of the coating material. Evaporation of alloys and compounds pose additional problems and can be overcome by varying the evaporation source and use of a plasma respectively. This will be discussed later.

In general, tables of recommended vapour sources for various coating materials are published by coating equipment manufactu-rers. A list is also included in the book by Maissel & Glang (1970).

3.2.3.1. Resistance Heated Sources:

The simplest are cheap resistance heated wires and metal foils of different types through which current is passed to produce joule heating (Fig. 3-3). They are usually made from the refractory

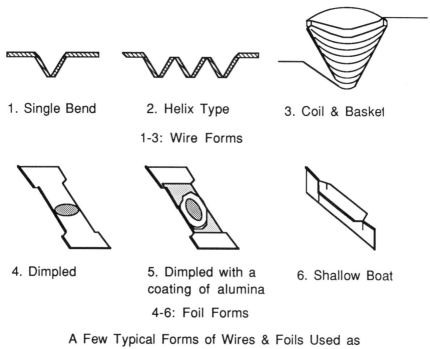

1. Single Bend   2. Helix Type   3. Coil & Basket

1-3: Wire Forms

4. Dimpled   5. Dimpled with a coating of alumina   6. Shallow Boat

4-6: Foil Forms

A Few Typical Forms of Wires & Foils Used as
Resistance Heated Sources

Fig.3-3

metals, W, Mo and Ta with low vapour pressure so as not to contaminate the deposit. Pt, Fe or Ni are sometimes used for materials which evaporate below 1000°C and are loaded with the evaporant material.

Considerable variation in the rate of evaporation from such sources occurs due to localized conditions of temperature variation, wetting, hot spots etc. So for a given thickness of film, the procedure is to load the source with a fixed weight of evaporant and evaporate to completion or use a rate monitor and/or thickness monitor to obtain the desired evaporation rate and thickness. Where reaction of the evaporant and metal source is likely, it can be circumvented by employing different oxides and other compounds which are more stable than the metals as the containing crucible. For instance, aluminium which reacts with refractory metals is contained in a boron nitride - titanium diboride boat through which current is passed.

For materials evaporating above 1000°C, which sublime to produce a sufficiently high vapour pressure before melting, the contact area between the evaporant and source crucible is held to a

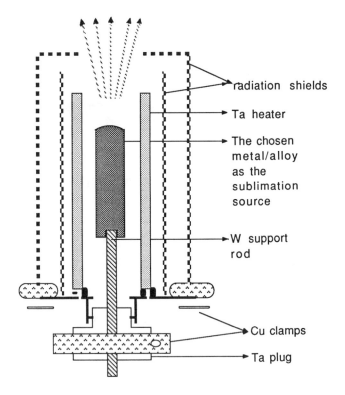

Fig.3-4: A Typical Sublimation Source for a
Refractory Metal or Vaporizable Ceramic

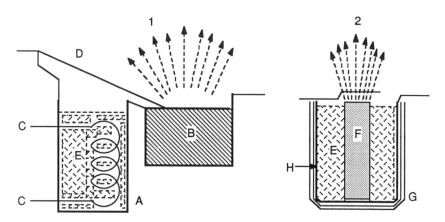

A - alumina crucible; B- tantalum box; C - tantalum wire heater;
D - reflecting hood; E - silica source; F - perforated tantanlum heater tube;
G - tantalum radiation shields; H - tantalum crucible

Fig.3-5: Silica Evaporation Sources; Optically dense.
1: Compartmentalised source; 2: The Drumheller source.

minimum (Fig. 3-4). In the case of the thermally stable SiO_2 which is usually in the form of powder, heating releases gases which cause ejection of particles which get incorporated into the film. Reflection of the vaporized material is another type of sublimation which overcomes the problem (Fig.3-5).

3.2.3.2. Radiation Heated Sources:

Reactive evaporants are held in oxide crucibles. Wire-coil heaters wrapped round the crucible are used to heat the crucibles.

3.2.3.3. Induction Heated Sources

Oxide or $BN-TiB_2$ crucibles can be induction heated. Water cooled RF coils of suitable number of turns and power input allow manipulation of the hot zone and the temperature. In cases where sublimation evaporation is required, copper crucibles can be water cooled and this is suitable for the evaporation of metals which react with the refractory oxides, nitrides etc., e.g. Ti, Be (Bunshah & Juntz 1966).

3.2.3.4. Electron Beam Heated Sources:

The electrons from an electron beam gun are directed to the evaporant which is usually contained in a water cooled copper hearth. This eliminates the problem of crucible contamination. Refractory crucible liners reduce heat loss along with the water cooling and so produce high coating rates. Electron beam heating has the advantage of a high power density thus enabling control over evaporation rates.

The two main types of EB-guns are those using electrons from a hot filment and those using electrons generated with a plasma.

(i) Thermionic Gun: The system consists of a hot filament cathode which emits electrons and an anode with a potential difference between 10-40kV. The assembly is placed in a vacuum chamber ($<5 \times 10^{-4}$ torr) so that negligible ionization occurs.

Basically the anode may be the evaporant in which case the gun is "work accelerated" (Fig. 3-6a); if the anode is located fairly close the cathode forming an intergral part of the gun then it is a "self accelerated" gun structure (Fig. 3-6b). The latter is now the most commonly used type; electrons leave the cathode surface, are accelerated by the potential difference between the cathode and anode, pass through the hole in the anode and continue onward to strike the work piece. The main advantage of thermionic guns is that they are easily focussed.

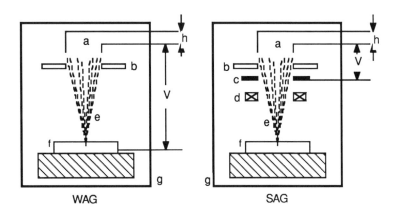

WAG - Work-accelerated gun; SAG - self-accelerated gun
a: filament (cathode); b: electrostatic focus plate; c: anode;
d: electro-magnetic focus coil; e: electron beam; f: workpiece;
g: vacuum chamber; h: filament heating voltage; V: High voltage

Fig.3-6a,b: Simple Models of Electron Beam Guns

There are 2 main types:

(a)The Transverse Gun: This uses a linear filament as cathode. In the most common type, the cathode is hidden from direct line-of-sight of the molten pool and the electron beam is bent by electromagnets and the beam can be scanned.

(b) The Pierce Gun: The cathode is either a disc shape filament for high power or a hair pin filament for low power. Fig. 3-7 shows the beam geometry. The gun is usually horizontal and the beam is bent electromagnetically through 90° into the crucible containing the coating material. Sometimes, the electron source is in a separately pumped chamber to keep its pressure below 1 millitorr, with a small orifice into the crucible chamber for the passage of electrons.

(ii) Plasma Electron Beam Gun: A glow discharge plasma can be defined as a region of relatively low pressure and low tempera-ture gas in which a degree of ionization in a quasi-neutral state is sustained by the presence of energetic electrons. When an electric field is applied to an ionized gas, energy is transfered more rapidly to the electrons than to the ions and the electrons can be extracted from the plasma to obtain an energy beam.

In hollow cathode guns a hollow cylinder with an aperture con-tains an inert gas at 10^{-3} to 10^{-1} torr which is introduced after evacuating the system. The cylinder forms the cathode. A glow discharge is struck between the cathode and earthed parts of the equipment at a particular pressure and voltage. Positive ions

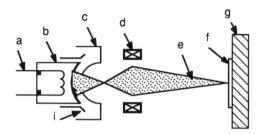

The gun is normaly horizontal and the beam is bent
at right angles to the work piece.

a: heater filament; b: cathode; c: anode; d: magnetic coils
e: electron beam; f: substrate; g: substrate holder; i: shield

Fig.3-7: Line Diagram of a Pierce Gun

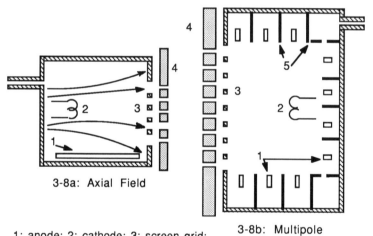

3-8a: Axial Field

3-8b: Multipole

1: anode; 2: cathode; 3: screen grid;
4: accelerator grid: 5: magneticaly
permeable pole pieces

Fig. 3-8a,b: Schematic of Industrial Broad-Beam Ion Sources

caused by collisions in the gas bombard the cathode releasing electrons which emerge from the aperture. Teer (1983) found that at high voltages and low argon gas pressures, the electron beam is fine and collimated by the action of the focusing effect of the positive ions, while in the reverse situation a "fanned" beam emerges from the aperture and an additional electromagnetic lens is needed for focusing. The values of the voltage, the gas pressure and the dimensions of the hollow cylinder and aperture are inter-related.

Plasma guns are increasingly used in vacuum operation systems because of their ability to operate at higher pressures. They are useful for gas scattering evaporation, reactive evaporation and ion plating. Hot hollow cathode (HHC) is another development in heating devices in this area. The term refers to a cathode with a more-or-less cavity geometry, such that a plasma region is enclosed by walls held at the cathode potential. Within the 'cavity' electrons are emitted and reflected from the cathode wall and a more sustained discharge in the arc mode is realised. This is considered as a major advantage for a coating device, because the huge amount of ions produced via electron impact result in high coating adhesion. Further extension of this technology to modify coating microstructures is considered promising (Kuo, Bunshah & Okrent 1986).

3.2.3.5. Broad-Beam Ion Sources:

The major obstacle for a wider application of a broad-beam ion source is due to operational problems, although it was brought into exploratory use as early as in the 1960s. The system is termed 'broad-beam' beacause of the use of many apertures in parallel to generate the ion beam. The multi-aperture arrangement allows much higher ion currents to be attained than is possible by single-aperture ion optics. Initially the broad-beam ion source was used for simple material removal, etching or deposition. More recent applications include reactive etching and deposition of compounds under various background gas media. Fig.3-8a,b show the schematic arrangement of axial field and multipole broad-beam ion sources used in industry (Kaufman 1986).

3.2.3.6. Arc Sources:

High current electrical discharges are suitable heat source for evaporation. Arcs have been used both in vacuum and reactive evaporation. The cathode-spot arc is a high current low voltage electrical discharge in which a considerable fraction of the conducting medium consists of ionized cathode material which is generated at areas known as cathode spots. Cathode spots are highly luminous regions of high current concentration on the cathode surface. Cathode-spot arc deposition achieves metallic plasma production concentrated in minute areas near the cathode surface which produce fully ionized energetic plasma jets directed away from the cathode surface. The plasma jets carry 7-12% of the arc current. By modifying aperture geometry and applying collimation the spatial distribution can be controlled (Lindfors et al 1986, 1987; Boxman & Goldsmith 1987).

The conventional arc evaporation process involves a random moving cathode spot. Control of the movement of the arc on the cathode has led to the steered arc evaporation technique (Boelens & Veltrop 1987). Magnetic enhancement of cathode arc deposition is claimed to achieve better adhesion (Sanders & Pyle 1987; Sanders 1988). The technique can be used to deposit a wide variety of coating compositions. The process allows control of ion concentration of the evaporating material and does not require added process gas, thus rendering it viable for UHV applications. Refractory metals can be deposited with a fine grain structure but care is needed for low-melting metals such as Al and Cu to avoid undesirable microstructures. TiO_2 and TiN films deposited by cathodic arc evaporation were found to be superior in quality – structure and adhesion. OES revealed that the excited Ti-ions present in a Ti-arc discharge have a higher velocity than the neutral Ti-emitters. Magnetic field application was used to focus the plasma stream, to enhance deposition rates, to accelerate ions and to modify film properties (Martin et al 1987).

3.2.3.7. Laser Beam Sources:

Although laser beams can be used as a heating source, problems arise since evaporation deposition on the laser window can block the path to the molten pool. Moreover the laser energy can be reflected from bright objects such as aluminium. Using laser sources for this mode of heating proves inefficient, but a number of adaptations in its use as a tool for deposition have been made in recent years.

3.2.4. EVAPORATION TECHNIQUES

The production of different coating morphologies and compositions by evaporation, not only depends on the vapour source but also on whether a plasma is used, or whether the substrate is biased. Fig. 3-9 shows the different evaporation techniques. A review by Bunshah (1981) outlines the important parameters. The technique of flash evaporation has been successfully used to deposit a variety of materials – metals, alloys, metal-dielectric mixtures and compounds. Experimental work has indicated that alloys can be deposited with component vapour pressures differing by even a factor of 5000, but not when one of the components is an oxide, e.g. $Ni-ThO_2$. Intermetallic compounds such as GaAs, PbTe, InSb etc., consisting of constituent elements with low m.p. and high vapour pressures are best suited to flash evaporation or sputtering; either of these methods allow careful stoichiometric conrol of the ratio of cations to anions. Several techniques of

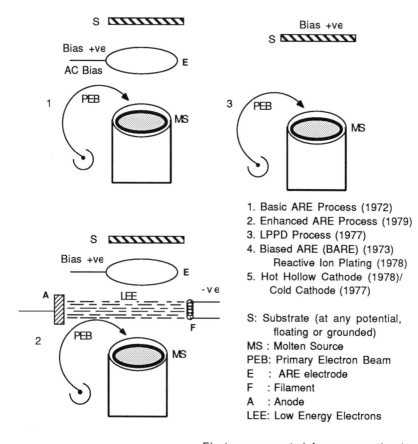

1. Basic ARE Process (1972)
2. Enhanced ARE Process (1979)
3. LPPD Process (1977)
4. Biased ARE (BARE) (1973)
   Reactive Ion Plating (1978)
5. Hot Hollow Cathode (1978)/
   Cold Cathode (1977)

S: Substrate (at any potential,
   floating or grounded)
MS : Molten Source
PEB: Primary Electron Beam
E   : ARE electrode
F   : Filament
A   : Anode
LEE: Low Energy Electrons

Electrons generated from a negative heated filament are drawn into the highly positively charged molten pool, thus maintaining it molten.

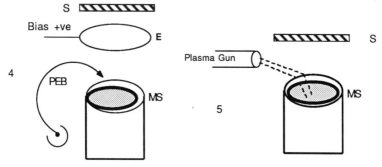

1. Bunshah & Raghuram; 2. Yoshihara & Mori; 3. Nakamura et al;
4. Bunshah & Beal; Kobayashi & Doi; 5. Vomiya et al: Zega et al

Fig.3-9: Basic ARE Process & Later variations

powder feed for flash evaporation are illustrated. Belt, disc and magazine powder-feeders, worm-drive with mechanical vibrator, electromagnetically vibrated powder dispenser etc., are some of the variations (Bunshah 1981).

Refractory compounds, apart from a few exceptions, have characteristically high m.p. and tend to form deposits over a wide range of stoichiometric compositions resulting in an extensive defect structure, e.g. the C/Ti ratio in a TiC deposit can vary from 0.5 to 1.0, with vacant carbon lattice sites. There are two types of evaporation processes for refractory compounds, viz. direct evaporation, where the refractory compound itself acts as the evaporant source, and reactive evaporation (RE) or activated reactive evaporation (ARE) where a metal or its compound with low valency is evaporated in the presence of a controlled amount of a reactive gas permitting a compound formation and subsequently a deposit, e.g. Si or SiO in O_2 to form silica, and Ti in N_2 to form TiN (Bunshah 1981).

3.2.4.1. Direct or Vacuum Evaporation:

This is the basic evaporation technique whereby a vapour source is vaporised by a high power density heat source in a vacuum chamber followed by deposition on a preheated substrate. The deposition rate can be varied. For high rate vapour deposition an electron beam heated source is used. Single elments can be deposited using a single vapour source. The choice of source depends on the melting point of the metal, its vapour pressure and the compatability of the molten metal with the source material.

In case of alloy coatings using a single source the rod fed electron beam gun is the most widely used technique. At first, the constituents in the rod with the highest vapour pressure are preferentially evaporated and as a result the molten pool becomes deficient in such constituents. This continues until an equilibrium state occurs where the composition of the rod and pool become the same as the vapour leaving the pool. The gun conditions must be kept constant in this method. Alloys with components whose vapour pressure ratios vary by less than 10^4 can be successfully deposited. However, in the deposition of MCrAlY coatings, it has been found necessary to add excess Y to the molten boat to obtain the desired composition.

Alloys can also be produced by an alternative method using multiple sources. The material evaporated from each source can be a metal, alloy or compound. Fine dispersed particles, e.g. Al_2O_3 can be codeposited to give stronger materials. Electron beam evaporators can coat steel strip 80 cm wide by 0.1 to 1 mm thick at 200 m/min using four 600 kW electron guns.

The Electron Beam PVD (EBPVD) process for batch process coating of gas turbine blades used a single rod fed EB evaporation source which continuously evaporated CoCrAlY (Hill & Boone 1982). Electron beam PVD CoCrAlYTa and NiCrAlYSi are more protective than the plasma sprayed versions but a high vacuum is needed, unlike for sputtering. 25 microns per minute is possible but there is a line-of-sight limitation. A detailed description of multiple coating operation is given. The application of EB-PVD for multi-blade coating on a large scale is described by Hill & Boone (1982) and several million gas turbine blades for aero-engines have been successfully coated. Power supplies of 100 to 200 KW are used. The thermodynamics of the distribution of elements in the pool and in the vapour has been studied (Krutenat 1972). The vapour cloud shape, as well as its density and composition are important factors. Waste deposits (overspray) are collected on a screen and re-cycled.

The capital cost of an evaporation unit is high and a large throughput is needed to keep costs low (Boone et al 1979). The equipment is complex due to the controls needed, e.g. to keep the molten pool in a constant state (Nimmagadda et al 1972) and to manipulate the specimens inside the vacuum (Scheuermann 1978; Halnan & Lee 1983) since the process is a line-of-sight one. When the vapour pressure varies by more than 100 times, e.g. MCrAlY, there is a difficulty of reproducibility of coatings (Gupta & Duvall 1984).

Plasma spraying does not have some of these problems but manipulation within the vacuum is still needed for coating complex shapes. In addition the blade needs polishing from the as-sprayed 6 micron surface finish to 1 micron for engine use (Restall & Wood 1984). A few other PVD techniques are available or are in development and avoid some of the above limitations. These are ion plating (Boone et al 1977), sputter plating (Patten et al 1976) and sputter ion plating (Hecht & Wright 1976; Dugdale 1977). Disadvantages of EB-PVD are occasional spits from the pool and columnar growth defects. Their presence is difficult to detect and degrades coating properties (Schmitz 1977).

The ratio of coating thickness in non-line-of-sight regions to line-of-sight regions is greatly improved by proper use of inert gas for scattering, but deposition energy is reduced and hence bond strength. One study found an improvement in coating quality with deposition rate (Hill & Boone 1982) but other results are the opposite and more research is needed on the kinetics of depositing atoms etc.

The disadvantages of the multiple sources method lie in controlling the deposition rate from each source and the relatively large distance between substrate and evaporants needed for blending vapour streams before deposition (38 cm for a 5 cm diameter source), Fig. 3-10. This decreases the deposition rate. By evaporating each component sequentially, multi-layered deposits which are homogenized after deposition are produced. But again this arrangement suffers from a low rate of deposition.

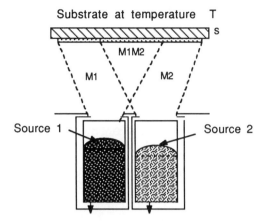

Substrate at temperature T

Composition of M1M2 variable
via temperature control

Fig.3-10: Twin-source coating technique
(evaporation method)

Surface alloying of mechanically activated strip steel using Al or Ni vapour deposition is facilitated by the high concentration of grain boundaries (rapid diffusion paths) (Schiller et al 1982). A rotating steel brush in vacuum gives such mechanical activation. Compounds are not successfully produced if direct evaporation is used. Usually on evaporation of the compound dissociation occurs and on recombination at the substrate surface, the resultant compound is not stoichiometric.

A flow sheet of processing steps is given in Table 3:4.

3.2.4.2. Activated Evaporation:

Use of a plasma gives the activated evaporation technique which has been reviewed (Bunshah 1981; Bunshah & Deshpandey 1985,1986).

3.2.4.3. Reactive Evaporation:

This process was developed more than 20 years ago (Conf.1974; US Patents - Brismaid et al; Auwarter) for the deposition of thin films of compounds at low deposition rates. The evaporation process is the same as in direct evaporation except for the presence of a reactive gas in the chamber. Due to the fact that the partial pressures of the reacting species are low ($<10^{-5}$ torr) and the mean free path therefore longer than the source-to-substrate distance, reaction between the metal atoms and the gas atoms occurs only on the substrate, e.g.

$$2Al(vapour) + 3/2\ O_2(gas) = Al_2O_3(solid\ deposit).$$

67

ELECTRON BEAM PVD PROCESS FLOW SHEETS

1.Receive Substrate

2.Scrutinise & Group

3.Prepare Surface: a.Degrease b.Grit blast c.Vapour clean
              (d. abrade/strip where required)

4.Weigh:(for reference on thickness and process control)

5.Load onto mounts and Fix

6.Mask

7.Coating Cycle Set Controls: a.Pre-heat b.Deposit c.Cool

8.Unload from fixtures

9.Unmask

10.Weigh

11.Remove overspray

12.Apply peening: (stress relief; texturing)

13.Diffusion heat treat; (substrate alloy solution)

14.Ageing heat treat (where specified)

15.Inspect; Document ; Despatch

16.Recycle if defective: Go to 2

3.2.4.4. Activated Reactive Evaporation:

This method is used to achieve high deposition rates of compounds (Bunshah & Deshpandey 1985; Bunshah 1974; Greco & Baldauf 1968). The partial pressure of metal vapour and gas atoms used is 5×10^{-4} torr or higher. At these pressures the mean-free-path is smaller than the source-to-substrate distance and collision of the reacting species occurs in the gas phase. This could lead to non-stoichiometric compounds; however, by activating the metal and gas atoms, the reaction probability on collision is increased leading to compound formation with the desired stoichiometry, e.g. $2Ti + C_2H_2 = 2TiC + H_2$, where the carbon/metal ratio of TiC can be varied (Bunshah 1974; Bunshah & Raghuram 1972).

Ti evaporated from a rod-fed source by an electron beam was reacted with C_2H_2 (at about 10^{-3} torr) between the source and a Ta substrate at 400 to $1000^{\circ}C$ (Jacobson 1981). The deposition rate was 1 to 3 microns/minute and the Ti:C ratio was 0.8 to 0.9. A very fine grain structure was obtained below $700^{\circ}C$ and coarse above, but of similar hardness. Evaporation of V-Ti in C_2H_2 to give VC+TiC was also studied. Ti-Ti$_2$N-TiN coatings for cutting tools have also been produced by ARE (Jacobson 1981).

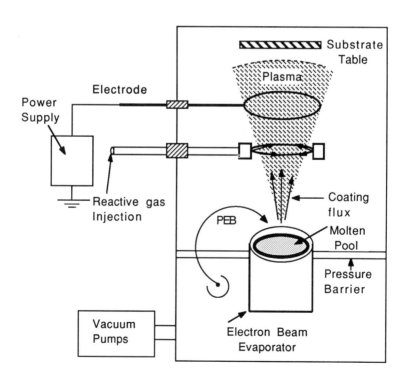

Fig.3-11: Schematic of Equipment for ARE

Fig. 3-11 shows a schematic diagram of this process. The molten metal, heated by a high acceleration-voltage electron beam (thermionic electron beam gun) has a plasma sheath on top of the pool. The low-energy secondary electrons from the plasma sheath are pulled upwards into the reaction zone of an electrode placed above the pool biased to a low positive dc potential (20-100V). The low-energy electrons have a high ionization cross-section thus ionizing or activating the metal and gas atoms.

Fig.3-9 (p.64) shows the variations in the basic ARE process which could be due to different vapour sources or different methods of generating a field, and also includes the use of a plasma electron beam gun instead of the thermionic eletron beam gun. This can be either a hot hollow cathode gun as used by Komiya et al (1978) to deposit TiC films or a cold cathode discharge electron beam gun used by Zega et al to deposit titanium nitride films (1977).

In the Biased Activated Reactive Evaporation process (BARE), the substrate is usually negatively biased to attract the positive ions in the plasma. This process was later called reactive ion plating by Kobayashi & Doi (1978). TiC on Mo by activated reactive evaporation using a hollow cathode is affected by substrate temperature and the gas pressure (Inagawa et al 1982). Stoichiometry can be achieved, necessary for use as a fusion reactor first wall coating (Mullendore et al 1981; Rao & Kaminsky 1981).

A plasma assisted deposition process designated RF Reactive Ion Plating was developed and used by Murayama (1975) to deposit thin films of In_2O_3, TiN and TaN. A resistance or electron beam heated evaporation source is used and the plasma is generated by inserting an rf coil electrode of aluminium wire in the region between the evaporation source and substrate.

The enhanced ARE process (Yoshihara & Mori 1979) is the conventional ARE process using electron beam heating with the addition of a thermionic electron emitter, e.g. a tungsten filament, for the deposition of refractory compounds at lower deposition rates as compared to the basic ARE process. The low energy electrons from the filament sustain the discharge. Many more papers have appeared during 1986-1988 in the several conferences listed, mainly technique modifications to suit component coatings in the fields of VLSI and surface mounting technology.

3.3. SPUTTERING PROCESS

3.3.1. THE PROCESS IN GENERAL

This is a momentum transfer process in which a fast particle, e.g. Ar^+, ejects an atom from a (usually) cathodic surface. The sputtering yield is the number ejected per incident particle (ion). Only about 1% of the bombardment energy gives sputtered (ejected) atoms and about 75% causes target (cathode) heating. Yield increases with mass and energy of the incident particle until it dissipates too deep, and it also increases (by about a factor of 2) with off-normal angles of incidence up to 70° and then decreases. Yield increase is disproportionately fast with dose, due to surface damage progressively reducing bonding; it decreases with gas pressure, due to backward scattering (Mattox 1982). Agglomeration of low yield materials can lead to surface cone deposits (Thornton 1982). Table 3:5 lists the process features. Table 3:6 gives a comparison of EB-evaporation, ARE and sputtering processes.

The substrate is placed near the cathode so that the sputtered atoms will coat it. Generally, sputter films have compressive stress whereas evaporative deposited films have a tensile stress, often near yield point, according to X-ray lattice studies. Internal stress in sputtered Ni films is compressive at low sputtering angles, and then becomes tensile with a tensile maximum at 45° (Shinzato & Kuwahara 1982). Increasing gas pressure may reduce stress by allowing crystallised films. Low stress is good for adhesion but compressive stress may reduce cracking propagation (Mattox 1982). There are few studies of hardness and yield strength.

TABLE 3:5

FEATURES OF A PVD SPUTTERING PROCESS

Type	1	2
Source of Power	DC or RF	Magnetron
Gas Medium	Argon & Nitrogen	
Operating Pressure	0.1 - 1Pa; 0.75-7.5 microns Hg	
Activating Species	Secondary electrons	Electrons in spiral paths
Target Material	Cooled or Heated Solid	
Equipment Features	Target can be inverted	
Type of Plasma	Low-intensity	
Drawbacks	Deposition Rate Limited	
	Small volume	Low target yield
Nomenclature		HRRS (High Rate Reactive Sputtering)

===

SOME CHARACTERISTICS OF THREE PVD TECHNIQUES

	Electron Beam Evaporation	Sputtering	Activated Reactive Evaporation
Rate of Deposition (microns/min)	25	9	1.2-3
Source/Substrate Distance		4-6 cm	20-25 cm
Electron Energy		50-100 eV	5-20 eV
Coating Deposition & Composition	Substrate is pre-heated in vacuum to 1000°C prior to exposure and held at that temp. when being coated.Using single source eva-ration of elements with considerable vapour pressure differences can be tolerated. Using multiple evapora-tion sources graded coat compositions can be achieved. Fine dispersed par-ticles can be co-deposited to give stronger materials.	Allows low temp. deposition of almost any kind of material and composition. But intermediate compound film synthesis is very difficult.	Substrate temp. 400-1000°C. High deposition rates and intermediate compound film sysnthesis is carried out well. A number of compounds and composites can be deposited,e.g.Ti-Ti_2N-TiN.
Structure	Fine, as-deposited structure at lower temp. Coarse grain & more substrate inter-diffusion at higher temp. Coat-ings deposited on rotating substrates often result in colunar grain with unbonded interfaces called "leaders". This can be over-come by increasing dep. temp. & by inc. substrate-to-vapour flux angle.	Suitable for the synthesis of ex-tremely fine-grain or amorphous stru-ctures, non-equi-librium phases, & non-reacting, multi-layered structures. Amorphous metals deposited are superior in corro-sion resistance.	Fine & coarse struc-tures as in EBPVD. Fine cavities may be included in fine-grain structures. Intermediate temp. deposits can yield bimodal structures, e.g. at 550°C Ti is interspersed with subgrain Ti_2N but with no cavities.
Post Coating Processes	Surface peening with 200 micron glass beads follow-ed by heat treatment results in complete closure of leaders.		Equilibrium phase composition structures remained unchanged during high temperature annealing.

To obtain reproducible coatings results in sputtering and other PVD processes, the following variables must be controlled (Mattox 1982; Thornton 1974): system geometry, initial vacuum, substrate/source distance, pre-conditioning, gas purity, pressure and flow rate, voltage and current, substrate temperature, time, system cleanliness, deposition rate. Process control is an area needing further research, including deposition rate monitoring and pressure and gas composition measurements.

The process has been reviewed extensively (Teer 1983; Thornton 1982; Stuart 1983; Bunshah & Deshpandey 1985) and its scope and limitations may be assessed. The film growth process in the sputtering technique differs from that of the ARE technique. Low deposition rates and a great difficulty in synthesising intermediate compound film is an inherent consequence of the reactive sputtering process mechanism.

3.3.2. SPUTTERING METHODS

3.3.2.1. Planar Diode Sputtering:

Fig. 3-12 shows a schematic system (Thornton 1982). The gas (usually Ar) is held between 2×10^{-2} and 1 torr. The voltage is between 1 and 5 kV, current about $1mA/cm^2$ and the cathode to substrate distance is about 5cm. Deposition rates are about 500 AU/minute (rather low) (Teer 1983, Thornton 1974) but can be increased by good target cooling (McClanahan et al 1974). Planar diode sputtering is suitable for thin complex coatings, for research and for small production volumes (Thornton 1982), its advantage being its simplicity. Use of two separated cathodes (or 'hollow cathode') improves efficiency (Stowell & Chambers 1974; Mah et al 1974) and coating growth features which relate to inter-grain shading are accentuated (Thornton 1974). An Ar ion source has been used to sputter C with a diamond structure onto steel substrates (Mirtich 1981).

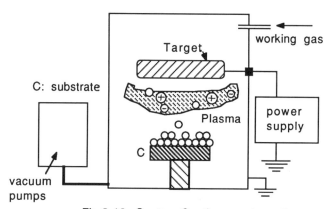

Fig.3-12: Sputter Coating - schematic

A new dc sputtering technique using Nb and W hollow targets in argon enabled the application of a high density power above $20W/cm^2$ at a low voltage, <1 kV and thus a high rate of deposition can be realized without a magnetic field (Yamamoto & Tanaka 1974). TiAlN of hardness more than double that of TiN has been deposited using a double cathode sputtering method (Munz 1986). A biplanar source for a high rate reactive sputtering of SiO_2 has also been reported (Rostworowski & Parsons 1985).

3.3.2.2. Triode Sputtering:

A heated filament and a 100 V positive plate are positioned as shown in Fig. 3-13, to increase gas ionization, resulting in ion currents of several amperes. The sputter rate is limited by the target cooling. Lower gas pressures of 10^{-3} torr (mean free path 5 cm) reduce backward scattering (Teer 1983). Deposition rates of 20000 AU/minute are reported (McClanahan et al 1974).

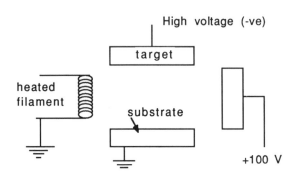

Fig.3-13:Triode Sputtering

3.3.2.3. Magnetron Sputtering.

Magnetron sputtering is perhaps the most widely investigated amongst sputter deposition methods since the application of refractory and high temperature wear resistant materials deposition found increasing industrial use (Sproul 1987; Mattox et al 1987; Schiller et al 1987; Emiliani & Richman 1987; Wert et al 1987; Konig 1987; Weissmantel et al 1986). Reactive sputtering is usually a slow process because the reactive gas poisons the target by interaction with it, and the deposition rate drops rapidly. Pulsing the reactive gas brought some improvement, viz. TiN deposition (Aronson et al 1980) with about 50% of the metal depositon rate. A closed-loop feedback control system with d.c. magnetron sputtering cathodes, using nitrogen peak height monitored by a mass spectrometer yielded better process control and eliminated pulsing (Sproul 1987; 1986). At fast pulsing rates (0.2 s on/off) TiN deposited nearly at the same rate as Ti.

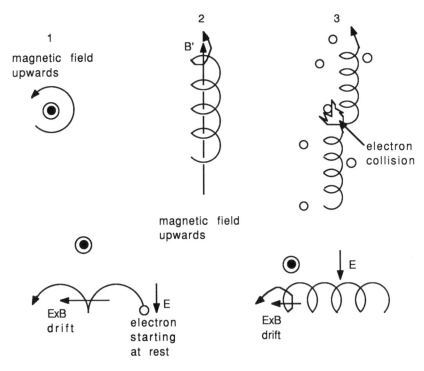

Fig.3-14: Electron Motion in Static Magnetic and Electric Fields.

In magnetron sputtering, electric and magnetic fields cause the filament's electrons to follow long non-linear paths giving many more collisions ionising more Ar. Deposition rates of 10000 AU/minute occur (Teer 1983; Thornton 1982; Gillet et al 1983). Teer (1983) also discusses the equipment design. Fig.3-14 – 3-17 show some of the salient features of the magnetron sputtering technique (Thornton 1982). It is possible to set up a large size scale-up of this equipment, e.g. continuous coating of 2m x 3.5 m glass sheets (Fig. 3-18, 3-19). Post cathodes in cylindrical magnetrons have been used to coat the insides of tubes up to 2m long (Penfold 1985). Hollow long cathodes will coat all points in it uniformly, thus being good for coating complex shape objects (Thornton & Hedgcoth 1975). High current densities of 10 micro-amps/cm^2 at pressures of 10^{-3} torr are typical. The electron trapping limits substrate damage by electron bombardment (Teer 1983). Magnetic substrates must be magnetically saturated (Thornton & Penfold 1978). Fig. 3-20 shows some current-voltage characteristics. Thornton (1974) used 10 cm diameter water cooled cylindrical hollow cathodes (targets) at about -800V, and sub-strates biased at -30 to -50 V relative to the anode. At a typical deposition rate of 1500 AU/minute the substrate heat load was 0.1W/cm^2, due to heat of condensation of coating atoms and plasma radiation. Using rod cathodes, 3 cm diameter and 30 cm

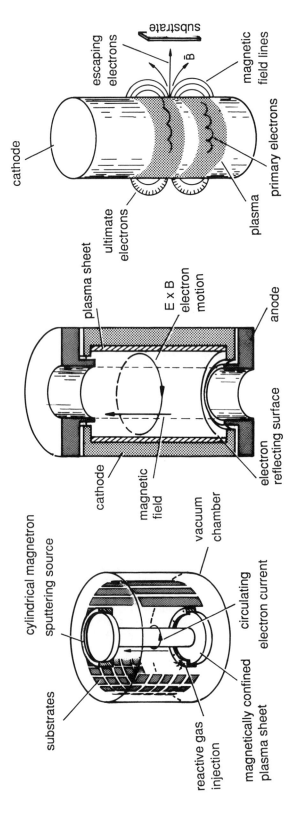

3-15: Cylinder-Post Type Showing
Substrate Alignment for
Multiple Coating.

3-16: Cylindrical Hollow Type;
Electrostatic End Confinement.

3-17: Cylindrical Type with
Magnetic End Confinement.

Fig.3-15, 3-16 & 3-17: Magnetron Sputtering Sources

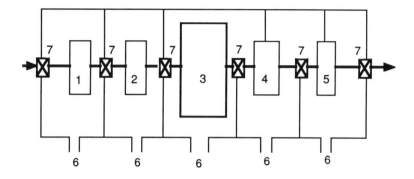

1. entrance interlock, 2. entrance buffer chamber, 3. coating chamber
4. exit buffer chamber, 5. exit interlock, 6. vacuum pumps, 7. valves

The in-line vacuum assembly is employed to achieve
production volumes with planar magnetrons.

Fig.3-18: Planar magnetron sputtering - Schematic

long, the field prevented plasma species from reaching the substrates (floating at almost anode voltage). Cathodes (targets)
were tubes of Al alloy, Cu, Ti, pressed and sintered Mo; Fe and
Cr were electroplated onto stainless steel mandrels. Substrates
were resistance heated glass, stainless steel and Ta. Liquid
nitrogen cooling can give substrate temperatures of -196°C.

A sputtering system of rectangular facing targets with a magnetic field normal to the targets was studied (Nakamura & Hosakawa
1983). Theoretical calculations of the optimum target-target
distance was 400mm and the target-substrate distance was 100mm
for 5 inch x 15 inch targets systems. Design advances in rota-
table cylindrical magnetrons have been reported (Wright & Beardow
1986). High sputtering efficiency of Ni with a dc planar magnetron has been achieved (Chang et al 1986). A triode magnetron
system has been formulated by adding a hot hollow cathode source
to a conventional planar magnetron system (Cuomo & Rossnagel
1986). Kuo et al (1986) review the advantageous features of HHC
in vacuum deposition. A planar magnetron sputtering process can
have variable throwing power and shadowing effects on substrates
of irregular shapes (Stowell et al 1985). Window & Sharples
(1985) discuss the increase in yield in magnetically enhanced
sputter deposition using a modified Penning geometry for the
sources. Fe, Nb, Mo, Ta, W and Pt have been deposited onto substrates at room temperatures at 1 Pa argon pressure (Window et al
1988). Falcone (1987) presents a new formulation of collisional
sputtering.

Helmer & Wickersham (1986) indicate the significant effect of
argon pressure in the system on the entire coating quality and
deposition efficiency. A more uniform thickness and improvement

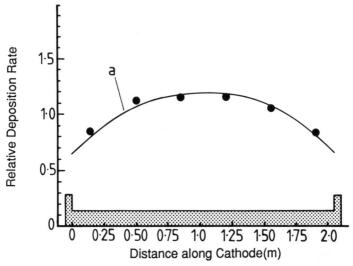

Fig.3-19: Deposition profile Parallel to the Cathode Axis
at Radius 0.86 m. (line - theoretical; points - experimental)
Cylindrical-Post Magnetron, Cr-cathode (2.08 m long, 0.13 m dia.)
Deposition conditions: 20A, 840 V, Magnetic field 50G(0.05T),
Argon at 1 m Torr (0.13 Pa).

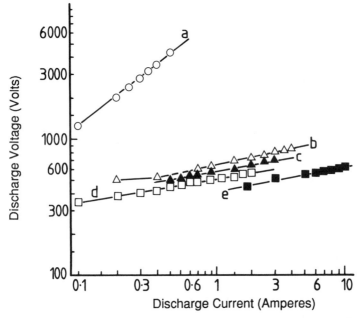

Fig.3-20: Curent-voltage Characteristics of Sputtering Sources
- Using argon as working gas & operating with Al targets.
a. Planar diode (6.5 Pa) vs Magnetron (all at 0.13 Pa)
[Planar -c:Ring-type; e:Rectangular-type; Cylindrical - b - Post; d - Hollow]

in deposition rates were realised with lower argon pressures. One to three layer coatings of uniform thickness and quality has been reported by using a unique movable post-cathode with a stationary substrate (Mattox et al 1987). Stabilization of the sputter process by means of a plasma emission monitor has been advocated by Schiller et al (1987). Plasma shield with aperture limitation or multiple apertures, or floated plasma shield arrangements have been demonstrated. The self-stabilizing operation of the discharge is suggested to be by a getter action similar to that controlling the operation of HHC . The target and substrate are surrounded by a closed wall to give a plasma-confining reaction chamber. Hoffman (1985) reports on experiments with a freewheeling fan blade in the path of energetic particles.

Sputtering has an advantage over both CVD and metalliding, which require process temperatures exceeding the steel substrate tempering temperature. Sputtering is also adaptable to tubular geometry and has been used for coating inside steel tubes; it generally operates at lower substrate temperatures. Atomic cleaning by intense ion bombardment at 0.6 microns/minute before sputtering Ta and Mo in low pressure Kr was succcessful. Adherent dense coatings resulted (McClanahan 1982).

A coating of 304 stainless steel with excess carbon was deposited by the magnetron sputtering technique. The substrate was float glass at room temperature. The deposition rates directly under the sources were: 0.6AU/second for C and 8.1 AU/second for 304SS (304SS has the composition: 18.1%Cr-7.6%Ni-1.6%Mn-0.9%Si-0.02%C-Fe). The chemical composition of the deposits was exactly the same as the 304SS target material except for the carbon concentration. The structure was however, ferritic bcc whereas that of the target material was austenitic fcc. With excess carbon the structure is changed dramatically; the crystalline-to-amorphous transition occurs with increasing carbon concentration at about 5 at% C. The crystallization temperatures of these amorphous structures increases with the carbon content, from 377°C close to the transition zone at 6 at% C to between 450-500°C at 15 at% C. A transition from ferritic to austenitic occurs at about 575°C.

Magnetron sputtered Cr in Ar (0.002 torr) + methane or nitrogen has microhardness up to 3500 HV (500g load), with fretting wear, fatigue and corrosion resistance better than electrolytic hard chrome. The power was 2 kW and target-substrate distance was 5 cm and 200 to 400°C substrates were negatively biased to -100 volts (-2 kV for pre-cleaning). Deposition rates were 0.5 microns per minute. Coatings were columnar with some surface defects. X-ray analysis showed supersaturation by -C and -N in Cr up to several wt%, with no nitrides or carbides (Gillet et al 1983). The carbon and nitrogen compound content induce better properties and the coating could replace electrolytic Cr in some uses.

The changes in characteristics of TiC coatings by a magnetron-sputtering process onto Mo was investigated (Shinno et al 1982). Without Ti addition, the deposited films were excessively rich in carbon. The stoichiometric or Ti-excess films deposited with Ti addition were more resistant to thermal cycling. The films containing large amounts of Ar deposited with Ti addition were apt to form blisters and exfoliate during the heating tests. Erosion resistant $NiCr_3C_2$ coatings on IN 718 and Ti-6Al-4V alloys have been characterized (Wert et al 1987). Hard and protective amorphous borides of W-Re and Mo-Ru have been sputter coated onto AISI 52100 steel (Thakoor et al 1985).

The deposition of hard TiN films on tools can be realized using high rate sputtering techniques (Munz et al 1982). The optimum coating process parameters are discussed and the principles of a high throughput production plant with cycle times of about 60 minutes is demonstrated. High rate sputtering of TiN below 550°C allows coating of tools after the final machining operation; higher temperatures required for CVD cause structural changes in the steel. A large scale production plant schematic is given. Nitrides of Hf, Ti, V and Zr have been extensively investigated (Konig 1987; Sproul 1985). Danroc et al (1987) report on the prominent part played by the pumping speed of the system and the gas inlet geometry in TiN deposition by reactive magnetron sputtering. NbN films by dc and rf reactive magnetron sputtering has been reported (Bhushan 1987).

Heat reflecting films on architectural glass are important in reducing energy consumption. High rate sputtering is employed for film systems of oxide/metal/oxide sandwich (Munz et al 1982). The deposition process must be carefully controlled especially to avoid corrosion of the metallic film during the deposition of the top oxide film; oxidation of metallic constituents in the film, e.g. Ag or Cu, causes higher absorption of the film system as well as a loss on long term stability.

Highly corrosion resistant films of amorphous $Co_{100-x}Nb_x$, $Fe_{80-x}Cr_xP_{13}C_7$, $Cr_{75}B_{25}$, $Ti_{75}B_{25}$ alloys were deposited by dc-toroidal sputtering on water-cooled copper substrates (Okamoto & Arata 1982). The effect of bias voltages applied to substrates, on the microstructure of rf sputtered Ni films was investigated by transmission electron microscopy and electron diffraction (Murakami et al 1982). At bias voltages of -100 V, the film is continuous and the NiO phase disappears. Deposition parameters for magnetron sputtering of ZrO_2 are described (Deshpandey & Holland 1982). Double arcs formed in plasmatrons are theoretically discussed (Ionin 1982).

3.3.2.4. RF Sputtering

Insulating substrates build up charge and cannot be sputtered by dc methods. Details are given in reviews (Teer 1983; Thornton 1982). A negative self-bias (about -40 V) appears on any surface capacitatively coupled to the Ar glow discharge, due to the mobility difference between electrons and Ar^+ ions (Butler & Kino

1963). This is because the electrons are more mobile and so more negative current would enter during the negative half cycle, but since the net dc flow must be zero for capacitative coupling the surface must assume a negative bias to stop such an effect (Thornton 1982). An applied (additional) negative bias may give structural benefits, e.g. prevention of a columnar deposit for W and Cr (Bland et al 1974). This bias causes a simultaneous ion bombardment as used in ion plating (section 3.3.2.5).

The schematic diagram of rf sputtering is similar to that for planar diode sputtering (Fig.3-12, p73), except that the dc is replaced by an rf power supply. The pressure can be much less for rf, due to much less loss of electrons from the discharge plasma and a higher percentage ionization due to oscillating electrons. Magnetron sputtering equipment can also be powered by rf, but some re-design may be needed (Thornton 1982). A continuous rf sputtering apparatus was developed (Naka et al 1982) for solar thermal collector pipes of 50mm outside diameter and 4m length were coated with a ZrC/Zr film system and produced at a rate of 900 mm/h. The system shows good optical characteristics and suffered no deterioration at temperatures up to 800K for 30 hrs in vacuum.

Reactive sputtering occurs when one or more of the coating species is gaseous, e.g. sputtering of Al in O_2 to form Al_2O_3, Ti to form TiO_2, Nb in N_2 to form NbN (Thornton 1982). Sputter deposition in UHV systems is one of the current developments (Fischer et al 1988). Diamond-like carbon films from CH_4 have been deposited in an rf system with separately controlled ion acceleration by a lower electrode, and the upper electrode controlled for dissociation and ion production. A pyrex cylinder inside a coil formed an rf oven. Highly dense carbon films, low in stress levels could be deposited by this technique (Nyaiesh et al 1985).

Advantages of reactive sputtering are:
- complex compounds can be formed from easy-to-fabricate metal targets,
- insulating compounds can be deposited with dc power,
- graded compositions are feasible.

Reactions occur on the cathode and then the reacted material is sputtered. They occur also at the substrate. The condensing film is a getter for the reactive gas. This gettering falls quickly as stoichiometry is approached. Factors affecting growth are reviewed in detail and discussed by Thornton (1982). Planar magnetrons achieve high coating rates. By injecting the reactive gas near the substrate and Ar near the target, it is possible to maintain high deposition rates for stoichiometric coatings. 1 micron per minute for Ta_2O_5 and 0.7 microns per minute for TiO_2 were reported with optical properties close to those of the bulk materials.

Pumping down a vacuum system quickly removes gases (which are physically adsorbed on surfaces, with bond energy <0.5 eV). Then water vapour is slowly removed (because its bond energy is about 1 eV) and this dominates chamber pressure during typical pumpdown times (<10 hours) at room temperature. The relation of this to factors such as the gettering area of sputtered material and consequent incorporation of gases into coatings is well discussed by Thornton (1982). These factors can lead inexperienced workers to conclude that Al metal cannot be sputtered.

Reactive rf sputtering of stoichiometric TiN for wear resistance is described (Yabe et al 1982). Optical and structural properties of metal-carbon composite materials produced by reactive sputtering of stainless steel in argon and acetylene for solar energy conversion have been described in detail (Harding 1982). Manufacturing of evacuated collectors with a selective coating of stainless steel-carbon film overlayed on sputtered copper and the applicability of magnetron sputtering for production of semiconductor films for photovoltaic devices is also described (Harding 1982).

3.3.2.5 Ion Beam Sputtering/ Sputter Ion Plating

(a) General: Ion beam sputtering allows independent control of energy and current density of the bombarding ions (Kaufman 1978). A sputtering target is held obliquely in an ion beam from an independent ion source. The substrate is placed suitably to receive the coating resulting in unique coating properties (see reviews in Harper 1978; Weissmantel 1980; Bland et al 1974; Machet 1983). Reactive ion-beam mixing, the Kr ion bombardments of $PtCl_4$ and $RhCl_3$ on a metal surface, is described (Padmanabhan & Sorensen 1982).

Sputter ion plating (SIP) is carried out in a vacuum chamber at $300^\circ C$ lined with coating source plates at −1 kV (which causes ion bombardment and consequent sputtering of material from them). The components for coating are held (in the volume enclosed by the source plates) at about −100V to ion polish their surface to promote a dense coating without significant re-sputtering. SIP is a very simple technique. A clean atmosphere (10 to 100 mtorr) is obtained using only a mechanical pump and a getter. The specimens can be cleaned by ion bombardment before the SIP begins. No manipulation of the specimens is needed. SIP has good throwing power due to the distribution of sputtered source material around the components and to the scattering effect of the low pressure gas (about 1 Pa). The nascent deposits on the substrates are ion-polished by low energy ions attracted to the surface by a small negative voltage (−100 volts). Coating rates are not high (typically 10 microns/hour) but adequate thicknesses (125 microns) can be deposited overnight onto a batch of turbine blades (Coad & Restall 1982). The costs of SIP for large numbers of turbine blades are about the same as for EB evaporation, LPPS and Ar-shielded PS. The costs of EB evaporation rise rapidly if the thoroughput is reduced, but SIP costs remain constant per item

(Coad & Scott 1982). Wear resistant TiN coatings have been deposited by the SIP process (Jacobs 1986).

Cd electroplating can be advantageously replaced by ion vapour-deposited (IVD) Al (Cd coatings on strong Al and steel alloys promote exfoliation corrosion). Al coatings can be used up to 500°C. The system is pumped down to 10^{-4} torr, backfilled with Ar to 10 microns and then a high negative potential applied between the parts being coated and the evaporating source. An ionising glow discharge bombards Ar^+ ions onto the substrate for cleaning. Al wire is then fed into resistance heated crucibles and vapour passing into the glow discharge becomes ionised. About 45 minutes is required. For coating small parts, rotating barrels were placed over the Al evaporating source. IVD Al coatings have good fatigue properties and so are used on aircraft.

(b) Other Process Features: The nucleation and growth of sputtered coatings is similar to vacuum evaporation, although the pressure is much higher. The sputtered atoms reach the substrate with about 10 to 40 eV kinetic energy, compared with 0.3 to 1 eV for vacuum evaporation. The higher gas pressure lowers the temperatures defining the zone boundaries in the Movchan & Demchishin diagram (1969), Fig.7-1, and so a higher temperature is needed to produce a dense structure (Teer 1983; Thornton 1982). This needs further research.

Alloys can be deposited if the target contains the constituents in the required ratio, e.g. cast alloy, mixed powders, strip and wires. Differences in sputtering yields are usually much less than the vapour pressure differences which severely affect vacuum evaporation of alloys. Some compounds can be sputtered from compound targets, but the stoichiometry may change (Teer 1983). Use of a reactive gas allows deposition of Al_2O_3, TiN and CdS, for example (called Reactive Sputtering). The rate is usually low because the reactive gas poisons the target surface. Using several targets sequentially creates a layered coating. The substrate temperature can be independently varied to get the best coating microstructure.

For CoCrAlY, 600°C or less gave cracked coatings. As substrate temperature was increased, the microstructure coarsened (Stowell & Chambers 1974). With Substrate at 0 V bias, the coating was coarse and columnar but with -30 or -50 V bias it was finer and less columnar. Hardness decreased with increasing temperature and X-ray showed phase changes occurring. At constant substrate temperature, floating bias gave the least texture; -50 V bias gave the strongest crystallographic texture and zero bias was intermediate. The -50 V bias gave the sharpest diffraction patterns (lowest internal stress level). In other systems compressive stress reached a maximum at about -250 V substrate bias and then decreased. Substrate biasing also affects the trapped gas (Ar) content, impurity content and bulk composition. Density increases with negative bias (Carmichael 1974; Bland et al 1974). Further discussions attribute stress relief at -15 V bias to the incorporation of about 30 at.% O and N into rf sputtered Mo films, causing an fcc structure. Floating bias (i.e. about -40 V) gives a bcc structure, negligible gas incorporation but compressive stress (Nowicki et al 1974).

Al alloy sputtered coatings (Vossen et al 1974), refractory coatings (Vossen et al 1974; Wagner et al 1974) and wear resistant coatings (Sproul & Richman 1975; Sproul 1987; Eser & Ogilvie 1978) have been described. Other properties of sputtered coatings are described in later chapters.

The bibliography given in this book includes further useful sources of information on sputtering (Chopra 1969; Maissel & Glang 1970; Lamont 1973; Francombe 1975; Westwood 1976; Thornton 1977; Vossen & Kern 1978; Chapman 1980; Green & Barnett 1982; Auciello & Kelley 1984; Sproul 1986; IPAT Proc. Vacuum 1987).

3.3.2.6. Plasma Assisted PVD Processes (PAPVD)

Table 3:7 summarises the features of evaporation and sputtering methods (Bunshah & Deshpandey 1986). Among the several forms of evaporation and sputtering methods discussed above, reactive sputtering (RS) and activated reactive evaporation use a plasma to enhance the deposition reaction and the term reactive plasma assisted physical deposition process (RPAPVD) covers these two processes. Reactive ion plating (RIP) is not basically different since it only involves positive ion bombardment from the plasma to impart modifications of the interface/growing film. In the direct evaporation and sputtering processes, the target is the same compound as the desired deposit. The configuration of steps involved may be summarised as: compound evaporation - compound dissociation into fragments - transport to substrate - recombination - nucleation and deposit/film growth. Complications set in because of problems with maintaining the required stoichiometry. Instead of compound fragmentation, constituent species are introduced in a reactive gas environment.

Theoretical studies on plasma discharges have been shown to be useful in optimising system lay-out for ion-assisted deposition systems; (i) Evaluation of cathode fall length with implications for field uniformity around components being coated and the glow penetration within holes, (ii) the ratio of cathode fall length to charge exchange mean free path which allows determination of the ion and neutral energy transportation characteristics, and (iii) ionization efficiency used to indicate the ion flux to total particle flux ratio at the cathode, are three assessments pointed out as of value in practical applications (Fancey & Matthews 1987; see also Mort & Jansen 1986).

Plasma volume reactions play an important role in film deposition by PAPVD. OES, MS and PMS methods have been used to analyse the substrate vicinity reactions (Deshpandey et al 1987). Cathode arc

TABLE 3:7

SURVEY OF PROCESS PARAMETERS & LIMITATIONS OF EVAPORATION & SPUTTERING PROCESSES

Process	Direct Evaporation	Reactive Evaporation	Activated Reactive Evaporation	Direct Sputtering	Reactive Sputtering	Magnetron Sputtering
Operating Pressure Range, torr	10^{-4}–10^{-5}	10^{-3}–10^{-4}	10^{-3}–10^{-4}	10^{-2}–10^{-3}	10^{-2}–10^{-3}	10^{-3}–10^{-4}
Substrate Temperature, $^{\circ}$C	500–1000	500–1000	room temp.–500	200–500	200–500	room temp.–500
Source-to-Substrate Gap, cm	25	25	25	5–10	5–10	5–20
Deposition Rate, Å/min	generally low; <300 for compound films	generally low 200	<500–10000	500 for compound films	up to 1000	up to 2000
Compound Dissociation	A major problem			limits dep. rates	Target poisoning	
Comments	Difficult to deposit fully stoichiometric film	High temp. reqd.; low dep.rates	High dep.rates; plasma control indep. pressure & source power.	High temp. due to energetic particle bombardment.	Interdependent reactive gas pressure, source power & surface state of target.	Magnetic confinement restricts substrate bombardment.

plasma deposition (CAPD) combines ion plating and use of plasma. CAPD-TiN provides tool life greater than arc or sputter coated TiN (Randhwa & Johnson 1987). Films of TiC and TiC_xN_{1-x} are also reported (Randhwa 1987).

An assessment of the scopes of reactive sputtering (RS) and ARE is presented by Bunshah & Deshpandey (1985). The ARE process is shown to have x12 to x30 higher deposition rates, and a larger range of products, viz. nitrides (Ti, Zr), carbides (Ti, Zr, Hf, Ta, Nb, V), oxides (Ti, Y, Al, V, Be, In-Sn, Si) and sulphides (Cu-Mo, Ti) have been synthesised. The RS method, on the other hand, appears to be more limited with only Ti and Zr nitrides, and oxides of Ti, Y, V, Si, In and In-Sn. The advantages of the ARE is indicated as due to the availability of an independent control of plasma conditions. The more updated versions published during 1987-1988 on magnetron sputtering and other sputtering devices appear to overcome many of their earlier limitations of target poisoning as well as deposition rates. A fresh assessment may be worthwhile.

3.4. ION PLATING & ION IMPLANTATION

3.4.1. ION PLATING

This technique has been used for more than two decades. Mattox, who invented the conventional ion plating in 1964, suggested a re-definition in 1982 that ion-assisted and plasma-assisted deposition be termed ion plating (Kuo et al 1986; Mattox 1982; Mattox 1964). The system is similar to the inert gas discharge used in sputtering but the substrate is the cathode. Better adhesion may result. High defect concentrations are put in the surface and physical mixing occurs of the coating and substrate surface. A higher 'throwing power' than vacuum evaporation methods is due to gas scattering, entrainment and re-deposition on the charged substrate. A comprehensive review and bibliography (1963-1980) on ion plating is given by Mattox (1982). Ion plating is the most used method for steel tool coating. Irregular shapes are coated with little variation in adhesion, thickness, composition and structure.

Ion plating with Al to protect U from oxidation has been successful in contrast to vacuum deposition which has been very unsuccessful due to the thin oxide which forms on U immediately after cleaning. In Ar glow discharge Al ion plating (Fig. 3-21), the specimen surface is sputter cleaned by positive ion bombardment and is then held contamination-free until and during the ion plating process. The high energy of ion collision with the surface heats it and promotes chemical reaction and interfacial diffusion (Bland 1968). Before ion plating U, the specimen was first deoxidised in 50% HNO_3 and electro-polished in phosphoric acid. It was then suspended from a water cooled cold finger in a

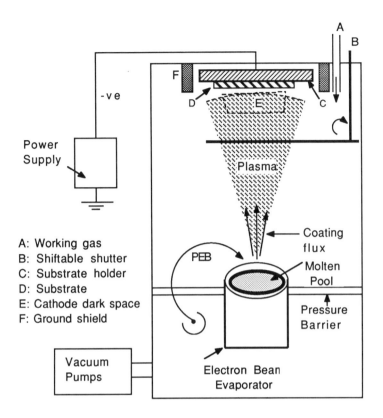

Fig.3-21: Ion Plating Apparatus: Schematic diagram

$<5 \times 10^{-6}$ torr vacuum chamber, above a W filament of the same area as the specimen. Al evaporation clips were put on the W. The filament was grounded and the specimen held at −5KV potential (drawing typically 0.5 mA/cm^2). Ar was leaked in to about 10 microns Hg pressure and the high voltage applied, to clean the substrate with Ar$^+$ ions. About 10 minutes later the filament was slowly heated to deposit the Al. Then the filament, Ar and high voltage were turned off, in sequence. An axial magnetic field can be used to increase the electron path length and thus the ionization (Mattox 1982; Bland 1968; Thornton 1980).

With a 6cm separation from substrate to filament and 25 mtorr pressure, the current density decreased from 1.2 mA/cm^2 to a reproducible and stable 0.53 mA/cm^2 when the surface had become sputter cleaned. Diffusion of the Al coating at 10^{-8} torr and 600°C was accompanied by an appearance change from typical white Al to a dull matt grey. UAl$_2$ and UAl$_3$ were found by X-ray diffraction, as well as Al. Some coatings did not diffuse in 2h while others took only 20 minutes, due to surface contaminants such as, (a) oxygen in the system due to poor vacuum, (b) W-Al compounds due to continued use of same filament.

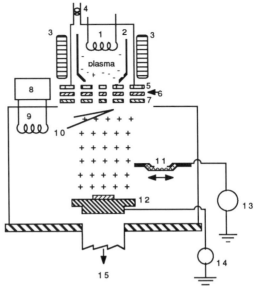

1. Heated Cathode 2. Anode 3. Magnet 4. Auomatic gas leak 5. Screen grid
(200 - 2000 V) 6. Extractor grid (-200 V) 7. Ground grid 8. Power supply
9.. Evaporator filament 10. Charge compenation filament 11. Movable shutter
with Faraday cup 12. Target bearing the Substrate to be coated 13. Current
density monitor 14. Target current monitor 15. To vacuum pumps
(A resistive heater is used for the source of depositing material)

Fig.3-22: Ion Gun Ion Plating System

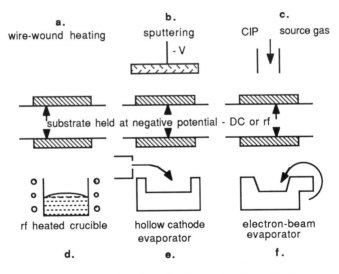

Fig.3-23: Vaporization Devices for Ion Plating

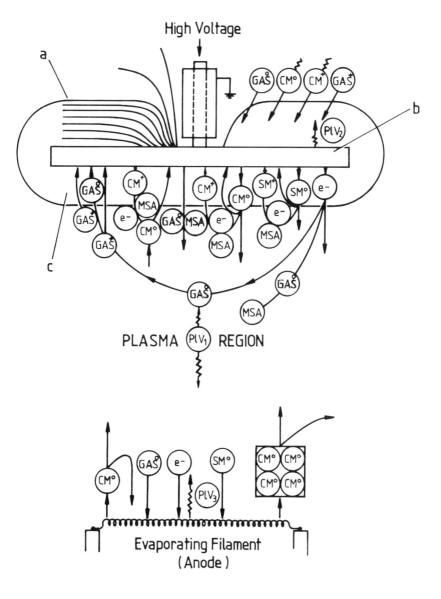

High Voltage

PLASMA (PlV₁) REGION

Evaporating Filament
(Anode)

a. Equipotential lines
b. Substrate (high
   voltage cathode)
c. Cathode dark space

e⁻	electron
MSA	Metastable Atom
SM⁺,SM⁰	Substrate Material
CM⁺,CM⁰	Coating Material

Fig.3-24: Schematic diagram of Typical Processes Occuring in
a DC Diode Discharge in an Inert Gas medium

Fig.3-22 (p.88) shows a vacuum ion gun ion plating apparatus, the maximum substrate diameter at present being about 25 cm (2 KeV, 1 mA/cm^2). The advantage is control of ion bombardment parameters independent of other deposition parameters but the disadvantage is absence of inert gas scattering. Fig. 3-23 shows various sources used to generate vapour for ion plating, Figs. 3-24 and 3-25 show the ion processes occurring and Fig.3-26 shows methods of enchancing the ionization. Ions accelerated to the cathode originate from the plasma edge, so the plasma density must be

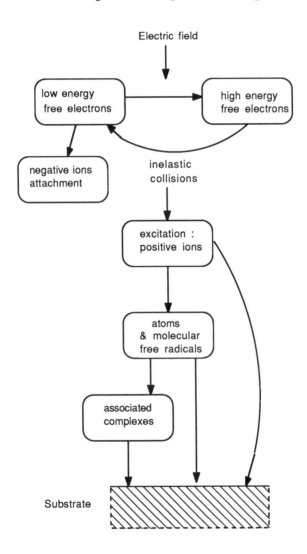

The Mode of Production of Active Species in Molecular Plasma-Schematic Representation

Fig.3-25

kept constant over the cathode (substrate) surface. Hybrid techniques follow a 10 micron thick Ti ion plated film by diffusion heat treatment and siliconizing.

Use of some O_2, N_2 or hydrocarbon gas allows the deposition of compounds, and the techniques are known as reactive ion plating (RIP) or activated reactive evaporation (ARE), already discussed, e.g. Ti in nitrogen deposits TiN (Teer 1983). In the reactive ion plating of TiN, composition and properties (wear resistance) can be regulated by controlling the colour temperature of the plasma flame (Oki 1982). For ion plating onto an insulator (glass, ceramic), a positive surface charge build-up must be prevented as this would repel the plating ions. The insulator is placed on a metal cathode backing plate and an rf (>500 V) field applied instead of DC. Positive ion plating then occurs in the positive half-cycles and the electrons in the plasma prevent charge build up (Mattox 1982). This is called rf ion plating. It can be used to deposit an insulating coating.

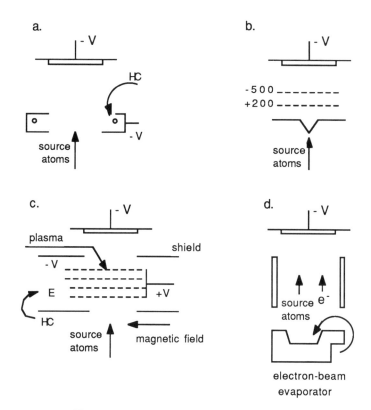

Fig.3-26: Ionization enhancement methods

Ion plating is good for wear and erosion resistant coatings, due to its good adhesion characteristic, e.g. HfN for low friction and TiC for deformation and bonding resistance. For an incompatible coating and substrate, ion plating can be used as a "strike" for electroplating. The very good adhesion of ion plated coatings (much better than vacuum evaporated coatings) is attributed to the high energy of deposition. Ionization energy effects modify the coating structure, similarly to the effect of substrate temperature in the Movchan & Demchishin model. The energy of the depositing particles is a critical factor in PVD (Teer 1983).

An electron beam gun can be used to heat the coating material source, the electron filament being kept in another compartment at higher vacuum. Cold cathode induction heated and sputter sources are also used. Feed systems (wire, rods) are described. Holders are described for obtaining a uniform coating on a complex part (Mattox & Bland 1967). The part in contact with the electrode is not coated and so a rotating mesh cage, in which the specimen is tumbled, is described. Insulating substrates can be coated by replacing the dc by rf. There is also a much greater ionization efficiency from rf than dc and rf discharges are also more stable than dc and can be held down to lower pressures (Teer 1983). Ionization can also be increased using a third electrode at +100 V and a grounded filament, as in the triode sputtering system (Enomoto & Matsubara 1975).

About 0.1 to 1% of the atoms are ionized in the glow discharge. At typical ion-plating pressures of about 0.01 torr the mean free path is about 5 mm and so many energy loss making collisions occur before the ion hits the substrate (Davis & Vanderslice 1963). Lower pressures are desirable to increase both mean free path and ionization efficiency. The good throwing power of ion plating is attributed to gas scattering effects rather than to ions following electric field lines to the substrate, but raising the gas pressure to further increase this gives poor structure and adhesion (Teer 1983). Ion bombardment during deposition supresses porous grain boundaries formed on unheated substrates. In the Movchan & Demchishin dense zone 3, vacuum evaporated coatings have larger grains than ion plated ones.

Fig. 3-27 to 3-31 show schematic diagrams of equipment and morphology for ion - sputtering, -plating and -bombardment.

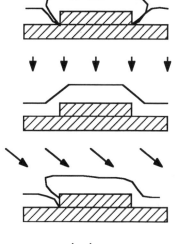

1. Covering pattern of evaporated or sputtered film in the absence of ion bombardment

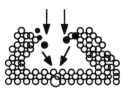

2. Ion bombardment normal to substrate surface; step coverage improved.

3. Ion beam incidental at acute or obtuse angles; step coverage non-uniform.

4. Forward sputtering technique influencing film microstructure by prevention of voids

Fig.3-27: Coating Coverage Modes in Evaporation & Sputtering

Advantages of Ion Plating:

- Good adhesion at low substrate temperatures.
- Good coverage of complicated shapes.
- Good structures and structure-dependent properties, at low substrate temperatures.

Disadvantages of Ion Plating:

- Must fix specimen to a high voltage electrode.
- Must work in a relatively high gas pressure.
- Masking difficult.
- Deposition energies are in a wide indeterminate range, hard to control.
- Not suitable for coating internal surfaces of tubes. (A recent report describes hollow cathode ion plating of large calibre gun bores (White 1986)).

A review of ion plating in German has appeared (Schintlmeister et al 1982). The theory of plasmas relevant to coating processes is reviewed (Thornton 1980). Coad & Restall (1982) review one of the conferences on ion plating processes. The development of magnetron ion plating and sputtering is an important area for research. The ICMC conferences (12th – Current), and the annual Am. Vac.Soc. meetings cited in the bibliography provide more detailed information on compounds and alloys deposited by PVD methods.

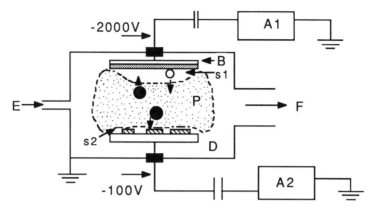

A1:radio frequency (rf) power supply to target; A2: Bias voltage
to substrate; B: target; D: substrate table & substrates; E: Gas inlet
F: to pump system; s1 & s2: cathode & anode dark space respectively;
P: plasma discharge: pressure in chamber - about 4Pa: dark circles
+ve charge; open circle -ve charge

Fig.3-28: Thin film deposition technique - rf sputtering system

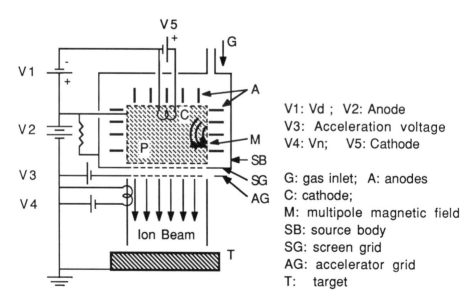

V1: Vd ; V2: Anode
V3: Acceleration voltage
V4: Vn;   V5: Cathode

G: gas inlet; A: anodes
C: cathode;
M: multipole magnetic field
SB: source body
SG: screen grid
AG: accelerator grid
T: target

Fig.3-29: Ion Beam Source - Broad Beam & Multiaperture

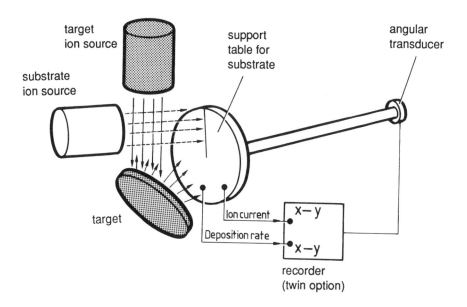

Fig.3-30: Schematic Diagram for Thin Film Deposition Using Controlled Ion Bombardment

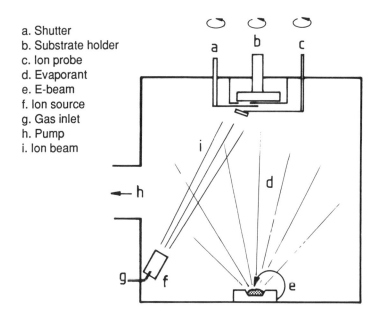

a. Shutter
b. Substrate holder
c. Ion probe
d. Evaporant
e. E-beam
f. Ion source
g. Gas inlet
h. Pump
i. Ion beam

Fig.3-31: Mounted In Beam Source for Ion Bombardment During Thin Film Deposition

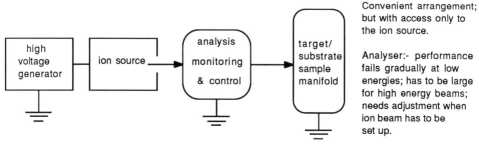

Convenient arrangement; but with access only to the ion source.

Analyser:- performance fails gradually at low energies; has to be large for high energy beams; needs adjustment when ion beam has to be set up.

1. Conventional accelerator

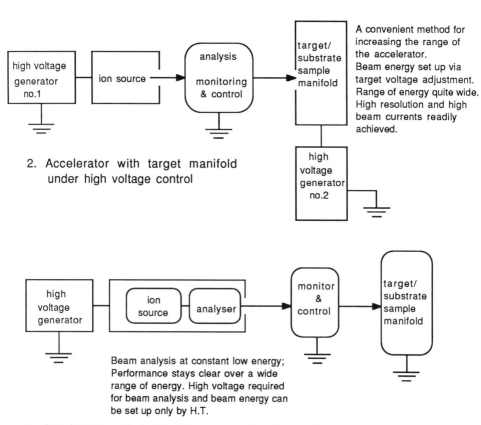

A convenient method for increasing the range of the accelerator. Beam energy set up via target voltage adjustment. Range of energy quite wide. High resolution and high beam currents readily achieved.

2. Accelerator with target manifold under high voltage control

Beam analysis at constant low energy; Performance stays clear over a wide range of energy. High voltage required for beam analysis and beam energy can be set up only by H.T.

3. Accelerator with beam analyser under high voltage control

Fig.3-32: Types of Ion Beam Acceleration & Control

3.4.2. ION IMPLANTATION

This technique was pioneered in the 1960s for the introduction of electrically active elements into silicon and other semiconducting materials. Subsequent development has led to more versatile applications in improving wear as well as thermal oxidation of metals and alloys.

In this process, chosen atomic species are ionised and then accelerated in an electric field to energies usually between 10 and 1000 keV in a moderate vacuum (about 1mPa). The ion source and extraction electrodes are designed so that the ions emerge as a well-defined beam. These ions slow down on striking the target and assume a roughly Gaussian distribution of rest positions. The peak position depends on the beam energy and its width on energy loss mechanisms. Although ion penetration is only in the region of 0.1 to 0.2 microns, the properties of materials can be considerably altered (Bennett 1981). The combination of vacuum coating with ion bombardment can extend the thickness of the treated layer (Andoh et al 1987). This will be discussed in a subsequent section.

Equipment variations are shown in Fig.3-32 (p.96), a schematic in Fig.3-33 and a typical installation in Fig.3-34. Fig.3-35 shows

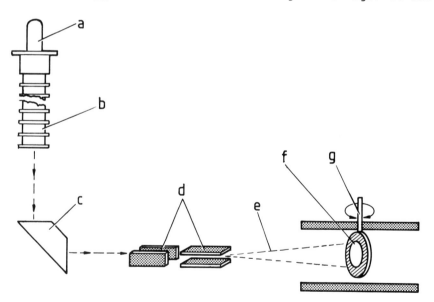

a. Ion source; b. accelerator, c. analysing magnet, d. beam scanner plates, e. scanned beam, f. sample, g. rotation system, h. 0-90° rotation

Fig.3-33: Van de Graaff 400 kV Accelerator Layout for Ion Implantation
- Schematic

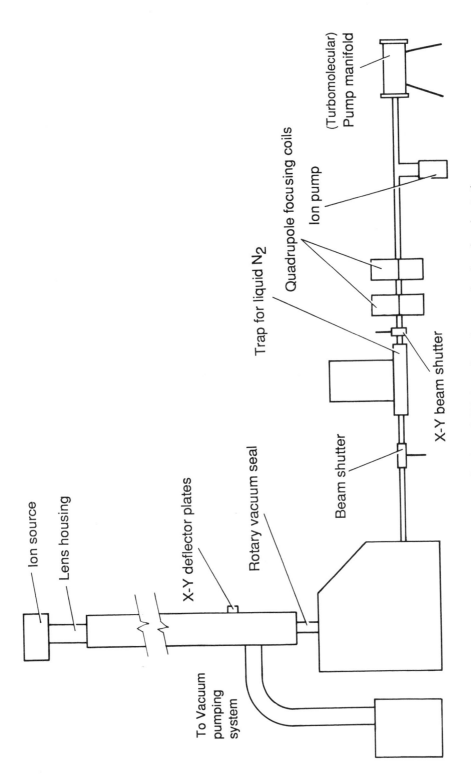

Ion source

Lens housing

X-Y deflector plates

Rotary vacuum seal

To Vacuum
pumping
system

Beam shutter

Trap for liquid N$_2$

Quadrupole focusing coils

Ion pump

X-Y beam shutter

(Turbomolecular)
Pump manifold

Fig.3-34: 500 kV Cockcroft Walton Accelerator for Ion Implantation

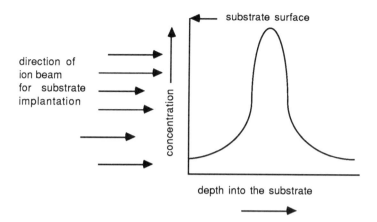

direction of
ion beam
for substrate
implantation

substrate surface

concentration

depth into the substrate

Fig.3-35: Gaussian Profile of Ion Concentration in the Substrate
on Ion Implantation

the Gaussian profile of the ion concentration in an implanted
metal surface. A new technique for monitoring nitrogen ion
implantation dose is described. This relies on colour changes
induced in anodised tantalum. Visual inspection of these dosi-
meters, by comparison with a set of calibrated standards, provides an accuracy of around 20% over the range $1x\ 10^{17} - 5\ x\ 10^{17}$
nitrogen ions/cm^2 (Gardner et al 1987).

Ion implanted metal surface does undergo radiation damage to the
metal lattice but is usually annealed out at or below the oxidation temperature used. Ion bombardment also causes sputtering
which has complex effects on oxidation. Thus the parameters of
ion dose, dose rate, metal temperature and annealing temperature
are especially informative in separating these effects from the
chemical effects of the implanted atoms on the oxidation rate.

These parameters, viz. ion dose, dose rate, substrate temperature
etc., will depend on the type of application. Ion implantation
rates are between 10^{10} ions/cm^2 (1 nA sec/cm^2) for semiconductor
doping, and 10^{19} ions/cm^2 (1A sec/cm^2) for surface chemical
conversion, e.g. $Si + 2O^+ = SiO_2$.

Many elements can be conveniently ionised in a discharge from the
vapour of a suitable compound. Multiply charged ions will attain
higher velocities (Dearnaley et al 1973). The effect of ion
implantation in standard and in ultra-clean conditions is reported (Le 1982) and metal ion implantation sources are described
(Wolf & Keller 1982).

Ion implantation has been reviewed by many, including Dearnaley
et al 1973, Bennet (1981), Wolf & Keller (1982), Hettche & Bath
(1983), Le (1982), Mcttargue & Yust (1984), Greene & Barnet

(1982) and Dearnaley & Grant (1985). Other books and conferences on the subject are edited by the following: Ashworth, Procter & Grant (1981), Ryssel & Glawischnig (1982), Teer (1983), Auciello & Kelly (1984), Williams & Poate (1984), Ziegler (1984).

A recent review by Dearnaley (1987) outlines the advantages of the process which are summarised as follows:

- High temperatures are not required for implantation; as a consequence thermal distortion of precision components is absent.
- The absence of an interface eliminates vulnerability to decohesion due to mechanical stress or corrosion.
- The surface finish is improved because sputtering erodes machining protrusions etc. Sputtering also induces complex effects on thermal oxidation (Bennett 1981)
- The implanted species is very finely dispersed, and this generally produces the most efficient effect of the additive eg. during corrosion.
- The process induces a considerable biaxial compressive stress in the bombarded surface, and in ceramics this can be particularly beneficial in closing surface microcracks. In the case of metals and alloys, annealing would eliminate these stresses.
- As an industrial process, it can be monitored and electrically controlled throughout (unlike a thermochemical treatment), and it can therefore be performed automatically.

Ion implantation is now well-established in the semi-conductor industry where it is estimated that over 2500 machines are used world-wide for the production of microchips. This application area remains a powerful driving force for the further development of ion sources and implantation systems. Recent conferences are cited below, and it is beyond the scope of this book to even summarise them here: Campisano et al (1986), Takagi (1988) and Hall (1988).

Several papers on ion implantation in High-Tc superconductors have recently been presented at the Ion implantation conference in Japan (Takagi 1988) and the sixth international conference on ion beam modification of materials held in Japan (Hall 1988).

Modified Ion Implantation Techniques:

The versatility of the ion implantation process for the protection of metals has been further extended over recent years by combining a vacuum coating process with ion implantation. This can ususally be carried out by straightforward adaption of the implantation equipment. By the continuous or sequential processes of deposition and ion bombardment, it is possible to build up the coating to a thickness of a micrometre or even more. Furthermore it is possible to choose an ion species that may react with this coating to convert it, partially or completely to a desired compound. At the same time, the interface between the coating and the substrate, always a plane of weakness, can be improved by ion beam mixing. Energetic ions, caused to pass through such an interface, can induce collisional mixing together with radiation-enhance forms of diffusive interpenetration to create a graded interface with excellent adhesion. Dearnaley (1984; 1987b) describes the various terms that have emerged due to the vast research programmes in the USA, Europe and Japan in the field of ion-assisted coatings (IAC) and describes the recent developments in this field. Although there is no consensus as yet some authors prefer to reserve the term ion-beam assisted deposition (IBAD) for the situation in which deposition and ion bombardment are carried out simultaneously. Other terminologies include ion vapour deposition (IVD), dynamic recoil mixing (DRM) ion beam enhanced deposition (IBED) etc. A review of (IAD) is given by Thompson et al (1988). A whole section in the recent conference "Ion beam modification of materials" Japan (Hall 1988) is devoted to ion beam mixing. Ion beam assisted deposition is now applied to the fields of lithography and etching. A series of papers were presented at the 5th international conference on ion beam modification of materials (Campisano et al 1986).

Ion implantation and laser processing of surfaces can improve the cracking tolerance of structural alloys (Hettche & Rath). Alumina remains crystalline if ion implanted with Cr, Ti or Zr and its hardness increases. It becomes less sensitive to fracture. But SiC becomes amorphous, softer and deforms without fracture when scratched (Mcttargue & Yust 1984). Fig.3-36 shows the abrupt transition from tensile to compressive stress in evaporated Cr films on Xe ion implantation.

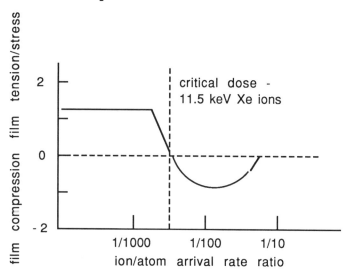

(note abrupt transition from tensile to compressive stress)

Fig.3-36: Stress generation in Cr films in Xe atmosphere

Tribological applications (Hirvonen 1985; Lempert 1988) and several corrosion and oxidation resistance studies on ion implanted alloys have been explored in recent years. Amorphous surface layers have been produced on steel with Ti and C-ion implantation. In the early stages of ion implantation inert gases such as Ar, Kr, Xe etc. were used, and later oxygen, nitrogen and carbon were experimented. A number of other metallic and non-metallic species have been used in ion implantation in recent years viz. Al, B, Ce, Co, Fe, S, Ti, Hf, Y etc., on Al-alloys, steels, and superalloys. The relevant literature will be mentioned in the appropriate sections (Chapters 7-9). The term ion-assisted coatings and ion bombardment are mentioned together with ion implantation, and sometimes replace the latter term. Large-area ion beams with higher intensities together with selected backfill atmospheres have been used to achieve reactive deposition coupled to ion implantation. Surface modifications to obtain a high degree of hardness, corrosion resistance or electrical conduction have been reported (Dearnaley 1987).

Although suitable for doping intergrated circuits, and considerable beneficial effects have been demonstrated in the development of resistance to oxidation and carbonification, the process equipment is very complex and its current production economics may prove a deterrent to the development of coatings for high temperature corrosion protection of sizeable components. Other coating methods would be preferred. But it can greatly improve sputtering coverage (see Fig.3-27, p93) and adhesion. With the development of large area ion beams many hybrid CVD and PVD processes may be envisaged.

CHAPTER 4

Chemical vapour deposition (CVD)

4.1. INTRODUCTION

4.1.1. GENERAL:

Chemical vapour deposition (CVD) is a process wherein a stable solid reaction product nucleates and grows on a substrate in an environment where a vapour phase chemical dissociation or chemical reaction occurs. It uses a variety of energy sources, viz. heat, plasma, ultraviolet light etc., to enable the reaction, and operates over a wide range of pressures and temperatures. CVD is a long-established, economically viable industrial process in the field of extraction and pyrometallurgy. Some of the well-known processes are:

$$\text{The Mond Process:} \quad Ni(CO)_4 \xrightarrow{\;150^{\circ}C\;} Ni + 4CO$$

$$\text{The Van Arkel Process:} \quad TiI_4 \xrightarrow{\;1200^{\circ}C\;} Ti + 2I_2$$

$$\text{The Kroll Process:} \quad TiCl_4 + 2Mg \xrightarrow{\;900^{\circ}C\;} Ti + 2MgCl_2$$

The method has also been used to form free-standing, simple and complex shaped articles from metals which are not very amenable to fabrication, e.g. W

$$WF_6 + 3H_2 = W + 6HF$$

Thick deposits of W and W–Re alloys have been obtained as 'vapo-forms' by CVD processes (Holman & Huegel 1974).

The technology of CVD took on new dimensions with a fresh view taken at the deposition aspects of the process. This transition of emphasis from extraction to deposition of the CVD phenomenon has made CVD an important sector of coating technology for producing new materials and coatings with improved resistance to wear, erosion and corrosion, good thermal shock resistance, and neutron absorption characteristics. Components from micro- to macro-sizes are now coated by CVD. In the deposition area of CVD, the lamp and electronics industry use many of the earlier processes to obtain longer service life and better performance for a number of products. The CVD-industry took a bigger leap when demands arose for new high temperature materials with greater high temperature corrosion resistance and strength in all the sectors using gas turbines and other propulsion hardware, in the military, science, engineering and aviation fields. In parallel, CVD technology has had rapid application in the semi-conductor industry, the many branches of fuel and energy – nuclear, fluidized bed, oil, solar and chemical industries, as well as the tool industry. Solid state devices, electrode materials, thermionic devices, erosion-, corrosion-, wear-resistant components, fission barriers and many more applications can be listed in CVD technology.

A general trend in new products produced since 1974 by CVD is summarized in Table 4:1. A summary of some CVD coating systems is given in Table 4:2 (Yee 1978).

4.1.2. THE STATUS OF INFORMATION ON CVD:

One of the greatest drawbacks in reporting on any operational industry is the lack of adequate and intelligible technical data freely available as published literature. As in any industry, the production wing of CVD has always outstripped the theoretical understanding of all the controlling parameters which govern the process. Also, products of commercial and strategic importance render all knowledge and development of these proprietary. By the time information actually becomes available other materials will have superseded them or they will have been researched well enough to obviate fundamental know-how as confidential. The actual, successful process parameters will, of course, never be revealed by the manufacturers.

It would be confusing and misleading to list a number of processes as CVD methods in as much as different high temperature coatings cannot be merely termed as diffusion coatings. The minor or major parameters which influence the actual achievement of the coating itself are those which facilitate categorization. Otherwise, conventional CVD, plamsa assisted CVD (PACVD), laser assisted CVD (LACVD), pack, vacuum pack, pack-slip, vacuum pack-slip, CCRS or RS may all be just CVD. Similarly, diffusion coatings can be obtained by first hot-dipping, spraying, cladding, fused salt electrodepositon, electrophoresis, plasma, or all PVD, and all the above "CVD" methods. What gives the distinction to any process?

The obvious factor which emerges is that a basic CVD process has a number of limitations, e.g. the types of reactions possible and available, and the minimum substrate temperature which has to be maintained. This results in restricting any eventual substrate/ deposit configuration and control. Improved techniques stretch the reactant – product spectrum and can also (i) lower the temperature, (ii) reduce the hazards and mismatch in the morphological aspects, viz. nucleation and growth, (iii) alter the physical-mechanical properties favourably, and (iv) provide better bonding.

CVD technology offers the following favourable aspects to semiconductor industry (Hammond 1979):

  (i) A wide range of silicon epitaxy thicknesses and resistivity;
 (ii) A low cost, highly productive method of depositing poly silicon, Si_3N_4 and high temperature SiO_2; and,
(iii)A proven technology for passivating silicon deviceswith low temperature deposition of silicon -oxide and -nitride.

A wide range of metallic and inorganic coatings can be produced by CVD. In many instances CVD coatings exhibit unique properties, which are difficult to produce by other methods. This survey attempts to collate CVD information reported up to mid-1988. The literature survey is not exhaustive but fairly representative.

4.1.3. EARLIER LITERATURE ON CVD:

Hirth and Pound (1967) give a summary of the status up to 1966. Alcock and Jeffes have discussed the application of thermodynamics to CVD processes (1967). Blocher has reviewed the structure/ property/process relationships of CVD (1974) and a more recent review (Blocher 1982) brings it up-to-date. Important sources of information are listed in the following references (given chronologially): National Acad. Sci. USA (1970); Blocher (1974), Nickel (1974); Holman & Huegel (1974); Tick et al (1974); Peterson (1974); Hammond (1979); EEC Petten Estbl.(1979); Duret & Pichoir (1982); Hayman (1984); Morrell & Nicholas (1984); Habig (1986); Headinger & Purdy (1987). All the international bi-ennnial conferences on CVD are good sources of reference (12th ICCVD in 1987).

TABLE 4:1

C V D Product Survey

Ref.No	Year	Deposit	Substrate	Deposit Temp °C	Process	1	2	3	4	5	6	7	Comments
Single Elements													
1	1967	single species			model								General guide on Thody.Diag.
2	1977	B	Earlier lit.	1050–1250	chloride/H_2	/x		/x	x/x	x/x			
	1977	B	Ta,Mo,W & graphite							/x			
3	1978	B	Ti,Ta,W & their borides	700–820	chloride/H_2	x/		/x					
4	1982	B	hard metals		fluidized bed	x/		/x					Ceramic-on-metal.
5	1974	C	hard metals		fluidized bed								Need stressed for more inter-assessable dep/prop. studies
6	1974	C	Inner wall of graphite tube		chloride/H_2	x/		/x	/x				Part of a review. Tube dia. 18mm.
7	1978	Nb	Inner wall of graphite tube		chloride/H_2	x/x		x/x	/x				
8	1973	Nb	Pt & W wire	500–1200	chloride/H_2			/x					Poor adherence. Part of a review.
8	1973	Si	cast iron	870–900	$SiCl_4$/H_2			/x					
6	1974	Si		900<T<1200									
9	1977	Si	Ni,Co,Mo Nim 105		$SiHCl_3$/H_2								
10	1977	Si	Si-coated graphite	1227	$SiHCl_3$/H_2	x/x		x/x	x/				Micro-CVD number assigned.
11	1982	Si	pure Ni	950–1080	$SiCl_4$/H_2	/x	/x			/x	/x		Prelim. for gas turbine appln.
9	1979	Si	Mo,Co,Ni, Nim 105, IN 738	990–1150	$SiCl_4$ or $SiHCl_3$/H_2	/x		x/x	x/x	x/x	x/		Surface process; mass transport rate controlling, not Thermodynamics.
12	1979	Si	Ni,Co,Inconel IN 738							/x	x//		
13	1971	Ta				x/							Good source for ref earlier than 1971.
14	1977	Ta	mild steel	900–1000	$TaCl_5$/H_2	x/x			/x		x//		Inner face of long tubes of mild steel coated.
7	1978	Ta	graphite tube			x/x		x/x	/x				Substrate dia. 18mm.

Table 4:1 (contd.) CVD Product Survey

Ref.No	Year	Deposit	Substrate	Deposit Temp°C	Process	1	2	3	4	5	6	7	Comments
15	1974	W	Mo;detachable substrates		fluoride/H_2	x/							Vapoformed W components.
6	1974	W			WCl_6 or WO_2Cl_2/H_2	/x	/x						Part of a review. Oxychloride results in a brittle deposit.
16	1968	W	Mo; quartz	600									
Compounds													
17	1982	$Al,B_{12}-B$	Mo & W filament	100-1500	Halides/H_2	x/x			/x	/x	//x		2 methods of preparation
18	1968	AlN	SiC, reactor wall	1030-1150	Al powder/N_2 = AlN; AlN/HCl,NH_4Cl	x/						///x	
19	1979	BP	Mo	880-1000		x/	/x			x/	//x		New substrate; BP usually dep. on Si, graphite & cemented carbide.
20	1975	BN	graphite	>1700		x/							Vapoform crucible production.
21	1982	IrO_2	quartz	1050		x/x				/x	x//		Reaction at 1230°C; growth at 1050°C; single crystal; electrode material.
8	1973	Nb_3Sn	Pt & W wire	500-1200	$NbCl_5-SnCl_4/H_2$								
21	1982	RuO_2	quartz	1050		x/x				/x	x//		Reaction at 1230°C; growth at 1050°C; single crystal; electrode material.
22	1977	SiC	SiC	800-1200	silane/NH_3	x/							Ceramic-on-ceramic.
23	1976	Si_3N_4	vitreous silica						/x				
24	1977	Si_3N_4		800-1200	SiF_4/NH_3	/x	/x		x/				2 methods of prodn., intermediate species characterized.
25	1978	Si_3N_4			silane/NH_3	x/	x/x/x						

Note: Each number lists 2 or 3 or 4 aspects of investigations. A symbol x//
indicates that the first aspect is reported & not the other two.
x/x/ means the first two aspects, //x means only the third aspect
& so on

Key to Column Headings-
1: mainly production purpose; equipment drawing
2: theoretical; coating characterisation; mechanism
3: thermodynamics calculations; analysis; deposition kinetics
4: mass transport; microstructure (S.E.M.)
5: substrate effect; X-ray of deposit
6: oxidation properties of coating; stress; hardness
7: optical; T.E.M.; storage; memory

107

Table 4:1 (contd.) CVD Product Survey

Ref.No	Year	Deposit	Substrate	Deposit Temp°C	Process	1	2	3	4	5	6	7	Comments
26	1978	Si_3N_4	radoms;optical fibres;turbine blades, vanes	1100-1550	silane/NH_3	x/					x//	x/x//	3 processes compared;CVD-Si_3N_4 lowest oxidation rate.
27	1978	Si_3N_4	Si wafer	1000		x/						/x/x	Thermo-ammonialysis.
28	1983	Si_3N_4	fused silica	1150,1250 1300	$SiCl_4$/NH_3, N_2	/x				/x	x/x		SiO_2 from reactor-wall interferes with dep.morphology; Si_3N_4 was at 1150 + at 1250 & at 1300°C
29	1974	SiO_2	9%Cr steel	300-430	silane/O_2+H_2 or CO_2	x/		/x	/x		x//		5-25 m dep.,non-porous, adherent enough to withstand temp. cyclic in advanced gas cooled reactors.
30	1978	SiO_2	9CrMo0.6Si-steel	325		x/					x//		For oxidn. resistance.
31	1982	SiO_2	20Cr25NiNb-steel	800-830	2-step-tetra ethyl oxysilane /steam	x/					x//		Brit.Petroleum process; for oxidn. resistance;step-1; pre-oxidn.; step-2; SiO_2 dep.
32	1977	TiB_2	uncoated & TiC-precoated cemented carbide	900-1600	chloride/H_2	x/		x//x					ceramic-on-ceramic process
33	1982	TiB_2		950-1100				/x					
34	1978	TiC	cemented carbide (WC-Co alloys)	1000	chloride/H_2		/x	/x	/x	x/			Influence of substrate on prop.
35	1980	TiC	cemented carbide	950-1100	halide/H_2	x/		/x					CVD model developed
36	1974	TiN	Fe	900-1100	chloride/	x/x	/x	/x	x//x	/x	/x		Dep. carried out in an ultrasonic field; improved adhesion; 160 m dep.
37	1977	TiN								/x			Vapoformed components.
15	1974	W-Re	Mo; detachable mandrels		fluoride/H_2	x/							
38	1979	ZrN	tubular quartz	1000-1200	chloride/H_2, Ar, N_2	x/x				/x			
39	1977	$ZrSi_2$	Zr	1000	chloride/	x/					x/		for oxidn. resistance in nuclear reactor.

Mixed Compounds

No.	Year	Compound	Substrate / Notes	Temp (°C)	Reactants	1	2	3	4	5	6	7	Remarks
40	1978	AlN–Si_3N_4	sapphire	600–1000	$AlCl_3, SiH_4/NH_3$	x/	y/x		/x	/x	x/x/	x//x	memory devices
21	1982	RuO_2^- IrO_2^-	quartz wall	1050		x/x	/x				/x	x//	reaction at 1230°C; growth of dep. 1050°C single crystals; electrode material
41	1974	Si–As–Te	Si,As & Te in individual boats /15H_2 + He										
42	1978	Ti–B–N	sintered TiN pellets	1050–1500	chloride/H_2, N_2	x/x	/x	/x		/x			released boron vapour reacts with substrate nitride pellet
43	1979	TiC–TiN TiCN	carbon steel	1050	$TiCl_4/H_2$, N_2, CH_4	x/x	/x	/x	/x	/x			ultrasonic field appln.; improved adhesion
44	1977	Ti–Zr–B	graphite disc	900–1400		x/x	/x	/x		/x	/x	//x	high metal chloride partial pressures used; m.p. & shock resistance detd.
45	1979	ZrC–C	semicircular graphite rod with Mo along its centre	1280–1300	bromide/H_2, Ar, CH_4	x/x	x/x		/x	/x	/x	/x	fission product barrier cost
46	1981	ZrC–ZrN & ZrC_xN_y	structural C–fibre carbon steel	950–1150	chloride/H_2, N_2, Ar, CH_4	x/x		//x		/x	x/x		
47	1988	C–C	structural C–fibre matrix	step1:400–1000 & >2000 CVD/CVI of C step2: 1000–1600 & 2500		x/x		/x	x/	x/	/x/x		Carbon–carbon composite; process & properties – hardness & strength discussed.
48	1988	SiC; also ZrC, ZrB_2, B_2O_3, Si_3N_4 etc.	Structural C			x/				x/	x/x/x		Review of ceramic coatings on carbon for oxidation protection

Key to Column Headings:-
1: mainly production purpose; equipment drawing
2: theoretical; coating characterisation; mechanism
3: thermodynamics calculations; analysis; deposition kinetics
4: mass transport; microstructure (S.E.M.)
5: substrate effect; X-ray of deposit
6: oxidation properties of coating; stress; hardness
7: optical; T.E.M.; storage; memory

Note: Each number lists 2 or 3 or 4 aspects of investigation. A symbol x// indicates that the first aspect is reported & not the other two. x/x/ means the first two aspects, //x means only the third aspect & so on

References in column 1 Table 4:1

1. Alcock C.B. & Jeffes J.H.E., Trans.I.M.M.,C, 76 C246 (1967).
2. Vandenbulcke L. & Vuillard G., J. Electrochem.Soc., 124(12) 1931 (1977).
3. Thebault J., Naslain R & Bernard C., J. Less-Common Metals, 57(1) 1 (1978).
4. Voigt K. & WestPhal H., High Temp. & High Pr., 14, 357 (1982).
5. Nickel H., J. Vac. Sci. Technol. 11(4), 687 (1974).
6. Blocher J.M., ibid., 11(4), 680 (1974).
7. Kehr D.E.R., High Temp. & High Pr., 10, 477 (1978).
8. Samsonov G.V.(Ed.), 'Protective Coatings on Metals', Vol.5, Consultants Bur. NY (1973).
9. Wahl G & Furst B., 'Materials & Coatings to Resist High Temperature Corrosion', Holmes D.R. & Rahmel A. (Ed.), Applied Sci.Pub., London (1977) p.333
10. Van den Brekel C.HJ., Phillips Res. Rep., 32, 118 (1977).
11. Subramanyam J., J. Mat.Sci., 17(7) 1997 (1982).
12. Hidebrandt U.W., Wahl G. & Nicoll A.R., 'Materials & Coatings to Resist High Temperature Corrosion', Holmes D.R. & Rahmel A. (Ed.), Applied Sci.Pub., London (1977) p.213.
13. ***
14. Beguin C., Horvath E. & Perry A.J., Thin Solid Films, 46, 209 (1977).
15. Holman W.R. & Huegel F.J., J. Vac. Sci. Technol., 11(4), 701 (1974).
16. Mehalchick E.J. & MacInnis M.B., Electrochem. Technol., 6, 66 (1968).
17. Motojima S., Takai K. & Sugiyama K., J. Nucl. Mat., 98, 151 (1981).
18. Chu T.L., Ing D.W. & Noreika A.J., Electrochem. Technol., 6, 56 (1968).
19. Morojima S., Ohutska Y., Kawajiri S. Takahashi Y. & Sugiyama K., J. Mat. Sci., 14(2) 496 (1979).
20. Anon., Metallurgist & Mat. Technol. p.292, June (1975).
21. Georg C.A., Triggs P. & Levy F., Mat. Res. Bull., 17 105 (1982).
22. Wu S.Y., Diss. Abs. Intl., 38B(1) 288 (1977).
23. Mellottee H., Cochet G. & Delbourgo R., Rev. Chim. Min., 13 373 (1976).
24. Lin S.S., J. Electrochem.Soc., 124(12) 1945 (1977).
25. Cochet G., Mellottee H. & Delbourgo M., J. Electrochem. Soc., 125(3) 487 (1971).
26. Galasso F.S., Vetri R.D. & Croft W.J., Am. Ceram. Bull., 57(4) 453 (1978).
27. Ito T., Nozaka T., Alakawa H. & Shimoda M., Appl. Phys. Lett., 32(5) 33 (1978).
28. Oda K., Yoshio T. & O-oka O., Comm. Am. Cram. Soc., C8 (1983).
29. Graham J., High Temp. & High Pr., 6, 577 (1974).
30. Dallaire S.& Cielo P., Met. Trans. B, 13B, 479 (1982).
31. Bennett M.J., Houlton M.R. & Hawes R.W., Corrosion Sci., 22(2) 111 (1982).
32. Besmann T.M. & Spear K.E., J. Electrochem. Soc., 124(5) 786 (1977).
33. Zeman F., Mayerhofer J. & Kulburg A., High Temp. & High Pr., 14 341 (1982).
34. Hara A., Yamamoto T. & Tobioka M., ibid., 10, 309 (1978).
35. Subrahmnayam J., Lahiri A.K. & Abraham K.P., J. Electrochem.Soc., 127(6) 1394 (1980).
36. Peterson J.R., J. Vac. Sci. Technol., 11(4) 715 (1974).
37. Takahashi T. & Itoh H., J. Electrochem. Soc., 124 797 (1977).

Table 4.1 Ref. Contd.

38. Motojima S., Kani E., Takahashi Y. & Sugiyama K., J. Mat. Sci., 14(6) 1495 (1979).
39. Caillet M., Ayedi H.F., Galerie A. & Besson J., 'Materials & Coatings to Resist High Temperature Corrosion', Holmes D.R. & Rahmel A. (Ed.), Applied Sci.Pub., London (1977), p.387.
40. Zirinsky S. & Irene E.A., J. Electrochem. Soc., 125(2) 305 (1978).
41. Tick P.A., Wallace N.W. & Teter M.P., J. Vac. Sci. Technol., 11(4) 709 (1974).
42. Peytavy J.L., Lebugle A. Montel G., High Temp. & High Pr., 10, 341 (1978).
43. Takahashi T. & Itoh H., J. Mat. Sci., 14(6) 1285 (1979).
44. Takahashi T. & Kamiya H., High Temp. & High Pr., 9, 437 (1977).
45. Ogawa T. Ikawa K. & Iwamoto K., J. Mat. Sci., 14(1) 125 (1979).
46. Takahashi T., Itoh H. & Fukao K., J. Less-Common Metals, 80, 171 (1981).
47. Buckley J.D., Ceram. Bull., 67(2) 364 (1988).
48. Strife J.R. & Sheehan J.E., Ibid., 67(2) 369 (1988).

Note: Please see Section 4.7. (p149-157) for further examples.

T A B L E 4:2

S U M M A R Y   O F   C V D   C O A T I N G   S Y S T E M S

Coating	Substrate	Chemical mixture	Substrate Temp, °C	Method	Ref	Note
B_4C	Tungsten	$BCl_3-CH_4-H_2$	1300	CCVD	109	
	Graphite	BCl_3-H_2	1500	CCVD	109	
$Cr_3C_2 \cdot Cr_7C_3 \cdot Cr_{23}C_6$	Steel	dicumene chromium	450–650	CCVD	34	
	Steel	$Cr(vapour)-C_2H_2-Ar$	20–500	ARE,RIP	69	
	Glass or ceramic	$Cr(AA)_3-H_2$	600	CCVD	110	AA = acetylacetonate
	Steel	$CrCl_2-H_2$	1000	CCVD	111	Gas inchromizing
	Steel	$Cr(CO)_6-H_2$	250–650	CCVD	2(p.360)	$Cr_3C_2-Cr_2O_3$ mixture
$(Fe,Mn)_3C$	Steel	$FeMn-MBF_4-Al_2O_3$	650–1200	PC	112	M = Na, K, NH_4
HfC	Steel	$Hf(vapour)-C_2H_2$	515	ARE	43	
	Tungsten alloys	$HfCl_x-CH_4-H_2$	1000–1300	CCVD	113,114	x = 2 − 4
Mo_2C	Steel	$MoF_6-C_6H_6-H_2$	400–1000	CCVD	15	C_6H_6 = benzene
	Steel	$Mo(CO)_6-H_2$	350–475	CCVD	2(p.362)	
NbC	Steel	$Nb(vapour)-C_2H_2$	540	ARE	43	
	Nb, W, Mo, graphite	$NbCl_5-CCl_4-H_2$	1500–1900	CCVD	115	
SiC	SiC	$Si(target)-C_2H_2-Ar$	>1150	RS	58	
SiC	Graphite, TiN	$CH_3SiCl_3-H_2$	1400	CCVD	82	see also literature cited in ref 82
TaC	Steel	$Ta(vapour)-C_2H_2$	550–1490	ARE	43,93	
	W filament	$TaCl_5-CH_4-H_2$	1100	CCVD	116	
TiC	Steel, Ta	$Ti(vapour)-C_2H_2$	450–1450	ARE	90	
	Steel, Ti	$Ti(vapour)-C_2H_4$	Not given	RIP	67	
	Various	$Ti(vapour)-CH_4$ (or C_2H_2)	870	RIP	68	
	Glass	$Ti(target)-CH_4-Ar$	300	RS	56	
	Cemented carbide, steel	$TiCl_4-CH_4-H_2$	>800	CCVD	5–8	see TiC literature in Refs. 5–8

Compound	Substrate	Reactants	Temperature	Process	Ref	Remarks
TiC–VC	Not specified	Ti–6Al–4V(vapour)–C_2H_2	560–1000	ARE	117	Deposits of TiC–VC solid solutions
Ti–Si–C	Graphite	$TiCl_4$–$SiCl_4$–CCl_4–H_2	>1000	CCVD	118	Coating contains mixed phases
VC	Steel	V(vapour)–C_2H_2	555	ARE	43	
	Steel	VCl_2–CH_4–H_2	1050–1150	CCVD	119	
	Tungsten	VCl_4–C_7H_8–H_2	1500–2000	CCVD	2(p.374)	C_7H_8 = toluene C_6H_6 = benzene
W_2C	Steel	WF_6–C_7H_8–H_2	400–1000	CCVD	15,16	
	Steel	WF_6–C_6H_6–H_2	600–1000	CCVD	120	
	Steel	$W(CO)_6$–H_2	350–400	CCVD	2(p.372)	
	Cemented carbide	WCl_6–CH_4–H_2	900–1150	CCVD	24	Coating is mostly WC
ZrC	Steel	Zr(vapour)–C_2H_2	540	ARE	43	
	Various	Zr(vapour)–CH_4 (or C_2H_2)	870	RIP	68	see also Ref.97
	Carbide–carbon composite	$ZrCl_4$–CH_4–H_2	1050–1500	CCVD	121	
	Alumina	ZrI_4–C_2H_6	900–1400	CCVD	122	Starting chemicals are Zr metal and CH_3
AlN	Silicon	$AlCl_3.3NH_3$–H_2	800–1200	CCVD	123	see also literature cited in Ref.123
	Metal–coated sapphire	Al(vapour)–N_2(orNH_3)	300–1200	ARE,RE	46	
	Sapphire	Al(vapour)–NH_3	1400	RE	124	
BN	Various	BBr_3–NH_3–H_2	>500	PACVD	45	RF discharge
BN	BN felt	BF_3–NH_3	1100–1200	CCVD	125	
	Various filaments	BCl_3–NH_3–H_2	1000	CCVD	37, 2 (p.381)	
CrN	Steel	Cr(vapour)–N_2–Ar	20–500	ARE, RIP	69	
Fe_4N	Steel, iron	N_2–H_2	520	CIP	71	Ion–nitriding process
HfN	W alloys	$HfCl_x$–N_2–H_2	900–1300	CCVD	113	x=2,3. see also Ref.94

AA* = acetylacetonate
ARE = activated reactive evaporation
RE = reactive evaporation
(Ref. nos. here refer to those in Yee (1978))

CCVD = conventional CVD
PACVD = plasma-activated CVD
RIP = reactivated ion plating

CIP = chemical ion plating
CSD = chemical spray deposition
RS = reactive sputtering

Table 4:2 (contd.) CVD Coating Systems

Coating	Substrate	Chemical mixture	Substrate Temp, °C	Method	Ref	Note
NbN	Various	$Nb(NEt_2)_5$-Ar, H_2 or N_2	500	CCVD	35	Et = ethyl
	Cemented carbides	$NbCl_5$-N_2-H_2	1000–1300	CCVD	94	
	Fused silica	$NbCl_5$-NH_3-H_2	900–1000	CCVD	126	
	Hastelloy or fused silica	Nb(target)-N_2-Ar	500	RS	127	see also Ref.8 (p.258)
Si_3N_4	Silicon	SiH_4-N_2H_4-H_2	550–1150	CCVD	31	see also literature cited in Ref.31 for SiH_4-NH_3 system
	Various	Si_3N_4(target)-N_2-Ar	<100	RS	53	Much Si_3N_4 literature cited in Ref.53
	Silicon	Si(target)-N_2	200	RS	54	
	Various	SiH_4-NH_3	<40	PACVD	49	
	Silicon	SiH_4-N_2	25–500	PACVD	48	
	Si-Mo alloys	Si(vapour)-NH_3 or NH_3 + SiH_4	100–300	RIP	70	
(Si,Al)N	Si-Mo alloys	Si + Al(vapour)-NH_3 (or NH_3 + SiH_4)	100–300	RIP	70	
Ta_xN	Cemented carbides, SiO_2	$TaCl_5$-N_2-H_2	800–1500	CCVD	94,128	x = 1, 2
	Glass, Al_2O_3	Ta(target)-N_2-Ar	100	RS	129	
TiN	Various	$Ti(NR_2)_4$-Ar, H_2 or N_2	250–600	CCVD	35	R = methyl, ethyl, propyl, butyl
	Cemented carbides SiC	$TiCl_4$-N_2-H_2	800–1400	CCVD	82,94,95	
	Steel	Ti(vapour)-N_2(or NH_3)	20	ARE, RE	43	
	Various	Ti(vapour)-N_2	870	RIP	68	
(Ti,V)N	Cemented carbide	$TiCl_4$-VCl_4-N_2-H_2	1100	CCVD	94	
ZrN	Various	Zr(vapour)-N_2	870	RIP	68	
	Cemented carbides	$ZrCl_4$-N_2-H_2	1150–1200	CCVD	94	
Al_2O_3	Copper, copper alloy	$AlCl_3$-H_2O-H_2	500	CCVD	104	

Material	Substrate	Reactants	Temperature	Method	Ref.	Notes
Al_2O_3	Cemented carbides	$AlCl_3-CO_2-$ (or $H_2O)-H_2$	850-1100	CCVD	105	
	Copper	$Al(AA*)_3-H_2$	350-500	CCVD	110	
	Silicon	$Al(CH_3)_3-O_2$	275-475	CCVD	130	
	Silicon	$Al(OC_3H_7)_3-O_2$	425-500	CCVD	130	
	Silicon	$Al(vapour)-H_2O$	20-300	RE	131	See also Ref.33
Cr_2O_3	Glass	$Cr(AA*)_3$ in butanol-Ar	520-560	CSD	41	
	Not given	$Cr(CO)_6-O_2-$ (or H_2O,CO_2)	400-600	CCVD	2 (p.390)	
Fe_2O_3	Glass	$Fe(AA*)_3$ in butanol-air	400-550	CSD	41	
HfO_2	Silicon	$Hf(AA*)_4-O_2-He$	450-700	CCVD	132	
In_2O_3	Glass	$In(AA*)_3$ in acetylacetone-air	470-520	CSD	41	
Nb_2O_5	Various	$Nb(OC_2H_5)_5-O_2-He$	450	CCVD	133	See also Ref.33
SiO_2	Quartz	$SiCl_4-O_2-Ar$	>1000	PACVD	50	
	Silicon	SiH_4-N_2O	25-1000	PACVD	45	
	Cr steel, silicon	SiH_4-O_2-Ar(or N_2)	300-450	CCVD	134-136	
SnO_2	Silicon	$SnCl_4-O_2$	600-800	CCVD	137	
	Glass	$SnCl_4$ in aqueous HCl-air	480-500	CSD	41	
	Various	$(C_4H_9)_2Sn(OOCCH_3)_2-O_2-H_2O$	420	CCVD	138	
Ta_2O_5	SiO_2, Al_2O_3	$TaCl_5-O_2-H_2$	600-900	CCVD	139	
	Various	$TaCl_5-O_2-Ar$	900-1300	CCVD	140	
	Various	$Ta(OC_2H_5)_5-O_2-He$	450	CCVD	133	
TiO_2	Various	$TiCl_4-O_2-He$	25-1000	PACVD	45	

AA* = acetylacetonate
ARE = activated reactive evaporation
RE   = reactive evaporation
(Ref. nos. here refer to those in Yee (1978))

CCVD = conventional CVD
PACVD = plasma-activated CVD
RIP   = reactivated ion plating

CIP = chemical ion plating
CSD = chemical spray deposition
RS = reactive sputtering

Table 4:2 (contd.) CVD Coating Systems

Coating	Substrate	Chemical mixture	Substrate Temp, °C	Method	Ref	Note
TiO_2	Various	Ti(vapour)-O_2	>450	ARE, RE	141	Various Ti oxides formed
TiO_2	Quartz	Ti(target)-O_2-Ar	>25	RS	55	See also $TiCl_4$-H_2O-O_2 system in Ref.113
	Various	Ti(OC_2H_5)$_4$-O_2-He	450	CVD	133	See also Ref.33
TiO_2	Silicon	Ti(OC_3H_7)$_4$-O_2-He	450-700	CCVD	132	
V_2O_3	Glass	V(AA*)$_3$ in butanol-O_2	450-510	CSD	41	
Y_2O_3	Steel	Y(vapour)-O_2	>25	ARE, RE	43,142	Y(thd)$_3$ = Y chelate of tetramethyl-heptanedione
	Quartz, glass	Y(thd)$_3$-O_2 (or O_2 + H_2O)	500	CCVD	36	See also Ref.33
ZrO_2	Silicon	Zr(AA*)$_4$-O_2-He	450-700	CCVD	132	
	TiN-coated cemented carbides	$ZrCl_4$-CO_2-CO-H_2	1000	CCVD	143	
MoB	Niobium	$MoCl_5$-BBr_3	1400-1600	CCVD	145	Solar furnace used
	Mo alloy	BF_3-B granules	1000	CCVD	146	Mo_2B also formed
NbB_2	Tantalum	$NbBr_5$-BBr_3	1500	CCVD	20	Solar furnace used
	Quartz	$NbCl_5$-BCl_3-H_2	950-1200	CCVD	147	
TaB_2	Niobium	$TaBr_5$-BBr_3-H_2	1500	CCVD	20	Solar furnace used
TiB_2	Tungsten	Ti(BH_4)$_3$-Ar	800-1100	CCVD	148	TiB coating obtained on Ti substrate
TiB_2	Steel, graphite	$TiCl_4$-BCl_3-H_2	>800	CCVD	106	
WB	Niobium	WCl_6-BBr_3	1400-1600	CCVD	145	Solar furnace used
Amorphous boride coatings	Various	BCl_3-H_2	950-1200	CCVD	149	See also Ref.148 for BCl_2-H_2 and metal borohydride systems

AA* = acetylacetonate
ARE = activated reactive evaporation
RE = reactive evaporation
(Ref. nos. here refer to those in Yee (1978))

CCVD = conventional CVD
PACVD= plasma-activated CVD
RIP = reactivated ion plating

CIP = chemical ion plating
CSD = chemical spray deposition
RS = reactive sputtering

Blochers's review in 1974 shows that apart from the well known extracted elements like Ni and Ti and vapoforms like W and W-Re, the deposition of some other elements and compounds had also been applied in some industries. Thus, Si, C, B, Ta deposition, vapoformed carbon or pyrolytic carbon, SiO_2, Si_3N_4, and SiC were well known in 1974. Processes like boriding, siliciding, and nitriding use CVD as one of the process-routes. Many new products by CVD have emerged since then and new experimental developments in CVD include plasma-CVD and laser-CVD. Free-standing shapes by CVD of graphite, rhenium, aluminium, boron nitride, boron carbide silicon nitride and zinc sulphide are in commercial production in addition to W and other earlier vapoforms (Hayman 1984). Reactive sputtering is a mixed PVD-CVD technique. All these may be termed hybrid CVD processes. A significant trend in CVD-industry is in the production of many refractory metal oxides, carbides, nitrides, borides and silicides both as individual compounds as well as solid solutions, mixtures and multilayered-bonded deposits (see Tables 4:1 and 4:2).

4.1.4. VARIATIONS IN CVD PROCESS:

CVD is an attractive method for applying ceramic coatings to composites because complex shapes can be coated without extensive equipment alterations, a wide variety of coatings can be applied, and coatings of unique properties can be produced. Chemical Vapour Impregnation (CVI) by CVD is a rapidly developing technology (Buckley 1988; Reagan et al 1988; Strife & Sheehan 1988; Headinger & Purdy 1987; Fitzer & Gadow 1986; Hayman 1984). The possible detrimental effects on substrate by high temperature CVD is avoided by applying plasma technology. PACVD and PECVD are in the forefront of developing CVD technology (Inspektor-Koren 1987; Dharmadhikari 1988; Mito & Sekiguchi 1986; Tsu & Lucovsky 1986; Supple & Stoneham 1985). Laser assisted (enhanced) CVD (LACVD, LECVD), (Shaapur & Allen 1987) and metal organic CVD (MOCVD) (Nakamura 1986) are other CVD processes which are being investigated.

4.2. THE CVD PROCESS

4.2.1. GENERAL:

Until recently it has been valid to categorize CVD as a thermochemical process in which the substrate was regarded as a collector of the deposit and the substrate/deposit interaction was both undesirable and unnecessary for deposit growth. For a CVD deposit however, the post-deposition state is a mandatory parameter to meet the service requirements; substrate/deposit interaction to some degree is expected. The ternary and mixed CVD products such as ZrC-C on a graphite-Mo substrate (Ogawa et al 1979), or Ti-B-N formed by boron released on a sintered TiN pellet (Peytavy et al 1978), illustrate this point. Yet another instance is where an intermediate layer is deposited between a substrate and the final deposit. Further, heat treatment and therefore diffusion, is often a step following the primary CVD process (or even during deposition, since the substrate is heated). Thus interaction to some degree does occur between substrate and deposit in CVD-produced coatings; it is also one of the ways deposit consolidation can be achieved.

4.2.2. CVD TERMINOLOGY & EQUIPMENT:

A CVD reactor is the apparatus-equipment configuration which carries out the CVD process. It is refered to as a 'reactor' which is confusing in general parlance where the term implies a nuclear reactor.

CVD can be a 'closed tube' process when the reactants and products are recycled in a closed system. The de Boer-Van-Arkel iodide process for Cr is a good example. Closed-tube CVD processes can be adopted in cases where (i) a reaction can be made reversible with a temperature differential, or (ii) where one (or more) of the products can be recycled without loss, to be re-introduced as a reactant. In the reaction, $CrI_2(g) = Cr(s) + I_2(g)$, I_2 reacts at one temperature with impure Cr, deposits Cr at another temperature and is recycled. However, most CVD processes are the 'open tube' type, where after deposition, the reaction chemicals are swept out of the reactor. If the expense justifies it some recovery may be made of the remnant products and reactants. In an open tube reactor, optimisation of deposition is very important to render the reactor efficient and safe for operation. Because the CVD reactants and products are often toxic, corrosive, hygroscopic, readily oxidising, inflammable, poisonous and have high vapour pressures, the post-deposition section of the plant must be efficient to render these chemicals harmless before disposal. Thus the open tube process is still, effectively, a closed system for safety.

CVD reactors are categorized on the basis of the temperature and pressure they operate under. A CVD reactor can have a cold wall or a hot wall depending on the experimental optimisation parameters. When the reactor wall surrounding the heated substrate is comparatively cool relative to the temperature of the substrate, it is a cold-wall reactor. The substrate is heated directly, and product deposition occurs mostly on the heated component. In a hot wall reactor the deposition product can nucleate both on the reactor wall and the heated substrate, though to different degrees, if a temperature difference prevails across the wall-substrate area and zone. The substrate may be independently heated or not, but the reactor itself is heated in order to facilitate the process. Table 4:3 gives CVD reactor groups w.r.t. the operating regime (Habig 1986; Hammond 1979).

TABLE 4:3

C V D   P R O C E S S   C A T E G O R I Z A T I O N

	Metal/Refractory	Semi-Conductor					
Pressure Range:	1–1000 mbar	10^{-4} – 10^{-1} mbar					
Pressure Category:	(Low to High)	Atmospheric Pressure		Reduced Pressure		Reduced Pressure & Plasma Enhanced	
CVD reactor wall temperature:–		Hot	Cold	Hot	Cold	Hot	Cold
Temperature		<——— Operational Status Ratings ———>					
High	$850 < T < 1200$ e.g. $TiCl_4, N_2, H_2$ -> TiN, HCl	<——— Single Crystal; 900 – 1300 ———>					
		NCP	DCP	NCP	LCP/I	NCP	NCP
Moderate	$700 < T < 850$ e.g. $TiCl_4, CH_3, CN, H_2$ -> $Ti(C,N), CH_4, HCl$	<——— Poly-Crystal; 500 – 1000 ———>					
		NCP	DCP	LCP/D	NCP	NCP	NCP
Low	(i) $300 < T < 600$ e.g. WF_6, C_6H_6, H_2 -> W_2C, HF	<——— Poly-Crystal; 200 – 500 ———>					
		NCP	DCP/D	LCP/I	NCP	NCP	DCP/I
	(ii) Plasma CVD e.g. $TiCl_4, N_2, H_2, Ar$ -> $TiN, HCl, (NH_3)$						

Note:
NCP – Not in significant Commercial Processing
LCP – Limited Commercial Processing
DCP – Dominant Commercial Processing
/D – Decreasing Usage; /I – Increasing Usage

(temperatures in °C)

In the achievement of an acceptable deposit for a particular application, the CVD process itself, the size and number of the substrates to be coated, and their distribution geometry play decisive roles. The type of heating, viz. resistance, induction etc., and the type of substrate holder viz. slab or carousel etc., have to be custom-designed. Since CVD can be applied to micro-chips in the semiconductor industry as well as to produce free-standing components or deposit inside large diameter tubes, there can be no one universally applicable reactor system. Some of the equipment used will be refered to in the sections below, but they are not to be regarded as process blue-prints.

4.2.3. CVD CONTROL PARAMETERS:

The principal variables which require control and monitoring are: pressure, temperature, reactant/product - activity/mass transfer and gas/vapour-flow dynamics. CVD processes are carried out from well below 1 atm, to high pressures. The total pressure of the reaction, the individual pressures of the reactant input and the product retrieval require control. The temperature is varied over the several zones of a CVD reactor.

The actual chemical reaction decides the heating mode and opera-ting temperature range. The reaction thermodynamics and kinetics must be defined; a temperature differential has to be maintained for vapour transport; endothermic or exothermic reaction effects have to be compensated suitably; the substrate may require sepa-rate control to induce product deposition, nucleation and growth, and to achieve a uniform deposit; mass and heat flow aspects have to be taken into account; and a control on the reactant/product removal at the post-deposition stage is necessary either for recycling or recovery and disposal.

The actual quantity of the gaseous products introduced, their movement along and through the reactor onto and past the sub-strate are inter-related. Although good mixing may be obtained under turbulent flow conditions, this would not be a desirable situation at the actual deposition site. Streamlining the gas flow is necessary in order to achieve reproducibility and unifor-mity. The boundary layer theory is advocated to provide an under-standing of CVD process dynamics (Hammond 1979). Optimising the gas flow for reactant supply to the substrate is crucial for achieving satisfactory deposition.

4.3. CVD REACTOR

The CVD reactor assembly consists of three components:

1. The reactant supply system.
2. The deposition system.
3. Reactant/product retrieval, recycle and disposal system.

A detailed discussion of the reactor assembly is available (Blocher 1982). Since each CVD configuration is individually tailored, only the main features of a reactor assembly will be given here.

4.3.1. THE REACTANT SUPPLY SYSTEM:

The most important fact to be remembered is that although a solid deposit is achieved on a substrate in a CVD process, the actual state of the reactant-product configuration is not always the vapour or gaseous state. Both the supply and retrieval stages must consider introduction, renewal and disposal of solid and gaseous components (and perhaps sometimes even liquid). Suitable measures are necessary to counteract condensation and blocking.

Solid reactants have to be held at the reaction temperature with optimum exposure of surface area. So the form in which they are used will be important. A reactant of the same weight will have greater surface area as it varies from a solid chunk to a pellet, sponge, wool or powder form. Sintering of powder or any other finer form of substance progressively reduces the effective surface area. Pellets will need to be in a perforated tumbling container. Reactants in liquid form would have to be either evaporated or be borne by neutral carriers or reactant gases at the required saturation levels. By controlling the volume and rate of flow of the carrier gas, knowing the vapour pressure of the liquid reactant and the temperature, the actual partial pressure can be controlled. Replenishment procedures must be worked out.

Gaseous reactants may be available individually or pre-mixed in tanks and their supply controlled by pressure gauges and flow meters. Gases have to be suitably dehydrated and purified, e.g. H_2 through Pd- Ag, A and He over Zr at $900^{\circ}C$, Cu turnings and Ti getters for de-oxidising, carbon stack etc., and diluted suitably with neutral gases to control their partial pressures. Other gases may have to be generated by _in situ_ chemical reactions, e.g. H_2S or SO_2, HCl or Cl_2 in dilute concentrations. The entire reactant supply has to be carefully balanced before introduction into the reactor system, predicted to reach the required partial pressures at the temperatures of reaction and deposition.

Preferential pick up of a more volatile reactant and separation of gases of different densities have to be compensated for. A known quantity of a volatile component can be injected by a "flash vaporizer". A minimum gas flow rate must be maintained (usually >60 ml/minute) to avoid thermal segregation of gases of different densities. A by-pass from the deposition chamber is essential so that the reactant composition can be monitored to the desired potentials prior to commencing the deposition. Careful control on the quality (purity) of the reactants is vital.

4.3.2. THE DEPOSITION SYSTEM:

(a) Substrate Preparation:

Deposition on a substrate should not be adherent and can have any morphology if the product is to be merely extracted via CVD. To obtain a vapoform it should not adhere to its mandrel but must have a good, and even multi-crystal growth. But for other CVD purposes deposits have to be not only adherent and strongly bonded, but also be compatible to the substrate in many other physical, chemical and mechanical aspects like thermal expansion, chemical resistance, prefered crystal orientation, hardness, stress control or release, and optical and electrical properties. The substrate/deposit adherence and growth are perhaps the two primary features to consider. The substrate,

1. must be free of organic or inorganic contaminants (oxides, grease, dust) which can affect adhesion or cause side-reactions with the highly reactive reactant-product configuration.

2. must be shielded from any marked attack by reactant components at least until the coating process has resulted in a thin layer of the deposit (some attack may be acceptable if it can improve the bonding).

3. must be covered with an intermediate layer if the substrate itself is highly reactive and can act as a reducing agent, or if it has poor resistance, e.g. Ti or Al substrate; or if the substrate is too weak in corrosion resistance, e.g. Fe-alloys; Fe-alloy/Ni/W and Fe-alloy/TiC/Ta sandwiches. Treating Ti with BCl_3-doped H_2 allows subsequent B deposition on it (Thebault et al 1978).

4. CVD processes are normally at higher temperatures than PVD. The coating/substrate configuration has to consider its compatibility at the temperatures it is going to be used at, and not merely the actual deposition temperature. A possible substrate/coating interaction, constituent inter-diffusion and changes in phases and micro-structure at working temperatures must be pre-considered.

Degreasing, pickling, grit or shot blasting (impact treatments), a quick exposure to a reducing agent, inert gas gettering, gaseous- or ion-etching, ultrasonic cleaning, vapour degreasing and electrochemical cleaning are a few of the common methods of substrate cleaning.

(b) Substrate Heating:

The means by which the substrate is heated, the manner in which the temperature is controlled over the entire substrate area, and the method by which an even deposit can be achieved over the whole of the area to be coated, form the temperature features incorporated in the deposition systems. The substrate size and geometry are two of the factors which decide the heating method and of course, whether it is a conductor or a non-conductor.

Heating can be achieved by

1. Direct contact of a heat source.
2. Direct heating.
3. Induction substrate heating (cold wall).
4. Indirect furnace heating, by radiation (hot wall reactor).

CVD reactions can be exothermic but mostly are endothermic. This could result in substrate cooling. On a production scale large volumes of gases are handled, which can induce convective cooling. Both these situations have to be compensated for. Substrate heating optimization involves achieving the correct temperature profile and coating exposure time.

Fibre-form CVD products pose special problems in substrate heating. Resistive heating using mercury contacts present contact problems if the coating is not very conductive and also health hazard problems; glow discharge requires reduced pressure. Continuous coating of moving wires has to circumvent breakdown due to sparking at contact points. A new contactless method has been devised in the authors' laboratory to overcome this problem (Hocking & Sidky 1988).

(c) Substrate Positioning:

The third parameter is the actual positioning of the substrate in the reactant-product stream. In a turbulent flow region (Reynolds number >2000) the reactant gases will have an ideal mixing condition, but a reduced dwell time of contact with the heated substrate for a solid product to form. The solid product, even when formed would be buffeted preventing adequate nucleation and growth on the substrate, and it can be swept out of range of the substrate or cause it to be uneven in supply. Turbulent gases also cause uncontrollable cooling effects and affect the substrate temperature profile, and are additionally disadvantageous where an intermediate reaction product is a necessary reaction step needing some extra-reaction time, e.g. $HBCl_2$ in boride production. Also, very large volume of gases will be needed which would be uneconomical.

The substrate/reactor geometry in a deposition system is thus designed to achieve a laminar flow. The input can be parallel, perpendicular or angular to the substrate depending on the substrate holder design. Movement (rotation or lateral etc.) of the substrate within this region also compensates to some extent any non-uniformity in the deposition. A jet design (one or 2 stage-type) is aimed at optimising variations arising from the depleting product/reactant conditions; and even a fan or blower is used to obtain gas circulation, e.g. chromizing of coils of steel sheets).

The relevant and basic aspect is that there should be a good thermal control and even contact and supply between the reactant /product system and the receiving substrate. It is easier to alter and manipulate the substrate section of the assembly where the substrate size is small and can be easily situated in the laminar flow region. The bigger the bulk of the substrate, the more is the necessity to actually move the reactant/product configuration. Thus fan blowers are on the other extreme end of the reactor requirements. Section 4.4 outlines some of the background principles.

4.3.3. THE RECYCLE/DISPOSAL SYSTEM:

Safety and economics of operation are the two main factors which decide the design of this segment in the assembly. The deposition efficiency, dictates the degree of retrieval, recycle and disposal procedures. Condensates have to be collected, toxic and corrosive items have to be rendered harmless (broken down, decomposed or converted) by auxiliary chemical (or physical) reactions, inflammables have to be combusted, dissolved or absorbed. Waste effluents disposal must be done with care.

4.4. CVD REACTORS: LINE DIAGRAMS

A selection of line diagrams of CVD reactors used in a few well known industrial processes as well as some experimental projects are collated in this section. This is a random selection and is only intended to give an idea of the variations in CVD reactor assembly.

Fig.4-1: Experimental CVD reactors (Takahashi & Kamiya 1977)
Fig.4-2: Barrel & Horizontal slab reactors (Juza & Cermak 1982)
Fig.4-3: Multi-boat reactant supply to reactor (Tick et al 1974)
Fig.4-4: Ti-Zr-B deposition reactor (Takahashi & Kamiya 1977)
Fig.4-5: AlB_{12}-B on Mo or W filament (Motojima et al 1981)
Fig.4-6: TiN by CVD (Peterson 1974)
Fig.4-7: Another TiN reactor (Takahashi & Itoh 1977)
Fig.4-8: CVD of TiB_2 (Besmann & Spear 1977)
Fig.4-9: Schematic for Ti-B-N deposition (Peytavy et al 1978)

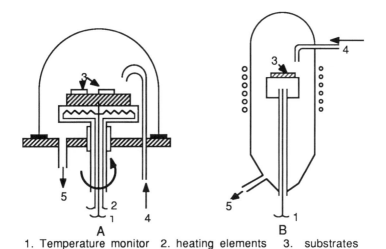

1. Temperature monitor 2. heating elements 3. substrates
4. inlet 5. exhaust

Fig.4-1a,b:Experimental CVD Reactors :
A: resistive heating   B: Induction heating

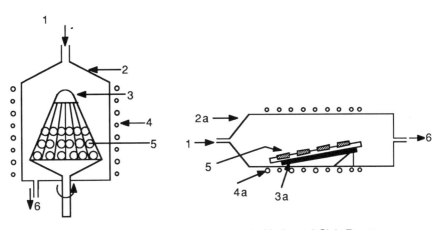

a. Barrel Reactor   b. Horizontal Slab Reactor

1. gas inlet (eg. H_2 + $SiCl_4$) 2. fused silica tube 2a. quartz tube
3. graphite heater 3a. graphite susceptor   4. rf heating coil
4a. induction coil or halogen lamps 5. Si wafers 6. gas outlet

Fig. 4-2: Schematic of Barrel & Horizontal Slab Reactors

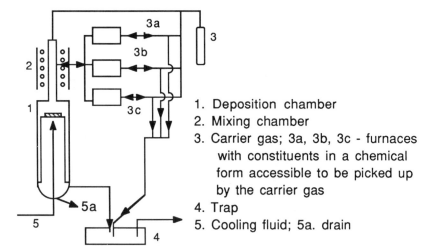

1. Deposition chamber
2. Mixing chamber
3. Carrier gas; 3a, 3b, 3c - furnaces
   with constituents in a chemical
   form accessible to be picked up
   by the carrier gas
4. Trap
5. Cooling fluid; 5a. drain

Fig.4-3: Schematic diagram of a CVD reaction system

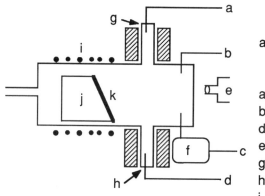

Fig.4-4: CVD of Titanium Zirconium
Boride-Schematic diagram

a-d: gas train with driers,
     flowmeters & oxygen
     gettering units;
a : chlorine + argon
b : hydrogen   c : Chlorine
d : argon
e : pyrometer f: BCl_3 generator
g : $TiCl_4$ generator
h : $ZrCl_4$ generator
i : Induction coil
j : susceptor
k : substrate

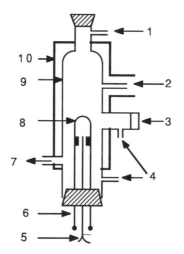

1. BCl_3 generator; Cl_2 +Ar carrier gas passed over B_4 C in a heated boat;
2. Cl_2 +Ar carrier gas passed over Al metal in a heated boat
3. optical window 4. Ar &/or H_2 5. chromel-alumel thermocouple
6. 2mm thick Mo conductors 7. gas outlet   8. Mo or W filament
9. Quartz reaction tube (26 mm i.d.) 10. Outer envelope encasing
nichrome heating elements

Reactor for Boride Composite on Al or W Filament
Fig. 4-5

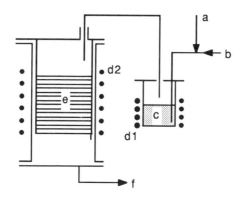

a: hydrogen   b: nitrogen; c: $TiCl_4$; d1: furnace at 25-50 °C;
d2: furnace at 900-1000 °C; e: carbide inserts
f: HCl out

TiN coating by CVD - Schematic diagram
Fig.4-6

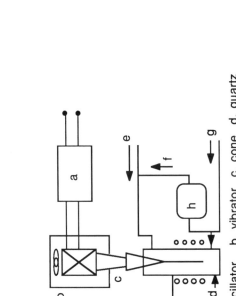

1. cooling coil 2. horn
3. silver solder 4. substrate
5. quartz reactor tube
6. window 7. mirror
8. pyrometer 9. $TiCl_4 + H_2 + N_2$
10. hydrogen

a. Oscillator b. vibrator c. cone d. quartz
reactor tube e. dried, hydrogen passed over
activated copper f. from Ti-tetrachloride
saturator h g. dried nitrogen over activated
copper

Fig.4-7a: Experimental CVD reactor for Ti coating   Fig.4-7 b: 'd' - The reactor tube (left)

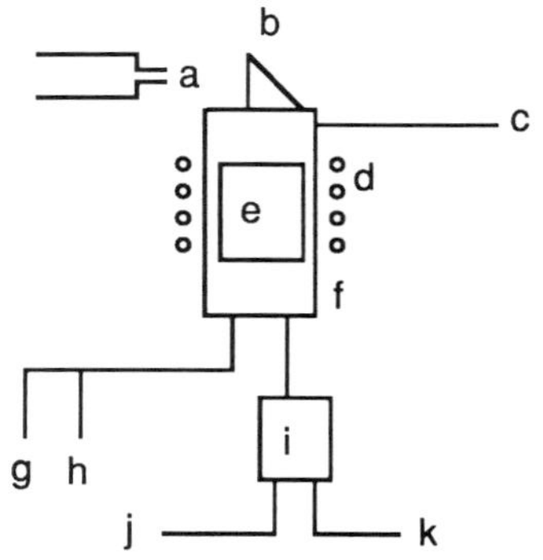

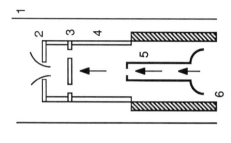

1. silica tube 2,3,4. rf-heated graphite assembly 5. vycor tube 6. mullite.

note: the graphite is surrounded by Mo shields and Mo foil covers the lid on top.

Graphite assembly in f - the fused silica reactor & gas flow pattern

Fig.4-8b

a. pyrometer b. prism c. vacuum pump & exhaust d. rf coil e. substrate f. fused silica reactor g. inert gas h. boron tri-chloride i. Ti-tetrachloride boiler j. hydrogen k. injector for $TiCl_4$

Fig. 4-8a

Schematic diagram of the CVD system for TiB_2

129

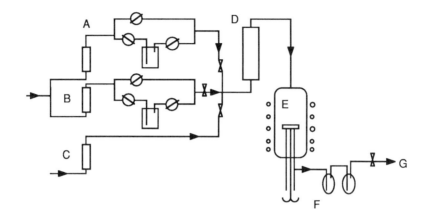

A. TiCl$_4$ evaporator   B. BCl$_3$ evaporator :- with H$_2$ as carrier gas;
C. N$_2$ line; Gases dried, metered and de-oxygenated before entering
A,B & D. D. Mixer unit E. HF- induction heated CVD reactor tube with
substrate temperature monitored.   F. Traps   G. to vacuum pump

Fig.4-9: Flow diagram of CVD apparatus for Ti-B-N Coatings

Fig.4-10: TiC by CVD (Hava et al 1978)
Fig.4-11: Deposition apparatus for ZrC+C (Ogawa et al 1979)
Fig.4-12: ZrN whisker by CVD (Motojima et al 1979)
Fig.4-13: ZrC$_x$N$_y$ deposition (Takahashi et al 1981)
Fig.4-14: Pyrocarbon from fluidized bed (Nickel 1974)
Fig.4-15: Coating of particles with pyrolytic carbon by CVD
          (Blocher 1982)
Fig.4-16: Si on Ni by CVD using SiCl$_4$ (Subramanian 1982)
Fig.4-17: Oxide crystals of Ru and Ir by metal vapour transport
          in an open tube reactor (George et al 1982)
Fig.4-18: Internal CVD of large substrate - mild steel tube coa-
          ted with Ta by CVD (Beguin et al 1977)
          (a second example of large substrate:- TiN on long
          steel tubes,10mm I.D.:tube to be coated is horizontal
          with a moving furnace (Itoh et al 1986)
Fig.4-19: CVD of MoSi$_2$ (West & Beeson 1988)
Fig.4-20: CVI/CVD :a.Thermal gradient design; b.Isothermal CVI
          (Buckley 1988)

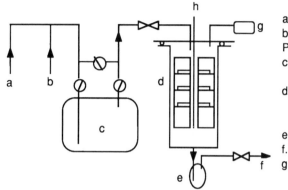

a. Hydrogen line
b. Methane line
Purified gases enter
c. TiCl$_4$ evaporator

d. Reactor with test
   substrates with
   thermocouple 'h' at centre
e. Liquid nitrogen trap
f. To vacuum pump
g. Vacuum gauge

Fig. 4-10: CVD of Ti-carbide

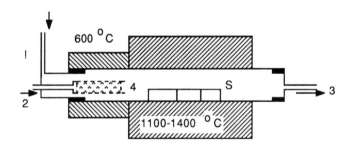

1. Argon+hydrogen+ methane; 2. Argon +bromine
3. Trap; 4. Zr; S. Substrates

Fig. 4-11: Deposition apparatus for ZrC+C

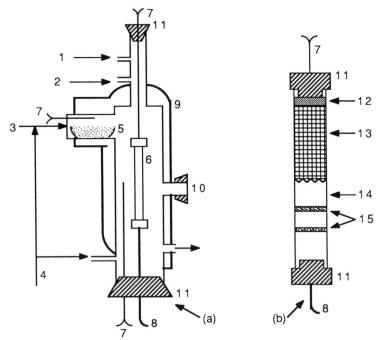

1. hydrogen 2. nitrogen 3. chlorine 4. argon (all gases
dried/purified/deoxygenated as appropriate)   5. boat with
Zr sponge 6. substrate assembly 7. a.c. thermocouple
8. electrode 9. outer manifold with heater element all
around the inner manifold   10. viewing window
11. graphite end cap 12. carbon powder 13. SiC heater
14. Quartz substrate 15. impurity painted zone

Fig. 4-12: (a) Apparatus for CVD of ZrN Whiskers
(b) Details of sample holder 6

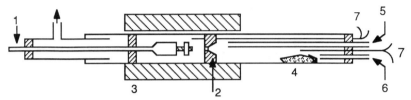

1. quartz holder carrying the substrate 2. graphite nozzle
3. furnace 4. $ZrCl_4$ powder boat 5. CH_4-N_2-H_2 6. argon
7. thermocouple

Fig. 4-13: CVD reactor design for $ZrC_x N_y$ Coating

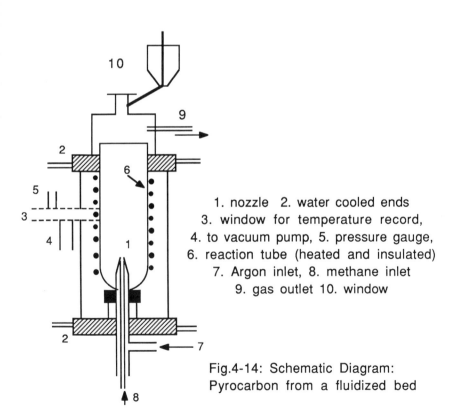

1. nozzle 2. water cooled ends
3. window for temperature record,
4. to vacuum pump, 5. pressure gauge,
6. reaction tube (heated and insulated)
7. Argon inlet, 8. methane inlet
9. gas outlet 10. window

Fig.4-14: Schematic Diagram:
Pyrocarbon from a fluidized bed

133

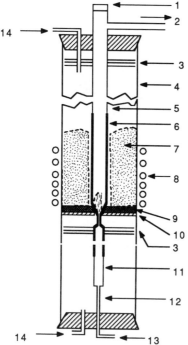

1. optical window
2. exhaust products
3. radiation shields
4. quartz manifold
5. vycor adapter
6. graphite reactor with nozzle
7. carbon black insulation
8. induction coils
9. carbon wool insulation
10. quartz plate
11. teflon tubing
12. rubber tubing
13. He+Hydrocarbon input
14. He purge

Fig. 4-15: Powder Coating with
Pyrolytic Carbon. An Experimental
Set up

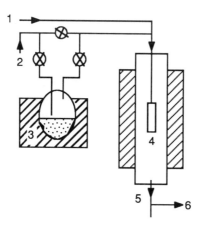

1. hydrogen
2. argon (gases dried & metered)
3. silicon tetrachloride bubbler in a
   low temperature bath
4. substrate in a resistance heated
   quartz reactor
5. To rotary pump & exhaust
6. to mineral oil bubbler

Fig. 4-16: Silicon Coating on Ni by CVD

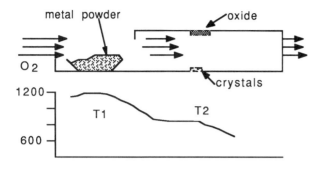

Fig. 4-17: Chemical vapour transport of Ru or Ir in open tube mode

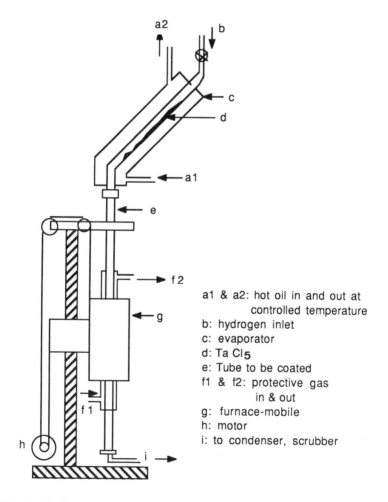

a1 & a2: hot oil in and out at
controlled temperature
b: hydrogen inlet
c: evaporator
d: Ta Cl$_5$
e: Tube to be coated
f1 & f2: protective gas
in & out
g: furnace-mobile
h: motor
i: to condenser, scrubber

Fig.4-18: Schematic diagram for CVD of Ta inside a long tube

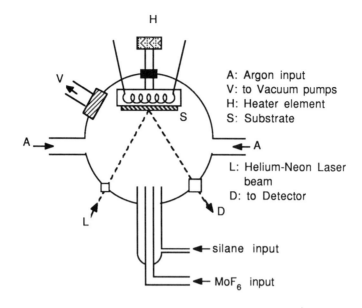

A: Argon input
V: to Vacuum pumps
H: Heater element
S: Substrate

L: Helium-Neon Laser
   beam
D: to Detector

← silane input

← MoF$_6$ input

Fig. 4-19: Reactor for CVD of MoSi$_2$

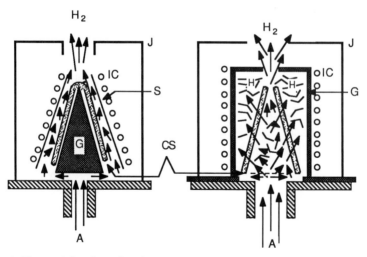

1: Thermal Gradient CVI/CVD   2: Isothermal CVI/CVD

A: Methane+Hydrocarbon+Carrier Gas; J: Outer jacket; S: Sleeve
IC: Induction Coil; CS: Carbon substrate; G: Graphite susceptor

Fig. 4-20: Carbon-carbon CVI/CVD

4.5. THE FUNDAMENTALS OF CVD AND CVD TECHNOLOGY

CVD occurs through a chemical reaction. It has to consider (i) the thermodynamics of the system which can drive the chemical reaction, (ii) the basic chemical kinetics to provide information on the reaction rate, and (iii) mass and heat transfer profiles which are influenced by the reactor and substrate size and design.

4.5.1. CVD THERMODYNAMICS:

The fundamental equation relationships are:

$$\triangle G^O = \triangle H^O - T \triangle S^O = - RT \, LnK \qquad and$$

$$K = \frac{\text{Multiple of product activities \&/or partial pressure}}{\text{Multiple of reactant activities \&/or partial pressure}}$$

$\triangle G^O$ = The reaction free energy change; $\triangle H^O$ = The reaction enthalpy; $\triangle S^O$ = The reaction entropy; K = The equilibrium constant and, T = The reaction temperature in Kelvin.

CVD reactions are complex and have to consider several equilibria and interactions and many data may have to be estimated or even experimentally determined. There are good sources for thermodynamic data (Bernard 1981; Colmet et al 1981; Vandenbulcke 1981; Hunt & Sirtl 1972; JANAF tables 1971; CRC Handbook of Chemical Data 1987; Kubaschewski et al 1979). Several computer programmes have been used in analysing complex reactions. SOLGAS for TiB_2 from $TiCl_4$, BCl_3 and H_2 (Besmann & Spear 1977) and EQUICA of Battelle for the Si-H-Cl system (Hunt & Sirtl 1972; Hall & Broehl 1967) are a few early examples of computer technology as a diagnostic tool in CVD fundamental work. Besmann (1986) gives programmes for Si_3N_4-BN coatings. Computerization is now an inherent aspect of process design, technology and production control (Headinger & Purdy 1987; Eckersley & Kleppe 1987).

Schafer (1964) formulated ten rules for predicting chemical transport reactions. Later, Alcock & Jeffes (1967) & Jeffes (1968) published the utility of G^O-T diagrams in assessment of vapour transport. The several points of interest are:

(i) The multiplicity of a system is the ratio of the number of moles of reacting gas to the number of moles of gas produced.

(ii) Vapour transport by variation of K, the equilibrium constant, requires a non-zero $\triangle H^O$ for the reaction and the maximum attainable transport is when K=1 and p, the partial pressure of the product = 0.5 atm.

(iii) For CVD to take place, $\triangle S^O$, the entropy change, must be negative for exothermic reactions and positive for endothermic reactions.

(iv) Exothermic CVD reactions are favourable with an overall diminution in volume while endothermic CVD reactions function best with increase in volume.

(v) A multiplicity of 1/1 is not ideal for vapour transport; the greater the ratio and difference the better it is for control.

(vi) The total pressure of the CVD system is influenced by $K = 10^n$ where K is the equilibrium constant and n is the multiplicity.

(vii) The productivity function is defined as

$$PF= p(product) \ [dp(product)/dK]$$

Comparisons of Fig. 4-21 (i) to (viii) showing thermodynamic productivity predictions, with the experimental and calculated data for TiB deposition in Fig. 4-22 (i)-(iv), illustrates the usefulness of fundamental data. Fig.4-23 shows the effect of turbulence on the thermal decomposition rate of iron carbonyl.

Thermodynamic equilibria for Si-H-Cl and Si-H-Br systems have been examined using JANAF tables; a higher enthalpy of formation for $SiCl_3(g)$ is suggested (Hunt 1988; also Woodruff & Sanchez-Martinez 1985). The thermodynamics of TiB_2 deposition has been given with effective detail (Besmann & Spear 1977). Thebault et al (1978) have studied the chemical compatibility of substrates in terms of free energy for Ti, Ta & W and their borides as substrates for CVD of boron from BCl_3-H_2 mixtures. Computer calculations used for prediction of the CVD of SiC are reported by Wu et al (1977). Computed thermodynamic values have also been reported for Si_3N_4-BN coatings (Besmann 1986). Fig. 4-24 shows the well-known free energy diagram for a few CVD reactions (Alcock & Jeffes 1967).

4.5.2. CVD KINETICS:

It is seldom that kinetics is considered on its own as it is closely linked to the thermodynamics of the system on one side and to the mass and heat transport on the other. Deposition kinetics quantifies the rate of growth of a CVD product in terms of variable total and partial pressures, activities and temperature of the system. Deposit thickness and distribution on substrate, diffusion - of the reactants and products at the substrate vicinity, reaction/deposit/product interaction, are all factors which cannot be clearly predicted, because of either a lack of experimental data viz. ceramic systems, or because of the variable geometry of CVD reactors for the deposition of one product.

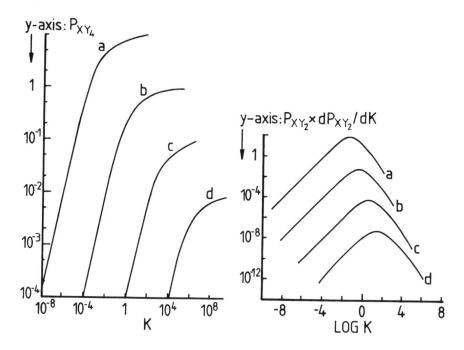

a. ΣP=10 atm. b. ΣP=1.0 atm c. Σp=0.1 atm d. ΣP=0.01 atm.

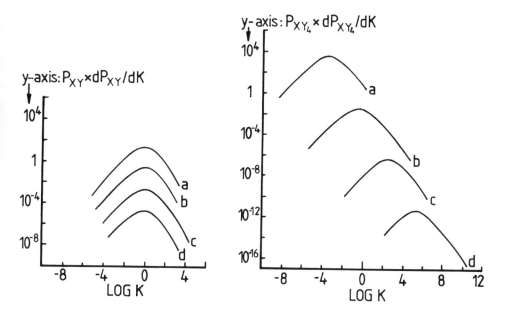

Fig.4-21 (i)-(iv): Product Partial Pressure as a Function of
the Equilibrium Constant at Various total Pressures P.
Reactions from Left to Right: 1/2; 1/4 & 4/1.

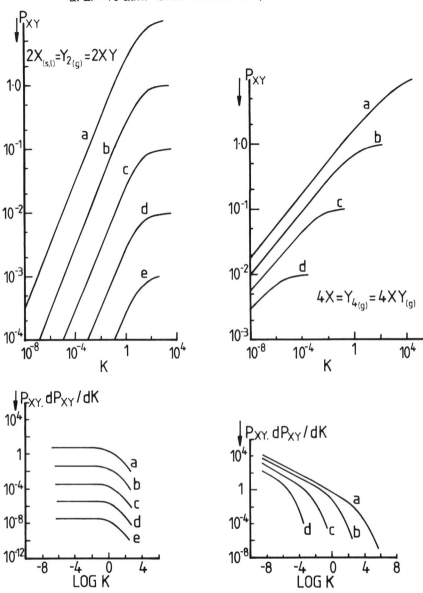

a. $\Sigma P = 10$ atm.   b. $\Sigma P = 1.0$ atm   c. $\Sigma p = 0.1$ atm   d. $\Sigma P = 0.01$ atm.

Fig.4-21 (v)-(viii): Product Partial Pressure as a Function of the Equilibrium Constant at Various total Pressures P. Reactions from Left to Right: 1/2; 1/4 & 4/1.

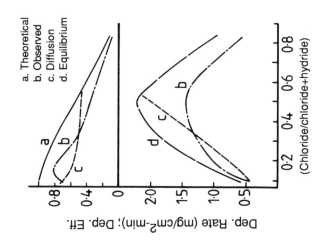

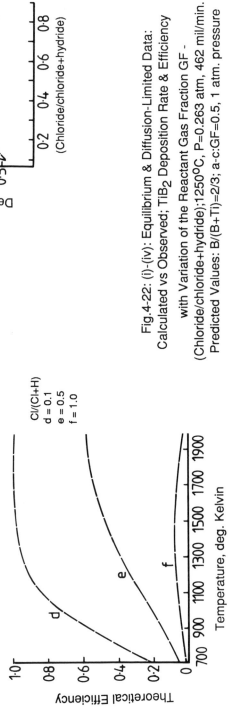

Fig.4-22: (i)-(iv): Equilibrium & Diffusion-Limited Data: Calculated vs Observed; TiB$_2$ Deposition Rate & Efficiency with Variation of the Reactant Gas Fraction GF - (Chloride/chloride+hydride);1250°C, P=0.263 atm, 462 mil/min. Predicted Values: B/(B+Ti)=2/3; a-c:GF=0.5, 1 atm. pressure

141

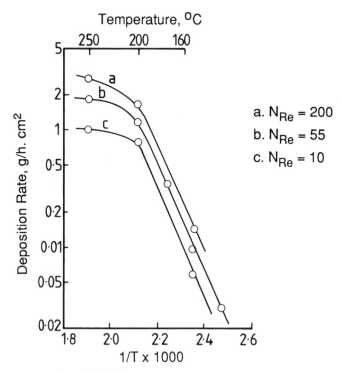

Fig.4-23: Thermal Decomposition of Iron Carbonyl
at 20 Torr Pressure. Effect of Temperature &
Reynolds Number on Fe Deposition Rate.

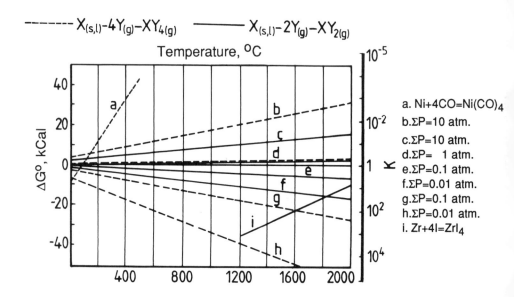

Fig.4-24: Equilibrium Constants from Vapour Transport Reactions
2/1 & 4/1 (most favourable values).

4.5.3. CVD & MASS TRANSPORT:

Mass transport of a CVD system links diffusion control, chemical kinetics and fluid-flow dynamics. Diffusion growth can be ascertained from chemical kinetics by,

Rate constant of growth = Aexp(-E/kT), where,

E is the activation energy, k is Boltzmann's constant, and T the temperature. With increase in substrate temperature the growth rate changes at a transition temperature from a predominant kinetic to a diffusion control, the latter being mostly under mass transport influence (Fig. 4-25) (van den Brekel 1977).

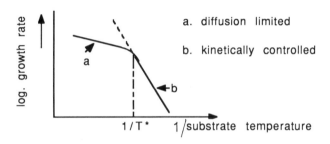

Fig. 4-25: Growth rate curve of a CVD process - schematic

Several mass transport models have been discussed by Blocher (1982). A few are listed: Hill & Boone 1982; Treyball 1955; Takahashi et al 1972; Diamuka & Curtis 1973; Vandenbuleke & Vuillard 1977; Subrahmanyam et al 1980; Juza & Cermak 1982).

A few significant assumptions in considering mass transport models can be set out as follows:

1. Kinetic control implies only a linear rate restriction.

2. Transport is considered in terms of diffusion and mass transport and therefore flux, across a boundary layer at the substrate/vapour interface.

3. Fluid flow - stagnant, laminar, turbulent, convection etc., is an important parameter to the above, since this determines the boundary layer.

Two mass transport treatments are given below:

(i) The transition temperature where diffusion takes over from kinetic control is process-dependent. van den Brekel (1977,a,b) has derived dimensionless CVD and CVD_{micro} numbers :

$$CVD = \frac{P_b - P_s}{P_s - P_{eq}} = \frac{k_D \, \delta \, T_o^2 \, \ln (T_s / T_b)}{D_o \, (T_s - T_b) \, T_s}$$

where, P_b and P_s = partial pressures of the reactant gas in the bulk and at the surface respectively,

P_{eq} = equilibrium partial pressure of reactant at surface,

k_D = mass transfer coefficient at the surface, cm/sec, (rate of coating thickness increase),

δ = thickness of boundary layer,

T_s and T_b = temperatures of surface and bulk gas stream respectively,

D_o = diffusion coefficient, cm^2/sec, at reference temperature To; $D_T = D_o \, (T/To)^2$.

Where the surface reaction is not at equilibrium, $P_s \neq P_{eq}$, and P_s/P_{eq} is the supersaturation at the surface which determines the ultimate rate of deposition as well as the morphology (temperature dependent).

In the simplest case, where $T_s \cong T_b$,

$$CVD) _{T_s = T_b} = \left. \frac{P_b - P_s}{P_s - P_{eq}} \right) _{T_s = T_b} = \frac{k_D \, \delta}{D_{T_s} T_s}$$

which is the equivalent of the Nusselt number (hd)/k in heat transfer where,

h = film coefficient for heat transfer, $cal/cm^2/sec/deg$ C
d = length parameter, cm,
k = thermal conductivity of gas, cal/cm/sec/deg C.

The equivalence of mass transfer and heat transfer has been clearly established (Treyball 1955) and the correlations of heat transfer with fluid flow parameters have been useful in establishing mass transfer coefficients.

In the equation for the CVD number, where the diffusional resistance is relatively large, $(P_b - P_s) > (P_s - P_{eq})$ and CVD >1. Where the surface reaction is slow, $(P_s - P_{eq}) > (P_b - P_s)$ and CVD <1, the process is surface reaction limited. By properly evaluating the CVD number, the equivalence or lack of equivalence of different CVD systems can be established using this tool.

144

Van den Brekel has been successful in explaining quantitatively the filling or absence of filling of surface grooves or depressions by CVD coatings (Blocher 1982).

A kinetic control process has CVD<1 and a diffusion limited process has CVD>1. To relate deposition to depressions in the substrate a CVD micro number is defined as CVD(micro) = (d/D)CVD where, d is the depth of the groove or depression and D is the boundary layer thickness.

A value of CVD(micro) = 10^{-4} is caused by surface control, while CVD(micro) = 10^3 results from diffusion control. The parameter is linked to the various partial vapour pressures calculated by thermodynamics (van den Breckel 1977a,b).

(It may be of interest to note that deposition and crystal growth influenced by polarization effects in electrolytes, are very similar to mass transport effects in CVD).

(ii) Hammond's Boundary Layer Theory (BLT) considers the simplest of the epitaxy reactors - the horizontal slab reactor (Hammond 1979):

BLT assumes zero gas velocity for some distance away from the substrate surface, and rises instantly to the average velocity. The zero velocity area is termed the boundary layer. The diffusion rate variations of reactants and products across BLT are:

1. directly proportional to temperature, as $T^{3/2}$,
2. inversely proportional to total pressure P_t, and,
3. directly proportional to the ingredient partial pressures, P_n

The BL thickness through which both reactants and products must diffuse through varies,

1. directly proportional to distance x, as $x^{1/2}$ and,
2. inversely proportional to the gas velocity v, as 1/v.

Tilting the substrate holder brings the surface more parallel to the $x^{0.5}$ BL; increasing the main gas velocity makes the BL thinner especially towards the downstream section; and, increasing the temperature at the rear of the substrate holder will compensate for the lowered partial pressure or depletion of the reactants. Tilting is advantageous for both horizontal and barrel reactors. Working at very low pressures, eg. 1/700 bar, where gas diffusion rates are very high, wafers can be aligned in a parallel pack vertical to the reactor wall. Increasing the total gas flow rate increases deposit thickness at the rear (downstream section) and increasing the total pressure increases thickness at the front. Rotation tends to average out thickness variations.

Boundary layers have also been interpreted in terms of a chemically frozen state as well as in local thermochemical equlibrium.

Laminar convective-diffusion boundary layers at high Reynolds numbers include thermal diffusion but no kinetic barriers. The theories have been used to interpret both CVD and Na_2SO_4 deposition (Gokoglu 1988).

A number of CVD reactor models have appeared in recent literature. Three dimensional effects in horizontal reactors for Si from SiH_4 (Moffat & Jensen 1988), a mathematical model for CVD of boron (Rebenne & Pollard 1985), Greg, a hot wall reactor for close spaced vapour transport deposition systems (Menezes et al 1985), models to meet needs of VLSI and MOCVD (Houtman et al 1986), vertical flow reactor model (Coltrin et al 1986), a versatile reactor for UHV applications based on a reusable metal sealing system (Hayes et al 1985), deposition thickness prediction (Bar-Gadda 1986; Middleman & Yeekel 1986) are a few of the many.

4.6. CVD REACTIONS : A SELECTION

(i) Thermal decomposition of metal carbonyls, hydrides, halides, and organo-metallic compounds, e.g.,

$$Ni(CO)_4 \xrightarrow{\quad 150 \ - \ 220^{\circ}C \quad} Ni \ + \ 4CO$$
    (H_2S as catalyst can lower the temperature to $100^{\circ}C$)

$$SiH_4 = Si + 2H_2$$

$$TiI_4 \xrightarrow{\quad 1200^{\circ}C \quad} Ti + 2I_2$$

$$CrI_2 = Cr + I_2$$

$$(C_8H_{10})_2Cr \xrightarrow{\quad 400 \ - \ 600^{\circ}C \quad} Cr + 2C_5H_{10} + 6C$$
(HI used as catalyst to suppress codeposition of C)

(ii) Reduction by hydrogen, or active metals like K, Na, Mg, Zn, Cd etc. The latter reactions are difficult to control, e.g.,

$$WF_6 + 3H_2 \xrightarrow{\quad 350-1000^{\circ}C \quad} W + 6HF$$

$$WCl_6 + 3H_2 \xrightarrow{\quad 500-1100^{\circ}C \quad} W + 6HCl$$

$$TaCl_5 + 5/2H_2 \xrightarrow{\quad 700-1100^{\circ}C \quad} Ta + 5HCl$$

$$SiHCl_3 + H_2 = Si + 3HCl$$

$TiCl_4 + 2Mg = Ti + 2MgCl_2$

$TaCl_5 + 5/2Zn = Ta + 5/2 ZnCl_2$

(iii) Disproportionation, e.g.,

$2GeI_2 = Ge + GeI_4$

$2WF_6 + 1/6C_6H_6 + 11/2H_2 = W_2C + 12HF$

$TiCl_4 + CH_4 = TiC + 4HCl$; (H_2 is used as a mixing component)

(iv) More examples of simple and complex compounds.

$$2AlCl_3 + 3H_2O \xrightarrow{600-700^{\circ}C} Al_2O_3 + 6HCl \text{ (uses tri-chloride)}$$

$$Al_2Cl_6 + 3CO_2 + 3H_2 \xrightarrow{800-1400^{\circ}C} Al_2O_3 + 3CO + 6HCl$$
(uses hexa-chloride)

$BCl_3 + 3/2H_2 = B +3HCl$

$$BCl_3 + NH_3 \xrightarrow{500-1500^{\circ}C} BN + 3HCl$$

$BCl_3 + PCl_3 + 3H_2 = BP + 6HCl$; ($N_2$ as diluent and carrier)

$CH_4 + H_2 = C + 3H_2$

$CrCl_3 + 3/2 H_2 = Cr$ (in Fe) $+ 3HCl$

$NH_4Cl \longrightarrow NH_3 + HCl$; $2HCl + Cr \longrightarrow CrCl_2(g) + H_2$

$CrCl_2(g) + Ni(s) \longrightarrow NiCl_2(g) + Cr(s)$;

$CrCl_2(g) + Ni(s) + H_2 \longrightarrow Ni-Cr(s) + 2HCl$

$Ga + AsCl_3 + 3/2 H_2 = GaAs + 3HCl$

$6xMoF + 6(1-x)WF_6 +3H_2 = 6Mo_xW_{(1-x)} + 6HF$

$SiHCl_3 + H_2 = Si + 3HCl$

$SiCl_3 + 3/2 H_2O+1/2O_2 = SiO_2 + 3HCl$

$$3SiF_4 (\text{or } SiCl_4) + 4NH_3 \xrightarrow{800-1400^{\circ}C} Si_3N_4 + 12HF \text{ (or 12HCl)}$$

147

$$TiCl_4 + CH_4 \xrightarrow[1000^{\circ}C]{\text{excess } H_2} TiC + 4HCl$$

$$TiCl_4 + 1/2\ N_2 + 2H_2 = TiN + 4HCl$$

$$2TiCl_4 + 2xCH_4 + yN_2 + 4(1-x)\ H_2 \xrightarrow{1050^{\circ}C} 2TiC_xN_y + 8HCl$$

$$TiCl_4 + 2BCl_3 + 5H_2 = TiB_2 + 10HCl$$

$$(1-x)\ TiCl_4 + xZrCl_4 + 2BCl_3 + H_2 \xrightarrow[x=1 \text{ to } 0]{900-1400^{\circ}C}$$

$$Ti_{(1-x)}Zr_xB_2 + 2HCl + 4Cl_2$$

$$ZrCl_4 + 1/2\ N_2 + 2H_2 \xrightarrow{1220^{\circ}C} ZrN + 4HCl$$

Equations cited above indicate the range of deposition which can be achieved by CVD, from pure metals and non-metals to complex compounds like $Ti_{1-x}Zr_xB_2$ whose stoichiometry can be manipulated by varying the reactant partial presssures and activities, and ammonialysis of $AlCl_3$ and silane to obtain mixtures of Al and Si nitrides, and the production of Ti carbide + nitride, boron nitirde and carbide are all interesting and practically useful CVD products.

(v) Ceramics production by CVD.

A number of inorganic and organic compounds have been used to produce ceramics by CVD. Metal organic CVD (MOCVD) (Besmann & Lowden 1988), Self-propagating high temperature synthesis (SHS) (Munir 1988) and chemical vapour impregnation CVD (CVI CVD) (Strife & Sheehan 1988, Buckley 1988; Reagan et al 1988; Fitzer & Gadow 1986) are all developments in producing ceramic materials in various forms.

The ceramics listed indicate their choice for high temperature corrosion protection by virtue of the high melting point of the elements and the strength and barrier layer properties of the resulting ceramic (Besmann & Lowden 1988).

The most common ceramics are:

Borides of :			Hf, Mo, Nb,		Ta, Ti,		W, Zr
Carbides of :	B, Cr,	Hf,	Nb, Si,	Ta, Ti, V,	W, Zr
Nitrides of : Al, B,		Hf,	Nb, Si,	Ta, Ti, V,	Zr
Oxides of : Al,	Cr,		Si, Ta, Ti,		Zr

(vi) MOCVD: Many organosilicon compounds have been used for producing silicon carbide and nitride ceramics by thermal decomposition, and are well suited for CVI/CVD. Whiskers, fibres, ceramic impregnation, and composite deposition are a few of the successes of silicon ceramics. Methyl and phenyl silicon chlorides, silazanes and polycarbosilanes have been investigated (Fitzer & Gadow 1986).

Liquid reactants such as methyltrichlorosilane and dimethyldi-chlorosilane are held in cylindrical or elliptic shaped pressure vessels into which the carrier gas is bubbled under controlled temperature. Segregation effects can be a hazard in deposition control. Heavy molecules in dilute concenrations carried by low atomic weight carrier gases are subject to thermal diffusion gradient and variations. It was found that such diffusion effects were more prominent with H_2 as carrier gas than with He (Holstein 1988). Headinger & Purdy (1987) present a process flow sheet for production level equipment.

The reactions or models listed are by no means exhaustive. More and more new reactions are under trial and may prove of indus-trial value. Models, process sheets and pilot plants are an inseparable segment of CVD development.

4.7. CVD- PRODUCTS & PROCESS ROUTES – A CROSS SECTION

A selection of CVD products and some salient process features are given in this section. Process features are not strictly confined in validity only to the product they appear in. Any similar system can be expected to be correlated. The main idea is to present the variety of CVD products and the extent of process technology.

4.7.1.

Al: Hot wall reactor and Ni substrate at 700-1100°C; $AlCl_3-H_2$ mixture held at 130°C max. passed over powder mixture of Al/Al_2O_3 (3:7) (Sun et al 1986).

Al: By MPCVD (Magnetron Plasma enhanced CVD) on Si wafer; $AlCl_3$-I_2 mixture; deposition at <100°C. In a planar magnetron with a closed path locus of the magnetic lines of force, cycloidal and helical electron motions are locally generated; The magnetic field applied to the plasma region confines the chemical reaction to the substrate surface region (Kato et al 1988).

B: Beta rhombohedral boron appears to be the thermodynamically stable form. Alpha rhombohedral and beta tetragonal form as metastable crystalline forms at low and high temperatures respectively. A cold wall retort and a hard metal substrate has been used (Voigt & Westphal 1982).

BC: 4 types of substrates: (i) Reactive attack with substrate compounds forming at the substrate/vapour interface; (ii) one reaction occurs at substrate/vapour interface and another in solid state; (iii) an 'overlayer' is formed at substrate; (iv) inert substrate. BCl_3-H_2-hydrocarbon at 1400K and 6.7 kPa total pressure, initiated Mo-boride formation on Mo as well as an overlayer of adsorbed hydrocarbons, the latter inhibiting surface reaction and deposit growth rate (Jansson et al 1987). Free standing shapes are formed over 1200-1400°C (Hayman 1984).

BN: Free standing shapes are produced by CVD at 1750-1950°C using BCl_3-NH_3 (Hayman 1984). With H_2 as additional reactant and 5-60 torr pressure deposition is carried out over 1200-2000°C (Matsuda et al 1986,1988). MOCVD of BN films at much lower temperatures are achieved using ammonia and tri-ethyl boron (Nakamura 1986). NH_3 and decaborane at 1:20 at substrate temperature 850°C yield thin, amorphous BN_x, with x<0.75 being a mixture of B and BN. It is a low pressure reaction (10^{-3} Torr) and termed melocular flow CVD (MF–CVD) (Nakamura 1985).

BP: Usually on Si, cemented carbide and graphite. BP is a semiconductor. Here it is deposited on Mo. MoP was found to be an intermediate layer. Microhardness increased with increase in B:P from 1.08 - 1.47. BP is put on Mo for oxidation resistance. BCl_3+ PCl_3 + H_2 + N_2 to yield BP deposit (Motojima et al 1979).

AlB_{12}(alpha)-B: Mixed phase with tetragonal boron using $AlCl_3$ + H_2 + Ar at 800-1400°C, or supplementing BCl_3 and reacting at 800-1300°C, on boron to aluminise it. Deposition rate increased with temperature and (BCl_3/$AlCl_3$) gas flow ratio and reached maximum at 1200°C. B/Al ratio in the deposit is independent of temperature >1200°C and BCl_3:$AlCl_3$ >1. Maximum microhardness of 3900 kg/mm^2 at BCl_3/$AlCl_3$ of 1-2 (Motojima et al 1981).

Carbon: Carbon-carbon and/or pyrolytic carbon via hydrocarbons are well known (Buckley 1988; Nickel 1974).

C: In high temperature gas cooled nuclear reactors (HTGRs), 500 micron diameter urania-thoria particle kernels are coated with a buffer layer of low density pyrocarbon and a layer of high density pyrocarbon or SiC which acts as a diffusion barrier for other fission products. Failure occurs by irradiation. Pyrocarbon coatings are deposited by CVD from a propylene + CO_2 mixture at about 1300°C, and the SiC by thermal deomposition of CH_3SiCl_3 in a H_2-

containing carrier gas at 1300 to 1700°C. SiC is the main barrier against solid fission product diffusion (Stinton & Lackey 1982; Wallura 1982; Nickel 1974). Carburizing is done in gases, salt baths, packs, or by vacuum carburizing. For the latter, components are heated in a vacuum at 1 mtorr at 1000°C and then methane is put in at 200 torr. C enters to the solubility limit (1.7% at 1030°C in Fe) (Child 1983).

Cr: On Nickel substrate is carried out in a hot wall reactor at 950-1050°C. NH_4Cl is the transport agent and H_2 the carrier gas at 300 ml/min; controlled at 180-280°C, the gas mixture passes over a mixture of Cr and alumina 80:20 wt%. Dissociation of NH_4Cl, $CrCl_2$ via HCl/Cr reaction occur. $CrCl_2$ then reacts with Ni with and without hydrogen to result in Cr solid and Ni-Cr surface alloying (Sun et al 1987).

Si: CVD Si on Ni forms a mixture of Ni_5Si and Ni_3Si above gamma-Ni. On alpha-Co it forms Co_3Si, Co_2Si. C73 is a directionally solidified Co-Cr alloy with Cr_7C_3 in the matrix. Plasma sprayed Ni was used as a sandwich between the siliconized substrate and matrix. Ni_3Si leading to Ni_5Si formed first, but the entire morphology changed on exposure to air at 1000°C for 1000h. This is discussed in greater detail in Chapter 9 (Fitzer & Maurer 1977). Very high purity Si is produced from $SiHCl_3$ in a bell-jar type reactor with the decomposition at 1100-1200°C. Recombination is possible at 325°C (Ingle & Peffley 1985; Bloem et al 1985).

The criteria for Si-deposition have been stated using cold wall and hot wall reactors (Wahl & Furst 1977) and the use of a high Al-content in the substrate as the criterion to promote a sound diffusion bond has been reported (Fitzer et al 1977), and other reports for improving adhesion (Betz 1977). A graded seal mechanism which causes a modification of the growth process and oxide pegging effects has provided parameters for better coating adhesion (Whittle & Boone 1981).

SiC: CVI/CVD of SiC is in the production forefront (Reagan et al 1988; Fitzer & Gadow 1986). A choice material for a number of high temperature applications, SiC has been produced by many CVD, and PVD methods (Brennfleck et al 1984; Motojima 1986). Conventional CVD (CCVD) uses a modified van Arkel-de Boer apparatus - a cold wall hot susceptor configuration. Mixtures of tetrachlorides of Si and C with H_2 are reacted between 1400 and 1900 $^\circ$C. Graphite and Mo susceptors are employed. A detailed discussion on SiC morphology is available (Pampuch & Stobierski 1976). MOCVD of SiC is another explored process route. Gruber (1977) describes SiC deposition on W wires to produce continuous SiC fibres. Methyltrichlorosilane pyrolysis with and without H_2 yields SiC deposition at 900-1200°C (Schintlmeister 1986) or 1100-1400°C (So & Chun 1988) respectively. Graphite, pyrocarbon fuel particles at 700-1150°C (Minato & Fukuda 1988) and cemented carbide are the substrates used. Cobalt silicide interlayer formation is avoided by using TiC or TiN as interlayers (Schintlmeister 1986). Free Si co-deposits with beta SiC under some conditions.

151

SiC is an excellent refractory material because of its outstanding high temperature strength, oxidation resistance, low expansion, good thermal conductivity, hardness, abrasion resistance and low density.

To produce high strength forms of SiC (and Si_3N_4), powders are densified by sintering, hot pressing or hot isostatic gas pressing with the aid of additives which function either by the production of a high temperature liquid phase, or by affecting solid state diffusion. Without these additives diffusion in SiC and Si_3N_4 is too low to allow sintering of the pure material to theoretical densities. However, it is very difficult in these materials to achieve the best possible high temperature mechanical properties and oxidation resistance, because of the additive presence.

CVD opens the route to produce purer high density materials, but strengths tend to be low, because the CVD products can have large columnar grains, some amorphous content which slowly converts to a higher density crystalline form (Godfrey & Pitman 1978), or internal stress. It is also difficult to achieve a constant deposit character over a large surface area; thus production of large and complex shapes are rendered difficult.

If the effects of flow and chemical reaction factors can be better understood, improved deposit character and fabrication capabilities could result. For example, it is possible to produce equiaxed forms of SiC in a fluidized bed with a moving substrate and CVD has become an important process for making ZnS laser windows. Considerable improvements have been made in CVD processes in the electronics industry, and similar benefits could result from research with the objectives of better strength via microstructure optimization, and fabrication capability by improving the spatial uniformity in the deposit. Such examples of basic research applied to CVD have been reported (Breiland et al 1986; Houtman et al 1986).

In ceramics fibre reinforcement plays a major role in strengthening and toughening the matrix. A method of CVD of SiC on fine W wires, which uses a novel method of contactless heating of the moving W substrate wire, has been developed in the authors' laboratory (Hocking & Sidky 1988). Previous techniques relied on resistive heating (where contacts can cause sparking at a resistive deposit and hence wire breakage, or present a health hazard (Hg contacts), or require a rough vacuum for glow discharge (UK 1969 Patents 1141840 & 1141551).

Most CVD methods for SiC growth use modified van Arkel de Boer apparatus with cold reactive gas mixtures at atmospheric pressure passing into a cold wall susceptor ($1400 - 2000^{\circ}$C) reaction vessel (Pampuch & Stobierski 1976). The gaseous reactants most often used are mixtures of Si- and C- bearing molecular species, perhaps with a trace of B-bearing species to aid sintering:

$$CH_3SiCl_3 \quad \text{--------} > \quad SiC + 3HCl$$

$$SiCl_4 + CCl_4 + nH_2 \quad \text{--------} > \quad SiC + 8HCl + (n-4)H_2$$

From such a reaction a straightforward deposition of SiC is envisaged but unfortunately SiC is readily produced containing an excess of either C or Si. The above reactants decompose near the hot susceptor and the stable species resulting at $1400 - 2000^{\circ}C$ are $SiCl_2$ and CH_4 and these probably deposit SiC at the hot interface. Other organosilicon precursors have been studied to relate precursor structure to SiC yield (Schilling & Williams 1983). If oxygen is present SiO_2 is likely to form in addition to SiC (Gusman & Tumakova 1971).

It was shown that at $1500 - 1800^{\circ}C$ with a Cl:H ratio in the gas of 0.1, only SiC should be deposited when Si:C mole ratios are only 1 to 1.3. Above 1.3 some Si is expected and below 1, some C; this allows control of stoichiometry of the SiC (Kuznetsov 1976). Thermodynamic calculations on the 4-component Si-C-Cl-H system show $SiCl_2 + CH_4$ to be the stable molecular species (Hamilton 1962; Arken et al 1968; Knippenberg et al 1973). At the gas/solid interface the following reactions are expected:

$$SiCl_2 + H_2 \quad \text{--------} > \quad Si + 2HCl$$

$$CH_4 \quad \text{--------} > \quad C + 2H_2$$

The effect of composition of the gaseous mixtures on the formation of the 2H polytype as well as the probable reasons for the common occurrence of stacking fault twins in the 3C polytype are discussed in detail by Pampuch & Stobierski (1976) in the review on the morphology of CVD SiC.

Si_3N_4: Si_3N_4 is usually prepared by injection molding of Si and nitriding, or hot pressing Si_3N_4 powder with a fluxing agent such as MgO or Y_2O_3. CVD Si_3N_4 exhibits a single phase alpha-Si_3N_4 but is often a mixture of amorphous and crystalline Si_3N_4. Pretreatment of the substrate with SiF_4+N_2 produces a more adherent coating on Ni-Co and Nb-base alloys. This step is not necessary for substrates like graphite, Al_2O_3, Mo, BN and Si_3N_4. However a high NH_3/SiF_4 ratio initially protects the substrates from reactant attack (Galasso et al 1978). Beta Si_3N_4 CVD coatings were obtained from $SiCl_4+NH_3+N_2$ mixtures at $1300^{\circ}C$ (Oda et al 1983). Reaction bonded silicon nitride is much advocated (Stinton et al 1988). CVD on quartz substrates (7mm o.d. and 100 mm long) using Si_2Cl_6, NH_3, H_2 gas mixtures deposited dense and amorphous layers of silicon nitride at $800-1100^{\circ}C$ and alpha silicon nitride at $1200^{\circ}C$ with N/Si ratio 1.33-1.77 (Motojima et al 1986).

Si-thermal nitriding has been done to obtain Si_3N_4 on Si wafers for metal-nitride-Si devices (MNS) for diodes and n-channel field effect transistors (FET). Si wafers are boron doped, degreased, boiled in H_2SO_4 and HNO_3, etched in HF and rinsed in deionized water, dried in N_2 and held in quartz tubes in inert gas e.g. Argon. NH_3 is introduced and nitridation done at 950-1300°C. Smooth Si_3N_4 is obtained 40A in 10h. Using SiH_4-NH_3 triples the rate (Ito 1978). Mass spectrometer analysis has been reported (Lin 1977). A low temperature method is to bubble NH_3 into cool $SiCl_4$ to form amorphous Si_3N_4 powder. Subsequent heating to 1200°C produces alpha Si_3N_4. The high temperature CVD method reacts at 800- 1200°C to yield a mixture of amorphous and crystalline Si_3N_4. Greater crystalline conversion is achieved when heated to 1400°C. This is because of the strong N-H bond which creates NH and NH_2 groups in the Si_3N_4 lattice. Using SiF_4, the most abundant species are SiF_2NH_2. In Si_3N_4 deposits vapour phase nucleation and interaction occur.

Si_3N_4-AlN from $AlCl_3$ + SiH_4 + NH_3 + N_2. (Metal-insulator-Si: MIS; metal-insulator-oxide-silicon : MIOS etc) e.g. M-Si_3N_4-SiO_2-Si. MAOS= metal-Al_2O_3-SiO_2-Si. These materials have charge storage capability. $AlCl_3$+N_2 and SiH_4+NH_3 are reacted first and the final mixing is done just prior to the products entering the deposition tube. AlN is polycrystalline and Si_3N_4 amorphous with mixed compound two-phase. A low NH_3/$AlCl_3$ ratio is desirable for an even and high deposition rate but not always because of property control. The purpose of these products is to use as a two-phase (AlN-Si_3N_4-Al_2O_3) dielectric to enhance storage property (Zirinsky & Irene 1978).

Si_3N_4-BN has been deposited by CVD, but again as amorphous silicon nitride and turbostratic BN (Hirai et al 1982). Neither of the above two nitride composites are useful as tribological materials. The need for a thermodynamic approach for modeling deposits is emphasised for a better understanding of process conditions. The silicon nitride-boron nitride system has been analysed (Besmann 1986).

SiO_2: He, N_2 or CO_2 + 0.07-0.5 vol% silane and 0.5-5 vol% O_2 on heated substrate. The order of the gas mixing was important. Gas with large O_2% met gas with high silane content. Reaction occurred even at room temperature producing dust in gas lines and also if the diluted gas contacted the bulk gas at 50°C above or below the substrate temperature. Therefore only the substrate was heated and not the gas. Silane+N_2 mixture were used for calibration. Below 300°C $(H_2SiO)_x$ formed. Flow rates were not kept constant because of incomplete coverage, but had to be alternated from Re 4000 to Re 12000. There appears to be not much success in this process; the British Petroleum process appears to be better. CVD kinetics for SIO_2 has been reported (Grabiec & Przyluski 1986).

The CVD of silica onto stainless steel for nuclear AGR-s hasbeen performed by pre-oxidising the steel in steam at 800°C for 2 hours followed by thermal decomposition of tetraethoxysilane at 800°C with a total pressure of 100 mbar (including carrier gas). The average silica thickness was 7 microns (Bennett et al 1984).

Mo: Mo coating on 304 stainless steel by CVD of molybdenum carbonyl has been carried out at substrate (deposition) temperatures in the range of 400-700°C and coatings with columnar grains and facetted surface have been prepared. It was found that both the size of columnar grains and surface hardness value of the coatings decreased with increasing substrate temperature. Post-deposition annealing of the coatings at temperatures above 800°C resulted in the formation of a diffusion zone at the interface between the deposit and the substrate. It is shown that the growth rate of the diffusion zone depends on the size of columnar grains. Recrystallization of deposited Mo was observed at annealing temperatures above 900°C (Yamanaka et al 1982).

Ta: Novel set up to coat large long tubes. 17mm id, 23mm od, 1500mm long. Moving furnace; N_2-5%H_2 maintain outside suface non-oxidising. Ta deposition obtained was pore free and provided resistance against acid corrosion (Beguin et al 1977).

TaSi$_2$: CVD carried out at low pressure. Process sequence requires deposition of undoped polysilicon followed by $TaSi_2$ in a single reactor. $TaCl_5$ reacts with SiH_4 to give Ta_5Si_3, and with Si to give $TaSi_2$. The Ta_5Si_3 reacts with polysilicon to form $TaSi_2$. The displacement reaction, silane reaction and interdiffusion reaction compete to influence deposit stoichiometry and interface roughness. Ar, H_2 and HCl are the carrier and reactant gases. The deposition is at 600°C and annealing at 900° (Williams et al 1986).

Ti: Internal CVD at 1050°C of long steel tubes 10mm i.d. $TiCl_4$-N_2-H_2 reactant mixture with argon carrier/diluent (Itoh et al 1986).

Ti-Zr-B: Deposition rate increased with increasing temperature but was insensitive to the total flow rate. High BCl_3 concentration caused porous, dendritic growth with higher deposition rate. High metal chloride pressures gave dense, uniform deposits (Takahashi & Kamiya 1977).

TiB$_2$: Mass transport modeling. Most efficient deposition at <1500K, under reduced pressures and in excess H_2. Below 1373K chemical kinetics controls. Changes in gas compositions change rate limiting mechanisms (Besmann & Spear 1977).

TiB_2 applied directly on cemented carbide tool bit shows a substrate reaction forming a CoWB ternary phase. When TiC is applied as an intermediate coat and a diffusion barrier this substrate interaction is absent (Zeman et al 1982). Erosion resistant TiB_2 on valve and pump components deposited over 750-1050°C using chlorides of Ti and B with H_2 as carrier and reducing gas (Caputa et al 1985).

TiBN: Deposition on TiN pellet using $TiCl_4$, BCl_3, N_2, H_2. $TiCl_4/H_2$, $N_2/TiCl_4$, BCl_3/H_2 ratios control. Ternary compound $Ti(B,N)_{1.5}$ with fcc phase can form. Also $TiB_{0.87}N_{0.28}$, the equilibrium cubic boronitride. But depending on the BCl_3/H_2 ratio, mixture or single phases of cubic boronitride +/- TiB_2 can be deposited (Peytavy et al 1978).

TiC: The carbon content of cemented carbide substrate has a large effect on the deposit parameters contrary to reports in earlier papers. Co, or TiC morphology has little effect. Substrate property influence on deposition rate examined. Authors disagree with earlier papers on TiC/substrate influence (Hara et al 1978). TiC nucleation and growth is strongly affected by diffusion of both C and Cr in substrate steel (Yoon et al 1987).

TiC on WC at 850-950°C using propane deposited at greater efficiency and over a larger temperature range at reduced total pressure (Kim et al 1986). Non-stoichiometric TiC, $Ti_{0.6}C_{0.4}$ on Mo substrates at 1900K and 1 atm pressure, exhibited a C/Ti ratio highly dependent on input molar fractions of methane and Ti-chloride. Ti-Mo-C phase diagram limitations prevented C/Ti=0.57 (Teyssandier 1988a).

TiN: The most investigated wear-resistant refractory in recent years, using $TiCl_4$, N_2, H_2. TiN microhardness 1600 - 2000 Vickers achieved by ultrasonic field CVD. Adherence and toughness improve. Little effect on growth rate. <220> orientation. 2-step mechanism involving the substrate Fe. Formation of $FeCl_2$ at <960°C yields isotropic TiN and at >960°C H_2 incorporation yields anisotropic TiN (Takahashi & Itoh 1977). Same reactants are used at 1050°C to coat inside long steel tubes, 10 mm i.d with a mobile furnace. More homogeneous coatings are obtained with the furnace moving along the outlet to inlet direction (Itoh et al 1986); (ii) at 900-1200°C in a quartz rotary bed CVD reactor (0-500 rpm) to diffusion treat Ti powder using an infra-red focusing furnace (Itoh et al 1988).

Thermodynamic diagram at 1900K and production of non-stoichiometric TiN on Mo substrate is reported (Teyssandier 1988b). Process kinetics is also reported (Jung et al 1986).

TiC-TiN & TiC_xN_y: 30-100micron films of TiC_xN_y mono- and double layer of TiC-TiN on steel (C 0.6-0.7%) in ultrasonic field. Growth rate, grain size and <220> prefered orientation decreased with increase in pCH_4. TiC_xN_y hardness was 1800-2600 VHN adhesion strength >120 kg. cm^{-2} (Takahashi & Itoh 1979).

ZrC-C: Argon saturated with Br_2 at 0°C flows on Zr sponge at 600°C. The ZrB_2 is introduced with CH_4 and H_2 at 2100 cm^3/min flow rate on substrates at 1280, 1330, 1360 and 1380°C. Methane influences the morphology significantly causing non-coherence when inadequate to blistering in deposit when in high concentration, and further to crystalline with columnar structure. Other factors which influence the morphology are the ZrB_2 concentration, and temperature. H_2 has little effect (Ogawa et al 1979).

ZrN: Pt, Fe and Mn chlorides as impurity dopants encouraged whisker growth. Au, Co, and Co gave pillars, polygons or plain deposit repectively (Motojima et al 1979).

ZrC_xN_y on graphite by $ZrCl_4$, CH_4, N_2, H_2 and Ar at 950- 1150°C. Homogeneous solid solution of ZrC_xN_{1-x} obtained by controlling CH_4/N_2 and deposit temperature (Takahashi 1981).

4.7.2. CVD PROCESS OPTIMIZATION

A general review of this nature is restricted, as mentioned at the outset, by inadequate information in published literature. The assessment given here is thus not to be regarded as conclusive.

(a) Theoretical Modeling:

The foregoing sections reveal clearly that the success and scope of CVD processes require a three-pronged approach, viz. process optimization by theoretical modeling linked to experimental analyses and developed to a pilot plant (Headinger & Purdy 1987). The mass transport and thermodynamic models quoted here have been diverse to an extent, and the few results published show their utility and limitations. Although a number of experimental results appear to have closer affinity to diffusion-controlled estimated results, the eventual product efficiency control parameters are still unclear. Mass transport predictions have been based on (a) Vertical jet models (Vandenbulcke & Vuillard 1977) and stagnation point flow (Houtman et al 1986), (b) Horizontal and barrel epitaxial reactors e.g.Si-CVD, (c) Non-specified directional flow eg.TiC (Subrahmanyan et al 1980), (d) Dimensionless CVD numbers, using diverse reactions to suit the models (i) B from BCl_3, (ii) Si from $SiCl_2$, (iii)TiC on cemented carbide, and (iv) Si from $SiHCl_3$ (van den Breckel 1977), and (e) Boundary Layer Theory (Hammond 1979; Gokoglu 1988) etc.

Experimental analyses show some drawbacks in fitting a model to an experiment (Cochet 1978). Reactions which are all gaseous are, perhaps, the easiest to assess. Where one or more of the configurations is/are liquid or solid, both in situ sampling and theoretical modeling become more complicated. Models have to be offered on a basis which can be implemented and realised experimentally. The experimental reactor design has to accommodate a substrate area and type over which the eventual deposit is required in its practical application. Advances in reactor design include in situ laser mass spectrometric analysis but the equipment cost could limit its scope.

(b) Product Characterization:

The intended application of the CVD product influences the process optimization. Conditions worked out for whisker and fibre forms will be incompatible to tribological and broad substrate deposition. Refractory coatings required in semi-conductor technology and in power and energy areas are diverse in thickness and area requirements. Reproducibility of deposit composition and stoichiometry, microstructure and morphology need good definition with respect to the substrate material, hardness, stress, thermal expansion characteristics, deposit stability and environmental behaviour. One of the most important unfavourable feature of CVD is the high reaction temperatures. Chapters 7 and 8 cover these aspects in greater detail.

(c) Adhesion & Substrate Compatibility:

These are considered separately because of their importance if a CVD product is meant to be a reliable protective coating. Only those CVD processes which are aimed at producing free-standing vapoforms need non-adherence. All other CVD materials regardless of size require adherence. Very few papers (about 4%) report the extent of substrate participation. Substrate interaction is clearly taken advantage of in the production of compounds such as TiC on cemented carbide (Hara et al 1978), TiC-TiN , Ti-B-N (Peytavy et al), ZrC-C (Takahashi et al 1980) and BP (Motojima et al 1979). Interaction to some extent is evident when intermediate layers are introduced, e.g. Si before Si_3N_4 (Galasso et al 1978), or are allowed to form, MoP on Mo before BP, and preoxidised steel for SiO_2 (Bennett et al 1982). Ultrasonics superimposition during deposition appears to have scored very distinctly on TiC-TiN on steel (Takahashi & Itoh 1979) and TiN on Fe (Takahashi & Itoh 1977), where >100 microns of the ceramic has been successfully applied. The extreme relevance of substrate/deposit compatibilty and adhesion contrasts with the lack of adequate data on the topics (<5%).

(d) Other Problem Areas:

Coating damage, its repair and coating separation is still a problem in the CVD industry. Preferential precipitation and/or decomposition of individual elements during the deposition of a compound or a mixture impede novel CVD processes, e.g. Ti and B precipitation in TiB_2 depostion, amorphous Si release in SiO_2 coating. But Al-B differential separation has been used to an advantage to produce AlB_{12}-B composite on Mo and W filaments (Motojima et al 1981).

Substrate and reactor interaction with the reactants is yet another continuing problem. An initial 'dilute' reaction followed by the more 'active' deposition seems to be one solution. This can only be solved on a process- and substrate-dependent basis. Repairing partially damaged CVD-coated products is unreported. The coating process gets advertised but the coating life on the substrate concerned is not documented. Clearly this needs to be specified. Some studies have been reported of the improved oxidation resistance of SiO_2-coated steels in advanced gas cooled reactors (Bennett 1982; Brown et al 1978).

(e) The Future of CVD Processes:

CVD technology has tremendous potential which has been proved long since, and its vast scope is obvious when carbon powders and fibres, Si wafers, steel tubes 17mm dia and 150 mm long, and steel claddings can be coated and free-standing vapoforms manufactured. It is an established industry to produce a variety of composite shapes – tubes, baskets, corrugated and flat panels, fibres and powders as it has been in metal extraction and recovery.

Materials in special shapes which can be produced only by CVD include W and W-Re vapoforms, and coatings such as $AlN-Si_3N_4$, Ti-B-N, TiC_xN_y, Ti-Zr-B, ZrC_xN_y etc., which cannot be formed otherwise. CVD Si is ideal for semiconductor technology, and CVI/CVD of SiC is unique in the composite industry. Optimizations with the aid of mass transport and thermodynamic models will help achieve further product efficiency.

CVD has replaced powder metallurgy in many instances where shape and size restrictions have been imposed. Since it is a method by which metals, non-metals, alloys and refractory ceramics can be deposited, its scope is limited only to the discovery of suitable chemical vapour transport reactions and the working parameters. MOCVD is one such development. Ru and RuO_2 films have been deposited on Si and SiO_2 substrates using Ru-acetyl acetonate, Ruthenocene and Tri-ruthenium dodecacarbonyl (Green et al 1985).

A further proof of its advancement, is in the development of plasma assisted and reaction assisted CVD processes. These are as yet in their development stage but show good promise. Both the methods have lowered the temperature restriction on CVD from 800 to 300°C. The advantage in combining amorphous and crystalline products has been proved by the superior performance of Si_3N_4 on Si-wafers as well as gas turbine components and hydrogen-containing Si. Alloys and non-crystalline materials and composites with superior corrosion and high temperature resistance and other unusual properties are possible now by CVD. More research and development are needed in this area.

Certain advantages of CVD over other coating methods may be listed (Blocher 1973):

(i) It is a versatile process which provides coatings of a variety of metals alloys and compounds not readily obtained by other means.

(ii) CVD can be carried out on a wide variety of substrates viz. powders (fluidized bed reactors), wires, fibres, wafers, sheets and substrates with complicated shapes and geometry.

(iii) CVD is not restricted as a line-of-sight process,

(iv) Relatively simple equipment is required and deposition is possible over a wide range of pressure and temperature.

(v) There is a choice of chemical systems and chemical reactions, so that, e.g. hydrogen embrittlement can be avoided.

(vi) CVD has a wide range of throwing power (deposition and coverage into shielded regions).

(vii) The chemical composition of the deposit may be controlled so that graded coatings and mixed coatings can be produced.

(viii) The coating structure and grain size may be controlled.

(ix) CVD coatings are dense and their purity may be controlled.

(x) Uniform coating on object of complex shape is possible, even when using the CCVD process.

The various advantages and disadvantages do not apply equally to all the different CVD techniques. The reactive sputtering process generally has lower deposition rates than CCVD. Unless special care is taken, reactive evaporation and reactive sputtering give less uniform coatings on objects of complex shape than CCVD. Thus, the unique deposition characteristics must be considered when choosing a particular CVD technique. PVD techniques employ a totally different reactor design. Combination CVD/PVD techniques are now receiving more attention.

There are many industrial applications of CVD coatings for the protection of metals against wear, erosion, corrosion, and high temperature oxidation. Some examples of the more recently developed overlay coatings which have found commercial applications are TiC, TiN, TiC_xN_{1-x}, Al_2O_3, Ta and Al. High temperature diffusion barrier coatings are finding increased application as are fibre-reinforced metal, alloys and ceramics viz. SiC composites (Archer 1984; Fitzer & Gadow 1986; Reagan et al 1988), and carbon-carbon composites (Strife & Sheehan 1988; Buckley 1988).

Most of the present commercial CVD protective coatings are produced at relatively high temperature (pack cementation or CCVD). These coatings are generally not suitable for substrate materials such as normal steels, copper, and copper alloys. However, a recent process for the deposition of tungsten carbide in the temperature range of about 300-700°C shows considerable promise in extending the application of CVD protective coatings to these materials. The tungsten carbide coatings are particularly suitable for protection against low-load abrasive wear.

It has been shown that the chemical spray deposition technique (Viguie & Spitz 1975) does not require cumbersome protection against air, especially for deposition of oxides and metals that do not form oxides.

Some of the high temperature limitations may be overcome with the use of graded coatings and intermediate layers. Progress is expected in the development of more low-temperature CVD proces- ses. A reduction of deposition temperature in these processes may be realized with the use of suitable chemicals such as organo- metallic compounds and reactive atoms, or with the aid of addi- tional excitation such as plasma activation. Magnetron enhance- ment and ultrasonics are two other methods attempted. This will result in wider application of CVD protective coatings particu- larly for substrate materials which are temperature-sensitive in their microstructure stability or constituent inter-diffusion causing adverse compositional changes and therefore, properties.

Low pressure CVD (LPCVD), especially for deposition of silicides is a noteworthy development. Commercial LPCVD reactors are repor- ted for $MoSi_2$ (West & Beeson 1988), boron-doped polysilicon (Maritan et al 1988), $TaSi_2$ (Reynolds 1988), TiS_x (Rosler & Engle 1984), W (McConica & Krishnamani 1986) and WSi_x (Bors et al 1983).

The effect of low deposition temperatures on the structure and quality of the coating is not yet understood; more research in this area is necessary. This is particularly important for low- temperature plasma-activated CVD processes, for although coatings can, in principle, be produced at relatively low temperatures by PACVD/PECVD, it is the properties such as hardness, cohesion and adhesion of the deposit which ultimately determine the success of many protective coatings (Yee 1978).

4.8. PLASMA ASSISTED CVD (PACVD); PLASMA ENHANCED CVD (PECVD)

4.8.1. INTRODUCTION:

PACVD is a fairly recent process developed over 1974-1978 mainly for obtaining thin films for application in the fields of micro- electronics, optics and solar energy. There is a brief note on the process in the VIII international CVD conference (Reinberg 1979; Hollahan & Rosler 1978; Rand 1979; Yasuda 1978; Hollahan & Bell 1974) and a review by Bonnifield (1982) and others. PACVD principles and process are updated by Inspektor-Koren (1987). The process has expanded into other applications, especially for thermal barrier coatings (Stinton et al 1988; Strife & Sheehan 1988).

In PACVD, solid deposition on substrates is achieved by initia- ting chemical reactions in a gas with an electric discharge. There are two methods. At atmospheric pressure the electrons, ions and neutral gas molecules in local thermodynamic equilibrium are "arced" to obtain equilibrated plasma. This is the thermal plasma method. The electrons, and to a lesser extent, ions are much more energetic than the bulk gas molecules in a low pressure glow discharge state. This leads to a cold plasma and is a non-equilibrium method. Most PACVD processes operate this method and not the equilibrium thermal plasma technique.

The main advantage of PACVD is that it enables relatively low substrate temperatures ($<300^{O}$C), has better covering power, adhesion and control. Instead of thermal energy, the energetic electrons activate the reactant gases. Most of its favourable characteristics stem from this. Substrate/coating pairs with thermal expansion mismatch, high melting point or high reaction temperatures can be adaptable to PACVD; e.g. deposition of Si_3N_4 on low melting substrate, such as Al. Substrates of high vapour pressure, or with structures which may distort, flow, diffuse or be subjected to chemical reaction, such as polymers which are unstable at high temperature can be accepted for deposition by PACVD. PECVD, PCVD are other terms found in use.

Another point in favour of PACVD is that often higher deposition rates are possible than by conventional thermal CVD. It has a greater tolerance level for the various operational parameters, although is more complex to set up. PACVD is reviewed and compared with conventional PVD and CVD, for wear and corrosion resistant coatings (Archer 1982; 1984). Their complementary characteristics are effectively combined in PACVD.

However, deposition of pure materials by PACVD is virtually impossible except for polymers; almost always non-desorbed gases get trapped in the deposit. This is one of its major limitations, but turned to an advantage sometimes, as in the case of amorphous Si containing H_2. Another drawback is the strong interaction of the plasma with the growing film. The high deposition rate also results in less control on uniformity and requires a careful alignment of the reactor assembly to counterbalance this.

Reactive rf sputtering of stoichiometric TiN for wear resistance is described (Yame et al 1982). Nitriding of steel for wear resistance is done by gas nitriding in ammonia, salt bath nitriding in cyanides and cyanates, or plasma nitriding in nitrogen and hydrogen (Child 1983). Plasma nitriding allows lower times or temperatures (minimum 380°C) because nitrogen ions are transfered uniformly into the surface and not mainly at grain boundaries (as in the gas process). Also, gas phase FeN forms from sputtered Fe and then deposits on the surface. Plasma nitriding is preferred over other processes because it depassivates steel surfaces prior to nitriding, gives low distortion, uses lower temperatures to minimise tempering amd allows selective nitriding using metal shields. Only steels which contain nitride-forming elements, e.g.tool steels, will harden on nitriding. Nitride precipitation causes lattice expansion and compressive surface stresses giving good fatigue properties.

Deposition parameters in magnetron sputtering PCVD of ZrO_2 are described (Deshpandey & Holland 1982).

4.8.2. THE PACVD TECHNIQUE:

4.8.2.1. General:

The system sets up a low pressure plasma sustained by a high frequency electric field. The ion and electron densities in such a plasma are 10^{10}-10^{11} particles cm^{-3}. The ambient gas temperature is usually in the range 25-300°C but the electron temperature can reach to 10^{4}K. A low gas/pressure of 10 torr is maintained in order to keep the high ion-to-gas temperature ratio. The rate of formation of species in the gas phase is related to the several parameters as follows:

Rate of formation of species:-

The rate coefficient in the gas phase (i) = Product of (ii) the electron density and (iii) the reactant density, where (i) is a function of reaction cross section and the average electron energy (the latter is not easy to measure). However, it is a function of the electrical field strength divided by the gas pressure, which are both measurable; (ii) and (iii) are respectively proportional to the power density and the product of the gas pressure and length of the experimental column.

Collisions occur between the adsorbed species on the solid substrate surface with the ions and electrons from the plasma. Ions have active kinetic energy and momentum of $10 - 10^{3}$ eV which when approaching are Auger neutralized and collision occurs. The impact results in lattice damage, sputtering, and ion-induced chemical reactions. Electrons affect almost similarly except that they have a lower flux and momentum. The effect of frequency on PACVD has been discussed (Konuma et al 1982).

PACVD process can be divided into 4 main sections:

1. Production of the plasma,
2. Chemical dissociation-decomposition by electron collision,
3. The reaction transport kinetics, and,
4. The surface chemical effect and deposition (Bonifield 1982).

4.8.2.2. The Plasma:

PACVD adopts similar glow discharge techniques as used in sputtering, fluorescent lamps and some types of gas discharge lasers. Articles by Inspektor-Koren (1987), Bonifield (1982), Koenig & Massel(1970) and Coburn & Kay (1972) are good sources of reference. See also (Coad & Scott 1982).

Radio frequency discharge is the most common method. At low frequencies (low kHz range) secondary electron emission to some extent is necessary in order to maintain the discharge. In the MHz range of frequency, sufficient density of electrons gain the energy to ionize the gas molecules to sustain the discharge.

Dissipation of electron energy is by inelastic collisions and molecular dissociation. Electron removal is when they neutralize or reduce ions mainly at surfaces.

PACVD contrasts with sputtering in that the gases it uses are polyatomic molecules with usually low ionization potentials, unlike e.g. argon, in sputtering. It also uses higher pressures 0.1-1 torr (0.01 torr for sputtering) to give higher collision frequecies and shorter mean free paths. The larger the molecules the lower the electron energies which can be a disadvantage sometimes. PACVD uses larger molecules in the deposition reaction. Unlike in sputtering, PACVD reactors maintain equal powered grounded and surface areas. Positive ions bombard all surfaces, grounded, powered or floating. The control parameters on the set up have been given earlier.

4.8.2.3. The Chemical Reaction:

PACVD is characterized by its ability to decompose reactants held in the discharge column in conditions where their normal disposition would have been to remain stable and unreactive, e.g. low temperature. It is the collisional dissociation by the energetic electrons in the plasma which is the major mechanism of decomposition. Ionization, dissociation followed by enhanced reaction rates are all the resultant effects, the rates being well above those occurring in normal chemical reactions with atoms and molecules even at higher temperatures.

It was mentioned earlier that PACVD products are rarely without inclusions. The reason will be clear with the following example:- The energy to dissociate one oxygen atom from N_2O is 40 kCal/mole but 127 kCal/mole for CO_2. Thus depositing SiO_2 using SiH_4 and N_2O yields a more homogenized SiO_2 than from CO_2 which will render it Si-rich. Also the bond strength of N-NO is 115 kCal/mole while $O-N_2$ is 40 kCal/mole. It is easier to decompose N_2O and the SiH_4-N_2 reaction will produce SiO_2 easier and purer with very little nitrogen inclusion.

This illustrates that dissociation energies and bond strengths can be used to gauge the reaction tendency, and its rate and quality to produce a PACVD deposit. Other parameters like equilibrium constant, free energy or entropy normally of great value under equilibrium conditions are rendered invalid for PACVD reactions which are in non-equilibrium states.

4.8.2.4. PACVD Kinetics:

The transport kinetics of PACVD is comparable to that of low pressure thermal CVD. However, the surface reaction, boundary layer and diffusion concepts can only be applied with severe limitations since the reactants are produced by discharge as dissociation fragments.

164

A slow, viscous, laminar flow concept can be applied but the mean free path of molecules in the gas, a small fraction of a millimeter, is too small to fit into this. Turbulence or convection conditions do not enter the substrate/deposit/reactant configuration. The dominant factor is the spatial distribution of the free radicals produced in the discharge. Not much information is available on the plasma discharge-adsorption-desorption chemistry.

4.8.2.5. Surface Effect:

Film formation proceeds by adsorption of neutralized radicals, their bonding and film growth. The substrate is not the only recipient. There is product ablation (wastage) as PACVD occurs on grounded or floating surfaces.

Substrate temperature does overcome such ablation and also minimises surface incorporations during film growth, e.g. Si_3N_4 formed from SiH_4-NH_3 at $25^{\circ}C$ has more hydrogen in it than that produced at $300^{\circ}C$. The higher temperature also improves its density and chemical stability.

Electron ion and photon bombardments can also produce local heating effects. Neither the electrons nor the photons can cause as much damage as ions can. The electrons have the same energy at the substrate surface as in the glow discharge. But the ions can bombard at high enough velocity to cause sputtering (see earlier section).

4.8.3. PACVD PRODUCTION TECHNOLOGY:

4.8.3.1. PACVD Reactor Design:

PACVD uses conventional systems for the flow, temperature and pressure control of its equipment. Several reactor designs have been tried on a laboratory scale; see Fig.4-26 (Bennett 1974), Fig.4-27 (Stinton et al 1988). Only the parallel-plate model is developed to production standards. Reactors are categorized on the basis of the gas flow patterns adopted, viz. radial flow or longitudinal flow. At Texas Instruments the Reinberg design (Reinberg 1973; Reinberg 1977) invented for radial flow is used (Bonifield 1982).

Gas flows from the periphery of a circular, grounded substrate plate, across the substrates and out at the centre of the reactor. As the initial gas is decomposed in the discharge, the direction of flow is such that the deposition occurs on decreasing areas of the substrate plate. Also, the flow velocity increases toward the centre of the reactor since the effective

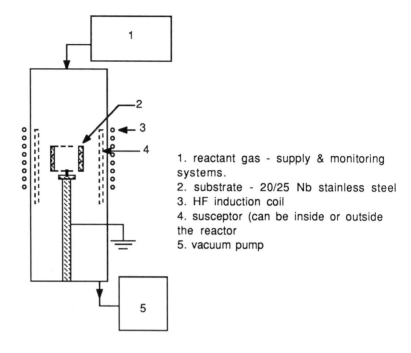

1. reactant gas - supply & monitoring systems.
2. substrate - 20/25 Nb stainless steel
3. HF induction coil
4. susceptor (can be inside or outside the reactor
5. vacuum pump

Fig.4-26: Schematic of a PACVD Reactor

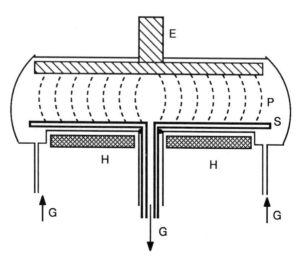

E: shielded rf power input electrode; P: plasma discharge; S: rotating susceptor carrying substrate to be coated; H: heater; G: gas in and out.

Fig. 4-27: Schematic of a PACVD Apparatus

channel area of the flow is decreased. As in a tilted susceptor CVD reactor, the higher flow velocity compensates for the reduced reactant concentration. These features, together with the cylindrical symmetry of the reactor, aid in maintaining deposition uniformity over large areas.

Flow effects are important for uniformity control, but discharge plasma conditions are the dominant effects. Egitto has published a detailed description of uniformity control in a radial flow reactor (Egitto 1980). The plasma density is also a function of the radial position in the reactor, increasing toward the centre, due to electron diffusion to the reactor walls. As a result, the reaction rate increases toward the centre compensating for reduced reactant concentration. The flow pattern can even be turned around from inside to outside, as in the Applied Materials radial flow reactor. Uniformity can still be obtained, but uniformity control differs from the outside-to-inside radial flow reactors (Rosler et al 1976).

Another major reactor design uses longitudinal flow between long parallel plates inside a quartz tube. The plates are alternately driven positive and negative by the rf voltage. Uniform depositions are achieved by maintaining a uniform discharge at low power. Apparently, operation at low power results in a depostion rate that is limited by the power density instead of the gas concentration. These and other reactor designs used in the semiconductor industry have been compared in the literature (Burggraaf 1980).

A multichamber plasma processing system for a research and pilot scale is described (Supple & Stoneham 1985).

4.8.3.2. Discharge Power Supply Unit:

Plasma CVD reactors vary in discharge power frequency from 50 KHz to 13.6 MHz (the industrial frequency band). Matching networks are needed to couple the rf power into the high impedance discharge. At high frequencies, matching networks consist of L-C networks with adjustable components – while low frequencies are matched through transformers. High frequency networks often require retuning with changes in power and pressure; low frequency impedance matching is less sensitive to discharge, but only a limited range of operating conditions is possible for a given matching network. If dc bias is desired on the powered electrode, the final coupling to the reactor must be a series capacitor. RF can be either to the substrate plate itself or to a counter electrode with the substrates floating or grounded. These factors determine the extent of ion bombardment seen by the film as described previously.

4.8.3.3. Vacuum Requirements:

Operating pressures are in the range of moderate vacuums, 0.1 to 5 torr. Although this is not a very high vacuum, high pumping speeds (2000-5000 litre/min) are usually required in order to achieve these pressures at reasonable flow rates. Rootes type blowers backed by vacuum pumps are most common. Many of the gases used or made in PACVD are flammable (SiH_4, H_2, CH_4, etc) or toxic. Scrubbers or "burn-offs" are required for safe handling of these gases in the exhaust of the vacuum pump.

4.8.4. PACVD MATERIALS:

1. Pyrocarbon (PyC):

Propylene-argon and methane-argon mixtures have been used to pyrocarbon coat thermally sensitive substrates. Non-equilibrium rf plasma at 0.5 MHz and 10 Torr was used at 300-500°C. Deposit properties were found similar to PyC obtained by high temperature CVD (1000°C) (Inspektor et al 1986).

2. Amorphous Carbon:

Nearly amorphous but "graphite-like" or "diamond-like" carbon can be produced from a variety of materials, usually butane, by PACVD. Exceptional hardness, high electrical resistivity and optical transmission are its special features. Significant ion bombardment of the PACVD film causes the "diamond-like" structure. The starting material apparently affects only the growth rate but not the properties of the deposited film. The ion-bombardment controls and influences the physical properties. A high rate is achieved by increasing the ratio of carbon:hydrogen in the gas. Low ion impact, low power and high pressure produce high hydrogen content polymers. High power, low pressure and increased ion bombardment results in hard, high resistivity (10^{12} ohm cm) "diamond-like" films. Further increase of ion impact and power and lower pressure yield "graphite- like" carbon films, with low resistivity (0.1 ohm cm).

3. Amorphous Silicon: a-Si:H

SiH_4 is the most common reactant for Si-PACVD. Reacted either on its own or diluted with inert gas, the Si obtained is amorphous and includes hydrogen. The material caught interest for application in solar cells due to its characteristically low density mid-band gap defect states, caused by the inclusion of hydrogen. Using $SiCl_4$ in H_2 (Bruno et al 1980) results in halogenised and hydrogenised silicon. Solid silicon/H_2 has been demonstrated (Webb & Veprek 1979). Polycrystalline Si from PACVD at 230°C (Iqbal et al 1980) and by thermal plasmas at 1100°C (Sharma 1980) have also been reported. Microcrystalline Si has been deposited from SiH_4, SiF_4,H_2 (Veprek & Maracek 1968; Matsuda 1983). Brodsky

(1978) has published a review on PACVD-Si.

4. Si_3N_4:

Silicon nitride was the first product to be deposited by PACVD, and has been reviewed by Reinberg (1979). Bonifield lists a number of references on Si_3N_4 films (1982). SiH_4 and NH_3 are the most common PACVD reactants, with SiH_4:NH_3 flow rate at 1:(1.4-6) as NH_3 decomposes slower than silane. Using N_2 lowers the hydrogen inclusion but the SiH_4/N_2 reaction is kinetically slower than the SiH_4/NH_3 reaction. Dissociating the N_2 before introduction to the main PACVD reactor overcomes this limitation. The dissociation energies (kCal/mole) are: 227 for N-N, 110 for $H-NH_2$, 90 for H-NH, and 79 for H-N. NH_3 dissociation has a total energy exceeding the N_2-breakdown but is lower on a step-to-step basis resulting in favourable overall kinetics. PACVD silicon nitride is non-stoichiometric and usually contains variable Si/N ratios, usually Si-rich with 15-30 at% H. The film is best produced under a mild compressive condition. Low rf frequency produces compressively stressed films. At high frequency (13.6 MHz) it can be tensile or compressive. High power, low pressure, high rf give compressively stressed films. It can be deduced that compressive films are formed under increased ion bombardment conditions during film growth. A careful control of power, pressure and anode-cathode spacing is required to control most properties viz. density, electrical resistivity and stoichiometry.

Amorphous Si_3N_4 produced by PECVD is better termed $Si_xN_yH_z$. The film stoichiometry is most affected by the silane/ammonia ratio and the rf power. They also influence the buffered-HF etch-rate of the films (Bohn & Manz 1985). Si_3N_4 by PECVD methods contains 10-30% at.% hydrogen. Using SiH_4 at 2% dilution in inert gas as one of the reactants reduces the hydrogen inclusion in the nitride film, but the product was inferior in performance to those produced by 'conventional' PACVD (Dharmadhikari 1988). Films produced by N_2-silane with argon (film-a) and helium (film-b) as carrier gases have been examined (Allaert & Van Calster 1985). Film-b was better for passivating purposes, while film-a was more uniform and less critical to deposition temperatures. Both the carrier gases influenced the hydrogen content markedly. The nitride is best denoted as $Si_xN_yH_zO_t$ and is deposited over 150-300°C. Induction heated PACVD (Mito & Sekiguchi 1986), Remote PECVD (RPECVD), Fig. 4-28 (Tsu & Lucovsky 1986) are other recent studies on Si_3N_4.

5. Other Si-Products by PACVD:

SiO_2, SiON, SiC are the other three Si-compounds deposited by PACVD. SiH_4 is the one common reactant to all. N_2O, O_2 or CO_2 are used for SiO_2:H, and these + NH_3 or N_2 yield $SiN_xO_yH_z$. The dissociation energies (in kCal/mole) are: 40 for $O-N_2$, 119 for O-O and 127 for O-CO. Near-stoichiometric SiO_2 requires a N_2O:SiH_4 of 1.7 but a CO_2:SiH_4 of 1:400. Using either N_2O or CO_2 results in low contamination as both, on dissociation, removing one

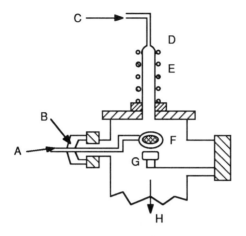

A: main reactant,e.g.silane input; B:feed through; C: input other reactants,
e.g.ammonia+nitrogen+oxygen; D: silica tube; E: RF coil; F:gas ring;
G:substrate holder+heater block; H:to pumping system

Fig. 4-28: Schematic Diagram for RPECVD

oxygen atom are left with a diatomic molecule. Si-rich films of
SiO_2 are composed of Si, SiO_2 and SiO_x, produced in silane/oxygen
at 350°C, at 20-90 W power and 2 Torr (Pan et al 1985).

SiC needs CH_4 with SiH_4 and can be produced across the Si-rich to
the carbon-rich spectrum.

6. Other PACVD Products:

Comprehensive information on PACVD products may be had from two
reviews by Hollahan & Rosler (1978) and Inspektor-Koren (1987).
A variety of reactants and conditions are possible to produce
polymers. It is not necessary to have a monomer for a PACVD
polymer. Radicals can be "chained" under PACVD conditions on the
substrate whose units may not be known to exist freely under
normal chemical conditions. Considerable coating work has been
carried out for use in VLSI devices.

Table 4:4 lists various other materials deposited by PACVD
(Inspektor-Koren 1987; Bonifield 1982).

TABLE 4:4

PACVD COATINGS & STARTING MATERIALS

PACVD Products	Starting Materials
Arsenic	AsH_3
Aluminium Oxide	$AlCl_3/O_2$
Aluminium Nitride	$AlCl_3/N_2$
Boron	BCl_3/H_2
Boron Carbide	B_2H_6/CH_4
Boron Nitride	B_2H_6/NH_3
	BBr_3/N_2
	$BCl_3/NH_3/Ar$
	$BH_3, B_2H_6/N_2/NH_3$
Boron Nitride (Isomorphous)	$B_3N_3H_6/Ar$
Boron Oxide	$B(OC_2H_5)_3/O_2$
Carbon (Isomorphous)	C_2H_2, CH_4, C_2H_6
Carbon (Pyrolytic)	C_3H_6/Ar
Germanium	GeH_4
Germanium Oxide	$Ge(OC_2H_5)_4/O_2$
Germanium Carbide	GeH_4/CH_4
Iron & Iron Oxide	$Fe(CO)_5$
Phosphorus Nitride	$P + N_2$
Silicon (Amorphous)	$SiH_4, Si(solid)/H_2$
Silicon (Microcrystalline)	$SiH_4, SiF_4/H_2$
SiC	SiH_4/C_xH_y
	$Si(CH_3)_4/Ar/H_2$
	$SiCl_2(CH_3)_2$
Si_3N_4	$SiCl_4/NH_3/Ar$
	$SiH_4/N_2/NH_3$
SiO_2	$SiCl_4/O_2$
	SiH_4/N_2O
	SiH_4/O_2
TiC	$TiCl_4, C_2H_2/CH_4/Ar/H_2$
TiN	$TiCl_4/N_2H_2$
TiO_2	$TiCl_4/O_2$

4.9. LASER - CVD

Laser technology has been adapted to CVD processing on small components. Possible applications of LACVD, LECVD or LCVD include (a) mask and circuit repair, (b) interconnects, (c) one- step ohmic contacts, (d) controlled area hard coatings, (e) generation of non-equilibrium materials and (f) materials with controlled grain-size. Several workers have tried using lasers to manipulate CVD processes, the main advantages being operating a clean heat source, minimum substrate distortion, rapid heating and cooling rates.

A few LACVD examples are given below:

Pulsed CW TEA CO_2 laser used to deposit metals (Ni and Fe form carbonyls) and dielectrics on quartz and stainless steel substrates; Ni/SiO_2, TiC/stainless steel. Blue-green lines of argon-laser to deposit Si and GaAs; and spots and lines of Ni, Fe, W, Al,Sn, TiO_2 and TiC (Allen et al 1981).

Transparent, conducting films of SnO_2 and In_2O_3 have been deposited using CO_2 laser (Tabata et al 1981).

$$(CH_3)_2SnCl_2 + O_2 = SnO_2 + 2CH_3Cl$$

Si coating from SiH_4 using pulsed or $CW-CO_2$ laser on glass, vitreous carbon, graphite and sapphire substrates has been carried out with the laser beam perpendicular or parallel to the substrate. A helium neon laser was used to monitor deposition rate. The intensity of the beam reflected by the depositing Si was measured by interferometry. The perpendicular configuration resulted in higher deposition rates than that obtained by the thermal technique. The parallel configuration enabled equivalent deposition at lower temperatures. Mass spectrometric analysis was also coupled to the parallel laser depositing equipment (Bilenchi & Musci 1981).

Very fine-grained films of TiC have been deposited by LCVD using a 1.4 kW CO_2 laser. Controllability of input energy and a small heat affected zone resulting in minimum distortion allow LCVD to be applied to a variety of selected area CVD applications (Mazumder & Allen 1979).

4.9.1. LASER CRYSTALS:

Amorphous and crystalline materials can be produced from laser interactions with a suitable gas medium. Diamond from methane is one such novel product (Fedoseev et al 1981). Hydrocarbon droplets can be laser decomposed to yield spherical amorphous, or irregular crystalline graphite or diamond, or metastable forms of carbon. An infrared laser was used to a power density of 5000 W/cm^2. Particles up to 0.1 microns were produced.

CHAPTER 5

Coatings by pack, slurry, sol-gel, hot-dip, electrochemical and chemical methods

5.1. INTRODUCTION

This chapter deals with five coating methods, the first two of which, may be regarded as particular extensions of CVD and high temperature diffusion. Sol-gel is an elctrophoretic process, hot dip is just molten metal dip coating. Electrochemical method includes plating at ambient and fused salt temperatures (Metalliding), and, chemical displacement and reduction coating is termed the 'electroless' method.

Molten salt/fused salt electrolysis is carried out to achieve 'Metalliding' of refractory metal deposits. Elctroplating of metals and alloys from aqueous, organic and non-aqueous electrolytes has been in use for a long time. Only those alloys and metals of interest to high temperature will be considered in this section. The sol-gel technique employs colloidal solutions as the starting medium which is sprayed, dried to a gel and then fired to obtain the coating. The slurry process uses larger particle sizes than the sol-gel, and with an activator brings about a reaction at fusion temperatures of the released product. Pack cementation is generally called the pack process and there are several variations of it, viz. vacuum-pack, slip-pack, pressure-pulse pack etc. The pack medium can be all-solid or part solid and part fluid (slip-pack).

5.2. THE PACK COATING

5.2.1. THE TECHNIQUE:

5.2.1.1. General:

Pack cementation and vacuum pack coating techniques can be gene-
ralised as methods in which a CVD process takes place with the
substrate surrounded by a mass of the depositing medium. 'Cementation' is a misnomer. The substrate is 'packed' in a 'cement'
consisting of a mixture of the master alloy (the source alloy), a
salt as activator and an inert filler. Normally, a pack is placed
in a heated 'retort' under an inert or reducing hydrogen atmosphere. The coating is carried out over a wide range of pressures
from a low, near-vacuum 1-20 torr to near-atmosphere i.e. 760
torr, in the enclosed 'retorts' (cf. 'reactors' for conventional
CVD). The substrate to be coated is surrounded by the pack;
various alignments within this principle, are possible depending
on the substrate requirements.

The first known 'cementation' process is that of Al on steel in
1914 (Allison & Hawkins 1914; Drewett 1969a, 1969b). However,
much attention was given to the process, and its variety and
development occurred during the mid-sixties to the late seventies
on Ni- and Co-base alloys, and iron alloys, when the protection
of high temperature gas turbine alloys became paramount, together
with rocket and space hardware, i.e. refractory alloys, mainly
Ta-, Nb-, Mo- and Cr-base. The technique itself in its principles, has changed remarkably little since 1914. The composition
and quality of the substrate-deposit configuration have been
modified in the research carried out in the last 10-20 years.

Even now, aluminium stands foremost amongst pack-coated deposits,
closely followed by Cr. Si and alloys of Al-Cr, Al-Si can be
coated as 1 or 2-stage packs (Brill-Edwards & Epner 1968). Most
of the literature refered to here is on Al, Cr-Al, Ni-Al and Fe-
Al systems. Pack coating is particularly suited to treat large
substrates either singly or in bulk, and can handle intricate
shapes as it is not a line-of-sight method. Much of the earlier
literature is available in the following references (Brill-
Edwards & Epner 1968; Goebel 1979; Duret & Pichoir 1983; Mevrel &
Pichoir 1987).

Table 5:1 gives a selection of process parameters used for pack
aluminizing, pack chromizing, vacuum-pack, slip-pack and slurry
methods.

T A B L E   5 : 1

P A C K   C O A T I N G S

Pack Processes at Ambient Pressures

Ref.No	Year	Deposit	Substrate	Pack mix	Process Temp.,°C	Time,h	Heat Treatment Temp.,°C	Time,h	Comments
1	1914	Al	Fe, steel	Powdered Al, or a ferro-aluminium alloy+Al_2O_3+ NH_4Cl+ a halide	800-950	2-24	815-980	12-48	A high Al mix can give a 25-150 μm thick coating with 50-60%Al on the surface which in stage 2 reduces 25-35% and a 0.6-0.1 mm coating.
2	1968	Al	Fe-,Ni-& Co-base	Proprietary to Chrom-alloy Am.Corp; HP, MP & LP packs for high, medium & low aluminium activity; *NH_4F_2 as energizer; HP & MP with NH_4I	480-980; *980; HP, MP at 650				Stainless steel retorts with heating rates of 14°/min. Detailed characterization & coating defect parameter studies.
3	1982	Al	plain carbon steel(0.17%C)	Al(-200,+325 mesh) or 10% of a 50wt% each of Fe & Al mix;+Al_2O_3,1-5% NH_4F-HF	750-900	2-20	900		Pre-aluminized mild steel retort; 150 μm thick(max.); surface Al conc 20% at all temp. except 750°C(<20%) kinetics and diffusion studies.
4	1983	Al	Fe(30ppm C)	Fe-50Al alloy powder (300 mesh)+NH_4Cl+Al_2O_3 in the ratio 49:2:49	750-1050	max.15			In argon; recrystallised alumina retort; interdiffusion studies.

Pack, Vacuum-Pack, Vacuum Pack-Slip Processes

Pack

Ref.No	Year	Deposit	Substrate	Pack mix	Process Temp.,°C	Time,h	Heat Treatment Temp.,°C	Time,h	Comments
5	1981	Al	Ni	(i) Al powder, calcined Al_2O_3 (at 1050°C in H_2 400 μm), CrF_2; 20g mix in retort, in flowing argon; pre-vacuum 1 hr for degassing.					The calcined alumina powder was further annealed at 860°C for 2 h before the activator was added; cylindrical alumina retorts. Morphology characterization.

Ref.No	Year	Deposit	Substrate	Pack mix					Temp.,°C	Time,h	Temp.,°C	Time,h	Comments
					Al	Ni	Al_2O_3	CrF_2 (wt%)					
5	1981	Al	Ni(contd.)	(ii)	19.9	28.9	49.9	1.3	760-1100	120 max			(i) High activity, (ii) low activity most probable pack mix as in ref(23) with pure Al
				(iii)	15.4	33.4	49.9	1.3					
				(iv)	32.8	67.2							
					48.8 =		49.9	1.3					
					16	32.8							
6	1982	Al(i)	IN100	Pack mixture not stated; heat treatment in Al_2O_3 powder					700	5	1050	10	
		Al(ii)	Without substrate alloy; Then IN100	Alloy powder; carbonyl Ni+atomized Al powder for an Al-30% Ni alloy; 25% of the alloy + 75% Al_2O_3 + NH_4Cl					1050	10			Kinetics & morphology studies.
									1100	5			
6	1982	Cr	IN100	Electrolytic Cr powder 25 & Al_2O_3 75wt% + NH_4Cl					1050	10			
		Cr-Al	IN100	2 steps; Cr step 1; Al step 2									
7a,b	1987	Al & Cr-Al	Fe and Fe-Cr-Mo alloy	Low activity Pack: 42Al-48Fe wt.%, with porous alumina wrap as a diffusion barrier. Master alloys: 95Cr-5Al for pure Fe & 90Cr-10Al for 2.25Cr-1Mo steel. Mixed activator: 1NaCl, 2 $AlCl_3$.									Hydrogen was found to support a marginally faster pack coating than argon. Kanthal-like compositions on Fe & Fe-Cr alloy. Tumbling pack eliminated diffusion zone around substrate. Yttria mixed as a fine powder into the cementation pack introduced Y into the coating, which gave better oxidation resistance to the alloy during thermal cycling. Process patent applied for.

Vacuum-pack

No.	Year	Materials	Source composition	Temp (°C)	Time (h)	Remarks
8	1968	Al,Cr,W TD nickel Mo & Ta	Pure metal, alloy or intermetallic compound (-8+30 mesh)+ a halide salt (0.1-2.0w/o); NaCl or NaF			
		Al	Cr-Al powder	760-1150	1-24	1-150 torr; max 8 mils coating i.e.
		Cr	Pure Cr or Cr-Mo,-W, -Ti etc	1040-1260	8-48	18mg/cm^2; 1-10 m pressure; 80mg/cm^2 max.
		W,Mo,Ta	-150 mesh metal powder	1040-1200		5-10mg/cm^2 max; sintering problems
9	1968	Nb, Ta, W, Mo		1260	7-12	1 -10μm pressures; 20-30μm/h or 30-50 μm/h leak rates; heating cycling 1-150 torr argon or helium

Pressure pulsing vacuum pack

No.	Year	Materials	Source composition	Temp (°C)	Time (h)	Remarks
10	1980	Al	Ni base alloys	Al source + AlF$_3$ as activator + inert filler	825-980		Retort must be heated rapidly & optimum temperature is about 1198-1223K to get more uniform coatings

Vacuum Slip-Pack

No.	Year	Materials	Substrate	Source composition	Temp (°C)	Time (h)	Remarks
9	1968	Cr-Ti-Si;	Nb-alloys	The slip-mix: metal or alloy particles (90w/o) (-250 mesh) + a halide activator (NaF or KF;1 w/o) + a binder 9 w/o, e.g. a lacquer, polyisobutylene; + a volatile liquid vehicle, e.g. Tulol, sufficient for required fluidity.	1090-1200	2-9	12-24 hr ball milling; spraying the bisque about 20 mils thick on the substrate.
8	1968	Cr-Ti	Nb-alloys	See previous entry			
9	1968	Si-W (WSi$_2$)	W & W-alloys Nb-, Ta, & Mo-alloys	See previous entry			
		W/WSi$_2$	Ta-	2-Step:- W on Ta followed by WSi$_2$			Step 1 by CVD, electrophoresis or slurry

Table 5:1 (contd.)

Ref.No	Year	Deposit Substrate		Pack mix	Process Temp.,°C	Time,h	Heat Treatment Temp.,°C	Time,h	Comments
			Slurry						
8	1968	Al; duplex:- Cr-Al & Cr-Al with Mo, Fe, or Ta or W; & Ta-Al	Ni- & Co-base alloys; TD Nickel	Al, -325 mesh or alloy or other mixtures of elemental powder additives; blended with a lacquer binder for 12-24 h.	1000-1100	2-4			In vacuum or in argon
11	1968	Si	Ta- & Nb-base alloys	Step 1: elemental powder mix ball milled with orgc. vehicle (ethyl cellulose) in secondary butyl alcohol, xylene & Stoddard solvent. Sprayed with paint spray guns. Step 2: Drying; Step 3: Sintering	1335-1515	0.5-15			In vacuum 10^{-5} torr. Sintered coat 0.0075-0.015 cm thick.
				Step 4: Si pack -200 mesh, no activator	1175-1230	7-8			Pack siliconizing in gettered argon 800 torr. Sintered slurry is porous but desirable to accommodate Si.
12	1968	Si- + Cr-Fe (20/20), Ti-Mo (20/10), and Cr-B₄Si (20/0.5)	Nb-alloy Ta-alloy Mo-alloy	Modifier elemental metal mix blended with lacquer; sprayed and air dreied.	½m.p. of deposit modifier alloy mix.		1370	1 at 10^{-3} torr	

Table 5:1   References:-

1. Allison G. & Hawkins M.W., GEC rev., $\underline{17}$, 947 (1914).
2. Brill-Edwards M. & Epner M., Electrochem. Technol.,$\underline{6}$(9-10) 299 (1968).
3. Sivakumar R. & Rao T., Oxid. Met., $\underline{17}$(5/6) 391 (1982).
4. Akuezue H.C. & Whittle D.P., Metal Sci., $\underline{17}$, 27 (1983).
5. Thevand A., Poize S., Crousier J.P. & Streiff R., J. Mat. Sci., $\underline{16}$(9) 2467 (1981).
6. Sivakumar R., Oxid. Met., $\underline{17}$, 27 (1982).
7a.Rapp R.A., Wand D. & Weisert T., "High Temperature Coatings", Khobaib M. & Krutenat C. (Ed.), Met. Soc. AIME (1987), p.131.
7b.Kung & Rapp R.A., Mat. Sci & Engg., $\underline{87}$,   (1987).
8. Gadd J.D., Fisch H.A., Kmieciak H.A. & Jones E.E., Electrochem. Technol., $\underline{6}$(11-12) 379 (1968a).
9. Restall J.E., Gill B.J., Hayman C. & Archer N.J., "Superalloys", Sims C.T. & Hagel W., Wiley NY (1980), p.45.
10. Gadd J.D., J.F. Nejedlik J.F. & L.D. Graham, Electrochem. Technol., $\underline{6}$(9-10) 307 (1968b).
11. Wimber R.T. & Stetson A.R., ibid., $\underline{6}$(7-8) 264 (1968).
12. Priceman S. & Sama L., ibid., $\underline{6}$(9-10) 315 (1968).

179

Fig.5-1 shows an example of the substrate-pack alignment (Blocher 1982; Thevand et al 1981). The choice of "retort" and/or the "can" material depends on the pack compositions, and the process is carried out at ambient pressures or under near-vacuum with controlled leak rates to balance the pack reaction pressure variations. As a "closed CVD" process, pack-coating poses a number of experimental problems. This is probably the cause for the more empirical approach taken and a poor characterization of the process. Some fundamental studies are attempted, which will be dealt with in Chapter 7.

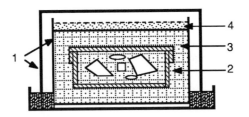

1. stainless steel   2. molybdenum can   3. pack mixture;
4. fusible silicate

(pack material (wt %):   Cr 3, Si 11, ammonium iodide 0.25,
alumina (bal)

Fig.5-1: Retort for Diffusion Coating of Non-Ferrous Metals
- Schematic Diagram

5.2.1.2. The Pack Mixture & the Substrate:

A typical, simple pack mixture consists of a fine powder of the depositing material (Al, Cr, or Si etc), an inert filler (e.g. Al_2O_3) and an activator (e.g. NH_4Cl or NH_4F). It is usual to subsidise the activator segment of the pack with another reactive halide (e.g. HI or HF, or NaCl etc). The concentration of the depositing material, e.g. Al, directly determines the activity of the Al vapour at the substrate and signifies the pack as a high-, medium- or low-activity variety.

The substrate must be thoroughly clean and free of oxide films or any other impurity, in order to ensure uniform deposition, bonding and diffusion on its entire surface. Elaborate cleaning procedures are normal as the flow sheets indicate. All operating conditions are worked out specifically to a substrate.

5.2.1.3. The Pack Reaction :

The substrate/pack interaction is a 3-stage process:

(i) chemical reaction within the pack; release of vapour of the metal/alloy to be deposited,
(ii) vapour condensation and nucleation on the substrate, and,
(iii) deposit growth plus substrate/deposit interdiffusion.

'Zero time' is the time taken by the pack to attain the furnace temperature. By then the substrate will have received a considerable fraction of the total deposit intended. A glance at Table 5:1 shows that the deposit material has, almost invariably, a high vapour pressure and the temperature range is 500- 1200^{0}C. The reaction itself may be carried out at temperatures when the deposit may be molten as it contacts the substrate. The use of an inert filler, and further, that of an alloy powder as a reactant, are both directed in preventing excess deposit activity and to prevent "flooding" on the substrate. The inert filler separating the droplets, also prevents them from seeping en masse on to the substrate. Failure in these adversely affects the vapour phase deposition and diffusion aspects of the pack process.

5.2.1.4. The Temperature Effect:

There are two stages where temperature influences pack deposits, viz. the reaction stage where the retort heating rate and the heat of reaction are in control, and the heat treatment step which consolidates the coating ready for use. Temperature control and effect is the absolute determinant of the reaction efficiency and deposition as well as the performance of the final substrate-deposit configuration. Like most other high temperature processes, deposit nucleation and growth are influenced by the reaction temperature, but it is diffusion and its effect that is of greater importance in pack reactions. Since the substrate is "packed" with the "deposit mix" the reaction time and temperature will affect the mechanical and physical properties of the component as a whole, and will also influence the morphology of one or more transition zones between the 100% substrate and the 100% deposit. In practice, very little, if any, of the surface layer is 100% deposit.

5.2.2. SPECIFIC PACK EXAMPLES

In the sections following, recent literature on specific pack coatings and substrates will be discussed. Work on gas turbine alloy coating has contributed a great deal to the conversion of an empirical process of the earlier period to a highly developed and standardized industrial mass production process. The pack process at ambient pressures has been improved by the vacuum-pack technique (Gadd et al 1968a). A combined vacuum slip-pack (Gadd et al 1968b) method has progressed further combining the advantages of the slurry method, and a pressure-pulse pack method claims yet another advantage in adapting to coat holes and channels, e.g. airfoils in turbine blades (Restall et al 1980). These processes, along with duplex and mixed coating processes are given in the following sections.

5.2.2.1. Pack Aluminizing:

Several ferrous and non-ferrous alloys can be pack aluminized. A low aluminium activity pack causes more of the substrate metal diffusion outwards and the high activity pack causes increased aluminium activity inwards. The choice is critical for Ni- base superalloys which have been studied more than the Co-base superalloys. Conditions optimized for one may not always be applicable to the other, especially in the homogenizing temperature aspects.

The phase diagrams of the M-Al systems are well-known, viz. Fig.5-2 and 5-3. Most of the phases formed in the pack aluminized Fe- & Ni-base alloys have been characterized, and are controlled by the pack activity and temperature.

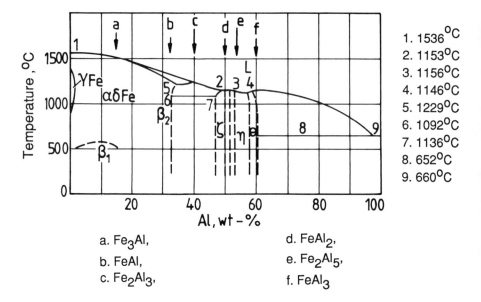

1. 1536 ^{o}C	
2. 1153 ^{o}C	
3. 1156 ^{o}C	
4. 1146 ^{o}C	
5. 1229 ^{o}C	
6. 1092 ^{o}C	
7. 1136 ^{o}C	
8. 652 ^{o}C	
9. 660 ^{o}C	

a. Fe_3Al,
b. $FeAl$,
c. Fe_2Al_3,
d. $FeAl_2$,
e. Fe_2Al_5,
f. $FeAl_3$

Fig.5-2: Phases in the Fe-Al System

5.2.2.1. (a) Fe-base Alloys:

Pack aluminized coating protects steels from oxidation and corrosion in hydrocarbon and sulphur-containing atmospheres. The pack
cementation process is ideal for small and/or intricate substrates. There are few studies of the factors controlling the coating
formation. The effects of pack activity and temperature on the
structure and kinetics of aluminide formation on EN3 steel has
been studied (Sivakumar & Rao 1982). The surface Al% was about
20% from 750-900°C pack temperatures. Weight gains and layer
thicknesses followed a parabolic law with time, indicating a
vapour-solid diffusion couple. The activator was $NH_4F.HF$ (1 to
5%). The role of the alumina powder is to prevent pack sintering
and prevent direct contact with Al droplets at the coating temperatures. The coating phases were $FeAl_3$ and Fe_2Al_5. Pack aluminizing of steel in an unalloyed Al pack gives non-uniform brittle
$FeAl_3$ and Fe_2Al_5 coatings. Smooth adherent and uniform FeAl is
obtained from a ferro-aluminium pack with $NH_4F.HF$ activator.
Alumina needles were found on the coated surfaces, due to the
reaction

$$2AlF_3(s)+3H_2O=Al_2O_3(s)+6HF.$$

At 750°C, the surface Al% rises with time until 16 hours and then
stays constant at about 20% Al. At higher temperatures, after 2
hours, the surface Al composition is constant with time.

Stainless steel aluminizing at 850 to 1000°C using a mixture of
ferro-aluminium and NH_4Cl powders gave an FeAl coating with 40%
Al. A diffusion inter-layer was also found (Ushakov 1973).

Plain carbon steel (0.17%C) aluminized in a pure Al pack forms a
two-phase structure of $FeAl_3$ and Fe_2Al_5; $FeAl_3$ is the Al-rich
phase. The coating proved to be too brittle. A better control on
the Al-activity was obtained with a ferro-aluminium alloy pack.
The deposit surface was found to be a mixture of well formed
grains and a number of randomly formed needles, the needle formation being due to the reaction quoted above.

FeAl was the only phase identified. The alumina needles from the
above reaction were confined mostly to the surface (Sivakumar &
Rao 1982). The activation energy was found to be 239 kCal/mole in
the temperature range 950-1100°C. Akuezue & Whittle (1983) report
the activation energy to be about 104.9 KCal/mole over the same
temperature range and an identical mixture for aluminized Fe
(30ppm C). Four phases were identified, viz. outermost beta-2
FeAl, alpha-delta Fe, beta-1 Fe_3Al and the original alpha-delta
Fe at the substrate (at 850°C, 10h).

The effect of carbon content in the ferrous substrate on the
relative diffusion of Fe and Al is not clear. It is an area where
further work is necessary as more Fe-base superalloys are being
considered in many high temperature energy-oriented applications.
Fundamental work carried out on iron at 900°C with a low activity

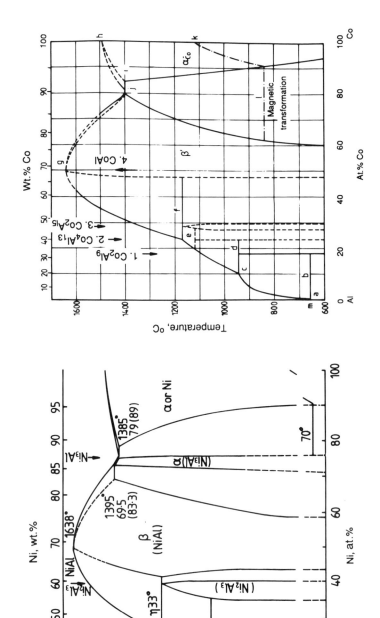

Fig.5-3b: The Co-Al System

Fig.5-3a: The Ni-Al System

pack of 42Al-58Fe wt.% with a porous alumina wrap as a diffusion barrier showed that hydrogen supported a marginally faster coating than argon (Kung & Rapp 1987). Rapp et al (1987) report on achieving Kanthal-like surface compositions on pure Fe and Fe-2.25Cr-1Mo alloy; yttria was also included in the pack cementation process in a separate attempt to dope Y in the pack coating.

5.2.2.1.(b) Non-Ferrous Alloys:

Pack aluminized Ni-base superalloys are the most-reported series because of their relevance to gas turbine application. Co-base superalloys, such as X-40, MA6000, etc. have also been studied, their pack aluminized morphology being somewhat different from those obtained on Ni-base superalloys (Mevrel & Pichoir 1987; N'Gandu Muamba & Streiff 1987).

The coating is applied by pack, vacuum pack, vacuum slip-pack, or slurry methods. All these methods were developed in the mid sixties-seventies. The alloy series, along with other superalloys and high temperature alloys is discussed under the various methods employed, the ultimate coating on the alloy being not very different. A good account of superalloy aluminizing is available (Gadd et al 1968a,b; Goward & Boone 1971; Pichoir 1978; Coutsouradis 1978; Goebel 1979; Strang et al 1981; Singhal 1983).

The essential morphology of a pack or slurry aluminized Ni- base alloy can be controlled by the Al-activity in the medium (pack or slurry). A high activity pack causes Ni_2Al_3 to be the predominant phase with the diffusion of Al inwards being greater. Most processes aim at the beta NiAl phase which is more suitable and resistant with favourable mechanical and oxidation properties.

Duplex coatings are quite common, e.g. Cr-Al or Si, Pt-Al, where the alloy is first chromized or platinized by CVD, electrodeposition, spraying or other methods, which is then followed by the pack treatment. Ternary coatings, e.g. Cr-Ti-Si (Gadd et al 1968a,b) and Co - Cr-Ni (Stetson & Moore 1975) have also been reported. The best known duplex coatings are Pt-Al, Cr-Al and Cr-Si.

Fig.5-3 shows the phases in an Ni-Al system. The process and the system have been fully reviewed (Chatterji et al 1976).

In determining the phase distribution morphology, the choice of the halide activator and even the filler, e.g. recrystallized or calcined Al_2O_3, can be critical apart from the actual Al source (pure or diluted as an alloy). The pack temperatures and heat treatment steps and their duration also have significant effects on the coating morphology. All these parameters will affect the relative diffusion of Ni from the substrate and the released Al from the pack. Multiple layers, their compositions with respect to Ni and Cr and their phase morphology determine the eventual performance of the coated superalloy.

Thevand et al (1981) have reported the control conditions for obtaining the Ni_2Al_3 phase, as well as the mixture of phases across the Ni-Al phase diagram. The cement activity passes through a maximum at 41 wt% Al, 1.3 wt% CrF_2 and 57.7 wt% Al_2O_3 for a Ni_2Al_3 coating. The coating is a dull grey when aluminized over 760-950°C, but is a shiny grey when treated between 950-960°C, the latter being due to a molten $NiAl_3$ formed underneath Ni_2Al_3, and above NiAl of the 3-layer diffusion band. A 100 micron thick coat of Ni_2Al_3 was found to provide a good coverage (treating temperature 760-1100°C), and a 180 micron coating resulted in fissuring at 760°C and catastrophic failure at 900°C. The aluminizing steps yielded equiaxed scale morphology which deteriorated with increase in temperature.

Aluminium can move at a relatively high rate through Ni_2Al_3 in the inward diffusion type of coating while in NiAl only Ni can diffuse significantly and does so at a relatively low rate. Most superalloy aluminizing exercises aim at the NiAl phase as this changes the substrate composition at the interface to a much more restricted degree, and thus causes little effect on the mechanical and physical properties which the superalloy component was designed for. However a columnar morphology needs to be avoided regardless of the overall phase distribution achieved. An example of surface compositions obtained on IN 100 after aluminizing or chrom-aluminizing at 1050°C for 5 h is given in Table 5:2 (Sivakumar 1982). It must be noted that the data quoted is not necessarily a recommended composition suited for industrial use.

Table 5:2

PACK COATING PARAMETERS

Type of coating	Typical surface comp. wt%	Total weight gain, mg/cm^3	Layer Thickness, Microns
High-activity aluminide	9Cr, 30Al, 47Ni	5.4	72
High-activity chromaluminide	15Cr, 24Al, 44Ni	5.8	65
Low-activity aluminide	2Cr, 23Al, 58Ni	5.0	55
Low-activity chromaluminide	6Cr, 20Al, 55Ni	5.4	55

By using the process of four-cycle vacuum pack cementation, a CrTi-Si coating with an internal layer of low silicide matching the Nb752 substrate and an external layer of improved oxidation-resistant disilicide was obtained (Huibo & Yuying 1982). The porous structure of the coating helps in reducing crack propagation. The proposed Si/CrTi ratio is between 0.4 and 0.8. Oxidation endurance tests on the coated alloy at $1300^{o}C$ exceeded 1000 hours.

Good protection against superalloy hot corrosion is given by a duplex silicon slurry & aluminide formed by a slurry or a spray of very pure Si followed by a 16 hour pack aluminizing at $1100^{o}C$ (Young CME 49).

Steel has been pack-boronized. When boron is diffused into steel (for wear resistance) at 900° to $1000^{o}C$ for 1 to 6 hours, a layer of iron borides forms, up to 150 microns thick (Child 1983). The steel is heated in a pack or paste of boron carbide + activators + diluent for 6 hours at $950^{o}C$, giving well-keyed (by diffusion) 150 micron thick layers of $FeB + FeB_2$.

5.2.2.1.(c) Pt-Al Duplex Coatings on Ni-base Superalloys:

A number of the Pt-group metals alloy with refractory metals like Si, Ti, Al and Cr (Ott & Raub 1976). Pt reacts with oxygen to form gaseous PtO_2 (Kubaschewski 1971) and is similar to Cr. Both Pt and Cr are principal undercoats or sandwich coats in a Ni-base/duplex coating configuration on Ni-base superalloys; their application has to take this feature into account because of its effect and influence on the coating performance. Pt has been deposited on Ni-base superalloys by sputtering (Jackson & Rairden 1977) and later by electrodeposition from fused salts (Wing & McGill 1981). Pt has also been deposited on turbine blades, prior to stabilized ZrO_2 as a ceramic barrier in order to improve the superalloy ceramic bonding (Anon. 1983). In another work, Pt was sputtered (max 25 microns) on an IN738 substrate and pack aluminized with a 20 g charge of pre-activated Al powder with Al_2O_3 in Inconel 600 retorts. Aqueous HF or NH_4F was used as the pre-activators. The furnace was at $1060^{o}C$, argon-filled, into which the retort was introduced after a 2 min residence in the cooler zone; zero time was 12 minutes (see section 5.2.1.3. above).

Addition of Pt to diffusion aluminide coatings can significantly increase the DBTT of the system. Lower surface platinum levels and a greater distribution of the $PtAl_2$ phase and Pt in solution over a larger coating zone brings down the DBTT in line with the unmodified high aluminium coating system (Vogel et al 1987). High and apparently structure- and composition-dependent residual compressive stresses are present in the coating at room temperature. The coating thickness and the elemental distribution are interdependent. A range of transition temperatures exists for the various structural forms in the Pt-modified coatings overlapping those for the higher aluminium content in the unmodified coatings. Three types of structures were identified, (Deb et al 1985) viz.

(i)   A continuous surface layer of $PtAl_2$ which can be produced by either HTLA or LTHA aluminizing and results from a minimum diffusion of the noble metal prior to aluminizing.

(ii)A 2-phase $PtAl_2$-NiAl occurs when a lower level of Pt is distributed over a wider coating fraction, and its distribution and volume percentage depend on the pre- and post-aluminizing heat treatments.

(iii)A single phase NiAl structure with Pt in solution forms with a greater platinum substrate diffusion with a lower level of Al established either during aluminizing or at the post-heat treatment stage.

At the instant when Al comes into contact with the platinized substrate, it is liquid. But the $PtAl_3$ phase formation renders it solid (m.p. $1121^{\circ}C$). The spectrum of Al-rich to Pt-rich phases extends through $PtAl_4$, $PtAl_3$, $PtAl_2$, Pt_2Al_3, PtAl, Pt_3Al_2, Pt_5Al_3, Pt_2Al and Pt_3Al. But on bulk Pt only the first four form. On an IN738 substrate the phases identified are $(Pt,Ni)_2Al_3$ at the platinized substrate end to $Ni_2Al_3+PtAl_3$, 28%(Ni, Co, Cr) in Pt_2Al_3 ($Ni_{1.25}Pt_{0.75}Al_3$) and $PtAl_2$ at the surface. $PtAl_4$ exists only below $800^{\circ}C$ while the others are stable at least to $1100^{\circ}C$. Thus at the reaction temperature, $1060^{\circ}C$, the first phase to render liquid Al on substrate to solid is $PtAl_3$. Thereafter Al-diffusion is rapid.

Al is assimilated more rapidly into Pt than the substrate IN738 probably due to (i) the apparent fast diffusion rate of Al into Pt and (ii) the high atom fraction of Al in $PtAl_3$ compared to that in Ni_2Al_3 or NiAl. With a 5 micron Pt coat, a 25 micron Al is sufficient to place all Pt in a 2-phase field. The presence of Pt retards Ni diffusion outwards and Pt is regarded as a diffusion barrier. This may not be quite the case. It could be termed more as a diffusion "diluent", presenting a Ni-Pt-Al ternary phase formation condition against a Ni-Al binary system (Jackson & Rairden 1977; Wing & McGill 1981).

Failure of platinized-aluminided superalloy arises from the large differences in thermal expansion at the Pt/IN738 interface or a poor initial bond. Also the outer Al-rich phases can prove to be brittle and not sustain a temperature cycle effect due to insufficient DBTT tolerance limits. This proved for some time a disadvantage for Pt-Al to be estabished as a coating against the more ductile MCrAlY coatings. However, tests in Rolls-Royce laboratories have contributed to a better understanding of the problem and Pt-Al duplex coatings have been used successfully in practice.

In general, a classification can be made as to those elements which can form diffusion barriers and those which show solubility in the Ni-rich NiAl phase (Fitzer & Maurer 1977).

Increasing solubility in Ni-rich NiAl

```
<-------------<-------------------<----------
Fe, Co, Pt      Ti, Cr, Ta, W         Nb, Mo
--------------->------------------->--------->
```

Increasing tendency to form diffusion barriers

In contrast to Fe-base alloys, Ni-base alloys spontaneously form Cr-rich interlayers during aluminizing which act as barriers against diffusion of Al from the outer NiAl layer into the solid solution zone. A total of 20 wt% additive consisting of Ti+Ta+Nb+Mo+W+Cr is found to reduce Al diffusion drastically.

5.2.2.2. Pack Chromising:

Pack cementation coating of Cr on steel uses typically 50% Cr powder, 1% NH_4Cl, 49% alumina. Higher surface concentration of Cr can be achieved if the substrate has small alloying additions. Pack chromizing gives 30 to 45% Cr in the surface. In use at 900°C, diffusion will reduce this with time. At lower temperatures both coating and substrate are bcc; at higher temperatures the Cr-rich coating will be bcc on an fcc substrate. Diffusion is faster in bcc lattices and so the interdiffusion will be mostly Fe into the Cr layer, equilibrating at about 13% Cr solubility in Fe, which may be too low for good protection (Haworth & Jha 1983). Adding a third element to raise the Cr solubility in the bcc phase is beneficial: 5% Ni raises the Cr solubility from 13 to 28%. A prior Ni coating can be diffused into the steel substrate before chromizing, for this purpose.

Many studies of CVD coatings on dense substrates have been listed but little is reported for sintered substrates. CVD of Cr on Fe-C sintered alloys gives a chromium carbide coating with Cr-Fe solid solution layers. The coating increases the density of the specimen surface and related surface structural properties. Atomized Fe powder + lubricant were sintered at 1180°C for 3 hours in H_2 and then pack chromized using Cr particles about 1 to 2 mm with 1.5% NH_4Cl activator and evacuated to < 1 Pa before backfilling with Ar and preheating for 5 minutes in H_2. Slow heating to 800°C initiated the series of reactions:-

$$NH_4Cl(s) = NH_4Cl(g) = NH_3 + HCl$$

$$2NH_3 = N_2 + 3H_2, \text{ and,}$$

$$Cr(s) + 2HCl = CrCl_2(g) + H_2.$$

The temperature was then raised to about 1000°C and H_2 (0.02 m^3/h) flooded around the pack box, for 1 hour. Quick removal and cooling in H_2 completed the pack chromizing (Audisio et al 1984).

Chromaluminizing is often carried out to combine the advantages of the protective action which Cr and Al can exert during oxidation and hot corrosion. The process is necessarily a minimum 2-

step pack cementation followed by post-diffusion heat treatment steps. The relative proportion and degree of distribution of Cr and Al are controlled by varying the pack formula amd time. Cr-enrichment can thus be manipulated to exist within the aluminide either near its surface or at the coating–substrate interface (Godlewska & Godlewski 1984).

5.2.3. VACUUM PACK PROCESS

The unique feature of a vacuum pack process is that no inert filler is employed unlike in the more conventional pack process. The metal, alloy or intermetallic granules (–8 to +30 mesh) form the major pack material. For high vapour pressure metals (e.g. Cr) halide activation is not required, but usually a halide activator (0.5-2 wt%) is included in the pack to promote coating element transfer to the substrate interface.

Fig.5-4 (p.191) gives a typical flow sheet for the vacuum-pack process, and Fig.5-5 shows the temperature profile in a vacuum pack retort-furnace assembly in near vacuum or under reduced pressure, 150 torr He (Gadd et al 1968a).

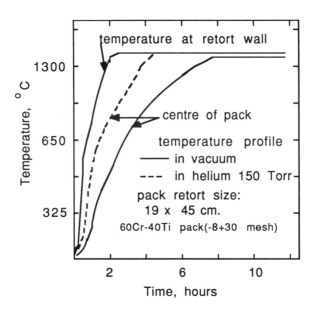

Fig.5-5 : Temperature distribution in a 19 x 45 cm pack retort in vacuum and in helium under reduced pressure. A smaller retort, 7 x 25 cm attains the helium profile at pack centre in vacuum itself.

Fig. 5—4

Flow Chart for (A) Vacuum Pack and (B) Vacuum Slip—Pack Processes

Receipt of Components to be Coated

Inspection

Cleaning: Grit/Sand Blasting (optional)
            Degreasing (solution; vapo—)
            Pickling (acid/ or alakali)
            Water—rinsing
            Drying

Coating Media Preparation:

A. Pack Media   B. Slurry
   Screening   Spraying Slurry
   Activator Mixing   Air Drying

Retort Loading

A. Loading into Retort   B. Parts & Support fixtures
                             assembled in retort

Charging Retort into the Furnace

Evacuation to pressure < 1 micron

Setting of Furnace Leak Rate

Furnace Heated to

A. $260^{o}C$ B. $420^{o}C$

Introduction of He or Ar (optional)

Heating Continued to Coating Temperature

Inert Gas Evacuation (if used)

Process taken through Isothermal Cycle

Process Termination

Cooling

Unloading

Component Cleaning
A. Simple Step   B. Removal of Bisque
                     (brushing or peeling)

191

The heat transfer in a vacuum pack is better than the conventional pack, as it is more conducting and has no inert filler, but there is still a zero time effect, and the larger the retort the higher the pack-furnace temperature gradient. Introducing He, A or H_2 reduces the gradient, H_2 being the best "normaliser". Chromizing by vacuum-pack requires more control of the control parameters because of the high vapour pressure of the pure Cr metal supplementing its halide salt pressure which would release more Cr. Thus when Cr-Al is put together in a single stage vacuum pack process, the resulting porous coating is 3-layered, with dispersed Cr in the outer-most NiAl, alpha-Cr beneath it and a gamma Ni-Cr solid solution sandwich with base substrate. It is the alpha-Cr which causes the porosity. The second drawback is the sintering of the pack particle Cr on to the substrate. Eliminating the activator from the pack, and relying entirely on the pure Cr vapour pressure, or a decrease in the Cr- activity by using Cr-alloy granules reduces porosity. Chromizing the substrate prior to aluminizing virtually eliminates the porosity because Ni diffusing through Cr gives a 2-band diffusion coating of Ni-Cr and Ni-Al with negligible alpha Cr (Gadd et al 1968b).

In general, "pre-deposition" before aluminizing is almost mandatory for Ni and Co based superalloys as well as fibre-reinforced and directional eutectics because of the particular problems encountered with them (Giggins & Pettit 1983). Fig.5-5 (p.190) indicates the beneficial effect of He in smoothing the temperature distribution in larger packs.

5.2.4. VACUUM SLIP-PACK PROCESS

Fig.5-4 includes the flow sheet for a typical vacuum slip-pack process (Gadd et al 1968a). It is an extension from the vacuum pack process and combines the advantages of a conventional slurry process method of coating. The essential departure from the vacuum pack method is the metal particle size, viz. -8 to +30 granules of vacuum pack vs. -250 mesh of the slurry method.

Vacuum slip-pack has 3 components, viz. fine metal particles blended with a binder + an activator + a volatile liquid vehicle. This is the 'slip' or 'slurry' or 'bisque'. Its usual composition is 90 wt% metal, 9 wt% binder and 1 wt% activator in a vehicle. The resulting slip is viscous but fluid enough to be applied uniformly by dipping, spraying, or painting. Spraying is perhaps the most practical and effective in achieving a reasonably uniform distribution, the minimum bisque thickness being 20 mils, the maximum not being too critical.

The coating temperatures are similar to that of the vacuum pack method but with a reduced process (step 1) time as the zero-time (see section 5.2.1.3.) is greatly reduced and heat transfer is superior. However, the restricted quantities of the metal-activator fraction in the slip presents the critical factor in near-vacuum conditions. It is usual to backfill the evacuated furnace with an inert gas in order to maximise the residence time of the reactive vapour species at the substrate/diffusion interface.

The particular advantage of this method is that coatings can be formed on pre-selected areas, instead of an entire component. Thus it is suitable as a local repair method for coating slightly damaged pre-coated components. Also, coatings can be made thicker at selected sections where greater coating sacrifice may occur, for instance, the leading edges of a gas turbine blade.

5.2.5. PRESSURE PULSE PACK & CCRS PACK METHODS

Two other pack methods are given in this section, viz. the pressure pulse pack (Restall et al 1980) and the controlled-composition-reaction-sinter (CCRS) methods (Stetson & Moore 1975).

(a) Pressure Pulse Pack

The pressure-pulse method utilises the direct CVD technique as the main auxilary with the pack serving as the metal source. Fig.5-5 illustrates this (Restall et al 1980). A Ni-alloy retort filled with the pack material is introduced into an argon-filled hot-wall reactor, the pressure in which is changed over an evacuation-refill cycle of 8 times/minute. Aluminizing is done over 825-980°C. The method is particularly suited to coat crevices and especially the cooling channels of the turbine blade. AlF_3 is employed as the pack activator in the aluminizing process because of its convenient vapour presssure, viz. 1 torr at 920°C and it generates AlF about 0.25 torr. During pulsing AlF enters the cooling channels and deposition of Al is achieved. It is important to prevent AlF_3 from concentrating at the substrate. The

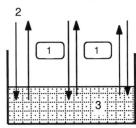

1. samples to be coated   2. Gas pressure
   and directions   3. aluminising pack

Schematic diagram of
the pack process
Fig.5-6

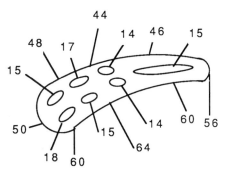

Variation of thickness (microns)
of an aluminide coating on
an aerofoil
Fig.5-7

process variables are as in the normal pack methods, but the critical parameter is the actual geometric disposition of the blades with respect to the pack, i.e. mass transfer is once again the influencing parameter. A typical optimum thickness distribution of a pressure-pulse aluminided aerofoil is given in Fig.5-7 (Restall et al 1980).

High pulse rates reduced the external surface to internal surface coat ratio, which is good, but it also resulted in the loss of the activator. A decrease in total pressure gave better "throwing power" but decreased the coating rate. The higher the temperature the more effective was the pressure-pulsed coating. It was also advisable to minimise the retort zero-time as much as possible; i.e. improved heat transfer conditions were necessary.

(b) The Controlled-Composition-Reaction-Sinter (CCRS) PACK

The Solar CCRS method (Stetson & Moore 1975) is essentially the TRW vacuum slip-pack method (Gadd et al 1968a) but using an aluminium modifier slip of either Co-20Ni-15Cr or Ni-15Co-15Cr. The 3-4 mil (0.08- 0.10mm) slurry is vacuum sintered at $1120^{\circ}C$ for 2 hours at $<10^{-4}$ torr. Aluminizing is done with a low activity Cr-Al pack at $1095^{\circ}C$ for 15 h.

The CCRS method produces deposits with a non-columnar structure unlike in conventional pack diffusion, claim Solar. It also is said to operate under a reduced cost giving it an additional advantage.

CVD-pack cementation has been used for MCrY coatings in a fluidized bed of MCrY powder of <40 microns diameter. The particles or the blade has a thin resin coating and only a few seconds is needed to pick up a coating of MCrY particles. The blade is then pack aluminized to densify and bond the coating to the blade, as MCrAlY. Complex shapes can be coated where line-of-sight PVD would be difficult and the cost is low (Stetson & Moore 1974; Stevens & Stetson 1976).

Similarly, aluminizing, chromizing or siliconizing a plasma sprayed coating can improve its hot corrosion properties. If the plasma coating is porous, vapour phase or pulsed pressure aluminizing is beneficial (Duret et al 1982; Restall 1984; Restall et al 1980).

Laser 'scanning' (laser-melting) can also densify and homogenise coatings.

5.2.6. FEATURES OF THE PACK PROCESS

Some of the drawbacks of the pack process, the defects inherent in pack-coatings, and attempts made to overcome these are considered here.

The most serious disadvantage in a pack process is the poor heat transfer in the furnace-retort configuration and then the pack substrate complex. Many of the pack coated components which appeared to have adequate resistance in simple atmospheric pressure tests are said to have proved to be inferior in low-pressure high temperature conditions encountered during Earth re-entry missions (see Fig.5-8a, 5-8b), (Priceman & Sama 1968). Brill-Edwards and Epner (1968) have reviewed the causes for unsound or inadequate pack-coat performance. Three main classes of defect were observed in the coatings on all the alloys coated (Fe-,Ni- & Co- base). These could be attributed to irregular vapour diffusion through the pack, selective reaction and nucleation at the substrate surface, or non-uniform solid-state diffusion through the coating. Irregularities in the physical and chemical structure of the pack led to macroscopic variations of coating thickness, while low processing temperatures and low aluminium supply rates promoted irregular nucleation and growth at high energy sites on the substrate surface. High deposition temperatures were conducive to more uniform nucleation and growth, whereas low aluminizing potentials encouraged sigma phase formation and entrapment of pack powder particles.

It has been established that irregular mass transfer plays a major role in the formation of chemical and physical discontinuities in aluminide cementation coatings on superalloys. Specific types of coating discontinuity have been related to irregular vapour diffusion through the pack, selective nucleation on the substrate surface, and non-uniform solid-state diffusion through the coating. The type of discontinuity formed in each case is similiar to that found in comparable coatings applied to Fe-, Ni- and Co-base alloys.

Irregularities in the chemical and physical structure of the pack promotes non-uniform deposition, particularly at low coating temperatures. Gross variations in coating thickness are also observed in systems where the transport efficiency of the aluminium halide vapour is low, such that nucleation occurs preferentially at specific high energy sites on the substrate surface. When coating progresses by predominant unilateral diffusion of the substrate elements outward, a chemical discontinuity develops which manifests as a zone deficient in the base metal element. This growth mode also results in entrapment of aluminium oxide particles in certain coatings applied to the Fe- base alloy.

Secondary carbide phases cause both chemical and physical discontinuities in the coatings on each of the three alloys. The nature of these defects is dependent on the location, reactivity, and orientation of the particles and interdiffusion between the matrix and coating. Under certain conditions, fragmentation and occlusion of the secondary particles occurs as the coating front advances. Most of the discontinuities observed can be accentuated or minimized by manipulating process parameters to obtain conditions of uniform and balanced material transfer through the pack, at the substrate surface, and through the solid coating deposit.

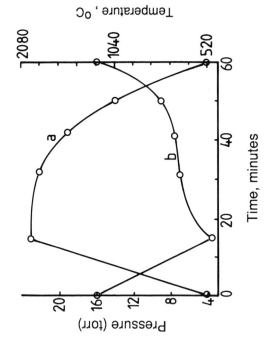

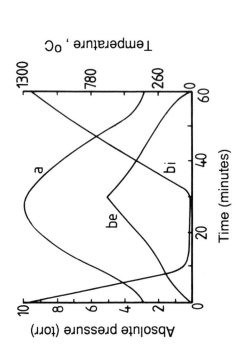

Fig.5-8a,b: Pressure & Temperature Profiles in Re-entry Simulation Tests.

A carefully optimized pack-process obviates most of the draw-backs. A thorough cleaning of the substrate is essential. Operating the process below ambient pressures with or without inert carriers helps better reactant control; other factors are the furnace-retort size ratio, and introducing a helium carrier at 150 torr (Gadd et al 1968a). The vacuum pack has better heat transfer characteristics as it need not use an inert filler like Al_2O_3. It can control the reactant pressure better by using an alloy or inert metallic granule to reduce metal activity.

Surface defects can be caused by sintered pack materials and particles trapped during the process. It is particularly noticeable if the substrate metal diffuses outwards more rapidly than the coating. A careful watch is also necessary on the mechanical and physical properties of the designed component. Thus the process and heat treatment temperature and the residence times become important.

A full characterization of pack-coats has rarely been made by NDT. It is essential that reproducibility of a coated component can be guaranteed based on experimental morphology tests. The process parameter, mass transfer and kinetics aspects require more clarification. Very little is known of the mechanism of element transfer in the vacuum pack process. Much less is known in the slip-pack and slurry methods. Optimization has been done on a largely empirical basis. The post-pack step is rarely described beyond specifying that the treated component removed from the pack is brushed free of the pack material. Micrographs show entrapped pack material but no mention is made of the detrimental effects of trapped filler or sintered material on coating performance. A B-grade diamond finish of the coated substrate has been mentioned (Thevand et al 1981). It is not clear how intricate shaped components can be surface treated at the post-aluminization stage. It is likely that many proprietary methods refrain from discussing the full process parameters.

One of the processes reported isolates the substrate from the powder mixture. For coating internal bores of turbine blade cooling holes, good throwing power is needed and complex halides of Al and and Na have this (Strang et al 1984). By evacuation and refilling, pulse pressure cycles of 10 per minute at $850^{\circ}C$, simultaneous surface and bore coatings of 50 and 25 microns were obtained (Refs. 17 and 18 in Strang et al 1984). Other bore coating processes include thermal decomposition of tri-isobutyl-aluminium at $200^{\circ}C$ (Ref.20 in Strang et al 1984) followed by heating at $1050^{\circ}C$, Ni deposition from nickel carbonyl at $200^{\circ}C$ and Cr from liquid dicumenylchromium at $400^{\circ}C$. Deposition of successive layers of Ni, Cr and Al allows formation of alloy coatings. More successful methods for bore coating are electroless Ni, and slurry with a halide activator for Cr & Al (ref.21 in Strang et al 1984). Masking of areas not to be coated during pack aluminizing can be done by painting these areas with a zirconia slurry. This is best in high activity packs at relatively low temperatures.

5.3. SLURRY COATING TECHNIQUE

5.3.1. THE PRIMARY METHOD:

This coating method, known as the slurry or slurry-fusion technique, can be used on its own for coating purposes without coupling
it to a subsequent vacuum-pack step, or it can be used for local
repair or augmenting purposes. The slurry-fusion technique is
different from the slip-pack method in that the initial metal mix
is of an even finer mesh, and the process is always carried out
at a temperature above the melting point of the slurry-composite-
elemental metals or alloy. It can be performed at ambient pressures but it is more common under vacuum or in low pressure inert
gas atmosphere.

The uniformity of the bisque on the substrate and a thorough
initial blending of the slurry are the first two essential features. Also, the substrate must be non-oxidised, blasted and acid
pickled. Failure in these will lead to non-adherence, non-uniform
composition and variable thickness in the heat treated deposits
(Gadd et al 1968b; Gadd et al 1968a; Priceman & Sama 1968; Wimber
& Stetson 1968). The advantage is that it avoids long heat-up and
cool-down times.

Slurries of Ni,Cr,B,Si (Fe,C) powders can be melted onto steel in
vacuum or inert atmosphere, giving wear resistant coatings. Fluxing forms boron silicates (oxides reduction) and good adhesion
results. Hardness and wear properties are reported (Knotek et al
1982).

Apart from metals and alloys, ceramic coatings have also been
applied (Newhart 1973), e.g. solaramic 55-8A, a proprietary coat.
Although it has proved very successful experimentally, three
primary causes have been given for its failure: (a) use of over-
age slurries which have not been adequately re-worked to eliminate agglomeration of the suspended frit, (b) improper curing or
baking of the applied slurry due to lax temperature control, and
(c) the presence of impurities, such as Al-compounds during the
baking (fusing cycle) which will alter the flow characteristics
of the bisque. The above three causes are also more or less valid
for all metallic slip mixtures.

5.3.2. DOUBLE SLURRY & ROLL SLURRY METHODS:

A double slurry method has been reported to obtain CrAl, CrAlY ,
CeCrAl. The second slurry application is the same to all the
above three compositions as it is the final aluminizing stage.
The first stage applies a Cr-rich slurry for the Cr-Al coating;
here Y and Ce are introduced respectively to obtain the latter
two coatings. The slurry compositions are as follows:

The coatings were applied successfully on Hastelloy X and IN 738 over a thickness range of 65-100 microns (Martinengo et al 1977).

TABLE 5:3

	First slurry stage				Second slurry stage		
	Composition, wt%				Composition, wt%		
	Cr	Al	NH$_4$Cl	Binder	Al	NH$_4$Cl	Binder
CrAl 1000°C,5h	61	18	8	13	83	4	13
CrAlY 1000°C,3h	64	18(Y 9)	9	9	83	4	13
CrAlCe 1000°C,3h	64	18(Ce9)	9	9	83	4	13

The 'Roll Slurry' method is a new method of diffusion coating of strip steel (Sugamo et al 1968) and consists of applying atomized aluminium powder to steel treated with an aqueous solution of polymetaphosphate (0.5 - 1.5 g/l). An addition of a fatty acid like capronic, 5×10^{-4} m-mol to 1g of Al powder (i.e. 0.5 mol/ton) improves powder adhesion to steel. Amines, esters, nitriles, alcohols and paraffins are found to be less effective. Essentially, a hydrophilic medium treated powder adheres less uniformly than a powder treated with a hydrophobic "glue". The applied powder is then compacted into a solid layer by rolling and the composite strip sintered and bonded by heat treatment.

This is an extreme form of a slurry method, applied very restrictedly to components which can be rolled satisfactorily. It does not seem to be recommendable for a wider selection of materials, i.e. those which are not malleable and to shaped substrates.

5.3.3. THE REACTION SINTER PROCESS:

Reaction sintering (Fitzer et al 1977) is an extension of the slurry method and was devised to counter the total loss of identity of the deposit which occurs during the diffusion process.

When a pre-alloyed powder of a particular Ni-Cr-Si composition is applied onto a Ni-superalloy substrate (by the slurry or plasma methods) the subsequent heat treatment results in a violent reaction between the substrate and the coat yielding a completely new morphology of reacted phases. The original coating material is totally absent. Although the overall heat treatment temperature is 1000°C the actual temperature measured during the exothermic reaction increases to 1450°C. A considerable fraction of

199

the substrate composition at the interface changes, a total loss of the original coating composition above it occurs and the reaction also introduces porosity.

The reactant phase band can be reduced remarkably when the slurry applied consists of $CrSi_2+NiSi_2$ mechanical mixtures. Plasma spray is less suited in this case since there is an in situ high temperature state built into the process which will cause the alloying of the mixture before it arrives on the substrate. The mechanical, unalloyed 2-component slurry deposit, when heat treated, expends most of the heat from the exothermic reaction within the Ni-Si, Cr-Si and Ni-Cr-Si reactions in the mixture itself and a much reduced substrate participation results in the targeted Ni-Cr-Si outermost coat. A thin diffusion zone, a much less affected substrate and a coating/substrate configuration almost free of porosity are thus achieved. Providing an inter-layer (e.g. Cr) can give even better results. The factors to be considered beforehand are: (i) relative thermal expansion compatibility of the substrate and the coating, and, (ii) the DBTT.

The authors report only Si-oriented coating compositions. This would be applicable to slurry Al, Cr-Al and Cr-Al-Si-Ni also.

5.4. THE SOL–GEL COATING TECHNIQUE

5.4.1. GENERAL:

This method is probably the most economical for producing high temperature ceramic coatings. The extent to which the technique has been adopted for high temperature applications is not yet clear. There is very little published information apart from a news report (Anon. 1980), two research papers (Nelson et al 1981; Spahn 1981) a patent (Steffens 1980; UK patent 2064370B) and reviews by Bennett (1983; 1984). The method was developed at AERE, Harwell, Oxon, UK in 1970 for particles, fibres and plasma spraying powders, and reported for coatings in 1981. The coatings have been evaluated as fuel claddings on steel.

Sols are more or less a further extension to the slurry discussed in earlier sections. The difference is, fine metal or alloy powders are applied in the slurry technique while oxides are applied by sols. The maximum point of departure is at the heat treatment stage. Slurry coatings are close-range CVD depositions which form a diffusion zone but sol-gel coatings are not.

5.4.2. THE SOL–GEL METHOD:

There are four stages in the sol-gel coating process:

(i) preparation of the sol, (ii) spraying of the sol on the substrate, (iii) drying the sol to a gel, and finally, (iv) firing to obtain a consolidated and adherent coating.

5.4.2.1. The Sol:

Colloidal dispersions are not new; their preparation is well characterized but not in the context of high temperature coatings. There can be aqua-sols or organo-sols, and these can be aggregate or non-aggregate in nature. Essentially, they are stable dispersions of particles in a fluid of colloidal units of hydrous oxides or hydroxides with particle sizes ranging from 20 angstroms to 1 micron. Non-aggregated colloidal units are less than 100 angstroms and such oxide-sols can form deposits of almost theoretical density even in low temperature firing (<850°C). The same oxide in aggregated colloidal units of 200-300 angstroms can form a porous layer of a high effective surface area. The degree of aggregation is controlled by the properties of the sol.

Sol-preparation can take two routes. Route 1 is hydrolysis followed by polymerization. Route 2 adopts precipitation and then peptization. Although only oxides have been considered for coating so far, it is possible that other ceramics may also have been attempted, e.g. Si_3N_4, AlN etc. It seems possible that cermets can be tried if they can be prepared as very fine aggregates. Oxide-sols of ceria, silica, zirconia, titania, thoria and alumina and gel-glass have been reported by the Harwell group.

Table 5:4 shows the formation of hydrous alumina sols.

TABLE 5:4

Route 1: Hydrolysis.

$$Al^{3+} + 3NO_3^- + H_2O = Al(OH)^{2+} + 2NO_3^- + HNO_3$$

$$Al(OH)^{2+} + 2NO_3^- + H_2O = Al(OH)_2^+ + NO_3^- + HNO_3$$

Polymerization

$$13Al(OH)^{2+} + 26NO_3^- + 27H_2O = [Al_{13}O_4(OH)_{24}(H_2O)_{12}]^{7+} + 7NO_3^-$$
$$+ 19HNO_3$$

Route 2: Preciptiation

$$Al(NO_3)_3 + 3NH_4OH = Al(OH)_3 + 3NH_4NO_3$$

Peptization

$$nAl(OH)_3 + xHNO_3 = [Al(OH)_3]_n^+ + xNO_3^-$$

==

The extent of aggregation of the sol and its stability depends on factors such as the nature of the precursor material and the ratio of the anion to metal oxide. Adsorption of H^+ on the metal oxide hydrates causes positive charging of the colloidal units. Anions in the solution associate with these and thus establish dipoles in the sol repelling each colloidal unit from its neighbours. The degree of aggregation depends on this inter-colloidal unit separation, and the anion to metal oxide ratio is a critical factor. A mole ratio of 0.32 for HNO_3/CeO_2 results in a gel density of 2.4 while a ratio of 1 yields a gel with 4.0 g cm^{-3} density.

5.4.2.2. Sol-on-Substrate, Drying & Firing:

Sols can be applied by immersion of the substrate, or, by spinning, electrophoresis or spraying. Any drying or thickening of the sol during spraying has to be compensated for, so that the aggregation optimized for the purpose remains so. The method of application of the sol depends on the shape of the substrate, which must be thoroughly cleaned and abraded at the outset.

The sol-to-gel transformaton is done by careful, gradual drying; an over-rapid drying can cause gel-crazing. At this stage the process is reversible. The gel can be "refloated" to become the sol by merely adding the vehicle, e.g. water for aquo-sols.

Consolidation to a coating occurs when the gel-coated substrate is fired to high temperatures (440-1200°C) where sintering can occur, but the heating times are short, usually 15 minutes. A single sol-application gives a 0.1- 2 micron thick coating which can be modified by controlling the rheology of the sol or by additives.

Unlike in slurry coats, a sol-gel on a substrate may not have a distinct diffusion zone at the substrate-coat interface. But the coating oxide and the growing oxide do get inter-dispersed. Also, there may be mixed phases of the solid oxide. For instance, sol-gel silica results in a mixture of beta-cristobalite with some alpha-quartz. TiO_2 forms coatings of density 4.07 and 4.11 g cm^{-3} when dried at 800 and 1000°C respectively, while the theoretical densities of anatase and rutile are 3.85 and 4.26 g cm^{-3}. CeO_2, on the other hand forms a single phase structured coating.

Table 5:5 lists the densification effects of four sol-gel coating products.

Except for their oxidation resistance data sol-gel coatings have not been characterized in the published literature. No information can be given here on micrographs of the cross section, adhesion and cyclic thermal effects, and stress studies on the substrate-sol-gel-coated system.

TABLE 5:5

Calcination Temp., °C	Density g cm^{-3}					
	Silica		Titania		Ceria	Zirconia
	Non-agg	Agg	Non-Agg	Agg	Non-Agg	Non-Agg
Dry Gel	1.8	0.5	2.6	1.5	4.3	3.6
400	1.5		3.2	1.5	5.2	3.9
600	1.5	0.4	3.4	1.6	6.0	4.8
800	2.1	0.5	4.1	2.6	6.9	5.2
1000	2.0	0.5	4.1	3.7	6.9	5.2
1200	2.0	1.8				
Theoretical density gcm^{-3}	2.2		4.3*		7.1	5.6
Melting pt, °C	1600		1840		2600	2715

(*Rutile; Anatase 3.9 g cm^{-3})

Agg – Aggregate; Non-agg – Non-aggregate

5.5. HOT DIP COATING PROCESS

Hot dipping is one of the oldest coating methods, which can be used to a limited extent for high temperature applications where the substrate can tolerate quite a wide variation in coating thickness and uniformity. The process is traditionally on ferrous alloys; zinc, aluminium, tin and sometimes lead, and a few alloys of these metals, are vat-melted. The substrate to be coated is floated through. Only Al is of relevance in the present context.

Hot dipped aluminized steel strip usually has 14 to 50 microns of Al on its surface. The process is usually carried out on a continuous feeding arrangement. The world production is estimated at 1 million tonnes per year, out of which 20 million tonnes are hot dip galvanized steel strips and sheets. Substrate pre-cleaning is carried out by hot reducing gas (N_2+H_2 at 800°C) as a last stage following the conventional steel cleaning steps. The Al melt bath is usually ceramic lined. Thermal cycling makes the steel less ductile but stronger. Brittle $FeAl_3$ forms and should be < 10 microns thick if ductility is required.

The heterogeneity of coatings on steel obtained by hot dipping has been studied (Samsonov 1973). General references on hot diping processes are: Lowenheim 1963; Heintz 1967.

Hot dipping of a thermal barrier coating may be mentioned here (Allam & Rowcliffe 1986). The eutectic composition Zr-25 wt.% Ni from the Zr-Ni system was selected as the coating alloy composition. 5 wt% ytrrium was added as stabiliser to the zirconia which would later form during final conversion step. Five subs-trates were selected, viz. experimental alloys: Co10Cr3Y, Co25Cr6Al0.5Hf, and Ni25Cr8AllY, and commercial superalloys MAR M509 and IN738. Coating was carried out as a 3-step process:

1. Substrates hot-dipped with the Zr-Ni-Y melt at 1027°C for 30-60 seconds in slowly flowing argon,

2. Annealing by holding coated samples at 1027°c above the melt for 3-5 hours to provide a metallurgical bond, and

3. Oxide conversion of the coated alloy by selective oxidation at pO$_2$ 10^{-17} in an H$_2$O/H$_2$/argon gas mixture of the appropriate ratio. Ni and Co were prevented from oxidizing by holding the oxygen pressure below their equilibrium value. Yttria formed at 1027°C was not in solution with zirconia and did not trans-form the monoclinic phase. Heating further to 1100° caused a complete transformation of the ceramic from monoclinic to cubic. This, however, did not produce any cracks normally associated with this structural transformation. Adhesion of the heat treated coating was good and the system was resistant to thermal cycling.

It is evident this is an alloy hot-dipping and ceramic thermal conversion process.

5.6. ELECTROCHEMICAL & CHEMICAL COATING METHODS

5.6.1. GENERAL

Five processes can be brought together in this section for high temperature coatings on the basis of a few common features:

   (i) They deal with one or more cations in solution.
   (ii)Deposition on a substrate is achieved by reduction (or oxidation) of solvated ions.
   (iii)With one exception, they all require input of electrical energy to produce coatings via redox reactions.
   (iv)They are well suited to apply coatings in bulk and/on large components.

The processes of relevance to high temperature applications are listed below:

(a) Composite deposition technology (electrolytic slurry medium),
(b) Aqueous solution electroplating,
(c) Electroless coating,
(d) Metalliding or fused salt electroplating and,
(e) Precious metal plating.

The fundamental principles of electrolytic reduction/oxidation, the practical operating parameters and their control and the overall depositing mechanism will not be described in this section. There are excellent sources of information, a cross-section of which is listed (Lowenheim 1963; Canning & Co Yearly Handbook; Metal Finishing Yearly Handbook; Mantell 1964; Bockris & Reddy 1971). It suffices to state that the main operating procedures have been well characterized and standardized, and deposition can be achieved to yield satisfactory coatings of controlled thickness and quality.

High temperature applications (300°C and above) of coatings by electrodeposition are in the following areas:

cutting and abrading (Simons 1951; Lukschadel 1978);
wear resistance (Williams 1966; Ishimori et al 1977; DuPont 1978; Samsonov 1976; Kedward 1969; Neckarsulm 1970; Kannan et al 1984; Carew 1983; Kannan 1985);
dispersion strengthening (Sauter 1963; Lakshminarayanan et al 1976);
low friction resistance and oxidation/corrosion resistance.

BAJ Vickers UK Tech. Information and DEF Standard 05-21, CAA, Rolls-Royce, SNECMA have applied composite coating in practice (Foster 1983). Composite deposition is a section of current relevance and will be dealt here. The bibliography provided covers a good section of the work on composites, as well as electroplating and chemical methods in general.

5.6.2. ELECTROPLATING & CHEMICAL COATING PROCESSES:

Three main process factors are involved:

1. Substrate preparation prior to coating is of vital importance as in any other deposition process. The section on adhesion and bonding elaborates these aspects (Chapter 7). The one disadvantage which may be mentioned here is that large areas and amounts of surfaces may need etching prior to plating to achieve adhesion. The depth etched into can be as much as 7.5×10^{-3} cm. Substrates containing additive radioactive elements, have to be be etched taking note of the hazards involved.

2. Electrolyte preparation, variations in its formulation, and maintenance form a critical section of the process. The electrolyte composition; the nature of the anion and cation - simple or complex, organic or inorganic; selection of the operating electrodes - anode (soluble or insoluble), any preliminary treatment required for the cathode; additional loading of particulates (for composites, and their nature and concentration) ; operating temperature ; pH ; stirring conditions or still electrolytes; barrel or tank operation ; addition agents, brightners, levellers, buffers and the operating current density, comprise the several factors which distinguish one process from another.

3. Substrate/coating compatibility, coating thickness, morphology, substrate/coating interface and bonding, coating uniformity, stability, strength, hardness, residual stress and other physical parameters, viz. hardness, abrasion and friction resistance, and corrosion resistance including the role as diffusion barriers, are the important requirements for high temperature coatings. Coatings produced have to be assessed in these respects.

5.6.3. COMPOSITE DEPOSITION TECHNOLOGY

5.6.3.1. Composite electrodeposition:

This is an extension of the conventional metal finishing process using aqueous electrolytes, the main additional feature being the loading of powders or fibres to the electrolyte. It can also be regarded as an electrochemical slurry technique with the process carried out at low temperatures in contrast to the pyro-slurry technique discussed in section 5.3. of this chapter.

The choice of particles for composite electrodeposition is wide ranging: oxides, carbides, silicides, or any other refractory powder, metallic powder, or organic powder e.g. PTFE , can be introduced into the electrolyte under consideration. With relevance to high temperature, the electrolyte choice is perhaps restricted to Ni-, Co- or Ni-Co- mixed electrolytes. Cr-plating has some limited use. The wider choice seems to be to introduce Cr in powder and granular form into nickel and cobalt plating electrolytes and the composite deposits are subsequently heat treated. Other refractory metal radicals can be added to deposit ternary alloys but these have not been explored in any detail with relevance to high temperature application.

Al_2O_3, ZrO_2, TiO_2, WC, SiC and similar refractory and/or abrasive resistant powders have been loaded into electrolytes as particle components. The references quoted earlier give a good cross section of a number of composites which have been prepared by electrodeposition and 'electro-inclusion'. Significant commercial development has been restricted to Ni-SiC (Neckarsulm 1970; Sauter 1963; Akzo 1978; Kedward 1979; Cameron 1981; Hart & Wearmouth 1979; Bazard & Bowden 1972; Hubbell 1978) and Co-chromium carbide (Foster 1983; Kedward 1969). A large number of aero engine components such as sleeves, stators and shroud assembly parts have been composite-coated.

Electrodeposition of overlay coatings should be studied as a simpler process than PVD. Co-deposition or occluded plating is possible for depositing oxides, carbides etc., in a metal matrix (Simons 1951; Lukschandel 1978; Kedward 1969). For depositing MCrAlY, 10 micron CrAlY powder can be co-deposited in a Co or Ni matrix by heavily loading the Co plating solution with CrAlY powder and using the barrel plating technique (Honey et al 1986; Restall & Wood 1984; Kedward et al 1979,1981).

The specimens are firmly aligned in a barrel covered with a membrane impervious to the particles but permeable to the electrolyte. The barrel containing the specimens and powder is horizontally immersed into a conventional electroplating bath and rotated at 4 rpm. Afterwards heat treatment bonds the coating to substrate and interdiffuses the Co or Ni matrix with the CrAlY particles. No porosity is found in the as-plated or heat treated coating. Surface finish is adequate for turbine use. The method claims excellent advantage on the basis of production costs but has not yet been scaled up for production. CoCrAlY coatings produced have been shown to be superior to plasma sprayed CoCrAlY and very definitely better than pack aluminized coatings (Honey et al 1986).

Electrophoretic composite coating has been briefly reviewed (Restall & Wood 1984; Restall 1979). Fine particles in a liquid dielectric, migrate in an electrostatic field and deposit on an electrode. High rates are possible (0.0025 cm in a few seconds), with good thickness control and throwing power, but heat treatment is essential to sinter and densify. Coatings have been tested in engines (Ryan et al 1974). Cold isostatic pressing is needed for MCrAlY powders (Holshire 1976).

Hybrid coatings have also been reviewed (Restall & Wood 1984). Electrodeposition or fused salt deposition (metalliding) of Pt group metals has been carried out to form 10 micron films which are later diffused in and the component aluminized. Better hot corrosion properties are reported than for plain aluminide coatings.

5.6.3.2. Composite Deposition Equipment:

Conventional electroplating tanks and auxiliary equipment are readily adapted and altered to handle composite electrolytes. The essential requirements are that the particles "loaded" into the electrolyte are well blended, non-agglomerated and kept suspended in solution in a uniform slurry. In order to prevent sedimentation, it is essential to pump air (or N_2) from perforated base plates of the electrolytic tank, to incorporate mechanical stirring and also to pump the electrolyte-slurry in and out of the working tank. The latter also permits sampling and adjustment of the electrolyte (Fig. 5-9a,b,c, p.208).

5.6.3.3. Composite Coating Mechanism:

The incorporation of particles within the matrix of a depositing metal on a cathode has been considered in several aspects. It is difficult to segregate and assess the physical volume and adsorption effect with an associated electrophoretic, i.e. charged particle, concept. The predominant factor appears to be the rate of arrival of particles at the cathode surface, coupled with the actual position of the cathode itself. Electrode geometry is a

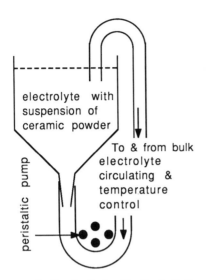

a. A Simple Cell with Air Agitation
and Slurry Circulation

b. Tank with Plate Pumper
& Slurry Circulation

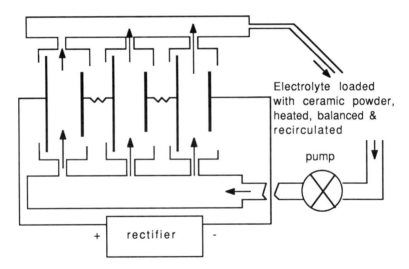

c. Triple cell arrangement in series and solution flow in parallel

Fig. 5-9 a,b,c: Electrodeposition of composite coatings :
schematic diagrams of cell and tank plating systems

critical factor. A surface inclined downwards from the vertical position (overhanging) will not receive a composite coat. This, and the fact that the volume/concentration of particles at the cathode is decisive for particle incorporation indicate that composites form on a cathode by physical incorporation of the particles by being there to be trapped and engulfed by a growing metallic crystal lattice. That there should be some physical bond via adsorption, is understandable, but there is no codeposition. This term implies simultaneous reduction of two different cation species, and is used in alloy plating when two or more metal cations are reduced at a cathode to yield alloy electrodeposits (Brenner 1960; Dunkerley et al 1966).

Guglielmi (1972) derived the following equation and proved its validity for titania and silicon carbide cermets.

$$\frac{C}{a} = \frac{Wi_o}{nFdV_o} \times e^{(A-B)OP} \times \frac{(I+c)}{K} \quad \text{where,}$$

C = concentration of the suspended particles, a = volume fraction of particles in the deposit, W = atomic weight of the electrodeposited metal, d = its density, n = its valency and F = the Faraday constant. i_o and A are constants in the Tafel equation w.r.t. the metal deposition relating current density and overpotential OP, viz. $i = i_o \times e^{AOP}$. V_o and B are similar parameters derived from the inert particle deposition. K is derived from Langmuir adsorption isotherm and depends essentially on the interaction intensity between particles and cathode. Copper-alumina cermets have also satisfied the above derivation (Celis & Roos 1977), and other workers have also considered the surface charge produced when the particle is adsorbed on cathode which later assists in mechanical bonding of the particle to the discharged metal (Foster & Kariapper 1973).

The percentage of the particles incorporated is influenced by the quality of the metal being deposited. Fine grain metal deposition conditions can incorporate a higher percentage of particles at higher current densities. In conditions where coarse grain or dendritic metallic deposits result, the particle content is low (Foster 1983; Carew 1983). Mobilisation of particles towards the cathode has been considered in terms of adsorption, and the effect of pH, temperature and agitation have been discussed by Hocking et al (1978), Foster (1983), Kedward and other workers (1976).

Tables 5:6-10 present selected data on composite coatings and electrolytes in general (Foster 1983; Hocking et al 1978; Carew 1983; Hocking, Vasantasree & Kannan 1985; Kannan 1985).

TABLE 5:6

A CROSS SECTION OF ELECTRODEPOSITED COMPOSITES

Matrix	Bath composition,	g/l	Dispersed phase
Copper	Copper sulphate	200	Graphite
	Conc. sulphuric acid	80	Molybdenum disulphide
Copper	Copper cyanide	100	Silica fibres
	Sodium carbonate	12	
	Sodium hydroxide	25	
Copper	Commercial pyro-phosphate soln.		Silicon carbide
Iron	Ferrous sulphate	240	Silica, fibres
	Ammonium sulphate	120	Alumina powder
	Potassium chloride	10	
Nickel	Nickel chloride	300	Alumina powder
	Boric acid	30	
Nickel	Nickel sulphate	200	Tungsten carbide
	Nickel chloride	180	
	Boric acid	40	
Nickel/ Cobalt	Modified Watts		Chromium silicide + Oxides (of Al,Ti,Hf & Zr)

==

TABLE 5:7

ELECTROLYTES USED & OPERATING PARAMETERS

Process	Electrolyte constituent	Operation limits, g/l	pH	Temperature, $^\circ$C
Nickel	Nickel sulphamate	330 - 370	4.5-4.9	50
	Nickel chloride	3 - 8		
	Boric acid	30 - 40		
Cobalt	Cobalt sulphate	430 - 470	4.3-5.0	50
	Sodium chloride	15 - 20		
	Boric acid	25 - 35		
Chromium	Chromic acid	240 - 260		20
	Sulphuric acid	2.4 - 2.6		

==

TABLE 5:8

Ni-Al$_2$O$_3$ CERMET DEPOSITION FROM WATTS NICKEL ELECTROLYTE

Ceramic content, g/l	Particle size	Current Density, A/dm^2	Deposit Quality
100 (low pH soln.)	5	5.9	very good dispersion; addition of urea was detrimental
100 (low pH soln.)	5	5.9	urea had adverse effects on dispersion, becoming worse with inc. of urea.
(High pH soln.) 100	5	4.3	very good dispersion,
100	400 mesh	3.9	poor dispersion
100	500 mesh	3.9	better than 400mesh
100	900 mesh	3.9	very good dispersion
300	30	4.3	good dispersion

Note:- low pH (1.5-4.5); High pH (4.5-6.0); (Foster 1983)

TABLE 5:9

METAL/CERAMIC COMPOSITION IN Ni & Ni—Co CERMETS

Cermet	Ceramic Conc. in soln.,g/l	Deposit Metal & Ceramic Content Weight %		
		Nickel	Cobalt	Ceramic
Ni-Alumina	10	Bal.	—	2.02
	100*			1.14
	100**			0.98
Ni-Zirconia	10	Bal.		1.84
	20	Bal.		3.92
Ni-Hafnia	10	Bal.		2.35
Ni-Titania	15	Bal.		3.33
Ni-Chromia	15	Bal.		6.80
Ni-CrSi$_2$	15	Bal.		3.80
Ni-(TiO$_2$+CrSi$_2$)	15	Bal.		3.60
Ni-Co-TiO$_2$	15	73.65	23.35	3.00
Ni-Co-Cr$_2$O$_3$	15	71.47	22.46	6.07
Ni-Co-CrSi$_2$	15	72.54	23.66	3.80
Ni-Co-(TiO$_2$+CrSi$_2$)	15	72.91	23.70	3.40

Note:- current density 2 A/dm^2,(except *1.61 and **3.23 A/dm^2 respectively). Watts-type electrolyte was used.

TABLE 5:10

MECHANICAL PROPERTY DATA ON ELECTRODEPOSITED CERMETS

Cermet	Hardness	Yield Strength MN/m^2	%Elongation
Ni	*187		
	118	93.0	6.26
Ni- $2.02Al_2O_3$	*275.4	68.4	
	247	67	
Ni- $3.33TiO_2$	*354		
	254	222.4	3.75
Ni- $6.80Cr_2O_3$	*409		
	295	284.0	1.50
Ni- $3.6(TiO_2+CrSi_2)$	*283		
	209	198.5	2.60
Ni- 24Co	*280		
	150	120.6	2.3
Ni- 23.4Co- $3TiO_2$	*383		
	219	206.6	2.0
Ni- 22.5Co- $6.07Cr_2O_3$	*462		
	285	264	1.2
Ni- 23.7Co- 3.8 $CrSi_2$	*302		
	207	196.4	1.6
Ni- 23.7Co-3.4(TiO_2+ $CrSi_2$)	*359		
	211	196.5	

===
*Hardness as plated; Other Values shown are of annealed samples

WEAR AND CORROSION RESISTANT FINISHES FOR ENGINEERING COMPONENTS

	Copper	Electroplated Nickel	Electroless Nickel	Chromium	Electroless Nickel + Chromium	Electroplated Nickel + Chromium	
Maximum working temperature (°C)	50	650	550	650	550	650	Short times at higher temp. possible
Non-toxicity	*****	****	****	****	****	****	
Covering complex shapes	***	***	*****	*	****	***	
Thickness range (µm)	12.5–500	12.5–500	12.5–500	12.5–500	12.5–100 + 25–50	12.5–500 + 25–50	Grinding needed over 200 µm
Wear 1. Hardness (HV)	60–150	200–300	450–500 900–1000 after heat treatment	850–950	850–950	850–950	Guide to abrasive wear resistance
2. Low friction anti-stick	**	**	***	*****	*****	*****	Guide to adhesive wear resistance
3. Resistance to impact	***	****	***	***	***	****	Thin coats and soft substrates prone to damage
Corrosion resistance	*	****	***	**	****	****	Ni at least 50 µm for corrosive environments
Typical applications	Build-up; lubricant in forming; heat sink; selective case hardening.	Build-up; under or instead of Cr in corrosive conditions printing surfaces	PVC moulding tools and dies; moving parts in process industries, glass, rubber moulds.	Moulds, tools, valves, rams, pistons, shafts gauges, dies saw blades.	High temp. anti-seize bolting; ball valves, shafts.	Marine crane rams and hydraulics; mine roof supports; print rolls.	
Specifications	None	BS 4758	DEF 03-5	BS 4641	DEF 03-5 BS 4641	BS 4758 BS 4641	

Star Code: ***** excellent; **** very good; *** medium; ** poor; * very poor.

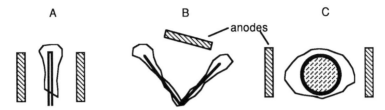

A   B   anodes   C

Note:- Uniform coating thickness is achieved by a careful location of anodes with respect to the cathodes. Mechanical and air agitation, movement of the substrate, e.g. by rotation etc. are control measures to overcome the non-uniformity of the deposition process. Free corners are always likely to be thicker. Confined corners, viz. acute angles and corners away from anodes result in reduced thickness. Circular substrates develop greater thickness at locations facing the anodes.

Fig.5-10: Thickness distribution of electrodeposited coatings

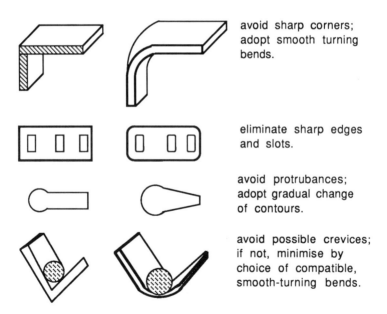

avoid sharp corners; adopt smooth turning bends.

eliminate sharp edges and slots.

avoid protrubances; adopt gradual change of contours.

avoid possible crevices; if not, minimise by choice of compatible, smooth-turning bends.

Fig.5-11: Design criteria to achieve reasonably uniform electrodeposits

5.6.4. AQUEOUS ELECTROPLATING

Industrial electroplating has been operating for a long time to provide decorative and protective finishes for use at ambient temperatures and a variety of environments. With reference to high temperature application their use is restricted to providing interlayers for diffusion bonding on large substrates of simple geometry. The main drawback in electroplating is the inability of achieving uniform depostion, viz. the electrolyte throwing power. Throwing power is substrate-shape dependent, as well as depending on the anode-cathode configuration, the current density and the electrolyte. Deposits are thicker at corners and edges, and thinner in crevices; poorer in holes and pores than the rest of the electrolyte-exposed substrate. The second drawback is that not all substrates can be readily coated, e.g. stainless steel in aqueous solutions; Al and Al alloys. Wear and corrosion resistance data for the various electroplated metals are summarised in Table 5:11 (p.213), and the main design features are given in Fig.5-10,5-11.

Taken in context with the other coating methods considered in this chapter, electrodeposition of only 4 metals seem to be of relevance to high temperature: Ni, Co, Cr and Pt. Electrolplating of Ni, Co and Cr is very well-established industrially and the limited use of Pt-coating is merely cost-oriented; Pt can also be coated from fused salts and more effectively. Many types of acid and alkali electrolytes are known for Ni and Co plating, namely, sulphate, chloride, sulphate-chloride, fluoborate, sulphamate, cyanide and pyrosphosphate, etc. The metals can be deposited at good current efficiencies, but have to be guarded against residual stresses. Plating conditions and post-deposition heat treatments are usually employed to control stress.

In spite of its poor efficiency, chromium electroplating from aqueous electrolyte is still the cheapest method for large-scale chromium deposition. Eletrolytic hard Cr coatings are much used in decorative and corrosion resistant applications but have the following problems:

  (i) Causes H embrittlement in steels
 (ii) Low deposition rates
(iii) Poor throwing power
 (iv) Poor cathode deposition efficiency (15%)
  (v) Pollution problems in plating bath.

Factors which must be considered before choosing interlayer deposition from aqueous electrolytes are:

(i) Substrate/coating adhesion and bonding e.g. for substrates of stainless steel, Al-, W-, Nb-, Mo- alloys etc. Note that a copper-strike acceptable for low- temperature use is not acceptable for high temperature application,

(ii)   The possibility of substrate oxidation while in contact with the aqueous solution,

(iii) Substrate attack by the electrolyte,

(iv)   Desirability of substrate wetting,

(v) _In situ_ incorporation of sulphur or phosphorous or other complex radicals in the deposit from the electrolyte ,

(vi) Hydrogen incorporation into the deposit when metal reduction efficiency is not 100%, and the subsequent embrittlement in the coating (Lowenheim 1961).

The advantages of aqueous electrodeposition are in:

  (i) bulk substrate handling,
 (ii) large sizes acceptance in plating tanks,
(iii) requiring very little surface finish on finished product,
(iv) easy control of deposit thickness and grain size (throwing power drawbacks can be compensated),
 (v) well bonded coatings,
(vi) speed and economy.

5.6.5. SPECIALISED APPLICATIONS OF ELECTRODEPOSITION

Al-C fibre composites are difficult to produce except by room temperature electrodeposition. (Embrittlement occurs at 500°C). Aqueous plating is also used for cladding Ni plated U cores, coating Mg and electroforming parabolic mirrors.

Co-Cr_3C_2 composites: heat treatment increases adherence. Wear resistant to 700°C, due to a 'glaze' of Co_3O_4 debris. The Cr_3C_2 promotes glaze retention.

Pt-Al coatings can be produced by pack aluminizing electrodeposited Pt. This is better than a simple aluminide coating. Simple aluminides degrade by oxidation and spalling, and by inward diffusion of Al. But Pt does not act as a diffusion barrier. Coating morphology and chemistry depends on the Pt layer thickness. Delamination can occur along phase boundaries.

Rh plating is used for electrical contact surfaces. Pt-group metals can be obtained by aqueous electroplating: the main problem is surface oxides not being removed by the pre-plating etch. Pt plating peels off, especially for thicker platings (half mil, which in fact is very thin). Thinner deposits are sound but too thin to give protection. The difference between aqueous and molten salt electrodeposition (discussed later) is that diffusion bonding (surface alloying) occurs in the latter case.

Si electrodeposition has been reported from a non-aqueous solvent (Gobet & Tannenberger 1988). Tetrahydrofuran was used as the electrolyte for adding $SiHCl_3$, $SiCl_4$, $SiBr_4$, $Si(CH_2CH_3)_4$ etc. Tetrabutyl ammonium bromide or $LiClO_4$ was used as the supporting electrolyte. Si deposits up to 0.25 microns were obtained at room temperature in a nitrogen atmosphere.

5.6.6. CHEMICAL COATING OR ELECTROLESS COATING

A "cold" coating technique intended for >300°C where a resin-matrix cannot be used, may be mentioned here (Kulkarni et al 1979). 20cm long carbon fibres in a bundle as 1000 filaments are heat treated in vacuum (10^{-4}-10^{-5} torr) at 700°C for 15 minutes to remove impurities and to activate the surface. They are then dipped in a metal salt solution with known amounts of glacial acetic acid (0.5-15 vol%). The latter improves the wettability. Displacing agents such as Mg, Al, Zn and Fe are added as granules or powder (100-200 microns). This addition displaces metallic Ni, Co, Cu, as the case maybe. Copper coated carbon fibres are treated in a 10% sulphuric + 5% orthophosphoric + 20% tartaric acid-stabilizer bath which prevents tarnishing. The coated fibres are washed and dried at 60°C. 0.5-7 micron thick coatings could be obtained.

Composites were then prepared by introducing the metal coated fibres into an aluminium matrix by vacuum infiltration techniques. The Ni, Co, or Cu coating dissolves in the molten Al and a fresh surface of the carbon fibres is well wetted by Al to result in a well bonded composite. The authors have not been quite accurate in applying the term 'cementation' for this process, which is in fact a displacement by reduction. The effect of Ni, Cu and Co dissolved in Al, on its wettability on carbon, is not clear. This is a hybrid hot-dip/electroless method.

The only high temperature metal in commercial production by electroless methods is nickel. "Electroless" is a trade terminology. The process mechanism is chemical decomposition, displacement and reduction. The significant advantage of the Electroless process is in its "throwing power". Very uniform coatings are possible by this method unlike in electroplating. Complex shapes and large-size components can be well-handled. The additional advantage is that it uses very little electrical energy except what is required for maintaining the electrolyte temperature (e.g. 85°C 95°C for a nickel elctroless coating solution). It also produces a quasi-crystalline, or near amorphous deposit depending on the control conditions.

Electroless Ni is less pure than electrodeposited Ni. Phosphorus is the main inclusion. Table 5:12 adapted from general data made available by Canning Co, Birmingham,UK, summarizes the procedure.

TABLE 5:12

Process & Characteristics of Electroless Nickel Solution
(NIFOSS-2000 by Canning)

Solution Composition:

Nifoss 2000 Base Slution N.21321   :– 100 ml
Nifoss 2000 Initial Additive N.21322:– 200 ml
Distilled or De-ionised Water   :– to 1 l

Operating Conditions:

Temperature   85 – 95°C;
pH   4.8-5.3

Process Control & Operation:

Polypropylene tanks are the most suitable containers
for the solution.   Direct heating is applied with
special immersion heaters;   solution is completley
stable up to the boiling point;   average life = 6
turn overs.

A plating rate of 15-20 microns per hour is achieved
at 95°;   this rate about twice that at 85°C.

Accurate   temperature   control is important to achieve
consistent deposit thickness;

Optimum pH = 5.0. Adjustment is done using 50% v/v
ammonia solution.

The nickel deposit obtained contains 7-10% P.
The deposit is corrosion resistant, with an as-plated
hardness of 600 HV$_{200}$ increasing to 1000 HV$_{300}$ on a
simple heat treatment. The deposit is bright and has
good levelling characteristics.

Most metals and semi-conducting materials can be coated
with electroless nickel. The deposit has low porosity
and high resistance characteristics and is useful on
components subject to wear.
==

5.6.7. FUSED SALT ELECTROPLATING OR METALLIDING

5.6.7.1. The Process:

Metalliding is the term coined for surface modifying and surface hardening processes by electrodepositing refractory metals or metalloids on cheaper or softer substrates from fused salt electrolytes. Electrodeposition at fused salt temperatures ($400^{\circ}C$-$900^{\circ}C$) also result in "cementation" and "diffusion bonding" as refractory metals and metalloids are discharged as particulates and not as coherent deposits. Substrates which are soft (Cu) are easily oxidizable. Fe-, Ni-, Co-, Mo-, W-, Nb-, Ti-, Cr etc., can be diffusion coated with Be, B, Al, Si, Ti, V, Cr, Mn, Y, Zr and rare earths in fused fluoride or chloride electrolytes. Improved component characteristics are noteworthy. Fused salt electroplating is attracting more interest in the fields of surface hardening and corrosion protection (Matiasovsky et al 1987).

To illustrate a few:

(i) Molybdenum which can burn away in air at ($650^{\circ}C$) can withstand twice that temperature for several hours when it is silicided;

(ii) A copper spring which can barely support a 10g weight, can easily accept a 50g weight when diffusion – alloyed with Be;

(iii) Steel can be aluminided, borided, chromized, silicided, and vanadided in molten salts;

(iv) Mo-B is a well applied bearing surface;

(v) Cr and Y can be co-diffused on to steel.

With the exception of Cr as a single metal and W and Mo as alloys, refractory metals mentioned above cannot be obtained in a coherent form either from aqueous or fused salt electrolytes. H_2 is preferentially discharged at the cathodes in aqueous electrolytes containing these metal radicals, and even in fused salt electrolytes they are mostly obtainable as powders or dendrites (Brenner 1960; Senderoff 1966).

Molten halide electrolytes are the most investigated. Of course the best known industrial production is that of aluminium extraction. Fluorides, chlorides, fluoride-chlorides have been reported mostly in literature although a few low temperature organic fused salts have also appeared in some research work. The only early claim at obtaining coherent coatings has been the Canadian patent of Mellors and Senderoff (1966). Except Ti, all other refractory metals have been obtained from pure fluoride electrolytes with the stipluated conditions: valency states of +3 for Cr, V and Mo; +4 for Nb, Zr and Hf; +4 or 5 for W and +5 for Ta; and, the conducting substrate must be more noble than the metal being reduced at it. The electrolytes operate over 750-850°C in a LiF-

KF main electrolyte with the refractory metal fluoride added at known concentrations.

Other works of interest are given in brief below:

1. Aluminiding of steel from fused $AlCl_3$-$NaCl_3$ (80-20 wt%), 175-205°, 18mAcm^{-2} (Nayak & Misra 1979)

2. Aluminium-carbon fibre composites from a $AlCl_3$ (3.4M), lithium aluminium hydride (0.4M), di-ethyl ether at 3-10°C, 50A/m^2.

3. A Cr-$CrCl_3$ mixture to obtain $CrCl_2$ (Ubank 1977) in solution with a NaCl-KCl (40-40) mole % at 1000°C, unspecified current density, to chromize carbon steel (Zancheva et al 1978).

4. Molybdenum carbide coatings have been deposited using FLINAK (LiF 29-24, NaF 11.7, KF 59.06), K_2CO_3 and Na_2MoO_4, in inert atmpsphere at 850°C, at 150 mA/cm^2. Current efficiencies were 45% at 10e$^-$ and 76% at 16e$^-$. Mo_2C deposited on a steel/nickel substrate as a dense and coherent coating (Topor & Selman 1988).

5. TiB_2 coatings increase tool cutting lives by up to 9 times. An 8 micron coating from a fused salt (KF, LiF and NaF eutectic, m.p. 453°C) has a Vickers hardness of 4060 (Kellner et al).

6. Tantalum carbide coatings on nickel substrates from a ternary (Li-,Na-,K-) fluoride eutectic containing K_2TaF_7 & K_2CO_3 at 750-800°C. Ta and C are simultaneously discharged at the Ni-electrode to result in abrasion resistant coatings stable up to 600°C (Stern & Gadomski 1983).

7. Adherent TaC has been deposited on Ni from a ternary (Li-,Na, K)F eutectic containing a few per cent of K_2TaF_7 and K_2CO_3. Ta and C simultaneously deposit on the Ni cathode where they react to form TaC at 750° to 800°C. These coatings resist abrasion and are stable to about 600°C (Oda et al 1983).

8. Dense, adherent coatings of Ta_5Si_2 have been plated from FLINAK with K_2TaF_7 and K_2SiF_6 at 750°C on a nickel cathode. The coating composition was between Ta_2Si and Ta_3Si and was independent of the melt composition. The deposits were stable in air up to 600°C with a hardness of 1200 kg/mm^2 (Stern & Williams 1986).

9. FLINAK eutectic was also used to deposit W_2C+WC at 750-1000°C, but the tungsten carbide was found unsuitable for wear resistance (Stern & Deanhardt 1985).

10. Boride layers have been developed on low alloy low-carbon steel, manganese steel and high alloy chromium steel. Electrolytically borided layer consisted of two phases FeB and Fe_2B. Thermochemical boriding produced an additional Fe_3B phase. On high alloyed chromium steel the Cr was found to act as a diffusion barrier, retarding the rate of boride layer formation and the FeB content increased (Chrenkova et al 1986). In a melt

mixture of $NaCl-KCl-Na_2B_4O_7$, chloride ions reacted with the sodium borate to form oxy-chlorocomplexes of boron or boron trichloride. Addition of a reducing agent such as B_4C or SiC released boron which dissolved in the melt. Boride layers of different phase compositions could be manipulated by controlling the boron activity which differed substantially between the two carbide additions (Makyta et al 1986).

5.6.7.2. Disadvantages of Metalliding:

Fused salt electroplating has not so far been utilised as a generally applicable deposition process for refractory metals and ceramics in line with processes such as CVD and PVD which have already established their versatility in being able to deposit all the above refractory metals and metalloids. The only exception is perhaps the Pt-group metals. The process conditions for fused salt electrolysis are too stringent, and difficult to control and maintain for a long time; a large scale operation which requires reproducibility and ease of maintenance may never be achieved. The electrolyte constituents have to be free of moisture and any oxide contaminants; the ambient atmosphere has to be a neutral gas like argon and helium which are expensive; and the electrolyte components are too readily reactive with metallic substrates or any other metallic components of the cell exposed even slightly to air. Specific applications do utilise the technique but to a very limited extent wherein its cost can be absorbed.

Although Cr can be deposited at a near 100% efficiency compared to the 15-30% efficiency achieved in aqueous electrolytes, a commercial scale fused salt deposition process is economically unfeasible. Very little industrial-scale applications have been reported. Small sized components can use metalliding because of the wide range of refractory metals which can be applied or recovered.

5.6.7.3. Utility of Metalliding:

Small-scale, small-size, specialized or strategically important components can be considered for technological development for coating by molten salt electrolysis if the cost justification can be made, in view of the fact that a one step diffusion coating can be achieved with greater ease. Metalliding deposits and heat treats the substrate simultaneously to produce sound diffusion bonded refractory products.

5.6.8. PRECIOUS METAL PLATING:

Only the Pt-group metals are relevant for high temperature coatings because of their high melting points and corrosion resistance. Platinum especially, has been used as a diffusion barrier layer in aluminiding Ni-base alloys and also with MCrAlY coatings. It is also used in the glass industry. Both aqueous and fused salt electrolytes have been used. The reviews by Cramer et al (1967) and Rhoda (1977) give an account of the operating conditions in some detail. Molten chlorides and cyanides have been used as base electrolytes to which the metal compound is an optional addition. Operating temperature range over 400-600°C. Deposits from cyanide melts appear to be bright and uniform. Notton describes the industrial plating of Pt from molten cyanide on Mo and Nb (1977). The electrolyte is H-free and Pt, Ir and Ru are commonly plated this way.

Stainless steel and Mo are plated with Ir to protect from air oxidation at 600°C but at 1000°C the attack is worse than un-plated steel. Mo was protected at 600°C and 1000°C. Ir coatings are used to protect graphite; 100 microns thick Ir is good to 1000°C in air, 120 microns of Rh protects Mo and W to 1330°C. Ir coats protect C, Ta, Nb, W, Mo from high temperature oxidation. A rate of deposition of 20 microns/hr and 270 microns thickness are easily obtained. Coatings thicker than 20 microns are pore-free, coherent, dense and columnar, harder than wrought Ir. Pull-off tests using epoxy (for Pt on Mo) always break the epoxy. Ti plated with 25 microns of Pt and boiled 1 day in 50 per cent HCl showed no Ti in the acid. Pt on Nb is also excellent. Electro-formed Pt crucibles are also very good. Fused-salt plated Pt is very ductile. The difference between aqueous and molten salt electrodeposition is that diffusion bonding (surface alloying) occurs in the latter case.

Pt from aqueous electrolyte (dinitro sulphate platinous acid) is highly stressed unlike the fused salt deposit, but rhodium can be plated satisfactorily up to 40 microns thick from sulphate solu-tions. Palladium can be plated better than rhodium (Hood 1976). Ru up to 14 micron thickness has been deposited on graphite at 5.4 mA/cm^2 with Ru concentration held at 0.26 wt.% (Bettelheim et al 1985).

Substrates particularly considered for Pt-group metal coating are the refractory group alloys of Mo, W, Ta, Nb and V which tend to form volatile oxides at high temperatures weakening their useful-ness as corrosion resistance materials. (Platinum itself forms a volatile oxide). The refractory metal alloy substrates are etched in fused salt electrolytes (anodic or cathodic as the case may be) in order to achieve satisfactory bonding of thick coatings of the Pt-group metals. Some of the operating conditions quoted in the literature are presented in Tables 5:13 to 5:15.

One of the main factors that controls the choice of plating platinum, iridium, or other related metals will be the cost. The offsetting factor will be that the scrap metal cost will be high and coating renewal should not be as expensive as an initial application. Once again, cost justification will allow these metals to be used in areas where diffusion barrier layers and corrosion resistant coatings are required and the components are small and strategically important.

T A B L E   5 : 13

SUMMARY OF ATKINSON'S MOLTEN CHLORIDE METALLIDING MELTS

Metal	Concentration of Metal in melt, wt%	Temp. °C	Current Density mA/cm^2	Cathode Efficiency,%	Anode Efficiency,%	Appearance of Deposit
Pt	3.6	500	3 to 6	68	60	Coherent, bright, coarsely crystalline
Rh	1.8	410 to 440	7 to 10	100	101	Partly smooth and bright, some sponge, some crystals
Pd	3.7	400	7	102	100	Non-adherent crystal leaves
Ru	1.8	400	9		47	Very little deposit crystals
Ir	3.6	500	3	70	97	Hard, white, coherent, some nodules

a. Impure soluble anodes were used.
b. Based on Pt
c. Formula not given, assumed to be as shown.
d. Values not given, estimated from total current used.
e. Based on Ir

T A B L E   5 : 1 4

I R I D I U M   E L E C T R O P L A T I N G   F R O M   M O L T E N   C Y A N I D E

Workers*	Atkinson	Dickson, Wimber, Stetson	Macklin, Owens, Lamar	Rexer Crisclone	Rhoda	Schlain	Withers Ritt
Main Support Electrolyte, wt%	NaCN, KCN, NaCN & KCN + or - NaOH(i)	70 NaCN & 30 KCN	70 NaCN & 30 KCN (ii)	70 NaCN & 30 KCN	NaCN	NaCN	70 NaCN & 30 KCN (ii)
Volume of Electrolyte cm^3	-------	70 (iii)	700 (iii)	200 (iii)	30 & 125	120 to 180	160 (iii)
Container	Porcelain or iron	Alumina	Alumina - quartz broke on freezing; melt decomposed with graphite	Alumina & graphite	Silica or porcelain	Silicon carbide quartz & mullite	Silicon carbide Mullite
Temperature °C	305 to 550	700 & 800	600	600	600	600	600
Atmosphere	Inert suggested	Argon in closed chamber	Argon in closed chamber	Argon in closed chamber with	Argon blanket	Argon in closed chamber	Argon blanket
Metal Concentration	-------	1.6 to 11	1	0.5 - not always given	0.3 described as equilibrium	0.19 to 5.7	0.43 to 0.53
Electrolyte Preparation	Metal immersion soluble anode or as cyanide complex	ac, soluble electrodes	ac, soluble electrodes	ac & dc, soluble electrodes	60 Hz ac, 200-300 mA/cm^2 (iii) 600°C, soluble electrodes	ac & dc, soluble electrodes	60 Hz ac, 10 mA/cm^2 500°C, soluble electrodes
Anode	Ir	-------	Ir	Ir	Ir	Ir	Ir
Cathode	-------	Ta	Graphite	Mo, W, & Ni-plated Ta & Cb	Cu, Ni & Pt	Mn	Cu, Mo, & 410 stainless steel

Anode Current Density mA/cm²	-------	-------	Probably < 10	15	3 to 100; black crust at > 200	2.5 to 30; optimum 10 passive > 100
Cathode Current Density mA/cm²	5 to 125	5 & 35 direct or P.R. 16 & 4	Probably 10	15	3 to 20	11
Anode Eff % (iv)	> Cathode Eff.	0.1 to 9.0	-------	-------	54.6 to 127 (v)	>>Cathode Eff.: up to 225 %
Cathode Eff % (iv)	-------	0 to 66	50 & 80 (may be based on +2)	9 to 20; 10 at 15 mA/cm²	25 to 69	0 to 75 with inconsistencies
Deposit Thickness μm	-------	Dendritic above 150 to 200	-------	up to 75	270	125 without difficulty
Deposit Characteristics (vi)	-------	Columnar	Columnar	Bright & smooth hardness (800)**	Smooth; nodular at 250 μm; hardness (727)**	Fine grain; 12 m protects Mo up to 1000°C
Comments	Source is patent, may make broader than justified claims	Electrolyte decomposes on "prolonged" standing at 600°C; good electrolyte is bright yellow or red	Electrolyte operated month with absolute exclusion of air. Air exposure causes irreparable deterioration. Good electrolyte is red at temperature.	Electrolyte may be frozen & reused. Throwing power appears adequate.	Only thin deposits possible if electrolyte is exposed to air.	

Notes: (i) "Convenient" mixture given as 60 NaCN & 40 KCN to which NaOH may be added. M.P. can be as low as 300°C. Examples cited: 66 NaCN & 34 KCN at 500°C; 30 NaCN, 20 KCN & 50 NaOH at 440°C; and 18 NaCN, 12 KCN & 70 NaOH at 305°C.

(ii) Exclusion of air incomplete. CNO & CO_2 probably present.

(iii) Not given, estimated from other information source.

(iv) Based on a valence of 13.

(v) Some anode decrepitation occurred.

(vi) All except Atkinson obtained dense, coherent, and adherent deposits.

*(Authors quoted by Cramer 1967); ** hardness (Knoop value)

225

T A B L E 5 : 1 5

E L E C T R O P L A T I N G F R O M M O L T E N C Y A N I D E S

Metal Plated	Pt		Pd	Rh	Ru
Worker	Rhoda	Schlain	Schlain	Schlain	Schlain
Main Support Electrolyte	NaCN	NaCN,NaCN & KCN NaCN & KCNO**	1/1 NaCN/KCN or 1/1 NaCN/KCNO*	NaCN,reagent grade vacuum dried	53% NaCN & 47% KCN
Volume of Electrolyte, cm³	30 to 125	180			
Container	Porcelain & Silica	Silicon carbide & quartz	Silicon carbide or fused quartz**	Mullite	Porcelain or silica*
Temperature, °C	600	500 to 600	450 to 475	600	560
Atmosphere	Argon blanket	------	Air***	Argon*	Argon
Metal Concn. mass %	0.3	0.28 to > 2	0.8 to 0.9	0.8 to 0.9	0.5
Electrolyte preparation	200 to 300* mA/cm² ac	Anodic solution at 10 to 40 mA/cm²	Anodic solution at 10 mA/cm²	Anodic dissolution at 10 mA/cm²	ac between soluble electrodes**
Anode	Pt	Pt or graphite	Pd	Rh	Ru
Cathode	Several	Several			
Anode Current Density mA/cm²	3 to 30	2.5 to 30	1.5 to 20	10	-----

Cathode Current Density mA/cm²	3 to 30	5 to 30	3 to 20	5 to 10	Current periodically reversed:10 mA/cm² forward for 60 s & 3 mA/cm² reverse for 7.5 s
Anode Eff %	------	98 to 150	------	100+**	------
Cathode Eff %	65 to 98	40 to 76	15 to 70	80**	30
Deposit Thickness, microns	250	125	125	200	50
Deposit Characteristics	Bright, smooth				
Comments	Attempts to make electrolyte from cyanide complex failed	Pt content of melt increases and Pt precipitates			

* estimated from indirect data
** air exposure during melting necessary

* Exposed to air during original melting
** Presumably graphite, mullite and alumina could also be used.
*** The reference seems to state air exposure is a constant need.

* Air exposure destroys electrolyte
** based on Rh?

* Graphite, mullite, and alumina should be suitable
** A platinum-ruthenium alloy electrolyte can be made by adding Ru-cyanide to a platinum electrolyte. Possibly a straight Ru-electrolyte can be made in the same manner

Coatings by laser surface treatment, rapid solidifications processing, spraying, welding, cladding and diffusion methods

6.1. IN THIS CHAPTER

Six coating methods are dealt with in this chapter, all of which are individual process methods, but are also more often used in conjunction. Diffusion is an inevitable surface/coating phenomenon except where diffusion barrier overlayers are applied, and a number of coating methods involve one or more post-coating heat treatment step wherein diffusion is the means for coating consolidation or modification of its morphology . Diffusion bonding (DB) as a coating method in its own right is discussed in this chapter. Cladding is self-explanatory. Welding has been a traditional 'cementing' or joining technique which has in recent years advanced rapidly as a surfacing technique. Both cladding and welding are classified as bulk-surfacing methods. The former uses the material to be coated in a solid form while solid and molten media in bulk are encountered in welding. Coating by droplet transfer using plasma-, arc- or detonation-gun (D-Gun) spraying is another form of bulk-coating method, with the material to be coated in a liquid or slurry form. Rapid cooling of molten alloys results in amorphous or 'vitreous' structures and the materials solidified thus exhibit high resistance to wear and corrosion. Rapid solidification processing (RSP) is a fast-developing coating technology. The laser has proved a very versatile processing tool to perform well-controlled, small area surfacing, welding, cladding, spraying and rapid solidification processes. This technique is covered at the outset in this chapter.

6.2. LASER COATING TECHNOLOGY

The application of lasers in coating technology has developed rapidly since 1980 although the first attempt at laser surface alloying was reported in 1964 (Cunningham 1964). Laser cladding by powder fusion progressed from research to development during 1974 to 1978 (Powell & Steen 1981). Now lasers are used as one of the main means of application of coatings, as a manipulated auxiliary in other depositon processes such as CVD, and PVD techniques, as well as in a number of surface modifying methods such as cladding, welding and spraying. The practical utility of lasers has stretched far beyond the area of coating technology itself.

A concise list of literature sources is appended to provide most of the preliminary information on the use of lasers in both thin film and bulk coating methods. They also provide a broad spectrum of the use of lasers in surface modification welding, drilling, cutting, machining and non-destructive testing of a number of metals and alloys: (Ferris et al 1979; White & Peercy 1980; Gibbons et al 1981; Bahun & Engquist 1964; Mukherjee & Mazumder 1981; Metzbower 1979; Draper & Mazzoldi 1986). The compatibility of lasers to refractories and cermets are undefined, although Kelly has succeeded partially in welding ODS-type Ni-base alloys (1979). Abbreviations very typical of laser processes are listed in the Acronyms section (Chapter 2; section 2.7.).

6.2.1. LASER CHARACTERISTICS

The laser is a singular source of heat produced by means of highly energetic, coherent, monochromatic beam of photons which can be manipulated to operate in focused or defocused states. It is capable of attaining very rapid rates of heating, and is clean, intense and localized. It can be produced either in a continuous form or pulsed. A laser beam incident on a substrate can heat, melt, vaporize or produce plasma depending on the manipulation of the controlling parameters. The principal sections of a laser are given in Table 6:1 and Fig.6-1 (Bass 1979). The overall portability and efficiency of the equipment and its repair and maintenance costs have to be considered in its eventual selection for use.

The main characteristics of the lasers utilized in LSA processing are given in Table 6:2 (Draper 1981). The wider application of lasers became possible when multi-kilowatt CO_2 lasers came into production in 1970. Now CW CO_2 lasers are available from a few watts to many kilowatts power; a CW Nd YAG laser of 400W is also in production.

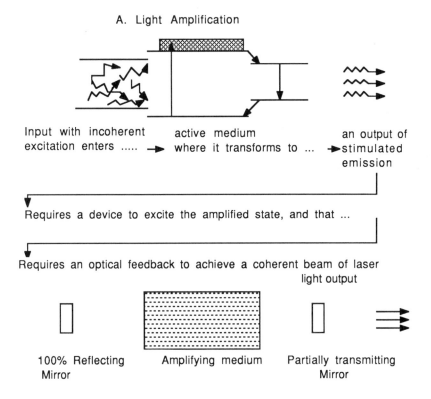

A. Light Amplification

Input with incoherent excitation enters ..... → active medium where it transforms to ... → an output of stimulated emission

Requires a device to excite the amplified state, and that ...

Requires an optical feedback to achieve a coherent beam of laser light output

100% Reflecting Mirror   Amplifying medium   Partially transmitting Mirror

Fig.6-1: Features of a Laser Device

TABLE 6:1

PRINCIPAL SECTIONS OF A LASER

Section	Factors in Design
Amplification of Light	Optical quality; Types of windows; Means of heat removal; Repair and efficency; Cost.
Excitation	Power supply: I stage amplifying state auxiliary supply for cooling, pumps etc; Weight, Performance, Efficiency; Cost.
Optical Feedback System	Reflector components: Mirrors, lenses, coatings modulators; windows; Mechanical and optical stability.

TABLE 6:2

LASER CHARACTERISTICS

Laser Source	Laser Wavelength (microns)	Power Output (W)	Log (Time, sec.) (Dwell or Pulse Time)	Repetition Rate, Hz
Continuous CO_2	10.6	500–20K	−1 to −5	cw
Pulsed Nd-YAG	1.06	200^	−2 to −3	100
Pulsed Nd-glass	1.06	10^6*	−2 to −3	1
Pulsed Ruby	0.69	10^6*	−2 to −3	0.2
Q-switched Nd-YAG	1.06 0.53	1^	−6.75^	$10^3 - 5 \times 10^4$
Q-switched Ruby	0.69	10^8*	−8.7 to −7.7	2

==
* peak value   ^ average value

6.2.3. LASER SURFACE TREATMENT (LST)

Laser surface treatment can be controlled to achieve alloying,
cladding, grain refining or transformation hardening a metal
surface without actually affecting the bulk of the metal itself.
This is mainly due to the localized nature of laser heating. LST
can be categorized into three main sections and its various
effects on a substrate may be shown as in Fig. 6-2 (Gnanamuthu
1979).

Fig. 6-2

Effects of Application of Laser Power on Materials

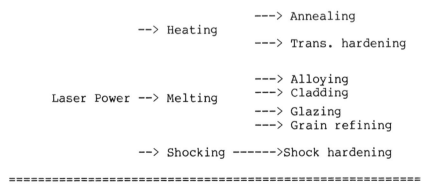

==

A laser beam can enchance surface properties to a controlled,
confined extent depending on the power and dwell time of the beam
and the thermal characteristics i.e. heating and cooling, of the

surface treated (Table 6:3). Surface treatment prospects by lasers were observed with pulsed lasers at first. A laser can be a giant-scale spot welder of one surface to another with very little seal or dilution zone, or it can cause good mixing of the substrate metal with the top coat to yield an alloy. Being inertialess, it has high processing speeds with very rapid stop and start facility. Q-switched laser processing offers quenching rates of 10^{10}K sec-1 and resolidification interface velocities of 1-20 m sec^{-1}. A material of poor oxidation or corrosion resistance but low cost can be modified with a surface alloy which can show improved resistance. Likewise can wear and abrasion resistance be improved. Laser grain refining eliminates or minimizes surface defects such as inclusions, intermetallic compounds and pores, and improves the grain structure. Laser rapid-heat, shock hardening or slow-heat annealing-hardening can improve the surface mechanical properties, relieve stresses and tendency to stress corrosion cracking.

TABLE 6:3

BEAM CHARACTERISTICS USED IN LASER SURFACE TREATMENT

Optical Telescope Parameter	State of Laser Beam	Beam Diameter/ Size; mm	Type of Laser Energy Profile	Application
f/7	Stationary	0.74	Gaussian	Alloying Cladding Glazing Grain refining
f/18	Stationary	1.9	Gaussian	Alloying Cladding Grain refining
f/125	Uni-directional oscillation	6.4 x 19	Rectangular Gaussian	Alloying Cladding Grain refining
f/125	Bi-directional oscillation	19 x 19	Square Gaussian	Alloying Cladding Annealing Transformation hardening
Beam Integrator	Stationary	15 x 15	Uniform square	Alloying Cladding Annealing Transformation hardening

The main limitation appears to be in the bulk area and size which can be conveniently handled by laser equipment. The advantage that a number of LST can be carried out in ambient atmosphere can be offset by the equipment and trained personnel cost. Its long-term advantages are, however, attractive.

6.2.3.1. LST Parameters:

Fig.6-3a,b show schematic diagrams of an overlay, a diffusion bond and LST surfaces and a typical concentration depth profile. In practice the substrate/coating interactions may not be as sharply divided and precise as in Fig.6-3a, but it shows adequately the effect on the interface. A simple overlay coating is hardly likely to exhibit any significant micro-structural change at the substrate/coating interface. LST is on the other extreme where a total surface modification can occur. Diffusion bonding, on the other hand, forms a diffusion zone of considerable thickness, with consumption of the original substrate and deposit materials. The advantage of LST on this is that the diffusion zone is very thin and very little substrate is lost in the dilution step. On the other hand, it can alter the total composition on the surface in a fraction of the time taken by diffusion bonding even to set up the interfacial zone. LST thus can 'create' or 'freeze' phases not captured to stability by any other means, because of the extremely rapid heating and cooling rates possible by laser heating.

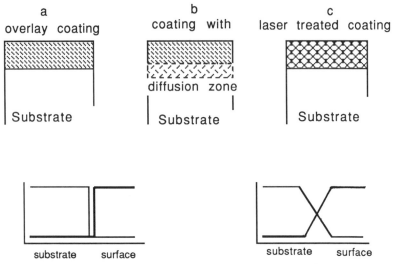

a	b	c
overlay coating	coating with	laser treated coating

diffusion zone

Substrate

substrate   surface

substrate   surface

Coating & Substrate concentration variation around the interface region

Fig.6-3 a,b: Schematic diagram of coating deposits & the effects of heat treatment & laser surface treatment

Fig.6-4 gives a schematic representation of the heat, absorption and reflection aspects of LST (Draper 1981). The temperature rise within the thermal diffusion layer is given by

$$T = \frac{(1-R)It}{C\rho(2Dt)^{1/2}}$$

where, R = reflectance of the surface, I = incident laser intensity, t = dwell time or pulse time of the laser output, D = K/C where K, C and ρ are respectively the thermal conductivity, specific heat and density respectively.

The incident laser intensity is absorbed by the metal within its electromagnetic skin depth, typically 10 - 100 nm and this is converted to heat. The heat rapidly diffuses away to depths directly related to the thermal diffusion length, i.e. $\sqrt{(2Dt)}$. The bulk of the solid then absorbs this heat. For T = 1000K at exposure times of 10 msec to 100 nsec quench rates can be 10^5 - 10^{10} ksec^{-1}. If metal loss is to be avoided, obviously the boiling point of the metal must not be exceeded.

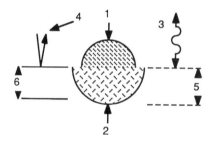

1. optical spot size diameter
2. effective melt pool diameter
3. Intensity of the incident laser beam and the pulse or dwell time
4. the normal spectral reflectance of the surface
5. melt depth
6. resultant solid coating thickness when overlapped

Important laser beam parmeters are: Thermal Diffusivity D, melting point T and vapour pressure P of the coating A and the substrate B

Fig.6-4: Laser beam parameters and the interaction of substrate - surface-beam configuration

Table 6:4 illustrates variation of alloy composition with speed and power of the laser beam (Draper 1981). Longer dwell times produce broader and deeper melt beads and more dilution in surface Cr-composition.

Much higher quench rates and regrowth velocities are possible with the 10-200 nsec laser pulses obtainable in Q-switched lasers. This would mean metastable surfaces even further removed from equilibrium than those produced by pulsed or relative motion continuous lasers are possible. The penetration depth is not more than 10000 AU. The technique would be particularly valuable in applying thin films of very expensive or strategic element surface alloy to a less expensive substrate. The problems in composition control are evaporation loss and "explosion" loss. Slower speeds (e.g. with CO_2 lasers) cut such losses but at the cost of greater alloying depths and hence need more material. A frequency doubled Nd-YAG laser appears to have offered a suitable compromise in alloying Au onto a Ni substrate giving a 50 atom% Au in a 1500 AU film (Fig 6-5) (Draper 1981).

TABLE 6:4

Velocity cm/sec	Melt Width, microns	Melt Depth, microns	Chromium %
38	326	99	11
50	259	51	23
63	276	37	31
75	167	20	58
88	202	42	35
100	133	14	80
125	133	18	60

===

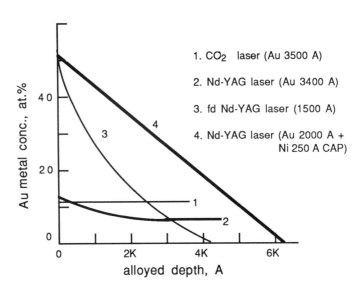

1. CO_2 laser (Au 3500 A)

2. Nd-YAG laser (Au 3400 A)

3. fd Nd-YAG laser (1500 A)

4. Nd-YAG laser (Au 2000 A + Ni 250 A CAP)

Close agreement found between calculated and RBS-spectra estimated values. The long melt time in 1has yielded a very uniform alloy composition by LST

Fig.6-5: Depth profiles of lasor surface alloyed Au-Ni alloys.

The interplay between the laser parameters and material properties influence the emerging microstructure and metallurgical properties of the laser-alloyed or laser surface treated material. A detailed discussion is given by Mordike & Bergmann (1984). Non-transformable steels gain a more homogeneous and finer grain structure but show no significant change in hardness. Transformable steels can exhibit a variety of microstructures and a considerable change in hardness. Stringent control conditions are needed for laser treating Ti-alloys. Most high temperature alloys are suitable for LST.

6.2.3.2. LST Parameter Estimation:

The foregoing brief account on surface treatment links four laser operational parameters: (i) incident laser beam power (ii) laser beam diameter (iii) absorptivity and (iv) traverse speed. The depth of penetration is directly related to the power density which is a function of the beam power and diameter. With the diameter constant, the higher the power the greater the penetration; it could be almost a linear relationship (Draper 1981; Locke & Hella 1974).

Mazumder (1981) indicates that the photon drag detector is probably the most accurate of all modes by which the beam diameter is estimated. Gnanamthu (1979) estimated beam intensity from burn patterns on a slab of polymethylmethacrylate. A Gaussian beam diameter defined as the diameter where the power has dropped $1/e^2$ of the central value is recommended as it contains at that value more than 80% of the total power.

The most material-dependent laser parameter is absorptivity (or emissivity, reflectance). All heat transfer calculations for laser processing are based on the energy absorbed by the work piece. The reflectance requirement for laser suface treatment is almost nothing to that required for laser welding (LW). The faster, cleaner, reduced LAZ (laser affected zone) is obtained with the capacity of the workpiece to cut reflectance losses. But in both LST and LW the total laser efficiency is absorptivity-linked. Defined variously,

$$A = 112.2\sqrt{\rho_r}$$

where, A = Absorptivity and ρ_r = electrical resistivity, emissivity $\epsilon_\lambda T = 1 - 0.547/\sqrt{B} - 0.009\sqrt{B}$, where $-B = 0.667 \rho_r(T)/\lambda$; $\rho_r(T)$ is the electrical resistivity (ohm-cm), T is temperature in deg. Kelvin for radiation wavelength λ (Bramson 1968).

Thus the estimated absorptivities are: when molten 2 and 3% for Al, Ag, Cu; <15% for Fe, Zr, 304 stainless steel and Ti-6Al-4V (at 300°C). This implies reflection losses are high. Reflectance losses can be reduced by applying an absorbant powder, or forming an anodized film (Arata & Miyamoto 1978), or by the use of a reactive gas; e.g. 10% O_2 in the shielding argon gas increased the welding depth almost 100% (Jorgensen 1980).

Metals are poor absorbers of infrared energy at room temperature, but above 10^6-10^7 W/cm^2 threshold, energy transfer can take place via a "keyhole", leading to much higher absorptivity. Multiple reflections within the keyhole can cut emissivity losses sufficiently to allow deep welding possible (Brown & Davis 1976). The threshold energy required by a laser beam to form such a keyhole is 1.5×10^5 W/cm^2, much higher than for an electron beam. This is why laser welding cannot handle thick materials unlike electron beam welding. However, laser welding of "difficult" materials is done with considerable success, e.g. Al- and Ti-alloys.

The laser traverse speed is a compromise between the melting and penetration the beam can achieve without causing damage and vaporization of the material being melted. Thicknesses greater than 3mm cannot be handled effectively for laser welding but is quite adequate for applying reasonably thick bulk coatings. The drawback arises from the attenuation of laser power caused by products formed in situ, e.g. plasma (Locke et al 1972a, 1972b). The traverse speed of the beam has to ensure total penetration and melting as the beam moves along. Electron beam welding scores over laser welding in its capacity to penetrate the work piece because the material absorptivity of the electron beam energy is not dependent on the keyhole reflections, its shape or extent.

Chande and Mazumder (1983) have illustrated a significant application of the statistical experimental design technique for composition control in LSA. A dimensionless parameter has been introduced (LASAP - Laser surface alloying parameter).

LASAP = $P/(pud^2)$, where P = incident power (W), p = shielding gas pressure at the surface, u = traverse speed (m/s) and d = beam diameter (m).

Another dimensionless parameter LAPP, (Laser-physical properties) is given.

LAPP = $Pd/(\rho \alpha u^2 d^2)$ where P, d and u are as defined above; and ρ & α are density and thermal diffusivity of the substrate; $(p/d^2)/\alpha$ = measure of the heat flux (the substrate's ability to dissipate this flux); $1/\rho u$ = measure of the mass flux in the melt pool, d/u = measure of time interaction and P/ud^2 = measure of the energy density.

LAPP quantifies the complex transport processes in LSA with respect to interaction time.

Processing conditions which lead to high alloy specie concentrations are accompanied by high cooling rates producing refined microstructures. Relatively wide beams (>1 mm dia) and long interaction times can produce zones of smooth flat surface profile (Christodolou et al 1983). When using alloying component elements or compounds which have high vapour pressure, surface

238

losses have to be taken into acccount.

6.2.3.3. Features of LST:

Some of the noticeable features in LSA and LST are:

(i) The small beam width: the actual area that can be treated is so small that several passes by the beam are needed to cover an entire area.

(ii) Also, the laser beam melts only a fraction of the longitudinal traverse area which results in ripple-like solidification patterns.

(iii) The melting effects in (i) and (ii) result in a striated, banded, wavy appearance on the finished substrate.

(iv) The two hazards in LSM are porosity and microcracking. The rapid heating and cooling traps bubbles induced by convection in the melt pool, and thermal shock can generate microcracks.

Optimization of laser parameters for LST has been a study-exercise to most workers using the laser (Chande & Mazumder 1981; Mehrabian et al 1979). Correlation of laser processing parameters (power density and interaction time) with the properties of the resultant surface has been investigated from the conventional to the statistical type of analyses. Metallurgical analyses include structure, composition and other properties. These range over conventional optical, transmission electron and scanning electron microscopy; x — ray diffraction; spectroscopy; EDAX and ESCA; hardness; corrosion/oxidation; wear; high temperature stability and heat treatment tests and so on.

6.2.4. LASER ASSISTED COATING TECHNOLOGY

6.2.4.1. Surface Treating in General:

A comprehensive cross section of LST literature up to 1980 is available (Draper 1981) and has been updated (Draper & Mazzoldi 1986). LSA- components are applied to the substrates as powder, paste (slurry), paints, or as sputtered, sprayed or electrodeposited layers.

Laser equipment in conjunction with other coating procedures has been variously tested. Laser cladding, laser spraying, laser-assisted CVD (LACVD), Laser assisted PVD (LAPVD), and laser modifying of plasma coatings have been reported. There are definite advantages in these processes and their economic feasibility has to be established.

1. Extensive laboratory studies exist on ferrous alloys, in modifying the surface characteristics, composition and morphology. For instance, application of a stainless steel composition, or altering the structure completely or partially to obtain austenite, austenite-martensite etc. (Christodolou 1983). Similar studies have also been done on Ti-alloys (Bayles et al 1981; Peng et al 1981; Moore et al 1979) and laser glazing on Zr-alloy stabilized a martensitic structure (Snow & Breinan 1979).

Carbon can be laser surface alloyed into pure iron and 1% C + 1.4% Cr steel (Walker et al 1984; Steen & co-workers). The substrate was precoated with DAG graphite and then up to 12 successive laser surface meltings were given, re-applying the graphite coatings between each treatment. Up to 6% C was alloyed in. Case layers of about 0.2 mm with a bainite structure has been recorded on 6 mm thick Armco Fe. A 15 kW CO_2 laser was used to carburize a graphite dag coated substrate. A maximum temperature of 1051-1058°C was reached over interaction times 1 and 20 secs (Canova & Ramous 1986).

2. Ni-10Cr and Ni-15Cr alloys were laser treated to melt depths of 170-175 microns at a power density of 1.7×10^9 W/m^2, with a speed traverse of 15-70 mm/sec. LST-treated healing Cr_2O_3 were found more resistant to cracking (Stott et al 1987). A general feasibility study on use of LST for fabrication of coatings resistant to high temperature degradation is given (Galerie et al 1987). SiO_2 coatings have been laser-fused onto Incoloy 800H samples (Ansari et al 1987).

3. Stellite coatings applied by TIG and oxy-acetylene torch were laser treated using high power Nd-YAG and CO_2 lasers. A CO_2 laser was also used to carry out laser coating. Laser modified stellite coatings were structure refined through to thin layers but their properties were not significantly improved. The oxyacetylene deposited coating, in fact, showed extensive crack formation on LST. The best results were obtained with the laser-applied stellite coatings (Tiziani et al 1987).

4. Dense crystalline ZrO_2 films have been deposited by pulsed laser evaporation (Sankur et al 1987). Deposition of conductors from solution with laser is reported (Auerbach 1985). Laser trimming of sputtered CrSi films covered with SiO_2 resulted in transforming the underlying CrSi with the top silica layer unaffected. Cr in CrSi is released on laser trimming to diffuse along leaving Si to oxidise (Masters 1985).

6.2.4.2. Laser Cladding:

Laser beams are particularly suited to apply cladding alloys with high melting points on low melting substrates. Wear, impact, erosion, and abrasion resistance, and high temperature corrosion resistance are the regime in which Co- , Fe- or Ni-base alloys are extensively used. Cast rods and pre-alloyed powders have been used for hardfacing. Self fluxing alloy powders, e.g. Ni-Cr-, Ni-base or Co-base alloy powder containing Si and B are flame sprayed and subsequently laser-treated. Oscillating and statio-nary laser beams have produced satisfactory, uniform, pore-free claddings.

Cladding materials range from alloy to ceramic, viz, Tribaloy, Ni-base, WC+Fe, TiC, Si and Al_2O_3. Shielding gases used are He, H_2, Ar, or oxygen as required (Gnanamuthu 1979). Vibro-laser cladding is an interesting variation in this section. Powell and Steen (1981) report that ultrasonic vibration (20 microns at 25kHz) during cladding reduced thermal stress-cracking and also porosity. The effects were monitored in situ by means of acoustic emission and later by optical macro- and microscopy, hardness and SEM studies.

A pulsed CO_2-TEA laser will melt the surface of coatings and thus seal porosity. A pulse length of 20 to 40 microseconds keeps the surface temperature below boiling. This is a useful treatment of plasma sprayed coatings which have a large fraction of open and closed porosity (lowers strength and corrosion resistance). A quick scan by a CW CO_2 laser is unsatisfactory for air-sprayed coatings as the expansion of gas-filled pores generates a spongy structure. A pulsed laser melts only the upper surface part of the coating. A plasma sprayed Ni coating was pulse laser treated with peak power pulses of 60 GJ/sm^2 and the open porosity dec-reased from 21% to 16% (Dallaire & Cielo 1982).

Detailed studies of laser cladding steel with stainless steel have been reported (Weerasinghe & Steen 1987). Tables 6:5 and 6:6 give a few representative data on laser cladding of various refractory and wear resistant materials and LST of steel.

TABLE 6:5

EXPERIMENTAL CONDITIONS AND RESULTS FOR LASER ALLOYING AISI 1018 STEEL

Alloying Elements	Cr	Cr, C	Cr, C, Mn	Cr, C, Mn, Al
Powder Coating Composition (wt%)	100 Cr	85 Cr-15 C	25 Cr-50 C -25 Mn	24 Cr-48 C -24 Mn-4 Al
Processing Conditions				
Laser Beam Size (mm x mm)	18 x 18	6.4 x 19	6.4 x 19	6.4 x 19
Oscillating/ Stationary Beam	Stationary	Oscillating (690 Hz)	Oscillating (690 Hz)	Oscillating (690 Hz)
Laser Power (W)	12500	5800	3400	500
Speed (mm/s)	1.69	21.17	8.47	8.47
Shielding Gas	Helium	Helium and Argon	None	None
Application Method Depth (mm) Width (mm)	Slurry 0.5 6	Slurry 0.75 25	Spray 0.025 25	Spray 0.125 25
Alloyed Casing Depth (mm) Width (mm)	1.95 21	0.38 15	0.13 15	0.66 15
Composition (wt%)	16.0 Cr, 0.7 Mn	43.0 Cr, 4.4 C,0.5 Mn	3.5 Cr, 1.9 C,1.3 Mn	0.9 Cr, 1.4 C, 1.0 Mn,0.5 Al
Hardness (R_C)	53	64	64	56
Major Microstructural Constituents	Martensite	Carbide M_7C_3	Martensite and Cementite	Martensite and Austenite

TABLE 6:6

E X P E R I M E N T A L   C O N D I T I O N S   U S E D   F O R   L A S E R   C L A D D I N G

	Tribaloy T-800 Alloy	Haynes Stellite Alloy No.1	Silicon	Tungsten Carbide and Iron	Alumina
Cladding Material	Tribaloy T-800 Alloy	Haynes Stellite Alloy No.1	Silicon	Tungsten Carbide and Iron	Alumina
Base Material	ASTM A387	AISI 4815	AA 390 Aluminium Alloy	AISI 1018	2219 Aluminium Alloy
Form of Cladding Material	Powder (Plasma Spray Grade)	Cast Rod (3 mm diameter)	Powder (44 μm size)	WC Granules (0.5 mm size) and Iron Powder (44 μm size)	Powder (0.3 μm)
Pre-heat Temperature, °C	20	250	20	20	20
Powder Application Method	Slurry	–	Slurry	Loose Powder	Loose Powder
Powder Depth (mm)	6	–	1	1	0.75
Powder Width (mm)	25	–	5	19	25
Oscillating/Stationary Beam	Stationary	Stationary	Stationary	Stationary	Oscillating (690 Hz)
Laser Power (W)	12,500	3,500	4,300	12,500	12,500
Laser Beam Size (mm × mm)	14 × 14	6.4 (diameter)	5 (diameter)	12 × 12	6.4 × 19
Shielding Gas	Helium	Hydrogen and Argon	Helium and Argon	Helium	Oxygen
Processing Speed (mm/s)	1.27	4.23	8.47	5.50	8.47

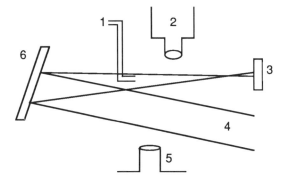

1. powder feed 2. Pyrometry system 3. graphite
4. laser beam 5. velocity system 6. mirror

Fig.6-6: Schematic set up for monitoring powder propulsion

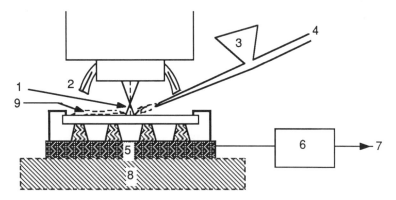

1. de-focused laser beam 2. water-cooled, reflector
copper dome 3. powder-feed hopper 4. Argon propellant
gas 5. 25 kHz ultrasonic vibrators set in rubber and housed
in a steel box 6. ultrasonic amplifier 7. to A.C. supply
8. hydraulic controlled X-Y table; X-direction - melt bead
running line; Y-direction - shift for overlapping laser tracks

Fig.6-7: Vibro-laser cladding - schematic diagram

(Steen & Powell 1981)

6.2.4.3. Laser Spraying:

Laser cladding and laser spraying are very related operations. A substrate may be clad by spraying when the material to be coated is in a powder/particle form. Cladding can also be done using sheet metal, and this can be later laser modified. Laser alloying is another term used in this context.

Fig.6-6 and 6-7 illustrate schematic arrangements for laser spraying and powder cladding respectively (Powell & Steen 1981). The coating powder is blown from a side, at an angle into a high power laser beam which heats the particles to their melting point and the molten droplets are then dispersed on to the substrate. Use of a highly polished, water-cooled copper dome above the work piece optimises the laser heating and reflection loss aspects to a better degree: (Powell & Steen 1981; Weerasinghe & Steen 1986; 1987). The process is not yet through to production levels (Steffens 1983), the disadvantages being the requirement of high laser power and reactions of the powder constituents in the ambient atmosphere. The latter aspect can be compensated for by maintaining a hovering envelope of a non-reactive gas. The cost effectiveness will be oriented to the choice of nitrogen, argon, helium or any other inert gases, together with the energy, equipment, and the strategic importance of the workpiece.

During boriding austenitic steels it was found that if the temperature at the point of impact of the laser beam rises above the substrate melting point, the resultant surface will be a mixed fusion composition. On boriding, the upper ferritic boride layer registered a hardness of 1400 HV sustained to a depth of 0.1 mm, decreased to about 850 HV over 0.1- 0.2 - 0.3 then fell sharply to 450 at 0.4 mm and 200 at 0.5 mm from the surface. (Sepold & Rothe 1979). Laser spraying of Ti and Cr has been reported (Yoneda et al 1988).

Hard metal matrix-ceramic composite layers for wear resistance have been laser-sprayed using a 10 kW, CW CO_2 laser beam oscillating with a frequency of 110Hz. 8-12 mm wide melt pools were formed on the IN 625 alloy samples. WC particles 45-75 microns were fed at rates of 0.35-0.5 cm^3/sec using helium at 35-85 kPa. Higher powder feed rates and lower gas pressures were necessary to inject the lighter TiC particles of 75-150 micron size. Relatively slow substrate velocities at 0.5-0.75 cm/sec were adopted to achieve 1-2 mm thick clad layer (Cooper 1986).

6.2.4.4. Laser Compaction & Laser Crystal Growing:

Coatings applied by thermal spray, plasma spray or vapour deposition can be laser-compacted to reduce porosity. The coating gets remelted either within the top thin layers or with a thin layer of the substrate depending on the LST parameters chosen. Optimum conditions were investigated on plasma sprayed IN 625 on type 316 stainless steel (Vasantasree & Hocking 1987). This results in an

245

improved bonding, and reduced porosity, and the coating also solidifies in a finer crystalline form (Sepold & Rothe 1979). A laser pulse length of 20-40 microseconds was found to fill up the pores without overheating the material above its boiling point. A pulsed CO_2 TEA laser was used on plasma sprayed Ni (Dallaire & Cielo 1982).

Amorphous and crystalline materials can be produced from laser interactions with a suitable gas medium. Diamond from methane is one such novel product (Fedoseev 1981). Hydrocarbon droplets can be laser decomposed to yield spherical amorphous, or irregular crystalline graphite or diamond, or metastable forms of carbon. An infrared laser was used to a power density of 5000 W/cm^2. Particles up to 0.1 microns have been produced. LST needs sophisticated handling and cost-justified outlay, but has a vast scope in coating technology.

6.3. RAPID SOLIDIFICATION PROCESSING (RSP)

Rapid solidification is a particular form of consolidation of a coated surface. Materials used for coatings are either rapidly solidified even as they contact the substrate in droplet form or as vapour, or, a solid coated surface is remelted and rapidly solidified. Many coating processes discussed in chapters 3-6 are suited to achieve RSP in addition to their individual capabilities for producing specific types and ranges of coatings. A separate identity is given to RSP in this chapter in view of the fact that it is comparatively new as a coating process. Current RSP investigations are both in materials and coatings development (Wadsworth et al 1988; Das & Davis 1987; Kear & Giessen 1984).

Section 2.5 in Chapter 1 has already introduced amorphous coatings which are produced by RSP methods. Controlled metallurgical properties on bulk materials by surface modification are manipulated using RSP principles. Amongst PVD methods, electron beam evaporation, sputtering and plasma enhanced PVD are capable of producing 'glassy' deposits. Plasma assisted CVD produces amorphous coatings. These two categories produce non-structural deposits from the vapour phase. Laser glazing and electron beam glazing are surface modification methods which almost eradicate surface microstructural features to produce frozen vitreous states which are normally metastable phases when mass solidified from bulk melts.

Rapidly solidified coating is achieved by creating a very large temperature gradient between the substrate and the contacting film or by rapidly cooling the molten droplet of the coating material originally in a powder form. Rapidly solidified material in powder or ribbon form is produced by rapid quenching by melt spinning on cooled copper wheels, or by rapid quenching of atomized liquid or by splat cooling (Johnson 1986). RSP powder is used in spraying techniques as coating material and it has also been consolidated to required forms by hot isostatic pressing (HIP) or hot isostatic compression (HIC) methods. The ultimate advantage in RSP is achieved via microstructure control and it is recognised by the absence of microstructure or by being amorphous.

Amorphous phases of a normally crystalline alloy have been produced by irradiation (Mori et al 1984; Brimhall et al 1983; Cahn & Johnson 1986). Hydrogen dissolution in an alloy has caused amorphization (Yeh et al 1983). The amorphous phase generates at the grain boundaries and grows into grain interiors as the concentration of dissolved hydrogen increases (Yeh & Johnson 1986). Solid state amorphization has been reported from seveal laboratories; elemental mixtures of metal powders when mechanically alloyed in a high energy ball mill result in alloys with microstructures typical of RSP products (Yermakov et al 1981;Schwarz & Johnson 1983; Koch et al 1983; Schwarz et al 1985; Politis & Johnson 1986). Amorphous mechanical alloys in bulk form as filamentary and lamellar composites have also been reported (Atzmon et al 1984; Schultz 1985). Slow ball milling has been shown to transform even an initially homogeneous intermetallic compound to the glassy state (Schwarz et al 1985).

6.3.1. RSP MATERIALS

1. The best known, simple amorphous coatings are Ni-P (also Ni-B) coatings produced by electroless methods. These have also been produced by CVD in an rf plasma on nickel substrates using meta carbonyls $Ni(CO)_4$ and $Fe(CO)_5$ and metalloid hydrides PH_3 and B_2H_6 (Bourcier et al 1986).

2. Foil source of a predominantly amorphous Ni-Cr-Fe-Si-B alloy $Ni_{63.5}Cr_{12.3}Fe_{3.5}Si_{7.9}B_{12.8}$ and a slab source cast of crystalline $Ni_{55.3}Cr_{16.9}Si_{7.2}B_{21.6}$ were used in a dual-beam ion sputtering system for producing amorphous alloys. The foil source produced amorphous coatings under all experimental conditions. Deposition rates were slower from the slab source and also the coatings were partially polycrystalline and amorphous (Panitz 1986).

3. Electron beam surface melting of Ni-Cr-Mo-V steel with traverse speeds 4.4 - 36.4 cm/sec resulted in grain refinement and increased hardness. The steel was austenitized at $980^{\circ}C$ for 1 hour, oil quenched, tempered at 620° for 6 hours and oil quenched, held for 4 hours at $400^{\circ}C$ for stress relief and then treated by electron beam melting and cladding. Cladding was done by pressing a uniform layer of 100 mesh size $Fe_{80}P_{13}C_7$ alloy powder, and electron beam melting it at 50-200 W beam power (for varying powder thicknesses) at similar traverse speeds. Hardness values were in the range of 880-1150 VPN for these amorphous alloy claddings (Mawella & Honeycombe 1984).

4. Electron beam glazing of high speed steel using a 1 - 15 KHz oscillating beam at 1 - 100 cm/sec has resulted in a 3-35 mm wide RSP. Homogenization of the substrate as a pre-treatment step was definitely advantageous to the subsequent beam glazing (Strutt et al 1984). Microstructure 'tailoring' by electron beam glazing has been achieved on hardfacing alloys, viz. Stellite 6 with and without TiC, TiB_2 particle strengthening (Kurup et al 1984).

5. Sputtered Si_3N_4-SiO_2 multi-layer amorphous sandwiched coatings show excellent resistance to laser damage when used as optical filters. Dendritic silicon nitride alternates with amorphous silica coating by selectively introducing N_2+Ar and O_2+Ar during the sputtering process. Control of the reactive gases allows the interface betwen layes to be chemically graded (Guenther 1986; Courtright 1987).

6. Co-sputtered 304 stainless steel and W on 304 stainless steel substrate resulted in an amorphous structure which conferred a superior resistance to stress corrosion cracking and pitting (Wang & Merz 1984). Refractory metals such as W, Ta etc. when co-sputtered or plasma sprayed along with stainless steel were found to stabilize the amorphous structure up to 400-600°C.

7. Thermal barrier coatings have been known to have columnar microstructure excellent for thermal shock resistance but poor for condensate corrosion. Plasma spraying and PVD microstructures provide additional defence by applying dense closeout top layers, or better as multiple sealing layers. On an IN 792 substrate a CoCrAlY bond coat was covered with a columnar sputtered ZrO_2-$20Y_2O_3$ initial layer. This provided a strain tolerant coating at the interface. A negative bias was applied to disrupt the columnar growth pattern, once the desired thickness was reached. This interrupted the columnar crystal growth and deposited a dense intermediate amorphous sealing layer. The process was repeated to result in three columnar segments with three sealing layers. The life time of the base coat was increased by a factor of three, the thermal shock resistance was high (survived five water quenches from 1000°C) and improved corrosion resistance to molten sodium sulphate (Prater 1987).

8. Laser glazing to obtain glassy alloys has been studied for Ti-Au, Ti-Co, Ti-Cr and Ti-Zr systems. All except Zr-Ti resulted in amorphous surface alloys (Von Allmen & Affolter 1984). Ultra-fast cooling at the rate of 10^{12} deg./sec by pulsed laser quenching has been investigated (Lin & Spaepen 1984). Laser glazing for wear and corrosion resistance has been done on Ni-base super-alloys (Yunlong et al 1984) as well as to obtain amorphous $Fe_{82}B_{12}C_6$ and $Fe_{82}C_{18}$ alloys.

9. Superalloy ribbons show a gradual progression from a dendritic to cellular and planar solidification in a melt spinning process (Huang & Laforce 1984).

10. Hardfacing alloys with excellent resistance to wear, oxidation and corrosion have been produced by rapidly cooled inert gas atomized powders and plasma spraying. The 'Toh-boride' system was investigated. Quaternary alloys of five compositions were prepared in the series Ni-Cr-Nb-B and Ni-Cr-Ta-B (Lugscheider et al 1987)

6.4. DROPLET TRANSFER COATINGS BY SPRAYING

Metal spraying coating was invented in 1909 (Kasperowicz & Schoop 1920). Table 6:7 is a summary of spraying processes developed since then. Fig.6-8 gives an idea of processes available and areas are identified for process variants still to be explored (Kretzschmar 1973). A major advantage of spray coating is in the production of graded coatings by gradual change in the feedstock. Different materials can be sprayed simultaneously and any coating composition is possible, although microhomogeneity may not be achieved (depending on solubilities).

Five spraying techniques are given in the next five sections: 6.5 to 6.9.:- Liquid metal-, Wire explosion-, Flame-, Detonation Gun- and, Plasma-spraying. Plasma spraying was carried out first at atmospheric pressures and later developed into the low pressure plasma spraying and vacuum plasma spraying methods. New advances such as the shrouded plasma spraying technique which offers the plasma coating route on a more economical basis for large-sized substrates and underwater plasma spraying are also reported. Plasma as a medium has been harnessed in many coating techniques. Plasma-assisted vapour techniques are PAPVD, PEPVD, PACVD and PECVD methods. On that basis section 6.5. may be considered to deal with plasma assisted molten media deposition.

TABLE 6 : 7

A SURVEY OF METAL SPRAYING PROCESSES

Type of Energy	Source of Energy	Atomising Gas	Material Form	Environment	Process	Practice Status
Electrical resistance heating	Induction heating	Argon	Molten metal in a crucible	4, (2)	Osprey (Molten metal spraying)	Effective
	Induction heating	3	Wire (a)	1	In-chamber induction spraying	Little used
		2	Wire	2		
	High Voltage	None	Wire	1	High Voltage Flame Plating (capacitor discharge)	Being developed
				2		
Electricity	Electric arc	3	Two Wires	1	Electric arc spraying	Widely used
		2	Two Wires	2	In-chamber arc spraying	Being developed
	Plasma	None	Powder	1	Plasma powder spraying	Widely used
				2	Plasma powder in a chamber	Being developed
	Fuel gas flame/3	None	Powder	1	Powder flame spraying	Little used
Chemical reaction	Oxy-fuel gas flame	None	Powder	1	-------- " --------	Not much used
		5	Powder	1	-------- " --------	Widely used
		-- " --	Wire	1	Oxyacetylene wire spraying	Very widely used
	Very rapid combustion	None	Powder	1	Flame plating	Being developed
		2	Powder	1, (2)	--- " -----	Not much used

Note:- 1. Normal atmosphere; 2. Shielding gas; 3. Compressed air; 4. Vacuum; 5. Compressed gas; numbers in parantheses indicate an alternate route. a: Wire can be single or multiple, or powder-cored; powder and wire can be 'fed' simultaneously.

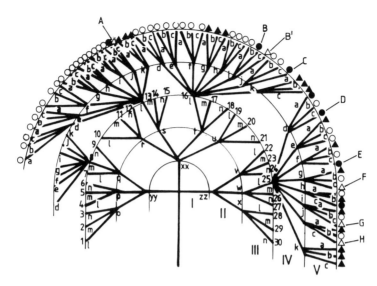

Fig.6-8

Energy Combinations & Process Variables Available & Explored for High Temperature Spray Treatment of Substrates

Segment No.	Process Variable	Segment No.	Process Variable	Segment No.	Process Variable
I	Type of Energy	II	Kind of Energy	III	Working Gas
IV	Type of Filler Material			V	Atmosphere

Section	Process Variable	Section	Process Variable	Section	Process Variable
I xx	Electricity	I yy	Beams & Waves	I zz	Chemical Reaction
II o	Ion Beam	II p	Laser Beam	II q	Electron Beam
II r	High Voltage Heating	II s	Induction Heating	II t	Arc
II u	Plasma	II v	Explosion-Combustion	II w	Oxy-fuel Flame
II x	Compressed Air-Fuel Gas Flame				
III l	Compressed Air	III m	Shielding Gas	III n	Without Air
IV d	Rod	IV e	Strip	IV f	Wire & Powder
IV g	Powder	IV h	Cored Wire	IV i	Two or More Wires
IV j	One Wire	IV k	Liquid		
V a	Vacuum	V b	Inert	V c	Air

Peripheral Notation (Clockwise)

A: High Frequency Spraying With One Wire

B: Arc Spraying in Shield Gas Chamber
B': Arc Spraying

C: Arc Spraying With One Wire
E: Flame Spraying With Wire & Enclosure
G: Flame Spraying With Two Wires

D: Flame Spraying With Ceramic Rods
F: Flame Spraying With Powder
H: Flame Spraying With One Wire

○ Possible Variant   △ Usual Variant   ● Tested Variant   ▲ Improbable Variant

Comments:- Clearly, the left hand segment was not much explored, viz. beams and waves (electron, ion, laser)/part electrical, in all media (air, gas, inert, vacuum) with filler materials and various working conditions. Beams and waves, on their own, are incapable of creating the coating material in a form viable enough to realise the end product as a **heavy duty** coherent deposit. PAPVD & PACVD developed as later techniques have enabled tribological deposits and many variations in multi-layer deposits.

6.5. COATING BY PLASMA SPRAYING

6.5.1. FUNDAMENTAL ASPECTS:

Plasma spraying technique has been practised for more than 25 years, but its fundamental technology is poorly established. Plasma generation, jet formation, particle injection, particle heat and momentum transfer, particle impact onto the substrate – its state and form, and, the post deposition substrate/deposit condition are the several aspects which have to be defined and understood (Pfender 1988).

Basically, the powdered coating material is carried in an inert gas stream into an atmospheric pressure or a low pressure plasma. Any coating thickness is feasible. Plasma generating equipment is available over a wide range from microplasma (1-10A) to mechanised plasma (100-400A) and the process (BS 4761, 1971) is well documented (Malik 1973; Houben & Zaat 1973; Anon. 1975; Tucker 1978; Anon. 1983; Kvernes 1983; Bennett et al 1984).

Frequent problems with plasma sprayed coatings are porosity and poor adhesion. The adhesion of plasma sprayed ceramics to metals is generally poor but can be considerably improved if a sprayed bond-coat or interlayer is used (e.g. Ni-Al, Mo). Special care is needed, as in electroplating, to clean and prepare the surface to be coated. Rough lathe turning is suitable for heavy coatings on large components (Steffens 1983) but unsuitable for aircraft components under stress. Grit blasting is also used for surface preparation and is claimed to provide the means for a strong mechanical bond (Wigren 1988).

The effect of quantified surface preparation parameters on adhesion of Al and Zn plasma coatings on steel has been studied. An electrochemical method for measuring true surface area, including the smallest irregularities, is given. The bond strength of arc sprayed Al is 3 times that of flame sprayed Al and Zn and of arc sprayed Zn. Further work is needed on quantifying the substrate surface properties which can determine coating adhesion and other relevant properties of powder materials (type, grain shape and size distribution). For low rust steel there is a linear relation between cleanliness (=reflectivity) and adhesion. Adhesion increases with increasing surface roughness (defined in a specific way). Arc sprayed Al has uniquely good adhesion with inferior cleanliness and topography, but flame sprayed Al and Zn and arc sprayed Zn need much stricter preparation (Bardal 1973). Sputter cleaning before spraying is worth investigation for possible improved adhesion.

6.5.2. ATMOSPHERIC PRESSURE PLASMA SPRAYING

A plasma torch has a water cooled Cu anode and W cathode. Typically, Ar (or any other carrier gas) flows around the cathode and through the nozzle anode. A d.c. arc is held and the plasma (ions and atoms) emerges at 6000-12000°C to about 1cm from the nozzle. Further out the temperature falls rapidly to <3000°C at 10cm, and still further out the temperature falls due to air entrainment (Nicoll 1984; Steffens 1983; Bennett 1983). Near the nozzle the gas velocity is 200 to 600 m/s while the particle velocity is only 20m/s. So the particle acceleration (due to frictional forces) is about 100000g. 18 micron particles reach 275m/s (maximum) at 6cm from the nozzle. Larger particles reach lower velocities, which gives them more melting time. Low density material particles reach higher speeds but also decelerate more. Fig.6-9 gives a schematic of a plasma spray gun head (Steffens 1983); Fig.6-10 shows a few electrode systems used for plasma discharge processes and Fig.6-11 gives an idea of plasma energies (Thornton 1980). Fig.6-12 shows a plasma coating torch with alternate powder injection locations.

The extremely high temperatures reached by the plasma can melt even very refractory powders into a molten mist (low mp materials, e.g plastics, can also be sprayed). The plasma jet is about 5cm long and nozzle to substrate distance is about 15cm. The carrier gas is Ar + H$_2$, or N$_2$ and H$_2$, flowing typically at 3.5 litres per minute (Dudco et al 1982). H$_2$ and N$_2$ raise the power level due to their high thermal capacity. Adding He allows an increased gas flow (up to 600m/s) and thus higher particle velocities, giving higher bond strength and coating density.

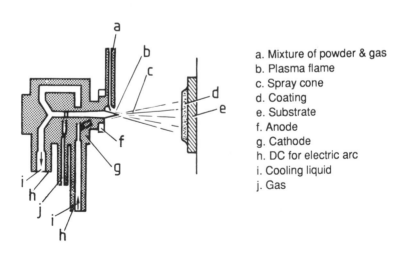

a. Mixture of powder & gas
b. Plasma flame
c. Spray cone
d. Coating
e. Substrate
f. Anode
g. Cathode
h. DC for electric arc
i. Cooling liquid
j. Gas

Fig. 6-9: Spraying Head - Cross Section of Plasma Spray Gun

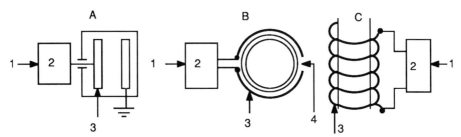

1. rf source 2. compatible network 3. electrode 4. dielectric walled tube
A. Planar electrode B. Clam shell electrode C. coil electrode

A selection of electrode systems used for plasma discharge in coating processes
FIG.6-10

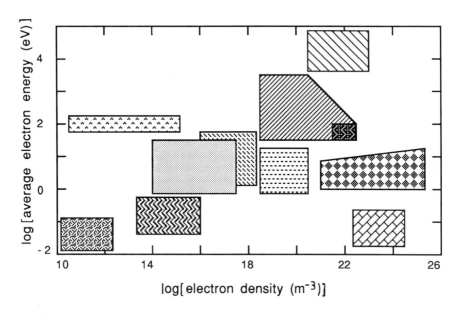

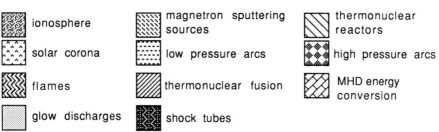

Electron Density & Energy Regime of Plasma from Various Sources
Fig.6-11

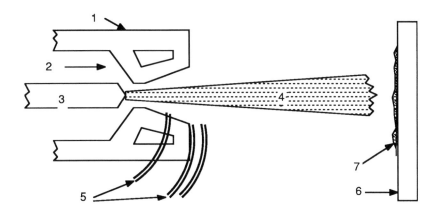

1. water-cooled anode   2. gas stream   3. centre cathode   4. plasma
stream carrying the injected powder via a carrier gas   5. powder
injection location alternatives   6. substrate   7. coating

Fig.6-12: Schematic diagram of a plasma coating torch and substrate

Interacting parameters influencing plasma coating:-
a. Plasma torch: Relative movement of torch and the torch-substrate distance;
b: Plasma: Gas composition, the heat content, temperature, air dilution and
   velocity of the plasma stream
c. Powder: Particle shape, size distribution and the individual constituent composition within the particles; dwell time in the plasma stream
d. Substrate: Temperature - initial and during coating, residual stress effects,
   quenching rates of the incident coating material

But at high gas flow rates about 80KW is needed to melt the
powder (typically about 700 amps). A water-stabilized plasma
torch has been described (Vesely 1973) in which a water jet
enters the plasma, liberating H_2.

A plasma is a very high energy state and thus transfers heat very
fast to the powder, reducing the necessary dwell time at high
temperature which minimises the oxidation of metal powders. Oxidation is also minimised by the inert nature of the heat source.
An unlimited area can be coated. Marchwood Engineering Laboratories (CEGB, UK) have devised a plasma spraying shielding box
which is attached to the torch and is contoured to half-fit on a
tube being sprayed. The exit, at the tube, is double-walled, to
prevent back-entry of air. Plasma spraying rarely heats the substrate > 320°C and it is usually < 150°C. Distortion of substrates
is indicated in Table 6:8. It is evident that plasma spraying is
the only process that does not introduce component distortion. A
high frequency plasma (instead of d.c.) does not contaminate the
coating with electrode materials and can be used with oxygen as
the carrier gas in order to generate oxides at a high rate by
injecting halide into the flame of an oxygen plasma (Gani &
McPherson 1980).

TABLE 6:8

A SURVEY OF HARDFACING PROCESSES

Process	Deposit		Component	
	Thickness mm	dilution %	Heating	Distortion
Gas Metal Arc (GMA)	2.5	30	medium	high
Gas Tungsten Arc (GTA/TIG)	1.6-5	5-10	medium	medium
Open Arc	2.5	30	medium	high
Oxy-acetylene	1.6-5	5	high	high
Plasma Spraying	1.25	nil	very low	nil
Plasma Transfered Arc (PTA)	0.25-6.4	5	high but local	medium
Powder Welding	0.08-0.2	5	medium	medium
Shielded Metal Arc	>= 3.2	10-25	low	high
Spray Fusing	0.8-2.4	nil	high but uniform	low
Submerged Arc (SAW)	>=3.2	20-30	low	high

Note on high degree of component distortion: In large masses distortion may not be marked but cracking might occur, especially with harder grades.

Optimization of plasma spraying processes has been attempted in recent years to decrease coating porosity and achieve better adherence. The deposition efficiency is strongly influenced by both particle size and distribution (Brunet & Dallaire 1987). The various parameters identified for plasma torch, plasma, coating material, and workpiece are as follows (Wuest et al 1985):

Plasma torch: relative movement and workpiece standoff distance;

Plasma: Gas composition, heat content, plasma stream temperature, velocity, and air dilution of plasma stream;

Coating powder: particle shape and size distribution, elemental distribution within the particles and dwell time in plasma stream;

Workpiece: temperature, residual stress control and particle quenching rates.

A nozzle-electrode set that had already been in spraying operation for 38 hours appears to have given deposition efficiencies >50% with the best optimised working conditions.

A laser two-focus velocimeter has been used to monitor particle velocity accurately at 45 kPa (340 Torr) (Smith & Dykhuizen 1988). A smaller particle size reduced coating porosity (Pawlowski 1987) and automated powder mass flow monitoring and control system is reported (Saenz 1988). Power output of 80 kW in the Mach 2 velocity range produced very dense coatings not obtained by normal air plasma spraying (Lugscheider et al 1987). Molten powder jets emerging from a twin injector system of powder feed were found to cross without interference with each other to produce two separate deposition areas. Partially molten particles produced globular deposits while fully molten particle flattened on to the substrate on contact (Mazars & Manesse 1987). An analysis of velocity and energy of Al, Al_2O_3 and W powders in an Ar plasma jet is given and operating conditions recommended for good coatings (Perugini 1976). 30% of the plasma energy is used in accelerating and melting the W particles with 5% for the melting.

To produce dense strong deposits most of the particles must be molten before impingement and must have sufficient velocity to splat into the irregularities of the previous splats. Powder-plasma interactions are a prominent aspect (Nicoll 1984). Many coatings contain easily oxidized elements, e.g. Al, Mo, Ti, and then plasma spraying at low pressure (20 mbar) or in Ar becomes advantageous. Ti coatings are produced if $TiCl_4$ is passed into an Ar plasma jet (Ohno 1979). The substrate must be at a high temperature or a Ti subchloride is deposited. Rare metals like Re and Pt are plasma sprayed (Kirner 1973). Table 6:9 gives an idea of the oxygen content in a few plasma deposited metallic coatings (Tucker 1978). Powder size distribution is controlled , e.g. 5 to 45, or 45 to 70 microns. This depends on surface finish or thermal shock properties, e.g. ceramic thermal barriers: 20 to 90 micron. A high pressure jet can restrict entry of fine particles to the hot zone.

Table 6:9

OXYGEN CONTENT OF PLASMA DEPOSITED METALLIC COATINGS

Coating material	Cu	Mo	Ni	Ti	W
Oxygen content (%) Conventional coating	0.302	0.710	0.456	>2.0	0.274
Coaxial gas Shielded coating	0.092	0.160	0.151	0.730	0.030

Bonding of a coating is due to: mechanical bonding (contouring around the surface topography and interlocking with it) metallurgical bonding (interdiffusion) and physical bonding (van der Waals forces). Particle reaction with carrier gases has to be prevented, e.g. Y in a MCrAlY particle reacts with H_2 in $Ar+H_2$ plasmas. H_2 should be avoided in cases such as this.

Changes, in the plasma, affect the microstructure of coatings. W and Cr carbides, show these changes. WC in a Co binder has wear resistant properties. During normal sintering production, Co diffuses into the WC and is precipitated out again during cooling, thus forming a bonded structure. But during plasma spraying the WC fuses and the Co in it has no time to escape on cooling to form a bonding matrix. Further, $WC+W_2C$ particles in a Co coating decompose and oxidise, losing carbon due to their solution in molten Co. The carbon then evaporates as CO and CO_2. WC + Co compositions prepared via powder metallurgy route are unsuitable for plasma spraying. Further work in this area is desirable. Table 6:10 gives some data on plasma deposited WC.

Table 6:10

EFFECT OF POWDER SIZE ON THE STRUCTURE OF PLASMA DEPOSITED WC

Powder size microns	Apparent density g/ml	Bulk density g/ml	% theoretical density	Porosity	Apparent hardness Kn500
Coarse (10-105)	0.5	8.7	60	17.1	538
Medium (10-74)	13.0	11.1	77	14.5	684
Fine (10-44)	14.2	13.0	89	8.5	741

Plasma spraying of carbides often does not give a homogeneous coating whereas ceramics such as alumina do (Basinska-Pampuch & Gibas 1976). $WC+W_2C$ form in an air plasma medium rendering the deposit superior in hardness and finer microstructure to that produced by vacuum plasma spray. However, the latter shows a better wear resistance because the brittle W_2C is absent (Tu et al 1985; Vinayo et al 1985). Complete fusion is difficult while plasma spraying high melting point materials, e.g. TiC, but sometimes this may be an advantage from the microstructural point of view. Partially molten particles are embedded in a network of x10 smaller grains which are recrystallized from the molten impacted fraction. Plasma melted TiC also undergoes high decarburization in both air and argon atmospheres (Fournier et al 1985).

Fully stabilized (20% yttria) and partially stabilized (8% yttria) zirconia with Si impurity level at 0.2-4%, air plasma sprayed and sintered, acquired up to x3.5 its original strength and x4 its original modulus. The improvement is correlated to a change from the as fabricated state of mechanical interlocked network to a chemically bonded system (Eaton & Novak 1987). A plasma sprayed bond coat of Co 32Ni 21Cr 8Al 0.5Y along with 8 wt.% yttria stabilized zirconia top coat sprayed on Hastelloy substrate has been tested for hot corrosion (Tjong & Wu 1987). LST has been a key modifying procedure for plasma sprayed coatings. LST of alumina and zirconia stabilized separately by yttria and magnesia produced significant structural modifications (Iwamoto et al 1988).

A study of Cr_2O_3 plasma spraying showed best results are obtained with 15 KW power, 30 1/minute of N_2 + 5 1/minute of H_2, 800g/hour of powder flow, 30 to 50 micron powder size, projection distance 12 cm, sand blasted steel targets at right angles. Spraying parameter variations have been discussed for chromia spraying (Boch et al 1976). Plasma coating of ceramics onto steel vessels used for holding molten Zn for galvanizing prolongs life (Vasquez et al 1976). Typical plasma conditions are 650 amps arc current, 35 volts arc voltage, 50 Hz frequency, 900 g/h powder feed rate, substrate temperature $200^{\circ}C$, 10 to 15 cm spraying distance.

Materials like ZrN, MoS_2 and the fluorine resistant coating CaF_2 can be plasma sprayed (Kirner 1973). ZrN decomposes slightly and MoS_2 forms some Mo_2S_3. Masking and materials selection is discussed and 80Ni20Cr plasma coatings are described for sputtering target use (Downer & Smyth 1973).

Ceramic composites by plasma spraying are well known. Metal-ceramic composites, both substrate reinforced and free forms have been fabricated (Quentmeyer et al 1985). Stainless steel mandrels are grit blasted with alumina plasma sprayed with 0.05-0.08 mm NiCrAlY. ZrO_2-$8Y_2O_3$ is then plasma sprayed to about 0.38 mm thickness to form the composite. The metal is subsequently dissolved in a 5/1 HCl/HNO_3 acid to obtain the ceramic composite free form. Zirconia-yttria ceramics have also been plasma sprayed directly on SiC. Fibre-reinforced ceramic cores have also been tried as substrates but poor adhesion and surface distortion have retarded their development. Cu and Pt have been electroplated through into the pores of plasma sprayed ceramic coatings (Hendricks et al 1985).

Plasma spraying is used for composite fabrication, e.g. B fibres in an Al matrix. A layer of the matrix as Al foil, or Al plasma-sprayed onto a mandrel, is wrapped with a spaced array of reinforcing filament which is then enveloped by spraying on more Al while the mandrel is rotated. A multilayer array has voids but these can be removed by hot pressure; or stacked monolayers can be prepared free of voids (Kayser 1973). A boron fibre reinforced pressure vessel can be made this way, a thin walled shell or inner vessel being used as mandrel. A ceramic matrix is also feasible. Carbon fibres need electroplating before incorporation.

6.5.3. NEW DEVELOPMENTS IN PLASMA SPRAYING:

Two new advanced techniques may be mentioned here, viz. Shrouded plasma spraying (SPS) and Underwater plasma spraying (UPS).

Shrouded air plasma spraying is a recent development in producing diffusion and thermal barrier coatings and bonding layers. Work on MCrAlY, MgO ZrO_2 and ZrO_2-7 wt.% Y_2O_3 have shown that substantial advantages may be realized using shrouded inert gas plasma torches operating in open air as against vacuum and low pressure plasma spray methods. The stringent requirement is that the starting powder must have very low oxygen content, vacuum melted, and argon-atomized to the required composition and sized to the uniquely designed plasma torches. The coating process flow chart is cut down by 50% by using shrouded plasma method and plasma spraying becomes more economical while retaining a high quality output (Taylor et al 1985).

Underwater plasma spraying offers much scope for the deposition of wear and corrosion resistant coatings on submerged substrates. Underwater processing drastically reduces noise, radiation and dust while giving a substantial improvement in quality. UPS was first carried out at shallow depths 300-500 mm (Schafstall & Szelagowski 1983) but is currently under development for processing at greater depths to meet demands from stationary and mobile offshore structures (Lugscheider et al 1987). Powders made from self-fluxing Ni-based hardfacing alloys, Colmonoy 5, 52 and 42 with metalloids boron and silicon to reduce oxide films on base metal have been used (Knotek et al 1975). Other materials such as oxides, intermetallics, high alloy steels and corrosion resistant alloys are under scrutiny. The Ni-based alloys sprayed by UPS were free of cracks, with homogeneous microstructures of low porosity. The spraying distance was a critical parameter amongst all the parameters for plasma spraying. For UPS it is greatly reduced. The plasma has been ignited both above (Karpinis et al 1973) and below water (Lugscheider et al 1987). Coatings 100-300 microns have been deposited depending on the operating parameters

6.5.4. LOW PRESSURE & VACUUM PLASMA SPRAYING

Some reactive materials cannot be sprayed in air. This problem is avoided by using low pressure plasma spraying (LPPS) which also increases particle velocity, giving pure and high quality coatings. The plasma jet is longer and so greater working distance is needed to avoid overheating. The LPPS jet velocity is up to mach 3, giving low porosity and good adhesion. The low particle dwell time requires powers up to 120KW to spray refractory materials.

The average coating rate is 25 microns per second over a circular 75 mm^2 area at 40cm distance, for most metal powders; for ceramics the rate is about half this. The density is typically 99% of the maximum possible. An advanced system description is given. Typical LPPS equipment are shown in Fig.6-13, 6-14 and 6-15 (Nicoll 1984; Steffens 1983; Muehlberger 1973).

The operating pressure is 50 to 70 mbar. The exhaust gas is filtered and cooled before reaching the vacuum pump. The inlet velocity distorts the arc about 50cm towards the substrate to be coated and additional power is needed to form the transferred arc which provides cleaning and rapid preheating of the substrate for coating, by transfering some of the arcing to the substrate. The powder is preheated and pressure fed to the nozzle. The large LPPS jet increases the heated surface area of substrate and small differences in spraying distance are acceptable. Higher powder losses and a lower energy density of the LPPS jet are sustained, compared with air plasma spraying. To avoid unwanted gas/metal reactions, powders and gases used are oxygen-free and the equipment is evacuated to 0.2 mbar before starting spraying. Successful LPPS coatings include the well-known Co(Ni)CrAlY used on gas turbine blades. Houben and Zaat (1973) give a mathematical analysis of LPPS torches.

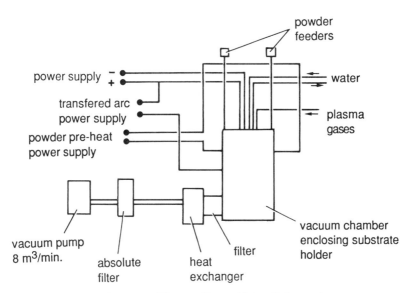

Fig.6-13: Low Pressure Plasma Spray Unit - Schematic

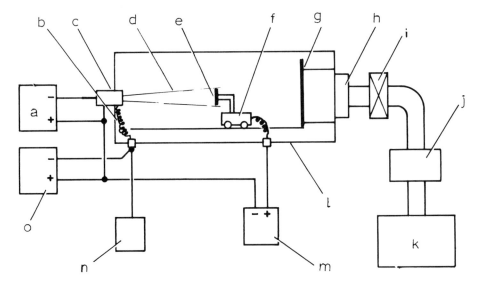

a. arc power supply; b.powder pre-heat tube; c. plasma generator; d. plasma powder
stream; e. substrate to be coated; f. trolley; g. heat shield; h. filter; i. gate valve;
j. roots blower; k. pump (mechanical); l. coating chamber; m. power supply (transfered
arc; n. powder feeder; o. powder pre-heat power supply

Fig.6-14: High Energy Plasma Coating - Schematic Diagram

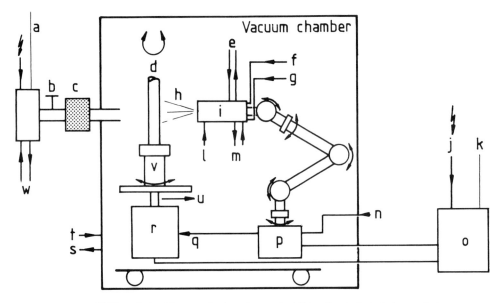

a. VPS control cable console; b. valve; c. dust filter; d. substrate to be coated;
e. water-cooled power supply; f. gas inlet to plasma torch; g. plasma trigger-
ing unit; h. plasma jet; i. plasma torch; j. mains supply; k. plasma control
cable; l. powder feed; m. torch connexion; n. argon cooling; o. robot control;
p. robot; q. workpiece holder+argon supply; r. worktable (rotary); s/t. cooling
water to spray chamber; u. power connexion channel; v. clamp

Fig.6-15: Vacuum Plasma Spray Unit - Schematic

In given plasma conditions there is likely to be an optimum powder particle size which will be melted and transported with enough momentum. Smaller particles may be vaporized and large particles may be incompletely melted giving a weak porous coating. The effect of powder size on adhesion of plasma sprayed coatings is discussed. Flow properties and feed systems are also described. Coatings 3 microns thick are obtained from metal powders 1 to 5 microns diameter. Finer powders do not necessarily give a surface smoother than 1 micron (result for Ni coating). Ultrafine particles tend to agglomerate in flight, giving lumpy coatings. No difference was found in coating adhesion of different particle sizes on grit blasted steel. Standard powders adhere better to glass and fine powders to ceramic and plastic; larger particles fuse to glass but degrade plastics and cause large thermal stresses with alumina substrates (Smyth 1973). Heat, mass and momentum transfer in plasmas (Perugini 1976; Vazquez 1976) and nitrogen absorption by liquid metals from plasmas which contain nitrogen are discussed (Basinska-Pampuch & Gibas 1976).

Powder quality is an important factor for high quality coatings and the following properties must be controlled: particle shape and size, size distribution, surface roughness, powder flow ease, composition and homogeneity, melting point, density, moisture content, contamination (Nelson 1973; Kvernes 1983). The gel process of powder production eliminates variations of density and homogeneity. The preparation of powders for plasma spraying is described (Nelson 1973), including the gel process for spherical particles of a specified size in the range 1 to 100 microns. The spherical form ensures a freely flowing powder even at <10 microns diameter and produces a very smooth coating surface. Particles finer than 5 microns tend to agglomerate in the plasma torch feed unit. Porous coatings if required are obtained from coarse powders, the maximum particle size being 0.1 mm. Big particles will penetrate to the flame centre but small ones will not enter the flame at all. Too small particles will oxidize or evaporate and too large ones will only partially melt and may initiate coating cracks. A small range of particle sizes is desirable. Some ceramic powders are hygroscopic and cannot be sprayed beyond a certain moisture level. Preheating before spraying is recommended.

The powder manufacturing process has a great influence on the chemical transformation which occurs during atmospheric plasma spraying. Cobalt coated WC powders and agglomerated sintered powders have been observed to undergo less decarburization during plasma spraying. WC+20%Co coated powder produced the best compromise phase composition and hardness in deposits. Mixed carbide formation was a feature, viz. Co_6W_6C. LPPS at 60 torr produced dense deposits of carbide grains in a cobalt matrix (Vinayo et al 1985).

Plasma spraying can be used for remote repair, e.g. of first vacuum walls of nuclear fusion reactors. But porosity and adherence are problems. Coarse spray particles are advantageous and surface tension lowering by oxygen etc., influences this. Plasma spraying combined with hot isostatic pressing (HIP) improves adhesion. Other combined methods include noble metal electroplating and pack cementation. Low pressure plasma spraying (LPPS) avoids breaking the spray particles by atmosphere collisions. This reduces the time for harmful gas-metal reactions and gives a higher particle velocity (McKelliget et al 1982; Asahi & Kojima 1982). High quality coatings result, e.g. CoCrAlY. Effect of Hf and Y additions has been examined. Plasma spraying Ti in N_2 gives TiN+Ti coatings. High density and hardness is obtainable. Mathematical models of plasma spraying allow parameters which may affect quality to be quantified.

Refractory powders like ZrO_2 can be plasma spray formed into a desired shape on a disposable mandrel or a mandrel precoated with a soluble material (Nelson et al 1973), e.g. yttria-stabilized ZrO_2 electrolyte coatings were produced by the gel process and plasma sprayed directly onto electrodes, or onto a water soluble mandrel to make tubes down to 100 microns thick. Other sprayed artefacts include crucibles for molten metals, furnace radiation shields and tubular resistance elements, mostly useable as sprayed and not needing sintering. The lowest gas permeability was obtained using uniform size spherical particles. The effect of particle size on porosity was determined. Highest density (95%) coatings were given by mixtures of spheres between 10 and 150 microns diameter, due to packing of spheres of different diameters. Loss of Y_2O_3, CaO or MgO cubic stabilizer from stabilized zirconia powder may occur during spraying and more must be added before spraying, which presents problems (Kvernes 1983).

Nearly stoichiometric silicon nitride films (up to 1 micron) were prepared by direct nitridation of silicon in an RF low pressure nitrogen plasma at temperatures below 700°C. The properties of the films were comparable with those of CVD silicon nitride films but uniform layers were however not obtained (Yoshida et al 1982). (Previously silicon nitride films could only be grown by a high temperature thermal nitridation process at temperatures ranging from 1000°C to 1400°C). Nitriding of titanium was performed by rf discharge at 5-20 torr and dc plasma jet at about 200 torr. The nitriding rate in the plasma jet was ten times larger than that in the rf discharge (Matsumoto et al 1982).

LPPS Ti coatings have been found inferior to LPAS (low pressure arc spray-ed/ing) coatings. LPAS coatings were lower in hardness and interstitial gas content but had higher bond strength with the substrate (Steffens et al 1985). LPPS of MCrAlY has found wide acceptance as overlay and bond coatings. The advantages claimed by SPS has been given earlier. NiCoCrAlYTa, 0.19 mm thick has been coated by LPPS on MA 6000 for burner rig tests (Smith & Benn 1987). For CoCrAlY the coating hardness rises with substrate temperature, to a maximum at 650°C and then falls. To cancel such effects, the component is heat treated after coating, e.g. 1 hour at 1100°C in vacuum, which also densifies (sinters) and interdiffuses the coating (Nicoll 1984).

$MoSi_2$ coatings deposited by LPPS on Mo suffer transverse cracks because of the difference in linear expansion. The use of inter-layers can overcome this problem; vacuum sprayed mullite (Al_2O_3 + SiO_2), at a substrate temperature of at least $1000^{\circ}C$ is found suitable. A special nozzle head was used where the powder injection ports are positioned at different locations and different angles in relation to the nozzle axis. Powders with high m.p. were injected at the throat of the nozzle and others downstream at the mouth of the nozzle. Interlayer compatibility to $MoSi_2$ invloves low solubility and low diffusion coefficient for Si and also no reaction with either Si or Mo. In this respect alumina at controlled thickness was found to be the best choice; zirconates, titanates, silicates and spinels were found unsuitable. Improvement in the Mo - mullite bond was achieved by avoiding large and intermediate splats of both mullite and Mo; very fine Mo and $MoSi_2$ spray are recommended. Sputter etching and coating parameters, in brief, are (Henne & Weber 1986):

Operating Conditions For

	Sputter	Coating
Burner Power, kW	30	40
Vessel Pressure, mbar	15	20
Plasma gas	$Ar+H_2$	$Ar+H_2$
Powder fraction		<20 microns
Spraying Distance		250 mm
Substrate Temp.,		$>/ 1000^{\circ}C$
Powder injection		near nozzle throat.

The multiple layer LPPS coat (by hypersonic spraying) on Mo thus was composed of a graded bond coating of Mo-mullite/mullite/mullite-$MoSi_2$ (enriched with GeO_2 for glass formation) (Henne & Weber 1986).

To coat very reactive single metals and alloys vacuum plasma spraying (VPS) is prefered even to LPPS. Mo, Nb, Ta, Ti, and Zr and their alloys can be best handled by VPS. Once again, the quality of VPS coatings largely depends upon the proper optimization of spraying parameters. Sandblasting, sputtering with a negatively polarized tranferred arc at 20 mbar, and, pre-heating the substrate yielded highly ductile and adhesive Ti and Ti5Al-2.5Fe coatings on austenitic and carbon steel substrates. (Lugscheider et al 1987). The substrates were pre-heated and held at $500^{\circ}C$. Typical temperatures can be as high as $850^{\circ}C$ during coating (Nicoll 1984). Computer-aided plasma spraying of turbine blades can help to give a constant distance from gun to substrate, rectangularity to substrate and constant scanning speed (Suzuki et al 1982).

TABLE 6:11

SOME CHARACTERISTICS OF DEPOSITION PROCESSES

	CVD	Electrodeposition	Evaporation	Ion Plating	Sputtering	Thermal Spraying
Metal Deposition	yes	yes – limited	yes	yes	yes	yes
Alloy Deposition	yes	quite limited	yes	yes	yes	yes
Refractory Compound Deposition	yes	limited	yes	yes	yes	yes
Production Mechanism of Depositing Species	chemical reaction	deposition from solution	thermal energy	thermal energy	momentum transfer	momentum transfer
Energy of Depositing Species	can be high with plasma-aided CVD	can be high	low 0.1 to 0.5 eV	can be high (1–100 eV)	can be high (1–100 eV)	can be high
Depositing Species	atoms	ions	atoms and ions	atoms and ions	atoms and ions	droplets
Deposition Rate	moderate 200–2500 A/min	low to high	can be very high up to 750 000 A/min	can be very high up to 250 000 A/min	low except for pure metals (e.g. Cu – 10 000 A/min)	very high
Throwing Power for: a. Complex Shaped Object	good	good	poor line-of-sight coverage except by gas scattering	good, but nonuniform thickness distribution	good, but nonuniform thickness distribution	no
b. Into Small Blind Holes	limited	limited	poor	poor	poor	very limited
Growth Interface Perturbation	yes (by rubbing)	no	not normally	yes	yes	no
Bombardment of Substrate/ Deposit by Inert Gas Ions	no	no	not normally	yes	yes or no depending on geometry	yes
Substrate Heating (by external means)	yes	no	yes normally	yes or no	not generally	not normally

A review of thermal spraying is given in Table 6:11 and is brief-ly discussed below (Tucker 1978):

6.5.5. THERMAL SPRAYING – ASPECTS OF COATING PRODUCTION:

6.5.5.1. Entry of Powder into Plasma Stream:

Plasma Coating Torches:- Very important. The ideal location is upstream of the anode, but powder adherence to the nozzle entry or throat and excessive heating occurs. So the powder entry is usually in the diverging position of the nozzle or just beyond the exit. Improvements consist of adjustments of point and angle of powder entry in accordance with its m.p. Alternatively, shock diamonds are generated and powder is introduced a short distance beyond the exit in the region of rarefaction in the plasma stream. Powder must be distributed uniformly in the plasma stream at a constant rate. A variety of powder dispenser designs are available including those based on auger, aspirated flow or fluidized bed.

Detonation Gun:- A pulsed flow of powder is required. Uniform distribution of the powder in the barrel is important and also of the constancy of amount of powder in each pulse.

6.5.5.2. Substrate Preparation:

Plasma Coating Torches:- Not only must all oxide scale and other foreign matter be removed, but all oils, machine lubricants etc., must be eliminated. Virtually all plasma coatings require a roughened substrate.

Detonation Gun:- Grit blasting may not be necessary because the particle velocity itself results in substantial surface rough-ening.

6.5.5.3. Substrate Temperature:

Plasma Coating Torches:- One of the major advantages of thermal spraying coats is that they may be applied to substrates without significantly heating them. As a result, a part can be fabricated and fully heat treated, without changing the substrate micro-structure or strength. During coating deposition a substantial amount of heat is transmitted to the part through the plasma gas and molten powder. Cooling air or CO_2 jets may be used. Normally the temperature does not exceed 150°C.

6.5.5.4. Coating Material:

Almost any material that can be melted without decomposing can be used to form the coating. For mixtures similar m.p. and b.p. and particle sizes should be used. Also, the difference between m.p. and b.p. of a single phase powder should not be too small, in order to prevent vaporization. Pre-alloyed powder may be better. Comparison of the relative heating rates is not as simple as comparing their m.p. Heat transfer in the plasma jet is primarily the result of the re-combination of the ions and association of the atoms in diatomic gases on the powder particle surfaces and absorption of radiation. Metals tend to heat much more rapidly than most refractory ceramics.

6.5.5.5. Shielding:

This is to stop or reduce the reaction of the powder with O_2 or N_2 from the air inspirated into the plasma stream after it emerges from the nozzle. This is done by an inert gas cloud that surrounds the effluent with argon.

6.5.5.6. Angle of Deposition:

Plasma Spraying Torch:- Plasma deposition is a line of sight process in which the structure of the coating is a function of the angle of deposition. Normally coatings with the highest density and bond strength are achieved at a 90° angle of deposition. This limitation may cause some problems in coating complex parts, particularly those with narrow grooves and sharp angles.

Detonation Gun:- The D-gun with its higher particle velocity can usually tolerate a wider deviation from 90°.

6.5.5.7. Size of Equipment:

Plasma Coating Torches:- The smallest torch available until 1980 could apply a metallic coating to a 3.1 cm or a ceramic coating to a 4.35 cm inside diameter at 90°. Another torch design with an outlet at 45° to the torch axis, can apply a coating to the inside of a closed cylinder 4.35 cm diameter. Other designs in plasma torches are available.

Detonation Gun:- The D-gun cannot fit into a cylinder or any other type of cavity. It can be used however, to coat an inside diameter to a depth about equal to the diameter, i.e. angle of deposition about 45°. The inherently high density and bond strength still allows superior coatings to be deposited at lower angles.

6.5.5.8. Masking:

Plasma Coating Torch:- For low velocity , long standoff plasma torches, tapes, oxide loaded paints or stop-off lacquers may be used. For high velocity, short standoff torches, glass fibre reinforced high temperature tape, adhesive backed steel, Al-foil, or sheet metal masking is necessary.

Detonation Gun:- Metal masking is required.

Nicoll (1987) has presented an updated review on thermal spraying equipment and quality control.

6.6. OTHER DROPLET TRANSFER COATING METHODS

6.6.1. COATING BY DETONATION GUN (D-GUN)

D-gun spraying is a form of thermal spraying which consists of heating and directionally propelling powder particles onto the workpiece from a combustion chamber by a stream of gas detonation products. It offers the following advantages:

1. D-gun equipment is relatively simple in design which is both dependable and has a reasonably long service life, unlike, for instance, plasma-arc devices.

2. Low porosity and high bond strength coatings are produced.

3. Substrate pre-treatment is not very stringent. The process is impurity tolerant.

4. With only a moderate substrate heating during deposition a strong bond can be achieved.

5. The powder velociy to temperature ratio has a wider choice and is flexible.

6. There are several means of controlling the thermal cycle of the coating being deposited unlike in other thermal spraying methods.

7. The method enables a high rate of growth of coating thickness.

Inherent deficiencies of the method are:

1. The spray coating process is cyclic and this impedes stabilizing and monitoring the production.

2. Environmental isolation and personnel safegauards are absolutely necessary because of the high noise level.

3. The detonation products are complex and reactive with a number of the components and constituents. CO_2, CO, H_2O, H_2, O_2, N_2 gases and radicals and atoms such as $-OH$, H, O, NO and N and other gases are produced. This situation imposes production res-

trictions and counter-measures.

4. Distortion of components is a definite possibility. Components to be coated have to be strong enough to sustain D-gun spray impact.

5. The flow from the gun is a two-phase stream. Workpiece to gun configuration is complicated for contoured substrates.

A comprehensive review on D-gun technique is provided by Kharlamov (1987) with references to work from the U.S.S.R.

Union Carbide Corporation (U.S. Patent 2714, 1955) developed a water cooled barrel about 1 metre long with inside diameter 2.5 cm. (Steffens 1983). A $C_2H_2 + O_2$ mixture is fed in with a charge of powder. Metered quantities of oxygen and acetylene are fed via poppet valves into the combustion chamber. Powder in a N_2 stream is metered from a heated pressurized vessel and admitted to the combustion chamber. A spark plug ignites the mixture. The explosion accelerates the powder to 720 m/s and melts it. The $4200^{\circ}C$ detonation flame temperature melts most materials. The particle velocity is much higher than with plasma spraying. There are 4 to 8 detonations per second with N_2 purges between each. A 25 mm circle of coating is deposited by each detonation, several microns thick and composed of many overlapping lamellae. The complete coating is made by overlapping many such circles (Tucker 1978). The flame can be oxidizing, carburizing, or inert by precise control. Carburizing conditions can be used to advantage (Price et al 1977). Inert gas shielding is essential for MCrAlY coatings, to stop oxidation during deposition. Fig.6-16 (Tucker 1978) shows a D-gun schematic diagram. A discussion of physical properties of D-gun coatings appears in Chapter 7.

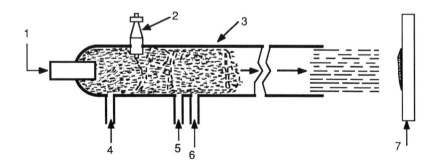

1. Powder injection 2. spark plug 3. gun barrel 4. nitrogen input
5. Acetylene injection 6. oxygen input   7. substrate

Fig.6-16: Schematic diagram of a detonation gun

All plasma and D-gun coatings have some open porosity which allows attack by corrosive environments and especially if a molten salt is allowed to freeze. Porosity is reduced by sintering, essential for MCrAlY coatings for gas turbine use. D-gun coatings have less porosity and may not need such sintering (Tucker 1978) as the particle kinetic energy is 25 times greater than in flame spraying. The D-gun hardness is 1300 V.P.N. for a WC/Co coating but only 700 for plasma spraying and less for flame spraying. D-gun bond strengths are very high, and greater than for plasma sprayed coatings (Steffens 1983). Detonation gun coating of Ni and other powders gives highly bonded and dense coatings (Samsonov 1973).

6.6.2. FLAME SPRAYING

Thermal barrier coatings are most often produced by spraying techniques. Flame spraying of ceramics, e.g. zironia + alumina for rocket nozzle coatings, alumina for MHD generating channels, zirconia for jet engine combustion chambers and gas turbines and Cr_3C_2 for gas turbines, is reviewed (Arata et al 1983). Present requirements include development of testing methods, improving adhesion and inter-particle bonding and reducing porosity.

A wire or powder feed of the coating material is fed through an oxy-fuel gas, e.g. C_2H_2 or C_3H_8 flame (Fig.6-17a,b; Tucker 1978). The wire is fed at a controlled rate which allows its tip to melt and to be blown off by the fast flowing compressed air. Wire flame spraying is viewed to be more economical than a powder feed, viz. Fig. 6-18. However, powder feed operation has been developed to more sophisticated levels in recent years and should prove its advantage over solid material feed (Crawmer et al 1987). Coating thicknesses range from about 50 microns upwards (Steffens 1983). Uses include corrosion protection at ambient and high temperatures, rebuilding worn parts and providing wear, abrasion and erosion resistance (Brit. Std. 03-011). Flame sprayed thermal barrier coatings are used in gas turbine engines. Inert gas can be used instead of compressed air, for shielding from oxidation, and the substrate can be independently heated, to obtain better adherence.

Flame sprayed coatings have a much higher percentage of porosity than D-gun coatings (0.25 to 1%). Advanced plasma torch coatings have porosities close to those of D-gun coatings. Flame sprayed coatings contain many round particles (i.e. not lamellar) which are not molten on impact (Steffens 1983). Porosity of flame sprayed coatings can be reduced by vacuum impregnation but only about 0.1mm is permeable by the resin. Atmospheric brushing with resin gives only 0.02mm permeation (Arata et al 1982). A new impregnation method is required, perhaps with a suitable molten salt. LSM is an attractive alternative for deposit consolidation.

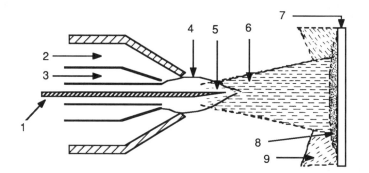

1. Feed wire introduced at central axis with wire feed coil-roller configuration 2. Input of compressed air   3. Mixture of oxygen & combustion gases 4. Flame   5. Wire tip   6. Spray cone   7. Substrate 8. Coating deposit 9. Loss of spray material

Fig.6-17a: Flame spraying process - wire form :
schematic diagram

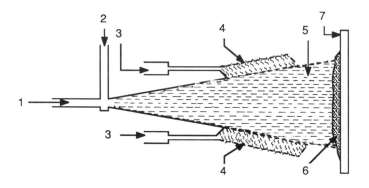

Main driving gas injector 2. powder feed nozzle 3.fuel gas injector channels 4. burning gases 5. spray cone 6. deposited coating 7. substrate

Fig.6-17b: Interior of a Flame Sprayer Employing Powder Feed

A Ni-Al composite wire generates extra heat in the flame due to exothermic reactions and the resulting coatings are more adherent and less porous than the usual flame sprayed coatings. Cine films have shown these particles becoming hotter as they leave the flame (Hebbert 1978). Details of the exothermic effects of Ni_3Al and Al_2O_3 formation are discussed (Knotek et al 1973; Houben & Zaat 1973). NiAl coatings on steel are described.

Flame sprayed yttria stabilized zirconia coatings on plasma sprayed bond coats of Co32Ni21Cr7.5Al0.5Y and NiCrAl, and arc sprayed 80Ni20Cr were tested for thermal shock (Steffens & Fischer 1987). Microstructure characterization procedures have been attempted and further quantitative correlation to the ceramic oxidation is required. Flame sprayed composites of nichrome and WC were tested for wear and corrosion. Four concentrations of carbide, viz. 85, 55, 15 and 5% were prepared. The wear rate was influenced by both the carbide agglometate and distribution in the nichrome matrix (Olivares & Grigorescu 1987).

6.6.3. ELECTRIC ARC SPRAYING

Two feedable wires are made electrodes for a d.c. electric arc and molten droplets are blown onto the substrate by a compressed air jet (Fig.6-18, Steffens 1983; German Std. DIN 8566). Although similar to flame spraying, the molten droplets temperatures are higher in arc spraying. High speed deposition of thick coatings are readily obtained on large components. Local interface welding is found for Al arc sprayed on steel and in other cases.

Arc spraying does not give as fine a spray as flame spraying. Nozzle design can be altered to optimize the spraying. Spraying wires must be accurately spooled and wear of wire guides causes problems. An attachment is described (Marantz 1973) to allow coating down the bores of long cylinders with diameters of 12cm (or more).

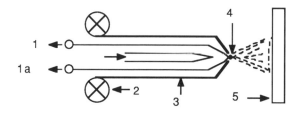

1. D.C. power source; 30 V, 300 A
2. Wire feed - drive and reel
3. Wire
4. Arc
5. Substrate

Fig.6-18: Principal features of an electric-arc spraying process

The adhesion of arc sprayed coatings is stronger than flame sprayed coatings, although the particle velocities are 100 and 180 m/second, respectively. The arc spray particle size is greater and temperature higher, causing local welding. But excessive heat increases shrinkage strains in the coating and burns out alloy constituents, which do not occur with flame spraying. The arc temperature is 6100K with 280 amps (Steffens 1983).

Pseudo alloy coatings are obtained by feeding different wires through each electrode. Thus a 2 phase alloy of Cu and stainless steel can be produced, giving a high wear resistance (due to the steel) and a high thermal conductivity due to the Cu (Steffens 1983). Electrical discharge machining (EDM) coating of steel by Cr, W, Ti and Ni is reported (Parkanski et al 1983). The high density of crystal lattice defects during spark discharge prevents removal of heat which then accumulates within the surface coating layers.

6.6.4. WIRE EXPLOSION SPRAYING

Faraday first obtained very good coatings by wire explosion over a century ago. Wire explosion spraying of W, Mo and piano wire onto Al and steel have adhesion 5 times stronger than by flame spraying (Suhara et al 1973), due to a welded zone between coating and substrate. This occurs because of high velocity and temperature. Applications include spraying Mo inside Al alloy cylinders of internal combustion engines, which is superior to the usual Cr plating. The inside face of glass tubes have been Al coated by this method. Smooth WC-Co coatings are obtainable.

In the optimum wire explosion (Suhara et al 1974,1973,1970) the wire vapour reaches the substrate, rather than just a spray of liquid droplets. This excludes air so that the subsequent shower of W and Mo can be wire explosion sprayed without oxidation. The droplets are spherical and only about 2 to 3 microns diameter. They travel at about 600 m/second. About 60% of the weight of the orignal wire is deposited on a hollow cylinder inner face. The inner surfaces of tubes of diameter as small as 1 cm can be sprayed. The coating thickness is 5 to 15 microns from a single spraying and can be built up by repetition. The coating is smoother and more adhesive than by flame and plasma spraying. The wear resistance of the W and Mo coatings is high and the friction small.

6.6.5. LIQUID METAL SPRAYING

The Osprey metal spray process can be used for building up very thick, e.g. 5 cm, metal coatings in a very short time (minutes). The coating metal is induction melted and atomized by an inert gas annular blast at the lower end of the melt crucible as it emerges through a hole into a reduced pressure area containing the substrate to be coated (Dunstan et al 1985). The droplets undergo in-flight cooling at 10^3 to 10^5 deg.C/second followed by a very fast deposition cooling and slow subsequent cooling. By controlling these stages a very thin liquid film is maintained on the surface of the coated substrate. Dendrites form in flight and shatter on impact, giving many nucleation sites and a fine equiaxed structure. One of the authors (M.G.H.) has seen this Osprey process and strongly recommends it for development and application.

6.7. COATING BY WELDING

6.7.1. GENERAL:

Surfacing by welding is one of the most widely practised bulk coating techniques. Cladding methods form the other group. Fig. 6-19 gives a comparative idea of the properties of major hardfacing alloys. The major welding processes currently used for surfacing are:

Gas, Powder, Manual Metal Arc (MMA), Metal- Inert Gas (MIG), Tungsten-Inert Gas (TIG) (which includes Plasma-Arc), Submerged Arc and Friction Welding.

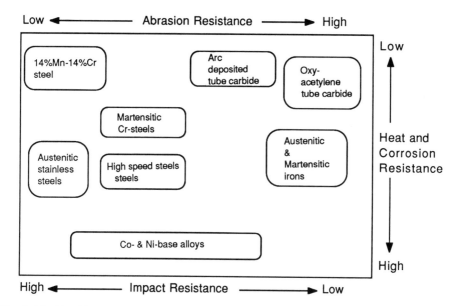

Fig.6-19: Position profiles of the major types of hardfacing alloys

The term 'surfacing' is generally used where coatings are thicker than 1 mm on the substrate, to greater than 8 mm. Several layers can be applied by welding techniques, the thickness restricted only by the mechanical and physical limitations specified for the application of the product material i.e. the coated substrate.

Bulk coating methods use coating material in the form of powder, sheet, strip, rod or paste, applied usually close to or above the melting point of the substrate-coating configuration. Amongst the bulk coating methods, welding, diffusion bonding and HIP methods result in an in situ diffusion bond unlike that obtained from overlay and other deposition methods. Bonding in overlay and other methods can occur by mechanical keying or over a period of time, by metallurgical bonding via thermal heat treatments, while bulk coating is wholly a metallurgical 'bond' formed during the liquid state with a good fraction of it containing the substrate material. Bulk coatings are applied to counter heavy wear and for use under conditions which impose mechanical and thermal shock. They have to be tough so as not to break under these conditions, apart from being metallurgically and chemically compatible to the substrate and resistant to the working environment. The thickness is a compromise for the wear and tear demanded during service and the utility not too dependent on significant dimensional variations as the coating wears. It is essential to achieve an optimised coating for the design service life, in view of the cost and bulk involved.

6.7.2. WELD SURFACING

A classification formulated by the British Steel Corporation may be considered representative of surfacing alloys normally used for coating by welding (Gregory 1980). The selection of the surfacing alloy involves three parameters at the outset, namely, the hardness, the purpose, and abrasion and/or oxidation resistance. Compatibility in characteristics with the substrate metal is the next important factor as only this ensures the integrity of form and performance of the finished component. In general, ferrous alloys are employed for abrasion resistance, nickel and cobalt alloys for oxidation and corrosion resistance, copper alloys for bearings, and, WC, chromium boride and similar compounds for wear resistance.

The process is known severally as hardfacing, weld-cladding, overlaying and by other proprietary names. Welding is particularly suited for local repair work of damaged bulk-coated components. It essentially involves melting the coating material with or without a flux and inert gas, on the substrate to be covered.

6.7.3. WELD SURFACING PARAMETERS

The foremost aspect is the dilution which results on welding. When molten weld metal contacts the substrate the substrate surface also melts and the resulting mixture is contained in the weld pool or the weld bead (Fig.6-20). Dilution is defined as $100A/(A+B)$, where A is the amount of weld metal and B is the substrate metal in the weld pool. On solidification there will be a zone in between the weld and the substrate which has a composition different from both the weld metal and the substrate. There will be an effective reduction in the original substrate thickness, but the total surfaced material will be thicker with the modified welded material bonded to the substrate. Dilution can vary from 10-40% and can be reduced to even 5% depending on the welding method used.

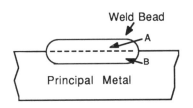

% Dilution = $(A \times 100)/(A + B)$

Fig.6-20: Dilution in a Weld Bead

The second parameter is the rate of deposition. The relative mass of coating dealt with in bulk surfacing far exceeds the normal rates of depositon. The latter are usually expressed in mg or g cm^{-2} hr^{-1} while surfacing is defined in kilograms per hour.

6.7.4. WELD SURFACING METHODS

Table 6:12 and Fig. 6-21 give some of the features of seven welding processes commonly used. Several variations on these basic themes are in practice. Oxy-acetylene methods are the slowest of all the methods causing also the minimum of dilution, as the flame temperature is much less than arc temperatures. The feed material is a rod or powder and lands on the substrate in a small pool or as molten droplets respectively, which causes restrained fusion and hence there is very little or virtually no loss of the substrate into the weld zone. A precise control and on-site welding is possible but the methods are unsuitable for large scale operations.

Very high temperatures are encountered in arc welding but TIG welding using a non-consumable W electrode is similar in scope to gas-welding in precision and operation, but somewhat higher in dilution and deposition rate. The plasma arc method improves on TIG- welding by employing two arc routes, the major arc current being the non-transferred arc passed between the W electrode and a water-cooled copper annulus around it, while the transfered arc passes between the electrode and the workpiece via the hot plasma issuing from the torch. Unlike TIG, plasma arc can cover large areas with precision and requires minimum finish. But it cannot lay down a coating more than about 1.5 mm thick.

TABLE 6 : 12

WELD-CLADDING

Process:	Oxy-acetylene	Powder	TIG	Plasma arc	MMA	MIG	Submerged arc
Also-known as:	Gas-welding		Tungsten-Inert-Gas; Gas-W-Arc; Argon arc	Variation of TIG	Manual-Metal-Arc Shielded metal arc; Stick-; Electric-arc	Metal-Inert-Gas; Gas-shielded-metal-arc-;CO_2-	
Weld-on-site:	Yes, portable	Yes, portable	Non-portable	Non-portable	Yes;versatile	portable	Non-portable
Type:	Wholly manual	Manual or mechanisable	Manual	Mechanisable	Manual	Manual or semi-mechanisable	Fully mechanised
Weld-material-form:	Rod	Powder	Rod	Rod or powder	Flux-covered rod electrode; tubular rod electrode containing flux	Continuous wire electrode	Continuous wire fed after a prior flux deposit
Dilution,%:	1-5	Virtually nil	4-10	5	10-30	10-30	10-40
Deposition rate kg/h:	1	0.5	2	3.5	1-7	1-10	30
Deposition thickness mm;	<1	0.05-3	2-4	<1.5	>2.5	>3	>4
Procedure:	Flame from gas source to provide heat. Welder co-ordinates rod-feed-rate into flame as weld-pool advances.	Powder fed via a hopper by gravity into the torch. Molten droplets contact substrate	Non-consumable W electrode Rod fed into the arc. Argon or similar inert gas (N_2, He) shields weld-pool	2-arc process	Very sturdy; vertical or overhead welding possible	CO_2 or argon shield. Flux cored wire, also flexible used as feed wire; can increase rate to 12 Kg/h without shielding gas.	Very good deposit achieved requiring little surface finish.

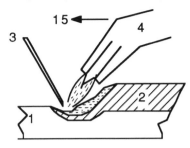

a. Oxy-acetylene hardfacing

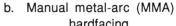
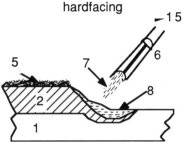

b. Manual metal-arc (MMA) hardfacing

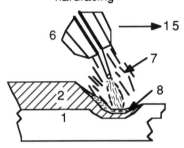

c. Metal inert-gas (MIG) hardfacing

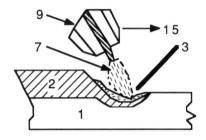

d. Tungsten inert-gas (TIG) hardfacing

e. Submerged Arc Weld Surfacing

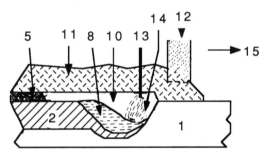

Fig.6-21: Weld Surfacing Techniques

Weld Surfacing Techniques : a: Oxy-acetylene; b. Manual Metal Arc (MMA); c. Metal Inert-gas (MIG; d. Tungsten Inert-gas(TIG); e. Submerged arc

a. 1.Parent metal, 2. Weld metal, 3. Filler rod, 4. Blowpipe nozzle (introducing oxy-acetylene gas mixture, 15. Welding direction

b. 1,2 ,15 - as in a; 5. Slag, 6. Consumable electrode (core wire with a flux cover), 7. Gas shield, 8. weld pool

c. 1,2,6,7,8,15 as in a,b above; the consumable electrode is fed through a contact tube which also carries the protective shielding gas supply

d. 1,2,3,7,8,15 - as in a,b, above; 9. Non-consumable electrode with a supply of shielding gas in a co-axial tube. 13. see 'e' below.

e. 1,2,5,8,15 - as in a, b above; 10. Excess weld metal, 11. Granular Flux, 12. Flux Tube feed, 13. Gas shield, 14. Arc submerged in the excess weld pool

MMA and MIG methods are more versatile and sturdy, capable of very high deposition rates. Vertical or overhead runs are possible and on-site operations. The flux-cored arc welding method can give thicker deposits and higher deposition rates using flexible feed wire packed with its flux and additives. A fully mechanised but strictly workshop method is submerged-arc welding. Granular flux is deposited ahead of a continuous wire and the arc submerges, under the flux resulting in a weld surface requiring mimimum finish. Friction welding or surfacing is performed by pressing a rotating metal rod with its axis at right angles to a laterally moving substrate.

A multiple strip weld surfacing schematic is shown in Fig. 6-22.

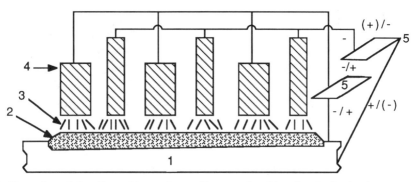

1. Parent metal   2. weld overlay   3. arc   4. strip sheet electrodes
passed through feed rollers   5. arc current source

Fig.6-22. Schematic of multiple strip weld surfacing

6.7.5. WELD COATED FINISH

Weld coated surfaces are usually wavy and slightly convoluted ending with a surface roughness of up to 2 mm difference between the maximum and minimum thickness. The wave nature is due to the electrode of a limited width having to pass to and fro on the work surface. Surface tension of the weld pool settling on the substrate which subsequently solidifies as the electrode passes on, causes an unevenness in the weld-coated surface. This surface roughness requires to be smoothed out by machining, grinding, polishing, or light abrading. For components like bulldozer blades or rock crushers this may not be necessary, but it is essential for gear wheels or valve seats.

Shrinkage effects during cooling can distort the coated component as stresses are created. Bending for instance, can be countered by offering the work pre-bent in the opposite direction along the same axis. Where thick layers or heavily restrained coatings are involved cracking is a possible or inevitable hazard. This can be compensated by applying an intermediate coat of soft-metal (ductile) and by intermittent applications with pre-heating of components especially when hard and brittle alloys are deposited.

280

6.8. CLAD SURFACING

Cladding is a broad term used which includes methods such as explosive impact and magnetic impact bonding, or hot isostatic pressing (HIP) or cladding (HIC), or mechanical bonding such as extrusion. This would introduce an overlapping between cladding and diffusion bonding when the above methods have to categorized.

Cladding methods can be classified on the basis of the speed with which substrate to coating material bonding can be achieved. Table 6:13 shows the three groups with the bonding they achieve and other details (Bucklow 1983). In general, the substrate is referred to as the backing plate and the coating material is called the flyer plate. Ferrous materials form the most handled clad components, with Ni-based alloys coming second. Other cladding alloys like Co alloys on steel are less used, chiefly in view of cost and availability. More detailed information may be had from the references appended (Bucklow 1983; Edmonds 1978; Bahrani 1978; Bahrani 1967).

Amongst the cladding methods listed, rolling and extrusion processes are perhaps the most widely applied. Explosive bonding was accidentally discovered in 1957; hot isostatic pressing (HIP) and electromagnetic impact bonding (EMIB) are comparatively new; diffusion bonding process spans the early 20th century ferrous technology to date on Ni- base and other high temperature alloys for special applications. Some of the outstanding features of the above processes will be given in the following sections.

6.8.1. ROLLING & EXTRUSION

This is the most economical of all cladding methods and the restriction in rolling and extrusion to sizes of components is only shop-capacity dependent. Both rolling and extrusion methods are practised for ferrous alloys with a wide variety of compositions and applications and for a number of Ni-based alloy/ferrous alloy combinations, Al-alloys and others have also been developed to meet increasing requirements for corrosion and oxidation resistance for ambient and high temperature applications in a variety of sulphur and chloride containing environments, and for high temperature uses ranging over energy, aviation and nuclear industries. Type 310 steel coextruded with 50Cr-50Ni alloy has been reported to have very good service under hot corrosion conditions in boiler systems (Meadowcroft 1987). Both rolled and extruded composite materials are only demand-limited, with the normal metallurgical limitations. Cladding materials are usually strip form, but powders can be employed quite effectively (Sugano et al 1968).

TABLE 6:13 (L.H.S.)

CLADDING PROCESSES

Group A
Ultra-rapid cladding

Process	(i) Explosive Cladding	(ii) Electromagnetic Impact-Bonding	
		(a) Hot	(b) Cold
Time-factor	Micro-seconds	Micro-seconds	
Pressure developed	not given	350 MPa	
Substrate shape	Simple; sheets	Simple; tubes or rings only	
Substrate size	15 m^2; max 35 m^2	<300 mm^2	
General requirement	Thickness >10 mm; and at least 2X cladding sheet thickness; sufficiently ductile	Sufficient strength to withstand magnetic discharge impact	
Cladding material	Sheet; reasonably ductile	Sheet or pre-sintered powder spray	
Bonding	Solid state, mechanical keying by shear flow, impact & scouring action, atomic contact & bonding	Components heated to recrystallisation temperature; high speed hot pressure diffusion bonding	As in explosive cladding – atomic keying as a result of a shock wave
Limitations	Comparatively large area – substrate/clad sheets as they are easy to handle; Explosive hazard; Some material deformation needs correction	Bulky equipment, expensive; small shapes restriction; components must have good electrical conductivity	
Advantages	Very fast; very strong & continuous bonds; unlimited combinations; very little surface preparation required	Safer than explosive cladding very fast	

==

TABLE 6-13 (R.H.S.)

CLADDING PROCESSES

Group B Medium-time cladding		Group C Slow cladding	
Rolling	(ii) Extrusion	(i) Diffusion-bonding	(ii) Hot Isostatic Pressing (HIP)
1-2 minutes	1-2 minutes	10 minutes to >10 hours	
		variable; 10-100 N/mm^2	100-200 MPa
Sheets	Tube; rods	Simple or complex	
Shop-capacity	Shop-capacity	Furnace-size-dependent	Reasonably small
Creep-resistant	Creep-resistant		
Sheet / sheet	Tube or ductile sheet	Sheet or powder	Sheet or powder
Mechanical	Mechanical	Metallurgical	Metallurgical
Several passes required to ensure sound bonding. Clad materials must be malleable	Shape restriction; only ductile coating materials can be handled	Expensive equipment and time consuming; uneconomical for large components; high technology components are usually considered for these two cladding methods	
Least expensive; several pairs can be rolled in a pack; also allows composite rolling	Economical	Well suited for very complex shapes and high technology components requiring close metallurgical control. HIP can be used for metal-ceramic bonding	

===

1. Quality check of the two component tube materials
2. Preparation of a 'sleeve' length of each component
3. Special cleaning of the tube inter-contact surfaces necessary for a faultless metallurgical bond. Accurate surface machining and a thorough removal of contamination are vital steps.
4. Inter-tube Gap controlled within tight limits
5.

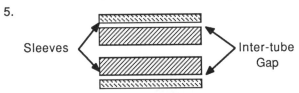

Sleeves   Inter-tube Gap

The sleeves are end-welded to prevent contamination during further processing

6. The composite billet enters a furnace to bring it to extrusion temperature.
7. Powdered glass is applied on the billet. At the extrusion temperature the molten glass acts as a lubricant.
8.

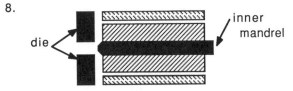

die   inner mandrel

An inner mandrel concentric with an outer die are aligned as the composite billet bearing the lubricant enters the extrusion press (3000 tons)

9. The billet undergoes a reduction as well as an elongation in the extrusion press.

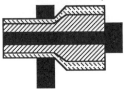

10. The extruded, clad tube is then cooled, and trimmed Further reductions can be carried out by reducing either cold or cooler temperatures.

Fig.6-23: Co-Extrusion Process Steps

Some of the special features in rolling and or extrusion are:

(i) A thin primary layer of electrodeposit usually ensures a better bond between the substrate creep resisting steel and the overlay clad material.

(ii) Cladding can be given with the substrate as a sandwich.

(iii) With a parting agent, usually Cr_2O_3, several composite pairs or trios can be packed together in a single thin container or "can" of steel and then rolled together. The can is sometimes evacuated or back-filled with N_2.

(iv) The exposed edges of the clad/substrate configuration are sealed by welding to prevent relaxation stresses from wrenching apart the rolled or extruded product.

Co-extrusion of clad tube is shown in Fig.6-23.

6.8.2. EXPLOSIVE CLADDING

This technique is also called explosive-welding. An accidental discovery in 1957 when a sheet metal got stuck to its die while an explosive experiment was carried out led to this method. An earlier report on the technique and its mechanism is available (Bahrani 1967; Bahrani 1978), and another has appeared in 1987 (Hardwicke 1987). The phenomenon seems to have been noted much earlier than 1957 in military circles when shells and metal fragments were seen to bond with metallic surfaces when impacted at certain angles.

Fig.6-24a,b illustrate schematically one of the modes of the process operation (Bucklow 1983). Bonding can be achieved by an oblique, or parallel high velocity collision between two plates to be joined. Shear and plastic flow are the instantaneous reactions as the shock wave speeds over the clad plate either at supersonic or sub-sonic velocities.

The Method of Explosive Cladding (Bahrani 1967; Hardwicke 1987):

The substrate plate is called the backing plate and the flyer plate is the cladding material. The backing plate must be at least twice the thickness of the flyer plate, preferably not less than 10 microns. Three basic requirements are as follows:

1. The flyer plate is spaced at a 'stand-off gap' parallel to the plate at a distance greater than its own thickness, or touching its edge to the substrate at an oblique angle. When explosion occurs the components must be brought together over this gap to collide progressively over the surface area. A collision front traverses this surface area.

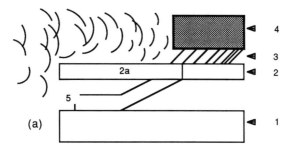

1. Backing plate 2. Flyer-plate 2a. Starting position of flyer plate - parallel to backing plate 3. Buffer 4. Explosive 5. Flyer-plate after deformation by detonated explosive. The two plates are inter-locked during the instant of contact.

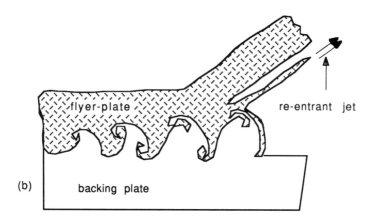

Melt flow and interlocking configuration at the impact region

Fig.6-24 a,b: Schematic diagram of Explosive Cladding

2. A protective (buffer) sheet is placed on the flyer plate (usually rubber) and the explosive is spread on it as a slurry or a sheet. The quantity of the explosive must be controlled to result in an optimum detonation velocity of, e.g. 2000 m/s for a thick plate. Slow explosions fail to secure a good bond and larger magnitude could shatter, distort, spall or melt the cladding. The velocity of the collision front must be below sonic velocity in the metals being joined. Shock waves travel through the materials at the velocity of sound. The subsonic velocity level of the collision front ensures that the shock wave precedes the bond being formed at the collision front. Failing this produces non-bonding or immature bonding in the composite. The explosion initiates this progressive shock wave which shear impacts the flyer plate to the backing plate. At such a high speed and pressure, the surfaces also get heated up and squirt out in a scouring action as if they are liquid. The surfaces of both contacting faces are pushed out along with any contaminants, e.g. oxide layers and atomically clean surfaces lock into each other.

3. The interfacial pressure at the collision front must significantly exceed the material yield strength to enable plastic deformation.

The collision angle of the flyer and backing plates determines the interface wave vortex morphology of the resulting composite. The volume of entrapped melt at the interface and the associated bond strength depend on the collision angle. A low collision angle results in an interlocking with a large vortex and a high degree of entrapped material as if it is a frozen melt. If the angle is very low the vortices link up and become continuous, instead of being discretely isolated, and the resulting bond is highly undesirable, since, effectively there are three layers of uncertain bond strength. The higher the collision angle, the more undulated the wave form will be with reduced vortices and entrapped melt. A very steep angle could result in a completly flat surface with little or no entrapped melt. The steep angle detonation configuration was thought commercially non-viable but has been achieved and patented (Szecket et al 1985; Inal et al 1985).

It seems that this process is only suitable for large masses of metal, but in principle ought to be applicable to any metal-metal bond as long as they are not very brittle, and have simple shapes such as sheets, tubes, and low convex-concave radius components - crucibles etc. Direct cladding of large areas, for instance 30 sq.m., have been produced and thin sheet composite cladders fusion welded together from 2 metre wide strips. Double sided clads, with the cladders bonded simultaneously have also been reported. Cup forgings and tube forgings have been streamlined. A better understanding of the process has enabled bonding of many metals and alloys. Laminar composites of alternating material layers of widely divergent mechanical properties have unique resistance to fatigue failure. Transition composite joints can be produced in tubular or flat form with two dissimilar metals. bonding of strictly identical materials (for multi-layer laminates) which are almost impossible to diffusion bond, e.g. Al-Li alloys, are very successfully explosion bonded. Imploded Zr-alloy to outer steel tubes have proved their worth in drilling tube service in sour gas oil wells.

The method has been found especially useful to bond normally 'mismatched' alloy bi-, tri- and quad-laminates, e.g. a V-20Ti bonded to Mo or Nb which itself was bonded to Cu (Kaminsky 1980); Pt, V, or Mo bonded to Ni-Cr-Fe-, or Fe-22Cr- alloys for high temperature use. Most of the clad materials listed by Gregory (1980) can be explosively bonded. A number of alloy combinations have emerged. Conversion rolling, formerly strictly reserved for stainless steel are now common for copper-nickel, all of Ti-base alloys, Hastelloys, many nickel alloys and Zr-clads. Zr-alloys which used Ti as interlayer, now are clad without it. Aerospace transition joints of Ti6Al4V to stainless steel, Al-alloys to Ti-alloys, multi-layer laminates of Al-Li alloys rank amongst the advances made by explosive cladding (Hardwicke 1987). Particular mention may be made of the straight waveless interface of Al to steel and Ti6Al4V to mild steel (Szecket et al 1985; Inal et al 1985).

6.9. DIFFUSION BONDING

A majority of coating processes include heat treatment as an intermediate or an ultimate step to ensure a sound substrate to deposit bond. Thus, almost all "non-overlay" deposition can be termed as diffusion bonds. This section discusses diffusion bonding employed as a primary process unlike its secondary role in other deposition techniques.

Diffusion bonding is a solid state method where component parts are pressed together and heated to form a bond by interdiffusion. Diffusion bonding requires high temperature data of all the components used in metallurgical technology for the concerned substrate-coating complex. Yet, the least characterized of all metallurgical parameters seems to be in the diffusion reaction area, e.g. diffusion constants for elements in multi-component alloys. Often, simplified or estimated values are extrapolated. This is perhaps, one of the important areas where futher fundamental research is needed.

Diffusion bonding is a slow process because a metallurgical bond by elements is achieved between two surfaces in contact by a solid-state diffusion process giving rise to a three-zone composite. Composition, microstructure, physical, chemical and mechanical properties are affected and controlled by the duration of heat treatment, the temperature and the pressure at which the process is carried out and the rate and type of cooling given subsequently. The physical, chemical and mechanical properties for the composite, such as hardness, stress/strain, oxidation, corrosion effects, microstructure etc., help in material selection and use. These are the parameters which decide diffusion bonding as a cladding process. Elliott and Wallach (1981) consider five main process parameters to be time, temperature, pressure, surface condition and environment. Unless otherwise stated diffusion bonding is carried out under vacuum in order to prevent inclusions and oxidation.

Diffusion bonding is also termed as solid-state welding. A few useful sources of information are appended: Tylecote 1968; Bartle 1969; Gerken & Owczarski 1965; Bartle 1978; Ohashi & Hashimoto 1976; Fielding 1978. Drewett (1969a,b) discusses diffusion coatings on steel on a wider context; Derby and Wallach (1982) present a theoretical model for solid-state diffusion bonding. Diffusion bonding is a suitable technique for joining materials which are difficult to join by conventional welding methods (e.g. Ti and its alloys) and to achieve bonding on a more commonly used material, e.g. steel.

6.9.1. PRIMARY CHARACTERISTICS OF DIFFUSION BONDING

Two material layers can be bonded when brought into elemental contact with an applied interfacial pressure at high temperature in the region 0.5- 0.8 Tm where Tm is the melting point in degrees Kelvin, of the material being bonded. Often additional layers in the form of coatings of a few microns, or loose shims are interposed to facilitate bonding or to act as diffusion barriers to specific elements.

The advantages of diffusion bonding are (Derby & Wallach 1982):

(i) Large areas can be joined.

(ii) Metal joining is possible under unconventional situations, e.g. space environment, or as part of a superplastic and bonding sequence.

(iii) Bonding is possible with minimum deformation.

(iv) Minimum microstructural damage and stress results while keeping a thin diffusion zone and smaller thermal gradients occur than in processes like fusion welding.

The disadvantages are:

(i) Poor adherence over large areas due to impurities such as oxides or grit at the interface.

(ii) Poor bonding due to interface separation via vacancy concentration and void formation during diffusion.

(iii) Unknown thermal effects on the properties of the bonding elements.

6.9.2. BONDING MECHANISM

Derby and Wallach summarise earlier approaches in their 1982 review in their model. The presence of an oxide film on the substrate is considered an inhibitor but not as rate determining to bonding since except for metals with insoluble oxides, e.g. Al, the oxide dissolves rapidly in the parent metal as bonding proceeds. The diffusion layer formed at the interface could be a solid-state bonding, purely mechanical, or a phase reaction in a solid-state exchange or a eutectic substrate-coating phase transformed to a solid microstructure on cooling. Surface roughness requires data specific to the material considered and generalization is not possible. The authors propose a broader based mechanism by considering diffusion bonding analogous to pressure sintering. The proposed six mechanisms of mass transfer are:

1. Surface diffusion from surface sources to a neck.
2. Volume diffusion from surface sources to a neck.
3. Diffusion along the bond interface from interfacial sources to a neck.
4. Volume diffusion from interfacial sources to a neck.
5. Power law creep deforming the ridge.
6. Plastic yielding deforming the ridge.

1 and 2 above are surface source related, driven by surface curvature differences across the surface of an interfacial void. 3 and 4 consider chemical potential along the bond line as driving force while 5 and 6 are gross deformation mechanisms driven by applied pressure with some surface tension effects. Five process parameters have been interwoven to apply the above six mechanisms, namely, temperature, pressure, initial surface roughness, initial surface aspect ratio and time.

Computer mapping yields the theoretical data which ironically offer limited verification for lack of adequate high temperature diffusion and creep data. Almond et al (1983) envisage a pressure-sinter mechanism in the diffusion bond of hard metal joints, lending partial support to the Derby-Wallach model. They consider an evaporation-condensation mechanism in the case of WC-Co wear resistant bonds. Further experimental work in high temperature creep, diffusion and most other physico-mechanical properties is needed.

6.9.3. ROLE OF INTERLAYERS IN DIFFUSION BONDING

The practical side of diffusion bonding has side-stepped technical difficulties in straight metal-to-metal bonding by using intermediate layers.

1. Applied as coatings, these prevent oxidation of the substrate prior to bonding (Crane 1967; Kammer et al 1969),

2. They assist micro-deformation and establish contact between faying surfaces (Bartle 1978; Hauser et al 1967),

3. They obviate bonding inhibition caused by contaminant films by increasing the rate of their dissolution,

4. They enhance the diffusion rates (Kharchenko 1969; Davis & Stephenson 1962; Lehrer & Schwartzbart 1961) and,

5. They can minimise Kirkendall void formation and prevent or reduce intermetallic phases at the interface (Crane 1967).

The above five causes have been inferred in the context of different diffusion bond pairs. Much more work is required to associate any of these with a proven mechanism (Elliott & Wallach 1981).

6.9.4. DIFFUSION BONDED MATERIALS

Aluminium on steels has been an industrial diffusion bond technique for nearly four decades. Aluminizing and chromizing on steels over forty years and later on Ni- base alloys over the last twenty years have been, consequently, the most investigated of all diffusion bond systems. High temperature and/or corrosion resistant materials such as Co-, Ti-, Mo-, and W- alloys have been investigated more recently. Materials listed by Gregory (1980) provide a reasonable cross section. Table 6:14 gives a list of elements which may or may not form good diffusion coatings on steel (Gregory 1980).

6.9.4.1. Aluminium on Steel:

The two impeding factors for good bonding in the Al- steel system are, (i) the oxide film which forms readily on Al- surface and (ii) brittle intermetallics, e.g. $FeAl_3$, Fe_2Al_5 which develop during the diffusion process at the interface. A number of circumventing parameters have been tried from pressure welding at one extreme which attempts to "squeeze" or break up the oxide film and affect plastic deformation, creep etc., to the other end where the oxide layer can be dissolved and the microstructure changed _in situ_ by interposition.

Comprehensive sources of reference are available (Drewett 1969b; Elliott & Wallach 1981). Pressure bonding, wedge bonding, butt-joint forming by twisting, force fitting tapered joints are a few of the mechanical deformation methods tried. Additions of Mg(0.3-2%), Si(1-6%) and Cu(3%) to Al have improved bonding, Si being the best, Mg being detrimental. Presence of 0.08% O_2 and N_2 in Fe is found to be beneficial, the probable effect being that a ceramic oxy-nitride barrier inhibits intermetallic formation.

TABLE 6:14

A SURVEY OF ELEMENT COMPATIBILITY TO FORM DIFFUSION COATING ON IRON

Elements Readily Compatible (good solubility)		Elements Incompatible (insoluble)		Elements of Unknown Ability or Untried	
Element	Size-factor	Element	Size-factor	Element	Size-factor
Al	1.15	Ag	1.16	Ce	1.47
Au	1.16	Ba	1.75	Ga	0.98
As	1.01	Bi	1.25	Ge	0.98
Be	0.91	Ca	1.58	Hf	1.27
Co	1.01	Cd	1.20	In	1.31
Cr	1.01	Cs	2.10	Ir	1.09
Cu	1.03	Hg	1.21	Os	1.09
Mn	0.95	K	1.86	Pd	1.11
Mo	1.09	Li	1.23	Pt	1.12
Nb	1.15	Mg	1.29	Ru	1.07
Ni	1.00	Na	1.50	Sc	1.22
Re	1.08	Pb	1.41	U	1.12
Sb	1.17	Rb	1.96	Y	1.46
Sn	1.13	Sr	1.73	Zr	1.26
Ta	1.15	Tl	1.45		
Ti	1.18				
V	1.05				
W	1.10				
Zn	1.17				

Several intermediate layers have been tried for Al- steel diffusion bonding. Ag (260-320°C), Cu(575°C and eutectic), Ag-Be (0.5-2%Be), Ni(0.2 mm) 625°C, Ni-Ag (6 micron electroplate – 5 micron foil) may be quoted as a few examples. Cu is a poor candidate layer because of its tendency to rapid oxidation at the diffusion bonding temperature. Ag and Ni have proved to be good, providing good bond strength.

6.9.4.2. Other Diffusion Bonded Materials:

Diffusion bonded WC-Co has been interlayered with Co, Ni, tool steel or mild steel at 1300°C (Almond et al 1983). Co and steel interlayers are found to be sensitive to the differential carbon contents encountered in the hardface-interlayer pair. At 1300°C a vapour phase evaporation/condensation mechanism is considered to play a vital role in diffusion bonding with creep and diffusion complementing it. The undesirable brittle eta-phase can be developed at low carbon contents at the interlayer/hardface metal junction. Choosing a high carbon content Fe, or Co or Ni interlayer prevents eta-phase formation.

6.9.5. HOT ISOSTATIC PRESS BONDING - (HIP) CLADDING

This is a relatively new method developed over the last decade and is particularly advantageous for complex shapes. Beltran and Schilling (1980) discuss the innovation in cladding at the General Electric Company for gas turbine components. A metal-ceramic HIP cladding method is reported by Allen and Borbidge (1983). Useful related literature can be found in these two sources. Patent literature (1980) and the powder HIP process (Nederveen et al 1980; Anon. 1982) have been mentioned by Bucklow (1983).

Ni-base and Co-base gas turbine alloys, MCrAlY-series, Ti- and Zr-alloys form some of the metal-metal HIP- processed components. It can also be applied to metal- ceramic pairs, e.g. Pt, Pd, Au, Ag or Fe, Co, Ni, Cu to Al_2O_3, ZrO_2, MgO, SiO_2 and BeO (Allen & Borbidge 1983). The process uses high temperature-high pressure equipment. The coating material is applied to the substrate by spot welding or local brazing a sheet, or by powder spraying. The substrate-coating pair is surrounded with a pressure-transmitting medium, e.g. glass chips later molten, in a deformable container and heated. Isostatic pressure is applied for various lengths of time; pressure, time and temperature are work-dependent. Sprayed deposits are consolidated and diffusion bonded, sheet material is forced by the pressure contact and diffusion bonded. The glass is then removed.

Powder ingots of Fe25Cr4AllY and Co25Cr3Al10Ni5Ta0.2Y alloys have been formed by HIP, with subsequent improved fabricability. Both the alloys contain intermetallic precipitates and dispersions and in cast or wrought form have low ductility. The components were argon-atomized, prealloyed powdered, and HIP-densified with a 20% improved tensile strength powder ingots (Beltran & Schilling 1980). The cladding process involves a preliminary substrate cleaning, a 12 micron Ni electroplate and then cladding, with the sheet (250 microns) fabricated from the powder ingot, and finally HIP diffusion bonding. The Ni interlayer prevents TaC formation at the interface due to interdiffusion between Ta in the clad material and C from the substrate. HIP temperatures and pressures ranged over 1093-1150°C and 6.9-103 MPa.

Two HIP-techniques are reported. The first where the work is totally immersed in a low viscosity ($<10^{-6}$ centipoise) soda-lime glass cullet in an outgassed, evacuated steel container, and heated to 1000°C in an argon pressurized autoclave. The second method side steps the difficult task of removing solidified glass at the post-process stage from cooling holes, crevices etc. This is achieved by a first step controlled vacuum brazing cycle to seal the substrate-cladding seams, e.g. a braze alloy Ni45Cr10Si is used to seal IN738/coating seams at 1150°C for 10 minutes, the temperature being slightly above the braze alloy solidus temperature of 1135°C. Direct pressure transfer occurs in the autoclave, between the substrate and cladding as the annular space will have been evacuated before seam sealing.

293

A metal-ceramic HIP process with Al_2O_3-Pt-Al_2O_3 sealing has been carried out using temperatures of $1100-1800^\circ C$, time 12 minutes to 10 hrs and pressures 0.13 to 10MPa. The standard operating conditions used were $1450^\circ C$, 0.8MPa, 4h. For maximum bond strength the conditions were $1700^\circ C$, >10h at 2MPa contact pressure. Note the low contact pressure contrasting the 100MPa used in the metal-metal HIP process. The authors do not consider mechanical keying as the bond mechanism since the bond strength does not continue to increase with increase in pressure. The bond is found to be very strong even at very high temperatures, with a 25% loss in strength at $1100^\circ C$ from that at ambient temperature. It is inferred that a HIP bond in metal-ceramic is a solid-state reaction bonding (Allen & Borbidge 1983).

Particle size and its distribution have been shown via an analytical deformation model, to influence the powder morphological changes and the kinetics of densification (Kissinger et al 1984). Samples of Rene 95 superalloy, graded on three particle sizes, viz. monosize 75-90, bi-modal 75-90 and 3-35, and commercial 100 microns, were canned and given the HIP treatment at $1120^\circ C$ over a time range 5-180 minutes at 10.3 or 103 MPa. It was found that component densification was achieved almost entirely by time independent plastic deformation of powder particles, that of smaller particles being the greater. The variations of HIP temperature and pressure, the effect of creep and superplastic flow etc., are yet to be investigated to clarify the role of the dominating plastic deformation behaviour.

HIP consolidation and the kinetics of densification of powder compacts need to take into consideration, the particle size, its distribution and powder morphological changes.

Cladding of Al onto steel requires 14×10^4 kN/m^2. at $400^\circ C$ or 21×10^4 kN/m^2 at $345^\circ C$. Subsequent annealing doubles the bond strength. Equipment for HIP is commercially available (Anon. 1985).

Hot isodynamic compression (HIC) may be mentioned in this section as a technique claiming an advantage over HIP in being more cost-effective while also providing improved metallurgical bonding. Both processes are new and quick conclusions of the advantage of one over the other may prove to be misleading. The HIC process has been applied to a low alloy steel substrate arc plasma sprayed with rapidly solidified platelet-powder mixtures of 57.5Ni23.5Mo9.0Fe10.0B prepared by spin-cast techniques. Particle size distribution (wt.%) was 65 of 100-270 mesh, 15 each of 80-100 and 270-325 mesh and 5 of >325 mesh. The spray coated steel was subjected to transformation treatments at two temperatures using hot forging techniques with a composite die as the transfer medium. This is described as the HIC treatment carried out at 827 MPa pressure over 0.5 seconds dwell time at $1010-1066^\circ C$ and subsequently normalized. Results based on EDS, microstructure, hardness and statistical evaluations have been used to show that HIC technique provides a better bond and very low porosity in the arc plasma sprayed coating than the HIP technique (Hays & Harris 1987).

6.9.6. ELECTRO-MAGNETIC IMPACT BONDING (EMIB)

EMIB employs magnetic energy to produce the impact which explosive cladding achieves by chemical combustion, but unlike it, EMIB cannot handle very large or thick components (not greater than 300 mm^2). It also requires the components to have a reasonable electrical conductivity, and only simple shapes such as rings, tubes, which are circumferentially continuous are acceptable. EMIB can act on heated or cold components. Hot EMI- bonded components have a diffusion zone at the interface while cold EMIB is achieved by mechanical keying. The substrate-cladding pair is introduced into the EMIB equipment, to be surrounded by a coil. A sudden surge of current through the coil produces an intense magnetic field which translates its force as a pressure on the work piece. In a few microseconds a pressure of 300 MPa can develop which impact seals the cladding onto the substrate (Fig.6-25). Gas turbine blades have bean sealed by this method (Obrzut 1972).

EMIB is expensive to install and its use is limited. But it is safe, quick and clean to operate within its materials limitations.

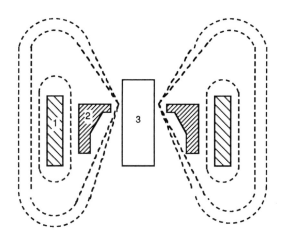

1. Compression coil 2. Field Sharper 3. Workpiece

Fig. 6-25: Electromagnetic Impact Bonding

CHAPTER 7

Physical and mechanical properties of coatings

7.1. INTRODUCTION

The brief survey made in this chapter is in three categories: (i) the physical aspects of coatings, viz. microstructure and tempe-rature-related properties, namely, thermal-expansion, thermal cycling, thermal conductivity and interdiffusion, (ii) physico-mechanical properties such as adhesion, stresses, strength, duc-tility, creep and hardness, and (iii) mechanical properties such as wear and erosion. Physical and mechanical properties are very inter-related and are strongly influenced by each other, and in turn also affect the chemical properties as will be evident in the following chapter.

There will be an inevitable overlapping of Chapters 7 and 8 with the mid-section of Chapter 9 which deals with characterization of coatings. The divisions of the three chapters are more for conve-nience rather than any validity for segregation of topics. Much of the information dealt with in this chapter should be useful for the assessment of coatings, dealt in Chapter 9.

7.2. EVOLUTION & MICROSTRUCTURE OF COATINGS

7.2.1. GENERAL

Coatings and thin films often exhibit different microstructure and properties from the bulk processed materials because many of the coatings processes can produce deposits in metastable states and non-equilibrium phases.

Factors which influence the formation of coatings are:

1. The substrate on which the deposit arrives,
2. The temperature of the substrate, and the temperature of the process,
3. The medium of deposition,
4. The rate of arrival and mass transfer, and,
4. The substrate/coating compatibility.

Coatings processes discussed in this book broadly divide into those which deposit from a vapour and/or charged medium, and those which arrive at the substrate as droplets, or attach to it in bulk. The routes by which they finally appear as coatings are thus vastly different. The deposit which arrives in a macroform of the material it will eventually be, needs mechanical keying to the substrate and interlinking to the arriving mass. But the deposit which arrives as a particle in a charged state, or as a vapour, needs to be neutralized, seek nucleation sites, then build up from an atom to unit cell and then onto a crystalline morphology or be amorphous as the conditions influence. The development of microstructures of the two types of deposition are different as are their modes of growth. There are further differences between those proceses which deposit through a 'vapour' medium, e.g. PVD, CVD, pack etc., and those which use an electrolyte or a gel or plasma.

7.2.2. THE ROLE OF THE SUBSTRATE

7.2.2.1. Nature of the Substrate:

The evolution of a deposit can start either by the formation of 3-dimensional nuclei on favoured sites on the substrate followed by lateral growth and thickness, or as a continuous film formed from the start with no island growth. The first mode occurs where there is poor interaction or bonding between the coating and substrate and the outward growth proceeds to produce in many cases a "columnar" structure. The second mode occurs under conditions where no oxide films are present or any other impedance, and good bonding between the coating and substrate metal occurs. The geometrical shape of the substrate significantly affects the coverage of the deposit in some of the processes, and coating discontinuity due to 'shadowing' is a common defect on contoured substrates for which adequate precautions must be taken. The types of bonding obtained between a superalloy IN 718 and a glass-ceramic is deeply graded with three zones of morphology (Madden 1987).

7.2.2.2. The Substrate Temperature:

Blocher (1974) & Bunshah (1981) have discussed the effect of temperature on deposit manifestation and mode of growth. Table

7:1 shows the pattern and influence of temperature and supersaturation on deposit structure.

TABLE 7:1

THE INFLUENCE OF TEMPERATURE & SUPERSATURATION ON THE SUBSTRATE OF CONDENSED MATERIALS
(Blocher 1974)

	Epitaxial Growth ↑	
	Platelets	
	Whiskers	Effect of
Effect of	Dendrites	Increased
Increased	Polycrystals	Temperature
Supersaturation	Fine Grained Polycrystals	
	Amorphous Deposits	
	↓ Gas Phase Nucleated "Snow"	

===

Increasing substrate temperature has the effect of increasing the mobility of the coating atoms on the surface thus promoting diffusion (Sherman et al 1974). The kinetic energy of the incoming atoms can be increased by partially ionizing the vapour flux and this can lead to epitaxial growth (Boone et al 1974). In this case, the effective surface temperature of the growing film is much higher due to ion bombardment. Movchan and Demchischin (1974, 1969) studied the variation of microstructure with deposition temperature for Ni, Ti, W, Al_2O_3 and ZrO_2 and proposed a 3 zone model of structures as shown in Fig.7-1.

At low temperatures, the coating atoms have limited mobility and the structure is columnar, with tapered outgrowths and weak open boundaries. Such a structure has also been called "botyroidal" and corresponds to Zone 1. If the substrate temperature (Tl) is increased above 0.3 of the melting point (Tm) (in ^{o}K) of the coating material, then although the structure (Zone 2) remains columnar, the columns tend to be finer, with parallel boundaries, normal to the substrate surface. Tl is 0.3Tm for metals and 0.22-0.26Tm for oxides. These boundaries are stronger than in the Zone 1 structure and contain no porosity. At a substrate temperature (Tl) above 0.45 of the melting point (Tm) of the coating material, the structure (Zone 3) shows an equiaxed grain morphology. The transition from one zone to the next is not abrupt but smooth. Zone 2 structures have been associated with enchanced surface diffusion and Zone 3 structures with bulk diffusion. Zone 3 is not seen very often in materials with high melting points.

Formation of thick coatings by atomistic deposition processes often gives a very columnar morphology of low density. In electrodeposition this morphology can be prevented by levelling or brightening agents which continually re-nucleate the growing film

(a) Zone 1: T= (0.1-0.4) T$_m$ (b) Zone T: T= (0.4-0.7) T$_m$

(c) Zone 2: T= (0.7-0.8) T$_m$ (d) Zone 3: T= (0.8-1.0) T$_m$

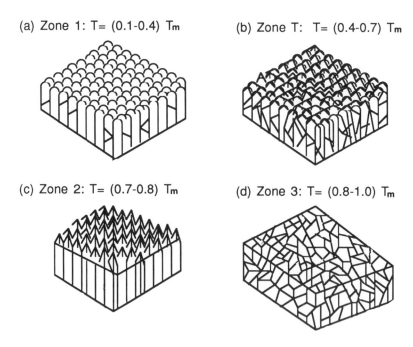

Fig.7-1: Structural Zones in Vacuum Condensates

(Movchan & Demchischin categorization; Zone T - Tucker Model)

(Bockris & Reddy 1970). In CVD (Holman & Huegel 1967) and electrodeposition the columnar structure can be prevented by mechanically disturbing the surface during coating (Kennedy 1968). Ion
bombardment during coating alters the density, appearance and
growth of thick metallic coatings produced by CVD (Culbertson &
Mattox 1966), sputtering (Bland et al 1974; Mattox & Kominiak
1972) and vacuum evaporation (Mattox & Kominiak 1972; Bunshah &
Juntz 1972) including electron beam evaporation (Bland et al
1974; Mattox & Kominiak 1972; Mattox 1973). High internal stresses and defect concentrations can sometimes cause unexpected
structures (Patten et al 1971).

Low rate low temperature PVD produces fine columnar grains of 250
Angstroms diameter (Nb and Sn), on a substrate at 570°C. Grain
boundary grooves (cavities) occupy about 15 volume% of the structure. Low rate, high temperature deposition (750°C) increases the
grain size about 10 times and gives no grain boundary cavities,
but the structure is still columnar. At a five times higher
rate, and high temperature deposition at 750°C, crystal grains
are twice as small and more uniform (Jacobson 1981).

Coatings of mirrors for solar energy heliostats by Ag vapour
deposition on glasses show strong preferred structural orientation effects, which are absent on chemically deposited films

which have much smaller crystallite sizes. Heat treatment of chemically deposited films to 350°C increases their crystallite size to that of vapour deposited films. Vapour deposited films are more uniform and defect-free (Shelby 1980).

Post coating mechanical peening densifies the surface and then a heat treatment promotes some interdiffusion for bonding. Methods of simultaneous spraying and peening have been investigated (Singer 1984). Heat treatment is usually at the solution temperature and does not disrupt the mechanical properties of the substrate.

7.2.2.3. Surface Roughness & Angle of Incidence of the Vapour Stream:

Since evaporation is primarily a line-of-sight process, the effect of any asperity or projection on the surface will be increased when the vapour stream is at a low angle to the substrate surface. Leaders are columnar defects which are poorly bonded to the rest of the coating and result from localized rapid growth due to shadowing effects caused by surface protrusions. Decreased initial specimen surface roughness can significanty reduce the number and size (width) of open columnar defects although they do still occur. Exposing surface height irregularities to varying angles of incoming vapour by compound specimen rotation results in improved as-deposited coating quality.

7.2.3. GAS PRESSURE

High gas pressure during deposition inhibits surface mobility and hence can be the cause of columnar strutures even at elevated temperature.

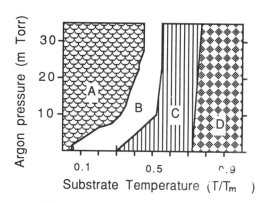

A: Porous structure; tapered crystallites separated by voids. (Zone 1)
B: Transition structure; densely packed fibrous grains. (Zone T)
C: Columnar structure. (Zone 2)
D: Recrystallised grain structure (Zone 3)

Zone 1: Metals $< 0.3T_m$;
Oxides $< 0.26T_m$
Zone 2: Metals $0.3 - 0.45T_m$;
Oxides $0.26 - 0.45T_m$
Zone 3: Metals & Oxides $> 0.45T_m$

Fig.7-2: Influence of Substrate Temperature & Argon Pressure on Condensate Structure

The model proposed by Movchan and Demchischin (Fig. 7-1) was extended by adding an additional axis to account for the effect of gas and was subsequently modified by Thornton (1975,1981) to account for an additional transition zone (Zone T, Fig.7-1,b). This was identified between Zones 1 and 2. Zone T is a columnar Zone 1 structure with small crystal sizes that appear fibrous and with boundaries that are sufficiently dense to yield good mechanical properties. Fig.7-2 shows the gradation of structure as related to argon pressure and substrate temperature. Coatings produced for laboratory scale studies and for industrial applications have been reviewed (Bunshah 1981).

7.2.4. CVD DEPOSIT MICROSTRUCTURE

In CVD processes the structure/property/process inter-linking is complex and specific to the deposit-substrate-process-purpose configuration chosen. Thus a successful CVD coating of carbon on particles is governed by a set of control parameters totally different from that required to achieve a carbon rocket nozzle by CVD. A good cross section of available information may be had in references cited : (Blocher 1974; Takahashi & Itoh 1977; Proc. International CVD Conf. 1976, 1978, 1980, 1981, 1983,1985,1987). It would be out of scope to deal with them here.

The main tools required for determining the desired structure are two-fold, viz. those which can provide the analyses of the reaction and those which can characterize the substrate/deposit configuration.

In general, a columnar deposit is weaker both mechanically and in environmental resistance when deposited in bulk on a broad substrate. But a prefered orientation may be desirable in particular cases, e.g. carbon with random a-b-growth but preferred orientation in the c-direction for increased thermal conductivity in rocket nozzles and re-entry vehicles and as a bowl liner in smokers' pipes. A prefered orientation for W increases its emitter efficiency to a maximum.

Davis (1980) discusses the formation of deposits by CVD, in particular, epitaxial films. A sustained growth of a heterogeneous film, energetically favourable requires a critical size for the nucleated, coalesced clusters of the deposit. Those which are smaller will decay or re-vaporize. In CVD processes supersaturation produces high yielding cluster radii of one or two atoms. Surface free energy and other classical thermodynamics parameters cannot be appropriate (being based on large statistical assemblies). An atomistic concept (Walton 1962) combining statistical mechanics and chemical bonding can be fitted. Surface sites with strong bonds are the more favoured nucleation sites. Since CVD is also a diffusion-limited process, stable clusters endeavoring to capture atoms in a 2-dimensional field have a much longer range than a corresponding 3-dimensional

302

nucleation (Lewis & Campbell 1967).

Low temperatures, high atom arrival rates, and atoms with co-valent or ionic bonds favour amorphous growth, e.g. the group IV elements of C, Si, Ge, SiC, and oxides of Si, Zr and Ti. Blocher's review (1974) discusses microstructural aspects of CVD in considerable detail. Useful sources for literature on the morphological aspects of CVD deposits are: (Davis 1980; Kern & Ban 1978; Walton 1962; Lewis & Campbell 1967)

7.2.5. PACK & PLASMA SPRAY DEPOSIT MICROSTRUCTURE

As a "close-up CVD" process, pack-coating poses numerous experi-mental problems. This is probably the cause for poor characteri-zation of the evolution of coating structures. The substrate must be thoroughly clean and free of oxide films or any other impuri-ty, in order to ensure uniform deposition, bonding and diffusion on its entire surface. Elaborate cleaning procedures are normal as the flow sheets indicate. All operating conditions are worked out specifically for a substrate. Because the pack surrounds the substrate adequate coverage is ensured, enhanced further by vacuum slip pack and other methods. The nucleation and deposit growth aspects are left 'unseen'.

Deposit arrival in a greater mass and its subsequent consolida-tion on the substrate is encountered in bulk coating methods. The microsructure of both plasma and flame sprayed coatings is a series of overlapping lamellae caused by the splat solidification of droplets. Gamma alumina forms at large substrate-to-torch distances, low currents and high gas pressures; otherwise, delta and alpha alumina form. Gamma alumina has inferior properties but may be transformed to alpha alumina by heating above $1100^{\circ}C$, although the lower density of alpha causes an increase in coating porosity. The proportion of alpha in flame sprayed coatings increases linearly with substrate temperature reaching 85% at $800^{\circ}C$, but this does not apply to plasma sprayed coatings. How-ever, pure alpha coatings are obtained if the substrate is held above $1100^{\circ}C$, but for low porosity it must be at $1450^{\circ}C$ (McPherson 1980). The effect of substrate temperature on porosity is discussed (Smyth 1973).

Metastable structures and amorphous solid solutions are frequent-ly observed, particularly in ceramic bulk materials. These are frozen in up to temperatures of $0.3T_m$. Thus materials with T_m of $2500-3300^{\circ}C$ can be used over $700-1000^{\circ}C$. Predicting the constitu-tion of a multi-component coating requires phase diagram and thermodynamic data. Deposition at low temperatures resulting in a less ordered structure followed by monitored heat treatment is an area open to further investigations. The system Ti-Al-N would be of interest in this context. Free energy calculations show that 0-60 mol.% AlN is a cubic mixed nitride when deposited at low temperatures and the metastable (Ti,Al)N solid solution does not decompose up to 800–1000°C. Similar studies on other systems have shown that:

TiC–TiB$_2$ form amorphous solutions with good adherence to substrate but lower strength than its crystallized state; the system also develops a multiphase layer when TiC and TiB$_2$ are deposited simultaneously, and crystallize slowly from 800–1000°C; TiC can form a multilayer deposit when sequentially sputtered with TiN, and WC with TiC and TiN can form gradient coatings Holleck & Schulze 1987).

Many hardfacing processes yield dendritic deposits; their size was found to decrease in the order of processes - oxyacetylene, TIG, PTA (Plasma Transferd Arc) and laser deposits (Monson & Steen 1986). The process by which discrete molten particles impact at random points within a surface area has been simulated using the Monte Carlo method and a computation model presented (Knotek & Elsing 1987).

A mathematical viscosity analysis of the impact of droplets on a cold substrate is given by McPherson (1980). The analysis shows tht flattening and solidification are separate events for lamellae < 1 micron thick; thicker lamallae are thickness controlled by the solidification kinetics. Most alumina droplets reach 2500°C but freezing occurs at 1600°C due to undercooling. The viscosity is between 0.15 poise (2500°C) and 5 poise (1600°C). Cooling rates are found by direct measurements or indirectly from dendrite arm spacing. The gun technique (D-gun) where a droplet at high velocity hits a cold substrate, gives splat cooling rates of 10^6 to 10^8 degrees/second. Heat transfer theory shows, for thin lamallae, that the cooling rate depends on the nature of the interface and not on the substrate conductivity, the heat transfer coefficient being about 10^5 watts/m^2 degree. The lamellae undercool and crystal nuclei grow at the limiting rate, liberating heat of fusion which supresses further nucleation. It can be shown that this leads to a columnar microstructure.

If the heat transfer rate (to the substrate) were such that the temperature decreased after the initial nucleation, then the nucleation rate should increase due to the extreme undercooling. Hence the solidification rate would be controlled by the nucleation rate, resulting in a very fine equiaxed microstructure (McPherson 1980). The (undesirable) columnar microstructure of plasma sprayed alumina suggests that the heat removal rate is less than the heat generation rate. Heat transfer to particles in a plasma has been discussed (Vardelle and Fauchais 1983). For cold substrates, lamella thickness increase gives proportionately longer solidification times, but at high substrate temperatures the reduced heat transfer reduces the degree of undercooling and solidification becomes controlled by the heat removal rate.

The fraction of partially melted particles rises rapidly with particle size and explains the occurrence of some alpha alumina in cold substrate coatings. It is suggested that transformation to alpha is significant at lamellae thicknesses >20 microns for cold substrates and >10 microns for $1000^\circ C$ substrates. For lamellae thicknesses resulting from <5 micron powder feeds, formation of alpha cannot be expected even for substrates heated to several hundred degrees. Flame sprayed alumina has more alpha content as lamellae are thick (about 10 microns) because the feed rod end must melt before it can be detached and so large particles are sprayed. The relation between alpha, gamma and delta alumina is discussed in detail by McPherson (1980).

7.3. THERMAL EXPANSION, THERMAL CYCLING

Thermal expansion is the commonest source of mismatch strains, especially important for ceramic coatings. The thermal expansion coefficient of ceramics is about $8\times10^{-6}K^{-1}$ and steel is about $14\times10^{-6}K^{-1}$ and a thin intermediate layer of oxide and metal can be used to reduce interface mismatch strains. Transformation strains depend inversely on the degree of stabilization in ceramics like ZrO_2 and may counteract some of the thermal expansion strain. Microcracking due to a phase transition or porosity may relieve thermal mismatch strains (Kvernes 1983). Ion plated coatings have excellent adhesion and resistance to thermal shock and vibration (Teer 1983).

Data on thermal expansion, especially on coatings of all categories are yet to be well documented. A useful source for nonmetallic solids may be cited here (Toulonkian & Ho 1987). Table 7:2 lists selected data on oxide and nitride coatings (Richerson 1982; Hancock 1987).

TABLE 7:2

THERMAL EXPANSION COEFFICIENTS OF OXIDE & NITRIDE COATING MATERIALS

Coating Material	$a \times 10^6/^\circ C$
Al_2O_3	8-9
CoO	15
Cr_2O_3	7-7.8; 9.6
MgO	12.9-13.9
$MgO-Al_2O_3$ spinel	9.1
NiO	14-17.1
SiO_2	3
ZrO_2	8-10
ZrO_2+MgO	10-12.2
SiAlON	3-3.2
Si_3N_4 (hot pressed)	3
Si_3N_4 (reaction bonded)	3
SiC	3.7-4.8

Thermal expansion coefficient values (a x $10^6/^{\circ}C$) for typical substrate alloys are in the range 9.7-19.2 for Ni-base super-alloys, 16.5-20.0 for stainless steel (25Cr 20Ni wt.%), 14.8 - 19.0 for stainless steel (18Cr 8Ni) and 12-14 for carbon steel. Thermal stability and thermal shock characteristics of carbides and nitrides are important in view of their barrier layer role on substrates of a wide variety, viz. on semi-conductors, graphite and refractory superalloys. AlN was found to be a suitable encapsulant for GaAs up to $1000^{\circ}C$ in air and $1400^{\circ}C$ in vacuum. It forms gamma-AlOOH at steam temperatures but is inert at room temperature (Abid et al 1986). Electron-beam thermal shocks imparted to TiC on graphite substrates were monitored to assess its application in fusion reactors (Brunet et al 1985). More work in this area is needed.

The possibility of adopting materials with negative thermal expansion has been explored, and the best way to achieve very low coefficients of thermal expansion seemed to be through composites. Many oxides have negative coefficient of expansion at low temperatures but few materials show negative values of this parameter at room temperature or above! During this study the authors found that although thermal expansion is an inherent thermo-physical property of a material, the observed values depended on the routes for sample preparation, and this has been attributed to errors creeping in via microcracking, grain boundary separation, plastic deformation and retardation in phase stabilization (Chu et al 1987).

Thermal cycling techniques and some results are discussed by Nicoll (1983). A few results on coated stainless steel and superalloys are shown in Fig.7-3. It is clear from the histogram (qualitative) that the superalloy substrate with stabilized zirconia coating where the substrate alloy was maintained at $20^{\circ}C$ or heated to $316^{\circ}C$ is far superior in thermal cycling conditions than the others shown in the diagram. Other substrate/coating systems tested for thermal cycling showed that superalloy substrates - Hastelloy, IN 738 with CoCrAlX (X = Y, Ce) and directionally solidified NiNbCrAl with Pt undercoat and NiCrAlY topcoat withstood thermal cycling very well, while steel substrates with a selection of coatings, e.g. Cr, WC12Co, 75CrC-25(80Ni20Cr), and silica and ceria 2-4 microns, were mediocre, and Al-alloys with cemented carbide, CrC-NiCr, NbC and AlN coatings were very poor as was Nb with Mo- and Nb- silicides.

Thermal barrier coatings have performed creditably in thermal cycle tests. Coating failure by thermal cycles depends on the applied stress on it (thermal and/or mechanical) and the strength of the substrate adhesion and interlinking to the coating. In a Rolls-Royce thermal cycling rig test on several coating systems ($1125^{\circ}C$ for 1 min. and forced cooling to $250^{\circ}C$ for 1 min), argon shrouded LPPS coated 8% yttria stabilized zirconia and $CaOTiO_2$, with MCrAlY bond coat proved to be the two superior systems with 7000 and >6100 cycles respectively. Coatings of magnesia and yttria stabilized zirconia with Ni/Cr alloy bond coat failed in

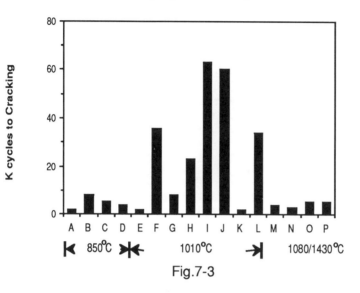

Thermal Cycling at Various Temperatures

Fig.7-3

Legend	Substrate	Coating
A	IN738LC	none
B	IN738LC	NiCrSi
C	IN738LC	SiB
D	IN738LC	TiSi
E	Superalloy1	$ZrO_2 20\%Y_2O_3$ (plasma sprayed)
F	As in E	As above but columnar
G	As in E	As in E, but electron beam coated
H-K	As in E but heated substrate	
H	at -34 °C	As in E
I	at 20 °C	As in E
J	at 316 °C	As in E
K	at 649 °C	As in E
L	As in E	As in E, but with 21%MgO (microcracked)
M	As in E	$ZrO_2.12\%Y_2O_3$/Ni16Cr6Al0.6Y Duplex coat
N	As in M	As in M
O	CrNiNb-st. steel	CeO_2 - by Sol-gel process
P	As in O	SiO_2 - by CVD

<1000 cycles; PVD coated 20% yttria stabilized zirconia fared better with 4000 cycles (Bennett et al 1987).

Thermal fatigue cracking depends on the coating-substrate thermal expansion mismatch (Strangman et al 1977). To minimise the sharp transition from metal bond coat (expansion 17×10^{-6}/deg.C) to ceramic (10×10^{-6}/deg.C), cermets with intermediate expansion have been used. But oxidation of the metallic particles in the cermet can cause strain and spalling and hence a two-layer coating system is preferred. But some type of monophase interlayer with the correct expansion and anti-strain features is desirable. With recent MCrAlY systems (M=Ni, Co), DBTT of 200°C and strain rates of 0.1%/second are achievable at low loads above 850°C. So at high temperatures such bond coats can deform to match the substrate and ceramic thermal barrier overcoat, so providing effective strain isolation for the latter. Oxidation, corrosion and ageing, in the bond coat may slowly increase its rigidity and so reduce its strain isolation potential.

Care must be taken to ensure that the values of coefficients of thermal expansion used are those for the phases actually present in the coating. Because of their lamellar microstructure, the thermal conductivity of D-gun and flame sprayed coatings is lower than that of solid, fully dense materials of the same composition. Their absorption characteristics may be very different as well because of their surface topology and in some cases slight shifts in composition (non-stoichiometry). One of the most common uses of coatings employing their thermal properties is for oxide thermal barriers. 500 cycles to 1000°C in air (9 minutes heating, 6 minutes cooling) cracks Ti-Si pack coatings on IN 738LC (Allen 1968). The thin trailing edge of a gas turbine blade heats and cools faster than the bulk and different expansion coefficients can cause cracking and hence corrosive attack and failure.

A successful coating in diesel engines for thermal shock resistance is MgO + ZrO_2 on a NiCrAlY bond coating (Kvernes 1983). MgO.ZrO_2 has good resistance to wear and flexural fatigue also. Ceramic thermal barrier coatings reduce substrate temperatures by 120 to 150°C. Control of porosity, segmentation, microcracking and residual stress, has given great increases in the number of cycles to failure compared with 1977 values (Spicer 1980).

Sol-gel and CVD silica and ceria coatings on 20Cr25Ni-Nb stainless steel (for gas cooled nuclear reactors) showed no effect on 825° to 625°C cycling for 860 days but some deterioration of the sol-gel silica coatings occurred after 9000 hours. The sol-gel ceria and CVD silica coatings were still effective after 15000 hours (Bennett 1983, 1981), no deterioration being caused by daily thermal cycling nor mechanical loads. No thermal fatigue or loading caused deterioration in 15K hours of CeO_2 (sol- gel coating) or SiO_2 (CVD), on 20Cr25NiNb stabilized stainless steel (Bennett & Price 1981).

CVD pyrolytic graphite is highly anisotropic and its thermal conductivity anisotropy is 300 (comparing radial and circumferential directions). Its accompanying thermal expansion anisotropy limits the ratio of thickness to bend radius in a curved section, beyond which delamination occurs on thermal cycling. Co-deposition with other materials to form 'alloys' modifies these effects and is an area for further research.

Self-healing plasma coatings are used in casting metal moulds which are subject to thermal cycling, doubling the life to 2000 castings and giving advantageous slower cooling rates (to avoid unwanted tempering). These coatings are also used on a 2-stroke diesel piston head, subject to thermal shock and sulphur corrosion from the oil. The piston showed no damage after 28000 hours and 230 million thermal shocks. Self-sealing coatings are also good on burners, combustion chambers and cracking reaction vessels, e.g. a cracking reactor of AISI-321 steel had a hot surface temperature of $2300^{\circ}C$ (ZrO_2 surface) but only $510^{\circ}C$ at the underside of the ZrO_2. Self-sealing coatings have an outer layer, e.g. ZrO_2, an intermediate layer, e.g. $Cr + ZrO_2$ whose porosity is reduced by its oxidation (which also causes mechanical and chemical bonding to the outer ZrO_2), and an inner layer, e.g. Cu-Ni. Recommendations are given (Perugini et al 1973).

The major factors affecting thermal cycling and high temperature resistance of plasma sprayed ceramic coatings are: thermal expansion mismatch between the coating and the alloy substrate, behaviour of the bond coating, interfacial phenomena during thermal treatment, phase transformations and composition changes (Herman & Shankar 1987).

Thermal fatigue degradation occurs due to phase transformations in FeAl coatings on Fe-Mn-Al steel (Guan et al 1981). Cr coatings on steel show spalling on high temperature thermal cycling, due to thermal expansion (Drewett 1969). Mo coatings on 304 stainless steel prepared by a CVD method have been subjected to two types of thermal cycling tests (Nicoll 1983). For uniform heating and cooling between 500 and $900^{\circ}C$, the coatings survived well up to 400 cycles whereas above $950^{\circ}C$ intergranular microcracking was observed and failure occurred as maximum cycle temperature was increased. Results of electron pulse heating tests indicated that the coatings were still intact until maximum surface temperature reached the melting point of the substrate. It is suggested that cracking in the coatings is very closely related to melting of the substrate (Yamaneka et al 1982).

7.4. THERMAL CONDUCTIVITY, THERMAL BARRIER COATINGS

A high thermal conductivity increases resistance to thermal shock, while a low value allows higher engine temperatures and adiabatic designs. Unspecified refractory coatings are claimed to greatly improve the emissivity of furnace walls, roofs and hearths, which gives fuel savings by re-radiation of heat back into the furnace (Anon. 1981). A very fine grain structure dec-reases the porosity of the walls which affects conduction and re-radiation. Coatings are applied by air spray gun. Refractory dusting and spalling is reduced. The emissivity of steel is modified by Al or Si coatings (Pikashov et al 1980).

Thermal and electrical conductivity of plasma sprayed and D-gun coatings deposited in an inert atmosphere is much higher than for conventional coatings, due to very little oxide formation during coating. The lamellar microstructure of D-gun coatings gives a lower thermal conduction than of the fully dense solid.

The thermal conductivity of Al_2O_3, Zr_2SiO_4 and Al_2O_3-MgO coatings on metals has been measured. Tensile stress in AlN coatings is due mainly to differential thermal contraction on cooling but the net stress is compressive (Zirinsky & Irene 1978) and increases with deposition temperature. Adding Si_3N_4 sharply reduces the net stress and puts it into the tensile region. High temperature radiative coatings of TiO_2 and ZrO_2 can be plasma sprayed (Goward 1970) onto refractory metals and resist thermal shock. Holding MgO-stabilized ZrO_2 above 1000°C allows MgO to precipitate out, causing ZrO_2 instability and its increasing thermal conductivity can rise to three times its previous value. This causes the hottest areas to degrade fastest and needs further research.

7.4.1. THERMAL BARRIER COATINGS (TBC)

Ceramic coatings were first tried in the late 1940s and 1950s, and in the 1960s rocket nozzles and gas turbine sheet metal combustor components were candidate substrates; much wider test performances on turbine section components were made in the 1970s and now they are in regular service high temperature components, in a variety of gas turbines and diesel engines (Miller 1987b). Thermal barrier coatings (TBC) as they are known today are compo-site coatings with a minimum 2-layer configuration of a metallic bond coat on the substrate with a ceramic top coat. The service performance, endurance and life time of TBC thus depend on the survivability of both the top- and bond-coats.

A coating graded in Al from 0% at the interface to 12% at the surface was produced by using a dual source of evaporation (Boone 1979). A dual target sputtering method has also been reported (Prater et al 1982; Patten et al 1979). A multi-source electron beam ion plating system designed specifically for graded coating production is another development for TBC (Nicholls & Hancock 1987). The equipment features EB sources, two rod-fed and four multi-hearth crucibles, which allow mixing of a coating in the vapour phase. Sputter cleaned components (using rf glow dis-charge) are given a strike layer of either the substrate alloy itself or a suitable underlay using one of the multihearth sour-ces and the rod sources take over to produce a graded interface.

Vacuum and power control allow more variations and rare earth element additions and other modifications are also reported to be possible.

Thermal barrier coatings fall into 2 types, thin (<0.05 cm) and thick (up to 0.625 cm). The thin barriers are used in gas turbine engines and diesel and petrol engine piston heads and valves. They are Ni-Cr, Ni-Al or MCrAlY with, e.g. ZrO_2 or $MgZrO_3$ as a thermal barrier on top. In earlier studies a continuous gradation from metal to oxide was considered for gaining on better thermal shock resistance. The metal part of the coating must resist oxidation (since the oxide layer is porous). This continuous gradation is more necessary with thick thermal barrier coatings. Thermal and mechanical properties of plasma and D-gun coatings are reviewed by Tucker (1981). As-sprayed coatings can have very low thermal conduction but on heating this can increase greatly due to sintering shrinkage. The performance and technology of TBC have been reviewed (Strangman 1985; Miller 1987a).

Cracks in a thermal barrier coating are stopped on reaching a more ductile NiCrAlY layer, for example, but this can lead to bad hot corrosion (Nicoll 1983). Yttria stabilized zirconia coatings have a high microcrack density next to their bond coat and were found inferior to MgO stabilized zirconia in thermal cycling. A martensitic transformation can occur and if the sprayed ceramic powder is a phase mixture the same nominal composition can give 2 orders of magnitude differences in rig test life. To get a single phase (homogeneous) powder for spraying, the sol-gel process is used and greatly increased performance results (Bennett et al 1984). Thermal cycling tests conducted on RB211-22B nozzle guide vanes have proved yttria stabilized zirconia coatings to be superior to the magnesia stabilized type (Bennett et al 1987). However, the latter appears to have had no bond coat as the former had. The results are, therefore, not conclusive.

Thermal barrier coatings have currently advanced to a point where they are in service in the turbine section of advance gas turbine engines. Fig.7-4a gives a schematic of the temperature profile across a metal substrate/bond coat/TBC (Bratton & Lau 1981). With an assumed hot spot gas temperature at $2280^{\circ}C$, an inlet pressure of 3.85 MPa, the coolant temperature and pressure at $538^{\circ}C$ and 40.4 MPa, the calculated surface temperature of an uncoated blade at the suction side is $1055^{\circ}C$. A ceramic layer 0.0127 cm thickness decreased the metal temperature by $189^{\circ}C$ and the thermal gradient across the ceramic was almost $400^{\circ}C$. A 50% reduction in the cooling air flow still held the gradient at $133^{\circ}C$ (Miller 1987b). The profile will be considerably modified when TBC configurations as envisaged on a multiple layer, Fig.7-4b (Prater & Courtright 1987) or different coating methods, e.g. PAPVD and EBPVD, to combine the advantages of both, with PAPVD improving the adhesion first followed by the more rapid EBPVD deposition (James et al 1987), are implemented. Whatever the coating method may be, one of the vital aspects is the surface state of substrate and the bond coat to ensure sound adhesion. The bondcoat has to be sufficiently rough to key the ceramic coat (Tucker et al 1976).

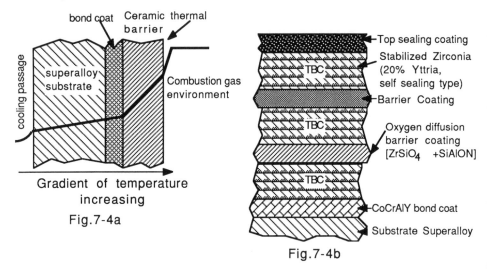

a: shows the temperature insulation offered by TBC;
b: shows the multiple sealant, diffusion barrier and
    bond coats alternated with ceramic layers with
controlled porosity to accommodate thermal shocks.

Schematic Single & Multi-Layer Thermal Barrier Coating
Fig.7-4a,b

Typical bondcoat compositions are shown in Table 7:3 (Dahman et al 1980):

Table 7:3

THERMAL BARRIER COATINGS – BONDCOAT & CERAMIC COAT COMPOSITIONS

Designation	Bondcoat Composition					Ceramic Composition		
	Co	Ni	Cr	Al	Y	ZrO_2	MgO	Y_2O_3
LTB 4	23	48	17	12	0.3	75	25	
LTB 5	22	48	20	9	0.5	75	25	
LTB 6	23	48	17	12	0.3	88		12
LTB 7	22	48	20	9	0.5	88		12
LTB 8	39	32	21	7.5	0.5	75	25	
LTB 12	39	32	21	7.5	0.5	93		6.5
LTB 13	39	32	21	7.5	0.5	93		6.5

Notes: 1. Bondcoat thickness is normally 180 microns (0.007")
and ceramic is normally 250-300 microns (0.01-0.012").
2. LTB 13 is processed to produce a controlled micro-crack density.

Application of TBC together with a bond coat is established, the survival of which is a major long-term stability factor for the coating unit as a whole. Many temperature-dependent factors have not been yet rationalized, e.g. bond coat DBTT, ceramic creep and ratchetting, sintering and phase transformations. Residence time at high temperatures and the frequency of temperature cycles, the coating thickness and quality are also to be made available for long-term utility and planning. TBC are known to tolerate very high heat fluxes as shown with a plasma torch test, which indicates that the thermal gradient prevalent in a gas turbine is not a major contributor to TBC failure. However, failure occurs through cracking of the TBC, regardless of the origin of the stresses. This area of improving bond coat properties and its compatibility to TBC is still unexplored. Ceramic microstructure and composition control, bondcoat property control, surface sealing and stress accommodation capacity, improvement in both oxidation and mechanical properties, and alternatives to plasma coatings and design aspects, are all to be looked at critically for a better optimization of TBC (Miller 1987b).

7.5. INTERDIFFUSION

Many surface treatment methods depend on diffusion of an alloying element into the surface of the material, e.g. surface hardening by carburizing, nitriding or boronizing of steel. C and N diffuse fast interstitially and nitriding can be done as low as $500^{\circ}C$ in ferritic steel. Diffusion of Al and Cr is slower and needs higher temperatures. 1% Al and 11% Cr keeps steel in its ferritic phase (fast diffusion phase) at all temperatures. Phase diagrams of refractory metals with Ni, Co and Cr show many intermetallic phases of low melting point and rapid interdiffusion prevents use of Ni and Co-base coatings on refractory metals. Interdiffusion between LPPS Ni23Co20Cr8.5Al0.6Y4Ta on two Ni-base superalloys CMSX-2 (5Co8Cr6Al6TalTilMo8W) and CoTaC 784 (10Co4Cr6.5Al4Mo4W-4Nb4Re0.4C) was found to be very limited at $850^{\circ}C$ but quite significant at $1100^{\circ}C$ but with different effects. In CMSX-2 W-rich gamma-prime phases, 40 microns thick, formed along with Ta-enrichment at the substrate-coating interface with outward diffusion of Ni,Cr and Co. But in Cotac 784, a gamma-prime depleted zone resulted above a coalesced gamma-prime zone which contained precipitates enriched with W, Re, Mo and Cr, and the diffusion reaction was found to be that of Ni diffusion outwards as in CMSX-2 but inward diffusion of Co and Cr (Veys & Mevrel 1987). The study indicates the relevance of interdiffusion data in specific coating/substrate systems at their operating conditions. Interdiffusion references have been listed, but more data are needed (Natl.Acad.Sci. 1970).

Coatings of Al, CrAl, SiAl and PtCrAl on IN738LC showed a needle phase extending into the IN738LC at $750^{\circ}C$ but NiCoCrAlY showed no such interdiffusion. This did not affect creep strength (Strang 1979). Two main processes occur with aluminide coatings:

(i) Outward Ni and inward Al diffusion, which increases coating thickness but decreases its outer Al content, (ii) Loss of Al due to alumina scale formation and spalling. These cause the beta-NiAl to become the less oxidation resistant gamma prime Ni_3Al and fingers of this phase finally penetrate through the coating causing rapid failure. As a basis for controlling the pack aluminizing process and the subsequent degradation of the coating, interdiffusivity measurements are needed. There is a strong dependence of D on composition in binary Fe-Al alloys. Good coating structure photographs and interdiffusivity values have been given. An equi-volume mixture of nitric, sulphuric and phosphoric acids is recommended as an etchant for the Fe-Al system (Akuezue & Whittle 1983). There is microstructural evidence of some interdiffusion of W, Mo and exothermically reacting NiAl coatings sprayed on steel, Ni or chromized Al (Coates et al 1968). The bond strength of ceramic coatings is generally attributed to interlocking, but some spinel formation is reported for Al_2O_3 on steel (Kofstad 1966) and Al_2O_3/TiO_2 on Al (Wagner 1965). But oxides on substrate surfaces or oxidation during plasma spraying, decrease bond strength (Perkins 1978). Diffusion bonding to fix WC tips onto machine tools has been studied (Almond et al 1983). Anomalous diffusion may accompany ion implantation due to vacancies and interstitialcies created by ionising radiation, which far exceed those produced thermally (Dearnaley 1973).

Interdiffusion in Si coatings on Ni produces several phases (Subrahmanyan 1982). In binary Fe-Al alloys interdiffusion shows strong dependence on composition (Akuezue & Whittle 1983). Interposing a diffusion barrier layer is unsuccessful. A high surface Al concentration is needed to form protective alumina films but this increases interdiffusion (Akuezue & Whittle 1983; Sivakumar & Rao 1982; Jackson & Rairden 1977; Glitz et al 1982). Pt acts not as a diffusion barrier but as a diffusion medium which allows Al to form a subsurface nickel aluminide (on Ni alloys) and a Pt_2Al_3 intermetallic outer layer which prevents outward Ni diffusion and thus allows unrestricted alumina formation (Wing & McGill 1981). W and Mo (solid solution strengtheners) are also prevented from escaping from the alloy; their escape would have an adverse hot corrosion effect. Pt_2Al_3 is preferred over $PtAl_2$ as its volume change is much less on forming PtAl (caused by alumina formation).

Inward Si diffusion, reacting with the interlayer, decreased the strength of $NiSi_2$, $CrSi_2$ and NiCrSi coatings on Fe, IN738, IN 100, Nimonic 90 and IN601 (Fitzer et al 1979). Considerable inward diffusion of a Ti-Si pack coating on IN 738LC occurs after only 500 cycles to $1000^{\circ}C$ in air. 1000 cycles to $1000^{\circ}C$ for CoCrAlY on IN738LC formed AlN and TiN under the coating (Nicoll 1983). Such brittle phases reduce alloy strength. The loss of Pt by its inward diffusion into the substrate can be estimated from $x^2 = 2 Dt$, where x^2 is the mean square unidirectional displacement of solute atoms in a solvent in time t. For a Pt layer on steel at $900^{\circ}C$ with $D = 10^{-10}$ cm^2/sec, t is some tens of hours for a 25 micron thick Pt layer and thousands of hours for a 250 micron Pt layer. Usually D is smaller the more refractory the substrate, and coating life on steel is much greater than on Mo, so Pt coatings on refractory metals last much longer. Loss of Pt also occurs at 900°C as volatile PtO_2 and 25 microns would last a few hundred hours and 250 microns for several thousand hours. For a 30Pt63Pd7Au alloy, the Pt life is 6 times longer (Kubaschewski 1971). Other reactions of Pt, with refractory oxides, are reviewed; metals can be brazed to refractory oxides using a Pt-containing brazing alloy (Exner 1982).

A 1 micron Al film reacts rapidly at 635°C with a thick Ti substrate forming $TiAl_3$, confining the Al to the surface region. At 900°C, Ti_3Al forms with little loss of Al into alpha Ti. But subsequent annealing at 900°C decomposes the Ti_3Al and releases Al into the alpha-Ti (Rao & Houska 1983). Considerable interdiffusion of Ti and Cr from a Nb-base alloy occurs into a Ni-base alloy at 1050°C in 30 minutes (Samsonov 1973), but Nb diffuses little. Ni diffuses considerably into the Nb alloy. Data for MSi_2 layer consumption, i.e. coating life, are given along with aluminide and beryllide coating life. No Al diffuses into W at 1950°C, making W a good diffusion barrier. N embrittles Cr alloys; aluminide coatings prevent this but substitute interdiffusion and a barrier is needed to immobilise the Al. C from Nb-alloy substrates diffuses into Ti-rich coatings and lowers the Nb alloy strength in hours at 1200°C (Natl.Acad.Sci.1970).

TiC coated molybdenum or graphite is considered as a promising material as diffusion barrier for application to fusion reactor first wall components. Thermal stabilities of TiC coatings on molybdenum and graphite prepared by ion-plating, magnetron sputtering, and chemical vapour deposition have been examined with particular reference to the interaction between the TiC layer and the substrate. A TiC coating was found to be markedly stabilized at elevated temperatures when it was deposited on the graphite substrate. This may be due to the diffusion behaviour of carbon from the graphite substrate. In contrast a TiC layer on Mo showed a large degree of weight loss above 1800°C on account of the Ti vaporization from the coating. This may be partly due to the formation of Mo_2C in the film-substrate interface. This tendency to lose weight at elevated temperatures was comparatively strong for the coatings prepared at low deposition temperatures by ion-plating and magnetron sputtering (Fukutomi et al 1982).

Sputtered yttria coatings 0.1 microns thick much reduce the reaction between metals and Si-base ceramics. Reaction bonded Si-SiC ceramics above 950°C react with Ni coatings by forming Ni-Si eutectics. Yttria coatings on SiC survive 26 thermal cycles between 1100°C to 25°C but yttria on Ni spalls after 4 cycles. But if yttria coated SiC and Ni were separately cycled and the spalled yttria cleaned off, then on assembly together and heating 100 hours at 1000°C the residual yttria gave considerable interdiffusion protection (although it could not be seen in the microscope). The barrier may be Y_2SiO_5 (Mehan et al 1983).

Bonding via interdiffusion is a developing technology. Bonding of Si_3N_4 to ZrO_2 without use of pressure but due to a solid state reaction and the presence of $CaSiO_3$ has been reported (Becher & Halen 1979). Reaction bonded ceramics have been known to give better performance than their hot pressed counterpart under certain hot corrosion conditions (Godfrey 1983). Pyrolytic deposition of Si_3N_4 avoids undesirable silicaceous second phase but gives either coarse columnar or 'amorphous' textures. A sputtered interlayer of yttria is necessary to prevent reaction of a Si_3N_4 coating with a superalloy substrate. A post-coating heat treatment is necessary to improve the Si_3N_4 properties, by causing outward grain boundary diffusion of undesirable grain boundary second phases to an external silica sink (Lang 1983). Oxidation at $1350^\circ C$ thereby increases the strength (Kreher & Pompe 1981; Evans & Heuer 1980). Additions of oxides such as CaO, SrO, Y_2O_3 and $MgAl_2O_4$ migrate to the grain boundaries, causing a loss of strength, forming silicates and causing weakness due to crystobalite formation. Less additives are possible if densifying is done by HIP. Further work is needed in this area.

Diffusion barrier layers are often required on carbon/carbon composites. Reviews on refractory coatings for graphite are available (Chown et al 1962; Criscione 1964; Schulz et al 1964; Samsonov & Epik 1966). Ir on graphite is reviewed (Criscione 1965a,1965b; Criscione 1967). Ir is almost impervious to oxygen diffusion up to $2100^\circ C$ and forms no carbide nor inter-diffusion. The expansion of Ir needs a graphite of high expansion. Coating methods include slurry and sintering, cladding, vapour deposition and electroplating (Natl.Acad.Sci. 1970). Two multiple layer systems have been described for providing compensations for thermal expansion and as barrier for oxygen and inter-diffusion:

(i) on the carbon/carbon substrate, CVD SiC is deposited which accommodates thermal expansion cracking, and this is top coated with CVD Si and B containing glassy coatings as a barrier to oxygen and carbon diffusion while also serving as crack sealers for the SiC undercoat;

(ii) The second type comprises of three layers, with a pyrolytic carbon coat on the carbon/carbon composite substrate, and CVD SiC is deposited on this, and in turn is coated with CVD Al_2O_3 (Stinton et al 1988).

7.6. ADHESION

7.6.1. GENERAL

The aim of any coating process is that the resultant workpiece functions as one component with superior properties for the environment for which it is designed. A coated material fails when the surface layer exposed to the environment separates completely or partially from the surface it is put upon. This can occur either due to an unsound contact between the substrate and the coating, or to an in-service breakdown arising from a surface environment reaction. This section is concerned with the first cause, and examines briefly, factors affecting bonding and adhesion.

Poor adhesion can arise at any of the three major stages in a coating process:

1. a. The substrate condition.
   b. The purity of the coating chemicals.
   c. The cleanliness of the equipment used.

2. The bonding between the substrate and the initial deposit; deposit growth.

3. The final compatibility of the substrate to the coating under the performing environment and temperature conditions.

Surface preparation of the substrate, its purity and that of the reactants, and their composition are the first-front influencing factors. Surface finish of the substrate is the last step prior to coating. All processes have this in common. Bulk coating processes, where the coating is itself a sheet, e.g. explosive cladding, need to surface treat two solid interfaces which interlock into each other.

Important parameters for adhesion and bonding are: substrate surface roughness, activation energies for surface and bulk diffusion and coating atom surface bonding energy. The last two properties are related and proportional to the m.p. and provide a basis for the Movchan and Demchischin structure zone model (fig. 7-1). A columnar structure (zone 1, low T/Tm ratio) is due to condensing a coating from a vapour arriving from one direction and with little subsequent surface diffusion. Inter-columnar voids become larger as the angle of vapour incidence increases. The three zones are well discussed by Thornton (1975, 1981). As T/Tm is increased, the columnar structure disappears but severe surface irregularities, e.g. debris can cause zone 1 (columnar) structures to persist into the Zone 2 and 3 regions (called columnar or linear defects). A smooth homogeneous substrate is essential. The use of ion bombardment (and its disadvantages) to suppress columnar zone 1 structures is also discussed by Thornton (1981). The adhesion of sputtered coatings is found to be better than those obtained by simple evaporation and condensation.

Coating adhesion depends on the bonding strength to the substrate and on the interface microstructure. The bonding can be chemical, Van der Waals and/or electrostatic . Chemical bonds are strongest (several eV), van der Waals arise from polarization (0.1 to 0.4eV) and electrostatic bonds are due to a double layer (0.1 to 0.4 eV). An 0.2eV bond should resist a high stress of $5 \times 10^8 N/m^2$ (72 kpsi) but adhesion failures often occur due to internal stresses in coatings exceeding this value and to microstructural flaws (Thornton 1981). Theories of adhesion are briefly discussed by Brewis (1982).

The thermodynamics of wetting of surfaces and its relevance to bonding is reviewed by Rance (1982). Three types of attractive forces operate at interfaces: primary bonding (covalent, electrostatic or metallic bonds, with binding energies of 40 to 400 kJ/mole), secondary bonding (Van der Waals forces, with energies of 4 to 8 kJ/mole) and hydrogen bond forces with energies of 8-35 kJ/mole. If intimate molecular contact between two materials were possible, then Van der Waals forces, even though they are the weakest of the bonding forces, would give adhesion far exceeding that observed (Tabor 1951). Attraction between two materials in molecular contact (about 0.5nm) is greater by a factor of about 10^4 than the attraction across a (typical) 10nm gap. 'Wetting' results in molecular contact. Metals have high surface energy, about 1 J/m^2 but all high energy surfaces readily adsorb contaminants (Rance 1982).

Wetting of steel is needed to obtain quality in coatings such as Zn and Al. Steels containing small amounts of strong oxide forming elements, e.g. Al, Si, have erratic adherence problems. Surface preparation methods may promote internal oxide formation. If the surface is hot pre-oxidised to form mainly a surface iron oxide which is then easily reduced before entering the liquid metal (Al or Zn), the surface is then readily wetted and a good coating results. If the surface is merely hot reduced before coating, adherence is poor because reducing conditions will favour alumina formation (Arnold et al 1977).

Interfaces are classed as: abrupt, compound-forming, e.g. intermetallics), diffusion (needs considerable solubility), pseudo-diffusion, e.g. co-deposition to form graded interfaces, or irregular, e.g. due to grit blasting or etching (Mattox 1978). Compound interfaces may be brittle and so should be kept thin. For diffusion, an intermediate layer soluble in substrate and coating can be used. Abrupt interfaces are liable to act as 'contamination' layers which weaken Van der Waals bonds. If the nucleation density is low, lateral growth gives interfacial voids with poor adhesion due to the reduced contact area and easy crack propagation. Sputter cleaning forms surface defects which enhance the nucleation density and minimises void formation; it also promotes diffusion via these surface defects. Sputter cleaning is necessary (but sometimes difficult) when using cylindrical magnetrons (Thornton 1981; 1978). The substrate is negatively biased and plasma ions then sputter material and contamination from the surface: 1 to 5 mA/cm^2, 100 to 1000 V, 5 to 20 minutes, removes 200 to 1000 A.U.; an rf bias is needed for insulator substrates. Irregular interfaces resist cracking since an interface crack would have to change direction or propagate through a stronger material (Mattox 1978). Plastic deformation due to the roughening operation is another factor that must be taken into account (Tucker 1974).

Oxygen-active metals often adhere well to glass and ceramics and no sputter cleaning is needed (Thornton 1981). Thus Ti, Cr, Nb, Ta and W are used as intermediate layers for depositing Cu, Ag and Au on glass or ceramic (Hollar et al 1970). The effect of oxides on bond strength is reviewed by Tucker (Tucker 1981). Sometimes an abrupt interface gives more adhesion than a co-deposited, graded interface, e.g. alumina on Hf (Vossen et al 1977). Usually adhesion is good at temperatures >0.3Tm if inter-diffusion is allowed by solubility (Natl.Acad.Sci. 1970). Hollow cathode ion plating of Ag onto stainless steel and Be gives excellent adhesion and a ductile fracture mode (Mah et al 1974). Fracture occurred in the Be, the Ag/Be interface being stronger. Adhesion of CVD coatings can be poor if the gases used attack the substrate (Bryant & Meier 1974). Adhesion test methods have been reviewed (Werber 1987; Tucker 1974; Davis & Whittaker 1967; Apps 1974).

The maximum thickness of a given coating is often limited by the increase in residual stress with coating thickness. Residual stress in a coating may change the apparent bond strength. Comp-ressive stresses induced during rf magnetron sputtering of Al_2O_3 film have been closely correlated to its adhesion, the magnitude of the stress depending on the film thickness, deposition pres-sure and sputtering power (Roth et al 1987). Arc deposited TiN on high speed steel substrates showed increase in adhesion by a factor of two when the substrate bias was changed from 0 to 100 V, and deposits were dense and very adherent when the substrate temperature was 200°C, and registered higher critical loads (>90N) when heated further up to 450°C (Erturk & Heuvel 1987). Highly ductile and adhesive Ti coatings have been obtained by vacuum plasma spray, with the process yielding deposits of low porosity. Substrate pre-treatment by sputtering and pre-heating resulted in very good adhesion on different substrates such as carbon and austenitic steels and Ti-alloys. Marginal variations were observed between as-deposited and annealed coatings (Lugscheider et al 1987).

A similar improvement in adhesion has been observed in TiN coat-ings by reactive magnetron sputtering on high speed steels. Adhesion increases with temperature reaching a maximum between 400-500°C, and is attributed to formation of FeO at the substrate interface, decomposing Fe_2O_3 and Fe_3O_4 as the substrate tempera-ture is raised. Ti as an interlayer increases adhesion but only up to substrate temperature 400°C. Above that TiC formation lowers adhesion. Carbide distribution in steel influences like-wise. Carbide as VC with lattice structure similar to that of FeO with only a slight mismatch, favours adhesion. In general, lower-ing of interfacial energy increased adhesion (Helmersson et al 1985).

Adhesion mechanisms have been reviewed as due to mechanical interlocking, alloying with the substrate, van der Waals forces, epitaxy, etc. (Tucker 1974; Van Vlack 1964; Malting & Steffens 1963). Other suggestions observed during corrosion studies such as oxide pegging (Choi & Stringer 1987) will not apply in this context unless a heat treated interlayer containing Y as a constituent is considered. For Ti as a thin film on oxide substrates, chemical interactions are needed in addition to Van der Waal's forces and electrostatic forces, as the metal has a high negative free energy of oxidation. However, the adhesion strength has been found to depend on the mechanical strength of the substrate as indicated by 'peel' data decreasing in the order - sapphire, MgO, quartz, and fused silica substrates (Kim et al 1987).

Most plasma deposited coatings need a roughened surface to get significant bonding (Grisaffe 1965). Plastic deformation due to the roughening operation may be more significant than the actual roughness, e.g. NiCr on steel (Epik et al 1966). Bond strength and recrystallization of a worked surface layer are correlated (Malting & Steffens 1963; Allsop et al 1961). For spray coatings where oxygen was present, spinels or 'oxide cementation' may be important (Epik et al 1966; Ingham & Shepard 1965). But oxide films on impinging particles would reduce adherence. Bond and mechanical strength achieved for Al and Ti are not possible for steel (McNamara & Ahearn 1987). Between CVD TiC and TiN on cemented carbides, the latter shows superior adhesion because it is less brittle; TiC shows interfacial fractures and considerable work hardening of the interface (Laugier 1986).

The ideal requirements of an adhesion test are that the test should yield results which are quantitative, applicable to a range of thicknesses and materials, economical with minimal machining, reproducible, sensitive, and suitable for routine use. A test based on indentation fracture is described and fracture mechanics analysis gives a quantitative measure of the interface toughness (Chiang et al 1980).

7.6.2. SUBSTRATE CLEANLINESS vs ADHESION & BONDING:

The extent and thoroughness of substrate cleaning are tailored to the initial state of the substrate and the needs of the coating process. Treatment given to a steel sheet with grease and mill scale is drastically different to that given to a wafer-thin or brittle or precision substrate, e.g. Si, or a gas turbine blade, a vapoform or an electronic component. There are several standard publications on the subject which should be consulted. Only a brief discussion is given here. Table 7:4 outlines broadly the type of surface and the various macro-cleaning/treating procedures adoptable. Cleaning by heating, sputtering, physical and chemical etching, vapour treatment and ultrasonics are methods adopted flexibly for substrates of all nature, shapes, and sizes (see section 7.8. for impact treatment of substrate).

TABLE 7:4

SUBSTRATE CLEANING OPTIONS

a. Pre-Coating Step

Surface Type	Treatment
Dust-covered	Suction
Solid dirt, loosely adherent	Brush (soft or wire); & suction
Solid dirt, firmly adherent	Abrasion or blasting with right size grit, or impact treatment with sand, metal shots, glass beads or ceramic granules.
Organic:grease, oil	Organic solvents wash
Inorganic:scales of known composition; viz rust	Abrasion; suitable acid and/or alkali cleaning with intermittent washing; Electro-cleaning, cathodic (or anodic)
Substrates which can be wetted	Washing with clean/or de-ionized water
Light scale	Acid/alkali etch/wash
Fine particle or debris	Ultrasonic cleaning
Final degreasing step	Vapour cleaning with organic solvents incorporating safety precautions; Solvents:Trichloroethylene, xylene, etc.,
Roughness a pre-requisite	Suitable peening.
Specific: metallurgically incompatible surface	Heat treatment to alter phase composition and/or microstructure or interlayer application to act as a diffusion barrier by suitable deposition techniques, viz. electrodepostion, spray, pack or vapour depositon.

b. Post-Coating Step

Electroplated finish	Soft-polishing to enhance reflectance or alter surface finish/light impact
Consolidation or extra step	Ion-implantation, Laser treatment; Sealing processes:- auxiliary sealing by CVD, PVD or amorphous material transformation - metallic or non-metallic.

It is necessary to remove millscale completely from steel before coating using hot H_2SO_4 or cold HCl causing H_2 evolution knocking it off mechanically while chemically etching the surface. Inhibitors (pickling restrainers) are added to reduce attack on the metal. Grit blasting is the most efficient mechanical method, using chilled iron or abrasive grit (but not shot, which embeds itself). Rust removal by phosphoric acid leaves a thin Fe_3PO_4 layer which gives some protection. Grit blasting is again the best mechanical method. A clean, grease-free surface will not show water droplets on it, when wetted. Guidelines for cleaning metals are available (Sykes 1982; Brit.Std. CP3012 1972; Moloney 1982).

Surface preparation can be a critical parameter for bond strength of sprayed aluminium coatings on grit blasted steel substrates. The type and condition of grit and the blasting angle influence bond strength markedly but blasting pressure or speed and nozzle-substrate distance has little effect. Blunted grit and low blasting angles reduce bond strength. The steel/sprayed Al bond is of a mechanical type ; alumina grit increases bond strength and moves the failure location from the interface to within the spray coat (Apps 1974).

An adhesive mechanism is proposed to account for the bonding of plasma sprayed Ni, Cr, Mo, Ta and W on aluminium and mild steel substrate. Part of the substrate gets molten giving rise to a boundary layer of an intermetallic compound and this bridges and links the substrate to the plasma coat (Kitahara & Hasui 1974). A similar adhesion mechanism is suggested for wire-explosion spraying which seems likely as both plasma and wire explosion can result in a high temperature small particle, high impact contact area of the substrate with the coating material (Suhara et al 1974).

7.7. BONDING

7.7.1. GENERAL

It is difficult to differentiate between bonding and adhesion except that the former implies substrate-coat linking on the micro-crystal, atomic, ionic scale while adhesion is a macro-term referring to a larger area-to-area bonding. Factors affecting bonding may be identified in the following areas:

1. Mechanical factors, e.g. slip and shear, in bulk coatings, and temperature effects, e.g. ductile- brittle-transformation-temperatures, and temperature cycle have to be noted. A Pt sandwich layer, for instance, is applied on a Ni-base alloy prior to coating stabilised ZrO_2 to operate above the DBTT of the composite.

2. Duplex coatings such as Cr-Al, Cr-Ti-Si, Cr-Al-Si are best given by applying the first single coat or alloy coat followed by

the top coat, e.g. siliconizing after Cr- Al, Cr-Ti or, Cr before Al.

3. Interlayers are mandatory where undesirable diffusion between substrate and coating has to be stopped. Interlayers will also compensate for Kirkendall void formation which if allowed to coalesce or concentrate, can lead to major non-bonded areas joints and junctions at the interface. Al is applied as an inter-layer to a Ti-SiC composite where the Ti substrate can form brittle silicides during service. (Rao & Houska 1983).

Solid state bonding of Al_2O_3 to austenitic stainless steel Type 316 has been achieved using a low pressure (3.1 MPa), low tempe-rature (1173-1473K) bonding of 0.5 micron thick Mo on the subs-trate steel followed by Ti. Ti/Mo/Ti interlayers have also been given (Hatakeyama et al 1986). Electrodeposited Ni forms the bond layer between Cu substrate and Pt (Ott & Raub 1986;1987). Ni-Cr-Al has been applied as the interlayer bond coat for 410 stainless steel prior to LPPS or air spray of 316 stainless steel (Eaton & Novak 1986).

The role of interfaces in sintering is reviewed (Anon.1982). Controlled precipitation of fine monoclinic needles of ZrO_2 in cubic ZrO_2 increase its toughness and adhesion (Kvernes 1983). A low Young's modulus decreases the chance of brittle fracture occurring during deformations. Ceramic coatings, e.g. ZrO_2 + MgO, have a lower modulus than sintered ceramic materials. Residual stresses cause coating failure. All melt coating processes result in significant residual stresses. Further work is needed, to investigate effects of residual stresses in spray coatings, to develop shear adhesion tests as coatings are more often used in shear than in tension, and on the effects of surface embrittle-ment and grit inclusions (Tucker 1981).

4. "Wetting" a powder-loaded vehicle is better achieved and more uniform when it is hydrophobic, when Al powder is mixed with it for steel aluminizing (Sugano et al 1968).

5. Aluminizing of Ni-base alloys is inherently linked with the Ni/Al diffusion phenomenon. A good documentation of the available literature prior to 1976 is given by Fontana and Staehle (1978). The relative advantages and disadvantages of Ni_3Al and NiAl phases must be noted.

6. Aluminizing of Fe-alloys is done in more than one way, viz. pack, CVD, slurry, cladding and hot dipping. When hot dipping or spraying of liquid metal is done, it is absolutely essential that the molten metal should wet the substrate. The presence of oxide layers generally impede adhesion. However Arnold et al (1977) demonstrate that an oxidizing preheater step followed by entry into a reduction furnace before the substrate enters the liquid metal, actually produces excellent adherence. The alloy should have more than the critical content of the alloying element needed to form a stable, external oxide layer. On Fe-alloys the

oxidizing preheater step induces a predominant iron oxide layer to develop in which the other oxides get dispersed. The reducing section of the process reduces the iron oxide and the surface is readily wetted by the liquid metal.

7. Vapour-phase and diffusion type coatings will not adequately coat faying surfaces at a spot-weld or rivetted joint. On such areas a mechanical fastener or closed fusion weld must be used. Fusion welding must be done before the vapour deposition (Gadd et al 1968).

8. CVD coatings exhibit poor adherence when the substrate has been permitted to be attacked by the highly corrosive reactants and/or product gas (Bryant & Meier 1974). For example:

Coating reaction
$$MF_{(g)} + 1/2\ H_{2(g)} = M_{(s)} + HF_{(g)}\ (1)$$

Metallic substrate reaction
$$Me_{(s)} + HF_{(g)} = MeF_{(s)} + 1/2\ H_{2(g)}\ (2)$$

If the free energy of reaction for (2) is less, i.e. more positive than that for (1), then substrate attack can be avoided. Substrates such as Mo-, W-, Ni-, and Ti with a high negative free energy for the fluoride reaction are likely to show poor deposit bonding in chloride-, fluoride-, carbonyl- and carbide-forming media. Factors which affect bonding and adherence in CVD processes are:

(i)   Free Energy of reaction,
(ii)   Reduced activity conditions whence the substrate undergoes alloying reactions,
(iii)   Unfavourable specimen geometry,
(iv)   Reaction in gas phase rather than on the substrate surface which results in a flaky deposit,
(v)   Formation of brittle intermetallics between substrate and coating,
(vi)   Substrate contamination prior to deposition, e.g. by the formation of a thin oxide film, and,
(vii)   Hydriding (or any other precipitate-forming reaction) of the substrate. Generation of hydrogen at the substrate interface is particularly detrimental, particularly in the case of metals which readily dissolve hydrogen. Dissolved atomic hydrogen tends to coalesce to molecular hydrogen and induce stresses resulting in cracking and disbonding due to embrittlement (Bryant & Meier 1974).

Bond coats used to key ceramic coatings show difference in adhesion according to whether the ceramic has been deposited by PVD or CVD (EB-PVD bond coats are usually 50 to 120 microns thick, with surface finish 0.2 to 1 micron). The initial ZrO_2 layer deposited on the alumina of the bond coat is dense and columnar and must be kept below about 2 microns thick to prevent stresses due to thermal expansion mismatch. The outer yttria-stabilized zirconia layer must have lower density for strain tolerance. Mechanical bonding between the ceramic top coat and the metal bond coat in a TBC is an accepted idea. These are experimental conclusions, but the area requires further investigation to clarify diverse bonding behaviour and quantify control parameters.

7.7.2. BOND STRENGTH

Tensile tests for assessing adhesion provide no fundamental data on bonding mechanism , but fracture mechanics methods with a controlled crack path do provide such data (Berndt & McPherson 1980; Ellsner & Pabst 1975; Pabst & Ellsner 1980; Becher & Newell 1977; Becher & Murday 1977; Bascom & Bitner 1977). The cohesive fracture toughness of plasma coatings is much less than that of the same bulk material, due to the lamellar structure. Bond coat interlayers succeed because their metallic lamellae allow plastic deformation; failure near a ceramic/metal interface occurs by brittle fracture between ceramic lamellae or between ceramic lamellae and substrate (Berndt & McPherson 1980).

Adhesion test methods are reviewed in Chapter 9 and reliable data may be generated (Stanton 1973; Bardal et al 1973). Modelling to quantify diffusion bonding of metals has been carried out (Guo & Ridley 1987). Three sub-processes of bonding are proposed: volume and interfacial diffusion coupled with creep, rigid collapse and surface diffusion. An alternative geometry is assumed for the shape of interfacial cavities. Effects of grain size and ratio of phases have also been considered. The time to achieve sound bonding has been predicted to find good agreement with data obtained on a Cu/Ti-6Al-4V experimental system.

Free energy data can sometimes predict adherence of CVD coatings, e.g. CVD reactions such as

$$2MoCl_5(g)+5H_2(g)=2Mo(s)+10HCl(g), \text{ for substrates}$$

with $\triangle G_f^O$ of the substrate's chloride $< \triangle G_f^O$ for HCl (both per g atom Cl), non-adherence is expected. This occurs in many systems (Bryant & Meier 1974) but the effect of alloying and gas pressures on activities for the $\triangle G^O$ calculation must be allowed for. Other factors must also be considered: specimen geometry, thermal expansion coefficient mismatch, surface contamination, and whether solid solutions form (allowing interdiffusion and hence good adherence).

Ion plated coatings have excellent adhesion, even if no alloying or diffusion is possible. Failure often occurs in the metal, not the interface. Bend, tensile, shear and scratch tests gave no failures, in contrast to vacuum deposited coatings. Adhesion is often due to interdiffusion, but ion plating energy (about 100 eV) is only enough to implant to a few atom diameters. Other mechanisms include atom mixing of sputtered substrate with depositing atoms and recoil implantation and ion-aided or radiation aided diffusion. It thus appears that diffusion of only a few atom diameters will give good adhesion. The mechanism is not clear but the adhesion is certainly due to the high energy of deposition (Teer 1983).

Ion bombardment during coating, e.g. sputtering, strongly affects adhesion, perhaps by increasing nucleation site density and so reducing voids (see fig.3-27, Chap.3) (Thornton 1981).

Most plasma coatings have bond strengths below 70 to 84 MPa, but most D-gun coatings and a few plasma coatings have strengths that exceed it. The mechanism of bonding of plasma deposited coatings in many respects is still in dispute (Tucker 1981,1974; Matting & Steffens 1963; Van Vlack 1964). Mechanical interlocking has been considered the most important mechanism by most investigators. It has been shown that bond strength increases with increasing surface roughness in both shear and tensile tests, although it may diminish again above 0.6 to 0.8 microns, rms (Grisaffe 1965; Marchandise 1965). Few published reports have taken into account the detrimental effects of surface embrittlement, peak blunting and grit inclusions when excessive grit blasting is used (Levinstein et al 1961; Leeds 1966).

Plasma sprayed coatings are formed by the impact (100 - 400m/s), deformation and solidification (in microseconds), of liquid droplets, forming overlapping lamellae. Adhesion of plasma sprayed ceramics to metals is poor but is improved by a sprayed bond coat interlayer, e.g. Mo or Ni-Al. The high bond strength between interlayer and steel is easily explained as due to the Ni-Al exothermic reaction and formation of an interdiffusion zone between coating and substrate, but the reasons why ceramic coats adhere so well to the bond coat interlayer needs an explanation (Berndt & McPherson 1980; Ingham & Shepard 1965).

Plasma coating adhesion is thickness dependent, due to internal stresses accumulating with increasing thickness until they exceed the bond strength. Internal stresses in fine powder sprayed coatings rise much faster than for coarse powders. For hard-facings, coarse powder coatings are limited to about 500 microns thick and fine to about 250 microns. This limitation is overcome by using sandwich coatings, e.g. Ni-Al or Mo wth the hard coat on top, or by electroplating Ni up to 1mm thick and spraying on top of it; Cr plate is not used as the plasma flame cracks it and because its high hardness prevents surface preparation and leads to poor adhesion. Sandwich coatings perform well under thermal cycling (Malik 1973).

Most of the factors that affect the bond strength of plasma deposited coatings also apply to D-gun coatings. Because of the unusually high velocity of the particles, they are actually driven into the surface of most metallic substrates. Some D-gun coatings are so well bonded that a crack starting in the coating may propagate into the substrate on cylic stressing. Some substrates require no grit blasting to achieve adequate bonding, since the coating process itself roughens the interface. This embedding/roughening process creates atomically clean interfaces between the coating and substrate over most of the coating area which facilitates chemical bonding and can be likened to the explosive bonding of sheets of metal. This undoubtedly plays a large role in forming the unusually high bond strengths of D-gun coatings. In some cases holding the substrate at constant temperature, e.g. 300°C during D-gun spraying can give improved adherence where a large thermal mismatch is present.

Heat treatment during coating may cause diffusion and improve adhesion by keying on the coating and by causing a gradual transition from substrate to coating. This may not apply to PVD and some CVD processes. Failure may occur within the coating (cohesive failure) or close to the substrate interface (adhesive failure). The adhesion of a coating is controlled by the deformation mechanism of the coating particles during fracture. Metal coating particles deform extensively and much energy is needed for crack growth and so coating adhesion, as measured by fracture toughness, is high compared with ceramic coatings which show brittle fracture.

Plasma sprayed 'self-sealing' composite coatings of a brazing material (Cu-Ni etc), as well as Cr and ZrO_2 are described; bend testing showed no cracks or coating detachment. Strength values are higher than for other types of powder plasma sprayed coatings and further research is needed on such self-sealing types of coatings (Perugini 1973). Arc-sprayed Ti, Nb, Zr and Mo coatings have very high adhesion (>8 kgf/mm^2) if produced in pure Ar, attributed to diffusion at the interface. For Ti on a structural steel the diffusion zone width was 20 microns, but insufficient time and temperature prevent the usual form of diffusion. Prior grit blasting gives a high defect concentration which may explain the anomalous diffusion. Adhesion and the diffusion zone width depends on the density, heat conductivity and specific heat (Muller 1973; Steffens & Muller 1973).

Adhesion of wire explosion coatings is excellent as in D-gun coatings, e.g. W on Al fails within the Al. W explosion cladding on mild steel caused the steel surface to sink several microns during the process due to deformation or melting which creates a 1 micron thick melted layer. The strong adhesion is shown to be due to the complete welding of both materials at this boundary, but the weld strength is less than that of a conventional weld due to voids etc. Further studies are needed on this problem. Adhesion tables of plasma sprayed coatings on ceramics and glass are available (Suhara et al 1973; Downer & Smyth 1973). W wire explosion sprayed coating on steel has 24 kg/mm^2 shearing stress of adhesion, which is the same as that of W itself, and the fracture propagates in the W coating. But W on Cu had only 10 kg/mm^2 (compared with 28 kg/mm^2 for solid Cu) and the interface fractured. W and Fe form a solid solution but W and Cu do not and the W/Fe interface thus has a 1.5 micron thick intermediate solid solution layer over 70% of its area which confers great strength (Suhara et al 1974). The adhesion of ion plated W and W-Re coatings is described (Schintlmeister et al 1982).

The microstructure of flame sprayed alumina consists of a series of overlapping lamellae produced by the splat cooling of impinging droplets. Porosities <10% can be achieved, with good adhesion (McPherson 1980). Metastable gamma alumina is formed by flame spraying, not the expected stable alpha form. Flame sprayed bond-coat interlayers of NiCrAl alloy are shown to improve the adherence of SiO_2-Al_2O_3 coatings on Ni-Cr and Co-Cr alloys (Kliminda et al 1980). Adhesion tests of CoCrAl, CoCrAlY and CoCrAlHf coatings have been reported (Whittle & Boone 1981).

Angular iron grit blasting gives good adhesion of arc and flame sprayed coatings. A surface roughness factor is defined to relate topography to adhesion (Bardal 1973). Some coatings plasma sprayed with transferred arc onto surfaces not grit blasted, show better adhesion than those sprayed without transferred arc. Their adhesion is beyond the testing range of DIN 50 (as available adhesives do not have strengths exceeding 70 N/mm^2). D-gun coatings have stronger adhesion (Tucker 1981).

The adhesion of flame-sprayed Al coatings on mild steel (irregular interface) is very dependent on surface preparation. Grit type and condition are important but variations of blasting speed and nozzle-substrate distance had no effect. With chilled iron angular grit, adhesive failure was at the interface, but with alumina grit it was within the Al coating. Other grit blasting variables are described (Apps 1974). Plasma air sprayed coatings of Ni, Cr, Mo, Ta and W on Al and steel were found to have melted to form intermetallic intermediate layers which improved adhesion (Kitahara & Asui 1974).

Sheets of Nb and Al_2O_3 pressed together above 1200°C, exhibited a bonding fracture resistance increasing linearly with temperature, varying from 0 at 1300°C to 4 MN/m$^{0.5}$ at 2000°C at a bonding pressure 5 N/mm^2 for 1 hour (Elssner et al 1978). Presence of some C, O and N (with Nb_2N formation) had no effect.

7.8. INTERNAL STRESSES, STRAIN & FATIGUE LIFE

7.8.1. GENERAL:

Internal stress is an important parameter in coating technology since it often relates to the maximum coating thickness which can be deposited without spalling, irrespective of the coating being produced by PVD, CVD or thermal spraying. It is known that even mild grit blasting will cause serious damage on ZrO_2 coatings. Semi-molten debris contracts on freezing and spalls off the coating. Thermal barrier coatings fail due to cracking. The excellent erosion resistant property of Si_3N_4 is profitably made use of in gas turbine technology (Godfrey 1983). However, more correlative studies are needed to explain how stresses generated and transmitted through coatings of various types and thicknesses may be controlled, and how coating production can form more uniform and stress-compensating structures. During a coating process, the reaction of the powder particles with their local environment in transit, particularly the extent of their oxidation, and, the stresses generated during coating and in the acting environment are very important in determining the properties of the coatings. Residual tensile stresses may be present in a substrate at a 1/6 stress gradient against a 2.75 gradient in a TBC upon it, with the coating in compression at the interface (Herman & Shankar 1987; Marynowski et al 1965).

7.8.2. FILM THICKNESS EFFECT & ION-IMPACT STRESSES

Internal stresses present in the substrate and coating have to be taken into account in order to work out a predictable model for the behaviour of a coated component. Internal stresses in thin films may be generated during its formation due to the difference in thermal expansion coefficients between the deposit and substrate materials. Atomic interaction, particularly at the nucleation stage at the substrate/film interface is another source, the effect of which attenuates as the film thickens. Another source is from reflected energetic neutral particles such as neutralized argon ions reflected from the target to produce an atomic peening effect on the bombarded film surface (and substrate). If such particles are in high flux then distorted structures, high defect concentrations and abnormal compressive stress levels can be generated. A working mode for magnetron sputtered alumina has been presented (Roth et al 1987).

Ion bombardment during deposition increases stress by displacing atoms from their lowest energy arrangement, especially for refractory materials. Sputter deposition using planar diodes allows contact of the plasma with the coating. Consequent ion bombardment occurs and electron (100 to 1000 eV) bombardment from the cathode heats and cleans the substrate and causes radiation damage and nucleation sites (Thornton 1981). The relatively high pressure in planar diode sputtering units unfortunately slows down the fast sputtered atoms to near thermal equilibrium with the gas by the time they reach the substrate (Westwood 1976). Triode sputterers can cause Ar trapping in coatings. Magnetron sputterers are generally free of bombardment effects (Thornton 1981). Few measurements exist on how hardness, yield strength etc., are affected by ion bombardment of films during deposition.

Evaporated and deposited films have a tensile stress and sputtered films have a compressive stress, often near yield point. But stress deduced from X-ray lattice parameters may be misleading as intergranular forces are excluded (Mattox 1981). Beam deformation measurements are more meaningful. The stress is also partly a thermal expansion coefficient mismatch and is mainly of this type for low m.p. materials. On reaching room temperature, mounds or depressions are created by compressive or tensile stresses (Thornton 1981). For sputtered films, higher pressure and temperature and additional sputtering from an oblique angle reduce the stress (Mattox 1981). But compressive stress may be beneficial by not allowing cracks to propagate. Thornton (1981) discusses the effects of working gas species and pressure, apparatus geometry and angle of incidence, on coating stress. Low pressure and angle promote smooth surfaces in compression; higher pressure and angles promote rough surfaces in tension. The compression to tension transition temperature increases with atomic mass ratio of coating to working gas. Effects of substrate bias on stress in sputtered coatings have been mentioned in Chapter 3. Residual stress increases linearly with coating thickness (Tucker 1974; Malik 1985). More studies are needed in which microstructure is related to mechanical properties.

7.8.3. STRESS MODELS:

A heat transfer model has been developed for plasma sprayed ZrO_2 and W coatings to predict effects of internal stress levels. Variation in residual stress distribution as a function of deposition rate and coating thickness was applied, but an extension is needed to include second-order factors such as microstructure, porosity and included oxides. A model is also formulated for sputter ion-coated W, where the dominant feature is the deposit microstructure influenced by the substrate bias voltage. Stresses are accommodated elastically at the substrate/deposit interface as the bias voltage increases and as the film thickens plastic flow occurs with increase in grain size (Rickerby et al 1987). Two new approaches are proposed to relate microstructure to surface stress and surface tension for solid vapour interfaces, considering them at thermal equilibrium. A finite crystal model with rigid planes illustrates that a modification of the bulk equation of state in a crystal slab of finite width can lead to such a difference, and a similar difference can be found between stress and potential density in a fluid in a periodic potential (Wolf et al 1985).

Amongst the in-service stress contributing factors, oxide inclusions play a prominent role. Oxidation results in a volume increase resulting in misfit strains which induce large hydrostatic stresses in the elastic medium around it. The effect of such hydrostatic stresses on diffusion kinetics, dislocation, fracture, pit formation, crack formation and propagation and finally coating failure is considered in another model (Louat & Sadananda 1987). Using oxidation and thermal expansion mismatch as parameters in another model based on an elastic environment has shown that ascribing total elasticity to a coating system, especially to TBC may lead to unrealisitic values, incompatible to observed data (Chang et al 1987; Miller 1987). The ceramic is far less elastic than the alloy bond coat which has different elasticity than the substrate. More work is clearly needed in this area.

7.8.4. STRESS RELIEF MEASURES:

Plaster (1983) has given a detailed account of impact treatment which is often employed for stress relief or compensation. Its permutations giving more than 21000 types of impact treatment, are possible from the following variables: (i) Abrasives (unlimited choice in solids liquids and vapours) (ii) Pressures (iii) Distance (iv) Angle of impact (v) Material quality and size (unlimited). The scope of impact treatment ranges from the destructive to the decorative. It can be used to blast or carve out a surface. It can produce anti-friction surfaces or surfaces of precise roughness, and it can even erase selectively, misprints which occur during bulk-printing operations to facilitate positioned overprinting of corrected matter. Impact generates stress, which can be detrimental to coating or may be turned into an advantage.

The advantages and beneficial effects of controlled shot peening as a pre-treatment to metallurgical coatings are discussed (Eckersley & Kleppe 1987). Shot peening is also known as impact prestressing, precisely because it induces compressive stresses which compensate stresses generated during the coating process or during service of a coated component. The impact of a high speed pellet creates a dimple of diameter 'd' with a 1/10th depression. The substrate surface stretches and the depth of stretching is about 'd'. The unstretched cone then exerts a compressive force in attempting to restore the surface to its previous state. A treated component may have an imparted compressive stress about 60% of its yield strength. It is argued that since most of the catastrophic failures originate from surface tensile stresses or residual stresses, a pre-stressed compressive state should be an effective compensator, and should be present to extend below all discontinuities caused by surface preparation procedures. The choice of shot for peening is directly related to the material to be treated. To generate a high residual compressive stress the material must be 'shot' with a medium as hard as itself. For high strength steel a hard shot is used; for non-ferrous metals glass, ceramic or stainless steel are used; thin sections are peened with light shots. The critical factors are that material distortion or deep damaging are avoided.

The several areas in a coating process where impact prestressing is beneficial are:

(i) Before grit blasting to offset stress risers created by angular grit. Trailing edges of jet engine blades tend to erode-corrode and/or distort if left untreated.

(ii) To offset negative side-effects of electrodeposited coatings of Cr, Ni etc. Plated Cr and Ni are brittle and hard and are not only susceptible to crack under load but can also reduce the substrate fatigue strength by almost 50%. Flame deposited coatings do not induce residual tensile stresses like plated Cr and Ni but are brittle and easily generate cracks which continue to propagate into the substrate. Impact prestressing of substrate can offset this condition. High speed steels, titanium and aluminium alloys are shot peened.

(iii) Shot peening is done on coatings when they are to be used under stress-generating conditions, e.g. aircraft landing gears, steam turbine blades etc. Plasma sprayed coatings are consolidated by shot peening, as are silver plated ball-bearings.

7.8.5. STRAIN & FATIGUE

Most metallic coatings have strain-to-failure of less than 1%. D-gun coatings have a higher modulus of rupture than comparable plasma coatings. Some coatings, particularly D-gun coatings are so well bonded that a crack generated in the coating may propagate into the substrate under cyclic stress. The magnitude of the residual stress due to rapid cooling is a function of torch (or gun) parameters, deposition rate, the relative torch to substrate surface speed, the thermal properties of both the coating and substrate and the amount of auxiliary cooling used. Coatings are normally under tension as a result of the residual stress, and this stress must be subtracted from the allowable fracture stress calculated from mechanical property tests of free standing specimens.

Strain and fatigue life results are discussed by Nicoll (1983). The effect of coatings on fatigue life varies (Bartocci 1967; Wells & Sullivan 1968). Some results are shown in Table 7:5. Studies show that if the strain-to-failure of a coating is not exceeded, then the coating will not affect the substrate fatigue strength, e.g. the roots of mid-span stiffeners on gas turbine compressor blades must not be coated, due to extreme fatigue sensitivity, and masking is used. Further research is needed to predict the effect of a specific coating on a given substrate (Tucker 1981). The longitudinal residual strain of 0.4 mm partially stabilized ZrO_2 on a 3 mm thick Al-alloy sheet with a 0.1 mm bond coat of Ni18Cr6Al, was markedly affected by the substrate temperature in a diesel engine combustion chamber; the higher the temperature the greater was the residual compressive strain. The temperature gradient across the coating was not significantly affected by surface cooling. The bond coat had no significant effect on the residual stress in the ceramic but it reduced the stress discontinuity at the substrate interface (Hobbs & Reiter 1988).

Grit blasting prior to coating affects fatigue strength. Coating thickness does not usually affect fatigue strength. Stainless

TABLE 7:5

STRAIN   FATIGUE   LIFE

	Substrate	Coating	Effect
1	Fe alloys	Al, CrAl,PtCrAl	Only very high strain fatigue life is reduced in air at $750^{\circ}C$; cycles to failure reduced 50% at $25^{\circ}C$. Falls as coating thickness increases.
2	Ni alloys	AlPt, Al, CrAl, SiAl CrAl/Pt, NiCoCrAlY	Same as uncoated ($900^{\circ}C$, 1500 cycles)
3	IN738	as in 2	as in 2
4	IN738LC	as in 1	as in 1
5	IN738LC	LDC2 (PtAl)	Allows double uncoated stress for a given life for precorroded specimens
6	Nb alloy sheet 20 mil thick	V-(80Cr20Ti)-Si	50% reduction at room temp.
7	Nb alloy XB88	(35W35Mo15Ti.15V)Si	Good at $1000^{\circ}C$
8	FX414	Al, CrAl, NiCoCrAlY	
9	?	Ni20Cr	Better than uncoated at $900^{\circ}C$
10	?	Ni20Cr10Al2Hf.1C	Slightly worse than uncoated substrate at medium stress but 100 times worse at high stress

steel coatings decrease the fatigue strength of low alloy steels and Al-alloys (Malik 1973). Al on stainless steel increases its hot fatigue strength (Drewett 1969); a sprayed Al coat reduces its creep resistance while a hot dipped coat increases it.

Both plasma and D-gun coatings consist of many layers of thin lenticlar particles but the latter have a higher density and modulus of rupture. The cooling rate may vary significantly with the substrate material and thickness of the coating. As a result of rapid cooling, some coatings have been found to have no crystallographic structure when examined by X-ray or neutron diffraction. Others may have a thin amorphous layer next to the substrate followed by crystalline layers. Many coatings form columnar grains within the splat in one or two layers perpendicular to the surface substrate. High local residual stresses due to the rapid quenching can occur and also non-equilibrium phases may be present. Both types of coatings have a strain-to-failure <1%. Non-metallic coatings are worse, e.g. a Cr_2O_3 coating on a hydraulic Al cylinder cracked due to the cylinder expanding under pressure. A substrate must be able to support a coating without yielding beyond the coating's strain-to-failure, e.g. a D- gun WC–Co coating gave ten times longer life to steel mill acid line roller guides, but spalling occasionally occurred due to the substrate yielding under impact from steel sheets. The problem was solved by raising the substate hardness to $55R_c$ (Tucker 1981).

CVD SiO_2 films are in compressive stress at $1000^{\circ}C$, decreasing with temperature. Silicon oxynitride films from 600 to 8000 A.U. thick are in tensile stress, independent of film thickness and deposition rate (Gaird & Hearn 1978), but parabolic with the nitrogen content. NiCrAlY+Pt coated superalloys are reported to have longer rupture lives and lower creep rates than the substrate material viz. Ni20Nb6Cr2.5Al (Strangman et al 1977). Nitrogen embrittlement occurs in hot gas atmospheres, of Al and Al–Cr coatings on nickel superalloys (Schmitt-Thomas et al 1981). Cracking occurs at the interface.

Hydrogen embrittlement of electroplated Ni and Co cermet coatings containing carbides of W, Ti or Cr, was measured by bending a specimen around a fulcrum at a controlled rate. The angle at which it breaks is compared with that for known specimens. Heat treatment at $200^{\circ}C$ (3hr) or $300^{\circ}C$ (1hr) gave full restoration of the fracture angle. Fatigue strength reduction at 1 million cycles is typically 60% for Cr electroplated steel and is only about 35% for Co+Cr_3C_2 cermet electroplate. The stress behaviour of W and W-Re ion plated coatings is described (Schintlmeister 1982). Internal stresses can be measured by a spiral contraction meter (Fry & Morris 1959). A value of 117 MN/m^2 tensile was found for optimum plating conditions which is better than for pure Ni electroplate and shows the stress reduction due to included particles (Kedward et al 1976). More work is needed on ways to stop crack propagation, e.g. by incorporating suitable spherical particles which will arrest cracks which reach them.

Stress levels induced by various surface hardening processes are shown in Fig.7-5. Bars A-G which includes CVD and conventional hardening treatments register lower stress levels than plasma nitriding and vacuum carburizing processes.

7.9. STRENGTH

Strength of hybrid type coated materials present a complex situation at high temperature. Ceramics are of particular interest in this context. They are notch sensitive and have poor impact resistance, have low fracture toughness, but are very resistant to high temperature erosion and corrosion. Duplex ceramic materials are a developing technology and much of their high temperature data as applied to coatings are obscure.

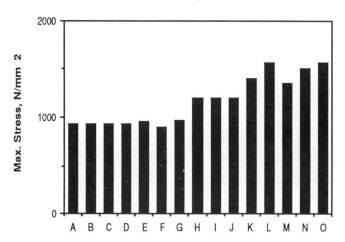

Surface Hardening Process Effects

Fig.7-5

Legend	Substrate steel		Surface Hardening
A	Cr-Mo, 722-M24}		
B	H13	}	VC Layer, TD process
C	M2	}	
D	As in A	}	
E	As in B	}	Hardened & Tempered
F	As in C	}	
G	M2		CVD
H	As in A	}	
I	As in B	}	Gas or salt-nitrided
J	As in C	}	
K	As in A	}	
L	As in B	}	Plasma nitrided
M	As in C		
N	As in B	}	Vacuum-
O	As in C	}	Carburisation

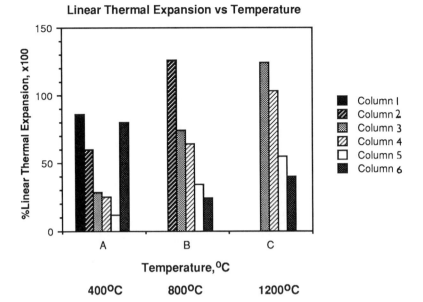

Fig.7-6: Thermal Expansion Characteristics of
Al-Alloys, Superalloys & Ceramics

Column	Material	Column	Material
1	Al-Alloys	4	Al_2O_3
2	Superalloys	5	SiC
3	Stabilized ZrO_2	6	Si_3N_4

Fig.7-6 to 7-8 compare ceramics, superalloys and carbon-carbon composites. (Richerson 1982; Buckley 1988; Hancock 1987). In general, strength at high temperatures can drop to almost half that at ambient conditions. Oxide ceramics, e.g. ZrO_2, HfO_2, Y_2O_3 and ThO_2 are protective carbon-carbon composite at >2000°C but are not strong, while the stronger SiC, Si_3N_4, and HfB_2 can be used around 1800°C, the latter with SiC for short times (Strife & Sheehan 1988). Table 7:6 indicates the effect of aluminide and silicide coatings on substrate strength.

Tensile and creep rupture testing data is relevant in the 0.1 to 1% extension range. Problems of extension measurement are overcome using television systems. PVD Cr by electron beam evaporation was used to produce 1 mm thick Cr films. A submicron columnar grain structure was produced over a wide range of deposition parameters. Brittle intergranular fractures occurred in tests at 25°C to 1000°C, related to intergranular defects reflected in density measurements and linearly dependent on the dissolved oxygen concentration (Anon.1983). Sputtered metal films show a sharp intrinsic stress reversal at low gas pressures, useful for diagnosis of coating quality (Hoffman 1982). Young's modulus

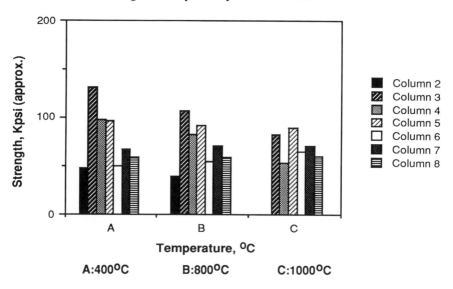

Fig.7-7: Comparison of Superalloy Srength with Ceramics

Column	Material	Column	Material
2	Cast X-40	5	Hot Pressed Si_3N_4
3	Cast IN 100	6	Reaction Bonded Si_3N_4
4	IN 713LC	7	Sintered SiC
		8	Reaction Sintered SiC

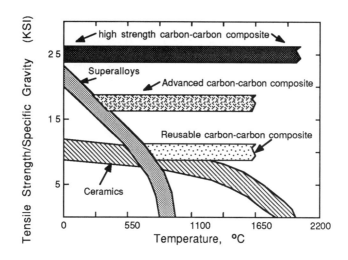

Fig.7-8: Comparative Data (approx.) on High Temperature Strength of Superalloys, Ceramics & Composites

(Buckley 1988)

measurements carried out on TiN, TiC, ZrN and HfN by plasma
reactive sputtering on type 316 stainless steel substrates showed
that porosity lowered the values; TiN was the best with 640 GPa
and HfN the lowest at 380 GPa (Torok et al 1987).

TABLE 7:6

EFFECT OF COATINGS ON TENSILE STRENGTH

Substrate	Coating	Effect
Steel	Al	slight decrease
Hastelloy-X	TiSi (CVD)	slight decrease
		(100-1000°C)
FX414	Al, CrAl	no effect
	NiCoCrAlY	
IN738	Al, CrAl, SiAl	no effect
IN738	CrAl/Pt	no effect
	NiCoCrAlY	
IN738LC	Al, CrAl, SiAl	slight decrease
	PtCrAl,	(20-900°C)
	NiCoCrAlY	
NiNbCrAl	NiCrAlY/Pt top	increase
(directional		
solidification)		
Nb alloys	CrTiSi	good to 1200°C
Nb alloys	Ti & TiV	decrease
Nb alloys	Nb15Ti10Ta10W2Hf3Al	decrease
Nb alloys	Nb5W30Hf5Ti3Re	decrease
Nb alloy sheet		decreased 25 %
20 mm thick	V-(80Cr20Ti)-Si	at 25 & 1200°C
Ta10W	Sn23Al5Mo	no effect
		5 hr at 1440°C

Coatings do not necessarily impair the substrate strength; CVD
Ti-Si coatings on Hastelloy X and various coatings on IN738LC
belong to this group. But if the cooling in the coating process
is too slow the impact strength is reduced, e.g. by allowing
carbide precipitation at grain boundaries (Nicoll 1983). The
strength of FSX414 (Co alloy) up to 1000°C is hardly affected by
pack Al, CrAl, and SiAl coatings, but coated IN738LC (Ni alloy)
was weaker at 700°C (Strang 1979). Creep rupture and high strain
fatigue properties of these superalloys were hardly changed by

coating. Metallography showed little interaction between coatings and FSX414 cobalt superalloy, but a needle precipitate (probably sigma phase) extended below the coating into the IN738LC substrate (except for NiCoCrAlY coatings).

CVD W from a WF_6+H_2 reaction mixture has higher yield strength if deposited at $500^{\circ}C$ (83 MPa) than at $700^{\circ}C$ (62 MPa), due to a finer grain structure. Further refinement of grain structure and higher strength is obtained by co-deposition of a material which causes re-nucleation, e.g. Re (Blocher 1974; Shim & Byrne 1972; Holman & Huegel 1967). The most effective grain refinement is caused by brushing the surface with a W brush during deposition (Holman & Huegel 1967,1974; Huegel & Holman 1971), which creates defects by local cold working and momentarily obliterates the gaseous boundary layer through which the reactants must diffuse (thus increasing the local supersaturation. Both effects aid re-nucleation. Similar grain refining occurs during CVD in a fluidized bed of W seed particles; these grow spherically (Blocher 1974) but hot isostatic pressing and rolling gives fine grained sheet (Oxley et al 1965a). Hence a Ta corrosion resistant coating thickness can be reduced by a factor of ten (Oxley et al 1965b).

Fluoride and chloride produced CVD W had equal tensile strengths at $1540^{\circ}C$ and $1700^{\circ}C$. Below this, chloride CVD W had lower yield strength but higher ultimate strength. The grain size of fluoride CVD W hardly changed during 5000 hours at $1700^{\circ}C$, but chloride CVD W showed poor grain growth resistance and an accompanying high DBTT (Bryant 1974). CVD pyrolytic graphite has a $25^{\circ}C$ flexural strength of 145 MPa which rises to 159 MPa for finer grained material (Blocher 1974).

Rupture strength reduction, e.g. in strong Ni-base alloys, is due to incompatibility between coating heat treatment and those needed to develop optimum mechanical properties in the metal. The coating heat treatment must be adjusted or post-coating heat treatments must be done. Of course, coatings can increase rupture life in corrosive conditions by preventing the reduction of substrate cross section which would occur with no coating.

Strang, Lang and Pichoir (1982) give a general review of aluminide coatings degradation effects on substrate creep and strength properties, ductility and fatigue. Coatings of $NiSi_2$, $CrSi_2$ and NiCrSi on Fe, IN100, IN738, Nim90 and IN601, show inward Si diffusion and interlayer reaction and causing strength loss. Specimens with Al in the interlayer had 50% strength loss while those with Cr had 10%. Heat treatment is beneficial (Fitzer et al 1979).

Fine grained fibrous Ni electroplate deposited at the high rate of 30 microns/minute with fast solution flow, using 150 A/dm^2, had a very high tensile strength of 130 kg/mm^2 (190 kpsi) (Safranek & Layer 1973). Fine grained elctrodeposited Al is 2.4 times stronger than the annealed, rolled metal. Strong electro-plated Cr is 6.7 times stronger than the annealed metallurgical counterpart (Safranek 1974). Strong Co electroplate is 4.7 times stronger than annealed wrought Co. This is due to fine grain sizes. Recrystallisation on heating reduces tensile strength to the normal values. Burnishing (rubbing during electrodeposition) improves mechanical properties (Stoltz et al 1974). The strength of electrodeposited Al and the effect of annealing has been studied (Addison et al 1972). Heat treatment caused severe embrittlement and porosity development due to gas evolution. H embrittlement and fatigue are discussed by Kedward and Wright (1978) for chromium carbide/Co cermet electrodeposits. A reduction in fatigue properties of fatigue sensitive substrates is usual when electroplated. Tensile stress also exists in platings, e.g. Fe, Ni, Co and their binary and ternary alloys.

ZrO_2 coatings on steel substrates have been tested for strength and fracture toughness using four-point bending and double cantilever beam techniques respectively. Strength values of 14.6 - 53 MPa and modulus 20 - 47 kPa were obtained for 0.375 cm thick ZrO_2-20%Y_2O_3 coatings on steel. Predicted values based on a model developed taking into account powder particle heating and its acceleration, as well as substrate residual stress and thermal shock agreed to within 72-88% of the measured data (Eaton & Novak 1986). ZrO_2 with 6 and 20 wt.% Y_2O_3 or 10 and 15 wt.% CeO_2 showed a greater toughness to cohesive fracture than adhesive fracture. For ZrO_2-CeO_2, the tetragonal structures coatings were tougher than fully stabilized coatings; for 10wt.% CeO_2 addition toughness varied from monoclinic > transformable tetragonal > nontransformable tetragonal; for 15 wt.% CeO_2 both tetragonal types > monoclinic > cubic; for 6 wt.% Y_2O_3 non-transformable tetragonal >> cubic > monoclinic; the highest strength level was registered for the 20 wt.% yttria ceramic (Heintze & McPherson 1988).

7.10. DUCTILITY, CREEP

The Ductile Brittle Transition Temperature (DBTT) is the lowest temperature at which a ductile 90° bend can be formed on a specified radius. If the as-coated alloy has 90° bend ductility at room temperature, bend tests done after oxidation exposure will quickly reveal ductility loss due to interdiffusion, entry of atmosphere contaminants and other effects not readily detected by visual inspection. Specifications for such bend tests are available (Natl.Acad.Sci. 1970) and are widely accepted. It is suggested that pile-up and work hardening theories include the inner grain structure in cases where alloys are hardened by a second phase. The grain boundary properties and structure are influenced by impurities or particle inclusions. Ultra-fine grain inclusions are found to provide ductility to high strength materials if surface preparation eliminates micro-cracks. In steady state creep calculations of complex alloys, introducing grain size influence (Hall-Petch stress) as one of the internal stress parameters provides rationalization in relating optimal grain size to maximising creep resistance. Controlling the grain boundary structure and grain size can lower crack growth rates, their initiation and propagation (Lasalmonie & Strudel 1986).

Sputtered coatings deposited at low substrate temperatures have many structural defects, while those deposited at high temperatures approach bulk properties. Simple metals with structures at the zone 1/zone 2 border (Fig.7-2; p.301) have high strength and hardness but little ductility. Dense zone 1 coatings have similar hardness but little lateral strength (Thornton 1981). In zones 2 and 3 the grain sizes increase with T/Tm and the strength and hardness fall towards bulk annealed values and ductility rises. In contrast ceramics deposited at low T/Tm have low hardness but approach bulk values at high T/Tm. Alloys have smaller grains and better phase dispersion than conventionally produced alloys (Westwood 1984), perhaps giving better corrosion resistance.

Electron-beam-evaporated Ni films vacuum deposited at < 0.3 times the melting point has uniform columnar half micron grains and is free of defects. It has the same hardness and yield strength as conventional Ni sheet and better creep ductility. Films condensed at 0-35 times the melting point had larger 2 to 3 micron grains. Hardness and yield strength were slightly lower but creep properties were erratic due to abnormal grain growth. Codeposition of alumina with Ni gave an equiaxed homogenous film with 0.3 micron grains, fully dense. Strengthening effects were observed with good hardness and hot creep resistance (Jacobson 1981). Alumina was on grain boundaries below 0.4 vol.%. $Ni+ZrO_2$ films were also studied.

Hardness, strength, ductility, creep and thermal conductivity can be controlled by microlaminar layers formed by alternate condensation from electron beam heated evaporation sources, e.g. Fe/Cu, Cr/Cu and Cu/Ni coatings from 0.1 to 300 microns. But Ni/Cu microlaminates are structurally unstable due to interdiffusion, forming a homogeneous solid solution. Microlaminates containing Cr can have twice the hardness and strength of pure Cr and also high ductility (pure Cr has no ductility). Superplasticity occurs for Fe/Cu microlaminates of 0.5 microns thickness, due to formation of islands of Fe in a Cu matrix. The steady state creep in Fe/Cu decreases sharply when the layer thickness equals the grain size (30 microns). Thermal conductivity is much reduced (below that of either component) for microlaminate layers, (see section on hardness) and these give possible thermal barrier coatings which have ductility, unlike ceramics (Movchan & Bunshah 1982).

Coatings can vary from brittle (MCrAlY, Al>12%) to relatively ductile (MCrAlY, Al<8%) and exhibit a range of DBTT. Reducing the Al and Cr contents reduces the hot corrosion resistance while increasing the ductility, and a compromise can be reached with overlay coatings (Nicoll 1983). Brittle coatings withstand cyclic engine conditions if the lower temperature is above the DBTT of the coating. The low-Al beta-NiAl phase can accommodate about 1%

341

tensile strain at 650°C with deviation from ductility occurring at 540°C; for high-Al beta-NiAl, the DBTT is about 815°C. The DBTT of MCrAlY overlays however, is from 350°C to 500°C and so these coatings can withstand a wider range of cyclic temperatures (Wing & McGill 1981). The 'strain to fracture' test gives the ductile-to-brittle transition temperature (DBTT) by observing the coated surface by telescope during a hot tensile test and measuring the strain at which the coating spalls or cracks. Plotting these values against temperature gives the DBTT; Fig.7-9 shows data on CVD SiB$_2$ coated on superalloy IN 738LC and Fig.7-10 compares the DBTT of overlay coatings and diffusion aluminide coatings. Some superalloy creep and fatigue test results are reviewed by Nicoll (1983).

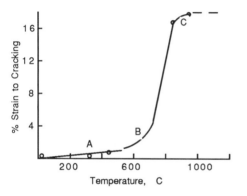

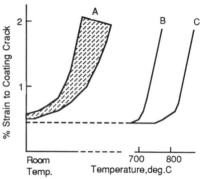

A:Elastic substrate, Coating delaminates; A: Ductile Overlay; B,C: Aluminide
B:Substrate begins to become plastic;   B-Low Al, Diffusion Coating
C:Substrate failure, Coating stays plastic C-High Al, As Coated

Fig.7-9: SiB$_2$Deposited by CVD on   Fig.7-10: Trend in DBTT of Ductile
IN 738LC Superalloy Effect of   Overlay & Aluminide Coatings
Temperature on Strain to cracking

A low DBTT is desirable (see Table 7:7). Adding Ti to an Si coating on IN738 increases the DBTT from between 450- 700°C to 850-1000°C. For Si-base coatings on IN-738LC, Nimonic 105 and IN-939, the DBTT rises with strain rate and with grain size. Presence of small particles lowers the DBTT. N embrittles Cr, H embrittles steels and Cu, and O embrittles Cu and W. It is desirable to reduce brittleness inducers such as Si and Al (Nicoll et al 1979). Decreasing the coating thickness can lower the DBTT. Cr has a DBTT just above room temperature, but N contamination unfortunately increases this to over 600°C. Coatings offer protection from this effect. Noble metal coatings have very low N solubility. Aluminides containing Fe and Ti and silicide coatings have also been evaluated (Natl.Acad.Sci.1970). Pt aluminides are less ductile than MCrAlY overlays and further development is

342

needed to improve this aspect (Wing & McGill 1981). The resistance to embrittlement during air oxidation of Cr-base alloys can be measured by the DBTT. Plasma sprayed Cr+10vol.% yttria was an effective coating. Other coatings studied were Li-doped yttria-stabilized chromia (Clark 1971).

Sol-gel ceria and CVD ceria and silica coatings on 20Cr25Ni- Nb stainless steel were oxidised with a load of half that calculated to rupture after 10000 hours at 825°C, in CO_2. The creep rate of the CVD silica coated steel was lowest and of sol-gel ceria was highest. This was consistent with theories of the oxidation-induced creep of the stainless steel; the ceria stopped oxide exfoliation and as the oxide is in compression it aided creep of the coated steel. Uncoated steel exfoliates its oxide, reducing its creep. CVD silica coated steel suffered almost no oxidation and hence had good creep properties (Bennett 1983).

Microhardness measurements show that arc-sprayed Ti coatings become hard and inductile if reactive gases are not excluded. But absence of oxygen forms droplets which do not easily wet the substrate, as oxygen (liquid oxide) reduces surface tension. So controlled amounts of O_2 are added. A thick-walled Cu tube arc spray coated with Nb can be reduced by drawing by about 20% without any coating damage or loss (Muller 1973). The high cooling rate of arc-sprayed Ti gives some metastable beta Ti in stable alpha Ti. Nitrogen causes a decrease in beta Ti, an increase in internal stress, solution expansion of the mixed crystal, a hardness increase and ductility and adhesion decrease (Steffens & Muller 1974). Spraying in pure Ar is essential. The ductility of D-gun and plasma coatings is limited. Even metallic coatings show a strain-to-fracture of <1% (Tucker 1974).

Structure and property relationships for electrodeposited coatings are discussed by Safranek (1974) and others (Sadak & Sauter 1974; Chen et al 1974; Stoltz et al 1974). As with sputtered coatings, fine grained coatings have much better strength etc., than columnar structures (which can be avoided by correct plating conditions). Proprietary brighteners give a banded or lamellar coating, also promoted by periodic current reversal. Cracking of highly stressed deposits often show a banded structure (Safranek 1974).

The effect of coatings on creep rate is shown in Table 7:8.

TABLE 7:7

DUCTILE-BRITTLE TRANSITION TEMPERATURE

Substrate	Coating	Effect
Ni alloys	NiAl, AlCr/Ta	DBTT is 700°C for low Al and 800°C for high Al
Cr5W.1Y	NiCr NiCrW	Ln & Y reduce DBTT. N increases it
Cr5W.1Y	Pt, Pd, Al	N increases DBTT
Cr5W.1Y	Cr_5Al_8 $(FeCr)_xAl_y$ NiCr, NiCrW	increases from 250° to 420°C
Cr5W.1Y	Ni30Cr, Ni30Cr20W same + 4 Al same + barrier layers of: W, Wl%ThO$_2$, W25Re, & W	After 100 hr at 1150°C the DBTT was 540°C. After 1260°C it was 870°C
IN 738LC	Si base coatings	25 to 1000°C Increases with strain rate Increases with grain size Decreases if small particles are present
IN 738	Si	DBTT is 500°C
IN 738	NiCrAlY/ZrO$_2$	DBTT is 700°C for NiCrAlY, no DBTT for ZrO$_2$
Nim 105	Si base coatings	25 to 1000°C Increases with strain rate Increases with grain size Decreases if small particles are present
Nim 105	Si	DBTT is 400°C
Superalloy	Co15Cr10AlY	DBTT is 250°C
Superalloy	Co22Cr12AlY	DBTT is 750°C

Note: A low DBTT is desirable, to prevent coating cracking on passing through this temperature.

The NiAl phase is brittle at 25^{C}C, and like all b.c.c. structures it has a brittle-ductile transition at around 720°C (or more, depending on Al content - NiAl has a wide stoichiometry range). Gas turbine temperatures are above this, but transients can cause cracking. More ductile coatings are needed. High Al "NiAl" coats are fine grained, precipitation strengthened and more resistant to thermal fatigue. After some use, outward diffusion of Ni leads to a pure NiAl phase with a weak columnar structure. Re prevents the ductile-brittle transition.

Table 7:8

EFFECT OF COATINGS ON CREEP RATE

	Substrate	Coating	Effect
1a	steel	CeO_2 (sol gel)	Reduces creep
1b	steel	SiO_2 (CVD)	Increases creep
1c	steel	hot dip Al	Reduces creep
1d	steel	sprayed Al	Increases creep
2	Ni superalloy	AlPt	Pack aluminising heating irreversibly increases creep
3	IN738LC	Al, CrAl, SiAl PtCrAl,NiCoCrAlY	No change at 750° & $850^{\circ}C$
4	FSX414	as in 3	as in 3
5a	Nb alloys	Nb15Ti10Ta10W2Hf3Al	Reduces creep
5b	Nb alloys	Nb5W30Hf5Ti3Re	Reduces Creep
5c	Nb alloys	CrTiSi	Good at $1200^{\circ}C$

Note: The effect of coatings on creep rate is generally small. The creep rate in SO_2 and in melts is decreased by a coating.

7.11. HARDNESS vs WEAR & EROSION RESISTANCE

7.11.1. BASIC ASPECTS

Hard materials may be categorized broadly into three classes based on the types of chemical bonding prevalent, viz. metallic, covalent and ionic. Transition metal borides, carbides and nitrides, e.g. TiC, VC, WC, TiB_2, TiN, are predominantly metallic in character; borides, carbides and nitrides of Al, Si and B are covalent; oxides of Al, Be, Ti and Zr are ionic. At a glance, all are in a broad group as ceramics, but a simple bonding model does not define these materials, nor is there a sharp division. The transition metal ceramics – borides, carbides and nitrides, considered to be some of the hardest and most chemically inert materials known, are ascribed with a complex bonding structure combining covalent, metallic and ionic bonding (Sundgren et al 1986b). All the compounds are distinctive with high hardness and high melting points, but they differ in other physical properties. The linear thermal expansion coefficient decreases in general from ionic to metallic to covalent hard materials. The first have the lowest modulus of elasticity. Tables 7:9 and 7:10 give a general trend in properties as applied to coated materials (Holleck 1986).

TABLE 7:9

TREND IN PHYSICO—CHEMICAL & PHYSICO—MECHANICAL PROPERTIES OF HARD COATING MATERIALS

Decreasing Tendency in Property	Trend in Chemical Bonding		
Melting Point	Metallic -> Covalent -> Ionic		
Stability	Ionic -> Metallic -> Covalent		
Thermal Expansion Coeff.	Ionic -> Metallic -> Covalent		
Hardness	Covalent -> Metallic -> Ionic		
Brittleness (Toughness)	Ionic -> Covalent -> Metallic		
Interaction tendency	Metallic -> Covalent -> Ionic		
Adherence to Metallic Substrates	Metallic -> Ionic -> Covalent		
Multi-Layer Compatibility	Metallic -> Ionic -> Covalent		

TABLE 7:10

TREND IN PHYSICO—CHEMICAL & PHYSICO—MECHANICAL PROPERTIES OF BORIDES, CARBIDES & NITRIDES

Decreasing Tendency in property	Trend in the Ceramics		
Melting Point	Carbides -> Borides -> Nitrides		
Stability	Nitrides -> Carbides -> Borides		
Thermal Expansion Coeff.	Nitrides -> Carbides -> Borides		
Hardness	Borides -> Carbides -> Nitrides		
Brittleness (Toughness)	Nitrides -> Carbides -> Borides		
Interaction Tendency	Borides -> Carbides -> Nitrides		
Adherence to Metallic Substrates	Borides -> Carbides -> Nitrides		

Factors which fix coated material properties are three-fold:

(i) Fabrication parameters - temperature of the substrate and the process, reactant-product interaction, and stresses created by thermal, ionic, particle and process impacts;
(ii) The Constitution of - the Substrate, Coating and the Substrate/Coating system,
(iii) Microstructure - grain size, grain orientation, grain boundaries; density (porosity)

Optimization of the properties of hard materials must then examine the fundamental relationships in the context of the high temperature regimes which are encountered in practice, on the following basis:

1. Bonding characteristics of the specific materials
2. Stoichiometry
3. Phase relationships and transformations
4. Anisotropy
5. Specific properties of miscibility - solid solution or intermetallics formation etc., within the coating candidate hard materials as well as the substrate.

Further criteria of hard material selection then depends on the interaction of its surface as a hard coating with the environment and the workpiece (if it is a tool finish), its own hardness, fatigue strength, fracture toughness and ability to accommodate stress, its adherence and interaction, if any, with the substrate and thermal expansion mismatch. Problems arise because good adherence at the substrate/layer interface cannot rule out surface interactions, and high hardness varies inversely with high toughness. Necessity for increasing hardness and strength means acceptance of decreasing toughness and adherence (Holleck 1986).

Tribological coatings have excited considerable interest in nuclear and other energy areas apart from the traditional machine tool industry because of their relative inertness to chemical reactions, lower thermal conductivity, resistance to erosion and ability to accommodate friction. The resistance they provide in nuclear applications to static adhesion, galling in vacuum environment, wear (wear scars can induce stresses), friction in the absence of conventional lubricants, static adhesion (of metallic surfaces in static contact) and chemical resistance has been reviewed (Lewis 1987). The general conclusion for hard coatings is that metallic hard materials appear to be very suitable and versatile except for their chemical response to environment, ionic hard materials are particularly suitable as surface finishes because of their high stability and low interaction tendency, but to get the best out of the coating/substrate system with an optimum wear resistance, only multiphase and multilayer coatings may be envisaged (Gurland 1988; Quinto 1988; Grunling et al 1987; Lewis 1987; Holleck 1986; Hillery 1986; Sundgren et al 1986; Sundgren & Hentzell 1986). Data processing for wear optimization has been enabled for more than 30 coating and surface treatment processes (Syan et al 1987; Kramer & Judd 1985).

7.11.2. HARDNESS & WEAR

Hardness of a material provides a measure of its ability to resist plastic deformation, and fracture toughness is a measure of its resistance to crack propagtion and fracture. Coatings for tool materials have to be dominantly hard in order to function as a tool without suffering undue deformation, but they also have to be sufficiently tough to withstand bending loads imposed by cutting forces, and to sustain shocks generated by interrupted cuts at high speeds and feed rates. With a compromise between hardness and fracture strength achieved, the service life is decided by the progressive wear and erosion. Hard coatings for other high temperature applications have to normalize their hardness and toughness with respect to the thickness required and the response of their thermal characteristics to environment and load.

The several forms of wear are (Gurland 1988; Hillery 1986; Child 1983):

(i) Adhesive wear (galling, scuffing, seizing) – which occurs particulary severely between mutually soluble pairs while in sliding contact, and often occurs in inert environments where even thin oxide films are absent to separate sliding components; and this could lead to micro-welding. As such welded areas formed at asperites (high points) shear and get plucked out during repeated sliding from the weaker (softer) surface, severely roughened surfaces result.

(ii) Abrasive wear which results from circular or linear sliding motions, normally with an acting load, of hard particles, cutting, impacting or ploughing into surfaces which are less hard;

(iii)Fretting wear characterized by low-amplitude loads and/or motion causing localized pitting (erosion); wear due to subsurface fatigue cracks is called delamination wear (Child 1983).

(iv) Corrosive wear which occurs as products from corrosion contribute to chemical and mechanical interaction to any of the three types given above, and,

(v) Diffusional wear which develops at high temperatures by selective constituent diffusion resulting in weakening areas and crater formation.

When the coating thickness is small compared with that of the substrate, its deformation will first follow that of the base material. At high temperatures interdiffusion, ageing and cyclic parameters and other factors interfere. Creep deformation is determined by the load and it is a volume effect, and fatigue causes locally concentrated damage (Grunling et al 1987). Under mild abrasive wear thin coatings fail due to localized detachment at the grooved intersections caused by the abrasive particles, and thicker coatings support contact stresses elastically and degrade by microchipping or a polishing mechanism. Severe abrasive wear causes thick coating failure by cohesive fracture. Under erosion conditions thick coatings survive angular particle impact for longer times while thin coatings perform well under blunt erodents. Internal stresses induce thin coat spalling when eroded or scratched (Rickerby & Burnett 1987).

Hardness is a good first approximation for adhesive and abrasive wear, for materials of the same type, but not for wrought materials (Tucker 1981). Earlier results of wear tests have been reviewed (Tucker 1974). Gregory (1980) reviews the wear resistance of materials and the improvement in wear resistance produced by surface welding. Hardness is related to wear resistance if comparable microstructures are considered. High wear resistance is not necessarily associated with low friction; soft materials may be good lubricants but do not wear well. A very hard smooth surface with small inclusions of a soft metal or a lamellar solid, e.g. MoS_2) is recommended, obtainable by ion implantation (Dearnaley 1973). Optimum coating thicknesses for best wear resistance are very thin, e.g. 2 microns for MoS_2 , due to the need to conduct heat away, which is difficult for MoS_2 in a resin binder coating. Areas of wear are usually limited and ion implantation can be applied to small regions of a bearing or shaft (without dimension change) to keep costs down. Ba ion-implanted into Ti-1Al-4V gives oxidation and fretting fatigue resistance; 40% Ba was still present after 10 million fretting cycles (Mattox 1981). When the matrix of a hardfacing alloy is Fe, Ni, or Co, the alloy will typically contain C, Cr, W and Mo. Higher amounts of these increase carbide formation and wear resistance. Good wear resistance is obtained from plasma sprayed alumina or chromia + silica.

Laboratory experiments have shown that implantation of a wide range of materials with gaseous ions, especially nitrogen, can cause a reduction in mild abrasive wear by factors typically 2-10, provided operating temperatures remain below 400°C. Work at A.E.R.E. Harwell, has focussed on analysing the mechanism of improved wear resistance as well as the production of prototype implantation equipment.

The latest development at Harwell (Gardner 1987) is the large "Blue Tank" ion implantation machine, currently the biggest in the world. This can treat workpieces up to 2 metres maximum dimension and 1016 kg weight using a bucket-type ion source capable of generating 35mA of nitrogen beam current over an 800mm diameter treatment area. This machine enables increased flexibility and reduced unit treatment costs for nitrogen ion implantation and an accessory monitors the nitrogen dose to an accuracy of around 20%. Examples of large items treated in the "Blue Tank" include racing car crankshafts, printing cylinders, crushing rollers, confectionery processing rollers and large plastic processing moulds.

A series of papers from Harwell deal with the various aspects of ion implantation and its role in improving adhesive and abrasive wear, resistance. Dearnaley (1985b) discusses the various categories of wear and shows that because wear rates under abrasive condition are very sensitive to the ratio of the hardness of the surface to that of the abrasive particles, large increases in working life are attainable as a result of ion implantation. Under adhesive wear conditions, the wear rate appears to fall inversely as the hardness increases and, it is advantageous to implant species which will create and retain a hard surface oxide or other continuous film in order to reduce metal-metal contact. In a review report from Harwell UK, Goode (1985) compares various surface treatments for metallurgical applications including modified ion implantation techniques. The possible mechanisms by which ion implantation can be used to improve wear and to strengthen metals and other materials have also been reviewed (Dearnaley 1986; 1987). These include,

(i) The pinning of dislocations by the segregation of interstitial species such as N, C or B for instance in steel (Dearnaley et al 1976),

(ii) The introduction of obstacles to dislocation movement consisting of fine dispersion of hard second phase precipitates, e.g. nitrides (Hutchings 1985),

(iii) Reduction in the coefficient of friction as a means of lessening subsurface stresses during sliding (Hubler et al 1985),

(iv) Modification of the work-hardening mechanism that takes place during wear so as to render the material more self-protective (Dearnaley et al 1985),

(v) Implantation of one or more species such that the surface is rendered amorphous, and so generally possesses a low coefficient of friction which may be combined with the presence of a high content of strongly-bound contituents (eg. TiC) (Follsdaedt et al 1983), and,

(vi) Ion bombardment in the presence of sufficient oxygen to cause its takeup, eg as an oxy-nitride in reactive metals, such surfaces having improved lubricity and freedom from galling (Oliver et al 1984).

The effect of temperature reached during nitrogen ion implantation of tungsten carbide and steel has been studied (Dearnaley et al 1985b). Implantation below $200^{\circ}C$ softens tungsten carbide, but at higher implantation temperatures significant hardening can be achieved.

Thermochemical and surface heat treatments are also used to combat wear, e.g. carburizing, nitriding, nitrocarburizing, boronizing, chromizing, CVD and PVD (thermochemical) and induction, laser and electron beam hardening (surface heat treatments). A hard 'case' of 5 to 10 microns is enough except for heavy loading (gears, bearings) if the core material is soft. Surface treatments or coatings allow the advantages of a composite: cheap but tough core and a hard surface (Child 1983). Earlier reviews on wear resistance are: (Lang 1963; Hurricks 1972; Donovan & Sanders 1972; Gerdeman & Hecht 1972). Wear resistance test methods are reviewed in Chapter 9.

Sections 7.11.3 to 7.11.8 consider hard materials studied and developed variously in the context of wear and erosion applications and production routes. TiN dominates to a large extent; again, a representative picture may be had, and it is out of scope here to cover the entire literature output.

7.11.3. HARDNESS vs APPLICATION

Before deciding which coating to apply for wear resistance it is important to assess the reasons for the wear. Wear may involve mechanical separation of small particles, corrosion, thermal shock or mechanical stress. If due to the scouring action of hard particles, a hardfacing alloy may be suitable. If due to oxidation, Co-C-Fe alloys etc., may be suitable; Co-base alloys are known for their low temperature and high temperature (1000°C) wear resistance (Coutsouradis et al 1987). An intentional oxide dispersion deposited under conditions that do not significantly oxidize the metal matrix is far superior to a coating heavily oxidized during deposition, both in wear resistance and mechanical properties.

Smoothest coatings do not always give the lowest friction, e.g. a nodular brush finish gives lowest friction in liquid Na (Tucker 1981). Self-welding of mostly static components can be prevented by coating, e.g. load pads on fuel ducts of a Na-cooled nuclear reactor - stainless steel, stripped of its oxide by the liquid Na, is self-welding but this is prevented by a D-gun chromium carbide-nichrome coating.

7.11.4. HIGH TEMPERATURE HARDNESS & HARDFACING

Requirements for hot wear resistance are (Child 1983):
- high hot hardness (resistance to thermal softening),
- structural stability (temper resistance),
- retention of adherent oxide films to act as lubricants,
- high thermal fatigue resistance, and,
- corrosion resistance.

The improved wear resistance of hard carbide layers is not at the expense of lower fatigue strength. For best fatigue properties, plasma nitriding or vacuum carburizing is needed. Oxide and fluoride films reduce high temperature friction, oxide films having temperature independent friction coefficients (Child 1983; Bhushan 1980).

Comprehensive reviews on tool wear are listed: Gurland 1988; Quinto 1988; Wolfe et al 1986; Sundgren & Hentzell 1986; Knotek et al 1986; Dearnaley & Trent 1982. Properties of types of hard-surfacing alloys are shown in Fig.6-19 (p.275). Hardness achieved by various processes is compared in Fig.7-11. Fig.7-12 shows the life increase of pins for Al die casting, due to Nb coating.

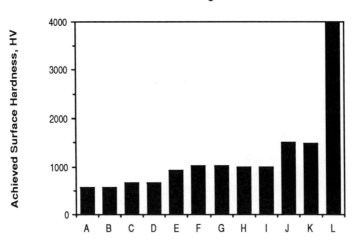

Surface Hardening Process Effects

Fig.7-11

Legend	Type of Steel	Surface hardening
A	Mild steel	Ferritic nitro carburising
B	0.4%C steel	Induction & flame hardening
C	0.5%C steel	As in B
D	Cr-Mo-steel }	
E	Cr-Mo-V-steel }	Nitriding
F	Cr-Mo-Al-steel }	
G	High alloy tool steel } }	
H	Most steels	Carburising
I	Carbon steel	Carbo-nitriding
J	Mild & Tool steels	Boronising
K	Carbon & Tool steels	Chromising
L	Tool Steel	Toyota Diffusion Process

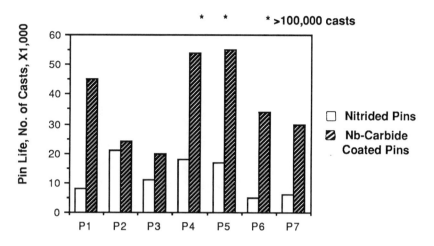

Increased Life of Coated Core Pins

Life increase of core pins for aluminium die-casting,
due to niobium carbide coating.

Fig.7-12

Friction and wear of alumina coatings with various pins is
discussed in pin-on-disc wear tests (Kohno & Kuwahara 1982).

7.11.4.1. Hardened Surfaces:

Chromised carbon steel is hard and brittle with good wear and
abrasion resistance and low friction coefficient. Chromised low
carbon steel is ductile (Drewett 1969). The main problem of
boriding steel by diffusion saturation is the brittle surface
layer produced and the method is little used. Combined diffusion
coatings based on B, Si, Al etc., with multiple saturation give a
high hardness, wear resistance, bonding and corrosion resistance
(Samsonov 1973). Addition of Ta and Nb increases the m.p. of an
Al coating which reduces interdiffusion and hot erosion. A nitrogen plasma of 10000°C for a few minutes surface hardens steel by
nitrogen and 10 microns below the surface the hardness was 5
times more than the untreated steel. Ti and Ta can be similarly
hardened (Lee & Ariyasu 1979). TiN is about 5 times harder than
hardened steel and has low friction against steel. X-ray diffraction and electron microscopy were used to investigate the hardening behaviour of ion-nitrided SUS304 austenitic stainless steel
as opposed to the usual ferritic nitriding steels (Urao et al
1982). Hardsurfacing but non-uniform wear was registered for high
dose ion implantation of carbon or boron on Ti-6Al-V alloys
(Bolster et al 1987).

Ni-Cr-B-Si hardfacing alloys are used for wear resistant coatings, e.g. in producing glass bottles (Knotek & Lugscheider
1974). They can be applied as a self-fluxing powder by welding or
by fusing in a furnace (slightly oxidising, or vacuum). Volume
diffusion of the B into the steel occurs, giving increased hardness as for boronized steel. The growth mechanism of Fe_2B, FeB,
and FeB_{1+x} by powder boride reaction with Fe indicated that
reactions proceeded mainly at discrete points of low boron activity with an inward diffusion controlled process resulting in
continuous layer growth. The outer layers are thus structurally
disordered and mechanically weak (Palombarini & Carbucicchio
1987). Plasma spraying can also be used for boriding; the consequent phase transformations are discussed (Pilous 1981). Laser
assisted boronizing of austenitic steel surfaces gives a higher
compaction of CVD coats and improved bonding (Steffens 1983). At
least 15 microns of FeB + FeB_2 surface layers on steel gives wear
resistance (Child 1983). For thicker layers (to reduce erosion),
monophase FeB_2 should be used, to avoid cracking. Boronized
surfaces wear 100 times less than untreated steel. Amorphous
borides $Ni_{20-21}M_{2-3}B_6$ where M is Ti, V, Nb or Ta have shown good
wear and corrosion resistance. They were plasma sprayed atomized
powders (Lugscheider et al 1987). Low wear and friction resistant
properties of ion beam deposited tribological BN (Miyoshi et al
1985) and CVD composites of $TiC-TiB_2$ (Hollech et al 1985) are
worth noting. Moulds and dies given CVD boriding (Ni-B) have
hardness of HV850 kg/mm^2 and wear better (Mullendore & Pope
1987).

Cemented carbide inserts, both coated and uncoated with TiC, TiN
or alumina, were used to cut EN8 and EN24 steels. Rake-face TiC
coatings wore by atomic diffusion and plastic deformation while
TiN coatings and uncoated WC-Co wore by atomic diffusion. Alumina
rake coatings wore by plastic deformation. Flank wear of coated
and uncoated tools was by atomic diffusion. Alumina and TiN
coatings resisted groove formation better than TiC. Grooves formed by oxidation and by fatigue cracking, both of which were
prevented by cutting in jets of Ar or N_2. Cutting temperatures
attained by the tool edge are about $1125°C$, for EN8 steel cut at
183 m/minute. Turbine components, e.g. the midspan shroud of the
blade is carbide coated. Fan and compressor blade dovetails and
disc slots are also carbide coated for fretting wear (Hillery
1986). Erosion-corrosion resistant coatings of TiN and Ti_5Si_3 or
$TiAl_3$ on Ti owe their hardness to the nitride, and their oxidation resistance to the silicide or aluminide. Fe-Cr-C hardfacing
alloys for high temperature use have been produced by a manual
metal arc welding technique. The hardface alloy consisted of
large primary M_7C_3 carbides in a eutectic mixture of austenite
and more M_7C_3. Hardfacing alloys usually depend upon 3 layers to
avoid dilution of the top layer by the base material, but thick
deposits lead to weld cracking. Diffusion barrier layers are
warranted for this purpose (Svensson et al 1986).

A baseline coating composition of a nickel alloy bonded Cr_3C_2
formulated with Ag, BaF_2 and CaF_2 was designed for wear resis-

tance at 900°C for aerospace and advanced heat engine applications (Sliney 1986). Very hard WC coatings with interlayers of Ti, Ta, W, Mo and Si between WC and stainless steel have been evaluated (Srivastava et al 1986).

Nitrides, particularly TiN and its combinations are one of the most widely investigated materials groups, both as applied and surface transformed coatings (Dearnaley 1987). Ion nitrided Fe-Ti alloys (1.07, 2.08, 2.58 wt.% Ti) increased in hardness with increase in Ti. Gamma Fe_4N was the surface layer with TiN as small dispersed precipitates (Takada et al 1986). Plasma nitrided Type 316 stainless steel also show Fe-rich surface layer with sub-surface Cr-rich nitrides. Peak hardness occurred at a 10 micron depth as ageing at 700°C increased the solubility of Fe_4N (Sundararaman et al 1987). Ion plated 3 micron TiN on cemented carbide showed tempering effects inducing lattice contraction (Perry et al 1988). A new material - Tribocor 532N is formed by surface nitriding a Nb-30Ti-20W wt.% alloy produced by conventional melting methods. TiN forms preferentially, but as the Ti level decreases Nb enters the reaction resulting in a fine structure of complex nitrides in a Nb-W-rich matrix. Nitriding at 1350-1900°C is parabolic with Ti showing the greatest mobility and affinity to nitrogen. Tribocor is developed for high temperature wear and corrosion resistance with greater hardness than WC-4.5%Co (Ziegler & Rausch 1986).

Frictional force and thermal shock caused the H 13 steel tool failure of ion-plated nitrides (Wiiala et al 1987). Ion beam deposited BN on non-metallic substrates for wear resistance showed good affinity to SiO_2 but not to GaAs or InP (Miyoshi et al 1987). ZrN deposited by cathodic arc plasma process outperforms TiN by a factor of two in cutting Ti-alloys and is marginally superior on other conventional steel substrates (Johnson & Randhwa 1987). The same ranking has been shown with (Ti,Al)N as an intermediate for ion plated Zr-, Ti-, Ti-Al- and Ti-Al-V-nitrides, with an interesting feature of substrate polishing prior to coating yielding equivalent wear resistance to doubling deposit thickness from 3-5 microns (Molarius et al 1987). High rate reactive sputtered nitride (and carbide) deposits have shown diverse ranking in wear resistance to the two above and also on different substrate steels. On 4340 steel the ranking order was TiN, TiC, ZrC, HfC, ZrN and HfN, and on 1045 steel it was TiN, HfC, ZrN, HfN, TiC and ZrC (Sproul 1987). A comparison of the overall defect levels of nitrides produced by all these methods could elucidate the reasons for divergence of wear results.

7.11.5. RUBBING SEALS

Compressor rub coatings, labyrinth seals and turbine gas path seals are yet to be developed to long-term acceptance and service levels. The incursion of a vane or blade tip into the seal can result in wear of either or both the components. The blade/shroud clearance problems have been addressed by the application of abrasive tips to the blades. 'Borazon', a cubic boron nitride has been useful but it has poor thermal stability and does not last long. MCrAlY alloys, which have been successful as overlay coatings have also served well as gas path seals. TBC such as ZrO_2 and its variations have ben tested for ceramic gas path seals, and in diesel engines (Hillery 1986; Levy & Macadam 1987; Keribar & Morel 1987). Above $250^{\circ}C$ or in vacuum, fluid lubricants are unsuitable. Solid lubricants are used, usually as thin coatings containing a binder, but adherence is often poor and failure occurs by peeling or by accumulation of soft wear debris. Ion implantation has resolved this problem to some extent by providing surface additives without the form of a coating which could peel off, and with no change in dimensions nor any surface roughening (for coating keying) required. Mo and S implantion gave a considerable decrease in friction. Pb, and Ag are good such lubricants in vacuum, but not in air due to oxidation (Dearnaley et al 1973).

Rubbing seals for ceramic gas turbine engine regenerators must provide a low leak seal between exhaust and inlet air while continuously rubbing the abrasive regenerator, in a high temperature corrosive atmosphere subject to repeated thermal shock. Low friction is needed to minimise rotation energy losses. Requirements are <25 microns per 100 hours for both matrix and coating and a friction coefficient of <0.35. Conventional solid lubricants (graphite, MoS_2) oxidise below $500^{\circ}C$. Plasma sprayed NiO-CaF_2 on Inconel 601 were studied at $760^{\circ}C$ (Moore & Ritter 1974). Ideally, this type of composite coating with a hard (NiO) matrix containing a small amount of soft (CaF_2) material, will have friction no higher than that of the soft material and wear no higher than the hard material.

Rubbing surfaces for liquid Na cooled nuclear reactors must resist liquid metal corrosion and irradiation. Chromium carbide in a 15 vol.% nichrome binder and Tribaloy 700 gave good results when applied as a D-gun coating, the only disadvantage being that it requires a line of sight with impact angles $>45^{\circ}$ (Lewis 1987; Johnson et al 1974). A composite powder of Ni5Cr3Al coating on bentonite core at a coating to core ratio 4:1 has been applied by thermal spray for high temperature abradable seals (Clegg & Mehta 1988).

Boron nitride deposited on 440-C bearing steel substrates by ion-beam extraction of borazine plasma was friction tested against various transition metals, viz. Ti, Zr, V, Fe, Ni, Pd, Re and Rh. The 2 micron thick BN included some oxide and carbide picked up during the tests, and oxygen adsorption was mainly responsible for increasing friction. The higher the affinity to oxygen the higher the friction (Miyoshi et al 1985). Oxides have been found effective in reducing sliding wear, but neither friction nor wear is eliminated completely (Glascott et al 1985). TiN sputter deposited and ion plated on bearing steels provided improved resistance to rolling fatigue for steels but gave varied performance on softer copper-alloy substrates (Hochman et al 1985). Ion nitrided steel surfaces register lower coefficients of friction but are columnar in structure (Spalvins 1985).

7.11.6. HARDNESS & EROSION RESISTANCE

Erosion may be considered as deformation wear of brittle materials and cutting wear of ductile materials. The maximum erosion wear of brittle coatings occurs at perpendicular impact angles and of ductile coatings it is maximised at oblique angles of 15-30° (Raask 1988; Hillery 1986). Large particle ingestion, typically at 90° affects the leading edge of a turbine airfoil. Further downstream in the compressor, particles tend to be smaller and are centrifuged towards the rotor periphery. Here the erosion affects the airfoil above midchord as well as the trailing edge at low impact angles. Erosion resistant coatings are important for gas path seals and for nuclear reactor applications (Hillery 1986; Nicoll 1983; Schwarz 1980). Erosion is combatted by hard coatings and its effect varies with particle size, velocity and angle of incidence, the last factor being the most crucial. It is thus obvious that no single coating can resist both low and high angle erosion impacts. Chromium carbide- and tungsten carbide-based coatings are being used in power recovery turbines and gas turbine compressor sections. A high temperature erosion testing apparatus built for their evaluation is described by Sue & Tucker (1987).

Erosion and corrosion occur together in coal gasifiers (Wright 1987; Meadowcroft 1987) and have been studied at 870°C with alumina and magnesia particles at 180 m/s in a burner rig (Barkalow & Petit 1979). Alumina-forming alloys resist erosion oxidation and hot corrosion better than chromia-formers. Hot corrosion-erosion is much worse. A comparison of the wear and erosion resistance of coatings is given in Figs.7-13a,b and 7-14a,b,c. Erosion-oxidation and erosion-hot-corrosion histograms are given for comparison in Chapter 8. Table 7:11 (Yee 1978) details the wear resistance of coatings.

T A B L E 7 : 11

ABRASIVE WEAR RESULTS OF SOME COATED AND UNCOATED MATERIALS[*]

Test material	Treatment or coating[**]	Thickness (μm m)	Hardness (HV)	Relative Volumetric Wear
Steel				
Ck15N	Annealed normally		180	165
100Cr6H	Hardened, tempered at 150°C		840	72
100Cr6V3	Hardened, tempered at 500°C		300	111
X5CrNi189	Annealed		275	124
34CrNiMo6	Quenched and tempered		330	108
X220CrMoV12H	Hardened, tempered at 150°C		810	48
Electroplating				
Ck15EH	Hard chromium	50	1050	23
CuZn40	Bright nickel	50	580	97
Ck15G	Dull nickel	50	560	106
Anodizing				
AlMgSi	'Ematel'	20	450	63
Bath Nitriding				
Ck15	'Tenifer'	8–10	640	73
Spray Coating				
NiCrBSi	Plasma, remelted		700–800	33
Mo	Flame sprayed		800	79
Cr_2O_3	Plasma		700–900	159
Al_2O_3	Plasma		850–950	127
WC–Co	Plasma		800–1000	106
Cr_3C_2–NiCr	Plasma		500	50
Chemical Vapour Deposition				
105WCr6	TiC(1050°C)	12	3200	3.5
C100	W_2C(550°C)	30	1900	4.8
100Cr6	Vanadized	30	2400	7.5
C100	Vanadized	30	2400	9
C100	Borided	100	1600	12
100Cr6	Borided	100	1600	14
Ck15	Borided	100	1600	30

===

*　From ref.159 of Hara et al (1978).

**　Vanadizing by pack cementation at 1100°C, 4 h; boriding by pack cementation at 900°C, 3 h.

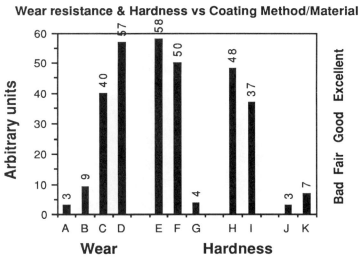

Fig. 7-13a

Legend	Coating Method	Material Or Coating
A	---	Hard Chrome
B	---	Wrought Steel
C	Plasma	WC-Co
D	Detonation Gun	WC-Co
E	----- " ------	91WC-9Co
F	----- " ------	85WC-15Co
G	Plasma	88WC-12Co
H	Detonation Gun	Al_2O_3
I	Plasma	Al_2O_3
J	Plasma	Ni
K	Plasma	58Co25Cr10Ni7W

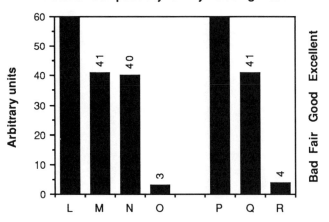

Fig. 7-13b

Legend	Coating Method/Material	vs	Coating Method/Material
L	Detonation Gun/WC-Co		GA Mechanite
M	----- " -----/WC-Co		440C Stainless Steel
N	----- " -----/WC-Co		Detonation Gun/WC-Co
O	----- " -----/WC-Co		Inconel-X
P	----- " -----/Al_2O_3		Haynes-25
Q	----- " -----/Al_2O_3		Hastelloy-C
R	----- " -----/Al_2O_3		Detonation Gun/Al2O3

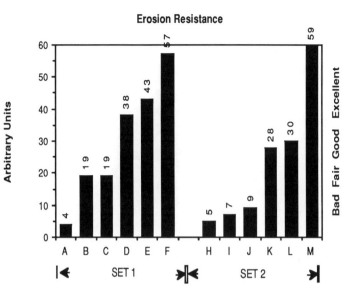

Fig.7-14a

Legend	Coating	Legend	Coating
A	CoCrAlZr	H	IN738 } Chromia-
B	CoCrAlTi	I	X-40 } formers
C	CoCrAlSi	J	MA754 }
D	CoCrAlHf		
E	CoCrAlY	K	IN738 }Alumina-
			(Al Coating) }formers
F	CoCrAl	L	CoCrAlY }
		M	Si_3N_4

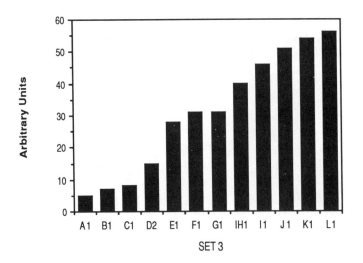

Erosion Resistance

SET 3

Fig.7-14b

Legend	Substrate	Coating
A1	TRW1900	--
B1	SM200	--
C1	SM211	--
D1	IN100	--
E1	IN717	--
F1	IN713LC	--
G1	PDRL162	--
H1	B1900	--
I1	PDRL162	Al
J1	IN100	Al
K1	B1900	Al
L1	IN713LC	Al

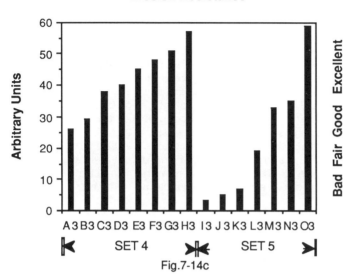

Erosion Resistance

Fig.7-14c

Legend	Substrate	Coating
A3	Haynes C9	none
B3	Hastelloy X	none
C3	Hastelloy X	Al
D3	Hastelloy X	Haynes C3
E3	Hastelloy X	CrAl
F3	Hastelloy X	Al-Cr-Si
G3	HastelloyX	Al-rich Al-Fe
H3	Hastelloy X	Al
I3	HA 188	none
J3	IN738	none
K3	X 40	none
L3	MA 754	none
M3	IN 738	NiAl
N3	IN 738	CoCrAlY
O3	Si_3N_4	none

7.11.7. HARD COATINGS - ROUTE RECORD

7.11.7.1. PVD Coatings:

The microhardness and tensile properties of Ni, Ti and W vapour condensed coatings were studied by Movchan and Demchischin and Bunshah. Zone 1 structures show high hardness, decreasing rapidly to zone 3 which has the hardness of the annealed metal. Increasing deposition temperature (on going from zone 1 to zone 3) gives larger grain size, low strength, higher ductility and lower hardness. Both hardness and yield strength varied as the inverse square root of the grain diameter (Hall-Petch relationship) (Bunshah 1974). Variation of microhardness with temperature for Al_2O_3, ZrO_2 and TiC, is quite different from metals. It falls from zone 1 to zone 2 but rises in zone 3. The high zone 3 hardness of TiC ranks it second to diamond (Raghuram & Bunshah 1972).

PVD wear resistant TiC, TiN and Ti(C,N) coatings onto low temperature substrates usually gives poor adherence and film morphology. A low N_2 pressure improves adherence, as does a thin Ti film (Pepper 1978; Brainard & Wheeler 1979; Matthews & Teer 1980), and vaporization in $C_2H_2 + N_2$ (Enomoto et al 1982). The hardness of BP coatings on Mo has been measured (Morojima et al1979) and decreased as the B:P ratio increased from 1 to 1.5. Very hard, wear resistant ceramic coatings are obtainable by activated reactive evaporation (ARE), reactive sputtering and ion plating. TiN coatings by reactive ion plating are used on twist drills, cutting tools, dies and moulds (Teer 1983). ZrN is used on cutting tools (Kirner 1973). Low temperature, <400°C, deposition of tribological 20% yttria-zirconia, TiB_2 and B-18 wt.%Si produced on steel, Cu- and Al-alloys showed good wear resistance. Excellent thermal and wear resistance were shown by zirconia coatings on copper and Al-alloys (Prater 1986). Cathodic magnetron sputtering of Mo alloyed with carbon or nitrogen, on 304L steel were found superior to hard chrome in corrosive, cyclically stressed and abrasive wear conditions (Danroc et al 1987).

Modified arc evaporated (Union Carbide) TiN coatings were found more erosion resistant than those produced by other PVD and CVD methods. A wide range of hardness values has been reported for TiN from 340-3000 HV which is felt to be primarily due to the crystallographic orientation and presence of a second phase such as Ti_2N (Sue & Troue 1987). Magnetron sputtered Cr-C and Cr-N coatings show much greater wear resistance than Cr (Gillet et al 1983). Solid solution hardening of (Hf,Ti)N was found when deposited by a high rate reactive sputtering process using a dual-cathode configuration (Fenske et al 1987). Amorphous alumina coatings on grey cast iron produced by rf magnetron sputtering have measured 300HV0.05 hardness and a tensile strength of 300 N/m^2 (Roth et al 1987).

Ion implantation of interstitial species like N and C has major effects on surface-related mechanical properties (Herman 1982;

Hale et al 1982; Zhong-yin et al 1982; Delves 1982; Roberts & Page 1982). The lattice distortion is analogous to the effect of shot blasting the surface and increases hardness. The relation between wear and oxidation in ion-implanted metals has been discussed in detail (Dearnaley 1982); ion implanted species on oxide grain boundaries or dislocations block fast diffusion paths. Ion implantation has succeeded where oxidation and stresses coexist. Ion implantation by inert gases acting on a sputtered TiB_2 film on stainless steel improves the erosion resistance to He^+ and D^+, which is of importance to the first wall of fusion reactors (Padmanabhan & Sorensen 1982).

Several workers report that PVD coatings give superior performance to their CVD counterparts. Claims such as this have to be treated with caution since a number of process and coating/substrate aspects have to be taken into account: Magnetron sputtering vs CVD - Konig (1987); Steered Arc Evaporation vs CVD - Boelens & Veltrop (1987); Plasma assisted activated reactive evaporation vs CVD - Bunshah & Deshpandey (1986). A state of art report of all CVD and PVD is made by Sundgren & Hentzell (1986). The implication of PVD via different deposition processes on tool wear models has been discussed (Kramer & Judd 1985; Kramer 1986). Results reported by Quinto et al (1987) would be of interest in this context. At room temperature, PVD microhardness values were significantly higher than the CVD products, but both registered the same values at $1000^{\circ}C$. TiN by PVD, IP, MS and AES; TiAlN, HfN and ZrN by PVD and MS and all by CVD were studied. TEM studies confirmed that the high residual compression growth stresses of all PVD coatings were associated with lattice distortion and fine grain size. Thermal expansion mismatch between coating and substrate was responsible to the low residual stresses in high temperature CVD coatings. Thin coatings, about 5 microns thick have also been examined (Quinto 1988). Bulk material structure and property relationships can be extrapolated to some extent for both PVD and CVD coatings, but substrate/deposit features will be unique to coatings only.

7.11.7.2. CVD Coatings:

CVD Ti, Cr and Al carbides and TiN coatings are used to increase wear resistance in many machining operations. TiC, VC, NbC and Cr_7C_3 are used on tools, valves, in die casting, metal cutters and in magnox reactors to reduce sliding drag (Child 1983). CVD carbides produce a continuous chip when cutting steel tools and segmented chips when cutting pearlitic cast iron. Tools coated with $TiN-Al_2O_3-TiC$ stay 200 degrees (C) cooler (Dearnaley et al 1986). CVD TiC on ball bearings shows very good wear resistance and has high hardness up to 35000 N/mm^2 (Boving & Hintermann 1987).

Codeposition of SiC with CVD pyrolytic C gives increased wear resistance, applicable to rocket nozzles (Blocher 1974). Hot hardness of 2150 kg/mm^2 has been measured for CVD SiC on graphite

365

(Morojima et al 1986). Siliciding of grey cast iron by CVD (H_2+$SiCl_4$) at 960° to $990^\circ C$ produced an adherent and wear resistant coating (Samsonov 1973). Below 960° the adherence was poor and above $980^\circ C$ the coating was thin on parts distant from the gas entry. A 30 micron CVD film of Ta is used on subterrene graphite heat receptors to resist hot rock temperatures of 1700 to 2000K (Stark et al 1974). Very high (10 to 20 GN/m^2) hardness at $1500^\circ C$ is reported for CVD Si_3N_4 (Niihara & Hirai 1982).

7.11.7.3. Electrochemical & Chemical Coatings:

Cr plating is the most extensive electrodeposited coating for wear resistance, usually 20 to 500 microns thick. The effect of changing the deposition conditions and heat treatment for 1 h at 400, 600 and $800^\circ C$ of the conventional deposit on abrasive wear has been assessed. Conventional deposits produced at $55^\circ C$, 4000 A/m^2 had hardness of HV 9.0 GN/m^2, and had a surface crack network. Crack-free deposits produced at $72^\circ C$, 1600 A/m^2 showed a hardness of 5.5 GN/m^2. The conventional deposit in the as-plated condition was a factor of several superior to those heat treated or crack-free (Gawne & Despres 1985).

Failure of a wear resistant WC-9%Co binder coating may be caused by surface fatigue causing spalling. This is prevented by a more impact resistant WC-15%Co coating, which is used on gas turbine engine compressor blades (Tucker 1981). 30% chromium carbide powder electrolytically codeposited in a Co matrix has excellent wear resistance (Kedward & Wright 1978). These coatings have been successful for gas turbine aircraft engine parts, at 15 to 125 microns. Worn Al alloy pistons can be salvaged by such 250 micron coatings (Kedward & Wright 1978). Wear resistance up to $700^\circ C$ is given by a Co + Cr_3C_2 cermet coating which provides a hard glaze of Co_3O_4 debris retained by Cr_3C_2 (Cameron et al 1979); wear resistance is also partly due to an hcp lattice. Tensile stress in AlN coatings is due mainly to differential thermal expansion on cooling but the net stress is compressive (Zirinsky & Irene 1978) and increases with deposition temperature. Adding Si_3N_4 sharply reduces the net stress into the tensile region.

Electroless Ni-P deposits are used as wear resistant coatings. The unlubricated wear characteristics have been examined (Ma & Gawne 1985). A standard rubbing test of like-on-like shows electroplated Ni+WC cermet coatings to be better than other Ni cermets but all the Ni based coatings had unacceptably high wear rates. Co+Cr_3C_2 was more wear resistant than Ni or Cr based cermets. Co forms oxides with lower friction than oxides of Ni, Cr or Fe. Moderate wear occurs at $200^\circ C$ and very low wear at $300^\circ C$ to $700^\circ C$. A hammer wear test is more arduous and assesses the coating cohesion and adhesion also. Co+Cr_3C_2 shows good wear properties, nearly matching plasma and flame sprayed Co+WC deposits and equalling plasma sprayed Co alloy X-40 (Kedward et al 1976). Excellent high temperature wear resistance (unlubricated friction wear tests) of Waspaloy has been reported with Co+Cr_2O_3

electrodeposited cermets at 300-700°C, the resistance attributed to a Co_3O_4 glaze layer (Thoma 1986).

7.11.7.4. Sprayed Coatings:

D-gun and plasma coatings are mainly used for wear resistance (adhesive and abrasive) (Tucker 1974). Considerations for wear resistant coating selection have been given (Tucker 1981). The effect of powder feed composition and size, in plasma spraying of wear resistant carbides, and results of friction tests are discussed. The use of NiO + CaF_2 (plasma sprayed from powders) for coating truck turbine engine heat exchanger seals is described Clegg et al 1973).

The mechanical properties of plasma and D-gun coatings are anisotropic because of their splat structure and directional solidification. Plasma sprayed ZrN on steel in argon atmosphere was hard (945-1045 VHN) but showed poor oxidation resistance and a sharp thermal expansion coefficient (Derradji et al 1986). The hardness of a detonation gun coating is generally higher than that of a plasma coating of the same composition. Hardness is usually reduced for a given material if the coating is applied in an inert atmosphere as compared to spraying in air. A typical wear test machine has rollers touching at different speeds, the roll diameter and weight loss being measured. This is used to select candidate materials for diesel engine testing. Plasma sprayed alloy coatings containing 20 to 30% free graphite gave promising results (Solomir 1973).

D-gun coatings are often much harder than plasma coatings of similar composition due to their higher density and cohesion. The high bond and cohesive strength of D- gun coatings allow their survival to impact, temperature and fretting (Barnhart 1968). Hardness varies with cooling rate if relative differences occur in various phases which form (Tucker 1974). Erosion (Sue & Tucker 1987) and fretting (Lindgren & Johnson 1987) wear tests have been reported on D-Gun coatings of Cr_3C_2-NiCr(A) and $Cr_{23}C_6$-NiCr(B) on steels and superalloys. The coatings were distinctly superior with bond strength greater than 6.9 x 10^4 kN/m^2 to those produced by plasma spray which showed 30% lower hardness due to porosity. The D-gun coatings had an apparent metallographic porosity of 1%. The two chromium carbides were similar in fretting wear but coating B was more resistant to spalling and did not affect the susbstrate creep-rupture properties. Coating A underwent structural changes to Cr_7C_3.

The microhardness of arc sprayed Ti Nb and Mo is low (ductile coating) if gas/metal reactions are avoided (Steffens & Muller 1974). Wear mechanisms of thermally sprayed hard facings such as Ni_3B and carbides are discussed (De-jie et al 1982). Flame sprayed 13% Cr steel coatings are used for wear resistance (Gagnet 1969). NiAl powder or Ni-graphite (75%-25%) are flame sprayed to give abradable seals (Messbacher 1975). The lower density (than plasma spraying) gives better coatings for this purpose. Grey cast iron has long been used for diesel piston rings as its graphitic structure prevents seizing (Solomir 1973). But at high power the wear becomes unacceptable and so hard Cr coatings are applied. But this needs good lubrication. Flame sprayed Mo (successful in petrol engines) corrodes.

Arc sprayed mixed metal coatings and pseudo-alloys (e.g. Cu + stainless steel) are wear, erosion and abrasion resistant and can be deposited at high rates to give thick coats on large components with bond strengths around 30 N/mm^2. Plasma sprayed Mo/Mo_2C coatings resist wear, Mo being the ductile matrix. An inert atmosphere is necessary to stop oxidation of the Mo_2C and loss of Mo ductility. Mo/Mo_2C composite coatings have low friction (Steffens 1983). Although plasma-deposited materials are softer than the wrought materials, they are much more wear resistant. Wrought alloys develop piles of debris but no debris occurs with plasma-deposited alloys as their structure limits the sizes of adhered wear particles. Addition of a small amount of alumina to aluminium bronze results in almost the same wear resistance as pure alumina, without increase in hardness (Tucker 1974).

7.11.8. WEAR vs MULTI-COMPONENT & MULTI-LAYERED COATINGS

It is evident from the discussion and information given in the preceding sections of this chapter that a considerable improvement in wear resistance can only be met by product optimization on a multiple-component, multiple-phase, multiple-layer concept. Ternary nitrides have been investigated and other systems and process methods may be expected to follow.

An increase of 200% of tool life has been achieved by diffusion boriding tool steel with Co-W-B followed by CVD TiC-TiCN or TiC-Al_2O_3 (Zachariev & Zlateva 1987). TiC-Al_2O_3 composite and TiC/Al_2O_3 layer deposited on stainless steel by direct electron beam evaporation showed greater abrasive and adhesive wear loss when compared with TiC by the ARE process (Memarian et al 1985). Steered arc evaporated TiN (6-7 microns thick) showed 42% lower flank and crater wear than TiN by random arc evaporation and 10 micron deposits by CVD. Steered arc evaporated TiHfN (2.5 microns thick) and TiNbN (4.9 microns thick) were both superior to the TiN (Boelens & Veltrop 1987). Magnetron sputtered TiHfN, TiZrN and TiVN also show a similar improvement in wear characteristics (Konig 1987). A two-fold increase in high temperature drill cutting both in terms of wear and oxidation is reported for TiAlN against TiN. Oxidation resistance extended from 550° for TiN to 800°C for TiAlN. The coatings were sputter ion plated (Munz 1986; Jehn et al 1987).

Innovations in multi-phase, multi-layer and multi-component wear resistant-deposits require reliable data on phase relationships of the metals and ceramics involved. Detailed discussions are presented by Holleck (1986), Sundgren & Hentzell (1986) and Konig (1987). Nitrides of similar lattice parameters and structures can be co-deposited more effectively and to give better resistance. Thus TiN can be combined with CrN, NbN, HfN, VN and ZrN but not with MoN, TaN or WN (stoichimetry not shown) (Konig 1987). Combinations of B_4C with LaB_6, TiB_2, ZrB_2, SiC and Al_2O_3 in binary and some ternary combinations have all resulted in 30-60% improvement in wear (Holleck 1986). Carbo-nitrides and oxy-carbonitrides of Ti and many other metallic and ceramic systems are listed (Sundgren & Hentzell 1986). A graded layer ceramic seal system developed from plasma spray has been a candidate aerospace coating. Inconel 713 substrate is given a NiCrAl bond coat followed by four layers of ZrO_2, viz. with 60% CoCrAlY and 15% CoCrAlY as the first and second layers on top, followed by 100% ZrO_2 and a porous zirconia as the topmost layer (Hillery 1986). Many more such sealing, thermal barrier, wear- and erosion-corrosion resistant coating composites are expected to be common in future high temperature coatings.

CHAPTER 8

Chemical properties of coatings

8.1. INTRODUCTION

The output of literature on alloy degradation is varied and voluminous, and it is difficult to do justice to all of this here; a reasonable cross-section is given in this chapter which reviews in brief, the influence of chemical factors on the durability and performance of coatings used for high temperature systems. It is difficult to determine any single package in explaining degradation, categorized by specific environment, mechanism or types of coating. The same coating can be applied by various processes, and each process is inherent with its own advantages and incorporated defects which can initiate coating failure later in service life. The chapter may be viewed in four main areas, divided into a number of sections for clarification. Coating/substrate/environment aspects in general are given at the outset, followed by the prevailing conditions for degradation of coatings in gas turbines – marine, power- and aero-engines, and in coal conversion systems, nuclear technology and tribology. The fundamental aspects of chemical reactions and examples in all aspects of chemical degradation, viz. general oxidation, hot corrosion, sulphidation, carburization, and erosion-corrosion form the third area. The fourth area examines the degradation of specific coating systems. References quoted in all sectors have been selected on a representative basis and should provide useful sources for further information.

The references listed below will be particularly useful in determining the direction and emphasis on aspects which need to be considered for high temperature investigations:

Conclusions published in the 1986 symposium on high temperature corrosion (Lacombe 1987); Workshop questionnaire and summary on high temperature corrosion (Rahmel et al 1985); Reviews on:-

- coatings in energy-producing systems (Stringer 1987; Fairbanks & Hecht 1987; Hancock 1987; Peichl & Johner 1986),
- the role of oxide scale point defects (Gesmundo 1987),
- electrochemistry of hot corrosion (Rapp 1987; Rahmel 1987),
- erosion (Wright 1987) and corrosion (Meadowcroft 1987) in coal combustion and conversion processes,
- oxide/metal interface and adherence - compositional fluctuations (Newcomb & Stobbs 1988), fracture mechanics (Hancock & Nicholls 1988), stresses and cohesion (Schutze 1988), improvement on adhesion (Stott 1988), protective oxides on gas turbine components (Rhys-Jones 1988).

8.2. COATING/SUBSTRATE/ENVIRONMENT vs DEGRADATION

8.2.1. GENERAL:

The deterioration of a coating and its eventual failure is primarily linked to:

(i)   The process by which it is applied,
(ii) Its metallurgical compatibility with the substrate, and
(iii) The joint response of coating plus substrate to the operating environment.

Coating failure associated with the coating process can arise in four ways:-

Coating composition, nucleation and build-up,
Coating thickness and uniformity,
Coating/substrate bonding - adhesion,
Coating defects, e.g. pores, cracks, disbonding and microstructure formed in as-processed condition and/or heat treated condition.

Chapters 3-7 review the process and physical aspects of high temperature coatings which control the above aspects. In brief, the coating composition is decided on the basis of the operating environment and the substrate on which it is applied. Deposition processes which are line-of-sight have to align substrate/target geometry to achieve uniform 'cover'. All processes resort to mechanical and electrical devices in moving the substrate and/or target to achieve uniformity in deposition, and not rely entirely on ionic, particle and physical mobility. Appropriate surface preparation and temperature control have to be exercised for ensuring sound adhesion, while controlled substrate exposure w.r.t. its geometry enables achievement of desired thickness, coherence and uniformity of the coated product. Ion bombardment or implantation for creating compressive stresses, or, negative bias to substrate to be sputtered are also used as a substrate pretreatment apart from chemical cleaning with a view to improving adhesion. Shot peening or grit blasting of the substrate or an interlayer of oxide ensures good coating nucleation and adhesion. Heat treatment as a post-coating step is often necessary to improve substrate-coating bonding.

Coating processes incorporate physical defects such as pores, voids and microcracks via nucleation, dislocations and crystal growth during the build-up of the deposit on the substrate; they can also cause interlayer structural changes, or substrate-coating interactions and interdiffusion leading to compositional and microstructural changes. The porosity in plasma and D-gun coatings is partially interconnected and so may have a strong influence on the corrosion rates of the coatings. Some D-gun coatings have sufficiently small pores to be unimportant in oxidation (Wolfla & Tucker 1978) but may not be ignored in hot corrosion where molten media are involved, and the porosity of plasma coatings have to have a post-treatment sealing of the top layer to render them acceptable. Heat treatment is not essential but may have to be considered. Columnar morphology is particularly unwelcome in hot corrosion environments regardless of how chemically resistant the coating may be. The evolution of barrier and multiple sealant layers compensate this physical inadequacy of an otherwise satisfactory coating.

The porosity of flame sprayed ceramic coatings can be greatly reduced by chromate impregnation, making these useable in hot corrosion conditions. Porous coats achieved by plasma spraying can be of advantage, as are microcracked and segmented ceramic coats like ZrO_2 (+ MgO, Y_2O_3 etc), for accommodating thermal shocks. However, microcracks can expose thermal barrier coats to selective leaching, e.g. of Y_2O_3 by molten salts (Pettit & Goward 1982). NiCrAlY(bond layer)/Cr intermediate coat/ZrO_2-Y_2O_3 (thermal barrier) exposed to Na_2SO_4 + vanadates caused total ceramic spallation in 100- 200 hrs with selective attack on Y_2O_3 while Cr was unaffected (Kvernes 1983). Thermal-mechanical (freezing) effects of condensed salts which penetrated porous coatings and disruptive chemical reactions between environment and coatings can both cause coating failure (Lau & Bratton 1983).

In the low temperature range, sealants can be applied for use up to 150°C (Tucker 1983). A new high temperature adhesive, LARC1-13 can provide high strength bonds up to 300° and can be used up to 600°C for short periods (St.Clair & Progat 1985). Its use as a pore filler appears promising. Bronze shaft sleeves handling saturated brine in a chlorine processing plant provided good service at 175°C when it was applied with an epoxy-sealed machinable metallic plasma undercoat and plasma deposited chromium oxide sealed with epoxy (Tucker 1983). Aluminium-bronze coatings on aluminum alloy substrates of aircraft landing gear bearings are corroded when the hydraulic fluid gets contaminated with salt water in marine conditions. A plasma coating virtually identical to the substrate and its subsequent sealing solved the corrosion problem.

At high temperatures however, plasma coatings have to be sealed by sintering and sometimes given supportive mechanical treatments (Tucker 1983). Laser sealing plasma coats of MCrAlY (M=Fe, Ni or Co) reduced the porosity to about 4 vol.% for dense coatings and about 10% for more porous coatings (Bhat et al 1983). The as-plasma-sprayed coatings revealed Al and Y segregated as thin splats. The Al_2O_3 developed on oxidation was discontinuous. The porosity caused premature degradation by internal sulphidation and reduced mechanical strength while the chemical inhomogenity would also contribute to the former. Plasma spraying with 5 wt % Al yielded only 1% Al in solution with Ni, while Al and Y segregated to thin splats. Laser treatment resulted in a four-fold increase of Al content in the top layer, while Y segregated to the coating/laser treated zone interface. Laser treatment also consolidated the outer 20-25 micron layer which provided a coherent continuous 5 micron layer of Al_2O_3 even during thermal cycling at 1100°C. However, laser treatment produced an outermost layer of slag which recorded a weight loss during thermal cycling. Slag formation can be prevented by laser surface modification in inert atmosphere or vacuum and by post-peening.

Laser fusion, laser irradiation or merely LST as it is termed, has been applied to aluminided and chromided superalloys and electrodeposited Cr (Galerie et al 1987), to PAVD SiO_2 coatings (Ansari et al 1987). Ni-10Cr and Ni-20Cr develop the protective Cr_2O_3 layer on identical mechanism; the oxide nucleates at the alloy grain boundaries which provide rapid diffusion paths for Cr and preferential oxide nucleation sites for internal oxide precipitates, and the healing layer is established initially at their intersection with the surface; stepwise extension into the grains promotes lateral growth, and thereof the cover of the healing layer. Ni-10Cr, at 1025°C, was found to be slower in developing the self-healing cover due to surface insufficiency of Cr. LST resulted in a change in surface microstructure, and provided rapid diffusion paths via retained alloy grain boundaries and twins which resulted in the 10Cr alloy evening up with the 20Cr alloy in self-healing (Stott et al 1987).

Burman and Ericsson (1983) have achieved plasma coating consolidation by HIP and EB. EB remelting prevented big oxide inclusions while allowing fast diffusion of Ni^{++}, and a 6% strain resistance prior to cracking. Although HIP treatment allowed only 3% strain release; it prevented Ni^{++} diffusion and retained oxide inclusions which could act as diffusion barriers. Both EB and HIP post-treatments of FeCrAlY plasma coats on Alloy 800 (Fe-base) and IN 738 improved oxidation by a factor of 10, the HIP treated surface developing a smooth oxide scale and the EB treatment yielding a nodular scale. Laser melting using a gas flow shield results in a high cooling rate to cause microcracks while EB is done in vacuum and can implement controlled cooling. The former has to be monitored to minimise slag forming. LSA and LSM have been improved with many modifications rendering them more versatile (Powell &

Steen 1981; Gnanamuthu 1979; Draper 1981, 1987; Bass 1981; Burley 1982).

Bond strength is very sensitive to variations of the plasma spraying process and to microstructural alterations during service life. A controlled precipitation of fine monoclinic ZrO_2 toughens up a cubic ZrO_2 at thicknesses greater than 0.5 mm (Kvernes 1983). The first order rule-of-thumb for thermal fatigue is that coatings have 1-2% ductility at the temperature of occurrence of maximum strain and should possess enhanced thermal fatigue resistance compared to elastically brittle coatings. Ceramic coats, especially, spall due to transient thermal stresses with failure occurring within the ceramic, near to but not along the ceramic/metal interface (Pettit & Goward 1982). A metal felt in between the metal substrate and the ceramic coat can help to accommodate shear stresses resulting from thermal expansion mismatch (Kvernes 1983).

Sol-gel-applied CeO_2 and SiO_2 coatings on preoxidized steel exhibit a low corrosion rate in CO_2 at 825°, the oxidation of steel itself having been reduced by a factor of 3 to 4 . The improvement is thought to be due to reduction in spalling during thermal cycling as an alpha quartz layer formed during the sol-gel process was incorporated in the growing oxide scale (Bennett 1983). Abrading the alloy coated surface prior to oxidation delays spalling (Hutchings et al 1981). ODS alloys in this respect are superior to ordinary superalloys (Lowell et al 1982), their lower thermal expansion causing reduction in spalling of the oxide layer. One method of combating corrosion fatigue and pitting of low pressure turbine blading (AISI alloys 403 and 630) and other surfaces washed by steam condensate is by coatings. The Southern California Edison Company have tested a number of nickel-gold electroplates (Kramer et al 1982). Coating selection for future prototype blade applications was based on a composite ranking of both laboratory and field specimen performance. The results indicated that fused powder Teflon, a nickel-cadmium diffused electroplate and ion vapor deposited aluminium all have significant benefits worth further investigation in field exposures.

8.2.2. EFFECT OF ION IMPLANTATION ON DEGRADATION

A drastic reduction in degradation of steels by oxidation drew attention to ion implantation as a surface modification method. The effect of ion implantation on high temperature oxidation has been extensively reviewed (Bennett & Tuson 1988; Bennett 1981). The surveys cover the oxidation behaviour of implants like Y and Ce in metals such as Ti, Zr, Cr, Fe, Ni and Cu as well as a range of stainless steels to Fe-Ni-Cr-Al alloys. Aluminized coatings are also discussed. In all cases beneficial effects could be obtained by selecting the right dopant as well as the optimum dosage. Several elements, metals and non-metals were ion implanted and some resulted in an increase in corrosion rate.

Radiation damage and effect of dosages of ion implantation were considered responsible for the onset of failure of the modified surface. Other workers also affirm that radiation damage and implant atomic radius are important parameters, along with solubility, compound and ternary oxide formation (Pons et al 1982).

Subsequent investigations indicate that any radiation damage to the lattice due to ion bombardment during implantation is annealed out at or below the oxidation temperature used. Ion bombardment also causes sputtering which has complex effects on oxidation. The physical effects of ion implantation can be best established by examining the effect of self ion implantation. For this purpose, variation of ion dose, dose rate, bombardment temperature and annealing temperature prior to oxidation, are particularly informative. Only when the role of radiation damage, sputtering etc., has been ascertained can the more complicated chemical influence of impurity atoms be distinguished.

Although the shallowness to which ions are implanted was predicted as a weakness in the technique, this has turned out to be its unique strength. Because the implanted element is confined initially to a near surface layer its subsequent redistribution has provided detailed and unique information about atomic migration mechanisms during oxidation. Also, the technique has provided further insight that could not have been obtained by any other procedure, that many elements, e.g. Y and rare earths on the oxidation of austenitic and ferritic stainless steels, acted through strategic location within the oxide film. Ion implantation also has the advantage of independence upon mutual solubilities which can provide mechanistic understanding of the role of specific elements in the thermal oxidation behaviour of metals and alloys. The implantation of Y in Ni-Cr alloys does not lead to grain structure change and this is an advantage for comparing the beneficial percentage with an alloy where change in grain structure occurs. Y implantation has been found to reduce oxidation and spalling of oxide in the stainless steel used for cladding nuclear fuel reactors (Bennett et al 1986). Doses as low as 2×10^{16} Y^+ ions/cm^2 have been shown to have a strong and long-lasting effect at temperatures up to 950°C. In contrast, most implanted impurities in Si were found to increase oxidation rate (Holland et al 1988).

The review by Bennett and Tuson (1988/89), addresses the improved oxidation characteristics of Cr_2O_3 and Al_2O_3 forming alloys in detail and reviews recent oxidation studies on ion implanted alloys particularly over the last five years. These cover studies in O_2 and CO_2, of Fe, Ni and Co base alloys implanted with Y, La, Ce etc.. The review gives a critical discussion of the possible mechanisms.

Ion implantation has established that the dramatic influence of many elements is derived from their incorporation into the oxide during the early stages of oxidation (Bennett et al 1976; Bennett 1983) and that a dispersion of reactive oxides is more effective than the corresponding metal (Hou & Stringer 1988). Another positive effect observed in 20Cr-25Ni-Nb stabilised stainless steel was that implantation of Y and Ce extended the scale growth induced by the low temperature oxidation process to higher temperatures, up to $1000^{\circ}C$ (Bennett et al 1987).

In all scales, the reactive element was located in a continuous band, probably acting as markers for the original metal surface. In some cases Y_2O_3 segregated at grain boundaries provided the source of Y in the band. Parallel observations confirmed that segregation of reactive elements also occurred for alloy additions. The most probable mechanism (Stringer et al 1972) is that the dispersed reactive element oxide particles on an alloy surface act as heterogeneous nucleation sites for the first formed oxides, thereby reducing internuclei spacings and hence the time for subsequent lateral growth. A discussion of the role of segregates in blocking diffusion along grain boundaries indicates that the most probable mechanism could be due to complex defect formation, by vacancy-rare earth association, increasing the activation energy for diffusion along segregated short-circuit paths (Nagai & Okabayashi 1981; Duffy & Tasker 1986).

Numerous models have been proposed to account for the enhanced adhesion of ion implanted materials. These have been reviewed in detail recently (Moon 1988). In any system one or more can apply and no fixed rule exists. The beneficial effect of implanted materials could be due to reduction in stress, oxide keying or pegging, chemical bonding of the scale-substrate interface, vacancy sink provision or impurity gettering. All these possibilities have been discussed in the review by Bennett and Tuson (1988/89).

Several of the mechanisms identified by numerous experiments in reducing corrosion rate are summarised below (Dearnaley 1987):

1. The production of a coherent film of material that resists diffusion of anions and cations,e.g. Al_2O_3 following high dose implantation of Al^+ (Bernabai et al 1980).
2. The blocking of short-circuit diffusion paths through the scale, e.g. at grain boundaries (Bennett et al 1985).
3. Modification of oxide plasticity,e.g. to allow the relief of growth stresses without cracking (Bentini et al 1980).
4. Modification of electronic conductivity and hence the defect population (the classic Wagner-Hauffe mechanism).
5. Electro-catalytic processes may occur which can influence reactions, e.g.in moist atmospheres, involving ingress of hydrogen or OH.

Oxidation of binary Ni-33 and 20Cr, IN 939, Rene N4, FeCrNiNb, and FeCrAl alloys with implanted Y, Ce, Ar, Pt, Zr, Al, Ca and Si have been scrutinised. The implanted dose, the test duration and the alloy condition, e.g. pre-oxidised, etc., are factors which influence the oxidation rate, and often without any

improvement or, when it does retard, it does so for a limited
duration. The protection is not comparable to that obtained by
the element incorporation as an alloy constituent in the coating
or the substrate (Srinivasan 1987). The effects of Al and B
implantation on Fe-oxidation were studied (Brown et al 1985;
Galerie et al 1982). Fe and Cr ion implantation into Fe had
detrimental effects on its oxidation rate at $400^{O}C$ at 0.1 torr.
Self-implantation caused increased oxidation due to more diffusion paths in the oxide and the benefit of Cr lasted only until
the 200 nm thick implanted layer was oxidised (Howe et al 1982).

La implantation to a dose of 10^{17} ion/cm^2 on 20Cr/25Ni/Nb-stabi-
lized stainless steel (also containing 0.9%Mn and 0.6%Si) was
oxidized in CO_2 at $825^{O}C$ up to 9735 h. The beneficial effect was
evident in scale adherence during thermal cycling. Microstructural changes were manifested in finer Cr_2O_3 grains and an intermediate region of La-rich spinel between the outer $Mn(Fe,Cr)_2O_4$
spinel and inner Cr_2O_3+spinel, beneath which was a continuous
band of Cr_2O_3 and internal SiO_2. The implanted La while oxidizing
also acted as a marker due to its low diffusivity. A major consequence of the improved adherence was that the steel was prevented
from pitting attack (Yang etal 1987).

The effect of Ni, Ar, C, Cr and Li ion implantation into Ni was
studied for oxidation resistance at $1100^{O}C$ in O_2. $NiCr_2O_4$, LiO_2,
etc., impeded scaling and later dissolved in the scale, doping
the NiO. A long term effect occurred in all cases, the oxidation
rate eventually becoming higher than that for pure Ni (Stott et
al 1982a). Unimplanted Ni-20%Cr and Cr-implanted surfaces gave
spalling scale on cooling, but Y and Ce implanted surfaces resisted spalling (Stott et al 1982b). Similar effects of Y and Ce on
steel are reported (Bennett et al 1982). Cr and Ni- implanted Ni
developed a NiO scale more facetted, more mismatched and generally smaller with more extensive stress during growth. The initial
slower rate gave way to much higher rates at longer times. A
doping effect on a semi-conductor NiO is ascribed as the cause
(Stott et al 1982c; Peide et al 1981).

Two ternary oxides were detected on Al-implanted Fe at $727^{O}C$
after 100-120 h. $FeAl_2O_4$ was ascribed to cause a blocking effect
(Pons et al 1982). But Fe-15Cr-4Al alloy implanted with Al in a
25 micron thick surface layer showed no significant influence on
its oxidation behaviour in air at $1100^{O}C$ (Smith et al 1987). An
Al implant dose <$1x10^7$ ions/cm^2 had no effect on the oxidation
rate of Fe-6 at.% Al alloy at $900^{O}C$. Ion mixing of Al and Al_2O_3
on SiO_2 improved adhesion, increasing with the ion dose, with the
best adhesion enhancement obtained by implants which produced
glassy phases with Si-O-Al bonding (Galuska et al 1988). Comparison of oxidation rates of ion implantation treatment and LST of
boron-alloyed iron showed that LST reduced the rate better. The
scale morphology seemed to be responsible with the implant oxidized sample showing FeB_2O_4 as top scale followed by FeO, Fe_3O_4
and Fe_2O_3 sequentially beneath. The LST sample scale analysed to
a top layer of Fe_3BO_5, Fe_3O_4+Fe_2O_3, $FeBO_3$ and B_2O_3 at the metal-

/scale interface (Pons et al 1986).

When 0.86% Y was alloyed with this FeCrAl, oxide spallation was reduced for at least 3271 h while Y with Al registered an improved behaviour for a limited duration (784 h). The beneficial effect of Y is inferred to be within the oxide film rather than the subscale (Bennett et al 1980). Y implantation was carried out on FeCrAl alloy with and without Y (0.19 wt.%) and Y added as a sulphide. Oxidation in air at $1050^{\circ}C$ showed that the rates of FeCrAl with and without yttrium sulphide were the same. FeCrAlY registered improved adhesion with convoluted scale formation. Pegged alumina scales on FeCrAl were cracked and non-adherent; the scale was adherent in the implanted alloy but with no change in the convoluted morphology. Y added to FeCrAl as an alloy normally forms flat scales but Y implantation at fluences of 10^{15} and 10^{17} ions/cm^2 did not achieve this. Helium bombardment on scales formed on implanted alloys exhibited local blistering (Smeggil & Shuskus 1986).

Fig.8-1a shows the parabolic oxidation rate constant of Al-implanted FeCrAlY was less by a factor of 140 in tests at $1100^{\circ}C$ (Bernabai et al 1980; Bennett 1981). Annealing Y-bearing Fe-Cr-steels is good and is thought to be due to either a lowering of the diffusivity of spinel-forming elements, or a degree of pre-oxidation of Y in the annealing pO_2. A similar mechanism is envisaged for the influence of Y implantation on Ni-Cr, Fe-Ni-Cr and Fe-Ni-Cr-Al alloys but does not appear to be well supported. Yttrium dissolution in the oxide layer changes the growth rate and composition and gives then an improved adherence. But the oxidation rate of the Fe-Ni-Cr-Al alloy is hardly affected; see Fig. 8-1b, 8-1c and 8-1d (Pivin et al 1980). A beneficial effect has also been reported for 20/25/Nb steel by Ce and Y in CO_2 at 825° and seemed to have originated during the early stages of subscale development, negating any initial detrimental influence caused by radiation during implantation. CeO_2 and Y_2O_3 grains modified the mechanism of growth of FeCr(MnNi) spinel with a thinner Cr_2O_3 underlayer. Outward cation grain boundary diffusion was arrested by the ingrained CeO_2 and Y_2O_3. The change caused in the microstructure resulted in increased adhesion. The formation of a continuous, weak, fracture-prone layer of SiO_2 was inhibited as scale modification occurred (Bennett 1984).

The effects of ion-implantation on Cr_2O_3-forming alloys appears to be variable, but often beneficial in air or oxygen environment. The same cannot be said for Al_2O_3-forming alloys. Most of the studies reported are on Fe-base alloys in air or O_2 atmospheres only.

Implantation of Y and Ce improves oxidation resistance of chromia former alloys but that of Al or noble gas results in little or no beneficial effect (Bennett et al 1980; Antill et al 1976). Pre-oxidation followed by yttrium implant in IN 800H has been shown to be beneficial (Kort et al 1986). The oxidation rate at $1020^{\circ}C$ was reduced by 45% for this alloy when Y-implanted. However, when

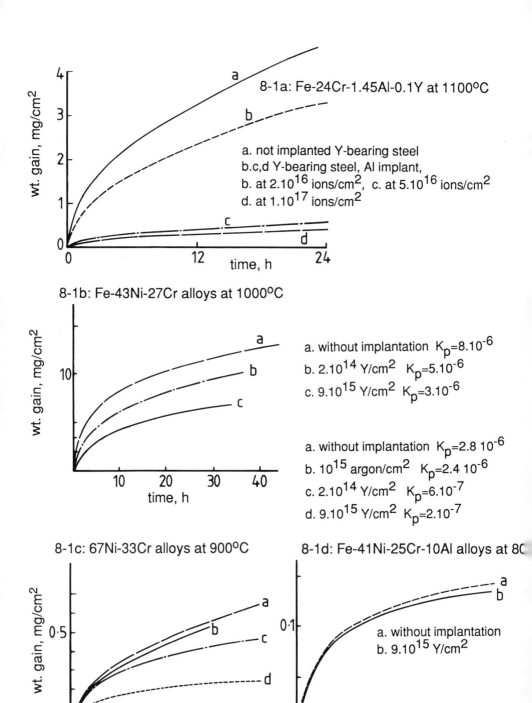

8-1a: Fe-24Cr-1.45Al-0.1Y at 1100°C

a. not implanted Y-bearing steel
b.c,d Y-bearing steel, Al implant,
b. at 2.10^{16} ions/cm^2, c. at 5.10^{16} ions/cm^2
d. at 1.10^{17} ions/cm^2

8-1b: Fe-43Ni-27Cr alloys at 1000°C

a. without implantation $K_p=8.10^{-6}$
b. 2.10^{14} Y/cm^2 $K_p=5.10^{-6}$
c. 9.10^{15} Y/cm^2 $K_p=3.10^{-6}$

a. without implantation $K_p=2.8 \cdot 10^{-6}$
b. 10^{15} argon/cm^2 $K_p=2.4 \cdot 10^{-6}$
c. 2.10^{14} Y/cm^2 $K_p=6.10^{-7}$
d. 9.10^{15} Y/cm^2 $K_p=2.10^{-7}$

8-1c: 67Ni-33Cr alloys at 900°C

8-1d: Fe-41Ni-25Cr-10Al alloys at 8C

a. without implantation
b. 9.10^{15} Y/cm^2

Fig.8-1 (a-d): Effects of Ion Implantation on the Oxidation
of Fe- & Ni-base Alloys.

exposed to H_2-H_2S at 560°C, Y-implantation has no effect on the alloy sulphidation rate without pre-oxidation. A 70% decrease occurs for the pre-oxidized, implanted alloy in the first 50 hours but the mass gain rate accelerates to twice the non-implanted alloy in 150 hours. Implantation of Kr shows a similar initially slow but accelerating later effect, but over much shorter durations and Al implantation has no beneficial or detrimental effect. A similar pattern of an inital decrease in rate (95%) and later acceleration (x3) in 200 hours is observed in H_2-H_2S-O_2 at 640°C (Polman et al 1987).

CoCrAlY overlay coating implanted with yttrium and oxidised in air at 700-1000°C developed voids which occurred in the initial stage of oxidation and was found to be dependent on the distribution of Y on the alloy surface. Implanted Y reduced the extent of interfacial void growth and increased the oxide film thickness. Implanted cobalt had no effect (Sprague et al 1983). Si and Y implants in CoCrAl and CoCrAlY alloys conferred resistance to exfoliation during thermal cycling (Smeggil et al 1984a, 1984b). Al implantation in NiCrAl and NiCrAlY alloys did not affect the alumina-forming characteristics or pattern. The oxide scale exfoliation of alumina formers did not seem to be automatically associated with any microstructural damage/changes brought about at the implantation stage. Implantation of $^{32}S^+$, however clearly resulted in cracked scale growth and exfoliation at 1050°C, 24 h in air. In the sulphur implanted NiCrAlY large (>1 micron) dark precipitates, of Y_2S_3 devoid of Ni, Cr and Al, were found 10 microns <u>below</u> the scale metal interface, while the actual implantation could not have been more than a few thousand angstroms deep (Smeggil et al 1985). This movement of sulphur inwards and Y segregation from Ni, Cr and Al can be explained in conjunction with the non-adherence of the alumina-former alloy, on the basis of the fact that yttrium which is normally available for oxide adherence is now not available at adequate levels at the metal-/scale interface as sulphur getters the subsurface Y. It could be inferred that at low sulphur potentials Y_2S_3 would form preferentially even before Cr; the implanted sulphur appears to force reaction with yttrium, preventing its outward diffusion when it reaches the required activity through internal diffusion. A similar argument has been put forward earlier (Abderrazik et al 1984).

From the above discussion it may be inferred that most of the degradation studies are in oxygen or air. Very few studies are available on hot corrosion (Bhat et al 1983). Ion implantation has been used to successfully reduce the corrosion/erosion type wear of burner orifices used in oil-fired power generating plants and thereby, increase their in-service life (Bennett 1981). The process has beneficial features recognized in nuclear technology, tribology and to a good extent in most carburization/oxidation environments but has yet to be explored for adapting to more aggressive environments.

Particular experimental areas requiring more comprehensive cover
are :

Comparison of the scale (at a point in time where thickness is
the same in both cases) of implanted and unimplanted substrates
to shed more light on the mechanism of scale adhesion.

The variation with scale thickness of the temperature drop required to cause spalling from the implanted substrate.

Mechanistic studies in sulphidising environments as well as mixed
environments.

Although ion implantation has become an accepted technique in
corrosion research, it has yet to be used practically for corrosion protection. The present scale up of implantation equipment
would make this economically viable. The crucial factor however,
in determining the practical applicability is the demonstrated
persistence of the beneficial influence of implantation on degradation.

8.2.3. DEGRADATION AFFECTED BY PROBLEM AREAS IN SUBSTRATE/COATING COMPATIBILITY

Several problems occur at the process stage which affect the
coherence of the resultant substrate-coating complex. Intermetallic compound formation, diffusion, interdiffusion, effect on
ductile brittle transformation temperature are the main influencing factors, which could create gradient-type changes in substrate and coating compositions in regions near the interface,
create stresses, undesirable phases and precipitates, all of
which could adversely influence the protective capacity of the
coating. Other effects are vaporization, selective attack, mechanical strain and creep effects, effects on microstructure and
phase formation, and themal shock and fatigue.

These are discussed below with a few examples:

Intermetallic Compound Formation:- Ir coatings on Nb,Mo and W
form intermetallics which reduce the original coating thickness.

Diffusion:- In Ti and TiV foils on Nb alloys, interstitial
compounds (which strengthen the substrate) diffuse to an external
sink (such as a coating containing Group IV elements).

In $MoSi_2$ coatings on Nb alloys, Ta10W, Mo alloys and W, oxygen
diffuses in and embrittles Nb and Ta alloys. Mo and W form volatile oxides. The composition and coherence of the coatings are
affected. $MoSi_2$ and $NbSi_2$ coatings on Nb show that Si diffusion
occurs from an outer silica layer through an $(Mo+Nb)_5Si_3$ interlayer and depletes the coating of Si. In $NiSi_2$, $CrSi_2$ and NiCrSi
coatings on Fe, IN738, IN100, Nim90 and IN601, inward diffusion

of Si reacts with an interlayer and decreases strength.

In coatings of Al, CrAl, MCrAlY etc., on IN100, nitrogen penetration forms carbonitride and stresses cause cracks. AlPt coatings on superalloys form an undesirable sigma phase due to inward diffusion.

Interdiffusion:- Coatings designed for hot corrosion resistance usually contain higher Cr and Al than the substrate component alloys. High temperature exposure over long periods increases the levels of both elements in the interdiffusion zone, which will lower the Cr and Al at the scale/environment interface and thus the renewal capacity, while also causing embrittlement at the interdiffusion zone. In 200 h at $1000^{o}C$ the following surface compositional changes (at.%) at the near surface region occur as 200 micron thick oxide scale forms on a NiCrAlYSi alloy (Grunling et al 1987):

0.4Si alloy:- Al 17 to 8.5; Cr 25 to 30; Ni 43 to 55;
4Si alloy :- Al no change: Cr 23 to 29.5; Ni 48 to 31;
changes in 2%Si and 4.6% si alloy were not as significant.

At intermediate temperatures the sigma phase formation is possible. Al appears to be considerably affected by temperature gradients with a 1% drop across 10, 2.8, 2 and 1.4 mm at temperature gradients of 0, 100, 200 and 300°/mm respectively. Increasing gradients and exposure time cause steeper profiles; grain boundary diffusion of coating elements could get enhanced which results in greater penetration depths. Decrease in low cycle fatigue limits with corrosion and interdiffusivity for LDC 2 coated IN 738 is illustrated (Grunling et al 1987).

Sputtered TiC applied as a diffusion barrier between Al and $CoSi_2$ breaks down at $500^{o}C$ to form Co_2Al_9 and $Ti_7Al_5Si_{12}$ at $550^{o}C$ (Applebaum & Murarka 1986). In Ni30Cr, and Ni30Cr20W with and without 3Al, or Mo barrier layer, or $W-1\%ThO_2$ barrier layer, or W25Re barrier layer, embrittlement occurs at $1150^{o}C$ due to interdiffusion and at $1250^{o}C$ due to nitrogen. In Al coatings on Ta and Ta10W, interdiffusion forms Ta-Al phases at $1150^{o}C$. NiAl coatings on Ni alloys degrade due to interdiffusion. Reactive diffusion occurring when exposed to high temperature over a period (termed as the soaking time) were found to induce a continuous change in the local compositions of both the coating and substrate, which in turn changed the growth of precipitates and surface oxidation. PVD-NiCoCrAlY on Udimet 520 developed the detrimental sigma-phase in this manner which affected both the microstructural stability as well as oxidation resistance. Tested on soaking up times over 500-2000 h, the system showed a sharp change between 1000 and 1500 h (Zambon & Ramous 1987).

It is proposed that three stages of interdiffusion be recognized during the degradation of a substrate/coating system. The stages can be tagged by measuring the solute loss from the coating and the variation of the surface concentration with time taking note of the coating thickness. In isomorphous systems with constant diffusivity, stages 1 and 3 are closely related to the coefficients of the square root diffusivity matrix [r] - an asymptotic approach to the final coating composition, and in binary and ternary systems any improvement in interdiffusion in stage 1 will be reflected likewise in stage 3. But higher order systems, including ternary show kinetic behaviour dependence clearly on the coating composition. Taking a Ni-Cr-Al system it has been demonstrated that decreasing %Cr decreases the Al-diffusion rate from the coating to the substrate. Changing %Cr in the range of 10 at.% results in a x10 reduction of the time taken to lose 50% Al from the coating (Thompson & Morral 1987).

Internal Oxidation:- In Hf20Ta coating on Ta and IrHfTa on W, oxidation occurs via coating defects at $1800^{\circ}C$. CrTiSi coating on Nb also undergoes similar degradation by oxidation in air via coating defects. In $TaAl_3$-Al coating on Ta, internal oxidation occurs in air at $1600^{\circ}C$.

Ductile Brittle Transition Temperature (DBTT):- Applying Cr_5Al_8 and $(Fe-Cr)_xAl_y$ coatings on Cr5W.1Y by the pack cementation coating process causes recrystallization of the substrate due to thermal effects and increases the DBTT from 250 to $420^{\circ}C$. Table 8:1 lists the DBTT of typical diffusion and overlay coatings determined at a 1% fracture strain (Nicholls & Hancock 1987):

TABLE 8:1

DUCTILE—BRITTLE TRANSITION TEMPERATURE OF DIFFUSION & OVERLAY COATINGS

Coating	DBTT, $^{\circ}C$; Range
Co-35%Al	970
Ni-35%Al	740
Ni_2Al_3	570- 710
Ni_3Al	730 - 900
NiAl	868 - 1060
Siliconized:-	
Nimonic 105	350
IN 738LC	560
IN 939	510
Co15Cr10AlY	250
Co18Cr9AlY	235
Co23Cr12AlY	740
Co27Cr12AlY	910
Ni38Cr11AlY	430

==

Miscellaneous Substrate/coating Effects:-

Chromalloy W-3 and PFR-30 on Nb alloys: Coatings vaporize; poor adherence, cracking, non-uniformities, occurring in air at $1420^{\circ}C$.

SiO_2 and CeO_2 on steels: Iron oxides destroy coating.

ZrO_2 (MgO or Y_2O_3 stabilised): Vanadates (from diesel fuel combustion) attack the MgO or Y_2O_3.

SiO_2 on 9Cr-steel: Cracks due to mechanical strain allows oxidation in CO_2.

Al on Fe alloy: Phase transformation occurs and Al-deficient FeAl phase starts to degrade in air at $750^{\circ}C$.

NiCrAlY/Pt coating: Coating thermal process affects substrate microstructure and creep strength.

ZrO_2-12%Y_2O_3/Ni16Cr6Al.6Y duplex coating on superalloys: Salt melts absorbed by coat porosity and then thermal shock failure follows; acid fluxing hot corrosion mechanism also occurs. The problems however, are not universal but dependent on the service arena.

8.3. ENVIRONMENTAL CONDITIONS FOR COATING DEGRADATION

Table 8:2 surveys the many types of coatings and process routes used in high temperature environments. Table 8:3 surveys the substrate/coating/environment interaction situation and effects. Chapter 1 has introduced the problems encountered in high temperature systems. Microstructure (Buhler & Hougardy 1980) and diffusion profile studies have provided the means for selection of coating component elements (Fitzer & Maurer 1979).

8.3.1. FLUIDIZED BED COMBUSTORS - COAL GASIFIERS

Fe-Cr-based alloys are the predominant structural materials and are also coating candidate materials. Ni and Co, in large proportions are detrimental to the coating or alloy service life and are not cost-effective for the industry, being more expensive than iron to produce and fabricate. The problem areas are summarised in Table 8:4 and Fig.8-2.

High temperatures and pressures prevail in the processing components for coal preparation and feeding systems, and the effluent streams vary from slag and liquid slurries to gas entrained with fly ash. Erosion-corrosion including wear is a problem of equal or greater magnitude than hot corrosion itself, in coal combustion and fluidized-bed systems (Raask 1988; Wright 1987).

TABLE 8:2

TEMPERATURE & ENVIRONMENT INTERACTION – THE RESPONSE OF THE ALLOY EXPOSED TO HOT CORROSION , EROSION & WEAR

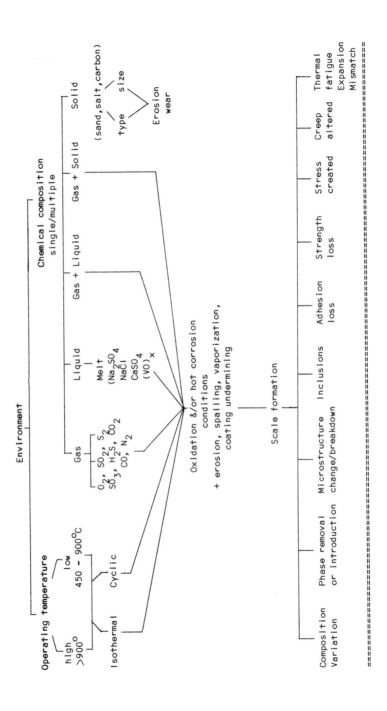

TABLE 8:3

CHARACTERISTICS OF COATINGS OBTAINED BY DIFFUSION PROCESSES(A) & BY OVERLAY(B)
(A Survey of interest to service in high temperature and hot corrosion environments)

Coating Process	Type	Average Coating Thickness, μm	Advantages (coded)	Coating Material
PVD – sputtering, evaporation, ion-implantaion, ion-plating.	B	75-100 (can vary)	(b),d	Ce, CoCrAlY, FeCralY NiCrAlY, NiCoCrAlY, $ZrO_2(Y_2O_3)$
CVD –inert/vacuum media	A,B	Variable	b	Al, Ni-Cr, Si, SiB, TiSi SiO_2
CVD – pack cement-ation	A	100	b	Al, AlPt, Cr, CoNiCrAlY, Si, SiAl
Spray – vacuum and air plasma, d-gun, flame, slurry laser spray, plasma sinter-ing.	A,B	150-400; can be varied	a,(b),c,d	Al-Si, CoCrAlY, FeCrAlY, NiCrAlY, NiCoCrAlY, Ni-Cr-Si
Cladding, Surfacing	B	Can vary	a	NiCrSi
Sol-gel	B	1	(c)	Ceria, SiO_2
Galvanic	A	100 can vary	(c)	Ce, Cr, CrFe

Code: a – suitable for repair work; b – good adhesion; c – suitable for thermal
barriers; d – Composition can be easily varied. Brackets indicate average
performance.
===

Conventional combustion systems (steam boilers), atmospheric and pressurized pressure fluidized-bed combustors (AFBC,PFBC), and some gasification systems use pulverized and sized coal. Coal-oil mixtures, coal-water mixtures are used which reduce emissions of oxides of nitrogen. Coal washing is done in some cases, to reduce sulphur emissions. Gasification reactor systems have compromised in efficiency to circumvent materials design problems required by the advanced systems proposed in the 1970s. The cool water integrated gasification unit is one such operational plant. Erosion-corrosion is a severe problem in coal liquefaction systems, and rapid corrosion of equipment in the distillation stage was found to be due to the concentration of amine hydrochloride fractions with a boiling point of $240^{\circ}-280^{\circ}C$ (Keiser et al 1981). WC-Co coatings have retarded the erosion problem but are expensive to renew very frequently. Borides, carbides and nitrides have also been tested for erosion control with Si_3N_4 giving the best erosion-hot-corrosion resistance (Wright 1987). Cladding and LPPS coatings are generally used.

Boiler Tube Failures & The Problem Spectrum

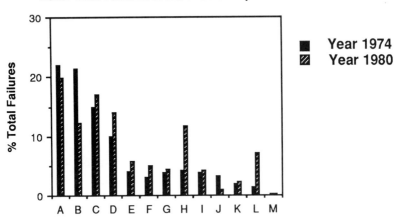

Fig.8-2

Column	Cause of Failure	Column	Cause of Failure
A	Mechanical Design	B	Weld Defects
C	Excessive Heating	D	Erosion of Sootblower
E	Corrosion due to water & Steam	F	Blocking Effects
G	Mixed Causes	H	Erosion due to Dust
I	Defects in Tubes	J	Wrong Materials
K	Mechanical Damage	L	Fireside corrosion
M	Other Operational Factors		

TABLE 8:4

CORROSION ASPECTS WHICH HAVE CAUSED NEAR–TOTAL SHUT–DOWN OF PULVERISED COAL–FIRED PLANTS

Corrosion Areas	Status
Dew-point Corrosion: Air heaters, ducting and stack locations	Alternative material for mild steel not yet in full operation; mechanism of attack reasonably clear; replacement economics under review.
Exfoliation: Oxide scales on ferritic & austenitic re-heaters and s/h steels; shut down caused via tube blocking and turbine erosion.	Spalling in austenitic steels predictable and under control; Ferritic spalling predictable but less clear for compensation.
Furnace using coals high in sulphide/chloride – damage of furnace wall tube.	Existence diagrams of the oxide-sulphide-chloride regimes vs temperature on many resistant alloys and mild steels are available. The mechanism of chloride effect and influence of carbon and sulphides are reported.
Heat flux effects	Corrosion rates predictable theoretically; plant data available; controlled investigations not widely reported.
Integrated action of stress-strain-oxidation at stress points/transition joints on gas and and steam sides.	Maximum threshold of effects predicted theoretically. Some pilot investigations available in published literature.
Multi-lamination and oxide-thickening of Superheaters and re-heaters on the steam-side. Superheater tube the coal used. corrosion.	Predictability reported; laboratory simulation not known to be reported. Damage caused by high sulphide/chloride coals documented. Corrosion mechanism yet to be clearly explained vis-a-vis the coal used.

==

The environment in coal gasification is predominantly reducing unlike in a gas turbine with oxygen potential varying over a large range of $10^{-1} - 10^{-14}$ in a cyclic pattern. The temperatures are not very high but in the region of 650- 900°C. Because steam and limestone are used, the environment has H_2/H_2O, $CaO/CaSO_4$ reactions as well as those of C/CO, CO_2, S_2 and some chloride.

Thus the reaction environment consists of S_2, SO_2, CO, CO_2, H_2S, H_2O, HCl, with carbon and unreacted fuel particles, with CaO and $CaSO_4$ creating erosion conditions (Meadowcroft & Manning 1982). In the U.K. furnaces operate around 160 bar at 350°C, and the heat exchanger units run at a metal surface temperature of 450°C in the evaporator tubes and 650°C in the superheater tubes. Co-extruded tubes are feasible if fabrication of the 50Ni-50Cr clad can be overcome, so that the niobium-stabilized Type 310 steel component can be replaced with another class of steel as the inner tube (Meadowcroft 1987).

8.3.2. THE GAS TURBINE ENVIRONMENT

The environment presented during the operations of a gas turbine is complex and variable but mainly oxygen-rich with sulphur and chloride as chemical contaminants and erodents such as salt particles and pyrolytic carbon. An appraisal of surface-oriented problems for gas turbine applications appears to be as shown in Table 8:5 (Pettit & Goward 1983).

TABLE 8:5

Problems	Aircraft	Marine	Utility
Oxidation	Small	Medium	Medium
Hot Corrosion	Medium	Large	Small
Interdiffusion	Medium	Large	Medium
Thermal Fatigue	Small	Medium	Large

Table 8:6 indicates the factors leading to degradation and the environmental conditions which are conducive to create them. Table 8:7 summarises the industrial turbine environment.

Gas turbines are run on a wide range of fuels from natural gas and high sulphur diesel to the highly refined aviation fuels. The ambient atmosphere is generally highly oxidizing with a high air to fuel ratio. The flame temperature reaches 2000°C. The full potential of the hot gases from the combustor confronts the blade section, with the thermal energy from combustion transformed to mechanical energy. The role of protective coatings is to maintain the integrity of the blades and guide vanes over extended periods. Cooling passages in the blades keep their temperature down. The first stage nozzle guide vane with a low design stress is the hottest at 900-1150°C and is subjected to severe thermal stresses and hot corrosion. The turbine rotor blade suffers a high creep stress because of the centrifugal force; it also undergoes thermal cycling and mechanical fatigue at 800-1050°C, the second stage vanes and blades operate about 100 degrees below that of the first stage.

LEVELS OF LOADING EXPECTED IN GAS TURBINES FOR AIRCRAFT APPLICATIONS

Type	Type of Loading	Effective Limits of Parameter	Detrimental effect
Mechanical	Centrifugal stress	170 N/mm^2	Cyclic strain in all temperature ranges
	Stress gradients: i. local ii. time-dependant (temperature & gas pressure variations)	e.g. + 30 N/mm^2	Formation of cracks in the coating
	Gas velocity	up to 600 m/s	Stripping of the coating erosion, mechanical removal Spalling of the coating
	Impact by external media		

[T, σ : lifetimes from 1000h to 5000 – 10,000h]

Type	Type of Loading	Effective Limits of Parameter	Detrimental effect
Chemical	Oxygen excess	12% by volume	Oxidation and corrosion depletion across surface or local attack
	Fuel contamination, e.g. sulphur	0.3w/o permissible 0.01w/o usual	Alloy impoverishment
	Air intake contamination, e.g. sea salt, industrial atmosphere	up to ~1 ppm	Roughening of the surface
	Pressure	up to 25 bar	
	Flow rate	up to 600 m/s	
Thermal	Temperature range: in combustion gas	up to 1400°C	Diffusion processes
	in component	up to 1050°C	Changes in structure
	local	up to tp 200°C/mm	
	Temperature gradients influenced by time	up to 100°C/s	Mechanical stresses

TABLE 8:7

SERVICE CONDITIONS IN HEAVY DUTY INDUSTRIAL GAS TURBINES

Pressure	:	$p = 10$ bar approx.
Max. Gas Velocity	:	300 m/s
Mechanical Stresses During Operation	:	up to $16 Nmm^{-2}$
Frequency Range	:	150 - 2000 Hz
Temperature Variations	:	$200^{o}C$/min.
Temperature Gradient (normal to surface)	:	$200^{o}C$/mm
Deposit Constituents	:	1 mg/m^3
		CO, CO_2, SO_2, SO_3, H_2O, O_2, N_2, CaO, MgO, K_2O, SiO_2, Fe_2O_3, Zn, V, SO_4

==

Turbines which operate under marine conditions ingest salt from the sea. In general, industrial aero-engines last longer than military craft. The high altitude aero-engines use highly refined fuels and pure air; coatings under these conditions are subjected to high temperature oxidation. Low altitude marine engines use lower quality diesel fuel in salty air under gas + solid conditions, an ideal condition for hot corrosion over a wide temperature range. V/STOL aircraft engines are subject to severe cyclic stresses leading to thermal fatigue. Turbines used in power generation use medium to low quality fuel and face hot corrosion by vanadium and sulphur products and erosion problems. An alumina coating for instance, may go through a damage and repair cycle with adequate reserve of aluminium activity in an oxidation-erosion environment. A hot corrosion environment could flux the Al_2O_3 compelling support by Cr_2O_3 with a duplex, triplex and multicomponent coating. It has been shown that attack from contaminants can be a high probability because both Na and K can build up to high levels of concentration if untreated potable water is used for water injection which later evaporates (Hsu 1987). Erosion and abrasion are caused by sand, salt or carbon particles (EUR 1979).

Na_2SO_4, NaCl, and in some cases, vanadic oxides present the additional threat of molten salt which changes high temperature oxidation into hot corrosion conditions. However catastrophic corrosion can occur if a molten reaction eutectic product can form in purely gaseous conditions (Simons et al 1955; Hocking 1961; Vasu 1965; Alcock & Hocking 1966; Hancock 1968; Viswanathan & Spengler 1970; Vasantasree 1971; Brown et al 1970; Vasantasree & Hocking 1976; Holder et al 1977; Mrowec & Werber 1978; Birks 1970; Wood et al 1978; Strafford 1976). The reactant environment is composed of O_2, SO_2, SO_3, Na_2SO_4, NaCl with sand, salt and pyrolytic carbon particles. The superalloys used in gas turbines are mostly Ni-Cr- and Co-Cr base, the Fe-Cr- base alloy series being used mostly under reducing conditions like in coal gasifiers, fluidized-bed combustion and in nuclear technology.

Degradation in gas turbines is predominantly temperature oriented, be it hot corrosion by chemical factors or corrosion induced by physico-mechanical factors, with cyclic conditions aggravating all types of corrosion. The temperature parameter is classified in three main ranges, where the low range is set over 600-750°C, the intermediate over 800-900°C and the high range above 1000°C. The regions 750°-800°C, and 900°-1000°C are overlap regions and are dependent on the specific alloy and coating system. All other working conditions are complementary. Most of the studies reported are carried out at atmospheric pressure but some high pressure rigs and high pressure laboratory studies have also been carried out (Conde et al 1982; Ma et al 1984; Ma 1983).

Interdiffusion plays an important role; coalescent Kirkendall voids reduce adhesion strength and cause coating failure by substrate parting; creep and thermal fatigue are common modes of turbine degradation. At high temperatures Al in an aluminide or an MCrAlY not only undergoes oxidation but it also diffuses inwards into the base metal (Smialek & Lovell 1974). At 1100°C, a CoNiCrAlYTa coating put through cyclic oxidation was shown to undergo interdiffusion with Co and Cr diffusing inwards while Ni diffused outwards (Mevrel & Pichoir 1984; Peichl & Johner 1986). In a NiCrAlZr/Ni22Cr system the Al was completely consumed in 15 h at the same temperature (Nesbitt & Heckel 1984). Linear porosities created by Kirkendall voids could reach as high as 70% in 400 h at 1100°C, and such porosity can lead to separation of coating from substrate even without any external influence (Peichl et al 1985; Gedwill et al 1982). Alloys such as MOO2 which contain W (9.5 wt% in this case) tend to form W-rich plate-like and cubic phases over quite narrow temperature ranges, e.g. found at 1050° but not at 1100°C in this case (Peichl & Johner 1986). MCrAlY coatings tend to develop precipitates of needle-like phases, often described as sigma-phases at the substrate/coating interface (Grisaffe 1972; Pichoir & Hauser 1980; Moorhouse & Murray 1979; Lang & Tottle 1981). These acicular precipitates induce embrittlement and introduce intergranular cracks proceeding to transgranular paths in the coating. Plate-like precipitates on aluminided IN100 showed nitrogen in Auger analyses at 95-101 ppm levels in the precipitates, against the 14-20 ppm in the alloy, which was tied up with Ti(C,N) and formed between 800° and 1000°C with a maximum at 950°C (Meisel et al 1980).

Localized low temperature hot corrosion occurred in CoCrAlY coating on a Stage 1 blade of LM2500 high pressure turbine after 600 h on a North Sea oil platform. Internal oxidation with a corresponding volume increase had generated enough mechanical stress to cause spalling. The influence of a molten salt attack was evident after 10000h. Spit defect incorporated during the EBPVD of a CoCrAlY caused lifting and spalling of coating on a similar turbine blade after 21500 h in service at intermediate temperatures. Other defects which can be EBPVD-process generated are leaders and flakes. Alloys with high chromium levels are reasonably resistant to intermediate temperature attack, but not lower chromium alloys which require protective coatings. Additions of Hf, Zr, Pt, Ce and Ta to MCrAlY-type coatings have improved the service-life. Superalloys are normally used in a coated state for high temperature operations because of the danger of losing Cr by formation of its volatile oxide CrO_3. Single crystal alloys, directionally solidified alloys and mechanically alloyed products have brought in considerable improvement in service life. 1%Hf incorporated as a laser-clad alloy with Ni, Cr and Al in the ratio 10:5:1 was found to improve oxidation resistance at $1050^{\circ}C$ of Inconel 718 when clad by underfocus lasing rather than an overfocus deposit. The NiCrAlHf alloy clad was uniform in the second phase with undissolved Hf particles. The reduction in oxidation rate was reasoned to be due to competitive reactions wherein the HfO_2 would act as sinks for excess vacancies, thereby reducing void formation and thus improving adhesion (Singh et al 1987).

Aluminiding with and without Pt, MCrAlY and thermal barrier layered coatings are further advances made in gas turbine alloy coating technology. Degradation in these coatings occurs by preferential loss of one phase, e.g. beta phase reduction in CoNiCrAlY produced by LPPS after 600 h cyclic oxidation at $1050^{\circ}C$ in a hot gas stream of 0.3 Mach and cyclic frequency of 1/h; the coating however, was still sound. An uncoated single crystal Ni-base superalloy MOO2 collected sub-surface voids within 400 h under the same conditions. A Pt-modified aluminide coating was observed to have the subsurface voids covering the whole of the gamma-prime depleted zone (Peichl & Johner 1986). Many overlay formulations of cobalt alloy compositions have also been made (Coutsouradis et al 1987).

A qualitative survey of gas turbine alloys hot corrosion resistance is shown in histograms, Fig. 8-3 to 8-7.

Gas Turbine Hot Corrosion Resistance, Isothermal, 700 °C

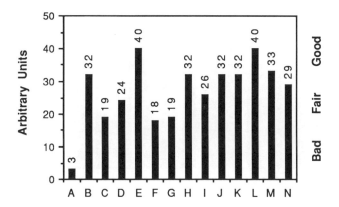

Fig.8-3: Hot Corrosion Resistance of Coated Superalloys

Legend	Substrate	Coating and Conditions
A	Superalloy	ZrO_2 ($pSO_3 = 10^{-2}$)
B	Superalloy	ZrO_2 ($pSO_3 = 10^{-3}$, no salt)
C	*	CoCrAlY
D	Superalloy	CoCrAlY
E	*	Co26Cr11AlY
F	*	CoCrAlY
G	Superalloy	CoCrAlY/Pt
H	Superalloy	Pt/CoCrAlY
I	*	NiCoCrAlYHf
J	*	NiCoCrAlY
K	*	NiCrAlY
L	*	NiCrSi
M	IN738LC	Cr
N	Nim80A	Cr

- -

Note: *indicates that the substrate used was not stated. Pre-annealing, NaCl content of test environment and other conditions cause big variations in hot corrosion rates. For these reasons it is very difficult to produce even semi-quantitative histograms like the ones displayed.

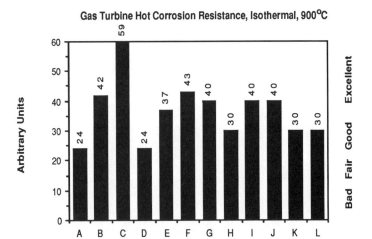

Fig.8-4

Legend	Substrate	Coating & Conditions
A	IN738	none
B	IN738	NiAl
C	IN738	57Co25Cr3Al10Ni5Ta0.2Y
D	IN738LC	none
E	IN738LC	Cr
F	IN738LC	Si
G	IN738LC	Si-Ti
H	IN738LC	CrNiSi
I	IN738LC	CoCrAlY
J	IN738LC	CrAlY
K	IN738LC	CrAlCe
L	IN738LC	PtAl

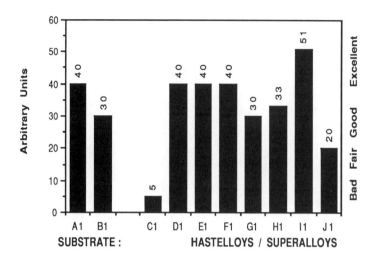

Gas turbine Hot Corrosion Resistance, Isothermal, 900°C

Fig.8-5

Legend	Substrate	Coating & Conditions
A1	Hastelloy-X	CrAlY
B1	Hastelloy-X	CrAlCe
C1	Superalloy	Ni16Cr6Al0.6Y/ZrO_2.12Y_2O_3 duplex coat
D1	Superalloy	Ni20Cr11Al0.4Y/ZrO_2.8Y_2O_3 duplex coat
E1	Superalloy	2CaO.SiO_2
F1	Superalloy	MgO-NiCrAlY Cermet
G1	Ni-Superalloy	PtAl
H1	Superalloy	CoCrAlY
I1	Superalloy	ZrO2 (pSO_3, no salt)
J1	IN700	CrAl

397

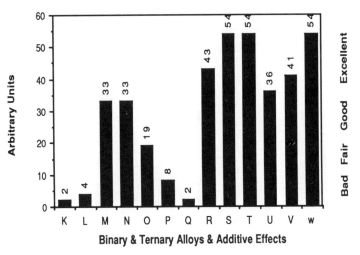

Gas Turbine Hot Corrosion Resistance, Isothermal, 900°C

Fig.8-6

Legend	Substrate	Coating / Conditions
K	B1900	none
L	Ni20Cr	none
M	Ni20Cr	Si^
N	Ni30Cr	Ni25Cr12Al1.1Y^
O	Co30Cr	Co22Cr6Al1.1Y^
P	Ni8Cr6Al	none
Q	Ni6Cr6Al16Mo	none
R	Ni15Cr6Al	none
S	Ni16Cr3.4Al	none
T	Ni16Cr3.4Al2Mo2W	none
U	Ni16Cr3.4Al2Mo2.6W	none
V	Ni16Cr3.4Al2Mo2.6W0.1C	none
W	Ni25Cr3.4Al2Mo2.6W0.1C	none

^only cases wherein coatings have been applied;
others have been tested as individual candidate
coating materials.

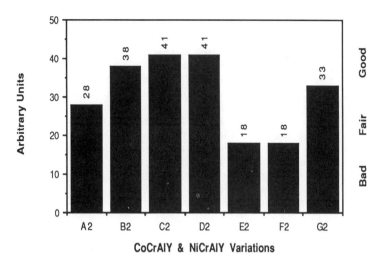

Gas Turbine Hot Corrosion Resistance, Isothermal, 900 °C

Fig.8-7

Legend	Coating /Single Material Tested *
A2	CoCrAlY*
B2	CoCrAlY*
C2	Co26Cr11AlY*
D2	CoCrAlHfPt*
E2	NiCrAlY*
F2	NiCoCrAlY*
G2	NiCoCrAlYHf*

- -

*Authors do not state the substrate

8.3.3. NUCLEAR TECHNOLOGY AND TRIBOLOGY ENVIRONMENTS

Corrosion problems in reactors are mostly encountered in the coolant and sealing systems. Carburizing and oxidation forms of degradation prevail which result in highly active oxide spallation. Carbon deposition interferes with heat transfer and causes erosion and crevice corosion problems. The reactant environment is generally composed of CO_2 + 2% CO with 300–350 ppm CH_4, 200–300 ppm H_2O, 200–300 ppm H_2. Carbonaceous deposition occurs from the catalysed degradation of a gas phase precursor derived from methane radiolysis, with the Fe and Ni constituents of the catalyst steel (Bennett et al 1984; Bennett 1982). Oxidation can be controlled and is not a serious problem but carburization is. Mechanical wear and abrasion is aggravated at high temperatures. Coatings vary from electroless Ni on steel for protection at 100°C in pressurised water reactors against boric acid, to aluminides and other coatings for much higher temperatures (Gras 1987).

Systems are tested in prototype nuclear processes for assessment. Be on steel, Monel, Mo, Cu and graphite (Dua et al 1985), and TiC on Mo (Fukutomi et al 1985) are used to reduce plasma contamination and power loss in Tokomak nuclear fusion reactors. Helium coolant in high temperature gas-cooled reactors can cause corrosion due to impurity inclusions such as CO, CO_2, H_2, CH_4 and H_2O. Incoloy 800, Nimonic 75 and Inconel 617 have been tested. An instability temperature T_i is tagged & found to be alloy dependent, particularly with activity of Cr. CO forms as a reaction product between the surface oxide and the carbon in solution and initiates the damage. T_i was found to be 907°C for IN 800H and 807°C for Nimonic 75. Decarburization occurs in the latter alloy at 850°C. Wet helium was found to suppress this reaction in IN 800H and reduce it in Nimonic 75 (Grimmer et al 1987). This temperature dependence is clarified further into three regions, and for Inconel 617, below a critical temperature of 830°C simultaneous oxidation and carburization occur in the helium coolant. Carburization reoccurs at 930°C with evolution of CO, thus severely limiting the alloy function. Between these two temperatures normal oxidation as in undiluted oxygen-containing gases, occurs (Christ et al 1987). Thermon 4972 is found to be similar to Inconel 617 in mechanical and creep rupture properties but with superior oxidation properties due to the formation of $Mn_{1.5}Cr_{1.5}O_4$, which prevented both further oxidation as well as a carbon transfer process (Huchtemann & Schuler 1987). Enhancement of oxidation by tellurization has also been recorded for an SUS 316 stainless steel fast reactor fuel cladding (Saito et al 1987). Significant reductions in weight gain and oxygen penetration of V-15Cr-5Ti (a blanket alloy for fusion reactor) was found when it was covered by vapour deposited Cr. The environment was He with 100 vppm H_2 impurity at 650°C. Without the protection of Cr, vanadium oxides cause severe degradation above 600°C (Tobin & Busch 1987).

The main concern of all aspects of degradation in nuclear technology is that the industry has to have a very long-term, fool-
proof solutions to all its problems in view of the prospective
risks involved when failure occurs. Physical degradation, and
tribological aspects predominate over chemical degradation; this
has been discussed in Chapter 7. The temperature range is wide
from 100° to $650^{\circ}C$ in which ferritic and austenitic steels as
well as low carbon steels are employed. Chemical degradation is
viewed in conjunction with the reducing effect of liquid sodium
coolant in fast breeder reactors, where Na can reduce the protective oxides on steel by forming sodium chromite. This leads to
the exposure of the bare metal to problems such as wear, fretting, static adhesion, galling and friction. 40 micron thick
aluminide coatings on IN 718 which are fabricated as slip rings
to fit recesses of grid plate holes have been successful. The
coated material withstood more than 3000 cycles between 530°-
$200^{\circ}C$ and damage was negligible even up to $627^{\circ}C$. D-gun coatings
are under review (Lewis 1987).

8.4. CHEMICAL DEGRADATION MECHANISMS

8.4.1. GENERAL:

The environmental conditions discussed in the previous section
show that the chemical factors controlling and influencing degradation are composed of several reactants under cyclic and isothermal conditions and mixed situations such as erosion and
stress. The coating which confronts them is often as nearly
multicomponent as the substrate itself. The mechanism by which a
multicomponent substrate/coating/reactant configuration leads to
coating degradation and termination of component service life is,
needless to say, complex.

The fundamental thermodynamic and kinetic principles which govern
the primary reactions are available in the references listed and
are beyond the scope of detailed conventional treatment here. The
judgement that has to be exercised in the context of practical
systems is in deciding how far results from an equilibrium state
or a simple system can be extrapolated. Many reactions occur in a
non-equilibrium state, the phase rule could be inapplicable and
mass transfer and heat transfer reactions have to make assumptions not always valid. Yet, fundamental studies are vital to an
understanding of the problem, especially in understanding diffusion phenomena. This section merely introduces work carried out
in the fundamental field with reactions involving only air and
oxygen for oxidation, followed by the more complex hot corrosion,
sulphidizing, sulphating and carburizing reactions. The influence
of NaCl which was recognized at a very early stage of gas turbine
research by investigators in the U.K. and in other countries in
Europe, is also considered. This chapter also includes a summary
of the outcome of a panel meeting for a co-ordinated attempt at

addressing fundamental aspects of high temperature corrosion which warranted further research (Rahmel et al 1985).

The general flowsheet for surveying the degradation mechanism starts with three main parameter blocks (Table 8:8):

TABLE 8:8

a. The Coating	b. The Environment	c. The Temperature
1. all metallic	1. all gas	1. low isothermal
2. all ceramic	2. gas+solid	2. high isothermal
3. metal & ceramic	3. gas+liquid	3. cyclic with iso-
	4. gas+liquid+solid	thermal residence periods

===

A few simple cases of chemical reactions with oxygen as the reactant are considered in the following three sections.

8.4.2. CASE 1: Bivalent metal + Oxygen = Metal Oxide

$$2M + O_2 = 2MO$$

$$\Delta G^o = RT \ln pO_2; \text{ or, } pO_2 = \exp(\Delta G^o / RT), \text{ where,}$$

pO_2 is the equilibrium oxygen pressure and ΔG^o is the free energy of formation for the reaction as written at T, the temperature in Kelvin. This is the baseline for a typical oxidation reaction. Most metals, precious metals excepted, form stable oxides at prevailing oxygen pressures at almost all temperatures. A few exceptions are elements such as Si, Mo, W and Cr, which can also form volatile oxides under certain operating temperatures and pressures. Fig.8-8,a-c, show the scale growth monitored as mass change viewed w.r.t. time.

(i) If the oxide scale is dense and protective, the scale growth will follow parabolic kinetics (Fig 8-8a) expressed as

$$dw/dt = k_p/w$$

where w is the weight change per unit area (can be expressed also in terms of oxide scale thickness, or thickness of metal lost/reacted or consumed), with time t, over an area A; k_p is the parabolic rate constant. The oxide layer, if coherent and completely covering, becomes a barrier between the coating it covers and the reactant, and with time, the oxidation rate levels off to a low value.

(ii) A non-protective scale will follow linear kinetics as in Fig.8-8b; the coating continues to oxidize until all of it is

402

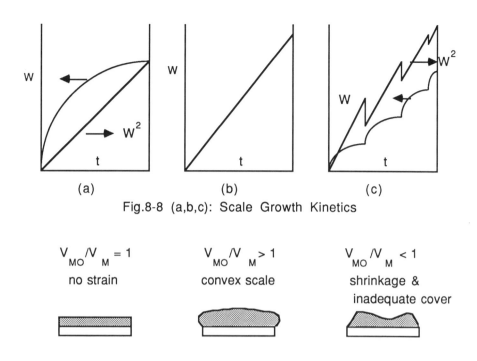

Fig.8-8 (a,b,c): Scale Growth Kinetics

$V_{MO}/V_M = 1$

no strain

$V_{MO}/V_M > 1$

convex scale

$V_{MO}/V_M < 1$

shrinkage &
inadequate cover

Fig.8-9 (a,b,c): Scale/Substrate relationship

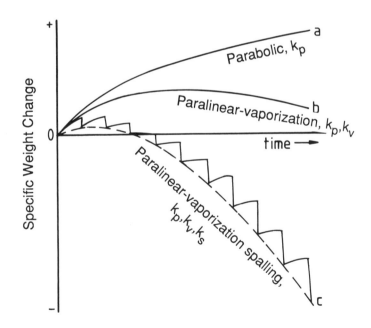

Fig.8-10: Typical Modes of Weight Change with Time
for Different Types of Oxidation Behaviour.

consumed.

Cases (i) and (ii) in Fig. 8-8 illustrate the simple configuration al+ bl at cl or c2 (using notation given in Table 8:8). When the configuration al + bl at c3 prevails, then the physical and mechanical stability of the oxide MO is subjected to the thermal stress-fatigue situation. The oxidation pattern shown in Fig.8-8c will then occur. The ratio of metal oxide molecular volume: metal atom volume influences the strain under which the metal oxide is stretched (or convoluted) laterally on the metal surface (Fig.8-9 a,b,c). The stress causes spalling even in isothermal conditions resulting in a make and break pattern. This is aggravated under thermal cycling conditions, c3. Configurations, especially a3 with b4 and c3, present the most aggressive conditions, introducing mechanical and physical factors to interact with chemical reactions, demanding scale adherence, strength, plasticity to accommodate thermal stresses and fatigue. Equations relating temperature and stress to predict the initiation of spallation have been proposed (Evans & Lobb 1984). Vaporization of one or more of the products along with constitutional changes arising from diffusion and phase related problems, when more than one metal is involved, push the substrate/coating endurance to extreme limits. Fig. 8-10 shows the mode of weight change with time when protective oxide, vaporization and spalling occur during oxidation.

8.4.3. CASE 2: Multiple elements + Oxygen = Multiple oxides:

The oxidation of each metal is governed by its activity in the alloy and the free energy of oxide formation; a metal oxide of higher negative free energy (lower equilibrium pO_2) can form at the dissociation pressure of an oxide with a higher pO_2 of oxide formation. Scale growth and oxide predominance is temperature dependent and is determined by the relative diffusion rate of the species, with the most stable oxide not necessarily being that with the most negative free energy.

More than one layer can develop in a scale. Thus:

$$(pO_2)^{1/2}{}_{(A, AO)} = [\exp(\Delta G^O_{AO}/RT]/a_A,$$

$$AO + B = BO + A$$

$$a_A / a_B = \exp[(\Delta G^O_{AO} - \Delta G^O_{BO})/RT], \text{ where,}$$

pO_2 = the oxygen potential, A and B are the two metals with the oxide of A reacting with B to form the oxide of B, and a = the activity of A or B as subscripted.

Ideally, all the metal oxides which are thermodynamically stable at the prevailing temperature and oxygen pressure should form in the same relative proportions as their activities at the coating-

/environment interface (not necessarily their activity in the alloy). A single oxide which is the most stable and can grow with lateral coverage predominates, or more than one oxide species which are more stable than the rest can form to produce mixed oxides. However, with time, diffusion, displacement as well as dissociation reactions interfere; and, the oxidation kinetics of just one or two of the constituents can predominate. There is very little basic thermodynamic data on spinel formation but spinels have been reported (Pettit & Goward 1982; Wood et al 1965, 1966, 1967, 1969, 1970a,b). Under isothermal conditions with protective scale growth, but for an initial depletion in the alloy constituents, particularly that of the predominant oxide, stable conditions will prevail. Under all conditions which lead to spalling or unlimited scale growth, there will be progressive element/s depletion at the alloy/scale interface to some depth into the alloy, with the alloy showing a deficit in the most predominant element that forms the oxide.

The operating temperature has a strong influence on oxide predominance. Thus a Ni-20Cr alloy at 700^OC is predominant in NiO while at 900° it is Cr_2O_3-rich. Ni-Cr-Al alloys can be NiO- , Cr_2O_3 - or Al_2O_3- formers as the %Cr and %Al in the alloy are increased. This mode of selective oxidation is manipulated for the benefit of achieving protective coating. The transient oxidation period over which selective oxidation occurs varies from alloy to alloy. Coating alloys are developed to have very short transient periods of oxidation of the order of seconds while structural alloys have an extended transient oxidation stage extending to a number of hours and days.

In service, thermally induced stresses are primarily responsible for the cracking and spalling of scales. Apart from the oxide thermodynamic stability, its plasticity becomes important. The most stable oxide again predominates during selective oxidation and with each subsequent scale repair step the transient oxidation period becomes longer. Eventually the alloy becomes severely depleted in the element that is selectively oxidised and a continuous scale growth of that oxide will cease. Degradation steps up as less stable oxides start forming incoherently, leading to eventual failure. Cr as an additive is an ideal example of this situation. Chromia, alpha-alumina and silica are used as effective barriers and coating alloys are designed for their selective formation. Chromia and silica however, are not stable as they can form gaseous products like CrO above 1000^OC, and SiO at low pO_2 as reaction products.

Four aspects are important in selective scale growth, retention and re-growth:

(i) The selective oxide must be capable of good lateral growth,
(ii) Diffusion processes must be slow once the 'barrier is up',
(iii) The alloy must have adequate activity of the metal for replenishment of the selective oxide which needs to re-form when damaged. [Coatings unlike substrate alloys are bound to

        fail in this aspect, because of their limited 'resources'],
(iv) Perhaps the most important - the oxide must have excellent
     thermal fatigue resistance.

Alpha-alumina formed on CoCrAlY offers an excellent example although there is much controversy about the actual mechanism by
which rare earth additions influence scale retention. Some workers argue on mechanical keying and pegging effects, i.e. influence on scale nucleation, development and morphology (Whittle &
Boone 1980; Whittle & Stringer 1980). Others argue on the rare
earth influence on oxide streses, the effect related to the
cationic mobility of Y in Al_2O_3 (Delarnay & Huntz 1982; Delarnay
et al 1980). Oxidized samples of CoCrAl and CoCrAlY were shown to
bend at room temperature (Giggins & Pettit 1975). Scale spalling
was far more pronounced in CoCrAl. The adherence of oxide on
CoCrAlY is argued not to be due to stress relief but because of
pegging. FeCrAl and FeCrAlY oxidation were compared, and found
that Y incorporated in alumina changed the growth mechanism,
reducing compressive stresses and the convoluted morphology (Wood
et al 1980; 1976).

The feature of prime importance for a coating is to guard against
thermal fatigue by good adherence. Prevention of void formation
or loss of contact between developed scale and the coating must
be ensured on one hand, while other means of maintaining scale
plasticity have to be exercised. Voids can manifest either by
diffusion and coalescence of Kirkendall voids and/or by gaps
created by mechanical lifting trapped by subsequent oxidation.
Additives are known to improve scale plasticity. Controversies
apart, a coating needs to develop an adherent scale which can
withstand thermal shock and not be incident to void coalescence
at either of the interfaces - its protective scale and itself, or
the interface at the substrate.

8.4.4. CASE 3:

Multiple Element + Multiple Reactants = Multiple Products

This is the most complex situation for coating degradation which
exposes Case 2 conditions to a variety of permutations and combinations.

(i)   Each gas reactant can compete for the same element to form
      its own reaction product;
(ii) Several elements can react with the same reactant;
(iii) Two different compounds can react on an interchange basis;
(iv) One compound and another element can react; and,
(v)   The most dangerous situation of all - Reactant deposition -
      /incorporation. Product formation and/or element distribu-
      tion can occur in a drastic change of phase of the reactant
      and/or the reaction products to result in catastrophic
      degradation.

Looking at the reactions:

A + X = AX ; A + Q = AQ ; AX + Q = AQ + X ; AX + BQ = AQ + BX ;
...... and so on

$$aX/aQ = \exp[(\triangle G_{AX}^{O} - \triangle G_{AQ}^{O})/RT]$$

The development of a single product layer, an oxide in particular, is still the best way to prevent further corrosion. And this product must be the one through which diffusion is slow. The case of thermal barrier ceramics is not considered in this context, since that introduces an aspect of inertness to the environment rather than the situation of a protective reactant product.

Oxygen-containing species such as CO, CO_2, SO_2, SO_3, and H_2O may be involved; or others such as H_2, H_2S, CH_4, He etc. with no oxygen radical may appear as reactants in a mixture or as impurities. Subscales of other products will form and can not be altogether prevented, as diffusional reactions occur with time. These products are stable although they cannot occur at the scale/gas interface. Repair of the stable scale subsequent to thermal fatigue for instance, will become a problem as the selective element has to diffuse through the partially damaged subscale-outerscale complex. Chromia former alloys with Cr_2N subscale or NiO top scale with Ni- sulphide subscale are good examples of this type of degradation. Although oxide barrier layers can be first formed at $pO_2 < 10^{-10}$, their repair and re-growth is not easy under reducing conditions with prevailing low pO_2 environments. Rapid degradation will then follow, e.g. in carburizing and sulphidizing environments. Diffusional growth of multiphase scales and subscales on binary alloys has been reviewed (Smeltzer 1987).

8.4.5. CASE 4: CYCLIC OXIDATION:

The ultimate test of a substrate/coating configuration is its response to the operating environment under cyclic temperature conditions. The physical, mechanical and chemical properties of the coating, the substrate and the scale become decisive factors, and the effect of cyclic conditions on the substrate/coating interface as well as the coating/scale interface indicate the survival capability of the system. Thermal and mechanical stress, fatigue, creep, thermal expansion effects, and the stability of the chemical reactants and products exercise a combined control on the coating performance and its protective oxidation characteristics.

The processes which lead to spalling of the oxide scale may be considered in terms of the relative strengths of the surface oxide and the oxide-metal interface. If the major source of stress is assumed to be due to the differential contraction strains produced during the temperature decrease, a critical temperature may be predicted. Oxide scale cracking or adhesion loss will result when the strain energy, W^*, per unit volume of oxide contained in the layer thickness, z, equates to the work required either for internal cracking, g_o, or for decohesion at the oxide-metal interface, g_f. The critical condition for oxide cracking is not explicitly dependent on its thickness, but the critical value of the temperature difference for interfacial decohesion decreases with increasing oxide thickness. Expressions for a strong or weak oxide/metal interface have been tested with 20Cr-25Ni-Nb-stabilized stainless steel (Evans & Lobb 1984a). Alloy depletion profiles prediced by selective oxidation under non-protective, i.e. spalling conditions, have also been presented (Evans & Lobb 1984b).

Cyclic oxidation histograms in Fig.8-11 - 8-13 give a qualitative assessment of coating/substrate combinations. Behaviour for 20 hour cycles may be quite different from that for 1 hour cycles, for the same material. Behaviour after 1000 hours of cycling may be quite different from that after 500 hours. Adhesion to the substrate is a decisive factor. Aerospace re-entry alloys (Ta-10W) are good for short times and higher temperatures. Different test conditions used by different investigators make it very difficult to draw up even a semi-quantitative histogram like those displayed in Fig.8-11 - 8-13. Fig.8-14 illustrates cyclic oxidation and corrosion effects (Giggins & Pettit 1980). The effect of cyclic oxidation on CoCrAl and NiCrAl alloys is shown in Fig.8-15 (Nicoll 1984). Additive effect on oxidation is illustrated in Fig.8-16, where Hf and Y added to CoCrAl show the more beneficial effect of Y and the critical level effect of Hf.

Alloy existence diagrams are of great value in interpreting the fundamentals of hot corrosion phenomena. The standard free energy change for the formation of compounds is plotted as a function of temperature, and isothermal diagrams are drawn for the compound stability in mixed atmospheres. All the compounds are plotted for the normalized gas pressures and reactant activities, so that the hierarchy of stability remains constant w.r.t. temperature. Superposed data indicate the shifts expected due to the activity changes of a particular species. However, it must be emphasised that equilibrium conditions mandatory in these diagrams expose them to the criticism that they cannot clarify practical situations which are not always in equilibrium. Superposed existence diagrams have shown their utility which may not be ignored. Two simple illustrations of reactions of Ni and Cr are included here (Fig.8-17a,b; Hocking & Vasantasree 1976). Many other complex diagrams showing variations due to changing activities and partial pressures are available in the literature (e.g. Perkins & Vonk 1979; Natesan 1983; Jacob et al 1979; Rapp 1987).

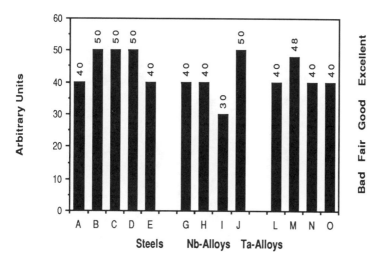

Cyclic Oxidation Resistance: 1000 °C in Air, 500 h

Fig.8-11

Legend	Substrate	Coating & Conditions
A	Steel	PSR1, 800 °C, 6000 h
B	20Cr25NiNb stainless steel	CeO_2 or SiO_2, CO_2, 825 °C, 15 000h
C	Cr5W0.1Y	Cr5Al8
D	Cr5W0.1Y	(Fe-Cr)xAly
E	Cr5W0.1Y	Ni30Cr20W4Al+Barrier (barrier is W-1ThO$_2$, W25Re or W)
G	Nb	VSi_2
H	Nb-alloy	Si20Cr20Fe
I	Nb-alloy	Si20Cr20Fe
J	Nb-alloy	Cr-Si-Ti
L	Ta10W	VSi_2
M	Ta10W	Sn23Al5Mo
N	Ta10W	$MoSi_2/ZrO_2$ (duplex coat)
O	Ta10W	WSi_2/ZrO_2

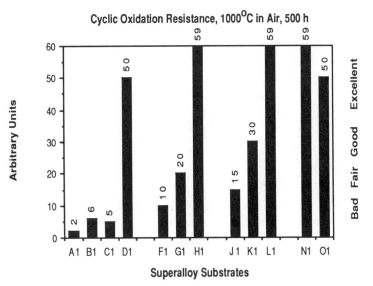

Fig.8-12

Legend	Substrate	Coating & Conditions
A1	MAR-M200	none
B1	MAR-M200	Al
C1	B1900	none
D1	B1900	Al
F1	IN939	none
G1	IN939	Si
H1	IN939	SiB
J1	IN738	+Si (solution)
K1	IN738	Si
L1	IN738	SiB
N1	Superalloy	ZrO_2 .12Y_2O_3/NiCrAl0.6Y (duplex coat)
O1	Réné80	Ni20Cr5Al0.1Y0.1C

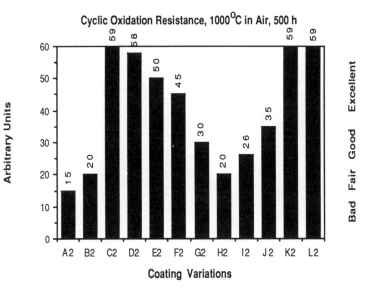

Fig.8-13

Legend	Substrate	Coating & Conditions
A2	*	CrAl
B2	*	CrTiSi
C2	*	CoNiCrAlY
D2	Co17Cr12Al0.8Y	
E2	Co25Cr6Al0.5Y	
F2	*	Co26Cr11AlY
G2	*	CoCrAlTaY
H2	*	NiCrAl
I2	Ni17Cr11Al0.5Y	
J2	*	NiCrAlSiY
K2	*	NiCrAlY + Pt(PVD-sputtered)
L2	*	NiCrAlY + Al (PVD +pack)

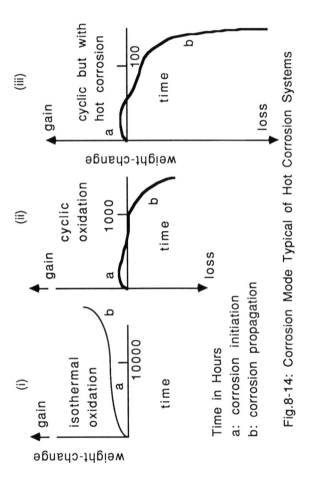

Fig.8-14: Corrosion Mode Typical of Hot Corrosion Systems

Time in Hours
a: corrosion initiation
b: corrosion propagation

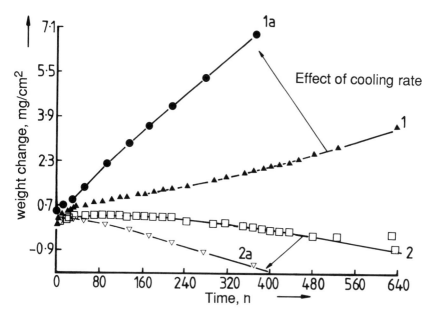

Fig.8-15: Cyclic Oxidation of CoCrAl & NiCrAl Alloys. Cooling Rate Effect.
(Nicoll 1984)

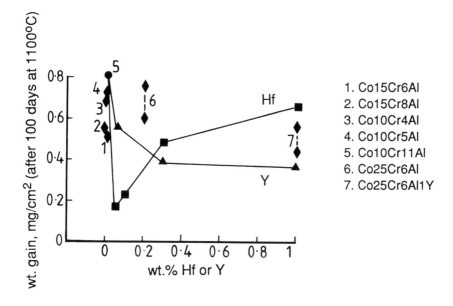

1. Co15Cr6Al
2. Co15Cr8Al
3. Co10Cr4Al
4. Co10Cr5Al
5. Co10Cr11Al
6. Co25Cr6Al
7. Co25Cr6Al1Y

Fig.8-16: Co-10Cr-11Al at 1000°C for 100 h in Air. Effects of Hf & Y.

413

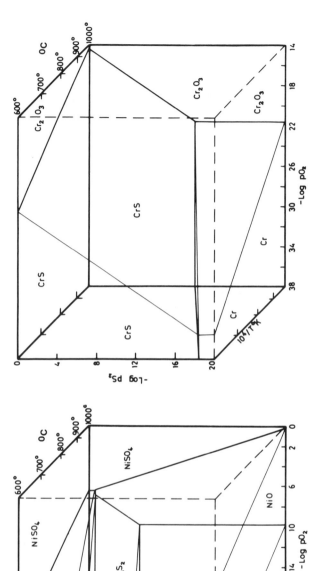

Fig.8-17a: Existence Diagram for the SystemNi-S-O Fig.8-17b: Existence Diagram for the SystemCr-S-O

600° - 1000°C

8.5. OXIDATION & HIGH TEMPERATURE CORROSION

Oxidation is often used to refer to corrosion in general, in addition to expressing degradation as a result of reaction with oxygen. Many of the early concepts, e.g. the classic Wagner model for diffusion and the Pilling and Bedworth ratio of scale to substrate molecular/atom volume ratio, evolved using oxides as examples. However, high temperature reactions have become quite complex and the simple framework indicated in the previous section has to be considered with respect to many systems undergoing corrosion simultaneously. The outcome of a reappraisal may indicate the several features which need to be addressed for scale growth, composition and stability (Rahmel et al 1985):

1. Adsorption, nucleation and initial stages of oxidation (Grabke 1985).
2. Complex defects in metal oxides and sulphides; their nature and transport mechanism (Hobbs 1985).
3. Lattice, line defect and grain boundary transport mechanism (Atkinson 1985).
4. Transport of gaseous species in growing oxide scales (Mrowec 1985).
5. Atomistics of scale growth at the scale/gas interface (Rapp 1985).
6. Lateral growth in oxide scales (Smeltzer 1985; 1987).
7. Adhesion mechanisms - the reactive element effect (Stringer 1985; Moon & Bennett 1987).
8. Key mechanical properties of oxidising components in maintaining scale integrity (Manning 1985).
9. Interaction between oxidation, sulphidation, carburization and creep (Ilschner & Scutze 1985).
10. Internal and intergranular oxidation of alloys (Yurek 1985).
11. Influence of impurities such as sulphides on the transport properties and nature of oxide scales (Wagner 1985).
12. Faults, fissures and voids in scales (Graham 1985).
13. Role of chlorides in high temperature corrosion (Hancock 1985).
14. Hot corrosion (Pettit 1985).
15. Role of oxides in abrasion, erosion and wear (Stott 1985).
16. Oxidation of compounds (Schmalzreid 1985).
17. New methods for studying high temperature corrosion, particularly in situ methods (Rahmel 1985).

It is beyond the scope of this review even to summarise the outcome of the above appraisal; coatings are an integral part of a high temperature system as more and more high temperature structural components are coated. Many of the available studies are conducted on individual candidate materials with a view to their utility as coatings, and later followed up by both corrosion and mechanical property studies in a substrate/coating mode. Information condensed herein must be viewed as such.

Oxide layers are not always protective although they predominate scale morphology in most respects. They are liable to damage, non-stoichiometry, mixed phases and stresses and present a complex diffusion pattern depending on scale morphology. Thus oxidation has to be viewed in many aspects w.r.t. scale growth, where plasticity is affected (Douglass 1969), where stresses generated in columnar scales lead to pit formation (Louat & Sadananda 1987), or lamellar stresses which occur in plasma spray coatings affect its adhesion and transport properties (Knotek & Elsing 1987). Defects in oxide scales influence transport properties, and at intermediate temperatures short-circuit diffusion paths can take over and become the predominant mode (Gesmundo 1987; Moon & Bennett 1987; Moon 1987; Atkinson 1985; Gesmundo et al 1985; Chadwick & Taylor 1984, 1982). The marked influence which elements like Y and Ce have has been argued on the basis of pegging (Stringer 1987; Pendse & Stringer 1985). But increasing evidence appears to come from effects of grain boundary diffusion and nucleation (Moon & Bennett 1987; Smeggil 1987). Oxides of Ce, La and Y exert a strong influence on oxide composition, adhesion and the scale growth direction of Cr_2O_3-formation; the effect is more pronounced on Ni-25Cr than the Co-25Cr (wt.%) alloys at 1000° and $1100^\circ C$, and in this case the 'peg' theory does not apply (Hou & Stringer 1987). A similar effect was observed on stainless steels doped with Y, Ce and La enhancing oxidation resistance while Hf and Zr offered no benefit (Landkof et al 1984).

Dispersed oxides of Mg, La and Y and element Y in plasma sprayed NiCrAl coatings at $1150^\circ C$ and $1225^\circ C$ showed that yttrium oxide gave the best oxidation resistance which was influenced by the size and distribution of the oxide particles (Luthra & Hall 1986). La was ascribed to provide a diffusion barrier layer as Cr^{3+} diffused 40 times slower in $LaCrO_3$ than in Cr_2O_3 in Cr-La alloys (Tavadze et al 1986). Y_2O_3 dispersoids were found to influence the microstructure rather than providing oxide nucleation sites for Ni-20Cr in low pO_2 of 10^{-19} to 10^{-24} at $1000^\circ C$. It reduced Cr-evaporation, the oxide grain size and porosity (Braski et al 1986). A small amount of impurities enriched at grain boundaries may greatly affect the deformation characteristics and influence the mechanical and transport properties of the growing scales (Kofstad 1985). Implanted Y in chromia-forming Co-45 wt.% Cr alloys exhibited a x100 reduction in oxidation rate at $1000^\circ C$ in pure O_2, with the mechanism of chromia growth changing from cation to anion diffusion, and Y-segregation at Cr_2O_3 fine-grain boundaries. A solute-drag effect is proposed as the mechanism (Przybylski et al 1988a,b; Przybylski & Mrowec 1984).

Embrittlement, break-away oxidation, internal oxidation are a few other detrimental effects. Hydrogen embrittlement is a well-recognised form of corrosion in aqueous systems. Water vapour can play a large part in high temperature embrittlement as also H_2 and O_2. Thus pre-oxidation, envisaged as a means of prolonging the initiation stage of hot corrosion may not always be beneficial. Ni undergoes oxygen-embrittlement at high temperature; the onset of embrittlement is dependent on the pO_2. Results on Inconel 718 seemed to indicate that the first stage oxidation process occurs at the grain boundaries before chromia formation influenced embrittlement (Andrieu & Henon 1987). An inverse phenomenon can occur in alloys prone to break-away oxidation. In 9Cr-1Mo steels, general oxidation occurs at 500-600°C and above 700°C Cr_2O_3 is the dominant oxide (Khanna et al 1986). Such preferential temperature-dependent scale development is not uncommon in Ni- and Co-base superalloys.

A side-step development of a 20Cr/25Ni/Nb steel precluding Al in the Fe-base alloy is reported. The wt.% steel composition was 19.9Cr , 24.6Ni, 0.7Nb, 0.6Mn, 0.56Si, 0.04C, balance Fe, which was first selectively oxidised in a 20% cold worked condition in a 50:1 H_2:H_2O for 2 hours at 800°C or 1hour at 930°C. The scales which developed were Fe- and Ni- free, with an average of 0.4 - 0.8 microns of Cr_2O_3. The alloy showed improved resistance to oxidation and carburization. Its sulphidation resistance is not reported (see Fig.8-19) (Bennett et al 1984). Implantation of Ce and Y reduced the oxidation rate consistently by more than 50% at 750°-950°C (Bennett & Tuson 1988/89). Alloying Ce (0.001-1.00 wt.%) and CeO_2 had a similar beneficial effect on Fe-(10-20)Cr alloys at 1000°C, pO_2 0.13 bar (Rhys-Jones 1987). Ceria dispersion added by jet injection to carbon steel was beneficial to oxidation of carbon steel (Tiefan et al 1984).

Ni- and Co-base superalloys generally are formulated to be preferential chromia- or alumina-forming variety. The degradation modes of MCrAl-coatings will be summarized later in this chapter. There are many studies on their oxidization behaviour (Wood & Whittle 1967; Wood & Hobby 1969; Wood et al 1970, 1971; Benard 1964; Kubaschewski & Hopkins 1961). Co effect on the oxidation of Ni-base alloys was found to lower the cyclic oxidation resistance on high-Cr alloys and was to be an optimum 5% to be effective in Al-containing alloys when tested in static air at 1000°, 1100° and 1150°C. The ratio Cr/Al decides the Cr_2O_3/chromite spinel (Cr/Al>3.5) or the Al_2O_3/aluminate spinel (Cr/Al <3.0), with the latter having a better resistance to cyclic oxidation. Refractory metal additions (Ta, Nb, W and Mo) were beneficial, with Ta the most effective. In all cases, any factor which promoted NiO formation resulted in scale breakdown (Barrett 1986)

Ti, as a light , high temperature metal is a prime candidate for aerospace vehicle systems. The metal oxidizes readily and also is prone to embrittlement (Shenoy et al 1986; Strafford et al 1983; Datta et al 1983; Strafford 1983; Datta et al 1984). It is found to manifest a moving boundary parallel to the interface as the two oxides change in proportion (Unnam et al 1986). Coatings of Al by EB, sputtered SiO_2, and CVD silicate via silane and borane, and mixed coatings of Al with the latter two were tested on Ti-6Al-2Sn-4Zr-2Mo foils and oxidized. Al+SiO_2 presented the best barrier layer performance with no effect on its mechanical properties (Clark et al 1988). Amorphous metals are not widely studied for their high temperature characteristics. Ti, Zr and Hf

417

affected void formation at the interface in a Ni_3Al-0.1B alloy. Hf addition was the most effective in promoting a protective oxide layer (Taniguchi & Shibata 1986). Studies on a rapidly solidified Ta-Ir alloy at 500 and 700°C are also reported. Ta was selectively oxidized, while the Ir coalesced into platelets of Ir-rich crystalline alloy oriented roughly parallel to the oxide-alloy interface. Although the unoxidized core remained in the glassy state, dissolved oxygen and the oxidation process had embrittled it (Cotell & Yurek 1986). Spectacular stratified oxidation layers develop on Ti (760°, 960°), Ta (900°) and Nb (450°, 600°C), in pure oxygen as demonstrated by Rousselet et al (1987).

8.6. HOT CORROSION

Hot corrosion has been variously defined but always involves the presence of sulphur species in the environment. Hot corrosion attack is recognized by the fact that a protective layer loses its resistance characteristics to shield the system it covers and this may lead to catastrophic failure. In general, the useful life of a coating in a corrosive atmosphere is very much a factor of time. There is a period over which it develops a protective scale; once the protective coating forms, its endurance to stress and temperature changes decides the subsequent breakdown of the scale. Coating degradation can be rapid or restrained, depending on the coherence, plasticity and shock proof qualities of the outer scale, the stability and compactness, free of voids of the intermediate scale, and, the stress and embrittlement accommodation and adherence of the scale/metal interface. Hot corrosion aggravates the scale breakdown mode and occurs in two recognizable stages, and it is particularly severe where a liquid phase is involved, irrespective of whether it is a fluid product or a fluid reactant.

Monitoring coating life thus can be considered in two stages:

Stage 1 - Scale initiation, retention and repair;
Stage 2 - Scale damage, weak repair, and rapid deterioration.

Stage 1 is termed the initiation stage and stage 2 is the propagation stage. Fig.8-14 (iii) (p.412) shows the two stages in environments ranging from mild to severe.

8.6.1. FACTORS AT THE INITIATION STAGE OF HOT CORROSION:

The overall factors which govern the onset of hot corrosion are (Giggins & Pettit 1979; Condé et al 1982):

1. Coating condition and its composition;
2. Gas composition and velocity;
3. Reactive particle/salt composition, its deposition rate and

existence state;
4. Temperature - isothermal/cyclic;
5. Non-reactive particle inclusion, deposition and impact;
6. Component geometry.

Hot corrosion is always a secondary reaction in the degradation process. It originates at the initiation stage, reduces the scale coherence and thus its protective life time there itself and follows it up with a catastrophic attack at the propagation stage. There have been several discussions on it which are available in the various references cited. Several contributory factors appear at the initiation stage; opinions differ about the definition, the factors of importance and the mechanism. One of the very common featurs is that a liquid phase is involved. Although in carburization this does not occur a liquid phase involvement is common with reactants where sulphur- or chloride media are present and also certain oxides, e.g. oxides of V. Gaseous oxide formation such as CrO_3, SiO, oxides of W , Mo etc., are also to be considered as promoters of catastrophic corrosion. The role of a coating then is to prolong the initiation stage, under isothermal and cyclic hot corrosion conditions by developing stable scales with adequate creep and rupture strength.

8.7. CARBURIZATION/OXIDATION DEGRADATION

A typical situation of carburization occurs in coal gasification and fluidized-bed combustion processes. Fe-base alloys are the most widely used in this field and hence documented (Hsu 1987; Tachikart et al 1987; Debruyn et al 1987; Ramanarayanan 1987; John 1986; Kofstad 1984; Terry et al 1987).

The 4-step kinetics involved in carburization may be seen as follows:

1. Transport in the gaseous environment by flow or diffusion. Uncombusted hydrocarbons, CO, CO_2 or CH_4 can induce carbide formation in the matrix;

2. Transfer of carbon to the metal matrix by phase boundary and/or reduction reactions which result in carbon atoms;

3. The dissolved carbon diffuses inwards;

4. Reaction of carbon with any or all of the alloy constituents which have the free energy for carbide-forming at the available carbon activity, is accompanied by diffusion of these constituents to the precipitate.

Oxide layers can be destroyed if graphite (or coke) deposits or gets trapped in the growing scale, or if reducing conditions prevail. High carbon activities are possible as the pCO/pCO_2 ratio links to the metal/oxide equilibrium. The metal matrix itself and the oxide that grows preferentially on it determine the growth of the graphite reductant. For instance, graphite grows much faster on Fe than on Ni, but in the presence of H_2S, the growth is accelerated on Ni, while on Fe it is retarded. Internal carbide formation in alloys such as Fe-Ni-Cr is called 'metal dusting'. Often, the formation of the same oxide can allow or arrest carburization. Naturally formed oxide on a Fe-12Ni-20Cr alloy caused local carburization of the alloy by impurities present in a N_2-H_2 atmosphere, while the material was fully resistant under the same conditions once the oxide was sandblasted. At higher than $1100^{O}C$, the protective Cr_2O_3 can undergo reduction by CO to form Cr_3C_2 and/or Cr_7C_3, especially if coke deposition is in significant amounts, and the oxide is buried under it. Internal carbide precipitation can also occur in CO-H_2-H_2O atmosphere.

In the absence of a protective scale, carbon ingress into a Fe-Ni-Cr alloy is by phase boundary reaction and diffusion controlled. The presence of sulphur retards the transfer of carbon, but to curtail the solubility and diffusion, a high Ni/Fe ratio is needed and additives like Si. If the oxide is coherent, dense and adherent carbon cannot penetrate since it has no solid solubility in oxides, but if there are fissures and cracks or pores, then carburization is possible. The integrity of the scale could be hampered by creep fatigue or thermal cycling. Any factor which induces stress is thus conducive to carburization. Additions of Nb, Ce, Si etc., would control in this case. Formation of higher oxides, spinels or mixed oxides lower the resistance to carburization. Fig.8-18 shows the oxide failure modes by carburization (Grabke & Wolf 1987; Ramanarayanan 1987; Hsu 1987).

The following carbon pick-up was recorded for various Fe-Base alloys in Argon-5%H_2-5%CO-5%CH_4 environment (Rothman et al 1984): (ranking in order of increase of carbon pick-up)

$925^{O}C$, 215 h:
Cabot 214 < Cabot 800H < Multimet < Cabot 600 < Hastelloy S < Hastelloy X < Inconel 610 < Haynes 230 < Inconel 617 < 310 stainless steel;

$980^{O}C$, 55 h:
Cabot 214 < Cabot 800H < Multimet < Hastelloy S < Haynes 230 < Hastelloy X < Cabot 600 < Inconel 601 < Inconel 617

The influence of selective oxidation on 20/25/Nb stainless steel at 650° and $700^{O}C$ is given in Fig.8-19a,b. On electropolished surfaces of chromia-forming Fe-Cr-Ni and Cr-Ni alloys, the oxidation layer appears to form non-uniformly, while cold worked samples undergo uniform oxidation in H_2-CH_4 with very low pO_2 of 10^{-30} at $825^{O}C$ with carbon activity at 0.8. Removal of alpha-chromia during nucleation of the carbide M_7C_3 occurred followed by internal carbide precipitation. (Smith et al 1985a,b). At low pO_2 heat resistant steels are completely under the influence of carbides wth M_7C_3 developing beneath Cr_3C_2 and are not affected

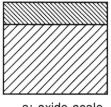

a: oxide scale

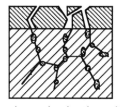

b. carburization via
cracked oxide layer

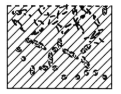

c. carburization with
no protective scale

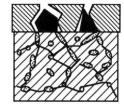

d: carburization with
graphite deposition

e: surface-converted
oxide-carbide

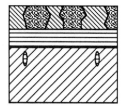

f. silica barrier layer

Schematic diagram of carburization with oxidation &
barrier layers

FIG.8-18

by brief periods of oxidation (Kinniard et al 1986). Breakaway oxidation and laminated structure morphology were observed at $600^{\circ}C$ on Fe-9Cr-1Mo steel in a high pressure CO_2/CO atmosphere (Newcombe & Stobbs 1986).

SiO_2 coatings by PAVD on IN 800H provided excellent resistance to carburization in H_2-CH_4 mixtures at $825^{\circ}C$, but were totally ineffective at $1000^{\circ}C$ where the SiO_2 was converted to SiC by a gas phase reduction. Partial degradation was found to occur even at 825° as Ti, Mn and Al additives in the alloy reduced the silica (Lang et al 1987). Silicide coatings are not protective on steels and if they are produced via a methyl-silane, carburization was found to set in at the coating stage itself (Southwell et al 1987). Ferritic steels with 6 wt.% Al showed good resistance to cyclic oxidation-carburization in H_2-H_2O-CO-CO_2 atmospheres with carbon activity 0.2, and Ti and Zr had very favourable effects in catalyzing the alumina phase transformation from the early theta- to the alpha-phase. They also increased the sintering rate and fracture toughness, while suppressing grain growth. None of these benefits were realized in the austenitic series tested with 4%Al, and Y also did not improve oxide adherence. Instead it caused grain boundary embrittlement. In creep tests the oxide layer cracked with subsequent intergranular oxidation and carburization (Wambach et al 1987).

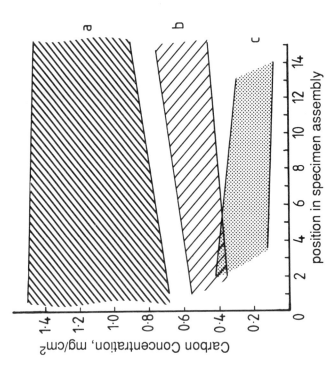

Fig.8-19b: a.20/25/Nb stainless steel; a. no treatment
b. selectively oxidised; c. selectively oxidised.

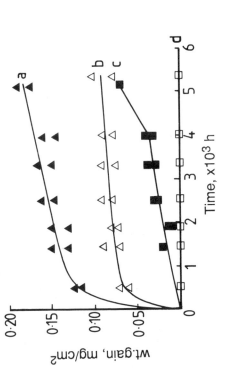

Fig.8-19a: a. 20/25/Nb steel at 700°C
b. selectively oxidised 20/25/Nb steel at 700°C
c. 20/25/Nb steel at 650°C
d. selectively oxidised 20/25/Nb steel at 650°C

Fig.8-19: Effects of Selective Oxidation. 20/25/Nb Stainless Steel
in $CO/CO_2/H_2/H_2O/CH_4$

In the above case the alloys were pre-oxidized prior to exposure to reactive atmosphere. Pre-oxidation has been found to be a deterrent, but not with sustained efectiveness. A coating of 63%Al-33%Cr-4%Hf on Incoloy 800 performed well at 980°C in coal char environment after pre-oxidation (Douglass & Bhide 1981). MA956, the mechnically alloyed product, yielded good mechanical stability of the alumina scale (Sheybany & Douglass 1988).

8.7.1. CARBURIZATION IN THE PRESENCE OF SULPHIDATION:

Carburization conditions are not encountered on their own in coal-derived atmospheres, but along with gases such as H_2-H_2S, and H_2-H_2O. The oxide layer which is perfectly resistant other-wise, is then exposed to conditions which render it unstable. In the previous section the 'inhibiting' effect of sulphur to carbu-rization was briefly mentioned. A number of environmental varia-tions have been studied: CO_2, CO, H_2 (introduced as forming gas), H_2S, with and without CH_4, chloride etc., with pO_2 variation brought in from the equilibrium CO/CO_2 and/or H_2/H_2O. Temperature appears to exercise a critical control on the degree of carburi-zation. At carbon activity 0.8 in H_2-CH_4, H_2S at 100 ppm at $pS_2=2.2x10^{-12}$ to $5.5x10^{-11}$, at 1000°C, carbide M_7C_3 on Fe-Ni-Cr alloys exhibited a preferred growth orientation in the [001] direction, and in commercial alloys surface carbides of M_7C_3 and $M_{23}C_6$ nucleated with MnS buried underneath, in contrast to sub-scale $M_{23}C_6$ and surface M_7C_3 embedded in surface alpha-Cr_2O_3 in sulphur-free atmospheres. An apparent reduction of 75% in weight gain occurs when pS_2 is introduced at $1.4x10^{-10}$, with significant decrease of internal carburization. However, sulphide scales are formed and the overall corrosion increases with corrosion at 950° greater than at 1000° and 1050°C (Barnes et al 1985,1986).

The inference from the above would be that although 'metal dus-ting' by internal carbide precipitation and external carbide formation can be arrested in the presence of sulphur, no ultimate benefit is obtained as one form of corrosion is exchanged for another. Internal precipitation reflects in material degradation by physical and mechanical factors such as creep and stress, and a marked loss in constituents will also result in chemical degra-dation. Surface scales formed by chemical reaction, on the other hand need to be plastic, stable and coherent; adverse factors will destroy them. Attack by sulphur manifests in catastrophic corrosion in the majority of cases, and such corrosion is often associated with a liquid phase. The effect of H_2-H_2S has been extensively studied and reported in literature. A few papers are listed here:- Weber & Hocking 1985; Majid & Lambertin 1985; Strafford et al 1983,1985; Floreen et al 1981; Gibb 1983; Grabke 1984; Norton 1984; Grabke et al 1980; Hemmings & Perkins 1977; Mrowec 1976; Mrowec & Werber 1975. The principal aspect of attack by sulphur-media is that one or more liquid phases are often involved.

8.8. THE LIQUID PHASE EFFECT

8.8.1. LIQUID PHASE FROM THE ENVIRONMENT:

In the marine gas turbine, sodium is picked up as NaCl, which reacts with sulphur and oxygen from the fuel + air mixture to form Na_2SO_4:

$$4NaCl + S_2 + 4O_2 = 2Na_2SO_4 + 2Cl_2.$$

The melting points of NaCl and Na_2SO_4 are 800° and 884^OC respectively. The two salts however, form low melting eutectics and can exist in a solid + liquid state over a range of composition and temperature (Fig.8-20a). The deposits which arrive at the hot turbine blade are composed of NaCl, Na_2SO_4, carbon (from the fuel) and perhaps dust/sand . They do not cover the entire blade but are deposited at discrete areas in the leading edge or at a middle concave (suction) region. The brunt of the deposition is taken by the first stage vanes and blades operating at 900 – 1150^OC. In coal gasifiers $CaSO_4+CaO$ provides a deposit and sulphidizing conditions and vanadium participates in the formation of molten oxides (Fig.8-20b). The eutectic $NaSO_4+NaVO_3$ forms at 610^OC.

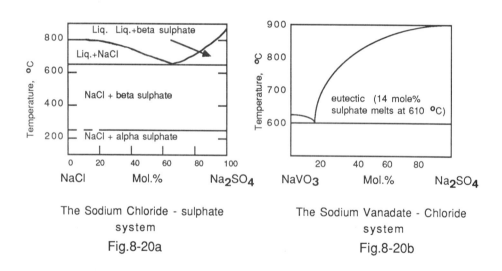

The Sodium Chloride - sulphate
system
Fig.8-20a

The Sodium Vanadate - Chloride
system
Fig.8-20b

8.8.2. LIQUID PHASE FROM ALLOY + GAS REACTIONS:

The gas enviroment in a turbine is O_2 predominant with a trace of sulphur which arrives as SO_2. The presence of metals can catalyse the reaction to give a mixture of O_2, SO_2 and SO_3. The metal / SO_2, SO_3, O_2 reaction mechanisms have been discussed elsewhere (Alcock et al 1969; Birks 1975; Kofstad & Akesson 1979; Seybolt & Beltran 1967). It suffices here to say that oxides, sulphides and sulphates form as reaction products under specific conditions. A few examples of the melting points of metal + metal sulphides of turbine and coal gasifier alloys are given below:-

Eutectic Ni + Ni_3S_2, m.p. = 645°C; Ni_3S_2, m.p. = 810°C;

Eutectic Co + CoS, m.p. = 879°C; CoS, m.p. = 1070°C;

Eutectic Fe + FeS, m.p. = 965°C

The more reducing environment in coal gasifier environments promotes Fe sulphide eutectics and in Ni-additive steels Ni-sulphides can appear as a liquid phase.

Liquid Phases from Alloy & Environment Reaction Product Interaction; a few examples :-

CoO or Co_3O_4 + SO_3 = $CoSO_4$; $CoSO_4$ + Na_2SO_4 ; m.p. = 565°C

NiO + SO_3 = $NiSO_4$; $NiSO_4$ + Na_2SO_4 ; m.p. = 670°C

$2NaVO_3$ + 2NiO = $Ni_2V_2O_7$ + Na_2O ; m.p. = 850°C,

$Ni_2V_2O_7$ + NiO = NiV_2O_8 ; m.p. >1000°C,

$2NaVO_3$ + NiO = $Ni(VO_3)_2$ + Na_2O ; m.p. 750°C,

$2NaVO_3$ + 3NiO = $Ni_3(VO_4)_2$ + Na_2O ; m.p. = 1210°C,

Al_2O_3 + $2NaVO_3$ = $2AlVO_4$ + Na_2O; $AlVO_4$ decomposes at 625°C.

Cr_2O_3 + $2NaVO_3$ = $2CrVO_4$ + Na_2O; $CrVO_4$ melts between 810° and 900°C.

8.8.3. THE EFFECT OF THE LIQUID PHASE:

The emergence of a liquid phase during a corrosion reaction will result in:

(i) drastic effects on diffusion control parameters, transport and mobility,
(ii) create electrochemical conditions - bimetallic cell situation between different metals and different phases,
(iii) dissolve protective reaction product oxides via acidic and/or basic fluxing,
(iv) facilitate fast transport of reactant species to the alloy /scale interface hitherto barred by coherent scales,
(v) physically undermine scale coherence by mass flow.

Products which vaporize, or react to form a vapour product and deposit in a more stable form on the cooler parts are well known. For instance in a chloride-containing environment reactive species can form chlorides which later oxidize. Mass spectrometric

Fig.8-21: Pre-Oxidised IN 738 Exposed to Sea Salt Melt at 750°C

studies by the authors (Hocking & Vasantasree 1976) showed formation of volatile TiS_2; needles of TiO_2 growing out of the Cr_2O_3
barrier layer may be expected to have formed via a vapour deposition mode (see Fig.8-21). Fig.8-22a,b show another needle-form-
ing system where liquid sulphide eutectic pipes up a column of
$Ni-Ni_3S_2$ to be rapidly covered by NiO (Hocking & Vasantasree
1976, 1982).

The catastrophic effect of Na_2SO_4 would have been confined to
temperatures just around $900^{\circ}C$ except for the chloride effect
first recognised in UK laboratories in the sixties. Much of
turbine hot corrosion was shown to fall into two categories – the
low temperature degradation in the region of $620-750^{\circ}C$ and the
high temperature corrosion over $850 - 950^{\circ}C$. Sulphides and sulphates largely responsible for hot corrosion of Ni - and Co-base
alloys are unstable or do not form beyond these temperature
regions. Aggravated corrosion occurs well above $900^{\circ}C$ if a floating potential of SO_2 is allowed to prevail in the reactant
(Vasantasree & Hocking 1976).

NaCl itself is completly converted to Na_2SO_4 within 3 minutes
(Conde et al 1977), but its continuous arrival and its stability
in solution with Na_2SO_4 at lower temperatures well below its
melting point mostly in a solid + liquid state and sometimes
liquid state is an important factor in low temperature hot corrosion. Co-base alloys are vulnerable to Na_2SO_4 itself as the CoO –
$CoSO_4$ – Na_2SO_4 reaction and solution progresses. Volatile chlorides are often intermediate products in hot corrosion, which
enhance the corrosion rate to result in more stable products and
also to form more stable reactants. Chloridation of Fe, Ni and
Fe-Ni alloys in pCl_2 10^{-3}, 10^{-5} and 10^{-7} with pO_2 range 10^{-11} to
10^{-15} indicates that the aggressive effect of the chloride medium
on Fe is greater between 800 and $1000^{\circ}C$, except for pCl_2 10^{-7}. A
limiting content of 50%Ni in the Fe-Ni system provided excellent
chloridation resistance; above $1000^{\circ}C$ Ni was inert in pCl_2 10^{-5}
(Strafford et al 1987).

Vanadic melts form the molten salt media encountered in low grade
fuel combustion and residual fuel oil ash hot corrosion (Johnson
& Littler 1963). The deposit-forming reaction may be of the type

$$wNa + xO_2 + yV + zS + aCl \text{ -----> } \text{deposit (Halstead 1973)}$$

in which Na, O_2, V, S and Cl are in the vapour phase, and can be
involved in several reaction systems.

8.9. FLUXING MECHANISMS IN HOT CORROSION

Thermochemical diagrams give a direction on corrosion in melts
(Pourbaix 1987) and the eletrochemistry of molten salt corrosion
is reviewed by several workers: Rapp (1987); Rahmel (1987);
Pourbaix (1987); Hocking & Sequeira (1982).

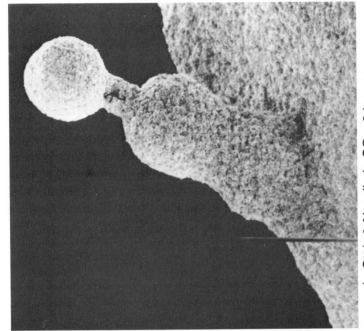

b: Stub & Needle in SO_2, 3 h exposure

Fig.8-22: Ni_3S_2 Core Within a Case of NiO Clusters. Needle & Stub Growth.
Ni-0.1%Cr Alloy

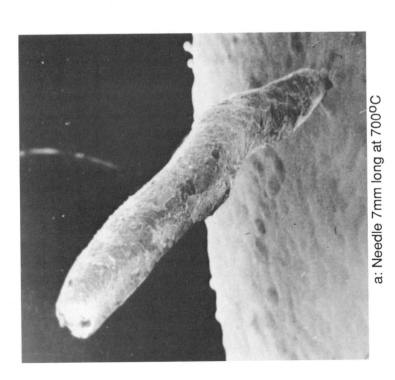

a: Needle 7mm long at 700°C

8.9.1. THE Na_2SO_4 MODEL:

The oxy-anionic Na_2SO_4 melt has been a model on which most of the acidic and basic-fluxing concepts have been developed. The concept can be extended to all molten salt metal product reactions as long as ion-exchange reactions can be applied validly to the reaction system. Thus sulphates, carbonates, nitrates and vanadates can be included in oxy-anionic reactions at high temperature (Johnson & Laitinen 1963; Cutler 1971). The primary factor in viewing molten salts with respect to hot corrosion is that of the availability of the melt and not its mass. An attack will be self-sustaining as long as the melt can participate in the exchange and remain as the intermediate means by which alloy component elements will eventually react to solid corrosion products. In the turbine operating conditions melt can form as and when the component particles are deposited. Products can remain in solution with the melt, form a eutectic, precipitate out or form a solid complex with the total mass of available melt.

The following reactions will clarify the various points noted above:

$$Na_2SO_4 \; \text{-----------} \rightarrow \; Na_2O + SO_3$$

$$SO_4^= \qquad\qquad\qquad O_2^= + \; SO_3$$

$$\text{Acidic} \qquad\qquad\qquad \text{Basic} \quad \text{Acidic}$$

$$1/2S_2 + (3/2)O_2$$

8.9.2. ACIDIC FLUXING:

Hot corrosion reactions occur where $SO_4^=$ participates with the S_2, SO_2 and SO_3 species from the gas or dissociated melt in converting the alloy to corrosion products either by chemical thermodynamic reaction or electrochemically transported as an ion for a subsequent reaction with the gaseous medium. Thus for an alloy AB,

(i) $A \; (alloy) + SO_3 + (1/2)O_2 = A^{++} + SO_4^=$

For a continuous solution of ASO_4 in Na_2SO_4, SO_3 and O_2 must be available, e.g. $CoSO_4 + Na_2SO_4$,

(ii) $\underset{\text{alloy}}{A^{2+}} + \underset{\text{melt}}{SO_4^=} + (1/2)O_2 = \underset{\text{solid}}{AO} + SO_3$

(iii) AO can remain in solution with Na_2SO_4 melt if there is a negative solubility gradient [note this cannot happen in a small mass or thin layer of melt].

429

(iv) $B \text{ (alloy)} + SO_4^= + (3/2)O_2 = BO_4^= + SO_3$

or

$A\text{(alloy)}B\text{(alloy)} + 2O_2 = A^{2+} + BO_4^=$ (solution in melt)

$$A^{2+} + BO_4^= = AO + BO_3$$

The melt remains as a via media; very small amounts of Na_2SO_4 can permit a substantial alloy-to-alloy metal oxides conversion.

In practical systems virtually all alloys are susceptible to acidic fluxing depending on the level of pSO_3 or the amount of V_2O_5 formed or deposited. The condition is prevalent when chloride is present and induces alloy depletion in gas turbine environment and in carburizing conditions when oxygen starvation occurs. Alloys containing Mo, W or V are very vulnerable because they can be auto-generative to acidic oxides, e.g. B-1800, IN 100, Mar M-200 etc. High-chromium alloys present a reasonably good resistance, as well as silica-formers. IN 738 and some Hastelloys are found to out-perform the above listed alloys. Refractory metal additions has to be restricted to avoid acidic degradation. Hot pressed Si_3N_4 performs well in acidic fluxing conditions (Giggins & Pettit 1979).

8.9.3. BASIC FLUXING:

Basic fluxing of the reaction product occurs when the alkali Na_2O or the $O^=$ part of the oxy-anionic melt participates in the reaction process.

(i) $A \text{ (alloy)} + O^= + (1/2)O_2 = AO_2^=$

Na_2SO_4 is converted to Na_2AO_2 and ceases unless fresh melt Na_2SO_4 is available.

(ii) $A \text{ (alloy)} + O^= + (1/2)O_2 = AO_2^= = AO \text{ (solid)} + O^=$.

Melt can act as a transport - precipitation reaction medium as long as it is balanced by an SO_3 supply; else, it will stifle the reaction when sufficiently basic. This means, that for this reaction to be possible there should be a negative solubility gradient of the oxide in the melt (Stroud & Rapp 1978).

Viewing melt fluxing in the context of protective oxide scale formation it may be generalized that:

Cr_2O_3 is more resistant under low pSO_3 conditions, i.e. basic conditions,

SiO_2 has minimal solubility in high pSO_3, i.e. it is good to resist acidic fluxing,

Al_2O_3 has a lower solubility at low pSO_3 than Cr_2O_3.

Degradation due to basic fluxing can be resisted effectively by promoting continuous scale growth of Al_2O_3 under Cr_2O_3. Since it requires an oxygen gradient for promotion, the best means of counteracting it is by formation of oxide scales which grow at a slow rate and need very low pO_2. In Ni-Cr-base superalloys it is better not to have any Al at all than Al in a low level since the chloride effect is particularly marked in low-Al Ni-Cr-Al-alloys. Deposits of carbon are observed to hasten the onset of basic fluxing as it creates local reducing conditions while a 5 micron surface top coat of Pt inhibits basic fluxing.

Susceptibility to basic fluxing occurs with Ni and Co and their alloy systems, binary e.g. Ni-Al and Co-Al, ternary Ni-Cr-Al, or multi-element systems where Cr and Al levels are lower than is required to form their stable oxides. Alloys with 20Cr or more and 10-12 Al with reactive elements such as Y have good resistance in basic fluxing media. If Al has to be lowered for mechanical purposes, then the preference is given to a CoCrAl system rather than a NiCrAl alloy. Alloy depletion caused by chloride reactions and carbon induced oxygen depletion are, once again, the contributory factors (Giggins & Pettit 1979).

8.9.4. STUDIES IN MELT SYSTEMS:

8.9.4.1. Sulphate-Chloride Systems:

Several workers have studied the molten salt as an electrochemical system and corrosion medium since the early studies on fluxing effects (DeCrescente & Bornstein 1968; Bornstein & DeCrescente 1969, 1971; Goebel & Pettit 1970, 1973). The sulphate – chloride provides a molten salt system over a wide range of temperatures relevant to gas turbine conditions. The binary eutectic of the sodium salt (Hocking & Sequeira 1980, 1981,1982) and the ternary system with Mg- and Ca- sulphates have been studied (Swidzinski et al 1978; Swidzinski 1980; Carew 1980; Hocking et al 1984; Rapp & Goto 1979; Gupta & Rapp 1982; Shores & Fang 1981; Shivakumar et al 1985; Kameswari 1986; Rapp 1987; Rahmel 1987).

Although alloy composition can be formulated for degradation by melts, it is more difficult to resist gas-induced acid fluxing. Both the temperature and a high pSO_3 (hence pS_2) favour this mode of degradation and attack is inevitable at low temperatures (700°C) or high, as long as a stable phase of liquid is generated. Cr_2O_3 and SiO_2 give the most favourable durations of resistance. Acidic and basic fluxing can succeed one another as is shown in an attack on B-1900 and IN 738 by Na_2SO_4 (Fryburg et al 1982, 1984). Alloy induced acidic fluxing is generated from elements like Mo, W and V and aggravated by chloride and carbon from the reactant environment.

Na_2SO_4 melt caused basic fluxing of alpha-alumina on alloy B1900 which was followed by a catastrophic attack and acidic fluxing by a Na_2MoO_4-MoO_3 molten phase that had formed underneath the oxide layer. In the case of Na_2SO_4 induced hot corrosion on IN738 at $975^{o}C$ the protective Cr_2O_3-TiO_2 scale was basic fluxed in the first 10 hrs. A long slow, linear oxidation rate ensued over the next 50 hrs and gave over to a rapid corrosion which decelerated as reactant availability was reduced considerably. MoO_3-WO_3 formed at the oxide - alloy interface during the slow stage; the fluxed Na_2CrO_4 and Na_2O $(TiO_2)_n$ reacted with MoO_3 to form the low melting Na_2MoO_4 and Na_2WO_4 which lowered the melting point of the MoO_3-WO_3 areas by fluxing and resulting in catastrophic attack. The difference from an earlier work (Giggins & Pettit 1981) is that the liquid phase is argued to be Na_2MoO_4/ MoO_3-WO_3 not the Na_2SO_4-acidified MoO_3-WO_3.

A number of solubility and stability studies of oxides, sulphates, sulphides and chromates in a Na_2SO_4 melt have generated activity diagrams w.r.t. to the Na_2O activity at $900^{o}C$ and above (Rapp 1987). Fundamental treatments for the corrosion potential and rate calculations can be referred to in the literature (Rahmel 1987). NaCl itself can enhance stainless steel corrosion at its melting point and above ($800^{o}C$) (Shinata et al 1987), but even in the solid state it can affect scale coherence (Hancock 1985). Because it can form low temperature eutectics with sulphates, sodium sulphate in particular, it is important to monitor melt effects at lower temperatures. The effect of NaCl has been examined by corroding alloys in melts of Na_2SO_4-NaCl and ternary sulphates of Mg-, Ca-and Na, to which additions of NaCl were made. These allow observation of melt attack at low temperatures in the range $650^{o}C$ and above (Swidzinski et al 1978; Swidzinski 1980). Products such as cobalt oxides and $CoSO_4$, form eutectics with Na_2SO_4 at $575^{o}C$ and lower at $535^{o}C$ with sodium and potassium sulphates for which the stability regions have been shown (Luthra & Leblanc 1987).

Tests at $850^{o}C$ show how the sulphate-chloride has significant detrimental effect on creep and fatigue behaviour of both IN 738 and Udimet 500 (Pieraggi 1987). Coated Hastelloy X and Haynes 188 fuel injector tips were tested in 95-5 sodium sulphate-chloride for cyclic oxidation at 982° and isothermal oxidation at $1150^{o}C$. NiCoCrAlY-YSZ (Yttria stabilized zirconia) showed poor adherence and spalling at 1150°. YSZ was the most resistant and Pt-aluminide the least; in the cyclic tests in the melt, the MCrAlY overlay was the best and the YSZ was the least resistant with the Pt-aluminide in-between. Vitreous ceramics failed in both the melt tests (van Roode & Hsu 1987). Low alloy steels (Kraft recovery boiler alloys) develop chromia scales which are not stable in the oxo-basic carbonate added sulphate-chloride melt at $800^{o}C$. Ni-rich alloys 600 and 800 presented the best resistance (Petit & Rameau 1987). A transition from ductile to local corrosion cracking occured in AISI 304L stainless steels at $570^{o}C$ in a NaCl-$CaCl_2$ melt, leading to intergranular fracture (Atmani & Rameau 1987). Chloride pollution of a (Na,K)- nitrite and nitrate melt mixture used as coolant in solar thermal electric power plant was detrimental to chromia scales on low alloy steels at temperatures as low as 450°C. The hottest part at 500°C was well resisted by AISI 316L stainless steels (Spiteri et al 1987).

Non-protective scales form on alloys due to several processes, and where a molten phase is involved exchanges via donation to the 'acid' or acceptance from the 'basic' oxide part of the melt will occur, neither of them requiring an actual dissoluiton and precipitation of the oxide product (Pettit 1985). Work quoted above gives a cross section of the sulphate-chloride effect on the stability of coatings developed at high temperature, over an entire range as low as 450°C and as high as 1100°C. The chloride effect is particularly detrimental to creep and fracture resistance of prospective scales which has been summarized with its corrosion effect (Hancock 1985).

8.9.4.2. Studies in Molten Vanadates:

Vanadic melts as oxy-anionic melts, are much more aggressive as the oxides can exist in multiple valency and form a number of oxides as products, decompose and re-react. The vanadate melting points range from 750° to 1210°C for the NiO-Na-vanadate reactions alone and are much lower in oxide melts. The vanadic melt attack is more self-sustaining and stable and therefore more harmful (Vasu 1964; Richard 1971). In Na-vanadate melts non-protective $CrVO_4$ forms and Al was detrimental in Na_2SO_4 + $NaVO_3$ melts for Ni-base alloys (Sidky & Hocking 1987). In a simple system of Ni in V_2O_5 at 900°C, the increase in corrosion by a factor of 3 has been explained as due to enhanced cationic diffusion along defective grain boundaries and pores of NiO in which (Ni,V) oxide is located (Chassagneux et al 1987). CoCrAlY coating alloy exhibited a retardation in sulphate – SO_3 attack at 700°C in the presence of V_2O_5. $Co_2V_2O_7$ developed as a surface layer seemed to act as a barrier layer for the inward diffusion of SO_3 and O_2. The effect is deemed to be temporary until the vanadate is decomposed and yields to the eutectic sulphate attack mechanism (Jones & Williams 1987). Vanadate attack occurred on Ni and MCrAlY on IN 600 with pre-formed vanadate deposits. Ni content at >75% produced a $Ni_3(VO_4)_2$ layer which retarded vanadic attack (Seirsten & Kofstad 1987). Aluminosilicates are easily fluxed by both sodium and vanadium oxides over a range of 600 – 1300°C, with the effect more pronounced at the higher end, while the effect of SO_3 was found to be higher up to 900°C (Mascolo & Marino 1987). Fig.8-23 illustrates vanadic attack on Ni-20Cr with and without protection of a Si coating.

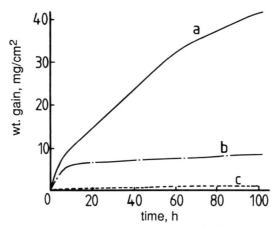

a: Uncoated Vanadic attack; b: Silicon Coating in Vanadic Slag; c: oxidation
with and without coating

Fig.8-23: Vanadic Corrosion & Oxidation Kinetics
for Ni20Cr With & Without Silicon Coating
at 900°C.

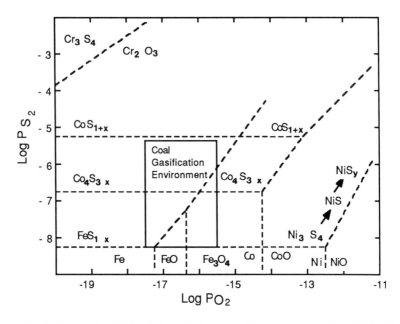

Fig.8-24 a: Sulphide-Oxide Existence Diagram for Fe, Ni & Co
at 871 °C , & the Coal Gasification Regime

8.10. HOT CORROSION IN S_2 & SO_2 ENVIRONMENTS

8.10.1. COAL GASIFIER CONDITIONS:

Sulphides and oxides predominate as corrosion products although the coal gasification environment appears to be mainly reducing with prevailing H_2/H_2S and CO/CO_2 potentials. The sulphur content of coal is identified to be in sulphatic, organic and pyritic forms, the latter two present in nearly x10 that of the sulphatic variety (Raask 1988). Sulphur is known to have a strong poisoning effect, and is disruptive to the formation of protective oxide layers on the surfaces of many metals (Oudar 1980; Grabke 1980; Strafford 1982; Hocking 1981). Adsorbed sulphur or oxygen at a constant available activity can inhibit carburization (and nitrogenation). The synergetic action of a sulphur-containing environment is its ability to sustain reactions to produce oxide, sulphide and sulphate products from the scale/environment surface to the metal/scale interface and subsurface internal sulphides. The presence of H_2S, H_2O, SO_2, SO_3, CO and CO_2 thus result in oxide/sulphide existence at various sulphur and oxygen potentials over a wide range of temperature. Fig.8-19 shows the existence diagram for sulphides and oxides of Al, Co, Cr, Fe, Mn and Ni at $871^{\circ}C$ (Perkins & Vonk 1979; Hemmings & Perkins 1977; Natesan 1983; Natesan & Delaplane 1978; Weber 1986). Fig.8-25 indicates the relative sulphide stability to oxide at $900^{\circ}C$.

Above 4% Al an alloy can form a continuous layer of alpha-alumina at $900^{\circ}C$, because the pO_2 needed is almost five orders of magnitude lower than most other constituents, but it has very poor adhesion. Formation of a protective Cr_2O_3 is possible at a given temperature and sulphur activity if the oxygen activity is maintained at least five orders of magnitude higher than that needed for the Cr_3S_4-Cr_2O_3 and CrS-Cr_2O_3 systems. Ni-base alloys with low Al% (<5%) are susceptible to sulphur induced degradation. MCrAl systems with >10-12% Al (M=Fe, Ni, Co) are in general resistant as are, TD-NiCr, X-40, Hastelloy X, 304 stainless steel and CoCrAlY systems (Giggins & Pettit 1979).

A number of investigations have been done on candidate coating and substrate materials which can be seen in references quoting co-workers with Birks, Condé, Hancock, Hocking, Hsu, Kofstad, Mrowec, Natesan, Nicoll, Perkins, Rahmel, Strafford and others. A brief account is given here on degradation of FeCrAl and additive alloys.

A series of studies have been done in the authors' laboratory in H_2, H_2O, H_2S, HCl, CO, CO_2 and N_2 mixtures over a wide temperature range from 450-1150$^{\circ}C$ (Hocking & Sidky 1987; Hocking & Weber 1986). Cast FeCrAl-base alloys with additives such as Si, Zr, Hf, Ta, Mn and Y, and mechanically alloyed FeCrAl with Y_2O_3 have been studied. In the cast alloy series the corrosion was higher in some alloys at $450^{\circ}C$ than at $850^{\circ}C$ where more compact oxide scales of Fe and Al formed and additive effect was more evident. At $450^{\circ}C$ multilayered scales with $Fe_{(1-x)}S$ and (Fe,Cr) spinels

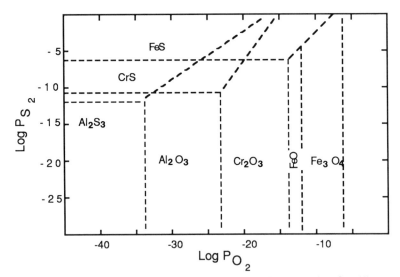

Fig. 8-24b: Fe-Cr-Al Existence Diagram in Gasifier
Environment at 1050 °C

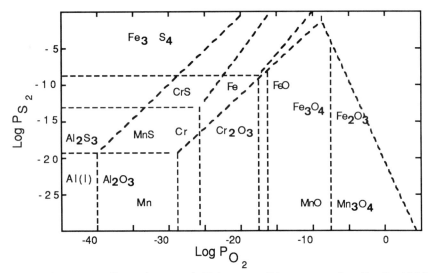

Fig.8-24c: Superimposed Existence Diagrams for Fe-Cr-Al-Mn
in Gasifier Environment at 871 °C

with traces of Al and Si formed together with alumina whiskers.

Gas potentials were at $pS_2=4.1 \times 10^{-5}$, $pO_2=2.4 \times 10^{-18}$ and $pHCl=9 \times 10^{-6}$ atm. ; scale spalling was greater at $450^\circ C$ than at $850^\circ C$ (Sidky & Hocking 1984). Pre-oxidation of these alloys were beneficial to a large extent, especially at $450^\circ C$, where no protective layer formed otherwise (Hocking & Sidky 1987). In Yttria-dispersed FeCrAl mechanical alloy series two distinct types of corrosion have been identified dependent on the pS_2 and the temperature. At $850^\circ C$, thin adherent alpha-alumina forms with a slight negative deviation from parabolic behaviour when pS_2 is 3.4×10^{-8} atm. But at pS_2 7.3×10^{-7} atm. severe sulphidation occurred with a duplex scale pyrrhotite $Fe_{(1-x)}S$ outer layer covering a multiphase spinel inner layer. The negative deviation in parabolic kinetics is reasoned to be due to alumina 'laths' in the inner layer reducing the cross section for outward cation diffusion. Alloys pre-oxidized in air at $1100^\circ C$ were shown to retard sulphidation by up to 500 h. At pS_2 8.1×10^{-6} atm. at $850^\circ C$, the protection due to pre-oxidation lasted up to 1500 h (Weber & Hocking 1985). FeCrAl with and without Mn have been examined in H_2-H_2O and H_2-H_2S atmospheres, and the formation of duplex and lamellar scale formation is illustrated. The effect of Mn is noteworthy (Colson & Larpin 1987).

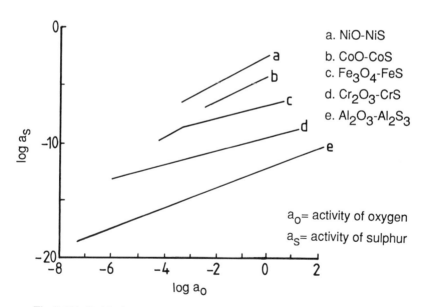

Fig.8-25: Oxidation-Sulphidation Equilibrium Boundaries for Al, Co, Cr, Fe & Ni at $900^\circ C$.

The solubility of S in oxides is detectable (Strafford & Hunt 1983; Alcock et al 1969), but proposals for the mechanism of sulphur transport for the formation of internal sulphides are wide and varied. When the outer scale cracks or opens up fissures under stress generated by any means, it is pertinent to expect direct ingress of the reactant to near the metal/scale interface. Both solute diffusion and diffusion through physical defects as transport media are likely to operate in Cr_2O_3 scale (Perkins & Vonk 1979). The mechanism for the Ni-alloy systems has been explained by Hocking (1979) in studies in SO_2-SO_3 atmospheres and is also discussed in the context of coal gasifier corrosion in a review by Hsu (1987). The sulphur effect in non-oxygen atmospheres has been elucidated by Strafford and co-workers (1969, 1981, 1983), and many others (see reference list for Mrowec et al, Smeltzer et al, Narita et al, Jacob et al). Iron and Chromium sulphides are shown to have large mutual solubilities, and do not form liquid phases under intermediate sulphur potentials unlike Ni- and Co-base alloys.

Cr and $Cr_{23}C_6$ are shown to have similar sulphidation kinetics and scale structure at 800°C in the range of pS_2 = $10^{-3.5}$ to 10^{-6}, except that the sulphidized carbide had an additional inner layer Cr_7C_3. Co, Fe- and Ni-Cr alloys show their sulphidation dependence on the Cr-content, but in Fe-Cr alloys suphidation is shown to be confined to grain boundaries as the chromium carbides sulphidize and not the Cr in the matrix (Narita & Ishikawa 1987). Addition of 10 wt.% Al to Fe-25Cr alloys showed two modes of early stage scale growth, viz. slow-growing alumina scales or faster growing sulphide scales depending on the pH_2O/pH_2S in the range 50 to 7. With time a shift in morphology was imminent (Yurek & Przybylski 1987). Corrosion with H_2/H_2S and H_2/H_2O mixtures with CO have been monitored over 100, 400 and 900 h at 600 and 800°C for a number of steels with 17-20% Cr, 8-26% Ni with major additives as Mn, Mo, Si and minor additives of Ti, Al, Ce etc., Ni-base alloys and ferritic stainless steels. Sulphur attack was observed in almost all of the alloys tested and the chromia stability depended on the sulphur potential and time. Grain boundary attack was evident (Debruyn et al 1987).

Porous sulphate deposits are more damaging in causing sulphidation than any assessment made on the sole basis of their composition because of the tendency to attain higher sulphur potentials at the deposit/oxide interface by creating possible entrapment channels. The deposition of $CaSO_4$ under fluidized-bed combustion conditions has been investigated with reference to the corrosion of heat exchanger alloys. The deposits appeared to contain 10-40% porosity, with the corrosion ascribable to the decomposition of the sulphate deposit, its porosity and the solid state reaction with the protective scale (Saunders & Spencer 1987).

Fig.8-26 and 8-27 show some representative line diagrams of sulphide corrosion product morphology (Hocking & Weber 1986; Hocking & Johnson 1988; Colson & Larpin 1987; Yurek & Przybylski 1987).

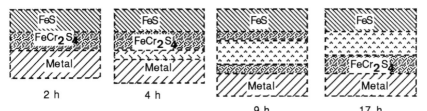

Fig.8-26a: Evolution of Fe- & Cr- sulphide compounds with time in Suphur Vapour in the 800 - 900 °C Range

CrS ⬚   ⬚ FeS+Cr$_2$S$_3$ (or Cr$_3$S$_4$)

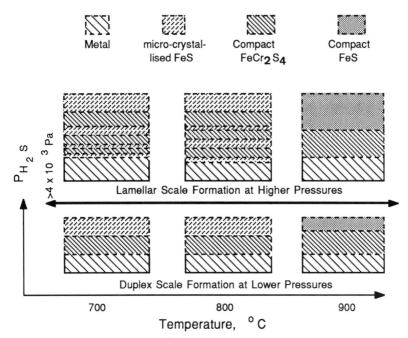

Metal   micro-crystal-   Compact   Compact
               lised FeS   FeCr$_2$S$_4$ FeS

P_{H_2S} >4 x 10^3 Pa

Lamellar Scale Formation at Higher Pressures

Duplex Scale Formation at Lower Pressures

700   800   900

Temperature, °C

Fig.8-26b: Sulphidation in H$_2$ - H$_2$S Gas Mixtures -
Schematic Diagram of Scale Formation

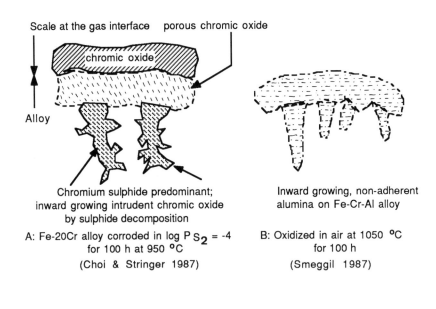

Scale at the gas interface   porous chromic oxide

Chromium sulphide predominant;
inward growing intrudent chromic oxide
by sulphide decomposition

A: Fe-20Cr alloy corroded in log $P_{S_2} = -4$
for 100 h at 950 °C
(Choi & Stringer 1987)

Inward growing, non-adherent
alumina on Fe-Cr-Al alloy

B: Oxidized in air at 1050 °C
for 100 h
(Smeggil 1987)

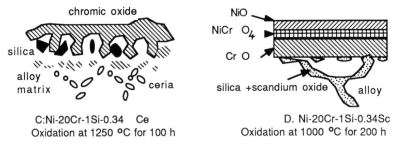

C:Ni-20Cr-1Si-0.34 Ce
Oxidation at 1250 °C for 100 h

D. Ni-20Cr-1Si-0.34Sc
Oxidation at 1000 °C for 200 h

(Y, La and Er produce shallower scallops in chromia, some mixed oxide
with Si and a compound MNi_y ,with Ni, M=Y ,Er , y=17;M=La,y=5)

(Saito et al 1987)

Fig.8-27: Schematic of Oxide Scale Growth

8.10.2. S_2, SO_2 EFFECT IN GAS TURBINE CONDITIONS:

8.10.2.1. General:

SO_2 and SO_3 are the principal reactants in a predominantly oxidizing atmosphere. The sulphur potential as pS_2 is considered because of the equilibria involved where sulphides are the stable end corrosion products with oxides and sometimes with spinels, while the intermediate product reaction involves reactions to form sulphate and decompose it under certain temperature and activity conditions. Investigations to simulate corrosion under gas turbine conditions range from testing a candidate alloy or coating in gas environment with and without sulphate and chloride media, to electrochemical polarization in molten salt media, and crucible tests, thin melt film tests, exposure to molten sea salt and burner rig tests with and without salt injection at the air intake. Effects of mechanical and thermal factors have been followed up with applied stress effects, vibratory rigs and thermal cycling. Thermodynamic calculations have greatly assisted product assessment. The state-of-the-art survey (Rahmel et al 1985; Lacombe 1986) may be mentioned in this context.

Some of the laboratories which have contributed to gas turbine hot corrosion research in the last twenty years are listed here by quoting the group leaders and/or principal workers:-

Alcock, Armijo, Barrett, Bennett, Birks, Booth, Bornstein, Carew, Conde, Coutsouradis, Cutler, Davin, Decresente, Douglass, Duret, Elliot, Erdos, Fairbanks, Foster, Fryburg, Gadomski, Galsworthy, Giggins, Goebel, Goward, Grabke, Hancock, Hart, Hed, Hocking, Jacob, Jones, Kedward, Kofstad, Kohl, Kubaschewski, Lambertin, Luthra, Lloyd, McGill, McCreath, McKee, Meadowcroft, Mevrel, Misra, Mrowec, Natesan, Nicholls, Nicoll, Pettit, Pichoir, Rahmel, Ramanarayanan, Rapp, Restall, Rhys-Jones, Romeo, Taylor, Saunders, Shores, Sidky, Smeggil, Stearns, Stephenson, Stern, Stott, Strafford, Stringer, Swidzinski, Smeltzer, Vasantasree, Wallwork, Wagner, Whittle, Wood, Worrell, Wright.

Many of their publications are listed and can be used for further reference. Individual references are not quoted in this section. Several more investigators who may have published in this area have not been specifically mentioned in the above list; again, the list is not exhaustive.

8.10.2.2. Hot Corrosion Reactions:

Ni and Co are the main matrix metals, and it is their preferential formation that has to be prevented in order to arrest hot corrosion reactions. Ni forms NiO, while Co can oxidize to CoO and Co_3O_4 and unless sufficient Cr is in the alloy and the temperature is sufficiently high, e.g. $900^{\circ}C$, these oxides will form preferentially because of the faster cationic diffusion. The formation of Ni (or Co) and Cr sulphide products beneath these oxide scales, to be distributed in the scale in a multi-layered morphology, as well as form internal sulphides preferentially with Cr has led to controversial explanations for sulphur transport through the growing scale. Considering the corrosion in just gaseous reactants, simple reactions such as

$$7Ni + 2SO_2 = 4NiO + Ni_3S_2 \text{ and,}$$

$$9Ni + 2NiSO_4 = 8NiO + Ni_3S_2$$

may be expected, permitted by the stability of the possible products and available Ni (or Co) activity within the scale. A similar approach is written for Co, Cr and all other additives. The reaction becomes a hot corrosion reaction because of the eutectic reactions of $Ni+Ni_3S_2$ (645°C) and Co+CoS (879°C), the liquid phases of which de-stabilize solid state transport, assuming that the oxide scale is free of physical defects generated by factors such as strain and stress. Both the oxides grow at quite a rapid rate, to considerable thickness and are rarely coherent at temperatures below 850°C. Scale formations at high temperatures are non-coherent, and the thicker the scale the worse its coherence, because of volume expansion and stress generation. Once a liquid product is formed a rapid rate of attack is imminent. This is the baseline of gas turbine hot corrosion. For alloys with less than 18-25% Cr and 10-12%Al preferential Cr_2O_3 and Al_2O_3 can be manipulated and these reduce the sulphidation effect for long periods unlike Ni and Co. Thus if hot corrosion is viewed to occur in two stages, the initiation and the propagation, Cr and Al are able to prolong the initiation stage. The controversies are in agreeing to a mechanism by which the prolonging of the initiation stage occurs and what happens during that time to trigger the later higher corrosion rates.

Sections 8.6-8.9 discuss the various effects of the sulphate and chloride on hot corrosion. Much of the information has been the outcome of gas turbine alloy research and will not be re-elaborated here. It suffices to point out that two temperature regions are recognized in hot corrosion with a low temperature region from 600-850°C and the high temperature from 850°C and above. Cobalt alloys are attacked by the formation of $CoSO_4$ and its eutectic formation with the Na- (and K in some cases) sulphate and stability diagrams for this system have been given.

Gas turbine alloys are multi-component with additives such as Ti, Al, W, Mo, Zr, Hf, Nb, Ta and the coatings formulated follow the same pattern in composition. Co-alloys which are more resistant than Ni-alloys below 850°C in gaseous reactants, are more liable to hot corrode in sulphate and sulphate-associated media. The effect is not as drastic in Ni- or Fe-alloys, but in general all the three groups of alloys have shorter survival records at low temperature in the presence of molten salts. Coatings with multi-element NiCoCrAl- compositions are thus formulated to compensate for the individual vulnerabilities with additive 'reactive elements' such as Y, Ce, Hf etc. Hf showed a critical minimum corrosion effect at 0.85-1.2%, below and above which the corrosion rate was higher in Ni-30Cr alloys. At 700°C at $SO_2:O_2=2:1$, ternary alloys are rated in the order Ni-20Cr-3Al > Ni20Cr-0.3Si > Ni-20Cr-5V > Ni-22Cr with the aluminium alloy as the best in reducing corrosion rate. Pre-oxidation contributes to a retardation in corrosion rate, or prolonging the initiation stage. Care has to be exercised in not depleting the metal content that is being preferentially oxidized, and to ensure that the scale developed is free of defects such as pores and fissures.

Breakaway corrosion is suggested to occur in Fe-Cr alloys in SO_2/O_2 atmospheres due to a critical microstructure development characterized by intrusions of relatively coarse chromia into the substrate matrix. The intrusions are reasoned to develop via the progressive oxidation of CrS which also causes the oxide intrusions to be porous and thus allowing access to gas to develop a reaction front close to the metal interface. Oxides of Ca, Ce, Y, La and Hf applied by conversion of nitrates on Co-15% and 25% Cr and Ni-25% Cr appeared to have considerable influence on the oxidizing properties of the nickel alloy but very little effect on the cobalt alloy.

Schematic diagrams given in Fig.8-28a,b,c, and Fig.8-29 elucidate scale formation typical of a number of alloy hot corrosion in $S_2/SO_2/SO_3$.

8.11. EROSION - CORROSION

The mixed-mode degradation of erosion-oxidation and erosion-hot corrosion is caused by deposition and impact of particles on the developed scale, resulting in scale damage under oxidation or hot corrosion conditions. The erosion aspect of the damage can set in by adhesion, surface-fatigue, delamination, fretting and gouging effects. Particle size, shape, velocity, rate of arrival, mass, physical state, chemical nature and retention control the extent of damage. Three velocity ranges are classified, viz. low impaction velocities <10m/s; medium range 10-100 m/s and the high velocity range >100 m/s. A uniform, smooth scale can be cracked or peppered on impact; the embedded particle may be non-reactive by itself but can change the local reactive conditions; or the particle can change its state from solid to liquid, or react to form a liquid. It also enables liquid retention as the surface is rendered rough. The resultant effect is that erosion reduces the useful lifetime of a coating by mechanical damage and more often by physical-chemical impairment of a protective scale. Any self-repair capacity of the coating attenuates as the constituent reserve depletes (Raask 1988; Santoro et al 1984).

Erosion-related problems in the field of coal-gasification have been well documented (Raask 1988). They may be identified as given in Table 8:9 and Fig.8-30 (Wright 1987).

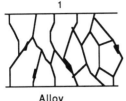 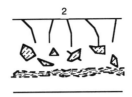 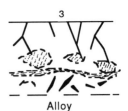

Alloy   Alloy   Alloy

1.protective alumina scale with
titanium oxide nucleating at the
grain boundaries, partially blocking
paths for oxygen.

2 & 3. Non-protective alumina with FeS islands above
a vestigial alumina layer; $Fe(Fe,Cr)_2 S_4$ above alloy
in 2, and in 3 an internal alumina nucleates along with
internal oxidation.

FeCrAlTi-alloy in Coal Gasification atmosphere in protective
& Non-protective Modes at (low and high P_{S_2})

Fig.8-28a

(Hocking & Weber 1987)

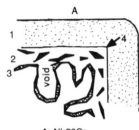 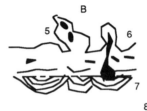 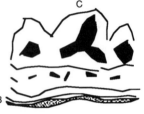

A: Ni-20Cr   B: Ni-20Cr-3Al   C: Nl-20Cr-4Ta

1. Compact NiO layer with $Ni_3 S_2$ islands; 2. Porous region with $NiO+Cr_2 O_3$ +
$Ni_3 S_2$ islands; 3. CrS layer in contact with alloy; 4. Retained sharp corner of
the original metal surface; 5. Multiple globule formation from $Ni_3 S_2$ reservoir
and subsequent oxidation shell with remnant sulphide inside; 6. Globule development;
7. Sulphide striations; 8. Multiple layer formation in high pressure hot corrosion

(Hocking & Vasantasree 1971; Hocking & Sidky 1981; Ma, Baker, Hocking & Vasantasree 1984)

Fig.8-28b: Hot Corrosion of Binary & Ternary Ni-20Cr- Alloys in SO_2 /O_2
Gas Mixtures - 2:1; 1:4; 1:39 (air) at 20 atm. Respectively, at 700 °C

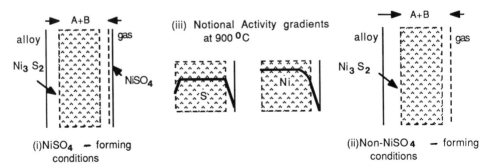

(i)NiSO$_4$ → forming
conditions

(iii) Notional Activity gradients
at 900 °C

(ii)Non-NiSO$_4$ → forming
conditions

Schematic of Sulphate-Sulphide-Oxide Stability in SO_2 -O_2 -SO_3
Gas Mixtures
Fig.8-29

Table 8:9

Type of Boiler/Turbine	Erosion Area
Conventional coal -fired boilers:	Fireside
Pulverized coal -fired boilers	: Fly-ash erosion of pulverizers, coal feed pipes,burner nozzles, superheaters and economiser tube banks.
Steam Turbines	: Flue gas de-sulphurization equipment.
AFBC and PFBC	: Heat transfer tubes, side walls proximate to the turbulent top surface of the fluidized-bed; fouling of the expander turbine

Alloys which are weak but ductile at the erosion conditions, and form tenacious oxide scales exhibit very good resistance to erosion-oxidation (Wright et al 1986). Erosion rates of mild steel by sand was found not to be linear with time at ambient temperatures at low particle velocities; morphologies observed indicated ductile erosion (Lloyd et al 1987). Fig.8-31 shows the relative slurry-erosion rates for valve material candidate ceramics and cermets (Wright 1987).

Degradation in sulphur bearing environments is generally much more severe, as sulphide eutectics tend to form. A sulphur + oxygen environment to a gas turbine alloy is as damaging as sulphur, or hydrogen sulphide is to fluidized bed combustion materials. It is the threshold pS_2 at the attack interface or corrosion front that will decide the sulphide product formation and not the ambient sulphur pressure. The initiation and propogation mode of sulphide formation has been shown to be dependent on temperature and thermodynamic stability (Hocking 1979). Further arguments on direct transport of SO_2 and SO_3 through fissured oxide layers and molten subscales, the temperature effect, eutectic formations and the relative importance of SO_2, SO_3 and sulphates have been put forward (Vasantasree 1971; Vasantasree & Hocking 1976; Sidky & Hocking 1987; Wooton & Birks 1972). The mode of degradation of cobalt and its alloys has also been developed based on similar arguments (Luthra & Shores 1980; Luthra 1982).

Erosion-hot corrosion in gas turbine environments occur as a result of ingested particulates in the air intake. Pyrolytic carbon and NaCl deposition are two of them. Coatings enhance erosion resistance by developing thin and slow-growing films

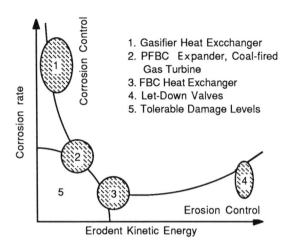

Fig.8-30: Schematic of Corrosion-Erosion Regime
(Wright 1987)

1. Gasifier Heat Excchanger
2. PFBC Expander, Coal-fired Gas Turbine
3. FBC Heat Exchanger
4. Let-Down Valves
5. Tolerable Damage Levels

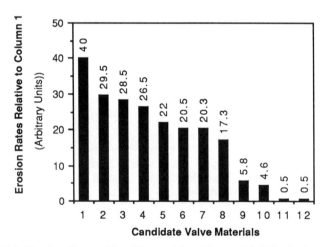

Fig.8-31: Erosion Tests with a Slurry of Coal Derived Solids+Anthracene Oil
(Wright 1987)

Column No.	Test Material	Column No.	Test Material
1	WC+Co+Cr(Std.)	7	B_4C
2	WC+Co	8	WC+Co+Cr(c)
3	B_4C+Co	9	CVD SiC
4	WC+Co+Cr(b)	10	Sintered alpha-SiC
5	WC+Co	11	CVD TiB_2
6	SiAlON	12	Diamond Compact

primarily based on SiO_2, Al_2O_3 and Cr_2O_3 or combinations of these. Erosion modes in superalloy aero-engine blades and helicopter blades by carbon and sand respectively have been presented (Restall 1976; Galsworthy et al 1982; Nicholls & Stephenson 1986). Erosion which is significant on thermal barrier coatings over 10000 h on surfaces parallel to flow may be seen within 3000 h on vane leading edges or wherever normal impingement may occur. For airfoil applications requiring smooth finishes and good erosion reistance PVD ceramics are thought to be suitable (Bennett et al 1987). A Monte Carlo model for erosion-corrosion as applied to a typical gas turbine environment is proposed taking into account, the particulates, protective oxide and the substrate alloy. The degree of damage is shown to depend on substrate temperature, scale composition and scale thickness. If a high velocity (150-300 m/s) is maintained the particle role is thought to be minor (Hancock et al 1987). Test conclusions were as follows:

1. Irrespective of the coatings beneath, top alumina scales formed on coated superalloys exhibit the same fracture behaviour under erosive conditions.

2. Quantitative predictions can be made on the influence of surface oxides on the stresses generated at the substrate/coating interface and the surface oxides show a considerable influence on the erosion response of the underlying coating.

3. High rates of strain of the order of $10^5 - 10^8$ /s are created at high velocity particle impact (150-300 m/s), and at these high strain rates all types of particles can be viewed on par because divergence in their mechanical behaviour is very insignificant.

4. At high velocity, similar impact damage is caused by particles from 50 microns to 1 mm, and the hertzian stress theory can be used to predict damage caused by smaller particles.

5. Tests conducted with single-impact erosion in the range 700°-1000^oC under low and high particle loading have yielded quantitative oxidation-erosion data agreeing satisfactorily with a Monte Carlo model of prediction.

The significance of the DBTT of a coating has been demonstrated by single impact tests on aluminized MAR-M002 and IN 738 with sea salt or pyrolytic carbon particles at impact angles 30° or 90°. Aluminide coatings provide the greatest deterrent to erosion at higher than 850^oC, above the DBTT of Al_2O_3 in marine gas turbine applications. But for the coal-fired gas turbine, components such as nozzle guide vanes and turbine rotor blades, erosion prevention has to be via gas clean-up procedures rather than material improvements (Restall & Stephenson 1987). Fig.8-32 shows the magnitude of loss in coating lifetime under erosion-oxidation and erosion-hot corrosion of aluminided IN 738, a typical gas turbine alloy (Barkalow et al 1984; Wright 1987). Alumina used as an erodent indicated a particle effect where 0.3 micron reduced

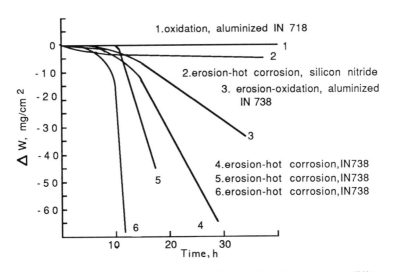

Fig. 8-32: Oxidation, Erosion & Hot Corrosion Interaction Effects
on Superalloy IN 738 in Burner Rig Tests

(Barkalow, Goebel &Pettit, 1980)

sulphate hot corrosion and 2 micron enhanced it, although without
sulphates it was non-erosive. Impacting at 245 m/s velocity, the
corrosion rates at $871_2^{O}C$ on gas turbine alloys was ranked at
0.05, 0.1, 3 and 7 $mg/cm^2/h$ for oxidation, hot corrosion, erosion
and erosion-hot corrosion respectively (Barkalow 1984). Ash impaction at 344 m/s gave the following ranking of erosion-hot
corrosion in decreasing resistance to wastage (Spriggs & Brobst
1982; McCarron 1984):-

At $737^{O}C$: CoCrAlY > FeCrAlY > IN 671 > FSX 414> Pt-Cr-Al> IN 738
         claddingcladding cladding alloy   coating   alloy

At $872^{O}C$: Pt-Cr-Al > CoCrAlY > FeCrAlY> IN 738 > IN 671 > FSX 414

Raask (1988) has compiled a comprehensive discussion on gas
turbine erosion in coal-energy systems.

Effects of coating ductility, bond strength and microstructure on
erosion -oxidation resistance have been evaluated for application
in power recovery turbines and aircraft gas turbine compressor
sections. On an IN 718 substrate, D-gun coatings of chromium
carbide-nichrome and WC with pre-alloyed Ni and Co metal matrices
were sprayed and tested at $760^{O}C$, with impact angles 30° and 90°;
the erodent was 27-micron alumina in a heated gas stream. Smaller
particles of coating powder gave better resistance. Chromium
carbide-nichrome was better than the WC composite; microstructural differences and greater oxidation resistance are believed
to be contributory (Sue & Tucker 1987). The erosion resistance

448

and hardness of sputtered $NiCr_3C_2$ increases with increasing deposition rate. The deposit is homogeneous at 25% bias above which microstructure is adversely affected and also reflects in poor erosion resistance. At 30° impingement erosion results in a ductile forging-extrusion mechanism, and at normal impingement brittle chipping occurs. Coating defects greatly enhanced erosion rates with rapidly expanding pinhole formation, stressing the importance of microstructure effects (Wert et al 1987).

PVD arc evaporated TiN shows a similar crystallographic influence to erosion resistance (Sue & Troue 1987). Ion plated TiN-Cr on 12%Cr stainless steel substrate showed that good resistance to cavitation-erosion is achieved by strong adhesive force at the bonded substrate/coating interface, along with high compressive residual stress, small grain size and preferred orientation. Substrate bias voltage was once again shown to have a decisive effect. WC-Co and TiC were inferior to TiN-Cr in many respects. Cavitation-erosion resistance is enhanced with a metallic interlayer between the coating and substrate (Odohira et al 1987). Thick coatings of PVD-TiN were more resistant to angular particle erosion, and thin coatings to blunt impact. Stresses in coatings cause aggravation by spalling in both cases (Rickerby & Burnett 1987). Ductile IN 617 coatings magnetron-sputtered and nitrided onto Ti-6Al-4V substrates exhibited good resistance to normal impact erosion by 15 micron SiC particles (Emiliani et al 1987). Plasma sprayed porous ZrO_2 ceramic eroded at $1287^\circ C$ with 27 micron alumina at 244 m/s showed plowing, fracture and tunnelling at low, medium and high incidence impact angles. The strength of the ceramic, the porosity distribution between intersplat boundaries, and discrete pores at a given porosity level could be used to predict erosion rate (Eaton & Novak 1987).

Histograms in Fig. 8-33 to 8-35 (cf. Fig.7-14a and 14-b) indicate chromia-forming alloys to be inferior to alumina-forming alloys with both being inferior to coated alloys. Al-rich/Al-Fe coating on Hastelloy X ranks almost equal to Si_3N_4, the most resistant. Fig.8-36 and 8-37 show erosion reistance of coated stainless steels with chromite powder as erodent (Qureshi et al 1987). Data with erodents which occur in practice, such as sand, pyrolytic carbon and sea salt have to be assessed against test erodents to obtain more realistic ranking and extrapolations. Further work is clearly needed to clarify alloy performance under well planned erosion-corosion conditions.

Sections 8.12 to 8.15 consider individual coating systems in three groups viz. aluminides, silicides and thermal barrier coatings.

Erosion-Oxidation Resistance

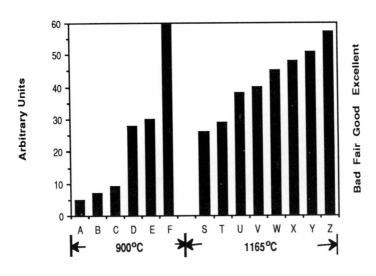

Fig. 8-33

Erodent:- Alumina particles; A-F: Tested at 900°C
S-Z: Tested at 1165°C

Legend	Substrate	Legend	Coating
A	IN738	U	Al
B	X40	V	Haynes C
C	MA754	W	CrAl
D	IN738 (with Al coating)	X	Al-Cr-Si
E	CoCrAlY	Y	Al-rich Al-Fe
F	Si3N4	Z	Al
S	Haynes C9		
T	Hastelloy X		

==
Where no coating is shown the material was tested alone for comparison.

Erosion - Oxidation Resistance

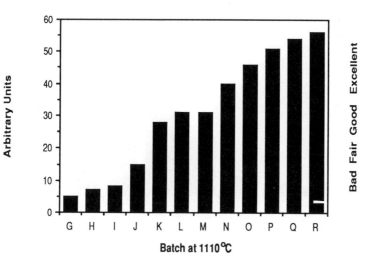

Fig.8-34

Erodent:- Alumina particles; G-R: Tested at 1110°C

Legend	Substrate	Coating
G	TRW1900	none
H	SM200	none
I	SM211	none
J	IN100	none
K	IN717	none
L	IN713LC	none
M	PDRL162	none
N	B1900	none
O	PDRL162	Al
P	IN100	Al
Q	B1900	Al
R	IN713LC	Al

=================================

where no coating is shown the material was
tested alone for comparison.

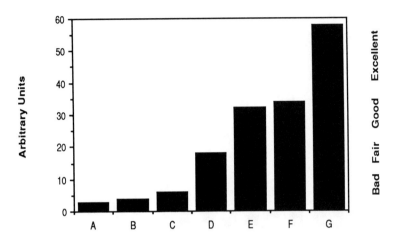

Fig.8-35

Legend	Substrate	Coating/Conditions
A	HA188	none
B	IN738	none
C	X40	none
D	MA754	none
E	IN738	NiAl
F	IN738	CoCrAlY
G	Si_3N_4	none

===
The erodent was alumina (particulate) at 871°C and the hot corrodent was a Na_2SO_4.$25K_2SO_4$ melt. Where only one material is shown, it was tested alone.

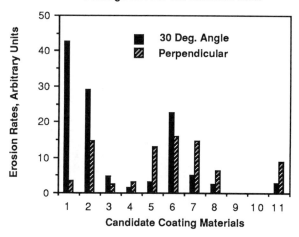

Fig.8-36: Erosion Rates of Coatings on AISI 403 Stainless Steel

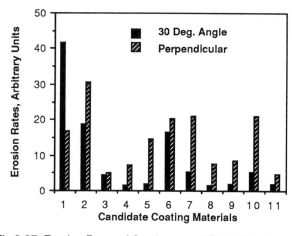

Fig.8-37: Erosion Rates of Coatings on AISI 422 Stainless Steel

Column No.	Test Material	Column No.	Test Material
1	Substrate Steel	7	WC+Co; High velocity Plasma Spray
2	Nitrided Case; Diffusion Process	8	XC+NiCrB; Clad Overlay
3	CrB+FeB(a); Diffusion Process	9	WC; CVD
4	Cr_3C_2; D-Gun	10	Co-Mo-Cr; Electrospark Dep.
5	TiN; PVD	11	CrB+FeB(b); Diffusion Process
6	NlCrBC; Plasma Spray		

8.12. DEGRADATION OF ALUMINIDE COATINGS

In this and the next two sections chemical degradation of specific coating systems is reviewed. All of them have been extensively investigated in all high temperature applications. A few new
methods developed and applied in recent years are also examined
in the context of their efficacy over conventional methods of
coating application in improving the performance of the coating
and the component.

A broad grouping can be made under three categories of coating:

Group 1 : Aluminides; chrome-aluminides; and Pt-aluminides; the
MCrAlX systems, M = Ni, Co, and/or Fe; X = Y, Hf, Ce, Ta, Zr etc.

Group 2 : Silicides (single or in combination)

Group 3 : Thermal barrier ceramics with overlays

This section deals with coatings in Group 1. Sections 8.13 and
8.14 consider silicides and thermal barrier coatings respectively.

8.12.1. GROUP 1. ALUMINIDES

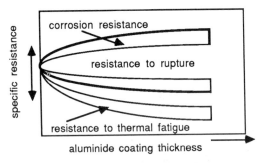

Fig.8-38: Aluminide Coatings - Thickness
Related to Corrosion Resistance.

The overall durability of any
coating can be assessed as
its specific resistance to
degradation by corrosion,
rupture and thermal fatigue.
Fig.8-38 shows a schematic
diagram of the resistance of
aluminide coatings to thermal
fatigue and hot corrosion.
Fig.8-39 shows the compositional changes occurring in
NiAl on Ni. Morphological
changes in CoAl on Co and
NiAl on Ni are shown schematically in Fig.8-40 (Fontana
& Staehle 1976; Nicoll 1984). Degradation of aluminide coatings
is largely dependent on the coherence and structure of the Al_2O_3
layer which is largely influenced by the available Al-activity
from the substrate, and the transition metal it is alloyed with
in the diffusion band.

Diffusion processes from the substrate always happen in conjunction with those occurring in the coating itself. At an early
stage Al in the aluminide coating is expected to diffuse into the
substrate. But as degradation proceeds the diffusion direction
reverses with Ni from the substrate entering the coating that is
being consumed. This situation is ideal for Kirkendall void
formation and causes physical dislodgement of coating if the

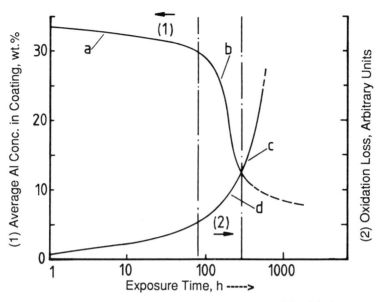

a. spallation induced loss   c. rapid oxidation of γ
b. enhanced loss due to martensitic   d. rapid oxidation of γ'
transformation

Fig.8-39: Oxidation Kinetics of Ni Coated with NiAl.
- Schematic of Variation in Al Level & Loss due to Oxide Penetration.

voids coalesce at the substrate/coating interface. Although Al_2O_3 is generally considered as an n-type oxide, it is argued that it is so only at low pO_2. Thus at $600^{\circ}C$, crystalline alpha Al_2O_3 grows beneath an amorphous film of gamma Al_2O_3 by inward diffusion of O_2. At $1300^{\circ} - 1750^{\circ}C$, Al_2O_3 is n-type up to $pO_2 = 10^{-5}$ and becomes p-type at higher oxygen pressures (Natl.Acad. Sci.1970).

Nickel aluminide is still the best aluminide system in the gas turbine field although aluminiding was first used to benefit steels. At the commencement of oxidation the non-protective gamma alumina forms and spalls easily. A more dense and adherent alpha alumina then forms and as long as this remains undamaged it confers protective kinetics on the substrate. However, a damage and repair cycle soon causes Al depletion which affects the diffusion mode and composition of the NiAl below the scale. Below the critical Al concentration the single phase NiAl now rendered Ni-rich tends to precipitate the gamma prime Ni_3Al phase which causes NiO to react with Al_2O_3 to form spinels and also NiO on its own develops porosity during growth, thus increasing mass transport. Appearance of the gamma prime phase thus marks the onset of the rapid oxidation stage. In a mixed gamma and gamma prime the gamma phase is more readily oxidisable and forms a voluminous, porous multi-phased non-protective scale.

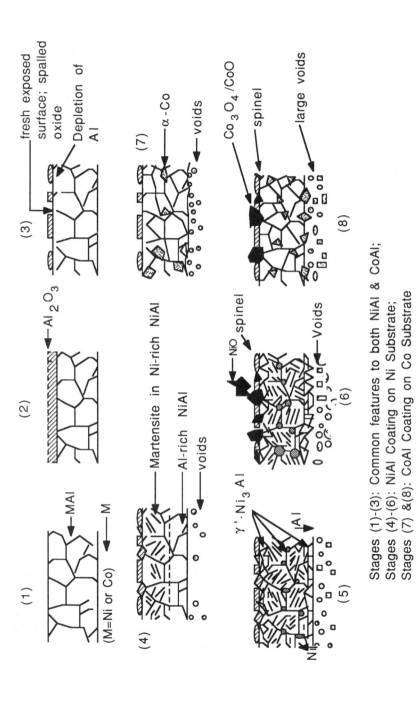

Stages (1)-(3): Common features to both NiAl & CoAl;
Stages (4)-(6): NiAl Coating on Ni Substrate;
Stages (7) &(8): CoAl Coating on Co Substrate

Fig.8-40

Morphological Changes in NiAl & CoAl Alloy Coatings on Ni & Co Substrates respectively

456

8.12.2. ALUMINIDE CASE HISTORY: SELECTED WORK

Carbon steels with varying Cr-content are being employed over a wide range of applications, and aluminide is an important surface coating. A high bulk Al is necessary for most carbon steels as diffusion levels the Al concentrations in the surface aluminide layer. A level of 18% Cr in the substrate leads to a lower level of Al on the surface for protection (Kindlimann & Greeven 1983). At 750-900°C although there may be some spalling of the outer layer the inner layer is more compact and offers protection.

Low carbon steels used in higher mass production than the more expensive structural steels resort to a logically cheaper method of aluminizing by a hot dip process. A Ti- modified low carbon steel, ALUMA-Ti, with aluminide coatings of 75 - 120 g/m^2 shifted breakaway behaviour beyond 2000 hours at 704°C, a far greater period than the projected useful life. Its cyclic behaviour is yet to be tested (Denner & Kim 1983). Pack aluminizing done on low carbon steels is yet to be optimized as diffusion of iron outwards is still a predominant feature of aluminide degradation. Application in the petrochemical industry is linked to the development of aluminided low grade steel to substitute for 18-8 stainless steels. A promising oxidation rate at 900°C has been achieved within an order of magnitude with uncoated low carbon steel and aluminided low carbon steel (rates of 3.0×10^{-2} and 3.6×10^{-3} $mg/cm^2/sec$ at 900°C) and at 700°C with low carbon steel and 18-8 stainless steel (rates of 0.14 and 6.1×10^{-14} $mg/cm^2/sec$ at 700°C) (Mitani 1983). Simultaneous aluminizing and chromizing has also been done on austenitic stainless steels and their degradation investigated (Miller et al 1988).

At 850 and 900°C aluminide layers less than 50 microns thick on EN-3 plain carbon steel oxidized in the first 24 hours at a slower rate than thicker coatings but over a 72 hours period the overall weight gains of coatings thicker than 50 microns were still low while the thinner coatings oxidized completly progressing deep into the substrate. A 75 micron thick coated EN-3 is more oxidation resistant than 304 stainless steel at 900°C (Kindlimann & Greeven 1983). Fe-17Cr alloys pack aluminized and oxidized at 800°-1000°C form quite thick alpha-alumina scales unlike Fe-Cr-Al alloys. Although the oxidation is diffusion controlled, the rate constant varies due to scale sintering (Majid & Lambertin 1987). Al as an alloy additive to Fe-Cr-alloys is found to be more beneficial, and with Ti the improvement even allows economy in the more expensive addition of Cr (Becquerelle et al 1987).

Molten $NaNO_3$ - KNO_3 is considered as an energy transfer and storage medium for solar central receiver applications. Molybdenum steels with 2-9Cr and 1-2Mo, and 316 stainless steel were aluminided by pack cementation. The coatings were found to survive 6000 hours at 600°C with a slow degradation occurring over the period. $NaAlO_2$ and $NaFeO_2$ were found to be the corrosion products (Carling et al 1983).

60 micron beta NiAl coatings on IN100, IN738LC and IN939 yielded a different morphology of the substrate coating complex as processed by Cr or Al pack, or Cr or Al vapour phase, wherein the latter samples were coated with 1 mg cm^{-2} Na$_2$SO$_4$ every 100 hours and hot corroded in oxidising conditions in air at 1 atm with hourly thermal cycling between 850°C and room temperature. Hot corrosion under reducing conditions was carried out at pO$_2$ = 10^{-15} in a N$_2$ - CO - CO$_2$ mixture; a small amount of H$_2$O is suspected. Air exposure cycled with a 2 hour exposure to the above mixture followed by nitrogen purging was adopted for any "redose hot corrosion".

Under oxidizing hot corrosion conditions the pack and vapour phase aluminides on IN738 recorded lifetimes of 1500 and 1300 hours as against 300 and 100 hour on IN100, and 400 hour each on IN939. Substrate influence is again evident. In hot corrosion under reducing conditions in Na$_2$S, NaCrS$_2$ and NaCrO$_2$ were observed as reaction products within an hour of reaction with the melt, while NaAlO$_2$ appeared after 70 hours. Eutectic Na$_2$S-Na$_2$O (m.p 682°C) is also formed as Na$_2$S reacts with traces of H$_2$O.

$$Na_2SO_4 + 4CO = Na_2S + 4CO_2$$

$$Na_2S + H_2O = Na_2O + H_2S$$

$$Na_2O + H_2O = 2NaOH$$

are the reactions proposed for Na$_2$SO$_4$ reduction. Al$_2$O$_3$ is more stable in molten Na$_2$S than Cr$_2$O$_3$ is, and in reducing conditions gets undermined by the molten reaction products (Steinmatz et al 1983; Sivakumar 1982). Aluminiding was first done on steels, but optimization has continued as its application has spanned over several grades in various environments and higher temperatures. In sulphidizing atmospheres at pS$_2$ of about 10^{-5} the alpha alumina scale thickness is grown over by sulphide scales in pure sulphur while in 10%H$_2$S-H$_2$ mixtures the sulphide is found nucleated under the alpha-alumina layer (Mari et al 1982).

Improved oxidation characteristics at 900-1000°C under both isothermal and cyclic conditions were achieved for an aluminided Ni-Al-Cr$_3$C$_2$ directionally solidified alloy by applying an intermediate layer of a Ni-20Co-10Cr- 4Al coat. A 10 micron Pt was more successful and alpha Al$_2$O$_3$ was formed and maintained; at 1100°C under thermal cycling it broke down but was capable of self-healing (Wood et al 1980).

Substrate influence on aluminides can be considerable. Ni-base alloy of composition Ni10Cr5Co12Ta4W1.5Ti withstood cyclic hot corrosion, the longest clocking 144 and 168 hours at 7 cycles. Mo, Re, Hf in the alloy were not beneficial to either or both low and high activity pack coated aluminides, while V strongly decreased the oxidation resistance (Smeggil & Bornstein 1983). This is shown in Table 8:10. Morphology and RBS studies on the Hf-effect on NiAl coatings on IN 738 and Rene 80 show that oxide

458

intrusions ('pegs') into the substrate increased with the Hf content 1-2 wt.% with scale adhesion for Rene 80 alloy increasing as follows:

Alloy added with 3.5Ti, 2Ti+1.5Hf and 3.5Ti+1.5Hf in high activity aluminide registered adhesion coefficients of 55, 82 and 95%;

Alloy added with 5Ti, 2Ti+1.5Hf, 3.5Ti+1.5Hf and 5Ti+1.2Hf in low activity aluminide had adhesion coefficients of 70, 69, 50 and 100 respectively.

Hf diffusion through the aluminide coating of low or high activity was a function of time. RBS data IN 738 with 1Hf analysed only Al and Ni initially and after 100 h at 1000°C Mo, Hf+W were identified (Muamba et al 1987). Titanium alone has been aluminided and oxidized at 850-1000°C with thermal cycling in static air. Interdiffusion effects have to be considered (Subrahmanyam & Annapurna 1986).

Low activity aluminide on IN738 and MAR-M002 was resistant to both O_2 and O_2-SO_2 environment at 700 and 830°C but in the presence of Na_2SO_4 (8 micro-gm/cm^2) in the latter gas mixture produced catastrophic corrsion discussed in sections 8.8-8.10 (Rhys-Jones & Swindels 1987). Substrate influence and diffusion effects

TABLE 8:10

HOT CORROSION TEST RESULTS

Alloy	High Temperature / Low Activity Coating		Low Temperature / High Activity Coating	
	Cycles to Failure	Time to Coating Failure (hrs)	Cycles to Failure	Time to Coating Failure (hrs)
MAR-M200	3	59	4	72
MAR-M200+Hf	3	72	1	0.08
Alloy 454	6	144*	7	168*
Alloy 231	7	144	3	53
Alloy 245	4	72	3	60
Alloy 250	7	168	3	66

* Failure determined by visual inspection of test specimen, not thermogravimetric results.
==

are further confirmed in the studies on aluminided gamma-phase Ni-base alloys in air at 1100°C. Ni-7Al, Ni-12Cr, Ni-23Cr and Ni-12Cr-7Al alloys used to monitor the additive flux at the substrate/coating interface indicated that ternary "cross-term" effects could be often large and determine the direction and magnitude of both Cr and Al fluxes. Oxide spalling was a common feature, with the 12Cr binary proving the best resitant to spalling. Inter-diffusion coefficients are derived (Fink & Heckel 1987). Mechanically alloyed and oxide dispersed Ni-base alloys (ODS MA-alloys) exhibit acute porosity effects. Diffusion aluminide coatings on all ODS MA-alloys showed porosity after 146 h at 1100°C. Higher Al levels in the substrate alloy reduced the porosity. The combined effect of Cr+Ta was also found beneficial (Benn et al 1987).

Beta NiAl implanted with Y in doses of 2×10^{14}, $\times 10^{15}$ and $\times 10^{16}$ ions/cm^2 with penetration to about 300 angstroms showed that scale growth on the implanted specimen was only by the outward diffusion of Al and did not have the inward diffusion of oxygen in addition as in the non-implanted sample. Scale adherence and reduced growth rate was attributed to the growth mechanism influenced by Y (Jedlinski & Mrowec 1987). Incorporation of Y into an aluminide by pack cementation did not result in the beneficial effect normally associated to Y addition to that alloy. It is not clear how sustaining implantation of Y is and how it compares with Y as a substrate alloy additive. Pack yttriumizing with or without pack aluminiding IN 738 resulted in an adverse effect on cyclic oxidation (Tu et al 1987).

The mechanisms which govern aluminiding impose particular limitations in composition and microstructure. Incorporation of a second and third element either as a pre-deposit/treatment or together with Al, combining the advantages of it being a simple process, unrestricted to line-of-sight offers scope for further exploration of aluminide coating formulae (Mevrel & Pichoir 1987).

8.12.3. CHROME - ALUMINIDE COATING DEGRADATION

Chromium is the principal metal linked with aluminiding, and most high temperature alloys rely on the oxides of Cr and Al for a consolidated resistance to degradation. The effect of many of the substrate components on the aluminide layer has been given in Chapter 5. The following solubilities in NiAl have been reported: Pt - up to 30% (Betz et al 1976), which with Fe and Co forms the first group with highest solubility. An average solubility of about 4 wt% is observed for Ti, Cr, Ta and W at 1000°C. Nb and Mo at 1 wt% in NiAl exhibit the lowest solubility. Thus:

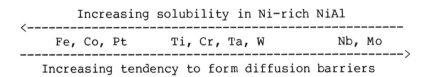

Increasing solubility in Ni-rich NiAl

<--

Fe, Co, Pt   Ti, Cr, Ta, W   Nb, Mo

-->

Increasing tendency to form diffusion barriers

The total element content in the substrate and in Ni_3Al and $NiAl$ and the thickness of the solid solution barrier layer after 50h at $1000^{\circ}C$ as Al diffuses inwards and the substrate elements dissolve in the Ni-Al complex, is reported (Fleetwood 1970). A barrier layer for resistance to degradation is found effective when Al diffusion towards the substrate is arrested and this is achieved when the cumulative content of Cr, W, Ti, Ta, Nb and Mo is more than 20 wt%. It is evident that in most alloys a high level of Cr is necessary to compensate for the absence of a low level of the other elements listed above. The chrome level can be increased by Cr in the substrate, a better degradation resistance by chromization before or during aluminization.

The degradation in a chrome - aluminide layer is influenced closely by the major element diffusion and degradation mode and the temperature effect on them. In general nickel-base alloys form Cr-rich intermediate layers during aluminizing unlike Fe-base alloys (Fitzer & Maurer 1979). Fe-25Cr were found to be the most resistant to sulphide-forming and sulphur containing environments at temperatures less than $800^{\circ}C$ where Ni-and Co- base alloys fail, the latter especially in Na_2SO_4 deposition conditions (Upadhya & Strafford 1983). Oxidation with thermal cycling and isothermal oxidation in exposure to fused Na_2SO_4 showed that the beneficial effect of Cr was to arrest spallation, and change in the coating microstructure was considered instrumental in restricting pitting of the chrome-aluminide diffusion coating. It is postulated that the Cr-enriched zone acts as a barrier to refractory metal diffusion, e.g. Mo, W, V, from the substrate and their subsequent oxidation (Godlewski & Godlewska 1987, 1986). An internal sulphidation zone where the spinel $(Cr,Al)_3S_4$ can exist with a broad range of Cr/Al ratio has been characterized (Erdos & Rahmel 1986).

A mathematical model has been developed to explore carbon diffusion through aluminided layers to form Cr_7C_3 (Marijnissen & Klostermann 1980).

8.13. DEGRADATION OF MCrAl-SYSTEMS

The extension of aluminided coating life for degradation is rarely achieved by Al alone. At its simplest, it is a dual metal system like Ni-Al, Co-Al, Cr-Al or Pt-Al and at its most complex a minimum of 4 and a maximum of 6 elements are formulated, usually of the general form MCrAl-Xl or MCrAlY-X2. These formulae are the most investigated; M = Ni, Co, Fe; Xl and X2 = Ce, Hf, Pt-group, Si, Ta, Ti, Y, Zr etc.; unusual additives like Mn, Zn and Cu have also been investigated. Schematic representations of degradation morphology of CoAlX and NiAlX are shown in Fig.8-41 and 8-42 respectively. Investigations on gas turbine coating systems have mostly concerned NiCrAlX1-X2 and CoCrAlX1-X2 compositions while the coal gas conversion systems have concentrated on FeCrAlX1-X2 systems. Fig.8-43 a,b show the oxide regions of

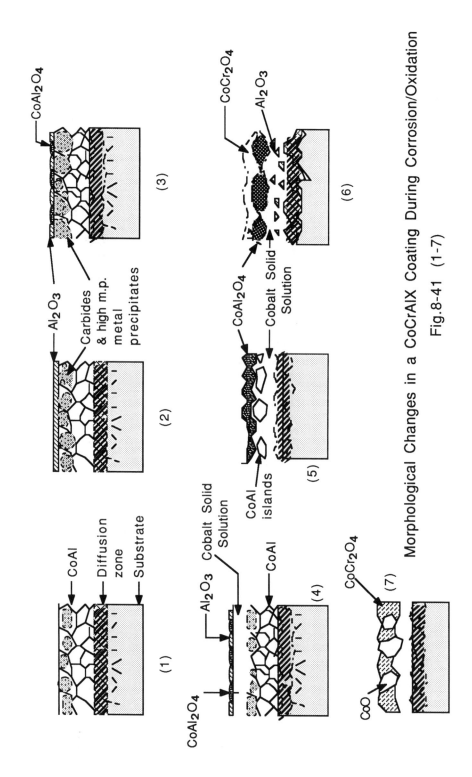

Morphological Changes in a CoCrAlX Coating During Corrosion/Oxidation

Fig.8-41 (1-7)

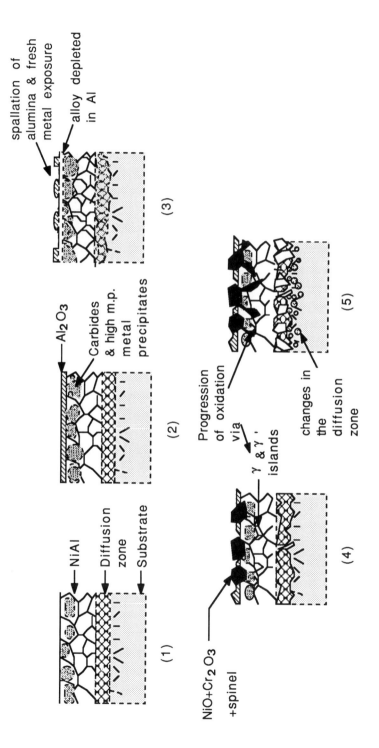

Fig.8-42: Morphological Changes in a NiCrAlX Coating During Corrosion/Oxidation

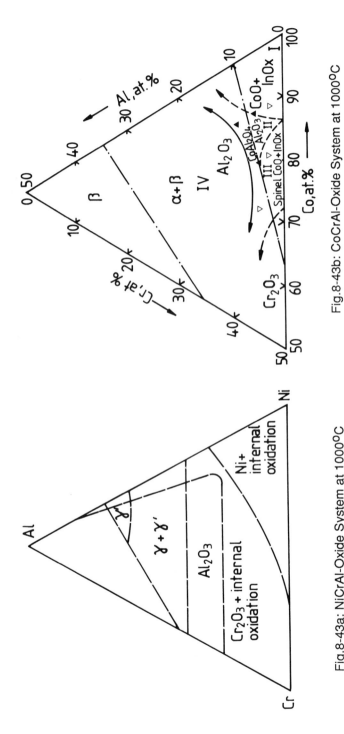

Fig.8-43b: CoCrAl-Oxide System at 1000°C

Fig.8-43a: NiCrAl-Oxide System at 1000°C

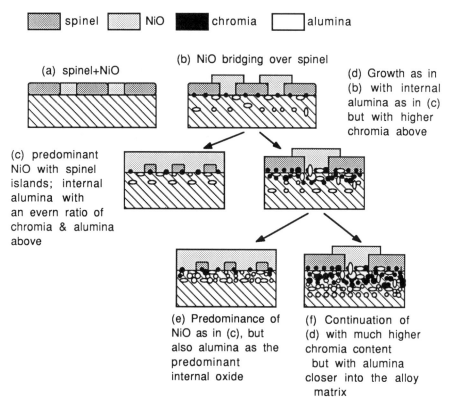

Schematic Representation of Oxides and Spinel Formation in NiCrAl Alloys

Fig.8-44

ternary NiCrAl and CoCrAl systems. Fig.8-44 and 8-45 give an idea on oxide and spinel formation across the substrate/coating/environment and the distribution modes of internal oxides/sulphides in MCrAl systems, M= Ni, Co or Fe. In sulphur-predominant environment sulphide products take a major spectrum in distribution (Nicoll 1984; Wallwork & Hed 1971).

Chromium is an essential additive in all aluminide systems as it influences the stability, coherence and continuity of the alpha-alumina layer sometimes as a top-scale and often as the scale immediately beneath the chromia layer. There have been several investigation groups working on optimization of the additive elements to MCrAl-systems. A full discussion of these results for even the last few years is beyond the scope of this chapter. MCrAl coatings mainly rely on their ability to form barrier layers of alpha alumina and chromia. Degradation of these coatings therefore means a breakdown in the scale coherence and in the ability to repair and sustain the barrier layer. The actual mechanism by which a breakdown in the scale occurs depends on the

Corrosion Mode, Group 1: Low Temperature

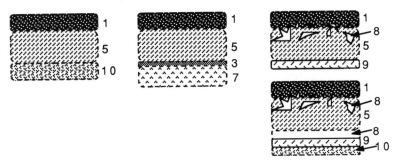

Corrosion Mode, Group 2: Intermediate Temperature

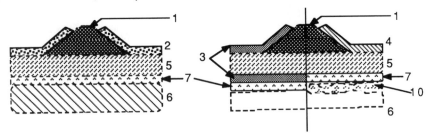

Corrosion Mode, Group 3: High Temperature

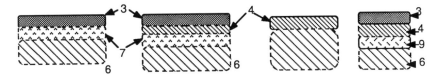

Schematic Representation of Scale Formation, Breakdown, Reformation and Compositional Changes Which Occur in NiCrAl, FeCrAl & CoCrAl Coatings.

Fig.8-45

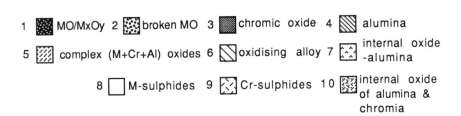

1 MO/MxOy 2 broken MO 3 chromic oxide 4 alumina

5 complex (M+Cr+Al) oxides 6 oxidising alloy 7 internal oxide -alumina

8 M-sulphides 9 Cr-sulphides 10 internal oxide of alumina & chromia

interaction of the several parameters involved: the environment composition, particulate effect/velocity and temperature. It would be futile to ascribe any one mechanism as a rule of thumb to even one of the M-Cr-Al systems. The majority of MCrAlY applications has been as overlay coatings with ceramic barrier coatings in the field of gas turbines. A broad picture of degradation is provided with reference sources.

The behaviour of some MCrAlX alloys with Cr levels of 20-35% and Al contents of 2-5% has been studied (Strafford et al 1981), where M is Ni, Fe, Co, Fe+Ni or Fe+Co; X is Zr or Hf , with 1.5%Si , 1-4%Ti and 3-4%Ti or 11%Ta. A main aim was to find a composition range with improved resistance to low temperature sulphur-induced hot corrosion while maintaining adequate performance at high temperature (900°C). Specimens coated with $NaCl+Na_2SO_4$ and in SO_2+O_2 mixtures showed the detrimental effects of high Ni contents at 700°C. Additions of 0.5-2% Si and of Zr and Hf delayed the onset of accelerated attack. In contrast, (FeCo)CrAlX alloys had low corrosion rates at 700°; negligible sulphidation occurred. Good scale adherence was found for alloys with both Ti and Si additions. Ta was very effective in reducing the sulphide content in the scale while enhancing chromia formation. At 900° corrosion was much less for all the alloys studied, due to decreased sulphide stability and formation. Additions of Ni and Co (reducing the Fe content) were beneficial at 900°. It should be possible to develop coatings with adequate performance at both 700 and 900°C (Upadhya & Straford 1983; Conde & Booth 1972; Hocking & Vasantasree 1978,1984; Conde et al 1984).

It is common knowledge that an Fe-18Cr-8Ni is a chromia former at room temperature. The situation for oxide formation is more complex at higher temperatures. The entire oxidation phenomenon is temperature dependent and easily responsive to additional reactants in apparently low levels of concentration. Oxide predominance of any one species, either Cr , or the main matrix metals will not occur. Mixed oxides, and multi-layer scales are more common as Fig.41-45 indicate. Fe-base alloys can accept up to 4 to 5 wt% aluminium in alloying before fabrication difficulties arise; the Fe-15-25Cr-4Al series is principally an alpha-alumina former (Bennett et al 1980). Unlike Fe- base alloys the Ni-and Co-base superalloy compositions form chromia only when not superseded by the faster-growing Ni- and Co- oxides respectively. The aluminium levels here have to be high to influence an alumina-forming capacity, although chromium helps considerably in lowering the critical Al level needed. The systems NiCrAl and Co-Cr-Al have been extensively studied and reviewed (Wright 1972; Wallwork & Hed 1971). The Ni-Al system requires 15 wt % Al to form Al_2O_3 while a Ni-10Cr alloy requires only 5 wt% Al. Co-30Cr and Co-25Cr-2Al are both mainly Cr_2O_3-formers while Co-15 to 21Cr-10 to 12Al form alumina preferentially at 1000°C (Irving et al 1977). Preferential alumina formation is, again, temperature dependent requiring temperatures above 900°C and more commonly manifested in the Co-Cr-Al compositions rather than the NiCrAl based coatings, the latter being more often chromia formers above 800°C.

The degradation patterns of the Ni-Cr-Al and Co-Cr-Al systems change drastically when the corroding environment changes from O_2 to SO_2/Na_2SO_4-NaCl environment, although the severity may be more curtailed unlike in the binary alloys. The mechanism has been discussed in the previous sections, viz. 8.6-8.10 (Bornstein et al 1973; DeCrescente & Bornstein 1968; Giggins & Pettit 1980; Hart & Cutler 1973; Hocking et al 1977; Johnson et al 1975; Lewis & Smith 1962; McKee et al 1978; Reising 1977; Sequeira & Hocking 1978; Strafford & Hampton 1978). Sea salt and sulphur seriously affect scale coherence and subscale composition (Conde & McCreath 1980; Hossain et al 1978) but in pure O_2 NiCrAl alloys develop more protective and adherent scales than CoCrAl alloys (Barrett & Lowell 1978; Stott et al 1971).

8.13.1. MULTIPLE ADDITIVE EFFECT ON MCrAl-DEGRADATION:

Yttrium is the only additive which has been very widely investigated in all the three systems, viz. FeCrAl, NiCrAl and CoCrAl, in both laboratory and rig - evaluation tests. Other additives have been tested in addition or as an alternative. Degradation of M-Cr-Al-X systems is mostly the degradation of the M-Cr-Al scale. The mechanisms for the remarkable beneficial effect yttrium has on scale retention especially under thermal cycling conditions, have been reviewed (Lacombe 1987; Stringer 1985).

The directions investigations need to take would be to see how metal-oxide bonds are established, how stress generation modes change, how scale plasticity gets affected at the metal/scale interface both for the substrate metal and the growing scale, how vacancy and void formations are affected, how preferential growth in directions towards and inwards into the substrate occur, and how do cracks propagate (Stringer 1985).

Additives such as Y, (and Ce, Hf, Zr etc.) are referred to as active additions. The observed effects as summarized from various investigations are (Lacombe 1987):

(i) Y influences the scale plasticity due to the alteration in scale microstructure,

(ii) Oxide pegs are grown which anchor scale to substrate,

(iii) Mechanical property differences are graded between the scale and the substrate,

(iv) Scale growth mechanisms are influenced by influencing the diffusion parameters, thus reflecting on the transport and growth of the barrier layer

(v) Pore formations are eliminated by vacancy coalescence effects, and,

(vi) Chemical links between the scale and substrate are improved.

(Whittle & Stringer 1980; Stringer et al 1972; Douglass 1970; Cathcart 1975; Tien & Pettit 1972; Giggins & Pettit 1975; Stringer et al 1977; Allam et al 1978).

Transport properties and microstructure studies made on doped alumina show that results from these cannot be interpreted for thermally grown alumina. The yttrium effect is likely to be several phenomena acting simultaneously, so that, although the pegging mechanism cannot be ruled out, the more prominent effect is felt to be in microstructural modifications which result by a decrease in either anionic or cationic diffusion and in the scale stresses caused by oxidation (Huntz 1987). Porosity at the metal/ scale interface is found to be present irrespective of Y as an additive; excellent adherence occurred in both NiCrAlY and FeCrAlY with small pegs; FeCrAl and NiCrAlHf with well-developed Al_2O_3 pegs had poor scale adherence, and it seems that adherence is not necessarily due to 'pegging'. While elemental Y is adherence promoting, Y added as sulphide is not; the adherence effect of Y added to a sulphur-containing alloy at levels adequate to getter the sulphur in alloy as well as any sulphur produced by the decomposition equilibrium of Y_2S_3, has been examined (Smeggil 1987). It appears that the nature and geometrical features of pegs are important from a scan of all the micrographs presented in literature. The mystery of Y still is to be solved acceptably!

The beneficial effect by Y is not realized unless its levels are kept low as an additive. Y is shown to produce a series of oxides, viz. $Y_3Al_5O_{12}$, $YAlO_3$, $Y_4Al_2O_9$ apart from two spinels – $Ni(Cr_{0.7}Al_{0.3})_2O_4$, $Ni(Cr_{0.45}Al_{0.55})_2O_4$, and oxides of Y, Al and Ni when its concentration is varied over 0-5.5 wt.% in triode sputtered NiCrAlY coatings on NiCrAl substrates, when oxidised in 1 h cycles in air at $1100^{\circ}C$. Convoluted surface oxides, with fractures, voids entrapped in the mid-layers and internal yttrium oxides, were identified (Choquet et al 1987).

Yttrium is added at a level not more than 1%, more usually in the range 0.3-0.8%. It forms intermetallic compounds YNi_9 and YNi_5 with Ni. A Ni9CrAl formed NiO as the main product at 800° and Al_2O_3 at $1000^{\circ}C$ and $1200^{\circ}C$ in O_2. Variation of Y over 0.005 to 0.7 wt% showed that 0.1-0.2 wt% was optimum for NiCrAl. CoCrAlY alloys consist of the dark beta-CoAl phase in a light matrix of the Co-rich alpha phase. The intermetallic Co_3Y localized the grain boundaries. Below 0.1 wt% this phase disappears suggesting a solubility limit below 0.1 wt%. One view is that Y provides enhanced diffusion of Al to the surface and increases the number of oxidation sites. An Al-Y double oxide forms giving a key-on effect and thus an increased oxide adherence (Kvernes 1973). In Ni10Cr5Al0.5Y, spallation was evident over long periods. It was postulated that $Y_3Al_5O_{12}$ forms at grain boundaries, Y_2O_3 releases Ni to form NiO or $NiAl_2O_4$ and yttrium annihilates voids leading to improved scale adherence (Kuenzly & Douglass 1974).

469

Alloy structures have an important influence on the degradation modes of multi-component alloys (and also for binary and ternary alloys), which becomes evident in systems such as NiCoCrAlYTa with 22-24 Co, 18-21 Cr, 7.5-9.5 Al, 0.7-0.8 Y and 3.5-4.5 Ta wt.%. Both the cast alloy and the LPPS-sprayed alloy had gamma, gamma-prime, and beta-aluminides, and sigma and M_5Y phases. The cast alloy also had Y_2O_3. Isothermal and cyclic oxidation in air at $850^{\circ}C$ showed that the cast alloy with its large grain size of gamma and beta phases promoted scale heterogeity with chromia mainly in the gamma phase and alumina in the beta phase. In contrast, the LPPS alloy favoured a compact, homogeneous alumina partially modified by chromia (Frances et al 1987).

Y and Hf decrease the corosion rate of NiCrAl-, NiCrAlTi- and NiCrAlZr- alloys but this beneficial effect does not extend in full over the hot corrosion environment in the range 600-800$^{\circ}C$ because of liquid product or liquid product-deposit formations. The medium is particularly damaging to CoCrAlY coatings (Smeggil & Bornstein 1978; Jones & Gadomski 1977). Earlier theories postulated appear to be on lines similar to those discussed above (Bolshokov & Fedorov 1956). Co22Cr11Al0.3Y deposited on IN 738 superalloy by EBPVD and implanted Y and Hf on Co22Cr11Al were both found to degrade badly at 700$^{\circ}C$ under hot corrosion conditions (Provenzano & Cocking 1987).

Using CoCrAlY or NiCrAlY as the main coating system, many permutations and combinations have been studied; Hf has been a prime additive element (Hocking & Vasantasree 1986; Swidzinski 1980); Zr, and Si follow close and several others have been tested, e.g. W, Mo, Mn, Ce, Zn and Cu. Overlay and underlay coating systems are well known. The degradation in all these involve interlayer - adhesion - compatibility to thermal cycling and stress, and of corrosion interdiffusion effects during hot corrosion, and high temperature oxidation/carburization, nitridation, etc. Some of the candidate coating systems and their corrosion performance may be cited here (Atkinson & Gardner 1981). The data in Tables 8:11 to 8:13 and Fig.8-46 to 8-48 were generated on a directionally solidified Ni-Nb-Cr-Al eutectic superalloy (Strangman et al 1977). At high temperature the substrate influence especially on interdiffusion is considerable. The hot corrosion test was with coated Na_2SO_4 for 20 hour intervals at 870$^{\circ}C$; the chloride effect is not registered. It must be re-emphasised here that degradation patterns are specific to the substrate/multilayer-coating/temperature/corroding environment system. Assessment can be broadly generalized. In the above study NiCrAlY / Pt and NiCrAlY physical vapour deposition coating systems exhibited the best combination of properties.

It was discussed in Section 8.8 that pure Na_2SO_4 melts at 884$^{\circ}C$, NaCl-Na_2SO_4 is solid-liquid at 620$^{\circ}C$, and a $CoSO_4$-Na_2SO_4 eutectic forms at 565$^{\circ}C$ from Co_3O_4, it being the stable oxide below 947$^{\circ}C$. CoCrAlY degradation occurs in the form of pits or what is called a discontinuous precipitation effect in the coating adjacent to the hot corrosion pits. Beta-CoAl is found to dissolve locally in

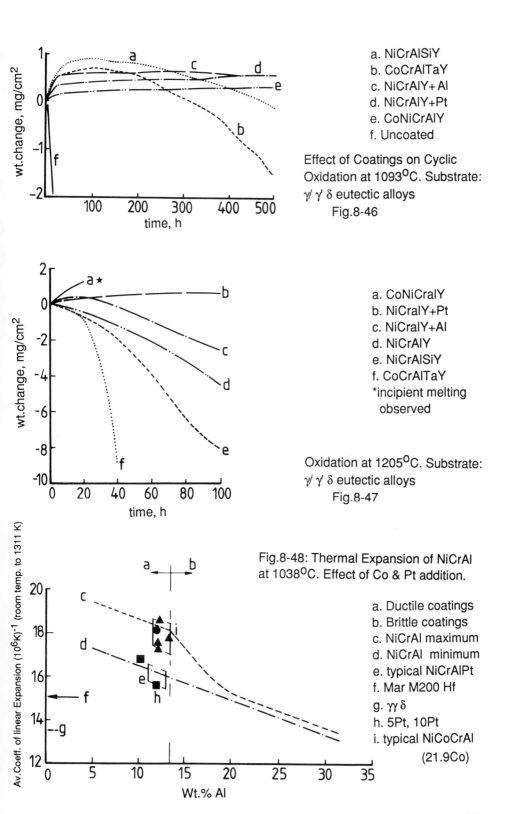

a. NiCrAlSiY
b. CoCrAlTaY
c. NiCrAlY+Al
d. NiCrAlY+Pt
e. CoNiCrAlY
f. Uncoated

Effect of Coatings on Cyclic Oxidation at 1093°C. Substrate: γ/ γ' δ eutectic alloys
Fig.8-46

a. CoNiCralY
b. NiCralY+Pt
c. NiCralY+Al
d. NiCrAlY
e. NiCrAlSiY
f. CoCrAlTaY
*incipient melting observed

Oxidation at 1205°C. Substrate: γ/ γ' δ eutectic alloys
Fig.8-47

Fig.8-48: Thermal Expansion of NiCrAl at 1038°C. Effect of Co & Pt addition.

a. Ductile coatings
b. Brittle coatings
c. NiCrAl maximum
d. NiCrAl minimum
e. typical NiCrAlPt
f. Mar M200 Hf
g. γγ'δ
h. 5Pt, 10Pt
i. typical NiCoCrAl
    (21.9Co)

471

the alpha-Co (Cr, Al) matrix immediately adjacent to such pits. On re-precipitation, the fine beta-CoAl lamellae are dissolved in the alpha-Co matrix all the way to the oxide pit (Smeggil & Bornstein 1981).

TABLE 8:11

A SELECTION OF COATINGS TESTED IN HOT CORROSION TRIALS

A. Overlay Coatings

Coating	Process	Thick-ness,μm	Coating	Process	Thick-ness,μm	Notation
Ni	E-pl	25	Co-a	EBVD	127	Ni+CoCrAlY
W	Sp.Dep.	13	Co-a	- " -	127	W +CoCrAlY
			Co-b	EBVD	127	CoNiCrAlY
			Ni-a	EBVD	63, 127	NiCoCrAlY
Ni-a	EBVD	63	Pt	Sp.Dep.	6	NiCoCrAlY-P
			Ni-b	EBVD	127	NiCrAlY
Ni-b	EBVD	63,127	Pt	Sp.Dep.	6	NiCrAlY-Pt
W	Sp.Dep.	13	Ni-b	EBVD	127	W+NiCrAlY
Ni-b	EBVD	127	Al	Pack		NiCrAlY-Al
			Co-c	Pl.Spry.	127	CoCrAlTaY
			Ni-c	Pl.Spry.	127	NiCrAlSiY

B. Aluminide Coatings

			Cr+Al	Pack	51-76	Cr-Al
Ni	E-pl	76	Cr+Al	Pack	51-76	Ni-Cr-Al

Note: E-pl: Electroplating; EBVD: Electron beam vapour deposition; Sp.Dep.: Sputter Deposition; Pl.Spry: Plasma Spraying; Pack: Diffusion coating by Pack Cementation.
Co-a: Co 18Cr 11Al 0.6Y; Co-b: Co 33Ni 18Cr 15Al 0.6Y; Co-c: Co 18Cr 11Al 5Ta 0.3Y
Ni-a: Ni 23Co 18Cr 12.5Al 0.3Y; Ni-b: Ni 18Cr 12Al 0.3Y; Ni-c: Ni 18Cr 12Al 2Si 0.3Y

The initiation of hot corrosion attack on Na_2SO_4-coated CoCrAlY at 700°C in SO_3-O_2 mixtures was observed to be due to hole formation in the oxide scale. It was thought to occur due to a reaction between yttrium oxide and Na_2SO_4, but this is not supported as holes develop in oxide scales formed on alloys not containing yttrium (Hwang et al 1983) A similar scale porosity was observed in NiO scale on a Ni- Mo-W-Al after 25 hours at 800°C in air (Smeggil & Bornstein 1981). It seems hole formation is a scale oriented phenomena and not concerned directly with any one base metal alloy Co- or Ni-. Some of the main differences in the hot corrosion of NiCrAlX and CoCrAlX alloys at 600-800°C and above 800°C may be summarised here. Given the same composition, in an

SO_2-O_2-SO_3 atmosphere, with a salt deposit of Na_2SO_4, NiCrAlX is attacked to a high degree only when the Ni- Ni_3S_2 eutectic is allowed to form, which will be when Cr_2O_3 and Al_2O_3 layers break down either to allow Na_2SO_4 and gas in or promote NiO growth outwards which leads to Ni_3S_2 nucleation within it (Hocking & Vasantasree 1981).

TABLE 8:12

CYCLIC HOT CORROSION TESTS ON COATED ALLOYS

Test Conditions: Substrate (Eutectic Alloys);
Coatings listed in Table below;
Duplicate test samples were salt-coated
with 0.5 mg/cm^2 Na_2SO_4 every 20 h and
oxidised for 13 cycles of 20-h at 871°C

Coating	Total Weight Change mg/cm^2	Alloy Performance
Ni+CoCrAlY	0.06, 0.04	Very low-degree attack
W+NiCrAlY	0.10, 0.60	Excessive attack at coating repair sites
NiCrAlY	0.17, 0.12	Post-test state of samples excellent
NiCrAlY-Pt	0.11, 0.03	As above; excellent resistance
NiCrAlY-Al	0.36, 0.43	Localized failures at coating defects
CoNiCrAlY	0.06, 0.02	Low-degree attack
CoCrAlTaY	0.64, 0.15	Samples covered with a bluish spinel oxide; reproducibility only fair
NiCrAlSiY	0.02, 0.01	Loss due to spalling and Cr-removal as soluble Na_2CrO_4
Cr-Al	1.47, 1.57	Selective attack at corners
Ni-Cr-Al	0.37 0.38	Localized failures at coating defects
Uncoated Alloy	4.90	Accelerated attack

===

473

TABLE 8:13

HIGH TEMPERATURE STRESS RUPTURE TESTS ON COATED & UNCOATED ALLOYS

Test Conditions: 1038°C; 151.7 MN/m^2, in air. Where 'Y', samples were furnace aged in argon at 1093°C for 500 h (except no.5 which was aged in air instead of argon)

A. NiCrAlY-Pt Coated Alloys:

Coating thickness, um	Aged, Y/N	Time to rupture, h	Affected thickness, Coating +/- oxidation affected substrate, um	Recalculated Stress level MN/m^2
107	N	206.2	27.9	160.0
107	N	206.1	25.4	158.6
39	Y	213.4	42.2	162.8
46	Y	287.3	47.0	164.1
95	Y	328.8	42.7	163.4

B. Uncoated Alloy

	N	106.2	67	169.7
	N	122.9	76	171.7
	Y	114.7	67.3	169.0
	Y	166.4	80.0	173.1

Note: Original stress level was calculated on the basis of gauge cross-section area of the uncoated alloy sample. The new stress level was then calculated on the cross section unaffected by either oxidation or diffusion with coating.

Unlike NiCrAlY, in CoCrAlY, breakdown in Al_2O_3 and Cr_2O_3 leads to Co_3O_4 formation which forms a $CoSO_4$ as a secondary product, and forms a eutectic with Na_2SO_4. It is this reaction that results in a major degradation of CoCrAlY and most other Co-base systems. At 1100°C CoCrAlY coatings showed very good corrosion resistance to V_2O_5-Na_2SO_4 and NaCl-Na_2SO_4 synthetic corrosion ash (Nakamori et al 1983). Ni-base alloys are not suitable , since they undergo catastrophic attack. Creep rupture and alternate load tests have shown that a high Co content is desirable for scale retention of aluminides, chromide-aluminides and MCrAlY-X coatings on IN100 as substrate (Schmit-Thomas & Johner 1983). Other additives like Hf, Ce, Zr and W, reduce the corrosion rate in general. Hf has been very rarely identified as a part of the scale product, but the other elements have been found as oxides incorporated in the Cr_2O_3 matrix or presenting an internal oxide precipitate band at the coating/scale interface.

MCrAlY coatings 100-150 microns thick, applied by LPPS onto superalloy substrates, heat treated for 4 h at 1115°C in vacuum were tested in O_2 potentials from 10^{-24} to 10^{-25} with pS_2 potentials ranging from 10^{-10} to 10^{-8} and carbon activities of 0.65-1.00 at 650 and 871°C. The degradation was on the lines discussed above; high aluminium NiCrAlY and CoCrAlY gave the best resistance over a wide range of pS_2 at both temperatures. NiCrAlLaY, CoNoCrAlY and NiFeCrSiBC were promising in the low to intermediate pS_2 (Natesan 1987).

The beneficial effect of yttrium on the scale adherence of Fe-Cr-Al alloys is similar to its role in Ni- and Co-base alloys except that more than one iron oxide and sulphide is involved in the diffusion process and in the oxide / sulphide scale degradation. The aluminium content has to be increased as the chrominum content rises and on addition of Ni more Al is needed. Thus a 20 wt% Cr requires 2-3% Al while with 20% Ni, 12-14% Al is required (Tomaszewicz & Wallwork 1978) and and it will be at the cost of the mechanical properties, viz low creep strength from 650-900°C (Wilhemson 1983). At $pS_2 = 10^{-4} - 10^{-6}$ in H_2S/H_2 over 700-900°C a 26.6 atom% Cr-Fe alloy developed a rapid growing (Fe,Cr) S and a slower growing $(Cr,Fe)_3S_4$ scale; at $pS_2 = 10^{-2}$ three layers are formed, a duplex scale at 10^{-5} and a single scale at 10^{-5} (Narita & Smeltzer 1984).

8.14. DEGRADATION OF SILICIDE COATINGS

Silicon is the third element chosen in hot corrosion systems for its ability to form a barrier layer to resist degradation, the protectivity relying on the formation of SiO_2. It offers the best melt fluxing resistance in hot corrosion environments at high pSO_3. The pattern of growth and retention of SiO_2 alone or as an oxide associate is influenced by the manner in which Si is introduced into the system, as an element additive to form an alloy by slurry or CVD methods, or by deposition or controlled oxidation to form SiO_2 or deposition as ceramics such as SiO_xN_y or Si_3N_4 or SiAlON. Although normally very stable, the degradation of SiO_2 can occur through cracking, peeling or spallation or vaporization loss as SiO (Pettit & Goward 1983; Goebel & Pettit 1975; Caillet et al 1979).

8.14.1. SILICIDE ON STEELS:

A pre-annealed thermal silica on steel shows a reduced reactivity in steam by a factor of x20 (Rigo et al 1982). In steam over 950-1100°C $SiO_2 > 8000$ angstroms was rated as a zero oxidation barrier. The thickness corresponding to stabilized reproductive index was taken as a measure of the unoxidized ceramic applied as SiO_xN_y or Si_3N_4 (Gaird & Hearn 1978). A layer of $MoSi_2$ on steel was found to give an oxidation resistance time increasing linearly with the increasing thickness with no cracks or peeling developed up to 10 microns thickness (Motojima & Fujimoto 1982).

A $MoSi_2$ layer of high stability could be achieved at low silico-nizing (CVD) temperatures when $SiCl_4$ was maintained at greater than 10 vol %. The reduced oxidation resistance at $SiCl_4$ < 5 vol% was probably due to the formation of Mo_3Si (Motojima et al 1982). A similar thickness-related oxidation resistance has been obser-ved on a CVD-SiO_2 on 20/25/Nb stainless steel for over 5975 hours in CO_2 at 825°C. A layer > 2 microns reduced cation diffusion and decreased oxide spalling and attack by a factor of 5 with a chromium content of 15-16% complementing this prevention of spal-ling. Provided an adherent, non-spalling silica scale was main-tained, pitting attack was also prevented (Bennett et al 1982).

The efficacy of silica layers on Fe-base alloys has been ascribed to a barrier effect, the diffusion of iron (Fe^{3+}) being 10^5 times slower in amorphous silica than in Fe_3O_4. The diffusion coeffi-cient in silica is also lower than that of Cr in Cr_2O_3 and of the same order as that of Fe in Fe_2O_3 at 1100°C, and SiO_2 films slow down formation of sesquioxides on high Cr-steels (Atkinson & Gardner 1981). This has been amply proved in the use of SiO_2-coated structural steels in the nuclear industry. Vapour depo-sited silica coatings give extented life to 20/25/Nb stainless steels with protection against both oxidation and carbonaceous deposition (see Fig.8.49, 8.50) (Bennett et al 1984). Plasma sprayed WSi_2 coatings proved satisfactory in mechanical testing but failed in corrosion and scaling tests because of porosity (Knotek et al 1987).

8.14.2. SILICIDE ON REFRACTORY METALS

Si deposited on titanium substrates via silane (SiH_4) formed Ti_5Si_3 which was found to decrease the oxidation rate over the temperature range 700-1020°C in O_2. Scale formation was found to be a mixed oxide (TiO_2 + SiO_2) layer at temperatures below 875°, while above 900°C, an outer TiO_2 and an inner, inward growing TiO_2SiO_2 developed (Abba etal 1982). A similar mixed oxide of NiO + SiO_2 is reported on silicided Ni at 950-1000°C (Subrahmanyam 1982). Silicides have also afforded over 600 hours protection at 870° and 1315°C on Ta-9.6W-2.4Hf-0.01C and a 1064 hours survival at 1315°C on 35Mo-35W-15Ti-15V alloys (Wimber & Stetson 1968). The coating lives of complex silicides were found to be shorter under reduced pressure (0.01-0.1 torr) than at higher (50 torr) and ambient pressures. Coatings 0.003 inches thick were applied by the slurry technique : Si20Ti10Mo on Ta alloys, Si20Cr0.5B$_4$ on Mo alloys and Si20Cr20Fe on Nb alloys. The coatings survived 553 pressure cycles at about 290°C and were ductile and at 1425°C survived in excess of 100 hour cycles (Priceman & Sama 1968).

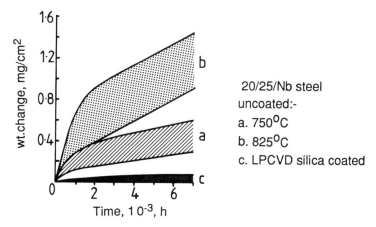

20/25/Nb steel
uncoated:-
a. 750°C
b. 825°C
c. LPCVD silica coated

Fig.8-49: Effect of LPCVD Silica Coating on
the Oxidation of 20/25/Nb Steel in
$2\%CO+300$ ppm CH_4+200 ppm H_2O+CO_2.
(scatter bands shown)

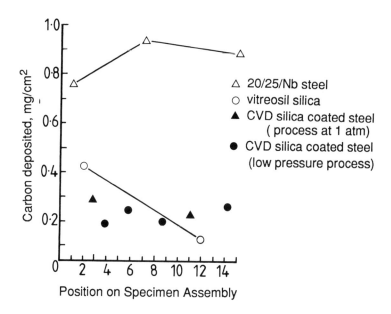

△ 20/25/Nb steel
○ vitreosil silica
▲ CVD silica coated steel
    (process at 1 atm)
● CVD silica coated steel
    (low pressure process)

Fig.8-50: Carbonaceous Deposition Inhibition by
LPCVD silica Coating. Substrate - 20/25/Nb Steel.

8.14.3. SILICIDE DEGRADATION IN GAS TURBINE ALLOYS:

Silicon-containing protective scales can be formed by Ni-5Si, Ni-11.5Si; a chromia-silica scale on Ni-40Cr-5Si and Ni-40Cr-11.5Si; and an alumina-silica scale on CoCrAlY-5Si and CoCrAlY-12.5Si.

Co-silicides melt above 1200°C and no liquid silicates form below 1380°C. But a solid state reaction between CoO and SiO_2 is said to occur by an unusual surface diffusion process; Fig.8-51 (Schmalzreid 1974). None of the refractory Cr-silicides melt below 1300°C but a Ni-50Cr with a protective Cr_2O_3 scale is pitted extensively when pressed in contact with a Si/SiC composite for 120 hours at 1150°C in air. A Ni-Cr silicide underlayer in the molten state at 1150°C is envisaged (Ni-5wt%Si solid solution and silicides Ni_3Si and Ni_5Si_2) (Mehan & McKee 1976).

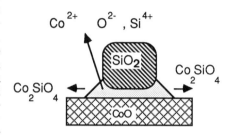

Cobalt oxide - Silica Interaction

Fig.8-51

An Al-11.7w%Si alloy forms a eutectic at 577°C and an $Al_2O_3-SiO_2$ reaction yields mullite at very high temperatures. Except at very low pO_2 a reduction reaction between Al and SiO_2 is unlikely. A Ni-30Al alloy (NiAl) showed Al_2O_3-scale fracture at 1150°C on 120 hours contact with Si/SiC, with a Si-rich phase penetrating the alloy. Contact reaction between IN718, a Ni-base superalloy, and Si/SiC showed severe pitting at 1150°C, and 1050°C, which was largely reduced at 1000°C and absent at 950°C. In all cases there was surface depletion of Si and the alloy matrix was seen to have complex mixed silicides some with Nb and Ti. A similar behaviour was observed for IN 738. Silicide phases $Ni_{16}Cr_6Si_7$, Co_2Si and Ni_2Si were identified along with Ti- and Nb- silicides. Degradation of Incoloy 800H was reduced by more than 50-80% by PAVD SiO_2 at $700-900^{O}$C in oxidation, carburization, and sulphidation. Laser fused SiO_2 have shown similar arrests, the coating having interacted with the substrate on application to result in a Cr:Si at 66:25 (Ansari et al 1987).

Silicide degradation in hot corrosion conditions must be viewed in the context of the above data. It is clear that Si diffuses inwards into the alloy substrate, with the diffusion penetration more marked at low pO_2. SiO_2 can be reduced by outward diffusing species like titanium which will then need a barrier layer to prevent this. Silicon gives a slight improvement in oxidation in air in the presence of NaCl in Ni and CoCrAlY systems but hot corrosion tests conducted in our laboratory showed scale spalling at both 700° and 900°C (Sidky & Hocking 1979, 1987a,b; Hocking & Sidky 1987a; Marriott et al 1980; Vasantasree & Hocking 1984). $MoSi_2$ is unsuitable for the protection of Ni-base alloys because of its low thermal expansion and also its high reactivity with the substrate, while it is so successful on Nb- and Ta- based substrates. A Ni-Cr-Si with about 8-10 at % Si and a low Cr content, about 10% in gamma-solid solution is found very beneficial (Fitzer et al 1979).

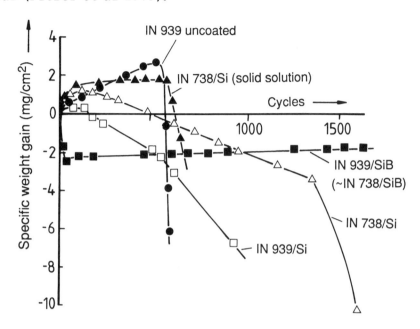

Fig.8-52: Effect of CVD Silicide Coatings on Cyclic oxidation. (Superalloy Substrates)

The phase stability of a vapour deposited silicide (Si on Ni) with a plasma sprayed chromium boride and their hot corrosion characteristics have been investigated. Siliconizing IN738 sprayed with Ni results in a gamma-Ni zone enriched with Co, Cr and Al, upon which develop the Ni_3Si and Ni_5Si_2 phases. Exposure to air at 1000° for 300 hours resulted in a breakaway of the coating which was clearly associated with the separation of the Ni-Si eutectic structure. A brittle Si-rich Ni_5Si_2 below which is Ni_3Si followed by a heterogenous gamma-Ni layer with voids and inclusions leads to the degradation; Fig.8-52. Degradation of the better substrate-coating configuration of C73-Ni-silicide followed a similar pattern but with improved performance. The directionally solidified eutectic superalloy consists of Cr_7C_3 fibres in a CoCr matrix. After 1000 hours exposure at $1000^{\circ}C$ the Ni content in the coating dropped from 90% to 50%; Cr and Co of about 10% were present in the porous zone; the carbide from the substrate yielded a lamellar carbide layer of the type $Cr_{23}C_6$ with large amounts of Co, Ni and Si; no Ni_3Si_2 was formed but a Co-rich (up to 20 at%) Ni_5Si_2 compound was formed. It is the lattermost layer which would determine the corrosion resistance over long periods (Hildebrandt et al 1979).

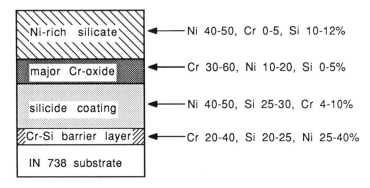

Fig.8-53: A Typical Morphology of Corrosion Product Formation of a Silicided Superalloy

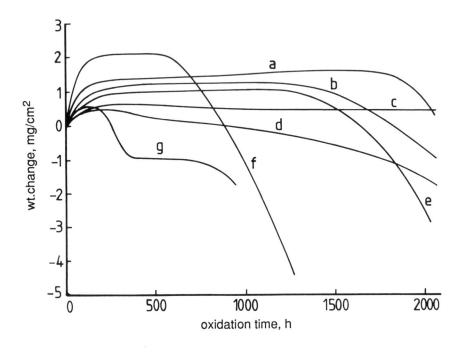

a. IN738 aluminised
b. ATS 340 aluminised
c. IN738 chromised
d. ATS340 chromised

e. IN100 chromised
f. IN100 aluminised
g. -SS on Nimonic 90

Fig.8-54: Silicide Coating 50Si-26Ni-24Cr wt.% on ATS 340 (equivalent to Nim 90). Oxidation effect over long exposure times in natural gas with excess air at 1000°C.

Much improved resistance based on the theory of providing a Cr-reservoir is achieved in a NiCrSiB coating on IN738LC. At 900°, during a 1000 hour exposure, the Cr-boride zone decreased as the oxide layer thickened, with the volume fraction dropping from 65-45% in 50 h reaching a stable condition after undergoing an Ostwald ripening process up to about 500 hours (Hildebrandt et al 1979). No data has been recorded on these coatings in sulphur-containing environments, but good oxidation resistance of 26Ni24-Cr50Si wt% coating in natural gas with excess air has been reported; Fig.8-53, 8-54 (Fitzer et al 1979).

Very good resistance to vanadic attack has been reported for silicon coated Ni-20Cr alloys at 900°C with a reduction of as much as 80% in the corrosion rate. No breakaway effects were observed for more than 600 h at 900°C in $80V_2O_5-20Na_2SO_4$ melts. The increased protection appears to be due to the development and retention of a Cr-rich barrier layer beneath a Si-rich surface with the slag showing little reaction with Si. Although Si coated by pack, vapour deposition, plasma spraying or ion plating methods was uniformly effective, ion plated coatings were found to be the most reliable with good adhesion, uniform composition and even thickness (Elliott & Taylor 1979). Fig.8-23 (p.434) shows the effect of a Si coat (Holmes & Rahmel 1979). The combined effect of molten sodium sulphate and vanadate is catastrophic in most cases as shown in studies on binary and ternary Ni-Cr(10-30 wt.%)-X alloys (X=Al, Si, V) and IN 738 (Sidky & Hocking 1987b).

A superior performance of ion plating as a coating method has been recorded for duplex coatings containing Si which have shown excellent resistance to attack by melts of $80V_2O_5-20Na_2SO_4$ at 850°C and 10% $NaCl-90\%Na_2SO_4$ at 850 and 950°C. The tests were cyclic with 24 hr residence and recoating steps. A Ni-base alloy (Udimet 500) showed a 50% reduction in weight loss while the coatings were more effective on the Co-based alloy X-45 with a reduction by a factor of 24 (Nakamori et al 1983). The effect of fusing conditions of a Si-15Cr-10Ti-10Zr coating on C-103 niobium alloy was investigated. 1300°C for 30 minutes gave the best high temperature oxidation resistance (Kun et al 1982).

8.15. DEGRADATION OF THERMAL BARRIER COATINGS

Failure of thermal barrier coatings could be initiated by poor bonding to the substrate matrix, or the coating parameters such as porosity, microstructure-microcrack distribution, thickness, phase distribution and cohesive strength, or the relative thermal expansion mismatch and residual stress of the substrate-coating system. Ceramic sintering and bond coat inelasticity could also contribute to TBC degradation. Bond coat pre-oxidation causes degradation by adhesion failure. The actual failure generally occurs within the ceramic layer near the bond coat, and is believed to be due to slow crack growth and microcrack link-up with the ceramic. Table 8:14 (and Tables 8:12 and 8:13) give a qualitative comparison of cyclic oxidation resistance. (Lowell et al 1976; Miller 1987; Kvernes & Forsyth 1987; Chang et al 1987; Strangman et al 1977).

TABLE 8:14

EFECTS OF THERMAL CYCLING ON NiCrAlY & NiCrAlZr ALLOYS COATED WITH $ZrO_2-Y_2O_3$

A. NiCrAlY/ZrO_2-Y_{23}

Heating Cycle Duration, h	1		7		20	
	Calc.	Exp.	Calc.	Exp.	Calc.	Exp.
No. of Cycles to failure	148	150	48	50	16	16
	110	120	35	35	15	15
	75	74	28	18	12	14
Time, h, to failure	148	150	300	300	320	320
	120	110	75	200	240	240
	70	75	160	115	220	280

B. NiCrAlZr/$ZrO_2-Y_2O_3$

Heating Cycle Duration, h	1		7		20	
	Calc.	Exp.	Calc.	Exp.	Calc.	Exp.
No. of Cycles to failure	49	50	13	17	6	8
	38	38	13	13	5	5
	-	30	15	9	5	5
Time, h, to failure	50	52	70	100	150	160
	40	40	75	75	90	90
	-	30	52	70	90	90

Note: the coating performance is superior on the NiCrAlY.

With the failure criterion defined as surface cracking visible under X10 magnification, an oxidation based model has been developed for predicting ceramic barrier coating life. The model is based on the cumulative effect of oxidation and thermal cycling on strains which promote crack growth. A fair agreement appears to have been achieved but the model fails where more complicated reactions are involved, and is also vulnerable to variations in coating characteristics during coating production. Table 8:14 indicates ceramic coated NiCrAlY to be superior to NiCrAlZr in thermal cycling tests (Miller 1984). Thermal fatigue failure is common in ceramic coats. The first order rule-of-thumb may be postulated to be that a coating should have 1-2% ductility at the temperature of occurence of maximum strain and should have enhanced thermal fatigue resistance compared to elastically brittle coatings (Pettit & Goward 1983). For Ti-Si coats on superalloys by CVD the DBTT and the minimum strain to spalling or cracking were a function of both the coating composition and thickness. CVD-Ti added to CVD-Si increases the DBTT from 450-700°C to 850-1000°C. Decreasing the thickness from 180 to 50 microns for Si-B increased the minimum strain at 600°C from 1% to 6% (Wahl et al 1981).

Ceramic coatings degrade by spallation due to transient thermal stresses aggravated by salts which induce hot corrosion. For non-oxide based ceramics such as silicon carbide and nitride, resistance depends strictly on the nature of secondary phases at grain boundaries due to the densificaton of additives, e.g. MgO, Y_2O_3, Al_2O_3 etc. (Billy 1987). Reaction-bonded Si_3N_4 and self-bonded SiC coatings were severely attacked in vanadic-sulphate melts at 820°-1100°C when exposed to a residual oil-fired environment in burner rigs (Brooks et al 1979). Tested in both crucible and burner rigs with melt compositions based on Na_2SO_4 with additions of NaCl, $NaCl+V_2O_5$ and $NaCl+Li_2O$ (950°, 200 h) reaction-bonded Si_3N_4 was found to be highly dependent on the overall environmental conditions for its corrosion. Excellent resistance was observed in oxidising and acidic media but very poor resistance in reducing and mildly alkaline environments. Several crystalline phases also occur on the ceramic surface during the corrosion reactions (Erdos & Altorfer 1979). Sintered SiC also suffered a similar attack in sulphate melts at 900°C, less in oxidising conditions but more in basic melts or slags especially with carbonaceous material. However SiC was inert in pure N_2, H_2 or H_2-H_2S mixtures at 900°C. Fig.8-55 shows the possible modes.

Pressure sintered Si_3N_4 exposed to air with Na_2SO_4 at 1000°C showed extensive oxidative corrosion and the $Na_2O-SiO_2-Y_2O_3$ phase equilibrium and is examined in relation to it since the nitride ceramic contained traces of $Y_2Si_2O_7$, YSi and $SiYNO_2$. The silica film formed during oxidation dissolved and reacted with basic fluxing to form sodium silicate (Riley et al 1987). Degradation by pS_2-environment at 900-1300°C with pS_2 at 10^{-6} and pO_2 at 10^{-15} show attack increasing via suphide products and volatile SiO (Datta et al 1987). Migrations of cations Mg^{+2} and Y^{+3} from the secondary phase material at silicon nitride grain boundaries are considered to contribute and interfere in the protective SiO_2 film growth (Riley & Andrews 1987).

Coatings of Si_3N_4, TiB_2 and Al_2O_3 proved to be effective barriers when free of fissures and pinholes (Hintermann 1981). Self-fluxing alloys of Ni-Cr-Si-B have great oxidation resistance. These are arc-sprayed from wire or powder stock and surface oxidation of the particles (which would prevent coalescence) is removed by the B and Si to give a fully dense fused deposit, also cleaning the base metal. If the substrate is Cr-rich, its oxidation occurs however, unless spraying and fusing are simultaneous.

483

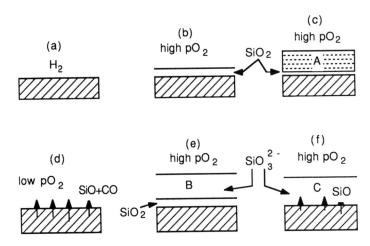

(a) Reduction environment; SiC is inert and suffers no material loss.
(b) A protective scale of silicon dioxide forms; could be termed as passivation.
(c) As long as the oxygen partial pressure is high, a passive layer of silica forms in the presence of an acidic or neutral melt.
(d) At oxygen pressures less than 10^{-16} SiC undergoes material loss via the formation of gaseous monoxides of silicon and carbon.
(e) Accelerated corrosion of SiC occurs in basic salt melts with soluble ionic silicates; a high oxygen partial pressure induces a layer of silica on the substrate.
(f) The highest degree of attack occurs in spite of a high oxygen pressure if an oxygen-depleted salt melt confronts SiC. Loss occurs through gaseous products as well as soluble silicates..

Fig.8-55: Effect of Gas &/or Molten Salt Environment on the Stability of SiC

Failure of SiC, Si_3N_4 (and also WC and Ni-Cr-B) by brittle fracture on erosion has been recorded on components used in fossil fuel energy proceses. The brittle fracture showed no relationship to hardness, the erodent being SiC particles (200 microns) at a velocity of 300 m/sec. The test conducted was at room temperature for 8-15 min with an erodent impingement of 5 gm/sec. Fine-grain structured, low porosity ceramic surfaces had the lowest erosion rates (Levy et al 1983). Attack on SiC under thin films of molten sodium carbonate and sulphate at 1000°C was found to occur at structural discontinuities with a crater-like material loss. The latter was correlated to bubble formation during the oxidation of SiC which created unprotected regions subsequently to be exposed to enhanced attack leading to pit formation. The simple oxidation reactions were (Jacobson & Smialek 1986):

$$SiC + 3/2\ O_2 = SiO_2 + CO$$

$$Si_3N_4 + 3O_2 = 3SiO_2 + 2N_2$$

Alloy compositions which have themselves been used to retard hot corrosion have been used as interlayers. A plasma-sprayed calcium silicate ceramic layer with a CoCrAlY or NiCrAlY interlayer was applied on to B 1900 and Mar M-509 and corroded at 1030°, 1100° and 1160°C under hourly cyclic conditions in air. The specimen life was a strong function of the temperature although the weight gain at the time of specimen failure was temperature-insensitive; Fig. 8-56 a,b (Miller 1983). Immersion in synthetic slag melts is used to rank on an accelerated test basis. Fig.8-57a shows tests on IN 738 coated with various bond coat materials (Bauer et al 1979). Creep rupture life of diffusion, overlay and thermal barrier coatings on IN 900 show the degradation of the uncoated IN 100 and the improvement gained by the bond coats and TBC (Fig.8-57b).

A good bonding interlayer betwen the ceramic coating and the substrate superalloys is vital for TBC survival. Titanium carbide and nitride are well-bonded to superalloys if the Ni_3Ti inter-metallic phase is controlled. An excess produces a rough deposit and thus a poor bonding. A marginal carburization is also a requirement for good bonding (Bertinger & Zeilinger 1981). Without an interlayer, ceramic-substrate interactions are inevi-table. MgO, Al_2O_3, S_iO_2 (as fused quartz), SiC and Si_3N_4 were all found to react with a Ni-based superalloy substrate, with the most severe being Si-SiC over $700^\circ - 1150^\circ$C, with conditions at the interface held at minimized pO_2. Si and C diffused inwards forming silicides and carbides with the degree of reaction on the ceramic side being similar to that of the substrate below 900°C, and less above 900°C (Mehan & Bolon 1979).

Chemical and thermal-mechanical interactions are cited for the failure of porous, plasma-sprayed ZrO_2-Y_2O_3 (8,15,20 wt%), ZrO_2-MgO (24.65 wt%) and Ca_2SiO_4 with NiCrAlY as a bond coat on U-720 and ECY 768 alloy samples. Fuels ranged from pure diesel GT no.2 to doped 1-100 ppm Na, 2-180 ppm V, 2-18 ppm P, 0.25-2.25 wt% S impurities. Partially stabilized ZrO_2 containing small fractions of monoclinic ZrO_2 out-performed fully stabilized ZrO_2. The 2-phase ZrO_2 could develop micro-fissures to relieve thermal stres-ses. In contaminated fuels graded coatings of NiCrAlY bond coat/ NiCrAlY-oxide graded zone/oxide overcoat were better than a dup-lex NiCrAlY bond coat/oxide overcoat. Higher impurity content, higher mass flow rate or gas velocity and thermal stress condi-tions were detrimental, especially in vanadium 180 ppm and sea salt 100ppm (Lau & Bratton 1983). Plasma sprayed magnesia-stabi-lized zirconia (MSZ) thermal barrier coating over a mixed MSZ+NiAl coating over a thin NiAl bond coat has been used in gas turbine combustion chambers but oxidation of the NiAl occurs. Better bond coats are NiCrAlY and CoCrAlY which resisted hot corrosion. The best ceramic overcoat was 8% yttria-stabilized zirconia, which did not peel or spall (199) in molten salt tests at 800°C in 11 ks or in 1000 cycle 400°C to 900°C burner rig tests. The substrate was Hastelloy X or Haynes 188 (Akikawa & Uenuna 1982).

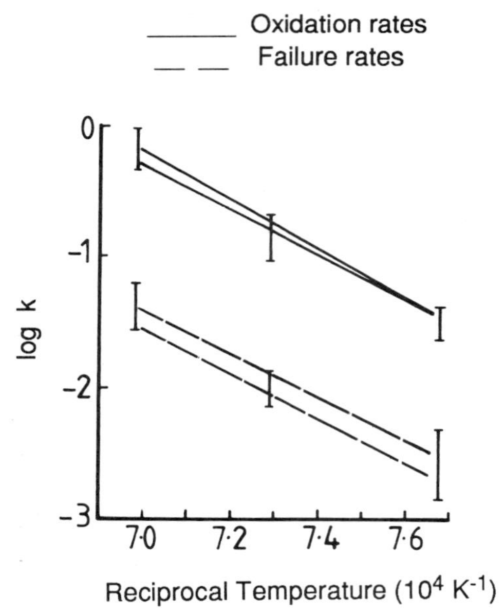

Oxidation rates

Failure rates

---- NiCrAlY

—·— CoCrAlY

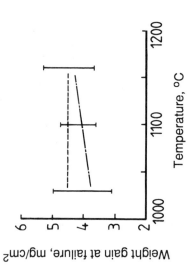

Fig.8-56a: Bond Coat Effects on Oxidation & Failure Rates. Arrhenius plots for Ca$_2$SiO$_4$/MCrAlY/MAR M509; (M=Ni or Co)

Fig.8-56b: Bond Coat Effects on Weight Gain at Failure for Ca$_2$SiO$_4$/MCrAlY/MAR M509; (M=Ni or Co)

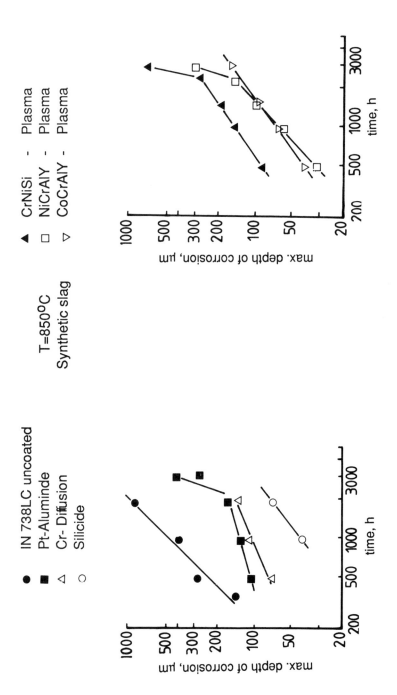

Fig.8-57a: Resistance of Coated IN 738LC to Hot Corrosion in Synthetic Slag
Immersion Tests (Bauer et al)

Single crystal ZrO_2 ceramic with CeO_2 and yttria as stabilizers degraded less than the sintered variety in sodium vanadate melts. Although ceria and yttria did not react with the vanadate in the solid state, the melt leached out ceria (Jones & Williams 1987). Improved yttria-stabilized zirconia ceramics were found to adapt very well to thermal cycling with sealing coats inserted into the columnar ceramic, but molten sodium sulphate was found to penetrate the ceramic. The bond coat/sealing coat resistance was thus proved to be a key factor in determining coating life (Prater & Courtright 1987). $Pt-ZrO_2$ and $Pt-Y_2O_3$ have been tested in methane at $500^{\circ}C$, hydrazine at $800^{\circ}C$ and in CO_2, H_2, NH_3, N_2, and steam at $1200^{\circ}C$ for 2000 h as potential space station resistojet materials (Whalen 1988).

Unlike in the oxidation-hot corrosion area, the failure mode of ceramic coats used for wear resistance is via micro-chipping. The combined effect of wear and hot corrosion conditions would be such as to allow access to the bond layers. Cemented carbides such as TiC, TiC_xN_y and TiN degenerate with abrasive excursions and impact. The ceramics fail by transverse crack development, TiC being superior to TiN (Venkatesh et al 1981; Cho & Chun 1981). The CVD temperature has to considered relative to the tempering treatments required for the steel substrate so that the projected hardness can be achieved in use (Ruppert 1981). Boron-based ceramic coats offer good wear resistance, the process being boriding succeeded by metalliding (electrodeposition from molten salts). B-Cr, B-V, B-Si and B-Cr-Ti have been used with B-V giving excellent wear (Chatterjee-Fischer 1981).

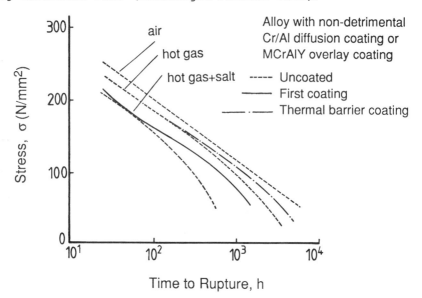

Fig.8-57b: IN 100 at $950^{\circ}C$. Creep Rupture Life under Imposed
Hot Corrosion Conditions.
(Scweitzer & Track)

8.16. BURNER RIG DEGRADATION DATA

Service-like tests carried out in burner-rigs have been helpful in ranking the performance of various types of coatings in a variety of environmental conditions. Table 8:15 gives a number of groups of coatings evaluated in this manner, the burner rig design is not necessarily identical. Specimens tested appear to vary from a simple form of cylindrical pin or are more specifically close to the component design. Fig.8-58a shows the schematic of a burner rig. Specimens are evaluated in several combinations, broadly along the following lines:

1. Coated and uncoated; coatings from various processes, variations in coating thickness, variations in substrate shapes, preparation and heat treatment,

2. Isothermal and cyclic temperatures,

3. Hot gas - in varied velocities with various potentials of pO_2, pS_2, SO_2, pSO_3, pH_2, pH_2O, pCO, pCO_2, pCH_4 etc.,

4. Air intake injected with combinations of sulphates, chlorides, vanadates, etc.,

5. Erodent additions in the gas stream, the particles being carbon, sand, salt (e.g. sulphate, chloride) etc.,

6. Hot gas spinning tests,

7. Creep and fatigue tests, and,

8. Temperature and load cycling conditions.

Conclusions drawn on coating performance on IN 100 from one such series of tests summarize as follows (Schweitzer & Johner 1985):

Diffusion Aluminides: High temperature/low activity (HTLA) aluminides show adequate hot corrosion resistance at <950°C for lifetimes <1200 h; High temperature/high activity (HTHA) aluminides tend to reduce creep life of the substrate alloy and is unsuitable for use on components likely to be subject high mechanical stresses. Cooling passages are protected by HTHA coatings. Re-aluminiding is possible with a small loss in the substrate alloy.

MCrAlY Overlay Coatings: LPPS or PVD coatings allow longer lifetimes of x3-x4 over the aluminides up to about 1050°C with tolerance to higher corrosion. Coatings, however, need to be on parts directly visible, but can be renewed without loss of substrate alloy.

Modified Aluminides: Coatings in this class such as Pt-aluminides - LDC 2, RT 22 etc., are preferable to MCrAlY on complex shaped components like nozzle guide vane segments.

489

TABLE 8:15

BURNER RIG TESTED COATING DEGRADATION DATA

Coating	Substrate	Temp., °C	Time, h	Investigators' Comments
NiCoCrAlYTa (LPPS)	MA 6000	925 1025 1100	500	Isothermal & Cyclic tests; (Smith & Benn 1987)
Aluminide & NiCoCrAlY	Ni-base Superalloys	900		Protective capability rests on the diffusion zone resistance rather than the additive layer. Pre-ageing at high temperature (1100°C) markedly decreases coating life. (Fryxell & Leese 1987)
Zirconia (PSZ & YSZ), CoNiCrAlY	IN 617			ZrO_2 with $8Y_2O_3$ with 80Ni20Cr bond coat found to be the best for thermal shock. NiCrAl bond coat was slightly inferior. Co32Ni21Cr7.5Al0.5Y was very poor. (Steffens & Fischer 1987)
Our Codes: A,B,C,D,E,F	LM2500	650-770	1077	Marine test conditions; sea salt ingestion levels 0.006 ppm. (normal level 0.002 ppm); Coatings D & C clearly superior, and also F in low temp. hot corrosion resistance. (Grossklaus et al 1987)
Co-Cr (Cr=40-43) Co-Cr-Si (Si=0.6,1,5) Co-Cr-Y(0.1)	IN 738 & Rene 80	732° & 871°	1000	At 732°C these experimental compositions were better than conventional CoCrAlY & Pt-aluminides; halved the corrosion rate, but at 871°C there was no advantage. (Luthra & LeBlanc 1987).
CrAl-I, CrAl-II, CrAl-III CrAl-IV RT22	IN 100, IN 738 & IN 939	800°- 900°		Molten sodium sulphate environment under aero-engine conditions; benefit of Pt and Cr and mechanism established.(Morbioli et al 1987)
Aluminide, Chromide CoCrAlY, CoNiCrAlY & NiCrTiAl		700°, 750°, 850° & 950°	500 500 500 3000	Pt-aluminides (RT22, LDC2) perform best under all conditions. Ranking given in Appendix (Nicholls & Hancock 1987)
Aluminide, Pt-Aluminide NiCoCrAlY	MAR M200 +Hf	1070°	300	Aimed for F100 I-stage turbine blade with inlet temp. at 1425°C; since 1070°C was the max. rig temp. test blades were not air-cooled; data comparable because max., cruise and idle temp. in practice were 1050°, 813° and 590°C resp. NiCoCrAlY overlay was the best; ranking given in Appendix (Mom & Hersbach 1987)

[A=PVD CoCrAlY; B=VPS CoCrAlHf; C=PVD CoCrAl+PackAl/Hf overcoat+E.dep.Pt; D=VPS CoCrAlHfPt+PackCr overcoat; E=HighCr PVD CoCrAlY; F=PVD YSZ(20%) over Al

490

For thermal cycling tests the test sample carousel 'g' can be lowered from the hot zone for quenching by air or inert gas.

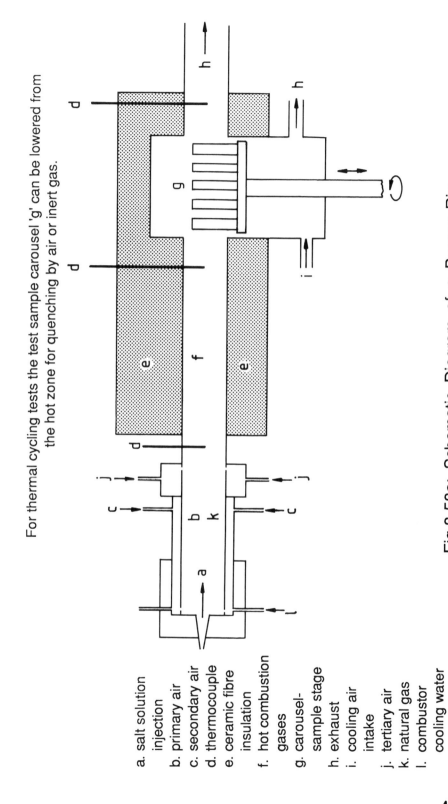

Fig. 8-58a: Schematic Diagram of a Burner Rig.

a. salt solution injection
b. primary air
c. secondary air
d. thermocouple
e. ceramic fibre insulation
f. hot combustion gases
g. carousel-sample stage
h. exhaust
i. cooling air intake
j. tertiary air
k. natural gas
l. combustor cooling water

491

Ceramic Barrier Coatings: Plasma sprayed stabilized $ZrO_2-7Y_2O_3$ appear to be highly recommended on cooled components reaching very high temperatures.

High Cycle Fatigue: Resistance to high cycle fatigue has to take note of avoiding carburization conditions.

Fig.8-58b - 8-62 present histograms of burner rig evaluations of several coated superalloy systems.

Attack on selective structures have been noticed in most of the coating systems (Goward 1986). These include oxidation of MC-type carbides often observed in the external surface of inward diffusion aluminide coatings, accelerated attack via acidic MoO_3 initiated by Na_2SO_4 on Mo-containing MC-areas, attack on Cr-denuded grain boundary areas caused by complex chromium carbides, attack on $Ni_3(Al,Ti)$ phases etc. In MCrAlY coatings the solid solution phase is more Cr-rich than the MAl phase and plasma sprayed MCrAlY is more resistant than the EBPVD-type. The difference is attributed to the manner in which Y is tied up in the matrix. In the EBPVD and cast MCrAlY complex yttride intermetallics form which oxidize to spiky Y_2O_3, which traverse the alumina-MCrAlY interface and provide initiation sites for an acidic-fluxing effect by Na_2SO_4 since yttria is more soluble than alumina in the molten sulphate medium. In plasma sprayed MCrAlY the Y is considered not to readily form the detrimental intermetallic.

High temperature systems are expanding and cheaper fuels are being used. More studies appear to be needed on hybrid substrate systems and in fused salt media with nitrates, carbonates and phosphates in conjunction with sulphates and chlorides of Ca, K, Mg, and Na; additive effects also need further clarification. Degradation studies of multilayer coatings are complex and need further investigations. Poor reproducibility in processing TBC systems is felt to be reflected in their degradation assessments (Nicholls & Hancock 1988). Metastable phase systems and their degradation characteristics are an open field.

Burner Rig Tests: Hot Corrosion of Coated Superalloys

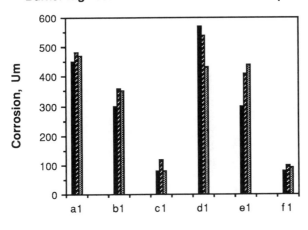

Fig.8-58b

Comparative Evaluation of IN 738 & IN 792 at 700°C in
Burner Rig Tests
(Condé 1984)

Legend	Coating	Substrate
a1	--nil--	}
b1	Pack Aluminised	}IN 738
c1	Elcoat 360 TiSi	}
d1	--nil--	}
e1	Pack Aluminised	}IN 792
f1	Elcoat 360 TiSi	}

==
Original Coating Thickness:-
Pack Aluminided samples: 90-100 microns
Elcoat 360 TiSi   : 80-100 microns

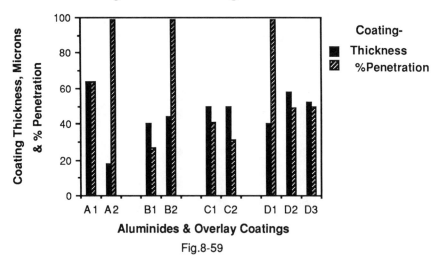

500 h Engine Test on First Stage Turbine Blades

Fig.8-59

A Comparative Performance :- Aluminides &
Overlay Coatings in a Gas Turbine Engine Test

Legend	Type of Coating
A1,A2	Conventional Aluminides
B1,B2	Modified Aluminides
C1,C2	Platinum Aluminides
D1-D3	CoCrAlY - Overlays

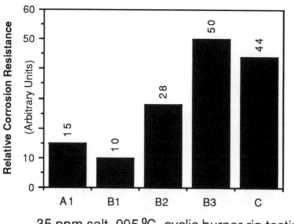

35 ppm salt, 995 °C, cyclic burner rig testing

Fig.8-60

Relative corrosion of coated alloys on different substrate
superalloys:

A1: Diffusion Aluminide on Cobalt-base Superalloy
B1-B3: Nickel-base Superalloy with,
B1: Diffusion Aluminide; B2: Pt-Aluminide B3:
Ni/CoCrAlY Overlay
C: NiCoCrAlY Overlay on a Nickel-base Superalloy with
no Co or Cr.

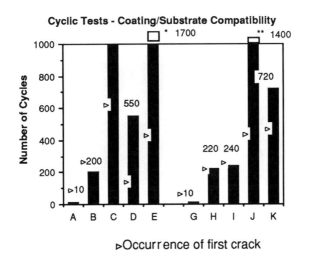

Cyclic Tests - Coating/Substrate Compatibility

▷Occurrence of first crack

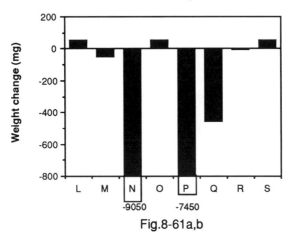

Hot Corrosion Under Cyclic Conditions

Fig.8-61a,b

A-E & L-O: Ni-base alloy type 1 Substrate.
G-K & P-S: Ni-base alloy type 2 Substrate.

Legend	Coating	Legend	Coating
A,G	none	D,J,N,R	Aluminised
B,H,L,P	Aluminide		NiCrAlSi
C,I,M,Q	CoCrAlY (PVD)	E,K,O,S	Pt-Al

Burner Rig, 500 h, 750°C, Thermal Cycling Conditions

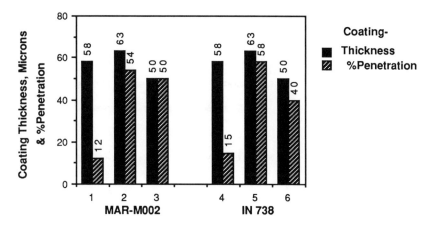

Burner Rig, 500 h, 850 C, Thermal Cycling Conditions

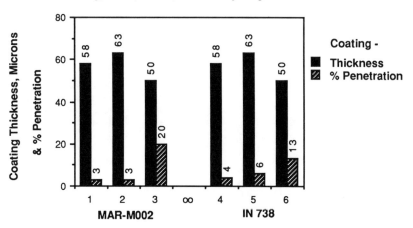

Fig.8-62

Burner Rig Tests Comparing Pt- & Ni-Aluminide Coatings

Legend	Substrate	Coating
1	Mar-MOO2	JML-1
2	--- " ---	JML-2
3	--- " ---	Ni-alumnide
4	IN 738	JML-1
5	--- " ---	JML-2
6	--- " ---	Ni-aluminide

===

Test Conditions: 24 ppm by wt. salt with 1.56 l/h SO_2 with air fuel ratio 30:1; hourly thermal cycling between room & test temp. over500 h. 127 micron NiCrAlY+Pt lasts >1000h at 1093°C; eutectic alloy creep life at 1038°C & stress at 151.7 MN/m^2 doubled when coated.

(Wing & McGill 1981)

COATINGS NOMENCLATURE, SUPPLIER & PROCESS DETAILS

Nomenclature	Coating	Thickness, microns	Supplier	Process
(van Roode & Hsu 1987)				
ATD 63	Ni-33Co-28Cr-8Al-0.3Y	50-100	Temescal	EB-PVD Overlay
RT-22A	Pt-Aluminide	45-55	CRT	Diffusion
PS 138	Pt-Aluminide	75-150	TMT	Diffusion
YSZ	NiCoCrAlY + ZrO_2O_3-$8Y_2O_3$	350-400	CRT	MCrAlY bond coat; Ceramic Overlay
Solaramic SN-5A	Ceramic	50-70	STI	Vitreous phase ceramic

(CRT:Chromalloy Research & Technology; TMT: Turbine Metals Technology; STI: Solar Turbines Incorporated)

(Mom & Hersbach 1987)				
PWA 73	Aluminide	65	PWA	Al-Si Pack; inward diffusion
PWA 273	Aluminide	50	PWA	Al Pack; outward diffusion
PWA 275	Aluminide	55	PWA	Gas phase process
PWA 270	NiCoCrAlY overlay	80	PWA	EB-PVD
RT 22A	Pt-Aluminide	90		Pt electroplate/Al Pack
BB	Rh-Aluminide	85		Rh electroplate/Al Pack

Ranking best-to-worst oxidation-corrosion resistance:
PWA 270 > RT 22A > BB > PWA 273 > PWA 275 > PWA 73
PWA: Pratt & Whitney Aircraft Co.Ltd.

(Morbioli et al 1987)
CrAlI: Pack cementation with 85Cr-15Al wt.%+alumina+NH_4Cl; 16 h at 1050^oC;
CrAlII: As in CrAlI but with thermal cycling
CrAlIII:Pack 70Cr-30Al wt.% + NH_4F; 3 h at 1150^oC;
CrAlIV :Pack Cr + NH_4Cl; 3 h at 1150^oC
RT22: Pt by electrodep. (8 microns) + diffusion treatment and aluminiding for 20 h at 900^oC.

(Nicholls & Hancock 1987)
RT22, LDC2: Pt aluminides
JML 1: Johnson Matthey Pt-Aluminide
LCO22: Plasma Sprayed CoNiCrAlY
ATD 6: EBPVD CoCrAlY
ATD 2:SIP CoCrAlY
Other coatings and coating combinations: conventional pack aluminides (Pk-Al), chromides(Pk-Cr), and chrome-aluminides(Pk-Cr-Al), pulse Al and SIP NiCrTiAl

Ranking best-to-worst in burner rig tests:
950^oC: RT22=LDC2 > Al > PWA 73 > Pk-Cr > Pk-Al (3000 h max.)
850^oC: RT22 > Pk-Al > LCO22 > Pulse Al (500 h)
750^oC: SIP NiCrTiAl > JML 1 > LCO22 + pulseAl > Al > LCO22 > ATD 6 > Pk-Cr > PkCr-Al (500 h)
700^oC: RT22 > Al > LCO22 > LCO22 + pulse Al > SIP NiCrTiAl > ATD 2
===

CHAPTER 9

Characterization, repair and functions of coatings

9.1. GENERAL

Testing and inspection of coatings may be categorized into four areas:

- Non-destructive testing (NDT),
- Standard Characterization (property measurement),
- Screening, and,
- Environmental endurance simulation

The last type of test has long been contentious, with different laboratories evolving different tests and ending with different ranking orders for coatings (Natl.Acad.Sci. Rep. 1970). Some attempts at standard testing have been made and results in literature data should be more comparable.

There are 9 sections in this chapter moderating on examination, measurements, characterization and modeling aspects of coating inspection, with some inevitable overlaps in the use of equipment necessary for the studies, and in each case the principal theme is emphasised.

The main aspects of testing are defect detection and description, and the application of testing systems in adverse conditions at high temperature. Care is needed for error allowance; for instance, radiation pyrometry can have 20 degree errors at $1000^{\circ}C$ which can make an order of magnitude difference in coating life. The specimen emittance must be known. Even 2- colour pyrometers have small errors due to emittance differences (Natl. Acad. Sci.Rep. 1970). Heating methods have a profound effect on test results, e.g. torch heating (as in some burner rigs) will cause equilibration at a surface temperature dependent on the heat flux through the specimen and thermal gradients in torch heating can reach several hundred degrees. This affects both substrate/coating interdiffusion and corrosion reactions. Isothermal tests can have quite different results. Furnace oxidation tests, torch testing, cyclic tests, burner rig tests, hot corrosion tests, thermal fatigue and hot abrasion tests, have been reviewed (Natl.Acad. Sci.Rep. 1970); formulae have been developed for life prediction of coatings (Strangman 1984).

Grinding wheel or abrasive powder wear tests of coatings can sometimes indicate the hardness or brittleness of coatings, rather than their wear resistance under the usual conditions of service (Stanton 1973). Acoustic (and Laser) microscopy for characterizing PVD films for adherence, hardness and wear resistance is a fairly new development; subsurface microstructures and flaws can be detected (Yamanaka & Enomoto 1982a,b). The following NDT topics are also discussed briefly : hot emittance, radiography, reflectivity, infrared inspection, dye penetration for cracks and pores, and back-emission electron radiography for large specimens (30 cm^2 sheets) and the application of ultrasonics - velocity measurement, holography, tomography, CS technique, pattern recognition areas, acoustic emission, acoustic microscopy and eddy current methods (Steffens et al 1983).

The overall success of a CVD process has to be assessed by monitoring the influence of the fundamental factors of the process and the optimum production of the deposit in the state it is required in. The several parameters to be considered are:

1. The theoretical interaction assessements based on thermodynamics, chemical kinetics, and heat and mass transport.
2. The deposit growth morphology.
3. The physical, chemical and mechanical data on the deposit.

Chemical sampling and spectroscopy are useful analytical diagnosers. A sampling system (Cochet et al 1978) and a mass spectrometric method for Si_3N_4 (Lin 1977; Ban 1973) are given; an in situ method using Raman and fluorescence spectroscopy is explained (Sedgwick 1977); four methods for pyrocarbon characterization are specified, viz. the OPTAF technique (optical microscope photometer to the obtain optical anisotropy factor); microporosity measurement; TEM and SEM; and oxidation resistance (Nickel 1974).

Deposit characterisation ranges from optical to scanning and transmission electron microcopy and x-ray studies. Mechanical tests of hardness, stress, adhesion, wear and erosion as well as tests on deposit stability, viz. resistance to environment, resistance to thermal changes and electrical parameters, are carried out depending on the ultimate purpose with which the particular CVD process is performed. The substrate itself influences the deposition in addition to the above test parameters.

500

Weight gain per unit area per unit time and cross section morphology of the substrate/deposit are two of the most reported pack characterization methods. Although many well known pack process groups have discussed the kinetics of the pack reactions it is felt that little has been achieved in establishing the actual mechanism. The weight cannot be defined with respect to any particular equilibrium or steady state conditions since a number of variations in the actual metal activity can occur:

(i)   the rate of pure metal or alloy evaporation in a mixed pack,
(ii) its transport through to the substrate interface, and,
(iii) the actual deposition, as the pack heats up and approaches zero time and resides in it for the specified period.

The aluminizing process for instance, can be fitted into a linear, parabolic or a linear-parabolic kinetics graph and equations can be generated to fit the data obtained (Jackson & Rairden 1977). There are very little direct and in situ measurements done to substantiate the arguments. A mass and heat transfer approach with in situ sampling and analyses of the reaction products have to be made to characterize the control parameters on a more predictive basis. A good effort has been made (Brill-Edwards & Epner 1968) but this is still an area which has not been well characterized (Kung & Rapp 1987).

9.2. MODELING & SIMULATION

Theoretical modeling and computer simulation is in increasing use and may be mentioned at the outset in this chapter. Statistical methods supplement these efforts and the Monte Carlo method is one such. It is a numerical technique in which a stochastic model matched to a given problem is constructed, and the corresponding random variables simulated with the help of random numbers; the difficulty is to establish the random processes associated with the problem and to model them with a suitable algorithm using random numbers. Diffusion equations, material transport in PVD and many others can come within this sphere of analysis (Binder 1979).

Monte Carlo methods and computer modeling have been done in many areas of materials science and material degradation. A Monte Carlo method has been used to interpret the lamellar structure of thermally sprayed coatings (Knotek & Elsing 1987), to simulate grain growth phenomena (Srolovitz 1986) and in high temperature degradation (Hancock et al 1987). Computer modeling has been reported for selection of expert systems for wear reduction (Syan et al 1987), for controlling shot peening parameters (Eckersely & Kleppe 1987), for microstructure evolution and degradation (Lee et al 1987), for assessing the influence of foreign element penetration for generating internal precipitation (Sockel et al 1987), and for establishing criteria for different growth modes during deposit initiation (Nieminen & Kaski 1987).

9.3. NON-DESTRUCTIVE TESTING (NDT) METHODS

9.3.1. ACOUSTIC METHODS:

A number of non-destructive methods are used and their classification has been based on the sound, light and electrical devices employed (Sharpe 1980). Ultrasonic, acoustic emission, thermal and holographic interferometry are discussed in detail (Steffens & Crostack 1980; McGonnagle 1981; Nicoll 1983; Ivanov 1984; Kishi 1985; Dunhill 1987). Acoustic emission is a purely passive method of testing. Sound waves generated by events in the specimen are recorded via a tuned frequency transducer (Sarkar et al 1970; Rawlings 1985). Acoustic microscopy is used to study PVD films; adherence, hardness and wear resistance can be characterised (Yamanaka & Enomoto 1982a); subsurface microstructures and flaws can be detected. (Yamanaka & Enomoto 1982b). Figures 9-1 and 9-2 show stress assessment by holography (Crostack & Steffens 1984).

The controlled signals (CS) ultrasonics technique uses a modulated carrier wave, the modulation feature allowing effectively about 900 times the acoustic energy of an unmodulated wave. The carrier wave (CW) frequency is set to suit the transducer used. Ultrasonic tomography can test the homogeneity of a coating. In this method the coating is tested ultrasonically from several directions and sound attenuation is measured. Equations are developed to distinguish porous areas. But the method has not yet been developed for practical testing (Steffens & Crostack 1982). Ultrasonic Testing is suitable for volume production, but interpretation of faults is difficult. $LiSO_4$ crystals cease to be piezo electric above $130^{\circ}C$ but lead metaniobate works up to $400^{\circ}C$ and oxygen purged lithium niobate works up to $900^{\circ}C$. Coupling pastes are available for use up to $550^{\circ}C$. It is easier to avoid direct contact and to generate the sound using an a.c. induction coil with an additional constant magnetic field. The transducer coil needs to be 0.5 to 1 mm from the hot surface however, and thus requires protection. $1000^{\circ}C$ with the transducer at $350^{\circ}C$ has been used (Crostack & Steffens 1984).

A laser pulse of 15ns can give a high energy (30MJ) pulse to the surface, causing fast expansion and a mechanical pulse. A large separation is feasible and the short pulses have high axial resolution making them suitable for thin coatings inspection. The attenuation and velocity of sound alter with temperature and attenuation differs for longitudinal and transverse waves. Signal to noise ratios are improved by using tuned frequencies in the transduceer, as in the CS method. Signal averaging and cross correlation methods can also be used to remove background noise and leave the required signal; see Fig.9-3 (Crostack & Steffens 1984). Pico-second acoustics has been applied as a NDT tool for thin film characterization (Thonsen et al 1987). Adhesion testing by acoustic emission is described for Ti(C,N) by PVD and TiC by CVD (Steinmann & Hintermann 1987).

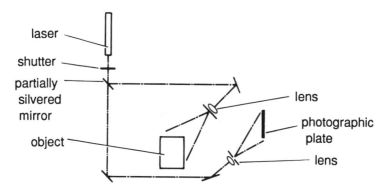

Fig.9-1a: Optical Holography - Schematic
of Apparatus

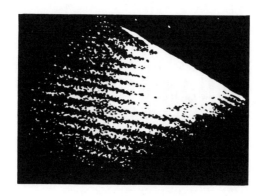

Fig.9-1b: Demonstration of Adhesion Defects as Revealed
by Optical Holography.
(sample heated 10° between holograms
to create stress to detectable level)

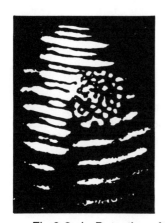 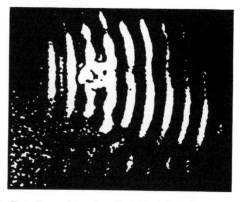

Fig.9-2a,b: Detection of Defects in Bonding of an Arc-Sprayed Coating;
0.63 MHz. Defect Size:- a: 14 mm; b: 7 mm.

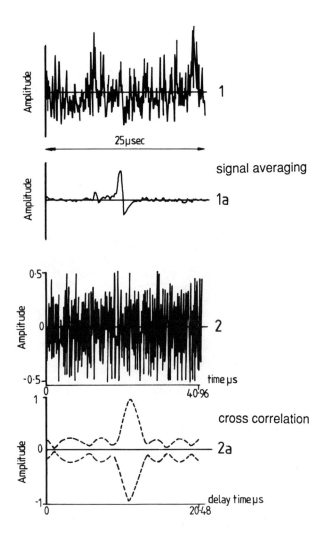

Tests with Ultrasound Combined with Optical Holography.
Suppression of Stochastic Noise in Acoustic Emission Tests.

Fig.9-3a,b

Acoustic emission has been used in several areas of coating assessment:

Fracture behaviour of plasma sprayed coatings in bend tests (Wen et al 1987); crack initiation and propagation of thermally sprayed coatings (Marquez & Olivres 1987); brittle oxide formation (Khanna et al 1987); weld cracking (Racko 1987); breakaway oxidation and internal cracking of scale (Jha et al 1986; Schutze 1985).

Phased arrays of transducers improve flaw detection (Gebhardt et al 1978). More accurate images are obtainable by acoustic holography, giving a 3-D picture of any interior fault (Schmitz 1977) but further development is necessary due to high temperature attenuation of high frequencies (5MHz). Resolution is about one wavelength. Optical holography allows an advantageously large working distance, and a laser image of the surface is combined with a reference image and then compared with the images when the specimen is stressed. Irregularities due to flaws will appear, if near the surface (as in coatings). Fig.9.1b and 9-2a,b show adhesion flaws thus detected. Large surfaces can be inspected by this method (several square metres).

9.3.2. THERMAL METHODS

Conventional NDT (non-destructive testing) using ultrasonic and dye penetrant methods have variable success on thermal spray coatings (Steffens & Crostack 1981). Thermal NDT for coatings have been reported (Green 1962; Crowe & Dawson 1975; Scott et al 1968) but only two included elimination of surface emissivity effects (Green & Day 1968; Green 1968). Thermal wave analysis is an advanced recent method (Green et al 1982; Green 1966; Green 1969). After a transient thermal injection from a gas jet heater, the surface temperature falls due to inward heat conduction which is modified by the coating-to-substrate bond, coating thickness variation, porosity, density and emissivity. An infrared scanner builds up a map of surface temperature which thus shows the condition of the substrate-to-coating bond. Emissivity effects can be removed by a ratio method (Green 1968). Surface reflexions can be avoided by screening. Proper choice of scanning speed and heater-to-detector distance can separate the factors of coating bond soundness and coating thickness; thickness has little effect on temperature in well-bonded regions (Green et al 1982). The method works for both rough and smooth coatings (Green et al 1985).

An automatic thermal impedance scanning (ATIS) has been developed for non-destructive evaluation (NDE) of thermal spray coatings on gas turbine components (Green et al 1987). The method is based on interpretation of surface infrared emittance data on heat injection to the test substrate. Computerized methods are adopted for signal processing and bond defect locations, while robots are employed to load parts into the system and scan under computer control. ATIS can be used to test thermal spray in the presence of bond coats and ceramic overcoats. Fabrication and coating process defects have been detected. Thermal wave interferometry has been applied to testing TBC ZrO_2, NiCr-carbide and WC coatings on stainless steel (Morris et al 1985; 1988).

In the thermoelectric method, a hot metal tip (Cu, W, stainless steel) is put on the coating surface and the emf generated depends on the composition below the point of contact. This method detects thin areas in coatings and failure sites. It is very useful for slurry coatings having thin areas at edges, or Cr-poor areas in Cr-Ti-Si pack cementation coatings. Considerable skill is needed and there are edge effect limitations. A pointed anvil micrometer technique is recommended as better. Conventional means of monitoring thermo-electric emf cannot be applied to gas turbines and internal combustion engines in operation. Thin film thermocouples have been developed for this purpose (see section 9.9.2.).

Miniature probes have been described to monitor high temperature gas streams (Vassalla 1966; 1967). Monitoring surface temperatures by means of thin film thermocouples can be hampered by thermo-electric instabilities which occur firstly due to compositional inhomogeneities and secondly by contamination from diffusion or environmental reaction. Differences in sputtering rate, material loss due to evaporation and cross contamination caused by evaporation and condensation of different vapour species contribute to variations in composition and hence thermo-electric instability. The major diffusing contamination is that of 3d metals from the Al_2O_3 base. Pt and Pt-Rh can act as catalysts because they present a large surface area being thin films, in cracking hydrocarbons, and their local temperature may be 50 degrees higher; unburnt hydrocarbons cracked thus cause inclusion of intergranular carbon precipitate which will not affect temperature measurement but deteriorates the thermal sensor by embrittlement effect (Budhani et al 1986). Thin film thermocouples 10 cm long, 3mm wide and 2×10^{-4} cm thick have been joined on to MCrAlY coatings on steel and superalloy substrates for temperature monitoring (Kreider 1986; Budhani et al 1986).

9.3.3. DYE PENETRATION

The NDT techniques of radiography, eddy current, thermoelectric measurements and dye penetration have been reviewed (Dunhill 1987; Natl.Acad.Sci 1970). Use of fluorescent dyes to find surface cracks and porosity can also be used to find cracks in the visually inaccessible inside surfaces of objects, e.g. inside drilled holes. The method is fast and cheap. As with the conventional dye method, fluorescent dye is spread over the surface to be inspected and wiped clean. Then a silicone rubber molding compound is applied to the inner surface and peeled away after curing. Dye that entered any inside surface crack is absorbed by the silicone and will glow under ultraviolet light.

9.3.4. EDDY CURRENT METHOD

Eddy current testing requires no contact and multiple frequencies allow interfering effects to be eliminated (Crostack & Steffens 1984). Penetration depth is small, useful for coatings inspection (except for insulating ceramic coatings). An improved low frequency eddy current method (Mih 1980) can detect cracks in inner layers of built up Al structures. Cracks are sensed as inductance changes by a search coil. Considerable skill is needed and there are edge limitations.

9.4. ADHESION

In Chapter 7, adhesion was considered as a physical property; in this section the test aspects and the relating factors are accorded relevance; attempts have been made to minimise overlaps. An ideal adhesion test should offer : quantitative values, applicability to a wide range of coating thicknesses and coating/substrate combinations, minimal preparation, reliability, sensitivity and routine usage. Many of the qualitative adhesion tests can also be considered as hardness tests; scratch and indentation are both likely to be taken in this category. Single-point contacts are used in both scratch adhesion testing and microhardness measurements, usually with a diamond pyramid for the latter and a Rockwell C120° diamond cone for scratch testing, both inducing plastic deformation in the substrate and coating. The elastic-plastic factor, prediction of the critical load causing the interfacial shear strain and finally failure, are developed in a model to enable inter-relating hardness and adhesion (Burnett & Rickerby 1987).

The widely used ASTM (1979) test is felt to introduce shear stresses. Work of adhesion mesurements can be obtained by a periodic cracking method, applicable to brittle films on a plastically deformable substrate. Thick film adhesion can be measured by an impact separation method, impacting the opposite face and allowing stress waves to loosen the coating; coating separation was recorded by cine camera (Brown & Ferber 1985). A recommended indentation method for adhesion determination has been correlated with acoustic properties; based on fracture mechanics it leads to quantitative results and requires only a standard hardness testing machine and optical microscope. The abilities of coatings to withstand machine finishing as a good gauge for adhesion strength has been suggested (Smart 1976).

A number of adhesion/bond strength related tests have been developed in the course of tests on coatings on tool materials and thermally sprayed coatings. Premature failure of coated tools has been related to the following: substrate material selection, metallurgy and heat treatment, tool manufacturing w.r.t. edge finish and presence of Beilby layers, and, the various steps of the coating process (Bergmann et al 1987). A tensile adhesion test has been used on thermally sprayed coatings (Ostojic et al 1988). Adhesion failure of thermally sprayed coatings are monitored by the development of cracks under mechanical loading with the crack propagation mainly along the lamellae boundaries.

Adhesion of Al to glass and Si can be measured by a topple test (Sato et al 1982) in which a rod is fixed with epoxy to a coating and the force required to topple it over is measured. Tests on coatings include: adhesion test using adhesive (epoxy), scratch hardness test, indentation test, hardness test, microscopic observation, wear abrasion, blast erosion, corrosion resistance. Use of adhesives to hold a coating to a tension bar can be misleading if the adhesive penetrates a porous coating and confers added strength. Adhesion testing avoiding use of adhesives involves placing 2 conic sections apex-to-apex, bolting together by an axial tie-bar (through an axial hole). The assembly is then sprayed, coating the contiguous conical parts, and then pulled (after removing the tie-bar) in a testing maching, until one of the conical substrates pulls away. Failure always occurred within the coating, in the case of arc-sprayed coatings, but for gas flame sprayed coatings at least 20% of the area was stripped clean to the original grit-blasted surface. The method is simple, reliable and has low scatter (Stanton 1973).

Another non-epoxy method for use where coating adhesion exceeds epoxy strength, has been given (Bourgoin 1973) but with insufficient detail for comprehension. A new acoustic emission parameter for adhesion is called the C-value. This shows differences in adhesion better than the conventional cross-hatch and pull-off methods (Hansman & Mosle 1981; 1982).

9.5. TESTS ON MECHANICAL PROPERTIES

Chapter 7 deals with a number of aspects of mechanical properties of materials in general and the effects of process routes, substrates, interfaces and environment on coatings. Information on conventional methods of testing tensile and yield strength, stress etc. is available in standard texts and monographs; a few features as applied to coatings are briefly considered here. Mechanical properties of coatings have been examined in the context of high temperature behaviour and may be accessed from the following: Grunling et al 1987; Hancock & Nicholls 1988; Nicoll 1987; 3rd & 4th US-UK Conf. on Marine Gas Turbines 1975,1978; Inst. Metals Conf. Series 1984 onwards.

The main difference in standard material test procedures and high temperature coatings are in assessing thick coatings and thin films in their service conditions - isothermal and cyclic. Mechanical properties are strongly influenced by variables which control microstructure (section 9.6.), and variables which control chemical degradation (section 9.8.), both of which have been discussed in considerable detail in Chapters 7 and 8. Properties which are examined under both isothermal and cyclic conditions under various environmental reaction situation are: residual strains and stresses, fatigue, creep and crack propagation, tensile stress, strength, stress rupture, fracture and decohesion (Grunling et al 1987; Hancock & Nicholls 1988; Schutze 1988).

Hardness measurement is widely used as a guide to the strength, wear and erosion resistance of a coating, particularly TBC systems. A number of PVD process systems incorporate hardness tests to assess porosity and defect properties of deposits. Scratch tests, cracking, chipping and flaking, and, partial and full indentation are a few of the several modes of monitoring hardness (Johansen et al 1987). Scratch adhesion and scratch removal tests are done (Hilton et al 1986). Almost all process reports on refractory nitrides, carbides etc. include hardness as a guide to deposit quality. The critical aspect of hardness and wear tests is the selection of the optimum load by which indentations or scratches are made on the test samples. Wear tests such as abrasive and ball wear scar tests are used as a hardness indicator, but the former is difficult to correlate for relatively thin coatings. The Palmqvist indentation method using a light load at all times is suggested (Mehrotra & Quinto 1985; Palmqvist 1957). Wear and erosion test equipment often appear to be indigeneous in design and results are comparative within the group tested. Development of tool materials in the light of detailed characterization has been reviewed (Gurland 1988).

Characterization of thin films by uni-axial and bi-axial (bulge) tension testing, elastic moduli measurements as well as hardness is reviewed; it can be hampered by coating plasticity. A nano-indenter which carries a load-unload sequence while continuously monitoring indentation depth is felt to offer more reliable data. The actual indentation is calculated from the 'plastic' depth and the final depth. The load vs. distance then gives the true values for correlation (Hardwick 1987). Thin film characterization using a mechanical probe is also described (Oliver et al 1987). Elimination of inaccuracy by extrapolation to zero thickness in hardness measurements of thin insulator coatings while measuring indentation via SEM because they have to be coated with conducting films, e.g. Au, is reported. Ultramicro-indentation hardness measurements, Young's modulus and internal stress developed for thin coatings and films are worth noting (Tsukamoto et al 1987; Wagendristel 1987; Hardwick 1987).

The importance of measuring residual stress and its nature after coating a substrate is revealed by the fact that a high speed steel substrate with a low stress registered a compressive stress of 900 MPa after CVD Ti(C,N) coating and when ion plated was greater than 4000 MPa, the substrate itself apparently unaffected (Perry & Chollet 1986). Thickness mesurements of amorphous metals Ni-B and -S plated by dual beam ion deposition were

509

measured by a stylus instrument after separated from their glass substrates (Panitz 1986). Stress models and fractographs for thermally sprayed coatings taking into account the involvement of bond coats have been attempted for their characterization (Rickerby et al 1987, Miller 1987).

In service conditions stresses are often not constant, varying (ir)regularly about a mean value, thus causing fatigue failures. At high temperature even a steady stress leads to creep failure and stress fluctuations affect the time to fracture. Fatigue effects are influenced by the mean stress, amplitude and frequency of the fluctuating stress and the temperature (Nicoll 1983). In testing, axial loading is ensured by coplanar knife edge joints and shock loading is avoided. Temperature, load, rupture time and elongation (of a scribed gauge length), are measured. Hydraulic, computer controlled mechanical test machines allow tensile, compression and bending stress cycles to be performed for fatigue and variable load creep fatigue and stress relaxation. Variations in results may be due to: composition, processing and heat treatment variations which affect microstructure, diversity of specimens (shape, size and surface finish), temperature and stress variations, and, coating variations. Creep measurement by different laboratories deviate due to the great stress sensitivity of high temperature alloys.

9.6. STRUCTURAL CHARACTERIZATION

A microscope is an integral part of almost all physical and mechanical testing and provides characterization procedures with X-ray diffraction as a strong supplement. Conventional equipment such as optical, SEM and TEM (replica, and direct - thin discs) with combinations of STEM and other types are used for sample examination. Augmentation by acoustics and laser principles have given optical and scanning microscopy a capacity for greater resolution and imaging. Interference layer metallography has provided an interesting tool in the study of hot corroded alloy samples (Buhler & Hougardy 1980). Thin film imaging and characterization by scanning acoustic microscopy (Lee et al 1987), surface topography, grain boundary and surface defects by scanning tunneling microscopy (Marchon et al 1987), cross sectional TEM (Sheng et al 1987; Veilleux et al 1987) and interfaces and thin films by TEM with optimum accelerating potentials of 300 -400 kV (Madden 1987), are some of the new techniques reported.

A schematic of the random defects in coatings is given in Fig.9-4. Microstructure evolution during and after the coating process, results of heat treatment, if any, and development and morphological changes in the working environment are monitored, and cross-section morphology and analyses are indispensable in assessing the coating-interface-substrate response. Production of amorphous phases by controlled microstructure is a developing technique; it has shown considerable control on arresting

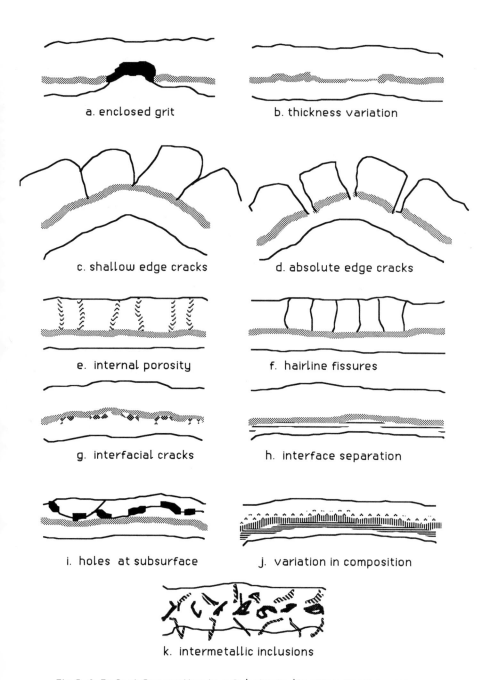

a. enclosed grit

b. thickness variation

c. shallow edge cracks

d. absolute edge cracks

e. internal porosity

f. hairline fissures

g. interfacial cracks

h. interface separation

i. holes at subsurface

j. variation in composition

k. intermetallic inclusions

Fig.9-4: Defect Generation in a Substrate/Coating Configuration

corrosion at temperatures up to 500°-$600^\circ C$ (Courtright 1987). Microstructure characterization has been invaluable in determining additive influence and limits for dispersion hardening, stress influence, matrix stabilization, intermetallic compound formation, solid solution strengthening, carbide forming, and environmental resistance. The survey on cobalt-base alloys is one such example (Coutsouradis et al 1987).

Correlation of microstructure tests to properties are both qualitative and quantitative. Porosity, inclusions, defects, phase distribution and grain size, for instance, reflect measurements made on density, tensile and ultimate strength, stress and strain, crack, deformation and fracture modes. To consider the effect of second phase particles on thermal and electrical conductivity, dielectric constant, magnetic permeability, linear and elastic property etc., it is necessary for a structural assessment to note the number of phases, their geometrical configuration, concentration, shape and orientation (Gurland 1988).

X-ray topography provides the direct means of characterizing the deformed subsurface zone and has been used to image abraded dendritic alloys (Newkirk 1959; Wilkens 1967; Armstrong et al 1980; Mitchell & Laufer 1980; Rappaz & Blank 1982, 1983, 1987). The technique is combined with X-ray microdiffraction by using a laser beam for analysis of heterogeneus materials (Rappaz & Blank 1987). Grazing angle incident X-ray methods have been reported for substrate/coating interface studies. These have a short controllable penetration depth and are well suited to probe surface and near-surface structures. Where a low density coating and high density substrate pair is involved the x-ray can be manipulated to penetrate the coating and be reflected by the substrate. Synchrotron X-ray sources are useful for this technique, and allow measurements of XRD, reflection and absorption (Heald et al 1985).

Combination of dispersive X-ray analyses together with scanning and transmission electron microscopy have been established techniques for some time. Scanning Auger microscopy is a subsequent successor technique with greater resolution and enables collection of more simultaneous structural and analytical data (Browning 1985; Gomati et al 1985; Smith et al 1986).

9.7. SPECTROSCOPY

Chemical analyses by spectroscopy has made rapid strides in high temperature work and almost always includes equipment for high resolution microscopy. The list of acronyms given at the end of Chapter 2 includes most of the old and newly developed techniques and several monographs are available (Psaras & Langford 1987). Glow discharge spectroscopy is fast, sensitive, accurate, simple and reliable and can be used for surface analysis, if the specimen can be attached to a vacuum cell (Berneron & Charbonnier

1982). The resolving power in depth profiling is similar to AES and SIMS. A 1 KV glow discharge causes ion bombardment and sur- face erosion which is fed (optically) to a multichannel spectro- meter for elemental analyses. Other long and complex methods of surface analysis such as Auger Electron Spectroscopy (AES), Secondary Ion Mass Spectrometry (SIMS), X-ray Photoelectron Spec- troscopy (XPS), ion scattering spectroscopy, Rutherford backscat- tering, nuclear reaction analysis, ion-induced X-ray emission and ESCA are difficult for field use. These methods are reviewed (McGuire et al 1981; Jacobson & McGuire 1983; Walls & Christie 1982). Tables 9:1,9:2 and 9:3 (McGuire et al 1981) and Table 9:4 (Brewis 1982) compare the techniques. Fig.9-5 (McGuire et al 1981) shows the relative sizes of areas analysed by these techniques.

Commonly used methods are optical and scanning electron microsco- py for surface studies. Transmission electron microscopy of int- erfaces has been explored. Selected area diffraction patterns show the orientation between different grains. In a ceramic coating, the interface between different phases can be coherent, semicoherent or incoherent. Coherent phases are usually strained and can be studied by TEM contrast analysis. Other aspects of analytical electron microscopy analysis are discussed (Thoma et al 1980; Hansman & Mosle 1982). TEM resolution is better than 10 Angstroms and selected volumes of 30 Angstroms diameter can be chemically analysed. Methods of preparing thin TEM transparent foils are described (Jacobson 1981; Jacobson & Mcguire 1983).

Photoemission with synchrotron radiation can probe surfaces on an atomic scale (Spicer 1980) but requires expensive equipment. Complex impedance measurements can separate surface and bulk effects, but problems of interpretation need to be resolved (Stratton et al 1980). X-ray and gamma radiographs, as used in weld inspection, can be used to inspect coatings for defects. The method is explained by Helmshaw (1982). Inclusions, cracks, poro- sity and sometimes lack of fusion, can be detected. Surface compositions of ion implanted metals have been studied by Ruther- ford backscattering spectra (RBS) (Eskildsen 1982; Baun 1982). In this non-destructive way a micro-analysis of the near-surface region is obtained. Interpretation is relatively easy. Assessment of radiation damage in ion implanted metals, by electron channe- ling, is described (Solnick-Legg et al 1980) using SEM. Nuclear reaction analysis (NRA) for the characterization of surface films is described (Padmanabhan & Sorenson 1980).

AES and XPS analyse the top of the surface only and erosion by ion bombardment or mechanical tapering is needed to analyse deeper regions. AES detects 0.1% of an impurity monolayer in a surface. Auger electrons are produced by bombarding the surface with low energy (1 to 10 KeV) electrons. In XPS the surface is exposed to a soft X-ray source and characteristic photoelectrons are emitted. Both AES and XPS electrons can escape from only 1 nm deep (those generated deeper cannot escape as such), which gives greater sensitivity for surface analysis (Walls & Christie 1982).

TABLE 9:1

Methods of Material Characterization by Excitation & Emission

Primary Excitation	Detected Emission	Method of Analysis: Name & Nomenclature
Photons a. Optical	Optical	AA: Atomic Absorption } IR : Infra-red } - Spectroscopy UV: Ultra-violet } Visible }
	Electrons	UPS: Vac. UV Photoelectron Spectroscopy - *Outer Shell*
Photons b. X-Rays	Electrons	XPS: X-Ray Photoelectron Spectroscopy - *Inner Shell*; also called ESCA ESCA: Electron Spectroscopy for Chemical Analysis
	X-Rays	XFS: X-Ray Fluorescence Spectrometry XRD: X-Ray Diffraction
Electrons	X-Rays	EPMA: Electron Probe Micro-Analysis
	Electrons	SEM: Scanning Electron Microscopy TEM: Transmission Electron Microscopy STEM: Scanning-Transmission Electron Microscopy SAM: Scanning Auger Microanalysis AES: Auger Electron Spectroscopy
Ions (+ & -)	Optical	SCANIIR: Surface Composition by Analysis of Neutral & Ion Impact Radiation
	X-Rays	IIXA: Ion-induced X-ray Analysis
	Ions(+/-)	ToFMS: Time-of-Flight Mass Spectrometry SIMS: Secondary Ion Mass Spectrometry IPM : Ion Probe Microanalysis ISS : Ion Scattering Spectrometry RBS : Rutherford Backscattering Spectrometry
Radiation	Optical	ES : Emission Spectroscopy
	Ions (+/-)	SSMS: Spark Source Mass Spectrography

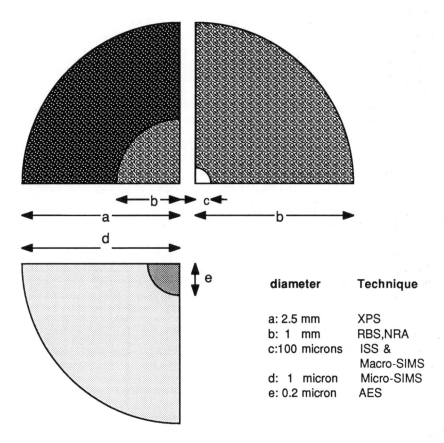

diameter	Technique
a: 2.5 mm	XPS
b: 1 mm	RBS,NRA
c:100 microns	ISS & Macro-SIMS
d: 1 micron	Micro-SIMS
e: 0.2 micron	AES

Fig.9-5: Schematic of Relative Sizes of Areas
Scanned by Spectrometric Analytical Techniques

It is most important to avoid surface contamination during preparation for surface analysis. Semi-quantitative in situ analysis by AES is reported (Bossebouef & Bouchier 1985). Nitrides and other compound refractory coatings are frequently analysed by AES and RBS methods. Depth and crater edge profiling have been done for dc magnetron sputtered and ARE samples of (Ti,Al)N, TiN and TiC coatings (Jehn et al 1987; Kaufherr et al 1987). 'Round robin' tests of characterization by including a range of analyses such as XPS, EPMA, XRS, AES, APMA and XRD are not uncommon. Amonsgt these XRD is felt to be unreliable (Perry et al 1987).

Ion spectroscopy is a useful technique for surface analysis (Baun 1982). Ion scattering spectrometry (ISS) uses low energy back-scattered ions (Buck 1975) and has a high sensitivity. Secondary ion mass spectrometry (SIMS) has the possibility of sputter removal of layers allowing depth profiling (Baun 1982). It can act as a stand-alone single system to solve surface analysis. Three dimensional SIMS of surface modified materials and examination of ion implantation is reported (Fleming et al 1987). Lattice vacancy estimation by positron annihilation is another approach (Brunner & Perry 1987). Transmission and scanning electron microscopy (TEM and SEM) are valuable techniques and replication techniques can non-destructively reveal surface features of specimens too thick for TEM (Cope 1982). ARE coatings of V-Ti in C_2H_2 gives wear resistant (V,Ti)C coatings. The hardness is related to: grain size, stoichiometry, free graphite, cavity networks. SEM and X-ray diffraction analysis could not explain the large hardness variations obtained by varying temperature and gas pressure but TEM revealed microstructural changes (Jacobson & McGuire 1983).

Beta backscatter and X-ray fluorescence have low sensitivity (0.5 cm^2/minute and 1 cm^2/hr, respectively). Thickness and uniformity of silica coatings on steel have been determined by X-ray fluorescence measurements of Si concentrations along the surface (Bennett et al 1984). Round robin tests for microstructure and microchemical characterization of hard coatings have included XPS, UPS, AES, EELS, EDX, WDX, RBS, SIMS, TEM, STEM and XTEM (Sundgren et al 1986). Field emission STEM has been applied for profiling Y across a spinel-spinel grain boundary (Bennett & Tuson 1988/89).

The cost of equipment and its maintenance, the cost for data collation and equipment availability appear to be the ultimate determining factors.

TABLE 9:2

SUMMARY OF VARIOUS CHARACTERISTICS OF THE ANALYTICAL TECHNIQUES

Characteristic	AES	XPS	ISS	SIMS	RBS	NRA	IXX
Sample Alteration	high for alkali halogen organic insulators	[-----low------]			[--------very low---------]		
Elemental Analysis							
Sens. Varia.	good	good	good	poor	fair	fair	good
Resolution	good	good	fair	good	fair	good	good
Detection Limits	0.1%	0.%	0.1%	10^{-4}% or higher	10^{-3}or higher	10^{-2} or higher	10^{-2} or higher
Chemical State	yes	yes	no	yes		no	
Quantification	[---with difficulty---] req. standards			very difficult; req. std.	[-------absolute---------] no standards		
Lateral Resolution	200nm	2mm	100 m	100-1 m	[-------1mm--------------]		
Depth Resolution	[-------atomic layer to--------]				10 nm	10 nm	none
Depth Analysis	[--------destructive-----------] sputter				[-non-destructive-]		very difficult

===

TABLE 9:3

OUTLINE OF SOME IMPORTANT TECHNIQUES TO STUDY METALLIC SURFACES

Technique	Abbreviation	Information	Comments
Optical microscopy	OM	Surface topography and morphology	Inexpensive but modest resolving power and depth of field
Transmission electron microscopy	TEM	Surface topography and morphology	Very high resolution but requires replication-artefacts can be a serious problem
Scanning electron microscopy	SEM	Surface topography and morphology – combined with X-ray spectroscopy gives 'bulk' elemental analysis	Resolving power >>optical microscopy. Preparation easier than TEM and artefacts much less likely
X-ray photoelectron spectroscopy*	XPS (ESCA)	Chemical composition – depth profiling	Especially useful for studying the adhesion of polymers to metals
Secondary ion mass spectroscopy	SIMS	Elemental analysis in 'monolayer range' – chemical composition and depth profiling	Extremely high sensitivity for many elements
Auger electron spectroscopy	AES	Chemical composition, depth profiling and lateral analysing	High spatial resolution which makes the technique especially suitable for composition–depth profiling
Contact angle measurement		Contamination by organic compounds	Inexpensive; rapid

*: Also called Electron Spectroscopy for Chemical Analysis, ESCA

TABLE 9:4

TYPES OF SAMPLES AND

TECHNIQUES NEARLY APPROPRIATE FOR THEIR ANALYSIS

Required Sample Analysis	Appropriate Technique
Depth profiling lower Z elements and thin films; trace or minor analysis of light elements; quantitative analysis	NRA
Depth profiling of higher Z elements and thin films; trace analysis of heavy elements in light matrix; quantitative analysis	RBS
Trace, minor, and major element analysis in thicker samples; quantitative analysis	IIX
Minor and major elements at the surface or interface of small samples	AES
Trace elements at surface or interface of medium to small size samples; analysis of insulators; sputter profiling of light elements	SIMS
Chemical state analysis; analyses of organics, insulators	XPS
Analysis or outer atom layer; analysis of insulators	ISS

==

9.8. DEGRADATION CHARACTERIZATION

Hot corrosion, carburization and oxidation degradation are at the outset characterized by material loss and/or scaling response in the environments configured to simulate those encountered in practice. Chapter 8 has included the spectrum of physical, mechanical and chemical tests including burner rig tests. The analytical methods discussed above form an inevitable segment for assessments. This section includes a few analytical methods of interest.

Table 9:5 gives a summary of corrosion monitoring and mechanistic study techniques (Castle 1987).

Assessment of coatings ranges from small scale laboratory tests to burner rigs. The specimen weight gain, weight loss, thickness loss, depth of attack determined by section-microstructure scanning, or individual element losses across the alloy profile can all be used for assessment against time at various temperatures and environmental conditions. Thermal cycling tests have no standardised procedure and vary from laboratory to laboratory, but they essentially aim at cooling the specimen in between isothermal residences at the corroding temperature. The system of SO_3 in conjunction with SO_2 and O_2 presents quite a problem in setting up apparatus for monitoring hot corrosion as SO_3 is liquid at ambient temperatures. Some designs of laboratory equipment are referenced: Vasantasree et al 1967; Mrowec et al 1977; Flatley & Birks 1971; Holthe & Kofstad 1980. Molten salt tests have ranged from crucible tests, direct and intermittent salt application at the test temperature and salt injected into the gas stream (Littlewood & Argent 1961; Grant 1979; Baudo et al 1970; Cramer 1979; Garfinkle 1973; Wonter & Strachan 1977; Kohl et al 1979; Johnson et al 1975). A two-zone temperature technique with the melt held at one zone and the specimen at the test temperature, e.g. Dean-Rig, claims close simulation to hot corrosion encountered in practice (Swindells et al 1980).

Electrochemical methods have often been used for hot corrosion assessment (Sequeira & Hocking 1977 a,b,1978; Vasantasree et al 1982; Shores 1975; May et al 1972; Sidky & Hocking 1987; Rapp 1987; Rahmel 1987).

Initial stages of oxidation have been studied by HVEM techniques (Baxter & Natesan 1985), but sulphur-bearing atmosphere are difficult to adapt for practical microscopic work. Marker studies have been very helpful but have to be monitored with care (Hocking 1962). Tracer studies have also been reported (Basu & Halloran 1987; Young & deWit 1986). Experimental set-up for SO_2-SO_3 studies vary from a thermobalance seal (Vasantasree & Hocking 1976; Vasantasree 1971; Vasantasree et al 1967) to argon/nitrogen counterflow measures and other heating tape arrangements (Birks et al 1969; Worrell et al 1972). Apparatus only suitable for gas-solid equlibration reaction is reported (Morin 1986). Moderating weight gain data with other auxiliaries may be briefly mentioned

TABLE 9:5

SUMMARY OF CORROSION MONITORING & MECHANISTIC STUDY TECHNIQUES

Film/Coating Thickness, m	Monitoring Techniques	Mechanistic Corrosion Study Techniques
Kinetics:-		
10^{-9}	Interference colour (1); Thin Layer Activation(2)	Quartz piezobalance(10);XPS/AES(11,12)
10^{-6}	Volumetric(3,4); Conductivity(5)	Microbalance(13-15); Volumetric measurement (3,4) AES depth profile(16)
10^{-3}	Ultrasound(6); Eddy current (5); Laser etch (7)	Thermobalance(13,14); Glow Discharge Optical Spectroscopy(8)
Local	External appearance	Metallographic examination(17)
Chemistry:-		
10^{-9}	Glow Discharge Optical Spectroscopy(8)	SIMS(18,19); Electron Diffraction(20,21) XPS/AES(22)
10^{-6}	Laser Optical Spectroscopy(9)	Laser Optical Spectroscopy(9); Raman Spectroscopy(23,24); XRD (25); EPMA (26); STEM (27)
10^{-3}	Beta-source X-ray Spectroscopy; Direct Analysis	Taper Section XRF (3); Spark Source Mass Spectroscopy (28)
Local		AES (22); Analytical TEM (27)
Morphology:-		
10^{-9}		Holography (29,30); TEM (31)
10^{-6}	Thermal resistance	Markers (32); RBS/RNR (33); SEM (34)
10^{-3}		Flexure & Vibration(35)
Local	General appearance	SEM (34)

Reference list for Table 9:5:- (Castle 1987)

1. MacSwain A.M., Proc. Phys.Soc., **72**, 742 (1958).
2. Finnigan D.J., Garbett K & Woolsey I.S., Corros. Sci., **22**, 359 (1982).
3. Scharfstein L.R. & Henthorne M., 'Handbook on Corrosion Testing & Evaluation', Ailor W.H. (Ed.), Electrochem. Soc., Princeton NJ, (1971), p 281.
4. Lange W. & Sceghenck H, 'Techniques of Metal Research', Rapp R.A. (Ed.), Interscience, NY, **4**(1), p 267.
5. Britton C.F., 'Corrosion', Shrier L.L., Butterworth, London (1976), Sec. 20.144-20.157.
6. Silk M.G., Int. Met. Rev., **27**, 28 (1982).
7. Klewe R.C., Lobley E.H., Parr M.K. Richards P.H. & Tozer B.A., J. Phys. E, Sci. Instrum., 5, 203 (1970).
8. Oshashi Y., Yamaoto Y., Tsunoyama K. & Kisidaka H., Surf. Interface. Anal., **1**, 53 (1979).
9. Tozer B.A., Optics Laser Tech., 8, 49 (1976).
10. Baker M.A., 'Progress in Vacuum Microbalance Techniques 1', Gast T. & Robins E. (Ed.), Heydon, London, (1972), p49.
11. Castle J.E., Surf. Sci., **68**, 583 (1977).
12. McIntyre N.S., 'Practical Surface Analysis by Auger & X-ray Photoelectron Spectroscopy', Seah M.P. & Briggs D. (Ed.), Wiley, Chichester (1983), p379.
13. Gregg S.J., 'Vacuum Microbalance Techniques 7', Masson C.H. & van Beckam H.J. (Ed.) Plenum, NY (1970), p107.
14. Rodin T.H., Adv. Catal., 5, 118 (1953).
15. Fergusson J.M., Livesey P.M. & Mortimer D., 'Progress in Vacuum Microbalance Techniques 1', Gast T. & Robins E. (Ed.), Heydon, London, (1972), p87, 101.
16. Hofmann S., 'Practical Surface Analysis by Auger & X-ray Photoelectron Spectroscopy', Seah M.P. & Briggs D. (Ed.), Wiley, Chichester (1983), Chp.5.
17. Wood G.C., 'Techniques of Metal Research', Rapp R.A. (Ed.), Interscience, NY, **4**(2), p493.
18. Brown A. & Vickerman J.C., Surf. Interface Anal., 6, 1 (1984).
19. Magee C.W. & Honig R.E., ibid., **4**, 35 (1982).
20. May J.C., Adv. Catal., **21**, 151 (1970).
21. Rhodin T.N., Corrosion, **12**, 123 (1956).
22. Seah M.P., 'Practical Surface Analysis by Auger & X-ray Photoelectron Spectroscopy', Seah M.P. & Briggs D. (Ed.), Wiley, Chichester (1983), p247.
23. Colthup N., Daly L. & Wilberley S., 'Introduction to Infrared & Raman Spectroscopy', Appl. Sci. NY (1981).
24. Castle J.E., 'Corrosion', Shreir L.L., Butterworth, London (1976), 1.241-1.265.
25. Erdos E., 'Analysis of High Temperature Materials', van der Biest (Ed.), Appl. Sci., London (1983), p189.
26. Reed S.J.B., 'Electron Microprobe Analysis', Cambridge University Press, Cambridge (1975).
27. Berneron R. & Charbonnier J.C., Surf. Interface. Anal., 3, 134 (1981).
28. Mulvey T. & Webster R.K. (Ed.), 'Modern Physical Techniques in Materials Technology', Oxfor University Press, Oxford (1974), p247.
29. Azzam R.M. & Bashara N.M., 'Ellipsometry & Polarized Light', North-Holland Amsterdam (1977).
30. Neal W.E.J., Appl. Surf. Sci., 2, 321 (1979).
31. Hren J.J., Goldstein J.I. & Joy D.C., 'Introduction to Analytical Electron Microscopy', Plenum, NY (1979).
32. Verma S.K., Raymond G.M. & Rapp R.A., Oxid. Met **15**, 471 (1971).
33. Zeigler J.F. (Ed.), 'New Uses of Ion Accelerators', Plenum, NY (1975).
34. Castle J.E., 'The Use of the Scanning Electron Microscope', Hearle J.W.S. et al (Ed.), Pergamon, Oxford (1972), p104.
35. Hancock P. & Hurst R.C., 'Advances in Corrosion Science Technology 4', Fontana M.G. & Staehle R.W. (Ed.), Plenum, NY (1974), p1.

to include thermodynamics of oxidation/sulphation/carburization reactions and ionic diffusion (Young & Gesmundo 1988), passivity and product solubility assessments by electrochemical techniques (Pourbaix 1987; Kruger 1988; Rapp 1987; Rahmel 1987), parabolic rate constant simulation and assessments (Pieraggi 1987, Denis & Garcia 1988), and time dependent gas pressure conditions for multi-layer growth at high temperature (Tomellini & Gozzi 1986).

Electronic absorption spectroscopy experiments (EASE) are demonstrated to be useful in continuous monitoring of dissolution rates of metals in melt in prescribed environmental conditions in the presence of a heat flux. Since each dissolved species will have a unique electron absorption spectrum its presence can be tagged as a function of time. Examples are dissolution of Ni and Cr from IN 738 in a eutectic $(Na,K,Al)SO_4$ melt at $600^{\circ}C$. Monitoring in crucible tests it was shown that presence of 20 mole% chloride nearly doubled the dissolution rate of the species; however, the duration of 260 h over which this was done was considered inadequate to obtain all the necessary information (Griffiths 1987). Mass spectroscopy (Hocking & Vasantasree 1978; Kohl & Fryburg 1978) has been used for characterizing vapour species forming during high temperature corrosion. SIMS and Laser-linked mass spectroscopy are subsequent developments. Raman spectroscopy is observed to be very useful for in situ analyses but unfortunately does not yet have as extensive a data base of reference spectra as exists for X-ray diffraction. Raman-scattering techniques offer a method of phase identification even above $1200^{\circ}C$ and can simultaneously monitor phase changes occurring in a molten salt film and the corrosion scale formation on the substrate. Phase sensitivities are very different in that it is difficult to observe alumina using Raman scattering but zirconia produces a strong signal. It is particularly well suited to analyse gas phases adjacent to the sample. As an optical technique Raman spectroscopy would be very useful for high temperature corrosion (Nagelberg 1985).

Rate controlling processes in multi-component materials will not conform to any fundamental mechanism governing scale growth. At high temperatures, a coating/substrate configuration would be hard to asssess on the basis of weight loss or weight gain. The use of electron microscopy in obtaining an insight into the diffusional patterns of component elements has been elaborated. Vertical segregation of a preferentially oxidized element to form the surface scale does not indicate all the other heterogeneous changes. Analyses of scale formation of Fe10Cr34Ni alloy and the breakaway oxidation of Fe9Cr are reported in this context with results from electron microscopy (Newcomb & Stubbs 1988). The utility of analytical electron microscopy is that it can be used to obtain morphological information with a resolution of less than 0.5 nm and also crystallographic and chemical analyses data with a spatial resolution of 20 nm. The corroded samples are prepared for TEM studies as plated Ni-sandwich sliced, polished to 150 microns, dimpled to 100 micron depth and ion milled for perforation (Reaney & Lorimer 1988). Preparation of H-shaped metal sections (1.5 cm x 5 mm) with a low speed saw for corrosion tests and subsequent fracture in liquid N_2 before TEM cross-section work is described (Desport & Bennett 1988).

The thin layer activation (TLA) technique has been used to measure material loss from wear, erosion and aqueous corrosion (Evans 1980; Conlon 1982; Finnigan et al 1982; Asher et al 1983). Its use in monitoring high temperature oxide spalling is demonstrated with 20Cr-25Ni-Nb stainless steels in CO_2 at 900°C (Asher et al 1987). High energy ion beams are used to produce trace quantities of radio isotope atoms within a solid surface layer. Loss of the surface layer is reflected in a subsequent scan for the isotope levels, whereby isotope loss can be linked to a depth of material loss.

9.9. PACK COATING CHARACTERIZATION

Some attempts have been made to study pack aluminide coatings under controlled experimental conditions. These include: thermodynamics and kinetics of pack aluminiding on IN 100 (Levin & Caves 1974); kinetics of gas transport in halide activated aluminizing packs (Kandaswamy et al 1981); relative infuence of transport in the gas phase in aluminiding (Neiri & Vandenbulcke 1983); influence of viscous flow on the kinetics of gas transport in aluminizing packs (Wang & Seigle 1987), and studies on simultaneous chromizing and aluminizing through controlled barrier layer (Rapp et al 1987). Changes in operation of the cementation equipment, e.g. a tumbling pack, which was found to eliminate the depletion zone around the substrate, and variations in the pack activator have also been reported (Kung & Rapp 1987).

9.10. COATING REPAIR & JOINING TECHNIQUES

9.10.1. REPAIR TECHNIQUES:

If the sputtering of a CoCrAlY coating is interrupted and then resumed, a visible interface is produced, but this vanishes on heat treating (4 hours at 1080°C) which causes healing by diffusion. The observation is relevant to coating repair capability (McClanahan et al 1974).

Activated diffusion brazing has been developed for crack healing. Many used turbine components can be replaced at a fraction of their new cost, with the repair technologies now available. The repaired parts are comparable with new parts in mechanical properties and reliability. Failures occur by tensile and bending stresses which propagate cracks, high cycle fatigue (due to improper operation, such as overspeed), low cycle thermal fatigue, hot corrosion, erosion, mechanical damage (due to

524

ingestion of a large object) (Gooch et al 1983).

Two main ways to reverse hot component damage are: welding and diffusion braze repair. Welding is limited to short or shallow cracks and superalloys are difficult to weld; distortion due to shrinkage is a problem and cooling holes can be blocked by weld spatter and burn-through. Even so, TIG is a viable method for turbine component repairs. Alternative methods have been long sought. In 1969 vacuum brazing was proposed. The advantages are:

-Economical (costs are 25% of a new part),
-Parts can be simultaneously braze repaired and solution heat treated,
-Parts can be batch processed to reduce unit costs,
-Repair is better than welding (distortion is avoided and tiny cracks, which would not have been seen during inspection, are healed by surplus filler),
-all cracks are uniformly treated, avoiding the conflicting stresses of welds.

Early work used NiCrBSi filler for cracks but these alloys form continuous boride and silicide chains which are hard and brittle and form ready crack paths. Ductile alloys were developed with microstructure free from such chains. B diffuses into the host metal but Si stays at the filled joint and so efforts were made to find fillers free of Si. Table 9:6 shows the development of such alloys. The powders were inert dry gas atomised to remove oxygen and improve wetting and flow. To increase joint strength, alloy powders of the airfoil material were blended in. The filler powder has a smaller grain size to give dense packing of the powder mixture. On melting the filler powder components diffuse into the other powder and the airfoil, giving a "casting". The more the B diffuses into the superalloy, the closer will be the joint properties to those of the base metal and good shear and stress rupture properties will be obtained.

Four of the filler alloys in Table 9:6 are Ni based and one is Co based. In filler compositions, Ta,Zr,Hf,Ge,B,Si,P and Ti are melting temperature raisers with both B and Si. Fe gives joint ductility, especially at $1260^{o}C$ and acts as a diffusion barrier to host metal elements entering the joint. Ta is a solid solution strengthener and (like Al) is a gamma prime former. Zr and Hf are also solid solution strengtheners. Zr can lower the m.p. by 40 to 65 degrees (C) causing joint embrittlement and lowering oxidation resistance. Hf improves oxidation resistance, Y and La stop boride chain formation and add to spheroidization. Major Cr additions increase hot corrosion resistance. The filler is essentially a superalloy with B additions giving a lower melting point. Some diffusion occurs in the brazing cycle but an extra 4 hours at $1080^{o}C$ gives better joint properties. Before use, any protective coating should be chemically removed, cleaned by glass bead peening, chemical cleaning to remove oxides from inside cracks and vacuum heat treat cleaned. Cleaning of cracks has been discussed and is especially important for airfoils containing

T A B L E 9:6

CONSTITUENTS OF A SELECTION OF Ni- & Co- SUPERALLOY SUBSTRATES USED FOR ACTIVATED DIFFUSION BRAZING

No.	Ni	Co	Cr	B	C	Fe	Ti	Al	Si	Ta	Mo	W	Zr	Y	La
1	Bal.	1.5	22.0	--	0.10	18.5	--	--	0.50	--	9.0	0.60	--	--	--
2	Bal.	9.5	14.0	0.015	0.17	0.2	5.0	3.0	--	--	4.0	--	0.03	--	--
3	Bal.	20.5	20.5	3.0	0.06	--	--	--	--	3.5	--	--	--	--	0.040
4	Bal.	9.5	14.0	2.6	0.06	--	--	3.3	--	3.0	--	--	--	0.02	--
5	Bal.	9.5	14.0	2.6	0.06	--	--	3.3	--	3.0	--	--	--	0.02	--
6	Bal.	--	13.5	2.8	0.04	--	--	4.0	--	3.0	--	--	--	0.02	--
7	10.5	Bal.	25.0	--	0.50	--	--	--	--	--	--	7.5	--	--	--
8	9.7	Bal.	21.3	3.3	0.05	--	--	2.5	2.0	4.7	--	--	--	0.02	--

significant Al and Ti (whose oxides cannot be hydrogen atmosphere reduced). The filler alloys can be tape or slurry applied and masking of areas not to be coated must be done by alumina with a volatile binder. The brazing is done in vacuum or atmosphere containing H_2 or in Ar. 15 minutes is enough to cause flow and close a crack. The solidus range for the filler alloys shown in Table 9:6 is 1045° to $1080^{\circ}C$ and the liquidus range is 1130° to $1228^{\circ}C$. Test cracks of 0.1 to 1.5 mm and deeper than 1.25 mm were metallographically examined and found well sealed (Gooch et al 1983).

A different purpose coating repair method which has been proposed is a slurry method where a suspension of Al and Si in an organic binder is sprayed onto the cleaned surface of the blade. The deposit is then cured at $340^{\circ}C$ followed by a high temperature heat treatment. Such coatings are available commercially as Sermalloy J or PWA47 (registered trademarks). It can be applied as a patch repair or a whole blade coating. The cost is more than pack aluminizing but only half that of a Pt-Al coating. Worn turbine airseals can be repaired by a diffusion bonded plasma coating of NiCoCrAlY. The wear path is first removed by grinding and then plasma spray used to build up worn surfaces to the required levels. Inert gas shielding is used to stop oxidation. Increased density and bonding is achieved with a heat treatment at $1100^{\circ}C$ (Union Carbide process).

New parts with reject coatings and old parts from engines needing new coatings can be chemically stripped with acids to remove only MCrAlY. The thin interdiffusion layer with PVD coatings is advantageous over pack aluminide coatings which have a much thicker interdiffusion layer (Hill & Boone 1982). Recoating requires stripping off the old coating (Strang 1984). To remove corrosion it may be necessary to strip below the original substrate surface, which reduces the mechanical strength of the residual component. Low activity pack cementation is an effective method. If the original coating is not removed, spalling risks for the new coating rise by a factor of 2. During stripping, grain boundary attack and preferential substrate dissolution must be avoided. Oxides and sulphides are hard to remove from cooling channels and hydrogen reduction may be needed, or use of fluorine or the fluoro-carbon cleaning process. Recoating rejection rates are about 5%, compared with 0.4% for new coatings rejects. Some blades in service have been recoated 12 times. Effective field repairs of some coating systems by infrared heater or gas torch have been described (Strang et al 1984).

9.10.2. COATED COMPONENT JOINING TECHNIQUES:

Joining methods for industrial gas turbines and other high temperature uses have been described in the literature (Snowden 1982). Hot pressing will join two metals if the mutual solubilities and diffusivities are sufficient. A thin oxide film will prevent bonding; oxygen or brittle intermetallic compounds may embrittle the interface. Hot rolling (20 to 50% single pass reduction) may allow successful bonding even if oxide films are present. The oxide is broken up. Hot swaging is also effective, e.g. for putting a Pt group metal tube onto another material. Embrittlement may still cause failure. Surface oxides and brittle intermetallics can both be avoided by explosive bonding. Flat plates with a small gap between receive a detonation from an explosive on top along one edge. The explosion goes across the sheet, peeling off a layer from the upper sheet which then scours the lower sheet face. This gives clean surfaces which bond by the explosive pressure. Oxide scale is blown out of the interface and no oxygen embrittlement can occur. Heating effects are too transient to form any brittle intermetallics (see Chapter 7) (Hood 1976). Fig.9-6 gives an illustration of weld-joining a clad plate.

Little work appears in the literature on joining of coated materials, but reviews of general joining are available (Elliott & Wallach 1981; Phelps 1982; Derby & Wallach 1982; Almond et al 1983). Best results would probably be obtained by a normal joining procedure (Elliott & Wallach 1981) followed by a coatings repair procedure (see coatings repairs section in this chapter). Metallization of AlN ceramics by electroless nickel gave better bonding on etching with NaOH (Osaka et al 1986). A 97% alumina has been metallized at $1000^{\circ}C$ using refractory metals applied as

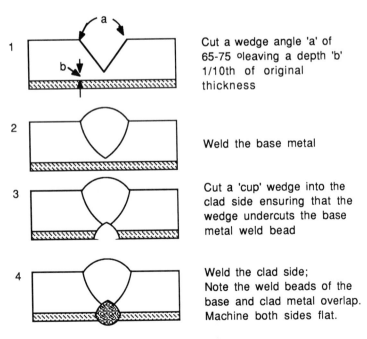

1 — Cut a wedge angle 'a' of 65-75 °leaving a depth 'b' 1/10th of original thickness

2 — Weld the base metal

3 — Cut a 'cup' wedge into the clad side ensuring that the wedge undercuts the base metal weld bead

4 — Weld the clad side; Note the weld beads of the base and clad metal overlap. Machine both sides flat.

Fig.9-6: Steps in weld-joining a clad plate

oxides or in solution. Seals were 85% as strong as those metal-lized by high temperature processes. A fine refractory metal powder, e.g. Mo or W is fired on to the ceramic surface in a reducing atmosphere at about 1500°C to produce a sintered layer which adheres to the ceramic. To facilitate wetting by the braze metal Ni is electroplated, followed by the metal braze (Tomlinson 1986).

The basic understanding in joining a conductor and an insulator is that, adhesion between metals is a maximum between metals when their electronegativity and atomic size are nearly the same. It is improved by deposition of a graded cermet at the film/subs-trate interface. The polarity of the interface and adhesion can be enhanced by applying a strong electrical field from the metal film to the insulator at high temperature, which is called field assisted bonding (Pomerantz 1983). Atomic scale roughness created by bombarding substrate with energetic ions can achieve a mecha-nical locking and is usually effective when carried out at the initial stages of film growth, the advantage being it will not only roughen the surface but also lead to embedding of the depo-siting species into the substrate thus creating a diffuse inter-face (Budhani et al 1986).

A thin film thermocouple is a fairly new development to meet the need to monitor surface temperature on the blades and other working components of a gas turbine (Budhani et al 1986) and internal combustion engine hardware like valves, valve seats, combustion chamber surfaces, cylinder walls and piston heads (Kreider 1986), with an aim to achieve better optimization of operation at higher temperatures and increasing fuel efficiency. A conventional wire thermocouple is unsuitable for structural and aerodynamic reasons on the gas turbine, and thermally grown alumina on the MCrAlY is not adequately insulating.

The 2 micron thin thermal sensing element made of a normal Pt/Pt-10%Rh couple has to be on an insulating surface on, typically, MCrAlY-coated superalloy blades and vanes, and faces four main problems: adhesion of the 2 micron insulating layer to a 120 micron MCrAlY overlay, adhesion of the sensor elements to the alumina, thermoelectric stability of the sensor elements on con-tamination, and, corrosion prevention of the sensor elements in the aggressive environment.

In the case of the gas turbine, the four crucial processing steps were MCrAlY surface pretreatment, application of Al_2O_3 on NiCoCrAlY and the thermocouple elements on the alumina, and, the final protective cover. Best adhesion was obtained on a polished, vacuum treated MCrAlY surface, applied with alumina deposition by ARE followed by thermal oxidation; polished, vacuum treated, thermally grown oxide was also promising. Polishing followed by thermal oxidation with or without ARE of alumina as an intermedi-ate failed to achieve good adhesion. Cr and Mo have been tried as interlayers and maximum adhesion of the interface was achieved by graded bonding of $Pt-Pt_xO_{1-x}-Al_2O_3$. The latter was carried out by alternating oxygen with argon during deposition. for the last step of corrosion protection, Si-O-N and Si_3N_4 graded structures were deposited by PACVD (Budhani et al 1986).

Thin thermocouple joining conditions appear to have been less problematical on FeCrAlY coating alloy on components for internal combustion engines. A thermal oxide, viz. alumina was grown on the 120 micron thick FeCrAlY, and the film showed outstanding adhesion and dielectric properties as well as stability in the hot gases environment, and was better than Cr_2O_3. Substrate polishing to -600 mesh grade or shot peening proved equally adequate. Thermal oxidation was carried out between 827° and 927^OC in controlled argon + $4\%H_2$ + air, wherein the hydrogen suppressed formation of oxides of Fe and Ni. For good adhesion onto sputtered alumina, thermal alumina had to be about 0.5-1 micron and was grown in 50 h at 902^OC (thickness of the sputtered alumina was monitored by a simultaneous deposition on a glass slide and measured by interferometer with thallium illumination). Pt and Pt-10%Rh was sputtered onto the oxide layer. Components of cast iron, low alloy ferritic steel 4340 and a high temperature austenitic 347 steel have all been joined with thin film thermocouples 10 cm long in an 'S' bend with leads (Kreider 1986).

To achieve adhesion and prevent spalling in cemented carbide tools coated with TiC, non-decarburizing is necessary. This prevents the formation of W and element diffusion from the Co matrix into TiC as a result of a relative increase in M_6C w.r.t. $M_{12}C$, the predominance of the latter contributing to adhesion failure (Bhat et al 1986). In contrast, adhesion is undesirable between copper forming tools and the workpiece, and this is achieved by coating MoN_x which has low galling tendency, by magnetron sputtering (Valli et al 1986).

Hot isostatic pressing (HIP) applies about 2000 atm. at high temperature, e.g. 1500^OC to compact powders, densify porous materials or to heal defects like cracks or shrinkage cavities. The process can be used to join Mo, Cr, Re and W which are difficult to weld. Weld zone recrystallization is avoided by using high pressures and moderate temperatures (0.3 to 0.7 T_m). The metals do not need to be ductile (in contrast, a minimum of 5% elongation is required for explosive bonding). For application of the pressure the couple is placed in a sealed steel can. Good HIP bonds have been made between Mo and Kanthal and between Mo and Inconel 601, thus combining the hot strength of Mo with the corrosion resistance of FeCr and NiCr alloys (Werdecher & Aldinger 1982).

9.11. FUNCTIONS OF HIGH TEMPERATURE COATINGS

High temperature coatings are required to provide resistance to the following effects:

- abrasion
- corrosion in aggressive media
- cyclic operating conditions (thermal and stress)
- erosion
- excessive heat flow
- friction
- interdiffusion
- oxidation
- wear

Coatings are also used as catalysts in high temperature processes.

Tables of coatings and their functions are given in Tables 9:7 to 9:9.

Key:	ER	Erosion	DB	Diffusion Barrier
	FR/WR	Friction & Wear	HC	Hot Corrosion
	OX	Oxidation	NI	Nuclear Industry
	TB	Thermal Barrier		

Note:- No individual references are indicated; several sources have been cited from literature; list may not be exhaustive.

TABLE 9:7

COATING COMPOSITIONS FOR REFRACTORY SINGLE METALS & ALLOYS

SUBSTRATE	COATING	FUNCTION
Aerospace alloys	Pt-Ir	OX

Al	Al_2O_3	OX, W
Al	TiO_2	OX

Al-alloys	RSP-amorphous metals & alloys	OX,WR
	WC-12Co 75CrC-25(80Ni20Cr) Mo, 80Ni20Cr	FR/WR
	Oxide dispersions Ion implanted N, O etc.	OX, WR
	RSP-amorphous metals & alloys	OX, WR
C fibre	Ni	OX
C-graphite	Pt,Ir,Re	OX
Graphite disc 25mm dia 2mm	Ti-Zr-B	NI
Semi-circular Graphite with Mo sheet in the middle area	ZrC-C	NI
Graphite tube fine grained, 18mm	Nb, Ta	FR/WR
Graphite, Si-coated (contd.)	Polycryst Si	OX

Cast Metals	Oxide dispersions ZrC_xN_y	WR WR/TB

Cu-Pd-Si, Cu-Zr	RSP - top coat	FR/WR

Cr particles	Al_2O_3	OX

Cr5W0.1Y	Silicides	OX
	Cr_5Al_8	OX

**COATING COMPOSITIONS FOR REFRACTORY SINGLE METALS & ALLOYS
(Table 9:7 contd.)**

SUBSTRATE	COATING	FUNCTION
Cr5W0.1Y	Ni30Cr, Ni20Cr20W, Ni30Cr20W, also with 3 to 5%Al, & barrier layers of Mo, W-1%ThO$_2$, W25Re, & W	OX, DB
	FeCrAlY, Fe-25Cr-14Al-1Y	OX
	Pt, Pd, Al	OX

Fe (cast iron)	Co + Cr$_3$C$_2$ cermet	FR/WR

	ZrO$_2$ Stabilised with 5wt%CaO or 8wt%CaO or 17wt%Y$_2$O$_3$ Best with an intermediate layer 0.2-1.2mm in 3 steps:
	    96%Ni 4%Al (wt)
	    80%Ni 20%Al
	    80%Ni 20%Cr.
		TB

FeBC & FeNiB	RSP - top coat	FR/WR
	NiCrBSi, Al$_2$O$_3$-TiO$_2$-Mo	TB

Hard metals	Borides	OX

Mo	Ir, Re	OX
	MoSi$_2$, SiO$_2$, NbSi$_2$, TaSi$_2$, TiSi$_2$, ZrSi$_2$, CrSi$_2$	OX
	BP	OX, TB

Mo & Mo filament (0.3-0.5 mm dia. & 50 mm long)	AlB$_{12}$ (thermoelectric converter for nuclear reactor)	OX

Nb	Ir, Re, MoSi$_2$, NbSi$_2$	OX

Nb alloys	Nb15Ti10Ta10W2Hf3Al (cladding) Nb5W30Hf5Ti3Re (cladding) Hf24Ta2CrlSi (cladding)	OX OX
	Ti foil, TiV foil	OX
	Si20Cr20Fe Si(35W35Mo15Ti15V)	OX OX

COATING COMPOSITIONS FOR REFRACTORY SINGLE METALS & ALLOYS
(Table 9:7 contd.)

SUBSTRATE	COATING	FUNCTION
Nb alloys	V-(80Cr20Ti)-Si, VSi$_2$ VCrTiSi, V-(80Cr-20Ti)-Si VCrTiAlSi MoTiSi, (95Mo-5Ti)-Si	OX
	MoCrFeAl, Mo-Cr-(Fe-25Cr-5Al)	OX
Nb-base alloys: CB742 D43	CrTiSi	OX

Ni	TaC, MoC, WC	ER,WR

Single Metals: Ni, Cr, Cu, Fe, Pb	Borides of Ta & Zr; Carbides of B, Si, Ta, Ti, W; Nitrides of B, Si; Also as fibres	OX,FR/WR

Quartz, Tubular	ZrN	FR/WR

Quartz wall of tube	Ru-, Ir-, & Ru-Ir oxides	OX

Single Metals: Ni,Cr,Cu,Fe,Pb	Oxides of Al, Ce, Si, Ti, Zr;	OX,FR/WR

Cermets   WR, OX

A	B	C
Cu-Ni;Ni-Cr;	Cr+/-Ni,Co,	Ceramic
Cr-Fe;Si-B-C;	or Ti	oxides
Cu-Zn		m.p.>1900°C

Binary mixes:
A 60-10% + B 40-90%
B 70-30% + C 30-70%

Ternary mixes:
A 60-10% + B 20-60% + C 20-30%

Sapphire	AlN-Si$_3$N$_4$	

Si wafer	Si$_3$N$_4$	OX, WR

SiC	SiC ;Si$_3$N$_4$; B	FR/WR

SiO$_2$ (Vitreous)	Si$_3$N$_4$	W

COATING COMPOSITIONS FOR REFRACTORY SINGLE METALS & ALLOYS
(Table 9:7 contd.)

SUBSTRATE	COATING	FUNCTION
Ta	Re, Hf20Ta	OX
	Al, Al20Cr, Al30Cr	OX
	TaAl3-Al	OX
Ta	B	in B-fibre production for chemical compatibility
Ta10W	Hf20Ta, Hf20Ta2O5, Hf20TaN	OX, ER
	Hf24Ta1.2Cr.7B.12Al (cladding)	OX
	ThO$_2$, HfO$_2$, with W diffusion barrier layer	OX, DB
	Sn23Al5Mo, Sn23Al7Mo	OX
	MoSi$_2$/ZrO$_2$, WSi$_2$/ZrO$_2$	OX
	Al, Al20Cr, Al30Cr, 50Sn50Al	OX
	Sn-Al-Mo, Sn-23Al-Mo	OX
Ta10W	HfTa, Hf-20Ta, HfTa2O5, Hf-20Ta2O5, HfTaN, Hf-20TaN	OX, ER
	VSi$_2$	OX
Ti	Ti$_5$S$_3$	OX
Ti	RuO$_2$, IrO$_2$	Catalyst
Ti	B	in B-fibre production for chemical compatibility
TiAl4V	Ba (ion implant)	OX
TiN sintered pellet	Ti-B-N	WR
W	Ir, IrHfTa, Pt, Re	OX
	AlB12 (thermoelectric converter for nuclear reactor)	OX
	SrZrO$_3$, Nd$_2$Zr$_2$O$_7$, Yb$_2$Zr$_2$O$_7$	OX, TB
	SiC or ZrO$_2$ over Ir	OX, TB
	ThO$_2$	OX, TB

535

**COATING COMPOSITIONS FOR REFRACTORY SINGLE METALS & ALLOYS
(Table   9:7   contd.)**

SUBSTRATE	COATING	FUNCTION
W & W-boride	B	in B-fibre production for chemical compatibility
WC 70%,TiC 7%, Ta(Nb)C 13%, + Co 7% (cemented carbide tool bits)	TiC, TiN	FR/WR
Cemented carbide (WC -Co alloys)	TiC	WR
Uncoated & TiC precoated Cemented carbides	TiB_2	WR
Zr	$ZrSi$, $ZrSi_2$,$Zr_2Si/ZrSi$	OX
Zircalloy-4	Sn	OX
Zr alloys	RSP-amorphous alloys	OX,FR
	Ion Implanted N, O etc.	OX,FR

===

TABLE 9:8

COATINGS FOR SUPERALLOYS & COMPONENT EXPERIMENTAL   ALLOYS

SUBSTRATE	COATING	FUNCTION
Ni	Cr, Li, Si	OX
Co	Cr, Li, Si	OX
Mo	Cr, Li, Si	OX
Ni-alloys	NiAl (has a wide stoich. range) AlCr/Ta Al/Pt	OX
	Si	OX
	Cermets	OX, ER
Ni-alloys (-10Cr, 20Cr)	Al, NiAl, Ni2Al3,CoAl, CrAl	OX

COATINGS FOR SUPERALLOYS & COMPONENT EXPERIMENTAL ALLOYS
(Table 9:8 contd.)

SUBSTRATE	COATING	FUNCTION
Ni-alloys	$NiSi_2$, $CrSi_2$, NiCrSi NiCrSi Ni-20Cr-10Si Low Si Ni-30Cr-10Si 26Ni-24Cr-50Si High Si	OX OX
Ni-alloys	ZrO_2/Ni-Cr, ZrO_2/Ni-Al ZrO_2/MCrAlY (M=Ni, Co, or Fe) (same but with $MgZrO3$ instead of ZrO_2)	OX, TB
Ni-alloys	CrC-NiAl	OX, ER
Ni-Al-Cr3C2	Al	OX
Ni-20Cr	CrC	FR/WR
Ni-20Cr	Co + Cr_3C_2 Cermet	FR/WR
Ni-20Cr	ZrO_2 Stabilised with 5wt%CaO or 8wt%CaO or 17wt%Y_2O_3 Best with an intermediate layer 0.2-1.2mm in 3 steps: 96%Ni 4%Al (wt) 80%Ni 20%Al 80%Ni 20%Cr.	TB
Ni-20Cr	CoCrAlY	OX, HC
NiNbCrAl	NiCrAlY; NiCrAlY/Pt	HC
NiNbCrAlY direc.solidified	NiCrAlY/Pt top layer	HC
Ni-base alloys	CoCrAlY, CoCrAlY/Pt	HC
	Si_3N_4	HC, TB
B 1900+Hf	Ni, Co, Cr, Al Hf, Mo, Ta	OX

	NiCrAlY bondcoat/ZrO_2-Y_2O_3
	    " /ZrO_2-MgO
	    " /ZrO_2-$CaCO_3$
		OX, TB

B 1925	PtAl	HC
Nim 80H	Cr	HC

COATINGS FOR SUPERALLOYS & COMPONENT EXPERIMENTAL ALLOYS
(Table 9:8 contd.)

SUBSTRATE	COATING				FUNCTION
Nim 90	$NiSi_2$, $CrSi_2$, NiCrSi NiCrSi Ni-20Cr-10Si Low Si Ni-30Cr-10Si 26Ni-24Cr-50Si High Si				OX

Nim 105	Si				OX
Nim 105	Si-base coatings				HC

IN 100	NiAl, Ni_2Al_3, CoAl, Al				OX
IN 100	Al, AlCr, NiAl, CoCrAl, NiCrAl, CoCrAlHf, CoCrAlY, NiCoCrAlY, NiCoCrAlYTa, NiAl+Cr+Ti				OX, HC
IN 100	$NiSi_2$, $CrSi_2$, NiCrSi				OX
IN 100	Ni-20Cr-10Si Low Si Ni-30Cr-10Si 26Ni-24Cr-50Si High Si				OX, HC
IN 100 CoCrAlY: (med.Cr)	Co 63 64 62 61.6 61	Cr 24.6 24.1 25.0 25.4 24.8	Al 12.1 11.8 12.0 11.9 11.6	Y 0.12 0.23 0.73 1.10 2.20	OX, HC
IN100 NiCrAlY	46 (Ni)	42.4	10.7	1.02	OX, HC

IN 601	NiCrSi				OX

Incoloy 800	Al, AlCr, AlSi				OX, HC

Hastelloy X	$NiSi_2$, $CrSi_2$, NiCrSi NiCrSi: Ni-20Cr-10Si Low Si Ni-30Cr-10Si 26Ni-24Cr-50Si High Si				OX
Hastelloy X	CrAl, CrAlY, CrAlC				HC
Hastelloy X	NiCrAl: Cr/Al =1:1; Ni10 NiCrAlY: Cr/Al/Y = 1:1:0.3; Ni10 NiCrAlCe: Cr/Al/Ce = 1:1:0.3; Ni10				OX, HC

538

COATINGS FOR SUPERALLOYS & COMPONENT EXPERIMENTAL ALLOYS
(Table 9:8 contd.)

SUBSTRATE	COATING	FUNCTION
IN738	NiSi$_2$, CrSi$_2$, NiCrSi	OX
IN 738 LC	SiTi, NiAl	OX
	NiSi$_2$, CrSi$_2$, NiCrSi	OX
	NiCrSi:	
	Ni-20Cr-10Si Low Si	
	Ni-30Cr-10Si	
	26Ni-24Cr-50Si High Si	
IN 738LC	Al; NiAl; CrAl:35%Cr - 65%Al	OX
IN 738LC	NiCrAl: Cr/Al =1:1; Ni10	OX, HC
	NiCrAlY: Cr/Al/Y = 1:1:0.3; Ni10	
	NiCrAlCe: Cr/Al/Ce = 1:1:0.3; Ni10	
IN 738LC	PtCrAl, NiCoCrAlY	OX, HC
IN 738LC	pack Al, Al, Cr, CrAl, SiAl,	OX, HC
	Cr-Al, Si-Al, Pt-Cr-Al,	
	CrAlY, CrAlC, CrAl/Pt	
IN 738LC	NiCoCrAlY, CoCrAlYPt, CoCrAlHf	OX, HC
IN 738LC	Si base coatings	HC
IN 738	CoCrAlY (High Cr, Low Al):	HC
	Co62.5 Cr29 Al7.1 Y1.35	
IN 738	NiCrAlY: Ni50 Cr41.5 Al8.1 Y 0.57	HC
IN 738LC	CrNi, 50Cr-50Ni	HC
IN 738LC	CoNiCrAlY:	HC
	Co39 Ni32 Cr21 Al7.5 Y0.5	
IN 738	NiCrAlTaC:	HC
	Ni10 Cr25 Al3 Ta5 Y0.2 C57	
IN 738	Experimental MCrAlX compositions:	OX, HC

Coating:	Co	Cr	Al	Y	Ni	Ta	Hf	Al$_2$O$_3$
CoCrAlYNiTa	56	25	3	0.7	10	5	-	-
CoCrAlYNiTa	45	25	10	0.5	10	10	-	-
CoCrAlYNi	32	22	9	0.3	38	-	-	-
CoCrAlYNi	32	13	23	0.3	33	-	-	-
CoCrAlY	63	23	13	0.7	-	-	-	-

COATINGS FOR SUPERALLOYS & COMPONENT EXPERIMENTAL ALLOYS
(Table 9:8 contd.)

SUBSTRATE	COATING	FUNCTION
IN 738	Experimental MCrAlX Compositions:	OX, HC

Coating;	Co	Cr	Al	Y	Ni	Ta	Hf	Al_2O_3
CoCrAlHf	63	23	13	–	–	–	0.7	–
CrAlNi	–	2	33	–	65	–	–	–
CoCrAlYAl$_2$O$_3$	63	23	13	0.7	–	–	–	0.5
LCO-5*	70	19	10	0.5	–			
LCO-7*	63	23	13	0.6	–			
LCO-22	39	21	7.5	0.5	32			
LCO-29	73	18	8	0.5	–			
LCO-37	44	30	3	0.5	23			
LN-11*	23	17	12	0.3	48			
LN-21	22	21	7.5	0.5	49			
LN-34	0.5	20	11	0.5	67	(& 0.5 Mo)		

* Proprietary alloys.

Substrate	Coating					Function
IN738	CoCrAlY: (High Cr, Low Al)	Co 62.55	Cr 29.0	Al 7.1	Y 1.35	OX, HC
IN 738	CoCrAlY: (low Cr)	Co 67	Cr 20	Al 11.7	Y 0.55	
		65.7	21.4	11.5	1.4	
		66.3	20	11.6	2.1	

IN 738	MCrAlY: Cr 18–39%; Al 4–16% (M=Co,orNi)	HC
	MCrAlHf: Co-19Cr –12Al-1Hf Co-19Cr –6Al-1Hf	HC

IN 792	AlCr, NiAlCr	OX
IN 792	Aluminides, Pt-Rh	OX
IN 792	MCrAlY: Cr 18–39%; Al 4–16% (M=Co,orNi)	HC
	MCrAlHf: Co-19Cr –12Al-1Hf Co-19Cr –6Al-1Hf	HC
IN 792	PtAl	HC
IN 792+3/4%Hf	Al, Cr, CoNiCrAlY: Co39 Ni32 Cr21 Al17.5 Y0.5	HC

IN939	Si-base coatings	HC

COATINGS FOR SUPERALLOYS & COMPONENT EXPERIMENTAL ALLOYS
(Table 9:8 contd.)

SUBSTRATE	COATING	FUNCTION
MAR-M-200+Hf (Ni alloy)	Ni, W, Co, Cr, Al, Hf, Nb, Ti	OX
	NiCrAlY bondcoat/ZrO_2-Y_2O_3 " /ZrO_2-MgO " /ZrO_2-CaCO3	OX, TB

Co-base alloys	TiB2	OX
	CoCrAlY, CoCrAlY/Pt, Si_3N_4	HC

CoTaC eutectic	NiAl, Ni_2Al_3, CoAl, Al	OX

Mar-M-004	NiAl, Ni_2Al_3, CoAl, Al	OX

MAR-M-509 (Co alloy)	Co,Cr,Ni,W,Ta,Zr (17)	OX
	NiCrAlY bondcoat/ZrO_2-Y_2O_3 " /ZrO_2-MgO " /ZrO_2-$CaCO_3$	OX, TB
Mar M002 & Mar M246	Al, Cr, CoNiCrAlY: Co39 Ni32 Cr21 Al7.5 Y0.5	HC

FX414 (Co alloy)	pack Al, CrAl, NiCoAl	OX
FX414	Al, SiAlFX414	
	CrAl, PtCrAl, NiCoCrAlY	OX, HC
FSX414	Al,Cr-Al,Si-Al,Pt-Cr-Al,NiCoCrAlY	HC

Rene 80	'Codep' PkAl; CoAl; CoCrAlY	HC
Rene 80	CoCrAlYPt; CoCrAl.AlHf; CoCrAlHfPt	HC
Rene 80	CoCrAlY/Pt	HC
Rene 80	Pt/CoCrAlY	HC
Rene 80	CoCrAlY:	OX, HC

(low Cr)

Co	Cr	Al	Y
67	20	11.7	0.55
65.7	21.4	11.5	1.4
66.3	20	11.6	2.1

COATINGS FOR SUPERALLOYS & COMPONENT EXPERIMENTAL ALLOYS
(Table 9:8 contd.)

SUBSTRATE	COATING					FUNCTION
Rene 80	CoCrAlY:					OX, HC
	(med Cr)	Co	Cr	Al	Y	
		63	24.6	12.1	0.12	
		64	24.1	11.8	0.23	
		62	25.0	12.0	0.73	
		61.6	25.4	11.9	1.10	
		61	24.8	11.8	2.20	
Rene 80	CoCrAlY:					
	(High	56.6	30.0	13.0	0.52	
	Cr & Al)					
Rene 80	NiCoCrAlY:					
	Ni41	17.5	35.4	6.0	0.3	

SUBSTRATE	COATING	FUNCTION
X-40	NiAl, Ni2Al3, CoAl, Al	OX
X-40	'Codep' PkAl, CoAl, CoCrAlY	OX, HC
X-40	NiCoCrAlYSi: Co32.9 Ni35 Cr26 Al2.9 Y0.23 Si2.8	OX, HC
X-40	FeCrAlY: Fe61 Cr25.8 Al12.4 Y0.63 Fe60 Cr27.3 Al12.2 Y0.59	
Superalloy	$2CaO.SiO_2$	OX, TB
Superalloy	$ZrO_2-12\%Y_2O_3/76.4Ni$ 16Cr 6Al 0.6Y (duplex)	HC, TB
Superalloy	$ZrO_2-8\%Y_2O_3/78Ni$ 17Cr 5Al 0.35Y (duplex)	HC, TB
Superalloy	MgO-NiCrAlY Cermet	HC, TB
Superalloy	MCrAlY	HC
Superalloy	ZrO_2 (+CaO or Y_2O_3) with inner NiAl or NiCr	HC, TB
Superalloy	Ni/Al, Ni/C, Co/WC, NiCr/ diatomite, $Ni/WTiC_2$, $NiCr/Cr_3C_2$, $NiCaF_2$, $NiOCaF_2$	OX, ER
Superalloy	NiCoCrAlY, Ni-18Cr-23Co-12Al-0.5Y	HC
Superalloy	Al_2O_3	TB, ER

COATINGS FOR SUPERALLOYS & COMPONENT EXPERIMENTAL ALLOYS
(Table 9:8 contd.)

SUBSTRATE	COATING	FUNCTION
Turbine airfoils	$ZrO_2-Y_2O_3$; ZrO_2-Mg	OX, TB

===

TABLE 9:9

COATINGS FOR STEEL

COATING	SUBSTRATE	FUNCTION
Fe	$Fe50Al$, $CrAl$, Al	OX
	$NiSi_2$, $CrSi_2$, $NiCrSi$	OX
	Carbon diffused through aluminide	FR/WR
	TiB_2	FR/WR
Fe/Ni-plated	Cr_7C_3	FR/WR
FeSi	TiN	FR/WR
Fe18Cr	$Fe18Cr8Ni/CeO_2$ & $/ZrO_2$	OX, ER
Steel wire	Ni	OX
Nailhead Carbon steel	$TiCx Ny$; $TiCx-TiN$	FR/WR
Carbon steel	ZrC_xN_y & $ZrC-ZrN$	FR/WR
Inner face of long mild steel tubes	Ta	OX
Structural steel	Ti, Mo, Zr, Nb, Ti6Al14V	OX
Fe-Mn-Al-Steel	Al	OX
9Cr-Steel	SiO_2	OX
Cr-Ni-Nb-Stainless steel	Y(ion implanted)	OX
20Cr25NiNb stabilised Stainless steel	CeO_2 (sol-gel) SiO_2 (CVD)	OX

COATINGS FOR STEEL
(Table 9:9 contd.)

COATING	SUBSTRATE	FUNCTION
316 Stainless steel	NiCrBSi	OX
Stainless steel	Co+Cr$_3$C$_2$ Cermets	FR/WR
	ZrO$_2$ Stabilised with 5wt%CaO or 8wt%CaO or 17wt%Y$_2$O$_3$ Best with an intermediate layer 0.2-1.2mm in 3 steps: 96%Ni 4%Al (wt) 80%Ni 20%Al 80%Ni 20%Cr.	TB
9CrMo0.6Si steel	SiO$_2$	OX
Fe2Cr2AllSi 0.5Ti	Al	OX
FeCrAl	Al$_2$O$_3$	HC
Coal gasifier alloys	MCrAlY MCrAlHf (M = Co, Fe or Ni)	HC
Steels used in Nuclear Industry	ZrC-C	prevents escape of nuclear fission products
	Al$_2$O$_3$	W, TB,OX
Steel	Al, Cr	OX
Steel	SiO$_2$, CeO$_2$, Cr$_2$O$_3$, Al$_2$O$_3$	OX
Steel	13-20% Cr	OX
Steel	MoSi$_2$-Mo-Ni	OX, TB
Steel	B	OX
Steel	ZrO$_2$ (24%MgO stabilised), ZrO$_2$ (20%Y$_2$O$_3$ stabilised), ZrO$_2$ (5%CaO stabilised)	OX
Steel	WC-9Co, WC-12Co, WC-15Co 75CrC-25(80Ni20Cr) Mo, 80Ni20Cr	FR/WR

COATINGS FOR STEEL
(Table 9:9 contd.)

COATING	SUBSTRATE	FUNCTION
Steel	W, Mo, piano wire (by wire explosion)	FR/WR
Steel	Cr_3C_2/Mo, TiNi alloy with 3% free C	FR/WR
Steel	TiC, TiN	W

==

CHAPTER 10

Strategic metal conservation, coatings achievements, and future

10.1. INTRODUCTION

The following two sections of this last chapter survey the role of high temperature coatings in conserving strategic metals, and the extension of working life achieved by reclamation, replacement and re-coating procedures. The present achievements of coatings are also assessed in these two sections and the last section sums up the current status and future efforts required to provide the much needed fundamental data together with research and development work necessary for further improvement in coating life and increasing efficiency.

10.2. COATINGS TO CONSERVE STRATEGIC METALS

Eight metals which are vital to the European and World economy are classed as strategic metals, viz. Cr, Co, W, Mn, V, Mo, Nb and the Pt group (Anon., Inst.Mech.Eng. 1981). The definition of strategic materials, which includes non-military uses, has been discussed (Archer 1980). Listings of strategic minerals have been compiled (Hargreaves and Fromson 1983). Table 10:1 shows prices and reserves available. Summarised information is provided below on strategic metals, their uses and sources (Anon., Inst. Mech. Eng. 1981).

Cr: This is one of the main alloying elements added for corrosion resistance and application of coatings can be expected to reduce the percentage required in substrates. The main sources of this metal are S.Africa, Zimbabwe and a small amount for Scandinavia (ferrochrome). USSR reserves are falling and USSR import from Africa may soon be necessary.

Co: This is also a vital anti-corrosion alloying element and coatings can be expected to reduce requirements in substrates. Zaire and Zambia are the main suppliers. Belgium is the main European importer and distributor. The main uses are in magnets, superalloys and cemented carbides.

W: This is used in high speed W-Mn steel and for WC tipped tools. It is imported mainly from Portugal, Thailand and East European countries, although the main reserves are in China, Canada, USSR an USA. There are W reserves in Austria and Turkey, and also in the UK, sufficient to allow exports, but these are presently undeveloped.

Mn: This is used in most steels and would be very difficult to substitute for. The main reserves are in S.Africa, USSR and Australia. Coatings applications cannot alleviate shortages of Mn or W.

V: This is used in alloy steels, high temperature alloys and catalysts. The main sources are USSR and S. Africa. Precipitates of vanadium carbide in steels confers hot strength and creep resistance and coatings cannot alleviate shortages of V.

Mo: This is mainly used in alloy steels and the major reserves are in N. and S. America. As its main use is to improve the mechanical properties of steels, coatings cannot alleviate Mo shortages. It would be very difficult to substitute for Mo in Nimonic alloys where it is also an important constituent.

Nb: This is an alloying element in steels and superalloys, added for grain refining in steels and for creep resistance. In its acid corrosion-resistance aspect its use could be reduced by applying protective coatings. The main reserves are in Brazil, USSR and Canada.

Pt, Pd, Rh, Ir, Ru and Os: The concentration of these rarely exceeds 10 parts per million even in the richest deposits. Chemical inertness, catalytic activity and high melting points are their important properties. None of these materials is likely to be reduced in quantity required, by the application of coatings.

Much of the literature on strategic materials is of limited availability. The survey published by Strang et al (1984) is a useful source of accessible references. Table 10:2 shows strategic materials imports. The USA, EEC and Japan depend greatly on imports but the COMECON countries are self-sufficient except for Co. Stockpiling of strategic metals in the USA is mainly intended for military use and not available for general use.

Substitution technology has been discussed but little specific attention has been given to using coatings technology for saving strategic metals. It is felt that industry has little incentive to invest in research and development just to find solutions which would only be used in an emergency. Table 10:3 gives various approaches to the strategic materials problem (Strang et al 1984). Use of all coating methods could be quickly increased in the event of a severe Cr shortage, but advances in coating processing methods should be pursued to develop the technology into a usable option for any emergency. The lead-time is very short in an emergency and so an information stockpile on substitution and conservation technology is also a stockpile of time and is of low cost.

Non-coatings approaches, such as joining very different materials can produce items with strategic metals only in the places where they are really needed. Machining losses can be reduced by contour forging, flow turning and powder metallurgy. Some parts can be made re-usable if only the coating needs to be replaced.

Cladding with stainless steel allows Cr reduction (where the Cr is used for corrosion or wear resistance. Electroless Ni plate and nitriding can be similarly used. Pack chromising on steel can replace ferritic stainless steels and pack aluminizing of steel gives oxidation resistance similar to Cr-steels.

Coating efforts to reduce the content of the strategic element Co in the hot section of gas turbines involves replacing Ni for Co in substrates and reducing the Co content of coatings. The coating developments which allow this are (Hecht & Halfpap 1984):-

1. Development of MCrAlY overlay coatings.
2. Pt-modified diffusion aluminide coatings.
3. Thermal barrier coatings.
4. Development of multiple layer coatings.
5. Wear resistant coatings.

CoCrAlY coatings with 67%Co gave a 3-fold increase in corrosion life of first stage blades and allowed higher strength Ni-base superalloy vanes to replace the Co-base aluminide-coated vanes; this reduced Co requirements by 36Kg per engine.

NiCoCrAlY coatings were developed for higher temperature use, giving higher ductility. They have the same hot corrosion resistance as CoCrAlY but contain 45% less Co. NiCoCrAlY coatings on new Ni superalloys containing no Co and Cr give corrosion resistance similar to conventional coated conventinal Ni superalloys (Strang et al 1984).

Pt aluminides do not give the corrosion protection of CoCrAlY but have contributed to Co saving. Pt is strategic but only $0.01g/cm^2$ is used in coatings (1/60 of the Co used in CoCrAlY) (Hecht & Halfpap 1984).

Co-base sheet metal alloys are preferred over Ni-alloys for their better creep/rupture strength and fatigue properties. To substitute Ni alloys for Co alloys, thermal barrier $MgO.ZrO_2$ coatings were developed to get about 60 C degrees substrate temperature reduction. There is a 7-fold reduction in Co usage for these alloys.

Use of thermal barrier coatings on turbine airfoils has developed considerably to enter the field of practical application, but the system requires further investigation for a fool-proof consolidation.

The best area for reducing strategic metals, apart from redesign, is component life extension.

TABLE 10:1

STRATEGICALLY SENSITIVE MATERIALS

ELEMENT	PRICE*	RESEERVE INDEX*** years	
Chromium	£4400/tonne, lump (UK) $3.750/lb, electrolytic (US)	£3650**	340
Cobalt	$7.20/lb (Europe) $6.65/lb (US) $11.30/lb as shots	£25000**	49
Tungsten	£13.50/kg metal powder	$6700**	47
Manganese	$1920/tonne, electrolytic $1810/tonne, flake	£780**	190
Vanadium (as V_2O_5)	$2.65/lb	£3700**	340
Molybdenum	$3.20/lb Mo (canned molybdic oxide)	£20000**	83
Niobium	$2.90/lb (ore with Ta-oxide)	£30000**	950

===
* Millbank (1987) (metal prices are not quoted where ore prices are indicated)
** Solid, pure, basic forms, London, January 1982
*** Reserve Index = Known World Reserves/Current Prodn. Rate (this figure does not take into account any future increase in the production rate).

TABLE 10:2

NET IMPORTS AS A PERCENT OF CONSUMPTION

MATERIALS	U.S	E.E.C	JAPAN	COMECON
Manganese	98	100	99	3
Cobalt	97	100	100	68
Bauxite	91	97	100	28
Chromium	91	100	98	2
Asbestos	85	90	98	1
Nickel	70	100	100	13
Zinc	57	91	74	9
Iron Ore	48	82	100	5
Silver	36	93	71	10
Copper	13	100	97	4
Lead	13	76	78	3
Phosphate	Export	99	100	23

Source: Bureau of Mines data (1977)

TABLE 10:3

WORLD PRODUCTION & CONSUMPTION*

MATERIAL	YEAR	PRODUCTION	CONSUMPTION
Al	1986	11938000	6585100
	1985	15428700	16138200
Ni	1986	446800	341000
	1985	762700	786800
W	1985	43382	46629
(ore &	1984	43699	48490
concentrate)			

* Tonnes; figures do not include data from communist countries (Millbank 1987).

TABLE 10:4

DISTRIBUTION OF R&D FUNDING FOR CRITICAL MATERIALS BY TECHNOLOGY GOAL

TECHNOLOGY GOAL	FUNDING IN $1000		
	DIRECT	RELATED	TOTAL
Substitution	9,648	13,960	23,608
New Sources	10,258	2,523	12,781
Reclamation	560	1,850	2,410
Life Extension	28,010	3,060	31,070
Conservation	1,585	2,400	3,985
Total, all technology	50,061	23,793	73,854

TABLE 10:5

WEIGHT % OF STRATEGIC ELEMENTS USED IN THE MANUFACTURE OF AN F100 GAS TURBINE ENGINE

ALLOY	Cr	Ni	Al	Ti	Ta	Cb	Co
Cobalt base:							
MAR-M 509	23.4	10	–	0.2	3.5	–	55
MAR-M 302	21.5	–	–	–	9	–	58
WI-52	21	–	–	–	–	–	64
Nickel base:							
B1900	8	65	6	1	4.3	–	10
IN100	9.5	61	5.5	4.7	–	–	15
MAR-M-200	9	61	5	2	–	1	10

TABLE 10:6

STRATEGIC ELEMENTS IN COBALT- & NICKEL-BASE SUPERALLOYS USED FOR CAST TURBINE VANES & BLADES
(Composition given in weight percent)

ALLOY	Ni	Cr	W	Nb+Ta	Co
Cobalt base:					
L605	10	20	15	--	53.5
Ha 188	22	22	14.5	--	41.4
Nickel base:					
Hast X	Bal	22	0.6	--	1.5
IN 625	Bal	21.5	21.5	3.65	0

TABLE 10:7

STRATEGIC ELEMENTS IN Co- & Ni-BASE SHEET ALLOY USED IN COMBUSTOR, AUGMENTOR & TURBINE NOZZLES (wt%)

Alloy	Ni	Cr	W	Nb+Ta	Co
Co-base:					
L605	10	20	15	–	53.5
Ha 188	22	22	14.5	–	41.4
Ni-base:					
Hast.X	Bal.	22	0.6	–	1.5
IN625	Bal.	21.5	21.5	3.65	O

10.3. PRESENT ACHIEVEMENTS & FUTURE REQUIREMENTS OF COATINGS

Chapters 1 and 2 gave a broad outline of the high temperature areas and systems which required coating applications. Further requirements and present achievements of coating applications in the following areas are indicated in this section:

- Aerospace
- Combustion engines
- Energy conversion (MHD, solar, H.T. fuel cells, coal gasifiers)
- Industrial (e.g. petrochemical, glassmaking, machine tools)

Further development work is needed for coatings forming dense glasses, complex oxides or spinels, and self-healing coatings for longevity (Kear & Giessen 1984; Das & Davis 1988).

10.3.1 AEROSPACE:

Engines operating at $1000^{\circ}C$ and rocket nozzles at higher temperatures need refractory and oxidation resistant coatings. A several hundred degree temperature reduction is achieved by applying ceramic thermal barrier coatings. Thermal barrier coatings are an important area needing further research. Abradable coatings are needed to restrict gas escape around turbine blading and through labyrinth seals, at 700 to $1000^{\circ}C$, e.g. Ni-Cr-Fe-Al and aluminium bronze, containing BN (Sickinger & Wilms 1980).

Improved coatings are needed for Nb alloys for use above $1500^{\circ}C$, for Mo and Ta alloys $1650^{\circ}C$ and for W alloys above $1820^{\circ}C$. Creep occurs in Si20Cr20Fe coatings on Nb alloys at $1465^{\circ}C$, although scratches are self-healing by slow oxidation and cycling does not affect ductility. $MoSi_2$ on Nb forms a protective SiO_2 layer and is a successful coating because it has the same thermal expansion as the substrate. $MoSi_2$ on Mo is good up to $1760^{\circ}C$. V-(80Cr20Ti)-Si coating on Nb alloys reduces the tensile strength of 20 mil sheet by 25% at $25^{\circ}C$ and $1205^{\circ}C$ and fatigue 50% at $25^{\circ}C$, embrittling the substrate, but the bcc coating is ductile and resists air oxidation at $1260^{\circ}C$. CrSiTi coatings on Nb alloys resist 150 oxidation cycles at $1444^{\circ}C$, better than VSi_2 coatings. W barrier layers between Ni30Cr and Ni30Cr20W + 3 to 5% Al coatings on Cr5W0.1Y prevented interdiffusion with resistance to cyclic oxidation over 600 hours at $1150^{\circ}C$. ThO_2 and HfO_2 with a W diffusion barrier layer on Ta10W resist air oxidation up to $2482^{\circ}C$. Cr_5Al_8 and $(FeCr)_xAl_y$ coatings on Cr5W0.1Y were also satisfactory (Allen 1968). Claddings of Hf-24Ta-2Cr-1Si on Nb alloys and Hf-24Ta-1.2Cr-0.7B-0.12Al cladding on Ta-10W extends oxidation life 25 hours at $1482^{\circ}C$; minor elements decrease oxygen diffusion by occupying octahedral sites in HfO_2.

$SrZrO_3$, $Nd_2Zr_2O_7$ and $Yb_2Zr_2O_7$ are structurally stable coatings on W, good in vacuum, with no vaporization or reaction of the coating with substrate or O_2. On Ta10W, $MoSi_2/ZrO_2$ and WSi_2/ZrO_2 showed good spalling resistance to cylic oxidation and lasted longer than 23Sn5Al or 7Mo coatings (Allen 1968c). 20Cr20Fe60Si and CrTiSi coatings on Nb alloys are good in subsonic air at low presssure at $1538^{O}C$ (Allen 1968a,c). Bend tests showed Ir on Nb is ductile but on W and Mo is brittle (Allen 1968a). SiO_2 coatings on refractory metals have limited self-healing at low temperatures due to cracking of the glass formed. Otherwise, glassy oxides flow and accommodate mechanical strain. But SiO is evolved at low pressure. Pt- Ir coatings on aerospace alloys form volatile oxides but are useful due to boundary layer effects retarding the loss. Ir is satisfactory on graphite. Modified silicides performed better than FeCrAlY, Pt, Pd and Al on Cr-5W-0.1Y and were the best coatings in a series tested on Ta and W alloys. Large and complex Mo shapes can be silicided but joining problems and a tendency for brittle fracture and dependence of strength and ductility on a strain hardened state limits use (Nat.Acad. Sci.1970).

Good erosion and fatigue resistance is shown by 35W35Mo15Ti15V and CrTiSi coatings on Nb alloys at $1205^{O}C$. High fatigue strength plus oxidation resistance is achieved by CrFeSi coatings on XB88 (Allen 1968a). V-(80Cr20Ti)-Si coatings on Nb alloys resist oxidation at $2300^{O}F$ but the ductile bcc coating reduced the tensile and fatigue strength of the substrate by embrittlement (Allen 1968c). Cr ion implanted steel bearings are used in U.S. aircraft to increase corrosion resistance (Anon. 1984)

10.3.2. COMBUSTION ENGINES:

10.3.2.1. Gas Turbines:

Requirements include overlay coatings processes for low cost quality coatings, coatings for erosion protection which are compatible with overlay coatings and coatings processes for internal blade passages. Requirements for both alloys and coatings are reviewed (Kear & Thompson 1980; Foroulis & Petit 1976; Perkins 1982; Goward 1986; Stringer 1987; Rahmel et al 1985; Lacombe 1987; Miller 1987; Lewis 1987; Gurland 1988; Wadsworth et al 1988).

Thermal efficiency improvement and corrosion resistance requires coatings for gas turbine components. Metallic coatings reduce corrosion and ceramic types act as thermal barriers to reduce static component temperatures. Problems in adherence and thermal shock resistance of ceramic coatings prevents their use on rotating blades (Herman & Shankar 1987).

Alumina coatings on aluminised superalloys are successful due to the thin and thus adherent film with good mechanical properties. Failure is due to Al diffusion into the substrate reduced by using an Sn-Al coating in which Sn flow fills defects. But vaporization can cause breakdown. Oxides and carbides reduce high temperature creep of Nb alloys, but Ti regresses this by reducing these oxides (Nat.Acad.Sci. 1970). Aluminided Ni and Co alloys

(Anon. 1984) and Al-Pt, Cr-Al, Si-Al, Cr-Al/Pt and NiCoCrAlY on IN738LC and FX414 Ni and Co alloys have increased oxidation life. Some deleterious sigma phase formed in the Ni alloy, but no coating/substrate interaction occurred for the Co alloy. 25Cr3Al-10Ni5Ta0.2Y57Co coated IN738 showed no penetration in 13000 hours with 225 engine startups but uncoated IN738 showed 180 microns intergranular penetration . Pack aluminised and chromised IN100 developed stress due to the coatings (Betz 1979). Pack aluminised CrAl, SiAl, PtCrAl, and NiCoCrAlY coatings on IN738LC and FX414 were unaffected in their tensile, rupture and fatigue properties if the normal heat treatment was done after coating (Strang 1979). Good thermal stress resistance was achieved on aerofoils by ZrO_2-MgO and ZrO_2-Y_2O_3 coatings on superalloys and direction-ally solidified alloys (Pettit & Goward 1983). An inner NiAl or NiCr layer improved fracture properties in 7000 hours tests at $900^{\circ}C$ (Mozar et al 1973). Otherwise, thermal cycling de-bonds such coatings. The presence of salt melts however prevents use of coatings such as ZrO_2-12%Y_2O_3 on Ni-16Cr-6Al-0.6Y intermediate layer on superalloys, due to absorbtion of salt in the porosity and subsequent thermal shock failure. Better resistance is ob-tained with glassy coatings like $2CaO.SiO_2$ or with a MgO- NiCrAlY cermet, on superalloys (McClanahan et al 1974). Pt in coatings for gas turbines has recently been reviewed (Anon. 1983).

NaCl vapour spalls the protective alumina layer on CoCrAlY coated superalloys (Jones et al 1975), although adherent, and crack-free fine grain structures are achieved (McClanahan et al 1974). NiCrAlY with a Pt top coat on a Ni-Nb-Cr-Al directionally solidi-fied alloy gave protection for 1000 hours at $1366^{\circ}K$ in a rig test, with very good interdiffusion resistance to $1478^{\circ}C$. Elec-tron beam coating gave the best oxidation resistance and a long stress-rupture life and thermal fatigue resistance were achieved (Goward 1970).

Six and seven element MCrAlY-type coatings containing elements with widely varying vapour pressures can be produced, e.g.NiCoCr-AlYHfSi by EB-PVD. With a multi-source evaporation unit, coatings with composition graded from substrate to surface have been produced, as well as multiphase structures containing oxide dis-persion. Zirconia coatings are important in the combustion zones of aero gas turbine engines. There is a need to develop thicker coatings or monolithic materials. Further work is needed on production of strain-tolerant ceramic thermal barrier coatings for combustion zones.

Good mechanical properties are reported for some coatings in the series Ni/Al, Ni/C, Co/WC, NiCr/diatomite, Ni/$WTiC_2$, NiCr/Cr_3C_2, Ni/CaF_2 and NiO/CaF_2, on superalloys (Clegg et al 1973). The corrosion resistance of Cr-Al and Al on Fe and Ni base alloys improves with thickness but the rupture and fatigue strength falls (Fitzer & Maurer 1979). Other studies on aluminide coatings on Ni and Co superalloys (Pichoir 1979; Duret & Pichoir 1987; Martinengo et al 1979) recommend the presence of Y. A Ta barrier layer under Al-Cr coatings prevents interdiffusion from degrading the coatings, but these coatings are damaged by carbide intrusions from unburnt fuel. Al-Pt on Ni alloys gives increased oxidation life (3 times better than NiAl) but some substrate interaction forms sigma phase (Ubank 1977; Grunling et al 1987).

10.3.2.2. Diesel Engines:

Reducing heat losses increases efficiency but raises temperatures. Thus thermal barrier coatings are needed on valves, piston crowns and cylinder heads. To improve efficiency the engine can be operated uncooled, viz. the adiabatic engine (Fairbanks & Hecht 1987), requiring thermal barrier coatings such as zirconia. At least 3.5mm is needed to obtain an 80% heat loss reduction. This is beyond plasma spraying capabilities and thinner zirconia on a metal fibre layer has been suggested (Kamo et al 1979; Kvernes & Forseth 1987).

If cylinder liners are made from a solid ceramic, the piston rings need a wear resistant coating with compatible thermal expansion. Multicomponent plasma coatings can be self-lubricating up to 870°C (Sliney 1979), e.g. 30% nichrome, 30% Ag, 25% CaF_2, 15% glass (to prevent nichrome oxidation). Ag improves wear properties at lower temperatures (start-up). Stabilized ZrO_2 on a NiAl or NiCrAlY bond coating, graded and up to 0.8mm thick have perfomed well on piston crowns of marine diesels using normal fuel. Low quality fuel contains V which reacts with the Y_2O_3 or CaO stabilizers in the ZrO_2 and causes disintegration. MgO stabilizer is more resistant (Kvernes 1979). Large diesel engine piston heads plasma spray coated with self-sealing ceramic resisted sulphurous combustion gas attack up to 1500°C under severe thermal shock and fatigue stress (Perugini 1976).

Braze/cermet/ceramic coatings on steel (AISI-4320) gave very good corrosion resistance with hot wear and thermal shock resistance and good tensile and shear resistance (Houben & Zaat 1973). Pt, Ir and Pd are recommended for engine exhausts. W, Mo and piano wire coatings by wire explosion on mild steel and Al gave 5 times better adhesion than flame sprayed coatings (Kirner 1973). Cr_3C_2/Mo and Ti-Ni alloy with 3% free C as coatings on piston rings on marine diesel engines showed improved wear resistance compared with grey cast iron (Solomir 1973). Stabilized ZrO_2 coatings on steel components had big tensile strength differences (Kvernes 1983). Kvernes has recently discussed coatings for diesel engines (Lang 1983).

WC-12Co, 75CrC-25(80Ni20Cr), Mo and 80Ni-20Cr coatings on steel and Al-alloys are used for abradable seals in compressors, turbines and bores of hydraulic cylinders; plasma sprayed coatings are better than hard Cr. Sandwich coats perform better under thermal cycling (Malik 1973).

10.3.3. ENERGY CONVERSION:

10.3.3.1. Coal Gasifiers:

Fire-side corrosion of power station boiler tubes, especially with high Cl content coal, is a problem around $640^{\circ}C$. Higher Ni and Cr content claddings on stainless steel tubes reduce this corrosion. Flame sprayed coatings have not yet been very successful. Plasma sprayed coatings are difficult to apply in-situ. The shut-down cost of a power station plant is up to 50000 pounds per day (Flatley et al 1980); coating development, therefore, is certainly justified (Meadowcroft 1987; Colson & Larpin 1987; Debruyn et al 1987; Restall & Stephenson 1987).

The low oxygen (10^{-15}) atm and high S (10^{-8}) activity in coal and oil gasifier and coal fluidised bed combustors will cause sulphidation of Ni based alloys (forming low m.p. eutectics) and Fe alloys like FeCrAlloy are preferred. High pressure hoppers and valves are used for discharging hot char and ash at $1100^{\circ}C$ (but metal temperature probably about $550^{\circ}C$) and valve seat inserts of stellite, silicon nitride or boron nitride are used. Need exists for other materials and coatings. Electrodeposited TiB_2, CVD SiC and B diffused into Mo or WC/Co are very erosion resistant and thermal cycling resistant up to $700^{\circ}C$ (Hansen 1979) if at least 60 microns thick.

Heat exchangers are needed for waste heat recovery if SO_2 removal is done by cooling the gases. Coatings could protect heat exchangers. Plasma sprayed and then laser fused CoCrAlHf coatings (Packer & Perkins 1980) and slurry fused CoAl and CrAlHf (Dapkunas 1980) have good corrosion resistance but would be very difficult to apply to large heat exchangers. Claddings of FeCrAl-MoHf (Perkins & Packer 1979; Lathem 1980), FeCrAlloy A and Super-FeCralloy , are probably the most suitable protection. Wrought MCrAlY and MCrAlHf on gasifier alloys, are better than cast versions (Packer & Perkins 1980).

Si on Ni20Cr alloy at $900^{\circ}C$ in oxygen plus $80\%V_2O_5 20\%Na_2SO_4$ for 600 h, gave an 80% reduction in corrosion (Elliott & Taylor 1979; Wahl & Furst 1979). Cr on Nim 80H and IN738LC improved corrosion resistance in engine and crucible tests (Bauer et al 1979). On the more suitable Fe-based coal gasifier alloys, MCrAlY and MCrAlHf (M=Co, Fe or Ni), alumina-forming coatings were best (with 8-10% Al); mechanical properties were better with wrought alloys than cast (Packer & Perkins 1979). Laser surface fusion improved uniformity.

Some problems are reviewed (Robinson 1981; Southwell et al 1987; Meadowcroft 1987). A maximum corrosion rate of 50 nm/h is needed for 100000 hours tube life. Reduction of furnace wall corrosion by using finer coal particles, better distribution to each burner and increased secondary air to raise pO_2, have only given marginal corrosion reduction. Materials improvement is thus necessary and coextruded tubes have been evaluated. 18% Cr austenitic steel suffers from stress corrosion cracking on the water-side, so coextrusion is used to prevent its waterside exposure and present a ferritic carbon steel core to the waterside. 30000 hours satisfactory service is reported (Hara et al 1978). Further work is needed on coatings (cladding) for coal gasifier environments.

Aluminide coatings on steel resist hydrocarbon and sulphide atmosphere corrosion and increase oxidation resistance. They are expected to have wide applications in coal gasification and liquefaction plants. Aluminized steel is better than stainless steel where oxidation-carburization is the main surface degradation.

10.3.3.2. Nuclear Reactors:

Advanced gas-cooled reactor (AGR): UO_2 pellets in $20Cr/25Ni/Nb$ stainless steel tubing are cooled by CO_2 at 40 atm., reaching 750 to $850^\circ C$. CVD SiO_2 on pre-oxidised $20Cr/25Ni/Nb$ stainless steel inhibits carbon deposition and spalling at $825^\circ C$, in 6000 hour tests (Bennett et al 1981).

Fast breeder reactor (FBR): Moving fuel handling components need protection only against wear and fretting corrosion in liquid Na. These components normally contact the Na only when it is cooled to about $250^\circ C$ (from the $600^\circ C$ operating temperature). Ferritic and austenitic steels are not significantly attacked by liquid Na at $600^\circ C$.

High temperature gas-cooled reactor (HTR): The He reactor atmosphere restricts oxide formation, so sliding movements (due to differential thermal expansion) cause galling or adhesive wear due to self-welding. Coatings are needed to stop this up to $950^\circ C$ and for the reactor life duration. Plasma sprayed, D-gun and CVD ceramic and cermet coatings have been tested (Engel & Klenmann 1981); ZrO_2 and TiC performed well in wear tests up to $950^\circ C$ but ZrO_2 disintegrated at long times due to phase destabilization. Further research is needed.

Sliding wear and corrosion resistance are areas where ion implantation has contributed but the very small penetration depth is a limitation and time dependence of corrosion protection needs study (Gebhardt et al 1978). The beneficial effects achieved by ion implantation has been updated (Bennett 1983; Bennett & Tuson 1988/89). HTRs also have the problem of radioactive tritium diffusing out through the heat exchanger tube walls (Van der Biest 1979). Oxide films on superalloys are good tritium diffusion barriers. Studies are in progress on the use of Al_2O_3 and SiO_2 coatings.

Fusion Reactors: Inner walls of the stainless steel or Cu plasma vessel receive high energy particle bombardment and high thermal loading. Material thus eroded contaminates the plasma and cools it, more effectively for high atomic mass contaminants. Low atomic mass coatings are under study: TiC, TiB_2, B_4C and B (by CVD); TiB_2, Be and VBe_{12} (by plasma spraying); SiC (by sputtering); C and Be (by vacuum evaporation). The hottest parts are water cooled and are typically Cu-Cr alloy explosively clad with Ni-Cr alloy; surface temperatures are about $500^{\circ}C$.

General: The corrosion problems encountered in general are summarized by Quadakkers 1987. Ti, Mo, Zr, Nb and Ti-6Al-14V coatings on structural steel gave improved hardness with good adhesion (Muller 1973). TiC, VC, NbC and Cr_7C_3 coatings give good wear resistance (Child 1983). Sn on Zircalloy-4 prolongs the time taken for mechanical breakdown which occurs due to insufficient plasticity of the ZrO_2 layer (Hauffe et al 1975). Other successful coatings include SiO_2, CeO_2 and ZrSi (Bennett 1983; Niklasson & Granqvist 1982).

10.3.3.3. Magneto Hydrodynamic Converters:

A very hot fast gas jet is directed between the poles of an electromagnet. Presence of some alkali metal ions makes the gas conductive and electrodes collect the electric current thus induced. Co, CO_2, H_2O and N_2 from burnt coal, plus some K_2CO_3 at $2500^{\circ}C$ produce problems in the ducting similar to those in slagging gasifiers. Refractory linings, or water cooling, may protect metalwork, but electrodes (water cooled Cu covered with stainless steel and an electrically conductive ceramic thermal barrier) have corrosion problems. Stabilised ZrO_2 is conductive above $1100^{\circ}C$ but near the metal ZrO_2-Inconel cermet is needed to increase conductivity. Plasma spraying a graded composition minimises thermal mismatch effects but is slow to give the required 4mm thickness.

10.3.3.4. Solar Energy Converters:

Photovoltaic, photochemical or photothermal effects may be used. Solar heat collectors focus energy on a coated tube having a high absorption in the solar spectrum and low infrared emissivity. Coatings are oxides or semiconductors of controlled thickness, morphology and composition. Coatings are needed with long term thermal degradation resistance and good adherence under thermal cycling. The theories required for the optical properties of cermet coatings for selective absorption of solar energy have been outlined (Niklasson & Granqvist 1982). The theories are then used to interpret the evaluated complex dielectric function of coevaporated $Co-Al_2O_3$ cermet films. Finally these data are used to modify surface coatings with optimised spectral selectivity.

Radiative cooling was demonstrated for practical use with Si_3N_4 coatings on a reflective substrate. The dielectric function of evaporated Si_3N_4 is reported and employed to evaluate the obtainable cooling power and temperature difference (Eriksson et al 1982). Reactive ion plating is used to deposit InN for high efficiency solar cells (1.8 eV energy gap) (Takai et al 1982a,b). For effective collection and retention a solar collector must absorb at wavelengths below 2 microns and not radiate above 2 microns. Such selective absorption is obtainable by several methods: Mattox & Sowell 1974; Koltun 1971; Drummeter & Haas 1964. Suitable coatings include Si, Ge and PbS. Thickness control gives destructive interference at the wavelength of the solar maximum; energy reflectance is also reduced by anti- reflection coatings or by porous coatings. Cu can be blackened by a $NaOH + NaClO_2$ mixture and steels can be blackened by other treatments (Mattox & Sowell 1974). Dendritic surfaces have been developed for high temperature solar collectors (Pellegrini et al 1979).

10.3.4. INDUSTRIAL (Petrochemical, glassmaking, machine tools):

Coatings achievements cover a wide industrial range. Ceramic coatings protect graphite crucibles from corrosion by liquid Al; metal casting mold life is doubled by self-sealing ceramic coatings which decrease carburizaion. But for the present discussion, the petrochemical, machine tool and glassmaking industries have been selected.

10.3.4.1. Petrochemical Industry:

Most problems occur in oil and gas fired tube furnaces (Swales 1979; Edeleanu 1980). Fireside temperatures are about $1150^{\circ}C$. For strength, high alloy cast austenitic steels are used, giving adequate corrosion resistance. Carburization occurs in ethylene production. Carbon (coke) deposits thicken and tube temperatures need raising to maintain heat transfer. On reaching $1100^{\circ}C$ cleaning is essential. Progressive carburization, called 'metal dusting' in extreme cases, occurs. Plasma spraying inside straight tubes with Cr, using a tubular plasma head in Ar, improves carburization resistance (Arcolin et al1980). Subsequent saturation with ceramic paint fills pores. Sulphidation occurs in processes using CS_2 and H_2S.

Fuel ash (from residual oil) fireside corrosion is prevented by duplex centrifugally cast tubes, with an outer wall of 50%Ni50%Cr. Refinery furnace tubes are used at $550^{\circ}C$ in desulphurization plant with Al diffusion coatings, but if too large it requires coating to be done in sections which gives a welding problem of how to protect the welded zones (McGill & Weinbaum 1979; Meadowcroft 1987).

Burner plates at 10^6 kcal/dm^2 at 1700°C in oxidising and carburizing conditions are protected by ceramic coatings. Combustors of hydrocarbon crackers have been protected from overheating and corrosion by 3 mm thick self-sealing ceramic coatings and soot scrapers in hydrocarbon crackers were found to withstand the hottest flame areas (Perugini 1976; Herman & Shankar 1987; Miller 1987).

Most petrochemical processes can work below 600°C and there is little interest in coatings designed for higher temperatures.

10.3.4.2. Glass Manufacturing:

Pt alloys containing 5% Au, are not wetted by molten glass but have insufficient hot strength. Adding Rh for strengthening lowers chemical resistance and causes embrittlement. A dispersion of 0.1% ZrO_2 in 5% Au Pt avoids these problems (Heywood & Benedeck 1982). Contamination of glass melts gives (unwanted) coloured glass and Pt coatings are widely used in the glass industry. Alternative cheaper coatings are being sought. Linings of mullite, Ca and Mg zirconates and silicides have been tried and plasma sprayed NiAl with some success (Nat.Acad.Sci. 1970).

For spinning of glass fibres for thermal insulation purposes, molten glass is passed downwards through a Pt sieve to spread out the stream and then into a fast rotating Rh-Cr-based spinner about 30 cm diameter and 10 cm deep. This cylindrical spinner has about 20000 small holes in its edge and strong external vertically-downwards directed flames force the emerging glass fibres downwards. (Centrifugal force forces the molten glass through the holes in the spinner). The spinner lasts only about ten days and coatings may offer an extended life and be of great production benefit in reducing the 'down time' of the plant. So far no coatings have been developed for this purpose.

Inclusion of Pt particles in glass laser rods degrades performance and is due to PtO_2 formation (as Pt solubility is very low). This precipitates and decomposes to Pt on cooling. Removal of oxygen from the glass furnace atmosphere should solve this problem (Kubaschewski 1971).

10.3.4.3. Machine Tools Industry:

When tools are reground, the coating left on the unground face (cutting face, for drills; flank for millers) continues to give a significant impovement over the uncoated tool (longer life between regrinds and faster cutting). Tool costs can be reduced by a factor of 10, biggest improvements being in tough materials (reinforced plastics, abrasive composites, high strength alloy steels). Better finish is reported, often removing the need for reaming and deburring. Cutting speeds are increased between about 20% and 400% and feed rates between 10 and 100% (Boston 1983;

Hansen et al 1979). Chapter 7 gives an updated outline on wear in hard coatings.

10.4. HIGH TEMPERATURE COATINGS: PRESENT & FUTURE APPLICATIONS

Salient points which emerge from the preceding chapters and sections are summed up in this and the last section.

Coatings have to withstand complex conditions; it is essential that research programmes in different laboratories should be coordinated on a well-phased programme to test different properties of the same coating on a stand alone and in a coating/substrate configuration and that the same type of test should be agreed for each property. It is lack of compliance with these which has made it very difficult to present coherent histograms and comparison tables which are outlined in the present survey. Normalization of units in assessing and expressing physical properties, and degradation data is essential on both sides of the Atlantic and would greatly help inter-laboratory data output. In 1985 an international workshop was held to assess the field of high temperature corrosion. An international code and standardization must now emerge for future work to avoid ambiguity and uncertainty of grading coating performance. High temperature coatings technology has developed adequately to allow a standardization of test procedures, a number of which have been evolved during the course of research in the last 30 years.

It is clear that no one coating can offer all aspects of resistance. Also, coatings must be developed for specific substrates as integrated systems and cannot be adequately developed in isolation. Duplex and multilayer coatings must be studied in a graded programme to enable assesment of the influence of each of the coating layers.

Further clarifications in fundamental studies such as thermodynamic data, phase diagrams, diffusion - its thermodynamics and kinetics, corrosion kinetics, heat and mass transfer, and modeling are needed. Mechanical property measurements require to be standardized, viz. hardness, adhesion, creep, stress and fatigue. The effects of thermal treatment of coatings on the mechanical properties of substrates is a field yet to be quantified meaningfully. Thermal and mechanical fatigue is an area that needs clarification. Cyclic tests devised for each application and environment have to be formulated and standard test procedures agreed upon on an international level.

Coatings and substrates systems should be developed as substitutes for those containing strategically sensitive materials. Directionally solidified and mechanically alloyed substrates, and rapidly solidified systems have yet to be much investigated.

Several new analytical tools are now available in the fields of microscopy and spectroscopy, but with very restricted and limited availability and are expensive for access. Research, while staying competitive should also aim to be collaborative in order to optimise time, money and team efforts, if the end result is to be the development of efficient coating systems over wide fields of application and a maximisation of its users.

10.4.1. GENERAL PROPERTIES ASSESSMENT:

A coating system has to have/be,

- High resistance to oxidation and/or hot corrosion.
- Ability to form alumina, chromia or other protective oxides.
- Resistance to embrittlement, e.g. from carburization.
- Resistance to forming low melting eutectics.
- Resistance to forming vaporizing products.
- Adherence.
- Ductility.
- Resistance to mechanical and thermal shocks and fatigue.
- Compatibility with base alloy in composition, thermal expansion and DBTT.
- Low porosity.
- Low coating/substrate interdiffusion.
- No detrimental element transfer or phase formation.
- Self healing if damaged.
- No degrading effect on substrate mechanical properties.
- Possibility of repairing damaged coatings and of welding.
- Multiple sources of coating materials and alternates if strategic.
- Overall cost-effectiveness.

(Note that each application has its own list of priorities)

10.4.2.. FUTURE WORK IN COATINGS FOR GAS TURBINES:

Corrosion and degradation are affected by several factors. The following aspects appear to need further investigation:

- effects of elements like Ta, Ti, Nb, Si, Mn, Zr, B, Cu, P, Zn and Pb. Effects of these must be linked to specific failure modes.
- beneficial effects of Si and Ta not yet clear; siliconising embrittles coatings; electron beam evaporation of Si on CoCrAlY appears promising.
- re-coating processes without much loss of substrate.
- effect of corrosive salts on thermal barrier coatings, e.g. ZrO_2 and additive oxides added to it for structure stabilization, e.g. CaO, MgO, Y_2O_3. Other oxide additives must be sought as alternatives, if possible, and tests conducted to observe melt fluxing effects over the turbine operating and cycling temperatures.
- development of better spalling-resistant thermal barrier ceramic coatings.
- research to stabilise oxide coats formed during corrosion, to

seek an alternative to alumina.
- coating/substrate mechanical interactions causing strain in aerofoils.
- study of the role of Y in MCrAlY to promote adherence in cyclic exposures. Investigation of other elements and alloy systems to check or confirm 'monopoly' role of Y. Strategic importance of Y to be assessed.
- study of cyclic corrosion loss mechanisms.
- study of the effect Cr, Al and Ti:Al.
- study of erosion-corrosion resistance.
- effect of salt contamination on creep behaviour.
- investigation of duplex and multi-layer coats.
- behaviour of coatings on directionally solidified, PM and MAP alloys
- composite system investigation
- rapidly solidified metallic system assessment.
- metal-ceramic bonding and temperature monitoring coated devices
- cooler component temperatures and steeper thermal gradients designed for higher engine efficiency have to be accommodated, and coatings for internal cooling passages developed.

10.4.3. COATINGS FOR SUPERSONIC & AEROSPACE HARDWARE:

- improved coatings needed for Mo and Ta alloys above $1650^\circ C$ and for W alloys above $1800^\circ C$.
- improved alloy development for Nb- and Zr- systems.
- coatings and composite systems for PM and MAP alloys.
- coatings and claddings match for the carbon-carbon composite.
- rapidly solidified material adaptation.
- metal-ceramic bonding and temperature monitoring coated devices

10.4.4. COATINGS FOR DIESEL ENGINES:

- more studies on plasma sprayed ZrO_2 coatings, and other types such as Mo and Cr carbides.
- denser and smoother wear resistant coatings for piston rings.
- coatings with controlled thermal conductivity.
- coatings for adiabatic diesel engines.
- erosion and fretting resistant coatings & seals.
- temperature monitoring coated devices.

10.4.5. COATINGS FOR COAL GASIFIERS:

FeCrAlY compositions have the lead as coating or cladding materials. Other candidate materials including those with dispersed oxides should be studied. A wider temperature range and more thermal cycling studies should be investigated.

Studies of component gas pressures and compositions in the fluidised bed combustor and studies which elucidate the role of deposits will support and clarify the development of coatings.

Studies to explain anomalies such as why Incolloy 800H is better in pressurised than in 1 atmosphere fluidised beds and other mechanistic studies are neccessary to provide clarifications for the development of future coating compositions.

10.4.6. COATINGS FOR NUCLEAR REACTORS:

- improved coatings for fuel cladding materials.
- improvements in borides and carbide coatings.
- further development in wear-resistant seals.
- further work on oxidation and carburization resistant barriers.

10.4.7. COATINGS FOR POWER PLANT SYSTEMS:

Development for-

- higher gas inlet temperatures.
- directionally solidified eutectics.
- refractory metal fibre reinforced superalloys.
- ceramic coatings like Si_3N_4 and SiC.
- higher ductility and toughness.
- develop coatings with chemical and mechanical compatibility with the substrate.
- combine ceramic coats and tailored cooling technology to allow $1500^{\circ}C$ to be exceeded with a capability of 100 N/m^2 stress to rupture in 10^5 hours.

10.4.8. COATINGS & SOLAR POWER:

Coatings are required for high temperature heat exchangers (gas cooled, liquid metal or salt cooled, steam cooled, oil cooled) and for mid-temperature solar absorbers. Coatings must tolerate daily thermal cycling. Bimetal claddings or overlays seem likely candidates; stainless steels or superalloys may have to be cost-justified. Absorber coatings of black Cr, Ni or Co with high selective absorbtion and low emissivity are required. Abrasion resistance would be a necessary feature.

10.5. COATING PRODUCTION METHODS: FUTURE WORK

10.5.1. EVAPORATION:

Process refinement is needed in EBPVD for :- avoidance of spits from the melt pool which degrade coatings, corrections for the occurrence of columnar defects, and, work on the kinetics of depositing atoms, its effect on coating quality, and laser sources for evaporation.

566

Electron beam coating requires further work on increase of deposition rate and cost reduction. Further studies on possible structural variations in thick ceramic coatings and effects of reactant pressure variation and substrate orientations are needed.

10.5.2. SPUTTERING:

Process control needs further research, including deposition rate monitoring and gas pressure and composition measurements. Magnetron ion plating and sputtering and sputter ion plating are important research areas. Contamination and interreaction effects during processing have to be minimised. More studies of hardness and yield strength of deposits, and non-stoichiometry effects in sputtering process are also necessary.

10.5.3. ION IMPLANTATION:

This is a rapidly developing field, but much of its beneficial effects are confined to oxidizing and carburizing media and in tribology. More developmental work is needed in other high temperature areas for application in aerospace alloys. More information is required on equipment and processing costs.

10.5.4. PVD IN GENERAL:

This includes ceramics, metals, alloys and cermets; and, thin films deposition.

The present uses of PVD coatings in industry on a wide spectrum are (Bunshah 1984):

- in microelectronics, viz. Al, al-Si, GaAs, SiO etc.,
- for heat insulation and decorative purposes, e.g. al on polymer films,
- Al on kovar for lead frame manufacture,
- as insulating films for capacitors, e.g. Al and Ta,
- refractory compounds such as MgF_2 in the optics industry,
- coating of metals and compounds for architectural glass,
- MCrAlY coatings for high temperature gas turbine applications in marine and non-marine environments,
- Al coating on steel strip as a replacement for tin-plate,
- Ni on perforated Fe-strip for Ni-Cd batteries (in USSR),
- Cr on glass for masks in photolithography (in Japan),
- Ti and other metal coatings on steel for corrosion resistance (in USSR),
- ultrafine metal powders and alloys for various applications, viz. magnetic memory devices,
- wear resistant coatings, and many others.

Future applications include further development in most of the areas discussed in the foregoing chapters including both high temperature and wear resistance.

In general, these will include,

- new alloys of fine grain size, high stength and toughness for ODS alloys, laminate and composite materials,
- thin corrosion resistant films of metals and alloys, e.g. Ti,
- abrasion resistant graded ceramic layers on tool materials,
- superconducting materials
- bio-medical materials and materials of controlled porosity,
- amorphous materials,

Investigations and further development are needed in,
- high speed high density LSI and VLSI circuits which have already made rapid progress,
- development of electron beam technology and x-ray lithography for patterning sub-micron geometries.
- stability of layered films and research on interdiffusion.
- improved computer control to achieve precision coatings.
- MCrAlY coatings can only be applied by EBPVD; more alloy compositions have to be explored both for initial application and renewal processes.

10.5.5. CHEMICAL VAPOUR DEPOSITION:

Coatings by the CVD (and other) process have not been normalised to standard specifications, unlike those applied by the aqueous electroplating industry. A working booklet on process parameters, terminology, service specifications and standards appears to be necessary. Standardization of test and inspection methods is also desirable.

In the CVD industry, coating damage, separation and repair are still a problem. Preferential precipitation and decomposition of individual elements impede novel CVD processes, although in some cases this has been turned to advantage. Substrate-container interaction is also a continuing problem.

PACVD and RACVD are particularly of interest since CVD coating temperatures can be lowered from about 800°C to about 300°C by these methods. However, more work is needed in both these areas in the development and characterization of microstructures and service life of coatings produced.

Ceramic-metal, crystalline-amorphous combinations have shown superior performance over single materials. Considerable work is needed to understand in situ and separate deposition patterns, heat treatment and diffusion effects on the physical, mechanical and chemical/corrosion properties of such materials. Duplex and triplex coatings have to be investigated in more detail.

Work on SiC and Si_3N_4 is needed to investigate low deposition temperatures and their effect on the coating quality and bonding, especially PACVD and PECVD methods. A method of bonding, applicable to silicon nitride is transient liquid phase bonding using oxy-nitride glasses which melt and wet Si_3N_4 (Loehman 1983). Application of this needs to be extended to achieve metal/silicon nitride bond. Silicon nitride has been bonded to ZrO_2 without use of pressure, via a solid state reaction and the presence of $CaSiO_3$ (Yee 1978). CVD of ZrO_2 to protect Si_3N_4 under well-bonded conditions will be worth a further investigation.

Heat and mass transfer studies have been sporadic and often worker-attitude oriented. Results are thus not comparable between one research group and another. More information is needed on erosion test procedure on post-coated materials, especially those produced by pack, slip and slurry processes. Research is needed to develop CVD below $800^{\circ}C$ for TiN, alumina and other wear-resistant coatings for steel.

Work records of coating thickness against coating service life in the working environment would be a useful data source for future planning, design, production and testing.

10.5.6. THE FUSED SLURRY PROCESS:

This process should be developed to get better and diverse coatings with B_2O_3 and modified SiO_2 by additions of oxides like Nb_2O_3. Aluminizing by slurry, hot dip and metalliding, is not well developed. Aluminized and chrome-aluminized Incoloy 800 are used for hydrogen production at 1073K by H_2SO_4 decomposition; studies on materials behaviour in these conditions are required.

10.5.7. LASER ASSISTED COATING:

All the laser assisted processes, viz. LACVD, LECVD, LPACVD, LAPVD, LEPVD, etc., need futher research to optimise conditions and obtain pore-free coatings. Use in CVD (LACVD) is an important area. Some problems are: the laser melting of highly reflective alloys e.g. Al, the relation of heat and mass flow to other process parameters, detailed studies of laser assisted alloying, achievement of uniform smooth deposits, effects of fast quenching. Work on laser assisted coating production methods has shown promise and deserves support for wider investigation. The mechanism and evaluation of plasma and keyhole formations need to be better understood and LSA parameters systematized better. Effects of rapid quenching on coating properties and during service achievement of more uniform and smooth laser deposits and safety handling procedures need to be looked at.

The prospects of employing lasers in the coating industry is linked directly to its cost effectiveness. It is ideal for clean, precise, miniature, controlled deposition and repair work. Small components are easily handled with negligible distortion and damage. Laser welding competes well with electron beam welding in the superalloy area and thin sheets (<3mm).

More work is needed in the following areas:

- Welding metals of high reflectivity eg Al alloys.
- Mechanism evaluations on plasma and keyhole formations.
- Heat and Mass flow-transfer characteristics links to other process parameters.
- A systematic study of LSA. Work so far has been more on the qualitative platform. Substrate purity/deposit effects, quality and form of deposit, powder particle size vis-a-vis the focussed or defocussed beam size need more study.
- Achievement of uniform and smooth deposition.
- More detailed substrate-deposit-process interaction and characterization.
- Effect of rapid quenching.

10.5.8. SPRAYING METHODS:

Plasma spraying: Coating adhesion aspects clearly need further research - e.g. the effect (on adhesion) of sputter cleaning before plamsa spraying. Research on problems in the plasma spraying of WC + Co and methods of reducing porosity of plasma sprayed coatings are required.

Low pressure plasma spraying needs developing for higher quality coatings as particle velocity can be increaseed. Nitridation of materials during plasma spraying has not been well clarified.

Liquid metal spraying: The Osprey process is excellent and builds up very thick coatings in very short times. Further investigation is desirable.

Thermal barrier coatings, in general, need further development work, both in production technology and characterization. Erosion -corrosion resistant thermal barrier coatings are needed to prevent overheating of metal parts in engines and heat exchangers, for which both coating formulation and deposition methods are needed. Thicker zirconia coatings and more strain-tolerant coatings are needed for combustion zones of aero gas turbine engines.

10.5.9. WIRE EXPLOSION:

This produces excellent coatings and deserves technical clarifications and further research.

10.5.10. SOL-GEL CERAMIC COATINGS:

Sol-gel coatings require more process clarification. More published literature on their utility as ceramic modified coatings on superalloy, ferrous and light alloys is necessary.

10.5.11. DIFFUSION BONDED COATINGS:

A well-controlled and reproducible diffusion bonded coating process requires a good documentation of fundamental diffusion data which are often lacking. The role of interlayers, and their influence on thermal and mechanical properties require much further research. More fundamental work is needed on diffusion of CrTiSi coatings on Nb alloys. Further work on the production of cermet, composite and duplex coatings is desirable, e.g. Al_2O_3-CoCrAlY.

10.5.12. AMORPHOUS ALLOYS:

This is a wide open field where coatings are concerned, surface modification methods with an aim to produce these metastable structures are proving their usefulness in high temperature applications. Microstructure modification in relation to amorphous alloy formation, thermal and mechanical effects, iso- and cyclic conditions, tribology applications and nearly all coating property parameters require definitive work in this context. Carbide, P- and B-systems, laser and electron beam glazing vs amorphous surface alloying require further study.

10.5.13. WEAR RESISTANT COATINGS:

- high temperature wear need controlled investigations and data.
- the effect of surface heat treatment on wear needs study.
- hot hardness is related to wear resistance and needs study. An international quantitative unit must be formulated.
- laser processed coatings, found good in laboratory studies, need to be developed in practice, e.g. for piston rings.
- the effect of oxidation above $500^{\circ}C$ on carbide and boride coating life needs study; many applications exist where the local transient temperature can cause oxidation, e.g. hot forging, extrusion dies, die casting, metal cutting, engine and valve parts and dry sliding in magnox reactors.
- further development of hard materials for use in corrosive environments (e.g. valves for coal gasifiers).

10.5.14. EROSION RESISTANT COATINGS:

Standardization of tests is essential if comparisons of results produced by different investigators are to be possible, e.g. erodents, particle size and hardness, velocity and frequency of particle impacts, coating thickness and temperature, mechanical properties of coating and substrate and environment prevailing.

10.5.15. MISCELLANEOUS:

Coatings are needed for use in the glass industry, especially for glass fibre spinning. Coatings research to save strategic metals is inhibited by little incentive for firms to spend funds on something which would only be used in an emergency (cut-off of strategic metals).

Much research on gas turbines has helped in an understanding of the problems and a recognition of the hazards involved which has provided a better start for material modification in coal gasifier systems. In both the industries, better substrate materials have been developed and tested but appear to be steadying off to give way to coating development and substrate conversion work.

Mechanically alloyed products have caught considerable attention but more data are necessary in their performance at high temperatures under various environments, loads and other service conditions and porosity problems. Laser and electron beam glazing and other surfacing finishes must be defined.

Although some noteworthy service records have been achieved with the current substrate/coating complexes, the danger must be recognised when such components are forced to perform under adverse conditions, such as obligatory use of an inferior fuel. Investigations are necessary on documenting service performance and projected life under such environments.

10.5.16. REPAIR & JOINING METHODS:

Repair & joining techniques for coated materials are always in need of further research. Some turbine blades in service have been re-coated 12 times, showing the importance of this aspect. The uncoated blade costs hundreds of pounds sterling and so is worth recoating many times if necessary.

Sandwich joints of $Si_3N_4/Zr,Nb,Zr/Si_3N_4$ have been made at $1150^\circ C$ to study crack growth due to stress. Below $900^\circ C$ adherence does not occur. Adherence, as a whole needs quantifying with a choice of specified tests. Research is needed on multilayer coatings such as the $MoSi_2$-Mo-Ni-steel substrate coating system which has excellent adherence due to the multilayer structure. Further work is desirable on coating of already corroded metals. Methods of coating particles, e.g. by sol-spray in a fluidized bed are promising and need further work. E.g. Al_2O_3 coated Cr particles. Research is needed on intermediate layers to prevent degradation of coatings by diffusion and other effects.

Few measurements exist of the work of adhesion or adhesive energy, which elucidate wear properties of, e.g. Co-cemented WC hard composites. The thermodynamics of this is described and reviewed (Levinstein et al 1971) and it is clear that interfacial energy measurements are an important fundamental area needing research for improved metal-ceramic coating systems. Porcelain enamelling is a well established technology for bonding a glassy coating onto steel. Theories of the adhesion are reviewed with the conclusions that there is mechanical keying by metal dendrites into the glass and also the glassy interface is saturated by FeO (which, when in this solution can not be reduced by the substrate Fe). Adhesion oxides (CuO or MnO) or Ni are often used as an interlayer. Studies of adhesion promoters are needed for all metal-ceramic coating systems. Accumulation of H_2 at the interface (from H_2O reduction) causes a problem which the presence of Ni solves by catalytically atomising it and so promoting its diffusion away into the steel; the role of adhesion oxides may be to stop the H_2O reduction. Research is needed in this area. Flame sprayed interlayers may improve adherence (Leeds 1966). The poor tensile behaviour of ceramics and reactions of SiC coatings with NiCrAl are reported at 700-1150°C. Interlayers are needed to prevent eutectic formation at above 700°C. Barrier layers are needed to prevent metal silicide formation (from silicon-containing coatings) at the metal/coating interface.

Ways of reducing oxide spalling are needed. Improved adhesion is vital for many types of coatings and methods for achieving sound adhesion have to be systematized.

10.6. COATINGS PROPERTIES – FURTHER WORK

Codeposition of CVD pyrolytic C with other materials can prevent thermal cycling spalling and this needs more detailed investigation. Fibre-incorporated CVD composites are materials with much potential applicaton. More work is required to prevent the hottest areas of MgO-stabilized zirconia coatings from degrading fastest due to MgO precipitation.

Improved diffusion barriers are in demand. Bondline effects and clarifications on how and why bondcoat interlayers adhere well to ceramics are needed, particularly where property monitoring during service is contemplated. Research is also needed on interfacial energy data. Studies of adhesion promoters are needed for all metal-ceramic coating systems. The adverse effects of surface embrittlement and inclusions on adhesion needs further data documentation.

Research is needed to predict the effect of specific coatings on given substrates and to develop a range of coating alloys, each optimised to combat a particular environment. Coatings forming dense glasses, complex oxides or spinels are needed, self-healing for long life. More studies relating microstructure to mechanical properties have to be undertaken.

A number of phase diagram, thermodynamic and diffusion data are very badly required and much support for this field of fundamental studies must come from both the industrial and governmental research areas, because of their potential use in developmental work.

10.7. TESTING METHODS – FURTHER WORK

Most NDT methods require reliable comparative data and ready reckoner tables for site testing and evaluations. Ultrasonic tomography needs development for practical testing. Acoustic and optical holography, thermal wave analysis and controlled signals (CS) ultrasonics teqniques, would benefit from more research. Research is needed to develop shear adhesion tests.

Reliable correlations between accelerated tests, predicted life, actual service record and performance have to be established. Raman spectroscopy requires further attention as a characterizing tool.

In conclusion, a number of new coating methods have evolved and older methods have been refined. New analytical tools are introduced. Even as this is book is being published three more conferences have been held. Coatings are a vital section in all phases of advanced technology. The development and evaluation of coatings are expensive and there is now a need, more than ever, to conserve resources, optimise test programmes and for a greater coordinated dissemination of information. The contribution that can be made by the University work force is of tremendous potential and value-for-money; the industry is urged to take a more active interest and accord support.

BIBLIOGRAPHY

GENERAL: An on-going source of coatings references is "Bibliography on the High Temperature Chemistry & Physics of Materials" published by M.G. Hocking & V. Vasantasree (Editors) for International Union of Pure & Applied Chemistry (IUPAC). Copies are available quarterly at £14 (US $24) per year (4 issues).

Address: M.G. Hocking, (IUPAC), Materials Dept, Imperial College, London SW7 2BP, UK.

NOTE: The Bibliography below is listed in **reverse chronological order.**

'Surface Modification Technologies', 3rd Intl. Conf., Metal. Soc. AIME & Centre Suisse d'Electronique et de Microtechnique S.A., Neuchatel, Switzerland, Aug. 28 to Sept. 1, (1989)

'Metallurgical Coatings', Intl. Conf., April 17-21 (1989), San Diego, California, USA

Proc. Inst. Metal Finishing Annual Conf., April 11-14 (1989), Brighton, UK.

'Technology & Markets for Inorganic Coatings in Western Europe', Briggs J., Frost & Sullivan UK, (1989)

'The Role of Active Elements in the Oxidation Behaviour of High Temperature Metals & Alloys', Conf. Proc., Dec. 12-13, (1988), Petten (N.H.), The Netherlands.

'Ceramic Composites & Coatings', IBC Technical Services Ltd., IBC Ho., Canada Rd., Byfleet, Surrey KT14 7JL, UK (1988).

'Advances in Coatings & Surface Treatment for Corrosion & Wear Resistance', Intl. Conf., 6th-9th Sept. (1988), Newcastle-Upon-Tyne, UK

'Ion Beam Modification of Materials', 6th Intl. Conf., Tokyo, Japan, 12-17 June, Hall S-K. (Ed.) (1988).

'Ion Implantation Technol.', 7th Intl. Conf., Takagi K. (Ed.), Kyoto,

Japan, 7-10 June (1988)

'Refractories, Their Properties & Applications', Course No. 8806-309, Center for Professional Advancement, Plestrinastraat 1, 1071 LC, Amsterdam, The Netherlands, June 14-17 (1988).

'Oxide/Metal Interface', Conf. Proc., Mat. Sci. & Technol., $\underline{4}$, (5) (1988)

'Hot Isostatic Pressing of Materials: Applications & Developments', Intl. Conf., 25th-27th April (1988), Antwerp, Belgium

9th Intl. Conf. on Vacuum Metallurgy featuring 15th Intl. Conf. on Metallurgical Coatings & 1988 Vacuum Metall. Conf. on Special Melting, 11-15 April, (1988), San Diego, CA, USA; see J. Vac. Sci. & Technol. issues in 1988.

'ASM Metals Handbook', Lyman T. (Ed.), 8th ed., vol. 1, (1988)

'Computerized Metallurgical Databases', Cuthill J.R., Cokcen N.A. & Morral J.E. (Ed.), The Metall. Soc. AIME, Cincinnati, Ohio (1988), pp 250

'Erosion Wear in Coal Utilization', Raask E., Hemisphere Pub. Comp., NY, (1988)

'Superalloys 1988', Reichman S., Duhl D., Maurer G., Antolovich S & Lund C. (Ed.), The Metall. Soc. AIME, Seven Springs Conference, Champion, PA, (1988) pp. 1000

'Surface Modification Technologies', Sudarshan T.S. & Bhat D.G. (Ed)., STM Distr. Ltd., Enterprise Ho., Ashford TW15 1XB, UK (1988).

'Tomorrow's Materials', Easterling K., Inst. of Metals, London UK, (1988)

34th Natl. Symp. Am. Vac. Soc. 1988, J. Vac. Sci. & Technol., A $\underline{6}$, (3), (1988)

'Metallic Corrosion', 10th Intl. Conf., Nov. (1987), Madras, Vasu K.I. (Ed.) CECRI, Karaikudi, India, (1987).

'What are the New Materials?', Metallurgia, $\underline{54}$, (11), 510 (1987)

'UK Corrosion 87', Proc. of the Conf., Brighton, Oct. (1987), Inst. Corr. Sci. & Technol., Vol. 2, (1987).

1987 Annual Intl. Seminars on Surface Engg. & Coating Technol., Davis Congress Centre, Switzerland, Sept28 -Oct. 2 (1987)

'National Thermal Spray Conf. & Exposition NTSC87', Orlando FL, Sept 14-17 (1987)

'Rapidly Quenched Metals', Proc. 6th Intl. Conf., Montreal, Aug. 3-7 (1987). Publ. in Mat. Sci. & Engg. $\underline{97, }$ $\underline{98}$ & $\underline{99}$, in 3 parts (1987)

'Surface Science ECOSS-9', Proc. of the 9th European Conf., Lucern, Switzerland, April 13-16 (1987), publ. in Surface Science, 189/190, 1 (1987)

'Metallurgical Coatings', 14th Int. Conf., San Diego CA, USA, March 23-27 (1987), publ. in Surf. & Coatings Technol., $\underline{32}$, & $\underline{33}$, (1987) & in Thin Solid Films, $\underline{153}$, & $\underline{154}$, (1987).

'Advances in Surface Treatments', Niku-Lari A. (Ed.), Pergamon, UK (1987)

'Recent Developments in the United Kingdom in Ion Implantation Equipment for Engineering Components', Contact AERE, Harwell, UK (1987).

'Advancing Materials Research', Psaras P.A. & Langford H.D. (Ed.), Nat. Acad. Press. (1987)

'Ion Plating Technology - Developments & Applications', Ahmed N.A.G., John Wiley & Sons NY, (1987)

'Metal Bulletins Prices & Data, 1987', Millbank P. (Ed.), Bull. Books Ltd. (1987) Park House, Worcester Park, Surrey, KT4 7HY, UK

'Thermal Expansion', 'Non-Metallic Solids', Vol. 3, 'Thermophysical Properties of Matter', Toulokian Y.S., Kirby R.K., Taylor R.E. & Lee T.Y.R. (Ed.), IFI, Plenum, (1987)

33rd Natl. Symp. Am. Vac. Soc. 1987, J. Vac. Sci. & Technol., A 5, (1987)

'Ion Plating & Allied Techniques', Proc. of the Intl. Conf., Brighton, (1987)

'Scanning Tunneling Microscopy STM86', Proc. of the 1st Intl. Conf., Surf. Sci., 181, 1 (1987)

'High Temperature Coatings', Proc. Symp. Metall. Soc. AIME, Orlando, FL, Oct. (1986), Khobaib M. & Krutenat R.C. (Ed.), Met. Soc. AIME (1987).

'CRC Handbook of Chemistry & Physics', Weast R.C. (Ed.), 67th Edn. 1986-1987, CRC Press, FL (1987).

'Ion Beam Modification of Materials', Proc. 5th Intl. Conf., Cantania (1986), Nucl. Instr. & Methods, B19/20 (1987).

'High Temperature Corrosion of Materials & Coatings for Energy Systems & Turboengines', Proc. of the 1st Intl. Symp., Univ. Provence Marseille, France, July 7-11 (1986), Mat. Sci. & Enggg., 87 & 88, (1987).

'ASM International Materials Week 1986', Orlando FL, USA, Oct. 4-9 (1986)

'Ion Beam Modification of Materials', Campisano S.U., Foti G.,Mazzoldi P. & Rimini E. (Ed.), Proc. 5th Intl. Conf., Catania, Italy, 9-13 June (1986).

'Metallurgical Coatings', 13th Intl. Conf., San Diego CA, USA, April (1986), J. Vac. Sci. & Technol., A 4, (6), (1986)

'Materials in Aerospace', Proc. of the 1st Intl. Conf., Royal Aeronautical Soc., London, April (1986)

'PVD 86', Proc. Intl. Conf., Darmstadt FRG, March 11-12 (1986)

'Prediction of Materials Performance in Plants Operating with Corrosive Environments', Proc. Intl. Conf., Cranfield Berks., Feb. (1986)

'Extended-Life Oxidation Protection for Carbon-Carbon Composites', McCormick E.S., AFWAL Contract F33615-85-C-5044 (1986).

'Ion Plating & Ion Implantation', Hochman R.F. (Ed.), Proc. ASM Conf., (1985), Publ. (1986).

'Modern Metallographic Methods for Solving Diffusion Problems', Hays C. & Anastas D., Metallography (1986)

'Plasma Deposited Thin Films', Mort J. & Jansen F. (Ed.), CRC Press (1986)

Proc. of the Marine Eng. Conf., Washington DC, May (1986), Soc. of Automotive Engs NY.

'Fundamental Aspects of High Temperature Corrosion', Proc. Symp., Shores D.A. & Yurek G.J. (Ed.), Electrochem. Soc., Pennington, NJ (1986)

'Rapidly Solidified Alloys', Proc. MRS Symp., Taub A., Giessen B.C. & Polk D. (Ed.), MRS, Boston (1986)

'Interaction of Molten Salts & Metals', Proc. Conf., York (1986),

32nd Natl. Symp. Am. Vac. Soc. 1986, J. Vac. Sci. & Technol., A 4, (3), (1986)

'Protective Coating Systems for High Temperature Gas Turbine Components', Proc. Conf., Royal Soc., Nov. (1984), Mat. Sci. & Technol., 2, 193 (1986)

'Ceramic Coatings for Heat Engines', Symposium C, in Conf. on 'Advanced Materials Research & Development for Transport', European Materials Research Soc., Nov. 26-28 (1985)

'Recent Developments in Surface Coating and Modification Processes', Seminar organized by the Tribology Group of the Inst. of Mech. Engg., 10 Oct. 1985.

Proc. of the 32nd Automotive Technol. Dev.Contractors Co-ordination Meeting, Soc. of Automotive Engineers, Warrendale PA, Oct. (1985)

'Coatings for High Temperature Oxidation Resistance', Proc. ASM Symp., ASM, Toronto, Oct. (1985)

'Laser Surface Treatment of Metals', Draper C.W. & Mazzoldi P. (Ed.), Proc. Mtg. Italy 2-13 Sept. (1985), NATO ASI Series E, Appl. Sci. No. 115, Martinus Nijhoff Dordrecht, (1986).

'Continuous Coating', Proc. Conf., London, (1985), Inst. of Metals (1985)

'Recent Developments in HIP Equipment', Metallurgia, $\underline{52}$, (8), 316 (1985)

NASA Thermal Barrier Coatings Workshop, May 21-22 (1985)

'Metallurgical Coatings', 12th Intl. Conf., Los Angeles, USA, April (1985), J. Vac. Sci. & Technol., A $\underline{3}$, (6), (1985)

'High Temperature Alloys', J.B.Marriott et al (Ed.), Elsevier (1985)

'Ceramic Coatings for Heat Engines', Proc. Materials Res. Soc. Meeting, Strasbourg, (1985), Les Editions de Physique, Paris (1985)

'High Temperature Alloys', Peichl L., Wortmann J. & Raetzer-Scheibe H.J. (Ed.), Intl. Conf., Petten Netherlands, (1985)

31st Natl. Symp. Am. Vac. Soc. 1985, J. Vac. Sci. & Technol., A $\underline{3}$, (3), (1985)

'IPAT 85 Ion & Plasma Assisted Techniques', Proc. Conf., Munich, (1985)

'Surface Modifications & Coatings', Proc. Conf., Toronto, Am. Soc. Metals (1985)

Proc. 28th Annual Meeting of the IPC, New Orleans, LA, (1985)

'Rapidly Quenched Metals', Proc. 5th Intl. Conf., Steep S. & Warlimont H. (Ed.), North Holland (1985)

'Materials for High Temperature Environments', Proc. Conf., EEC, Petten (1985)

VIII Congress European de Corrosion, Nice, 71-1 (1985).

Proc. Conf. Heidelberg, Sept.(1984)

Proc. 9th Intl. Cong. on Metallic Corrosion, Toronto, NRC, Canada, (1984), 4 Vols.

'Protective Coatings & Their Processing - Thermal Spray', Nicoll A.R., CEI Course on High Temperature Materials & Coatings, Finland, June (1984)

'The Environment High Temperature Oxidation & Hot Corrosion', Nicoll A.R., CEI Course, on High Temperature Materials & Coatings, Finland, June (1984)

'International Workshop on Critical Issues Concerning the Mechanism of High Temperature Hot Corrosion', Bad Honnef FRG, May (1984), Rahmel A., Wood G.C., Kofstad P. & Douglass D.L. (Ed.), Oxid Met, $\underline{23}$, (5/6), (1985).

'9th Intl. Conf on CVD', Robinson M.C.D., Van den Brekel C.H.J., Cullen C.W., Blocher Jr. J.M. & Rai-Chaudhury P. (Ed.), J. Electrochem. Soc. Publn. (1984)

'Development of Oxidation-Resistant Structural Carbon-Carbon Composites for 3500°F Turbine Engine Applications', Dodson J.D. & Washburn R.M., AFWAL-TR-84-4010 Final Rep., Oct.(1984).

'Erosion-Corrosion by High Velocity Particles in Fluidized-bed Combustion Systems', EPRI report Corr. Sci.-3396, Palo Alto CA, (1984)

'High Temp. Materials Corrosion in Coal Gasification Atmospheres',

Norton J.F. (Ed.), Elsevier Appl. Sci. Publ., London (1984)

'Inorganic Fibres & Composite Materials', Pergamon, Oxford (1984)

'Ion Bombardment Modification of Surfaces', Auciello O. & Kelley R., (Ed.), Elsevier (1984)

'Ion Implantation & Beam Processing', Williams J.S. & Poate J.M. (Ed.), Academic Press, NY (1984)

'Ion Implantation Science & Technology', Ziegler J.F. (Ed.), Academic Press, NY (1984)

'Magnetic Glasses: Production, Properties & Applications', Anantharaman T.R. (Ed.), Trans. Technol., c/o Karl Distributors, Rockport MA, USA (1984)

'Materials Substitution & Recycling', Hecht R.J. & Halfpap D.S., Strang A., Lang E. & Pichoir R., AGARD - CP - 356 (1984)

NATO Advanced Workshop on Coatings for Heat Engines, Italy (1984)

'Rapidly Solidified Metastable Materials', Kear B.H. & Giessen B.C. (Ed.), Mat. Res. Soc. Symposia Proc., vol. 28, North-Holland, NY (1984)

5th Internat.Symp.on Superalloys, Seven Springs (1984), Gell M. et al. (Ed.)

Proc. Mat. Res. Soc. Symp., vol.37 (1984), Mat. Res. Soc. Pittsburgh PA, (1984)

'Protective Coating Systems for High Temperature Gas Turbine Components', Proc. Conf., The Metals Soc., London (1984)

Proc. of the 9th Intl. Cong. on CVD, Pennington NJ, (1984)

Proc. of the Intl. Cong. on Metallic Corrosion, 4 vols., Toronto, (1984)

Proc. Conf., McCafferty E., Clayton C.R., & Oudar J.(Ed.), Electrochem. Soc., (1984)

'Intl. Symp. on Light Metals: Science & Technol.', Proc., Varnasi, India, Nov. (1983)

'Recent Developments in the Use of Electrodeposition Technology for Engineering Coatings', Inst. Met. Fin. Conf., London, Sept. (1983)

'Coatings and Surface Treatments for Corrosion and Wear Resistance', Proc. Conf., Newcastle upon-Tyne, April 1983, Strafford et al (Ed.), Ellis Horwood (1984)

'Corrosion in Fossil Fuel Systems', Proc. Conf., Wright I.G. (Ed.), Electrochem. Soc., Pennington, NJ (1983)

'A Survey of Convective Instabilities in Silicon CVD Systems, Semiconductor Silicon -1983', Huff H.R. & Burgess R.R. (Ed.), Electrochem Soc., 258 (1983)

'Behaviour of Joints In High Temperature Materials', Gooch T.G., Hurst R., Kroeckel H. & Merz M.(Ed.) Appl Sci., (1983)

'Coatings for High Temperature Applications', Lang E.(Ed.), Applied Sci., NY (1983)

'Corrosion Resistant Materials for Coal Conversion Systems', Proc. Conf., London, May (1982), Meadowcroft D.B. & Manning M.I.(Ed.), Elsevier Applied Sci Pub., (1983)

'Vacuum Technology, Thin Films & Sputtering', Stuart R.V., Academic Press NY (1983)

'Evaporation & Sputter Techniques', Lang E. (Ed.), Appied Sci., (1983)

'Frontiers of High Temperature Materials II', Benjamin J.S. & Benn R.C. (Ed.), Incomap, London (1983)

'Fundamental Aspects of Corrosion Protection by Surface Modification', McCafferty E., Clayton C.R. & Oudar J. (Ed.), Symp., Electrochem. Soc. (1983)

'Glassy Metals: Magnetic, Chemical & Structural Properties', Hasegawa

R. (Ed.), CRC, Boca Raton, Florida (1983)

'High Temperature Corrosion', NACE-6, Rapp A. (Ed.) NACE, Houston, (1983)

'Plasma Processing', Maltal G.S., Schwartz G.C. & Smilonsky G. (Ed.), 1983 Symp., Electrochem. Soc. (1983)

'World Index of Strategic Minerals Production Exploitation & Risk', Hargreaves D. & Fromson S., Gower Publ Co NY (1983)

'High Temperature Protective Coatings', 112th Mtg., Atlanta GA (1983), Singhal S.C.(Ed.), Met Soc AIME (1983)

'Organosilane Polymers, V, VI, VIII', ONR Contract N00014-81-C-0682, Tech. Rept., (1983)

'EUR 5623 EN Review of Technological Requirments for High Temperature Materials', Joint Res. Centre, Petten, The Netherlands, (1979)

Metallurgia, 49, (4) 160 (1982)

Metal Prog., 121, (1) 38 (1982)

'Advanced Ceramic Coating Development for Industrial/Utility Gas Turbine Applications', Anderson C.A., Lau S.K., Bratton R.J., Lee S.Y., Rieke K.L., Allen J. & Munson K.E., NASA-CR-165619 (1982).

'Deposition Technologies for Films & Coatings Development & Applications', Bunshah R.F. (Ed.), Noyes, Park Ridge, NJ (1982)

'Emergent Process Methods for High Technology Ceramics', Davies R.F., Palmour H. & Porter R.L. (Ed.), Plenum Press NY, (1982)

'High Temperature Alloys for Gas Turbines', Brunetaud R.(Ed.), Reidel (1982)

'Industrial Radiology', Helmshaw R., Appl. Sci., London (1982)

'Ion Implantation Techniques', Ryssel H. & Glawichnig H. (Ed.), Springer Verlag (1982)

'Suface Analysis & Pretreatment of Plastics & Metals', Moloney A.C., Brewis D.M.(Ed.) Appl Sci., (1982)

'Surface Modification Concepts for Enhancement of the High Temperature Corrosion Resistance of Gas Turbine Superalloys', Burley N.A., Tech. Alert CME42, Dept Trade & Ind UK (1982)

Metal Sci., 16, 451 (1982).

Proc.7th ICVM Conf., Tokyo (1982)

Proc. 3rd Intl. Symp. Metall. & Mat. Sci., Roskilde, Lilholt H. & Talreja R. (Ed.) (1982)

'Advanced Materials for Alternate Fuel Capable Heat Engines', Proc. 2nd Conf. EPRI RD 2369SR (1982)

'The Chemistry of Materials at High Temperatures', Symp., Harwell, Sept (1981) AMTE (M) TM 82203 Jan (1982)

22nd Corrosion Sci. Symp., Newcastle - Upon - Tyne Sept (1981)

Proc. 7th Intl. Conf. on Surface Technology', Berlin, VDI Verlag (1981)

Proc. 8th Intl. Cong. on Metallic Corrosion, Mainz, Sept (1981)

'Ion Implantation into Metals', Ashworth V., Procter R.P.M. & Grant W.A.(Ed.), Proc.3rd Intl. Conf.on 'Modification of Surface Properties of Metals', Manchester, UK, June (1981)

'Evaporation, CEI Course on Deposition Technol. & Their Applications', Bunshah R.F., May (1981)

'High Temperature Corrosion', Proc. Conf., San Diego, March 2-6 (1981), Rapp R.A. (Ed.), NACE, Houston, TX (1981)

'Lasers in Metallurgy', Conf. Proc. Metall. Soc. AIME, Chicago, IL, Feb (1981), Mukherjee K. & Mazumder J.(Ed.) (1981)

'Criteria for Selection of Coatings', Perkins R.A.(Ed.), Seminar on Coatings for High Temp. Applications at Joint Research Centre,

Commission of Eur. Communities, Petten Esablishment, Netherlands (1981)

Proc. Intl. Conf. on Metallurgical Coatings, Am. Vac. Soc., (1981)

'Films & Coatings for Technology', McGuire G.C., Jacobson B.E. & Bunshah R.F.(Ed.) (1981)

'Intl. Advances in Non-Destructive Testing', McGonnagle W.J. (Ed.), Gordon & Breach Sci. Pub. (1981) vol. 7

'Ion Plating Technology in Films & Coatings for Technology', Jacobson B.E. & Bunshah R.F.(Ed.) Vol. 1 Sweden (1981)

'Laser & Electron-Beam Solid Interactions & Materials Processing', Gibbons J.F., Hess L.D. & Sigmon T.W.(Ed.), Proc 1980 Mat.Res.Soc., Elsevier North-Holland NY (1981)

'Material Science Research', Evans A. (Ed.), Plenum Press NY, (1981), vol. 14

'Plasma Processing', Dieleman J., Frieser R.G. & Mathal G.S. (Ed.), 1981 Symp., Electrochem. Soc. (1981)

'Strategic Metals and the UK', Inst. Mech. Eng., The Materials Forum (1981)

'The Properties & Performance of Materials in the Coal Gasification Environment', Hill V.L. & Black H.L. (Ed.), Am. Soc. of Metals, Metals Park, Ohio (1981)

Proc. Intl. Conf. on the Science & Technol. of Zirconia, Heuer A.H. & Hobbs L.W. (Ed.), Am. Ceram. Soc. (1981)

Proc. 8th Conf on CVD, Blocher J.M., Vuillard G.E. & Wahl G.(Ed.), (1981)

Strang A., Lang A. & Pichoir R., ONERO Tech. Publ. 1981-94 (1981)

'Behaviour of High Temperature Alloys in Aggresive Environments', Proc. Intl. Conf., Petten, Netherlands, Oct.1979, Kirman J., Marriott J.B., Merz M., Sahn & Whittle D.P.(Ed.), The Metals Soc U.K. (1980)

4th Intl. Symp. on Superalloys, Seven Springs (1980)

Proc. 31st Meeting of Intl. Soc. Electrochem., Venice (1980)

'Laser & Electron Beam Processing of Materials', Proc. Mat. Res. Soc., Acad. Press (1980)

'Superalloys', Miller R.A., Levine S.R. & Hodge P.E., ASM, Metals Park, OH (1980).

'Superalloys', Sims C.T. & Hagel W., Wiley (1980)

'Surface & Interfaces in Ceramic & Ceramic - Metal Systems', J.Pask & Evans A.(Ed.), Plenum NY (1980)

'Atlas of Interference Layer Metallography', Buehler H.E. & Hougardy H.P., Metallography sub. Committee, Deutsche Gesselschaft fuer Metallkunde, D-6370, Oberursel, FRG (1980)

'Glow Discharge Processes, Sputtering & Plasma Etching', Chapman B. Wiley NY (1980)

Corrosion - 80 Conf, Chicago, NACE (1980)

'Environmental Degradation of High Temperature Materials', Proc. Conf., Isle of Man, March (1980), Vol.1 & 2, Lathem E.R., (1980)

'The Current Status of Coatings for High Temperature Applications', Pichoir R. & Hausser J.M., Inst. of Metallurgists UK, Series 3, 1, (13) (1980)

Proc 9th Intl Metal Spraying Conf., The Hague, Netherlands (1980)

'Research Techniques in NDT', Vol. 6, Sharpe R.S., Academic Press, London (1980)

US Patent 4,218,007, 19th Aug (1980)

'Availability of Strategic Minerals', Proc. Natl. Symp., London Nov (1979) M.J. Jones (Ed.), IMM (1980)

'Advanced Materials for Alternative Fuel Capable Directly Fired Heat Engines', Proc. Conf. Maine, Dec., Kamo R. et al (Ed.), (1979)

'Materials for Coal Conversion & Utilisation', 4th Ann. Conf., Perkins R.A. & Packer C.M., Gaithersberg Oct (1979)

'Gas Turbine Materials in a Marine Environment', Proc. of the 4th US/UK Conf., Annapolis, June (1979) 2 vols.

'Combustion Engines', Proc 13th Intl. Cong., Strang A.(Ed.), Vienna, (1979), CIMAG (1979), Sektion HB2/GT33

'Applications of Lasers in Meterials Processing', A Metzbower E.(Ed.), Proc Conf ASM, Washington DC (1979)

'Laser-Solid Interactions & Laser Processing', Ferris S.D., Leamy H.J. & Poate J.M., Proc.1978 Mat.Res.Soc., AIP Conf Proc., (1979) No.50

'Materials Problems in Fluidised-Bed Combustion Systems', Perkins R.A. & Vonk S.J., EPRI-FP-1280, EPRI, Palo Alto, Calif. (1979)

'Metallurgical Thermochemistry', Kubachewski O. & Alcock C.B., 5th Edn. Raynor G.V. (Ed.), Pergamon Press Oxford UK (1979)

EUR 5623 EN, (1979) Review of Technological Requirements for High Temperature Materials R&D, Petten, Netherlands (1979)

'Advanced Materials for Alternate Fuel Capable directly Fired Heat Engines', Proc. of the 1st. Intl. Conf., EPRI 790 749 (1979)

'Corrosion/Erosion of Coal Conversion System Materials', Proc. Conf., Berkeley CA (1979)

Reinberg A.R., Ann. Met. Rev. Mat. Sci., $\underline{9}$, 341 (1979)

Proc. Intl. Conf, Petten, Netherlands, North-Holland Publ., March (1978)

'Adhesion Measurements of Thin Films, Thick Films & Bulk Coatings', ASTM STP 640, Mittal K.L.(Ed.), ASTM (1978)

'Gas Corrosion of Metals', (Engl. Translation), Mrowec S. & Werber T., Foreign Sci. Pub., Warsaw (1978); Natl. Bur. Std., Washington DC (1978), Avail., US Dept. Commerce, Springfield VA 22161

'High Temperature Alloys for Gas Turbines', Appl.Sci., (1978)

'High Temperature Metal Halide Chemistry', Proc. Symp., Hildebrand D.L. & Cubicciotti D.D. (Ed.), Electrochem. Soc., (1978)

'Plasma & Detonation Gun Techniques & Coating Properties', UCLA Course on Deposition Techniques & Applications (Sept.), Tucker R.C. (1978)

'Thin Film Processes', Vossen J.L. & Kern W.(Ed.), Academic Press, NY (1978)

2nd Portugese Congress on Chem. (1978)

'Advances in Welding Processes', Conf., Harrogate (1978)

Du Pont, Ind. Finishing, 2, 2, 42, (1978)

'Corrosion-Erosion Behaviour of Materials', Proc. Symp., Natesan K. (Ed.), American Inst. of Mining, Metallurgical & Petroleum Engineers, NY (1978)

Proc. Electrochem. Conf., Pittsburgh, Abstracts only (1978)

Proc. Intl. Conf. on Metallurgical Coatings, San Francisco, April (1978)

Hemmings P.L. & Perkins R.A., EPRI FP-539 Proj. 716-1 Interim Rep., Dec. (1977)

22nd. Natl. SAMPE Symposium & Exhibition, San Diego, Calif., April 26-28 (1977)

'Advances in Extractive Metallurgy', Jones M.J. (Ed.), Inst. Min. & Metall., (1977)

'Ion Plating & Applied Techniques', Dugdale R.A., CEP Consultants, Edinburgh (1977)

'Materials & Coatings to Resist High Temp Corrosion', Holmes D.R. & Rahmel A.(Ed.) Appl Sci., (1977)

'Thermodynamics Phase Stability Diagrams for the Analysis of Corrosion Reactions in Coal Gasification/Combination Atmospheres', Hemmings P.L., Perkins R.A., Lockhead Co. LMDC D558238 (1977)

Proc. Conf. on Refractory Oxides, Odeillo (1977)

Proc. Electrochem. Soc. Conf., Georgia, U.S.A. (1977)

Proc. 7th Intl. Vacuum Cong., Vienna, Dobrozemsky (Ed.) (1977)

Proc. 6th Intl. Conf. on CVD, Electrochem. Soc., NJ (1977)

Proc. of the 3rd CIMTEC Intl. Meeting on Modern Ceramic Technology, Rimini, Italy, May (1976)

'Advances in Corrosion Science & Technology', Vol.6, Fontana M.G. & Staehle R.W.(Ed.), Plenum NY (1976)

IEEE Int Electronic Devices, Meeting, Washington, DC (1976)

'Ion Implantation, Sputtering & Their Applications', Acad. Press (1976)

'New Trends in Materials Processing', Bunshah R.F., Am.Soc.Metals (1976)

'Physical Vapour Deposition', Hill R.J.(Ed.) Pub. Airco Temescal (1976)

Proc. Conf. Thermal Spraying, Miami, Fl (1976)

'Properties of High Temperature Alloys', Foroulis Z.A. (Ed.), Electrochem. Soc., New Jersey, (1976)

'Properties of High Temperature Alloys with Emphasis on Environmental Effects', Proc. Symp., Foroulis Z.A. & Pettit F.S. (Ed.), Electrochem. Soc., NJ (1976)

Proc. 3rd Intl. Symp. on Superalloys, Metall. & Manuf., AIME Claitor's Publ. Div., (1976)

'Gas Turbine Materials in Marine Environment', 3rd US/UK Conf., Bath, UK, Hall B. (Ed.) (1976)

'Scientific & Industrial Applications of Small Accelerators', Proc. 4th Conf. Denton (1976), IEEE, NY (1976)

'Metal-Slag-Gas Reactions & Processes', Proc. Conf. Toronto, May (1975) Foroulis Z.A. & Smeltzer W.W.(Ed.), Electrochem.Soc.(1975),

'Application of Infrared NDT Techniques to Hardface Coating Bond Inspection', Crowe J.C. & Dawson J.F., BNWL - B- 427 (1975)

'Growth of Epitaxial Films by Sputtering', 'Epitaxial Growth', Part A, Mathews J.W.(Ed.), Acad. Press NY (1975)

'High Temperature Corrosion of Aerospace Alloys', Stringer J., AGARD No. 200, NATO (1975)

'Methods of Surface Analysis', Czanderna A.W.(Ed.) Elsevier (1975)

'Stress Effects and the Oxidation of Metals', Cathcart J.V. (Ed.), Met. Soc. AIME NY, (1975)

Proc. 25th Intl. Meeting, Soc. Chimie. Phys., Barrett P. (Ed.) Elsevier, Amsterdam (1975)

NASA CR134735, Felten E.J., Strangman T.E. & Ulion N.E., Oct (1974)

'Gas Turbine Materials in Marine Environment', Conf. Zurich, March 1974, ASME 74-GT-100, Fairbanks J.W. & Michlin I. (Ed.) (1974)

'Techniques & Application of Plasma Chemistry', Hollaham J.R. & Bell A.T. (Ed.), Wiley NY (1974)

'The Nimonic Alloys', Betteridge W. & Heslop J. (Ed.), 2nd. edn. , Arnold, London (1974)

'Chemical Transport Reactions', Schmalzreid H., Academic Press NY (1974)

Proc. 4th Intl. Conf. on CVD, Princeton (1974), Wakefield G.F. & Blocher J.M. (Ed.) Electrochem. Soc., (1974)

'Deposition & Corrosion in Gas Turbines', Hart A.B. & Cutler A.J.B.(Ed.),Halstead Press, UK (1973)

'Selected Values of the Thermodynamic Properties of the Elements', Hultgren R., Desai P.D., Hawkins D.T., Gleiser M., Kelley K.K. & Wagman D.D., Am. Soc. Metals, OH (1973)

Proc. 3rd Intl. Conf. on SiC, Miami, (1973)

Proc.16th Ann Conf, Soc. Vac. Coaters, Paton B.A., Movchan B.A. & Demchishin A.V.(Ed.), Chicago (1973)

Proc. 7th Intl. Metal Spraying Conf., Perugini G., Galante F. & Maraccioli E., London (1973)

'Protective Coatings on Metals', Vol.5, Samsonov G.V. (Ed.), Consultants Bureau, NY (1973)

'Properties of Electrodeposits as Materials for Selected Applications', Symp., Battelle Columbus Lab., OH, Nov. (1973)

'Deposition & Corrosion in Gas Turbines', Conf., CEGB, Sudbury House, London, Dec.(1972), Hart A.B. & Cutler A.J.B. (Ed.), Applied Sci. (1973)

'Arc Plasma Technology in Materials Science', Springer -Verlag NY (1972)

'Temperature, Its Measurements & Control in Science & Industry', Plumb H.H. (Ed.), Instruments Soc. of America, (1972)

'Cleaning & Preparation of Metal Surfaces', British Standard C.P. 3012 (1972) Brit Std. Inst., London (1972)

Metals & Ceramics Information Center, Report MCIC-72-007, Battelle, Columbus, OH (1972)

Proc. 3rd Intl. Conf. CVD, Hinsdale IL, Glaski F.A. (Ed.), Am. Nucl. Soc.(1972)

Proc. Gas Turbine Conf., NAVSEA/NAVAIR (1972)

JANAF Thermochemical Tables, Stull D.R., Prophet H. (Dir), NSRDS, US Dept. Comm. NBS.37 (1971)

'Sprayed Unfused Metal Coatings for Engineering Purposes', British Standard 4761 (1971), Brit. Std. Instn., London

Proc 2nd Intl. Conf. on CVD, Blocher J.M. & Withers J.C.(Ed.) Electrochem Soc., (1971)

'Handbook of Thin Film Technology', Maissel L.I. & Glang R.(Ed.), McGraw-Hill (1970)

'High Temperature Metallic Corrosion of Sulphur & its Compounds', Foroulis Z.A.(Ed.) (1970)

'Modern Electrochemistry', Bockris J.O.M. & Reddy A.K.N., Plenum, NY(1970),

6th Metal Spraying Conf., Paris (1970)

'Flame Spray Coatings for Wear', Soc. Automotive Engrs., Mid Year Meeting, Chicago IL, May (1969)

'Thin Film Phenomena', Chopra K.L., McGraw-Hill NY (1969)

DMIC Rev, Allen B.C., Battelle, Columbus OH, Nov 8 (1968a)

DMIC Rev, Allen B.C., Battelle, Columbus OH, April 17 (1968b)

DMIC Rev, Allen B.C., Feb 2 (1968c)

DMIC Review, Allen B.C., Feb (1968) Battelle Memorial Inst, Columbus OH

'Infrared Radiation: A Handbook for Application', Cramson M.A., Plenum NY (1968)

'The Contribution of Metallic and Ceramic Coatings to Gas Turbine Engines', Contributed by the Gas Turbine Div, ASME, Gas Turbine Conf. & Products Show, Washington DC (1968)

13th Refractory Composites Wkg. Gp. Mtg., AFLMTR - 68 - 84, Priceman S.

& Sama L., (1968)

'Corrosion of Alloys at High Temperatures in Atmospheres Consisting of Fuel Combustion Products and Associated Impurities', Hancock P., HMSO, London (1968)

'Hot Corrosion Problems Associated with Gas Turbines', ASTM STP 421 (1967)

'The Solid Phase Welding of Metals', Tylecote R.F., Edward Arnold (1968)

'A Computer Program for the Rapid Solution of Complex Chemical Equilibrium Problems', Hall E.H. & Broehl J.H., 8th Qtly Report to CVD Res GpII, Battelle, Columbus OH, April 15 (1967)

Congress on High Temperature Coatings, Leningrad, Izdaltelstro Nauka, Moscow (1967)

'CVD of Refractory Metals, Alloys & Compounds', Proc. Conf., Schaffhauser A.C.(Ed.), Am Nuclear Soc., (1967)

'Coatings of High Temperature Materials', Part 1, Hausner H. (Ed.) Plenum, NY (1966)

'Oxide Dispersion Strengthening', 2nd. Boston Landing Conf., AIME, NY June 27-29, (1966)

'High Temperature Oxidation of Metals', Kofstad P., John Wiley, NY (1966)

'Some Observations on the Interface Between Plasma Sprayed Tungsten & 1020 Steel', Leeds D.H., Defence Documentation Center, AD-803286 (1966)

Proc. 8th Conf. on Tube Technol., IEEE Conf. Record (1966)

'Metco Flame Spray Handbook Vol III Plasma Flame Process', Ingham H.S. & Shepard A.P., Metco Inc., Long Island NY (1965)

'Encyclopaedia of Eletrochemistry', Mantell C.E.(Ed.), (1964)

'Physics of Thin Films', Hass G. & Thun R.(Ed.), Academic Press, NY, Vol 2 (1964)

'Chemical Transport Reactions', Schafer H., Academic Press NY (1964)

Proc. Intl. Conf., Marchwood, UK, May (1963), Johnson H.R. & Littler D.J.(Ed.), Butterworths, London (1963)

'Condensation & Evaporation', Hirth J.P. & Pound G.M., Macmillan NY, (1963)

'Modern Electroplating', Lowenheim F.W.(Ed.), Wiley NY (1963)

'Temperature, Its Measurements & Control in Science & Industry', Herzfeld C.M. (Ed.), Reinhold NY (1962)

Proc I Intl. Cong. on Metallic Corrosion, London, (1962), Butterworths (1962)

Proc. Symp. on Special Ceramics, Popper P.(Ed.) Academic Press NY (1962)

'A Text-Book of Inorganic Chemistry', Partington J.R., 6th ed., Macmillan & Co Ltd., London (1961)

'Electrodeposition of Alloys', Brenner A.(Ed.), (1960)

'Vacuum Deposition of Thin Films', Holland L., Wiley NY (1956)

'Mass Transfer Operations', Treyball R.E., McGraw-Hill NY (1956)

'Applications of Coatings in Advanced Technologies', Perkins R.A., Joint Res. Centre, Petten, Netherlands

'Electrodeposition of Thick Coatings of Platinum & Palladium on Refractory Metals from Aqueous Electrolytes', Cramer S.D., Kanahan C.B., Andrews R.L. & Schlain D., US Dept of the Interior

'Fabrication & Processing Technol. for Improved Creep Tolerant Properties of Structure Alloys', Hettche L.R. & Rath B.B.,

'Materials Substitution & Recycling', Coad J.P. & Scott K.T.,AGARD-CP-

356, Conf.Proc., No.356

'Metallic Materials, Specification Handbook', ,Ross R.B. (Ed.), E & FN Spon Ltd

Canning & Co, Handbook on Electroplating & Metal Finishing

DIN 8566 (German Standard) Metallspritzdrahte Lichtsbogen Spritzdrahte

'Durability & Fracture', Proc. 2nd Irish Conf., Limerick, Ireland, Bolton J (Ed.)

SNECMA DEF Standard 05-21, CAA, Rolls - Royce,

Tech. Alert (Patent counsel, Marshall Space Flight Center, Alabama USA), Orders to Tech Alert, Room 206, Ebury Bridge house, Ebury Bridge Road, London SWIW 8QD

'Flame spraying', UK Standard 03-011, Ministry of Defence

REFERENCES

NOTE: For names with an umlaut, e.g. Schroder, see both this and Schroeder. For names with prefixes like El-, Ben-, De-, Van-, please see both the prefixed and the unprefixed names.

Abba A., Galerie A. & Caillet M., Oxid.Met., 17, (1) 43 (1982)

Abderrazik G.Ben, Millot F., Moulin G., Huntz A.M., J Am Ceram Soc 68, 302, 307 (1985); also see J Mat Sci 19, 3173 (1984)

Abderrazik G.Ben, Moulin G., Huntz A.M., Young E.W.A. & de Wit J.H.W., Solid State Ionics 22, 285-294 (1987)

Abid A., Bensalem R. & Sealy B.J., J. Mat. Sci., 21, 1301 (1986)

Addison C.W.E., Harris S.J. & Noble B., Trans Inst Met Finishing, 50, 171 (1972)

Ahmed N.A.G., 'Ion Plating Technology - Developments & Applications', John Wiley & Sons NY, (1987)

Akikawa N., Uenuna T., Miki S. & Mutoh M., Proc 7th Intl Conf on Vacuum Metall (ICVM), Tokyo, p.330 (1982)

Akuezue & Whittle D.P., 'Coatings for High Temp. Applications', Lang E. (Ed.), Applied Science (1983)

Akuezue H.C. & Whittle D.P., Metal Sci., 17, 27 (1983)

Akzo N.V., Brit.Patent 1,511,109 (1978)

Alcock C.B. & Jeffes J.H.E., Trans IMM(C),76, C246 (1967)

Alcock C.B., Hocking M.G. & Zador S., Corrosion Sci., 9, 111 (1969)

Alexander J.D., Driver D., 'Proc. of the 1st Int. Conf. on Materials in Aerospace', April (1986), Royal Aeronautical Soc. London, p. 168 (1986)

Allaert K., Van Calster A., Loos H. & Leguesne A., J. Electrochem. Soc., 132, (7), 1763 (1985)

Allam I.M. & Rowcliffe D.J., J. Vac. Sci. & Technol., A 4 (6) 2652 (1986)

Allam I.M., Whittle D.P. & Stringer J., Oxid.Metals, 12, 35 (1978)

Allen B.C., DMIC Review Feb (1968) Battelle Mem. Inst, Columbus OH

Allen B.C., DMIC Rev, Battelle, Columbus OH, Nov 8 (1968a)

Allen B.C., DMIC Rev, Battelle, Columbus OH, April 17 (1968b)

Allen B.C., DMIC Rev, Feb 2 (1968c)

Allen S.D., Trigubo A.B. & Chiu Y., Proc.8th Conf on CVD, Blocher J.M., Vuillard G.E. & Wahl G.(Ed.), (1981) p 267

Allen R.V. & Borbidge W.E., J Mat Sci., 18, 2835 (1983)

Allison G. & Hawkins M.K., GEC Rev., 17, 947 (1914)

Allsop R., Pitt T. & Hardy J.,

Metallurgica, 63 (377) 125 (1961)

Almond E.A., Cottenden A.M. & Gee M.G., Metal Sci., 17, (4) 153 (1983)

Amicarelli V., Borghetic G. & Cento L., 'Mn Diffusion Coating of Steels', Met. It., 71, (11), 471 (1979)

Anantharaman (Ed.), 'Magnetic Glasses: Production, Properties & Applications', Trans. Technol., c/o Karl Distributors, Rockport MA, USA (1984)

Andersen A., Hastan B., Kofstad P. & Lillerud P.K., Mat. Sci. & Engg., 87, 45 (1987)

Andoh Y, Ogata K., Suzuki Y., Kamijo E., Satou M. & Fujimoto F., Proc. 5th Int. Conf.on Ion Beam Modification of Materials, Cantania, 1986, Nucl Instr. & Meth. B19/20, 787 (1987).

Andrieu E. & Henon J.P., Matls Sci & Eng 88, 191 (1987)

Anon (1974) J. Vac. Sci. Technol., 11 July/Aug (1974)

Anon (1975) The Metallurgist & Materials Technologist, June (1975) p 292

Anon (1978) Aviation Weekly & Space Technol., 20th Nov. (1978).

Anon (1980) Ed. Report, Metallurgia, 47, (9) 459 (1980)

Anon (1980) Metallurgia, 47, (9) 459 (1980)

Anon (1980) Report, Metallurgia, 47, (9) 459 (1980)

Anon (1981) Ed.report, Metallurgia, 48, (8) 372 (1981)

Anon (1981) Report, Metallurgia, 48, (8) 372 (1981)

Anon (1981) Metal Construction 13, (1981) p232

Anon (1981) Inst. Mech. Eng.

Anon (1982) Metal Prog., 121, (1) 38 (1982)

Anon (1982) Metals Sci., 16, 451 (1982)

Anon (1982) Report, Metallurgia, 49, (4) 160 (1982)

Anon (1983) Metallurgia 50, (7) 299 (1983)

Anon (1983) Pt. Met. Rev., 27, 30 (1983)

Anon (1984) Naval Res Rev (US), 36, 31 (1984)

Anon (1985), 'Recent Developments in HIP Equip.', Metallurgia, 52, (8), 316 (1985)

Anon (1987), 'What are the New Materials ?', Metallurgia, 54, (11), 510 (1987)

Ansari A.A. et al, Mat. Sci. & Energy, 88, 135 (1987)

Antill J.G., Bennett M.J., Carney R.F.A., Dearnaley G., Fern F.H., Goode P.D., Myatt B.L., Turner J.F. & Warburton J.B., Corr. Sci., 16, 729 (1976)

Applebaum A. & Murarka S.P., J. Vac. Sci. & Technol., 4, (3) Part 1 p.637 (1986)

Apps R.L., J. Vac. Sci. & Technol., 11, (4) 741 (1974)

Arata Y. & Miyamoto I., Technocrat, 11, (5) 33, May (1978)

Arata Y. et al, Proc.7th ICVM Conf., Tokyo (1982), (1983) p 313

Archer A.A., Proc Natl Symp on Availability of Strategic Minerals, London Nov (1979) M.J. Jones (Ed.), IMM (1980) p1

Archer N.J., Proc 5th Intl Conf on CVD, Blocher J.M. et al, eds, Electrochemical Soc, Princeton NJ, p.556 (1975)

Archer N.J. in "Ceramic Surfaces & Surface Treatments", ed Morrell R. & Nicholas M.G., Brit Ceram Proc No 34, p.187-194 (Aug 1984)

Archer N.J., Proc.7th ICVM Conf., Tokyo (1982), p 224

Archer N.J. & Yee K.K., Wear, (1976)

Arcolin C. et al, Proc 9th Int Metal Spraying Conf, The Hague May (1980) p94

Arken D.J., Bromberger G.C. & Lewis T.J., J.Appl.Chem., 18, 348 (1968)

Armstrong R.W., Boeltinger W.J. & Kurijama M., J. Appl. Crystallogr, 13, 417 (1980)

Arnold J.L, Dunbar F.C. & Flinchum C., Met Trans B,8B, 399 (1977)

Aronson A.J, Chen D. & Class W. H, Thin Solid Films,72, 535 (1980)

Asahi N. & Kojima Y., Proc.7th ICVM Conf., Tokyo (1982), p 305

Asher J., Bennett M.J., Hawes R.W.M., Price J.B., Tuson A.T., Savage D.J. & Sugden S., Mat. Sci & Engg., 88, 143 (1987)

Asher J., Conlon T.W., Tofield B.C. & Wilkins N.J.M., UKAEA Rep. AERE-R10871 (1983)

Ashworth V., Grant W.A., Procter R.P.M. & Wellington T.C., Corr. Sci. 16, 393 (1976).

Ashworth V. Procter R.P.M. & Grant W.A., Treatise on Materials Sci. & Technol. 18, 175 (1980).

Ashworth V., Procter R.P.M. & Grant W.A.(Ed.), 'Ion Implantation into Metals', Proc. 3rd Intl. Conf. on Modification of Surface Proper ties of Metals, Manchester June (1981)

Atkinson A. & Gardner J.W., Corr Sci., $\underline{21}$, 49 (1981)

Atkinson H.V., Oxid. Met., $\underline{24}$, (3/4), 177 (1985)

Atmani H. & Rameau J., J. Mat. Sci. & Engg., $\underline{88}$, 221 (1987)

Atzmon M., Veerhoven J.R., Gibson E.R. & Johnson W.L., Appl. Phys. Lett., $\underline{45}$, 1052 (1984)

Auciello O. & Kelley R., (Eds.), 'Ion Bombardment Modification of Surfaces', Elsevier (1984) p.158

Audisio S., Knoche R. & Mazille H., Powder Metall., $\underline{27}$, 147 (1984)

Auerbach A., J. Electrochem. Soc., $\underline{132}$, (1), 130 (1985)

Auwarter M., US Patent 2,920,002

Ayers J.D., 'Lasers in Metallurgy', Proc Conf Metall Soc AIME Chicago, Feb 1981, Mukherjee K. & Mazumder J.(Ed.) (1981) p 115

BAJ Vickers UK Tech. Information

Baardsen E.L., Schmatz D.J. & Bisaro R.E., Welding J., $\underline{52}$, 227 (1973)

Bahrani A.S., Proc Roy Soc A, $\underline{296}$, 123 (1967)

Bahrani A.S., Surfacing J, $\underline{9}$, (1) 2 (1978)

Bahun C.J. & Engquist R.D., Metals Engg. Qtly, $\underline{4}$, 27 (1964)

Bailey F.P. & Borbidge W.E., 'Surface & Interfaces in Ceramic & Ceramic - Metal Systems', J.Pask & Evans A. (Ed.), Plenum NY (1980), p 525

Balog M. et al, J Cryst Growth $\underline{17}$, 298 (1972)

Ban V.S.(1973), see Ban et al, J Electrochem Soc $\underline{122}$, 1387, 1389 (1975) and J Japan Assoc Crystal Growth $\underline{5}$, 119 (1978)

Ban V., J Crystal Growth $\underline{31}$, 248 (1975)

Bardal E., Molde P. & Eggen T.G., Brit Corros J., $\underline{8}$, (1) 15 (1973)

Bardal E., Proc.7th Intl.Metal Spraying Conf., London (1973), p 215

Bar Gadda R., J Electrochem Soc $\underline{133}$, 2123 (1986)

Barkalow R.H. & Pettit F.S., Proc Conf Corrosion / Erosion of Coal Conversion System Materials, Berkeley CA (1979), p 139

Barkalow R.H. (1980), see Barkalow (1984) & Proc 4th US/UK Conf on Hot Cor of Gas Turb Matls for Marine Appl, Annapolis,$\underline{1}$,493 (1979)

Barkalow R.H., 'Erosion-corrosion by high velocity particles in fluidized-bed combustion systems', EPRI report Corr. Sci.-3396, RP 979-4, Palo Alto CA, (1984)

Barnes J., Corish J. & Franck F. & Norton J.F., Oxid. Met., $\underline{24}$, (1/2), 85 (1985)

Barnes J., Corish J. & Norton J.F., Oxid. Met., $\underline{26}$, (5/6), 333 (1986)

Barnhart R.E., 'The Contribution of Metallic and Ceramic Coatings to Gas Turbine Engines', Contributed by the Gas Turbine Div, ASME, Gas Turbine Conf. & Products Show, Washington DC (1968)

Barrett C.A. & Lowell C.E., Oxid Metals, $\underline{12}$, (4) 293 (1978)

Barrett C.A., 'A Multiple Linear Regression Analysis of Hot Corrosion Attack on a Series of Nickel Base Turbine Alloys', NASA Technical Memorandum 87020, July (1985)

Barrett C.A., NASA Tech Mem. 87297, Metall. Soc., Am. Inst. Mining, Metall. & Pet. Engrs., New Orleans, March (1986)

Bartle P.M., Metal Constn & Brit Welding J., $\underline{1}$ (5) 241 (1969)

Bartle P.M., Conf. Advances in Welding Processes, Harrogate (1978), paper No 50

Bartocci R.S., ASTM STP 421 (1967), p 169

Bascom W.D. & Bitner J.L., J.Mat Sci., $\underline{12}$, 1401 (1977)

Basinska-Pampuch S. & Gibas T., Proc. 3rd. Intl. Mtg on Modern Ceram. Technol., Rimini, Italy (1976) p 201 & 212

Bass M., 'Applications of Lasers in Materials Processing', Metzbower A.E.(Ed.), Proc Conf ASM, Washington DC (1979) p 13

Bass M., 'Lasers in Metallurgy', Conf Proc Metall Soc AIME, Chicago, IL, Feb (1981), Mukherjee K. & Mazumder J.(Ed.) (1981), p 13

Basu S.N. & Halloran J.W., Oxid Met $\underline{27}$, 143 (1987)

Baudo G., Tamba A. & Bombara G., Corrosion-NACE, $\underline{26}$,(7) 193 (1970)

Bauer R., Grunling H.W. & Schneider K., 'Materials & Coatings to Resist High Temperature Corrosion', Holmes D.R. & Rahmel A.(Ed.), Applied Sci., NY (1979), p369

Baun W.L., 'Suface Analysis & Pretreatment of Plastics & Metals', Brewis D.M.(Ed.) Appl Sci Publ., (1982), p 45

Baxter D.J. & Natesan K., Oxid. Met., 24, (5/6), 331 (1985)

Bayles R.A., Meyn D.A. & Moore P.G., 'Lasers in Metallurgy', Proc Conf Metall Soc AIME Chicago, Feb 1981, Mukherjee K. & Mazumder J.(Ed.) (1981) p 127

Bazzard R. & Bowden P.J., Trans IMF 50, 207 (1972)

Becher P.F. & Murday J.S., J Mat Sci., 12, 1088 (1977)

Becher P.F. & Newell W.L., J Mat Sci., 12, 90 (1977)

Becher P.F. & Halen S.A., Am Ceram Bull., 58 582 (1979)

Becquerelle (see Bequerelle)

Beguin C, Horvath E. & Perry A.J, Thin Solid Films, 46, 209 (1977)

Beindt C.C. & McPherson R., 'Surface & Interfaces in Ceramic & Ceramic - Metal Systems', J.Pask & Evans A. (Ed.), Plenum NY (1980), p 619

Beltran A.M. & Schilling W.F., 'Superalloys 1980' Sims & Hagel (Ed)

Benjamin J.S. & Benn R.C. (Eds.), 'Frontiers of High Temp. Materials II', Incomap, London (1983)

Benn R.C., Deb P. & Boone D.H., 'High Temp. Coatings', Khobaib M & Krutenat R.C. (Eds.), Proc. Symp. Orlando Florida, (1986), Metall. Soc. AIME (1987) p.1

Bennett A., Toriz F.C. & Thakker A.B., Surface & Coatings Technol., 32, 359 (1987)

Bennett M.J., 'The Role of Ion-Implantation in High Temperature Oxidation Studies', Proc NACE Intl Conf on High Temp Corr., March (1981)

Bennett M.J., Houlton M.R. & Hawes R.W.M., Corr.Sci., 22, (2) 111 (1982)

Bennett M.J., Nucl.Energy, 21, 63 (1982)

Bennett M.J., High Temperature Corrosion, NACE, 6,145 (1983).

Bennett M.J., 'Sol-gel & Vapour Deposition', 'Coatings for High Temperature Applications', Lang E. (Ed.(Appl Sci., (1983) p 169

Bennett M.J., 'Coatings for High Temperature Applications', Lang E. (Ed.), Applied Sci., NY (1983), p 169

Bennett M.J., J. Vac. Sci. & Technol., B12, (4) 800 (1984)

Bennett M.J. & Price J.B., Proc. 8th Intl. Cong. on Metallic Corrosion, Mainz, Sept (1981), p 1026

Bennett M.J. & Tuson A.T., To be published in Matls. Sci. & Eng. (1988/89).

Bennett M.J., Antill J.E., Carney R.F.A., Dearnaley G., Fern F.H., Goode P.D., Myatt B.L., Turner J.F. & Warburton J.B., Corros. Sci., 16, 729 (1976).

Bennett M.J., Bellamy B.A. Knights C.F. Meadows N. & Eyre N.J., Mat. Sci. Eng. 69, 359 (1985).

Bennett M.J., Bishop H.E., Chalker P.R. & Tuson A.T., Mat. Sci. and Eng., 90, 177 (1987).

Bennett M.J., Dearnaley G., Houlton M.R., Hawes R.W.M., Goode P.D. & Wilkins M.A., Corr. Sci., 20, 73 (1980)

Bennett M.J. et al, 22nd Corr Sci Symp., Newcastle-Upon-Tyne, Sept (1981)

Bennett M.J. et al in V.Ashworth, Grant W.A. & Procter R.P.M.(Ed.) 'Ion Implantation into Metals', Pergamon Press (1982), p 264

Bennett M.J. et al, Nucl.Technol., 66, 518 (1984)

Bennett M.J. et al, Proc Conf., Heidelberg, Sept.(1984)

Bennett M.J., Horsley G.W. & Houlton M.R., Proc. Conf., McCafferty E., Clayton C.R., & Oudar J.(Ed.), Electrochem Soc., (1984)

Bennett M.J., Houlton M.R. & Dearnaley G., Corros Sci., 20, 69 (1980)

Bennett M.J., Houlton M.R. & Hawes R.W.M., Corrosion Sci., 22., 111 (1982)

Bennett M.J., Richerky D.S., Thin Solid Films, 154, 403 (1987)

Bentini G.G., Berti M., Carnera A., Della Mea G., Drigo A.V., Russo S.Lo, Mazzoldi P. & Dearnaley G, Corros. Sci. 20, 27 (1980).

Bequerelle P., Hubert M., Savage B., Bavy J.C. & Bowergain P., Mat. Sci & Eng 87, 137 (1987)

Bergmann E., Vogel J., Simmen L., Thin Solid Films, 153, 219 (1987)

Bernabai U., Cavallini M., Bombara G, Dearnaley G. & Wilkins M.A., Corros. Sci., 20, 19 (1980).

Bernard C., Proc.8th Intl. Conf., Blocher J.M. & Vuillard G.E.(Ed.), p 3 (1981)

Berndt C.C. & McPherson R., 'Surfaces & Interfaces in Ceramic &

Ceramic - Metal Systems', Mat Sci Res., Vol.14, J.Pask & Evans A.(Ed.), Plenum (1980), p 619

Berneron R. & Charbonnier J.C., Proc.7th ICVM Conf., Tokyo (1982), p 592

Bertinger R. & Zeilinger H., Proc.8th Intl Conf on CVD (1981), Blocher J.M., Vuillard G.E. & Wahl G.(Ed.), (1981)p 664

Besmann T.M. & Spear K.E., J Electrochem Soc 124, (5) 786, 790(1977)

Besmann T.M., J. Am. Ceram. Soc., 69, (1), 69-74 (1986)

Besmann T.M. & Lowden R.A., Am. Ceram Soc Bull.,67,(2), 350 (1988)

Bettelheim A., Broitman F., Mor U. & Harth R., J. Electrochem. Soc., 132, (7), 1775 (1985)

Betteridge W. & Heslop J. (Eds.), 'The Nimonic Alloys', 2nd. edn. , Arnold, London (1974)

Betz G.(1977), see Betz et al in "Sputtering by Particle Bombardment", Berich R. (Ed.), Springer, Berlin, p.11 (1983)

Betz W., 'Materials & Coatings to Resist High Temperature Corrosion', Holmes D.R. & Rahmel A.(Ed.), Appl. Sci., NY (1979), p 185

Betz W., Huff H. & Track W., Werkstoff Z u Korr,7, 161 (1976)

Bhat D.G., Cho T. & Woerner P.F., J. Vac. Sci. & Technol., A4 (6) 2713 (1986)

Bhat H., Herman H. & Coyle R.J., 'High Temperature Protective Coatings', Proc. Symp. Atlanta, GA, March (1983), Singhal S.C.(Ed.), Metall.Soc.AIME, p 37

Bhoem J., Oei Y.S., de Moor H.H.C., Hanssen J.H.L & Giling L.J., J. Electrochem. Soc., 132, (8), 1973 (1985)

Bhushan B., Metal Finishing, June (1980) p 71

Bhushan M., J. Vac. Sci. & Technol., A 5, (5), 2829 (1987)

Bilenchi R. & Musci M., Proc.8th Conf on CVD, Blocher J.M., Vuillard G.E. & Wahl G.(Ed.), (1981) p 275

Billy M., Mat. Sci & Engg. 88, 53 (1987)

Binder K. & Herrmann H.J., Z Phys (B) 35, 171 (1979)

Binder K., Baumgart A., J Chem Phys 71, 254 (1979)

Birks N., Proc.25th Intl. Mtg. Soc. Chimie. Phys. ,Barrett P.

(Ed.) Elsevier, Amsterdam (1975)

Birks N., 'Properties of High Temp. Alloys', Foroulis Z.A. (Ed.), The Electrochem. Soc., New Jersey, (1976) pp.215-260

Birks N. & Tattam N., J Sci Instr 2, (1969)

Biswas D.R., J. Mat. Sci., 21, 2217 (1986)

Biswas D.R., J. Mat. Sci., 21, 2217 (1988)

Bland R.D., Electrochem.Technol, 6 (7-8) 272 (1968)

Bland R.D., Kominiak G.J. & Mattox D.M., J. Vac. Sci. & Technol., 11(4) 671(1974)

Blocher J.M., Auto Eng, No.730543, May (1973), Mach Des., 43, (16) 58 (1971)

Blocher J.M. (Ed.), Princeton Pub Electrochem Soc., p 563 (1973)

Blocher J.M., J. Vac. Sci. & Technol., 11, (4) 680 (1974)

Blocher J.M., 'Chemical Vapour Deposition', Jacobson B.E. & Bunshah R.F.(Ed.) 'Films & Coatings for Technology', Sweden (1982)

Blocher J.M., CEI Course, Jacobson B.E. & Bunshah R.F.(Ed.), Sweden (1982)

Bloem J., Oei Y.S., Moor H.C.de, Hansen J.H.L., Giling H.J., J Electrochem Soc 132, 1973 (1985)

Boch P., Fauchais P. & Borie A., Proc. 3rd. Intl Mtg on Modern Ceram. Technol., Rimini, Italy (1976) p 208 & 191

Bockris J.O.M. & Reddy A.K.N., 'Modern Electrochemistry', Plenum, NY(1970), p 22

Boelens S. & Veltrop H., Surf. & Coatings Technol., 33, 63 (1987)

Bohn P.W. & Manz R.C., J. Electrochem. Soc., 132, (8), 1981 (1985)

Bolshakov K.A. & Fedorov P.I., Zh.Obgch.Khim, 26, 348 (1956)

Bolster R.N., Singer I.L. & Vardiman R.G., Surf. & Coatings Technol., 33, 469 (1987)

Bonifield T.G., 'Films & Coatings for Technology', Jacobson B.E. & Bunshah R.F.(Ed.) (1982)

Boone D.H., 'Protective Coatings for Use in High Temperature Combustion Zone Environments', Course on 'Films and coatings for Technology', Jacobson B.E. & Bunshah R.F., (Ed.), Soderkoping, Sweden, Vol 2 June (1981)

Boone D.H., Shen S., & Matialu S., Proc. '4th Conf. on Gas Turbine Materials in a Marine Environment,

Annapolis, USA (1979) p.735

Boone D.M., Lee D. & Schafer J.M., 'Ion Plating & Allied Techniques', CEP Consultants, Edinburgh (1977), p 141

Boone D.H., Strangman T.E. & Wilson L.W., J. Vac. Sci. & Technol.11, 641 & 645 (1974)

Bornstein N.S., DeCrescente M.A. & Roth H.A., Met Trans, 4, 1799 (1973)

Bors D.L., Fair J.A., Monnig K.A. & Sarahwat K.C, Solid State Technol., 26, 183 (1983)

Bosseboeuf A. & Bouchier D., Surf. Sci., 162, 695 (1985)

Boston M.E., Metallurgia, June (1983), p248

Bourcier R.J., Nelson G.C., Hays A.K. & Romig A.D., J. Vac. Sci Technol., A4 (6) 2943 (1986)

Bourgoin P.B., Proc.7th Intl.Metal Spraying Conf., London (1973), p185

Boving & Hinterman Thin Solid Films 152, 253 (1987)

Boxman R.L. & Goldsmith S., Surf. & Coatings Technol., 33, 153 (1987)

Bracke P., Schurman H., Verhoest J, 'Inorganic Fibres & Composite Mat.', Pergamon, Oxford (1984)

Brainard W.A. & Wheeler D.R., NASA TP 1377 (1979)

Braski D.N., Goodell P.D., Caltcant J.V. & Kane R.H., Oxid. Met., 25, (1/2), 29 (1986)

Bratton R.J. & Lau S.K., 'Advances in Ceramics', in 'Proc. Intl. Conf. on the Science & Technol. of Zirconia', vol. 3, Heuer A.H. & Hobbs L.W. (Ed.), Am. Ceram. Soc. (1981) p.226

Braun M., Hultberg S., Brown A., Svensson B.M. & Hogmark S., "Corrosion Studies of Ion Implanted Iron and Carbon Steel", Nuclear Instruments and Methods in Physics Research, North-Holland, Amsterdam, B19/20, 259-262 (1987).

Breiland W.G., Coltrin M.E. & Ho P.L., Appl. Phys., 59, (9) 3267 (1986)

Brenner A.(Ed.)'Electrodeposition of Alloys' (1960)

Brennfleck K. et al., 'Proc. of the 9th Int. Cong. on CVD', Electrochem. Soc.,NJ, (1984), p. 649

Brewis D.M.(Ed.) Surface Analysis & Pretreatment of Plastics & Metals, Appl. Sci., London (1982) p1

Briggs J., 'Technology & Markets for Inorganic Coatings in Western Europe', Frost & Sullivan UK (19)

Brill-Edwards M., Epner M., Electrochem Technol., 6 (9-10) 299 (1968)

Brimhall J.L., Kissinger H.E. & Charlot L.E., Rad. Eff., 77, 237 (1983)

Brismaid S. et al, US Patent 2,784,115

British Standard 4761 (1971) 'Sprayed unfused metal coatings for engineering purposes', Brit Std Instn, London

British Standard C.P. 3012 (1972), 'Cleaning & Preparation of Metal Surfaces', Brit Std.Inst., London (1972)

Broc M., Fauvet P., Olivier P. & Sanvier J., Mat. Sci. & Engg., 87, 219 (1987)

Brodsky M.H., Thin Solid Films, 50, 57 (1978)

Brooks S., Ferguson J.M., Meadowcroft D.B. & Stevens C.G., 'Materials & Coatings to Resist High Temperature Corrosion', Holmes D.R. & A.Rahmel (Ed.), Applied Sci., NY (1979), p 121

Brown A.M., Graham J., Saunders K.G. & Surman P.L., Corrosion Sci., 18, 337 (1978)

Brown C.O. & Davis J.W., ' IEEE Int Electronic Devices', Meeting, Washington, DC (1976)

Brown C.T., Bornstein N. & DeCresente M.A., 'High Temp. Metallic Corrosion of Sulphur & its Compounds', Foroulis Z.A.(Ed.) (1970)

Brown D.W., Musket R.G. & Weirick L.J., J. Vac. Sci. & Technol., A 3, (3), 583 (1985)

Brown S.D. & Ferber M.K., J. Vac. Sci. & Technol., A3, 2506 (1985)

Browning R., J. Vac. Sci. & Technol., A 3, (5), 1959 (1985)

Brunet C. & Dallaire S., Surf. & Coatings Technol., 31, 1 (1987)

Brunet C., Dallaire S. & St.Jaegnes R.G., J. Vac. Sci. & Technol., A 3, (6), 2503 (1985)

Brunner J. & Perry A.J., Thin Solid Films, 153, 103 (1987)

Bruno G. et al, Thin Solid Films, 67, 103 (1980)

Bryant W.A., J. Vac. Sci. & Technol., 11, (4) 695 (1974)

Bryant W.A. & Meier G.H., J. Vac. Sci. & Technol., 11 (4) 719 (1974)

BS see British Standard or UK Standard

Buck T.M., 'Methods of Surface

Analysis', Czanderna A.W.(Ed.) Elsevier (1975) p 75

Buckel W. & Hilsch R., Z Phys 138 109 (1954)

Buckley J.D., Am. Ceram. Soc. Bull., 67, (2), 364 (1988)

Bucklow I.A., 'Coatings for High Temperature Applications', Lang E.(Ed.) Appl Sci., NY (1983) p 139

Budhani R.C., Prakash S., Bunshah R.F., J. Vac. Sci. & Technol., A4 (6) 2609 (1986)

Bühler H.E. & Hougardy H.P., 'Atlas of Interference Layer Metallography', Metallography Sub-Cttee. Deutsche Gesellschaft fuer Metallkunde, D-6370, Oberursel, FRG (1980); (Eng. trans. avail.)

Bunshah R.F., J. Vac. Sci. & Technol.11 (4) 633 & 814 (1974)

Bunshah R.F., 'New Trends in Materials Processing', ASM (1976), p200

Bunshah R.F., High Temp.High Press.10, 187 (1978)

Bunshah R.F., 'Evaporation', CEI Course on Deposition Technol. & Their Applications, May (1981)

Bunshah R.F. (Ed.), 'Deposition Technologies for Films & Coatings Development & Applications', Noyes, Park Ridge, NJ (1982)

Bunshah R.F., Surf & Coatings Technol 27, 1 (1986)

Bunshah R.F. & Deshpandey C.V., J. Vac. Sci. & Technol., A 3, (3), 553 (1985)

Bunshah R.F. & Juntz R.S., 'Berrylium Technology', Vol.1, Gordon & Breach Sci. Pub.(1966), p 1

Bunshah R.F. & Juntz R.S., J. Vac. Sci. & Technol., 9, 1404(1972)

Bunshah R.F. & Raghuram A.C., ibid., 9 1385 (1972)

Burggraaf P.S., Semiconductor Intl. March (1980) p 23

Burke J.J., Lenoe E.N., Katz R.N., "Ceramics for High Perf Applics, II", Brookhill Publ Co, Chestnut Hill, MA, p.397-443 (1978)

Burley N.A., 'Surface Modification Concepts for Enhancement of the High Temperature Corrosion Resistance of Gas Turbine Superalloys', Tech.Alert CME42, Dept Trade & Ind UK (1982)

Burman C. & Ericsson T., 'High Temperature Protective Coatings', Proc. Symp. Atlanta, GA, March (1983), Singhal S.C.(Ed.), Metall. Soc. AIME, p 51

Burne D.H., Deb P., Purvis L.I. &

Rigney D.V., J. Vac. Sci. & Technol., A 3, (6), 2557 (1985)

Burnett P.J. & Rickerby D.S., Thin Solid Films 154, 403 (1987)

Butler H.S. & Kino G.S., Phys. & Fluids, 6, 1346 (1963)

Byworth S., Rolls Royce Mag., (29), 21 (1986)

Cahn R.W. & Johnson W.L., J. Mat. Res. (1986)

Caillet M., Ayedi H.F., Galerie A. & Besson J., 'Materials & Coatings to Resist High Temperature Corrosion', Holmes D.R. & Rahmel A.(Ed.), Applied Sci., NY (1979), p 387

Cameron P.B., Thesis Ph.D, CNAA (UK) (1981)

Cameron B.P., Foster J. & Carew J.A., Trans Inst Metal Finishing, 57, 113 (1979)

Campbell D.S., 'Mechanical Properties of Thin Films', Handbook of Thin Film Technology, Maissel L.I. & Glang R.(Ed.) ,McGraw-Hill NY (1970)

Campisano S.U., Foti G.,Mazzoldi P. & Rimini E. (ed.), Proc. 5th Intl Conf. on 'Ion Beam Modification of Materials", Catania, Italy, 9-13 June (1986).

Canning & Co, Birmingham, Handbook on Electroplating & Metal Finishing (periodical issues)

Canova P. & Ramous E., J. Mat. Sci., 21, 2143 (1986)

Caputa A.J., Lackey W.J., Wright I.G. & Angelini P., J. Electrochem. Soc., 132, (9), 2274 (1985)

Carew J.A., Univ Of London Thesis (1983)

Carew J.A., Vasantasree V. & Hocking M.G., Hot Corrosion of Electrodeposited Cermets', Electrochem. Soc. Conf., Pittsburgh, PA, USA (1978)

Carling R.W., Bradshaw R.W. & Mar R.W., 'High Temperature Protective Coatings', Proc. Symp. Atlanta, GA, March (1983), Singhal S.C.(Ed.), Metall. Soc.AIME, p 261

Carmichael D.C., J. Vac. Sci. & Technol., 11, (4) 639 (1974)

Castle J.E., Brit Corros J 22, 77 (1987)

Cathcart J.V. (Ed.), 'Stress Effects and the Oxidation of Metals', Met. Soc. AIME NY, (1975)

Cellis J.P. & Roos J.R., J Electrochem Soc 124, 1508 (1977)

Chadwick A.T.. & Taylor (1982), see next column

Chadwick A.T. & Taylor R.T., 'Proc of the 9th Intl. Congress on Metallic Corrosion', (NRC Canada), **3**, 381 (1984)

Chadwick A.T. & Taylor R.T., Solid State Ionics, **12**, 343 (1984)

Chande T. & Mazumder J., 'Lasers in Metallurgy', Proc Conf Metall Soc AIME Chicago, Feb (1981), Mukherjee K. & Mazumder J.(Ed.) (1981) p 165

Chande T. & Mazumder J., Met Trans B **14B**, 181 (1983)

Chang G.C., Phucharoen W. & Miller R.A., Surf. & Coatings Technol., **30**, 13 (1987)

Chang S.A., Skolnik M.B. & Altman C., J. Vac. Sci. & Technol., A **4**, (3), 413 (1986)

Chapman B., 'Glow Discharge Processes, Sputtering & Plasma Etching', Wiley NY (1980)

Chassagneux, Thomas G. & Soustelle M., Mat Sci & Engg, **87**, 379 (1987)

Chatterjee-Fischer R., Proc 8th Intl Conf CVD (1981), p 491

Chatterji D., DeVries R.C. & Romeo G., Ch.1, 'Advances in Corrosion Science & Technology', Vol.6, Fontana M.G. & Staehle R.W.(Ed.), Plenum NY(1976)

Chatterji D., Hampton A.F., Graham H.C. & Davis H.H., Proc. Conf., 'Metal - Slag - Gas Reactions & Processes', Foroulis Z.A. & Smeltzer W.W.(Ed.), Electrochem.Soc., (1975), Proc.Conf., p 391

Chen Q.M., Chen H.M., Bai X.D., Zhang J.Z. & Wang H.H., Nucl. Instr. & Meth. 209/210, 867 (1983)

Chiang S.S., Marshall D.B. & Evans A.G., 'Surfaces & Interfaces in Ceramic & Ceramic - Metal Systems' Mat. Sci. Res., Vol.14, Pask J. & Evans A.(Ed.), Plenum (1980), p603

Child H.C., 'Coatings for High Temperature Applications', Lang E. (Ed.), Appl.Sci. NY (1983), p 395

Chisholm R.U., Electrodep. & Surface Treatment, **3**, 321 (1972)

Cho J.S. & Chun J.S., Proc 8th Intl Conf on CVD (1981), p 573

Choi S.H. & Stringer J., Mat. Sci. & Engg., **87**, 237 (1987)

Chopra K.L., 'Thin Film Phenomena', McGraw-Hill NY (1969)

Choquet P., Indrigo C. & Mevrel R., Matls Sci & Eng **88**,97 (1987)

Chown J., Deacon R.F., Singer N. & White A.E.S., Proc Symp on Special Ceramics, Popper P.(Ed.) Academic Press NY (1962)

Chrenkova M., Fellner P., Matiasovsky K. & Kristin J., Chem. Papers, **40**, (2), 153 (1986)

Christ H.J., Kunecke U., Meyer K. & Sockel H.G., Mat. Sci. & Engg., **87**, 161 (1987)

Christodolou G., Walker A., Steen W.M. & West D.R.F., Metals Technol, **10**, 215 (1983)

Chu T.L., Ing D.W. & Noreika A.J., Electrochem. Technol., **6**, 56 (1968)

Chu S.K., Saka N. & Suh N.P., Mat. Sci. & Engg., **95**, 303 (1987)

Chuzkho R.K., Balakhovskiy O.A. & Vorotnikov A.I., 'Temp. Effect on hardness of Ta_2B_2 Diffusion Coatings', Metallov0ed Term Obrab Met., (1), 57-9 (1978)

Clark J.W., NASA Report, CR-72763 (1971)

Clark R.K., Unnan J. & Wiedemann K.E., Oxid. Met., **28**, (5/6), 391 (1987)

Clayton C.R., Nucl. Instr. & Meth. 182/183, 865 (1981).

Clegg M.A., Silins V. & Evans D.J.I., Proc. 7th Intl. Metal Spraying Conf., London (1973), p62

Clegg M.A. & Mehta M.H., Surf. & Coatings Technol., **34**, 69 (1988)

Coad J.P. & Scott K.T., 'Materials Substitution & Recycling' AGARD-CP-356, Conf.Proc., No.356 (1982)

Coad J.P. & Restall J.E., Metal Technol, **9**, 499 (1982)

Coats A.W., Dear D.J.A. & Penfold D., J Inst Fuel, **41**, 129 (1968)

Coburn J.W. & Kay E., J Appl. Phys., **43**,4965 (1972)

Cochet G., Mellottee H. & Delbourgo R. J Electrochem Soc., **125**, 487 (1978)

Cochrane H., 'Temp., Its Measurements & Control in Sci. & Ind.', Plumb H.H. (Ed.), Instruments Soc. of America, (1972) p. 1619

Colin C., Boussuge M., Valentin D. & Desplanches G., J. Mat. Sci., **23**, (6), 2121 (1988)

Colmet R. et al, Proc.8th Intl. Conf., Blocher J.M. & Vuillard G.E.(Ed.) p 17 (1981)

Colson J.C. & Larpin J.P., Matls Sci & Eng **87**, 11 (1987)

Coltrin M.E., Kee R.J. & Miller J.A., J. Electrochem. Soc., **133**, (6), 1206 (1986)

Condé J.F.G., 'The Chemistry of Materials at High Temperatures', Symposium, Harwell, Sept (1981) AMTE (M) TM 82203 Jan (1982)

Condé J.F.G. (1983), private communication

Condé J.F.G. & Booth G.C., Conf.on Deposition & Corrosion in Gas Turbines, CEGB, Sudbury House, London, Dec.(1972)

Condé J.F.G. & McCreath C.G., ASME 80-GT-126 (1980)

Condé J.F.G., Birks N., Hocking M.G. & Vasantasree V., 'Aspects of the Aetiology and Mechanisms of Hot Corrosion in Marine Gas Turbines', Proc. 4th U.S./U.K. Conf. on Gas Turbine Materials in Marine Environment, Annapolis U.S.A. (1979)

Condé J.F.G., Erdos E. & Rahmel A., Proc Int Conf on High Temp Alloys for Gas Turbines, Liege, (Oct 1982), publ Reidel, Dordrecht, p.99-148

Cooper K.P., J Vac Sci & Tech A 4, 2857 (1986)

Cope B.C., 'Suface Analysis & Pretreatment of Plastics & Metals', Brewis D.M.(Ed.) Appl Sci Publ., (1982), p 95

Cotell C.M. & Yurek G.J., Oxid. Met., (5/6), 363 (1986)

Conlon T.W., Contemp. Phys., 23, 353 (1982)

Courtright E.L., Surf. & Coating Technol., 33, 327 (1987)

Coutsouradis D., 'High Temperature Alloys for Gas Turbines', Appl. Sci., (1978) Ch.8

Coutsouradis D., Davin A. & Lamberigts M., Mat. Sci. & Engg., 88, 11 (1987)

Cramer S.D., J.Electrochem.Soc., 126, (6) 891 (1979)

Cramer S.D., Kanahan C.B., Andrews R.L. & Schlain D., 'Electrodeposition of thick coatings of platinum & palladium on refractory metals from aqueous electrolytes', US Dept of the Interior (1967)

Cramson M.A., 'Infrared Radiation : a Handbook for Application', Plenum NY (1968).

Crane C.J., et al, Weld J, 46, (1) 23s-31s (1967)

Crawmer D.E., Bartoe R.L. & Kramer J., Surf. & Coatings Technol., 33, 353 (1987)

Criscione J.M. et al, 'High Temperature Protective Coatings for Graphite' AFML-TDR-64-173, part 1, AD 604-463 (1964)

Criscione J.M. et al, 'High Temp. Protective Coatings for Graphite' AFML-TDR-64-173, Part II AD 608092 (1965a)

Criscione J.M. et al, 'High Temperature Protective Coatings for Graphite' AFML-TDR-64-173, Part III AD 479131 (1965b)

Criscione J.M. et al, 'High Temperature Protective Coatings for Graphite' AFML-TDR-64-173, Part lV AD 805438 (1967)

Crostack H.A. & Steffens H.D., 'Behaviour of Joints in High Temperature Materials', Gooch T.G., Hurst R., Kroeckel H. & Merz M. (Ed.), Appl. Sci.,p 217 (1984)

Crowe J.C. & Dawson J.F., 'Application of Infrared NDT Techniques to Hardface Coating Bond Inspection', BNWL - B- 427 (1975)

Culbertson R. & Mattox D.M., Proc.8th Conf.on Tube Technol., IEEE Conf. Record(1966)

Cunningham F.E., Thesis, 'The Use of Lasers for the Production of Surface alloys', MIT, June (1964)

Cuomo J.J. & Rossnagel S.M., J. Vac. Sci. & Technol., A 4, (3), 393 (1986)

Cutler A.J.B., J. Appl. Electrochem., 1, 19 (1971)

DIN 8566 (German Standard) 'Metallspritzdrahte Lichtsbogen Spritzdrahte'

Dahman U. et al, 'Surfaces & Interfaces of Ceramic & Ceramic - Metal Systems', Metal Sci., Vol.14, J.Pask & Evans A.(Ed.) Plenum (1980), p 391

Dallaire S. & Cielo P., Met Trans B, 13B, 479 (1982)

Danroc J., Aubert A. & Gillet R., Thin Solid Films, 153, 281 (1987)

Danroc J., Aubert A. & Gillet R., Surf. & Coatings Technol., 33, 83 (1987)

Danyluk S. & Diercks D., in 'The Properties & Performance of Materials in the Coal Gasification Environment', Hill V.L. & Black H.L. (Eds.), Am. Soc. of Metals, Metals Park, Ohio (1981) p.155

Dapkunas S.J., Proc Conf Environmental Degradation of High Temperature Materials, Isle of Man, March (1980) Vol 2, p6/1

Das S.K. & Davis L.A., Mat. Sci. & Engg., 98, 1 (1988)

Datta P.K., Strafford K.N., Gray J.S., Gill C. & Fordham R.J., Proc. Conf. 'UK Corrosion '87', Brighton, Oct. (1987), p. 337 (1987)

Datta P.K., Strafford K.N. & Dowson A.L., 'Environment-Mechanical Interaction Processes & Hydrogen Embrittlement in Ti', Intl. Symp. on Light Metals: Science & Technol., Varnasi, India, Nov. (1983)

Datta P.K., Strafford K.N. & Dowson A.L., 'Some Observations on the Role of H_2 & O_2 in the Fatigue & Fracture Behaviour of Pure Alpha Ti', Proc. of the 2nd Irish Conf. on Durability & Fracture held in Limerick, Ireland, (Eds.) Bolton J

Davies R.J. & Stephenson N., Brit Welding J., 9, (3) 139 (1962)

Davis R.F., 'Surface & Interfaces in Ceramic & Ceramic-Metal Systems' J.Pask & Evans (Ed.), Plenum (1980)

Davis W.D. & Vanderslice T.A., Phys Rev, 131, 319 (1963)

Davis D. & Whittaker J.A., Metall Rev., 12, 16 (1967)

Dearnaley G., Nucl. Instr. & Meth. 182/183, 899 (1981).

Dearnaley G., in 'Ion Implantation into Metals', V.Ashworth, Grant W.A. & Procter R.P.M. (Ed.) Pergamon Press (1982), p 180

Dearnaley G., "Adhesive, Abrasive and Oxidative Wear in Ion Implanted Metals" Harwell report AERE-R 11428 August (1984).

Dearnaley G., "Adhesive and Abrasive Wear Mechanisms in Ion Implanted Metals", Nuclear Instruments and Methods in Physics Research, North-Holland, Amsterdam, B7/8, 158-165 (1985B).

Dearnaley G., Mat. Sci. & Engg., 69, 139-147 (1985C).

Dearnaley G., Surface Engg., 2, (3) 213-221 (1986).

Dearnaley G., "Ion Implantation Part II: Ion Implantation in Non-electronic Materials", Nuclear Instruments & Methods in Physics Research, North-Holland, Amsterdam, B24/25, 506-511 (1987).

Dearnaley G., Surface & Coatings Technol.,33, 453-367 (1987B).

Dearnaley G. & Grant W.A., "Ion Implantation", Seminar (Tribology Gp.) Inst. of Mech. Engg., Oct. 10, 1985. publ. in "Recent Developments in Surface Coating and Modification Processes"

Dearnaley G. & Hartley N.E.W., Proc. 4th Conf., Sci. & Ind. Appl. of Small Accelerators, Denton (1976) (IEEE, NY,USA) p20

Dearnaley P.A. & Trent E.M., Metals Technol., 9, 60 (1982)

Dearnaley G., Freeman J.H., Nelson R.S. & Stephen J., 'Ion Implantation', North Holland (1973)

Dearnaley G., Goode P.D., Minter F.J, Peacock A.T. & Waddell C.N., J. Vac. Sci. & Technol., A 3, (6), 2684 (1985)

Dearnaley G, Minter F.J., Rol P.K., Saint A. & Thompson V., "Microhardness and Nitrogen Profiles in Ion Implanted WC & Steels", Nuclear Instruments and Methods in Physics Research, North Holland B7/8, 188-194 (1985b)

Dearnaley P.A., Thompson V. & Greanson A.N., Surf. & Coatings Technol., 29, 157 (1986)

Deb P, Boone D.H. & Streiff R., 'Proc. ASM Symp. on Coatings for High Temp. Oxidation Resistance, ASM, Toronto, Oct. (1985)

Deb P., Boone D.H. & Streiff R., J. Vac. Sci. & Technol., A 3, (6), 2578 (1985)

Debruyn W., Casteels F. & Tas H., Mat. Sci. & Engg., 87, 169 (1987)

DeCrescente M.A. & Bornstein N., Corrosion (NACE) 24, 127 (1968)

De-Jie G. et al, Proc 7th ICVM Conf, Tokyo, p.352 (1982)

Delarnay D & Huntz A.M., J Mat Sci 17, 2027 (1982)

Delarnay D., Huntz A.M. & Lacombe P., Corros.Sci., 20, 1109 (1980)

Delves B.G., 'Ion Implantation into Metals', V.Ashworth, Grant W.A. & Procter R.P.M. (Ed.) Pergamon Press (1982), p 126

Denis A. & Garcia E.A., Oxid. Met., 29, (1/2), 153 (1988)

Denner S. & Y-W.Kim, 'High Temperature Protective Coatings', Proc. Symp. Atlanta, GA, March (1983), Singhal S.C.(Ed.), Metall. Soc. AIME, p 233

Dennison J.P. & Kisakurek S.E., Met Trans B, 8B, 237 (1977)

Derby B. & Wallach E.R., Metal Sci., 16, (1) 49 (1982)

Derradji A. & Kassabji F & Fauchais P., Surf. & Coatings Technol., 29, 291 (1986)

Deshpandey C. & Holland L., Proc. 7th ICVM Conf., Tokyo (1982), p 276

Deshpandey C.V., O'Brien B.P., Doerr H.J. & Bunshah R.F., Surf. & Coatings Technol., 33, 1 (1987)

Desport J.A. & Bennett M.J., Oxid. Metals, 29 (3-4) 327 (1988)

Dharmadhikari V.S., J. Vac. Sci. &

Technol., A 6, (3), 1922 (1988)

Diamuka J.P. & Curtis B.J., 'A Survey of Convective Instabilities in Silicon CVD Systems', Semiconductor Silicon -1983, Huff H.R. & Burgess R.R. (Ed.) The Electrochem Soc., 258 (1973)

Dieleman J., Frieser R.G. & Mathal G.S. (Ed.), 'Plasma Processing', Symp., J. Electrochem. Soc. (1981)

Doi Y. & Doi A., Proc.7th ICVM Conf., Tokyo (1982), p 195

Donovan M. & Sanders J.L., Tribology, 5, 205 (1972)

Douglass D.L., Oxid Met,1, 127 (1969)

Douglass D.L., Bhide V.S. & Vineberg E., Oxid Met 16, 29 (1981)

Douglass D.L., Gesmundo F. & de Asmundis C., Oxid. Met., 25, (3/4), 235 (1987)

Downer & Smyth R.T., 7th Intl Metal Spraying Conf., London (1973) p 199

Draper C.W., 'Lasers in Metallurgy', Proc Conf Metall Soc AIME Chicago, Feb 1981, Mukherjee K. & Mazumder J.(Ed.) (1981) p 67

Draper C.W. & Mazzoldi P. (Ed.), 'Laser Surface Treatment of Metals' NATO ASI Series E, Appl. Sciences, No. 115, Martinus Nijhoff Dordrecht, (1986), Proc. Mtg. Italy 2-13 Sept. (1985)

Drewett R., Corr Sci,9, 823 (1969)

Drewett R., Anti-Corrosion Methods & Materials, Apr/Jun/Aug. (1969a)

Drummeter L.F. & Hass G., 'Physics of Thin Films', Hass G. & Thun R.(Ed.), Academic Press, NY, Vol 2 (1964) p305

Du Pont, Ind. Finishing, 2, 42, (1978)

Dua A.K., Agarwala R.P. & Desai P.B., J. Vac. Sci. & Technol., A 3, (6), 2612 (1987)

Dudko Y.D. et al, High Temp., (Engl.Transl.) 20, (1) 120 (1982)

Duffy D.M. & Tasker P.W., Phil. Mag A, 54, 759 (1986).

Dugdale R.A., 'Ion Plating & Applied Techniques', CEP Consultants, Edinburgh (1977), p 177

Dunhill A.K., Powder Met., 30 (3) 149 (1987)

Dunkerley F.J., Leavenworth H.W. & Eichelman G.E., 'Oxide Dispersion Strengthening', 2nd. Boston Landing Conf., AIME, NY June 27-29, (1966)

Dunstan et al (1985): error, see Leatham & Brooks, Mod Devels in

Powd Met 15-17, 157 (1985); Evans et al, Powd Met 28, 13 (1985)

Duret & Pichoir (1982), see below

Duret C. & Pichoir R., 'Coatings for High Temp. Applications', Lang E.(Ed.), Applied Sci., NY (1983)., p 33

Duret & Pichoir (1987): error, see Mevrel & Pichoir, Matls Sci & Eng 88, 1 (1987)

Duret C., Davin A., Marjnissen G. & Pichoir R., 'High Temperature Alloys for Gas Turbines', Brunetaud R.(Ed.), Reidel (1982), p 53

Dust M., Dep P., Boone D.H. & Shanton, J. Vac. Sci. & Technol., A 4, (6), 2571 (1986)

Duwez P., "Prog in Solid State Chem" 3, Pergamon Press (1966)

Duwez P., Willens R.H., Clement W., T Appl Phys 31, 1136 (1960)

EUR 5623 EN 'Review of Technological Requirments for High Temperature Materials', Joint Res Centre, Petten, The Netherlands, May(1982)

Easterling K., 'Tomorrow's Materials', Inst. of Metals, London UK, (1988)

Eaton H.E. & Novak R.C., Surf. & Coatings Technol., 30, 41 (1987)

Eaton H.E. & Novak R.C., Surf. & Coatings Technol., 32, 227 (1987)

Ebata Y. & Kinoshita M., 'Methods of Adhesion of Ceramics', Ceramics Japan, 18, 17-22 (1983)

Eckersley J.S. & Kleppe R., Surf. & Coatings Technol., 33, 443 (1987)

Edeleanu C. & Littlewood R., Electrochem. Acta 3, 195 (1960)

Edeleanu C., Proc Conf Environmental Degradation of High Temperature Materials, Isle of Man, March (1980) Vol 1, p4/8

Edmonds D.P., Report ONRL-TM-6452 (1978)

Egitto F.D., J Electrochem Soc., 127, 1354 (1980)

El Gomati M.M. et al., Surf. Sci., 152, 917 (1985)

Elliot P. & Taylor T.J., 'Materials & Coatings to Resist High Temperature Corrosion', Holmes D.R. & Rahmel A.(Ed.), Applied Sci., NY (1979), p353

Elliott S. & Wallach E.R., Metal Construction, 13, 167, 221 (1981)

Ellssner G. & Pabst R.F., Proc Brit Ceram Soc., 25, 175 (1975)

Ellssner G., Diem W. & Wallace J.S., 'Surface & Interfaces in Ceramic & Ceramic - Metal Systems'

J.Pask & Evans A.(Ed.), Plenum NY (1980), p 629

Ellssner G., Jehn H. & Fromm E., High Temp.-High Pr., $\underline{10}$, 487 (1978)

Emiliani M. & Richman M.H., Surf. & Coatings Technol., $\underline{33}$, 267 (1987)

Emiliani M., Richman M.H., Brown R. & Gregory O., Surf. & Coatings Technol., $\underline{33}$, 267 (1987)

Engel R. & Kleenmann W., 22nd Sci Symp., Newcastle-Upon-Tyne, Sept. (1981)

Enomoto Y. & Matsubara K., J. Vac. Sci. & Technol., $\underline{12}$, 827 (1975)

Enomoto Y., Yamanaka K. & Mizuhara K., Proc.7th ICVM Conf., Tokyo (1982), p 209

Ensinger W., Meyer A., & Wolf G.K. VIII Congress European de Corrosion, Nice, 71-1 (1985).

Epik A.P., Sharivkes S.Y. & Astakhov E.A., Porosh Met, $\underline{39}$, 43 (1966), Sov. Powder Metal. Metal Ceram, $\underline{39}$, 210 (1966)

Erdös E. & Altorfer H., 'Materials & Coatings to Resist High Temperature Corrosion', Holmes D.R. & Rahmel A.(Ed.), Applied Sci., NY (1979), p 161

Erdoes E. & Rahmel A., Oxid. Met., $\underline{26}$, (1/2), 101 (1986)

Eriksson T.S., Hjortsberg A. & Granqvist C.G., Proc. 7th ICVM Conf. Tokyo (1982),p 696

Erturk E. & Heuvel H.J., Thin Solid Films, $\underline{153}$, 135 (1987)

Eser E. & Ogilvie R.E., J. Vac. Sci. & Technol., $\underline{15}$, 401 (1978)

Eskildsen S.S., 'Ion Implantation Into Metals', Ashworth V., Grant W.A. & Procter R.P.M.(Ed.), Pergamon (1982),p 315

EUR 5623 (1979), "Review of Tech Req. for High Temp Matls R & D", EEC, Petten, Netherlands (1979)

Evans (Ed.) Mat Sci Res Vol 14, Plenum NY (1980), p 379

Evans H.E. & Lobb R.C., Corr. Sci., $\underline{24}$ (3) 209-222 (1984a)

Evans H.E. & Lobb R.C., Corr. Sci., $\underline{24}$ (3) 223-236 (1984b)

Evans R., Wear, $\underline{64}$, 31 (1980)

Evans & Heuer, J Am Ceram Soc $\underline{63}$, 241 (1980)

Evans & Heuer, Proc Intl Conf on Sci & Tech Zirconia, Heuer A.H. & Hobbs L.W. (Ed.), Am Ceram Soc (1981)

Exner H.E., Metal Sci.,$\underline{16}$, 451 (1982)

Fairbanks J.W., Hecht R.J., Mat. Sci. & Engg., $\underline{88}$, 321 (1987)

Falcone G., Surface Science, $\underline{187}$, 212 (1987)

Fancey K.S. & Matthews A., Surf. & Coatings Technol., $\underline{33}$, 17 (1987)

Fedoseev D.V. et al, Proc.8th Conf on CVD, Blocher J.M., Vuillard G.E. & Whal G.(Ed.), (1981) p 284

Felten E.J., Strangman T.E. & Ulion N.E., NASA CR134735, Oct (1974)

Fenske G.R., Kaufherr N. & Sproul W.D., Thin Solid Films, $\underline{153}$, 159 (1987)

Ferris S.D., Leamy H.J. & Poate J.M., 'Laser-Solid Interactions & Laser Processing', Proc.1978 Mat. Res. Soc., AIP Conf Proc., (1979) No.50

Fielding J., Conf.Advances in Welding Processes, Harrogate, (1978), paper No 26

Finnigan D.J., Garbett K. & Woolsey I.S., Corros. Sci., $\underline{22}$, 359 (1982)

Fischer H., Peter G. & Koprio J.A., J. Vac. Sci. & Technol., A 6, (3), 2109 (1988)

Fishteyn B.M, Nikolaev V.E. & Kozlov V.G., 'Boronising of 40X Steel in Powder Mixt. by High Frequency Heating', Metalloved Term Obrab Metals, (1), 51-3 (1977)

Fitzer E. & Gadow R., Am. Ceram. Soc. Bull., $\underline{65}$, (2), 326-35 (1986)

Fitzer E. & Maurer H.J., 'Materials & Coatings to Resist High Temp Corrosion', Holmes D.R. & Rahmel A.(Ed.) Appl.Sci., (1977), p 253

Fitzer E. & Maurer H.J., 'Formation of Diffusion Barriers in Aluminide Coatings on Ni Alloys', Arch. Eisenhuettenwesen, $\underline{49}$, (9), 95-100 (1978)

Fitzer & Maurer (1979), see below

Fitzer E., Nowak W. & Maurer H.J., 'Materials & Coatings to Resist High Temp Corrosion', Holmes D.R. & Rahmel A.(Ed.) Appl Sci., (1978), p 313

Flatley T. & Birks N., J Iron & Steel Inst., $\underline{193}$, 523 (1971)

Flatley T. et al, Proc Conf Environmental Degradation of High Temperature Materials, Isle of Man, March (1980) Vol 2, p6/15

Flatley T., Latham E.P. & Morris C.W., paper 62, Corrosion - 80 Conf, Chicago (1980)

Fleetwood M.J., J. Inst. Met., <u>98</u>, 1 (1970)

Fleischer R.L., J. Mat. Sci.L, <u>7</u>, 525 (1988)

Fleming R.H., Meeker G.P. & Blattner R.J., Thin Solid Films, <u>153</u>, 197 (1987)

Flink P.J. & Heckel R.W., 'High Temperature Coatings', Khobaib M. & Krutenat R.C.(Ed.), (1987), p21

Floreen S., Kane R.H., Kelly T.J. & Robinson M.J., 'Frontiers of High Temperature Materials I', Incomap, London (1981)

Follstaedt D.M., Yost F.G., Pope L.E., Picraux S.T. & Knapp J.A., Appl. Phys. Lett. <u>43</u>, 358 (1983).

Fontana M.G. & Staehle R.W.(Ed.), 'Advances in Corrosion Sci. & Technol.' Vol.6, Plenum NY (1976).

Foroulis Z.A. & Pettit F.S.(Ed.), Proc. Symp. 'Prop High Temp Alloys With Emphasis on Environmental Effects', Electrochem Soc., Princeton NJ (1976)

Foster J. & Kariapper A.M.J., Trans. IMF, <u>51</u>, 727 (1973)

Foster J., 'Recent Development in the use of Electrodeposition Technology for Engineering Coatings', Inst.Met.Fin.Conf., London, Sept (1983)

Fournier D., Saint Jacques R.G., Brunet C. & Dallaire S., J. Vac. Sci. & Technol, A<u>3</u> (6) 2475 (1985)

Frances M., Vilasi M., Mansoour-Gabr M., Steinmetz M. & Steinmetz P, Mat. Sci. & Engg.,<u>88</u>, 89 (1987)

Francombe M.H., 'Growth of Epitaxial Films by Sputtering', Epitaxial Growth, Part A, Mathews J.W.(Ed.), Acad. Press NY (1975)

Fraser D.B., 'Thin Film Processes', Vossen J.L. & Kern W.(Ed.) Acad.Press NY (1978) p 131

Freche J.C. & Ault G.M ., Proc 3rd Intl.Symp.on Superalloys, Metall. & Manuf., AIME Claitor's Publ. Div (1976)

Freeman R.J., 'Temp., Its Measurements & Control in Sci. & Industry', Herzfeld C.M. (Ed.), Reinhold NY (1962) p. 201

Freller H. & Haessler H., Thin Solid Films, <u>153</u>, 67 (1987)

Fry H. & Morris F.G., Electroplating & Metal Fin., June (1959)

Fryburg G.C., et al, J.Electrochem.Soc., <u>129</u>, 517 (1982)

Fryburg G.C., Kohl F.J. & Stearns C.A., 'Chemical Mechanisms and Reaction Rates for the Initiation of Hot Corrosion of IN-738', NASA Technical Paper 2319, July (1984)

Fryt E.M., Wood G.C., Stott F.H. & Whittle D.P., 'Influence of prior Oxidation on the Oxidation of Dilute Co-Cr alloys in Oxygen', J. Oxid Met., <u>23</u>, (1-2) 77 (1985)

Fryzell R.E. & Leese G.E., Surf. & Coatings Technol., <u>32</u>, 97 (1987)

Fukutomi M. et al, Proc. 7th ICVM Conf., Tokyo (1982), p 711

Fukutomi M., Fujitsuka M., Shikama T. & Okada M.,J. Vac. Sci. & Technol., A <u>3</u>, (6), 2650 (1985)

Gadd J.D., Fisch H.A., Kmieciak H.A. & Jones E.E., Electrochem. Technol., <u>6</u>, (11/12) 379 (1968a)

Gadd J.D., Nejedlik J.F. & Graham L.D., Electrochem.Technol., <u>6</u>, (9-10) 307 (1968b)

Gagnet W.J., 'Flame Spray Coating for Wear', Soc Automotive Engrs., Mid Year Meeting, Chicago IL, May (1969)

Gaird A.K. & Hearn E.W., J. Electrochem Soc., <u>125</u>, 139 (1978)

Galasso F.S., Veltri R.D. & Croft W.J., Am Ceram Soc Bull., <u>57</u>, (4) 453 (1978)

Galerie A., Ashworth V., Grant W.A. & Procter R.P.M.(Ed.) 'Ion Implantat ion into Metals', Pergamon Press (1982), p 190

Galerie A., Pons M. & Caillet M., Mat. Sci. & Energy, <u>88</u>, 127 (1987)

Galsworthy J.C., Restall J.E. & Booth G.C., Proc Conf on High Temp Alloys for Gas Turbines, Brunetaud et al (Ed.), Pub. Reidel, Dordrecht, p.207 (1982)

Galuska A.A., Uht J.C., Adams P.M. & Coggi J.M., J. Vac. Sci. & Technol., A <u>6</u>, (2), 185 (1988)

Gani M.S.J. & McPherson R., J. Mat. Sci., <u>15</u>, (9) 1915 (1980)

Gardner P.R., "Recent Developments in the United Kingdom in Ion Implantation Equipment for Engineering Components", (1987).

Gardner P.R., Dearnaley G. & Rosenbaum S., Surface Engineering, 3, (3) 203-209 (1987).

Garfinkle M., Met Trans., <u>4</u>, 1677 (1973)

Gass H. & Hintlemann H.E., Proc 4th Intl Conf On CVD, Princeton, Wakefield G.F. & Blocher J.M.(Ed.) Electrochem Soc.(1974)

Gates J.E. et al (1976), US Patent No 3951612

Gawne D.T. & Despres N.J., J. Vac. Sci. & Technol., A <u>3</u>, 2334 (1985)

Gebhardt W., Bonitz & Woll H., Berichte Fh.G.,3-78, 560 (1978)

Gedwill M.A., Glasgow T.K. & Levine S.R., Thin Solid Films,95,65 (1982)

Georg C.A., Triggs P. & Levy F., Mat Res Bull., 17, 105 (1982)

Gerdeman D.A. & Hecht N.L., 'Arc Plasma Technology in Materials Sci.', Springer-Verlag NY (1972)

Gerken J.M. & Owczarski W.A., Bull., W.R.C.109, Oct (1965)

Gesmundo (1985), see next ref

Gesmundo F., Mat. Sci. & Engg., 87, 243 (1987)

Giarda L & Pizzini S., 'Dep. of Monocryst. Si from the Gaseous Phase', Riv. Combustibili, 33, (3-4), 87 (1979)

Gibb W.H., 'Corrosion Resistant Materials for Coal Conversion', Meadowcroft D.B. (Ed.), Applied Science (1983)

Gibbons J.F., Hess L.D. & Sigmon T.W.(Ed.), 'Laser & Electron-Beam Solid Interactions & Materials Processing', Proc 1980 Mat. Res. Soc., Elsevier North-Holland NY (1981)

Giggins C.S. & Pettit F.S., WPAFB-Contr.No.F33615-72-C-1072 (1975)

Giggins C.S. & Pettit F.S., Oxid. Metals, 14, (5) 363 (1980)

Giggins C.S. & Pettit F.S., PWA Report, 5364, June (1975)

Giggins C.S. & Pettit F.S., 'Films & Coatings for Technology', Vol 2, Jacobson B.E. & Bunshah R.F.(Ed.), p 1-140 (1979)

Giggins C.S. & Pettit F.S., PWA Report No.FR-11545 (1981)

Giggins C.S. & Pettit F.S., 'Coatings for High Temp. Applications', Lang E. (Ed.) Appl. Sci. (1983)

Gillet R., Aubert A. & Gaucher A., Metals Technol., 10, 115 (1983)

Glang R., 'Handbook of Thin Film Technology', Maissel L.I. & Glang R.(Ed.), McGraw-Hill (1970), p1

Glascott J.G., Stott F.H. & Wood G.C., Oxid. Met.,24(3/4), (1985)

Glitz R., Notis M.R. & Goldstein J.I., Met Trans, 13B, 1921 (1982)

Gnanamuthu D.S., 'Applications of Lasers in Materials Processing', Metzbower E.A.(Ed.), Proc Conf ASM, Washington DC (1979) p 177

Gobet J. & Tannenberger H., J. Electrochem. Soc., 135, (1), 109 (1988)

Godfrey D.J., Paper no 740238, Proc Auto Eng Cong, Detroit, Feb (1974), Soc Auto Engrs

Godfrey D.J., Proc Brit Ceram Soc 26, 1-15 (1978)

Godfrey D.J., Materials & Design 4,759 (1983)

Godfrey D.J., chapter in "Nitrogen Ceramics", Riley F.L.(Ed.), p.647

Godfrey D.J. & Taylor P.G., Special Ceramics 4, Popper P.(Ed.), Brit Ceram Res Assoc, Stoke on Trent, p.265 (1968)

Godfrey D.J., Pitman K.C., Proc. Brit. Ceram. Soc., 26, 225 (1978)

Godlewska E. & Godlewski K., Oxid. Met., 22, (3/4), 117 (1984)

Godlewski K. & Godlewska E., Oxid. Met., 26, (1/2), 125 (1986)

Godlewski K & Godlewska E., Mat Sci & Engg 88, 103 (1987)

Godulyan L.V., 'Concentration of Deposited Metal in Coatings by Molten Salt Electrolysis', Zasch Metallov., 15, (2), 237-9 (1979)

Godulyan L.V., Andreev Yu.Ya. & Isaev N.I., 'Corr. of Cu-Ni Diffusion Coatings prepared by Molten Salt Electrolysis', Zasch Metallov., 15, (1), 128-9 (1979)

Goebel J.A., 'Advanced Coating Development for Industrial Utility Gas Turbine Engines', Proc.1st Conf.on Advanced Materials for Alternate Fuel Capable Directly Fired Heat Engines, US Dept Energy (1979)

Goebel J.A. & Pettit F.S., 'Metal-Slag-Gas Reactions & Processes', Proc. Conf. Foroulis Z.A. & Smeltzer W.W. (Ed.), Electrochem.Soc., (1975), p 693

Gokoglu S., J. Electrochem. Soc., 135, (6), 1562 (1988)

Gomati M.M.El et al, Surf Sci 152/153, 917-924 (1985)

Gooch T.G., Hurst R., Kroeckel H. & Merz M.(Ed.) 'Behaviour of Joints In High Temperature Materials', Appl Sci., (1983)

Goode P.D., "A Comparison between Surface Treatments for Metallurgical Applications", Harwell report AERE -R 11643 April (1985).

Goode P.D., Peacock A.T. & Asher J., "A study of the Wear Behaviour of Ion Implanted Pure Iron", Nuclear Instruments and Methods, North-Holland, Amsterdam, 209/210, 925-931 (1983).

Goryachev P.T., 'Carbideless Diffusion Coatings on Grey Iron', Metalloved Term Obrab Metals, (7), 44 (1978)

Goward G.W., J Metals, _22_, 31 (1970)

Goward G.W., Maatls Sci & Tech _2_, 194 (1986)

Goward G.W., Boone D.H. & Giggins C.S., Trans Qtly, _60_, 228 (1967)

Goward G.W. & Boone D.H., Oxid. Metals, _3_, 475 (1971)

Grabiec P.B. & Przyluski J., Surf. & Coatings Technol., _27_, 219 (1986)

Grabke H.J, 'High Temperature Material Corrosion in Coal Gasifier Atmospheres', Norton J.F. (Ed.), Elsevier (1984)

Grabke H.J.(1985), see in Rahmel et al (1985), p.253

Grabke H.J. & Wolf I., Mat. Sci. & Engg., _87_, 23 (1987)

Grabke H.J., Moeller R. & Schnas B., 'Behaviour of High Temperature Alloys in Aggressive Environments', Metal Soc., London (1980)

Graham J., High Temp.-High Pr., _6_, 577 (1974)

Graham J., High Temp. & High Press., _6_, 577 (1974)

Graham M.J. (1985), see in Rahmel et al (1985), p.301

Grant C.J., Brit.Corros.J, _14_, (1) 26 (1979)

Gras J.M., Mat. Sci. & Engg., _88_, 47 (1987)

Greco V.P. & Baldauf W., Plating, (1968) p 250

Green D.R., Nucl Sci & Eng., _12_, 271 (1962)

Green D.R., J. Appl. Phys, _37_, 3095 (1966)

Green D.R., Materials Evaluation, _25_, 231 (1969)

Green D.R. & K Day C., ASTM STP 439, (1968) p 4

Green D.R., J. Appl. Optics, _7_, 1779 (1968)

Green D.R., Schmeller M.D. & Sulit R.A., Oceans 82, IEEE

Green D.R. et al, 'Thermal Spray Testing', (1985) CME 126, Tech Alert Room 373, Ashdown House, 123 Victoria st, London SW1E 6RB

Green D.R., Voyles J.W. & Prati J.H., 'High Temp. Coatings', Khobaib M. & Krutenat R.C. (Eds.), Metall. Soc., AIME (1987)

Green M.L., Gross M.E., Papa L.E., Schnoes K.J. & Brasen D., J. Electrochem. Soc.,_132_, 2677 (1985)

Greene J.E. & Barnett S.A., J. Vac Sci & Technol.,_21_, 285 (1982)

Gregory E.N., Metal Construction, _12_, (12) 685 (1980)

Griffiths T.R., 'Proc. Conf UK Corrosion '87', Brighton, UK, Oct. (1987) p.235

Grimmer H., Grman D., Iniotakis N. & Zimermann U., Mat. Sci. & Engg., _87_, 189 (1987)

Grisaffe S.J., NASA Tech Note, NASA TN D-3133, (1965)

Grossklaus W., Ulion N.E. & Beale H.A., Thin Solid Films, _40_, 272 (1977).

Grossklaus W.D., Katz G.B. & Wortmann D.J, in Proc. of a Symp. on 'High Temp. Coatings', Orlando, Florida, Oct 7-9 (1986), Kbobaib M. & Krutenat R.C. (Eds.) Publ. by The Metall. Soc Inc. (1987), p.67

Grüber P.E., 'SiC Filament Reinforced Composite Materials', 22nd. Natl. SAMPE Symposium & Exhibition, San Diego, Calif., April 26-28 (1977)

Grünling H.W., Schneider K. & Singheiser L., Mat. Sci. & Engg., _88_, 177 (1987)

Guan H.Y., Huang Y.S., Lou H.Y. & Bai L.X., Proc.8th Intl. Cong., 'Metallic Corrosion', Mainz, Sept (1981), p 1099

Guenther K.H., J. Am. Opt. Soc., _3_, 20 (1986)

Guglielmi N., J. Electrochem. Soc., _119_ (8) 1009 (1972)

Guo Z.X. & Ridley N., Mat. Sci. & Technol., _3_, 945 (1987)

Gupta D.K. & Duvall D.S., 5th Intl. Symp., 'Superalloys', Seven Springs (1984), Gell M. et al (Ed.), p 711

Gupta D.K. & Rapp R.A. (1982): see idem, J Electrochem Soc _127_, 2194 & 2656 (1980)

Gurland J., Intl. Mat. Rev., _33_, (3), 151 (1988)

Gusman I.Ya & Tumakova E.I., DRIC Transl. 2570 (1971)

Habig K.H., J Vac Sci & Tech A _4_, 2832 (1986)

Hahn B.H., Jun J.H. & Joo J.H., Thin Solid Film, _153_, 115 (1987)

Hale E.B. et al, 'Ion Implantation into Metals' V.Ashworth, Grant W.A., Procter R.P.M.(Ed.), Pergamon (1982), p 111

Hall E.H. & Broehl J.H., 'A Computer Program for the Rapid Solution of Complex Chemical Equilibrium Problems', 8th Qtly Report to CVD Res GpII, Battelle, Columbus OH, April 15 (1967)

Halmshaw R.,error, see Helmshaw R.,'Industrial Radiology', Appl.

Sci Publ., London (1982)

Halnan W.K. & Lee D., High Tempe-rature Protective Coatings, 112th Mtg., Atlanta GA (1983), Singhal S.C.(Ed.), Met Soc AIME

Halstead W.D., 'Deposition & Corr. in Gas Turbines', (Conf. held in 1972), Hart A.B. & Cutler A.J.B. (Ed.), Applied Sci. (1973)

Hall S-K. (Ed.) "6th International Conference on Ion Beam Modifica-tion of Materials", Tokyo, Japan, 12-17 June (1988).

Hamburg G, Cowley P., Valori R.J., Am Soc Lub Engrs (7), 407 (1981)

Hamilton D., J. Appl. Phys., 31, 112 (1962)

Hammond M.L., Solid State Technol p.61, December (1979)

Hancock P., Corrosion of Alloys at High Temperatures in Atmospheres Consisting of Fuel Combustion Pro-ducts and Associated Impurities', HMSO, London (1968)

Hancock P. (1985), see in Rahmel et al (1985), p.305

Hancock P., Mat. Sci. & Engg., 88, 303 (1987)

Hancock P. & Nicholls J.R., Matls Sci & Tech 4, 398 (1988)

Hancock P., Nicholls J.R. & Steph-enson D.J., Surface & Coating Technol., 32, 285 (1987)

Hansen J.S. et al, US Bur Mines Report R1 - 8335, (1979)

Hansman & Mosle , Adhesion, 25, 332 (1981)

Hansman & Mosle , Adhesion, 26, 18 (1982)

Hara A., Yamamoto T. & Tobioka M., High Temp.-High Pr., 10 309 (1978)

Harding G.L., Proc.7th ICVM Conf., Tokyo (1982), p 677

Hardwick D.A., Thin Solid Films, 154, 109 (1987)

Hardwicke R., Metals & Materials, 3, (10), 586 (1987)

Hargreaves D. & Fromson S., 'World Index of Strategic Minerals Prod-uction Exploitation & Risk', Gower Publ Co NY (1983)

Harper J.M.E., 'Thin Film Proces-ses', Vossen J.L. & W Kern (Ed.) Acad. Press NY (1978) p175

Harris S.J., Mat. Sci. & Technol., 4, 231 (1988)

Hart A.B. & Cutler A.J.B.(Ed.) 'Deposition & Corrosion in Gas Turbines', Halstead Press, UK (1973)

Hart C. & Wearmouth R., US Patent 4,153,453, May (1979)

Hasegawa R. (Ed.), 'Glassy Metals: Magnetic, Chemical & Structural Properties', CRC, Boca Raton, FL (1983)

Hatakeyama F., Suganama K. & Oka-moto T., J. Mat. Sci., 21, 2455 (1986)

Hauffe K. & Martinez V.in Foroulis Z.A. & Smeltzer W.W.(Ed.), 'Metal - Slag - Gas Reactions & Proces-ses', Electrochem. Soc.(1975), Proc. Conf. Toronto, May (1975)., p355

Hauser D., Kammer P.A. & Dedrick J.H., Weld J., 46, (1) 11s-22s (1967)

Hava A., Yamamoto T. & Tobioka M., High Temp - High Pr., 10, 309 (1978)

Haworth C.W. & Jha R., J Mat Sci Letters, 2, 701 (1983)

Hayashi C., Proc.7th ICVM Conf. Tokyo (1982),p1

Hayes T.R., Evans J.F. & Rusch T.W., J. Vac. Sci. & Technol., A 3, (4), 1768 (1985)

Hayman C., 'Ceramic Surfaces & Surface Treatments', Morrell R. & Nicholas M.G. (Ed.), Brit. Ceram. Proc., No. 34, Aug. 1984, p.175

Hays C. & Anastas D., 'Modern Metallographic Methods for Solving Diffusion Problems', Metallography (1986)

Hays C. & Harris D.H., Surf. & Coatings Technol., 30, 83 (1987)

Headinger M.H. & Purdy M.J., ibid, 33, 433 (1987)

Heald S.M., Tranquada J.M., Chen H. & Welch D.O., J. Vac. Sci. & $$ Technol. A3 (6) 2432 (1985)

Hebbert R.A., Metals & Materials, 1, 38 (1978)

Hecht R.J., P&W Rep., FR12170 (1979)

Hecht R.J. & Halfpap D.S., 'Mat-erials Substitution & Recycling', AGARD - CP - 356 (1984) p 18-1

Hecht R.J. & Wright R.J., Proc 3rd US/UK Conf on Hot Corr of Gas Turb Matls for Marine, Bath, UK, p.11-6 (1976)

Heintz P., Trans Ind Inst Metals, Dec (1967) p 231

Heintze G.N. & McPherson R., Surf. & Coatings Technol., 34, (1), 15 (1988)

Helmer J.C. & Wickersham C.E., J. Vac. Sci. & Technol., A 4, (3), 408 (1987)

Helmersson U., Johansson, B.O., Sundgren J.E., Hentzell H.T.G. &

Billgren P., J. Vac. Sci. & Technol., A 3, (2), 308 (1985)

Helmshaw R., 'Industrial Radiology', Appl Sci Publ, London (1982)

Hemmings P.L. & Perkins R.A., EPRI FP-539 Proj. 716-1 Interim Rep., Dec. (1977)

Hemmings P.L., Perkins R.A., 'Thermodynamics Phase Stability Diagrams for the Analysis of Corrosion reactions in Coal Gasification/combination Atmospheres', Lockhead Co. LMDC D558238 (1977)

Hendricks R.C., McDonald G. & Mullen R.L., J. Vac. Sci. & Technol., A3 (6) 2456 (1985)

Henne R. & Weber W., High Temp. & High Press., 18, 223-232 (1986)

Herman H. & Shankar N.R., Mat. Sci. & Eng., 88, 69 (1987)

Herman H., 'Ion Implantation into Metals',V.Ashworth, Grant W.A. & Procter R.P.M.(Ed.) Pergamon Press (1982), p 102

Hettche L.R. & Rath B.B., 'Fabrication & Processing Technol for Improved Creep Tolerent Properties of Structure Alloys', Ref CME 10

Heywood A.E. & Benedek R.A., Pt. Met. Rev., 26 (3) 98 (1982)

Hildebrandt U.W., Wahl G. & Nicoll A.R., 'Materials & Coatings to Resist High Temperature Corrosion', Holmes D.R. & Rahmel A.(Ed.), Applied Sci., NY (1979), p 213

Hill R.J. & Boone D.H., 'Review of Technical Development & Production Experience in Gas Turbine Overlay Coatings', Proc.7th ICVM Conf., Tokyo (1982), p 338

Hill R.J.(Ed.) 'Physical Vapour Deposition', Airco Temescal (1976)

Hillery R.V., J. Vac. Sci. & Technol., A 4, (6), 2624 (1986)

Hilton M.R., Middlebrook A.M., Rodrigues G., Salmeron M. & Somorjai G.A., J. Vac. Sci. & Technol., 2797 (1986)

Hintermann H.E., Proc 3rd Intl Conf CVD, Hinsdale IL, Glaski F.A. (Ed.), Am.Nucl.Soc.(1972) p 352

Hintermann H.E., Proc.8th Intl Conf on CVD, (1981),p 473

Hirai T., 'CVD of Special Ceramics', Bull. Japan Inst Met, 17, (4), 313-9 (1978)

Hirai T. & Goto T., J. Mat. Sci., 16, 17 (1981)

Hirai T., Goto T. & Sakai T., 'Emergent Process Methods for High Technology Ceramics', Davies R.F.,

Palmour H. & Porter R.L. (Ed.), Plenum Press NY, (1982), p. 347

Hirth J.P. & Pound G.M., 'Condensation & evaporation', Macmillan NY, (1963)

Hirvonen J.K., J. Vac. Sci. & Technol., A 3, (6), 2691 (1985)

Hobbs L.W. (1985), see in Rahmel et al (1985), p.259

Hobbs M.K. & Reiter H., Surf. & Coatings Technol., 34, 33 (1988)

Hochman R.F., Erdemir A., Dolan F. & Thom R.L., J. Vac. Sci. & Technol., A 3, (6), 2348 (1985)

Hocking M.G., 'Interaction between SO_2 and Refractory Oxides', Proc. Conf. on Refractory Oxides, Odeillo (1977)

Hocking M.G., Rev. Int. Hts. Temp. Refract. 16, 177-182 (1979)

Hocking M.G. & Sidky P.S., Corr. Sci., 27, (2), 205 (1987)

Hocking M.G. & Sidky P.S., 'Hot Corrosion of Ni-based Ternary Alloys & Superalloys for Gas Turbine Applications. II: Mechanism of Corrosion in SO_2+O_2 Atmospheres', Corrosion Sci. 27, 205-214 (1987).

Hocking M.G. & Sidky, P.S.,'Proc. of the Conf. UK Corrosion '87', Brighton, Oct. (1987), Inst.Corr. Sci. & Technol., vol. 2, p.197

Hocking M.G. & Sidky P.S., UK Patent Appln. No.8811893 (1988)

Hocking M.G. & Sidky P.S., 'Effect of Pre-oxidation & Thermal Cycling on the Corrosion Behaviour of Fe-Cr-Al-based Alloys in a Coal Gasifier Atmosphere', Proc. U.K.Corrosion Conf., Brighton (1987) p197

Hocking M.G. & Vasantasree V., Proc. 3rd. UK/US Conf. on Gas Turbine Materials', Bath, UK (1976)

Hocking M.G. & Vasantasree V., 'Mechanism of the High Temperature Corrosion of Ni-Cr Alloys in SO_2+O_2', Corr. Sci. 16, 279 (1976)

Hocking M.G. & Vasantasree V., 'Mass Spectrometric Studies of Hot Corrosion', Proc. 3rd U.K./U.S. Conf on Gas Turbine Materials, Bath. (1976)

Hocking M.G. & Vasantasree V., 'Studies on some Basic Mechanisms of Hot Corrosion', Proc. 4th U.S/U.K. Conf. on Gas Turbine Materials in a Marine Environment', Annapolis, U.S.A. (1979)

Hocking M.G. & Vasantasree V., Corro. Sci., 21, 731 (1981)

Hocking M.G. & Vasantasree V.,

603

'Effect of Temperature on Hot Corrosion of MCrAlY Alloys in SO_2+O_2', Proc. 8th Intl. Congress on Metallic Corr., Mainz, Germany (1981)

Hocking M.G. & Vasantasree V., 'Effect of Silica and Platinum on Thermal Decomposition of Na_2SO_4', Corr. Sci. 21, 731-2, (1981)

Hocking M.G. & Vasantasree V, unpublished work (1985)

Hocking M.G. & Vasantasree V., 'Evaluation of Additive Effects on Ni-Cr- and Co-Cr- Alloy Hot Corrosion', Materials Sci.& Tech. 2, 318-321 (1986)

Hocking M.G. & Vasantasree V., 'High Temperature Corrosion of Ni-Cr-Al-Y Alloys in Molten Sodium Sulphate + Chloride Mixtures', Proc. Conf. on Interaction of Molten Salts & Metals', York (1986), p.284-291

Hocking M.G., Vasantasree V., Swidzinski M.A.M. & Carew J.A., Proc Electrochem Conf. GA (1977)

Hocking M.G., Vasantasree V. & Wai A.H., Proc. of Conf. on 'High Temp. Corrosion', San Diego, March 2-6 (1981), Rapp R.A. (Ed.) Publ. NACE, Houston, TX USA

Hocking M.G., Vasantasree V. & Wai A.H., 'Electrochemical Assessment of Hot Corrosion of MCrAlY Alloys' Proc. 31st Meeting of Intl. Soc. Electrochem., Venice (1980)

Hocking M.G., Vasantasree V. & Wai A.H., Hot Corrosion of MCrAlY Alloys', Proc. Intl. Conf. on High Temp. Corr., San Diego, U.S.A. (1981)

Hocking M.G., Vasantasree V., Swidzinski M.A.M. & Carew J.A., 'Effect of NaCl on Sulphate Hot Corrosion: Mass Spectrometric, Hot Corrosion and Chemical Analyses', Electrochem. Soc. Conf., Georgia, U.S.A. (1977)

Hocking M.G., Swidzinski M.A.M., Carew J.A. & V.Vasantasree, Proc. Electrochem.Conf., Pittsburgh, (1978)

Hocking M.G. & Vasantasree V., unpublished work (1987).

Hoffman D.W., Proc.7th ICVM Conf., Tokyo (1982), p 145

Hoffman D.W., J. Vac. Sci. & Technol., A 3, (3), 561 (1985)

Holder J.C., Lanauze R.D., Rogers E.A. & Thurlow G.C., 'Ash Deposits & Corrosion due to Impurities in Combustion Gases', Engineering Foundation Conf., Henniker, New Hampshire, June (1977)

Hollaham J.R. & Bell A.T. (Ed.), 'Techniques & Application of Plasma Chemistry', Wiley NY (1974)

Hollahan J.R. & Rosler R.S., 'Thin Film Processes', Vossen J.L. & Kern W(Ed), Acad. Press, NY (1978)

Holland L., 'Vacuum Deposition of Thin Films', Wiley NY (1956)

Holland O.W., White C.W., Pennycook S.J. & Fathy D., "Thermal Oxidn. of Ion Implanted Si", 6th Intl. Conf., Ion Beam Modfn. of Mat., Hall S.K.(Ed), Tokyo, 12-17 June (1988)

Hollar E.L., Rebarchik F.N. & Mattox D.M., J Electrochem Soc., 117, 1461 (1970)

Hollech H., Kuhl Ch. & Schulz H., J. Vac. Sci. & Technol., A 3, (6), 2345 (1985)

Hollech H., J. Vac. Sci. & Technol., A 4, (6), 2661 (1986)

Hollech H. & Schulze H., Thin Solid Films 153, 11 (1987)

Holman W.R. & Huegel F.J., J. Vac. Sci. & Technol., 11(4) 701 (1974)

Holman W.R. & Huegel F.J., Proc Conf CVD of Refractory Metals, Alloys & Compounds, Schaffhauser A.C.(Ed.), Am Nuclear Soc., (1967), p 127, 427

Holmes D.R. & Rahmel A.(Ed.), 'Materials & Coatings to Resist High Temperature Corrosion', Applied Sci., NY (1979), p.333

Holshire J.O., AFML-TR-76-230 (1976)

Holstein W.L., J. Electrochem. Soc., 135, (7), 1788 (1988)

Holthe K. & Kofstad P., Corrosion Sci., 20, 919 (1980)

Honey F.J., Kedward E.C. & Wride V., J. Vac. Sci. & Technol., A4 (6) 2593 (1986)

Hood C., Pt Met Rev., 20, (2) 48 (1976)

Hornbogen E., Klein K.H. & Schmidt I., J Mat Sci Letters, 1, (3) 94 (1982)

Horvath E. & Perry A.J., Thin Solid Films, 46, 209 (1977)

Hossain M.K., Noke A.C. & Saunders S.R.J., Oxid. Metals, 12, 451 (1978)

Hou P.Y. & Stringer J., Mat. Sci. & Engg., 87, 295 (1987)

Houben J.M. & Zaat J.H., 7th Intl Metal Spraying Conf., London (1973) p 11

Houben J.M. & Zaat J.H., Proc.7th

Intl. Metal Spraying Conf., London (1973), p 77

Houtman C., Graves D.B. & Jensen K.F., J. Electrochem. Soc., <u>133</u>, (5) 961 (1986)

Howe C.I. et al, 'Ion Implantation into Metals', Ashworth V., Grant W.A. & Procter R.P.M.(Ed.), Pergamon (1982), p 211

Hou P.Y. & Stringer J., Oxid. Met., <u>29</u>, 45 (1988).

Hsu H.S., Oxid. Met., <u>28</u>, (3/4), 213 (1987)

Hsu L.L., Surf. & Coatings Technol., <u>32</u>, 1 (1987)

Huang S.C. & Laforce R.P., 'Mat. Res. Symp. Proc.', Kear B.H. & Giessen B.C. (Ed.), North Holland, vol. 28, (1984) p.125

Huang T.T., Peterson B., Shores D.A. & Pfender E., Corros. Sci., <u>24</u>, 167 (1984)

Huang T.T., Richter R., Chang Y.L. & Pfender F., Metall. Trans. A, <u>16A</u>, 2051 (1985)

Hubbell F.N., Trans IMF 56 (1978)

Hübler G.K. & McCafferty E., Corr. Sci. <u>20</u>, 103 (1980).

Hübler G.K., Smidt F.A., Nucl. Instr. & Meth. B7/8 151 (1985).

Hüchtemann B. & Schuler P., Mat. Sci. & Engg., <u>87</u>, 197 (1987)

Huegel F.J. & Holman W.R., Proc 2nd Int Conf CVD, Blocher J.M. & Withers J.C.(Ed.) Electrochem Soc., (1971) p 171

Huibo P., Yuying W., Proc.7th ICVM Conf., Tokyo (1982), p 524

Hultgren R., Desai P.D., Hawkins D.T., Gleiser M., Kelley K.K. & Wagman D.D., 'Selected values of the Thermodynamic Properties of the Elements', Am. Soc. Metals, OH (1973)

Hunt L. & Sirtl E., J Electrochem. Soc., <u>119</u>, 1771 (1972)

Hunt L.P., J. Electrochem. Soc., <u>135</u>, (1), 206 (1988)

Huntz A.M., Mat. Sci. & Engg., <u>87</u>, 251 (1987)

Hurricks P.L., Wear <u>22</u>, 291 (1972)

Hutchings R., Loretto M.H., & Smallman R.E., Metal Sci., <u>15</u>, (1) 7 (1981)

Hutchings R., Mat. Sci. Eng. <u>69</u>, 129 (1985).

Hwang S.Y. et al, 'High Temperature Protective Coatings', Proc. Symp. Atlanta, GA, March (1983), Singhal S.C.(Ed.), Metall.Soc. AIME, p 121

Ilschner B. & Scutze M. (1985), see in Rahmel et al (1985), p.287

Inagawa K., Watanabe K. & Itoh A., Proc.7th ICVM Conf., Tokyo (1982), p 118

Inal O.T., Szecket A., Vigueras D.J. & Pak H., J Vac Sci Tech A <u>3</u>, 2605 (1985)

Ingham H.S. & Shepard A.P., 'Metco Flame Spray Handbook Vol III Plasma Flame Process', Metco Inc., Long Island NY (1965)

Ingle W.M. & Peffley M.S., J. Electrochem. Soc., <u>132</u>, (5), 1236 (1985)

Inomoto Y., Yamunaka K. & Mizuhara K., Proc.7th ICVM Conf., Tokyo (1982), p 209

Inspektor A. et al, J Vac Sci Tech A <u>4</u>, 375 (1986)

Inspektor A., Plasma Chem Plasma Process <u>1</u>, 377 (1981)

Inspektor A. et al, Proc 6th Conf on CVD, Jerusalem (1987), ed Forat R., Iscar Ltd, Israel

Inspektor-Koren A., Surface & Coatings Technol <u>33</u>, 31 (1987)

Inst. Mech. Eng., 'Strategic Metals and the UK'. The Materials Forum (1981)

Ionin V.E., Proc 7th ICVM, Tokyo, p.373 (1982)

Iqbal Z., Webb A.P. & Veprek S., Appl Phys Lett, <u>36</u>, 164 (1980)

Irving G.N., J.Stringer & Whittle D.P., Corrosion NACE, <u>33</u>, (2) 56 (1977)

Ishimori S. et al, Metal Finishing Soc.Japan, <u>28</u>, 508 (1977)

Ito T., Appl.Phys.Lett., <u>32</u> (5) 330 (1978)

Ito T., Nozaki T., Arakawa H. & Shinoda M., Appl Phys Lett., <u>32</u>, (5) 330 (1978)

Itoh H., Kato K. & Sugiyama K., J. Mat. Sci., <u>21</u>, 751 (1986)

Itoh, H., Watanabe N. & Naka S.,J. Mat. Sci., <u>23</u>, (1), 43 (1988)

Ivanov V.I., NDT Int., <u>17</u>,(6) 323-328 Dec (1984)

Ivanov V.K., 'Ti-Nitride Coating on SiC Fibre interaction with Complex Alloy Matrix', Fis Metallov i Metall, (3), 153-5 (1978)'

Iwamoto N., Proc.7th ICVM Conf., Tokyo (1982), p 283

Iwamoto N., Umeskai N., Katayama Y. & Kuroki H., Surf. & Coatings Technol., <u>34</u>, (1), 59 (1988)

Jackson M.R. & Rairden J.R., Met. Trans., <u>8A</u>, (11) 1829 (1977)

Jackson M.R. & Rairden J.R., Met. Trans., <u>84</u>,(10) 1697 (1977)

REFERENCES

Jacob K.T., Rao D.B. & Nelson H.G., Oxid Met <u>13</u>, 125-55 (1979)

Jacobs M.H., Surf. & Coatings Technol., <u>29</u>, 221 (1986)

Jacobson B.E., 'Microstructure of PVD-Deposited Films Characterised by Transmission Electron Microscopy' in 'Films & Coatings for Technology', CEI Course, Sweden (1981), Jacobson B.E. & Bunshah R.E.(Ed.) Sweden (1982)

Jacobson B.E. & McGuire G.E., 'Coatings for High Temperature Applications', Lang E.(Ed.), Applied Sci., NY (1983), p 217

Jacobson N.J. & Smialek J.L., J. Electrochem. Soc., <u>133</u>, (12), 2615 (1986)

James A.S., Fancey K.S. & Matthews A., Surface & Coatings Technol., <u>32</u>, 377 (1987)

Janssen M.M.P. & Rieck G.D., Trans AIME, <u>23</u>, 1372 (1967)

Jansson U., Carlsson J.O. & Stridth B., J. Vac. Sci. & Technol., A <u>5</u>, (5), 2823 (1987)

Jarrett T.C., 'The Science of Metallography & Metallurgy', in 'Proc. 28th Annual Meeting of the IPC, New Orleans, LA, (1985), Paper IPC-TP-540

Jata K.V. & Hubler G.K., J. Vac. Sci. & Technol., A <u>3</u>(6) 2677 (1985)

Jedlinski J. & Mrowec S., Matls Sci & Engrg <u>87</u>, 281 (1987)

Jeffes J.H.E., J Cryst Growth <u>3</u>,13 (1968)

Jehn H.A., Hofman S. & Munz W-D, Thin Solid Films, <u>153</u>, 45 (1987)

Jelacic C. & Petrovski P., 'Kinetics of Al & Al-Si Alloys Nitridation in N_2 & NH_3', La Ceramica, <u>31</u>, (4), 17 (1978)

Jena A.K., Chaturvedi M.C., J. Mat. Sci., <u>19</u>, 3121 (1984)

Jha B.B., Raj B. & Khanna B.S., Oxid. Met., <u>26</u>, (3/4), 263 (1986)

Johansen O.A., Dontje J.H. & Zenner R.L.D., Thin Solid Films, <u>153</u>, 75 (1987)

John R.C., J. Electrochem. Soc., <u>133</u>, (1), 205 (1986)

Johnson D.M., Whittle D.P. & Stringer J., Corrosion Sci., <u>15</u>, 649 (1975)

Johnson D.M., Whittle D.P. & Stringer J., Werkstoffe u Korros., <u>26</u>, (8) 611 (1975)

Johnson H.R. & Littler D.J.(Ed.), Proc.Internat.Conf.Marchwood, UK, May (1963), Butterworths, London

Johnson K.E. & Laitinen H.A., J. Electrochem Soc., <u>110</u>, 314 (1963)

Johnson P.C. & Randhwa H., Surf. & Coatings Technol., <u>33</u>, 53 (1987)

Johnson W.L., Progress in Mat. Sci., <u>30</u>, 81 (1986)

Johnson R.N., Schrock S.L., Whitlow G.A., J. Vac. Sci. & Technol., <u>11</u>, (4) 759 1974)

Jones D.W., Trans. Brit. Ceram. Soc., <u>84</u>, 40 (1985)

Jones R.L. & Gadomski S.T., J. Electrochem Soc., <u>124</u>, 1641 (1977)

Jones R.L., 'Metal - Slag - Gas Reactions & Processes', Foroulis Z.A. & Smeltzer W.W.(Ed.), Electrochem.Soc.(1975), Proc. Conf. Toronto, May (1975)., p762

Jones R.L. & Williams C.E., Mat. Sci. & Engg., <u>87</u>, 353 (1987)

Jorgensen M., Metal Construction, <u>12</u> (2), 88 (1980)

Jung T.C., Bao C.E. Fang M.H., Trans IMM, <u>95</u>, C63 (1986)

Juza R. & Cermak J., J Electrochem Soc., <u>129</u> (7) 1627 (1982)

Kamata K., 'Y_2O_2 & ZrO_2 Films by CVD', Yogyo Kyokai Shi, <u>90</u>, (1), 46 (1982)

Kameswari S., Oxid. Met., <u>26</u>, (1/2), 33 (1986)

Kaminsky M., Thin Solid Films, <u>73</u>(1) 117 (1980)

Kammer P.A., Monroe R.E. & Martin D.C., Weld J., <u>48</u>, (3) 116s-124s (1969)

Kamo R. et al, Proc Conf on Advanced Materials for Alternative Fuel Capable Directly Fired Heat Engines, Maine, Dec (1979) p832

Kandaswamy N., Seigle L.L. & Pennisi F.J., Thin Solid Films, <u>84</u>, 17-27 (1981)

Kannan K.R., Ph.D. Thesis, Univ of London (1985)

Kannan K.R., Vasantasree V. & Hocking M.G., 'Hot Corrosion of Ni-based Cermet Coatings', Proc. Conf. Coatings and Surface Treatments for Corrosion and Wear Resistance, Newcastle upon-Tyne, Apr. 1983, Ellis Horwood (1984), p.263

Kant R.A. & Sartwell B.D., J. Vac. Sci. & Technol., A <u>3</u>, 2675 (1985)

Karpinos D.M. et al, Porosh Metal <u>4</u>, (1973)

Kasperowicz W. & Schoop M.U., 'Das Electro-Metallspritz-Verfahren von Scoop' M.U., Charl Marhold Velagsbuchhandlung, Halle', (1920)

Kato A., Hirota Y., Maeda T., Tokonuga Y. & Nishi H., 'Sintering

Behaviour of B-Al$_2$O$_3$ Powder by Spray Pyrolysis Technique', Ceramurgia, _12_, (3), (1982)

Kato T., Ito K. & Maeda M., J. Electrochem. Soc., _135_, 455 (1988)

Kaufherr N., Fenske G.R., Busch D.E., Lin P., Despandey C. & Bunshah R.F., Thin Solid Films, _153_, 149 (1987)

Kaufman H.R., J. Vac. Sci. & Technol., _15_, 272 (1978)

Kaufman H.R., J Vac Sci Tech _4_, 764 (1986); ibid _3_, 1774 (1985)

Kayser H., 7th Intl Metal Spraying Conf., London (1973) p 5

Kear B.H. & Thompson E.R., Science, _208_, 847 (1980)

Kear B.H. & Giessen B.C. (Ed.), 'Rapidly solidified Metastable Materials', Mat. Res. Soc. Symp. Proc., Vol. 28, North-Holland, NY (1984)

Kedward E.C., Metallurgia, 225 (1969)

Kedward E.C., Cobalt _3_, 53 (1979)

Kedward E.C., Metallurgia, June (1969), p 225

Kedward E.C., Addison C.A. & Tennett A.A.B., Trans Inst Metal Finishing,_54_, 8 (1976)

Kedward E.C., Addison C.A., Honey F.J. & Foster J., Brit Patent 2,014,189 (1979), US Patent 4,305,792 (1981)

Kedward E.C. & Wright K.W., Plating & Surface Finishing, August (1978) p 38

Kedward et al (1987), see Honey F.J., Kedward E.C. & Wride V, J Vac Sci Tech A _4_, 2493 (1986)

Kehr D.E.R., High Temp.-High Pr., _10_, 477 (1978)

Keiser B.A., J Phys Chem _73_, 4264 (1979)

Kellner J.D. et al, CME 15, 'Titanium Diboride Electrodeposited Coatings' p 56 & Burnley N.A. Chart Mech Engr (CME)42, 'Surface Modification for Enhancement of High Temperature Corrosion Resistance of Superalloys',

Kelly T.J., 'Applications of Lasers in Materials Processing', Metzbower E.A.(Ed.), Proc Conf ASM, Washington DC (1979) p 43

Kennedy K., Trans. Internat. Vac. Metall. Conf., Am. Vac. Soc., NY (1968), p 195

Kennedy K.D., US Patent 3,560,252 Feb (1971).

Keribar R. & Morel T., Surf. & Coatings Technol., _30_, 63 (1987)

Kern W. & Ban S., 'Thin Film Processes', Vossen J.L. & Kern W.(Ed.) Academic Press, NY (1978)

Khanna A.S., Jha B.B. & Baldevraj, Oxid. Met., _27_, (1/2), 95 (1987)

Khanna A.S., Rodriguez P. & Gnanamoorthy J.B., Oxid. Met., _26_, (3/4), 171 (1986)

Kharchenko G.K., Auto Weld, _22_, (4) 31 (1969)

Kharmalov Y.A., Mat. Sci. & Engg., _93_, (9), 1 (1987)

Khobaib M. & Krutenat R.C. (Ed.), 'High Temperature Coatings', TMS-AIME, (1986) pp 209

Kim D.G., Yoo J.S. & Chun J.S., J. Vac. Sci. & Technol., A _4_, (2), 219 (1986)

Kim Y.H., Chang Y.S., Chou N.J. & Kim J., J. Vac. Sci. & Technol., A _5_, (5), 2890 (1987)

Kim Y.K., Pryzbylski K. & Yurek G.J., 'Proc. of Symp. on Fundamental Aspects of High Temp. Corrosion', Vol. II, Shores D.A. & Yurek G.J. (Eds.), Electrochem. Soc., Pennington, NJ (1986) p. 259

Kindlimann L.E. & Greeven M.V., 'High Temperature Protective Coatings', Proc. Symp. Atlanta, GA, March (1983), Singhal S.C.(Ed.), Metall. Soc. AIME, p 217

Kinniard S.P., Young D.J. & Trimm D.L., Oxid. Met., _26_, 417 (1986)

Kirchener H.P. & Seretsky J., NASA Rept CR 134644 (1974); see also idem, "Strengthening of Ceramics", publ Dekker, NY (1979)

Kirner K., Proc.7th Intl.Metal Spraying Conf., London (1973), p 190

Kishi T., Z. Metall., _76_, (7) 512-518 (1985)

Kissinger R.D., Nair S.V. & Tien J.K., 'Proc. of the Mat. Res. Symp.', Kear B.H. & Giessen B.C. (Eds.), North-Holland (1984) p.157

Kitahara S. & Hasui A., J. Vac. Sci. & Technol., _11_ (4) 747 (1974)

Kliminda P., Lingstuyl O., Lavelle B. & Dabosi F., 'Surface & Interfaces in Ceramic & Ceramic - Metal Systems', Pask J. & Evans A.(Ed.), Plenum NY (1980), p 477

Knippenberg W.F., Verspui & Von Kemenade A.W., 'Proc. 3rd Intl. Conf. on SiC, Miami, (1973), p.92

Knotek O., Bohmer M. & Leyendecker T., J. Vac. Sci. & Technol., A _4_, (6), 2695 (1986)

Knotek O., Elsing R. & Heintz H.R., Surf. & Coatings Technol.,

30, 107 (1987)

Knotek O. & Elsing R., Surface & Coating Technol., 32, 261 (1987)

Knotek O. & Lugscheider E., J. Vac. Sci. & Technol., 11 (4) 798 (1974)

Knotek O., Leyendecker T. & Jungblut F., Thin Solid Films, 153, 83 (1987)

Knotek O., Lugscheider E. & Eschnauer H.R., Proc.7th Intl.Metal Spraying Conf., London (1973), p72

Knotec O., Lugscheider E. & Eschnauer H., Stahleisen Bucher, Verlag Stahleisen, Dusseldorf (1975)

Knotek O., Reimann H. & Lohage P., Proc.7th ICVM Conf., Tokyo (1982), p 484

Kobayashi M. & Doi Y., Thin Solid Films, 54, 57 (1978)

Koch C.C., Cavin O.B., McKamey C.G. & Scarbrough J.O., Appl. Phys. Lett., 43, 1017 (1983)

Koenig H.R. & Maissel L.I., IBMJ Res & Dev., 168, March (1970)

Kofstad P., 'High Temperature Oxidation of Metals', John Wiley, NY (1966) p 269

Kofstad P., Proc. Intl. Cong. on Metallic Corrosion', Toronto, (1984), vol. 1, p.1 (1984)

Kofstad P., Oxid. Met., 24, (5/6), 265 (1985)

Kofstad P. & Akesson G., Oxid. Metals, 13, (1) 57 (1979)

Kohl & Fryburg (1978), see refs below

Kohl F.J., Santoro G.J., Stearns C.A. & Fryburg G.C., J.Electrochem Soc., 126, 1054 (1979)

Kohl F.J., Stearns C.A. & Fryburg G.C., 'Metal-Slag-Gas Reactions & Processes', Foroulis Z.A. & Smeltzer W.W.(Ed.),Electrochem.Soc., (1975), Proc. Conf., p 649

Kohno M. & Kuwahara K., Proc.7th ICVM Conf., Tokyo (1982), p 216

Koltun M.M., Geliotekhnika, 7, 70 (1971)

Komiya S.N.Umezu & Narusawa T., Thin Solid Films, 54, 51 (1978)

König U., Surf. & Coatings Technol., 33, 91 (1987)

König U. & Wolf G.K., Surf. & Coatings Technol., 33, 501 (1987)

Konuma M., Takagi H. & Nagasaka M., Proc.7th ICVM Conf., Tokyo (1982), p 560

Kört J.H., Fransen T. & Gellings P.J., Appl. Surf. Sci., 25, 237 (1986)

Kovalev V.N., Lyapin A.A. & Chursin M.M., 'High Intensity Cathode Sputtering', Teplofiz Vyso Temp, 16, (2), 418-9 (1978)

Kramer B.M., J. Vac. Sci. & Technol., A 4, 2810 (1986)

Kramer B.M., J. Vac. Sci. & Technol., A4 (6) 2870 (1986)

Kramer B.M. & Judd P.K., J. Vac. Sci. & Technol., A 3, 2439 (1985)

Kramer L.D., Kratz J.L. & Ortolano R.J., Proc.7th ICVM Conf., Tokyo (1982), p 508

Krecher W. & Pompe W., J Mat Sci 16, 694 (1981)

Kreider K.G., J. Vac. Sci. & Technol., A4 (6) 2618 (1986)

Kretzschmar E., 7th Intl Metal Spraying Conf., London (1973) p 43

Krier C.A. & Gunderson J.M., Met Eng. Q., 5, (5) 1 (1965).

Krier C.A., 'Vapor Deposition', Powell C.F. (Ed.), Wiley NY, p.512

Krivoruchko V.M., 'Thickness of $MoSi_2$-electrical Transport', Izv. Acad. Nauk. SSSR Neorg. Mat., 17, 355-6 (1981)

Krüger J., Int. Mat. Rev., 33, (3), 113 (1988)

Krutenat R.C., Proc.Gas Turbine Conf., NAVSEA/NAVAIR (1972) p 136

Kubachewski O., Alcock C.B. & Raynor G.V. (Ed.), 'Metallurgical Thermochemistry, 5th Edn., Pergamon Press, Oxford UK (1979)

Kubaschewski O., Pt Metals Rev., 15, (4) 134 (1971)

Kubaschewski O. & Hopkins, "Oxidation of Metals", Butterworth, London (1961)

Kuenzly & Douglass D.L., Oxid. Metals, 8, (4) 227 (1974)

Kulkarni A.G., Pai B.C. & Balasubramanian N., J Mat.Sci., 14 (2) 592 (1979)

Kun G. et al, Proc.7th ICVM Conf., Tokyo (1982), p 516

Kung S.C. & Rapp R.A., Surf & Coatings Technol 32, 41 (1987)

Kuo Y.S., Bunshah R.F. & Okrent D., J. Vac. Sci. & Technol., A 4, (3), 397 (1986)

Kurup M., Tanqir A. & Strutt P.R., 'Proc. Mat. Res. Soc. Symp.', Kear B.H. & Giessen B.C. (Eds.) North Holland (1984) p. 93

Kustas F.M., Misra M.S. & Tack W.T., J. Vac. Sci. & Technol., A 4, (6), 2885 (1986)

Kuznetsov (1976): please see ref 4 in Pampuch & Stobierski (1976)

Kvernes I., Oxid.Met, 6, (1) 45 (1973)

Kvernes I., 'Coatings of Diesel Engine Applications' in 'Coatings for High Temperature Applications' Lang E.(Ed.), Applied Sci., NY (1983), p 361

Kvernes I., 2nd Conf on Advanced Materials for Alternative Fuel Capable Directly Fired Heat Engines, Monterey CA

Kvernes I. & Forseth S., Mat. Sci. & Engg., 88, 61 (1987)

Kvernes I., Solberg J.K. & Lillerud K.P., Proc Conf on Advanced Materials for Alternative Fuel Capable Directly Fired Heat Engines, Maine, Dec (1979) p233

Lacombe P., Mat. Sci & Engg 96, 1 (1987)

Lad R.J. & Blakely J.M., Surf. Sci. 179, 467 (1987)

Lai G.Y., J. Met., 14 (1985).

Lakhtin Y.M., Kogan Y.D. & Borodin V.A., MADI, Consultants Bureau NY, 978 1977)

Lakshminarayanan G.R. et al, Plating 5, 63, 35, (1976)

Lamont L.T., J. Vac. Sci. & Technol.10, 251 (1973)

Lane P. & Geyer N.M., J Metals, 18, 186 (1966)

Lang E.(Ed.), 'Coatings for High Temperature Applications', Applied Sci., NY (1983)

Lang E. & Tottle L., COST 50-II, CCR-1, (1981)

Lang E., Bennett M.J. & Knights C.F., Mat. Sci. & Engg., 88, 37 (1987)

Lasalmonie A. & Strudel J.L., J. Mat. Sci., 21, 1837 (1986)

Lathem E.R., Proc Conf Environmental Degradation of High Temperature Materials, Isle of Man, March (1980) Vol 2, p6/50

Lau S.K. & Bratton R.J., 'High Temperature Protective Coatings', Proc. Symp. Atlanta, GA, March (1983), Singhal S.C.(Ed.), Metall. Soc. AIME, p 305

Laugier M.T., J. Mat. Sci., 21, 2269 (1986)

Le K.O., 'Ion Implantation into Metals', Ashworth V., Grant W.A. & Procter R.P.M.(Ed.), Pergamon Press (1982), p 277

Lee C.C., Matijasevic G., Cheng X. & Tsai C.S., Thin Solid Films, 154, 207 (1987)

Lee S. & Ariyasu T., J High Temp Soc., (Japan), 5, (4) 185 (1979)

Lee S.R. & Szekely J., J. Electrochem. Soc., 133, (10), 2194 (1986)

Leeds D.H., 'Some Observations on the Interface Between Plasma Sprayed Tungsten & 1020 Steel', Defence Documentation Center, AD-803286 (1966)

Lees D.G., Oxid. Met., 27, (1/2), 75 (1987)

Lehrer W. & Schwartzbart H., Weld J., 40, (2) 66s-80s (1961)

Levine N.S.R., & Caves R.M., J. Electrochem. Soc., 121, 1051-1064 (1974)

Levinstein M.A., Eisenlohr A. & Kramer B.E., Weld J, Weld Res Suppl., 40, 85 (1971)

Levy A., Bakker T., Scholz E. & Azadabeh M., 'High Temperature Protective Coatings', Proc. Symp. Atlanta, GA, March (1983), Singhal S.C.(Ed.), Metall. Soc.AIME, p 339

Levy A. & Macadam S., Surf. & Coatings Technol., 30, 51 (1987)

Lewis B. & Campbell D.S., J. Vac. Sci. & Technol., 4, 209 (1987)

Lewis H. & Smith R.A., Proc I Internat Congress on Metallic Corrosion, Butterworths, London, (1962).

Lewis M.W.J., J. Vac. Sci. & Technol., A 5, (5), 2930 (1987)

Lewis (1988): error, see Lewis M.W.J. (1987)

Lin C.J. & Spaepen F., 'Mat. Res. Symp. Proc.', Kear B.H. & Giessen B.C. (Ed.), North Holland, vol. 28, (1984) p.75

Lin S.S., J Electrochem Soc., 124, 1945 (1977)

Lindfors P.A., Mularie W.M. & Wehner G.K., Surf. & Coatings Technol., 29, 275 (1986)

Lindfors P.A., Mularie W.M. & Wehner G.K., Surf. & Coatings Technol., 31, 303 (1987)

Lloyd D.M., Rogers E.A., Oakey F.E. & Pittaway A J., Mat. Sci. & Engg., 88, 295 (1987)

Littlewood R. & Argent E.J., Electrochem.Acta, 4, 114 (1961)

Locke E.V. & Hella R.A., IEEE J Quantum Electronics, QE-10, (2), 179 (1974)

Locke E.V., Hoag E. & Hella R., Weld J, 51, 245 (1972a)

Locke E.V., Hoag E. & Hella R., IEEE J of Quantum Electronics, QE8 132 (1972b)

Louat N. & Sadananda K, Surf. & Coatings Technol., 32, 169 (1987)

Lowell C.E., Deadmore D.L. & Whittenberger J.D., Oxid.Metals, 17, (3/4) 205 (1982)

REFERENCES

Lowell C.E., Grisaffe S.J. & Levine S.R., NASA TM X-73499 (1976)

Lowenheim F.W.(Ed.), 'Modern Electroplating', Wiley NY (1963)

Lugscheider E., Eschnauer H., Hauser B. & Agelten R., Surf. & Coatings Technol., 30, 29 (1987)

Lugscheider E., Hauser B. & Bugsel B., Surface Coatings & Technol., 30, 73 (1987)

Lugscheider E., Krautwald A., Eschnauer H., Wilden J. & Meinhardt H., Surf. Coating & Technol., 32, 273 (1987)

Lugscheider E., Lu P., Hauser B. & Jager D., Surf. & Coatings Technol., 32, 215 (1987)

Lukschandel J., J Trans Inst Metal Finishing, 56, 18 (1978)

Luthra K.L., Tech.Report G.E.No 82CRDO81 & 82CRDO82, April (1982)

Luthra K.L. & Briant C.L., Mat. Sci. & Engg., 88, 348 (1987)

Luthra K.L. & Hall E.L., Oxid. Met., 26, (5/6), 385 (1986)

Luthra K.L. & LeBlank O.H., Matls Sci & Eng 87,329, (1987)

Luthra K.L. & Shores D.A., J Electrochem Soc 127, 2202 (1980); also see 4th US/UK Conf on Gas Turb Matls in Mar Envir 1, 525 (1979)

Lyman T. (Ed.), 'ASM Metals Handbook', 8th ed., Vol. 1, (1988)

Ma U., Ph.D. Thesis, Univ of London (1983)

Ma U. & Gawne D.T., Galvanotechnik, 76, (11), 1650 (1985)

Ma U., Baker E.H., Vasantasree V. & Hocking M.G., 'High Pressure Hot Corrosion of Ni-Cr-Ta Alloys, Proc. 9th Intl. Cong. on Metallic Corros., Toronto, June (1984) 2, 64-73

Machet J., 8, 341 (1983)

Madden M., Thin Solid Films, 154, 43 (1987)

Mah G., McLeod P.S. & Williams D.G., J. Vac. Sci. & Technol., 11, (4) 663 (1974)

Maissel L.I. & Glang R.,(Ed.) 'Handbook of Thin Film Technol.', McGraw-Hill NY (1970)

Majid El.Z. & Lambertin M.,'High Temperature Sulphidation of Aluminized Alloys: Protective Properties of Alumina Layers', JRC Petten, The Netherlands, Oct (1985)

Majid El.Z. & Lambertin M., Oxid. Met., 27, (5/6), 333 (1987)

Makyta M., Chrenkova M., Fellner P. & Matiasovsky K., Z. Anorg. Allg. Chemie, 540, 169 (1986)

Malik (1985), see Malik A.U. & Whittle D.P., Intl Cong on Met Corros, Toronto (1984) 1, 73 (1984)

Malik M.P., 7th Intl.Metal Spraying Conf., London (1973) p 257

Maltal G.S., Schwartz G.C. & Smilonsky G. (Ed.), 'Plasma Processing', 1983 Symp., Electrochem. Soc.

Malting H.A. & Steffens H.D., Metall., 17, 583, 905, 1213 (1963)

Mantell C.E.(Ed.), 'Encyclopaedia of Eletrochemistry', (1964).

Manning M.I. (1985), see in Rahmel et al (1985), p.284

Marantz D.R., 7th Intl Metal Spraying Conf., Tokyo (1973), p 53

Marchandise H., 'The Plasma Torch & its Applications', European Atomic Energy Community, EUR 2439 f, (1965)

Marchon B., Ferrer S., Kaufamnn D.S. & Salmeron M., Thin Solid Films, 154, 65 (1987)

Mari P.A., Chaix J.M. & Larpin J.P., Oxid. Met., 17, 315 (1982)

Marijnissen G.H. & Klostermann J.A., 'Behaviour of High Temperature Alloys in Aggressive Environments', Proc. Internatl. Conf., Petten, Netherlands, Oct.1979, Kirman J., Marriott J.B., Merz M., Sahn (Ed.) (1980)

Maritan C.M. et al., J. Electrochem. Soc., 135, (7), 1793 (1988)

Marquez & Olivares (1987): see Mora-Marquez & Lira Olivares (1987)

Marriott J.B., Merz M., Sahn P.R. & Whittle D.P.(Ed.), The Metals Soc U.K. (1980), p 243

Marriott J.B., Merz M., Sahn P.R. & Whittle D.P.(Ed.), The Metals Soc U.K. (1980), p 813

Martin P.J., Mckenzie D.R., Netterfield R.P., Swift P., Filipczuk S.W., Muller K.H., Pacey C.G. & James B., Thin Solid Films, 153, 91 (1987)

Martinengo P.C., Carughi C., Ducati U. & Coccia G.L., 'Materials & Coatings to Resist High Temperature Corrosion', Holmes D.R. & Rahmel A.(Ed.), Applied Sci., NY (1977), p293

Marynowski C.W. et al, Elechem Technol 3, 109 (1965)

Mascolo G. & Marino O., Mat. Sci. & Engg., 88, 75 (1987)

Masters R.G., J. Vac. Sci. & Technol., A **3**, (2), 324 (1985)

Mathews A. & Teer D.G., Thin Solid Films, **72**, 541 (1980)

Matiasovsky K., Fellner P., Chrenkova M., Lubyova Z. & Silny A., Chem. Papers, **41**, (4), 527 (1987)

Matsuda A., J. Non-Cryst. Solids, **56-60**, 768 (1983)

Matsuda T., Nakae H. & Hirai T., J. Mat. Sci., **23**, (2), 509 (1988)

Matsuda T., Uno N., Nakae H. & Hirai T., J. Mat. Sci., **21**, 649 (1986)

Matsumoto O., Hayami E., Konuma M. & Kanzaki Y., Proc.7th ICVM Conf., Tokyo (1982), p 576

Matting H.A. & Steffens H.D., Metall., **17** (6) 583 (13), 17,(9) 905 (1963)

Mattox D.M., Electrochem. Technol., **2**, 295 (1964)

Mattox D.M., J. Vac. Sci. & Technol., **10**, 47(1973)

Mattox D.M., 'Adhesion Measurements of Thin Films, Thick Films & Bulk Coatings', ASTM STP 640, Mittal K.L.(Ed.), ASTM (1978) p 54

Mattox D.M., 'Ion Plating Technology in Films & Coatings for Technology', Jacobson B.E. & Bunshah R.F.(Ed.) vol. 1 Sweden (1981)

Mattox D.M., 'Deposition Technologies for Films & Coatings', Bunshah R.F. (Ed.), Noyes (1982), Chapter 6

Mattox D.M. & Bland R.D., J.Nucl Mat, **21**, 349 (1967)

Mattox D.M. & Kominiak G.J., J. Vac. Sci. & Technol., **9**, 528(1972)

Mattox D.M. & Sowell R.R., J. Vac. Sci. & Technol., **11**(4) 793 (1974)

Mattox D.M., Cultrell R.E., Peebles C.R. & Dreike B.L., Surf. & Coatings Technol., **33**, 425 (1987)

Matuschka A.G.V., 'Borieren', Vevey Delta Verlag AG, Switzerland (1979)

Mawella K.J.A. & Honeycombe R.W.K., 'Proc. Mat. Res. Soc. Symp.', Kear B.H. & Giessen B.C. (Eds.) North Holland (1984) p. 65

May E.R., Zetlmeisl M.J., Bsharah L. & Annand R.R., Ind. Eng. Chem. Prod. Des. Res., **11**(4) 438 (1972)

May W.D., Smith N.D.P. & Snow C.I., Trans Inst. Metal Finish, **34**, 369 (1957)

Mazars P. & Manesse D., Matls Sci & Eng **88**, 21 (1987)

Mazumder J., 'Lasers in Metallurgy', Proc Conf Metall Soc AIME Chicago, Feb (1981), Mukherjee K.

& Mazumder J.(Ed.) (1981), p 221

Mazumder J. & Allen S.D., SPIE **198**, Laser Appln. in Mat. Processing, 73 (1979)

McCafferty E., Clayton C.R. & Oudar J. (Ed.), 'Fundamental Aspects of Corrosion Protection by Surface Modification', 1983 Symp., Electrochem. Soc. (1983)

McCarron R.L., 'Corrosion & Erosion in PFBC Environments', EPRI report Corr. Sci.-3363, Palo Alto CA (1984)

McClanahan E.D., Busch R., Fairbanks J. & Patten J.W., Gas Turbine Conf., Zurich, March 1974, ASME 74-GT-100

McClanahan E.D., Proc.7th ICVM Conf., Tokyo (1982), p 481

McConica C.M. & Krishnamani K., J. Electrochem. Soc., **133** (12) 2542 (1986)

McGill I.R. & Weinbaum M.J., Metal Progress, Feb (1979), p26

McGonnagle W.J. (Ed.), 'Intl. Advances in Non-Destructive Testing', Gordon & Breach Sci. Pub. (1981) vol. 7

McGuire G.C., Jacobson B.E. & Bunshah R.F.(Ed.) 'Films & Coatings for Technology (1981)

McKee D.W., Shores D.A. & Luthra K.L., J.Electrochem Soc., **125**, 411 (1978)

McKelliget J., El-kadaah N. & Szekely J., Proc.7th ICVM Conf., Tokyo (1982), paper 6-3

McNamara D.K. & Aheam J.S., Int. Mat. Rev., **32**, (6), 292 (1987)

McPherson R., J Mat Sci., **15**, 3141 (1980)

Mcttargue C.J. & Yust C.S., J Am Ceram.Soc.**67**, 117 (1984)

Meadowcroft D.B. & Manning M.I.(Ed.), 'Corrosion Resistant Materials for Coal Conversion Systems', Elsevier Applied Sci Pub., (1983), Proc. Conf., London, May (1982)

Meadowcroft, D.B., Mat. Sci. & Engg., **88**,313 (1987)

Mehan R.L. & McKee D.W., J. Mat. Sci., **11**, 1009 (1976)

Mehan R.L. & Bolon R.B., J. Mat. Sci, **14**, 2471 (1979)

Mehan R.L., Jackson M.R. & McConnell M.D., ibid, **18**, 3195 (1983)

Mehrabian R., Hsu S.C., S Kou & Munitz A., 'Applications of Lasers in Materials Processing', Metzbower E.A.(Ed.), Proc Conf ASM, Washington DC (1979) p 213

Mehrotra P.K. & Quinto D.T., J. Vac. Sci. & Technol., A $\underline{3}$, (6) 2401 (1985)

Meisel H., Johner G. & Scholz A., Praktische Metallographie, $\underline{17}$, 261 (1980)

Mellottee H., Cochet G. & Delbourgo R, Rev.Chim.Min.,$\underline{13}$,373 (1976)

Mellors & Senderoff, J Electrochem Soc 113, 60 (1966) and see Senderoff (1966)

Memarian H., Budhani R.C., Karim A.A., Doen H.J., Destpandey C.V., Bunshah R.F. & Doi A., J. Vac. Sci. & Technol., A $\underline{3}$, 2434 (1985)

Menezes C., Fortmann & Casey S., J. Electrochem. Soc., $\underline{132}$, (3), 709 (1985)

Messbacher A., Wt-Z Ind Fertig, $\underline{65}$, 619 (1975)

Metal Finishing Handbook, Annual, Inst of Metal Finishing

Metzbower E.A.(Ed.), 'Applications of Lasers in Materials Processing', Proc Conf ASM, Washigton DC (1979)

Mevrel R. & Duret C., 'NATO Advanced Workshop on Coatings for Heat Engines', Italy (1984)

Mevrel & Pichoir (1984), see below

Mevrel R. & Pichoir R., Mat. Sci. & Engg., $\underline{88}$, 1 (1987)

Middleman S. & Yeckel A., J Electrochem Soc 133, 1951 (1986)

Mih D.T., 'Improved Low Frequency Eddy Current Inspection for Cracks', CME 39 (1980). Orders to Tech Alert, Technol.Rep Centre, Orpington, Kent BR5 3RF

Mikoni P. & Green A.K., J. Vac. Sci. & Technol.,11, 849,940 (1974)

Millbank P. (Ed.), 'Metal Bulletins Prices & Data', (1987) Bull. Books Ltd. (1987) Park House, Worcester Pk., Surrey, KT4 7HY, UK

Miller D.M., Kung S.C., Scanberry S.D. & Rapp R.A., Oxid. Met., $\underline{29}$, (3/4), 239 (1988)

Miller R.A., 'High Temperature Protective Coatings', Proc. Symp. Atlanta, GA, March (1983), Singhal S.C.(Ed.), Metall.Soc.AIME, p 293

Miller R.A., J. Am. Ceram Soc., $\underline{67}$, 517 (1984)

Miller R.A., 'High Temperature Corrosion-Resistant Coatings', Streiff R. (Ed.) (1987a)

Miller R.A., Surf. & Coatings Technol., $\underline{30}$, 1 (1987b)

Minato K. & Fukuda K., J. Mat. Sci., $\underline{23}$, (2), 699 (1985)

Minkevich A.N., 'TiC Coatings on Steels & Alloys', Metalloved Term Obrab Metals, (6), 36-40 (1979)

Mirtich M.J., 'Ion Beam Deposited Protective Films', Ref N81-19221, p 17, Tech.Alert, (1981) (see Mih above)

Misra A.K., Oxid. Met., $\underline{25}$, (3/4), 129 (1986)

Mitani Y., 'High Temperature Protective Coatings', Proc. Symp. Atlanta, GA, March (1983), Singhal S.C.(Ed.), Metall.Soc.AIME, p 251

Mitchell C.M. & Lauffer E.E., Wear 61, 111 (1980)

Mito H. & Sekiguchi A., J. Vac. Sci. & Technol., A $\underline{4}$, 475 (1986)

Miyoshi K., Buckley D.H. & Spalvins T., J. Vac. Sci. & Technol., A $\underline{3}$, (6), 2340 (1985)

Miyoshi K., Buckley D.H., Pouch J.S., Alterwitz S.A. & Sliney H.E., Surf. & Coatings Technol., $\underline{33}$, 221 (1987)

Moffat H.K. & Jensen K.F., J. Electrochem. Soc., $\underline{135}$, 459 (1988)

Molarius J.M., Kirkonen A.S., Harju E. & Lappalainen R., Surf. & Coatings Technol., $\underline{33}$, 117 (1987)

Moloney A.C., 'Suface Analysis & Pretreatment of Plastics & Metals' Brewis D.M.(Ed.) Appl Sci Publ., (1982), p 175

Mom A.J.A. & Hersbach H.J.C., Mat. Sci. & Engg., $\underline{87}$, 361 (1987)

Monson G., Ph.D. Thesis, Univ. of London (1986)

Monson G. & Steen W.M. (1986), private communication (see also ref above)

Moon D.P., Thesis Univ of Oxford UK (1987)

Moon D.P., To be published in Surf Interface ANAL. (1988/9).

Moon D.P. & Bennett M.J., 'The Effects of Reactive Element Oxide Coatings on the Oxidation Behaviour of Metals and Alloys of High Temperatures', AERE R12757, AGR-FPWG Aug (1987), p.1051

Moore G.D., Ritter J.E., J. Vac. Sci. & Technol., $\underline{11}$, 754 (1974)

Moore P., Kim C. & Weinman L.S., 'Applications of Lasers in Materials Processings', Metzbower E.A.(Ed.), Proc Conf ASM, Washington DC (1979) p 259

Moorhouse P. & Murray K.E., COST 50/II UK, Proj. 11, (1979)

Mora-Marquez J.G., Lira-Olivares J Thin Solid Films, $\underline{153}$, 243 (1987)

Morbioli R, Steinmetz P. & Duret C Mat. Sci. & Engg., $\underline{87}$, 337 (1987)

Mordike & Bergmann (1984), see Bergmann H.W. & Mordike B.L. in Z Werkstoffe Tech 14, 228 (1983)

Mori H., Fujita H., Tendo M. & Fujita M., Scripta Metall., 18, 783 (1984)

Morin F., Oxid. Met., 25, (5/6), 351 (1986)

Morojima S. et al, J Mat Sci., 14, (2) 496 (1979)

Morojima S., Iwamori N, Haltori T, Kurasawa K, ibid, 21, 1363 (1986)

Morrell R. & Nicholas M.G., 'Ceramic Surfaces & Surface Treatments' Brit. Ceram. Proc., (34) (1984)

Morris J., Patel P.M., Almond D.P. & Reiter H., Surf. & Coatings Technol., 34, (1), 51 (1988) and NDT Intl 18, 17 (1985)

Mort J. & Jansen F. (Eds.), 'Plasma Deposited Thin Films', CRC Press (1986)

Movchan B.A. & Bunshah R.F., Proc.7th ICVM Conf., Tokyo (1982), p 96

Movchan B.A. & Demchischin A.V., Fiz. Metall, Metalloved, 28, 653 (1969)

Movchan B.A. & Demchischin A.V., Phys Met Metallogr., 28, 83 (1969)

Movchan B.A., Demchischin A.V. & Kulak L.D., J. Vac. Sci. & Technol., 11, (4) 640 (1974)

Motojima S., Fujimoto A. & Sugiyama K., J. Mat. Sci. Letters, 1, 19 (1982)

Motojima S., Iwamori N. & Hattori T., J. Mat. Sci., 21, 3836 (1986)

Motojima S., Kani E., Takahashi Y. & Sugiyama K., J.Mat.Sci., 14, (6) 1495 (1979)

Motojima S., Ohutska Y. et al, J Mat Sci., 14 (2) 496 (1979)

Motojima S., Takai K. & Sugiyama K., J.Nucl.Mat., 98, 151 (1981)

Motojima S., Yagi H. & Iwamori N., J. Mat. Sci. Lett., 5, 13 (1986)

Motijima S., Yoshida H. & Sugiyama K., J Mat Sci Letters,1, 23 (1982)

Mozar Th., Francke W. & Tichler J.W., Proc 7th Intl Metal Spraying Conf, London (1973), p.151

Mrowec S., "Properties of High Temperature Alloys", Foroulis Z.A. & Pettit F.S.(Ed.), Electrochem Soc, NJ (1976)

Mrowec S., see in Rahmel et al (1985), p.266

Mrowec S. & Przybylski, 'Transport Properties of Sulfide Scales and Sulfidation of Metals & Alloys', J. Oxid Met., 23, (3-4) 107 (1985)

Mrowec S., Stoklosa A. & Damielewski M., Oxid.Met., 11, 355 (1977)

Mrowec S. & Werber T., 'Gas Corrosion of Metals', Natl. Bur. Std., Washington DC (1978), Avail. US Dept. Commerce, Springfield VA 22161

Muamba J.M.N.,Streiff R. & Boone D H., Mat Sci & Engg.,88, 111 (1987)

Muehlberger E., 7th Intl Metal Spraying Conf., London (1973) p245

Mukherjee K. & Mazumder J.(Ed.) 'Lasers in Metallurgy', Proc Conf Metall Soc, AIME, Chicago, Feb (1981)

Mukherjee S.K. & Upadhyaya G.S., 'Oxidation Behaviour of Sintered 434L Ferritic Stainless Steel-Al_2O_3 Composites with Ternary Additions', Oxid Met., 23, 177 (1985)

Mullendore A.W., Whitly J.B. & Mattox D.M., Thin Solid Films 83, 79 (1981)

Mullendore A.W. & Pope L.E., Thin Solid Films, 153, 267 (1987)

Muller K.N., Proc.7th Intl.Metal Spraying Conf, London (1973) p165

Munir Z.A., Am. Ceram. Soc. Bull., 67, (2), 342 (1988)

Münz W.D., J. Vac. Sci. & Technol., A 4, (6), 2717 (1986)

Münz W.D., Reineck S.R. & Hatig K.W., Proc.7th ICVM Conf., Tokyo (1982), p 633

Münz W.D., Reineck S.R. & Kienel G., ibid., (1982), p 641

Murakami S., Baba S. & Kinbara A., ibid, (1982), p 658

Murayama Y., J. Vac. Sci. & Technol., 12, 818 (1975)

Na S.J. & Bae K.Y., Surf. & Coatings Technol., 31, 273 (1987)

Nagai H. & Okabayashi M., Trans Japan Inst Metals, 22, 101 (1981).

Nagelberg A.S., J Electrochem Soc 132, 2502 (1985)

Naka K. et al, Proc.7th ICVM Conf., Tokyo (1982), p 689

Nakamura K. et al, J Japan Inst Metals 38, 913 (1974)

Nakamura K., J. Electrochem. Soc., 132, (7), 1757 (1985)

Nakamura K., J. Electrochem. Soc., 133, (6), 1120 (1986)

Nakamura K. & Hosokawa N., Proc.7th ICVM Conf., Tokyo (1982), p 610

Nakamori M., Harada Y. & Hukue I., 'High Temperature Protective Coatings' Proc.Symp.Atlanta, GA, March (1983), Singhal S.C.(Ed.), Metall. Soc. AIME, p 175

Napier J.C. & Metcalfe A.G., Soc Auto Eng, paper 770342 (1977)

Narita T. & Ishikawa T., Mat. Sci. & Engg., 87, 51 (1987)

Narita T. & Smeltzer J., Oxid. Metals, 21, (1/2) 39, 57 (1984)

Narita T. & Ishikawa T., Mat. Sci. & Engg., 87, 51 (1987)

Narita T. & Sme., J. Mat. Sci., 21, 2493 (1986)

Nat. Acad. Sci., 'High Temperature Oxidation-Resistant Coatings', Washington D.C. (1970).

Natesan K. in 'High Temperature Corrosion', NACE-6, Rapp A. (Ed.) NACE, Houston, (1983) p.336

Natesan K., Corrosion, 41, 646 (1985).

Natesan K., Mat. Sci. & Engg., 87, 99 (1987)

Natesan K. & Delaplane M.B., 'Corrosion-Erosion Behaviour of Materials', Natesan K. (Ed.), Am. Inst. Min., Metall. & Pet. Eng. NY, (1978), p.1

Natesan K. & Delaplane M.B., Proc. of Symp. on Corrosion-Erosion Behaviour of Materials', Natesan K. (Ed.), American Inst. of Mining, Metallurgicial & Petroleum Engineers, NY (1978) p.1

Nayak B. & Misra M., J Appl Electrochem., 9, 699 (1979)

Neckarsulm N.S.U., Brit.Patent 1,200,410, July (1970)

Nederveen M.B. US Federal Sci. & Tech. Inf., Springfield VA 22161

Nederveen M.B. et al, Proc 9th Intl Metal Spraying Conf., The Hague, Netherlands (1980)

Neiri B. & Vandenbluck L., J. Less Common Metals, 95, 191-303 (1983)

Nelson R.L. et al, Thin Solid Films, 81, 329 (1981), US Patent 2,023,453A

Nelson R.N., Woodhead J.L. et al, 7th Intl Metal Spraying Conf., London (1973) p 96

Nesbitt J.A. & Heckel R.W., Thin Solid Films, 119, 281 (1984)

Ness J.N.& Page T.F., J. Mat. Sci., 21, 1377 (1986)

Newcomb S.B. & Stobbs W.M., Oxid. Met., 26, (5/6), 431 (1986)

Newcomb S.B. & Stobbs W.M., Mat. Sci. & Technol., 4 (5) 384 (1988)

Newhart J.E., 'Corrosion 73', NACE, Houston(1973) paper 113

Newkirk J.B., Trans. AIME, 215, 483 (1959)

N'gandu Muamba J.M., Strife R. & Boone D.H., Mat Sci & Eng 88, 111 (1987)

Nickel H., J. Vac. Sci. & Technol., 11, 687 (1974)

Nicoll A.R., 'Protective Coatings & Their Processing - Thermal Spray', CEI Course on High temperature Materials & Coatings, Finland, June (1984)

Nicoll A.R., 'The Environment High Temperature Oxidation & Hot Corrosion', CEI Course, on High Temperature Materials & Coatings, Finland, June (1984)

Nicoll A.R., 'Coatings for High Temperature Applications', Lang E. (Ed.) Applied Sci., NY (1983) p269

Nicoll A.R., Surf. & Coatings Technol., 30, 223 (1987)

Nicoll A.R., Wahl G. & Hildebrandt U.W., 'Materials & Coatings to Resist High Temperature Corrosion', Holmes D.R. & Rahmel A.(Ed) Applied Sci., NY (1979) p233

Nicholls J.R. & Lawson K., Metallurgia, 51(7) 267(1984)

Nicholls J.R. & Hancock P., Ind. Corr., 5, (4), 8 (1987)

Nicholls J.R. & Stephenson D.J., paper at HT Corros of Superalloys Conf, Roy Soc London, Feb 1986; publ in Mat Sci & Tech

Nieminen J.A. & Kaski K., Surf. Sci., 185, L489 (1987)

Niihara K. & Hirai T., Proc.7th ICVM Conf., Tokyo (1982), p 180

Niklasson G.A. & Granqvist C.G., ibid., (1982)., p666

Niku-Lari A. (Ed.), 'Advances in Surface Treatments' Pergamon(1987)

Nimmagada R. & Bunshah R.F., J Vac Sci Tech 12, 585 (1975)

Nimmagadda R., Raghuram A.C. & Bunshah R.F., ibid, 9, 1406 (1972)

Nishiyuama Y., 'Present Status of Heat Resistant Materials for Gas Turbines', Tetsu To Hagane, 69, 1257-65 (1983)

Nomaki K., 'Glass & Ceramics Metal Bonding', Ceramics Japan, 15, (6), 411-417 (1980)

Norton J.F. (Ed.) "High Temp Matls Corros in Coal Gasifier Atm", Elsevier Appl Sci Publ (1984)

Notton J.H.F., Pt Met Rev., 21 (4) 122 (1977)

Nowicki R.S., Buckley W.D., Mackintosh W.D. & Mitchell I.V., J. Vac. Sci. & Technol., 11,(4), 675 (1974)

Nyaiesh A.R., Kirby R.E., King F.K. & Garwin E.L., J. Vac. Sci. & Technol. A 3, (3), 610 (1985)

Obrzut J.J., Iron Age, <u>209</u>, (16) 92 (1972)

Oda K.,Yoshio T & Oka K.O., Comm. Am.Ceram.Soc., Jan, C-8 (1983)

Odohira T. et al, Surface & Coatings Technol <u>33</u>, 301 (1987)

Ogawa T., Ikawa K. & Iwamoto K., J.Mat.Sci.<u>14</u> (1) 125 (1979)

Ohashi O. & Hashimoto T., J.Jap Weld Soc., <u>45</u> (6) (1976)

Ohno S., J.High Temp Soc of Japan, <u>5</u>, (5) 218 (1979)

Okamoto I. & Arata Y., Proc.7th ICVM Conf., Tokyo (1982), p 650

Oki T., ibid., (1982), p 110

Olivares J.L. & Grigoresen I.C., Surf. & Coatings Technol., <u>33</u>, 183 (1987)

Oliver W.C., Huchings R. & Pethica J.B., Met.Trans A15A, 2221 (1984).

Oliver W.C., McHargue C.J. & Zincle S.J., Thin Solid Films <u>153</u>, 185 (1987)

Osaka T. et al., J. Electrochem. Soc., <u>133</u>, (11), 2345 (1986)

Ostojic P. & Berndt C.C., Surf. & Coatings Technol., <u>34</u>, (1), 43 (1988)

Ött D. & Raub C.J., Pt Metals Rev., <u>20</u>, (3) 79 (1976)

Ött D. & Raub Ch.J., Pt. Met. Rev., <u>30</u>, (3), 132 (1986)

Ött D. & Raub Ch.J., Pt. Met. Rev., <u>31</u>, (2), 64 (1987)

Ött D. & Raub., Surf. & Coatings Technol., <u>32</u>, 153 (1987)

Oxley J.H. et al, Nucl Appl, <u>1</u>, 567 (1965a)

Oxley J.H., Powell C.F. & Blocher J.M., US Patent, 3,178,308 (1965b)

Pabst R.F. & Ellsner G., J Mat Sci., <u>15</u>, 188 (1980)

Packer C.M. & Perkins R.A., "Behaviour of High Temperature Alloys in Aggressive Environments', Proc. Intl. Conf., Petten, Netherlands, Oct.1979, Kirman J., Marriott J.B, Merz M, Sahn P.R.(Ed) (1980)

Padmanabhan H.R. & Sorensen G., 'Ion Implantation into Metals', Ashworth V., Grant W.A. & Procter R.P.M.(Ed.), Pergamon Press (1982), p 352 & 361

Padmanabhan K.R. & Sorensen G. (1980), see Thin Solid Films <u>81</u>,13 (1983)

Palmqvist S., Jernkontorets Ann, <u>141</u>, 300 (1957)

Palombarini G. & Carbucicchio M., J. Mat. Sci.L, <u>6</u>, (4), 415 (1987)

Pampuch R. & Stobierski L., 'Proc. of the 3rd CIMTEC Intl. Meeting on Modern Ceramic Technology', Rimini Italy, May (1976), p.180

Pan P., Nesbit L.A., Donse R.W. & Gleason R.T., J. Electrochem. Soc., <u>132</u>, (8), 2012 (1985)

Panitz J.K.G., J.Vac.SciTechnol., A<u>4</u> (6) 2949 (1986)

Parkanski N.Ya., Kats M.S., Goldliner M.G., Gittlevich A.E., J Mat Sci <u>18</u>, 1151 (1983)

Partington J.R., 'A Text-Book of Inorganic Chemistry', 6th ed., Macmillan & Co Ltd., London (1961)

Pask J. & Evans A.(Ed.), - Metal Systems', Plenum NY (1980), p 503

Pasolti G., Ricci M.V., Sachetti N., Sacerdoti G. & Spadoni M., 'Nb$_3$Sn Diffusion Layers: their Super Conducting Properties', Nuovo Cimento, <u>35B</u>, (2), 165 (1976)

Patnaik P.C. & Smeltzer W.W., 'Sulfidation Properties of Fe-Al Alloys at 1173K in H_2S-H_2 Atmospheres', J. Oxid Met., <u>23</u>, (1-2) 53 (1985)

Paton B.A., Movchan B.A. & Demchischin A.V, Proc.16th Ann Conf, Soc Vac. Coaters, Chicago (1973), p251

Patten J.W., Hayes D.D. & Fairbanks J.W., see US/UK Confs on Gas Turbines in Marine Envir and refs below

Patten J.W., McClanahan E.D. & Johnston J.W., J.Appl.Phys., <u>42</u>, 4371(1971)

Patten J.W. et al (1976), see Thin Solid Films <u>63</u>, 121 (1979)

Patten J.W., Moss R.S., Hayes D.D. & Fairbanks J., Proc. of the '1st. Intl. Conf. on Advanced Materials for Alternate Fuel Capable directly Fired Heat Engines, EPRI Conf. 790 749 (1979) p.459

Pawlowski L., Surf. & Coatings Technol., <u>31</u>, 103 (1987)

Peichl L. & Johner G., J. Vac. Sci. & Technol., A <u>4</u>, (6), 2583 (1986)

Peichl L., Wortmann J. & Raetzer-Scheibe H.J., 'Int. Conf. on High temperature alloys', Petten Nether-lands, (1985)

Peide Z., Stott F.H., Proctor R.P. & Grant W.A., Oxid.Met <u>16</u>, (5/6) 409 (1981)

Pellegrini G., Piatti G. & Schneiders A., Met Int., <u>71</u>, 154 (1979)

Pendse R. & Stringer J., 'The influence of Alloy Microstructure on the Oxide Peg Morphologies in a Co-10%Cr-11%Al Alloy wth & without Reactive Element Additions', Oxid

Met., 23, (1-2) 1 (1985)

Penfold A.S., Telic.Corp., Pvt. Comm.(1985)

Peng T.C., Sastry S.M.L. & O'Neal J.E., 'Lasers in Metallurgy', Proc Conf Metall Soc AIME Chicago, Feb 1981, Mukherjee K. & Mazumder J.(Ed.) (1981) p 279

Pepper S.V., NASA TM - 78838 (1978)

Percy A.J. & Chollet L., J. Vac. Sci. & Technol., A4, 2801 (1986)

Perez M. & Larpin J.P., Oxid. Met., 24, (1/2), 29 (1985)

Perkins R.A., Proc Internat Conf, Petten, Netherlands, North-Holland Publ., March (1978) p 213

Perkins R.A., 'Applications of Coatings in Advanced Technologies', EEC, Petten (1982)

Perkins R.A., 'Criteria for Selection of Coatings', Seminar on Coatings for High Temp. Appl. at Joint Research Centre, Commission of Eur.Communities, Petten Estbl. Netherlands (1981)

Perkins R.A. & Packer C.M., 4th Ann Conf on Materials for Coal Conversion & Utilisation, Gaithersberg Oct (1979) p2-50

Perkins R.A. & Vonk S.J., 'Materials Problems in Fluidised-Bed Combustion Systems', EPRI-FP-1280, EPRI, Palo Alto, Calif. (1979)

Perry A.J. & Cholet L., J Vac Sci Tech A 4, 2801 (1986)

Perry A.J. & Chollet L., Surf. & Coatings Technol.,34(2) 123 (1988)

Perry A.J., Strandberg C., Sproul W.D., Hofmann S., Erneberger C., Nickerson J. & Chollet L., Thin Solid Films, 153, 169 (1987)

Perugini G., Proc.3rd CIMTEC 3rd Intl Meeting on Modern Ceramic Technologies, Rimini, Italy, May (1976) p 191 & 212

Perugini G., 'Ceramic Self Sealing Coatings for High Temp. Surfaces', Cermurgia, 4, (1), (1978)

Perugini G., Galante F. & Maraccioli E., Proc.7th Intl.Metal Spraying Conf, London (1973) p143

Peterson J.R., J. Vac. Sci. & Technol., 11, (4) 715 (1974)

Petit L. & Rameau S.J., Mat. Sci. & Engg., 87, 38 (1987)

Pettit F.S., see in Rahmel et al (1985), p.311

Pettit F.S. & Goward G.W., 'Gas Turbine Applications', 'Coatings for High Temperature Applications' E.Lang (Ed.), Appl.Sci.,(1982)

Pettit F.S. & Goward G.W., 'Coatings for High Temperature Applications', Lang E.(Ed.), Applied Sci., NY (1983), p 1 & 341

Peytavy J.L., Lebugle A., Montel G. & Pastor H., High Temp.-High Pr., 10, 341 (1978)

Pfender E., Surf. & Coatings Technol., 34, (1), 1 (1988)

Phelps B., Metallurgia, 49, (1) 42 (1982)

Pichoir R., "High Temperature Alloys for Gas Turbines", Coutsouradis et al, eds, Applied Science Publ, London, p.191 (1978)

Pichoir R., 'Materials & Coatings to Resist High Temperature Corrosion', Holmes D.R. & Rahmel A (Ed), Appl. Sci., NY (1979) p271

Pichoir R. & Hausser J.M., 'The Current Status of Coatings for High Temperature Applications', Inst. of Metallurgists UK,(1980), Series 3, 1, (13)

Pieraggi B., Mat. Sci. & Engg., 88, 199 (1987)

Pieraggi B., Oxid. Met., 27, (3/4), 177 (1987)

Pikashov V.S., Erinov A.E. & Klevchishkin V.A., High Temp (Engl Transln) 18, (4), 575 (1980)

Pilling N.B. & Bedworth R.E., J. Inst. Metals, 29, 529 (1923)

Pilous V., Werkstoffe u Korrosion, 32, (8) 338 (1981)

Pivin J.C., Roques-Carmes C., J Chaumont & Bernas H., Corr.Sci., 20, (8/9) 947 (1980)

Plaster H.J., Metallurgia, July (1983) p 302

Politis C. & Johnson W.L., J. Appl. Phys., 60, 1147 (1986)

Pollard F.H. & Woodward P., Trans Faraday Soc., 46, 190 (1950)

Polman E.A., Fransen T. & Gellings P.J., Mat. Sci. & Engg., 88, 157 (1987) Roth Th., Kloos K.H. & Broszeit E., Thin Solid Films, 153, 123 (1987)

Pomerantz, US Pat. 3904783 (1983)

Pons M., Caillet M. & Gallerie A., Corr. Sci., 22, (3) 239 (1982)

Pons M., Caillet M. & Gallerie A., 'Ion Implantation into Metals', Ashworth V., Grant W.A. & Procter R.P.M.(Ed.) Pergamon (1982), p 201

Pons M., Caillet M. & Galerie A., J. Mat. Sci., 21, 4101 (1986)

Pons M., Galerie A. & Caillet M., J. Mat. Sci., 21, 2697 (1986)

Portnoy K.I., 'NiAl Coaitngs', Porosh Met., (2-206), 33-9 (1980)

Pourbaix M., Mat. Sci. & Engg., **87**, 303 (1987)

Powell J. & Steen W.M., 'Lasers in Metallurgy', Conf. Proc. Metall. Soc. AIME, Chicago, IL, Feb (1981), Mukherjee K. & Mazumder J.(Ed.) (1981), p 93

Prakash S., Budhani R., Doerr H.J., Deshpandey C.V. & Bunshah R.F., J. Vac. Sci. & Technol., A **3**, (6), 2551 (1985)

Prater J.T., Surf. & Coatings Technol., **29**, 247 (1986)

Prater J.T., 'Proc. Intl. Conf. Metallurgical Coatings', March 23-27 (1987), San Diego, CA, Surf. Coatings Technol., **32**, 389 (1987)

Prater J.T. & Courtright E.L., Surface & Coatings Technol., **32**, 389 (1987)

Prater J.T., Patten J.W., Hayes D.D. & Moss R.W., Proc. '2nd Conf. Advanced Materials for Alternate Fuel Capable Heat Engines', EPRI RD 2369SR (1982) p.7/29

Price M.O., Wolfla T.A. & Tucker R C,Thin Solid Films,**45**, 309(1977)

Priceman S. & Sama L., 13th Refractory Composites Wkg Gp Mtg, AFLMTR - 68 - 84 (1968)

Priceman S. & Sama L., Electrochem Technol., **6** (9-10) 315 (1968)

Provenzano V & Cocking J L, Surf. & Coatings Technol., **32**, 181(1987)

Przybylski K. & Mrowec S., 'Proc. Intl. Cong. on Metallic Corrosion', Toronto (1984), Vol.1, p47

Przybylski K., Garratt-Reed A.J. & Yurek G.J., J. Electrochem. Soc., **135**, (2), 509 (1988a)

Przybylski K. & Yurek G.J., J. Electrochem. Soc., **135** 517 (1988b)

Psaras P.A. & Langford H.D. (Ed), 'Advancing Materials Research', Nat. Acad. Press (1987)

Pushparaman M., Varadarajan G., Krishnamoorthy S. & Shenoi B.A., Electroplating & Metal Finishing, May (1974) p 10

Quentmeyer R.J., McDonald G. & Hendricks R.C., J. Vac. Sci. & Technol., A**3** (6) 2450 (1985)

Quinto D.T., J Vac Sci & Technol., A **6**, (3), 2149 (1988)

Quinto D.T., Wolfe G.J. & Jindal P C,Thin Solid Films,**153**,19 (1987)

Qureshi J., Levy A. & Wang B., in 'Proc. Symp. on High Temperature Coatings', Florida, Oct. (1986), Khobaib M., Krutenat R.C. (Ed.), Metall. Soc. AIME, (1987)

Raask E., 'Erosion Wear in Coal Utilization', Hemisphere Pub. Comp., NY, (1988)

Racko D., Mat. Sci. & Technol., **3**, (12), 1062 (1987)

Radomycelskiy I.D., Vlasyuk R.Z. & Furman V.V., 'Electrophoretic Coating Structure of Cr Carbide on Steel', Porosh Met., (4-196), 27-32 (1979)

Raghuram A.C. et al., Thin Solid Films, **20** 187 (1974)

Raghuram A.C. & Bunshah R.F., J. Vac Sci & Technol.,**9**,1389 (1972)

Rahmel A. (1985), see in Rahmel et al (1985), p.326.

Rahmel A., Mat. Sci. & Engg., **87**, 345 (1987)

Rahmel et al (1985): see papers presented at workshop below

Rahmel A., Wood G.C., Kofstad P. & Douglass D.L., Intl. Workshop on 'Critical Issues concerning the Mechanism of High Temp. Corrosion' Oxid.Met., **23**, (3/4) 251 (1985)

Rama Char T.L., in "Electrodeposition of Alloys", Brenner A.(Ed), 1960

Ramanarayanan T.A., Mat. Sci. & Engg., **87**, 113 (1987)

Ramanarayanan T.A., Ayer R., Petkovic-Luton R. & Lesa D.P., Oxid. Met., **29**, (5/6), 445 (1988)

Rance D.G., 'Suface Analysis & Pretreatment of Plastics & Metals' Brewis D.M.(Ed.) Appl Sci Publ., (1982), p 121

Rand M.J., J. Vac. Sci. & Technol **16**, 420 (1979)

Randhawa H., Thin Solid Films, **153**, 209 (1987)

Randhawa H. & Johnson P.C., Surf. & Coatings Technol, **31**, 303 (1987)

Rao A.S. & Kaminsky M., Thin Solid Films, **83**, 93, (1981)

Rao D.B., Jacob K.T. & Nelson H.G, Met Trans A, **14A**, 295 (1983)

Rao D.B. & Houska C.R., Met Trans A, **14A**, 61 (1983)

Rapp R.A., Mat. Sci. & Engg., **87**, 319 (1987)

Rapp R.A. & Goto K.S., "Molten Salts" Symp, Braunstein J. & Selman J.R.(Ed), Pittsburgh, Electrochem Soc, NJ (1979)

Rapp R.A., Wang D. & Weisert T., 'High Temp. Coatings', Khobaib M & Krutenat R.C. (Eds.), Proc. Symp. Orlando Florida, (1986), Metall. Soc. AIME (1987) p.131

Rappaz M. & Blank E., Wear, **84**, 387 (1983)

Rappaz M. & Blank E., J. Mat Sci.,

22 896 (1987)
Rappaz M. & Blank E., 'Proc. 3rd Intl. Symp. Metall. & Mat. Sci.', Roskilde, Lilholt H. & Talreja R. (Eds.) (1982)
Rawlings R.D., 'Introduction to Acoustic Emission', Private Communication May (1985)
Reagan P., Ross M.F. & Huffman F N, Adv Ceram Mat., 3, 198 (1988)
Reaney I.M. & Lorimer G.W., Mat. Sci. & Technol., 4 (5) 391 (1988)
Rebenne H. & Pollard R., J. Electrochem. Soc., 132, 1932 (1985)
Reiley T.C. & Nix W.D., Met Trans A, 7A, 1695 (1976)
Reinberg A.R., Ann. Rev. Mat. Sci., 9, 341 (1979)
Reinberg A.R., US Patent 3,757,733 Sept 1 (1973)
Reinberg A.R., J Elec Mat, 8, 345 (1977)
Reising R.F., Corrosion - NACE, 33, (3) 84 (1977)
Restall J.E., Proc 3rd US/UK Conf of Gas Turbine Materials in Marine Environ, Bath (1976), ed Hall, B., session 5
Restall J.E., Metallurgia 46, 676 (1979)
Restall J.E. & Wood M.J., Symp. on Protective Coatings for High Temperature Gas Turbine Components, Metals Soc UK, London, Nov.(1984)
Restall J.E., 5th Intl Symp on Superalloys, Seven Springs (1984), Gell M. et al (Ed.), Met Soc AIME, p 723
Restall J.E., Gill B.J., Hayman C. & Archer N.J., 4th Intl Symp on Superalloys, Sims C. & Hagel (Ed), Seven Springs (1980), p405
Restall J.E. & Stephenson D.J., Mat. Sci. & Engg., 88, 273 (1987)
Reynolds G.J., J. Electrochem. Soc., 135, (6), 1483 (1988)
Rhines F.N. & Mehl R.F., Trans AIME 128 185 (1938)
Rhoda R.N., Plating 64, 48 (1977)
Rhys-Jones T.N., 'Proc. of the Conf. UK Corrosion '87', Brighton, Oct. (1987), p. 307
Rhys-Jones T.M. & Swindels N., Matls Sci & Eng 87, 389 (1987)
Rhys-Jones T.N., Nicholls J.R. & Hancock P., 'Proc. of the Int. Conf. 'Prediction of Materials Performance in Plants Operating with Corrosive Environments', Cranfield , Berks., Feb. (1986)
Richerson D.W., "Modern Ceramic Engineering", Dekker, NY (1982)

Rickerby D.S. & Burnett P.J., Surf. & Coatings Technol., 33, 191 (1987)
Rickerby D.S., Eckold G., Scott K.T. & Buckley-Golder I.M., Thin Solid Films, 154, 125 (1987)
Rigo S. et al, J Electrochem Soc 129, 867 (1982)
Riley F. & Andrews P., 'Proc. of the Conf. UK Corrosion '87', Brighton, Oct (1987), p. 347 (1987)
Riley F., Gaultier G. & Fordham R.J., ibid., p. 325 (1987)
Roberts S.G. & Page T.F., in "Ion Implantation into Metals", Ashworth, Grant & Procter (Ed) Pergamon (1982), p.135
Robinson M.C.D. (1981), see below
Robinson M.C.D., Van der Brekel C.H.J., Cullen C.W., Blocher Jr. J.M. & Rai-Chaudhury P. (Eds.), '9th Intl. Conf on CVD', 1984 Symp., Electrochem. Soc. Publn.
Rosler R.S., Benzing W.C. & Baldo J, Solid State Tech, June (1976) p 45
Rosler R.S. & Engle G.M., J. Vac. Sci. & Technol., B 2, 733 (1984)
Ross R.B. (Ed.), 'Metallic Materials, Specification Handbook', E & FN Spon Ltd
Rostwolowski J. & Parsons R.R., J. Vac. Sci. & Technol., A 3, (3), 491 (1985)
Roth Th., Kloos K.H. & Broszeit E, Thin Solid Films,153,123 (1987)
Rothman M.F., Lai G.Y., Antony M.M. & Miller A.E., 'Proc. of the Int. Cong. on Metallic Corrosion', Toronto, (1984), Vol.1,p66(1984)
Rousselet C., Lattaud L., Ciosmak D., Bertrand G., Lamelle C. & Heizmann J.J., Mat. Sci. & Engg., 87, 145 (1987)
Ruppert W., Proc 8th Intl Conf on CVD (1981), p 489
Ryan K.H. et al., Proc NAVSEC Conf., p 237 (1974)
Ryssel H. & Glawischnig H.(Ed.) 'Ion Implantation Techniques' Springer Verlag (1982)
SNECMA DEF Standard 05-21, CAA, Rolls - Royce
Sadak J.C. & Sautter F.K., J.Vac. Sci.Technol., 11, (4) 771 (1974)
Saenz J.A., Surf. & Coatings Technol., 34, (1), 89 (1988)
Safranek W.H. & Layer C.H., Symp on Properties of Electrodeposits as Materials for Selected Applications, Nov (1973) Battele Columbus Lab OH

Safranek W.H., J.Vac.Sci.Technol., 11, (4) 765 (1974)

Saito M., Furuya H. & Sugisaki M., Mat. Sci. & Engg., 87, 211 (1987)

Sakai M. & Watanabe K., 'Compatibility of Ni-Coated SiC Fibres', J Japan Inst Met, 45, 624 (1981)

Salamah M.A. & White D., 'Surface & Interfaces in Ceramic & Ceramic - Metal Systems', J.Pask & Evans A.(Ed.), Plenum NY (1980), p 467

Samsonov G.V.(Ed.) 'Protective Coatings on Metals', Pub: Consultants Bureau NY, Vol 5 (1973)

Samsonov G.V. et al, Poroshkovaya Metallurgia, 7, 163 (1976)

Samsonov G.V. & Epik E.P., 'Coatings of High Temperature Materials', Part 1, Hausner H.(Ed.) Plenum NY (1966)

Samsonov G.V., Haydash N.G. & Chartokolenko P.P., 'Ti-Al Diffusion Coatings on Carbon Steel', Metalloved Term Obrab Metals, (4) 73-5 (1978)

Sanders D.M., J. Vac. Sci. & Technol., A 6, (3), 1929 (1988)

Sanders D.M. & Pyle E.A., J. Vac. Sci. & Technol., A 5, 2728 (1987)

Sankur H., DeNarale J., Gunning W. & Nelson J.G., J. Vac. Sci. & Technol. A 5, (5), 2869 (1987)

Santoro G.J. et al., NASA Tech. Paper 2225, March (1984)

Sarkar N.G., Frederick J.R. & Felbeck D.K., Met. Trans., 1, (10) 1979 (1970)

Sartwell B.D., Baldwin D.A. & Singer I.L., J. Vac. Sci. & Technol., A 3, (3), 589 (1985)

Sato K., Matudas N., Baba S. & Kinbara A., Proc.7th ICVM Conf., Tokyo (1982), p 164

Saunders S.R.J. & Spencer S.J., Mat. Sci. & Engg., 87, 227 (1987)

Sautter F.K., J Electrochem Soc., 110, (6) 557 (1963)

Schafer H., 'Chemical Transport Reactions', Acad.Press NY (1964)

Schaftsall H.G. & Szelagowski P., DVS Rep 80, (Deutscher Verband fur Schweisstech., Dusseldorf) (1983)

Scheuermann G.R., US Pat. 4108107 (1978)

Schiller S., Heisig U. & Goedicke K., Proc.7th ICVM Conf., Tokyo (1982), p 600

Schiller S., Hoetzsh G., Foerster H & Reschke J, Proc 7th ICVM Conf. Tokyo (1982), p 127

Schiller S, Heisig O & Goedick K, Proc 7th Intl Vacuum Cong Vienna,

Dobrozemsky R. (Ed.) (1977) p1545

Schiller S. et al., Surf. & Coatings Technol., 33, 405 (1987)

Schilling C.L. & Williams T.C., 'Organosilane Polymers V, VI, VIII', ONR Contract N00014-81-C-0682, Tech. Rept., 83-1 (1983)

Schintlmeister W. & Pacher O., J Vac Sci & Tech 12, 743 (1975); Metall. 28, 690 (1974)

Schintlmeister W., Wallgram W. & Gigl K., High Temp. & High Press., 18, 211 (1986)

Schintlmeister W., Kailer K.H., Wallgram W. & Pacher O., High Temp - High Pr, 14, 199 (1982)

Schmalzried H., 'Solid State Reactions', Acad Press NY (1974) p 165

Schmalzreid H., see in Rahmel et al (1985), p.322

Schmitt K.-Thomas H.G. & Johner G., Proc., 8th Internat. Congress on Metallic Corrosion, Mainz, Sept (1981), p 724 (& see next ref)

Schmitt K.-Thomas H.G., Johner G. & Meisel H., Werkstoffe u Korrosion, 32, 255 (1981)

Schmitz V., Fh.G-Berichte 1 (1977)

Schuetze M., Oxid. Met., 24, 3/4, 199 (1985) (see also Schutze)

Schultz L., 'Proc. 5th Intl. Conf. on Rapidly Quenched Metals', Steep S. & Warlimont H. (Eds.), North Holland (1985) p.1585

Schulz D.A., Higgs P.H. & Cannon J.D., R & D on advanced graphite materials, 34, Oxid Resist Coats for Graphite, WADD TR 61-72 (1964)

Schütze M., Matls Sci & Tech 4, 407 (1988) (see also Schuetze)

Schwarz R.B., Petrich R.R. & Saw C.K., J Non-Cryst Solids 76, 281 (1985)

Schwarz R.B., Petrich R.R. & Saw C.K., J. Non-Cryst. Solids, 76, 281 (1985)

Schwarz R.B. & Johnson W.L., Phys Rev Lett 51, 415 (1983)

Schweitzer K.K. & Johner G., J. Vac. Sci. & Technol., A 3, (6), 2525 (1985)

Scott C.W. et al, "Non-destruc Insp of Thin Plasma-Srayed Ceram & Cement Prot Coats for Coal Convers Equipt", ORNL/TM-6210 (1968)

Scott K.T., 'Plasma Sprayed Ceramic Coatings', in 'Ceramic Surfaces & Surface Treaments', Morrell R. & Nicholas M.G., eds, Brit Ceram Proc, No 34, Aug (1984) p195

Sedgwick T., Proc 6th Intl Conf on CVD, Electrochem Soc, (1977), p59

Seirsten M. & Kofstad P., Mat. Sci. & Technol., _3_, 576 (1987)

Senderoff S., Metall Rev _11_, 97 (1966)

Sepold G. & Rothe P., Laser 79, Munich (1979), p.257

Sequeira C.A.C. & Hocking M.G., Brit Corros J _12_, 158 (1977)

Sequeira C.A.C. & Hocking M.G., 'Potentiometric Measurements on Oxygen Electrodes in Molten Na_2SO_4', Electrochim. Acta. _22_, 381, (1977a)

Sequeira C.A.C. & Hocking M.G., 'Corrosion of Ni in Na_2SO_4+NaCl Melts: Potential-time and Topochemical Studies', J. Appl. Electrochem., _8_, 145, 179, (1978)

Sequeira C.A.C. & Hocking M.G., 'SO_3 Gauge for Monitoring Acidity of Sulphate Melts', 2nd Portugese Congress on Chem. (1978)

Sequeira C.A.C. & Hocking M.G., 'Polarisation Measurements on Solid Pt-molten Na_2SO_4+NaCl Interfaces', Electrochim. Acta. _23_, 381, (1978b)

Sequeira C.A.C. & Hocking M.G., 'Corrosion of Nimonic-105 in Na_2SO_4+NaCl Melts', Corrosion (NACE) _37_, 392, (1981)

Sethi V.K., Ganesan P., Robinson-Wilson D., Puentes E. & Sherman S.K., Corrosion-86, paper No. 92 (NACE, Houston, 1986)

Seybolt A.U. & Beltran A., 'Hot Corrosion Problems Associated with Gas Turbine', ASTM STP421, Am. Soc. Testing Mat., _21_ (1967)

Shaapur F. & Allen S.D., Surf. & Coatings Technol., _33_, 491 (1987)

Sharma K.R., Solid State Technol., 143, April (1980)

Sharpe R.S., 'Research Techniques in NDT', Vol. 6, Academic Press, London (1980)

Shatinskyi V.F., 'Diffusion Coatings on Mo Alloy by Liquid Metal Solutions', Izv. Acad. Nauk SSSR Metally, _1_, 7-13 (1978)

Shelby J.E. et.al., 'Surfaces & Interfaces in Ceramics & Ceramic - Metal Systems' Mat Sci.Res., Vol.14 Pask J. & Evans A.(Ed.), Plenum NY (1980), p 579

Sheng T.T., Thin Solid Films, _154_, 75 (1987)

Shenoy R.N., Unnan J. & Clark R.K., Oxid. Met., _26_, 105 (1986)

Sherman M., Bunshah R.F. & Beale H A, J Vac Sci & Tech _11_, 1128 (1974)

Sheybany S. & Douglass D.L., Oxid. Met., _29_, (3/4), 307 (1988)

Shim H.S. & Byrne J.G., J Cryst Growth _13/14_, 410 (1972)

Shimada M. et al, 'Fabrication & Characterisation of Si_3N_4 Without Additives by Hot Pressing' J. Soc. Matls. Jap., _31_, 967-72 (1982)

Shimotori K. & Suzuki T., 'Ceramic Thermal Barrier Coat for Gas Turbines' Ceramics Japan, _16_, (3), 163-169 (1981)

Shimotov S. & Aisaka T., 'The trend of MCrAlx Alloys', Tetsu To Hagane, _69_, 1229-41 (1983)

Shinata M., Takahashi F. & Hashima K., Mat. Sci. & Engg., _87_, 39 or 399 (1987)

Shinatu Y. & Nishi Y., Oxid. Met., _26_, (3/4), (1986)

Shiney H.E., J. Vac. Sci. & Technol., A _4_, (6), 2629 (1986)

Shinno H. et al, Proc 7th ICVM, Tokyo, p.618 (1982)

Shinzato S. & Kuwahara K., Proc 7th ICVM Conf, Tokyo (1982), p.172

Shiota I. & Watanabe O., J Mat Sci _14_, 1418 (1979)

Shivakumar R., Oxid Met _17_, 27 (1982)

Shivakumar: see also Sivakumar

Shiver W.C., McHargue C.J. & Zinkle S.J., Thin Solid Films, _153_, 185 (1987)

Shores D.A., Corrosion-NACE _31_, 434 (1975)

Shores D.A. & Fang W.C., J Electrochem Soc _128_, 346 (1981)

Sickinger A. & Wilms V., Proc 9th Intl Thermal Spraying Conf, The Hague, (May 1980), p.35

Sidky P.S. & Hocking M.G., 'Corrosion of Ni-Base Alloys in SO_2+O_2 Atmospheres at 700^oC and 900^oC', Proc. Conf. on Behaviour of High Temperature Alloys in Aggressive Environments, Petten, Netherlands (1979)

Sidky P.S. & Hocking M.G., 'Hot Corrosion of Ni-based Ternary Alloys & Superalloys for Gas Turbine Applications I: Corrosion in SO_2+O_2 Atmospheres, Corrosion Sci. _27_, 183 (1987)

Sidky P.S. & Hocking M.G., 'Hot Corr. of Ni-based Ternary Alloys & Superalloys for Application in Gas Turbines Employing Residual Fuels', ibid, _27_, 499 (1987)

Sidky P.S. & Hocking M.G., 'Corrosion of FeCrAlY and MA956 Alloys for Coal Gasifier Plant', Proc.

9th Intl. Cong. on Metallic Corr. Toronto, June (1984) 4 254-262

Silton M.R., Middlebrook A.M., Rodrigues G., Salmeron M. & Somorjai G.A., J. Vac. Sci. & Technol., A 4 (6) 2797 (1986)

Somorjai G.A., J. Vac. Sci. & Technol., A 4 (6) 2797 (1986)

Simons A., U.S. Pat. 2571772, Oct (1951)

Simons E.L., Browning G.V. & Liebhafsky H.A., Corrosion, 11, (12) 12 (1955)

Sims C.T. & Hagel W.C.(Ed), 'The Superalloys', Wiley NY (1972)

Singer A.R.E., Conf.on Rapid Solidification of Al Alloys, Metals Soc., London Dec (1984) Unpub

Singh A., Lessard R.A., Knystautas E.J., Matls Sci & Eng 90, 173 (1987/8)

Singh P. & Birks N., J. Oxid. Met., 13, 457 (1979)

Singh P. & Birks N., J. Oxid. Met., 19, 37 (1983)

Singhal S.C. (Ed.), 'High Temperature Protective Coatings', Proc Symp, Atlanta,GA, Mar 1983, Metall Soc. AIME, Warrendale PA, (1983)

Sivakumar R., Oxid Met 17, 27 & 391 (1982)

Sivakumar R., Sagar P.K. & Bhatia M.L., Oxid. Met., 24, (5/6), 315 (1985)

Sivakumar R. & Rao E.J., Oxid Met 17, 391 (1982)

Sivakumar: see also Shivakumar

Sladek K.J. & Gilbert W.W., Proc 3rd Intl Conf on CVD, Glaski F A (Ed), Hinsdale III, Am Nucl Soc (1972)

Sliney H.E., NASA TM 79113 (1979)

Sliney H.E., Surf & Coatings Tech 33,243 (1987) and J Vac Sci Tech A 4, 2629 (1986)

Smart R.F., 'Proc. Conf. Thermal Spraying', Miami, Fl (1976) p.195

Smeggil J.G. & Bornstein N.S., 'High Temperature Protective Coatings', Proc.Symp.Atlanta, GA, March (1983), Singhal S.C.(Ed.), Metall.Soc.AIME, p 61

Smeggil J.G. & Bornstein N.S., R78-914387-1, United Tech. Res. Center Rep., (1978)

Smeggil J.G. & Bornstein N.S., NASA CR-165464, July (1981)

Smeggil J.G., Matls Sci & Eng 87, 261 (1987)

Smeggil J.G. & Peterson G.G., Oxid. Met., 29, (1/2), 103 (1988)

Smeggil J.G. & Shuskus A.J., J.

Vac. Sci. & Technol., A 4, (6), 2577 (1986)

Smeggil J.G. & Shuskus A.J., Surf. & Coatings Technol., 32, 57 (1987)

Smeggil J.G., Finkenbusch A.W. & Bornstein N.S., Thin Solid Films, 119, 327 (1984b)

Smeggil J.G., Paradis E.L. & Shuskus A.J., Thin Solid Films, 113, 27 (1984a)

Smeggil J.G. et al, J. Vac. Sci. & Technol., A 3, (6), 2569 (1985)

Smeltzer W.W., Matls Sci & Eng 87, 35 (1987)

Smialek J.L. & Lowell C.E., J. Electrochem. Soc., 111, 800 (1974)

Smith D.J., Bursill L.A. & Jefferson D.A., Surf. Sci., 175, 673 (1986)

Smith G.D. & Benn R.C., Surf. & Coatings Technol., 32, 201 (1987)

Smith H.R. & Hunt D.C., Trans Vac Met Conf, Am Vac Soc, 1965, p227

Smith M.F. & Dykhuizen R.C., Surf. & Coatings Technol,34, 2531 (1988)

Smith P.J., Beauprie R.M., Smeltzer W.W., Stevanovic D.V. & Thompson D.A., Oxid. Met., 28, (5/6), 259 (1987)

Smith P.J., Van der Biest O. & Corish J., Oxid. Met., 24, 47 (1985a) and 24, 277 (1985b)

Smyth R.T., 7th Intl Metal Spraying Conf, London (1973), p.89

SNECMA DEF Standard 05-21, CAA, Rolls - Royce

Snow D.B. & Breinan E.M., "Applications of Lasers in Matls Proc", Mezbower E.A., ed, Proc Conf ASM, Washington DC (1979), p.229

Snowden W.E., "Behav of Joints in High Temp Matls", Gooch T.G., Hurst R., Kroeckel H. & Merz, eds, Appl Sci Publ (1982), p.651

Sockel H.G., Christ H.J. & Christl W., Matls Sci & Eng 87, 119 (1987)

Soddy F., Proc.Royal Soc., London, 78, 429 (1967)

Solnick-Legg H.F. et al, 'Ion Implantation into Metals', Ashworth V., Grant W.A. & Procter R.P.M.(Ed.), Pergamon (1980) p328

Solomir J.G., Proc.7th Intl.Metal Spraying Conf, London (1973) p194

Southwell G., MacAlpine S. & Young D.J, Mat Sci & Engg, 88, 81 (1987)

Spahn H., Proc.7th Intl Conf on Surface Technology, Berlin (1981) VDI Verlag, p 427

Spalvins T., J. Vac. Sci. & Technol., A 3, (6), 2329 (1985)

Spicer W.E., 'Surfaces & Interfaces in Ceramic & Ceramic-Metal Systems', Pask J. & Evans A.(Ed.) Plenum (1980) p 51

Spiteri P., St.Paul P., Picand G., Lefebre H. & Tremillon B., Mat. Sci. & Engg., 87, 369 (1987)

Sprague J.A. et al, 'High Temperature Protective Coatings', Proc. Symp. Atlanta, GA, March (1983), Singhal S C (Ed), Metall Soc AIME, p 93

Spriggs D.R. & Brobst R.P, ,'The Effect of PFBC particulate on the High Velocity Erosion-Corrosion of Gas Turbine Materials', in Proc. of the Conf. on 'Corrosion-Erosion -Wear of Materials in Emerging Fossil Energy Systems'

Sproul W.D. & Richman M., J. Vac. Sci.Technol., 12, 842 (1975)

Sproul W.D., J. Vac. Sci. & Technol., A 3, (3), 580 (1985)

Sproul W.D., J. Vac. Sci. & Technol., A 4, (6), 2874 (1986)

Sproul W.D., Surf. & Coatings Technol., 33, 73, 133 (1987)

Srinivasan V., Proc. of a Symp. on 'High Temp. coatings', Orlando, Florida, Oct 7-9 (1986), Metall Soc Inc. (1987) p. 153

Srivastava P.K., Vankar V.D & Chopra K.L., J. Vac. Sci. & Technol., A 4, (6), 2819 (1986)

Srolovitz D.J., J Vac Sci Tech A 4, 2925 (1986)

St ClairT.L. & Progat D.J., CME 38, Tech Alert (1985); see Mih for address

Stacey M.H., Mat. Sci. & Technol., 4, 227 (1988)

Stanton W.E., Proc.7th Intl.Metal Spraying Conf., London (1973), p 157 Stark W.A., Wallace T.C., Witteman W., Krupka M.C.

Stark et al, J Vac Sci Tech 11, 802 (1974)

Stecher P. & Lux B., J Less-Common Metals, 14,(1968)

Stecher P., Lux B. & Funk R., Entropie 21 May - June

Stecura S., Am Ceram Soc Bull., 56, (12) 1082 (1977)

Steen W.M. et al, see Weerasinghe & Steen

Steffens H.D., 'Coatings for High Temperature Applications', Lang E.(Ed.), Appl Sci, NY (1983) p121

Steffens H.D., AGARD-LS-106 NATO (1980), paper 5

Steffens H.D. & Crostack H.A., 9th Intl Thermal Spraying Conf., The Hague, May (1980), p 120.

Steffens H D & Crostack H.A., Proc 7th ICVM Conf, Tokyo (1982) p360

Steffens H.D. & Crostack H.A., Proc. Intl. Conf.on Metallurgical Coatings, Am.Vac Soc., (1981)

Steffens H.D., H-A Crostack & Beczkowiak J., 'Coatings for High Temperature Applications', Lang E (Ed), Appl. Sci, NY (1983), p193

Steffens H.D. & Muller K.N., J. Vac. Sci & Technol,11, 735 (1974)

Steffens H.D. & Fisher U., Surf. & Coatings Technol., 32, 327 (1987)

Steffens H.D., Estruk E. & Busse K.H., J. Vac. Sci. & Technol., A3 (6) 2459 (1985)

Steinmann P.A. & Hintermann H.E., J. Vac. Sci. & Technol., A3 (6) 2394 (1985)

Steinmetz P. et al, 'High Temperature Protective Coatings', Proc. Symp. Atlanta, GA, March (1983), Singhal S.C.(Ed.), Metall. Soc. AIME, p 135

Stern K.H. & Gadomski S.T., J Electrochem Soc., 130, 300 (1983)

Stern K.H. & Deanhardt M.L., J. Electrochem. Soc., 132, (8), 1891 (1985)

Stern K.H. & Williams C.E., J. Electrochem. Soc., 133, (10), 2157 (1986)

Stetson A.R. & Moore V.S., Aerospace-Finishing March (1975) p 67

Stetson A.R. & Moore V.S., 'Gas Turbine Materials in Marine Environment', Fairbanks J.W. & Machlin I.(Ed.), 1974

Stetson A.R. & Moore V.S., Aerospace-Finishing, March (1975) p 67

Stevens W.G. & Stetson A.R., AFML-TR-76-91 (1976)

Stinton D.P. & Lackey W.J., Ceram. Bull., 61, (8) 843 (1982)

Stinton D.P., Bessman T.M., Lowder R.A., Am Ceram Soc Bull 67, 350 (1988)

Stoltz D.L., Rahe A.H. & Hren J.J., J.Vac.Sci.Technol., 11, (4) 781 (1974)

Stott F.H., see in Rahmel et al (1985), p.315

Stott et al (1971), see Wood, Fountain & Stott (1980)

Stott F.H., Bartlett P.K.N, Wood G.C., Matls Sci & Eng 88,163 (1987)

Stott F.H. et al, see Matls Sci & Eng 87, 267 (1987)

Stott F.H., Bartlett K.N. & Wood G.C., Mat Sci & Engg,88,163 (1987)

Stott F.H., Bartlett P.K.N. & Wood

G.C., Oxid Met, <u>27</u>(1/2) 37 (1987a)
Stott F.H., Chong F.M.F. & Stirling C.A., 'The Effectiveness of Preformed Oxides for Protection of Alloys in Sulphidizing Gases at High Temp.', 9th edn., Intl. Cong. on Metallic Corrosion, (1984) p.1
Stott F.H., Wood G.C. & Hobby, 'Ion Implantation into Metals', V.Ashworth, Grant W.A. & Procter R.P.M.(Ed.) Pergamon Press (1982a), p 231
Stott F.H. et al in V.Ashworth, Grant W.A. & Procter R.P.M.(Ed.) 'Ion Implantation into Metals', Pergamon Press (1982b), p 245
Stott F.H., Peide Z., Grant W.A. & Proctor R.P., Corrosion Sci., <u>22</u>, 305 (1982c)
Stowell W. et al., J. Vac. Sci. & Technol., A <u>3</u>, (3), 572 (1985)
Stowell W.R. & Chambers D., J Vac Sci Tech <u>11</u>, 653 (1974)
Strafford K.N., 'High Temp. Corrosion of Alloys Containing Rare Earth or Refractory Elements: A review of Current Knowledge & Possible Future Developements', in High Temp Technol., 307-318 (1983)
Strafford K.N. & Hampton A.F., J Less Comm Mets <u>21</u>, 305 (1970)
Strafford K.N. & Chan W.Y., Proc 8th Intl Conf on Met Corros, Mainz, p.1476 (1981)
Strafford K.N. & Harrison J.M., Oxid Met <u>8</u>, 347 (1976)
Strafford K.N. & Hunt P.J., Intl Conf on High Temp Corros, San Diego, Rapp R.A.(ED), p.380 (1983)
Strafford K.N. & Hunt P.J., 4th US/UK Conf on Gas Turbine Materials in Marine Envir, Annapolis <u>1</u>, 221 (1979)
Strafford K.N., Datta P.K., Nagaraj B., Jenkinson D. & Dowson A.L., 'The Corros. Degradation of Ti in Gaseous Atmospheres at Elevated Temperatures & the Associated Influence on Mechanical Properties', Paper presented at Inst. of Metals, London, UK (1986)
Strafford K.N., Mistry P., Hocking M.G. & Vasantasree V., 'The Influence of Base Composition & Addition Elements on the Corrosion Behaviour of MCrAl-type Coating Alloys', Proc. 4th US/UK Conf on Gas Turbine Materials in a Marine Environment, Annapolis (1979).
Strang A., 'Proc 13th Int Cong. on combustion engines', Vienna, (1979), CIMAG (1979), Section HB2/GT33
Strang A., "Behav of High Temp Alloys in Aggr Envir", Proc Intl Conf, Petten, Netherlands, Oct (1979); Metal Soc, UK (1980), p.595
Strang A., Lang A. & Pichoir R., ONERO Tech. Publ. 1981-94 (1981)
Strang A., Proc 4th US/UK Conf on Gas Turb Matls in Mar Envir, Annapolis <u>2</u>, 273 (1979)
Strang A. (1984), see next ref
Strang A., Lang E. & Pichoir R., 'Materials Substitution & Recycling', AGARD-CP- 356 (1984) p 11
Strangman T.E., Proc Workshop on Gas Turb Matls in Marine Envir, Bath (II), (1984)
Strangman T.E., Thin Solid Films, <u>127</u>, 93 (1985)
Strangman T.E., Boone D.H., 'Gas Turbine Materials in Marine Envir' 4th US/UK Conf. <u>1</u>, 655 (1979)
Strangman T.E., Felten E.J., Benden R S, NASA CR-135103, Oct (1976)
Strangman TE, Felten E J, Ulion N E, Am Ceram Soc Bull <u>56</u>, 700(1977)
Stratton T. et al, 'Surfaces & Interfaces in Ceramic & Ceramic - Metal Systems', J.Pask & Evans A.(Ed.), Plenum (1980), p71
Strife J.S. & Sheehan J.S., Am. Ceram Soc Bull, <u>67</u>(2)369 (1988)
Stringer J, Corr Sci, <u>10</u>, 513 (1970)
Stringer J., in 'High Temp. Materials Corrosion in Coal Gasification Atmospheres', Norton J.F. (Ed.), Elsevier Appl. Sci. Publ., London (1984), p.83
Stringer J. (1985), see in Rahmel et al (1985), p.280
Stringer J., Wilcox B.A. & Jaffee R.I., Oxid. Met., <u>5</u>, 11 (1972)
Stringer J., 'High Temp. Corr. of Aerospace Alloys', AGARD No. 200, NATO (1975)
Stringer J., Mat. Sci. & Engg., <u>87</u>, 1 (1987)
Stringer J., in 'Proceedings of Corrosion in Fossil Fuel Systems', vol. 83-5, Wright I.G. (Ed.), Electrochem. Soc., Pennington, NJ (1983), p.1
Stringer J., Allam I.M. & Whittle D P, Thin Solid Films,<u>45</u>,377(1977)
Stroud W.P. & Rapp R.A., 'High Temp. Metal Halide Chemistry', Hildenbrand D.L. & Cubicciotti D.D.(Ed.) Proc. Symp. Electrochem. Soc., (1978), p 574
Strutt P.R., LeMay J. & Tanqir A.,

'Proc. Mat. Res. Soc. Symp.', Kear B.H. & Giessen B.C. (Eds.) North Holland (1984) p. 87

Stuart R.V., 'Vacuum Technology, Thin Films and Sputtering', Acad. Press (1983)

Stull D.R., Prophet H. (Dir), 'JANAF Thermochemical Tables', NSRDS, US Dept Comm NBS.37 (1971)

Subrahmanyan J., J Mat Sci 17, 1997 (1982)

Subrahmanyan J.& Annapurna J., Oxid. Met., 26, (3/4), 275 (1987)

Subrahmanyan J., Lahiri A.K. & Abraham K.P., J Electrochem Soc 127, 1394 (1982)

Sudarshan T.S & Bhat D.G.,(Ed.), "Surface Modification Technologies", STM Distr Ltd, Enterprise Ho, Ashford, TW15 1XB, UK, (1988)

Sugano G., Mori K. & Inone K., Electrochem Technol 6, 326 (1968)

Sue J.A. & Troue H.H., Surf. & Coatings Technol., 33, 169 (1987)

Sue J.A. & Tucker R.C., Surf. & Coatings Technol., 32, 237 (1987)

Suganuma K. et al, 'Solid State Bonding of Ceramics', Ceramics Japan, 18, 112-121 (1983)

Suhara T. et al, J. Vac. Sci. & Technol., 11 (4) 787 (1974)

Suhara T. et al, 6th Metal Spraying Conf., Paris (1970), paper B3

Suhara T., Kltajima K., Fukuda S. & Ito H., Proc.7th Intl.Metal Spraying Conf, London (1973) p179

Sugiyama K. et al, J Electrochem Soc 122, 1545 (1975)

Sun W.P., Lin H.J. & Hon M.H., Met. Trans., 17A, 215 (1986)

Sun W.P., Lin T.H. & Hon M.H., Met. Trans. B, 18B, 617 (1987)

Sundararaman D., Kuppuswami P. & Raghunathan V.S., Surf. & Coatings Technol. 30, 343 (1987)

Sundgren et al, J. Vac. Sci. & Technol., 2770 (1986)

Sundgren J.E. & Hentzell H.T.G., J. Vac. Sci. & Technol., A 4, (5), 2259 (1986)

Sundgren J.E., Rockett A., Greene J.E. & Helmersson U., J. Vac. Sci. & Technol., A 4, (6), 2770 (1986b)

Supple R.W. & Stoneham E.B., J Vac Sci & Technol, A 3, 504 (1985)

Suzuki T., Baba H., Takada H. & Shimotori K., Proc.7th ICVM Conf., Tokyo (1982), p 323

Swales G.L., Proc Int Conf Petten, Oct(1979); Met Soc, UK (1980) p.45

Swensson L.E., Gretoft B., Ulander B. & Bhadeshia H.K.D.H., J. Mat. Sci., 21, 1015 (1986)

Swidzinski M.A.M., Ph.D.Thesis, Univ of London (1980)

Swidzinski M.A.M., Vasantasree V. & Hocking M.G., 'Hot Corrosion of 4 Ni-base Alloys in 2:1 :: $SO_2:O_2$ Atmospheres with and without salt coatings', Electrochem. Soc. Conf., Pittsburgh, U.S.A. (1978)

Swindells N., Raper K. & Van H. Gameren, 'Behaviour of High Temperature Alloys in Aggressive Environments', Proc Intl Conf, Petten, Netherlands, Oct.(1979), Metal Soc., UK (1980), p 351

Syan C.S., Matthews A. & Swift K.G., Surf. & Coatings Technol., 33, 105 (1987)

Sykes J.M., 'Surface Analysis & Pretreatment of Plastics & Metals' Brewis D.M.(Ed.), Applied Sci. (1982), p153

Szecket A., Inal O.T., Vigueras D.J. & Rocco J., J Vac Sci Tech 3, 2588 (1985)

Tabata O. et al, Proc.8th Conf on CVD, Blocher J.M., Vuillard G.E. & Wahl G.(Ed.), (1981) p 272

Tabor D., Rep Progr Appl Chem., 36, 621 (1951)

Tachikart M., Davidson J.H., Almanet F. & Beranger G., Mat. Sci. & Engg., 87, 63 (1987)

Tagaki K. (Ed.), "7th Intl. Conf. on Ion Implantation Technol." Kyoto, Japan, 7-10 June (1988)

Takada J., Ohizumi Y., Miyamura H., Kuwahara H. & Kikuchi S., Oxid. Met., 26, (1/2), 19 (1986)

Takada J., Ohizumi Y., Miyamura H., Kuwahkawa N., Proc.7th ICVM Conf., Tokyo (1982), p 610

Takagi K.: error, see Tagaki K.

Takahashi T., 'CVD of B_3N_2 on Ni', Yogyo Kyukai Shi, 89, 63-68 (1981)

Takahashi T. & Itoh H., J. Mat. Sci., 14, (6) 1285 (1979)

Takahashi T. & Itoh H., J Electrochem Soc., 124 (5) 797 (1977)

Takahashi T. & Kamiya H., High Temp-High Pr, 9, 437 (1977)

Takahashi T., Itoh H. & Fukao K., J Less-Common Met, 80, 171 (1981)

Takahashi R., Sugarawa K. et al, J Electrochem.Soc., 119, 1406 (1972)

Takai O., Ebisawa J. & Hisamatsu Y., Proc.7th ICVM Conf., Tokyo (1982a)., p129

Takai O., Ebisawa J. & Hisamatsu Y., Proc.7th ICVM Conf., Tokyo (1982b)., p137

Takashio H., 'Ceramics & Glass to

Metal Bonds', Ceramics Japan, 15, (6), 4427-434 (1980)

Tamarin Y.A. & Dodonova R.N., Zashita Metallov 13, 114 (1977)

Tamura I., J Mat Sci, 21, 2493 (1986)

Taniguchi M., Wakihara M., Uchida T., Hirakawa K. & Nii J., J. Electrochem. Soc., 135, 217 (1988)

Taniguchi S. & Shibata T., Oxid. Met., 25, (3/4), 201 (1986)

Taylor T.A., Overs M.P., Gill B.J. & Tucker R.C., J.Vac.Sci.Technol., A3 (6) 2526 (1985)

Tavadze F.N., Mikadze O.I. & Keshelava N.P., Oxid Met 25 335 (1986)

Tech. Alert (Patent counsel, Marshall Space Flight Center, Alabama USA), Orders to Tech Alert, Room 206, Ebury Bridge house, Ebury Bridge Road, London SWIW 8QD

Teer D.G., "Evaporation & Sputter Techniques", in "Coatings for High Temp Applications", Lang E., Ed, Applied Sci Publ, NY (1983), p.79

Tempest P.A. & Wild R.K., 'Formation and Growth of Spinel and Cr_2O_3 Oxides on 20% Cr-25% Ni-Nb Stainless Steel in CO_2 Environments', J. Oxid Met., 23,

Tentarelli L.A., White J.M. & Buck R.W., Rept AD-632-362, (1966), Sperry Rand, avail from Clearinghouse for Federal Sci & Tech Info, Springfield VA 22161, USA

Terenteva V.S. & Zhorov G.A., 'Radiation Capacity of Si-Ti-Cr Plasma Coatings on Nb-Base', Teplofiz Vyso Temp, 16, (2), 299-303 (1978)

Terry B.S., Wright J. & Hall D.J., in ' Proc. of Conf. UK Corrosion '87', Brighton, Oct. (1987), p. 211 (1987)

Teyssandier F., Bernard C. & Du-Carrior M., J. Electrochem. Soc., 135, (1), 225 (1988a)

Teyssandier F., Bernard C. & Du-Carroir M., J. Mat. Sci., 23, (1), 135 (1988b)

Thakur A.P., Lamb J.L., Williams R.M. & Khanna S.K., J. Vac. Sci. & Technol., A 3, (3), 600 (1985)

Thebault J., Naslain R. & Bernard C., J.Less-Common Metals, 57, (1) 1-20 (1978)

Thevand A., Poize S., Crousier J.P. & Streiff R., J Mat Sci., 16, (9) 2467 (1981)

Thomas G., Gronsky R., Krivanek O.L. & Mishra R.K., 'Surfaces & Interfaces in Ceramic & Ceramic -

Metal Systems', Mat Sci., Vol.14, Pask J. & Evans A.(Ed.) Plenum (1980), p 35

Thoma M. (1980): see Matls Sci & Eng 90, 327 (1988), and next ref:-

Thoma M., J. Vac. Sci. & Technol., A 4, (6), 2633 (1986)

Thompson D.A., Oxid. Met., 28 (5/6) 259 (1987)

Thompson G., Stelmack L. & Thurman C. "Rev. of Ion-Assisted Dep - Res. to Prod.", 7th Intl. Conf., Ion Implantation Technol., Kyoto, Japan, 7-10 June (1988).

Thompson M.S. & Morral J.E., Proc. Symp. on 'High Temp. Coatings', Orlando, Florida, Oct 7-9 (1986), Publ. by The Metall. Soc Inc. (1987), p.55

Thonsen C., Maris H.J. & Tanc J., Thin Solid Films, 154, 217 (1987)

Thornton J.A., J.Vac.Sci.Technol. 11, (4) 666 (1974)

Thornton J.A., 'High Rate Thick Film Growth', Ann Rev Mat Sci., 7, 239 (1977)

Thornton J.A., 'Plasmas in Dep. Proc.', Dep. Technol. & Appln. Course, Univ Calif. June (1980)

Thornton J.A., 'Coating Deposition by Sputtering', "Films & Coatings for Technology', CEI Course, Sweden (1981), Jacobson B.E. & Bunshah R.F.(Ed.)

Thornton J.A., US Pat.,4126530 (1978)

Thornton J.A., "Deposition Technologies for Films & Coatings", Bunshah R.F., ed, publ Noyes, Park Ridge, NJ (1982)

Thornton J.A., Thin Solid Films 107, 3 (1983)

Thornton J.A., J Vac Sci Tech A 4, 3059 (1986)

Thornton J.A. & Hedgcoth V.L., J.Vac.Sci.Technol., 12, 93 (1975)

Thornton J.A. & Penfold A.S., 'Thin Film Processes', Vossen J.L. & Kern W.(Ed.), Acad.Press NY (1978) p 75

Thornton J.A. & Hoffman D.W., J Vac Sci Tech A 3, 576 (1985)

Tick P.A., Wallace N.W. & Teter M.P., J. Vac. Sci. & Technol., 11, 709 (1974)

Tiearney T.C.Jr. & Natesan K., J. Oxid. Metals, 17, 1 (1982)

Tiefan L., Jianian S. & Xun L., 'Proc. of the Int Cong. on Metallic Corrosion', Toronto, (1984), vol. 1, p. 53 (1984)

Tien J.K. & Pettit F.S., Met Trans

REFERENCES

1587 (1972)

Timoney (1978), see "Survey of Technol Requs for High Temp Matls R & D, section 1, diesel engines", publ EEC, Petten, EUR 7660EN (1981)

Tiziani A., Giordano L., Matteassi P. & Badan B., Mat. Sci. & Energy, 88, 171 (1987)

Tjong S.C. & Wu C.S., Surf. & Coatings Technol., 31, 289 (1987)

Tobin A.G. & Busch G.A., Proc. of a Symp. on 'High Temp. Coatings', Orlando, Florida, Oct 7-9 (1986), Publ. by The Metall. Soc Inc. (1987), p.143

Tomaszewicz P. & Wallwork G.R., Rev High Temp Matls 4, 75 (1978)

Tomellini M. & Gozzi D., Oxid. Met., 26, (5/6), 305 (1986)

Tomlinson W.J., Surf. & Coatings Technol., 27, 23 (1986)

Topor D.C. & Selman J.R., J. Electrochem. Soc., 135, 384 (1988)

Torok E., Perry A.J., Cholet L. & Sproul W.D., Thin Solid Films, 153, 37 (1987)

Toulonkian Y.S. & Ho C.Y., 'Thermal Expansion', Non-Metallic Solids, vol. 3, Thermophysical Properties of Matter, Toulonkian Y.S., Kirby R.K., Taylor R.E. & Lee T.Y.R.(Ed), IFI, Plenum (1987)

Townsend P.D., Kelly J.C. & Hartley N.E.W., 'Ion Implantation, Sputtering & their Applications', Academic Press (1976)

Treyball P.R.E., 'Mass Transfer Operations', McGraw-Hill NY (1955)

Tsu D.V. & Lucovsky G., J. Vac. Sci. & Technol., A 4, 480 (1986)

Tsukamoto Y., Yamaguchi H. & Yanagisawa M., Thin Solid Films, 154, 171 (1987)

Tu D., Chang S., Chao C & Lin C, J.Vac.Sci.Technol, A3, 2479 (1985)

Tu D.C, Lin C.C, Liao S J & Chou J.C, ibid, A 4, (6), 2601 (1987)

Tucker R.C., J. Vac. Sci. & Technol., 11, (4) 725 (1974)

Tucker R.C., 'Plasma & Detonation Gun Deposition Techniques & Coating Properties', UCLA Course on Deposition Techniques & Applications, Sept (1978)

Tucker R.C., 'Films & Coatings for Technology', Jacobson B.E. & Bunshah R.F.(Ed.) Sweden, June (1981)

Tucker R.C., 'Plasma & Detonation Gun Deposition Techniques & Coatings & Coating Properties', 'Films & Coatings for Technology', CEI Course, Sweden, Jacobson B.E. & Bunshah R.F.(Ed.) (1981)

Tucker R.C., 'Coatings for High Temperature Applications', Lang E. (Ed.), Applied Sci., NY (1983)

Tucker R.C, Taylor T.A. & Weatherly M.H., 'Proc. 3rd conf, Gas Turb. Mat. in Marine Env. (1976)

Turnbull, Trans AIME 221, 422 (1961)

Twentyman M.E., J Mat Sci 10, 765 (1975)

Twentyman M.E. & Popper P., J Mat Sci 10, 777 & 791 (1975)

Twentyman M.E. & Hancock P., 'Mat. Sci. Res.', Evans A. (Ed.), Plenum NY, (1981) Vol. 14, p535

Tylecote R.F., 'The Solid Phase Welding of Metals', Edward Arnold (1968) p334

Ubank R.G., Rev Int Htes Temp et Refr 14, 21 (1977)

UK Patent GB 2,064,370B

UK Standard 03-011 Ministry of Defence, 'Flame spraying'.

US Patent 4218007, 19th Aug (1980)

Unnam J., Shenoy R. & Clark R.K., Oxid. Met., 26, (3/4), 231 (1986)

Upadhya K. & Strafford K.N., 'High Temperature Protective Coatings', Proc.Symp.Atlanta, GA, March (1983), Singhal S.C.(Ed.), Metall. Soc.AIME, p 159

Urao R., Yamagata K. & Yoshida H., Proc.7th ICVM Conf., Tokyo (1982), p 584

Ushakov B.A., 'Protective Coatings on Metals', Vol 5, Samsonov G.V. (Ed.), Consul. Bur. NY (1973) p103

Valli J., Makela U., Hentzell H.T.G. & Raub C.J., Pt Metals Rev., 20, (3) 79 (1976)

Valli J., Makela U. & Hentzell H.T.G., J. Vac. Sci. & Technol., A4 (6) 2850 (1986)

Van der Biest et al, "Behaviour of High Temperature Alloys in Aggressive Environments', Proc Intl Conf Petten, Netherlands, Oct.1979,

Van Vlack L.H., 'The Metal-Ceramic Boundary,' 1964 Metals / Materials Congress, Philadelphia PA, Tech Rep No p (10-1-64)

Van den Brekel C.H.J., Philips Res Reports, 32, 118, 134 (1977a,b)

Vandenbulcke L., Proc.8th Intl. Conf., Blocher J.M. & Vuillard G.E.(Ed.) p 32 (1981)

Vandenbulcke L. & Vuillard G., J Electrochem Soc., 124, 1937 (1977)

Vandenbulcke L., J Less Comm Met 95, 191 (1983)

Vardelle M. & Fauchais P., Ann Chim Fr., 8, 383 (1983)

Vasantasree V. & Hocking M.G., 'Electrochemical Studies of Hot Corrosion of Ni-Cr Alloys', Proc. 3rd U.K./U.S. Conf. on Gas Turbine Materials in a Marine Environment, Bath (1976)

Vasantasree V. & Hocking M.G., 'Kinetics of the High Temperature Corrosion of Ni-Cr Alloys in $SO_2 + O_2$, Corr. Sci. 16, 261, (1976)

Vasantasree V., Pantony D.A. & Hocking M.G., 'A Gas-tight Thermobalance-load Seal', J. Sci. Instrum. 44, 791 (1967)

Vasantasree V., Thesis, Univ of London (1971)

Vasantasree V., Hocking M.G., Internal Report, Unpublished work (1984)

Vazquez A.J. et al., Proc.3rd Intl Mtg on Modern Ceram Technol., Rimini, Italy (1976) p 228

Van De Voorde M.H., Mat. Sci. & Engg., 88, 341 (1987).

Van Roode M. & Hsu L.L., Surf. & Coatings Technol., 32, 153 (1987)

Vassala F.A., ARL Rept 66-0115 (June 1966) & 67-0181 (Sep 1967)

Vasu K.I., Thesis, Univ. London (1964)

Vasu K.I. (Ed.)'10th Intl. Conf. on Metallic Corrosion', Nov. (1987), Madras, India

Veilleux G., Saint Jacques R.G. & Dallaine S., Thin Solid Films, 154, 91 (1987)

Venkatesh V.C. et al, Proc 8th Intl Conf.on CVD (1981), p 617

Venkatraman S., Nair M.R., Kothari D.C. & Lal K.B., "Anodic Polarization Behaviour of Ion Implanted Al in 3.5 wt% NaCl Electrolyte", Nuclear Instruments & Methods in Phys. Research North-Holland, Amsterdam, B19/20, 241-246 (1987)

Veprek S. & Maracek V., Solid State Electron, 11, 683 (1968)

Vesely V., 7th Intl Metal Spraying Conf, London (1973), p.137

Veys J.M. & Mevrel R., Matls Sci & Eng 88, 253 (1987)

Viguie J.C. & Spitz J., J Electrochem Soc 122, 585 (1975)

Vinayo M.E., Kassabji F., Gujonnet J. & Fauchais P., J Vac Sci Tech, A3 (6) 2483 (1985)

Vincenzini P., 'Present Status & Trends of Research on Reaction Bonded & Hot Pressed Si_3N_4 for Structural Appln. in Advanced Hot Engines', Ceramurgia, 8, (4), 181 (1981)

Vineberg E.J. & Douglass D.L., Oxid. Met., 25, (1/2), 1 (1986)

Viswanathan R. & Spengler C.J., Corrosion 26, 29 (1970)

Vogel D., Newman L., Dep P. & Boone D.H., Mat. Sci. & Engg., 88, 227 (1987)

Voigt K. & Westphal H., High Temp.- High Pr., 14, 357 (1982)

Vössen J.L. & Kern W.(Ed.) Thin Film Process, Acad Press NY (1978)

Vössen J.L., Schnable G.L. & Kern W., J. Vac. Sci. & Technol., 11, 60 (1974)

Vössen J.L., O'Neil J.J., Mesker O.R. & James E.A., J. Vac. Sci. & Technol., 14, 85 (1977)

Von Allmen M. & Affolter K., 'Mat. Res. Symp. Proc.', Kear B.H. & Giessen B.C. (Ed.), North Holland, vol. 28, (1984) p.81

Vuorinen S., Niemi E. & Korhonen A.S., J. Vac. Sci. & Technol., A3 (6) 2445 (1985)

Wadsworth J., Nieh T.G. & Steppers J.J., Int. Mat. Rev., 33, (3), 131 (1988)

Wagendristel A., Bangert H., Cai X. & Kaminitschek A., Thin Solid Films, 154, 199 (1987)

Wagner J.B. (1985), see in Rahmel et al (1985), p.296

Wagner C., Corrosion Sci., 5, 751 (1965)

Wagner R.S. et al, J. Vac. Sci. & Technol., 11, 3 (1974)

Wahl G. & Furst B., 'Materials & Coatings to Resist High Temp Corrosion', Holmes D.R. & Rahmel A.(Ed.) Appl.Sci., (1979), p 333

Wahl G. & Furst B., 'Materials & Coatings to Resist High Temperature Corrosion', Holmes D.R. & Rahmel A.(Ed.), Applied Sci., NY (1979), p333

Wahl G., Schmaderer F., Metzger M. & Nicoll A.R., 8th Conf. CVD, (1981) p685

Waits R.K., in 'Thin Film Processes' Vassen J.L. & Kern W. (Eds.) Acad. Press NY (1978) p.131

Wakefield G.F. et al., p 173, Proc 4th Intl Conf CVD, Princeton, Wakefield G.F. & Blocher J.M.(Ed.) Electrochem Soc., see also p 177 (1974)

Walker A, West D R F & Steen W M, Metals Technol., 11, 399 (1984)

Walls J.M. & Christie A.B., 'Surface Analysis & Pretreatment of

Plastics & Metals', Brewis D.M.(Ed.) Appl Sci. (1982), p13

Wallura E., High Temp.-High Pr., 14, 477 (1982)

Wallwork G.R. & Hed A.Z., Oxid. Met., 3, 171 (1971)

Walton D., J Chem Phys., 37, 2182 (1962)

Wambach K., Peters J. & Grabke H.J., Mat. Sci. & Engg., 88, 205 (1987)

Wang R. & Merz M.D., Corrosion, 40, 272 (1984)

Wang T.H. & Seigle L.L., 'High Temp. Coatings', Khobaib M. & Krutenat R.C. (Eds.), Met. Soc. AIME Symposium, Florida (1986) Publ. (1987) p.39

Waterman N.A., Metallurgia, 54, (11), 512 (1987)

Weast R.C. (Ed.), CRC Handbook of Chem. & Phys., 67th edn., 1986-1987, CRC Press FL (1987)

Webb A.P. & Veprey S., Chem Phys Lett, 62, 173 (1979)Weber J.K.R., Thesis Univ of London UK (1986)

Weber J.K.R., Thesis, Univ. London UK (1986)

Weber J.K.R. & Hocking M.G., 'Hot Corr. of MA-965 Composite Alloy in Coal Gasifier Atmospheres', Proc. Conf. on Materials for High Temp. Environments, EEC, Petten (1985)

Weber J.K.R. & Hocking M.G., 'An evaluation of Fe-Cr-Al Alloys in Simulated Coal Gasifier Environments from 850°C to 1050°C', "High Temperature Alloys" (Ed. J.B.Marriott et al, Elsevier 1985), p.125

Weerasinghe V.M. & Steen W.M., Met. Constr., 19, 581 (1987)

Weissmantel C., Ackermann E., Bewilogna K., Hecht G., Kupfer H. & Rau B., J. Vac. Sci. & Technol., A4 (6) 2892 (1986)

Weissmantel C., Thin Solid films, 72, (1980)

Wells C.H. & Sullivan C.P., Trans ASM, 61, 149 (1968)

Wen L.S., Zhang H.T., Zhen X.K., Guan K., Liao B. & Cao S., Surf. & Coatings Technol., 30, 115 (1987)

Werber T., Mat. Sci. & Engg., 88, 283 (1987)

Werdecher W. & Aldinger F., High Temp - High Pr., 14, 183 (1982)

Wert J.J., Baker D.M. & McKechnie T.M., Surf. & Coatings Technol., 33, 245 (1987)

West G.A. & Beeson K.W., J. Electrochem. Soc., 135, (1), 1752 (1988)

Westwood W.D., 'Glow Discharge Sputtering', Progress in Surface Sci., 7, 71 (1976)

Westwood W.D., 'Sputtering - Notes from AVS Short Course', Avail. Am. Vac Soc., 335 E 45th st, NY 10017

Whalen M.V., Pt. Met. Rev., 32, (1), 2 (1988)

White C.W. & Peercy P.S.(Ed.) 'Laser & Electron Beam processing of Materials', Proc 1979 Mat. Res. Soc., Academic Press, NY (1980)

White J.R., J. Vac. Sci. & Technol., A 4, (6), 2855 (1986)

Whittle D.P. & Boone D.H., 'Surfaces & Interfaces in Ceramic & Ceramic - Metal Systems', Mat Sci Res., Vol.14, J.Pask & Evans A.(Ed.) Plenum, (1980), p 487

Whittle D.P. & Boone D.H., Proc. 8th Intl. Cong. on Metallic Corrosion, Mainz, Sept (1981), p 718

Whittle D.P. & Stringer J., Phil. Trans. Royal Soc. London, A 295, 309 (1980)

Wigren J., Surf. & Coatings Technol., 34 (1) 101-108 (1988)

Wiiala U.K., Kivivuoriso J., Molarius J.M. & Sulonen M.S., Surf. & Coatings Technol., 33, 213 (1987)

Wilhemson P., 'Corrosion Resistant Materials for Coal Conversion Systems', Meadowcroft & Manning, (1983)

Wilkens M., Can J. Phys., 45, 567 (1967)

Williams J.S. & Poate J.M.(Ed.), 'Ion Implantation & Beam Processing', Acad.Press NY (1984)

Williams R., Electroplating & Metal Finishing, 19, 92, (1966)

Williams D.S., Coleman E. & Brown J.M., J. Electrochem. Soc., 133, (12), 2637 (1986)

Wimber R.T. & Stetson A.R., Electrochem Technol., 6 (7-8) 264 (1968)

Window B. & Sharples F., J. Vac. Sci. & Technol., A 3, 10 (1985)

Window B., Sharples F. & Savvides N., J. Vac. Sci. & Technol., A 6, (3), 1935 (1988)

Wing R.G. & McGill I.R., Pt. Met. Rev., 25 (3) 94 (1981)

Wolf G.K., Nucl. Instr. & Meth. 182/183, 875 (1981).

Wolf B H & Keller R, 'Ion Implantation into Metals', Ashworth V., Grant W.A. & Procter R.P.M.(Ed.), Pergamon Press (1982), p 302

Wolf D.E., Griffiths R.B. & Tang

L., Surf. Sci., 162, 114 (1985)
Wolfe G.J., Petrovsky C.J. & Quinto D.T., J. Vac. Sci. & Technol., A 4, (6), 2747 (1986)
Wolfla T.A., Unpublished Data
Wolfla T.A. & Tucker R.C., Proc. Intl. Conf.on Metallurgical Coatings, San Francisco, April (1978)
Wonter D.G. & Strachan A.M., 'Advances in Extractive metallurgy', Jones M.J.(Ed.) Inst Min & Metall, (1977)
Wood G.C. , Norwegian J. Chem., Mining & Metallurgy, 29, 209-213 (1969)
Wood G.C., Oxid. Metals, 2, 11-57 (1970)
Wood G.C., 'Oxidation of Metals', Chapter in 'Techniques in Metals Research, Vol. IV', Ed. Rapp R.A., Interscience, NY (1970) pp 493-569
Wood G.C. et al, Oxid Met 14, 217 (1980) and 10, 163 (1976)
Wood G.C. & Hobby M.G., Oxid. Metals, 1 23-54 (1969)
Wood G.C. & Bastow B.D., Corr. Sci., 18, 297-313 (1978)
Wood G.C. & Whittle D.P., Scripta Met., 1, 61-64 (1967)
Wood G.C., Davies G.A., Ponter A.B. & Whittle D.P., J. Electrochem. Soc., 113, 959 (1966)
Wood G.C., Stott F.H. & Hobby M.G., Oxid. Metals, 3, 103 (1971)
Wood G.C., Whittle D.P., Evans D.J. & Scully D.B., Acta Met., 15, 1421-1430 (1967)
Wood G.C., Wright I.G. & Ferguson J.M., Corr. Sci., 5, 645 (1965)
Wood G.C., Wright I.G., Hodgkiess T. & Whittle D.P., Werkst u. Korr., 21, 900-910 (1970)
Wood G.C., Fountain J.G. & Scott F.H., Oxid Met 14, 47 (1980)
Woodruft D.W. & Sanchez-Martinez R.A., J. Electrochem. Soc., 132, (3), 706 (1985)
Wooton M.R. & Birks N., Corros Sci 12, 829 (1972)
Worrell (1972), see Luthra K.L. & Worrell W.L., Met Trans A 9, 1055 (1978)
Wortmann D.J., J. Vac. Sci. & Technol., A 3, (6), 2532 (1985)
Wright I.G., Metals & Ceramics Info Center, Battelle, Columbus, OH, Rept MCIC-72-007, June (1972)
Wright I.G., Mat. Sci. & Engg., 88, 261 (1987)
Wright I.G., Nagarajan V. & Stringer J., Oxid. Met., 25, 175 (1986)
Wright M. & Beardow T., J. Vac.

Sci. & Technol., A 4, 388 (1986)
Wu C.S. & Birks N., J. Electrochem. Soc., 133, (10), 2185 (1986)
Wu S.Y., Diss Abstr Intl 38B, 288 (1977)
Wuest G., Keller S., Nicoll A.R. & Donnelly A., J Vac Sci Tech A 3, 2464 (1985)
Wynblatt P. & McCune R.C., "Surfaces & Interfaces in Ceramic & Ceramic-Metal Systems", Pask J. & Evans A., eds, Plenum, NY, (1980), p.83
Yabe K.et al, Proc 7th ICVM Conf Tokyo (1982), p 246
Yamaguchi T., Harano K. & Yajima K., 'Surfaces & Interfaces in Ceramic & Ceramic-Metal Systems', Pask J. & Evans A., eds, Plenum, NY (1980)
Yamamoto H. & M Tanaka, Proc.7th ICVM Tokyo (1974) p 602
Yamanaka K. & Enomoto Y., J Appl Phys, 53, 846 (1982)
Yamanaka K. & Enomoto Y., Proc.7th ICVM Conf., Tokyo (1982a), p 157
Yamanaka K. & Inomoto Y., J. Appl. Phy., 53 846 (1982b)
Yamanaka S. et al, Proc.7th ICVM Conf., Tokyo (1982), p 703
Yamanaka S., Kawagoe Y., Son P. & Miyake M., Proc.7th ICVM Conf., Tokyo (1982), p 500
Yame K. et al, Proc.7th ICVM Conf., Tokyo (1982), p 246
Yang C.H. et al (1987), see Mat. Sci & Engg 69, 351 (1985)
Yasuda H., 'Thin Film Process', Vossen J.L. & Kern W.(Ed.), Academic Press, NY (1978)
Yee K.K., Intl Metals Rev., 23, (1) 19 (1978)
Yeh X.L. & Johnson W.L., 'Proc. MRS Symp. on Rapidly Solidified Alloys', Taub A., Giessen B.C. & Polk D. (Ed.), MRS, Boston (1986) p.63
Yeh X.L., Sammer K. & Johnson W.L., Appl. Phys. Lett., 42, 242 (1983)
Yermakov A.Y., Yurchikov Y.Y. & Barinov V.A., Phys. Met. Metall., 52, 50 (1981)
Yoneda M. et al., J. High Temp. Soc. Japan, 14, 142 (1988)
Yoon S.G., Kim H.G. & Chun J.S., J. Mat. Sci., 22, (1), 2629 (1987)
Yoshida T. & Akashi K., 'Co-condensation of Metal Vapours', J Japan Inst Met, 45, 1138-45 (1981)
Yoshida T., Takata R. & Akashi K., 7th ICVM Conf, Tokyo (1982), p.568

Yoshihara H., Mori H.,J Vac Sci Tech 16, 1007 (1979)

Young D.J. & Gesmundo F., Oxid. Met., 29, (1/2), 169 (1988)

Young E.W.A. & de Witt J.H.W., Oxid. Met., 26, (5/6), 351 (1986)

Young S.G., Chartd Mech Engr (CME) 49, 'An Exptl Study of Si Slurry/ Aluminide Coatings for Super-alloys', p.24 (1987)

Younger P.R., J Vac Sci Tech A 3, 588 (1985)

Yunlong G., Zhuangqi H., Wei G. & Changshu S., 'Mat. Res. Symp. Proc.', Kear B.H. & Giessen B.C. (Ed.), North Holland, vol. 28, (1984) p.99

Yurek G.J. (1985), see in Rahmel et al (1985) p.292

Yurek G.J. & Przybylski K., Mat. Sci. & Engg., 87, 125 (1987)

Zakhaniev Z., Marinov M. & Zlateva R., Surf. & Coatings Technol., 31, 265 (1987)

Zambon A. & Ramous E., Mat. Sci. & Engg., 88, 123 (1987)

Zancheva L., Hillert M., Lange N., Seetharaman S. & L-I.Staffansson, Met Trans 9A,(7) 909 (1978)

Zancheva L. et al, Met Trans A, 9A, 909 (1978)

Zega B., Kornmann M. & Amiguet J., Thin Solid Films, 45, 577 (1977)

Zeman F., J.Mayerhofer & Kulmburg A., High Temp.-High Pr., 14, 341 (1982)

Zhibing M., Ruizeng Y, Liang G., Mat. Sci. & Technol., 4, (6) 540 (1988)

Zhong-yin X. et al in V.Ashworth, Grant W.A. & Procter R.P.M.(Ed.) 'Ion Inplantation into Metals', Pergamon Press (1982), p 117

Ziegler J.F. (Ed.), 'Ion Implantation Science & Technology', Acad Press NY (1984)

Ziegler P.F. & Rausch J.J., Surf. & Coatings Technol., 29, 259 (1986)

Zirinsky S. & Irene E.A., J Electrochem Soc., 125 (2) 305 (1978)

Zmiy V.I., 'B-Silicide Coating Diffusion Structure on W & Mo', Zasch Metallov., 14, 232-4 (1978)

Zmiy V.I., 'WSi$_2$ Diffusion Layer', IZV Acad. Nauk. SSSR Neorg. Mat., 17, (5) 916-7 (1981)

Zotov Yu.P., 'Mo$_5$Si$_3$ Coating, Reactivity with Mo Alloy', Izv. Acad. Nauk SSSR Metally., 2, 193 (1979)

Zuca S., 'Electrolytic Dep. of Al-Si Alloys', Metallurgia, 20, (7), 412-4 (1978)

INDEX

LILY MCGEE SOLVES THE MYSTERY IN A QUIET ENGLISH VILLAGE

DONNA DOYLE

PUREREAD.COM

CONTENTS

DEAR READER, GET READY FOR THE TRIP OF A LIFETIME TO A QUIET ENGLISH VILLAGE...

WILL YOU SOLVE THESE MYSTERIES?

Lily McGee has had enough of Hollywood!

After an altercation with the director of a movie she was working on, she decides to leave it all behind and live in a small English village to try her hand at writing murder mysteries.

Lily soon finds that her cozy crime-writing skills become crime-fighting skills as her sharp mind and amateur sleuthing ability is called upon to solve an array of strange happenings in little Didlington!

Starting with her first case, in which she has to prove her own innocence...

Turn the page and let's begin...

∾

A HUGE HOLLYWOOD
MYSTERY IN A QUIET
ENGLISH VILLAGE

PROLOGUE

Lily McGee rushed through the movie studio, dodging errant props and other people along the way. Her honey brown hair was pulled back in a tight ponytail, but strands had come loose and were falling around her heart-shaped face. Her cheeks were rosy, her eyes sharp, and her heart was pounding as she felt the pressure of time crushing her. Panting for breath, she reached the shadowed set and ran toward the director, Howard Boxall. Huge cameras stood behind the director, watching the set from different angles, while others on the crew looked on. Howard was standing there with his star, Rick Chambers. The two men were complete opposites of each other. Rick was the handsome, classic Hollywood archetype of a leading man while Howard was a squinting, wispy haired, narrow faced auteur who had made a career of being difficult, and brilliant. Currently, Rick had the script in his hand and was glaring at Howard, looming over him like death itself, not that Howard was intimidated. The director had his neck arched back and hands on his hips, fury pouring out of his eyes.

"You can write this stuff Howard, but you sure as shooting can't say it!" Rick fumed.

"You'll say whatever I want you to say. What, do you think because you're on the cover of magazines, the world's prettiest man, that you know better than me?"

"I know what the people want Howard. That's why my box office takings are gold and yours are... well... let's just say there's a reason why the studio wanted me on this picture. Your best days are behind you."

Howard narrowed his eyes. "I've had more good days than you've had witless teenagers fawning over you, and let's not forget that genius lasts forever while looks fade. You should be lucky you can make money while you can."

Rick threw his head back and laughed, shaking his head. Lily watched on meekly. Confrontations like these had been a regular occurrence on the film set and certainly wasn't anything like what she expected when she had taken the job as Howard's personal assistant. She assumed it would afford her the opportunity to work closely with a visionary director and benefit from his creative wisdom, instead she had been faced with a petty man whose cruelty surpassed anything else. Even the set had not been as exciting as she had hoped. It was mostly filled with surly faces. The glitz and glamor of Hollywood was as much of an illusion as anything else on the screen.

"What do you want?" Howard asked sharply, his gaze sweeping away from Rick.

"I just came with your coffee," Lily replied gently. She was almost thirty, yet when Howard spoke to her she felt like a misbehaving child.

"Finally," the director barked as he snatched the coffee from her. "I don't know why it took you so long to get back to me. He took a hearty gulp and then grimaced. "What the heck is this? I didn't ask for it cold," he spat in disgust. "I've had it with this place. What happened to the good old days?" he growled as he threw the coffee away. Lily assumed that he meant to throw it on the floor, but if that was the case then his aim was amiss because it hit Lily, right in the chest. The cold coffee (which wasn't as cold as Howard claimed it to be and left an uncomfortable warmth seeping over her chest) burst over her, leaving her shocked. Howard and Rick stormed away in different directions, leaving Lily to stare into the blank cameras that saw everything, sopping wet and forlorn.

Joanie rushed up to her, blonde tresses bouncing with every step, pink lipstick glowing like a neon sign.

"Oh my gosh Lily, are you okay?"

Lily's lips were a thin line and her tone was flat. "I've had it Joanie. I'm out of here. Lily McGee's love affair with Hollywood is over."

Lily had no idea where she was going or what she was going to do next. All she knew was that she needed a new start.

1

Lily McGee threw open the wide windows of her cottage and breathed in the sweet air of the English countryside. A small rock wall separated the grounds of her cottage from the winding lane that led to the heart of Didlington St. Wilfrid. Rolling fields rose toward the horizon, framed by trimmed hedges and populated by fluffy sheep. There was not a hint of smog in the sky. When she closed her eyes she was immersed in silence. Exhaling softly, she sighed with relief. The busy, bustling world of Hollywood was far away, and with it so too were all of her stress levels. She brought a slice of toast that was coated with jelly, although she told herself that she was going to have to get used to calling it jam, to her lips. The taste was rich and luxurious. The sky was pale blue, the scenery idyllic. She felt blessed for being able to be in a cottage like this. The rent was similar to what she had paid for a squat, blocky apartment in Los Angeles. Here she had the freedom of space, the luxury of solitude, and she didn't have to put up with the bumps and thumps of her neighbors. Nor was she plagued with the sound of police sirens. This village was quaint, quiet, and exactly what Lily needed to recharge her batteries after being drained by the rigors of Hollywood.

After finishing her breakfast she went to the small study where she opened her laptop and settled down for a day of writing. As blessed as she felt to live here, her savings would only last for so long, so she needed to find a way to earn money. She turned to doing something she had always wanted to do; write cozy mysteries. This village seemed like the perfect place to get a lot of writing done since it was quiet and free of distractions, but as she settled into her seat and stared at the blank document in front of her she realized that she needed something else; she needed inspiration.

Lily rested her fingers against the keys, tapping out a rhythm in the hope it would conjure an idea from the depths of her mind. She looked all around the room, as though the magic idea was lurking in a corner somewhere like a spider, and then she gazed out of the window. Clouds idly drifted by and they seemed to take form before her eyes, but only in the shapes of animals. She leaned back and pinched the bridge of her nose, trying to figure out where to start and where to end, and what to put in between. The more she stared at the document the more foreboding it seemed, as though it was a white abyss that threatened to swallow her. She muttered under her breath and rubbed her eyes.

"At this rate it'll only take me about, oh, fifty years to write one novel."

She tapped her fingers against the table and then shook her head. Whatever she needed wasn't going to happen in this one room. Rising, she ran a hand through her hair and made her way out of the cottage. Closing the door behind her, she looked at the etching of a bird on a plaque and the words 'Starling Cottage'. There were a number of cottages dotted around the village, all owned by Lord Huntingdon, who was renting them out. He must have had an interest in birds as they were all named after different types.

A winding path led to a small road that was framed with colorful flowers. The scent in the air was sweet and she strolled along, hoping that she would catch sight of something that would inspire her. The creative process had always been something that intrigued her. It was like alchemy in a way, creating something from nothing, and she had hoped that by working in Hollywood she would glean some sense to help her own process. Unfortunately all the people she had known were difficult and saw her as a nuisance more than anything else. There wasn't much she missed about the life, apart from Joanie. Friends had been few and far between, and for all her faults Joanie had been there when it mattered.

After she had been walking for a little while, the aroma of freshly baked bread drifted toward her and pulled her toward the bakery, which was owned by a woman called Marjorie Porter. She was a plump woman with a ruddy face, a cheery disposition, and a bob of curly hair. The picture of her husband, the dearly departed Henry, hung on the wall. This bakery had been a frequent stop of Lily's since she had moved in, although she had to control herself because she knew that she would balloon up if she didn't. It was easy to stay slender in Hollywood, what with the long and draining working hours, squirreling away something to eat only when chance allowed. The hours were long here, and there were plenty of opportunities to indulge her appetites.

"What can I get for you today Lily?" Marjorie asked. Lily was still getting used to the personal touch. People around here had gotten used to her quickly because she was new, and because she was from far away.

"I'm not sure. I just came in here to look really," Lily said, and her mouth began to salivate when she saw the currant buns, the éclairs, and the iced cakes that Marjorie had baked. "Actually I told myself that I need to be more disciplined around you. I'm already beginning to get a little too comfortable," she patted her belly and laughed. Marjorie chuckled as well.

"I don't agree with that. I think you need a little meat on your bones if you're going to attract a husband."

Lily offered a wan smile. "That's not really at the top of my list of priorities."

"You say that now, but you never know when you're going to meet the right person. I'm surprised you never settled down with one of those famous actors since you lived in Hollywood."

"Believe me, they might be handsome, but they're rarely charming," Lily said. The truth was that she hadn't exactly mixed with the higher echelons of Hollywood, but the people of this village believed otherwise.

"Well you just keep your eyes peeled because you never know when you're going to meet someone. And don't be afraid to enjoy life. The worst thing you can do is miss out on all the small pleasures." She said this as she reached in and brought out an iced bun with some tongs, placing it in a paper bag. Lily hadn't remembered ordering it, but she found that it was exactly what she was in the mood for.

When she left the bakery she pondered Marjorie's words. She certainly hadn't given much thought to romance as the possibilities out here were few and far between. After some car crashes of romance back in Hollywood it was refreshing to not have the pressure of it all on her mind. For the time being she wanted to focus on herself and her books. If she was going to meet someone then it would happen in its good time. For now she was just going to enjoy a bun.

While she was walking she continued to admire the scenery, which was so vivid and colorful it seemed unreal. She was used to the grey, drab color of city blocks and fences. This place only had the lightest human touch, and it was all the better for it. Lily was interrupted from her reverie by a fluffy black dog that bounded up to her. His pink tongue lolled out and his eyes were as brown as the trunk of an old tree.

"Hello boy," she laughed as the dog lifted itself on its hind legs and placed its paws on her waist. He was a medium sized dog, but in just the few moments she had interacted with him, Lily could tell that he had a lot of personality. He was sniffing toward her fingers, which were still dusted with sugary icing from the bun she'd eaten. She pulled her hand away and laughed as the dog stretched itself to its maximum length in an effort to reach her fingers, as though the sweet taste was the only thing that mattered in his life. Most of his fur was black, but there was a white patch running from his right eye down his neck, leaving a tinge of white on his long, floppy ear.

"Zeus, get back here," a stern voice barked through the air, and was followed by a whistle. Lily looked up and saw a stocky man walking toward her. He had a salt and pepper beard, he wore a tweed jacket, and a flat cap.

"What a lovely dog," Lily said as the man drew closer.

"What would you know about it?" the man snapped. Lily was taken aback by the disgruntled manner of the man. Most people had been pleasant when she encountered them and it had been a nice change from the city, but it seemed as though ill manners had found their way into this idyllic place as well.

"Excuse me?" Lily asked.

The man stopped in his tracks and clicked his hands, commanding Zeus to stay by his side. The dog sat there happily.

"I don't know what you want here, but you're not welcome."

"What I want? I'm sorry… but have you gotten me confused with someone else?"

"No, I haven't, not as long as you're that American writer who is coming along to make fools of us all?"

"I'm a writer yes, but I don't know where you've heard this. I don't want to make a fool out of anyone."

"That's what they all say," he grunted, shaking his head. "I know your type. You come here thinking that we're simple folk, and you want the whole world to laugh at us. Others might look at you and treat you like you're some kind of visiting royalty, but we're just a tourist attraction to you. We're real people, with real lives, and I'm not going to have you make us look like idiots."

Lily was left aghast. Before she could apologize and reason with the man to try and understand what had provoked this intense outburst, he was already marching away with his dog in tow. She wasn't sure she had ever seen more of a difference between an owner and their pet before, and she was left wondering what had caused the chip to appear on his shoulder.

The interaction brought back memories of her encounter with Howard and how he had berated her. It made her wonder what would happen if Howard had come to a place like this rather than her. Surely the irascible man would have clashed with the local population. A smile twitched on her face and suddenly she forgot all about the unpleasant meeting with the stranger. Inspiration had struck! She finally had her idea, and she raced back to Starling Cottage, flinging herself into her chair where her fingers typed so fast they became a blur.

An arrogant Hollywood director would come to a quiet English village to shoot a movie and end up infuriating one of the locals. It drew on her own experiences, and was something that she could sink her teeth into. After all, she figured that she might as well get *something* out of her Hollywood career.

2

L ily had been writing consistently for about two weeks. The words flowed from her mind and she was proud of herself for writing so much. She took a little too much joy in writing the murder scene as she worked through some lingering frustration with Howard. She didn't *mean* it to resemble him as much as it did, but it just happened that he fit the character she wanted to write about. She grinned as she turned over different plots in her mind and developed the suspects. It seemed so fitting, and she liked that the story had some meat to it. In her mind it wasn't simply a mystery, rather it was a way to tell the story of the modern age, of huge cities clashing with a simpler way of life. In truth it was something that she was caught between as well. She had spent all her life in the city, and while moving to Didlington St. Wilfrid was a nice change of pace she wasn't sure if she would want to spend her entire life here. Still, there was time for all that when and if she got her book published. For now she needed to focus on the writing of it.

To reward herself for such good work she found herself popping into Marjorie's bakery again, and was greeted with a smile. After selecting a few sweet treats to take home with her, Lily decided to ask Marjorie about the ill-tempered man she had encountered two weeks earlier.

Marjorie arched her eyebrows and sighed. "Ah yes, that sounds like Pat."

"Well what's his problem?" Lily asked.

"Who knows with Pat? He's not exactly the type to talk about his feelings, although if you ask me not many men learn that skill. He lives up on the old farm and keeps to himself. It's just him and Zeus you know, for as long as I can remember. I think he just doesn't like outsiders. I wouldn't worry about him though. Most people here like a bit of excitement and you've certainly provided that!"

"I have?" Lily asked quizzically, surprised that she had had such an impact on the small village.

"Oh yes," Marjorie drew in a gasp. "Everyone is talking about you. We rarely get anything exciting happening around here. I think you should give a talk at the community hall. I'm sure you must have so many amazing stories to tell. It's not as though people here get to travel that much. Some of us dream of leaving but, well..." she sighed and trailed off. Lily wondered where Marjorie's dreams would have taken her, and if it was her love that had rooted her to this place. Lily's eyes drifted to the picture of Marjorie's husband and wondered what it would be like to love someone so much you were willing to stay in one place.

"I'm sure I'm not all that exciting," Lily said dryly. In Hollywood nobody had ever paid any attention to her, and she had never been anything special. It was strange to think that people were talking about her behind her back and creating stories about her. "But I don't have anything to worry about from Pat or anyone else, do I?"

"Of course not," Marjorie chuckled. "Pat is Pat. He'd get angry at a passing cloud if it suited him, but he doesn't mean any harm. As for anyone else, well, they're just curious, that's all. It's not often that something unusual happens in our little corner of the world."

Lily wasn't too sure that Pat's anger should be taken for granted. It can't have done his disposition any good, but she hadn't arrived here to become a therapist so she thanked Marjorie for the kind words and departed from the bakery, carrying the sweet treats with her. She had been around many celebrities during her time in Hollywood, and it had never occurred to her that she would end up becoming one, but she supposed to these people she was a symbol of a life that they could only dream of. Little did they know that the dream was really a nightmare, and they were much better off here.

When Lily returned to the cottage she settled at her desk again and began to think of ideas for various other characters while eating the things she had picked up from the bakery. Marjorie really was a magician, and she wondered what other secrets were hiding in Didlington St. Wilfrid.

A while later, Lily was interrupted from her thoughts by a knock at the door. It took her a few moments to respond because she wasn't expecting anyone at all. She furrowed her brow, wondering if some interested villager had worked up the courage to visit her and possibly learn something about Hollywood. But when she opened the door she saw someone she hadn't been expecting at all.

"Lily! It's been so long. Oh I've missed you so much!" Joanie exclaimed. She walked in, the air sweeping around her. Her smile was wide, her statuesque beauty somehow not quite fitting the quaint humility of the cottage. Her platinum blonde hair glowed, her lipstick shone, and she looked as though she belonged in another world. Joanie flung her arms around Lily's neck and squeezed her tightly, all while Lily was nonplussed.

"Look at this place, it's so cute! Oh my gosh it's like something out of a movie," Joanie moved away from Lily, leaving a suitcase standing in the doorway. She tilted her head back and gazed around in awe at Starling Cottage. Lily had to pinch herself to make sure that she hadn't fallen asleep.

"Joanie what are you doing here?" she asked in disbelief.

Joanie turned and flashed a wide, dazzling smile.

"I came here because I missed you of course! Life isn't the same without you Lily and I wanted to pay you a visit, make sure you're not getting yourself in any trouble. Of course, to do that I had to bring work with me."

Lily arched an eyebrow and her chest tightened with anxiety. "What do you mean you brought work with you?"

Joanie sidled toward Lily, draping an arm around her shoulder. "Well you know I couldn't really afford to take a vacation. Howard needed another location for the movie and I thought why not shoot it in a small English village? We're here to scout. There are only a few of us you know, but I thought it would be the perfect opportunity to see you again!"

"Howard is here?" Lily asked, her face turning ashen.

"Yes! Howard, Rick, Doug the camera guy, and a few other people. It's going to be like a big party!" I thought that I might stay with you to make things more fun. The others can all go somewhere else. It looks like you have the room," Joanie wandered through the cottage and made herself at home, murmuring appreciation for the place. Lily's throat was dry as she leaned against a wall and pressed her hand against her forehead. A band of tension suddenly knotted around the front of her face. She had come here to get away from Howard and Hollywood and everything that happened there. What had Joanie done?

Lily heard some commotion outside and walked down her garden path to see what was happening. There was a huge coach hissing as it tried to eke its way down a narrow lane, and Pat was there, shaking his fist at the vehicle. Lily rolled her eyes, knowing that this was only the beginning. Her peaceful idyll had just been disrupted and it was only going to get worse.

3

"**G**et away from here! What do you think you're doing bringing this monster down this lane, it's clearly not going to fit!" Pat yelled. His expression was twisted into a scowl and his cheeks were crimson. Zeus was circling him, swept up in the excitement. The windows of the coach were tinted and it looked a foreboding sight. The doors opened and Howard emerged. Lily groaned and started toward the two of them before war could break out, although she feared she was already too late.

"Get out of my way!" Howard yelled, storming toward Pat.

"What right do you have coming down here with this thing?"

"I have every right. I'm making a movie," Howard shot back, as though that explained everything. They were yelling at the top of their lungs and Lily grimaced as the peace was shattered. Zeus stood beside Pat, looking at the proceedings with intrigue. Howard was his usual brash self, caring only about himself and his movie.

"I'm sure that we can settle this matter without having to wake the entire village," Lily said upon reaching them. The two men looked at her with withering stares.

"You should keep out of it as well. You're as bad as them," Pat said.

"I have nothing to say to you," Howard added. "As far as I'm concerned you're dead to me. I only want loyal people around me, not people who are going to walk away at the first sign of stress."

His words cut Lily sharply. She gasped, for she hadn't left at the first sign of trouble at all, not that Howard would believe her.

Howard turned back to Pat. "Now get out of our way. We have every right to stay here. We've paid the fees to camp in the field. You should be fortunate that we picked this forgotten little place to make our movie. It's not as though this place has any significance otherwise," Howard spoke in a scathing tone, and then he turned to Lily. "It's no wonder why you wanted to retreat here. You get to hide away in this little corner of the world without having to do anything difficult. That's the problem with your generation. You don't know the meaning of hard work."

Before Lily could reply two figures had emerged from the coach. One was the star, Rick Chambers, while the other one was the camera man, Doug.

"Come on Howard, let's just get back on the coach," Rick said, turning to Pat. "I'm sorry sir, it's been a long flight and a long way to get here. We're all a little tired and stressed. This village is lovely and we're honored that you've allowed us into your home. We'll be gone before you know it, I promise, and then life can get back to normal."

His voice was as smooth as honey and Lily had to give him credit for being able to defuse the situation. He could be charming when he wanted to be, which she supposed was the gift of acting. After all, she had seen his temper flare at Howard as well. Just being around them for a few moments was enough to make a band of tension run around her forehead, and she was glad that she had made the decision to walk away.

"Come on Howard, let's go," Doug said, putting his arms around the director. Doug was a big, husky man, and like most people he towered over Howard.

"Get your hands off me," Howard said, shrugging away Doug's efforts to get him back on the coach. "The last thing I need is you treating me like some old fool. Just point the camera where I tell you. That's the only reason you're here," Howard snapped bitterly as he climbed the steps back into the coach. Doug rolled his eyes and Lily pitied him. Meanwhile, Rick was shaking Pat's hand and wore a warm smile on his face before he turned away. The moment he did, and the moment he was certain that Pat had been placated, that smile fell from his face and he scowled. Lily was convinced that there was going to be fireworks on that coach.

The coach hissed and trundled down the narrow road, moving slowly to avoid scraping the small stone walls that stood on either side of the path. Unfortunately there were a few plants and bushes that were squashed by the wide tires. Pat rolled his eyes and glared at Lily, but he didn't say anything else before moving on. She watched the coach as it drove toward a field, where it stopped and all the crew filtered out to begin building their camp that would be their home for the duration of the shoot.

She sighed as she returned to Starling cottage. Joanie was there waiting for her.

"Well, I'm glad you brought this storm with you," Lily said tersely.

Joanie wrung her hands. "I'm sorry, but I thought it would be a good way to get a free trip over here to see you! It's not like it's going to last forever either. You know what Howard is like, he likes to get his movies done quickly. I'm sure it's going to be over before you know it and then you can carry on with whatever you want to do here," she said.

Lily found it impossible to be angry with Joanie, but she hoped that Joanie spoke the truth. The last thing she needed was to have a film crew around the village interfering with the peace and quiet. She had come to Didlington St. Wilfried to escape Hollywood, not to bring it with her.

Still, interacting with Howard again gave her fuel for her story. The man inspired her in all the worst ways and the ideas began swimming in her mind. He managed to irritate almost everyone he met, and writing about his demise was strangely satisfying. She wrote long into the night, losing herself in her story until the hours were small.

Lily awoke to the sound of heavy knocking on her door. She groaned and rubbed her eyes, for sleep had been fleeting. Yawning, she walked to the door and wondered who could be calling for her at this time of the morning. She dreaded to think it might be Howard. Even though she had quit Hollywood he would probably demand that she continue working for him because that's the kind of man he was. However, she was surprised to see a police officer standing before her. He was a tall man with a ruddy complexion and a thick mustache sitting above his upper lip. His uniform was stretched tight around his rotund body, and his voice was deep and rasping.

"Are you Miss McGee?" he asked.

"I am," Lily replied, narrowing her eyes as she wondered what this was about.

"I'm Constable Dudley, Bob Dudley and I'd like to ask you a few questions if I may," he said.

"Of course. What is this about?"

Constable Dudley sucked in a breath and lowered his voice. "I'm afraid to tell you that there was a murder last night," he said.

"A murder?" Joanie replied, squealing loudly. Lily's eyes were wide with shock. Constable Dudley's eyes were wide as well, but not because of the murder, but because of Joanie's appearance. She was dressed in a robe which showed off her long, elegant legs and it must have been a sight that Constable Dudley had never seen before because he was quite distracted.

"What exactly happened? Who was murdered?" Lily asked, trying to get the Constable's attention back on the matter at hand.

"Ah, yes," Constable Dudley was flustered, but he managed to tear his gaze away from Joanie. "This is the reason why I'm here. The victim is Howard Boxall, the director of the film. I believe you knew him. I'm trying to ask anyone that knew him for any insight about the man. As you can imagine we don't get many murders around here, so I'm trying to get the case wrapped up quickly. Can you tell me about your relationship with him?"

"Howard was murdered? Lily, this sounds just like your story!" Joanie gasped.

Lily had been blinking slowly, trying to process the news that Howard had died. When Joanie said what she said, Lily glared at her.

"Excuse me? What story?" Constable Dudley asked.

"It's nothing really," Lily said, chuckling lightly because it was all a silly misunderstanding. "I left my career in Hollywood to come here and write murder mysteries. It's just a coincidence that I happened to base the first book on my own life."

"I see," the constable made a few notes in his little book. "And in this book it's the director of the film who died?"

"Yes," Lily admitted. "But it really is just a coincidence."

"And why did you choose the director? Did you have some kind of animosity toward him? Is that why you left Hollywood?"

Lily arched her eyebrows. Constable Dudley certainly didn't waste any time in getting to the heart of the matter. He reminded her of a battering ram, and was clearly eager to crash his way through to the truth.

"When you speak to more people I think you'll find that everyone had animosity with Howard. He was that type of person. He's a genius and his films have won many awards, but he's not a pleasant man to work with. He's very demanding and he has a volatile temper, but I promise you it wasn't I who killed him."

"I see," the constable said. He didn't sound convinced by her declaration of innocence, and she supposed it was the exact thing a guilty person would say. "The murder took place during the night. Could you tell me what you were doing?"

"I was writing until quite late, around midnight I think, and then I went to bed," Lily said.

"And can you corroborate this Miss...?" the Constable said, turning to Joanie.

"Joanie Dawson," Joanie strode across the room toward the door and took the Constable's hand, flashing him a warm smile. His cheeks glowed an even brighter red and he muttered something unintelligible.

"Well Miss Dawson, were you in the house as well?"

"I was," she began, and then looked awkward, "but I'm afraid I was asleep so I didn't hear Lily come to bed. I take sleeping tablets you see because otherwise I simply can't get my mind to turn off and so I'm always out like a light. I'm sorry I can't be more help officer but I can assure you that Lily isn't the kind of person to do this. She's my best friend, but I'm sure you'll find out who the real murderer is. I just hope this isn't some kind of vendetta against people in the movie industry. What if we're targeted next," Joanie gasped.

"I'm sure you'll be safe. I'll get to the bottom of this soon enough," Constable Dudley puffed out his chest as he spoke.

"Where exactly did the murder take place?" Lily asked.

"In one of Lord Huntingdon's houses. I'm not sure what this director of yours was doing up there at that time of night, but I suppose he had his reasons."

"And how was he murdered?"

"Well that's the strange thing," Constable Dudley scratched the back of his neck. "It seems as though history is repeating itself."

"What do you mean? I thought you said that there weren't any murders around here?" Lily asked, suddenly intrigued.

"I meant nowadays. But in the past there was a murder, a grisly one at that. It involved one of the Lord's ancestors. A terrible thing. One of the stable hands was in love with the Lord's daughter at the time, but she was promised to another. The betrothed husband was found dead in the same room as your director, also from a head wound, again, the same as your director."

"So you don't know exactly what killed him yet?" Lily asked.

"All I know is that it was something heavy. I still have a lot of questions I need to ask, and a lot of people I need to speak to. I appreciate all your help and, well, I'd advise you not to leave the village until the investigation is over." The constable spoke with a warning tone and Lily was able to infer that she was under suspicion, which was a frightening thought. She thanked him and the constable went about his business, while Lily closed the door and turned to face Joanie.

"I can't believe Howard is really dead," Joanie said with a look of shock upon her face. Now that Constable Dudley had left they were given the opportunity to properly process their grief, although Lily did have a few other things on her mind.

"I know. So many people have threatened to kill him over the years, but someone finally did it," Lily said.

"I'm sorry for being unable to lie to the cop for you. I thought it was better to be honest," Joanie offered a small smile, but Lily told her not to worry.

"I didn't do it so I'm not upset, although I don't like the fact that I'm under suspicion. If they can't find the actual murderer then I might be suspect number one. After all, it's not a leap of logic to suggest that I still hold enmity toward him, given I left Hollywood because of him. It would have been a shock to me to see him again, and we've already had another altercation. I think I'm going to have to try and find the real murderer before people start believing that I had anything to do with this." The more Lily thought about it the more afraid she was. In her heart she knew she was innocent, but that wouldn't matter to anyone else. She was writing a story where the director was dead and that meant all eyes were going to be on her, but if she didn't do it, then who did?

"Where are you going to start?" Joanie asked.

"I don't know, but the fact that this murder resembles a local legend makes me wonder if it was actually anyone on the film crew at all. After all, they all know how Howard is and if they haven't killed him yet why would they start now? I don't think we can discount the possibility that it's a local, and I think I know which one I want to talk to first."

4

Lily was reeling from the news that Howard had died. Despite having a difficult relationship with him she never would have wished death upon him, especially not in such a grisly way. Still, she had to be pragmatic about the situation and acknowledge that Howard was the type of man who would create more enemies than friends, and not everyone was going to be willing to give him a pass for his genius. She had certainly not foreseen herself becoming involved in investigating an actual murder. Writing one was difficult enough, but she felt personally invested in this one because she knew the people involved, and now she was about to have a difficult conversation.

Lily made her way across a field toward an old farmhouse. A single wisp of smoke rose from a chimney. A pigsty was attached to the house, and elsewhere sheep roamed about the grass, grazing happily without a care in the world. The house was weathered and she imagined it must have been standing here for generations. She rapped her knuckles against the door and waited for it to open. Within moments Pat was there, with Zeus by his side as always. The contrast between the two greetings she got was huge. Zeus was happy to see her, nuzzling her hand and licking her fingers, while Pat stared at her with narrowed eyes.

"What do you want?" he asked gruffly.

"I just... I just wanted to speak to you about the murder that happened. Has Constable Dudley been to see you yet?"

"Of course he has, not that the fat headed fool has any idea what he's doing. He couldn't solve the crime if the murderer came right up to him with bloody hands and confessed. But what has it got to do with me?"

Lily flashed a smile, trying to put Pat at ease. She clasped her hands in front of her and spoke in her friendliest tone. "Well you may not know this but I'm actually writing a mystery story of my own and I suppose I'm trying to see how people have reacted to the news so that in my story I can make sure the reactions are genuine, and you being someone who recently had an altercation with Howard... well... I thought you might have some insight into the matter."

Pat barked a laugh. "You Hollywood types are all the same. Why say one word when a hundred will do? I know what this is about. You're getting bored and the moment something exciting happens you want to get involved. You think it's so simple don't you? The local man gets into an argument with an American and then the American turns up dead. It makes sense, doesn't it? It's all neat and tidy and you wouldn't have to worry about anything else. Believe me, I have better things to do with my time than going around killing people I get into arguments with. All I want is to be left alone. I don't know who killed him and I don't particularly care. All I hope is that it means the lot of you will clear out and go back to wherever you came from. Now I'd appreciate it if you didn't come by here again. I don't particularly like opening my door and being accused of killing someone."

Before Lily could say anything else Pat had slammed the door in her face. She cringed at the way she had gone about things and certainly hadn't meant to accuse him outright, but she did wonder if his protestations were done to protect him. Pat didn't strike her as a daft man. He might well have seen an opportunity to kill a man and use the rest of the film crew as a shield. After all, there were many likelier killers who know Howard more intimately than Pat did. Anger was a volatile emotion and it could lead people to do dramatic things. Pat hated his village being overrun with strangers, and since he lived alone, his negative emotions would have been allowed to fester in his heart. She wasn't sure why he would frame the crime to reflect something that happened to Lord Huntingdon's family, but she wasn't ready to dismiss him as a suspect either. She found his comments about Constable Dudley interesting too. Was Pat the killer and thought he was smarter than the local law, or was he right and Constable Dudley did verge on the side of incompetence. If that was the case then Lily had to work hard to ensure that she was not the one who ended up as the prime suspect.

Her next stop had to be the camp where the crew was set up.

The mood around the camp was somber. People were sitting around with long faces, sharing stories of Howard. There was a lot of uncertainty as well. The equipment stood unused, the cameras were pointed to the ground and covered with tarpaulin. The director was the lynchpin of the production, especially out here away from the studio. He had the vision and the command, and now everyone was going to be aimless. Lily smiled awkwardly at her old colleagues and friends, but made a beeline for Rick's trailer. He was the one who would know what was going on, and he was also a suspect himself. In fact most people in the camp were suspects considering what Howard had put them through.

She reached his trailer and knocked on the door, although there was no answer.

"He's on a call to the producers, trying to find out what the heck is going to happen now," Doug said. The camera man was standing by a chest filled with equipment with wires wrapped around his hands. "I'm just trying to keep busy."

Lily nodded. "How is everyone?"

"Pretty shaken up, as you can imagine. It's pretty crazy what's going on. I'm not sure any of us can actually believe it."

"I know how you feel. What actually happened? What was he doing up there?"

Doug shrugged. "I don't know. We went up to the house to scout it out yesterday afternoon. I guess he wanted to have another look at it at night. You know what he's like. Sometimes he gets an idea in his head and he just goes and follows it. It's not like anyone can tell him no."

"Yeah, I remember," Lily agreed. "Did you go with him the second time?"

Doug shook his head. "Just the first. Last night I was asleep. I don't know what happened around here. If anyone did go with him they're not talking. People were pretty tired due to the jet lag and setting up camp though. I don't know if anyone would have been aware of anything happening. People are pretty freaked out though, you know, at the thought that one of us is a murderer."

"You seem pretty calm."

"I'm just trying to keep busy. I don't really think it's anyone here. I think it's someone local. You saw the way that guy reacted yesterday. I wouldn't be surprised if Howard encountered someone else who was out late at night and got into an argument with them. He was lucky yesterday that he had Rick and me to pull him away. It's always the way with these things, isn't it? He was in the wrong place at the wrong time."

Lily nodded. Perhaps it was just that simple, although the cogs in her mind were turning. Was it possible that Howard had gone off on his own in a strange town to scout a location and ran into someone

"I'm just keeping busy, you know, to take my mind off things," he said. Lily angled her head to try and get a better look at the camera, but as she did so Doug turned his body to prevent her from seeing it properly. "What are you still doing here? I would have thought it would be too depressing to be around a place like this. You should head back to your cottage, get some rest."

"I think I'd rather be here," Lily said. "There's something that I've been wondering about Doug... you told me that everyone was asleep. Wasn't there anything that disturbed people during the night?"

"Not that I know of."

"That's funny Doug, because I was just talking to Rick and he said that there was a rooster crying out for a while that woke everyone up. But if you didn't hear it then..."

"I guess I'm a heavy sleeper."

"Or you weren't here at all," she said. As soon as the words slipped out of her mouth she knew that something was wrong with Doug. The bland expression changed on his face and he suddenly looked like a monster. Before she could move he had burst across the tent toward her and grabbed the bundle of wires he had carried with him earlier. She twisted away, trying to run, but his arms were already around her. The wires dragged around her throat and he started pulling tighter and tighter, and the world was growing darker and darker...

"You shouldn't have poked your nose in here. You should have left and kept your distance. This isn't your world anymore. It's not your problem." Doug spoke through gritted teeth as he tried to choke the life out of her. Lily gasped and gagged and flailed everything she had to try and escape. In one brusque movement she managed to stomp on his foot and for a moment his grip was released. It was only one brief moment, but she made it count. She screamed at the top of her lungs for help. Doug's hand came over her mouth almost immediately, suffocating her, and he pulled the wires even tauter.

She always feared that Hollywood would be the death of her, and she hated to be proven right. The strength slipped from her body and she began to lose the will to resist.

Then, just as he had done in so many movies, Rick Chambers came charging in, along with others from the crew who had heard the scream. They wrestled Doug off her and she fell to the ground, gasping for air and clutching her throat.

"What the heck is going on here?" Rick asked.

"I don't know! She went crazy! I think she was the one who did it. I think she killed Howard!" Doug exclaimed. Lily's eyes went wide with panic. She coughed and spluttered, trying to protest her innocence, but it was so hard to speak. Her neck blazed with pain, but she pointed to the camera and managed to utter a single word, directing Rick's eyes to the device.

He frowned for a moment as he tried to understand her, and then he followed her guidance to look at the camera. The moment he did so, he recoiled in shock.

"There's blood on here! There's blood!" he cried, and Doug roared in anguish.

EPILOGUE

"It's all been rather exciting here, hasn't it?" Marjorie said, placing a few sweet treats in a bag. Lily went to pay, but Marjorie shook her head and insisted she do no such thing. "I couldn't ask you to pay, not after what you did."

"I didn't really do all that much. I only asked a few questions," Lily said.

"Nonsense! You put yourself in harm's way. You're very brave Lily. I daresay I couldn't have done what you did. And you solved the crime before Constable Dudley managed to. What was it like? Did you find out why he killed the director?"

Lily nodded. The moments after Rick had seen the blood on the camera had been fraught with tension. Doug knew the game was up and he had confessed to everything. "Doug has been working with Howard for a long time, and because he's a cameraman he's seen everything. I think he just had had enough. He said that Howard dragged him out of bed last night to go and scout the location under the moonlight, which Doug didn't like because he was trying to sleep. Apparently Howard was running his mouth and insulted Doug, and Doug let his anger get the better of him. He hit Doug with the camera, and the impact was hard enough to kill him. Doug was trying to clean the evidence off the camera before anyone could find out what had happened, but he was too late."

"So there was nothing to the fact that it happened in the same manner as the old story?" Marjorie asked.

"Just coincidence I think, although I'm going to want to hear more about that story next time I come in here. I've only heard the basics."

"You're welcome any time Lily, any time at all," Marjorie said.

Lily thanked her for the sweet treats and returned to Starling Cottage, where Joanie was waiting for her. Lily handed her the bag and Joanie smiled with glee as she smelled the delicious treats.

"Are you sure you don't mind me staying for a while? I think this is a sign that I should take a break from Hollywood too," Joanie said.

Lily was slightly annoyed at having her peaceful idyll broken, but she couldn't be too despondent about her friend asking to stay. She would be glad for the company. The film crew was packing up already, eager to leave the sorry business behind them. Once it was revealed that a member of the crew was responsible for Howard's death there was no way the film could continue, so the peaceful town of Didlington St. Wilfried was left alone again. The flash of excitement was over.

"I promise you that it's not like this all the time," Lily said. "Most of the time it's perfectly relaxing. I think it's time I get back to writing my book," she said, and bid Joanie farewell. Lily retired to her study where she sat down in front of her computer and reviewed what she had written so far. She ended up changing much of the story out of respect for Howard. For all the hostility that had existed between them she hadn't thought he deserved to die, and she didn't want to appear as though she was taking advantage of his tragic death. But she was sure that in her book as in real life justice would be done, and she felt a sense of pride in the role she had played.

She was also glad that she had managed to solve the crime quickly, as Constable Dudley had revealed to her in a joking manner that she had been his prime suspect. It was a chilling thought to know how close she had come to being arrested, and she vowed to never take inspiration from real life again, or at least to have an iron clad alibi if she did.

A little while later she was disturbed by Joanie shouting from the lounge, telling her that someone was at the door. Lily rolled her eyes and wondered what this was about. Surely there couldn't have been another murder!

But as she emerged from her study she saw a strapping man standing in the doorway. It was as though he had been plucked from one of her dreams.

"I'm Blake Huntingdon," he said, "after the recent events I thought it best to check on my father's property." His voice was a low, rumbling baritone. Lily and Joanie were both swooning. Life in Didlington St. Wilfried was surprising indeed. She never knew what was going to turn up on her doorstep.

A PERFECT POISONED PASTRY IN A QUIET ENGLISH VILLAGE

1

"I'm glad that excitement is all over with," Frank Roberts said as Marjorie deposited some of her sweet treats into a bag and handed them to him. Frank was a man of average height with a face as round as the moon, and a smile as wide as the heavens above.

"I don't know about that; it was nice to have a change of pace around here," Marjorie replied.

"I moved here for a change of pace! The best thing about Didlington St. Wilfrid is that nothing happens. The beauty lies in the peace."

"It seems a lot of people are coming here for a change of pace," Marjorie said.

Frank arched an eyebrow. He knew who she was talking about. "How is our newest star in the making? I have to admit I'm not entirely convinced that an American fits in well here."

Marjorie gave him an admonishing look. "Lily McGee is one of the sweetest women I've ever met and if it wasn't for her all that terrible business with the murder would never have been solved. She's a wonderful woman and you would be blessed to know her."

Frank raised his hands, surprised at the force of the words. But Marjorie was the kind of woman who defended her friends with all the strength she could muster.

"I stand corrected," Frank said with a wry smile. "I would be a fool to ignore your judgment. Perhaps our paths will cross. After all, it's a small village."

"And if she takes a liking to you perhaps she'll put you in one of her books," Marjorie wore a teasing smile.

Frank shuddered and grimaced. "I don't know about that. I'm not sure how I'd feel about someone putting words into my mouth or making me do things that I have no control over, even if it's not really me they're controlling."

"I think you're thinking too much into it. Lily has already said she's going to base a character on me in her next book," Marjorie let out a little squeal of excitement.

"Well I'm sure that the character will be charming and beautiful," Frank said warmly.

Marjorie's cheeks glowed. "Get on with you," she said, waving a playful hand toward him. "Charlotte will be wondering where you are."

Frank sighed and arched his eyebrows. "Yes, that she will," he said with a sense of burden weighing upon him, the burden that came with marriage. He stood there for a moment and his gaze drifted to the pictures hanging on the wall, depicting her late husband. "Can I ask you something Marjorie? Did you and Henry ever have times when... well... when you wondered if you should have gotten married in the first place?"

The smile fell from Marjorie's face, replaced by a gleaming sorrow in her eyes as they turned liquid. Frank immediately felt ashamed for causing such a reaction in her.

"I think marriage is a journey, and on that journey there are always moments of doubt, but it's important to remember that those moments do not define the journey. Are you and Charlotte...?" she trailed off at the end, unable to bring herself to ask the question directly, but Frank knew what was implied.

He nodded and ran his hand along his bare chin, suddenly looking as though he had aged a decade in a matter of moments. "There are times when I look at her and I just think that she's not the woman I married," he said.

"I'm sure things will get better," Marjorie said. "And while you're having these problems you know that I'm always here for a chat, but you should probably get back to her. The only way you're going to get close to her is to be with her."

"Yes, you're probably right," Frank said. He thanked Marjorie and left the warm, cloying heat of the bakery. The door closed behind him, and he left Marjorie gazing at the picture of her husband. Frank thought it was a tragedy that a marriage should have ended so swiftly, but he wished his own wasn't plagued by trouble. Recently Charlotte had been... difficult. It was hard to describe to be honest, and he wasn't even sure how long it had been going on. The years had a way of blurring into each other. One blink and suddenly a boy was a man, with fewer years ahead than he had behind him, leaving ragged hopes and dreams for someone else to pick up.

He heard the bark of a dog as Zeus approached him. Frank crossed the road, never having been the kind of man who appreciated the company of canines. The appearance of Zeus also meant that Pat wouldn't be too far behind, and the less Frank had to do with Pat the better. In the dim light of the evening he looked across the road. Pat glared at him as he glared at everyone, for no apparent reason. Frank's skin crawled. Pat was the kind of man who carried threat with him, and Frank was glad when Pat continued on his way. Frank found himself walking more rapidly to his house, which happened to neighbor Pat's farm.

His small cottage was a cozy place, like something out of a fairytale. He paused for a moment before he entered, taking a deep breath and promising himself that tonight would be a special night for himself and Charlotte. Something caught his eye on the doorstep, a smooth envelope that looked as though it should have been pushed through the letterbox. He bent down to pick it up and was surprised to find that there was no address written on the envelope and no stamp. He unsealed the envelope and pulled out a single sheet of plain paper. The letters were large, and it appeared as though the ink had been slashed across the page.

"Your sins will find you out. Do you really have a death wish?"

Frank stared at the words, unsure of their meaning, but fear curdled in the pit of his stomach. He scrunched up the letter and shoved it in his pocket, ignoring the crawling sensation that ran up and down the back of his neck.

The smell of roasted meat and thick gravy greeted him as he entered. Charlotte was standing in the kitchen with her arms folded across her chest and her eyes as cold as ice. Her features were sharp, and her black hair formed a midnight veil across her face.

"You're late," she said, the words cutting through the air. Frank's heart sank. So it was going to be one of *those* nights.

"Yes, I just stopped off at the bakery to pick us up some dessert," Frank said, lifting the bag he had brought back from Marjorie's. At the mention of it Charlotte rolled her eyes.

"Of course you did. Any excuse to go and spend time with *her*," Charlotte spat.

Frank sighed and pinched the bridge of his nose. "Let's not do this again Charlotte, please."

"We wouldn't have to do it again if you didn't keep making a point of going to see her."

"She owns the bakery Charlotte, besides, this place is so small it's not like I can actually avoid anyone," his voice was strained, but he managed to keep himself from shouting.

"You could if you tried. I've seen the way she looks at you. You're this handsome stranger who came from the glamorous city. She's lonely, and lonely women do desperate things. You don't think I see what she's doing? She's plying you with all these sweet treats, fattening you up like a turkey for Christmas."

"Charlotte, you're really overreacting. We're just friends. We have a little chat when I buy something from her bakery. There's nothing going on, believe me. I'm married to you. I brought this treat back so that we could share it."

"I don't want to eat anything that's been touched by her hands," Charlotte said bitterly. Frank's head dropped, for he didn't know what to do. His stomach growled so he went to the table and began eating. Charlotte dropped down to the table as well, joining him in the meal. It was a strange meal though, silent and tense and Charlotte never took her gaze off him. There were so many things he wanted to say, yet he wasn't sure how to begin. It seemed as though anything he said was going to turn into an argument anyway, so perhaps it was better not to say anything at all.

But then, Frank had never been the kind of man to avoid a problem until it was too late.

"I think we need to talk Charlotte, about us."

"I knew it," she said. "I knew there was something going on with you and that baker."

"There's not Charlotte, and I don't like how you can even believe that there is."

"You can't blame me," Charlotte said, and the look that flashed in her eyes made Frank swallow a lump in his throat. For a moment he was filled with shame.

47

Frank chose his next words carefully, keeping his voice measured, speaking in between chewing the tender meat and the delicious vegetables. "We came here to move on Charlotte. We were supposed to find our happy ending here, away from the distractions of the city, but we're not going to be able to do that if we keep living in the past. We have to look to the future and to the life that we could have together."

Charlotte's voice was rasping when she replied. Her hands trembled and glistening lines trickled down her cheeks. "You're acting as though this is my fault Frank. I've given you everything. I've tried *everything*. I came out here for you. I left my friends and my family for you and this place, but nothing has changed really, has it? We can change all the things that are around us, but we can't change ourselves."

Frank was about to say something else, but she left her meal half eaten and left the room, choking on her sobs. Frank thought about going after her, but he wasn't sure anything he said would actually do any good. It might be better for Charlotte to have some time alone.

Frank finished the food on his plate and then decided to leave the washing up until a little later. He took his sweet treat into the lounge and settled into a plush leather chair, telling himself that he shouldn't let this argument ruin his enjoyment of the night. He still had to get his happiness after all, although in a way that had been the problem that had brought their lives to this quaint village. But Frank wanted to be a better man. He wanted to be a good husband, it was just that sometimes it was *so* hard.

As he opened the bag and pulled out the éclair he thought about Marjorie, how sweet and friendly she was. She had a kind of homespun beauty that would not have been remarkable at all in the city, but it was a beauty that was imbued with a kind of strength, a strength that Charlotte could never possess. The sugary, creamy treat bathed his tongue in a wild storm of sweetness and delight, and it was almost enough to soothe his aching heart. He wondered how his life would be different if he had chosen another woman to be his wife, and he hated himself for it.

The sweetness lingered on his tongue as he sat there in the shadows of the evening, reflecting on Marjorie's words. He wondered how much truth there was to them, and how he would know when the difficult moments actually became the journey. But for now he still had the wherewithal to try and fix his marriage. They had uprooted their lives to come here so that they could be together and find their happy ever after, but Frank wasn't naïve enough to believe it would be so easy. He prepared to pull himself up and go and see his wife, but then a strange sensation overwhelmed him. His chest tightened and the world grew hazy, as though it was spinning around him. Then there was a sharp pain in his torso, which made him clutch his stomach. He leaned forward and coughed, looking with concern at the dark blood and black bile that stained his hand. It felt as though he was on fire and when he tried to call out for help, nothing but a rasping whisper escaped his lips. He lunged forward and tried to crawl toward the staircase, to get to his wife, but all his strength left him. A hacking cough brought up yet more bile, and he felt the kind of dread that came with any man's final moments.

His hand reached toward the door, toward his wife, but he knew that nobody could save him now. Tears flowed from his eyes, mixing with the blood and the bile that lingered on his lips, and then the darkness closed in. Then, there was nothing.

2

Lily McGee opened the curtains to her bedroom and basked in the glorious view of the English countryside. Even though she had been living here for a little while now, she knew she was never going to take this for granted. The rising, rolling hills were covered in soft moss and beautiful flowers. The trees were like battalions of soldiers guarding the natural world. There was a fine mist through which the sun was filtered, making everything glow in golden rays. Her heart was soothed and she woke up feeling relaxed every morning rather than feeling stressed and pressured as she had been in Hollywood. No longer did she have to wake up with the sight of towering buildings and billboards advertising yet more things for people to spend their money on. She didn't have to hear the cacophony of cars rushing by or crowds of people stampeding through the streets. She was in another world, and she couldn't have been happier.

Sauntering down to have her breakfast, she found that Joanie was already awake. When Lily moved over to England she hadn't anticipated sharing her house with anyone, but Joanie had arrived along with her old production company that was shooting a movie. While the production had moved back to Hollywood after the scandalous murder, Joanie had decided to stay. However, even though she wasn't living in Hollywood anymore, Joanie was still glamorous with her golden hair, gorgeous smile, and effervescent personality. She was the kind of person that was made for a blessed, popular life, while Lily had always thought of herself as being made for a place like Didlington St. Wilfrid , a place that had quickly become her spiritual home.

"Morning sleepyhead," Joanie said between bites of toast. "I hope that you're not using your new career as an excuse to be lazy. These stories are never going to write themselves."

"I know that, believe me, but I already have an idea for the next one. I figure I might as well get something written while the first one is off for editing, so I stayed up late last night."

"Don't you feel strange about writing your story when it almost came true?" Joanie asked.

It was a question that Lily had wrestled with a great deal while she had been writing the book, for the events of her story had mirrored the real life tragedy that had occurred. "I did think that out of respect to Howard I should have rewritten certain things, but at the end of the day I did come up with the story before it happened. I don't think Howard would mind anyway. It's not like he ever let anything get in the way of a good story."

"No, you have a point there. Now, tell me, have you seen that handsome prince again?" Joanie's smile was wicked, and Lily's skin went warm. She turned her back on Joanie as she poured herself some coffee, not wanting her to see the crimson blush that bloomed upon her cheeks.

"He's not a prince," she said.

"He could have fooled me. He certainly looks like one," Joanie said.

The man in question was Blake Huntingdon, the son of the man whom Lily rented her house, Starling Cottage, from. The senior Huntingdon was a Lord with a distinguished history in the area and held a lot of land in his name. What this made Blake, Lily wasn't sure. He had appeared on her doorstep out of nowhere, inquiring about the events that had happened on his father's estate. He was a tall, handsome man with eyes like sapphires, a thick mane of hair, a square jaw, and an imposing figure. His low baritone voice was made even more attractive by the refined English accent, and Lily had been floored by him the moment they had met. During her time in Hollywood she had encountered many actors and models, but there was a sincerity about Blake's attractiveness that appealed to her. Those actors had always been pretending to be something they weren't, but Blake was the real deal. His blood was noble, his wealth excessive, and his looks were such that he seemed to have been sculpted carefully by God. Just thinking about him was enough to make her stomach queasy.

"Well he's not a prince, and I shouldn't think that I'd see him again. He strikes me as a busy man," Lily said, saying these things more for her own benefit than Joanie's. After all, it wouldn't do her any good to let herself get carried away by fantasies that were never going to come true. "I'm sure that he's left again for London now. I doubt he would spend the majority of his time here. How would he be able to do business in a place like this?"

"I'm sure he could find a way. A man like that could find a way to do anything," Joanie said in a playful tone. "I saw the way you were looking at him Lily. I think if you see him again you should ask him out."

"I could never!" Lily said. "I'm just a common American girl. There are probably rules about that sort of thing. Besides, given that he is who he is I should imagine that he has no end of women to accompany him."

"Perhaps, perhaps, but you can never tell what's in a man's heart. You might just end up being exactly who he has been waiting for. It's not as though there are many other eligible men around here anyway."

"Who even says I want a man," Lily turned around and flashed Joanie a snobbish glance. It seemed as though everyone was concerned with her ability to settle down and get married, as though that was the only choice a woman had in life. For all the progress that had been made in the world there was still this backward notion that the only thing a woman could accomplish was to be married and have a family. That wasn't to say that Lily *didn't* want these things, but if they happened then she wanted them to happen because she had chosen them, not because she had been forced into the decision.

"Lily, everyone needs someone. You've been single for far too long. I know what happened last time hurt you but…"

Lily held up her hand. "We don't need to go there again. And trust me Joanie, I'm not swearing off romance and I haven't been burned so badly in the past that I'm never going to entertain the idea again. But right now I want to concentrate on my career and just enjoy life for a while. I've spent so long running around like a headless chicken for directors who don't appreciate me that I'm still getting used to having all this time to myself. I'm sure that I'll meet someone eventually, but it's not as though I'm a spinster. Anyway, I could ask you the same thing. Have you decided how long your break from Hollywood is going to last?"

Joanie sighed and pushed an errant strand of golden hair behind her ear. "I'm not entirely sure. This place does have a certain charm to it and I'm not sure that I'm ready to leave yet. I was thinking about taking up photography again. It's been too long since I've picked up a camera, and the landscape out here is wonderful. It reminds me that the world is beautiful. I'm waiting on some equipment to be delivered."

"Don't take this the wrong way, but I never quite pictured you in a place like this," Lily said. "You seem far too glamorous for this place."

"Well, I'm a multi-faceted woman," Joanie smiled widely. "Besides, sometimes there's nothing better than removing yourself from a place for little while. It makes people wonder what happened, and then when the return comes there's always a celebration. When I return to Hollywood it's going to be a grand occasion." Her words were punctuated with a light chuckle, which made Lily smile too.

As she sipped her coffee she looked in the cupboard where she kept her sweet treats, only to be dismayed when she found the cupboard empty. "I'm going to pop over to Marjorie's. I find it very difficult to start writing without one of her treats."

"Then you'd better make sure to include her in the acknowledgments of your book," Joanie said. Lily laughed and nodded.

As she left Starling cottage, Lily breathed in the fresh, sweet air of this village. It was so unlike the world she had left that it almost seemed as though she had been lost on a movie set. But there was no mistaking the scent of baked bread and sweet flowers that drifted on the air, or the scene of winding old roads that had been there for centuries. It felt good to be strolling to a bakery for a home baked treat rather than rushing around in a crowded coffee shop to get some processed drink and a stale donut that did more harm than good. She had noticed that she had put on a little weight since she had arrived in Didlington St. Wilfrid , which she didn't mind. Outside of Hollywood she was free of the pervasive, almost cultish obsession with thinness. She remembered the looks of derision she used to get whenever she scarfed down an éclair or treated herself to something nice, and being in this place was a reminder that it didn't truly matter. The only thing that mattered was her own happiness, and she liked that her body had filled out to a healthy degree. Her skin was less sallow and she had been sleeping much better so the shadows under her eyes had disappeared. She hadn't realized until she'd left Hollywood how much tension and stress she had been carrying in her entire body. It was as though Didlington St. Wilfrid was a bath, one which she had slipped into with ease.

Because of her good mood she wore a wide smile on her face as she entered the bakery, and there was a strut in her step.

"Morning Marjorie! I'd like my usual please," she exclaimed, before her words fell flat. She wasn't the only one in the bakery. Constable Bob Dudley was there as well, and there was a grim look on his mustachioed face. Marjorie was standing behind the counter, looking sickly. Both her hands were on the surface shaking.

"What's going on?" Lily asked, her tone suddenly turning serious.

Constable Dudley waited for the door to close before he answered. He seemed a little relieved when he saw that Lily had entered, for it was she who had solved the last mystery when he had still been poring through the possibilities, although she accepted that it was her familiarity with the film crew that had come in handy on that occasion.

"I'm afraid there's been a death, and a very serious accusation against Marjorie. Frank Roberts has died, and his wife has accused Marjorie here of killing him," Constable Dudley said. There was a small whimper from Marjorie as the words were uttered, and the color drained from Lily's face.

"Tell me what happened," she said in a determined voice.

3

―――――――――

"I was woken up by a call from a very distraught woman, Charlotte, who said that her husband had died. Naturally I raced over there to investigate and found him on the floor in their lounge. She had placed a blanket over him, for his dignity I suppose. I didn't know the man myself, but Charlotte insisted that he was healthy and he hadn't complained of any pains or anything like that. The only thing that was near him was a bag from this bakery. Charlotte seemed to imply that there was something... something between her husband and Marjorie, and she believes that Marjorie poisoned him because Frank wanted to end things," Constable Dudley said, hemming and hawing through some of the words because speaking about romance clearly made him uncomfortable.

"I don't know why she's lying, but she is. I would never do anything like that," Marjorie gasped. She was white as a sheet and her eyes were red and raw.

"Of course Marjorie, of course, but I am required to take things seriously and to investigate every suspect. Until we have evidence to the contrary I do have to ask you some questions. I can't let my personal feelings get involved," Constable Dudley said.

"But I didn't do it. I couldn't... the very thought..." Marjorie sputtered. Lily's heart went out to the woman. She knew well what it was like to fall under suspicion, for the same thing had happened to her when the director had been murdered. Lily had been required to clear her own name, but Marjorie wasn't the kind of woman who could put herself through such a thing. She was kind and sweet and had been nothing but a friend to Lily ever since Lily had arrived in Didlington St. Wilfrid , and yet Lily knew the Constable was right.

"He has to do this Marjorie. I know it hurts, but it's the evidence that matters," Lily spoke in a sympathetic tone, but it didn't seem to help Marjorie any. The look of betrayal on her face was stark, and it broke Lily's heart.

"Then find the evidence that proves I didn't do this, because it's out there. I would never have done such a thing. Now if you'll excuse me I have to close the bakery while I... while I compose myself," she said, turning abruptly and disappearing upstairs in a flurry of emotion, leaving Lily and Constable Dudley standing in the bakery.

"I don't want to believe she was capable of something like this, but I can't rule out the possibility either," the Constable said in a low voice, ensuring that Marjorie wouldn't be able to overhear them.

"I know. I think she knows it too, deep down, she's just in shock at the moment. Is it possible that it could have been some kind of accident?" Lily asked.

"I think there was definitely some kind of poison involved by the state of his mouth. It was a disgusting sight. I have to say that things aren't usually like this around here. I'm not used to dealing with murders," Constable Dudley said, "would you mind helping me again? I need to take a statement from Marjorie, but I don't think she's quite in the mood at the moment. She might be more inclined to talk to you."

"I'll see what I can do," Lily said. She hadn't been expecting to become involved in another murder investigation either. The fictional crimes she wrote were enough for her, but there was something inside her that couldn't simply stand by while real people were being hurt. There was some kind of truth behind this, and Lily wanted to know what it was. But first she had to speak to Marjorie.

After closing the bakery Lily crept upstairs, each step creaking as she did so. Marjorie lived above the bakery, and the smell from the bakery rose, creating a warm, comfortable feeling. There were a few rooms jutting off a corridor, which made up a wide apartment. It was plenty of room for one person and might well have been cramped for two, but somehow she expected that Marjorie and Henry had made it work. The décor was what she would have expected from what she knew of Marjorie; with lots of pictures of animals displayed, as well as various forms of embroidery. She found Marjorie sitting in a chair, hunched over, shoulders shaking as she sobbed.

"Marjorie, do you mind if I come in for a moment?" Lily asked gently. Marjorie didn't say anything in reply, but she merely nodded and sniffed.

"Would you like something to drink?" she asked when Lily sat down, determined to be a good host despite being accused of a heinous crime.

"No, I'm fine, thank you. I just want to talk to you about what's happened. You know that neither of us believe that you actually did this, but when solving a crime everything needs to be explored. We would be doing the victim a disservice if we ignored something because of our own assumption," Lily said.

Marjorie wiped her red eyes and sniffed back her sorrow. "I know all that," she said, "but it doesn't make it any easier. Just the thought that anyone could think that I was capable of such a thing I... I'm not quite sure how to handle it."

"I'm not sure there is a right way. All we can do is stay true to ourselves and hope that the truth comes out."

"You will make the truth come out, won't you Lily? You solved the last one and you can solve this one too. It's not that I don't have faith in Constable Dudley, but you're worldlier than he is. You've been out there where there are real crimes like this. You have to help me," Marjorie said.

"Of course Marjorie, of course I will, but in order to do that I'm going to need to ask you some questions, okay?" Lily said gently, reaching out to place a reassuring hand on Marjorie's arm. Marjorie nodded and composed herself so that each word wasn't being lost amid a torrent of emotion.

"Can you tell me what happened last night? Did he buy something from you?"

Marjorie nodded. "Frank has been a regular of mine ever since he and Charlotte moved here. They arrived not long before you."

"And what was he like? I take it you got on well?"

Marjorie nodded. "He was a friendly man, and he seemed excited to be here. He was very complimentary of my baked goods; he said that he hadn't tasted anything like it, and he said that if I moved to a city I'd make a killing. I told him that I didn't want to be anywhere but here though."

Lily nodded. "And last night?"

"Yes, he came in and we spoke for a little while before he left."

"Was there anything different about him? Did he seem like he was in pain or in distress?"

"No, well, not physically anyway."

"What do you mean?" Lily asked.

"Well..." Marjorie paused for a moment and looked directly at Lily. "I do always prefer to keep conversations private between people, but I suppose there's no point now that he's dead, is there?"

"Anything you tell me might be able to help," Lily prompted.

Marjorie smoothed out her dress and continued. "Well he did ask me about my marriage to Henry, and if there had ever been a point where I wondered if we should have been married. We never spoke much about his home life, but I don't know... when he spoke about Charlotte his face didn't light up the way it usually does when two people are in love."

"So you think they were having problems?"

Marjorie shrugged. "Every married couple has problems now and then. It's the nature of sharing your life with someone."

Lily didn't know anything about that. She rolled her bottom lip under her teeth for the next question she needed to ask was an important one, and not one that would be welcomed at all.

"Marjorie, you know that I'm your friend and I wouldn't ever want anything bad to happen to you, but right now I need to ask you a question and I want you to answer me honestly. I need to know if there was anything going on between you and Frank. I promise I won't judge you if..."

"Of course not!" Marjorie said, drawing back in shock. "And I would appreciate it if you never insinuate anything like that again. I'm a good Christian woman and I would never take up with a married man." She softened, and a sad look came upon her face. "Besides, I don't have it in me to love anyone apart from Henry, God rest his soul."

Lily came from a world that was seedy and sinful, where tales of adultery and lust abounded. But Didlington St. Wilfrid was a pure place, where people were virtuous and where morals were still something to be cherished. In her other life Lily might not have believed Marjorie when she said that she was still devoted to her deceased husband. In Lily's experience many people weren't devoted to their spouses when they were still alive, but Marjorie was different. This place was different, and although there wasn't any evidence yet, Lily was convinced that Marjorie was not responsible for Frank's death.

That being said; who was?

"I'm sorry for bringing it up Marjorie, but you know I had to ask. Hopefully I won't have to bother you with this again."

"It's such a pity though, he was such a nice man," Marjorie said.

Lily had no doubt that was true, but even nice men had secrets.

4

Lily told Constable Dudley about what Marjorie had told her. "I don't for a minute think that Marjorie had anything to do with this. Is there any chance we could get a toxicology report so we know what exactly has gone into his body?" she asked.

"We can, but it'll take a while. I can drive the body up to the hospital myself, but getting them to do anything for us is... well... a small town like us isn't exactly high on their list of priorities. There are always other things coming in from the city that are more important," Constable Dudley said.

"I'm sure if you keep on them they'll have no choice but to do their job," Lily said. Constable Dudley puffed out his chest and looked a little prouder at this.

"Well, I don't mind doing that if you wouldn't mind speaking to the widow."

"Of course," Lily said. "Did you ask her any questions already?"

"Only the rudimentary things. She was distraught when I spoke to her and very angry. She was so convinced that Marjorie was responsible that she practically pushed me out of the house to make me go to the bakery."

"I want to know why she was so convinced that something was happening. Could Frank have lied to her about it?" Lily asked.

"I don't know why he would want to do that."

"Some men get bored and find grim ways to amuse themselves. Did you know him well?" Lily asked.

Constable Dudley shook his head. "I only saw him a few times. They tended to keep to themselves, which I suppose is their right. They're not like you and your, ah, friend," he said, his face reddening when he mentioned Joanie in passing. Lily stifled a smile. It wasn't the appropriate time to be amused by the fact that Constable Dudley had a little crush on Joanie, but she couldn't stop herself from feeling that way anyway.

"There is something else that was found on the body," the Constable said, producing a letter that had been scrunched up. He unfolded it and handed it to Lily, who read it and was horrified at the words.

"Why didn't you tell me about this sooner?" she asked.

"There's not a name or an address written on it, so there's nothing to point to who could have written it. I don't think we're going to get much information. Besides, the fact that it was scrunched up suggests that Frank wasn't too concerned about it."

"And the fact that he's dead suggests that he should have been," Lily said with an icy tone to her voice.

Constable Dudley harrumphed and frowned. "If you can get anything out of it then be my guest, but for now I'm going to focus on the chain of events that led him to the bakery." Since they both had their own things to take care of they went their separate ways. Lily was almost beginning to feel as though she should have a deputy's badge. Her gaze fell to the letter. It raised yet more questions. Who wrote the letter? Why did they threaten Frank?

Why did Frank stuff it in his pocket rather than tell anyone about it?

It reaffirmed her determination to see Frank's wife. Charlotte might be able to throw some light on Frank's secrets, because it seemed as though he had plenty lurking in the shadows.

Lily walked up the rising hill, casting a glance across to Pat's farm, which had a foreboding gloom about it. She wondered what it must have been like to neighbor his farm. She turned along a fork in the path and followed it to a cottage that was much like her own. The front garden was filled with a variety of crops, and she found a woman kneeling before a flower bed, kneading the soil with her fingers. She looked up when she saw Lily approach.

"Excuse me, but are you Charlotte? Charlotte Roberts?" Lily asked.

"I am," Charlotte replied.

"I'm here at the behest of Constable Dudley. I have a few questions if you wouldn't mind speaking to me. Of course I know it's a difficult time, but it would be a great help," Lily said.

Charlotte sniffed and sighed and leaned back, placing her hands on her thighs. Her skin was pale and her eyes were puffy and bloodshot, and although she wasn't wearing a veil it was clear that she was in mourning.

"I suppose that I'm going to have to get it over and done with sooner or later, and if it helps putting that woman behind bars then it'll be worth the pain," Charlotte pushed herself to a standing position and wiped some dirt off her hands. She led Lily into the cottage, and headed to the kitchen. Charlotte washed her hands and offered Lily a drink. Lily welcomed some water and looked around at the jars of herbs and spices sitting on shelves.

Lily produces the threatening letter and asks Charlotte if she knows what it might mean, looks around as if someone might be watching them. She asserts that maybe someone else had witnessed the philandering of her wayward husband and Marjorie.

"Why are you so sure that Marjorie poisoned him?" Lily asked.

"It makes sense, doesn't it? They spent a lot of time together. She was in love with him, and when she realized that she couldn't have him she didn't want anyone else to have him either. She knew he was coming. He was a regular. I can just imagine her now, smiling that sweet smile of hers, knowing that he was going to come home and have that and die. She wanted to steal my husband away from me in any way she possibly could, and now she succeeded." There was a small sob that accompanied the words. Lily held her tongue, for her natural inclination was come to the defense of her friend, but such a statement was not going to make Charlotte inclined to keep talking to her.

"How long have you suspected that something has been going on between your husband and Marjorie?" Lily asked.

Charlotte laughed at this, although it was a laugh without humor. "Oh probably not long after we arrived. I know her type and I know what a woman like her wants. I can imagine she gets bored seeing the same old faces around here, so when someone new comes into town there's a sense of the exotic, isn't there? I bet it was like an adventure for her, especially considering she is a widow."

"And you think Frank would have encouraged this?" Lily asked. In her experience there were always two people involved in adultery, although more often than not only one was blamed. Charlotte paused and twirled a lock of hair in between her thumb and forefinger. Her eyes glazed over and her voice was detached.

"My husband was many things. One of the things that first attracted me to him was that he was outgoing and flirtatious. At first I thought it was a quality he reserved for me, but he's gregarious with most people. I don't think it would have been difficult for Marjorie to get the wrong impression and read into the fact that more was there."

"But you don't think there could have been more? I'm sorry for the indelicate question and I certainly don't want to speak ill of your husband, but do you have any evidence that he and Marjorie were having an affair?"

Charlotte tilted her head in a way that made her hair fall across half her face, obscuring her right eye. "Only the kind of evidence that a wife can see. It was in the way he touched me, the way he smiled when he spoke about her, it was in a million things, but he never came out and said it, no."

"So you don't think that anyone else could have learned about his philandering ways?" Lily asked, sliding her hand into her pocket where she felt the crumpled letter.

Charlotte shrugged. "I suppose it's possible, why do you ask?"

"Because of this," Lily said, and produced the letter. Charlotte blanched as she took the letter and ran her hand up and down the back of her neck.

"What is this?" she asked.

"It was found in your husband's pocket. Are you saying you knew nothing about this?"

"He didn't tell me about it... no. I suppose it's another thing he was hiding from me," Charlotte didn't attempt to hide the acid in her words as she went to the window, turning her back on Lily. "I always thought I would feel safer out here, but somehow all this space... it makes you wonder who is hiding in the shadows."

"Do you have any idea who could have written this letter?"

"I don't know."

"Has anyone tried to blackmail your husband?"

"I don't know."

"Have you noticed anything strange outside your cottage? Have you heard any strange noises at night? I'm only throwing out possibilities here, but I'm wondering if someone approached Frank and threatened him. If Frank wasn't willing to play along then perhaps he was killed. Do you have any idea who could have done this?"

"I don't know. The only person I know he spoke to was Marjorie. Maybe she wrote the note as well. It's hardly the most romantic love letter though, is it?" Charlotte said with a weak smile. Lily was getting more and more irked by Charlotte's lack of information, although she supposed she shouldn't get too frustrated with the widow.

Charlotte then changed tack immediately, which took Lily by surprise. "I've heard about you, you know. It caused a great stir when you arrived. What do you think of it out here? Have you adjusted well from your former life?"

"I think so. There are certainly things I miss, like the convenience of having everything within a short walk of my apartment, but I like the peacefulness here. I like that there are no distractions."

"I suppose for a writer like you that's a big thing. The thing about it though is that if you want a distraction hard enough then you'll find one. Are you married?"

"I'm not," Lily said, her throat tightening uncomfortably as she spoke.

Charlotte smiled, although it was a smile formed of sympathy and pity rather than anything joyful. "Then you won't understand what it's like to be married, to tie yourself to another soul and promise things to each other that some people simply can't promise. We were supposed to make a new beginning out here. Frank was always prone to distraction, but out here I thought things would be different. Instead it's just all come to an end..." she sighed sadly.

It was hardly a glamorous advert for getting married.

"I'm sorry that this happened," Lily offered. "Was there anything else your husband was complaining about, or anyone else that he might have been… distracted by?"

"Not that I can think of. Just Marjorie. There's really nothing else I can tell you Lily. You need to go and speak to her. She's the only one that knows the truth. She's the only one that knows what happened."

Lily nodded and got a feeling that she wasn't going to get anything more out of Charlotte, or at least nothing useful. It was a difficult case because clearly Marjorie didn't do it, and the only other person so far with a possible motive was Charlotte herself, but why would she kill her husband if he wasn't actually cheating on Marjorie? It was one thing to suspect Charlotte, but quite another to prove it as well.

"That's an impressive collection of herbs and plants you have," Lily remarked, nodding toward the shelves that were packed with glass jars.

"Yes, the one benefit of being out here is that I have access to so many of these plants. It brings every meal to life," Charlotte said. Lily walked toward the door, but before she left she turned and looked at Charlotte again.

"Tell me, did Frank have anything else to eat last night?" she asked.

"I don't believe so. We had an argument and I went upstairs to be by myself. He might have made himself dinner," she said without batting an eye. Lily cast a glance toward the sink and saw enough pots and pans there for a meal to feed at least two people, perhaps more, and it only made her suspicions rise. How to prove it was another thing entirely though.

"Constable Dudley will be in touch when he has some more news. I am sorry for your loss. I didn't know him, but he sounded like a good man."

"He had his moments," Charlotte said. Lily couldn't tell if the woman was joking or not.

When Lily left the cottage she shuddered and tried to shake away the feeling of being in the same room as a potential murderer. If she was willing to discount the possibility of Marjorie being the murderer then it only really left Charlotte as the viable suspect, but how was Lily supposed to get any evidence for it? It could well be the case that neither woman was to blame and Frank had killed himself inadvertently by whatever meal he had concocted, but if Charlotte was to blame then would she stop at Frank?

As she walked along the path by Pat's farm something caught her eye. She stopped and inspected it more closely, for it was the same plant that had been in one of Charlotte's jars. There was no mistaking its forked leaves and mottled green color. She climbed over the short wall and approached the plant, bending down to inspect it. She was distracted by Zeus' yapping and the yell of Pat approaching.

"Get away from there! Get off my land!" he yelled, gesticulating wildly with his arms. Zeus came bounding up to her and had nothing but affection for her, making the man and the dog seem quite the odd couple. Zeus licked her hands and jumped around, while Pat stormed up. He whistled for Zeus to come to heel, which the dog obliged, and Lily found Pat glaring at her. "What are you doing here? I knew you were trouble, the lot of you, all these people coming from cities thinking you have the run of the place. Do you not see the wall? You don't know what you're messing with either," he said.

Lily's ears pricked up at this and she tilted her head to the side. "What do you mean 'other people'? And how did you know I was here?"

Pat sighed and shook his head. "Because I know how to defend my land," he said, gesturing to a small rise in the ground. When Lily looked carefully she could see the glint of a camera lens.

"Wait, do you have recordings of people who have been here before? Has someone else been trespassing? And what exactly is this plant?"

Pat narrowed his eyes. "You always have a lot of questions, don't you? Last I checked I didn't have to share any information with you."

"There's been a murder Pat, and I'm trying to figure out what happened. And no, maybe you don't have to share any information with me, but I'm sure that Constable Dudley would be interested to learn that you have cameras set to spy on people."

"What I do on my land is my business, and I'm not spying on anyone. It's to protect me."

"But Constable Dudley might not come to the same conclusion. It would be such a shame to have him poking around everywhere," Lily said, knowing that she had a trump card. Pat gnashed his teeth and drew his lips into a thin line.

"Come with me," he growled. Lily followed him back to his cottage.

Pat's farm was vast, but the inside wasn't at all what she expected. There was a grid of monitors displaying various angles of his farm lands, as well as a radio in one corner of the room. Filing cabinets formed a line on an opposite wall, and she found herself wondering what kind of man Pat was to have a set up like this.

However, that mystery would have to wait for another time.

"You're going to need to learn to be more careful. This place might seem harmless, but there are still plenty of things that can kill you," he said. "What you were reaching for I call the Widow's Kiss. It's deadly poisonous, unless you know how to cook it just right. I'm not having people come up here and take something that doesn't belong to them, especially not when it might kill them."

"Who else was up here?" Lily asked, her heart suddenly racing.

70

"Someone who shouldn't be up here, a stranger, but I have my suspicions about who it is," he scowled, "and I'm sure that they won't bother to do it again."

"Do you have a tape of this?"

Pat huffed and cycled through the footage of the past couple of days, and during one night it showed clearly that Charlotte came to fetch the Widow's Kiss. That was all Lily needed to confirm her suspicions and she told Pat to keep the tape safe, for it was likely the key to solving the murder. There was something else that Pat told her as well, something that made her see the case in an entirely new light.

5

It was dark and silent. The house creaked as though it was breathing. There was a noise that would have been imperceptible to anyone who was asleep. Stairs moaned under the weight of feet, and then a door opened. As soon as it did Lily switched a lamp on and the room was illuminated with light. Charlotte stood there with a knife in her gloved hands, a look of utter shock on her face. The shock gave way to anger, and she went to walk back down the hallway.

"I don't think you want to be doing that," Constable Dudley said in a stern tone. Charlotte's gaze danced from the Constable to Lily to Marjorie, who was standing in the corner of the room.

"You," Charlotte hissed, "you did this. You took him from me."

"She didn't do anything of the sort Charlotte. There wasn't anything going on between Marjorie and your husband," Lily said. "But I have you, on a video tape, stealing the poisonous plant from your neighbor. I know you cooked something for him, his last meal. You knew he was going to bring something home from Marjorie's bakery and you wanted to blame it on her."

"No... we would have eaten the same thing. Only he would have eaten something from her bakery. I wouldn't have touched it! It must have been her, otherwise I would be ill as well," she said.

Lily merely shook her head and sighed. "I'm sure someone with your expertise would have found a way to protect yourself."

Charlotte's tongue darted out to lick her lips. "But what about the note?" she said quickly. "Someone else was clearly angry at Frank." She sighed and her shoulders slumped. "Perhaps I have been wrong all this time and Marjorie wasn't to blame. What if there was another lover? Frank could have found someone else, someone who was jealous of myself *and* Marjorie. Or perhaps it was another of Marjorie's admirers. She's clearly a flirt, flaunting her sweet treats for the entire village to enjoy. I'm sure my husband wasn't the only man she had set her sights on." Charlotte's tone turned venomous and Lily couldn't quite understand why she had placed such a target on Marjorie's back. It didn't help the unsettling, gnawing feeling that itched the back of her mind.

"The thing is Charlotte that you're accusing Marjorie, or some unknown lover, of being the jealous woman, when in fact you're the only who was jealous. As it turns out I know exactly who wrote the letter, and it wasn't intended for Frank at all. It was left for you because you were trespassing and stealing from Pat's field. He left the note to warn you away, but Frank happened upon it instead." Lily had proven that Charlotte was a thief. She didn't think it was long until she proved Charlotte a liar either, and then perhaps a murderer. "Tell me, why exactly did you move here from the city? What were these distractions that you spoke of?"

Charlotte's face twisted into a scowl and her nostrils flared. "We left because he cheated on me. He promised it would never happen again, that this place would be a new beginning. I thought he was right. I figured there would be far fewer people to catch his attention, but even then he couldn't help himself. I knew it was only a matter of time and I wasn't going to be humiliated again. I couldn't do it. I couldn't live a life knowing that my husband wasn't going to be faithful to me. I was his *wife*. He was supposed to love me and only me, but I wasn't enough for him. I would never be enough for him…" she said, her voice turning into a sad moan.

"He did want to make things work between you Charlotte. He told me just as much. Sometimes you need to give a little faith to the other person, a little trust, even when they might not deserve it," Marjorie said.

Charlotte looked at her and began to blink fretfully. The sorrow rose within her and tears began to flow down her cheeks.

"He… he wanted to make it work?" she asked, her words dropping out of her mouth like stones as the dreadful realization washed over her. "I thought he… no… I thought he didn't care." The knife dropped from her hand and she leaned against the wall, suddenly losing all her strength. Constable Dudley stepped toward her and grabbed her wrists, placing her under arrest for the murder of her husband. He took her downstairs, and Lily was left with Marjorie.

"It's such an awful, terrible shame," Marjorie said.

Lily looked at her with shock. "The woman came here to kill you. She was convinced that you were having an affair with her husband, and you still feel sorry for her?"

Marjorie sighed. "The good Lord teaches us to rise above ourselves. She's a woman in pain, and she let her own fear get the better of her. It's such a shame. If they had just trusted each other and spoken to each other then maybe things would have been different. Maybe they would have been able to find their happy ending here."

"It doesn't seem as though they were suited to marrying each other at all."

"Marriage takes a lot of work," Marjorie agreed. "But there has to be a certain connection there. Charlotte must have lost that when Frank betrayed her trust. If I'd have known then maybe… maybe I could have made a difference."

"Or Charlotte would have been even more convinced that you wanted to steal Frank away from her. It seems to me that she was just waiting to believe the worst about him. In her mind she was convinced that he was cheating on her again, and she was willing to kill for the sake of her dignity."

"I'm just glad that you managed to get to the heart of the matter quickly Lily. If things had lingered then I might well be dead," Marjorie shuddered at the prospect.

"Yes, I'm glad too. I suppose in a crime like this it's most natural to look to the person closest to the victim. It helps that it could never have been you," Lily said, smiling.

"I'm reassured to hear you say that. For a moment I thought you might have actually believed it was me."

Lily shook her head. "When I looked at the case objectively and just had the information in front of me then yes, logically there was a possibility that Charlotte was telling the truth and you had been having an affair. But it was only a possibility when I failed to include my own judgment of you. I know you're not that type of person Marjorie. In fact I do believe you're the sweetest, kindest person I have ever known."

"Well that means a great deal coming from you Lily," she said.

"Before I go, there is something I want to ask you about Pat. Have you ever been in his home? He has all these monitors to watch over his land, and so many filing cabinets as well. I thought he was just a farmer. What exactly does he do?"

Marjorie shrugged and shook her head. "I'm afraid I don't know. Pat is the type to keep himself to himself. I suppose the only way you're going to find out is to ask him yourself."

Lily let out a scornful chuckle at this prospect. "I'm not sure that's ever going to happen. He only helped me with this under duress, and probably because he was angry at Charlotte for coming to steal some of his crops. I don't think he's going to be forthcoming with me, but I'm not sure I can shake this memory from my mind. There's definitely something about him, but I can't put my finger on what it is. I just get this feeling that there's more to him than meets the eye."

"Well if there's anyone who can get to the bottom of the mystery it's you Lily, but if I were you I'd let him be. I'm sure that nothing good can come from meddling in other people's affairs. Besides, you never know, there might be some handsome bachelor coming to stay in the village now that there's a house available."

"I'm not sure that anyone is going to want to live in a house where a murder took place," Lily said, "and I'm not looking for a husband at the moment. I have far too many things to concern myself with, such as a lot of writing I need to catch up on." She indicated that she was going to leave and walked past the hallway where Charlotte had dropped the knife. Lily picked it up and carried it out with her as Marjorie walked her down to the bakery exit. There were low lights illuminating the chilled cabinets that kept some of Marjorie's goods fresh, and it made the darkness outside even more stark.

When Lily was about to leave she saw a figure looming at the door and the door opened with a jangle. Out of instinct she raised the knife and pointed it at the stranger who was coming in uninvited.

"I hope you're not planning to use that on me," a familiar voice said. She knew who it was as soon as she heard his voice, and then he came into the light.

"Oh my gosh, I'm so sorry," she said quickly, the words rushing out of her in one quick breath as Blake appeared. He smiled effortlessly and his engaging eyes completely captured her attention. Usually the only thing she could smell in the bakery was the sweet treats, but upon his appearance there was something else in the air; a musky smell of cologne. "I didn't mean to do that. It's, well, it's been a long day," she said.

"Blake Huntingdon, well I never. It's been a long time since I've seen you here," Marjorie said. To Lily's surprise, Marjorie walked forward and hugged Blake tightly. Lily found her mouth opening, agog at what she was seeing. Blake seemed as though he should be in a mansion or a castle somewhere, not in a humble bakery.

"I know, it's been far too long. I hate that work has kept me so busy that I've had to take time away from this place," Blake said.

"You two know each other?" Lily asked.

"Of course, Marjorie's treats are the best I've ever tasted. I've been to so many different countries, but none of them can hold a candle to what you have to offer. I know it's late, but I was hoping that I might be able to get an éclair or something," Blake said.

"It is very late. What are you doing walking around the village at this time of night? There are dangerous people out there you know," Marjorie said.

Blake chuckled. "I'm sure I can take care of myself. Actually I'm still on Australian time and my body hasn't quite adjusted yet, but I'm sure it will get there. Besides, there's something nice about walking around the village when it's dark and there are only the moon and the stars for company. It feels as though you're the only person in the world, and it's all rather humbling," he said.

Lily felt herself warming to him, for she had felt the same thing.

"Well, each to their own. I prefer to be tucked up in bed with a cozy blanket wrapped around me," Marjorie said as she went to one of her chilled cabinets and pulled out an éclair for Blake to enjoy. Lily licked her lips, so Marjorie got her one as well, acting like a grandmother who was feeding her grandchildren far after their bedtime had passed. "How long are you staying this time? I'm sure your father would appreciate having you home for a while."

"That's why I've returned actually," Blake said, the tone of his voice turning grave. "Father has taken a turn for the worse and, well, I think I should be here with him. He still has a lot that's he's managing and you know what he's like; he'll never let himself rest, even though he needs to. He's too old to be running around like he used to, but try telling him that and he only wants to prove you wrong. I figure that at least if I'm here I can take some of the burden away from him. Besides, it'll give me a chance to refamiliarize myself with the village. Despite everywhere I've lived this place has always been home to me, and of course it will allow me to get used to the changes that have happened over the years, and the new people who have arrived."

As he said this his gaze lingered on Lily. She felt her skin growing warm again. It was becoming a habit whenever she was around him.

"I'm sure you'll have a blast. Lily is an interesting woman. She actually just helped solved another murder here, and Lily there's nobody who knows Didlington St. Wilfrid better than Blake here. It's in his blood. I'm sure he'd be happy to show you around."

Lily glared at Marjorie to stop her transparent attempts at trying to make something happen between Blake her. Blake merely laughed a gentle laugh. "It would be a pleasure to give you a tour, and I'd love to hear about this new crime you've solved. I had a feeling there would be something special to find at home," he said. With that he popped the last of his éclair in his mouth and smiled at Lily.

A FREAKY FAMILY MURDER IN A QUIET ENGLISH VILLAGE

1

It was another beautiful day in Didlington St. Wilfrid. The sky was azure, and it looked idyllic as Lily tried to swallow her nerves. She told herself that she was being silly because there was no reason to be nervous at all, and yet as she approached the church she had to consciously ignore the feelings that were swimming in the pit of her stomach. The church was centuries old, with an impressive spire reaching high into the sky, while the vast roof came sweeping down. Fragments of stained glass made up the windows, while in the distance Lily could see rows of tombstones standing like dominos, each one of them marking the grave of someone she would never know.

There was an aura in the air as she walked into the church, a sense of power and majesty that she hadn't experienced anywhere else, not even on the movie sets where millionaires and geniuses played around with the hopes and dreams of a generation.

The wide pews were wooden, and the church was filled with almost everyone in the village. Lily offered a weak smile to people she had seen out and about, while other faces she had not become acquainted with yet. Marjorie was near the front of the church, surrounded by her friends. She beckoned for Lily to join her, but

Lily preferred to sit in the back. Her gaze darted around, rising up to the impossible high ceiling, drifting over the artwork and religious symbology. It had been a long time since she had been in a church, and although she had never fully stepped away from her faith over the years, it had definitely been neglected in the whirlwind that was Hollywood.

Lily slunk into a wooden pew and settled down, holding the hymn book in her hands. The cover was faded and the pages were thick and discolored, but the words that were printed on them were eternal. She wished that Joanie had accompanied her, but Joanie hadn't been interested at all. It was another adjustment to make after moving here from Los Angeles. Where that community had been fragmented this one was close knit, and the fulcrum was the church.

The vicar, Donald Brown, was a trim man with soft features and honey brown hair that was thinning over his scalp. His white robe seemed a size too big for him, but when he spoke his blessed words filled the vast space within the church, as though they had a power entirely their own. But just before he started speaking, something caught Lily's attention. Footsteps pattered the old stone floor and someone came to stand next to her. She stiffened with tension when she realized it was Blake Huntingdon, the man who made her tremble, and her stomach curdle. Her cheeks flushed red as he smiled, whispering a greeting to her. She managed to mumble something in reply as the vicar began his sermon, and Lily was grateful for the distraction.

There was something about Blake that was unlike any man she had ever seen before. She supposed it helped that he was a part of the English nobility; the son of a Lord with all the distinguished heritage that came with that. He was impossibly handsome as well, with slick hair, smooth features, and eyes that were as deep as the ocean. The air was filled with the sweet fragrance of his cologne too, and it took Lily a few moments to realize that her heart was beating a little more rapidly than usual.

The vicar's sermon was about trust and faith. He read some passages from the Bible and the congregation rose a few times during the sermon to sing hymns. Lily's voice was hesitant at first and she mostly just mouthed the words, unsure if it was respectful or not to sing when she had not made a point of visiting church for such a long time. However, by the last hymn her voice was full and it swirled around with the chorus of the others, creating a beautiful sound that reached the heavens. It felt good to be a part of something that was bigger than herself again, to feel that swell of faith that was as warm as honey and strong as steel. It lanced through her with all the shock of lightning, and she was quite taken aback by its power.

When the service was over, Blake turned to her and spoke in a low voice.

"I have to admit that I wasn't sure I'd see you here," Blake said.

"Why?"

"Well, most people from big cities don't seem to have time for God in their lives anymore."

Lily blushed. "I have to admit that I haven't attended church as much as I should have in the past, so I thought I would make amends. Besides… with all that has happened recently I thought it was the right thing to do."

Blake nodded in understanding. "It has been a horrible time. But as long as there are good people like you in the world there's still hope for us all."

Lily beamed at the compliment, and became lost in Blake's eyes. It took everything she had to be able to keep her mind straight and focus on speaking.

"I'm a little surprised that you're here as well. I would have thought someone like you would have been too busy for a place like this."

"I don't think you can ever be too busy for church. I make a point to come here every time I'm back. I used to come here every week when I was a child. It was a good time, although I used to be a terror. I always hated sitting still, and when I was a kid it seemed as though this place was just made for running up and down, I mean, just look at all the space!" he said, smirking. Lily could well imagine him as a child running around as the adults looked on shaking their heads at the audacity of a child.

Before the vicar approached them, Blake's expression grew serious and he mentioned that he wanted to speak with her about something.

The vicar was making the rounds with his congregation and shook hands with Blake.

"I heard that you were back in the village. It's a pleasure to see you again," Donald said.

"The pleasure is mine. That was a lovely sermon Vicar, you always know how to keep my attention."

Donald smiled and Lily wondered where Blake got this gift to be able to improve the mood of everyone around him.

"Well, I try. I suppose your father couldn't make it?"

"I'm afraid not," Blake said. Lily noticed that the smile fell from Blake's face. "I did offer, but he preferred to stay at home."

"I must make it a point to visit him," Donald said, and then turned to Lily. "I don't believe we've been introduced. Welcome to the church," he said, offering Lily his name. Lily shook his hand and noticed that it was surprisingly strong for such a soft spoken man. His eyes lit up with recognition when she told him who she was, and when he heard her accent.

"Oh Lily! Yes I have heard so much about you. I have to thank you for what you've done for this village. What's happened here has been awful, just awful. When I heard what happened I prayed for someone to find the truth, and thankfully you did. We are blessed to have you here."

"Thank you," Lily said, shifting uncomfortably on her feet as she wasn't used to receiving such effusive praise. "I'm glad you've managed to find time to spend in my humble church. It must be quite different in America."

"It is, although to be honest it's been a while since I've stepped into one."

"Well, better late than never, and God is never going to turn away one of his children. Thank you for coming today and I hope that I will see you again, but for now I must go and attend to others." He bowed his head and strode away, engaging another member of his congregation in a conversation.

"He must know everyone in the village," Lily remarked.

"Everyone who comes here, anyway," Blake said, but he seemed to have something else on his mind. "Lily, I was hoping that you might help me with something."

The tone of his voice changed and it was unmistakable to Lily. "What's wrong Blake?" she asked.

He glanced around, as if to make sure that nobody was listening in on their conversation. He lowered his voice too as an extra precaution. "It's my father. His health has been deteriorating recently and I'm worried that someone might be taking advantage of him. We have a lot of land holdings and a lot of investments, but I've recently noticed that some money has gone missing and there are some irregularities in the accounts. Since you've proven yourself adept at solving mysteries I was hoping that you would help me get to the bottom of this one as well."

"I'm sorry to hear about your father Blake, but of course I'll help. I'm not sure how much help I can be, what happened before was mostly just luck, but I'll try."

"You're too modest Lily. It wasn't luck that caught those murderers, it was you. I'll be in touch," he said, and seemed relieved that she had offered her help. As she watched him walk away her mind started to wander toward the impossible, but she shook her head to try and dispel the thoughts. There was no point in thinking about those kinds of things since they were from two different worlds. She was an American in exile still trying to find herself, while he was a noble English gentleman with wealth and heritage behind him.

But still, it never hurt to dream...

2

Blake strode away from the church, walking through the village, enjoying the exercise and the scenery of the verdant fields and colorful flowers. The air was sweeter than in the city where he usually conducted his business, and he was glad to be home. This place would always inspire fondness in his heart no matter how much time he spent away from it, but it seemed as though he was going to have to spend more time here than usual since his father only seemed to be getting worse. It was bound to happen; Lord Huntingdon was an old man and as much as Blake wanted to believe they were an exception to mortality, deep down he knew that they weren't. It was difficult though to see his father, who had always been a vigorous man, fade away before his eyes.

But there were other things that caused fondness within his heart, like Lily. He found himself smiling whenever he thought about her. She was so unlike anyone else he had ever met, and she certainly didn't seem as though she belonged in the quaint and quiet village of Didlington St. Wilfrid, and yet in so many other ways she *did* belong. She may have been American and she may have worked in Hollywood, but she had a good heart and she cared about people.

That counted for a lot these days, and he was happy that God had brought her to him.

He walked into his house and found his father sitting in a chair reading a newspaper. "It's terrible, just terrible," his father muttered.

"Are you okay Dad?" Blake said.

Lord Huntingdon looked up. His eyes were cloudy and his lips parted. He still had a full head of hair, but it was thin and as white as a winter's sky.

"Yes, yes I'm okay, but the Aussies are at it again. I don't think our boys are ever going to beat them. I remember when I was young and I thought I could be the one to take the Ashes from them. Oh you should have seen me Blake; I was the star of the village team. Of course I had other duties and, well, dreams have to give way to that. But I would have shown them our English grit," he curled his hand into a ball and had a look of defiance on his face. "Still... there good people though. Fine cricketers, you can't take that away from them. I just wish our boys were better. It's a real problem these days you know, people just don't take pride in our nation anymore. If they knew how important the Ashes were they would never give up so easily," he sighed and folded the paper onto his lap. "Where have you been anyway? Did you take Bingo for a walk?"

Blake felt his heart break when his father mentioned Bingo, the old family pet that had been Blake's best friend when he was as boy. "Dad... Bingo has been dead for years."

Lord Huntingdon looked shocked for a moment, but then blinked and spoke in a bluster. "Well of course he has. I know that. You don't need to tell me. I was just remembering, you know, people only live as long as you remember them."

"Yes father," Blake said, sighing inside. "There's something else I want to talk to you about though. I'm a little worried that someone is stealing from us. Have you noticed anything strange recently?"

"Stealing?" he looked around. "Looks like everything is here to me."

"No, I mean from the accounts."

"No no, I'd know if there was something amiss. Still... if anyone is stealing then we'll get them son. Nobody is going to take anything from us! Get the accounts and I'll have a look through them."

"I don't think that will be necessary Dad. I'll look through them myself, I just wanted to see if you had noticed anything."

"No, I haven't, but I'll be sure to keep my eyes open. Nobody has ever been able to get anything past me before, and they're not going to begin now." For a moment he looked like the man Blake had looked up to when growing up, the tall, towering man who was like a force of nature. But just as quickly that image faded as Lord Huntingdon looked down at his paper again and rolled his eyes.

"Have you seen this Blake? Looks like our boys are failing down under again. If I was a young man I'd put myself forward. I could do a sight better than this lot. I used to play a bit of cricket when I was younger," he said, as though he was reading the information for the first time. Blake's eyes misted over as he nodded and humored his father by having the same conversation again, before he turned to move out of the room. When he left he noticed that his father's assistant, Jenny was standing there.

Blake frowned. "Afternoon Jenny, I wasn't aware you came in on Sundays."

Jenny was a small slip of a woman, with curled black hair and elfin features. "I don't usually, but, well, your father needs a bit more help than usual," she said, cringing a little when she spoke. "I know it might not be my place, but I couldn't help but overhear what you were talking about. I have to admit that I'm a little worried about him."

"So you've noticed it too then," Blake murmured.

Jenny nodded. "He goes round and round in circles sometimes, and there are times when he'll forget that I've made him his tea and then he'll blame me for letting it get cold. He's well most of the time, but then sometimes he's like this and…"

"I know. Thank you for looking after him. Next time he gets like this come and find me. I'm planning to stay here for a while longer to keep an eye on him," Blake said.

"I'm sure he'll appreciate that. He'll like having you around. He's always talking about you when you're not here," Jenny said, smiling in between licking her lips. Blake nodded and turned to walk away, but Jenny called him back. This time she whispered.

"Is it true what you spoke to him about? Do you think someone is stealing from him?" she asked.

Blake didn't like the fact that she had continued listening well past the point when any decent person would either have announced themselves or taken themselves away, but given her close proximity to her father she could be a good source of information.

"I'm afraid so. I need to investigate it more fully. I hate the idea of someone taking advantage of him when he's like this."

"Me too sir, me too. I see him at his best you know and he's so charming and lovely, and then he seems to just lose himself for a little while. I hope you don't think I'm speaking out of turn, but I think I might know who you should look into."

"Who?" Blake asked.

"George. I've heard things, you know how word spreads in a place like this. Apparently he likes to gamble a bit and, well, you know what kind of people he might meet when he's gambling. I think that's where you should start. He's a nasty man too. I've bumped into him more than once and the way he looks at me… he's one to take liberties sir, if you don't mind me saying," she said. Her cheeks flushed a deep shade of crimson and the tone of her voice rose to where it was sharp and jagged. She calmed herself almost immediately afterwards though and apologized for her outburst.

"I'm so sorry sir. I don't mean to speak ill of people behind their back but, well, I just hate the fact that someone your father has trusted could be treating him like this. It's just not right to treat a man in his condition like this. It's better that you're home though. You'll make everything right sir, I'm sure of it," she said, composing herself before she went in to see to his father.

Blake watched on as Jenny entered the room and smiled at Lord Huntingdon. He seemed delighted to see her and Blake thought it was a good sign that he recognized her, for it showed that he wasn't completely living in the past. However, within a few moments Blake got that sinking feeling again when he noticed that Lord Huntingdon was talking about the cricket games once more, going over the same beats as he had done with Blake. Jenny was kind about it though, listening and speaking to Lord Huntingdon in a soothing manner. He was glad that she was there to help him through these little episodes he had, and a part of Blake was consumed with guilt by the fact that he hadn't been here to take care of his father. Well that was about to change, and he was going to make amends. He turned away just as Jenny went to get Lord Huntingdon's medication, for Blake had a name now. It was time to get this investigation started properly.

3

Blake had come to find Lily again and updated her on what he had learned from the assistant. Lily's heart went out to Blake as it was clear how much this was affecting him. She wanted so badly to help him.

"I think there's one person in this village who will know of anything suspicious happening," Lily said.

"Who is that?" Blake asked.

"The vicar. I noticed how he spoke to everyone today. If there's a rumor I'm sure he's heard it."

"Isn't there something like vicar-parishioner confidentiality," Blake said.

"If there is then I'm sure I'll still be able to find out something from him. I doubt he'd want your father to be in any danger. Would you like to come with me?"

"I would like to, but I'm going to have to pass. I need to look through the accounts properly to see exactly what has been taken and how long it's been going on for. I feel really bad. I should have been here for him. I shouldn't have let myself get so distracted with all my business elsewhere."

It was clear that he was being hard on himself. "You were only living your life Blake, you can't feel guilty for that. It's not as though you knew how bad he was getting anyway. The important thing is that you're here now and you'll take care of him. You're a good son, and I'm sure Lord Huntingdon knows that as well."

Blake nodded, although she wasn't sure how much of her kind sentiment he was taking on board.

"But when I find out whoever has been stealing from him..." Blake said, unable to finish the thought. His face was like thunder though and rage poured from his dark and brooding eyes.

"When *we* find out we'll go to the police," Lily reminded him, emphasizing her words to show that Blake wasn't in this alone. She worried that he might do something drastic. She had seen all too well what lengths people would go to when their emotions were raw, although she hated to think that Blake would ever be capable of such a thing. It was important for her to find out who was stealing from Lord Huntingdon as soon as possible, for she feared what would happen to Blake if this matter lingered.

Lily returned to the church. It was quiet and eerie, almost as though if she listened hard enough she would be able to hear the secrets of the world. The heavy wooden door was open and a shaft of light from outside crept in through the doorway. The stained glass windows were colorful and danced with brightness, while the hallowed air had a great aura about it. Her footsteps echoed against the stone floor, and she was afraid to make a noise for she did not want to disturb the peace. She made her way to a small office at the back of the church, walking past the depictions of Christ and saints and disciples looking down upon her with their caring, pitying eyes.

She knocked on the door. The Vicar was sitting at his desk with a sheet of paper in front of him. He looked up and smiled when he saw her.

"Lily! Welcome, please come in and sit down," he said, gesturing to a seat. He offered her a glass of water as well, pouring it from a jug before she had a chance to reply. "What brings you here today? I was hoping that I would see you. It must be quite a shock to come from America to a place like this."

"It has been an adjustment yes, although I think I'm settling in quite well. Everyone here, well, most people have been friendly and it's refreshing to be in a place like this. I don't think I realized how much I needed to leave L.A. until I did leave."

"I can imagine, and it's brought you here as well, to this church. I'm sure God is pleased that you have found your way back to him."

Lily shifted uncomfortably in her seat. Recently it had been her putting people on the spot with unsettling questions, and she did not like being on the receiving end of them. "Yes, well, I thought I'd just try it out to see how it was."

"And what did you think? Shall I expect to see your face regularly on Sunday?"

"I'm not sure… I think so. It's still… it's still a little strange to me. I mean… it's been so long and I don't want to be disrespectful when I haven't been a practicing Christian for a while."

Vicar Brown tilted his head and looked at her reassuringly. "Given that God is eternal do you really believe that He cares about a few years here or there? Besides, even though I love this church and believe that it is important to build a house where we can gather or worship, the real place of worship that matters is in your heart. I'm sure even over the years you have turned to God in your darkest moments. That is what's important, that you have this faith and that it has never left you. God knows what is in your heart and He's not going to judge you for being absent. Every child leaves home at some point or another, and eventually they almost always return."

"Yes, I suppose they do," Lily said.

"Either way I'm glad to have you here. It's always pleasing to see my congregation grow. And I noticed you were speaking to Blake as well; he's a fine man. I'm glad to see he is sticking around a little bit longer, although it's a shame what's happening with his father," the Vicar said, his face falling into sorrow, the skin between his eyebrows crinkling sadly.

"I know, actually that's what I came to speak to you about," Lily said. The Vicar arched his eyebrows in surprise. Lily hoped he wasn't offended that she hadn't come here for his religious wisdom. "This is a delicate matter so I hope that I can speak to you in confidence," she said.

"Of course, there are only three of us here," he said, and then explained that he was talking about the two of them and God. Lily nodded, realizing that she should have understood what he was talking about.

"Well it seems as though Blake suspects someone of stealing from his father, and the likeliest suspect seems to be Lord Huntingdon's accountant, George. Since you're such a central part of the community and you know practically everyone in the village I wondered if you could provide any insight into the matter? Perhaps you've heard someone mention something? Anything could be illuminating, no matter how insignificant it seems."

The vicar leaned back in his chair and pressed his finger to his mouth, tapping his lips idly as he raised his gaze to the ceiling. He hummed as he thought and Lily waited patiently, already wracking her brains for another avenue to the truth in case the vicar was a dead end. Thankfully though Donald snapped his fingers.

"I don't know if this is important or not, and in fact it might well not matter at all, but I was taking a stroll through the village yesterday when I heard a bit of a commotion. There was a stranger having an argument with Pat. I'm not sure what it was about, but the man most definitely did not belong in the village. Perhaps that's somewhere you could start?"

Lily pursed her lips at the mention of Pat's name. Plenty of mystery surrounded the man, and for someone who prided himself on keeping to his solitude he seemed to get involved in plenty of different situations. She had a feeling there was more to his story as well, remembering the elaborate setup of security cameras she had seen in his home. She assumed it would remain a mystery though, for she didn't imagine Pat being forthcoming with the truth.

"Thank you Vicar, you've been very helpful," she said, and rose from the chair.

"When you speak to Pat please tell him that there's always a place for him in church, and I'm always here if he ever wants to talk."

"I shall do that," Lily replied.

"The same is true for you as well Lily. I'm sure that after what you've been through you could use someone to speak to now and then. I'm glad that you've found your faith here in our small village, and I look forward to seeing you next Sunday."

Lily paused for a moment before she left the office, letting the vicar get back to writing his sermon. She swallowed a lump in her throat as she thought about the vicar's offer and decided that she would take him up on it one day. However, for now she had other matters to attend to.

Pat's house was foreboding as usual. Her gaze drifted to a neighboring cottage, which was still empty after the terrible events that had occurred there, where a jealous wife had let her feelings get the better of her. The hatred of the woman flashed in Lily's mind, and a great shudder passed through her body. She pushed the unsettling feeling aside and tried to focus on the matter at hand. Murder was a macabre thing to think about, and unfortunately given her career as a mystery writer it was on her mind more often than not.

She walked up the well-tended path and knocked on the door. Smoke was rising from the chimney, indicating that Pat was home rather than walking Zeus out in the field. She heard stomping footsteps from behind the door and then it opened just enough for Pat's face to be visible. When he saw that it was Lily on his doorstep he scowled and rolled his eyes.

"You again," he muttered. "What is it this time? I've told you before I don't like you coming up here. I did you a courtesy before, isn't that enough? That's the problem with inviting people into your home; you do it once and they think they can come and go as they please."

Zeus, Pat's dog, rushed up and tried to push through Pat's legs. His pink tongue lolled out and he looked as though he was smiling at Lily. So far the biggest mystery in the village had been how someone as coarse as Pat could have such a friendly dog in Zeus. In her experience pets reflected the nature of their owner, but these two seemed the complete opposite of each other.

"I'm not coming and going as I please, I'm just trying to help Lord Huntingdon and his son. I'm sure you must have some affinity for the family?"

"They don't bother with me and I don't bother with them," Pat said, eyeing her suspiciously. "Say what you want to say so we can get this over with and I can get back to my affairs."

Lily had hoped that after helping her with her last murder investigation, Pat might have been more amicable toward her When she had first arrived in Didlington St. Wilfrid he had made no secret of the fact that he despised outsiders. As Lily had spent more time in the village she had come to the conclusion that he actually just hated everyone. Even so, she couldn't let that stop her from getting the information she needed.

"I need your help again Pat. It seems as though someone is stealing from Lord Huntingdon and I'm trying to track down who it might be. The Vicar mentioned that you had an argument with someone yesterday?"

Pat glowered, and she wasn't sure whether it was because she had mentioned the vicar or because of the argument.

"I did."

"What was the argument about?"

"I don't know. The man was looking for the doctor and someone named George. I told him we didn't have a doctor and that he should clear off because we didn't want him around disturbing our peace. He didn't look like the kind of man who should be here anyway. We've had too many strangers coming here for my liking."

Lily ignored his bluster and tried to get to the heart of the matter. Her ears pricked up at George's name, although she had no idea what he might want the doctor for.

"Did he say anything else? Was there anything else that you picked up on?" Lily asked.

"Just this," Pat said, reaching into his pocket. He extracted a crumpled and torn piece of paper and handed it to Lily. "He was looking at this as though it was important. He didn't like the fact that I took it," Pat grinned with pride.

Lily unfolded the paper. The creases made it difficult to read the words, and the fact that it was torn in certain places didn't help either. None of them jumped out at her and for the most part it seemed to be a random jumble of letters. She sighed and placed it in her pocket, deciding to look at it later. If it was a code then she would need to be clear of mind when she looked at it.

"Well thank you for your help Pat. I'll let you know if it comes in any use."

"Don't bother," he said.

"Oh, and the Vicar asked me to tell you that he's always there if you ever want to go and visit him."

A dark, humorless laugh erupted from Pat's mouth. "The next time I'll be anywhere near that place is when I'm buried."

The force of his words took Lily by surprise. She had rarely heard someone speak with such vehemence when approached with kindness.

"How could you say something like that? He seems like a nice man. All he wants to do is help."

"You should know better than anyone that some people are beyond help Lily. God doesn't want anything to do with me, and I don't want anything to do with Him either. I made my peace with that a long time ago and the best thing people in this village can do is LEAVE ME ALONE." He placed emphasis on each of these words as he uttered them, and then slammed the door before Lily had a chance to respond. She furrowed her brow and wondered what had happened in Pat's past to make him feel like this.

"I haven't been able to find George anywhere. It seems like he's skipped town," Blake said. Lily had gone to Blake after she had seen Pat to tell him what she had learned.

"If that's true then it certainly places him under suspicion," Lily said.

"I just wish I could find him to talk to him."

"Have you spoken to your father about it?"

Blake shook his head. "I haven't wanted to, but I think I need to. Would you come with me?" he asked.

Lily accompanied Blake upstairs to where Lord Huntingdon's bedroom was. Lily was surprised to see the man sitting on the edge of the bed with his head clasped in his hands. He sounded as if he was weeping.

"Dad? What's wrong?" Blake said, rushing into the room.

"I... I..." Lord Huntingdon said, before his gaze fell upon Lily. His mouth hung open and he gasped.

"Gracie... my Gracie is that you?" he asked.

Lily tilted her head and looked at Blake, who averted his eyes from her. "No Dad, it's not mother. This is my... friend, Lily," he said, his voice catching a moment before he called her his friend. Lily was flattered and surprised that Lord Huntingdon had taken her for Blake's mother, and she found herself looking away.

"Mother has been dead for a long time father," Blake said gently, lowering himself to the floor as he took Lord Huntingdon's hands. "What are you sad about?"

"I... I'm not sure. Something happened. Something bad, but my head is... my head is so fuzzy and it hurts to think. It's all in there, but it's so jumbled and I..." he clamped his eyes shut and held his head in frustration, cursing himself and stomping his feet on the floor. It was distressing to see and Lily wished she could do more to help.

"Dad, Dad it's okay," Blake cooed. "Just calm down, okay? I'm right here to help you and listen to you. We'll get through this together," he said, and then glanced down. He reached over and picked something up that had been obscured from Lily's field of vision. "What have you been doing Dad?" Blake asked, holding up a pair of muddy boots. The mud was dried and crusted on the soles of the boots, but it didn't look more than a day old.

The moment Blake showed Lord Huntingdon the boots, the old man groaned and threw his head back.

"It's true, it's true!" he cried.

"What's true?" Blake asked.

"The murder... I don't remember it, but I must have done it. Oh... how could I? God forgive me," he said.

Blake turned to look at Lily, and his eyes were cold with fear.

4

L ily wasn't sure what to say to Blake. Lord Huntingdon was babbling incoherent words. They streamed from his mouth and he kept shaking his head, as though he was remembering a terrible dream. Blake was holding his father's hand and speaking to him clearly, trying to get Lord Huntingdon to speak about the memories.

"Was it George?" Blake asked over and over again, but Lord Huntingdon was twisting and writhing as though a demon was in his mind. Lily was amazed that Blake was able to remain so calm when his father may have just confessed to committing murder. Lily's heart sank, for yet another murder had blighted the small town of Didlington St. Wilfrid.

There were footsteps approaching from outside. Lily turned and saw a small woman standing in the doorway. Her face was white as a sheet and Lily assumed this was the woman Blake had mentioned, Jenny, the assistant.

"I heard Lord Huntingdon cry out and I came to see what the matter was. What is going on?" Jenny asked.

"I need you to do something for me Jenny. I need you to go and call the constable."

"The constable?" Jenny asked, her eyes widening in fear.

"Yes," Blake said. His voice was curt, emotionless. Lily wondered how he could control himself so well. "It seems as though there may have been a murder. I need them to search the grounds for a body."

"A body..." Jenny gasped, and then she began to cry. Her head dropped and her voice trembled. "I didn't think he was going to do it. I didn't believe he could..."

"Do what?" Blake asked.

"The Lord... before I left last night he said that he was going to teach George a lesson. You know what he's like though. I didn't think he was actually going to do anything like this. He was just angry. He..." she lost all composure and let her head fall into her hands. It was a grim revelation as it seemed to confirm Lord Huntington's condition.

"Why didn't he say anything when I spoke to him about it this morning?" Blake asked.

"I suppose he didn't remember," Lily said, her voice hollow and empty. This was not the way she had wanted this case to turn out. She could already see the toll it was taking on Blake and his father, and she was sure that if the resident Lord was found guilty of murder in this small village the scandal would never be forgotten. That wasn't even to say what it would mean for Blake's heritage, or the property that Lord Huntingdon owned. Lily's home, Starling Cottage, was one such property. This had ramifications for everyone.

"Jenny, you need to go and call the constable," Blake said. "I'm going to stay here with my father."

~

Constable Dudley came a short while later and indeed confirmed that a body had been found on the grounds of the estate, a body that had been identified as George, the accountant. Constable Dudley cleared his throat and rocked back and forth on his heels, unsure about the way to conduct himself.

"I am sorry Lord Huntingdon, but I need you to answer a few questions," Constable Dudley said.

"Is there any point?" Blake asked, showing his emotion for the first time since his father had confessed. The words erupted from him, rushing on a wave of anger. "The man can barely remember anything!"

"He might remember enough," Constable Dudley said solemnly.

"Blake," Lily said gently. "I know it's difficult, but Constable Dudley does have to do his job."

Blake relented at her words and stepped back, allowing Constable Dudley to approach the Lord. The constable was very apologetic, but even so Lord Huntingdon didn't seem to be aware of what was happening. He held out his palms in front of him and shook his head, staring into space muttering over and over again that he couldn't remember what had happened, only that he had killed the man.

"My mind is so fuzzy. I wish... I wish that I could be more help. I wish that I could remember, but it's all so difficult..." he groaned.

"We're not getting anywhere here," Blake said.

"No, we're not," Constable Dudley said, scratching his chin. He looked to Jenny and exhaled deeply.

"Would you mind telling me again exactly what you know?" he asked.

Jenny sniffed and nodded. Her voice continued to tremble. "I was working late at night. I've been working here more than usual recently because of Lord Huntingdon's condition," she said, averting her gaze as she spoke these words. "He was upset about something and he was angry at George. I didn't think much of it because he has been more irritable lately. I simply gave him his medication, which I hoped would calm him down. Then today I overheard Blake speaking about the stolen funds. I assumed that George might know something about it, but I never thought he would be dead! I'm sorry for not telling you the entire truth sir," she turned to Blake, "but I didn't want you to think badly of your father. He didn't want to appear like a lesser man to you. He kept telling me over and over again that he wanted you to be proud of him, and I thought he was in control of things. I thought the medication would help. I never thought that this would happen."

The girl looked distraught and Lily felt pity for her since she had only tried to help her employer. Still, there was another element to the case that hadn't been spoken about yet.

"It seems that I don't have any other choice but to arrest Lord Huntingdon," Constable Dudley said.

Blake erupted in anger once again. "You can't arrest him! Do you think he'll survive in a prison cell? How can you expect to take a man in his condition from his home? He's barely conscious of what's happening around him."

"I know it's hard, but we have measures to keep him safe and we'll look after him. The law does not make exceptions Blake. He's the prime suspect and we need to take him into custody."

Blake laughed and threw up his hands. "What possible good is that going to do? Do you think he's going to flee? Do you think he's really capable of murder?"

Constable Dudley steeled himself and looked Blake directly in the eye. His voice was terse, weighed down with all the history of his career. "In my experience anyone is capable of anything if they're pushed hard enough. It would be naïve to think otherwise."

Blake opened his mouth, about to speak again, before Lily interrupted. She knew if she didn't intervene that Blake might say something he wouldn't be able to take back.

"There's something else that we haven't addressed yet, the stranger in the village," Lily said. "He was looking for George, and the doctor for some reason."

"It might be that this has nothing to do with the case," Constable Dudley said.

"No, but we should look for him anyway. If he knew George then he might have some more information."

"It was probably just one of the people George was talking to about his gambling. He was in debt. Maybe George got desperate and came back here to get more money from Lord Huntingdon. If he knew someone was in the area," Jenny said. The words came out in a torrent, and the sorrow had given way to something akin to fear.

"But why would he want to see the doctor as well?" Lily wondered aloud.

"Perhaps the doctor is a gambler too," Jenny offered.

Constable Dudley scoffed at this. "Dr. Hartley would never do such a thing. The man is beyond reproach and he's done more for this village than anyone else. I won't hear a word said against him!" He calmed for a moment as a thoughtful look came upon his face. "Perhaps you are right Lily. If we can't find this man then we should go and speak with Dr. Hartley. He might be able to shed some light on the situation."

As he said this Jenny wailed and threw herself against the wall. "Are you sure the body is George's? I can't cope with knowing that he might still be out there. He was a nasty man," she said, quickly becoming hysterical. Lily was surprised at this turn of emotion, but the act of murder hit people in different ways.

Blake told Jenny to sit with his father while they went to speak with Dr. Hartley. Lily could tell that Constable Dudley wanted to tell Blake to stay with Lord Huntingdon, but there was nobody that could tell Blake what to do at the present moment. The trio left and walked outside. When they did they saw a stranger standing a ways down the path.

"Who is that?" Constable Dudley asked, for none of them recognized the man. At the sight of three people approaching him, the stranger turned on his heels and sprinted away. Blake leaped into action, with the blustering, heavyset form of Constable Dudley not too far behind. Blake accelerated beyond the constable, his long legs eating up the ground. He cut through the field and vaulted over a stone wall. Lily found herself impressed with his athleticism. The stranger was slowed by the twisting paths. This was Blake's home, his land, and he was the master of his domain. He looked like a hunting lion, chasing down his prey, and only one outcome was inevitable.

The stranger gave up and Blake grabbed him by the arm, leading him back toward the house. Constable Dudley stopped running as well, puffing and mopping sweat from his brow. His cheeks were ruddy and he leaned against a helpful tree, his skin glistening.

When Blake returned he pushed the stranger toward Constable Dudley.

"Who are you and what are you doing lurking around my home?" Blake asked.

The man kept his lips tightly shut. They formed a thin line on his narrow face. He had a mess of hair and his eyes were beady.

"Speak to me! Did George owe you money? Did you kill him?" Blake yelled.

At this the stranger's eyes widened. "George is dead?" he gasped. The shock was genuine. Blake calmed himself at this, and he seemed troubled.

"I think you should tell us how you know George and what you're doing here," Constable Dudley said now that he had caught his breath.

"Maybe this will help," Lily said, and thrust the crumpled bit of paper toward the stranger. "What does this mean? I haven't had a chance to look at the code yet."

"It's not a code," the man said, seeming to grow more at ease. "My name is Tony and I'm George's brother. He got in touch with me when he became concerned about Lord Huntingdon. He wanted me to look into it."

"But what does Dr. Hartley have to do with this?" Constable Dudley asked.

"I need to speak to him about medication he was prescribing. That's what's on the paper. It's a list of tablets," Tony said.

Lily looked at the paper again and suddenly it all made sense to her. Medicines usually had unusual groupings of letters together, and without knowing the context it had been natural to assume there was something else going on.

"George was getting suspicious of someone who was working in the house. He said that she was getting these tablets and giving them to Lord Huntingdon. He thought if I asked the questions then she wouldn't learn someone was on to her, but I suppose if he's dead then she did," he looked downcast as the news that his brother was dead sank in. It was as though the whole weight of the world was hefted onto his shoulders. Yet another family had been torn apart by murder.

"Did George tell you a name?" Lily asked, although she had a sneaking suspicion she knew which name he was going to say.

"Jenny," he said.

Lily and Blake looked at each other and fear lined their faces. She had been left alone with Lord Huntingdon, and if she had killed once then she might well kill again.

5

T hey rushed back to the house to find that Jenny was in the process of leaving. She had a bag and stormed out of the estate just as Lily was rushing up. Jenny barreled into Lily and almost knocked her to the ground. Blake caught Lily in one of his strong arms before he turned and grabbed the bag, hauling Jenny back. She screamed as she lost her balance and toppled back, and when she realized that Blake and Lily were towering above her she knew there was nowhere to run.

Where before her face had been a picture of sorrow and fear, it was now bitter and vengeful. Her eyes were filled with rage and she glared at them as they surrounded her.

"What did you do? What have you done with my father? Did you kill him?" Blake asked, his voice rippling with tension.

"The old fool is fine," Jenny spat.

"Even though you gave him all this?" he asked, grabbing the paper from Lily and thrusting it in Jenny's face. "We know what you did Jenny. How could you? I thought you were looking after him. I trusted you," Blake said. The betrayal was plain on his face.

"Trusted me?" Jenny said, laughing as Constable Dudley and Tony joined them. "You only trusted me because you paid me, and not enough at that. Do you know how humiliating it is to come into a place like this and be reminded of wealth every second of the day? What good is it going to do to an old coot like him? That money could transform my life, and all he's doing is sitting on it. I'm paid a pittance, and you're all here sitting on your thrones surrounded by so many expensive things. I just wanted to even the playing field a bit. George did as well."

"So you were working with him?" Lily asked.

Jenny nodded, and Tony grimaced. "We had a good thing going as well. The old man didn't know what was going on. George had everything under control, and then you had to come back," she stared at Blake. "George got scared."

"He told me he thought you were going too far," Tony said.

Jenny dismissed the concerns with an idle wave of the hand. "He kept telling me that it was too risky, that Lord Huntingdon was going to end up dead. I told him there was nothing to worry about."

"How did you get the tablets from the doctor?" Lily asked.

"I just asked him to prescribe some for my grandmother," Jenny said, and laughed cruelly. "Everyone here is such a soft touch. You give them a sob story and they believe it. He was all too happy to help. It almost worked perfectly as well. It would have, if George hadn't gotten scared."

"He didn't get scared enough in time to save his life," Tony said.

"No, I thought I could pin the murder on the old fool and get away with all the money. You should have seen his face when he actually started to believe that he had done it. George should have just kept his mouth shut and everything would have been alright." She scowled and shook her head, as though she was the one who was the victim. Lily was struck by the cruelty of the woman and how remorseless she was. One man had been killed while another had been plied with drugs and accused of murder, and all Jenny cared about was that she had missed out on a windfall of cash.

Constable Dudley arrested her and then turned to Tony, saying that he was going to have to come to the station to answer some questions about George's part in all this. Tony hung his head as he walked away.

"I feel bad for him. He came here because his brother asked for help and now he finds out that not only has his brother been killed, but that he was a thief," Lily said.

"I guess sometimes we can't know what the people closest to us are capable of. It makes me wonder what other secrets are hiding in this village," he said as he turned away, not wanting to speak any more about the matter.

They returned to Lord Huntingdon's room. The old man was still sitting on the bed, mumbling to himself.

Blake went to his father and kneeled on the floor before him. "Dad, it's okay. You didn't do it. I don't know what Jenny told you, but she was lying. You're not responsible for the murder, and nobody is going to steal from you again. We'll get you back to your old self, okay? I promise. I'm not going to go anywhere either. I'm going to stay right here and be by your side and everything will be alright again. I made a mistake in leaving you here. This is where I belong, I promise," Blake said. It was a tender sight to see the son nursing the father like this, and it made Lily think about her own family ties, ties that had been broken a long time ago. She felt a pang of pain in her heart, before she stepped away.

EPILOGUE

After leaving Blake, Lily had gone to Dr. Hartley to ask him about the pills. He had seemed horrified that he had been deceived and promised that he would take stricter measures when deciding who to prescribe medication to. It turned out that Jenny had been very cunning in the tablets she chose for they caused fogginess, memory loss, and made people very prone to suggestion. She had been dosing Lord Huntingdon with these drugs for a long time, but Dr. Hartley assured Lily that once he had been weaned off the medication he would return to normal and regain normal cognitive functions. She was relieved at this, and was eager to share the news with Blake.

They stood outside the church on another crisp Sunday morning. Lily had given Blake some space for the week as he tended to his father, and she was glad to see that he was attending church. They had a few moments before the service started.

"How is your father doing?"

"He's getting there, although he still has moments where he can't quite remember things as well as he used to. I know Dr. Hartley said that he's going to get back to normal, but I'm worried that some scars are going to be left. He's woken up a few times in the night crying out, and he keeps saying that he 'sees him'. Jenny did a real number on him. I have to keep convincing him that he didn't kill anyone."

"It's horrible to think how our minds can be warped by these things," Lily said.

"I know, and it's all my fault."

"You can't blame yourself for that Blake."

"I can. I was the one who hired Jenny in the first place. I should have seen the warning signs. I should have been here to watch over him. I shouldn't have just left him to be taken care of by other people. That's not what real families do."

"You did what you had to do Blake. And we got there in the end. He's not going to be hurt anymore, and I'm sure that soon enough he's going to be back to the man that he always used to be."

Lily tried to be reassuring, but it was clear from Blake's expression that he wasn't ready to believe her words. In truth she wasn't sure how accurate her predication was going to prove to be either.

"The thing is I'm not sure that I'm the man I thought I was. There was a moment there when I actually believed my father was a murderer. Can you believe that? I thought he was guilty. I should have had more faith in him. I should never have let myself doubt him."

"You were just following the evidence Blake. At the time it did seem like he had some part to play."

"No. I jumped to conclusions and let his condition affect my judgment. I saw those boots and thought the worst, when all that happened was that he had gone to do some late night gardening. It was all so simple and innocent, but I thought that he was a killer. I know he never would have thought the same of me. I'm not a good son."

Lily wasn't sure what to say to him to make him feel better. She placed a hand on his shoulder and turned toward the church. "Well, if it's faith you're lacking then I know the right place to find some. Come on, let's go inside. You might not have faith in yourself Blake, but I do," Lily said.

They walked into the church and settled into their seats. Lily continued to glance at Blake, hoping that he would be able to resolve the anguish in his heart. It was easy to expect the worst of people, even when you should have known better. Jenny had managed to put on a convincing act, but even she had shown the evil within her, in the end. Before Blake could move on he had to forgive himself, and sometimes that was the hardest thing of all to do.

But something he said earlier struck her as she gazed at the congregation, letting her mind wander while the vicar delivered his sermon. Blake had made a remark about other secrets in the village. As she looked at every face in turn she realized there was so much she did not know about them, and her literary instincts started to tingle as she wondered what mysteries each of them might hold within their hearts. But the one that was on her mind the most was Pat. He was averse to spending time with anyone, even allowing God into his heart, yet Zeus was such a loving soul.

It was clear that people were filled with contradictions. Blake was a loving son who suspected his father was capable of murder. George was a thief who had the conscience to warn someone when the crime was getting out of hand. Lily herself was an outsider who was only just beginning to feel as though she belonged. She was a woman who had let her soul drift in the ether for too long, but now that she was back in a church singing her heart out to the tune of the old hymns, she felt as though she was getting back in touch with herself.

Her time in Didlington St. Wilfrid had not been without incident so far, but she had faith that as long as she trusted in the wisdom and the love of God then everything would turn out for the best. She sang her heart out and felt tears welling up in her eyes as she felt the warm love flowing through her, giving her the strength to know that she had a place in this world, and for the moment she was taken with the idea that her place was by Blake's side.

A SINISTER STORMY CRIME IN A QUIET ENGLISH VILLAGE

1

"The sky has really cleared up," Lily said, arching her neck back to look up at the azure sky. It was back to being the idyllic blue shade, endless and wondrous, as opposed to the crackling storm of the previous night.

"I thought the world was going to break last night," Joanie said, running a hand through her golden hair, still looking as glamorous as ever even though she has spent a little time away from Hollywood. Lily's friend had only meant to stay in Didlington St. Wilfrid for a short amount of time, but as of yet there hasn't been any plans to leave. Lily didn't mind though. It was nice to have the company, even if Joanie could be a little thoughtless at times. She could also be prone to exaggeration, and her latest statement was a case in point, Lily thought.

"I'm sure nothing like that would have happened. This place is too beautiful to be destroyed by anything, let alone a storm."

"If you say so," Joanie said, "I wonder how your new beau coped up there in his mansion."

Lily's cheeks burned crimson as Joanie's flicked with a mischievous gleam. "He's not my new beau," Lily said, clenching her jaw.

Joanie held her hands up in mock surrender and arched her eyebrows. "I never said it was a bad thing! In fact I think it would do you good. It would be a change of pace, and he can't be worse than the last one."

Lily grimaced and rolled her eyes. "I'm sure Blake has better things to do than get involved with an American writer. He has his hands pretty full right now anyway helping his father recover."

"I'm sure the two of you have been spending plenty of time together. There's no harm in it Lily. How far removed is he from the throne anyway? It's every girl's dream to marry a real live prince."

"I haven't asked him," Lily said, and wished the questioning about Blake would end. Lily's mind had been at peace since she arrived in Didlington St. Wilfrid, which was exactly the result she wanted to achieve when she moved there from Los Angeles. There were moments where in the back of her mind she could still hear the rush of traffic, only to realize she was dreaming. As for Blake, well, he had proven himself to be a good friend and wonderful company, but Lily wasn't so arrogant to think that she, some commoner, could steal the heart of an English noble. She wasn't exactly sure where Blake ranked as it was all so confusing what with their lords and earls and dukes and things, in fact she wondered quite how anyone could make head or tail of it. It was a more tangled mystery than the ones she wrote.

"You haven't? What do you two talk about when you're together then? You must do something fun when you visit the church."

"If you want to know so badly then perhaps you should come along with me next time. I'm sure the vicar would be happy to add another to his flock," Lily said.

Now it was Joanie's turn to have a reticent look on her face. She swept her hair across her face and for a moment she looked like a queen. "My relationship with God does not need a church," she replied. "In fact I feel such an awe just being out here that I don't understand why anyone would want to be cooped up in a stuffy old building."

"Wait a minute, when we were back in L.A. I'm sure you told me that you didn't have any place for God in your life."

"Well, at the time I didn't. But being here has given me time to think and time to appreciate everything around us. It's far easier to think about God existing in a place like this that is so… so beautiful than it is in L.A. Besides, people can change, as I remember it you never made a point of going to church back then either."

Lily squirmed a little and knew that she had been caught. It had certainly been a change for her, and going to church had been a little strange at first. It had soon become normal though and now she quite looked forward to Sunday mornings, and seeing Blake regularly was only half the reason why. Somehow having Joanie refer to her old life made Lily uneasy. When she thought back to her time in America it was as though she thought about a different person. The Lily who lived in Los Angeles and the Lily who lived in Didlington St. Wilfridwere two very different people as far as she was concerned. One was stressed all the time, frazzled because of work and had absolutely no time for herself. The other was relaxed and content, happy to be removed from a wellspring of tension.

Deciding to change the subject and ignore any further talk about their lives in America, Lily shifted topic. "Have you taken up photography because of your newfound appreciation of nature or has that happened because you have been looking at the world through a new lens, pardon the pun."

Joanie offered a small smile. "I suppose it's a little bit of both. I just hope my equipment hasn't been damaged by the storm."

Joanie had set up a number of cameras throughout the village and the fields surrounding it, seeking to capture photos from various points during the day and night to create a catalogue of images that showed Didlington St. Wilfridfrom different perspectives. It was a thoughtful and worthwhile endeavor, and one that Lily was looking forward to seeing the results from. It was something she had not expected from Joanie. As much as Lily loved her friend, there was no denying that Joanie was usually the flighty sort, the type to do something on impulse without thinking too much about it. That was pretty much how she had ended up in Didlington St. Wilfridin the first place, but Lily was glad of her company. Joanie was perhaps the one shard of her previous life that Lily actually wanted to hold onto.

Life in Didlington St. Wilfridwas practically perfect. Of course the murders and schemes were terrible, but thankfully life in the village had been uneventful for little while now. Lily hoped it would continue. All she needed from life was good company and good inspiration to fuel her books… then her mind turned to Blake and she wondered if perhaps there was room for something else to happen as well, although she quickly dismissed the idea with a swift shake of the head.

They were heading to Marjorie's bakery to pick up a few sweet treats when they saw the door open and someone walk out.

"Who is that?" Joanie asked. The woman was of medium height and had curled black hair that came to the base of her neck. She had soft features and a kind smile, and she walked briskly holding a basket of goods that she had purchased from Marjorie.

"That's the one who moved into the house near Pat's after that terrible affair," Lily said, thinking back to the horrid situation where a wife had killed her husband because she thought he was being unfaithful. "Her name is Emma I think, or Emily. Something like that."

Joanie shuddered. "I can't imagine why anyone would want to live *there*. The place must be cursed."

"I'm sure that something terrible has happened in every building if it's old enough."

"Even in the cottage?" Joanie asked with wide eyes.

Lily chuckled. "Of course! Think of what people are like; the odds are that eventually someone is going to do something horrible. If you don't want to live in a cursed building then you're never going to be able to live anywhere."

"Do you know what happened in our home?"

"I don't," Lily said, "but I'm sure someone as intrepid as you could find out if they were so inclined.

"I'm not sure I want to know."

"Anyway, my point is that the only way to break curses is to keep moving forward and doing the right thing. It's better that someone new should move in there and create a fresh history. I've been meaning to welcome her to the village actually. I thought we could both go, considering that we're new as well. It would sure help to make a good impression."

"If you want to," Joanie said. "I put one of my cameras up near that way so we're going to have to pass by anyway."

"You put one of your cameras up near Pat?" Lily asked, gasping a little.

Joanie rolled her eyes. "I still don't think he's as bad as you say."

"It's not a matter of him being bad or not, it's just that I feel like I've been making progress with him recently and if he finds out that my friend has been putting cameras near his home he's not going to be happy. He likes his privacy."

"People who like their privacy only have something to hide. Besides, it's not near his home and I'm not interested in capturing him coming out of the bath or anything like that. I only want to capture the beauty of nature."

"Well, still, I don't think he's going to be happy and I don't much care to have another run in with him."

"I think you should just leave him alone. That's clearly what the man wants."

"Yes, but there's something else about him, something he's hiding. I'm convinced he's not the man he wants us to think he is," Lily said. Joanie waved a hand dismissively through the air as they waltzed into Marjorie's bakery. The sweet aroma of freshly baked goods lingered in the air and if there was ever Heaven on earth, Lily was convinced this was it.

"We just saw that newcomer leave. What's she like?" Lily asked, and confirmed that her name was indeed Emma.

"Oh she's quite lovely," Marjorie said. "She said she's here for a new start. She wouldn't say directly, but I think she's suffered some tragedy. There was a look in her eyes that you only see in people who have lost something."

Lily immediately averted her eyes just in case Marjorie could glean anything that Lily didn't want her to know, and then stated her intention of welcoming Emma to the village.

"I'm sure she'll appreciate that. I can't imagine what it was like being up there in that house alone last night when that storm was raging. It's nights like those that really make you understand the fury of God," Marjorie said. "I can never sleep through a storm. I always have to put the light on and do something to distract myself. We haven't seen a storm like that for a long time."

"It was impressive," Lily said.

"I just hope my cameras survived so that they captured some good pictures," Joanie said. Lily could tell that Joanie was anxious to examine the condition of her equipment, so she cut short the conversation with Marjorie and left the bakery.

Before they could reach any of Joanie's equipment, Lily was interrupted by a high pitched yapping. She turned to see the bounding form of Zeus running toward her, the impact of his paws sending up a spray of water as he seemed to hit every puddle that he possibly could. Lily wasn't sure if this was intentional or not, but the playful nature of the dog couldn't be denied. It was such a sharp contrast to the grizzled manner of Zeus' owner, Pat. This man was the village's local grump, for lack of a better term. Although he had come to Lily's aid during a couple of her investigations it had been with a begrudging spirit, and he had made it plain that he didn't care for anyone else. It had always struck Lily as strange that Zeus would be so outgoing and friendly when Pat was anything but, and it made her wonder about who Pat truly was, and if she would ever find out.

2

Zeus came up to the two women with his tongue lolling out and pawed at them, immediately getting their coats wet. They were both glad they were wearing their anoraks. Lily petted Zeus and rubbed behind his ears, a place that usually made him grunt with appreciation, but on this occasion he seemed frantic. He whimpered and yelped and twisted around, almost as if he was pulling Lily in a certain direction. She glanced around to try and see why he was so agitated, thinking perhaps the storm had affected him in some way. From what she knew animals were quite attuned to the weather. Pat was never usually far from Zeus and she was surprised that she hadn't heard the man's curt cry yet, and indeed even more surprised when Pat was nowhere to be seen.

"What's wrong boy?" Lily asked. "Where's Pat?"

At the mention of Pat's name, Zeus jerked his neck back and the pawing became more intense.

"I think he wants us to follow him," Joanie said. Lily frowned. Although Pat was mostly a mystery to her, there were some things she knew for certain. One was that he could probably have gotten into an argument in an empty room, and the other was that he never let Zeus stray too far from his side.

As soon as Lily began moving, Zeus scampered away a few feet in front, and then turned to see if she was following. Once he became sure of this he moved more swiftly, and Lily felt more uneasy. If Zeus was this agitated then it could be something that might have happened to Pat. For all his faults she didn't want to hear of anything bad happening to him, for she was convinced he was a good man deep down. She wondered if anything had happened to him during the storm. One of the things she loved about Didlington St. Wilfridwas that the community was such a close knit one, but people could also be isolated. In the dead of night as darkness shrouded the world it was easy for people to be alone, and if any dark fate befell them it wouldn't be uncovered until the pale light of day illuminated them.

Lily doubled her pace to reach Pat's farm. The field was waterlogged and rainwater was still dripping from the drains. A huge puddle spanned the width of the path that led to Pat's house. Zeus ran to the door and scratched. Lily's pace quickened and she took a leap across the puddle. While she had never prided herself on her athletic prowess, she made the leap and felt a sense of pride swell within her.

Joanie, in contrast, sidled around the puddle with delicate footsteps, muttering under her breath that her boots were being ruined.

Lily tried the lock and found the door impossible to open. Zeus was pointing toward the rear of the house, and Lily followed.

"I'll wait here for you to come and open the door," Joanie said, wrapping her arms around herself to combat the morning chill. Lily wouldn't have expected anything else from her friend, so she followed Zeus around to the rear of the house The dog stopped at a window and pointed his nose toward it, his black coat silky and beautiful in the morning light. Lily approached the window carefully, afraid of what she might find. The window was open halfway, and she assumed this was how Zeus had gotten out in the first place. Just as she was about to climb in she looked up and saw two legs poking out from behind a table, and she recognized the boots. Suddenly the matter had taken on more urgency. If Pat had had a fall then time was of the essence. Lily battled with the window and squeezed herself through, glad that she was keeping up with her exercises. Given how many treats she allowed herself from Marjorie's bakery she might well have been unable to make it through the window otherwise.

She pushed herself through and landed inside, almost losing her balance upon impact. She could hear Zeus bounding around outside and whimpering loudly, so she was caught between going to check on Pat and opening the door. She rushed to the door and unfastened the multitude of locks that Pat had attached, then flung it open so that Joanie could enter. Even though Joanie had been waiting there, Zeus came rushing around as soon as he heard the door open. He was a black blur that surged past Joanie's legs and almost skittled the woman over. He ran directly to Pat, and Lily followed him.

"Pat? Pat are you okay?" Lily called out, but there was no response. Zeus nudged the man's legs and lower torso. Lily walked around the table and then gasped when she saw Pat there. The man had fallen alright, but not because of any medical condition. Pat was sprawled on the floor in a pool of dried blood. The expression on his face was blank, and Lily's heart was chilled.

Moments later Joanie joined her and exclaimed at the sight, shielding her eyes. Zeus continued to nuzzle his master, who was unresponsive. Lily's eyes were filled with tears as she dropped to her knees and pulled the dog into her, cuddling him as much as she could for the thought of Zeus being alone was heartbreaking, but it seemed as though trouble in Didlington St. Wilfridwas raising its ugly head again.

"Call the police," Lily screamed, "and an ambulance!"

3

"What's happened here?" Constable Bob Dudley said as he barged into the house, shaking off his wet feet as he apparently had not managed to avoid the puddle on the path at all, and instead had barreled through. His manner changed immediately as he realized that Joanie was there with Lily. A strange smile came upon his face and his cheeks reddened. He had never quite been able to keep himself calm when around Joanie, which wasn't a surprise to Lily as Joanie was a stunning woman, absolutely stunning. Still, seeing Constable Dudley act like this reminded her about how she acted when she was around Blake, and wondered if her attraction to the upstanding Englishman was as obvious as Constable Dudley's toward Joanie.

However, his reaction was short lived as he moved into the room and laid eyes upon Pat. Lily stood by as paramedics urgently tried to revive the hapless self-proclaimed protector of Didlington. Zeus had stretched himself out and had his head in his paws. He blinked slowly and when Lily looked at him she could sense the sorrow in the dog's heart. If anyone ever claimed that animals did not have feeling then she knew all she needed to do was show them a dog who had lost their master. Was there a sorrier sight in all the world?

It also reaffirmed her belief that there was good in Pat's heart somewhere. If Zeus could love him this much then surely Pat couldn't have be as bad as all that. She silently prayed for his survival and recovery.

Those attending to the body moved back for a moment so that the Constable could see the wound. He recoiled and frowned.

"A gun?" he exclaimed, as the frantic preparations were made to move the body to the ambulance and speed Pat to the nearest hospital.

"Is it such a surprise?" Joanie asked nonchalantly. Constable Dudley answered without looking away from the body.

"It is around these parts, especially a wound like this. We're not like your country. It's far more difficult to get a gun," he said.

"I'm sure I've seen people carrying them in fields," Joanie said.

"Rifles yes, to keep poachers away and to threaten any predators that come by to try and take liberties with the flock. But this wound was made by a pistol, and I can't think of anyone in the village who carries such a weapon, at least not to my knowledge. The more that time passes the more I'm seeing secrets in the world. I'm not sure that I like it," Constable Dudley sighed. He stepped aside as the body was carefully placed on a stretcher and carried from the building. It seemed that faint signs of life were still showing. Pat's too stubborn to die, Lily thought, and again prayed silently for his recovery and also that the culprit would be caught and brought to justice. She was ready to become the answer to her own prayer and was already determined to find out who committed such a heinous crime.

"I would ask you to help me find a list of suspects, but the people who have had run ins with Pat is about as long as the list of people who live in the village," Constable Dudley continued.

"But enough to kill him?" Lily asked.

Constable Dudley curled his lips and clasped his hands together, wringing them tightly. "I'm not sure, I'm not sure."

"Could he not have done it himself?" Joanie asked.

"No," Constable Dudley and Lily said in unison. Constable Dudley seemed amused by this and inclined his head, encouraging Lily to continue. She supposed he wanted to hear her estimation of the events.

"There's one obvious reason why; the gun is nowhere to be found. It would have been in his hand if he fired it himself. Besides, as ornery as Pat could be I never got the sense that he wanted to kill himself."

"I didn't either. He took far too much joy in making everyone else miserable to take his life."

"You've known him for longer than I have, don't you have any idea who might have done this?" Lily asked.

Constable Dudley scratched his head and blew out his cheeks. "I wish I did Lily, I truly wish I did. But it looks as though we're going to have to do this the old fashioned way. We're going to have to ask around and see if anyone had any arguments with Pat recently, anything that was out of the ordinary."

"Or we could try something different," Lily suggested. Constable Dudley arched his eyebrows, intrigued, as Lily explained. "During another investigation I came here for Pat's help. I think he only helped me to stop me irritating him, but he told me that he had security cameras set up all around his fields so that he could see when anyone was approaching his farm. I thought it was a sign of paranoia, but perhaps he did have reason to be worried. Anyway, he allowed me to watch some of them and that's how I knew that Charlotte had been on his farm, where she gathered the poisonous plants. His cameras might have captured whoever came to him last night. I doubt anyone would have known about it, so this case might be closed as soon as it's been opened!"

"We can hope," the constable said, and gestured for Lily to show him the way.

They walked through Pat's home, and it was unsettling to walk through the home of a man who had just been attacked and left for dead. Although Pat's home was devoid of many of the usual personal touches, there were still signs of life that indicated the time of the attack was very recent. There were things like a television guide with programs ringed throughout the week. There were dirty bowls in the sink, and some leftover cake he had bought from Marjorie's shop. A puzzle book rested on a coffee table with a crossword half-filled. It was a pitiable sight in light of Pat's present plight, and one that Lily tried to put out of her mind as she went into the room with the security cameras.

She settled down in front of the equipment and tried to remember how Pat had worked it. It took some doing, but she managed to find the recording from the previous night. The evening was bleak, but at the point of the recording the storm had not yet set in. Lily fast forwarded through a lot of empty space where nothing happened, and she wondered how Pat had stomached going through all of this footage when so much of it was empty. People had strange hobbies, she thought. Then a figure appeared, moving erratically and eerily through the sped up footage. Lily wound it back to see who had visited Pat, and there was a gasp from all of them when they saw Vicar Brown. Lily paused the footage on his face, and it was unmistakably the good vicar.

"I can't believe… no no, a man of God would not be capable of such a thing," Constable Dudley said. "There has to be more than this. Make it go forward more," he said, wagging a finger at the device. Lily did as she was asked and tried to ignore the lump that had formed in her throat, but they didn't even get to see the Vicar leave. She only wound it on for a few moments when the picture went fuzzy and it cut out, leaving nothing but static blackness behind.

"What happened? Where did it go?" the Constable exclaimed.

"I don't know. Maybe the storm affected the equipment," Lily said.

"Speaking of which, I should really go and check mine," Joanie said, taking her leave. Lily suspected that she had just been waiting for an excuse to leave the scene of the crime as such things were not as fascinating to Joanie as they were to Lily.

"I suppose we should go and speak to the vicar," Constable Dudley said, although he was made uncomfortable by the thought.

"Perhaps I could go and speak to him on your behalf while you put together a list of other suspects. As you said, there are going to be plenty of people who had a grievance with Pat," Lily offered. The constable seemed relieved at this.

"Come by the police station when you're done," Constable Dudley said. He put his hands on his hips and rose, shaking his head. "My father never told me the job would be like this," he muttered under his breath. Lily spared a thought for him. Being the chief of police in a small village like Didlington St. Wilfridshould have been a sleepy, easy task, with most of his duties taking care of interpersonal disputes, not terrifying crimes like attempted murder. But Constable Dudley had acquitted himself well so far, and Lily just couldn't stop herself from becoming involved.

As they went to leave there was something she had almost forgotten though, Zeus. He pricked up his ears and whimpered at her, looking at her through his beautiful dark eyes. Constable Dudley was going to take care of the long job of seeking other suspects, so it only seemed fair that Lily should take care of Zeus. She didn't have the heart to leave him behind.

"Come on boy," she called out, slapping her thigh. Zeus leaped up and ran to her side, circling her legs. Lily couldn't help but smile and hoped that she would be able to bring justice for the dog's master. Before she left she checked around to see if Pat had any treats left for Zeus. There was a packet on the coffee table, next to the crossword book. She picked up the packet and held one in her hand. Zeus' warm tongue ran across her palm as he gobbled up the treat, and he was nudging her hand and licking her to beg for more, but as Lily's gaze had idly drifted across the crossword book there was something that had caught her attention. When she had glanced at it before she thought Pat had been in the middle of solving it, but actually what he had written didn't make sense with the clues. There were two words that stood out to her, 'The Butcher,' and she wondered if that might have had something to do with the crime. She filed it away for later, for nothing could be ignored where murder, or in this case attempted murder, was concerned.

4

"Ah Lily, how are you this morning? I'm surprised to see you today, it's not Sunday yet after all," the vicar called down from a ladder. He was right at the top, standing there with a bucket.

"What are you doing up there?" Lily asked.

Vicar Donald Brown sighed and shook his head. "It's the storm you see, whenever it happens the wind blows a fierce gale and the drains get blocked. If I don't see to them now we'd all be like Noah by the time Sunday comes and I wouldn't want that. It's nice of you to come by and check on the church though," he said.

"Yes… well… that's not exactly why I'm here," she said.

"Oh no?" The vicar called out, getting his gloved hands mucky as he plunged them into the blocked drains. Lily winced as he pulled out a mass of wet leaves and mud. It was slimy and gross, and the damp smell was strong enough to reach her, all pungent and musty. "Well I'm happy to help with anything you need, although if it's a spiritual matter then perhaps you'd like to wait inside. I'll be down shortly," he said.

"Actually I'm not sure it's something that can really wait," Lily said. "It's about Pat. Someone tried to kill him."

As soon as Lily spoke the words the Vicar almost lost his balance. The ladder wobbled, but he managed to steady himself, although his face was pale.

"Dead? But… but that's not possible," he said.

"No, vicar, he's still holding on but in critical condition," Lily said, and then spoke the words that she hated to speak. "And I know that you went to see him last night."

The expression of the vicar changed when he realized what she was implying. His mouth grew very small and he nodded. He pulled off his gloves finger by finger and then left them in the bucket as he descended the ladder. He looked shaken, and Lily hoped this was due to shock, not to guilt or the realization that Pat was still breathing.

"I always find that a cup of tea can help even in the most trying circumstance. I suppose you must find that quintessentially English of me," Vicar Brown said gently as he brought a cup of tea to Lily as well. She noticed how his hands trembled. He sipped his drink moments after he sat down and she watched him patiently, while Zeus sat by her feet. "I suppose I should have known that something was amiss when I saw you with Zeus. Pat would never have let anyone else take care of his dog."

"I'm sorry if I startled you with the news, but I don't think there's ever a good way to tell anyone."

"No, I suppose there isn't."

"Did he say anything to you last night about being in danger? Was he worried? What did you go and see him about?" Lily asked.

"He didn't mention anything. I went to see him because I wanted to reach out to him and make him know that I was more than willing to speak to him if he wanted to. It does pain me to know that someone is struggling so badly with things. I felt that the warmth of God's love would be able to give Pat some respite and I thought if he wasn't going to come to me then the least I could do was go to him."

"And did he appreciate the visit?"

"Does Pat ever appreciate any visits?" the Vicar said with a smile, although it quickly faded when he realized that he was potentially the prime suspect in an attempted murder case.

"By the way, I did tell him what you asked me; I made it clear that you were open to him coming back to church, but he was adamant that he had no business being there. Do you mind me asking why he had such a drastic reaction to the church?"

Vicar Brown sighed before he spoke. "I suppose there's no harm telling you now that he's in such dire straits, especially if it might help find his nemesis. I did try to protect his privacy. It was the least I could do really, and I'm not sure that it's going to help you figure out who killed him."

"It might, we never know what might be important."

The Vicar nodded. "Well, it's all to do with his sister you see. Pat moved away from the village for a number of years and only returned when his sister fell ill."

"What was she ill from?"

"The kind of thing that people don't recover from," the Vicar said with sadness in his eyes. "She died and I don't think Pat has ever forgiven the world for taking her away. He never really shared much with me, but last night I tried to tell him that God was listening, but Pat wasn't having it. He said that if God was the kind of being who would give his sister so much pain, while leaving someone like Pat unharmed then he didn't want anything to do with religion. I could tell that I wasn't going to get through to him, despite knowing that I could have helped Pat if he had just opened his heart to the love of God. I suppose he just wasn't ready to forgive God for taking his sister away."

"And after his sister died Pat moved back here?" Lily asked, to which the Vicar nodded. "And do you know what he did while he was away from the village?"

"I'm afraid I don't. Pat was never forthcoming with personal details about his life," Vicar Brown said. "Whatever secrets he has, there hidden with him in that hospital bed. Lord, have mercy on the man."

Vicar Brown leaned forward and placed his head in his hands, rubbing his temples.

"What was he like when you left?"

"He was his usual self. He told me to never come to his house again. I could sense a storm was coming in so I wanted to make it back to the vicarage before I got caught in the rain."

"And you didn't see anyone else around the farm when you left?"

The vicar thought for a moment and then shook his head. "I'm sorry that I can't be more help. Perhaps I should have been more insistent. If I had stayed then maybe he would still be arguing with anyone who crosses his path. Maybe I could have helped fight off whoever attacked him," he said.

"You shouldn't blame yourself," Lily said, thinking that it was more likely Vicar Brown would have been beaten up or shot along with Pat. She bid the vicar farewell, a little annoyed that speaking with him hadn't shed any light on who might have committed the crime. It did help explain something about Pat though, and why he was always so miserable. Grief affected people in different ways. For some people it made them throw themselves into the world and form connections with all kinds of different people, while for others it left them isolated and solitary, forever scared to make that kind of connection again.

Lily visited the butcher on a whim, just in case the word Pat left in the crossword had been a warning. The butcher's shop was located at the end of a line of shops, with a row of joints and sausages displayed in the window. The smell in the shop was hearty and rich, and a man stood behind the counter with dark stains on his white apron. In his hand he clutched a sharp knife. He slammed it down again and again as he cut the meat, a fine display of savagery that gave her chills. It seemed to Lily that if this man was going to kill anyone it would have been with a knife, but perhaps he wanted to throw people off the scent. She looked at him warily, while Zeus came alive at the smell of all the meat.

The butcher, who was called Adam Locke, heard Zeus before he looked up. "Alright Pat? I'll be with you in a moment, I'm just putting together this order for Mrs. Hughes," he said.

"Actually Pat isn't here," Lily said, and greeted him with a disarming smile. Adam was a youngish man with a moon face and strawberry blonde hair.

"Oh, morning Miss McGee, how are you doing today? Where's Pat?"

"Actually I was going to ask you the same thing. I wondered if you'd had any recent dealings with him. I know he can often get under people's skin..." Lily said, hoping to lead the butcher into a trap.

Instead, Adam laughed and shook his head. "I know he's like that with a lot of people, but I've never had a problem with him. In fact he's one of my best customers. He always knows what he wants and he pays his bills on time."

Zeus left Lily's side and bounded up to the counter, placing his paws on the glass and panting heavily. "Yes, yes, don't worry Zeus I have the usual treat for you," he said, and turned away to fetch a bit of unused meat. He tossed it in the air. Zeus leaped up and caught it, gobbling it down in a moment of animal majesty. Adam chuckled to himself.

"Is something wrong with him? I've never known Pat to let anyone else walk Zeus before. I hope he's not ill, although I'm sure the storm would have gotten to many people."

"Actually it's not the storm that got him, and he's not ill," Lily said, approaching the counter and lowering her voice. "Pat's in hospital, he was shot last night by an unknown assailant."

Adam stopped what he was doing and looked at her in utter shock. "What? But... but I just saw him the other day. What... what happened?"

"Someone broke into his home and tried to kill him," she said, not willing to share too many details in case Adam was about to give himself away, besides, she knew how quickly gossip could travel around a village as small as Didlington St. Wilfrid.

Adam dropped his knife to the counter and hung his head. "That's... that's terrible," he said. Lily watched him closely and saw that he was blinking back tears. A wave of emotion swam over him and was draped over him like a cloak, and her instincts told her that he wasn't the man responsible for Pat's gruesome situation.

But if not him then who was, and why had Pat made a note of 'The Butcher'? Whoever had visited Pat's farm must have done so under the cover of night. For someone so concerned with security like Pat, Lily assumed that he might have had other methods to protect himself and his home, so she decided to return to his house and see if she could find any other clue regarding who might have wanted Pat dead.

5

When Lily walked into the farm house she was glad to see the Constable Dudley had clearly been on the scene making sure everything was properly logged and tagged, although Zeus was a little puzzled. He pawed at the area where Pat had been laying, and seemed to be looking for his master. Lily searched the cupboards and found his food, doling out a portion for Zeus into his bowl. The dog's nose led him to the bowl and he wolfed down his food. At least that would distract him for a short time, Lily thought. She went to the coffee table to double check the crossword and again there was no mistaking it; the words read 'The Butcher', and Pat didn't seem the type of man to write something like that at random. She rummaged around a few cupboards, but didn't see anything out of the ordinary. It reached a point where she was beginning to suspect that she wouldn't be able to get to the bottom of this at all. Pat's secrecy was harming her chances of discovering the truth, and she found herself getting annoyed with him again even though he was lying close to death just a few miles away.

After Zeus finished eating he joined Lily and nudged her legs, moving past her to a part of the wall. He scratched and scratched and Lily was puzzled, for it wasn't as though he could get through the wall to the outside.

"I'll take you for a walk in a minute Zeus, I promise, I'm just trying to figure out what happened to your master. If only you could talk, perhaps you could tell me something I need to know," she said, turning her back on him to go and look at something else. It was then that Zeus began to bark and the scratching intensified to the point where Lily couldn't ignore it, and then she wondered if perhaps Zeus was trying to tell her something.

She inspected the wall more closely and saw that there was a small button, which opened a secret compartment. She smirked and patted Zeus on the head, praising him for helping her. Then, she pulled out boxes and boxes of documents and leafed through them all, amazed by what she found. Her eyes were agape and she could barely believe what she was seeing, but it was all there in black and white. She had finally uncovered the mystery of Pat, and it was stranger than she ever would have believed.

But if this would lead to her finding out who tried to harm him then she was going to need help, and there was only one man who she could possibly think might have information pertaining to this part of Pat's life. She gathered a few of the documents she wanted to show, and then placed the others back in the secret compartment, swallowing a lump in her throat as she realized that killing Pat might not have been the end goal of the person who invaded his home. They might have wanted information, and if the information was hidden they might still be watching his home carefully. As Lily left she felt the fine hairs on the back of her neck stand on end, hating the unsettling feeling that maybe she was being watched.

Lily breathed a sigh of relief when she reached Blake's home. When he opened the door she burst through and pressed herself against the wall, her eyes darting in all directions.

"It's good to see you too Lily. What's going on?" Blake asked, peering out of the door to see if anyone was chasing her. Lily had run from Pat's house and worked up a bit of a sweat, as well as panting heavily. She got Blake to close the door and then dragged him to an office where they could be alone. As soon as she did so she revealed the documents. Blake was busy petting Zeus, and there was a question on his lips when he looked at her, although Lily was too busy with words of her own.

"I need your help Blake. Pat was left for dead in his own home and I just discovered that he used to be a spy, or perhaps he still is. He worked for the government. I found these documents and more in his home. They were hidden. I'm worried that someone from his past is back to kill him, or maybe they're looking for something and they're going to try again. I thought that you might know someone who knows about this sort of thing." The words fell from her mouth in a torrent and Blake took a few moments to process everything that was happening. He blinked slowly and then arched his eyebrows.

"Wait, Pat's dead?" he asked.

Lily nodded. "Not quite, he's in terrible way in the hospital with gunshot wounds, and he was a spy!"

"I suppose that explains why he led such a secretive life, and why he didn't like outsiders coming into the village," Blake said after a sharp intake of breath.

"He might have suspected them of coming after him for whatever he did when he was a spy. There has to be something in those documents that gives us a clue as to who wanted him dead, but maybe you know someone who actually worked with Pat or could have an interest in what happened?"

Blake stroked his chin. "I probably can get word to someone, although they're very busy and I'm not sure when I would hear back from them. Let me make a few phone calls though and I'll see what I can do."

Lily was relieved at this, but she didn't much like the idea of a spy operating in Didlington St. Wilfrid, and she didn't want to wait too long in case they struck again. There was an attempted murder to solve after all, and if she didn't strike while the iron was hot then she might never get to the bottom of it. For all the differences she had with Pat he still deserved justice, and it didn't seem right that he should be lying in a hospital bed fighting for his life without the truth of how he got there being revealed.

Then she thought about someone who might be able to shed light on the situation, if the storm hadn't taken its toll. She bid a quick farewell to Blake, leaving the documents in his care, and she and Zeus strode the village again, this time back to her home, Starling Cottage.

~

"Joanie, please tell me that your cameras still work," Lily said as she burst through the door.

Joanie was sitting with her feet up, reading a magazine. "I'm just waiting for the photos to develop. The equipment seems fine, how are you getting on? Have you caught Pat's attacker yet?"

"No, and I think I'm going to need your help to do so," Lily said, and explained everything she had learned so far. Joanie listened with wide eyes and shook her head, while Zeus made himself familiar with the cottage. He nosed and nuzzled his way into every nook and cranny, and at least he was keeping himself busy after enduring such turmoil.

"If you keep this up you're going to have to retire as a writer and become a full time detective," Joanie said.

"I hardly think so. It's not as though I seek these things out. It's all just been a matter of circumstance so far. I moved out here to have a quiet life, remember."

"And how is that working out for you?" Joanie asked. Lily merely gave her a cold look and then went to fix herself some food. Zeus followed her all through the house, never leaving her side.

She waited hours while Joanie set up a makeshift dark room, in the basement, to develop her film. It was agonizing and she paced around and around, wondering whether the pictures were going to be developed first, or if Blake was going to hear back from his contact in the government first. She wracked her brain, trying to think how a spy would slip into the village and slip out again without being seen. The storm would have provided good cover, and the rumble of thunder would have covered up the sound of the gunshot. Whoever committed this ghastly crime knew what they were doing, and it remained to be seen whether they had finished their task with Pat, or if this was just the beginning.

Eventually when twilight set in the photos were beginning to take form. Lily and Joanie waited patiently for the images to reveal themselves. The first lot were amazing photos that captured the beauty of the land and the raw power of the storm, but there was nothing that pointed to who snuck into Pat's farmhouse. But then Joanie developed a few images that showed a shadowy figure walking out in the slashing rain. It was difficult to tell who it was at first, but then in one image a flash of lightning illuminated the figure, and there was no doubt who it was. Lily and Joanie both glanced at each other.

"I need to get Constable Dudley," she said.

6

Lily and Zeus raced through the village to the police station, which was a small, historic building and had housed the police force since the village had been built. It was manned by Constable Dudley usually, but he wasn't anywhere to be seen. Lily pressed her lips together and looked around, wondering where he had gone. Someone popped their head out of a nearby window and asked if she was looking for the Constable, to which Lily nodded.

"He thought you might be coming along. He said that he's gone to ask the neighbor if she saw anything, if that means anything to you," they said. Lily thanked them and then doubled her pace, running as quickly as she could to get to the cottage before it was too late. She ran up the winding road with Zeus by her side and when she reached the cottage she burst through the door, worried that she was going to stumble on another grisly scene. Instead, she found the constable sitting at a table with Emma sharing some tea and biscuits.

"What is the meaning of this?" the Constable asked, rising from his chair.

"It's okay Bob," Emma said calmly. "I'm sure that there's nothing amiss. It's Lily, isn't it? Marjorie said that I should seek you out as a fellow newcomer to the village, and Bob here has spoken so kindly of you. You must have so many stories to tell, please, join us for some tea," she said, but Lily stared daggers at her.

"Don't say another word. Constable Dudley, you need to arrest this woman," Lily said. Zeus strode into the room as well and growled, the hackles on his back rising, his teeth bared.

"Lily, I know that a murderous crime has happened, but there's no need for this," he said.

"A murder!" Emma exclaimed, looking shocked.

"An attempted murder," corrected Constable Dudley.

"And you are the woman who attempted it." Lily said.

Constable Dudley looked as outraged at the thought as Emma herself. "I don't know what you're talking about," Emma said.

"I have evidence that you went to Pat's farm last night during the storm," Lily said in an accusing tone. Emma didn't miss a beat.

"I must confess that I find myself troubled by storms. I hate being alone in them and the fear of being in solitude was greater than being swept away by the rain. I saw lights on in my neighbor's house and thought he might like some company, that's all," she said.

"And what happened when you asked him?"

"Well, he turned me away. I thought he might take some pity on a young lady out by herself in the middle of the storm, but he practically slammed the door in my face. I returned here," she said.

"Well, that seems to fit with what we know about Pat. Perhaps whoever shot him came in after that," Constable Dudley said, but Lily wasn't finished.

"It's a fine story, but I know it's a lie. You didn't seem scared at all. In fact you seemed to be on a mission, which is why you had this pistol," Lily said, and presented a photograph that showed the gleam of a metal gun in Emma's hand. She showed the picture to the Constable and his expression changed. There was no denying the nature of the image. Emma didn't seem fazed at all. She took a bit of a biscuit and then sipped her tea as though nothing in the world had changed at all.

"This is rather damning. I'm afraid I'm going to have to ask you to come to the station with me so that we can ask you some more questions," Constable Dudley said.

"Of course, of course. I'm sure you have a job to do and I promise not to hold it against you," Emma said. Then she turned her gaze to Lily and as the icy stare bored into her, Lily felt a chill run down her spine. She didn't have to hear Emma tell her that she was going to hold this against her, for Lily already knew it to be true.

∼

"How can you deny this picture exists?" Lily asked with an aggravated sigh. She and Constable Dudley had been questioning Emma for what seemed like hours, and Emma wasn't giving them anything.

"I do not deny the picture exists," Emma said calmly, "I merely deny that I shot the man."

"But this shows you holding a gun!"

"It shows me holding something that looks like a gun. And even if it was a gun that does not mean I used it on him."

"So you're saying that there's a third person involved?" Lily asked.

Emma shrugged, "I am merely pointing out possibilities. It is you who has already concocted a fantasy where I tried to kill a man. I am a patient woman, but even my patience has its limits. Unless you have clear evidence that I was in his home and interacted with the man, I do not see how you can keep me here. I'm sure you don't mean anything by it Constable, you have been fair to me and I bear no ill will against you, but I find this treatment highly insulting."

Lily was more convinced than ever that this woman was responsible, but she wasn't being forthcoming with a confession and Lily wasn't sure how she was going to get one out of her. Emma was calm and collected. Usually the suspect was the one getting flustered, but this time it was Lily. She wasn't used to being in this position and a band of tension ran around her head.

While Lily was trying to figure out a different approach there was a noise outside as a car pulled up. Constable Dudley looked at her in confusion, but Lily's heart was lifted when she saw Blake enter the station flanked by another man in a suit. This man was older than him with a wide mustache and a thin frame. Lily should have been concerned by the fact that her smug smile was reflected by Emma.

When Blake entered, Lily assumed that light was going to be shed on the case and this woman was going to finally face justice.

"Constable Dudley, Miss McGee," the stranger said, nodding to each of them in turn. Lily noticed that Blake wasn't looking at her. In fact he had a kind of sick look on his face. "My name is Lord Avery Fauntleroy and I thank you for taking care of this matter, but I am now assuming responsibility. This woman has been remanded into my custody so you do not have to trouble yourself with it any longer," he said in a clipped tone. Emma rose from the other side of the table, but Lily was incensed.

"Wait, why are you taking her away? She tried to kill our friend, and upstanding citizen of Didlington St. Wilfrid. I know she did it. You can't just take her away like this. Where are you going? Don't we get a chance to hear her confession?"

"Unfortunately I am not at liberty to share that information. Needless to say she will be dealt with as we deem fit. Again, I thank you for your endeavor and I hope that life can return to normal for you," he said. With that he was gone. He led Emma out to the car and opened the door for her, and then it sped away, leaving the village behind. Lily stood there dumbfounded, wondering what on earth had just happened.

"I think we need to have a conversation," Blake said.

"What in the world was that about?" Lily asked. Her skin was burning and her fists were clenched into tight balls by her side. Anger poured from her eyes, and Zeus was mirroring her mood. He paced around, still with his teeth bared, ready to attack anything he thought was a threat.

"That was Lord Fauntleroy. He responded to my enquiries about Pat. They were very interested in learning what happened to him."

"So Pat was a spy then?"

"Yes, and quite a high ranking one from what I understand. I believe that a number of world governments wanted him dead, and unfortunately one of them found him and tried to do just that."

"So Emma is a spy too?"

Blake nodded his head slowly. Avery was kind enough to share a few details with me. Her name is Emilia Ivansky, also known as "The Butcher," he said, and Lily realized that Pat had indeed tried to leave a warning, although she would have had no idea what that warning meant. "She's a deadly assassin and is a highly prized target for our government's espionage services."

"Well, at least she'll see justice for all her crimes, including this one," Lily said, only relaxing for a moment because the uneasy look on Blake's face hadn't changed. "What other bad news is there?" Lily said.

Blake ran a hand along his jaw. "The government is going to use her to cut a deal so that they can regain some of their captured operatives."

Lily felt as though she had been hit by a hammer. "So she's going to get to go back to wherever she came from and she's not going to be prosecuted with attempting to murder a British citizen?"

"I'm afraid not. Our government believes the lives of our operatives are more important than justice in this instance."

"Well I don't agree."

"I can't say that I do either," Blake agreed, "but unfortunately we're not a position to make that decision.

"So nobody cares about Pat? Nobody cares about what happened here?" she said, shaking her head in disbelief.

"That's not strictly true. We care. I know Pat wasn't the most well liked man in the village, but everyone here would never wish him dead. We're all rooting for his recovery and from what I hear, he's fighting in true Pat-like fashion. He's not out of the woods yet, but by God's grace and the skill of his carers we should soon be seeing Pat harassing the locals back here very soon. I understand that this isn't how you wanted this to turn out with Emilia, but I can imagine her superiors aren't going to be happy about the fact that she got caught, and it's not just anyone who could catch a spy. I think in this situation it's best to focus on what you can control, otherwise it's easy to feel as though life is spiraling out of your grasp," he said, although his words didn't feel comforting at the moment. There was so much rage and anger building up inside Lily's heart. All she wanted to do was scream, not that it would make any difference either.

Then she felt Zeus pressing against her legs and licking her hand. With his master gone for now he was going to need someone to look after him, and it seemed as though he had picked Lily for the task. She would certainly be fond of his company and had no qualms with taking him into her home, knowing that Pat would hopefully be back on his feet. All of this just went to show that there is always more to people than is apparent to the eye. Pat it seems is a man who defended his country and grieved his sister, and perhaps one reason why he had pushed people away was so they would not be tainted by the secrets he held.

She reached down to stroke Zeus' head and tried to make sense of it all. It seemed so wrong that Emma, or whatever her name was, should end up being traded as a political pawn when she had tried to kill a man in cold blood. It made Lily wonder if this was the same kind of difficulty that Pat had to deal with when trying to marry his faith with the fact that his sister had died. There were going to be long nights ahead for Lily as she tried to come to terms with this chain of events. She bid farewell to Blake and went back to Starling Cottage, shutting herself in her study where she could be alone with her thoughts and her stories.

There was some solace to be had as she penned another mystery story, because at least in her world she could ensure that villains received the justice they so dearly deserved. A bright sky hung over Didlington St. Wilfrid, but clouds hung over Lily's heart as she thought about a man only days before she had thought nothing more than a grumpy and unthankful curmudgeon. Truth really is stranger than fiction she thought, as she tapped out another line in her latest mystery.

A BONY BURIED SECRET IN
A QUIET ENGLISH VILLAGE

1

Lily awoke to a sharp yapping sound, a sound that she was still getting used to. Suddenly she felt an impact on the bed as Zeus jumped up. He pounced on her and his warm, wet tongue licked her face. She squealed at this show of adoration from the adorable dog, who was panting excitedly, eager to begin the day. His dark fur was soft and her hands wrapped around the animal, feeling the rapid beat of his heart and looking into his deep, chocolate brown eyes. Although she wasn't sure it was physically possible, Lily was sure that Zeus was smiling. He tilted his head toward her and then whimpered a little, nuzzling into her and placing his paws on her chest. The weight of the dog was not inconsiderable, but it was the comfortable kind of weight that made her feel a part of the world, that gave her something to embrace, and she realized it was the kind of weight she had been missing.

While her time in Didlington St. Wilfrid had been wonderful she was beginning to feel a little lonely. Of course Joanie lived with her to offer her company, but it wasn't quite the same as having that special person in the world with whom she could share all of her innermost secrets and the blessings of love. A cloud threatened to form in her mind as she remembered the last time she had been close to anyone, and she shuddered, not wanting to repeat the mistakes she had already made. Sometimes it seemed far easier to be alone than allow someone else into her life, and sometimes it was easier to welcome a dog into her heart than a man.

"Good morning to you too Zeus. Are you feeling well today?" she asked, scratching the back of his neck and under his chin. The dog moved his head around, following the motion of her hand, and seemed to be quite relaxed. "You want a walk, don't you?"

Zeus snuffled and then let out a loud bark, which she took for a 'yes'.

It was difficult to not feel pity for Zeus after all that had happened. Lily rose from bed and yawned, stretching out her arms. She tied her hair into a ponytail and pulled on some comfortable clothes that would protect her from the chill of the early morning. Zeus bounded around her feet, almost tripping her up as she brewed herself a hot drink and had some toast.

"I know, I know, don't worry we'll be going out in a moment," she said, for Zeus kept moving between her and the door, almost going crazy with a desire to leave for his morning walk. Usually it would have been Pat doing this, dear Pat who was filled with complications. When she had first arrived in the village, Lily had thought him an intractable man with nothing but hate in his heart, but she had learned there was far more to him than that.

Before he returned to Didlington St. Wilfrid he had been a spy for the government, and Lily found his house to be full of secrets. Apparently his reputation was notable. A Russian spy, known as the butcher, had come to the village and attacked him, intending to kill. By some miracle and the grace of God Pat had survived and was now recovering in hospital. It would be a long road until he was healthy again, and until that time Lily had decided to take over the care of Zeus. It had always puzzled her how such a brash and rude man could have such a friendly dog, as usually pets reflected the nature of their owners. She chastised herself for not seeing the obvious truth; that Pat's dark mood was an act he indulged in to keep people away, thus keeping them and himself safe.

She wondered if he would be the same Pat as before, or if something in him would change. Before this he had nothing but loathing in his heart for God because of the tragedy he had suffered, but now he had been saved. Would this make any difference? Would it bring him back into the light?

Lily certainly hoped so, because it didn't seem right that a man like Pat should settle for being lonely. He had given so much to this country and clearly there was a part of him that cared for other people, and she hoped that she would see this side more. But, for now, she had to make sure that Zeus was looked after in Pat's absence. It actually helped to take her mind off a few things as well, for she was still hurting after the resolution to the last mystery. The Butcher had been taken away, with a sneering smile on her face, by a government official who was going to use her to trade with the Russians to get one of their own spies back. Even now it made Lily grit her teeth and stirred anger within her heart for it seemed unfair that justice could not be served. That woman should have been punished for what she had done to Pat, but instead she was able to return to her home country and suffer no consequences at all.

Lily crunched on the toast and chewed angrily, before she grabbed the leash and fastened it around Zeus' neck. A walk would do the both of them good, she thought.

Lily emerged from Starling Cottage as dawn settled upon the land. The sky was a breathtaking mix of colors, and the plants around her glistened with morning dew. There was a freshness to the air, the sign of a new day. She breathed all the sweetness in and felt a wave of peace settle over her. Her heart was calmed and she could embrace the glory of God and the beauty of the world He had created, and for a moment at least she could be at peace.

"Come on then Zeus," she said, and walked along the path through the village.

As it turned out, it was more that Zeus walked her than she walked Zeus. The dog was a curious animal, always darting this way and that to poke his nose into a flower or sniff along a wall to investigate. In a way he reminded Lily of herself. When she was a child she had incessantly pestered her mother with questions about the world, questions that her mother was ill equipped to answer. She sighed a little as she thought about her family, and sorrow filled her heart.

One look at Zeus though and she was happy again. It was impossible to be anything but happy around the dog, for he was such a bright and positive presence and she loved the way he didn't seem fazed by anything. He even found a stray butterfly and tried to chase it. The butterfly fluttered away and out of Zeus' reach, but he continued chasing it, looking at it until it rose high enough to escape his vision. Zeus stared into space momentarily, confused about how it could have disappeared, before going on to find something else he was interested in.

Zeus took her through the winding path of the village and up toward the fields. Some of these were private property, but not everyone was like Pat and paranoid about others stepping onto their land. People in Didlington St. Wilfrid were mostly open and had a grand sense of community, so she wasn't worried about anyone turning away. A lot of the land was owned by Lord Huntingdon as well, and he had opened them to the public. She gazed in the distance at his manor and wondered if she might see Blake on her early morning walk. She supposed she would probably have to wait until the next church service for that, and her lips curled into a smile at the thought.

She reached an open field and since there were no roads or other people around, she decided to let Zeus off the leash. She bent down and cupped his head in her hands. "Okay Zeus, you have to behave for me, okay? No running off, and when I call you I expect you to come back," she said in a firm tone. Zeus stared back at her silently, and she wished she could know what was going on in the dog's mind.

She unclasped the leash and immediately Zeus expressed how much he enjoyed the freedom by bounding around and yapping loudly. His tail wagged so fast that it was almost a blur and he ran away as Lily strode quickly to keep up. He was a small black mass, like a tiny storm cloud that raced across the field with absolutely nothing keeping him tethered or slowing him down. He careened around, almost seeming like he would be a danger to himself. Trees dotted the side of the field, although a low wall opened the field up to the road beside it, which was wide enough for cars to pass. Of course, in Didlington St. Wilfrid traffic was a rare sight. It had taken her quite a while to get used to the lack of sound around her, especially at night, for L.A. had been a hubbub of noise. Here there was time to think and time to ponder, but that wasn't always necessarily a blessing.

"Go careful Zeus," she cried out, wondering if he even heard her or took notice of her words. He ran along the field and Lily wondered if he liked exploring somewhere new. Pat always had such a regimented way about him and he only ever took the same path around the village. Of course now that she knew what she knew about him she could understand why he might want to be sure of his surroundings, but it must have been difficult for Zeus to not have the entire village open to him. He seemed to be making up for lost time though as he was scurrying around, his paws thundering against the ground as he ran in circles and played with himself. Lily moved toward the trees in the hope of finding a stick that she could fling around, but when she looked up she was surprised and horrified to find that she had lost sight of Zeus.

He was no longer a blur in motion, but as she scanned the field she breathed a sigh of relief as her gaze settled upon him. Pat would never forgive her if something happened to his beloved dog, and Lily would never forgive herself either. Zeus seemed to be preoccupied with something. He had his head close to the ground and his paws were digging at the ground as frantically as a rabbit. Lily worried for a moment that he had found some small animal to hunt, so she abandoned her search for a stick and walked toward Zeus.

He was making firm progress. Dirt was being flung behind him as he descended deeper and deeper into the ground. Lily realized she must have been lost in thought for longer than she assumed given how low the dog had become.

"Zeus! Zeus what are you doing?" she cried out, but the dog was so preoccupied with what he was doing that he didn't listen. He just kept digging.

Just as she was approaching Zeus, and hoping that nobody would be angry with her for allowing Zeus to dig this massive hole, he emerged and turned around, holding something in his mouth. His paws padded along the ground as he approached Lily and dropped it at her feet, looking utterly proud of what he had accomplished. His pink tongue lolled out of his mouth and he was waiting for praise, although Lily was far too preoccupied with what Zeus had just dropped to remember to praise him.

She fell to her knees and a lump formed in her throat. He had brought her a bone, and Lily had seen enough medical dramas to know that this was not the bone of an animal. It was long and thick and it caused an unsettling chill to run down her spine. She was glad it was morning and not the depths of night, for this was something gruesome and horrible that had the stink of death about it.

"What have you found here Zeus?" she whispered, daring to reach out to touch it, almost wishing that it would be some illusion, but no, her fingers ran against the bone and her stomach churned. Her gaze turned toward the hole. A big part of her didn't want to look, but she knew she had to. She summoned her courage and kept one hand on Zeus, which gave her an extra boost of strength. She peered into the hole and then let out a cry as she saw what she had dreaded to see. There was a plastic bag that had been ripped, and from which Zeus had presumably taken the bone. Staring back at her were the hollow eyes of a skull among an assortment of bones. Lily pulled Zeus closer to her, as though he could ward off the evil spirits that lurked around this place.

2

Lily was sitting in the police station with her hands wrapped around a cup of tea, although the warmth of the drink didn't help to soothe the uneasy feeling in the pit of her stomach. Zeus lay at her feet, having no idea of the importance of what he had discovered. Her face was pale and she couldn't stop thinking about the fact that the bones had belonged to a body, a body that had been a person, a person who had had a life. So many questions ran through her mind and none of them had easy answers, and none of the answers were particularly good ones either. They all led down dark paths, and it was clear to Lily that she was becoming embroiled in another mystery.

The door to the police station opened and Constable Dudley appeared. He poured himself a cup of tea and then mopped his brow as he sat at the desk.

"I certainly didn't think the day was going to begin like this," he said.

"Neither did I. Do you have any idea who the bones belonged to?" Lily asked.

"I'm afraid not. All I know is that they must have been a fairly recent killing because of the bag they were stored in. I'll have to send them away for forensic analysis though to determine the cause of death. I did see a nasty crack in the skull, but I don't know if that happened before or after the body was placed in the ground." He sighed again, pinching the bridge of his nose.

"So you really don't have any idea who this could be? I mean, isn't there anyone who has suddenly gone missing from the village?"

"It's funny you should say that." Constable Dudley rose from his desk and went to a filing cabinet. He pulled out a drawer and rifled through a number of files before pulling out a selection of them.

"Here are some missing persons," he said, spreading them out before Lily. Her eyes widened at the number of them. She opened them and looked at the pictures. Most of them were young.

"Who are these people?" she asked, her words little more than a gasp.

"They're people who have been reported as missing, people we've never been able to find."

"Haven't you been worried that they've been killed?"

The Constable had a pained look on his face. "It was determined by my predecessor that most of these people simply ran away from home in search of a more exciting life. I'm sure it hasn't escaped your attention that Didlington St. Wilfrid is not the most lively place in the world, and while the people who live here enjoy the peace and quiet that can't always be said for the children who are born here. A lot of them see this place as a prison and want to explore the wider world, but feel they're kept here by tradition and ties to their parents. Unfortunately it's fairly common for families to have some… misunderstandings shall we say, and for the children to leave as soon as they come of age. Frankly it's always made more sense to me than them being killed, but now… well…" he rubbed the back of his neck and looked concerned, as well he should have been. There could have been a murderer living in the village who had gotten away with the crime for far too long.

"Well we need to get to the bottom of this Constable. We need to find out who this person was and then we need to find out how they died."

"I know," he sighed. "I suppose we'll have to start by talking to the families of these people and see if they've heard from any of them, at least then we can rule out people who are still alive. It's not going to be easy reminding them of their past though."

"No, it's not, but it's something that just has to be done. It'll help if we split the workload," she said, taking half of the folders. Constable Dudley smirked.

"Are you sure you want to help with this? You've already done so much for the village Lily. I really can't ask you to help again."

Lily offered him a faint smile. "It actually helps to keep me distracted. With recent events I find my thoughts turning to things that I would rather not think about. At least this will keep me busy," she replied, and given how valuable she had proven her skills to be already the good Constable was not going to refuse her offer.

Lily spent the morning going around to a few houses with Zeus. Her reputation around the village preceded her somewhat and her exploits had taken on a life of their own, so people were pleased to see her, at first. However, when she explained why she was there their faces fell and the past came back to haunt them.

"I'm sorry for this," she said to all of them, "it's just that some bones were found this morning and we're trying to understand who has died."

Lily couldn't even say for certain that it was murder, although it certainly seemed like it given the evidence at hand. It was hardly a respectful burial and seemed more designed to hide the body than anything else. Her instinct told her that something was afoot here, although she tried to be as positive as possible with the families of the possible victims, as she didn't want to declare anything to be true when she couldn't be certain.

There were a couple of families who smiled with relief when she told them that the bones had been buried for a while, as it ruled out their children, for they had reconnected with them over the years and had spoken to them recently. Lily was glad to hear this, although she felt a pang in her heart. How long had it been now since she had spoken to her own mother? Some would say too long, while others would say it hadn't been enough time.

Still, she pushed that thought from her mind as she focused on the task at hand. It was pleasing to know that some of these families had managed to forgive their children for moving away and leaving the village behind, but unfortunately not all of them could say the same thing. Some of them had not heard from their children ever since they left. Tears glistened in their eyes and Lily felt absolutely horrible for bringing up the possibility that their child had actually died a number of years ago, without being able to offer any closure. Doors were closed to her and she left houses filled with sorrow, but the fire in her heart burned with a desire. This time she was not going to let the killer get away with it. This time she was going to see that justice was done.

It was nearing the middle of the day when she returned to the police office to speak with Constable Dudley. They shared their progress, and as it turned out his experience was much the same as hers. They were able to filter out about half of the cases as the potential victims were still alive, but that still left plenty in the way of possibilities, and few directions in which to turn.

"I have to admit Lily that I'm struggling to see where we go from here. There are still a few of them. How are we supposed to figure out which one is dead, even if it is any of these?" he asked.

"We'll find a way. I'm not letting anyone get away with this. Someone has died, and for all this time the person who has killed them has been able to live their life without suffering from the consequence of their action. They've taken something... taken the most valuable thing that exists in this world and they haven't felt any repercussions at all. I'm not going to allow this to happen again. I'm not going to let them get away with it," she snapped. Anger poured from her eyes and she could feel her blood boiling beneath the surface, as though she was a volcano about to erupt.

She and Constable Dudley were in deep discussions about how to proceed with the case when the door to the police station slammed behind them. Their eyes turned to see a woman standing there, looking about as angry as Lily felt. Her face was narrow and her eyes were sunken with sorrow. She had curled blonde hair, and striking features. Her feet thundered against the floor as she came into the room and stared directly at Constable Dudley.

"I've heard about what's happened. You found him, didn't you? You found Luke. I've waited so long for this moment. I told you that he had been killed, but you didn't listen to me, did you? It's all your fault this has happened! It's all your fault!" she cried, her voice cracking with the emotion of her words.

3

L ily stared in stunned silence at this outburst. The woman was trembling and her eyes were red, ringed with tears.

"What's going on? Who are you and who is Luke?" Lily asked.

The woman suddenly seemed to realize that there was someone other than Constable Dudley in the room, but it didn't soften her at all. She took a deep breath. Her lips became a thin line stretching across her face, but her eyes continued to exude hatred and rage.

"Luke was my boyfriend," she said, the words stuttering as they had to battle through raw emotion. "And when he disappeared I told the police that I thought he had been killed, but they fobbed me off and told me that he had run away. Well I knew he hadn't. I knew it! And all this time he's just been lying there rotting in the ground forgotten," she said, and utterly collapsed into tears. They streamed down her face and she looked as though she was going to crumble. Lily got up and pulled out a chair for her, helping her to sit down. Lily also poured her some water.

"What's your name? I'm Lily," she said.

"Lindsay," the woman replied.

"Okay Lindsay, look I don't know what happened before, but if this is Luke then we're going to find out what's happened to him. I'm sure that Constable Dudley wouldn't let anything like this happen."

Lindsay's voice hardened. "He would. They all would. None of them cared, they just brushed him away without looking into it properly. I told them all about what happened with Luke and they just ignored me. All these years I've been praying that I was wrong, and now I realize that I was right all along, but it doesn't help him, does it? I never even got to have a funeral for him. I never got the chance to say goodbye," she said, collapsing into tears again.

Lily gave Constable Dudley a questioning look while she placed her hand on Lindsay's back and tried to comfort her. So far Constable Dudley had proven himself to be a decent enough police officer. There were times when he could be a bit bumbling, and moments when he seemed overwhelmed with the viciousness of the crimes that were being committed in his small village, but she would never have believed he was capable of such heartless negligence. That was a conversation that would have to wait until later though, for there was much Lily wanted to learn about Lindsay and Luke.

"What exactly did you tell them? I know it didn't help then, and I'm not sure why, but I can promise you that I'm not going to rest until I find the person responsible. If it is Luke who has died then we will get justice for him," Lily said. This seemed to help Lindsay to relax. She sniffed back her tears and wiped her eyes. "Tell me what you think happened to Luke and how you knew him."

"We were always close. We were always good friends, and as we grew older that friendship turned into something more," she said. The words were halting as Lindsay was still clearly choked up with emotion. It must have been difficult for her, Lily thought, to suddenly have these intense feelings burst forth again even though they should have been left in the past. "We kept talking about the future and how we were going to leave the village together and

168

make something of ourselves. It's not that this place isn't nice, it's just that well… you know, we were young and we wanted to be special. We used to sit under the stars and dream about what everything was going to be like. The more we talked about it the more real it became and instead of just pretending we actually began planning it. He came to me one night and said that he was ready to tell his parents that we were going to leave. He said that he didn't care where he ended up, as long as it was with me."

"And how did his parents feel about it?" Lily asked gently.

Lindsay's expression became hard again. "They weren't happy. Of course they weren't happy. Luke was their only son. He was the one who was going to inherit the business. They had it all planned out for him. Luke used to hate it, the way his life was already mapped out. He wanted to be a free spirit you know, but his parents kept telling him that he needed to stay and take over the business. He didn't even like it. There was nothing more boring to him, and he thought it was unfair. Sometimes he wondered if they had even wanted a child at all, or if they just wanted an heir to the business." She shook her head and wrung her hands together in her lap. "But that didn't stop him. He was willing to turn away from it all for me, but then he disappeared. We were supposed to leave together and *they* said he left alone," Lindsay glared at Constable Dudley as she said this. "But he would never leave without me. We had a promise to each other. We were going to have a life together and he never would have turned his back on that. So for them to say that he ran away is just… it doesn't make sense!"

"And you told this to the police at the time?"

Lindsay nodded. "I did, but they didn't want to know. I told them that I thought his parents wouldn't have been able to handle the fact that we were going to leave. His father had an awful temper… the only possibility I can think of is that Luke was killed by Allan. But they never listened. They said since there wasn't a body and I had no evidence there was no point in pursuing the case, and that Luke was probably alive and well, he had just left me without telling me.

But he wouldn't do that! He'd never leave without me!" she said, and cried again.

"Okay Lindsay, okay, look, right now we don't even know if this is Luke, but I'm going to follow this up and investigate it myself. We're going to find out the truth of this," Lily said.

Lindsay nodded, and then her gaze turned spiteful again. "I hope you do find the truth, because you have a lot to make up for. God has not been kind to me, and He's not going to be kind to you either," she spat at Constable Dudley, before rising from her seat and leaving.

~

Lily sat there in stunned silence, processing everything Lindsay had said. Constable Dudley had his head bowed and looked pensive.

"Is there any truth to what she said?" Lily asked, almost hating that she had to ask the question, but there wasn't much use trying to deny the matter.

Constable Dudley inhaled deeply, making his broad body seem even bigger than it was. "Unfortunately, yes," he said.

"What? How could you let this happen?"

"It was years ago, before I earned these," he said, pointing to the rank emblems on his shoulders. "Constable Vickers was in charge at the time. He was a hard man, a traditional man. We had a bit of a problem with the youth of the village who had their heads turned by the cities around us and all the excitement they could promise. He didn't think they should leave the village, and he didn't take kindly to people coming up to us and doing our jobs for us, at least that's how he saw it."

"And how did you see it?" Lily asked, unable to hide the tone of accusation from her voice. Perhaps it was unjust of her to act like this, and even Zeus pricked up his ears, as though he wasn't used to her speaking like this, but she had been put on edge by the possibility that justice had been evaded by a guilty party once again.

"I saw it as he was the Constable and I was below him, so I had to follow his lead," Constable Dudley said tersely. "He told me that if we chased after every missing person or every broken heart then we'd never do any actual policing, so as far as I was concerned the matter was closed." He paused for a moment before he spoke again. "If I was the man then that I am now I would have stood up to him and told him that we had to pursue and investigate to the very end, but I was young and he was intimidating. I didn't think I had the right to question him, and I don't think I would have made much of a difference anyway."

"But do you recall much of the case?"

Constable Dudley shook his head. "She would have spoken to Constable Vickers directly."

"Then I think we should have a talk with this Constable, although after we have a word with Luke's parents," she said. "Did they say anything when you went to see them this morning?"

Constable Dudley shook his head. "No, they just looked sad like all the others."

4

Lily and Constable Dudley, accompanied by Zeus, walked through the village and came to the house where Luke's parents, Allan and Barbara lived. She cast her gaze past the house and noticed that the field where Zeus had found the body was not too far. Granted, the village wasn't huge and there were other houses in close proximity as well, but it struck her as tragic that their son's bones had been so close to their house without their knowledge. Or had there been some knowledge?

The door opened and Allan stood there, tall and intimidating with his glare and his dark hair, although the black was flecked with grey. His eyes narrowed when he saw the Constable and Lily at his door. Lily tried to have an open mind, tried to remember that Lindsay wasn't necessarily telling the truth, but Lily couldn't help thinking; was this man a murderer?

"What do you want? Why are you back here? Are you here to upset us again? Barbara has been crying ever since you left. Why are you bringing up the past? We've spent so long trying our best to move on and yet you keep coming back with this. Luke left us a long time ago and he's not coming back. If he was he would have tried to get in touch. We don't need to be reminded that our only son turned his back on us."

"I understand that, but we have some… new information, We'd just like to ask you some questions, please, we just need a little more detail to help us in our inquiries," Constable Dudley said.

Allan narrowed his eyes and for a moment Lily thought that he was going to slam the door in their faces. He had every right to, and such an action would not be enough to prove his guilt. Lily recognized Allan though. He and his wife attended church every Sunday. They always sat together and had a solemn look on their faces. Now she understood why.

"Please Allan, I know this must be difficult, but someone has died and if we're going to be able to lay their soul to rest then we need to investigate every angle. I'm not saying this is Luke, but if it is then we need to ask you some questions. I promise it won't take long, and if it's too traumatic for Barbara then we don't have to talk to her right now."

"Of course it's too traumatic for her. Do you know what it's like for a mother to lose contact with her only child?" he asked. Lily felt a sharp pang of emotion in her gut, but pushed it away. "Fine, ask your questions, but I'm not happy about this," he said. He glanced inside and then left the house, closing the door behind him. He walked to a garage nearby that was filled with all kinds of mechanical contraptions and machinery, which Lily assumed was from nearby farms.

"Looks like you're busy. What actually is it you do here?" Lily asked.

"I fix things. My family has fixed things for generations, until now at least," he replied dryly. "Is that all you want to know?"

Lily glanced at Constable Dudley, who cleared his throat before speaking. "Actually there was something else we wanted to ask you about. It's about your work actually. It seems that Luke wasn't exactly happy at the thought of following in your footsteps. He had dreams of his own, yes?"

Allan let out a dry chuckle. He looked down at his feet and folded his arms across his broad chest. His sleeves were rolled up, revealing thick lines of hair that came slashing across his forearm. "I think I know where this is going," he said. "Did a certain woman come and see you? Lindsay?"

Lily's brow crinkled, but she wasn't about to overplay her hand. "Why would you say something like that?"

"Because it sounds like something she'd do. I'm sure she told you how Luke hated us for wanting to push him into the family business and how we were such monsters to actually try and give him a future and teach him about his tradition and his legacy. What, has she tried to claim that we had something to do with his death?" Allan asked. He pursed his lips and nodded. The look that Constable Dudley and Lily gave him was enough to tell him everything he needed to know.

A dry laugh emanated from his mouth. He shook his head and sighed, shifting his weight between his feet. "Look, before you come around here with all this stuff, bothering me and my wife, you should really get your facts straight. Lindsay is a bitter woman who can't handle the fact that Luke left her."

"But she claimed that Luke loved her. She said they were going to run away together and begin a new life."

Allan sighed again. "I don't know when it happened, but Luke didn't exactly become the son I thought he would be. We tried to do our best by him and raise him right, but he wasn't interested in the lessons we had to give him. Lindsay wasn't the only person he was dating, and I wouldn't be surprised if he told the same things to this other girl. I'd love to stand here and tell you that Luke was a darling boy and who never did anything wrong, but unfortunately he had a bad habit of putting himself first. He didn't care about his family's tradition, and so I don't think he cared about Lindsay as much as he said he did. I'm sure that he just left with a smile on his face and wherever he is now I'm sure he doesn't think about us as much as we think of him. Barbara and I go to church every week, and every week she prays that he's going to come back and see us again, but he never does. You coming around here only makes it worse. Just leave us alone now," he said, and moved to walk away. Before he left Lily called out and asked him if he knew the name of this other woman that Luke had been seeing.

"Anne Lawrence," he replied.

~

"Well, that didn't go how I expected it to go," Lily said.

"What do you think?" Constable Dudley asked.

"I think they have been through a lot and they probably don't appreciate all the pain being dragged up again. I've seen them in church a lot. I always wondered what had been on their minds. I suppose now I know. I still think we need to talk to them again, especially Barbara. I wonder if Luke mentioned that he was thinking of leaving, or if there was a note or anything to suggest that he did in fact run away? I don't know… I can't shake the feeling that something is wrong here."

"I think we should talk to Anne. If she was in a relationship with Luke as well," Constable Dudley grimaced as he said this, as though he found it difficult to comprehend the fact that someone might have dated more than one person at once, "she might have some information about what happened to him. Maybe she's even heard from him over the years."

With that, Lily, Constable Dudley, and Zeus headed to their next destination.

5

Anne Lawrence was a pretty young woman with a soft voice and lips that seemed permanently upturned into a smile. She was a teacher at the local school, and was waving the children goodbye when Constable Dudley and Lily approached. Anne wore a red cardigan and a blue dress. She seemed humble and quiet, not the kind of woman to become involved in a love triangle at all. When Constable Dudley told her what they were there to talk to her about Anne let out a soft whimper and then led them into the school, taking them into the small staff room where she offered them some tea and biscuits. She patted Zeus on the head, and Zeus licked her hand in response. This gave Lily the impression that Anne could be trusted.

"I haven't thought about Luke in a while... it's so sad what happened to him," she said.

"Can you shed any light on the matter?" Lily asked. "What exactly was your relationship to him?"

Anne blushed a little and from the expression on her face it was clear that a happy memory had just flashed through her mind. "Luke and I were close. I used to see him riding on his bicycle every day.

He always smiled at me, and eventually he started bringing me treats from Marjorie's. Then one day we got to talking and it became quite a regular thing."

"So were you two romantically involved?" Lily asked. Constable Dudley coughed and spluttered, almost spilling his tea at the direct question. Lily glanced at him, wondering if he had ever delved into the realms of romance. She had noticed how flustered he became around Joanie, which admittedly was quite a common occurrence with her, but that was a mystery that would have to wait for another time.

Anne blushed again and sipped her tea. "As I said… we were close. It was never really anything serious you understand. Luke was always the kind of man who was more of a dreamer than anything else. He was never going to stay in one place for too long, and he never had the kind of heart that was going to stay in one place. I just used to enjoy the conversations we had. I was young and probably a little too naïve, and he was unlike anyone else in the village. He made me see the world in a different way and when you were around him he made you feel as though anything was possible." She had a wistful tone and the look in her eyes suggested that she was returning to a time when her life was filled with excitement. It was quite the contrast to the look that had been in Lindsay's eyes, a look of rage and hurt.

Lily placed her cup back onto the table and leaned forward. "Anne, were you aware that Luke was involved with someone else? Lindsay?"

Anne's gaze flickered and she pushed an errant lock of hair behind her ear. She looked away. "Yes, I did. At least, I knew they were good friends. I suppose at the time I tried not to think about it. I knew that Luke and I weren't doing anything serious. We were close and we enjoyed the odd kiss," the smile returned to her face and her cheeks were aglow, "but I knew he wasn't going to settle down with me. I was just hopeless around him. I just couldn't look away, and I know that probably makes me an awful person, but it was just the way Luke could make you feel… when I was with him it felt as though I was the only girl in the world. But then when he left I knew the reality of the situation and I suppose I have always had quite a pragmatic way about me. I only let myself get carried away to a point. Besides, Luke always used to talk about leaving the village behind and I knew I was never going to leave, so we never had a future anyway."

"Why wouldn't you want to leave?" Lily asked.

"Because this place is just the most perfect place in the world," Anne said with a wide smile, and a disbelieving stare, as though she couldn't imagine why anyone would leave. "That was one of the things that Luke and I used to talk about. He could never understand how I could be happy staying in one place for my entire life, and I couldn't understand why he would want to leave somewhere that was so perfect."

"Did he ever talk about Lindsay with you, or his relationship with his parents?" Constable Dudley asked.

Anne's face fell again. "He did. Things with Lindsay… I don't know. He seemed conflicted. The thing with Luke is that I don't think he meant to hurt anyone, but he had this way about him that almost made it inevitable. He said it was something he got from his father, and he was always so afraid of getting angry. He said it was why he liked being around me, because I calmed him down."

"So he told you that he was planning to run away with Lindsay?"

"He said that he wanted to run away, and he said that Lindsay wanted to go with him, but he also told me he wasn't sure about it. I think he was afraid of making the wrong decision and being tied to it forever. In his mind if he left with Lindsay then he would have to stay with her forever, and if he was going to do that then he might as well stay in the village. I could tell he loved her though. He was always so passionate when he spoke about her. I really wish he could have spoken that way about me… like I said he was a complicated man. Sometimes I think he was his own worst enemy. He always felt like he should rebel against anyone who tried to tell him what to do, and from what I understand Lindsay was starting to get pushy, you know, asking him when it was going to happen and what they were going to do when they left."

"So when he did leave did you ever think anything bad had happened?" Lily asked.

"I knew that Lindsay was upset. I tried to stay out of it. I don't think she knew about me and so I didn't want her to be angry with me. I just assumed that Luke had had enough and had left by himself," Anne said.

"Weren't you upset that he hadn't said goodbye or left a note?" the Constable asked.

"Of course I was, but it didn't surprise me, that's just the way Luke was. I always used to think he would come back one day when he realized he wasn't going to find what he was searching for out there, but if Lindsay *was* right…" Anne trailed away, unable to think about the truth of what she was about to utter.

"Do you think Luke's father would have been capable of killing him?" Lily asked. Anne looked toward her lap and remained silent. "It's okay, I know it's not nice to think of these things. We're really just trying to understand what might have happened."

"Luke did tell me that Allan could get loud and angry quickly. It was one of the reasons why he said he wanted to get away from this place, because he worried it was going to bring out the worst in him. I never witnessed him getting angry though. There were times when I could tell something was affecting him, but he always managed to bring it back. As for Allan, like I said, I was never actually Luke's girlfriend so I didn't meet his parents. I suppose it wouldn't be beyond the realm of possibility though, if Luke was being honest with everything he told me."

Lily and Constable Dudley thanked Anne and then left the school, walking back through the village toward the former Constable's house.

"I'm not sure if that was helpful or not. The more we learn about Luke the more I'm unsure of his character," Constable Dudley said.

"Because he was dating two women at a time?"

"Exactly! A relationship like that should be treated with dignity and respect. I can't understand anyone who would do such a thing as this."

"Let me ask you a question Constable, have you ever been in love?"

Constable Dudley spluttered and frowned. "I don't think that's an appropriate question for the time being," he said harshly, but that was all Lily really needed to know. She nodded and suppressed a smile.

"I see. Well, the thing about love is that it can make you do strange things. I am wondering something though... Anne said she had never spoken to Lindsay about this, but what if Lindsay found out about it? If she was convinced that she and Luke were going to leave together and then found out that he was being cozy with another girl she might have been angry. Maybe all this time her guilty conscience is beginning to tell."

Constable Dudley sighed. "In my experience this is exactly where love leads; to somewhere complicated and messy."

Lily looked solemn for a moment and gave a soft nod. "I suppose it does, and yet it still draws so many people in, and we're all still searching for it," she said, thinking about the past and the possible future, wondering if she could make herself so vulnerable again.

Constable Dudley rapped against the door of the fine, sturdy house. There were a number of neatly trimmed plants in the front garden and a lion's head knocker gleamed in welcome. The door opened and a man was standing there, looking barrel chested and full of life.

"Bob!" he exclaimed, and welcomed the man with a firm hug that left Constable Dudley off balance and ill at ease.

"Sir," Constable Dudley said.

"Don't be silly Bob, call me Stan. I've been retired long enough now. You don't have to stand on protocol with me," Stanley Vickers said. He turned his head to look at Lily, and his smile widened. "And you must be the famous Lily McGee, it's really a pleasure to meet you, please come in. I've just made some tea, and I'm wondering if you might do me a favor Lily?" he asked.

"I'm sure I can oblige," Lily said.

They walked inside to a tidy cottage. Classical music was playing in the background, which Stan turned off when they walked into the living room. Pictures of his family were dotted around the room, as well as pictures of boats. "Mary will be sorry she missed you," Stan said. His back was turned to them as he reached into a bookshelf and pulled out a book. "I wondered if you wouldn't mind signing this for me," he said, and handed Lily a copy of her first book.

"Oh, yes, of course! I wasn't aware you were a fan," she said, quickly scribbling her name on the first page.

"Oh I am indeed! I find it quite fascinating having lived in this village for so long. You know, if you ever would like to pick my brain I'm sure I could provide you with a number of exciting threads. I did always think I should write a book about my experiences as the constable of Didlington St. Wilfrid, but I find writing to be laborious. Still, I enjoyed reading your story and I'm looking forward to the next one," he said, taking the book back and beaming with pride.

He settled into his chair and the sunlight that poured into the cottage through the windows reflected on his round glasses. He wore a scruffy beard, and it appeared as though his hair had relocated from his scalp to his chin.

"I am sorry for intruding on you like this sir- Stan," Constable Dudley said, correcting himself. The word sounded strange on his lips though, as if he wasn't quite used to the idea of being so casual with his former superior yet.

"It's not an intrusion, not at all! Believe me I could use the company. I usually have Mary here, but she does need her time away from me. It's funny really, all those years I was a police officer she kept moaning at me to take more time off and to spend more time with the family, and then when I finally retire she tells me that I need to get out more and that I'm driving her crazy! Women are always such remarkable creatures, always filled with such contradictions," he said. Lily tilted her head to the side and was about to pull him up on his comment, when Constable Dudley spoke instead.

"Well actually this isn't quite a social call. It's about the job," he said.

"Oh?" Stan asked, leaning forward.

"Yes, actually it's about a case you worked on a number of years ago, one of those missing persons cases."

"Oh yes, one of those people who thought life was better elsewhere," he shook his head with derision and huffed. "It still baffles me how anyone can walk through our streets and breathe in our fresh sweet air and want to be anywhere else. People today don't know how good they have it. One day they're all going to come running back praying for us to take them back into our arms, and we'll probably do it too because we're foolishly sentimental, but that doesn't make it right. I still remember all those sad parents who had to face the reality that their children had forsaken them. It really makes me sad how far the standards in this country have fallen."

"We're here about one of these people in particular," Lily said sharply, finding his manner quiet insufferable. "Luke Jenkinson, do you remember him?"

182

Stan searched his memory for a moment, but then shook his head. "I'm afraid I don't. To be honest I don't remember many of their names. As far as I was concerned if they didn't want to be a part of this village then I didn't need to waste my time with them."

"We think he was killed, and we might have even found his bones," Lily said, hoping that the shocking statement would at least elicit some sympathy from Stan. Instead he merely tutted and rolled his eyes. Lily continued, "in fact someone came to see us today and told us that she raised this possibility with you, but you dismissed it. Her name was Lindsay, perhaps that rings a bell?"

Stan again searched his mind and sighed, nodding sadly. "Yes, yes she was quite distraught. I told her at the time that if she had any evidence concerning the matter then she should bring it to my attention, but she kept saying that the only evidence she needed was the promise he had made to her. We all know that that isn't enough for anything, and we only had limited resources. I couldn't very well hunt down this man when it seemed clear he ran away. The poor parents though, I know how hard Allan works and it was that boy's duty to learn from his father and take over the trade. That's what's wrong with the country today. Our history is being lost because the children don't have any respect. Their heads are filled with this idea that they can do absolutely anything they want to do, when really all they need to do is buckle down and start to think about something other than themselves."

Lily was amazed at his tirade and felt sorry for all the people who had had to deal with him while he was in charge of police affairs in the village. "So you don't think there's any truth to what Lindsay said then? You didn't feel the need to investigate his parents or the area around their house, or ask if anyone had seen anything suspicious?"

The moment she asked this, Stan furrowed his brow and glared at her. He had been nothing but pleasant so far, yet now he had changed. A dark, rigid shroud had fallen over him and he seemed utterly different to the man who had just been sitting before her.

"You may not be aware because you are an amateur sleuth, but police officers in England always investigate every possibility to the fullest extent. His parents were good God fearing people. Luke was the problem. I used to see him cycling around the village with that silly dreamy look on his face. I can't count the times he almost knocked me over when he should have been focused on his surroundings. Frankly the village is better off without him. It was clear he didn't want to be here, so no, I don't think that there was anything to her accusations. She just seemed heartbroken and was looking for a way to convince herself that her boyfriend hadn't just left her here. It was a simple enough solution. I know you have a very inventive mind Miss McGee, but if you begin to see mysteries everywhere you look then you're going to miss the actual crimes that are taking place. I know this isn't what you want to hear, but it seems evident to me that Luke has left this village behind, and we should leave him behind too."

"I can't do that," Lily said defiantly. Stan seemed taken aback by the response. She sensed that he was one of those old school men who were used to getting their way because they had always had the power and the innate respect of their role in the world. She had met enough of them in Hollywood, directors and producers who demanded things to be a certain way just because they could. Lily had spent enough time kowtowing to them and even now it made her skin crawl. When she left Hollywood she had vowed that she was never going to let anyone treat her that way again, and she certainly wasn't going to let that happen here.

"I want to know everything about the case. You were there when it happened. You might have been told something that seemed inconsequential at the time, but that actually has some significance now. There's a body that was buried in that field and is now just a bag of bones. Someone put it there. Someone has died. It might be Luke, it might be someone else, but if we can find out what happened to Luke then maybe-" she cut short her sentence when she saw the look on Stan's face. Everything had dropped, and he suddenly didn't seem so threatening.

"What field?" he asked quietly, his voice barely a whisper.

"Excuse me?" she asked.

"What field were the bones found in?" he asked. Lily told him the location and he nodded slowly, taking a deep breath. "There was something else. A man came to me, a witness who said that he had been taking a walk in the night and saw something there, near the trees."

Lily was agog. Even Constable Dudley looked at his former superior with disbelief.

"And you didn't think this was worth further investigation? You never mentioned this to us!" Constable Dudley said.

Stan scowled. "It wasn't conclusive. There was no point of identification and it was dark, so there were no identifying features either. I can't go chasing around everyone who goes around at night. It's not my business why someone would want to dig a hole in a field."

"Unless they're burying a body," Lily said coldly.

"Who was this man?" Constable Dudley said.

"I don't remember his name. It will be in the file somewhere though. I did make a note of it. I did do my job," Stan said defensively.

"Not well enough," Lily said, and stormed out of the cottage. Constable Dudley was at her heels, and Zeus plodded along beside her.

"I can't believe that man was in charge of investigating crimes here. How could he have someone come to him and say that they think their boyfriend has been killed, then have someone tell him they saw something suspicious in the night and not do anything about it? And you actually looked up to him?"

Constable Dudley rubbed his chin. "He, ah, he seemed more impressive when I was younger. I'm beginning to see that maybe I shouldn't have paid attention to everything he tried to teach me."

"No, I don't think you should have. He doesn't even seem to care about people. Think of how long Luke might have been suffering? And what about the others? They might be dead now, but we have no way of knowing. How many other crimes have gone unsolved because he couldn't be bothered to investigate them?"

Her words were caustic, and everything within her body raged. There was too much injustice in the world for her liking, far too much. Bile rose in her throat. She hadn't felt this angry since she had been in Hollywood and been bawled out by a director, but this was far worse. People's lives were at stake here. Someone was responsible for the most heinous crime, and they were getting away with it. Someone in this village knew the truth, and Lily was determined to find out who they were and then make them pay.

When they returned to the police station someone was waiting for them. Zeus perked up as Lindsay was standing outside, clutching her bag tightly, and she stared daggers at them.

6

"Have you found anything yet? Have you gone to speak to Luke's parents? What's been happening?" Lindsay asked.

"We have been making inquiries, but there have been no major breakthroughs in the case as of yet," Constable Dudley said in the most official tone of voice he could muster. "If you would return home we will let you know when there have been some developments."

"I can't go home," Lindsay said. "How do you expect me to sleep or think or do anything when this is hanging in the air? I thought I could cope. I thought I had finally reached a point in my life where I could move on with things, but then this happened and it just brings it all back. I need to know what's going on. Please, my heart has broken a million times over the years and I just can't handle all of this stress," she said.

When she acted like this Lily felt pity for her and it was difficult to believe that she could be the one responsible for Luke's death. Yet Lily had been duped recently. People could be manipulative and harsh and prey on the good, trusting nature of others. Lily had to erect walls around her heart to protect her from being so vulnerable. Lindsay was convincing, yes, but until Lily knew the truth she couldn't take anything for granted. However, there were more questions she wanted to ask Lindsay, so she accompanied her into the police station and asked for a few moments of her time while Constable Dudley went to examine the file for the name of this potential witness.

Lindsay seemed a little more relaxed as she settled into her chair.

"I'm sorry about all this, I truly am, it's just that it's swirling around my head and I can't escape it. I keep telling myself that none of this is going to make a difference because it's not going to bring him back."

"But it is going to make a difference, to you. It's going to bring you closure," Lily said.

Lindsay nodded and pressed her lips together. "I just don't want to come across as some kind of crazy lady. It's just that if there's a chance to find out what happened to him… I really need it."

"I know," Lily said. "Can I ask you a question? Why did you stay in Didlington St. Wilfrid if you were always so adamant about leaving? I mean, if I recall correctly it wasn't just Luke's idea, was it, it was yours too?"

Lindsay nodded. "Yes, it was, and I did think about it. Maybe I should have. It might have helped me move on, that's for certain." She inhaled deeply, and her entire body swelled and then deflated. "At first I wanted to stay because I wanted to fight and figure out what happened to Luke, but when the police abandoned the case I knew there was no hope. I had to try and tell myself that maybe Luke was still out there and maybe he had left. It was hard. The only hope I had that he was alive meant that I had to believe he would leave me after everything he promised me. I knew that if I left I was never going to see him again, because I never could have tracked him down in the outside world. My only hope was that one day he would miss me and come back to me, and eventually take me away just like he promised. So I made a vow to stay here for him. Besides, I also thought that if I left then I would start to forget him and I think he deserves more than that. Everyone else has already forgotten him," she said in a small voice, and gentle sobs slipped between her lips.

Lily listened silently and with sympathy. There was information she had of course that countered what Lindsay said, and it was always difficult in these situations to know if it was the right thing to share the truth or not. Right now Lindsay, as far as Lily knew, had this idea in her mind of Luke being a loving boyfriend who would have swept her away and pursued happiness with her if he hadn't been killed. But was that actually what was going to happen, or had Luke felt trapped in his relationship with her and had actually been planning to leave by himself all this time?

And had Lindsay known?

It was difficult to believe that this woman who was crying so profusely in front of Lily was putting on an act, but Lily could not allow herself to make assumptions, even if it meant she had to be cruel. Stan Vickers had made baseless assumptions while he had been Constable, and because of him victims had gone undetected and killers had been allowed to roam free.

So it was with a heavy heart that Lily took a deep breath and spoke in a quiet voice. She wondered if Lindsay was able to sense that what she was about to say might be devastating.

"Have you ever considered that Luke might not have been as wonderful as you think?" Lily asked.

Lindsay's eyes flicked up. "What do you mean?"

Lily spoke slow and deliberately. "I did speak to Luke's father. Allan seemed to think that you might have pointed the finger at him because you were angry that Luke left without you."

Lindsay's eyes went wide and then she laughed, although it was a brittle laugh without any humor at all. "And you believe him? Of course I'm bitter, but not because of that."

"Did you know that Luke was... involved with someone else?" Lily asked, watching Lindsay closely for her reaction. Lindsay's jaw tensed and her entire body tightened. A cold hardness flashed in her eyes, making them look like ice for a moment. She swallowed hard, and looked down.

"Of course I knew," she said in a small voice. "It's a small village. Not much can be hidden here, but she was just a girl. It was never anything serious. Luke was... he was troubled. I knew he loved me, but he was never the best at dealing with his emotions. One day he broke down and he told me about her. He felt so guilty. He hated himself and he started hitting a tree. His knuckles started to bleed. I was so scared for him I had to jump on him and tell him to stop before he did serious damage. I told him it didn't matter to me. I knew he loved me. When we were together there was this feeling between us that you couldn't fake. It was real, and I just knew that once we got away from this place everything would be perfect. Luke could have done something he was really passionate about and have some structure in his life. We shared things that nobody else could and this other girl was just... she was something to keep him occupied. I'm sure Allan would like to paint me as some crazy girlfriend who didn't know what Luke was up to, but I knew all about it."

"And he said that you would love to paint him as an angry father who was disappointed in his son," Lily said. "In fact it's easy to imagine that you might have been more upset when you found out the truth than you are now."

The two women stared at each other. Lindsay licked her lips slowly and ran her hands along her thighs. "I can't actually believe you're going to sit here and imply that I'm capable of killing Luke. I loved him. In a way I always will. I'm never going to lose that part of me and if you had ever truly loved anyone you would understand that feeling as well. I'm going to go now because this just feels like I'm going through the same thing as before. You can change the people, but the attitudes are always the same. I just hope that you do find the truth here, but right now you're looking in the wrong place. It's his father, I promise you that," Lindsay said before she walked out of the police station and left Lily feeling rather low. Nausea churned in her stomach at the fact that she had basically accused Lindsay of murder. She also sighed as she reflected on what Lindsay had just said.

No, Lily had never truly been in love. The only man she had come close to loving had hurt her deeply, and she wasn't sure if she could ever reach that point again where she was so vulnerable.

She reached down and patted Zeus, her fingers in his thick fur. "Oh sometimes I wish I was a dog like you Zeus. I have a feeling life would be far simpler if all I had to worry about was running around and eating treats. It's just not that easy being a human. Sometimes life is very messy."

It wasn't too long before Constable Dudley returned with the file and the name of the man who might be able to shed light on this case.

"How did your chat go?" he asked.

"Not well. I think I might have put my foot in it."

"Do you think she might have been responsible?"

"My heart tells me no, but after what happened I'm not sure what I can trust anymore. I always used to think my instinct was enough, but after what happened to Pat... we need evidence."

"Then let's go and get some."

＠

Lily and Constable Dudley had been welcomed into the home of Steven Chambers with no trouble at all. Like most of the people in

Didlington St. Wilfrid he was always happy for company and had a cup of tea seemingly at the ready. His house was filled with ornaments of cats. He had a gentle voice and manner, with wispy fair hair that was streaked with red, and that Lily suspected would once have burned as a ginger shade.

"What can I help you with Constable?" he asked, smiling widely.

"Well actually it's about a case that you might have been questioned about a number of years ago," the Constable said, and proceeded to tell Steven about the discovery of the bones and the resulting investigation. "It has come to our attention that you did actually see someone in the field, it was listed as suspicious activity?"

"Well, I certainly thought it was suspicious at the time," Steven said, "although nothing ever came of it."

"What happened?" Lily asked.

"Well, you see I've always had this trouble with insomnia. It's plagued me ever since I was a child and when I find it hard to sleep I tend to go for a walk. The night air seems to calm me, and usually when I return to bed I'm out like a light," he said, snapping his fingers as he spoke. Lily waited patiently for him to continue, wishing that he would hurry up and arrive at the point she thought was most pertinent. "So anyway, I went outside and usually I don't see many people around. It's always quiet at night as well, so the sound travels far. Well, I suppose it's always quiet here isn't, anyway, yes, so I was out walking and I went by a field and I heard this heavy breathing and the sound of movement. It was most unusual, and although I've never been a brave man I thought it was only right that I should go and see if anyone needed help with anything. I thought someone may have been in danger."

"Who was it?" Lily asked, unable to keep herself from butting in.

"Well that's just the point of it. I didn't get a chance to see. It was dark and he was hidden in shadows. The fact that he was a man is just about all I can tell you, a big man at that, but when I shouted to him to see if he needed any help he yelled at me and bellowed at me to get away. I was hardly going to argue with him, so I turned tail and returned home."

"I see," Lily replied, her heart sinking as the information wasn't as valuable as she had hoped. They bid farewell and walked outside. Lily's head was low and she dug her hands into her pockets, pursing her lips.

"I thought he was going to be more helpful. I thought he might have seen the man's face," she said.

"I agree. It is disappointing," Constable Dudley replied.

"What's disappointing is the fact that your old boss never caught onto this. How could he have a witness and just ignore the testimony? If he had investigated the area he might have found boot prints or dug into the hole and found the body all those years ago or at least had some clue about who the man was... I mean... the chances are that it's Allan," Lily spoke with a hint of reluctance in her voice, and Constable Dudley was quick to pull her up before she jumped to a conclusion.

"We don't know that for certain. I know it's plausible, but there are a number of men who could fit that description. Without evidence we're not going to be able to get anywhere."

"I just don't know what we're supposed to do now, or what I'm supposed to do. We have all these pieces of a puzzle, but they don't seem to fit together quite right. I'm worried that the murderer is going to go free," she said, resisting the urge to say 'again'. Although Pat hadn't died, he had come about as close as anyone had to death.

"Go home and get some rest for the night Lily. Come to the station tomorrow and we'll think about what we can do next. It's been a long day. It's not as though anything is going to change through the night," he said. Lily hated to admit that he was right, but she couldn't think of what else to do so she bid the Constable farewell and led Zeus home to Starling Cottage. She made herself some dinner and filled Zeus' bowl up with a hearty serving of food, and when she had eaten Zeus came to sit beside her, resting his head on her lap. She idly stroked the dog and rubbed her temples as she pondered the case. Tension throbbed behind her eyes and she wished she could see the truth behind the mystery.

She had a man who longed to leave the village, under pressure from his parents to take over the family business. This man was in love with a woman, Lindsay, who he had promised to leave with, but at the same time he had been seeing Anne.

She groaned, annoyed that people were so filled with contradictions.

Joanie came up from the dark room where she had been developing some photographs. She saw Lily sitting on the sofa, looking melancholy, and joined her. Lily sighed as she explained what had happened during the day.

"You like getting yourself into trouble around here, don't you?" Joanie asked with a wry smile.

"I don't intend to, it just seems to happen. I just can't turn my back when someone is in trouble, but I'm not sure if I'm being much help here."

"I suppose it is an old case. How much help can you really be?"

"There are people hurting because of this, and someone is responsible for burying this body. I have to find out who. I can't let them get away with it, not after last time."

Joanie let the words hang in the air before she responded. "You need to let that go Lily. You can't fight against the government. It wasn't your fault what happened to Pat."

"I should have seen it sooner," Lily said sharply, surprising even herself with the force of her words. "I should have known she wasn't who she seemed."

Joanie gave her a sympathetic look. "Lily, I know that you've enjoyed getting involved in these little mysteries, but at the end of the day she was a trained spy and you're a writer. You shouldn't feel bad that she came out on top."

"Maybe I'm not as good as I think I am. Maybe I should just stop interfering and let Constable Dudley get on with his job."

"Don't be like this," Joanie said, reaching over to squeeze her hand with reassurance. When Joanie leaned back, Zeus pricked up his ears and looked up, as if to ask where his stroke was, so Joanie scratched behind his ears, and this left him satisfied. "You've made a big difference in the lives of the people of this village and you got to Pat in time to save his life. You're not going to be able to solve everything Lily, and maybe this is another one that you're just going to have to let go. I mean, how many years was the body buried for? Some things just get lost over time, that's all."

"But I just hate thinking about people being in pain."

"Well… do you have any idea about who might be responsible?"

Lily sighed, and nodded. "I think it's the father. It all makes sense. He wanted Luke to follow his footsteps in the family business, he has a temper, and it fits with the man Mr. Chambers saw. It wouldn't surprise me if he used Luke's wishes to leave as a cover so that nobody would suspect anything. It doesn't help that the police at the time were useless," she pinched the bridge of her nose in the hope that it would relieve her tension, but it didn't seem to have any affect. "I just don't see how you can be a chief of police and not bother to investigate the crime because it's easier to believe someone just ran away. Think of how many other crimes slipped through his fingers? Think of how many other people suffered because of him…"

"So tell me more about this man, the boy's father," Joanie said, trying to focus Lily's thoughts.

"He's a hard man, you know, the kind who doesn't put up with anything he doesn't want to put up with. I definitely got the sense that he was hiding something from me."

"Then it seems you have your man."

"But what good is that when I can't do anything with it? There's no evidence to suggest that he actually killed his son. For all we know he's telling the truth and Luke is out there in the world somewhere and this body is someone completely different. I just don't know what to do next, and then there's Lindsay. I promised her that I was going to find out what happened to Luke and now I'm afraid I'm going to have to break that promise."

"I don't think you'll have to break any promise Lily. One of the things I've always admired about you is the fact that you're not afraid to follow your heart, and your instinct leads you to the right place. It led you here, after all," Joanie said, smiling. "I'm sure that you'll get to the bottom of this somehow. The answer will come to you, it always does. But you're not going to get anywhere sitting here feeling sorry for yourself. Go to bed and attack the problem again tomorrow."

Lily sighed and rose from the couch, making Zeus move too. "You're probably right. Thanks for the chat Joanie. I do appreciate it," she said, offering her friend a smile.

"No problem, I'm always here for you Lily," Joanie said, and Lily went off to her room.

It was strange, she thought, when Joanie had first arrived in Didlington St. Wilfrid Lily had been subdued and a little annoyed. She had come to England to escape her life in America after all, and the last thing she needed was someone following in her footsteps and ruining the peace she had found. But although Joanie did have her quirks, she was a good friend, and on this night especially Lily was grateful for her company as she didn't like feeling alone. She had always been prone to having thoughts whirl around in her mind

196

and if Joanie hadn't been there with her kind words Lily might well have fallen into despair.

But Joanie was right, Lily's instinct never usually failed her and there was always a way to the truth in these mysteries. She wasn't ready to let anyone get away with this. Someone had died, and everything in Lily's bones was telling her that it had something to do with the parents.

She let out a low, humorless laugh. It always came back to the parents.

Her talk with Allan had been brief, but she thought it was time for her and the good Constable to go and pay him another visit again the following day. If he was responsible for this then he was likely feeling confident that he had gotten away with it, but Lily was ready to bring him back to earth. There were always consequences, and she wasn't about to let him get away with it.

7

L ily awoke to the feeling of Zeus nuzzling into her palm. She rubbed her eyes as she rose and looked at the time, surprised that it was a couple of hours later than she usually woke up. Back in America she had always been running herself ragged for the directors and producers in Hollywood and her sleep schedule had suffered because of it. She had been like a rope that was frayed at both ends, but since arriving in Didlington St. Wilfrid she had managed to fall into a healthy routine of sleep that left her feeling refreshed and energetic. Usually she would spend the hours of the morning writing diligently, working on her next murder mystery. However, sometimes other matters took greater importance.

She washed her face and looked at herself in the mirror, pulling her hair back into a tight ponytail and clasping buttons of a blouse. She smoothed down her clothes, puffed out her chest, and held her head up high. She was not going to allow Allan to talk down to her or to bully her away. If he had something to hide then she was going to make sure it was uncovered. Everything pointed to him. It made sense. She wasn't going to let anyone deceive her again.

"Come on Zeus," she said, feeding the dog some breakfast before fetching his leash and preparing him for a walk. She left Starling Cottage and walked briskly toward the police station, ready to greet Constable Dudley for the day. Her heart went out to the Constable. She had noticed him squirming when in the presence of his old superior, and his face had turned rather pale when Mr. Vickers had rebuked them as they left. She assumed there was likely a healthy dose of guilt weighing upon the Constable's soul as well. Although he had only been following orders at the time, he still had a part to play in this travesty. Perhaps if he had spoken up against Vickers something might have changed... but probably not. Men like Vickers didn't like being challenged. He would have followed through with his own way of doing things regardless.

Still, this case probably meant as much to Constable Dudley as it did to Lily.

As she walked through the village toward the police station there was a commotion that caught her attention.

"He didn't love you!" someone cried, and Lily recognized the voice to be Lindsay's. Her heart sank and she increased her pace, running toward the origin of the noise, which happened to be the school. Zeus panted beside her and seemed to enjoy the run. People emerged from their homes, drawn to the loud shouts as well.

Lily arrived to see Lindsay looking like a ghoul. Her hair was tousled and her face was white, with two dark shadows under her eyes, suggesting that she had endured a night without sleep. Her hands were flailing about in the air, and she cut an intimidating figure in front of Anne, who was practically pinned against a wall. However, the soft spoken teacher was not cowering, rather there was an intense glare in her eyes.

"He didn't love you either!" she called back. "That's the thing, Luke didn't love anyone. He loved the idea of love. He loved the idea of things more than actually doing anything."

"Don't pretend like you know him better than me. I was his girlfriend. I was the one he spent time with, the one he was going to run away with. I was the one he loved. You were just something to kill time," Lindsay retorted.

"I think the fact you're here tells me that you know he didn't love you as much as you hoped he did. You were never a fairytale Lindsay. I'm sorry that he treated you this way. I knew what the deal was. It's a shame that you didn't. You know what he told me? Luke worried that he could never be what you wanted him to be. He was so afraid of letting you down and breaking your heart. You wanted to put him in a box. You wanted to define him, make him into a husband and a father and all these things when Luke didn't want that. All Luke wanted was to be free. You're just as bad as his father."

"Don't you dare say that," Lindsay uttered, her words as loud as a thunderclap. Without any hesitation she slapped Anne. When her palm met Anne's face there was a loud crack that whipped through the air. Anne's face twisted around, and her hair whipped with it. She let out a soft whimper and raised her hand, protecting herself, but it was too late.

The slap almost seemed to have a sobering effect on Lindsay. It was as though she had been lost in her own world during the argument, but now she stood there and looked around. A crowd had gathered, and all of them looked at her with accusing and suspicious eyes. Murmurs rippled through the crowd. Anne used the opportunity to steal away, running into the safety of the school, still holding her face.

Lily took it upon herself to stride forward and take Lindsay's arm, dragging her away from all the prying and judgmental eyes. It was a harrowing sight though, and it became clear that Allan wasn't the only one who had a temper. Lily was beginning to wonder if her instinct was as good as she thought.

~

Lily took Lindsay to the nearest safe place she could think of, which happened to be Marjorie's bakery. Marjorie greeted them with a smile, although when she saw the look on Lily's face and the state of Lindsay her smile fell almost immediately, replaced with a look of concern.

"Marjorie, do you mind if I use your living room for a few moments?" Lily asked, using a tone that suggested she would explain later. Marjorie was a kind soul and Lily had built up a lot of goodwill, so Marjorie nodded and brought up a plate of sweet treats behind them, before leaving them to themselves. Lindsay seemed deflated and had lost all sense of the anger that had flared within her. She sat in a chair, looking small and hunched over, and far older than her years.

Lily went to the kitchen and poured her a glass of water, which Lindsay sipped when it was handed to her. She didn't touch the plate of treats Marjorie had left though.

"Do you want to tell me what that was about?" Lily asked.

"I... I just couldn't stop thinking about them last night. All these years I've been telling myself that Luke never really cared about her and it wasn't anything serious, but then I wondered what if it was? What if all this time I was never really his love, just something to pass the time with? What if... what if what we had wasn't as special as I thought it was?"

"From what I've learned about him, Luke seemed like a complicated man. It can't have been easy to love him."

A wistful smile curled upon Lindsay's face, but it was quickly replaced by the sorrow that haunted her eyes. "It wasn't. But it still felt right. When we were together it felt as though we could conquer the world. We were going to have wonderful lives... but what if Anne was right? What if he was afraid he couldn't live up to what I wanted from him? What if he thought I was trying to control him? That was the thing he hated most of all, you know, people telling him what he had to do and who he had to be."

"If he hated it that much then I don't think he would have been in love with you just because he thought it was right or because he thought you were telling him to. Only you know what happened between the two of you, and if it felt real then the chances are it was real. I thought I was in love before, but now when I look back on it I can see that it wasn't right and it wasn't what I wanted. Time gives us a lot of wisdom, and if you can still look back at your time with him and miss him then I think it shows that he meant a lot to you, and no doubt you meant a lot to him as well."

"Thank you," Lindsay said, and offered her a grateful smile.

"But it still doesn't change that what you just did was wrong."

"I know," Lindsay sighed. "I will apologize to her. I know it's not her fault that Luke spent time with her. I just... I couldn't sleep last night and this morning I wanted to go and speak to her about him. I thought maybe if I spoke to someone who remembered him in a similar way that I remembered him it might make me feel closer to him, but instead it just... it got out of hand."

"That's certainly one way of putting it," Lily said.

"I suppose it doesn't help my case, does it?"

"Your case?"

"Well, you implied yesterday that I'm under suspicion too. I can't imagine this show of anger would have helped me any."

Lily pressed her lips together. It was true, she did not want to discount anyone from the running when there was so little evidence to help her reach the truth, but she also didn't want her fear of being deceived to get the better of her and live in a world that was filled with paranoia.

"I don't believe that you are responsible Lindsay. I know it must have been difficult to hear yesterday. Perhaps it was even cruel of me, but I had to examine every possibility. But there is something I must confess to you. I'm worried that I'm not going to be able to live up to the vow I made yesterday. I will try to the best of my ability, but so many years have passed and you were right; the handling of the crime by the police was negligent, but it means there's not much we can look into. I have my suspicions of who was responsible, but I'm worried that I'm not going to be able to prove it."

"I understand," Lindsay said, her face falling.

"I wanted to ask you if you don't think it's time to move on? I know that you feel tied to this place because of Luke, but it seems to me that it might be doing you more harm than good. It doesn't do anyone any good to live in the past Lindsay."

"I know, and I've thought about leaving so many times before. I just... I just can't leave him. Not when there is unresolved business."

"But what I'm saying is that things may never be resolved, at least not in the way you want them to be. Would he really want you to stick around in a place where you're not happy, because you don't want to move on from him?"

"No, no I don't think he would," Lindsay said, laughing as she spoke.

"Then it might be better if you to try and live that life you wanted with him. Live it for him, and live it for yourself as well. He might have died here, but it doesn't mean that you have to as well."

"I'll think about it," Lindsay said, nodding. She sniffed back some tears and breathed deeply. "I want to thank you Lily. You've done something that nobody else has ever done. You actually listened to me. All these years people have been saying that I'm crazy for believing in this conspiracy. They've always looked at me like I'm a fool, deluded, unable to see that he left me behind. I had nobody before this to actually sit with me and listen to me."

Lily furrowed her brow. "I'm a little surprised this didn't bring you and his parents together. I imagine they can't have been fond of you, but you were still the closest one to their son, and much of who he was would have been left behind with you."

"Well, they never much liked me. They blamed me for filling his head with ideas of leaving, Allan especially. It didn't help that I accused him of murder. I keep out of their way. Sometimes when I see Barbara in the village I think about talking to her, but she always seems on the verge of tears on the rare occasions I see her, and I don't much feel comfortable talking to Allan by myself."

"No, I can't imagine you would."

"I suppose you're right. If he did kill Luke they're going to take it to their graves. I suppose even though I might never get the proof I need, I can at least believe that he never left without me. I just wish I could have spoken to him one more time."

"What actually happened the last time you saw each other?"

Lindsay wiped away glistening tears from her cheeks. "It was in the evening and he said that he was tired of talking about leaving and just wanted to do it. He said he was going to go home and tell his parents that he wasn't going to be who they wanted him to be. I was excited and ran home as well. He said that once he was done with them he was going to come by my house and help me get ready, and then we were going to go out on our great adventure."

"But he never showed up."

"No," Lindsay said, shaking her head with sorrow. "Allan said that Luke had just left and he must have left me behind. But I didn't believe him. I never have. Maybe if Constable Vickers had believed me we might know more."

"Well, I think I need one more conversation with Allan before we put this matter to rest. Constable Dudley has sent the body away in the hope that they can find some identifying markers to confirm if it is actually Luke or not."

"I'm sure it is," Lindsay said.

Lily was too.

Once Lindsay had calmed down, Lily escorted her downstairs and out through the bakery and told her to go home and get some rest. Lindsay was hardly in a position to argue. Sometimes, Lily thought, it wasn't always the truth that could help, but just someone to listen.

Even so, the truth was always worth seeking.

As a writer Lily always approached mysteries with a pragmatic viewpoint. She looked for patterns and things that seemed plausible. Real life of course was not always as orderly as a well plotted story. Sometimes there were random moments of chaos that skewed everything that would have been scoffed at if she had written it. Real life could be unsatisfying sometimes like when a spy got traded back to their country of birth rather than punished for an attempted murder. She worried that the same thing was happening here. It just made sense that Allan was responsible for his son's death, as grim as that was to face.

With Zeus by her side, she went to speak to Allan again, this time without Constable Dudley. She hoped that without the Constable there she might at least be able to elicit some information from Allan that he may not necessarily want to divulge. The advantage of being a writer is that it didn't put people on edge as much as seeing a police officer did.

It wouldn't be the first time that she would be faced with a murderer. As she walked to his home her heart rate increased and an uneasy feeling swam in the pit of her stomach.

"I need you to be on guard Zeus," she muttered, glad that she had the dog to protect her.

She found Allan in his workshop. His sleeves were rolled up and sweat dripped from his brow. He hammered metal, and the noise sent reverberations through her bones.

"Hello Allan. I have a few more questions for you," Lily called out upon arrival. Allan glanced up and grimaced when he saw her.

"I told you to leave me alone. I know you've got a reputation for poking your nose into places, and while most people here have welcomed you I don't much care for outsiders. I'd appreciate it if you just leave."

"I'm afraid I can't do that. I need to talk to you about Luke."

Allan had brought his hammer back, but paused before he struck the metal he was holding. He turned and sighed, still with his hands curled around the hammer. Lily's gaze couldn't help but be drawn to it. It was a brutal weapon, well capable of killing someone in the right hands. A lump formed in her throat.

"I don't know who you are, but you can't just come around here and talk about my son like that. I know you have it in your head that he's dead, but as far as I'm concerned he left us a long time ago. The truth is that he never appreciated what his mother and I did for him, or the home we gave him. We've had to live with this ever since he left, wondering if we did the right thing, wondering if we loved him enough. Well the truth is that he had his own problems. It doesn't make it any easier though. He was a part of us, but we've had to move on. Do you know how hard it is to love each other after something like this? That woman in there is the light of my life. Ever since I met her I knew she was going to be my life and that I would do anything for her, and when we had Luke it was the happiest we had ever been. Then he had to grow up and turn into..." he scowled, unable to finish the sentence. "All you're doing is bringing up bad memories. Is that really what you came to this place to do? Just leave us alone. Don't you think we've suffered enough?"

"I'm just trying to figure out what happened to him. I thought you would want to know, considering he's your son and all. If he was killed don't you want to know who killed him and why?"

Allan's face turned hard. He grit his teeth so firmly that Lily was surprised they didn't all shatter. "It doesn't change what happened.

We're just trying to get on with our lives," he said.

"Then perhaps your wife will feel differently," Lily said, turning to move to the house. The moment she mentioned Barbara, Allan's temper flared and he stormed toward her, threatening her with the hammer if she stepped a foot in his house. But Zeus was there, standing between them, growling and baring his teeth. Lily moved to the house. There had to be someone who cared what happened to Luke, and she worried that if she left Barbara with Allan and he *was* the murderer then it might happen again.

She flung the door open and a woman screamed. Barbara was a wisp of a woman and cowered when Lily entered the house.

"Don't hurt me!" she cried.

"I'm not going to hurt you. I don't mean any harm. I don't know if your husband told you who I am, but I'm trying to investigate the death of your son. We found a body in a nearby field and I'm—"

Lily barely had the chance to finish what she was saying before Barbara sighed. It looked as though she was on the verge of tears.

"I suppose it was only a matter of time before you arrived," she said despondently. She took a seat at the table and closed her eyes for a moment, exhaling deeply. Before either of them could speak, heavy footsteps came thundering behind Lily and Allan arrived in the house, his face stricken. He was still holding the hammer and Lily's heart burst in fear for Zeus, but the dog came in after Allan, still growling. Lily commanded him to heel.

"I told you to leave us alone!" Allan said, rushing to stand beside his wife. He placed his hands on her shoulders and leaned in. "What have you asked her? What's happened?"

"Nothing yet," Lily said, although she had the distinct impression that there was indeed something she needed to know.

"It's time Allan. It's been too long. We've lived with this secret for too long now..." she said.

Allan shook his head.

"I think you had better tell me what's going on," Lily said. "Or do I need to bring Constable Dudley back?"

Before Barbara could say anything Allan was speaking again. "I hate you," he spat. "All you had to do was leave us alone, leave us in peace. Do you not think we've suffered already? You want the truth? Fine, I'll give you the truth. Yes, it was me. I did it. I killed Luke. He came back and he told us that he was going to leave. I didn't take kindly to it. I tried to remind him that his place was here, with us, and that he had responsibilities. He didn't want to listen. All through life all he ever wanted to do was make himself happy. He never cared about anyone else. He never cared about us. So we fought and I... I pushed him and that was that. He hit his head and he died. And I know you're going to think that I'm a monster, but all I wanted was the best for him. I tried to love him, and he kept pushing me away until he pushed me too far," Allan said. The words came out in a torrent, as though he had been holding them inside for so long it was inevitable they were going to escape. He fell to his knees and took Barbara's hands in his. He reassured her that everything was going to be okay, but she was in tears. It was a surprising scene of tenderness and love from a man who had murdered his own son, once again proof that people were filled with contradictions.

"I think that you had better—"

Lily began, but Allan was already ahead of her.

"I know the way to the station," he said gruffly, and walked past her. Barbara was in tears, and Lily had no choice but to leave her alone.

Sometimes the truth came at a price, a very high price, but justice had to be unwavering and sometimes cruel. It had to be absolute, but still, Lily didn't feel proud as she stood there looking at this woman who had her life torn apart, sitting in emptiness and silence and despair.

8

Lily was in the police station, poring over the other missing person files, wondering how many of them were as simple as runaways and how many had suffered some horrible injustice. She burned the faces into her memory for she doubted that she would be able to bring justice to each of them in the same way as she had for Luke, but the least she could do was remember them and acknowledge them, which is more than Vickers had done. Lily knew it was more important than ever for her to help where she could.

As she looked at the faces she wondered if they had been lonely when they ran away, or misunderstood. She knew what that feeling was like. If her own life had turned out differently she may well have been in one of these files herself. Knowing what Luke's father had been capable of made her think of her own family. She shuddered as the past reared its ugly head again, and felt goosebumps rise upon her skin. She had to remind herself that she was far away in a place where they could never find her, and where she never had to be worried by them again.

Allan had already been taken away to be charged. It was yet to be decided whether it was going to be murder or manslaughter. All this time Lindsay had been right. All this time Allan had been allowed to walk free and attend church every Sunday. She wondered if he had ever believed he would achieve the forgiveness he sought. And he left behind a wife, a wife who had lost both her son and her husband. Lily couldn't imagine what that was like, to have this dream life crumble and slip through your fingers. The tragedy of it all must have been overwhelming and she honestly didn't know how Barbara was going to cope. But at least Lindsay could have some closure to the matter and perhaps find some solace elsewhere in the world. For many people, including Lily herself, Didlington St. Wilfrid was an idyllic place filled with nothing but gentle bliss, yet for others it was a place of sorrow, a place where dreams went to die.

Luke had certainly felt that way, and when he had tried to escape he had paid the ultimate price.

She looked at the files and tried to think of a place to start on the next one, hoping that it might turn out with a happier ending than this current case had. She wished that Vickers could suffer some kind of consequence for his negligence though. Since he was already retired it seemed as though he was going to be able to get away with his prejudice and baseless assumptions.

There was a knock at the door of the office.

"Miss McGee? Could I have a word?" a trembling wisp of a voice asked.

Lily turned, and to her surprise found Barbara. "Of course, please sit down. What can I do for you?" she asked, pushing the case files to one side. She noticed Barbara take a long glance before she did so though. Barbara nodded and sat down, placing her bag by her feet. She wore a long green coat that made her body look narrow. Her face was plain and sallow, and she seemed a little lost, almost as though she had no idea where she belonged anymore.

"I needed to speak to you about… about what happened."

"Of course. You know, I'm sure Constable Dudley has the name of a counselor around here somewhere to help you with your grief."

"I've been dealing with grief for a long time," she said. "It's just that... there's something else that you need to know. Something that I need to tell you. It's about Allan."

"Okay," Lily said, taking in a sharp breath, for she worried the worst. Was Barbara about to tell her that there were other victims?

"He's innocent," Barbara said. Lily gave her a pitying look.

"I know it must seem difficult to believe that he's capable of such a thing, but all the evidence points to him and—"

"No, you don't understand," she said, the tone of her voice becoming sharper. Her hands were clasped together, and the skin was as pale as snow. "I *know* he didn't do it."

"If he didn't do it then who did?" Lily asked gently. She felt sorry for the woman who was clearly trying to make sense of all this by denying the truth. However, Lily soon received an answer that she did not expect.

"I did."

The words were simple and short, and it took Lily a moment to process what she had just heard. She leaned forward and furrowed her brow.

"I'm sorry Barbara, did you just say what I think you said?"

Barbara nodded. Her lower lip trembled and she started to twitch.

"I think you might need to tell me what exactly happened," Lily said in a low voice.

"It was... it was so terrible you see. We tried so hard with Luke, so hard to teach him what was important and what mattered in life. Allan has always been so proud of his history and his legacy and all he wanted was for Luke to share in that joy. But Luke was... he was a difficult child. He kept saying that he wanted to leave the village and we tried to explain why it was important that he stayed, but he never listened to us. He just couldn't seem to see it from our point of view. I hated all the arguments. It was never what I wanted from family life. I always thought we were going to be happy, but we were always so angry with each other. I think somewhere along the way we forgot how to talk to each other and we just resorted to shouting.

Neither Allan nor I believed that he was actually going to leave. We thought it was just a phase and that in the end he would come to his senses and take his place in the family, but then he came back and told us that he was going to leave with Lindsay and there was nothing we could say to stop him. By this point I had had enough. Allan was usually the one to talk to him about it. Men have a way with these things. But I was about to lose my son. It didn't matter what he had put us through, I didn't want him to leave. I couldn't cope with it. His place was with us, in our home. So I asked him to stay. I begged him. I was practically on my knees and still he said that he was going to leave, that nothing could change his mind. So I... I told him that if he wanted to go so badly then he should just go because he clearly didn't want to be my son and I... I pushed him."

Her face twisted in horror and she sobbed pitifully. Her entire head bowed and her shoulders shuddered. "I can still see it now. I pushed him out of the door, but he crashed down and his head hit the raised stone level. It was a crack as loud as thunder and then... then he was just lying there. Blood was underneath him and I fell by his side. I hugged him and said that I was sorry, but he wasn't listening. He couldn't listen he was...he was gone," she choked on sobs again as she relived the horrifying moment. Lily was stunned.

"What happened then?" she asked.

"Allan told me not to worry. He said that he would take care of it. I was a mess. I had just killed my son, but Allan said that it didn't matter. What mattered was that we stayed together and kept our marriage alive. He said that we had promised to love each other no matter what, and he was never going to forsake that vow. So he took care of the body and I cleaned up the blood and I always thought someone was going to come and speak to us. Allan told them that Luke had left, just like Luke was planning. I knew the truth though and I kept thinking someone was going to ask us more questions, but they never did. They all just... believed us. So our lives went on, but it was far from normal. Even if nobody else knew, I did. I went to church every week and prayed for forgiveness, both from God and from Luke.

You can't understand how hard it was, and how much Allan had to suffer. I took his child away from him. He could have left me so many times over the years, but he never did. He stood by me. He loved me. He treated me with far more respect than I deserved, and even at the end he is still willing to take the fall for me. I saw you, you know, when you came to our house the first time. I spoke to Allan about it afterwards and I told him that it was time to come clean, that it's the right thing to do. He kept saying that it wouldn't do any good, that it would just ruin our lives when it had only been an accident. I couldn't stop myself when you came in. I thought it was a sign from God that it had been enough time now, but then he protected me again, just like he's always protected me. But I can't let him be punished for this. I can't let him suffer the consequences alone. He's not to blame. I'm the one who killed Luke. I should be punished. I should be the one to protect my husband," she said, raising her head to reclaim some small measure of dignity.

Lily exhaled deeply and nodded, feeling numb at this revelation.

"Okay Barbara. I'm sure that can be arranged. You will have to tell this story again though."

"I know, and I'm prepared to finally tell the truth. I've been keeping it to myself for so long. It's only right that Luke should have this final dignity. I robbed him of so much. Do you know that I never went to visit the place where Allan buried his body. I wanted to, to pay my respects, but I was so worried someone would find it suspicious. I used to just sit in his room and sometimes I would pretend that he was out there in the world somewhere, following his dreams. Sometimes I was even able to convince myself that it was true…" she sniffed again. "But at least this will make everything right. At least Allan won't have to bear the brunt of this. I know that it's going to be difficult for him to manage alone, but he has done so many things for me over the years I can't let him do this. That's the thing about love and marriage, it has to be give and take. You have to help each other and trust each other. I know what I did was evil and awful, but if I can save Allan then maybe I can live with myself. Please don't judge him for it Lily. He only did what he thought was right. He was only trying to protect me."

Lily listened and was dismayed that she had to break this woman's heart one final time.

"Barbara, you do understand that just because you confessed to the crime doesn't mean that Allan is going to be able to return to his life?"

Barbara recoiled with shock. "But why? He didn't do it. I did!"

"That may be so, but he still helped you cover up the crime. He lied to the police, he buried the body, he conspired with you to hide. You're both going to have to face the consequences of what you've done," Lily said.

"But… but I came here to save him," Barbara said, stammering without understanding the nuances of the law. Lily rose from the table and called out to one of Constable Dudley's officers to arrest Barbara, before she left the station and went to see Lindsay. When speaking with Barbara it was easy to fall into the trap of pity and feel sorry for them, but they were entirely selfish in their sorrow, assuming that they had a monopoly on that emotion and Luke's love. The thing about parents is that so many of them thought a child was there to do their bidding and live life in the way they saw fit, to fill a role that had been picked out for them. That wasn't true though. A child should be free to make their own choices. Lily had been fortunate enough to escape a life like that.

Luke hadn't.

But there was one other person who truly loved Luke, and who could at least be at peace now that she knew what happened. Lily sat with Lindsay and told her everything. Lindsay cried now that she finally knew the truth, and then confessed that she was ready to leave Didlington St. Wilfrid and move on with her life. The village had been a salvation for Lily, but to others it was a place they needed to leave.

When the analysis of the bones was finally performed it was confirmed that the body belonged to a male of Luke's age, of the same height, and so an identification was assured. Allan and Barbara were taken away from Didlington St. Wilfrid so they could be charged with manslaughter and criminal conspiracy. This lie, this secret had been allowed to fester for years because it had not been properly investigated. As Lily stood atop a hill with Zeus beside her she couldn't help but wonder how many other crimes were lurking beneath the surface of this pretty, blissful village.

At least this time she had managed to find justice though. It would not always elude her, but it was scant comfort after all the pain she had witnessed. She whistled to Zeus, who bounded back to her and jumped up against her chest. She hugged the dog and closed her eyes, blinking away tears that had formed. She had recognized something of herself in Luke, but she was beginning to grow tired of being alone. She kissed Zeus on the head and then flung a stick in the air, watching him chase after it, wishing that something as simple as a stick was enough to make humans happy.

A CHILD GOES MISSING IN
A QUIET ENGLISH VILLAGE

1

L ily McGee was wearing a pale blue dress that complemented her eyes. It was the prettiest thing she had worn since she had left Hollywood, and it wasn't something that she was quite comfortable in yet. Her dear friend Joanie had taken great delight in taking her into the nearest city to go searching for a dress for the occasion, and was quite in her element. The dress was an elegant piece, seeming to cascade down Lily's body like waves, which was quite fitting as the party was being held in Blake Huntingdon's beach house, which stood upon the shore near Didlington St. Wilfrid. The cliffs rose up sharply along the beach, and the sand here was golden, with the water rushing onto it. There was a small dock where fishing boats sailed out, although at this time of the evening there were no fishermen out on the waves. Indeed, the sea was quiet and dark and endless, and quite beautiful. Lily stared at it as she stood on the porch, losing herself to the soft rhythm of the waves, finding it fascinating how the sea and the sky blended together in such a way that it became impossible to tell where one of them ended and the other began.

She held a glass of wine in between her fingers, although she only took small sips because the bubbly liquid rose straight to her mind and she never liked being robbed of her faculties. The breeze was cool, and it made goosebumps rise upon her arms. Joanie had styled Lily's hair too, which came down in soft ringlets, and she had even touched her face with makeup. Since arriving in Didlington St. Wilfrid, Lily had been quite content to wear casual, comfortable clothes, so it was taking her some time to get used to wearing a dress like this, again.

"You know, you're going to have to come inside at some point," the dulcet, smooth voice of her host, Blake Huntingdon said as he stepped out onto the porch. Here was a man who looked like he belonged in Hollywood; wearing a tailored suit that nicely fit his broad shoulders, his chiseled looks as gorgeous as any star, and those piercing eyes that seemed to gaze right into her soul. Lily felt her heart flutter whenever she was around him, but he was a member of the English nobility and she was just some American girl who found herself in a quaint English village. There was no sense in getting ideas above her station.

"Do I really have to?" she asked, turning to face him leaning against the porch.

"Well, the party is being held in your honor. I think it would be remiss of you to avoid speaking to your avid fans," Blake said with a warm smile, gesturing toward the wide glass doors that led into the living room. The beach house was a vast place, with many rooms and places to hide, although Lily didn't seem to be allowed to flee to any of them.

"I did tell you that I didn't expect this," Lily said.

Blake held his arms out wide and tilted his head to the side. "But you're Lily McGee! Your books have caused quite a stir and your own personal reputation as a solver of mysteries has created a mystique around you. You've become quite the star and I find myself speaking about you often with my friends," he said.

Lily blushed when she was told that Blake spoke about her, but her brow furrowed and she still showed evident dislike for the occasion.

"Well, I thought I had left all of this behind in Hollywood. I came to Didlington St. Wilfrid to live a quiet life. I don't seek out these mysteries, you know, I just happen to become embroiled in them for some reason. It's all rather unfortunate."

Blake arched an eyebrow and gave her a knowing glance.

"Why are you looking at me like that?" she asked.

"Because I know there's a part of you that enjoys it. You wouldn't be getting involved if you didn't secretly like it, and what's more is that you're good at it. You've done a lot of good in the world Lily; that's not something to be ignored."

"Yes, well, I suppose so," Lily said, although she still found it difficult to accept his kind words. Then again she had never been good at receiving praise. She even found this party to be quite over the top. All she had done was write a couple of books, and while such a thing was not to be scoffed at, the words had come easily to her and so she did not think of it as any great accomplishment. After all, there were so many books printed a year, that she didn't think hers stood out as something special. Others disagreed though, and her mystery stories had gathered quite a following. Fame was never something she had sought after, especially after seeing the effect it could have on people in Hollywood, so she was content to write her stories in Starling Cottage and then send them out into the world where people could enjoy them as they saw fit.

"You know you have. Have there been any mysteries lately? I was very disappointed that you left me out of the last one. You know I'm always here to help whenever you need it," Blake said. Lily got the impression that he was quite a curious man and liked to poke his nose into other people's affairs. It may have come with the territory though; his family had always had governance of the region so perhaps Blake saw anything that happened in Didlington St. Wilfrid as his own business.

"Well everything happened so quickly I didn't have much of a chance to involve you, but I'll be sure to remember you for next time. And no, actually, it has been mercifully quiet. You know I would much rather not have to get involved in these mysteries. If people would just stop killing each other the world would be a better place."

"You should put that quote in your next book," he said, and Lily smiled. Blake took a sip of wine. "Have you heard anything about Pat yet?"

"He's still recovering. He's having to go through some therapy to learn how to move again. The bullet did a lot of damage. I think if it had been an inch either side he would be dead. It's going to be a long path back for him, but I'm hopeful that he'll be able to return soon."

"So Zeus is still in your care then, where is he tonight?"

"Joanie is looking after him. I was tempted to bring him in case I needed defending from anyone here," she said, looking past his shoulder at the throng of people in the beach house.

Blake laughed. "They're all harmless. Trust me, do you really think I would have invited anyone who would make the night difficult?"

"I'm not sure, there are still many things I don't know about you."

"Well then, we'll have to do something about that, but right now my friends are waiting to meet the star author. I have invited them on the promise that they could meet you and I'm sure you would not want to make a liar of me, would you?" he asked in such a charming way that Lily found it difficult to resist, and she thought how easy life must be for him when he possessed such charm. She sighed and shook her head, knowing that she was going to be unable to escape her fate now that she was here. It had been impossible to refuse his invitation, although now that she was here she wished she had.

"I thought you would have been used to parties like this considering your time in Hollywood," Blake said under his breath.

"Those parties are exactly why I developed a distaste for them," Lily replied. There was a deeper story there, but Blake was not going to get it this evening. He walked into the main living room where the murmur of conversation bubbled as freely as the wine. Blake introduced Lily, and when he did so, there was an excited hush. Lily had been among famous people before, but she had never been the focus of everyone's attention. She smiled politely.

"Perhaps you'd like to say a few words?" Blake said, putting Lily on the spot and she hated him in that moment. She lifted her glass of wine.

"Well, I suppose I should say thank you to you all for enjoying my books and I hope that none of you find them too true to life," she said. The witty comment brought about a titter of laughter and Lily took another sip of wine, finding herself becoming more relaxed. After this Blake took her through the crowd, introducing her to the guests. She was presented to so many people she knew she would forget most of their names. They were all well to do, eloquent people dressed in expensive gowns and suits and walked around with an air of prestige. It was far from the type of people she had gotten used to knowing during her time in Didlington St. Wilfrid. In fact she had to stifle a laugh as she imagined how Pat would respond to these people. No doubt he would have a few choice words, and everyone would surely leave shocked.

She also found herself wishing that she had sent Joanie in her stead. Lily's face began to ache from smiling and she found herself having to have the same conversation over and over again. She was always asked where she got her ideas from and how difficult it must have been to write, and of course many of them had ideas for books but never had the discipline to sit down and write one from beginning to end. And then of course the conversation turned from grisly fiction to grisly reality as the recent spate of murders in the village became the focus of attention. Lily had to listen in polite silence as people debated the issue and wondered if there was anything that could be done. None of these people actually had any idea about what caused the issue though. Lily would have spoken up, but she knew how much this party meant to Blake, and she didn't want to see it end in controversy.

While other people spoke she idly cast her gaze around the beach house, taking in everything she could in case she wanted to use it in a future novel. People milled around, moving from room to room, engaged in their own conversations. They seemed to be in their own bubble, completely removed from the world as Lily knew it.

There was a moment where she was able to get Blake by himself.

"Do you truly enjoy parties like this?" she asked in a quiet voice, so as not to offend anyone in attendance.

"Of course, why wouldn't I?"

"Well… these people don't really seem to know how the world works. I mean, could you imagine any of them living in the village or going to church? I'm not sure any of them have done a hard day's work in their life."

"Not of manual labor, but there are different types of work. I know they're a little odd, but they're all good people. I've known them for years and I know that if I ever need a hand with anything I can count on them, and that means a lot. They all sent well wishes when they heard what had happened to my father. They just take a little getting used to, that's all."

Lily nodded and tried to not allow herself to fall into judgment. "Of course, they are your friends and I'm sorry. I suppose I'm just used to people in Hollywood not being very nice. The more power they got the more of their humanity they lost."

"Well I can assure you that nothing like that is happening here. I can vouch for each and every one of these people," Blake said. Lily decided to take his word for it, and to shrug this chip off her shoulder. It wasn't fair to all of these people to judge them and compare them to the snakes and rats that had inhabited Hollywood.

It may not have been her ideal evening, but she wore a smile anyway and tried to make the best of it, speaking to a number of people as the night went on. Music played in the background as some people danced, and she was delighted when Blake reached out a hand and beckoned her to the dance floor. His strong arms wrapped around her back as they swayed around the room, falling into the music's rhythm. The lights sparkled and he spun her around. The sensations was dizzying and she actually started laughing.

"See, this isn't so bad, is it?" he said.

"No, I suppose it's not. But don't expect me to make a habit out of this. I much prefer cozy evenings to this kind of thing," she said.

"Well, perhaps we can share one of them next time," he replied.

Lily blushed, unsure how much of what he said was genuine and how much had been fueled by the wine he had imbibed. Blake had become a good friend and they had worked together on a number of mysteries. They saw each other at church every week and it was always a pleasure to see him, but could she belong in this world? Did he want her to be a part of his life in the way that she thought she might want to be a part of his life?

The thoughts were dangerous. It was easy and perhaps natural to flee from them after what had happened in her past. Thoughts jabbed at her mind like thrusting daggers and she blinked to try and force them away, to push them back into the past where they belonged. Yet it was so difficult. They kept gnawing and clamoring to come to the forefront of her thoughts. That life was left in America. Why did the thoughts and feelings have to come with her?

She tried to focus on the dance, on Blake's smiling face and warm body and musky aftershave that left the air tingling with excitement. She was breathless and happy with him, and as she danced she thought she might well be able to soar away and float into the sky, joining the stars in their eternal bliss.

That was until there was a scream. It was high pitched and ran through the beach house, and brought the entire party to a standstill. Thumping footsteps crashed down the stairs and then a distraught woman stood there, horror etched upon her exquisite face.

"My son... Evan is gone!" she cried.

2

Everyone stopped dancing. The music still played, but it was faraway now. Lily breathed deeply to try and focus her mind. The wine had caused a dreamy haze to settle in her thoughts, but the scream had been sobering, and she knew that she was needed. People moved to the stairs. Blake was at the forefront, pushing past people to reach the woman, Cecilia Cox. She was a delicate woman, as thin as a pole with porcelain skin. She trembled and leaned against the wall, tears flowing from her eyes.

"It's okay Cecilia," Blake said. "Come with me and sit down. What happened? What do you mean Evan is gone?"

"He's gone I... I checked upstairs where I left him. He was sleeping before, but now the bed is empty and I don't know where he is. He could be anywhere? Oh my darling little prince! What happened to him?" she said, the sorrow falling from her mouth in helpless sobs. A few of her friends crowded around her to offer their support and get her some water, while Blake reassured her that he couldn't have gone far.

Some of Blake's friends had children, so he had offered them rooms to use so that they could still attend the party. The crowd gathered closely together, but it seemed as though some people hadn't noticed what was happening. Lily watched as Blake went outside to fetch a group of men who were enjoying cigars. The smoke rose into the air as he told them what had happened, and then one barreled past him into the room. This man was Heath Cox, Cecilia's husband. He made a beeline for his wife and stood towering over her.

"What's going on?" he asked in a callous tone. Lily noticed how he didn't make any effort to comfort his wife, not even a gesture.

"It's Evan," Cecilia sobbed.

Heath frowned deeply and then looked at everyone else. "Well what is everyone standing around for? Come on, let's find my son. He has to be around here somewhere," he spoke in a brittle tone, and everyone immediately followed his orders. They spread out around the house, searching every nook and cranny for the boy. Lily followed other worried parents upstairs who checked on their own children, all of whom were still present. Guilty relief was on their faces as they hugged their children, while Evan remained absent.

Lily walked into his room. The window was closed, and the bed sheets had been disturbed, meaning that Cecilia wasn't lying when she said she had put Evan to bed. The nightlight had been switched off, and the curtains had been drawn. There was nothing to suggest that any struggle had taken place. The door did not have a lock, and there were toys and other belongings scattered around the room.

"What are you thinking?" Blake asked, standing in the doorway.

"I'm not sure yet," Lily said. "There is always the chance that he awoke by himself, heard the sound of the party, and went to investigate. He might have wanted to see what all the fuss was about. You know what kids can be like." This was said more of optimism than anything else because the alternative was too difficult to think about. The pain of thinking that someone could do anything cruel to a child... it was too much to bear.

"We can hope that's all that's happened," Blake said.

"I should speak to Cecilia to see if she noticed anything important," Lily said.

"I'll give Constable Dudley a call. I suppose it won't be any harm for him to come here, just in case the worst has happened."

"You might want to give Joanie a call as well and get her to bring Zeus here; he might come in handy," Blake said. The dog did have a tendency to be able to sniff out danger, and if the worst happened they were going to need all the help they could get. But for now Lily needed to talk to the parents, for it Evan was missing then she needed to understand who might have taken him, and why.

Cecilia and Heath had moved to a more private room. Cecilia was sitting in a chair as though she had been draped on it like a blanket, while Heath was pacing about the room, shaking his head and muttering under his breath. He paused when Lily walked in.

"I hope you don't mind the intrusion. I thought you might like to talk," Lily offered.

"Come here to get some fodder for your next book?" Heath said sharply, although he apologized immediately. "I'm sorry, that was uncalled for. I'm just on edge at the moment."

"Has there been any word?" Cecilia asked.

"I'm afraid not, but people are still looking around the house for him. Is Evan the type of boy to get into trouble like this? Does he often wander out of bed?" Lily asked.

"Never," Heath said.

"He's a good boy," Cecilia wrung her hands together. "He never does anything bad. He is quite and sweet and just the loveliest boy in the whole world. He knows that when it's bedtime he has to stay in bed."

"So you don't think that he might have woken up and left the room of his own accord to see what was happening at the party? I remember when I was young I always wanted to know what my Mom was up to when she had people over," Lily said, pushing another memory back. There were times when curiosity should never be indulged.

Cecilia shook her head. "No, he never has done anything like that and he never would."

"She's right. To be honest sometimes I think he's too good. It's not natural for a boy to be so well behaved. There always needs to be some trouble to get into, but not Evan, not yet anyway. He does as he's told and we've never had any problems with him."

"If what you say is true, and I'm sure you have no reason to lie, then it means that something more nefarious has happened. I know it's going to be difficult to think about this, but we must try and think about the possibility if we are to find your son. Is there anyone who might want to take him?" Lily hated asking the question. A lump appeared in her throat and her chest tightened at the very notion of something untoward happening, but it was a question she needed to ask.

Cecilia and Heath glanced at each other.

"There's nobody," Cecilia said. "I don't know what kind of monster would have done this, but I would do anything to see my son again. He has to be close! He just has to be!" Cecilia cried, and once again burst into tears. Her head fell into her arms and her entire body trembled with sorrow. Heath stood there, looking at her, and Lily had the impression that he was a hard man.

"I need to do something to be useful. I'm going to look for him," Heath said. Lily thought Cecilia was a pitiful sight, and she thought that Heath was heartless for not even offering any comfort to the poor woman. But people always dealt with these things in different ways. Some people collapsed due to the pressure, while others wanted to fight back and make themselves feel as though they were doing something useful.

"Did you notice if anything was missing from Evan's room?" Lily asked before she left. Cecilia shook her head, but she did not raise it. Lily left then and whispered to a woman outside that she should go in and keep Cecilia company. Cecilia certainly was a woman who should not be left alone.

People had been swarming about the house and peering into every corner of every room. The other children were awakened and in their sleepy state they were asked if they had seen Evan or heard anything, but none of them had. Questions abounded about how a child could have disappeared without anyone noticing. If he had been kidnapped then how was it that someone had walked into the beach house without anyone seeing? It didn't help that there were so many people, an entire crowd, and Lily knew it was all too easy to get lost in a crowd.

When it became clear that Evan wasn't going to be found in the house they had to think about expanding their search to outside, which proved quite a problem as the outside area was far bigger than inside the house. Bright lights appeared as shafts in the distance as cars appeared. Joanie got out first, along with Zeus, who bounded up to Lily and licked her on the face.

"It's good to see you too boy, but we have a job to do right now. I need your help," Lily said, and quickly updated Joanie on what was happening. Joanie gasped and shook her head, and was quick to offer her help.

"Oh, by the way there was a parcel left for you. I put it in the kitchen," Joanie said. Lily wondered who had left her a parcel, but that would have to wait until later. Blake was doing a good job at organizing everyone, and Heath was calling people to his side to go and search near the cliff edge when the other car door opened. Lily expected to see the familiar face of Constable Dudley climb out. She had formed a good working relationship with him, although it was always a sorry affair when she had to renew their partnership. However, on this occasion she was surprised because two slender legs appeared, followed by a lithe body. Auburn hair was tied into a ponytail, and sharp eyes peered beyond a delicate nose. The officer walked with purpose and radiated an aura of control.

"Who are you?" Lily asked, mystified because she didn't recognize the officer.

"I'm Chloe Hargreaves. I'm the new Constable," she said, giving Lily a withering look.

3

"The new constable?" Lily spluttered. "But what about Constable Dudley?"

Constable Hargreaves rolled her eyes and barked out a sardonic laugh. "Well, after the information that came to light about the negligence of the police in Didlington St. Wilfrid, it was deemed inappropriate for him to continue serving as the constable while an investigation takes place. So I'm here now, and I can assure you that nothing like that is going to happen on my watch. Now I've told you who I am, who are you?" she asked.

Lily's stomach churned as she was well aware she had played a crucial part in this information being brought to light. It was unfair though, because Constable Dudley hadn't been the key offender, that had been his old superior, a man now retired. Lily dreaded to think that her meddling had gotten Constable Dudley in trouble, even if it had brought justice to killers who had been hiding in plain sight for too long.

"I'm Lily McGee," she said. A look of understanding passed across the Constable's face.

"Ah yes, I've heard a lot about you. I suppose I shouldn't be surprised that you've found yourself in the middle of something again. I must make my intentions plain Miss McGee, while I do appreciate your efforts in helping bring criminals to justice I don't usually like to employ consultants on the cases I'm working, so if you could stay back and keep to your role as strictly an observer I think that would be best for everyone. I know the policing here hasn't been up to standards, but I am going to change that, so there won't be any need for outside help," she said brusquely. Before Lily could respond, the Constable was already calling toward Blake, bringing out her notepad and a pen.

It was curious to see the way her demeanor changed as soon as Blake entered the picture. Officer Hargreaves went from a brash and abrasive person to a softer, gentler tone. She tilted her head and even wore a smile. Lily always liked to think that jealousy was below her, but she felt a flash of green pass in front of her eyes as the police officer asked Blake all the questions that Lily had the answers too.

Well, Lily wasn't about to be told that she couldn't help someone in need even if this constable didn't appreciate her skills. She had helped people before and she would help people now. Everyone split off into different groups to scour the local area. Some walked toward the beach, hoping to find the boy playing harmlessly in the sand. Others went into the forest, where the chances of finding him were slim, while still others, and this included Lily, walked along the edge of the cliffs where the beach narrowed below until it disappeared entirely.

Heath and his friends shone flashlights around, calling out Evan's name. The noise was enough to drown out the sound of the tide crashing against the cliff. The shafts of light danced around. Zeus sniffed the ground. Lily kept a tight hold on his leash. In the darkness she didn't want to risk him losing his footing and plummeting down the edge of the cliff.

As she walked along she tried to put thoughts of this new constable out of her mind. She and Constable Dudley had enjoyed a good working relationship and he had been a fine officer, if not a bit bumbling. Still, what had happened in the past had been terrible and sometimes a life of decency could not make up for that. She would have to see him to speak with him though. Hopefully this relief from duty was only a temporary one.

The shadows of the night swirled into the sky and nobody was responding to the calls to Evan. People yelled until their throats were hoarse. Lily peered into the darkness, but already she was beginning to lose hope. There were so many dangers in the night that it wouldn't have been beyond the realms of possibility for the boy to fall to his doom. The sea crashed endlessly, as if providing a reminder that it was always there. When she reached the cliff's edge, she peered down and was gripped by a swirling sense of vertigo that made the world spin around her. She composed herself and took a step back, still maintaining a fixed gaze on the sea below. It was as dark as wine and foreboding. To think of a child screaming as he fell, a scream that would be engulfed by the sounds of the water. And then a hand reaching out for rescue, only to clutch nothing but empty air before it was pulled below like the rest of him. She could well imagine him thinking about his parents, crying out for them, trying to depend on them like he always had, but with nothing in reply.

It was enough to break her heart and bring a tear to her eyes.

"I see something!" a man shouted. People rushed to the cliff edge, huddled together, all the flashlights focused on a central point, the beams of light creating a spotlight that showed something that could not be ignored. There was a scarf flapping in the wind, caught on the jagged rocks. It was lonely and morose, and an even more pitiful sight than the weeping mother.

"It does look like his scarf," Heath said. He clenched his jaw and his face was like stone. The light from his flashlight wavered as his hands trembled. "But it doesn't make sense. He wouldn't have walked off like this."

Lily was left wondering if his parents did not know their son as well as they thought. Children could always be surprising and do the unexpected. That was one of the most wonderful things about them, but sometimes the unexpected led to the unpalatable. While the others returned to the beach house, Lily remained on the cliff edge for a moment. She looked back, picturing the journey of the boy.

He would have awakened in the middle of the night and wriggled out of bed, then made his way downstairs and out of the beach house. Instead of being intrigued by the party that his parents were attending he preferred to go outside where it was quiet, perhaps because he did not like noise. Maybe he had seen something in the daytime that had caught his attention and he wanted to see it again, or he just liked the view. He would have made his way out here, unaware how the darkness was going to play tricks on his mind. He might have stumbled, or been unbalanced by a strong gust of wind. It was a cruel accident to send him falling into the unforgiving water, but what had brought him out here in the first place? Most children would have gone to the beach because that was far more of a fun area, but as their parents had admitted; Evan was not like most children.

She stared into the abyss below and wondered if the only reason she doubted the event was a wish that he might still be alive. Sometimes even she could not give into her pragmatic mind.

4

ily returned to the beach house. Given that the scarf had
been found the vast majority of the other search parties
had returned, although word was yet to reach some of
them. Cecilia and Heath were sitting in front of Constable
Hargreaves, while other people were standing together, mourning
and comforting each other, trying to cling onto something that
made sense in this senseless world.

"I'm afraid that we may never know exactly what happened to your
son," Constable Hargreaves said.

"He wouldn't have done this. He wouldn't have walked away like
this. There has to be something else. He has to be somewhere! We
need to keep looking. He's out there. He's probably scared. He
always does get scared," she said, her voice rising and falling with
trembling emotion.

"We will keep searching, but I'm just trying to tell you that you
might have to prepare for the worst," the constable said. Although
Lily didn't have the best first impression of the constable, she didn't
envy her task right now. However, the situation was made more

complicated when someone spoke up and said they had some information.

The woman's speech was a little slurred as she had perhaps had too much wine during the evening.

"I remember something, earlier, there was a servant speaking with the boy. A girl with short hair, a young thing you know, perhaps someone might try asking her," she said.

"And you didn't think to tell anyone this earlier?" the constable asked. The woman shrugged and the hazy look in her eyes spoke volumes. Constable Hargreaves huffed and asked Blake to bring in the servant that had been named as suspicious. Lily frowned at this development. It had of course been servants who had almost caused the death of Blake's father, Lord Huntingdon. Perhaps another servant had gotten inspiration from them and decided to strike back at the social elite by herself. There was no better time for it, not with them all gathered here. However, Lily wondered what anyone might gain by taking a child from his bed? It wasn't as though harming a child was going to cure the world's ills.

Blake walked in with his servant, who was daunted by the eyes that were upon her. Constable Hargreaves took her into a private room. Lily moved forward as well.

"May I join you?" she asked.

"I think I can handle this myself," the constable replied, and turned away from Lily immediately.

Lily frowned, irked that she could not be privy to the questioning as she had been previously.

"I don't know what her problem is," she muttered, crossing her arms.

"I imagine she just has her way of doing things," Blake said. "I'm sure she'll get used to you eventually. After all, I imagine most police officers would appreciate having someone like you help them. I bet she just wants to prove that she can do one by herself before she involves you."

Lily wasn't sure she liked the way Blake jumped to her defense, but she breathed deeply and tried to compose herself. After all, it wasn't her job and she had no divine right to expect the police to show her courtesy. Even so, it still stung. She patted Zeus' head and waited with the rest of them.

"I certainly hope it's not a member of my staff. If it is then I'm going to have to do a thorough review of my hiring process," he said.

"I suppose it's only a matter of time before we find out," Lily said, casting a gaze at all the other people in attendance. "The problem is that there are so many people here anyone had the opportunity to take advantage, although why I'm not sure."

"I suppose that's always the question. The simple truth might be the only thing in this instance though. It's not hard to believe he might have gone out walking of his own accord. I used to do the same thing when I was a child."

"It might well end up being the case," Lily said, although she hoped it wasn't. If the child had been kidnapped then at least there was a chance he would still be alive.

Speculative conversations were whispered between everyone in attendance. Lily could tell that Joanie was itching to speak to people, although it was hardly the time or place so she quelled her socialite tendencies. It took a little while before the door opened and the servant walked out, weeping fretfully. She had her arms crossed and her face was red. She was utterly bereft, and as soon as she was free she strode away, almost collapsing under the weight of her sorrow.

"What did you do to that poor girl?" Lily asked as Office Hargreaves emerged from the room moments later. The police officer seemed perturbed by the question. Her eyebrows were high as she answered, and her tone was one that suggested she did not care one jot about Lily's opinion.

"I questioned her about the matter, and I am sufficiently satisfied that she did not have anything to do with the boy's disappearance," Chloe replied.

"But you brought her to tears!" Lily said, not understanding how someone could be so unsympathetic.

"Sometimes in the course of an investigation there is collateral damage," Chloe said, and that seemed to be as much of an explanation as Lily was going to get. Chloe then moved about the crowd, asking them to move on as there was little more that could be gained in the night, but that the search would continue in the morning. She asked a few people to stay for further questions, including Blake, while Lily was summarily dismissed as though she had nothing to offer at all.

Lily found herself walking back to the car with Joanie and Zeus. Zeus hopped in the back seat, while Lily got in the passenger seat.

"Well, I can't say I like her that much," Joanie said. "Poor Bob. I hope he's coping well. He always took such pride in being a police officer."

Lily ignored the way Joanie referred to Constable Dudley by his first name. The two of them had a strange friendship and Lily thought it better to not be involved. She wondered if Joanie realized the way Constable Dudley felt about her. She could be oblivious to these things sometimes, even if Constable Dudley wasn't doing his best to hide it.

"I know, I'm going to have to speak to him soon. I fear that by involving myself in these affairs I've only made it worse."

"You know that's not true. I'm sure that in time this one will see what you can offer as well. She can't keep Lily McGee on the sidelines for too long!" Joanie said.

"I appreciate your confidence, although I think this time it's misplaced." Lily sighed.

"So, apart from the awful matter of the disappearing child, how was the party?" Joanie asked on the short drive back to Starling Cottage.

"Actually it wasn't as bad as I feared. Everyone was friendly and Blake was a wonderful host."

"I imagine he was," Joanie said, wearing a certain knowing smile that Lily hated.

She rolled her eyes before she spoke. "You don't need to look like that Joanie. Nothing happened."

"No? While you were off searching for the boy I took the opportunity to speak to Blake. He said that you danced together."

"We may have danced together for a song or two," Lily admitted. Joanie let out a sharp squeal. If she hadn't been driving she would have clapped her hands.

"You need to ask him out properly Lily! I can tell there's something between the two of you, and surely you don't think he would have thrown a party like that for just anyone?"

"He's grateful to me for how I helped him and his father, and we share a common bond at church. We are just friends. I'm sure he would tell you the same thing if he was asked."

Now it was Joanie's turn to sigh. "Sometimes you are hopeless Lily. You have the perfect man right in front of you. If you're not careful he's not going to be standing in front of you for very long. You need to let things in the past go. Blake isn't going to be the same as Tommy was."

Lily shuddered at the name and her stomach formed a tight knot. She gazed out of the window as the fields of the countryside went flashing by, but she didn't respond to Joanie's words. They reached Starling Cottage and bid each other goodnight, for Joanie retired immediately to her room. Lily gave Zeus some food and then saw the parcel that had been left for her, waiting on the kitchen table. She frowned, for she had not ordered anything, but when she unwrapped it she saw that it was a packet of sweet buns, so she assumed that Marjorie had had some extra goods left over. Lily tasted one and sighed happily, enjoying the flavor as it lingered against her tongue, trying to not think of all the bad things in the world. It was so difficult though when a child could go missing, and when she could not escape the scars that marred her heart.

The morning arrived, but Lily did not feel refreshed. Even though she had not had copious amounts of wine, she was not used to drinking it and so it did have an effect on her head. She had some tea and a hearty breakfast to try and clear her mind, while thinking about what she was going to do next. Rationally she knew she did not have any right to poke her nose into a case, especially when the police officer in charge made it clear that she did not want any help from an outside source, but Lily felt it wrong to not use her given talents to help where she could. God had led her to this small, quaint village, surely He would want her to help the people who were in need?

She intended to return to the beach house to see Blake and ask if any new information had presented itself. It seemed logical to assume that someone had seen something as there had been so many people in the house. Surely one of them would have seen a child running about the place, or being carried away.

However, first she had to go and see Constable Dudley. He lived in a small cottage, near the police station, that was trim and neat. He opened the door, and it was strange to not see him in his police uniform.

"Ah Lily, good morning," he said, yawning, and then invited her in for some tea.

They were sitting at his kitchen table when Lily looked at him with sympathy. "I'm so sorry Bob," she said.

"What for?"

"For everything. There was an incident at the beach house last night and I met the new constable. I had no idea that this was happening. Why didn't you tell me?"

Bob scratched behind his ear. "Well, I didn't really know how to broach the subject. I was hoping that they would just let the matter rest and then I could get on with things, but then she turned up and said that I had to be placed on leave. I suppose it's not the worst thing in the world. I get to catch up on all the things I need to catch up on, although once you've done that then the days do get rather long. It's surprising how life can seem without work..." he said, trailing off. Lily thought there was a lot of truth to that, and she suspected he was feeling more subdued than he initially seemed.

"But I don't understand why you have been placed on leave. Surely this is nothing to do with you?"

"Well, I was present on the cases. They're looking into whether I was a willing participant, or if I just looked the other way," he sighed, hanging his head, and he looked as though he had aged a decade in a matter of seconds. "I always had a feeling things were wrong and that I should have stood up and said something, but that's just not the way we were taught to do things. I was always told to obey orders and toe the line, so that's what I did. Maybe it's right that I pay the price though. I was a part of this, albeit unwittingly."

"I don't think that's right Bob. Sometimes it's not the people's fault, it's the system that's in place. If you didn't feel you were given the right support then how were you to know what to do? I think your years of service after that should show that you're a fine officer, and if you ever need a personal testimony you can count on me for one," she said, nodding firmly.

"I appreciate that Lily, I truly do. So what did you think of my replacement?" he asked.

Lily arched her eyebrows. "She's certainly an acquired taste. She made it clear that she doesn't want any outside help either. You'd have thought that when there's a missing child involved people would want all the help they could get."

"She's from the city. They're always told they can do no wrong. I can only imagine what she thinks of us out here," he said, shaking his head.

"Well I hope she isn't here for very long. The village will be better once you're back in the police station."

"Thank you Lily, but I'm afraid I might never get back there. The things that happened in the past… someone has to pay, and I'm the only one still here." He spoke hopelessly, and Lily felt bad for him. While Bob wasn't the man responsible, he was indeed someone his superiors could make an example of if they deemed it necessary. Sadly politics played a part in every element of life, and it may not have helped his favor that he had relied on the aid of an American author either. Perhaps that was why Chloe Hargreaves was so against Lily helping her.

Either way, Lily wasn't going to let that stop her from doing all she could to help this missing child. After seeing Bob she took Zeus with her to the beachfront again, which was still a hub of activity.

Blake smiled as she approached, and he bent down to pet Zeus with both hands.

"Looks like you still have quite a crowd," Lily said.

"Well, a lot of them thought they should stay to help. I did tell you they're good people," he said.

"Yes, or perhaps one of them wants to stay to make sure something remains hidden," Lily said.

"Always with the conspiracy," Blake teased. "I'm glad you're here though because there's someone who wants to speak to you," he said, promptly leading her through the house and along to the porch, where a man was waiting for her. He stood with his back to the house, staring out at the ocean. In the daylight it was far different than the night. The sea danced under the light of the morning sun. It looked fresh and beautiful. It was almost impossible to believe that it could be so dangerous, or that it could have pulled a child into its merciless depths.

5

"Lily, you remember Sam," Blake said. Lily remembered his face, but the name had been lost in the tornado of other names. She feigned a smile and nodded.

"It's good to see you again, although I'm sure we'd all prefer to meet under different circumstances," Lily said.

"Definitely," Sam replied.

"So what did you have to tell me?" Lily asked

There was an uncomfortable look on his face. "Well I don't want to get anyone in trouble, but it's something I saw last night." Sam paused for a moment. Blake gave him an encouraging look, and Lily waited with bated breath for any morsel of information that might shift the tide toward truth. "Evan and Cecilia were arguing last night."

"Arguing? About what?" Lily asked. She had noticed some tension between the married couple, so it was interesting to note that they had been having a full blown argument. "I'm not sure. I couldn't overhear them, I just saw that things were getting pretty animated. He was flinging his arms in the air and she looked as though she was going to burst into tears. I tend to keep my nose out of such business, you know, but maybe I should have gone to see what was going on," he said, a little mournfully. Lily knew the feeling of regret well. There were always things she could have done differently. The wisdom of hindsight often came with a bitter sting.

"I'm sure you wouldn't have been able to change anything, but thank you for telling me. Have you told Constable Hargreaves?" Lily said.

"Not yet," Blake said. "I thought we'd come to you first."

"We should tell her," Lily said reluctantly, although part of her wished she could have kept the information to herself. An uncomfortable knot twisted in the pit of her stomach as she thought about seeing Constable Hargreaves again. While being involved in these murder investigations had taken a toll on her, at least Lily had felt useful, far more useful than she ever had in Hollywood. Back there she had been consumed with work for every waking hour of the day, and some sleeping ones too, yet what had she really accomplished? No lives had been changed because of what she had been doing, and the world wasn't a better place for it. It was as though she had wasted her life, and she was keen to make up for lost time. But Constable Hargreaves did not feel the same way at all.

But still, there was a right way of going about things, and Lily wasn't about to keep valuable information from the police. She and Blake walked through the house and found Constable Hargreaves questioning other people who had been at the party. The mood was a somber one, a stark contrast to the joviality that had been present the night before. Everyone had a grim look on their faces. She noticed that the children were never far from their parents. No doubt the adults were afraid that their children might disappear too. The likely truth was that Evan had plummeted from the cliff and his soul had been lost at sea, but without a body there was always a chance that he was still alive.

"Constable, could we have a word with you a minute?" Blake asked, his resonant tone as smooth as honey. As soon as Constable

Hargreaves heard his voice, she turned around and flashed a smile at him, her face changing from the scowl that Lily was used to seeing. She played with a long strand of hair, twirling it around her fingers as she tilted her head back to look up at Blake. Constable Hargreaves wasn't the tallest woman, but she projected an aura of authority that made her intimidating. While her eyes sparkled with delight at seeing Blake, the mood dissipated as soon as she realized Lily was there too.

Lily didn't much care for the way she was looking at Blake either. Although she and Blake were just friends, there was something comfortable about their relationship that Lily enjoyed. It was petty of her to feel this way and she tried to rise above it, especially when there were more important things to concern herself with, but deep down this knot twisted in her stomach.

"Morning," Constable Hargreaves said toward Lily, a greeting that was devoid of all the charm of her greeting toward Blake. "I'm surprised to see you here."

"I'm a concerned citizen," Lily said.

"Lily has proven herself to be helpful on more than one occasion. She helped my father when he was being taken advantage of. I trust her implicitly," Blake said. Lily's heart was warmed at the words of praise, and she also had no small amount of enjoyment from the narrowing of Chloe's eyes and the pinched expression around her mouth.

"I just became aware of some new information," Lily said, and then proceeded to share what Sam had told her. Constable Hargreaves put her personal dislike of Lily aside for a moment and nodded. "That matches with what a few of the other people have told me. There's something else I've learned as well; our grieving mother was absent from the party for a time. I'm starting to think she may not have been as open with us as she claims."

"What about Heath?" Lily asked.

"As far as I can tell he never left the party. It seems he's more social than his wife and was reveling in the occasion. But I think we need to go and have another conversation with Cecilia," Chloe said, relishing the opportunity. Lily flashed a worried glance in Blake's direction as she feared for Cecilia's sanity. This time Blake and Lily followed Chloe and the constable did not try to deny them the chance to be present when Cecilia was being questioned. Perhaps this was because she was caught up in the excitement of following a lead, but Lily suspected it was because she wanted to spend more time with Blake, and at the moment it seemed as though Blake's friendship with Lily was providing a shield around her that allowed her to engage in the case.

Cecilia was alone in a room, still absent from the rest of the group. There were shadows under her eyes. Tears glistened and she looked pale. She trembled, as though all the horrors of the world had been visited upon her. Lily thought that was quite apt. She couldn't imagine what it would be like to lose a child. The role of a parent was to protect a child at all costs, to love them and care for them and if anything happened... well... it would make the parent quite distraught.

But sometimes there were parents who were not meant to be parents, who actively harmed their children. As heartless as it seemed, when Lily looked at Cecilia she had to ask whether the woman was merely a grieving parent, or if she had done something reckless and was wracked with guilt.

"Has there been any news?" Cecilia asked, her voice barely above a whisper. Her eyes were filled with hope, but that was quickly dashed by Chloe.

"I'm afraid not. But I'm not here to talk about Evan at the moment. I'm here to talk about you," the constable said.

"Me?" Cecilia asked, shocked.

"Yes, it seems as though you and your husband were arguing last night. Have you been having problems in your marriage?" the constable asked.

Cecilia averted her gaze and began gnawing on her lip uneasily. Her hands crept together in her lap, as though being drawn together like magnets. She began to pick at the skin around her cuticles.

"Everyone argues sometimes. It's just the nature of marriage," Cecilia said.

"What were you arguing about?" Chloe pressed.

"I don't remember."

"You don't remember, so arguing is a common occurrence between you and your husband?" Chloe asked.

"No, I mean—" Cecilia said, but Chloe cut her off.

"Because I certainly remember all the arguments I've had with boyfriends. It was just last night as well. Are you sure you don't remember anything? It seems strange to me that you would have an argument, at a party no less, and not remember what it was about."

"I just don't know," Cecilia hung her head and spoke in a small voice.

Chloe decided to change tack. "I've also learned that you haven't been entirely honest with me Cecilia, and if there's one thing that makes me suspicious of people it's lies. You'd think that with your son missing you would want to tell me everything, and yet you've been keeping something to yourself. You left the party for some time last night, didn't you? I've spoken to a few people who saw you walk upstairs, presumably to check on your son, and then they saw you leave quietly, closing the door behind you. Now, why wouldn't you tell me something like that?"

Cecilia shifted uncomfortably in her chair. "I didn't think it was important."

"You didn't think it was important?" Chloe's breath caught in her throat as she spoke.

"It was only for a moment, just to catch my thoughts. I needed some fresh air," Cecilia said.

Chloe pursed her lips and leaned forward a little. "I'm not sure that's the case. I think I know what happened. I think you had an argument with your husband and maybe it was one argument too many. You thought enough was enough. You went upstairs to get Evan and smuggled him out, pushing him out of the door in front of you. Of course he would ask you where you were going and all you'd want to do is get out of there. You would have told him that you were going away from Daddy, but he didn't want to leave Daddy. So he started making some fuss and trying to break away from you. Maybe you tried to stop him from calling out. Maybe you tried to drag him away, but it was dark and you didn't realize how close to the cliff edge you were. Then you saw him fall and you realized there was no taking back what you had done, nowhere to hide, so you came back in here and now you're playing the role of the grieving mother when you know exactly what happened to your son," the words poured from Chloe's mouth in a steady torrent, rushing out of her like marching soldiers running toward an enemy trench.

"No!" Cecilia cried, riding a wave of sobs. "I would never do that! I would never hurt Evan! I just left to clear my head. I left him in bed I swear. I wouldn't... I couldn't..." she said, and collapsed into tears once again.

Lily glared at Constable Hargreaves, shocked at the treatment of this woman.

"Can I speak to you outside?" Lily said in a stern voice. Chloe rolled her eyes, but allowed Lily to get her wish. Blake remained with Cecilia, calming the woman down after the ordeal.

"What was that about?" Lily asked in harsh words.

"That was me trying to find the truth of what happened," Chloe replied.

"Yes, but she might have just lost her son, surely you could be a little gentler?"

"I'm not here to be gentle Lily. I'm here to solve a potential crime. I can't think about the feelings of those who might be guilty."

"But you've seen what she's been through. Now she has to deal with knowing that people think she might have done it. That's not fair."

"It's not fair that a child might have died either. I'm here to get to the truth, no matter the cost. If this is the way Constable Dudley was going about things then it's no wonder things have been such a mess here. It's fortunate I've come along when I did. There's a lot of things that need to change around here, a lot. I'm going to find out what happened to Evan because it's my job. If people's feelings get hurt along the way then that's just what has to happen," Chloe said brusquely.

Lily was stunned at the hard way she spoke, as though these crimes were just puzzles to be solved without involving real people. Her heart ached for Cecilia and for whoever else was going to be accosted by Chloe and trampled on in her search for the truth. It was as though she was a bull rampaging through the case, intimidating people until one of them cracked, but surely that wasn't the right way to go about things.

Chloe moved off and spoke to other people, while Lily waited for Blake to emerge from the room.

"How is she?" Lily asked, concerned about Cecilia.

"She's pretty broken up about it, but I think she's okay, considering everything."

"I can't believe Chloe would do something like that. She had no evidence, she was just putting together a theory."

"I suppose it makes sense to see how Cecilia would react," Blake said.

"You agree with her?!" Lily asked in disbelief.

"I didn't say that, I just mean I can understand her methodology. She certainly has a different way of going about things."

"Yes. The sooner that Constable Dudley can get back to work the better. I'm not sure she's the best fit for this place."

"Well, you might have to get used to working with her for a while. I know how long these investigations can take."

Lily shuddered at the thought and shook her head. The sooner Chloe left Didlington St. Wilfrid the better.

〜

It seemed as though Chloe was not willing to give Lily any further time as the police officer continued making the rounds, meaning that Lily was left to her own devices. After having been involved in multiple criminal cases she didn't feel as though she could just go back to Starling Cottage and get on with her life. She wanted to be involved, especially if there was something nefarious going on. From what happened last night Chloe seemed to think that this was an accident, and Lily doubted that anything she had heard so far had changed her mind. But there was still the husband to speak to. Lily was following Chloe's lead, which was a little awkward, but she hoped that she could find something that Chloe had missed, and in this way she could insure that Constable Dudley would return to his rightful place as the protector of this village.

Heath was standing on the porch, as he had been for a lot of the time last night. He was impressively tall, and as he stared out at the sea there was a pensive air to him. his face was like stone, and as with Cecilia there were shadows under his eyes.

"Excuse me Heath, do you mind if I have a word with you?" Lily asked.

Heath turned, startled, for he had been lost in his own thoughts. "I don't mind Lily, it's better than standing here and staring out at the ocean," he said.

"Yes, well, I don't mean to pry, but I was wondering if I could ask you a little bit about what happened last night," she began.

"I just told the constable, so I suppose there's no harm in going over it again. Just as long as you change my name when you write the book," he said with a knowing smile. There was a kind of effortless charm with him that Lily sensed was dangerous. It was the same kind of thing that Blake had, and she wondered if it was a product of wealth or just something that handsome English men possessed.

"I promise," Lily said with a smile. "I hope this isn't too personal, but is everything alright with you and Cecilia? It's just that I noticed you and she reacted to things in a very different manner."

A slight coldness appeared in his eyes. "We always have," he said. "When something happens I prefer doing something about it. She gets overwhelmed easily."

"I'm just a little surprised that you're not with her, comforting her."

Heath hung his head. Perhaps he hadn't expected her to get so personal so quickly. "Sometimes it's just hard to be around her when she's like this. I can't do anything to fix it, you know? And it makes me feel worse. But Cecilia... I love her, but she never does herself any favors. She always lets herself fall into despair."

"Is that what your argument was about?"

"No, no I've long given up trying to change that about her. We were arguing about her brother, my brother-in-law."

"Her brother?" Lily asked.

Heath nodded. "Louie," he said darkly, almost spitting the name.

"Why were you arguing about him?"

"Because he's the bad apple of the family. Every family has one, except that he keeps coming back no matter how hard I try to get him to stay away. He usually leaves for a time when he finds some scheme to make himself rich, but when that inevitably fails he comes crawling back to her, begging her for money to keep him going, and she always obliges him. Which wouldn't be a problem, except the man is a leech and he's never going to stop, and it's not her money that she's using to help him. It's mine, and frankly I'm tired of it. He came by again recently and I told her in no uncertain terms that I'm not going to allow her to help him again. Of course she thinks that family is the most important thing and she should help him if she can, but it's not going to do her any good. It's better for everyone if he just stays away. Sometimes I think he just comes by to get some easy money. He has all the fancy suits and a flashy car you know. He wants to flaunt things, but he doesn't want to do the hard work to pay for it himself." A smile crept over his face. "He loves that car more than anything. I've often thought of taking that ugly red thing away from him and sending it to the scrap yard. I've basically paid for it from all the handouts that Cecilia has given him over the years. He pleads poverty and yet he drives a Cadillac he got from some collector. I don't know what people must think of him when they see him driving around the English countryside like that. But I'd love to see his face when it got crumpled," Heath said.

Lily found it a little strange that Louie hadn't been mentioned before. "Do you happen to have a contact number for him?" Lily asked.

"Sure thing," Heath said, and then flicked through his phone to give her the information. "I'm not sure if you'll be able to get a hold of him. After I told him to leave I think he's gone elsewhere. He's probably out of the country at the moment, looking for some other scheme to get some money," Heath said. Lily keyed in the number. "Listen, I want to thank you for all your help last night, and today. I know this isn't what you wanted from your party. I just wish things had gone differently."

"It's okay. I'm sorry too. I know this is going to be hard for you. I hope that you and Cecilia can manage to work through this."

"We'll see," he said, "but if I know her she's never going to give up hoping that he's still out there somewhere. The more I look out to sea the more I think that he must have just wandered off alone. It's not like him, but then sometimes we all do things that are out of character, don't we?" he said.

Lily wondered if there was any hidden meaning to his words. For a moment she wondered if he was relieved that the attention was being pulled away from a murder, but then remembered that he had been at the party all night, so was beyond suspicion. It did seem more and more likely that Evan had been the victim of some unfortunate accident, but then there was this intriguing kernel of information about the brother-in-law, and although Lily was reluctant, she went to share the information with Chloe again.

"Yes, I'm aware of it," Chloe said when Lily presented her with the phone number. "I do know how to question people. I know you've been working with Constable Dudley so you're perhaps of the opinion that the police in the village are incompetent, but you don't have to buzz around trying to tie up any loose ends I might have missed," Chloe said. "In fact I've already spoken to the man on the phone and he seems to have a solid alibi. I was just about to leave to check it out myself, but I think, as tragic as it is, all the evidence is pointing to the fact that Evan left the house of his own accord and fell off the cliff. It's sad that we're never going to know what was going through his mind. Maybe he just wanted to see something outside, or maybe he heard his parents arguing and wanted to get away from them. But I can't spend any more time on this when there doesn't seem to be any clear motive or any suspects. The only guilty party here is gravity," Chloe said.

Lily still thought it was a little too soon to be drawing conclusions like this, but she wasn't about to try and change Chloe's mind. In fact she was glad when Chloe left the scene, which gave Lily the opportunity to go and speak with Cecilia again.

She slipped into the room and closed the door behind her. "Cecilia, I just wanted to say that I'm sorry for everything you've been through. We're still working hard to get to the bottom of this. I just wanted to ask you about something though."

"Questions, questions, is there ever going to be an end to these questions? You're not going to be like her, are you? I don't want to be yelled at again," Cecilia said, fear lining her face.

"No, I'm not. I promise. I just wanted to ask you about Louie."

"Louie? What's he got to do with this?"

"Well it's just that Heath told me he was the cause of some tension in your marriage."

Cecilia sighed and looked to the ceiling. "He is, but then Louie has always been trouble. But he's family, what can you do? I try to help him where I can, but Heath tells me that I shouldn't. Maybe I am being taken for a fool, but he's my little brother. I can't just ignore him. We're the only ones left you know. Without me he has nobody. I can't just turn my back on him. Heath has never been able to understand that. But you do, don't you?" she asked, her eyes filled with hope. Lily could tell that after all she had been through Cecilia was just trying to cling to anyone who might understand her, to find some common ground that could push back the loneliness for a short while.

So Lily lied to her and told her that she did understand, when deep down she had never had much love for her family either. There was a long history there and memories that Lily preferred to remain buried deep within her heart, so she steeled herself against the rising emotion and left Cecilia, hoping that Heath would come by to comfort his wife soon.

6

Lily was sitting at her desk, staring at a blank page. The cursor on her laptop flickered, almost taunting her. She had hoped to lose herself in a story, but her thoughts rolled around in her mind, thinking about the tragedy of Evan and how helpless he must have felt as he fell down into the abyss. It was almost as helpless as she felt now. She had first come to Didlington St. Wilfrid when she needed to escape from an overwhelming and stressful life in Hollywood. The quietness and peace here was alluring and for much of the time it was enough to comfort her, but the mysteries that had been interspersed into her life here gave her a sense of purpose. She found that she liked helping people and she didn't want that to stop. She had always been of the opinion that if you had the ability to help people then you should, and she didn't like being refused by the new Constable.

Perhaps there was still a way for her to continue in this same vein even without Constable Hargreaves's opinion, but she wasn't sure how.

She typed out something, and then immediately deleted it because she didn't think it was up to her usual standards.

Then she got to thinking about Blake and family. Back in Hollywood she had barely had time for romance, and the romance she did have didn't leave her with the best impression either. Blake would surely want someone who could be a good wife and fit into his world, and even give him a family one day. Lily just wasn't sure she was that person. There were so many unknowns, so many things that she was still unsure of. It was beginning to feel as though her life was becoming as messy as it had been in Hollywood, and the only common denominator was her.

To clear her head she decided to go out for a walk with Zeus. His behavior never failed to make her smile as he didn't have a care in the world. She wondered if deep down he was worried about Pat though, but of course she couldn't know the true feelings of his heart.

As she walked along she thought about Evan and who was truly responsible for what happened to him, if anyone. Were Cecilia and Heath to blame for being negligent? Should they have checked on Evan more? She thought back to her own childhood and the way her mother had treated her. Lily had been forced to grow up quickly, taking on more responsibility than a child should. Her mother was still out there somewhere, probably making trouble in someone else's life. Still, Lily's heart sank at what she had lost, and at what she had been prevented from enjoying in life. Sometimes she wondered what her life would have been like if she had learned how to trust other people more freely, rather than relying on herself for most of the time. Maybe if that had been the case she would be able to fall in with other people and know how to act with them, taking Blake as the example.

Just as she was thinking of him, and just when she had been heading to Marjorie's to thank the baker for the sweet surprise she had left, Lily's phone rang and her heart skipped a beat when she saw that it was Blake who was calling her. She wondered if he was going to apologize for her party having turned into a crime scene. It was just like Blake to want to make it up to her, although she hoped it wouldn't be in the form of another party. One of them was enough, but an intimate dinner would have been something Lily could have enjoyed.

"Morning Blake," she said.

"Morning Lily. Listen… there's something happening here that you might want to know about. It seems as though we have an uninvited guest," Blake said. Lily found herself intrigued and raced back to Starling Cottage, throwing herself into her car.

She pulled up to the beach house and groaned inwardly when she saw the police car sitting in the drive way as well. There were far fewer cars than there had been last night, as people began to return to their lives, no doubt shaken by the events that had happened. But one car did catch her attention; a bright red Cadillac.

It seems as though Louie had made himself known.

Blake was standing at the doorway with a beleaguered look on his face as Lily entered to find Heath standing on one side of the room with Cecilia on the other. Another man, who she assumed was Louie, stood beside Cecilia and had his arm around her. With the resemblance to each other there was no doubting that they were brother and sister. Chloe was standing in the middle, her gaze darting between the two men. Zeus growled a little, sensing the hostility in the air and moving to guard Lily from anything that might threaten her.

"It's time for you to leave Heath. You've never been good for Cecilia. I tried to convince her of it, you know, I tried to talk her out of it, but she's always loved you too much. I know you had something to do with this. What, did you get bored of being a father? Or did you just want out of the marriage? I know that you've been cheating on her," Louie said.

Heath was remarkably composed for a man who had just been accused of cheating on his wife. He wore a wry smile. "It amazes me how you can lie just as easily as breathing Louie. You shouldn't be here. I thought you were elsewhere in the world, finding your fortune, but then you've never been any good at that, have you? You're nothing without Cecilia and I really wish that you loved her as much as you claim you do, because then you might actually leave her alone. Can't you see that we're grieving at the moment? We need to be alone. We need time."

"No, she needs time away from you. I'm tired of this Heath. I'm not going to have her suffer any longer. Come on Cecilia," he said, gently coaxing her away. "I have a hotel nearby. I'm not going to have you suffer here anymore. We can finally be free of this parasite," Louie said. Cecilia was so weak she didn't seem to offer any resistance, and Lily was left wondering if there was any truth to what Louie was saying. Heath snarled and scowled.

"Get away from my wife Louie. You might be her brother, but you've never cared about her. I don't know what you're doing here, but it's time for you to leave," Heath said.

Lily noticed movement by her side. Chloe ducked out of the room, pulling her phone out. Lily thought it strange that Chloe should decide to leave at this moment when things seemed to be heated, and it led to her questioning the constable's commitment. Perhaps she thought that getting involved in a domestic dispute was beneath her.

Blake started forward. "Let's all calm down. I'm sure we can talk about this."

"No, we're beyond talking. If Louie is here and wants to sort things out then we can sort things out. How about we start talking about settling the debt you owe me? I'm thinking we can start by giving me your car. I've always wanted to turn it into scrap," Heath said.

Louie's face twisted and he started to scream.

"Hang on a minute!" Lily cried, yelling loudly to get the men's attention. They both stopped and looked at her. "Let's not forget what this is really about. We're here to try and talk about Evan and find out if anything happened to him. Louie, when you said that Heath wants to leave the marriage was there any truth to that? Is there anything else that we should know, things that Cecilia or Heath might have been keeping from us?" Lily asked. Heath looked a little betrayed, while Cecilia just looked forlorn, as though she wanted the whole thing to be over.

"I'm sure there is, but I don't know what he's told you exactly. I don't know anything about Evan though. I've always liked the kid, but I've never been able to spend much time with him because I'm not exactly welcome," Louie said.

"And with good reason," Heath said, which led to another loud argument. It seemed almost inevitable that the two men were going to come to blows. Lily glanced helplessly at Blake, who was trying to stand in between them and stop them from getting too heated. Lily thought it was a shame that Evan's death should be marred with this family tension, and it seemed as though the hatred between Heath and Louie ran far deeper than any of them had implied. All she wanted was to tell them that they should use this opportunity to come together and not be so stupid, because they were a family and if they weren't going to come together now then when would they? But she also knew there were likely things that she was not aware of. While Louie was a liar perhaps he hadn't been lying about everything. After all, Heath hadn't been quick to comfort his wife and had spent most of the time keeping his distance from her. Given the tension in their marriage it wouldn't have surprised Lily to learn that he was cheating on her, and perhaps that was one of the reasons why Cecilia wanted to maintain a relationship with her brother. What if Louie was the only way out of her marriage, Cecilia had?

While it wasn't exactly a crime, Lily still felt as though she should get involved somehow, but before she could figure out what to do

Chloe returned to the room, striding in, holding Evan's scarf. She didn't seem to pay any attention to the argument that was raging and stood in front of Louie, unfazed.

"Do you recognize this?" she asked.

Louie took the fabric and turned it over in his hands. "Given the size I assume it's Evan's? I'm not sure why you're showing me though. Unfortunately I wasn't here in time to help him. I'm here to help my sister before it's too late. I know that Heath is responsible for this somehow. If you knew what he gets up to…" Louie trailed away.

Chloe arched her eyebrows. "Indeed, well, the thing is Louie that I know what you have been up to," she said, and Louie fell silent. "After Heath told me about your character I did look into you, and while your record shows some petty misdemeanors, and it's quite clear that you are a dishonest character, I was ready to discount you. After all, as you said on the phone you were at a show and you even sent me a picture of the ticket. Of course, that event began well before Evan disappeared, and I decided to do some asking around. It turns out that some people saw an unusual car last night. Now, of course this isn't enough to place suspicion on you as it was too dark to read the plate number, and while it would be extremely uncommon for there to be two Cadillacs in the area in such a short space of time it's not beyond the realm of possibility. So it seems as though you don't have anything to do with this crime at all, and that indeed you are here because despite your dishonest nature and your involvement in bending the law when you see fit, you might have some kernel of honor inside you and you want to help your sister in this time of distress," Chloe paused. Lily wondered where she was going with this, and Louie had a smug look on his face, as though Chloe was stating the obvious.

"But then you said something that I thought I should look into. This hotel you're staying at, well, there aren't many hotels in the area so I made some calls and surprisingly there's nobody with your name staying in any of them. Now why would you want to use an alias?

Well, then I gave them your description and there was one hotel that did recognize you, except that you've been staying there for a few nights now, and that got me wondering why you would have been in the area before you even knew your sister was upset?" Chloe turned to Lily. "Lily, would you mind bringing Zeus here please," she said.

Lily duly obliged as the color faded from Louie's cheeks. He looked around, but was backed against a wall. Zeus came up to him and sniffed the fabric, and then sniffed the man. He barked loudly and pawed at Louie, who fought to try and get the dog off him.

"I don't know what you're doing but this is police harassment!" Louie cried. Meanwhile Cecilia pulled away from him and Heath strode across the room, grabbing Louie by the scruff of the neck.

"What have you done with my son?" he thundered. Lily and Chloe glanced at each other and knew what they had to do. Blake stayed behind in an effort to keep the peace, while Zeus joined Lily. All animosity that existed between Lily and Chloe was pushed aside for the moment as a child was in need.

7

L ily, Chloe, and Zeus poured out of the car when they reached the hotel. They crashed in through the doors and the receptionist was about to lambast them, when Chloe flashed her badge. She ordered the receptionist to tell her which room Louie was staying in (although obviously Chloe had to give the name of the alias he was using).

"Has anyone cleaned the room?" Chloe asked.

"No, he left strict instructions that the room was to be left undisturbed," the receptionist said. She pulled a key from a rack behind her and led them upstairs, through the winding halls toward the room. She unlocked the door and opened it, allowing Lily, Chloe, and Zeus to move in. The room was sparse and empty. At first Lily was disheartened and wondered if Louie had actually killed the child, but then Zeus ran straight to a closet door and began scratching and sniffing at it. The door was locked and it took both Chloe and Lily yanking together to open it. Behind the door they could hear muffled whimpers. Lily strained her muscles. Sweat prickled on her brow as she summoned all the strength she could muster, and together with Chloe they were able to yank the door open. The receptionist didn't seem too keen to hear the sound of wood splintering, but she looked shocked when she saw who was in the dark closet.

The scared face of a small boy peered out. His hair was tousled and his eyes were ringed with tears. His hands and feet had been bound together, and Lily's heart immediately broke for him. Chloe pulled out a Swiss army knife and cut through the rope, while Lily pulled the gag away from his mouth.

"Are you Evan?" she asked, and the boy nodded. "It's okay Evan, we're going to get you back to your parents, okay? You're safe now," Lily said. She turned to Chloe and nodded her appreciation, a sign of begrudging respect. They gave Evan some water and a little bit of food, and Chloe gave him a quick check to make sure that he hadn't been harmed, other than the biting rope burns that were red around his wrists and ankles. Chloe showed him her badge to prove that he could trust her, and he seemed to relax when he realized that Louie wasn't with them. Zeus helped to calm him as well. The dog walked up to the boy and rested his head on Evan's lap. Evan petted him and soon enough he was breathing calmly.

"I know you've been through a lot Evan and we'll get you back to your parents soon enough, but I'd just like to ask you a few questions about what happened, if you feel up to answering them?" Chloe asked. Evan nodded. "It was your uncle who took you, wasn't it?"

Evan nodded again.

"How did he come to you?"

"He was in my room. I was asleep and then I woke up. He said that he was there to take me home because I wasn't to be at the party any longer. I didn't think much of it, although he said that it was a game and we had to be quiet so nobody saw or heard us, but he was holding me hard and hurting me, and when I told him he just told me to be quiet. I've always been told not to go away with strangers and even though he's my uncle I thought I should still ask so I went to go into the party and he pulled me back. And then I tried to cry out and he put a hand around my mouth. He pulled me outside and it was dark and cold. I wriggled and managed to get free, and I just ran. But he chased me. I couldn't get away from him. I was looking behind me," terror gripped his face, "I fell. He grabbed me and pulled me back, but I lost my scarf. Then he carried me and brought me here."

"Do you know why he brought you here?" Chloe asked. Evan shook his head. "Well you've been a very brave boy and I'm sure your parents are going to be proud of you. We're going to take you back to them now, okay? Your uncle is there, but I promise he's not going to hurt you."

"Zeus won't let him," Lily added with a smile. At the mention of his uncle, Evan cowered in terror, but he seemed relieved when he was promised that no harm would come to him. Chloe led him out of the hotel. Lily helped him get into the back seat of the car, where Zeus sat with him. They made their way back to the beach house and pulled up to the sight of Louie yelling loudly, trying to get into his car. Blake and Heath were standing in the way, while Cecilia was apart from them, looking distraught. Chloe pulled up and got out, immediately going over to help Blake and to arrest Louie, while Lily was left to open the door and bring Evan out. He was a little reticent considering all the noise that was happening, but as soon as he saw his mother his face lit up and he ran over to her. Cecilia reacted as though she was in a dream. She saw Evan running toward her and opened her arms wide, running toward him. She wrapped him up in her arms and whirled him around in the air, and suddenly all the tension that had been building up inside her over the last twelve to fifteen hours vanished as she squeezed her son. It was a happy reunion, and it brought a tear to everyone's eye.

Heath suddenly saw this and ran over to his family as well, and if there had been any doubts about the way he felt about his son they were vanquished by his reaction. He kissed his son's head and looked as though he was on the verge of tears. He turned back to glare at Louie and his words were ominous.

"What were you thinking? I'm going to kill you!" Heath yelled.

Chloe seemed to understand that the threat was borne of anger rather than having any strong intention. She remained calm as she clasped the handcuffs around Louie's wrists, not making any effort to make it comfortable for him.

"Yes, I would like to know what you were planning too," Chloe said. "It would be a good idea for you to cooperate now."

Louie hung his head. "I didn't mean any harm to the kid. I just wanted to speak to him. I wanted to convince him to speak to Cecilia on my behalf and convince her to give me the money I needed. She kept saying that Heath wouldn't let her, but I knew if Evan asked her she would change her mind. But I never got to ask him. All I wanted was to pull him outside, but then he had to go and run away and try to make noise and I knew that people would start to think the worst. So then I threw him in the trunk of the car and figured that I might be able to send a ransom note in a few days."

"And you thought you could show up here and turn suspicion onto Heath," Chloe said. Louie nodded and looked forlorn.

"I had every right to the money. If they had just given me the money in the first place then we wouldn't have had to go through any of this. I wasn't going to hurt him, I just wanted my fair share. We're family. They're supposed to take care of me," Louie said. Lily felt an echoing sensation in the back of her mind, remembering a time in the past when her mother had said something similar.

Cecilia marched up to him and jabbed a finger in the air. "I can't believe you did this Louie. He's my son! That's it. I want nothing more to do with you. Heath has been right all this time. I've been too soft on you. You're just no good Louie, no good at all. I never want to hear from you or see you again!" she cried out, and then fled back to her husband and son. Louie grumbled as he was led away and made one final plea, but this time it wasn't about Cecilia or money, it was about his car.

It was doubtful he would ever see the vehicle again.

Chloe drove away, leaving the scene of the crime behind.

Lily and Blake walked up to the family, along with Zeus, who was being petted by Evan.

"Thank you so much," Cecilia said. She wiped her eyes. It was impossible to not feel pity for her after the rollercoaster of emotion she had been through. "Both of you," she said.

"Think nothing of it. I'll just have to make sure I arrange for some security next time I have a party," Blake said.

"I wouldn't blame yourself. He's a slippery one. I'm just glad he's out of our lives. We can go back home now and everything is going to be better," Heath said, giving a meaningful look to his wife. Her smile faltered. They made one final thanks before gathering their things and driving away. Lily and Blake watched the car disappear into the distance, and then suddenly they were alone. Blake invited her in for a drink and a bite to eat.

They were sitting in the kitchen at a table, where the wide windows showed the beautiful horizon and the dancing sea. It was a much more comforting sight now when Lily didn't have to think about a child being lost in its maelstrom.

"Well I'm glad Evan was returned safely," she said.

"Me too. It's been a long fifteen hours though. I can't imagine what Heath and Cecilia have gone through."

"Yes, although it does make me wonder about them…" Lily said, wondering if she should elaborate on her thoughts. Blake looked at her with open eyes, waiting for her to continue. "I just mean that there's clearly something not right between them. I know they looked happy at the end there, but I can't forget the way Heath seemed so far removed from Cecilia all through the ordeal. It makes me wonder if there was some truth to what Louie was saying."

"You think you can trust anything he says?"

Lily shrugged. "I'm not sure, but I don't think we can just dismiss everything he said. There definitely seemed to be some tension between them."

"I suppose all relationships are complicated," Blake said.

"Yes," Lily said, annoyed that another thought, she would rather not think about, gnawed at the back of her mind.

"Speaking of complicated relationships, how do you think you're going to get on with the new constable now that you've spent more time with her?"

"I don't know. She's clearly capable at her job, but there are certain things she does that just rub me the wrong way. I suppose we'll have to see how it goes. I still firmly believe that Constable Dudley will be reinstated before too long, despite what you think," Lily said.

Blake held his hands up in mock surrender. "I hope so too, I just know how long things like this can take. So are you going to take a step back from helping out on cases?"

"I hope not. I've been thinking a lot about it, and this is the first time in a long time where I've actually felt useful."

"You didn't feel like that in America?"

Lily shook her head slowly. "Not at all. I had no time to myself at all and I was a complete mess. I lost myself in my work. That's why I came over here."

"And now you have found yourself?"

"I think so. At least I'm starting too," she said, smiling uneasily.

"Well that's always a good thing. There's never anything wrong with a little self-discovery. I've been thinking a lot myself about certain things as well."

"Oh yes?" Lily asked, her ears pricking up at this.

"Yes, and actually seeing Heath and Cecilia with Evan there does get me thinking about the future. So much of my life has been taken up by work and now that I've taken a step back to spend more time here I've been thinking about what could fill the gap. I know you're busy with writing and things, but have you given any thought to parenthood?"

Lily averted her gaze and blushed slightly. "I'm not exactly sure. It's only now when I've really had the chance. I suppose there are certain things that need to happen first."

"Oh of course, but you must know whether you'd be suited to motherhood or not, yes? It's always been expected of me to have a child, you know, being a part of the English nobility. He hasn't said it for a while, but my father is disappointed in me for not marrying yet. I suppose I've always had this feeling that I have all the time in the world, but the more I reflect on things the more I know that's not the case. I think it might be time to begin making some changes in my life," he said, and gazed at her intently. Lily felt herself getting hot as she blushed a bright red, and wasn't sure how to respond. Was he really saying what she thought he was saying, or was she reading too much into it?

"I'm sure there must be many women who would be well suited to being your wife."

"Yes, well, thankfully we don't live in the era where marriages are forced upon those of a noble class," Blake said with a smile. Lily wasn't sure if it was just a joke, or if he was intimating something else, something that she didn't dare dream about.

They smiled and continued chatting idly. The conversation never reached a dangerous edge again, but Lily was left wondering if she should have taken that leap and confronted her feelings, but there were still things about her that Blake did not know, things that might make him look at her differently.

Still, at least the family had been reunited, and at least she did not have to mourn a child. A little while later Lily called Zeus to the car and returned to Starling Cottage, where she spent time with Joanie and then lost herself in the new story she was writing. Meanwhile, in the back of her mind, she was thinking about how she was like a hook dragging the past along with her, for although she was far from America and her old life she couldn't quite escape it at all, and upon witnessing the reunion of Heath, Cecilia, and Evan, she wondered about what a reunion with her mother would be like.

It left her feeling uncomfortable and uneasy. There was a lot that woman had to answer for, but should Lily be like Cecilia and disown her mother completely, or should she make some effort to contact her again? After all it had been years. Perhaps her mother had changed...

8

A couple of days had passed since Louie had been arrested and charged. Lily had buried herself in her work and had barely come up for air. It was like that sometimes. Being in the midst of a story was akin to being in the middle of an ocean, completely surrounded by this other environment and without any desire to leave it behind. Time flew and became this elastic, malleable thing, while the only breaks she took were to eat and take Zeus for a walk, although Joanie also helped with this. However, she was taken away from her story when she was invited to a small ceremony in the city hall, where she, and more specifically Zeus, had been asked to attend.

The city hall was a small building that Lily had had no cause to visit, as of yet. It was filled with friendly faces and those she had only seen in passing. Chloe was there of course, sitting in the front row. The only person who she didn't see was Constable Dudley. The poor man must have been so ashamed by what was happening that he didn't dare show his face. She couldn't blame him, but she hoped he would not retire into his shell completely and she made a note to go and see him again. Perhaps she could take Joanie with her, because the Constable had a soft spot for Joanie and no doubt it would brighten his day.

The Mayor was greeting people as they came in and beamed when he saw Lily. He was a stocky man with thick sideburns that grew into whiskers, and reached down to his lips. His eyes were bright and his cheeks red. He wore a gold chain, signifying that he was the mayor, and seemed to revel in the lively atmosphere. He clasped her hand in his and shook it heartily.

"Lily McGee, well it's a pleasure to meet you finally! I've been meaning to invite you to share a cup of tea. You've caused quite a stir in our little village," he said.

"Thank you Mr. Mayor."

"Oh please, call me Steven," he beamed. "I must have an audience with you soon so that you can tell me everything that has been happening. I have a keen mind for this sort of thing myself you know, but who would have thought all these terrible things have been happening right under our noses! Still, all's well that ends well, eh? Well, help yourself to a drink or something and we'll get underway soon, and we'll have a special treat for Zeus here," he said, leaning down to pet the dog. Zeus smiled widely and his tongue lolled out. He too reveled in the social atmosphere.

Lily and Joanie took a seat near Chloe. She craned her neck around and saw Blake in the background, although he was deep in conversation with someone she did not recognize. The Mayor cleared his throat. His booming voice filled the hall.

"Thank you for coming here tonight. I know we have suffered through many strange times recently and it seems as though our little village has been quite the hub of some undesirable activity! When I was told of what happened recently my heart went out to that poor boy. I think it's clear that we must all remain vigilant about our surroundings to prevent anything like this happening again, and I'm glad to say that one member of our village was able to sniff out trouble," he took a moment to chuckle at his joke. "Now then, I'd like to welcome Zeus up onto the stage," he said. There was a huge round of applause as Lily led him onto the stage toward the Mayor, who extolled Zeus' abilities and said that he was one of the stars of the village. He pinned a small medal onto Zeus' collar, and Lily thought that Pat would be proud when he eventually returned.

Zeus was fussed over tremendously and everyone seemed to adore the dog. Although he couldn't have completely understood what was happening. Zeus was smiling with his eyes and he must have known deep down that this was all about him. The people poured around him, petting him and letting him lick their hand, and he loved every minute of it.

Lily, though, didn't, and she ended up beckoning Joanie over to offer her some respite.

Lily walked away and saw Chloe standing to the side of the room, watching everything with a pensive stare. Although Lily wanted to ignore Chloe, the new constable wasn't going anywhere and Lily thought she should at least try and insure they had a cordial relationship, otherwise she would never be able to help on a case again.

"Good evening," Chloe said when she saw Lily approach. "I'm glad Zeus is getting his due."

"Yes, well, I think it helps everyone to have an animal to rally around. I don't think he's the only one the Mayor should be thanking though. That was some good policing you did there. I'm impressed."

Chloe paused for a moment, as though she was going to say something and then thought better of it. In the end she settled for thanking Lily for the compliment.

"I know we got off on the wrong foot," Lily said. "I don't want you to think, that me getting involved is disrespecting your work as a police officer. I just want to help, really."

"I know," Chloe said. "And I can see that people here appreciate the help as well. Everywhere I turn there are people talking about you and the way you've helped the village since you arrived. You're something of a celebrity around here."

Lily cringed at the term as she had had her fill of actual celebrities in Hollywood, and had never been enamored with them.

"Well, I just wanted to say that I'm glad I could help you bring the case to a close, and that I am more than willing to help you again, if it's something that suits you," Lily said in an effort to be as diplomatic as possible.

Chloe inclined her head and did not dismiss the idea immediately, which Lily took as a victory. "While it's somewhat unconventional, I can certainly see the appeal that you have here and I would not be doing my job if I did not use every resource that was available to me. We'll see Lily, we'll see."

That was about as good as Lily knew she was going to get for now, and it at least left the door open to her helping on another case, although she doubted her working relationship with Chloe was going to be as smooth as it had been with Constable Dudley. Still, if she impressed Chloe then perhaps Chloe's opinion of her would change over time.

However, Chloe then said something that shook Lily to her core. "By the way, you seem to be friends with Blake. Can you tell me a little more about him?"

"What do you mean?" Lily asked.

"Well, you know, is he involved with anyone? I just thought that if I'm going to be here for a while then I might as well have a little fun while I'm going about it." A wry smile crossed her face as her gaze drifted across the room to settle on Blake. Lily's throat tightened and she wondered if she had missed her chance. Chloe was determined, confident, English, and she had a good job with the police. If Blake was looking for someone to fit into his world then he could do far worse than Chloe.

"I don't believe so, no," Lily admitted forlornly.

"Good, then I think the prospect of staying here has just increased. I have to admit Lily that I wasn't looking forward to coming here when I received the assignment, but I'm beginning to see the perks this place has to offer," Chloe said, and then moved away, walking with purpose and confidence. Lily's heart sank and she wondered if she had just fooled herself into believing that she belonged here. Blake was a good friend, but was that all she was? Was that all she wanted?

She knew she was going to have to figure it out, and quickly, otherwise she might end up missing out on something before it had a chance to begin.

9

A fter her conversation with Chole, Lily found her tolerance for enduring a social occasion was ebbing away. She thought about speaking to Blake, but had no idea what she was going to say to him. Her mind was hazy and she just wanted to be alone, so she went and asked Joanie if she could stay with Zeus for the rest of the evening, for it seemed cruel to take the hero of the hour away from all of this attention and affection. Joanie didn't mind at all, because like Zeus she enjoyed the attention of the public as well and threw herself into conversations, turning strangers into friends within mere moments of meeting them. Lily envied her a great deal as she had never found it as easy to get along with other people.

She walked through the village, enjoying the peace that came with the late evening. Most of the houses were dark as people were at the hall, while others like her had left earlier and now their homes glowed with the soft illumination sneaking through the windows. Stars glittered in the sky above, and Lily continued to marvel at just how many of them there were. After living in L.A. she was used to seeing a bleak sky, one that was rendered dull by so much light and pollution. Here the moon was larger than life, almost as though it shouldn't have been real. She considered herself lucky to be in Didlington St. Wilfrid, but at what point was writing going to be enough?

When she reached Starling Cottage she noticed that a package awaited her on the doorstep. At first she was a little surprised, before remembering that it was like the one she had received earlier. As she grew closer the sweet scent of a baked treat reached her nose and immediately made her feel better. She chuckled to herself because she assumed it was Marjorie's way of telling her it had been too long since she stopped by. Lily took it inside with her and unwrapped the contents. It came with no note, but the contents were delicious. Lily stayed there for a short while before she decided that although she didn't want to be in a crowd, she didn't want to be alone either. Clearly Marjorie missed her, and in truth Lily missed Marjorie as well.

Lily knocked on the door to the bakery. Marjorie was busy baking things for the following day, so she was still downstairs although the shop was closed.

"Good evening Lily, is anything the matter?" she asked.

"Well I just know that you've been missing me recently, so I thought I would stop by."

Marjorie chuckled. "Well someone certainly has a high opinion of herself, but I suppose it has been a while since we've had a chance to catch up. I heard about what happened at Blake's beach house. What a terrible thing. Anyone who would think to hurt a child..." Marjorie shook her head and closed her eyes. "I just can't imagine what was going through their head. At least he's caught now, and I see that you had plenty to do with that."

"Well, not as much as usual. In truth I was always one step behind the new Constable."

"Ah yes," Marjorie said. "Bob was in here earlier and I was asking him about her. She doesn't seem like the shy type, although I suppose you can't be in that boys' club. I assume she's good at her job then?"

"She is, in fact so good I'm not sure she needs my help at all. It was Zeus that she needed."

"Well, it might be good for you to take a step back from all this grisly crime stuff. You spend enough time writing about them, I'm sure the last thing you need is to be faced with them day in and day out, besides I'm sure there are plenty of other things that you could spend your time on, like a certain handsome man who just threw you a party."

Lily laughed off the comment. She wasn't sure she wanted to talk about Blake right now, especially when she knew that Chloe had set her sights on him, and Chloe seemed the type of woman to always get what she wanted. "I'm not sure about that. Blake and I are just friends."

"If you say so. For what it's worth I think you two make a cute couple. Or is there someone back in America who has stolen your heart?"

"Definitely not," Lily said, shuddering at the thought. Her gaze drifted toward the picture that hung on Marjorie's wall, of her dearly departed husband. Lily had come along too late to know him, but by the way Marjorie spoke about him so fondly Lily felt as though he was an old friend. "How did you know it was right to marry?"

"Well," Marjorie sighed, washing her hands in the basin after cleaning up what she was doing. "I suppose in our day it was a little different. It was always expected that people should get married sooner rather than later, and after the war people had this desire to bring some good into the world. We'd seen and heard about so much death that we all wanted to celebrate and bring a little life to balance things out I suppose. I never had many grand thoughts of love, and there certainly weren't as many movies depicting all this amazing romances. All I wanted was a man I could trust and a man who would take care of me, and I found that," she smiled wistfully. "At the end of the day I think you can think about these things too much and they become overcomplicated. All you really need is someone who can be a friend, someone who respects you, and someone who is willing to put your needs above everyone else's. That's what love is at the end of the day, and when you feel the same way about that person it's the most wonderful thing in the world. If you ask me people are looking for something so perfect that it never exists, and all these high expectations only leads to people being unhappy."

Lily wondered if that was what had happened to Heath and Cecilia. "I wish it was as easy as that nowadays."

"It is, it's just that people make it complicated. Have you tried talking to Blake about how you feel?"

Lily turned away from the picture and gave Marjorie a helpless look. "Well that's just the problem Marjorie, I'm not sure how I feel."

"Then that's what you need to figure out. How is he supposed to know what to do if you don't know where you stand with yourself? Now I can tell you from experience that living alone isn't the worst thing in the world. I carry on day after day and I'm quite happy with my little shop and speaking to people who poke their heads in now and then, but life is always better when there's someone by your side whose shoulder you can rest your head on, and who can make you laugh with just a look. You're young, you should be sharing life with someone special. If you don't make your youth count then you're only going to look back with regret when you reach my age. If you end up waiting too long then you might find that the chance has gone," she said. But then she sighed and looked at Lily with reassurance. "But I'm sure you'll end up fine. You've got a good head on your shoulders and there's no need to get yourself into a daze with this. I'm sure things will happen as the good Lord intends, and you'll have all the happiness that life can bring you."

"I suppose," Lily said, trying to push the thoughts away so they didn't completely overwhelm her. "By the way, I was surprised that you weren't at the meeting tonight. I'll have to bring Zeus by to show you his medal."

"Yes you will, and unfortunately there was just too much around here to do. No rest for the wicked," Marjorie chuckled, but Lily knew that wasn't true. She didn't think there was a purer soul in all the world than this woman's. "I took some time out of my day today to go and see Pat. I'm glad he's on the mend."

"Yes, hopefully it won't be long before he's back. Although I do wonder what he's going to be like."

"Well I think we just might see a change in attitude. At least I would assume so, then again this is Pat so I have no idea."

They continued chatting for a little while, with Lily telling Marjorie all about the disappearance of the Evan and how the case unfolded. She stayed there for a good while before she yawned and also saw tiredness within Marjorie's eyes. Then it became clear that it was time to leave. Just before she was walking out of the door though, Lily had one last thing to say to Marjorie.

"Oh by the way, I do appreciate the things you've been leaving for me at Starling Cottage, but next time you want to see me just pop in for a cup of tea, you know you're always welcome," she said gaily, and then turned to walk out of the door. However, Marjorie seemed puzzled.

"What things?"

Lily turned slowly. "You know, the baked treats you've been leaving on my doorstep."

"Lily, I haven't left you anything," Marjorie said. Lily's throat ran dry. She feigned a smile and said that she must have made a mistake, but in her mind she asked herself who was leaving these surprises for her, and why. It was odd that someone should be doing this without leaving a note, which suggested they didn't want her to know, but why should they want their identity to remain a secret? The thought of someone coming uninvited to her door made her skin crawl, and she suddenly felt vulnerable without Zeus beside her. The quiet darkness of the village was peaceful, but it also meant there were more places for people to hide. She found herself walking briskly back to Starling Cottage. Her heart pounded as her gaze darted into the shadows, trying to sense any movement. She was filled with the unsettling sense that somebody was watching her, but who was watching her, and how long had they been doing it for?

By the time she reached her home she fumbled with her keys and slammed the door behind her, breathing heavily. She went to search the container for clues, but there was nothing to suggest who had left the treat there. Another mystery had presented itself, and this time Lily was the one at the center of it.

CORPSE IN A CRYPT IN A
QUIET ENGLISH VILLAGE

1

"Come on, it's not that bad," Belinda said, rolling her eyes as she skulked through the undergrowth, pushing back fronds. The evening was settling in and the sky was slashed with purple hues, making it look as though a celestial artist had taken a paintbrush and smeared it across the sky. Belinda wasn't focused on that though. Her blonde hair was tied back in a tight ponytail. She wore jeans and a long sleeved top, which served to protect her from the thorny bushes of the forest.

"It is. I really think we should go back," Henry said, scowling as he tried to disentangle himself from thorns. His overalls had gotten caught more than once and it was beginning to get on Belinda's nerves.

"I don't know how you've lived near the forest for this long and never learned how to move through them. You need to pinch the branches and push them back, not just barrel through them," Belinda said.

"I must have missed that lesson at school," Henry grumbled, making long loping strides, as though he was afraid that the ground would swallow him up. Belinda supposed that she shouldn't judge him too harshly for that, as his right leg was still covered in mud from where he had stepped into a puddle, a puddle that had been far deeper than he had anticipated. "I still don't get why you want to be down here. Couldn't we just have stayed in and done a jigsaw puzzle?"

"A jigsaw puzzle?" she snorted and shook her head. "Come on Henry, we can't spend our lives inside. We live out here with all of nature around us and you want us to sit indoors? We have the winter for that. Life is happening all around us. You're only ever truly alive when you're exploring the world."

"I beg to differ. I'm quite happy watching it on the TV."

Belinda rolled her eyes. "I don't know where your sense of adventure has gone, but it's not going to do you any favors. One day I'm going to leave this place and I'm going to see the world. Don't you want to know what there is out there one day?"

"Yes, but I don't need to explore some muddy forest in order to do so," Henry said. "I don't even know what you expect to find. There's not going to be anything in these old crypts."

"That's what you think. There's always some kind of treasure in these things. Besides, even if we don't find anything it will still be cool to see. I mean, who gets to go in these crypts?"

"There's probably a reason for that," Henry said, shuddering a little.

"You don't believe those old stories, do you?" Belinda asked as they broke through the trees and came to the entrance of the crypts. The buildings were old, ancient, and were overgrown with weeds and moss. There was a damp smell coming from the old stone, and the dark entrance loomed with intrigue and danger. Henry swallowed a lump in his throat as he stared at the entrance.

"Every story has an element of truth to them. What if they are true? What if people have stayed away from here because there is a ghost?"

"There's not a ghost in there," Belinda said, "and even if there is it probably just wants some company. Come on, the sooner we get in the sooner we'll be out. Or do you want to stay out here and wait for me?"

Henry took one look around the obscured glade as the light faded. Didlington St. Wilfrid was only a stone's throw away, and yet because it was hidden around the other side of the hillock it may as well have been in another world entirely. Henry's home, his parents, his comfortable bed all seemed like things from a dream, and he wondered if he would ever see them again. He gulped and shook his head. "I'll come with you," he said, deciding that his chances were better with Belinda than by himself.

Belinda led the way, pushing through cobwebs. Something scuttled in the distance, and Henry jumped.

"It's just a bug. Are you going to be scared of a bug now?" Belinda said as she whipped out her phone and activated the torch. A beam of light glowed within the crypts, illuminating the old cracked stone. There were etchings and symbols carved inside, and the tunnels spread out in a labyrinth.

"This is amazing," Belinda gasped in awe, moving forward slowly.

"I can think of another world for it," Henry said, crossing his arms and rubbing himself to fight against the chill that whistled through the crypts. Belinda walked through the tunnels as though she knew exactly where she was going, while Henry made a mental note of which twists and turns they had taken, and he kept an eye out for ghosts as well. After a while they reached an opening. The light from the phone filled the small chamber, and Belinda stopped in her tracks. They had come across a few chambers before this one, but those had all been empty.

This one wasn't.

Henry grumbled as she stopped, for he hadn't been looking in front of him and had clattered into her, meaning that his teeth had knocked together. She shushed him and pointed into the chamber. There was a stone altar, and upon this was a figure clad in a robe.

"Let's go," Henry whispered, tugging at Belinda's sleeve. Belinda pushed him away.

"He could be in trouble."

"I don't like it. We should get an adult."

"Adults don't know anything. He might be injured. Come on."

"I don't know…"

"Didn't you learn anything in Sunday School? Did Jesus leave the Samaritan behind? Sometimes fear is just the Devil trying to steer us off the righteous path," she said. Henry began to argue, but it was too late. Belinda was already moving forward into the chamber. He heard another sound in the distance, but when he questioned it Belinda told him it was just a bug.

It sounded bigger than a bug though. Fear curdled in his heart and he wondered how Belinda was able to face this without feeling any dread at all. She crept closer to the altar, so he crept closer too, and as she shone her light upon the body they both cried out in horror. There was a horrible gash in the middle of the man's chest, and scarlet blood oozed out. His face was ashen, and the look in his eyes was one that neither of them were ever going to forget. Abject terror was displayed upon his face, and even this was enough to fill Belinda with fear.

"I think you're right. I think we should go and get an adult…" Belinda said, her voice quivering with fear. They backed away from the chamber into the shadows, and soon enough the crypts were filled with the sound of footsteps thundering through the tunnels. They panted and Henry twisted his head, looking to make sure they were going to emerge in the forest. He worried that they were going to running through these crypts forever, or that they would run into the person responsible for killing that man. Henry knew this had been a mistake, he just wished that he had been able to persuade Belinda not to follow the whims of her adventurous heart.

2

L ily walked briskly through Didlington St. Wilfrid. Her eyes darted about warily. Zeus was by her side, plodding along with the usual smile on his face. His round, dark eyes intimated a depth of emotion that many people would have denied an animal possessed, but here he was standing resolutely beside his mistress, the woman to whom he had been entrusted while his master was in the hospital. So far Zeus had proven himself adept at helping in crimes, although in this instance Lily didn't need him to help catch a murderer or a kidnapper, she needed him to protect her. He had even recently received a medal for his services to the community, a medal that sat proudly in Lily's office, tucked away in its ceremonial box.

As the night descended upon the small village Lily pulled her jacket more tightly around her in an effort to keep the warmth close to her body. When she exhaled there was a faint wisp curling in the air, but it was quickly left behind as she rushed toward the police station, just like the trail of steam left by an old train as it shunted down the line.

Fear trickled down the back of her neck, as cold as ice. It was strange... since arriving in the village she had been privy to some dark and gruesome things, yet she had not felt fear seize her with such a tight grip before. She supposed it was because she had always solved the crimes. There had always been a face to put to the atrocities committed. The mask of doubt had been lifted and the truth was always evident to see, even if justice had not always prevailed. However in this matter there was nothing but doubt, there was nothing but mystery, and the shroud of the unknown draped the world in danger and suspicion.

Lily quickened her pace as she reached the police station. The lights were still on. She wished she could have been greeted by the friendly, affable face of Bob Dudley, but he was still suspended from duty pending an investigation into police negligence. Replacing him was a prim woman called Chloe Hargreaves. She and Lily had developed a cordial working relationship, although the new constable had made it plain that she did not welcome Lily's help in cases as Constable Dudley had. What's more is that she had made her romantic attraction to Blake clear as well, and Lily was at a loss about what to do.

At present she was unable to think clearly because of the matter at hand, and unfortunately there was nobody she could turn to other than Chloe. She entered the police station and found Chloe hunched over her desk, scribbling notes into a manila folder. Lily closed the door behind her and Chloe lifted her head, her smile fading when she realized who had come to visit.

"Good evening Miss McGee," Chloe said briskly. "I'm sure that whatever you came here with is important, but I was just about to finish up for the night. I was hoping that there would be less paperwork out here in the country, but I was wrong. I suppose it's alright for you being a writer; you can finish work whenever you like." She chuckled lightly, although it wasn't hard for Lily to detect the bitterness behind her eyes.

"I still have deadlines to meet," Lily said, but quickly abandoned that line of conversation as she did not want to get dragged into an argument with Chloe. If she was to continue helping with mysteries then she was going to have to remain on the polite side of Chloe, as much as it pained her. She could have focused on her writing of course, and perhaps that was the sensible option, but since she had arrived in the village she realized she had a gift. She liked helping people, and she wanted to continue using her talents for the benefit of the community. "I actually wanted to come here to talk to you about a potential crime."

Chloe sighed a little and placed the pen on the desk, closing the folder she had been making notes about. "Lily, I understand that you helped out a lot with the missing boy and I know people around here think of you fondly. Believe me, this is nothing personal, it's just that I have a reputation to maintain and it doesn't make the police look good if they're turning to civilian consultants all the time, especially when there are members of this police force that are being investigated for negligence. I also don't particularly like the idea of putting you in danger. It really is for your own benefit to steer clear of these crimes and let me handle them. That being said, if you do come across a crime let me know and I will fill out a report and investigate it in due course, to the full extent of the law." There was a beleaguered tone to her words, and while Lily did not disbelieve her entirely, she did doubt the fact that Chloe said there was nothing personal about this. Chloe had needled Lily from the moment she had arrived here.

"I'm not actually here with a case that I want to consult on. I'm here because I... I think I might be in danger," Lily said. A lump appeared in her throat and it didn't disappear even when she swallowed. Chloe's expression changed, sensing that Lily was being earnest. There was a partition that separated the entrance to the police station with the reception desk. A small gateway was opened by Chloe, and she brought Lily through to the desk where she had been working. She was not entirely devoid of manners and offered Lily a drink. Lily decided to have some water.

"So what's this about?" Chloe asked.

"Well, it might sound a little strange, but I have been receiving these packages," Lily said.

"Packages?"

"Yes... packages filled with baked goods. You know, cupcakes, sweet buns, and things like that."

"Right," Chloe said, her pen hovering above the page as she studied Lily with a curious look.

Lily cringed with embarrassment. She knew how it sounded, but she could not deny the fear that curdled in her stomach. "It's just that I don't know who sent them. I thought it was Marjorie at first, but when I asked her she said she had no idea what I was talking about. And I've been trying to think who might be tempted to do something like this, but I just don't know who it is." Lily grew more flustered as she continued to speak and heat rose to her cheeks.

"And there's no note left with them, nothing to suggest where it might have come from or who made it? There isn't any message at all?"

"None," Lily said. "And that's what I find so strange about it. Surely if this was a nice gesture someone would want to take credit for it? I just hate the idea that someone is creeping around my cottage leaving these things for me. I feel like I'm being watched all the time." Lily tilted her head from side to side and scratched the back of her neck, feeling uncomfortable.

Chloe leaned back and tapped the pen against her lips. "I'm sure that this is a difficult time for you, but I'm not sure there's much I can work with here. Are you sure there hasn't been anything threatening in these packages? You haven't been feeling ill after you've eaten anything from them, have you?"

"I haven't eaten anything from the second package, but no I didn't feel any ill effects from the first."

"Well, I admit that it's unusual, but are you sure this isn't just the work of a fan? It wouldn't be the first time someone has made an anonymous donation. Perhaps this is just their way of saying thanks for writing some good books, or they're thanking you for the help you've provided to the police. Zeus might get a medal, and you might get some cakes."

"I really think it's something more than that. I can't put my finger on why, but—"

Chloe cut her off. "I think you might have been writing about too many murders. I'm sure that if there was someone who meant you harm they would do something other than make you baked goods. I'm not sure it means anything more than that, but I will file your complaint and make some enquiries. Unfortunately I don't have the manpower to station anyone to watch over your cottage, but it might be an idea to install some lights near the path and the flowerbeds. Sometimes prevention is better than a cure after all. I'd keep watch as well and if you see anything suspicious don't hesitate to call me, I just don't think there's much here that I can work with. I know you must be afraid, but it's not as though these baked goods are stale," Chloe said, laughing. Lily hoped it was an effort to put her at ease rather than to ridicule her. Lily inhaled deeply and pressed her lips together. Perhaps she had been over the top with her suspicions. After all, it wasn't as though this was Los Angeles, a place peppered with sinful and nefarious people.

At least she had told the police anyway, so if anything else happened there would be a record of it.

She was about to leave, and her mind was already thinking about ways to make Starling Cottage more secure, when two people burst through the door of the police station. They were a young boy and girl, in their early teenage years. The boy wore glasses, while the girl had her hair tied back in a ponytail. Looks of terror were on their faces, and they were panting deeply. As soon as they entered the station they slammed themselves against a wall, peering out of the window. Lily and Chloe glanced at each other.

"Can I help you?" Chloe asked, rising from her seat to address them.

"T-there's been a murder. There's a body in the crypts!" the boy said, his voice wavering and uneven. It sounded as though it was at the point of breaking, and the first question that Lily had was what these two teenagers, people who were little more than children, were doing lurking in the crypts. The second question on her mind regarded what they had found.

Had there really been another murder in this small village?

"Slow down and tell me exactly what happened," Chloe said in as calm a voice as possible. She fetched two glasses of water as the teenagers sank into their seats. It didn't take Chloe long to coax out their names, and while Lily hadn't directly been invited to take part in the case, she wasn't about to leave now. Zeus walked up to Belinda, who petted him, while Henry took one look at the big dog and shifted an inch or two away from Zeus. Zeus was harmless to anyone of a good nature, but some people weren't to know that and his size could be intimidating.

Belinda told Chloe and Lily about their ordeal, while Henry interjected with the occasional comment. Although both of them were terrified it was clear that Belinda was the one with the more rational mind.

"And you didn't see anyone else in the crypts, or near them?" Chloe asked after Belinda had finished telling her story.

Belinda shook her head.

"It could have been the ghost," Henry said, gnawing on his nails.

"The ghost?" Chloe asked. She looked to Lily, who shrugged. She hadn't been in Didlington St. Wilfrid for much longer than Chloe had, and was not up to speed with all the local legends.

"It's not the ghost," Belinda said sharply, scolding Henry. When she saw the confusion on the faces of the two women she offered an explanation. "There's an old story that we're all told as children that a ghost walks in the crypts. Apparently it's some dead priest who was killed by heathens so he wanders the crypts and protects the dead from anyone who intrudes. There are always stories of people who have been wandering around there at night and see him wandering around, and it's said that sometimes you can hear his ghostly wail if you listen hard enough."

"I heard it tonight," Henry said.

"That was just the wind," Belinda said. "It's obviously just a story meant to keep us away from the crypts because the adults think they're too dangerous or something. Everyone knows that ghosts don't exist."

"I'd assume that the crypts are dangerous if you're stumbling across a dead body," Chloe said, her words pointed. Belinda shifted in her seat. "What were you doing there anyway?"

Belinda shrugged her shoulders. "Just wanted to have a look really. There's not much to do around here."

"I think there are plenty of other things to keep two creative young minds occupied than going into some crypts. Perhaps I should think about making them a protected area," Chloe said, rising. "Now I'm going to investigate, so I'm going to need you to take me back to the body."

"You want us to go back there?" Henry asked, his eyes wide with fright. Lily could well understand his fear.

"You can go home if you like, but I need one of you to show me what happened."

"I'll take you Belinda said," and Henry seemed quite relieved at this. He breathed a sigh of relief, although Lily didn't think that Belinda would forgive him so easily. As they rose, Lily rose too, and Zeus walked toward the door. There was a tense moment as Chloe stared at Lily. She looked around the police office. The lack of other officers was notable. If there was a killer on the loose then it didn't serve her well to go into a murder scene with only a teenager as backup.

"Perhaps you and Zeus would like to join me?" Chloe asked.

Lily didn't have to be asked twice. She wondered what it said about her that she was more willing to venture into a crime scene than return home where someone unknown may well have been waiting for her.

They reached the crypt. Belinda had her phone out, while Chloe carried a proper flashlight. The beam cut through the darkness. Shadows spread across the crypts and they did look foreboding. Lily was sure she had seen something similar on a set of a horror movie she had once worked on. There was a whistling wind through the trees, and in the darkness an owl hooted. They entered the crypt. Belinda tried to retrace her steps, but the tunnels were twisting and uncertain like a rabbit's warren. Lily was glad to have Zeus by her side, and wasn't sure how she was going to cope when Pat returned and took Zeus under his care again. It was strange how much she had come to rely on Zeus in such a short time, but she supposed she had a habit of making big changes in her life. After all, she had settled into this village well enough.

The crypts were drenched in history. The stone was as old as the earth, and it was amazing to think of all the people who had walked through here, who had mourned the dead here. Lily also wondered how much truth there was to the ghost story Belinda had told them. Had this crypt been the home to more than one murder? It was certainly an appropriate place for it, far removed from the village while still being accessible by most people, and it was hidden away from the world.

They rounded a corner and Belinda's pace slowed.

"It's in there," she said in a small voice. Chloe and Lily moved forward slowly as the altar came into view. A man's body was draped over it. A leg hung over the edge of the altar, and his hands lay limply on either side of him. Lily and Chloe walked in opposite directions around the body. Lily's eyes were first drawn to the deep gash in his chest, an impact that cut through his flesh and must have stabbed directly into his heart. It wouldn't have taken him long to die. Both Lily and Chloe made their own observations of the dead, and then Lily's path took her around to the man's head, and there she gasped with even more surprise than when she had seen the dead body in the first place.

This man was not unknown to her.

"I know it can't be a ghost, but what if it's something else?" Chloe said. "Belinda, have there actually been any deaths here, or were there just stories of a ghost?"

"I think there were deaths. I remember my parents mentioning a few things, but that was long before I was born. It might have just been part of the story though," Belinda said.

"We should probably go and speak with someone who knows more about local history. I'll have to get someone here to pick up the body and take it away," Chloe said. They left the crypts. Chloe made a call for someone to come and examine the scene of crime and take the dead body away so that it could be identified. Since Lily didn't know the name of the man she couldn't offer much help, but she couldn't fail to remember his face. It was the man she had seen Blake talking to at Zeus' award ceremony.

If Blake knew him, that made him a suspect, and she needed to talk with him as soon as possible.

3

Before Lily could talk with Blake, she and Chloe went to Bob Dudley's house. Lily could sense the tension bristling around Chloe, although she did a good job of hiding it. The door opened. Bob smiled when he saw Lily, but his face fell when he realized that Chloe accompanied her.

"What do you want?" he asked, holding his door like a shield. Lily explained the situation and as soon as she mentioned that a body had been found in the crypts there was a sudden look of realization upon his face. "And what have you come to me for?"

"Lily and I are not familiar enough with the history of the village. I was hoping that you might be able to tell us if there is any truth to this story of a ghost, and if there have been murders in the crypts before," Chloe said.

Bob welcomed them into his home and offered them some tea. It was still strange for Lily to see him out of his police uniform, and she hoped the situation would change as soon as possible. It was never easy to tell how quickly the wheels of bureaucracy were going to spin though, and she feared she might have to endure Chloe's company for a while yet.

"There were murders, back in the 80s. I was a lad then. The stories of the ghost have been around a lot longer than that though. I remember being told them as a boy. It always used to keep me up at night. I think they did more harm than good because it was a long time before I could sleep with the light off," he said with a light chuckle.

"And the actual murders?" Chloe asked, not wishing to delay talking about the things that really mattered.

Bob cleared his throat and the pleasant look upon his face vanished. "Yes, well, actually there were a string of murders that took place in the crypts. It sounds much like what happened to the victim here. They were stabbed in the heart."

"And who committed these murders?" Chloe asked. Lily's mind was already racing with thoughts of a copycat killer, but why would someone act out a murder like this after so many years? And what did Blake have to do with it, if anything?

"Well, that's the thing. Nobody was ever caught," Bob said, scratching the back of his head.

Chloe snorted with laughter and rolled her eyes. "Why am I not surprised?" she huffed.

Bob frowned at her. "There was never any evidence, and if anyone knew anything they weren't speaking. The thing is that the crypt was used by this... well, for lack of a better term, a cult."

"A cult?" Lily and Chloe said in unison. They exchanged an embarrassed glance with each other.

"More like a brotherhood really. I've never known much about it. I just remember hearing stories of robed figures who met in the crypts. They were always linked to the murders, but nothing ever stuck. By the time I joined the force," he stared pointedly at Chloe, "the murders had stopped for a long time. They were all cold cases and there weren't any leads to follow. I'm pretty sure the crypts became deserted. I can't imagine anyone would want to spend any serious amount of time there after knowing what happened," Bob said.

"It seems as though someone has taken up murdering again. Perhaps it's some strange ritual. The victim was placed in a ritualistic position, and the heart has often been at the center of ceremonies," Chloe said. She turned back to Bob. "Do you happen to know anyone who is a member of this... brotherhood?" she asked, stopping short of calling it a cult.

"I don't I'm afraid. Like I said, the murders were before my time. I've never had cause to look into them," Bob said. Chloe and Lily bid him a good night and left his home. When they stepped out into the air Chloe muttered under her breath.

"That was a waste of time. I bet he knows more than he's letting on. I know how these brotherhood things work. It wouldn't surprise me if he's a member. Maybe he's covering for someone."

"He wouldn't do that. He's a good man. All he cares about is the truth," Lily said, wanting to rise to the defense of Bob Dudley.

"That's for the investigation to decide," Chloe said bitterly. "Well, I suppose I'm going to have to try and dig up records of this cult, and see who was murdered. Perhaps there's a link with tonight's victim. I was hoping to get some sleep tonight, but I suppose that's going to have to wait." She sighed and rubbed the bridge of her nose. Lily could well imagine that Chloe had not anticipated there being such work for her in the small village when compared with the city. Lily thought about letting the woman rest and keeping the nugget of information about Blake to herself, but as much as there was some antagonism between the two women, and as much as Lily was jealous that Chloe was attracted to Blake, she still could not bring herself to conceal any part of the investigation.

"I don't think you're going to need to dig up any information. I might know someone who is involved," she said. When she spoke

Blake's name Chloe's eyebrows arched. Lily felt horrible in case she had betrayed the confidence of someone she had come to think dearly about, but she was also afraid of what it might mean in the future.

What was she going to do if Blake wasn't the man she thought he was?

4

Blake greeted them wearing casual clothes, although his version of casual was still smarter than the average person. Upon seeing him Lily felt a stirring of emotion within her, but she was also wary that her emotions could betray her. Taking a deep breath to compose herself, she offered a thin smile. Blake looked at both of them with a little surprise, clearly wondering why they were together. There was an unspoken question in his eyes, which Lily could not answer just yet. He already know about the tension between Lily and Chloe.

"Good evening, what can I do for you? Would you like to come in for a drink?" Blake asked with a relaxed smile.

"Actually we'd like to come in for some questions. There's been a murder," Chloe said. The color drained from Blake's face and his jaw clenched. Was that a sign of guilt, or just a sign that he was taken aback by Chloe's blunt manner? Idle thoughts swam through Chloe's mind and she wasn't sure who to trust or what to think.

Lily went back to the mantra her main characters used in her stories.

Focus on the facts. People lie. Thoughts can be wrong. The facts are never wrong and they will lead to the truth.

"I suppose you had better come in then," Blake said. Blake lived in a spacious house that was far bigger than Starling Cottage, and likely far bigger than anything Lily had seen before. She knew he was descended from a rich and noble family, although she had no idea how much he was actually worth. Since his father had been ill Blake had been spending more time in the village, although given the way he lived it didn't seem to have made any difference to his income. They stood in the lobby of the house. A wide staircase led up behind him, where the stern visage of one of his ancestors looked down upon them. If Lily squinted she could see the resemblance between them, the genetic qualities that were passed down from generation to generation and formed a chain that stretched through history. Blake had the luxury of being able to point at all these people and see a tapestry of his life. She wondered what it was like to have such a legacy trailing behind him. She knew little of her own ancestry. Her mother had never bothered to tell her anything useful, and her father had never been around to tell her anything at all.

She shook the errant thoughts from her mind and focused on Blake. Chloe had just been telling him about the murder victim, so Lily jumped in.

"I don't know his name, but I saw him with you at the ceremony. You were shaking his hand," she said, trying to keep the sense of accusation out of her voice, although it was difficult.

Blake tilted his head as she spoke. His eyes narrowed a little, and she cringed with shame. The last person she wanted to suspect of murder was him, but she had a job to do. She hoped that he would be able to forgive her. His lips formed a thin line as he nodded and hung his head, crestfallen.

"Albie," he sighed, shaking his head.

"Albie?" Chloe asked.

"Albert Young," Blake clarified. "He's an old friend of the family. He was in the area and came to say hello."

"We think he might have been here for more than that. We know about the string of murders that happened here in the 80s, and that a cult might have been involved. What do you know of the cult?" Chloe said.

Lily watched Blake's reaction slowly. The moment Chloe mentioned the word 'cult', Blake's eyes flickered and he gulped. But then he smiled, and this took both women by surprise.

"I think "cult" is a strong word," Blake said.

"So you know about it?" Lily asked, unable to keep the terse shock from her voice.

Blake did not answer straight away. Instead, he turned and walked to a coat rack from which many jackets hung. He rifled through them and rummaged in one of the pockets, pulling out a box. He handed it to Lily, who opened it. Inside was a large ring made of gold. It was etched with the Latin symbol for Omega and was surprisingly heavy.

"I would say it's more of a brotherhood than a cult. It's called the Omega Society," Blake said.

"And you're a part of this?" Chloe asked.

"No," Blake said, dismissing the idea with a shake of the head. "I think its heyday was in the past. I always thought it was just a way for men to get together and spend some time together, like a club."

"But why Omega?" Lily asked.

"I think they wanted to be like Mensa. It was supposed to be an elite and exclusive club where only the finest members of society could join. I think they hoped they would be able to make big decisions that would help influence the development of the country, maybe even the world. When you get so many men who are in a position of power in one place then I suppose anything is possible," he said, gazing at the ring as he took it back from Lily.

"But you didn't want to be a part of this?" Chloe asked.

"As far as I understand it, the brotherhood never really reached the levels of influence it wanted to. I think people just got interested in other things."

"So why is Albie back here with this ring then?" Chloe asked.

"I think he's always been passionate about it. He's been through some hard times. He got divorced not long ago and his son died. I imagine he's lonely and he wanted to start it again so that he had something to cling onto in his life. He came to me and said that he wanted to bring in a new generation of members, and he wanted to start with sons of the original lot."

"And did you agree?" Lily asked.

Blake let out a small laugh and shook his head. "No. I told him I'd think about it, but honestly the thought of skulking around in robes and taking part in odd rituals isn't my idea of a good time. I think you have to have a certain mindset to be able to enjoy things like this. I didn't want to dash his hopes so quickly though. You should have heard the way he spoke about it. It clearly meant a lot to him. I did intend to spend some time with him though, just to remind him that he's not alone in the world and he does still have people who care about him. I suppose it's too late for that now though," Blake trailed off, as though it was only just sinking in that the man had died.

But something he said had caught Chloe's attention. "Did you just say that your father was a part of the cult, sorry, brotherhood," Chloe said.

Blake stopped suddenly, realizing that he had implicated the man. His head dropped. There wasn't anything he could do to take back what he said now.

"Yes, he was, back in the day. But he hasn't been associated with them for a long time," Blake said.

"As far as you know," Chloe said.

Blake's tone suddenly took on a different tone. "If you think my father could be capable of this then you're mistaken? He's already been accused of one crime. I don't want to put him through that again."

"I'm not accusing him of anything. I merely want to get as much information as possible," Chloe said, managing to keep her cool. Blake softened and took a deep breath.

"I know, I know, you're just doing your job. I'm just in shock I suppose. Albie was a good man. He was harmless really. He only wanted to have some friends," Blake said.

"Well it may have been one of these friends who killed him," Lily said. The three of them shared a grim look with each other.

"Can you let me know of anyone else who was associated with Albie? I'm sure a murder investigation will be a righteous cause to divulge the names of people associated with this Omega Society," Chloe said.

"I'll draw up a list of names for you," he said. "I don't know what Albie was doing in the crypts alone though. He should have known better than that."

"We'll get to the bottom of it," Chloe said.

They thanked Blake for his time and left. Lily breathed a sigh of relief, although she wished she could have stayed with Blake and told him that they didn't really suspect him of anything. She was impressed with Chloe though, who didn't seem fazed at the thought of questioning a man she was interested in. Perhaps she wasn't as interested in him as Lily had first thought.

"Well, it seems as though this case just got a little more complicated," Chloe said as they walked away from Blake's home.

"But at least I'll be saved the rigors of looking through all the records."

"Yes, but one thing does puzzle me though; the ring. Albie didn't have one," Lily said.

"So you're thinking a robbery?"

"It must have been. Clearly the society was important to him, and if so then I don't think he was likely to leave a ring behind. Perhaps whoever took it knew how important it was to him?"

"Yes. I don't think they would have stolen it for the value of the item. I doubt anyone would have followed him into the crypts for that."

"I suppose what we really need to know is who else he asked to join this cult. If he was trying to recruit new members perhaps he recruited the wrong person."

"Or there was a former member who took umbrage to the fact he was trying to revive the society in the first place," Chloe added. Neither thought was a particularly pleasant one, and the two women decided to leave the case for the day. There was no invitation from Chloe for Lily to continue helping her, but she didn't say anything that would imply the opposite either. Lily took that as a sign that she should keep thinking and keep speculating about the true nature of this mystery. She didn't want to admit it to herself, but there was another dark fear curdling in her mind. If Albie was targeted because he was a member of the Omega Society, then what about Blake? If the murdered saw him and Albie together then he might be next.

Icy fear trickled down her spine, and she shuddered as she peered into the darkness before retreating to the safety of Starling Cottage. Zeus was quiet, indication that nothing was amiss, but she still could not shake the feeling that someone was watching her from the shadows.

5

Morning arrived with a rosy light and a gentle warmth. Lily rubbed her eyes and yawned, stretching out as Zeus paced in circles at the end of her bed, getting himself worked up into a frenzy.

"Yes Zeus don't worry, I know you're hungry and you're aching for a walk," she said, flinging away the sheets, leaving them a mess. It was a bad habit she maintained from her time working in Hollywood. There she had always been needed and there had never been a spare moment to do something as simple as making the bed. She caught a glimpse of herself in the mirror as she washed her face and took a moment to gauge how much she had changed since she had arrived in England. Her body had been lean before, to the point where the shape of her collarbones and cheeks had been apparent. Even then she was always judged for eating anything more than a stick of celery. But here she had put on a little weight, which rounded out her features and gave her a rosy glow. She actually looked happy, and she couldn't remember the last time she had felt this way.

After Zeus had gorged himself on his breakfast she patted her thigh and he came bounding up to her. She opened the door, fear twitching inside her. She almost didn't dare to look in case something had been left for her, but the doorstep was empty. It was only then that she realized she had been holding in her breath, and now let it out in one deep exhalation. The village seemed different in the morning and there were fewer shadows in which people could hide. She tried to tell herself that this place was safe and Chloe was probably right in that it was just some misguided fan or someone who meant well, but it all still niggled in the back of her mind.

She put the thoughts away though and ventured outside, taking joy in watching how Zeus wandered around the world with such an innocent look in his eyes. To him there was nothing but adventure and excitement. Even though he saw the same walls and the same plants every day he still seemed amazed by them, always sniffling and snuffling as he explored them in the hope that he would find something new. He never did, but that hope was always there, and it was something that Lily admired.

She strolled past Marjorie's bakery and waved through the window, before making her way to the church. It wasn't a day when services were being held, but the reverend usually had some club or another going on later in the day. She walked into the hall and found him setting up a circle of chairs. He smiled as she walked in.

"Lily!" he exclaimed. "It's wonderful to see you. Are you here for the book club? It would be wonderful to have you sit in, especially as we're reading your novel," reverend Brown said. Lily offered him a weak smile, and tried to squash the embarrassed feeling that rose within her.

"I don't think that would be a good idea."

"Oh don't be silly. I'm sure everyone would be nothing but complimentary, and it would be wonderful to have the author here so we could ask you questions," he said, his eyes sparkling with hope. Lily could hear someone puttering around in the kitchen. A kettle whistled.

"I like to have all the questions answered in the text itself. I am a mystery writer after all, it would be bad form if I just gave all the answers away," she said. The Reverend looked a little dismayed, but he didn't press the issue. "I actually don't have time anyway," Lily continued, "there's been a murder and I was wondering if you might have any knowledge on the subject."

"A murder?!" Reverend Brown shook his head sadly and gasped with shock, placing his hand on his heart.

Lily gave him a quick rundown of the important points of the case. "I wondered if you knew anything about this society? Some people call it a brotherhood and some call it a cult. I thought that if anyone had a different faith you might be aware of it."

The Reverend nodded and had a sad look in his eyes. "Yes, unfortunately people do get confused sometimes and seek faith in the oddest places. I do remember my predecessor warning me about some odd behavior in the past, but I've never been privy to it. Of course, they're not likely to come here anyway. It's a shame that people who are looking for meaning in their lives don't come to me first. I suppose they think they know better than to trust in something that's been around for thousands of years," he said with a chuckle. At this point he was interrupted by someone who emerged from the kitchen. Her name was Jolene. She was in her sixties, although she looked a little younger. She had not dyed her hair and had instead let the natural grey grow all through her head of hair. She carried a cup of tea and expressed a little shock at the fact that Lily was there.

"Here you are Reverend, oh I'm sorry I didn't realize you had company. Would you like anything?" she asked, smiling sweetly at Lily

"I'm fine, thank you, I was just stopping by to ask the reverend about something."

"Oh Reverend Brown is always so willing to help. Well, you just let me know if you want a cup," she said, and turned to the Reverend.

"I've almost got everything ready for the book club. Do you think that we should have the plain biscuits or the ones with chocolate?"

The Reverend pondered the question for a moment. "I think the ones with chocolate, let's treat ourselves, shall we? We only live once after all!" he said. Jolene tittered with laughter and disappeared back into the kitchen.

"I'm surprised she remained so calm around you. She's a big fan," Reverend Brown said. "Perhaps you could stop by sometime and talk to her? She's been through a bit of a rough patch the last few years."

"Oh, what happened?" Lily asked.

"It's sad really. Her husband died of a heart attack. He was a successful businessman in the area, but then other people muscled in and the way the economy turns... well... I suppose there's only so much a man can take. It was too much for him and he collapsed. There were some years between them, but even so it seemed too early for him to leave us. She was absolutely devoted to him, and I still see the sadness in her eyes."

"I suppose it's good that she can help out here. It probably takes her mind off things."

"I'm not sure about that. The two of them were always more than willing to help out with the church, and they never missed a Sunday service. I think rather than coming here to distract herself from thoughts of him she comes here to be close to him. I can imagine his spirit lingering here, for this was the most important place in his life. Don't tell her I said this, but there are times when I've even caught her whispering to herself, talking to him."

"Are you sure she's okay?"

"Oh yes, it's all harmless," the reverend said, dismissing the concern with an idle wave of the hand. "We've all done it now and then. Anyway, as to this business of yours I'm afraid I don't really have much to contribute. It all happened before my time. I did like to think that people came to their senses and realized there is only one true path to everlasting life and faith, but I suppose if people never made mistakes then we would never know how blessed we are to find the Lord's glory."

"I suppose not," Lily said with a wan smile. "Was the Reverend before you not concerned that a group may have been committing heathen practices on God's land though?"

"I don't think he was ever concerned about that. As far as I understand it he thought it was just an excuse for a group of men to get together and drink. They probably just wanted some time away from their wives, you know what men are like. The murders were a tragedy of course, and there may have been a few members who did believe in an… alternative faith, but I'm not sure if all of them did. It didn't last very long though, did it? It goes to show that nothing can compare to the one true Lord," he said, gesturing to an image of Jesus that looked down upon the congregation.

Lily thanked the Reverend and left, sad that she had not managed to make any progress on the case.

Her walk with Zeus took her back to the police station and she stopped in to find Chloe staring at a piece of paper.

"Blake gave me a list of names," she said. "I've been cross referencing people on this list with people who were in the village at the time of the original murders, and who are still in the village now. I don't think this is going to be a copycat killer. Albie wasn't here for long. Anyone who saw him must have known what he was trying to do. If we can speak to people who were in the village at the time, then we might be able to have some more information. I can't find out if the original people who were killed had their rings on them or not."

"Are there many names."

"Not many, but there are two I want to check out first of all. One of them is Blake's father, and the other is the man who was the butcher at the time."

Lily nodded and followed Chloe as she left the station.

"I don't want to jinx this, but are you sure you're okay with me tagging along?" Lily asked, a little perturbed as to why Chloe was suddenly at ease with Lily joining her on a case when she had been so adamantly opposed to the idea just a short while ago.

Chloe smirked. "Well, you see, I've been thinking about it a lot and I started to realize that I could actually take advantage of it. I've witnessed how people in the village react to you and gravitate to you. You have this disarming way about you. Perhaps it's because you're American, or perhaps it's because you've become something of a minor celebrity around here, but people tend to be more open with you. And while you're talking to them I can spend more time observing them, which helps me get more information. I think I was a little too harsh in dismissing your influence before. It's going to help me do my job better, so I'm willing to have you tag along with me," she said.

"Oh," Lily said, unsure how to react. She had hoped that Chloe would have been impressed with her ability to solve crimes and her instincts at getting to the heart of mysteries instead of just using her as a distraction. However, Lily wasn't someone who defined herself by pride so she wasn't going to let this stop her from being involved in the investigations of these crimes. At least she would still be front and center, and would be able to have an impact in the lives of the people of the village.

They arrived at the butcher's. The smell of meat drifted out of the door. Sausages and racks of uncooked meat were displayed on the wall behind the butcher, who was busy chopping new joints. Behind a glass display there were steaks and packets of mince and chicken fillets laid out, with prices written on neon signs. The butcher wore an apron and a white smock, while he brought his heavy carving knife down with a thunderous impact. Lily saw the meat being severed behind the glass, and all she could think about was the impact that had killed Albie. She shuddered at the memory of the gruesome blood spreading out across his chest and the gaping hole that had been left behind. As the butcher raised his butcher knife again she saw the blade gleam before it went crashing down. His gloved hands were flecked with stray bits of meat and fat, and she wondered how different it was to kill a man than an animal.

"Morning," he said, "I'll be with you in a moment. I might even have a treat for the hero of the hour as well," he added with a smirk. Zeus seemed to know that the butcher was referring to him, and the dog barked with hope. His mouth hung open and his tongue lolled out. He sniffed the air. Lily had no doubt that he would have had a field day with all this meat if he wasn't being supervised.

"Actually we're here to ask you some questions, about a murder," Chloe said, and then threw a glance to Lily, as if indicating that she should take over. Lily didn't exactly appreciate being thrown into the firing line, but she stepped forward as the butcher placed the knife on his counter. He leaned over the glass display.

"A murder? I suppose I should have guessed given that the two of you are here," he said gruffly.

"Yes, well, we'd actually like to know about your father. I don't suppose you'd know if he's available to answer a few questions for us?" Lily asked.

"Sure, actually he's here now," the butcher said, and called out to the back. His bellowing voice rang in Lily's ears. It took a few moments, but eventually his father emerged. It was easy to see they were father and son. They both had the same shaped heads and stocky builds. The father, a man named Len, should have been retired really, but Lily got the impression he was the kind of man who had spent his life working and didn't know how to do anything different. It was an attitude that was not as prevalent nowadays, but he embodied the spirit of the working man, having found a passion and stuck to it all these years, even passing it on to a new generation.

"What's this about?" he asked, wiping beads of sweat from his brow. Despite his age his muscles were still knotted, and under a rolled up sleeve Lily could see trails of a tattoo.

"There's been a murder. They want to talk to you," the butcher said.

"A murder? What do you want to talk to me for?" Len asked, squinting at the two women. He saw Zeus and smiled. He reached in and gathered a couple of chunks of meat, and walking out from around the counter and he leaned down to pet the dog. Zeus walked up to him and eagerly wolfed down the meat, licking his chops when he was done.

"Because we think you know the victim," Chloe said, getting visibly frustrated that Len wasn't giving the matter his full attention. "Albie Young."

"Albie?" Len said, and then shook his head. "I haven't heard from him in years."

"No, but we think he was going to get in touch with you. Are you willing to talk about the Omega Society?" Chloe asked. She and Lily both watched his reaction carefully, but like Blake he merely smiled.

"That old thing? Well, sure, although I'm not sure what I can tell you. I haven't been a member for... oh... years now."

"Really? So it's not the kind of thing where you're a member for life?" Lily asked.

"I suppose so, but it changed over the years. It used to be fun and a way to get away from it all, but there were a few people who took it too seriously for my liking. Albie was one of them. He always had these grand plans for the society. I actually stopped going when I lost the election to become the Grand Master. He had a lot of drive though, but I didn't like where he was going. It's probably a good thing though, if I had been Grand Master then I might be the one you're asking about!" he said.

"You don't seem that upset," Chloe said, and gave a meaningful look toward Lily, suggesting that she should follow this line of questioning.

"Of course I'm upset. But when you get to my age you get used to people dying. It's just a part of life. I'll toast to him later, don't you worry about that. It's the least I can do to pay respect to a fallen brother."

"So you still consider him a brother even though you haven't been a part of the society for a long time? Do you ever miss those days?"

Len sighed. "I suppose I do, in a way, but as I said it lost its shine a little bit."

"Were you angry at Albie for taking the society in a direction you did not like?" Lily asked.

Len shook his head. "It's just the way things go. I don't know what he ended up doing with it after I left. I suppose if he's back here then he couldn't have done much for it," he said.

"We're also wondering if he was trying to recruit new people to the society. Perhaps the sons of former members," Chloe said, allowing her gaze to drift toward Len's son. "Did he get in contact with you?"

The butcher shook his head. "No, he didn't," he said curtly.

"Would you be able to tell us what you were doing last night?" Chloe asked.

The butcher sighed. "I was here, like I am most nights, getting things ready for the day," he said, and then made it clear in no uncertain terms that he had nothing more to say to Chloe, and that she was going to need to come back with more evidence if she wanted to directly accuse him of murder.

6

Although Len had managed to provide Lily and Chloe with some more information regarding the cult, it wasn't anything that helped them get closer to the possible suspects. Chloe did not like the butchers, and thought they had a good motive and the means to actually commit to the kill. There was certainly going to have to be more investigation done into the two men to see if they had alibis, or if they needed to be brought in for further questioning. Lily wasn't sure though; Len seemed to have put that life behind him, and given his age would he really be worried about Albie going about his business? Perhaps the son might have been angry to be approached by Albie, but angry to the point where he was willing to kill? It was a stretch, but at the moment there were few other suspects to think about, aside from Blake's father. She closed her eyes, for she didn't want to think either Blake or his father had anything to do with this, but until there was strong evidence proving otherwise she had to keep an open mind.

Sometimes the hardest thing about investigating these mysteries was having to believe the worst of people she knew and trusted.

Every instinct in her body told her that Blake had nothing to do with this, but she knew she couldn't simply trust them. Instincts could be wrong. Facts couldn't be, yet sometimes friendships could be ruined when suspicion was involved. She was glad when Chloe said they should end things for the evening and go to see Lord Huntingdon tomorrow, because there was somewhere that Lily wanted to be.

She had received a message earlier declaring that someone special was returning to the village. Zeus was getting excited, although she wasn't sure if Zeus was simply picking up on her own feelings. She put together a small bag of supplies and then went to Marjorie's bakery, exchanging swift words before Lily continued on her way and walked up the winding path to a house she hadn't been in for a long time. As she looked at the grim door she remembered her first welcome to the village; it had not been a kind one. An angry man had warned her away. Later on she would learn that the angry man, called Pat, had secretly been a spy for the English secret service, and had retired here to live out his life after his sister had died from an illness. It was not so easy to leave his past behind though, as a Russian agent, nicknamed 'The Butcher', had come to kill him. She had failed, although only just. Pat had been in the hospital for a long time now, and Lily was nervous about the state he was going to be in. It crossed her mind that in that case, the butcher had fallen under suspicion then because of the agent's codename, and she wondered if his irascible nature was due to being approached by her and the police.

Zeus perked up as he approached the door, back to familiar surroundings again. While Lily had looked after him, it was no comparison to his own home. She petted his head and knocked on the door, calling out to Pat that she was there. She heard a muffled voice emerge from beyond the door, telling her to come in. The door was unlocked, and this openness took her by surprise as Pat had never been the type to welcome anyone into his house. The only times Lily had been invited in were when she had insisted. The

321

place was as neat as she remembered, although this time instead of Pat coming to greet her he was on his couch, lying on his back, propped up with a fluffy pillow. His skin was sallow and his face was gaunt, while there was an awful scar running along his scalp. He offered her a weak smile.

"I would get up and get you a drink, but I've never been a good host," he said. Then his expression changed completely as Zeus bounded into view. Lily smiled, her heart warmed by the sight of a man and his dog being reunited, although there was slight envy in her heart, even though as she knew she had only been borrowing these feelings for a time. Zeus leapt onto his master, and it was to Pat's credit that he did not display any sign of pain. Zeus lapped Pat's face and whimpered happily, his tail wagging as he welcomed his master back.

"Oh I've missed you too boy," Pat said, unable to stop smiling. The difference between Zeus and Pat's demeanors had always struck Lily as odd. Zeus had always been friendly and sought out other people's company, while Pat had been gruff and reclusive. However, now she saw that Pat could be kind and caring when he had a reason to be. He cuddled Zeus and petted him and kissed him, and it was clear how much he had missed Zeus. Lily gave them a moment as she went to the kitchen and fetched them both some drinks, while unwrapping the sandwich and other treats she had bought for Pat. She placed them on plates, which she then placed on a tray, and carried them back into the living room, setting them down on the coffee table.

"Yes, yes, I haven't forgotten you either, Zeus," she said as she tossed a doggy treat up in the air. Zeus leaped up and caught in his mouth, chewing on it noisily before returning to his master's legs and curling up on them, happy to be back where he belonged. "He's missed you," Lily said.

"I've missed him as well," Pat said, blinking back tears. Lily had never seen him this emotional before. His voice was horse and raspy, and he looked tired, so tired. "I wanted to thank you Lily, for everything, not just looking after Zeus. I know that if you hadn't found me when you did I wouldn't have made it, and she wouldn't have been caught," he said, referring to the Russian agent.

"I'm just glad you're back with us Pat. I know we've never exactly seen eye to eye, but I've never meant you any ill, and we've all been worried about you," Lily said, handing a plate to him. Pat took a bite of one of Marjorie's buns and closed his eyes, acting as though he had just tasted a slice of paradise.

"This is much better than hospital food," he said, and sighed happily. "So what has been happening since I've been in hospital?" he asked. Lily was surprised as she hadn't expected him to engage in any lengthy conversation. She had assumed he would ask her to leave once she had returned Zeus, but perhaps this incident had caused a change in character for him.

She quickly told him about the cases she had worked on, and that Constable Dudley was currently being investigated. When she spoke about Chloe he wore a smirk.

"You don't like her, do you?" he said.

Now that she knew he was a former spy his insight did not take her by surprise. "I'm not sure yet. She just seems to have an edge about her, that's all, and I'm not sure if she puts more focus on her career or on the crimes. I'd hope it was the latter, but I'm not sure."

"Well, with you around I'm certain that justice will be done."

"I hope so. We're in the middle of a case now, but I'm not sure where it's going to lead," she said. Pat asked her about the case, and so she told him. As soon as she mentioned the Omega Society he nodded, confirming that he knew about it. Lily asked him to elaborate.

As he spoke he idly stroked Zeus. "I heard whispers about things from my sister. My superiors wanted me to become involved incase there was anything nefarious happening. There were quite a few prominent members of society being asked to join, but as far as I could tell it was just as harmless as any gentleman's club. It seems to me that men always need some excuse to go off and spend some time together. Now, it wouldn't surprise me if there have been some deals made because of connections made in this society, but that's nothing more than people making use of connections. I didn't find there to be anything noteworthy, so if you're looking for a motive then I'd suggest it's personal."

"Which means it's likely someone who knew Albie," Lily said.

"Yes," Pat confirmed.

Lily sighed and rested her hands in her lap. She dreaded speaking to Lord Huntingdon in case it revealed truth she did not know. Yes, it was a poor attitude for someone like her, but she couldn't help herself.

"Pat, have you ever been faced with a case where you didn't want to know the truth?" she asked.

"More often than you think," he replied with a smirk.

"How did you cope with it?"

He thought about it for a moment. Zeus' ears pricked up as Pat stroked them. "The truth is going to be the truth no matter what your opinion is. It's not going to change based on what you want to happen, so the best thing is simply to accept it. Anything else just leads you to being hurt," he said.

Lily nodded and cast her gaze around the room. It flicked toward the room adjoining the living room, where Pat kept a lot of his recording equipment. An idea flashed in her mind.

"Pat, do you think I might be able to borrow some of your surveillance equipment?" she asked.

"Of course, but if you're thinking about lugging that to the crypts I have to say I think it's a bad idea."

"I wasn't thinking about the crypts actually, it's more of a personal matter," she said. Of course Pat inquired as to what it was, and she

324

told him about the strange packages that had been left on her doorstep. She felt a little silly to confess them, especially after Chloe's reactions, but Pat had a far different opinion.

"I think you're right to be worried," he said. "People don't do things like that for no reason, and like it or not you're a prominent figure now. Not only are your books bringing you acclaim, but you're also becoming noted among the people of the village. It's why I tried to keep to myself and maintain a low profile. The more people who know about you, the greater the chance is that one of them is going to be crazy."

Lily's eyes went wide with fear. "Crazy?" she asked.

"I wouldn't be worried that anyone is going to come after you or anything like that. All in all I think leaving baked goods is low down on the list of hostile things, but it doesn't do any harm to be careful. You have to ask yourself why is this person bringing them to you, and why are they being secretive about it? I'm sure that someone of your talents will find it easy to solve the mystery before too long," he said. "And of course, if you need any help then just let me know because I am in your debt Lily." He added this part with a smile, and Lily could see the gratitude in his eyes. It was a far cry from the fury and darkness that lived in him before. He looked to be a new man, with a new lease on life.

"I really am," he continued, and his voice broke under the weight of emotion. He reached out to her and she took his hand. His skin was calloused and rough, but there was a lot of tenderness in that touch. "I would have died without you, and nobody would ever have been able to bring her to justice. I know that I have been... difficult, and I apologize for that. I thought it was safer for everyone involved if I closed myself off from the world. I thought I would be safer... but I wasn't. I'm here now because of you Lily, and I know I'm never going to be able to repay you for that."

Lily found herself becoming overwhelmed with emotion. "Pat... it's okay, honestly. I get wanting to shut the world away, and I can only imagine what you must have thought when you saw me coming into your cozy English village. I'm just glad I was able to make a difference, and I'm glad that you're able to be back here. You may not like it, but you are a big part of this village, and I hope that you're going to be around for a long time."

"I plan to be."

"I'm also honored to be with you as well, after knowing what you did for the country, and the world I suppose. And I wanted to say that I'm sorry for what happened to your sister as well. I think I would have liked to have known her."

"I'm sure she would have liked you," Pat said. "She always knew how to get on well with people, more than I did anyway." He wore a sad smile and she could see his mind drifting into sad things.

"There's something else I want to give you," she said, and pulled out Zeus' medal. "He earned it for how he's helped me these past few months."

Pat's face lit up with pride. "You got this boy?" he said, stroking Zeus' head once again. "I'm so proud. Lily, thank you. I couldn't have asked for anyone better to look after him."

"Honestly it's been my pleasure. He's been a good companion. I'm sorry to see him leave, but I know he's glad to be back by your side. I think he's been pining a little for you these months."

"Well, there is something I wanted to ask you actually," Pat said. "The doctors have warned me that I need to take it easy for a while and I shouldn't walk distances that much. I'm not going to be able to walk Zeus as much as he needs. I was wondering if you wouldn't mind continuing to walk him while I recover? And if you wouldn't mind coming by to bring him his food and things. I hate to impose like this, but I—"

"It's no imposition," Lily said without letting him finish, "I'll be happy to help. After all, that's what friends are for," she flashed him a smile and Pat returned the smile, although it was uneasy, as if he wasn't used to smiling. When they first met Lily never would have believed that she would call this man friend, but now she was glad to be in his company, and grateful for the opportunity to get to know him a little better. Her thoughts turned to Albie, and she wondered if he would still be alive had someone taken the time to be his friend.

Lily stayed with Pat for a little while longer before he got tired. When he began to move, Zeus immediately ran circles on the floor and headed toward the door, expecting to be taken for a walk. Pat groaned. It was clear that he wanted to do his duty as the dog's owner, but that he was incapable as of yet.

"I'll take him Pat, don't worry. You just focus on getting the rest you need and getting better," she said.

"It feels like all I've done for the past few months is rest," he muttered, and she chuckled a little because it felt like the old Pat was shining through. Lily promised she would be back soon and then she would pick up some surveillance equipment as well, in the hope that it would finally shed some light on who had been leaving things on her doorstep. She whistled to Zeus, who was accustomed to obeying her now, and she left Pat's house, feeling better than when she had walked in.

She had been walking Zeus for a little while when her phone rang. She picked it up, her heart skipping a beat when she saw that it was Blake.

"Lily, are you busy at the moment?"

I'm never too busy for you Blake, she thought, but refrained from saying these words out loud. "I'm just walking Zeus. Pat is home by the way, and he's in good spirits," she was tempted to tell him all about her conversation with Pat, but she sensed the worried tone in his voice.

"Would you mind coming over? Something happened that has my father worried. He received a letter. I've already spoken to Chloe, and she's on her way," he said.

Lily's heart sank at the mention of Chloe, but of course he should involve her because she was the chief of police in the village. "I'll be right there," Lily said, and then walked at a brisk pace, not wishing to delay.

7

When Lily arrived Chloe was already standing in the study. Lily had been shown through by a servant. Chloe, Blake, and Lord Huntingdon himself were standing around the desk, staring at the letter. Lily knocked on the door lightly to announce her presence. The three of them turned, and Chloe's face was a scowl.

"What are you doing here? I never gave you authorization to join us," she said, a harsh tone lining her words.

"I asked her," Blake said, his voice so strong and mellifluous. "I wouldn't want to go through this without her."

"I agree," Lord Huntingdon said. He walked over to Lily and took her hand, pressing his palms against hers and smiling, bowing his head. "It's a pleasure to see you again, although I fear we keep meeting in these terrible circumstances. Just once I would like to meet alongside a glass of sherry or port," he said, shaking his head, his grey whiskers shaking as he moved. Lily allowed herself a sly smile as she welcomed this endorsement from the Huntingdon men. If Chloe was going to try and generate opposition to Lily, she was not going to be successful here.

"Take a look at this," Blake said, pointing to the desk. Chloe reluctantly moved out of the way, allowing Lily to pass. She leaned over and looked at the note. It was a short note with just a few words etched upon the page in scraggly, jagged lines.

Come to the usual place where we share our secrets under the moon. I have news. We must protect each other. Come alone, for I fear they are onto us. We must fight this betrayal together. There are so few of us left. Come. The Omega Society demands it.

Lily read it out loud, and as she did so Lord Huntingdon placed his hand over his forehead and let out a groan.

"I always knew this was going to bring me trouble one day. We never meant any harm by it, you know, it was just a way for us to have a little fun," he said.

"Do you have any idea who could have written this?" Lily asked.

"I don't, I really don't. Most of us gave it all up a long time ago. You know, we grow up, have families, and this sort of thing falls by the wayside. Albie though… I suppose it always did mean more to Albie than most," Lord Huntingdon said.

"Len told us that Albie was taking things in a direction that he didn't agree with. Did you feel the same?" Lily asked.

Lord Huntingdon stroked his beard and exhaled deeply, nodding along. "Yes, yes, he wanted us to recruit more members and to have more influence. It started off as people in the village you know, and then it slowly spread to more people we knew until there were a lot of us. Albie was never the most influential in the real world, but in the Omega Society he thought he could lord it over us all. He was the one who took it most seriously, and over the years there were a few attempts where he tried to get us all together again."

"Did it work?" Chloe said. She seemed to have put her dislike for Lily to the side for the moment in order to focus on the crime.

"I don't know. It didn't on my end, I can tell you that. Perhaps I should have tried harder though, it's just that whenever he reached out to me all he wanted was to talk about the society, and I left that part of my life behind."

"So it's safe to assume that he reached out to other people. Perhaps one of them is behind this?" Lily said. "Do you have a list of names we could go through? Perhaps you know more than Len?"

"We don't need names," Blake said sternly, before his father could respond. "We can go to the crypts now. Whoever has done this is there, waiting for us."

"That's what worries me. If they're waiting for us then they're going to have a weapon ready," Lily said.

Blake pressed his lips together. "When I say 'us' I mean they're waiting for my father. They don't know that we're here looking at the note. We can catch them and stop this once and for all. We can get justice for Albie," he said.

Lily wasn't sure about the plan. She didn't much like walking into a place when she knew a murderer was going to be standing there, but at least it would afford them the opportunity to try and catch whoever was committing these crimes. She glanced toward Chloe.

"I wouldn't usually endorse something like this, but it does seem as though we have a good opportunity. I don't think we can afford to miss out on an opportunity where the killer is ready to reveal themselves," she said, "and I think we might need Zeus' skills as well." Lily wasn't sure if Chloe was just saying this to make her aware that Lily was only there because she came as Zeus' partner or if she was reading into things, but Lily wasn't going to let that stop her.

"This is terrible, just terrible," Lord Huntingdon said. He leaned against the desk and looked every one of his years. Lily had already walked to the crypts once and knew how difficult the terrain could be.

"Wait, are we sure about this? If they're expecting Lord Huntingdon they're going to be suspicious when we turn up. What if they're going to be watching the crypts, making sure that he's coming alone? If all of us show up they might escape before we even get a chance to see them, and I don't particularly want to put you in danger Lord Huntingdon."

"I should think not!" Lord Huntingdon said with a huff, but he pushed out his chest and seemed to summon some of the vigor he had lost over the years. "I wouldn't want anyone else to take the risk though. They want me, well, they're going to get me. We'll teach them a lesson," he said bullishly.

"That won't be necessary Dad. I've already been thinking about that," Blake said gently. "I can disguise myself. It'll be dark, and even darker in the crypts. I'll go ahead, with you, Chloe, and Zeus following behind me. If there's any trouble then you won't be too far back where you can't leap in. We won't have to worry about anything at all."

Lord Huntingdon put up a display of manly bravado so that his courage would never be in question, but it was clear that this was one mission where he was not needed. Blake didn't seem to be in the mood for discussing the situation as he left the room as soon as he had made his point, presumably going upstairs to changed into his father. Lord Huntingdon sighed as he looked at the note again.

"He is a good lad, always willing to do whatever he needs to protect his family. You should have seen how angry he was when the note arrived. As soon as he realized I was being threatened he snapped to attention," Lord Huntingdon said.

"Did you see who delivered the note?" Lily asked.

The Lord shook his head. "It appeared on the doorstep, as though an angel had left it there," he said.

"While we wait for Blake, is there anything else you can tell us about the Omega Society that might help us prepare for what we find in the crypts? Was there anyone you suspected was responsible for the murders that happened in the past?" Chloe asked.

Lord Huntingdon trembled at the mention of the crimes of the past, and he sank into his seat, as though his strength had deserted him. "A terrible time, a terrible time indeed. That's when we knew this had all gotten out of hand. We had many meetings you know, deciding what to do. Albie wanted to continue and work together to find out who was killing us, but the rest of us knew when to get out. The thing is we looked to each other and none of us could think of any reason why any of us would want to kill each other off. There didn't seem to be any motive at all, and as for other people... well... what happened in the Omega Society stayed in the Omega Society. As soon as we stopped meeting we stopped being killed. Albie should have stayed well enough away. I suppose he thought enough time had passed..." Lord Huntingdon said, his words trailing away. Perhaps he realized that his own days were dwindling, Lily thought.

"I think when we leave you should lock up the house and go to your room for the evening. Make sure that a servant you trust is with you, and listen for any strange noises. Call the police station if you suspect anything. I'll let them know that they might hear from you. I wouldn't even try calling us because there's not any signal in the crypts," Chloe said. Lord Huntingdon looked at her with shock.

"Do you think they'd come for me here?" he asked.

"I'm not willing to put it past them. This note was meant for you. They might expect us to go rushing away to the crypts, leaving you here, unguarded. We have to try and defend against every possibility," she said.

Lily nodded toward her, trying to show that she appreciated how Chloe was trying to cover every angle.

It wasn't long before they heard Blake's footsteps come down the stairs. They stepped out of the study and saw Blake standing there. He wore one of his father's long jackets and stooped over, giving the impression of shoulders that had been rounded by the years. He also wore one of his father's flat caps and tucked his chin underneath a turned up collar, hiding the fact that he did not have a bushy white beard like his father. In his right hand he held a cane, and when he walked he hobbled along.

Lord Huntingdon clapped. "It's like looking in a mirror! There's the future that awaits you my boy!" he said, seeming to take great enjoyment from the sight of his son pretending to be him. Lily couldn't say that the image of Blake as an old man was a particularly appealing one, but she appreciated the lengths he was willing to go to in order to protect his father.

"Shall we go to the crypts then?" he asked, mimicking his father's speech as well, which brought another howl of laughter bellowing out of Lord Huntingdon's mouth. Lily only wished she could find something to laugh about on this night, because they were about to go and confront a murderer.

8

As they made their way to the dark forest that led to the crypts, Blake walked normally. Lily cast her eyes around, and Zeus knew that he had to be on guard. His ears pricked up, ready to bark at any sudden noise. Chloe's training kicked in and she peered into the darkness. Night had fallen onto the village, and in the night anything could happen. Stars twinkled above, silent witnesses to all the dark deeds that happened during the night. Lily only hoped that on this night they would be able to prevent a murder. She had no idea who could be waiting for them, and if things went wrong then any one of them could die. Branches snapped under their feet and itching thorns caught their clothes as they pushed their way through to the crypts. They stood hidden by the trees when they saw the foreboding entrance to the ancient buildings that held more than bodies, they held secrets too.

"I'll go in first. Don't follow until I'm inside. The murderer might be waiting out here to sneak up on me from the rear. If you don't see anything then come inside after me. I'll cry out if I'm attacked," Blake said.

"What if you're attacked before we can get to you?" Chloe asked.

Blake lifted up his cane. "I've taken one or two defense classes in my time. Believe me, no murderer is going to get the drop on me. I'll be fine," he said, and then he strode out toward the crypts. Lily's heart was in her mouth as she watched Blake. Faint mist began to curl around the crypts, as though some unearthly spirit knew that he was coming into danger and sought to conceal it from the world. A lump formed in her throat as he submerged himself into the yawning abyss of the crypt entrance, losing himself in the darkness. As he disappeared she feared that she would never see him again, and suddenly she worried that she had not made the most of her time with him. There were so many things she wanted to tell him, so many things she wanted to confess, but fear had shackled her tongue. Now he was walking into the crypts, ready to face a murderer, and he might never return.

"You have to admire him, what a man," Chloe said, exhaling deeply after she spoke, with a sense of awe in her voice. Lily pressed her lips together tightly and swallowed bitter bile that rose up at the back of her throat.

"We shouldn't leave him alone for too long. I haven't seen anyone sneak in after him," Lily said, ignoring her comment. She was eager to follow Blake to insure that he would not fall foul of this vile murderer, and Chloe didn't try to stop her.

With Zeus by their side, Lily and Chloe crept across the dark hillock. The mist was rising with every moment, as though some specter was stretching its phantom tendrils across the land. Lily peered into the shadows, and whenever she saw movement her heart jumped. Chloe's flashlight beamed light in a narrow beam, but the darkness was encroaching and no amount of light seemed to be able to penetrate. Lily breathed deeply as fear threatened to overwhelm her mind. She kept telling herself that this was still the Lord's world, and that He would never let evil reign in these crypts. She breathed deeply and with a renewed sense of purpose she moved forward, determined that this would not overcome her.

The crypts were a slick, squalid place filled with a heavy atmosphere. It sounded as though whispers were all around them, although Lily knew rationally that it was just the wind creeping through the old labyrinthine corridors and the roughly hewn rocks. Water dripped in certain places, and occasionally there was a clash of claws against rocks as some rat or other rodent ran around in the darkness, no doubt annoyed that humans were trespassing upon its territory.

"I can't believe that anyone would want to meet in a place like this. What was the Omega Society thinking?" Chloe muttered under her breath. If she was afraid then she was hiding it well. In this instance her police training certainly gave her an advantage compared to Lily's amateur status. The temperature in the crypts was cold, made worse by the stones that ran all around them. The ceiling was low as well, which was alright for Lily, but she imagined that Blake must have been rubbing his head against the ceiling.

Suddenly there was a loud cry emanating from the darkness. It was sharp and raw, and there was no doubting it was Blake's voice. Chloe and Lily immediately broke into a sprint, following the sound, and Lily hated herself for letting him go into this place alone. It had been a foolish plan and she should have made sure that he was safer. She should have gone with him because there was safety in numbers. That was a lesson she had learned a long time ago, and it was one that she shouldn't have let slip now. If anything happened to Blake she would never forgive herself. It was bad enough what happened to Pat... but not Blake.

Gasps slipped through Lily's lips and tears swam in her eyes as they reached the chamber. She remembered how Albie had looked as he lay on that slab, his chest gouged and gashed with the lifeless, hopeless look in his eyes. Was she about to see Blake in the same position? She almost held back, but the corridor opened into the chamber and Chloe's flashlight filled the room. Lily breathed a sigh of relief as she saw that the slab was empty, but pressed against the side of the room was Blake. His cane had clattered to the ground and his clothes were disheveled.

"They came from nowhere, from the shadows. When I cried out they ran," he said, and pointed in the direction of another corridor where darkness loomed large. Blake was rubbing his arm and Lily could only guess that the attacker had gone to kill him, but somehow Blake had scared them off. Time was of the essence though. The killer evidently knew these crypts far better than anyone else, and if they had enough time they could escape and slip into the night, disappearing into the shadows again, and then the investigation would have to start from scratch.

Lily, Chloe, and Zeus left Blake behind. He groaned and picked up his cane, before trailing after them. The narrow beam of the flashlight didn't seem to make any difference against the darkness. It was impenetrable, bleak, and oppressive. It was as though it absorbed everything around it and that nothing could escape it, and she imagined it was like the killer's heart. Lily clenched her jaw. She and Chloe walked slowly, wary of anyone jumping out of the crypts at them.

"This is impossible. We're never going to get out of here," Chloe said as they reached an impasse. They had been twisting and turning and still seemed no nearer to catching the killer. They strained their ears, but no sound greeted them, and Lily became afraid that somehow they had passed the killer without knowing it and they had gone back to finish the job with Blake. She tossed a glance behind her, but there was only darkness in front and darkness behind, darkness all around. It was as though the world she knew was gone and this was all that was left, just ending blackness.

And then Zeus began to snuffle. He stretched his paws forward and his ears pricked up. He stood at the entrance to one corridor that seemed to narrow completely. Chloe shone her flashlight into it, and it only showed a wall. There was a corridor to the left of her that seemed more promising.

"I wonder what he can sense," Lily said.

"Probably just a rat. Come on, it's a dead end," Chloe said.

"I think Zeus might be onto something. If he thinks we should go this way then we should trust him. I certainly can't sense anything, and I don't think you can either. Maybe he knows something we don't," Lily said.

"But there's just a wall there. If we make the wrong choice then we're going to lose the killer for certain."

"I'm going to bet on Zeus. If you want to head down that corridor by yourself then go ahead," Lily said, "but we're only wasting time debating this." Chloe glanced down the tunnel beside her into the pitch blackness and decided that as much as she did not like working with Lily, she wasn't about to go off alone in these crypts. Lily followed Zeus, and she dearly hoped that this would not be the time when it was wrong to follow a dog's instinct.

They came to the wall, which looked like a dead end, but there was a narrow opening on the left side. It was a tight squeeze, and it looked as though the stone was weathered, and much of it had been deliberately removed. When they had worked their way through, Zeus hared (ran with great speed) down the corridor. The whispering wind around Lily became louder and it was as though a thousand voices blurred into one, forming a cacophony that filled her mind. She fought against them though, reminding herself that it was just the wind, that these bodies that were contained in the crypts were all in Heaven now, their mortal forms long since forgotten as their souls had risen to the ethereal plane.

There was an opening before her and her heart leapt as she saw a different kind of darkness. It was opaque and hazy, and it took her a few moments before she realized it was the mist from outside. At first she was elated to be out of the crypts, but then she was concerned that the killer had already escaped. Air burned in her lungs and her muscles cried out in pain as she pushed herself to her limits. Chloe was beside her, panting loudly, her footsteps crashing against the damp surface of the crypts. As they emerged and fought their way through the light mist they saw a robed figure sliding and slipping down the sloping hillock. Zeus barked loudly, and the figure stumbled.

"We've got them," Chloe said with a burst of excitement. The shot of adrenaline gave Lily a second wind, and the two women battled with the uneven terrain. Gravity pulled them, carrying them with momentum, pulling them toward the killer. They moved faster, closing the distance. The village was in the background, looking peaceful and still, although Lily had no time to enjoy the view as she was preoccupied with catching the killer. She and Chloe reached out at the same time and caught the killer's robe, pulling the hood back. A knife was swung haphazardly through the air. Chloe caught the killer's wrist and twisted it. There was a yelp of pain and the knife fell to the ground, and soon after the killer landed there as well. Chloe drove a knee into the back.

Lily looked at the killer's unmasked face and furrowed her brow, trying to understand it all.

"Get her off me!" the killer cried. "You're hurting me."

"And you did more than hurt one person, at least," Chloe spat, before hauling the killer to their knees and binding their wrists in handcuffs.

Lily stared at the killer, while Zeus stood by her side and growled.

"Jolene, why did you do this?" Lily asked, staring at the woman she had only seen recently at the church, helping Reverend Brown prepare for the book club. She had seemed so timid and so lost.

Jolene gasped, struggling to catch her breath. It was incredible to think that someone of her age could commit such a horrible act, but Lily supposed anything was possible if someone was caught by surprise.

"I had to do it. It was the only way," Jolene spat.

340

"The only way to do what?" Lily asked.

"The only way to make it right," Jolene replied.

"I'm sorry Jolene, but I don't understand. How is killing anyone ever going to make anything right?"

"Because other people die and justice is never done. My husband died because of them," Jolene's tone was harsh and laced with bitterness. She couldn't have been more unlike the woman that Lily had seen in the church.

"How did they kill your husband?" Chloe said in a flat tone. "From what we understand there was never anything criminal about the Omega Society. They were just men who wanted a place to get together, to spend time away from their real lives. It was never anything more than that."

"It was," Jolene spat. "It was to some of them. I never liked them in the first place. They had strange beliefs, and how dare they use such a special place to conduct their meetings? It's a sacred place. There's no respect for the dead, none at all I tell you. And they might have told you that they never meant any harm, but that's not the truth. They tried to get my dear old Keith to get involved in their club, but he refused them. He was too pure a soul for that. He never would have done anything like that. And what did they do to him? They refused to do business with him. They conspired with each other to cripple him. He had big plans. We both did. We were going to be successful. He was going to expand into cities and we were going to see the world, but they never let him. They stopped him before he could do anything. And it ruined him. He was never the same after that. If he had taken the gamble himself and lost then he could have settled for that. A man can live with his own mistakes, but when that chance is taken away from him through no fault of his own, at the whim of others... no, no that can't stand. So when I saw *him*," she spat the word as if it was a curse, "I knew that he was trying to ruin other people. I wasn't about to let him have his way with the world when Keith had his world destroyed. There's no fairness to it at all. Where's the justice?" Jolene asked.

"Justice is something you're going to become very familiar with," Chloe said.

"Why didn't you kill Blake?" Lily asked, a little surprised that Jolene had fled the scene rather than finishing what she had started.

"Because he wasn't the one I wanted. I thought they had stopped this silly club. I was willing to leave them alone, but then they had to begin again didn't they? As though nothing mattered. I know they were in on it. Keith isn't here, but they are, and why should they be allowed to be happy? It's not fair. It's just not fair..." she said, trailing away. Chloe arched her eyebrows in Lily's direction as she began to take Jolene away.

"I suppose that's another case you've helped me on," Chloe said. "Maybe we do work well as a team after all."

Lily wasn't sure if she was being serious or not, and before she had a chance to reply Blake caught up with them. He was walking normally again and stood at his full height rather than stooping. Lily was about to ask him if he wanted to join her for a late night snack, but Chloe beat her to it.

"Would you mind coming to the station and filling out a report?" she asked. Blake glanced at Lily and shrugged. Lily had to stand there and watch as Chloe and Blake walked away, and once again she wondered if she had missed her chance.

"Come on Zeus, let's get you home. Good job again, Zeus, the village is a lot safer with you around. I'm starting to think you're its guardian," she said. Zeus looked back at her with his dark eyes, and he seemed to smile, with his tongue lolling out of his mouth. She walked him back to Pat's house and told Pat all that had happened, but in the back of her mind she worried about what was happening between Chloe and Blake.

9

After Jolene had been arrested there was an investigation into her allegations that her husband had been threatened and bullied by members of the Omega Society. Sadly it seemed as though the blame was misplaced. Keith had unfortunately been the victim of some bad investments. He had put a lot of stock into a certain product that failed to work as well as he thought it would. It had been a speculative gamble, and one that had not paid off. After that his business had been a humble one, and he had lived out his life feeling that he was a failure, always ruing the opportunity he had missed. Jolene had been unable to admit that her husband had basically sacrificed their future for a business opportunity, so she had pinned the blame on the Omega Society. An investigation at her house had turned up a collection of rings that matched the ones the Omega Society wore. It seemed as though she killed the ones she thought were to blame a long time ago, and only now did she still seek revenge.

She had succeeded in one thing though; the death of the Omega Society. With Albie gone there was nobody else who was going to champion its cause or try and enlist other members. The elder members like Lord Huntingdon and Len had long left that life behind, and their sons had no intention of carrying it on. Lily felt foolish for ever suspecting that Blake had anything to do with this, and she looked forward to seeing him again so that she could apologize, and perhaps tell him a few other things that had been on her mind. There were many feelings that had been swirling within her heart, and she longed to open up to him and share with him why she was usually so guarded. The thought of it made her heart beat rapidly though, and hurt feelings from the past flashed in her mind.

She shook her head to dispel these thoughts, and told herself that if she was ever going to feel the things she wanted to feel then she was going to have to take a risk.

It was Sunday morning and Lily walked to church. It felt strange to not have Zeus by her side every moment of the day, and she wondered if she would ever get used to not having a dog. At least she would still get to spend time with him though.

As she approached the church her nerves increased. She had decided that today was the day when she was going to tell Blake that she wanted to be more than friends. She greeted people at church, and the place was abuzz with murmurs about the latest murder. Most people admitted they did not know how Jolene could have done these things because she seemed like such a sweet woman. They were all a little frazzled at knowing a murderer had been in their midst, but they looked toward Reverend Brown to guide them to a place where they could feel more comfortable with themselves, and not look with fear at the people around them.

Reverend Brown stood behind his pulpit and looked at his flock as he addressed them. Lily sat beside Blake, picking at the skin around her nails due to nerves.

"Masks. Sometimes we wear them without even knowing, and very often we look at them without suspecting. There are times when wearing a mask is important, such as when we need to protect ourselves. But it's equally important that we do not form the habit of seeing a mask everywhere we look. We need to look beyond what people show us, directly into their eyes, because then we will see the truth. I know it has been a traumatic time to learn that Jolene did this terrible thing. I am ashamed that I did not see it myself seeing as I spent so much time with her. I suppose it is a reminder that we must always remain vigilant, and that even when we think we can trust someone there is always a way for the devil to get into their hearts. I would like us all to speak more openly about our feelings, especially in times of trauma like this. We cannot allow hatred and bitterness and other feelings like that to define our existence. The only way to truly fight against them is to be a community, to take care of each other. We must be available to each other and patient with each other, and we must approach each other with kindness and empathy rather than judgment. We have all been afraid in life. We have all felt that darkness coming around us, but we must remember that there is always a light within us that can push it back. That light is faith and love and the Lord working through us. Jolene lost sense of that light," he paused for a moment and allowed his head to drop, as though in some way Jolene's failing was his failing too. "But we cannot allow her failure to be repeated. We can be stronger. We must always remember that no matter how dark it gets there is always the opportunity for light to shine, and to emphasize this point I would like to welcome a special guest to speak to you."

Reverend Brown stepped away and held out an arm, pointing to a side room of the church that he used for private meetings with the members of his flock. Murmurs arose as people speculated about the identity of his mysterious speaker, and an amazed gasp followed as Pat appeared. The local curmudgeon was in a place he had sworn he would never enter again, around people he had often argued with and moaned at. He walked slowly, and used a cane to help him keep his balance. Zeus was by his side, holding his head proudly, as though he approved of what his owner was doing.

345

He reached the podium and clutched it, using it to support himself. He was clearly still struggling, and Lily was proud of him for taking this step. She had secretly been hoping that he would find his way back to his faith after all this time. The murmurs died down as people waited to hear what he had to say. Unlike Reverend Brown's smooth, effortless voice that filled the church to the rafters, Pat's voice was uneven and rough, and there were moments where he mumbled, so people had to strain to hear what he was saying. But there was no denying the emotion in his voice.

"It's strange to be back here after such a long time," he began. "I used to come here a lot. I used to believe that the world was a good place, and then my sister died." He bowed his head to pay his respects and it was clear he struggled to cope with the emotion that was swimming inside him. "I lost my faith after that. I didn't understand how someone so pure and so sweet could be taken from the world when there were other people who survived... people like me... people who were bitter and broken. I lost a part of myself when she died and I wasn't sure I would ever get it back again. I've seen things in this world that I'll never be able to forget, and when I came back to live here all I wanted was to push people away. Most of you obliged me," he said. A titter of laughter followed this. "But one person didn't. She appeared from nowhere and I thought she was going to make my life miserable. She knocked on my door in the dead of night and she interfered in my business, but I'm proud to call her a friend now. Lily McGee showed me that while we're here we have to try and help people. None of us are perfect, but what's important is that we keep trying our best to be better. What the Reverend said there about light and darkness, well, I'm not sure I've ever heard anything that sums me up more perfectly. I've been in darkness for such a long time, so long that I couldn't see anything else. And then I almost died. The Lord spared me, and I believe He spared me for a reason. I see now the error of my ways. I don't want to shut myself away any longer. I know it's not what my sister would have wanted. I just... I want to be a better man and I'm here... I'm here to..." he was clearly struggling with his emotions a great deal. Words were choking in his throat and he bowed his head. Reverend Brown came up to him and placed a hand upon his shoulder, squeezing it tightly as if to remind Pat that he wasn't alone. He whispered something in Pat's ear, and Pat nodded.

Pat took a deep breath before speaking again. "I'm here to ask you for your help. I know I haven't always been the easiest person to get along with, but I hope you can forgive that. I hope that you can see I'm trying to be better, and that you will help me become better. I can feel the light in me shining again, the first time in what seems like forever, and it feels good. It feels so good," he said, and the smile on his face was one of genuine wonder. He lifted his gaze to the heavens and seemed overwhelmed with divine inspiration. The crowd was silent for a moment as they processed what they had just heard, before they rose to their feet and gave a standing ovation to Pat, who had embraced God's love once again. They cried out that they would help him, and that they would help each other. Reverend Brown never usually liked these types of outbursts in his services, but he allowed this one to continue without interruption as he knew how important it was to Pat. In fact it took over the entire service and people began to mingle among themselves, taking the time to ask each other about their lives and their wellbeing.

"I did not see that coming," Blake said, impressed at Pat's revelation.

"It's wonderful, isn't it? I thought he might do something like this after I saw him. He seems like a different man, as though he's let go of a lot of things that were bothering him."

"I suppose having a near death experience will do that to someone, although I hope we don't all have to go through something like that to get a better understanding of ourselves."

"I don't think we do. Sometimes all it takes is a conversation."

"So Pat did things the hard way? That fits for him," Blake said, chuckling to himself. "Do you think this new version of Pat is going to stick, or do you think he's going to go back to old habits?"

Lily allowed her gaze to wander to the front of the church where Pat was being surrounded by people. Usually it would be the kind of scene where he would be barging past them and telling them all to leave him alone, but here he was taking the opportunity to speak with each of them in turn, all with Reverend Brown smiling behind him like a proud father. "I think this Pat is here to stay. I'm sure there will be moments where he stumbles though, but that's like anyone. It's difficult to leave old habits behind completely," she said.

"I suppose that's true," Blake said.

Lily felt her stomach tighten as images of the past flashed through her mind, of tears and broken hearts and nights that seemed as though they were never going to end. There were some habits she had to force herself to change though, and there was no time to change like the present. Blake began to move away, intending to speak to Pat himself, but she pulled him back. She braced herself against her nerves and kept telling herself that she did not have anything to worry about.

"Blake, I was wondering something… we've known each other for a while now, but I feel like we've never really had the opportunity to sit down and get to know each other properly," she said, feeling that the words were clumsy as they slipped out of her mouth. As a writer she could always put elegant words into the mouths of her characters, but she could not do the same for herself. Emotion always seemed to interfere with the cadence of her words. "I thought now that things are settling down you might like to grab some dinner tonight? I could use the company now that Zeus is back living with Pat," she added as a joke and a light laugh to try and show him that she was at ease, although she wasn't sure she succeeded.

"That sounds like a wonderful idea Lily," he said with a twinkling smile, although her hopes were quickly dashed like a ship upon the rocks. "But I already have plans for tonight. Constable Hargreaves asked me to dinner. I think she feels bad that she had to treat me and my father as suspects in the case. But we'll definitely get together at some point," he said, so casually. Lily wore a wistful smile and nodded, trying not to show how much the words stung her. He walked away, not known how cruel he had just been. She sighed and pinched the bridge of her nose, wondering how she could have been so foolish. She was just an American writer after all, and Blake was the son of a Lord. Why would she even think that she had a chance of romance with him? She chastised herself for being so foolish. The only small mercy is that Blake hadn't laughed at her.

She left the church thinking of a life in which Blake had said yes. Was there a world out there in which she could be his equal, in which he would see her as being a viable partner for him? It felt as though she was too late. Once Chloe sunk her claws into him she would never let go, Lily was certain of that, and she regretted not taking the opportunity to ask Blake on a date earlier. It seemed as though she wasn't free of fear yet. Because of past experiences she was still reluctant to put her heart on the line, but the longer she waited the longer she was going to have to live alone. Now that she was free of the distractions of Hollywood she knew she wanted something meaningful in her life, but after her last relationship had been so disastrous she wondered how she was ever going to find love. Some mysteries just weren't easy to solve.

A FRIGHTFULLY FOGGY MYSTERY IN A QUIET ENGLISH VILLAGE

1

L ily McGee was sitting in her office in Starling Cottage. A bookshelf stood behind her, filled with all kinds of stories, not just the mystery ones she is used to writing. In truth she had her fill of mysteries from real life. She had moved to the cozy village of Didlington St. Wilfred, tucked into the heart of the English countryside, to get away from the hectic chaos of Hollywood. However, during her time in England she has been witness to a number of murders and has helped the police capture the culprits. Not all of the murders had been successful, and not all of the mysteries had been revealed, but overall she could be satisfied with what she has accomplished. But now she was sitting in front of her computer, staring into space. The story was ready to be written in her mind and her fingers hovered over the keyboard, but she looked as though she had been frozen in place.

There was still much on her mind, distracting her from her work.

First, there was the matter that her heart was concerned with; how the interim constable Chloe Hargreaves had made romantic advances toward Blake Huntingdon, son and heir to the Huntingdon fortune. He carried his name well and oozed nobility and charm. Lily's breath had been taken away by how handsome he was from the first time they had met. A part of her had longed to go on a date with him, but she felt out of place. She was just a common American girl who had found a modicum of fame as a mystery writer while he socialized in elite circles with other members of the English elite. On paper it didn't seem as though they would be a good fit for a long term relationship, yet every time they were together there was a spark in Lily's heart and she always found her mood lifted. Blake had proven himself to be a caring and wonderful person with a sharp sense of humor, but Lily wasn't sure if there was a genuine attraction between them or if he was just being friendly.

She supposed it did not matter any longer because he was out having dinner with Chloe. Lily had missed her chance.

The second matter on her mind was the strange affair of the mysterious parcels being left on her doorstep. There was nothing inherently sinister about the content of the packages, for they contained sweet baked goods, but the fact that there was nothing to indicate who had left them there was strange, and the idea that someone had been sneaking around the grounds of Starling Cottage left an unsettled feeling. Lily glanced around warily, feeling vulnerable in her own home. She had borrowed some security equipment from Pat to hopefully catch the perpetrator and solve this mystery once and for all. It was a source of frustration to her that in real life she did not have the omniscient vantage point of a writer, and could only make do with crumbs of information.

She had a grim feeling she was going to have to wait longer for confirmation though, because a thick fog had settled over the English village. Lily sighed, realizing that she was not going to be able to get any work done that night, and went to her curtains. She pulled them open to see if the fog had cleared, but it had only gotten thicker. It obscured the usual lovely sights of the English countryside, hiding the pretty flowers and the quaint buildings, and it gave Lily a chilling feeling.

At least she wasn't alone in the house though. Her glamorous friend Joanie lived in Starling Cottage as well, and Lily was glad of her company.

"Are you done with work for the evening?" Joanie asked as Lily left her office and walked into the living room. Joanie was sitting on the couch with a book opened on her lap.

"I think so," Lily sighed as she fell into a comfortable chair. "I just can't seem to get my mind focused at the moment."

"It's no wonder with everything you have going on. Have you heard from Blake yet?"

Trust Joanie to strike directly at the heart of the matter, Lily thought, especially when it was to do with a man. "No, I haven't. Chloe is enchanting company though and I'm sure that he would have had a good time."

Joanie rolled her eyes. "You really need to stop with this self-deprecating view you have of yourself. There's no reason why Blake wouldn't be interested in you. If you like him then you should ask him out."

Lily folded her arms across her chest and swallowed a lump in her throat. "There's no point. He's gone out with Chloe now. I've missed my chance."

"Sometimes I think you like being alone," Joanie said, shaking her head. "I think perhaps I should take a counseling course if I'm to help you. I might as well considering I can't take any photos until this fog lifts. I do love England, but if there's one thing I could change it would be the weather. Say what you will about L.A., but at least I could predict what would be a good outfit to wear."

"I don't like being alone," Lily said bluntly, enough to get Joanie's attention.

"Is this about—"

She was unable to finish her sentence before Lily snapped at her. "It's not about *him*." The truth was it was about so many things. Sometimes it seemed as though the biggest mystery in Lily's life was to do with her romantic interludes. The last relationship she had was better left forgotten, although when she delved into her past it didn't seem like a mystery as to why she found it difficult to connect with people. She had never known her father, and her mother was a spiteful woman who resented having a child to look after. Most of the time she simply ignored the responsibility, and other times, especially when she was drunk, she would actively taunt Lily.

Lily could remember it vividly, the hatred and envy in those bloodshot eyes, the venom in the alcohol stained breath; *Nobody will ever love you Lily. They don't love girls like us. They just use us and throw us aside. One day you'll learn how useless it all is.*

There was always a cackling laugh that followed the slurred words, words that had taken a deep root within Lily's heart and had always been with her. The sad thing was that her mother's words had come to pass. Lily was still alone, and she did not see that changing any time soon. Perhaps that was the way things were supposed to be though. Other, more deserving people were meant to be happy, people like Blake and Chloe.

She was just here to solve murders.

Lily had drifted into dark thoughts of the past while Joanie continued talking, although her words were vague and indistinct within Lily's mind. It was only when she suddenly stopped and uttered a sharp question that Lily's attention was seized from the shadows of her past into the sudden future.

"What was that?" Joanie said.

Starling Cottage wasn't an inherently spooky place to live, and neither was Didlington St. Wilfred, although Lily's imagination had been stirred by ghoulish things during the recent mystery surrounding the ancient crypts. The fog certainly did not help though, nor did the fact that Lily's suspicion was raised by the strange parcels. There was a clumsy noise outside the door. Lily leaped to her feet. What she wouldn't have given to have the faithful dog Zeus by her side, but he was back with his master, Pat, who had made a return to the village after a horrible injury that had almost cost him his life. She ran to the door and flung it open, howling a warning as though she was a banshee.

"Who's there? Show yourself!" she cried, but her words drifted away into the distant fog. The thick and heavy mist swirled in front of Starling Cottage, making it impossible to see very far into the distance. Lily took a step into the chilled air, but stopped herself from entering the fog for fear that she would get lost, even in this small area. The light from inside Starling Cottage glowed brightly and illuminated the fog, but even though she searched for movement she could see nothing. She might as well have been blind. Her words echoed out through the night. There was no response. Her gaze fell from the fog to her feet, where there was another parcel. A lump formed in her throat. She bent down slowly, carefully, picking up the plain cardboard box and carrying it inside.

"It's another one?" Joanie asked. Lily nodded with a terse look on her face. She had already been to Chloe about this, but unfortunately there wasn't anything Chloe could do. It was a mystery that Lily was going to have to solve by herself. Her best guess was that it was an enthusiastic fan who had tracked her down, but it seemed strange there would be no note or anything indicating who had created these treats. Lily set the box onto the coffee table and opened it. There was a range of biscuits dusted in sugar, simple treats that were nothing compared to ones the local baker, Marjorie, could make.

"Are you going to try them?" Joanie asked.

"No," Lily said, running a hand through her hair, scratching her scalp as she did so. She had eaten treats from the first box that had been sent to her house when she thought they had been a gift from Marjorie, and thankfully they hadn't been poisonous. She wasn't willing to take a chance with the subsequent treats though. These would end up in the trash like the others. Before that happened though she carefully examined them to see if there was a clue that might lead her to some revelation, but no matter how intensely she scrutinized them they were still just biscuits.

Lily gnawed on her lower lip and tried to ignore the uneasy tension that knotted in the pit of her stomach as she rose, throwing the box in the trash with a good deal of anger. It was incessant, this feeling that anyone could be outside, that someone was hunting her. She rushed to another room where she had set up the feed from the security cameras that Pat had allowed her to borrow. Joanie was half a step behind her.

"Do you think we might be able to see who it is?" Joanie asked.

"I hope so," Lily said, running the camera feed back so that she could peer into the recent past. It didn't seem to help though. The fog swirled and barely moved.

"There!" Joanie cried, pointing at the screen. Lily quickly paused the tape and squinted in an effort to see what Joanie had seen, but what she could make out was just an amorphous shape, formless and vague. If she tilted her head and looked at it a certain way then it might have been a figure of a person, but it could also just have been a shadow of the fog. There was nothing that she could be certain of, and there was nobody she could identify from the shape either.

She sighed heavily, her head drooping down as she was no closer to finding the person who was tormenting her.

"I just don't know who it is Joanie. I have no idea who it could be. What if it's someone I helped catch, or what if it's—"

Joanie reached out and placed a hand on Lily's. "It's not going to do you any good to speculate like that. Come on, you know it's not going to do you any good. You're always the one to say look at the evidence. I know there's not much evidence to look at, but we'll get it in time. They're going to make a mistake at some point and we'll figure out who they are. Once this fog clears they won't be able to hide again and we'll see them clearly."

Lily blinked slowly, trying to hide the tears that formed in her eyes. "It's just that with this and Chloe and Blake… and I haven't even been able to take Zeus for walks since the fog has been like this either. For the past few months I've felt better than I ever have before, but now it's getting to the point where I'm wondering if I'm always going to have the same troubles. I'm beginning to think that it might be time for me to leave England."

Joanie leaned back, her eyes wide with shock. "Surely you can't mean that Lily? You've been loving life over here! Think of all the good friends we've made and everything you've accomplished! I don't think you would have found the inspiration you needed to start writing back in America."

"I know, and I've really enjoyed helping the police solve mysteries, except that Chloe doesn't really want me helping her. I'm just starting to feel like I don't belong here. I never intended to stay here for the rest of my life anyway. I only ever wanted a quick break before I returned. Maybe it's time for that," she said.

Joanie gazed at her with sympathy in her eyes, while Lily tried to avoid looking at her friend. It was a giant thing to admit, but things had been grinding her down recently. All the murders she had helped solved had taken their toll on her. There was only so much one woman could take without having any lightness and freedom to balance out the darkness, and the fog was suffocating Lily to the point where all she wanted was to scream.

2

The fog cleared the following day, to some extent at least. Now, instead of being like a blanket draped over the world, it was akin to a veil that provided a hint of mystery. It at least allowed people to go about their business though. Lily peeked outside and immediately looked at her doorstep, just in case a parcel had been left for her. The deliveries had been erratic and at differing times of day so there wasn't even any pattern she could begin to work back from. She thought some fresh air would do her good though as the fog made her feel cooped up, and if she was going to go through with her decision to leave England then she wanted to soak in the atmosphere as much as she could while she was still here.

She intended to pay Marjorie a visit, wanting to see how the baker was coping with the fog. She was alone, her husband having died a number of years prior to Lily arriving in Didlington St. Wilfred. It was a shame, as Lily would have loved to have met him. While she was on her way to the bakery though, her attention was captured by a loud scream in the distance. Everyone who was outside were struck by the sheer terror in the voice and ran to the source, with

Chloe Hargreaves also walking briskly away from the police station to arrive at the scene of the crime.

They were on the outskirts of the woods, just beyond the main area of the village. The grass was long here, but people often used this area as a path toward the farms beyond the village. In the summer it was a lovely walk, with birds chirping in the trees and rabbits hopping around in the undergrowth. The fog prevented all of that though, giving it an eerie sheen that was difficult to ignore. The sight that greeted Lily and the others was nauseating. Some of the people had already turned away, hands to their mouths, regretting the way they had indulged their curiosity. There were gasps and murmurs, and at least one person had staggered away doubled over, having been brought close to vomiting by the sight.

Lily stared at the body, intrigued. It was like staring into a mirror. The body had been hanged from a tree, a thick rope around the neck. The face was pale and sad, the eyes empty as so many of the eyes that Lily had gazed into had been. The arms were spread out, with rope tied tightly around the wrists, leaving deep red marks. The legs had been left free to dangle. Lily could almost picture the grisly scene, the woman's legs writhing and flailing, attempting to catch onto anything that might offer her salvation, only to find nothing but empty air. She looked to be around the same age as Lily, with the same color hair as well. Her life had been snuffed out while Lily's continued to move forward, but Lily did not feel more deserving of life than this woman was. She would have had hopes and dreams as well, would never have planned for her life to end in this manner. Lily could already rule out suicide as there was no way the victim could have tied herself up like this. Someone had done this to send a message, quite literally, because one had been pinned to the body.

It was one of the other curious onlookers who first spotted it and walked forward, reaching out to touch it.

"Don't do that!" Lily cried out, pushing her way to the front of the crowd. "You're going to contaminate the crime scene." She spoke in a warning tone. The person who had gone to grab the note recognized her and acquiesced to her wisdom, for everyone in Didlington St. Wilfred knew of Lily's reputation as a sleuth.

"I think it's been contaminated enough already," Chloe said bitterly as she appeared on the scene. Her usually neat hair was in disarray and her pale cheeks were rosy. Beads of sweat trickled down the sides of her face and she heaved in a deep breath, placing her hands on her hips, having run directly from the police station. "Look at all these people. If there were any boot prints to tell us who had been here we're not going to find them." She scowled and shook her head at the sight of the crowd and the mess they were making of the ground. She seemed more erratic than usual. Envy flared inside Lily's heart at the thought of Chloe having been out on a date with Blake, but this was not the time to ask Chloe about that. Lily wasn't sure she wanted to know anyway.

"What happened here? Who found this?" Chloe asked, turning to the crowd. A woman stepped forward, and while she was describing how she had gone out for a morning walk and come across the body to Chloe, Lily took a closer look at the victim. It was uncanny how similar her features were to Lily, and it gave Lily a chilling feeling that she could well have found herself in a similar situation if things had gone differently. There had been a few occasions where she had been cornered by a threatening person, or where her life had been in danger. So far the odds had always been in her favor, but for how long would her luck keeping coming up trumps like this?

Her gaze moved from the woman's face to the note that had been pinned to her chest, the pin having been driven through her clothes and flesh, with blood having become crusted where the metal had pierced the skin.

'Living in Sin means Dying is a Virtue,' the note read. It had been scrawled across plain paper, handwritten, the ink having bled into the paper. It was a chilling message, but Lily had to ask herself who it was for.

Suddenly there was a cry from the crowd. "Hayley!" a man pushed through the people standing around and ran up to Chloe, falling prostrate before the hanged woman. A look of dread and sorrow was on his face. He reached up to the victim, as though he could still save her, but it was far too late for that. "No," he croaked, "no it can't be. Hayley. Hayley what happened?"

There was no response from the victim though. She was silent. It was up to the living to try and understand the whispers of the dead.

3

———

Lily had followed Chloe to the police station. Once Chloe had learned that the man who appeared was called David, and he was the boyfriend of the victim, Hayley, Chloe had some of the other officers prepare to get the body down so that it could be transported to the police station, and warned the crowd to stay back and give them a place to work. She also told them to be careful and to return home. Another murderer was in Didlington St. Wilfred, and it could have been any of them. It was in fact quite common for murderers to return to the scene of the crime, taking macabre pride in what they had done. Before Lily had left she had glanced at the people in the crowd, trying to see if any of them had smug smirks on their faces, but she was unable to see anyone who seemed noteworthy.

She had taken to tagging along whenever Chloe had a case like this to solve. At first Chloe had been opposed to Lily helping out, but over time her protests had fallen silent. While Chloe wasn't about to say how much she appreciated Lily's help, Lily assumed that she was welcome as Chloe wasn't the type to stew in silence. So Lily once again found herself in the police station with the steam from a cup of tea rising in front of her. David was slumped in a chair, still looking distraught. He stared into space. Lily knew that look well. It was the look of a person who had not yet accepted reality. He was a tall man with lean features. His glasses had a tendency to slide down the bridge of his nose and he absently pushed them up, a gesture that he must have performed dozens of times a day. He wore a loose fitting shirt and slim fitting jeans, with boots that were scuffed, albeit clean.

"David, now that we're here, could you tell me a little bit about your relationship with the deceased?" Chloe began.

"She… was my girlfriend. My partner," he said, choking on emotion, clearly still coming to terms with the idea of referring to her in the past tense. Lily tried to catch his eye and offer him a sympathetic smile, but he wasn't moving his gaze.

"I see, and were you two living together?"

"No, we were thinking about it, but we were just waiting for the right place to pop up. I always said I wanted to wait for the right time, but I guess there is no right time. I should have just done it. I should have just…" he trailed away, his breath falling to a sigh and he lowered his head, pressing his fist against his forehead while clamping his eyes shut, evidently frustrated that he had wasted time. Lily glanced at Chloe, wondering if she was going to cut the interview short, but that wasn't Chloe's style.

"David, I know this has been a shock for you this morning and that this is very difficult for you. I don't want to put you under any undue stress, but in my experience the sooner we get information the better our chances are of finding out whoever did this. I'm going to ask you some questions and I'd like you to answer to the best of your ability. If it becomes too much for you or you want a break then just tell me, okay?" she said.

David nodded his head.

"Right, so do you know what Hayley was doing being out last night?" Chloe asked.

"She was coming to see me," he said. His voice was weighted down by guilt. "I know it was stupid. I never should have asked her to come out in the fog. It was just... you know how it gets. With the village being so dark it seemed like the kind of night where we should have been cozying up together."

"Why didn't you go to hers? Surely it would have been gentlemanly of you to make the effort of walking?" Chloe asked, Lily thought with too much of a hint of accusation. David thought so too as his eyes narrowed and he replied in a terse tone.

"Her place is a mess right now, and it's too small. She works from home so she likes the opportunity to leave. We've always used my place as a place to hang out, and she's not the type to be shy. Maybe I should have insisted that I come and get her, but she said she was capable of looking after herself."

"And you didn't think anything was amiss when she didn't arrive?" Chloe asked.

"I just thought she had changed her mind. Sometimes she does that. It was maddening about her. Used to drive me crazy, but I loved her for it as well. It meant that life was always unpredictable. I just figured she had gone outside, realized how bad the fog was, and then decided to head back into her house. She sometimes gets tired and falls asleep easily. It's happened before. I shouldn't have assumed though. I should have gone out there. I should have—"

"Given the fog last night it's unlikely you would have managed to find her even if you had been looking," Lily said.

"But I could have at least tried. I love her and I didn't even go look for her. I should have sensed that something was wrong. How could I just have gone to sleep while she was being killed? I..." he lost the strength to speak and put his head in his hands again. Lily's heart went out to him. It was hard enough losing someone, but even harder when you felt there was something you could have done to prevent it.

"David," Chloe continued, using his name to try and get him to focus on her voice. "The questions I need to ask you now are going to be difficult and they might even seem unfair. I promise I'm not trying to upset you, I'm just trying to get as much information as possible, okay?" she said. David nodded.

Chloe took a sharp intake of breath. Lily braced herself because she knew what was coming. "I know that you said you were waiting for her. Is there anyone that can confirm this?"

"No, I was alone. The only person who could have confirmed it was Hayley," he said, and then his voice seemed to catch when he realized what Chloe was intimating. His expression shifted and the sorrow gave way to shock and indignation. He furrowed his brow and sat up, glaring at Chloe and Lily.

"What is this?" he asked in a harsh tone. "I've just lost my girlfriend and you're trying to say that I had something to do with it? Are you crazy?! What kind of monster are you?" he said, his tongue lashing like a whip. Chloe took it all in stride though.

"I told you that I was going to have to ask you difficult questions, David. I don't mean to cause offence."

"You might not mean it, but you're throwing plenty of it around," he said sullenly, folding his arms and slumping back in his chair.

"Was there anyone else important in her life? Anyone else that she might have gone to see last night or that might have known she was coming to see you?"

"Anyone else? No there was nobody else! She wasn't like that. She was just... she was just normal. She liked her work and then she liked spending time with me. That was it. She just... she always wanted more. She wanted to get married and I kept telling her that we needed to wait and save up more money. It always felt as though we were waiting for our lives to begin and now it's never going to happen. Now it's all too late. What am I going to do? How am I going to cope without her?" he asked. The anger had been replaced by complete and utter desperation, but neither Lily nor Chloe could offer him an answer to the question.

"I'm very sorry for your loss David. Thank you for your time. I'll be in touch if there's anything else I need to know, and if you think of anything that might be helpful just come by the station. Someone will be happy to talk to you," she said. David looked numb as he lifted himself out of his chair and walked away, in a kind of stupor. It can't have been easy for him, Lily thought, and she wished she could make it easier. Even finding the killer never really offered closure. People often thought it would, but it didn't bring the victim back.

"So what do you think about that?" Chloe asked, sipping her tea as she stretched her neck in one direction and then the other.

"I think he's a very upset young man who has just lost the love of his life," Lily said.

"So you don't think there's a chance he might have done it?"

"Well, he had the opportunity. He knew where she was going to be, and the fog gave him a shield so nobody would have seen him. I suppose it's possible he could have raced from his home and intercepted her, then killed her and returned, using the fog as an excuse."

"It does strike me as strange that he wouldn't have offered to walk with her. I would hope that if I had a boyfriend he would offer to come and walk with me."

"As he said, she might have been stubborn and insisted on coming to him alone," Lily said, feeling a wincing pain inside at the thought of Chloe having Blake as her boyfriend. "But he did seem utterly distraught. I would be hesitant to have him as the prime suspect. I think we need to look elsewhere."

"But where?" Chloe asked, gazing to the ceiling. "With this fog it's difficult to glean any clues. Nobody saw any movement. Nobody saw anyone during the night. It's as though we've all been rendered blind."

Lily nodded, thinking about the way the security cameras had been proven useless as well. She thought about mentioning the newly delivered parcel to Chloe, but then thought better of it. Chloe's stance wouldn't have changed now as there was still nothing sinister in the packages, plus there was a murder to solve, and that had to take precedence.

"Something he said does make me wonder if there's more to this than meets the eye," Chloe said. "David mentioned that he had been dragging his heels when it came to progressing in their relationship. What if she was exploring other options?"

"I think we'd need to go and see her house for that," Lily said. "There is also a possibility that it's just someone else, someone who was an opportunist."

"Opportunist in the fog?" Chloe asked, furrowing her brow. "I think to be close enough to kill Hayley, the murderer would have had to follow her."

"That's not what I meant. I meant that he might have been waiting for the fog, waiting for something to shield him and make it easier for him to be hidden. There is something else as well, although I'm not sure how important it is to the case."

"What's that?" Chloe asked.

"Well, it's just the fact that she looks like me," Lily said.

Chloe frowned and then closed her eyes, bringing forward the image of the dead woman. "Oh yes," she exclaimed, "I suppose she does. Well, I'm sure it's just a coincidence. Would you like to go to her house and look for any clues?" Chloe said. It was a big gesture for her to want to include Lily in the investigation, but in truth Lily wasn't in the mood. There were so many things playing on her mind that she could not concentrate, and was worried that she would miss something vital.

"Actually I think it would be better served being in your hands."

"Are you okay?" Chloe asked, a hint of concern creeping into her voice. "I'm not sure I ever would have expected to hear Lily McGee refusing the opportunity to go and investigate for clues."

"I know," Lily offered a weak smile, "I'm fine, I'm just feeling a little under the weather I suppose, and there are some errands I wish to run before the fog sets in again."

Chloe nodded with understanding and then Lily walked away from the police station, feeling a little guilty that she wasn't able to give this matter her full focus as she had been able to do with some of the other cases. There was something gnawing in the back of her mind that would not rest. It was entirely maddening, and the wispy fog that lingered around the village did not help at all. She found herself feeling frustrated, and dreaded the thought of returning to Starling Cottage. Guilt swam in her heart as she knew she should have wanted to remain with her dearest friend, but she found herself going to see someone else aside from Joanie, someone who might be able to shake the cobwebs from her mind.

4

L ily arrived at Huntingdon Manor and had been shown in by one of the servants. The walk had been brisk, and the fog was transparent enough where she could still see for a vast distance. She imagined that Joanie was using the opportunity to capture some beautiful still life imagery of Didlington St. Wilfred under the gaze of the fog. The way it shifted around the landscape gave it a sense of life, and there were moments when it seemed as though she was walking through a cloud, as though the heavens had descended upon the earth. If only that was the case, she thought, for then there would be no more of these gruesome murders.

Lily had been shown into one of the sitting rooms. A stuffed deer's head protruded from the wall, which she hoped was fake. Its eyes glinted in the soft electric light of the room. A bookcase stood in the corner, while a liquor cabinet was beside it. Lily wondered which one was used more frequently, although she had her suspicions. As she looked around the manor house she wondered if Chloe had been brought back here after their date, and then pushed the thought aside, feeling as though it was beneath her. There was a murder to solve as well, and she wanted to put as much effort into that as she possibly could, just like she had all the others. They deserved to have her full attention because Hayley was a person who had lost her life. Lily was still alive. She had the time and the ability to investigate the murder, to bring justice to the person who had killed her, but with everything going on in Lily's mind she just couldn't be sure about what to do.

Thankfully Blake entered the room. His mere presence was enough to soothe the anguish in her soul for a time. The air became alive with the scent of his cologne and she smiled at his friendly demeanor, getting the feeling that she would never tire of seeing the sparkle in his eyes.

"Lily! I wasn't expecting you today, how are you?"

"I'm well," she said, "although there has been a murder."

"Another one? Well, I'm sure you're hot on the trail of whoever committed it."

"I'm not sure about that," Lily said with a self-deprecating laugh and a roll of the eyes. Blake poured himself a whiskey and offered Lily one, although she refused. He joined her, taking a seat in the green leather chair and leaned forward, wearing a concerned look upon his face.

"That's not the usual tone I'm used to hearing from you," Blake said.

"I don't feel quite usual at the moment. It's probably just to do with the fog," she said, seeking to dismiss the feeling with the wave of a hand.

"Are you sure? You know that you can talk to me about anything Lily. We've been through enough to enjoy that luxury at least, or if you really want some words of wisdom I can fetch my father for you," he added the last part with a smile.

"I think I'll be fine with you Blake. I just wanted to speak with you for a little while. I know Joanie is at home, but it can get so lonely in Starling Cottage with the fog hanging around me. I just wanted to be able to take my mind off things for a little while. There have been so many murders around here and it's all becoming too much for me."

"Well, they say that the Lord never gives us more than we can handle, but there are always times when we need to rely on other people as well. You know you can always count on me Lily," he said. His smile was enough to melt her heart and his eyes were like two deep pools that she longed to drown in. "If you need to take your mind off things then I'm your man, let me see, what can we talk about..." he tapped his lower lip and the skin in between his eyes crinkled as he thought about what to say. Lily knew exactly what they could talk about.

"You could tell me how your dinner with Constable Hargreaves went," she said innocuously, trying to act as though it was a matter of little consequence even though it really meant a great deal to her. She knew her envious thoughts were sinful and she was trying to be a nobler person, but there were times when the world was trying and with everything weighing down her soul it felt as though she was as tattered and frayed as an old rag doll.

Blake threw back his head and laughed, which wasn't the reaction she had expected.

"It was actually quite a good night. I didn't expect her to be such charming company. I actually think you two might be good friends if you hadn't started off butting heads. She told me how hard it was for her to rise in the ranks of the police. You know, even though we live in a modern world there are still some institutions that have outdated views on things and sadly she found it difficult. I think she saw being posted here as a punishment, but it was quite enlightening to hear her story."

"It sounds as though you got on very well then," Lily said as tension rose through her body.

"Oh yes, by the end of it I think we became good friends," he said. Lily's ears pricked up at this.

"So there was nothing... more to it then?" she asked, gently hinting at a romantic nature.

Blake shook his head. "While I do admire her bravery and her spirit, she seems intent on returning to the city as soon as her posting here is finished, and I have had enough of the city. When I came back to Didlington St. Wilfred I was filled with a sense of belonging, something that I had not felt in a long time."

"I felt the same thing when I arrived here, but recently... I don't know..."

"What do you mean?" Blake asked.

Lily gnawed on her lower lip and wasn't entirely sure what to say in response. How much of herself was she willing to share with him? Every time she seemed to be getting close she kept hearing her mother's voice in the back of her mind, nagging her again and again that she wasn't good enough and she wasn't worthy of love.

"I've just been thinking of home lately," she said.

"Ah," Blake said, as though that explained everything. "I suppose it must be difficult for you to be so far from home, even though you're enjoying yourself here. Nothing ever feels quite the same as home, does it? Not even when you feel comfortable in a place. It's been the same for me over the years. There has always been a sense that I've been pulled back here. It's funny really, when I was younger I would have done anything to get away from here. But over the years I've realized that the things I'm looking for are here in the village. Perhaps you could take a small trip to refresh your memory of home?"

"Perhaps. I have been thinking of getting away for a while. Which is strange because this was supposed to be my way of getting away. I just feel so ill at ease at the moment, and I'm not sure what's changed. I guess it must be me, because the place hasn't changed," Lily said. The more she spoke the more embarrassed and ashamed she felt. She had come here with the intention of asking Blake about the possibility of her staying in Huntingdon Manor. With the increased staff she was certain that nobody would be able to sneak in, and she doubted anyone would track her here. However, if she was going to be so pitying and so whining she feared that Blake would never see her in a romantic light, and if he wasn't impressed with someone like Chloe, who had been prosperous enough to rise through the ranks of a male dominated police force, then how was he going to be impressed with her when she was at this crossroads? He needed a woman who knew her own mind, who was confident in what she wanted and could offer him certainty and an intrepid spirit. Lily could only offer him the opposite right now. It had been kind of him to offer her friendship, but surely there was no way he could be romantically attracted to her, and she certainly wasn't going to humiliate herself enough to find out. Seeking refuge with him would only serve to torment her, so she ended up deciding to not make that request of him.

In fact she wanted to show him that she was brave and strong, so after they had spoken some more she declared that she was going to leave before the fog set back in.

"Are you sure about that? I can drive you if you'd like. I wouldn't fancy being caught in the fog," Blake said.

"I'm sure it will be no problem. I can walk at a quick pace when I need to. I shall be at home before you know it, and I'm sure you have important matters to attend to here," Lily said, rising and moving out of the room before Blake could say anything. When she walked into the open air she was faced with a thicker fog than she anticipated, but was too proud to go back on herself and ask Blake for a ride home. Instead she tucked her coat around herself tightly and dug her hands into her pockets, and then ventured off to Starling Cottage.

~

Within moments of walking Lily realized it had been a mistake to refuse Blake's offer. She was so lost in the mire of her own mind that she had allowed her instincts to become twisted. The fog had been a mist before, but with every passing moment it was thickening into a cloud and it was getting more and more difficult to see in front of her. It was as though cataracts were forming in her eyes. The more she squinted and strained her eyes the darker things became. Among the fog there were shifting movements. Didlington St. Wilfred had always been considered a cozy village, but with this thick layer of fog around her it was terrifying and eerie. Strange noises whispered and rode the soft wind, shadows danced, and this did nothing to help Lily's paranoia. It felt as though grim pairs of eyes were watching her, peering through the fog, somehow able to see her even though she could barely see anything else. Her pace increased and her shoes clapped against the ground. There were dim lights shining in the fog, the lights of homes in which people were safe and comfortable. They were like the shining beacons of lighthouses, guiding ships into safety, and Lily was happy to use them like this.

The fog was not yet thick enough to prevent her from seeing everything, so she was able to make her way to Starling Cottage and breathed a sigh of relief when she walked through the door and was greeted by Joanie's smiling face again. She closed the door behind her, knowing that the fog was going to become denser, and that someone out there was going to be watching for her.

"Where have you been all day?" Joanie asked. From the imprint on the couch, Lily could tell exactly where Joanie had been.

"There was a murder," Lily said, and then spent a great deal of time explaining things to Joanie. Joanie gasped and shook her head.

"I thought this fog was bad enough, let alone knowing that someone is killing people. What were you thinking walking back from Blake's Lily? You should have gotten a ride from him."

"I know, I just… I wanted him to see that I was strong."

"Oh who cares about that. What matters is that you're safe," Joanie said. Lily was surprised at the force of the rebuke, and sat there like a scolded child. "Look, you really need to sort yourself out with Blake. I know things haven't been good for you in the past with romance, but that doesn't mean you can just lose yourself. I think you either need to tell him how you feel or stop thinking this way about him, because it's only going to cause you more heartache."

"But how can I tell him how I feel? I'll only make a fool of myself."

"You don't know that Lily," Joanie said. But in her heart Lily did. She kept hearing her mother whispering to her. *No good man is ever going to love you. Only the bad ones are going to want you.*

"As for this murder, maybe it's better that you sit this one out. If you're not feeling your best there's no reason why you should put yourself in a vulnerable position. Let the police do their job. I'm sure Chloe will be eager to solve a case on her own anyway."

"I have no doubt about that, but I feel a responsibility to Hayley to find the killer. I have skills and a talent that should be used. I can't just ignore them because I'm not in the right frame of mind."

"If you're not in the right frame of mind then you're not really going to be able to help them, are you?" Joanie said in a matter of fact way. "Now, I think you should make yourself some dinner and then run yourself a hot bath. You're holding a lot of stress inside you Lily, and that's not going to be good for you. Get a good night's sleep and maybe things will look better tomorrow. I hope they will anyway. The sooner this fog lifts the better. I have an itchy trigger finger just waiting to take some photos," Joanie said.

On a ranking of Joanie's ideas, the idea of a good meal and a hot bath was up there with the best. The frothing bubbles caressed Lily's skin and the scent of lavender filled the air. As she sank into the bath the knots in her muscles began to loosen, and all the tension began to drift away from her body. She dried herself and then slipped into the comfortable sheets of the bed, not even daring to look outside as the fog would only serve to sadden her. Unfortunately she could not stop errant thoughts from coming to her mind completely. When she closed her eyes she began to think about the case, about how scared Hayley must have been walking through that fog. Or perhaps she hadn't been scared at all. Perhaps she had been excited and happy at the thought of seeing David again, of speaking about the future and the home they were going to make together. She may have been so excited that she didn't even care about the fog, everything in her life moving along at a happy clip, making her feel invulnerable, because of course bad things only ever happened to people who weren't feeling on top of the world.

She would probably have been lost in her own thoughts, probably had not heard the slight movement beside her. It may have been a cracked twig, or a heavy breath. But then suddenly she would have felt the hands upon her, dragging her to the tree. One of them would have clamped around her mouth, silencing her screams. It wouldn't have mattered if she had screamed because nobody was around to listen. Nobody except the one who wanted her dead. Hayley would have been frightened beyond measure, and a part of her would probably have been in shock, telling her over and over again that somebody would come along to save her, that somebody would come along and stop this because she had too much to live for. In a way she hadn't even begun to live at all. She had to marry David, move in with him and bear his children. It was all written in the future. It was all going to happen... until that future was ripped away from her by a savage and cruel act. Until she was humiliated by being strung up amid the trees, forced to wait until the fog cleared until light could be shone on this grave and deadly act.

As Lily thought all these things her stomach churned. It sickened her that someone could steal away someone's life like this, that they could plunge the future into darkness. What were Lily's problems compared to this? How selfish could she be when there was a murder to be solved? Nothing she was going through mattered when Hayley had been killed. There were more important things in life, and she needed to focus, to do what God had put her on this earth to do. She thought about what Blake had said, about belonging, and Lily knew where she belonged. She could help people in a way that few others could, and it would have been wrong of her to take herself out of the game.

5

The following morning, Lily awoke with a renewed sense of purpose. Her malaise still lingered in the depths of her soul like the fog around Didlington St. Wilfred, but it wasn't going to stop her from bringing justice to the world. After having some breakfast she walked to the police station with a determined frame of mind, eager to find a way to the truth.

"Good morning," she said, nodding to Chloe as she marched in. "Did you find anything at Hayley's house?"

Chloe arched an eyebrow. One corner of her mouth curled in a smile. She was holding a piece of paper and folded her arms across her chest as she regarded Lily with a cool smile. "I was wondering how long it would take you to come here," she said. "And no, I didn't find anything of interest. She seemed to be as devoted to David as he was to her, which also creates a problem for him as he was the only one who knew where she was going to be."

"You're right. It can't have been an opportunistic murder, unless we're going to assume that the killer kept the note on them just in case. It's more likely that the killer was watching Hayley, lying in wait for the perfect moment," Lily said, annoyed with herself that she hadn't picked up on that the previous day. Her mind really had been fractured.

"Yes, so I think I might speak with David again, because I really don't have any other leads to go on. I tried looking at the crime scene, but there were so many people around it was impossible to know what marks were there before the morning came and what marks there were after. I've searched for the phrase as well, but it doesn't appear to be an extract from a passage. I was hoping it might be from a book or a song, just *something* to give us a hook."

"I know. It's not as if we can search the whole of the village for the culprit. There is something you should know, something I didn't tell you yesterday. I don't know if it's going to be involved in this case or not. I received another parcel yesterday morning, left in the fog. I didn't see who it was."

"I see," Chloe's lips formed a thin line. "And what was in it this time?"

"Biscuits," Lily replied, feeling a little silly as she revealed this information.

"Biscuits," Chloe repeated.

"I don't know if they were poisoned or not. I didn't risk eating them. I just... you know, considering that Hayley and I look alike I thought it might be worth mentioning, in case they're related."

Chloe swept a hand through her hair, pushing it aside from her face. "I understand that this is strange, but I don't think it's logical to assume the cases are related. We need more evidence of that before we start using that as a motivational force. I think for now we're going to have to treat them as separate incidences, but I will look into this mysterious fan or whatever it is you have," Chloe shook her head a little as she said this, clearly amused and disturbed by the crime in equal measure.

"I know, I just wanted you to be aware of all the information. Let's get back to the murder at hand. So far we know that Hayley left during the night, when the fog was at its thickest. If someone was watching her then they may have been plotting this for a while. David said she worked from home, so I guess that means we can rule out coworkers."

"We can, which means that it's probably someone who saw her going about her day to day life, and given the size of the village we can't rule anyone out," Chloe said.

"I really hate it when the murderer turns out to be a neighbor. Everyone in the village seems so friendly."

"There are always bad apples, no matter how small or big the place is. I think I'm going to check to see if there have been any unsolved murders of women who look like Hayley before. This may not be the first time the killer has struck. There might be other occasion, or perhaps people have been killed in the fog before. There have to be records somewhere."

"It would be easier going to speak to Bob about it. It'll save you more time than searching through the files," Lily said, trying to put in another good word for the constable, who had been relieved from duty pending an investigation into police negligence. The case had taken place when Constable Dudley had been a young officer and bullied into following orders by a brash and arrogant chief, but the investigation was still ongoing and Bob's fate had yet to be decided.

Chloe scowled at the mention of his name. "The only thing I can learn from him is what mistakes I shouldn't make if I want to do my job properly."

"It wasn't his fault Chloe, it was the chief he was following."

Chloe glared at her. "Then he shouldn't have followed the orders. There's such a thing called personal integrity Lily. Some of us are able to stand up for what we believe in." She spoke harshly, her tongue lashing like a whip. Lily was taken aback and was about to ask her what had caused such an outburst, when the door to the police station opened and Lily's mouth dropped open, because a ghost stood before her.

"I need your help," Hayley said, at least from a cursory glance Lily certainly thought it was Hayley, except that was impossible. As the woman moved into the station Lily focused on her and began to see the differences; the shades of the eyes, the tilt of the head, the curled ringlets in the hair. But the similarities were remarkable, and she saw a lot of herself in the woman as well, as did the woman. When their eyes met they studied each other, as though they were staring at their reflection.

"What's going on? How can we help you; I'm Officer Hargreaves and this is Lily," Chloe said.

"I'm Jo," the woman said. It took her a few moments before she was ready to peel her gaze away from Lily. She was clearly shaken and staggered to a seat, where she ran a hand through her hair and pushed it away from her face. "I think someone is following me. No, I *know* someone is following me. I can feel it. It's everywhere I go. I can just feel their eyes on me and I can't bear it any longer." As she spoke the words crashed out of her mouth as though they were being fired like bullets. She gnawed on her lower lip and her leg was shaking as well.

"Slow down Jo, it's okay, you're in a safe place here. Nobody is going to hurt you. Now what's this about someone following you. Have you seen them?"

Jo shook her head. "I haven't seen them, no, but I know they're out there. I can feel them. They're in the fog, hiding. I keep looking for them, but I can't see them. Then I heard about what happened yesterday. I just... I can't cope with it. I can't cope with this fog. It's bad enough to walk through it as it is. And when I saw who had been killed I just freaked out."

"Did you know the victim?" Chloe asked.

383

Jo looked up and furrowed her brow, shaking her head as though the question was foolish. "No, but she looks exactly like me. You must know how I feel," she said, turning her gaze toward Lily.

Lily swallowed a lump in her throat and nodded. "I felt the same thing when I was walking home last night, as though all these eyes were upon me. It's not a nice feeling," Lily said, reaching around to scratch away a chill that crept down her neck.

"Well I felt more than eyes. I saw someone," Jo did.

"You did?" Chloe's ears pricked up. "What did he look like? Do you have a description?"

Jo shook her head. "It was in the fog. I didn't see anything clearly. I can't see anything clearly in this fog. I just felt him behind me, you know, like his presence. I was so scared to turn around, you know, it was almost like when I was a kid. I used to hide under my covers and I had this fear that monsters were waiting for me to turn around, only able to attack me if I looked at them. If I didn't know they existed then they couldn't hurt me. I wanted it to be the same with this guy, but when I turned around he... he was there."

"Did you get a good look at him? Anything would be helpful? I know it's hard but think back, was there any sense of what he was wearing or the color of his eyes or hair? Think Jo, please," Chloe said.

"I'm trying," Jo said. The effort was plain on her face and Lily's heart went out to her. Lily was also afraid, because this was the second woman who looked like her. Chloe might try and say that the mysteries weren't connected, but *someone* had been leaving mysterious parcels at Starling Cottage, what if it was this man and he was getting frustrated that Lily wasn't responding in the way he wanted? This might be another cry for attention.

"I'm sorry," Jo said, shaking her head after a time, "there's just nothing. I don't know what he looked like. I wish I did, believe me. I wish I could tell you exactly what he was like, but I can't. All I know is that he was there and now I'm afraid he's going to be there again the next time I go out."

"How did you get away from him?" Lily asked softly, curious to see why the man, and presumed killer, had allowed the woman to escape.

Jo sniffed back tears and took a moment to compose herself. "It was only because of the fog. I walked normally at first, hoping that he wouldn't sense that I had spotted him. I couldn't keep doing that though. I started to run back to the village. I think the fog helped me. I made it to a friend's house and stayed there for a while until my husband came to pick me up. I thought I should come to you and tell you. I just can't wait for this accursed fog to lift. It takes my mind to all these dark places when I look outside. I can just imagine him standing there, waiting for me to be alone again, watching everything I'm doing…" her words trailed away into soft sobs. Lily certainly knew how she felt.

"You're safe now," Chloe said reassuringly, although Jo did not seem reassured. She continued to wring her hands in her lap and she looked around continually, as though she was afraid a shadowy figure would appear within the police station. Lily understood though. Paranoia was a difficult thing to shake and it was easy to believe that someone could find them, when that person was yet to be identified. It could have been anyone.

"When your husband came to find you, did he see anything suspicious?" Chloe asked.

"I'm not sure," Jo said.

"I think it would help us to talk to him," Chloe said. Jo nodded and mentioned that he was just outside, waiting in the car for her as she hadn't wanted to come to the police station alone.

～

385

Jo's husband, Will, walked in. He was a tall man. At first glance he appeared to be handsome, with a square jaw, a thick mane of hair, and deep set eyes. He dressed well too. Lily smiled at him as he entered, but was unnerved as his gaze lingered upon her. Lily was unsure if this was actually happening or if this was just a trick of her mind. She was uncertain if she was actually seeing this, or if it was just something that was borne from her own paranoia. Will leaned in and whispered in Jo's ear, before taking a seat himself.

"What can I help you with?" he asked.

"I just thought you could tell us your experience of what happened. Did you see anyone suspicious?" Chloe asked.

Will adjusted his position in his seat and wore a thoughtful look. "When Jo called me I went as quickly as I could. I arrived and I did see a shadowy figure in the distance, but I couldn't make out anything."

"Not even if they were male or female?" Chloe asked.

Will shook his head. "I'm afraid not. They must have been wearing a hoody or something. I suppose I should have chased them, but I was concerned about my wife. I just wanted to make sure that she was alright. I'm sorry. If I had thought more clearly then, I might have been able to get my hands on them. It was so foggy I wouldn't even be able to tell you if it was a man or a woman!" he said. After every few words he kept glancing toward Lily. There was a nervous laugh, and he kept gnawing his lip as he spoke. Perhaps it was just because Lily was in a nervous state already, but it set alarm bells ringing in her mind.

"Are you well Will? You seem to be a little unnerved," Chloe said. Lily breathed with relief when Chloe asked this question, as it confirmed that not all of Lily's instincts had been compromised.

Will ran long, slender fingers through his hair. When Lily looked at him more closely she almost sensed that he was trying not to look at her almost as much as he gazed at her. He smiled and again laughed nervously. "Sorry, I'm sorry, it's just that I'm a private man you see and I'm not used to all of this attention. I only really want to make sure that my wife is okay." He leaned toward Chloe and lowered his voice. "I would also prefer it if any... revelations in the case would pass through me before Jo, to protect her from anything unsavory, you understand. She has been shaken by this enough already and I would not wish to put her under any further distress," he said.

"Of course," Chloe said sharply. "I believe that's all for now. Thank you for coming in," she said, turning to Jo, "and I'll be in contact if I need anything else."

Jo and Will rose. Will reached out to support his wife, although before he left he turned around again to cast one last smile at Lily. His eyes gleamed with something that made her shudder, and she was quite glad to see the back of him.

"Well, I'm not sure what to make of that. It's one thing to be protective of his wife, but when he spoke he acted as though she wasn't even there! It's a shame that she can't remember more though. I wish she was more at ease here. She doesn't have anything to be afraid of in this station," Chloe said.

"I know how she feels though. I felt the same when I was walking home alone last night. The way she described it..." Lily said, trailing away. She was lost in her own thoughts, so she didn't see Chloe's incensed look.

"What do you mean you walked home alone?" Chloe asked.

Lily stood there open mouthed, unable to offer a good defense. Was she supposed to tell Chloe that she wanted to appear brave and strong to Blake? What a childish thing it seemed.

"I thought you were smarter than this Lily," Chloe chastised. "From now on you're not to go walking about by yourself, especially not at night. It was speculation before, but after seeing Jo I think we can definitely say that the killer has a type that he goes for, and you match it perfectly. Get back to Starling Cottage, hunker down, and don't go anywhere or let anyone in. We'll find out who this person is before too long, I promise you that. They might be able to hide in the fog, but the fog isn't going to last forever," Chloe said.

"It might last long enough though," Lily said. Up until this point she had tried to put on a brave face, especially in front of Chloe because of the rivalry that had existed between them, but things were too solemn for such a trivial matter. "I just don't know what I'm supposed to do. I don't feel safe in my own home. I can't concentrate on anything because I keep having this feeling that someone is out to get me, and I know you don't think there's anything sinister in these parcels that are being left at my door, but it still means that someone is lurking around. I just don't know what I'm supposed to do," she said. It was unusual for her to feel adrift like this.

"I'll tell you what you're going to do. You're going to stay somewhere safe and keep your head down until this matter is resolved. We may not have always seen eye to eye Lily, but I'm not going to let anyone hurt you. We can't deny the truth here. Someone is stalking women who look like you."

"Do you think someone wants me dead?"

Chloe sighed a little. "I'm not sure. We can't dismiss the theory. It might just be that you fit his type and it's unrelated, but given the parcels... I don't know. Either way you're in danger. Is there somewhere you can go to be safe?" she asked.

Lily thought about all the friends she had made during her time in Didlington St. Wilfried. There were a number of people she could count on. Pat and Zeus would have defended her, but Pat was still recovering from his illness. Marjorie would definitely have taken her in, but Lily didn't want to risk bringing trouble to her door. The obvious answer would have been Blake, but Lily felt strange about being vulnerable around him, and she thought it would be safer to remain in the heart of the village rather than near the outskirts, so she was left with one choice, a man she had trusted many times before; Bob Dudley.

As soon as she mentioned the name Chloe rolled her eyes, but Lily was certain of her choice and Chloe escorted her through the city to Bob's house.

6

"I really can't thank you enough Bob," Lily said after Bob had agreed to let her stay with him. The tension was evident between him and Chloe, but for Lily's sake they did not argue. Lily called Joanie to bring across an overnight bag from Starling Cottage, and as soon as Joanie arrived Bob became flustered. He disappeared to make a drink, and it was at this point when Chloe took her leave. She promised Lily that she would do everything she could to solve this case and find the killer so that Lily didn't have to hide away for too long. She also left Lily with a warning to be careful, glaring beyond Lily's shoulder toward the kitchen, where Bob was located. Lily thanked her, even though she didn't need a warning about Bob.

She had a quick look through the overnight bag and hugged Joanie, thanking her for being such a good friend.

"Think nothing of it," Joanie replied with a quick smile. With her golden hair and striking figure, Joanie was the farthest thing from Lily's appearance so wasn't as threatened by the killer as Lily was. They took a seat in Bob's living room and waited as he brought out some cups of tea with a plate of biscuits. It was clear from his manner that he was unused to playing the role of host, and it was somewhat charming.

"Are you sure you don't want to stay here as well Joanie? I don't want to impose on Bob, but I'd hate for you to get hurt in case someone comes to Starling Cottage looking for me," Lily said.

"It would be no imposition at all," Bob said, although he seemed a little flustered. He always did around Joanie.

Joanie smiled and dismissed the suggestion with a shrug and a tilt of the head. "I'm sure I'll be fine. Starling Cottage is nice and quiet. The person leaving the parcels hasn't made any threatening motions to break in yet, and if they do I'll be sure to raise enough of a stink that the whole of the village will hear me and come running. I still have a scream mastered from the horror movies I filmed when I was trying to become an actress," Joanie said with an assured smile. Their conversation continued until the tea had been drunk, and then Joanie thought she had better be going before the fog grew too heavy, leaving Lily and Bob alone.

"I wish I had her courage. She never seems to be afraid of anything. She just takes everything in her stride," Lily said.

"She is a remarkable woman," Bob replied, and there was something more than a sense of respect in his voice. It was something that had been there from the first moment Bob had laid eyes on Joanie. Lily hadn't needed to be an observant sleuth to detect it; it was the kind of reaction most men had toward Joanie as her beauty was rare. "I have never seen anyone like her."

"I'm sure you haven't," Lily murmured. Her mind turned to thoughts about the case, but Bob was distracted.

"There is something I've been meaning to talk to you about Lily. I'm not sure it's the right time for this, or if indeed there is a right time. But I've had a lot of time to think about my life since I have been suspended and it's made me realize that there's something missing. The days here are so endless and quiet and I really wish there was a way to bring a spark into my life. And then I think of Joanie and, well, I know that I am only a humble man with humble ways and I'm sure that I can't compare to the caliber of men who exist in Los Angeles, but there is just something about Joanie that I cannot ignore and I wondered if you had any advice for someone who might have romantic intentions toward her," he said, wringing his hands and blushing fiercely.

Lily smiled. It was a welcome distraction from her grim thoughts about the case. "Joanie likes people who are confident and take the lead. She likes people who are direct. I'm sure you'll do fine Bob, but good luck. You're going to need it because she's a handful," Lily said with a wry smile. Bob chuckled.

"I'm sure she is," he said with a knowing look in his eyes.

"Speaking of your time off work, have you heard anything about the investigation?" Lily asked.

Bob's demeanor changed immediately. He went rigid with tension and lines of stress appeared on his face. "I don't know for certain. I have been assured that it should be over soon, and I was relieved to hear that. I don't know what's taking them so long. It should have been an easy matter to resolve. I suppose the only thing I should fear is that they want to make an example out of me."

"I'm sure they won't be that petty," Lily said. "You weren't the one to blame. You're not responsible. We all know that you didn't do anything wrong, and everyone in the village would swear by it, I'm sure."

"Not everyone," Bob said with an arched eyebrow. Lily knew who he meant; Chloe, who had made her mind up about Bob once she had been sent here as his replacement. If Bob was found guilty and fired from his job then it was natural to assume that Chloe would be his permanent replacement, which meant that Lily was going to have to deal with her for even longer, and it meant she might try and woo Blake again. Although Lily had managed to strike up an efficient working relationship with Chloe, she always got the sense that Chloe was putting up with her begrudgingly rather than actively welcoming her assistance.

"Well, I hope that justice is served sooner rather than later. It's not been the same working the cases without you," Lily said.

"I'm sure it hasn't. I've heard some stories about my replacement, although she seems to be getting results."

"She has. I think it took her a little time to get used to the way things are handled here, but she is a capable officer and I'm sure she has a bright future in the force."

"I'm sure she has. I bet she just sees this place as a stepping stone as well. That's the problem when you take police officers away from their homes, they don't have a good understanding of the region and everything is seen as a way to get another step up the ladder. I've never wanted anything more than to protect this town, and if they take it away from me I don't know what I'd do..."

"Perhaps you could help me. We could team up and become McGee and Dudley, private investigators."

"I think Dudley and McGee has a nicer ring to it," Bob said, and laughed heartily.

Lily unpacked and settled into the small spare room before she descended the stairs again as evening settled in. Bob was busy with a jigsaw puzzle and asked Lily if she wanted to help him, although she said she had another puzzle to work on. He then asked if she wanted any help, but she said she wanted to work through it herself at first, but that she would call for him if she needed to discuss anything. Right now she needed to put down some information and work this through using her logic rather than by being fueled with fear. It was easy for her mind to run away with itself and she hated herself for it, wishing that she was more in control of her emotions like Joanie, or perhaps even Chloe. Then again it wasn't as though women who looked similar to them were being stalked and killed.

Lily sat down beside a lamp with a notebook and a pen. She worked on the assumption that she was in danger and the killer was looking for her, and began to think of anyone who might want her dead. She thought about the note that had been left on the body as well, and racked her brain to think if she had heard it anywhere before. She wished that she was in Starling Cottage so that she could quickly search through the books she had written to see if she had used that phrase, although nothing came immediately to mind. One by one she wrote down names of the people she had caught during her time in Didlington St. Wilfred, wondering if any of them had escaped and now sought vengeance against her. Her growing reputation as a cozy mystery author also leant itself to attracting some strange characters who were obsessed with death and blurred the lines between fact and fiction. There were other people as well who had nothing better to do than to leave inappropriate comments on her social media pages (pages that she had at the insistence of her publisher, who thought it was better for her chances of engagement). Was it possible that someone had found out where she lived and was now bringing this deplorable behavior toward her? It made her skin crawl to think that her success as a writer could also bring her so much sorrow, and even that an innocent woman had been killed because of it. All she had wanted was to bring a little bit of joy to the world and offer people an escape from the drudgery of modern life, but some people always sought to twist things into unrecognizable, awful things.

It was just a shame that the people who had consciences were always the ones to pay the price.

She was so lost in thought that when she heard a noise from outside. Panic flared and she almost leapt out of her seat. She called out to Bob, who had heard it too. It was probably just the wind, Lily thought, but in the back of her mind fear gnawed at her, and it was impossible to deny that someone might have been watching her all the way from the police station, tracking her movements to Bob's house and was now waiting to hurt her. Moving from Starling Cottage may not have made a difference at all.

Lily's face was white and her fingers dug into the arm of the chair as Bob placed a finger to his lips. He held a cricket bat in his hands. His burly figure was imposing, and there was no better person for Lily to have protecting her, but even so she was still scared. She waited carefully to hear anything, and was primed to leap up and flee through the back door. In her mind she was already plotting the direction to the police station, confident that she could find it even in the fog. However, when Bob called to her, Lily heard the surprise in his voice. It was absent of fear.

She went to the door and her entire emotion was changed when she saw Zeus sitting there, his pink tongue lolling out. His eyes sparkled when he saw her and he came bounding up to her, licking her playfully. She hugged him and buried her face in his fur, only realizing how much she had missed him when she now saw him. She kissed and petted him, before realizing how odd it was to see him by himself. As she petted him she found a note attached to his collar.

She unfolded it and saw that it was from Pat. The note claimed that he might have some information about the killer.

"You don't think it's a trick, do you?" Bob asked.

"Not at all. Zeus would never let someone unfriendly put a note in his collar," Lily said. "It seems we have a little trip to make," Lily said.

"Are you sure you want to leave the house? I thought the whole point of this was that you were going to stay safe?" Bob said.

"I will be safe with you and Zeus by my side. Besides, if Pat has information then we need to learn what it is as quickly as possible. The killer might strike again tonight. I don't want another woman to die because I was too afraid to go out in the dark," she said.

Bob grumbled a little about having to go out into the fog, but he took her point and fetched a flashlight, as well as putting his jacket on. The two of them followed Zeus outside. In truth Bob's flashlight didn't make much different as it didn't parse the fog, only illuminated patches of it, but it did make Lily feel safer. Zeus stayed close to her, as though the dog could sense her fear as well. She was still filled with the sense that someone was watching her, but this time it was alleviated by the fact that she had Bob and Zeus with her.

"I can't see a thing," Bob complained as they walked at a much slower pace than usual. With Zeus' nose guiding them though, navigating the fog wasn't as difficult as it would have been otherwise. Lily continued to be on her guard as they walked through the village, searching the fog for shadows, although she saw nothing. Zeus made no warning signs either, so his nose wasn't detecting danger. They made their way to Pat's and Lily was relieved when she was inside another house. Pat greeted Bob with a friendly handshake. Lily stifled a chuckle when she saw the look on Bob's face. Not everyone was used to Pat's now amiable demeanor yet.

Zeus walked back to his master and sat at Pat's feet.

"So what's this information then Pat?" Lily asked.

Pat smiled at her, amused at her ability to always get to the heart of the matter. "I'm glad you made it here without any trouble. I have been humbled by the news of what has been happening in this village. In my former life I always liked the fog coming down upon the world as it gave me a shroud in which to work. I imagine this killer is the same, but unfortunately for them they're not the only ones who know how to watch."

"What do you mean?" Lily asked.

Pat smiled wryly. "I've always been a cautious man, doubly so ever since a mortal enemy moved into a house nearby without me realizing. That house has remained empty ever since then, and yet I have detected signs of life there, small flickers of movement within the windows. It seems as though someone is making use of the house and trying to keep it a secret, someone who would rather not be found. It's rather curious, wouldn't you agree? Of course it may not be linked to this crime spree at all, but then again…"

"Then again what's more perfect for someone who likes to hide?" Lily finished the thought for him and glanced toward Bob, who was in fierce agreement with her.

"I think you should take Zeus with you as you investigate. I would love to come with you, but I am still not well enough to move around outside like I'd need to in this fog," Pat said. Lily could feel the frustration within him. It must have been awful to be shackled to his home when before he had known the freedom. The only benefit was his change in disposition, but Lily hoped he would regain his full mobility soon so that he could fully be a part of the Didlington St. Wilfred community. But for now there was someone who was living within the boundaries of the village who wasn't supposed to be there, and Lily was going to find them.

She whistled for Zeus to join them and he rushed to her side, well used to assisting her in these types of matters. Now that Lily was in control of the situation again and acting like the hunter rather than the hunted the fear dissipated from her heart. Her eyes gleamed and she felt like her old self again, hating this killer for taking that feeling away from her, even if it had only been briefly.

7

In a sense this house had been cursed. Lily had been privy to two crimes that had been located in this house, and now there was potentially a third. Zeus led them through the fog, while Lily kept herself low, creeping along even though the fog offered her cover. Bob was the same as they came to the front yard of the house. The plants had withered and died. Dry leaves crunched under their feet. Lily cursed inwardly as she made noise, but she hoped it was imperceptible to whomever was inside. They crept around the back. She knew well enough that a third night might mean the end of another woman's life, and if they could find something that would shed some light on the identity of the culprit then she could bring an end to this reign of terror, and the fog would be a mere inconvenience rather than a harbinger of death.

They reached the back door. Lily had not sensed anything from Zeus yet. Bob reached out and he tried the door handle. It creaked, which might as well have been as loud as a banshee's scream out here, but nothing reacted. The door swung open and the curling fingers of the fog stretched inside. Bob moved in first, along with Zeus. The dog stood in the middle of the room, sniffing the air as though he was inspecting it. Lily slipped in behind them, pressing her back to the wall. The house was dark and filled with shadows. There was a musty, damp smell, and the house was not filled with things that a home should be filled with. It was hollow, like a skeleton, and Lily wondered if it would ever be given another chance to be filled with joy and laughter again. In some ways she thought she was like this house, empty of love, filled with sorrow. She pushed the thought aside, wondering if she was just using these macabre things to distract herself from what she truly wanted in life.

Once she and Bob were convinced that nobody else was inside the house they began to poke around to see if there were any clues as to the identity of the killer. Bob swung his flashlight around. The beam illuminated the house. There was a plate on the table filled with scraps of food, and then she heard Zeus snuffling in the corner of the room. Bob followed the dog with his flashlight. Zeus was sniffing into a pile of clothes that had been left near the wall. Then the dog lifted his nose to the sky and raced toward the front door. Lily followed immediately, the chase being on.

She flung open the door and Zeus was already away, running while growling. Lily felt her heart thumping in her chest as she followed him, trying to remain within sight of him while also being cautious about not running into anyone she wanted. She had no idea how far Bob was behind her, although she caught sight of the shadow of his flashlight hitting the fog, making it gleam, as though it was made of gold. Lily had no idea where she was in relation to Didlington St. Wilfred, but as long as she was with Zeus she knew that nothing bad could happen to her.

Suddenly Zeus stopped and yelped, but before Lily could react she heard a heavy set of footsteps behind her as a powerful set of arms wrapped around her. She immediately struggled, but the hand that clamped around her mouth suffocated her scream, and the other arm was a vice like grip. She flailed her legs and tried to stomp on his foot, but he had them too far away. Panic flared and her eyes bulged. All she could see was the fog, and she worried it was going to be the last thing she ever saw.

"How could you betray me like this?" he spat. Lily felt the hot anger of his voice spilling over her cheeks. She could feel the fury in his grip as he tried to choke her and drag her away. It was the same thing that had happened to Hayley. Lily could feel exactly what Hayley must have been going through, the abrupt fear of suddenly having a man grabbing her, of taking her away from the world. The only difference was that Hayley had been alone when she had been taken, while Lily had allies.

Zeus came in, nipping at the man's heels, yapping and growling away. Bob was soon upon them as well. The light flashed in Lily's eyes, blinding her and the killer for a moment, before Bob was pulling him away from Lily. She staggered away, clutching her throat and gasping for breath. In the fog she could see Bob's huge frame slamming his fist into the man's chest and sending him falling to his knees, with Zeus baring his sharp teeth for good measure.

"I give up! Stop! Stop!" the killer pleaded. Lily swallowed a lump in her throat, although it ached badly after what the man had done.

"Let's get you back to the station," Bob said with a sense of satisfaction. It had been a while since he had caught a criminal, and Lily knew it must have been a big deal for him.

8

They were back at the station. Chloe's face had been filled with surprise when Bob had returned with the man in tow. She was forced to give Bob a begrudging look of respect. Lily sank into a chair and had some water, composing herself before she and Chloe questioned the suspect.

"Do you have any idea who he is?" Chloe asked before they began.

"I don't," Lily said, and that worried her most of all. What did the man mean by saying that she had betrayed him? Could he have been a fan who thought her books had taken a direction he did not like, or was there something else afoot?

Lily called Joanie to come to the station as well, and when she arrived Lily noticed how Bob blushed. While Chloe was doing the paperwork for the arrest, Joanie asked them what happened and Lily was not shy in praising Bob. He was a little flustered again, but seemed to remember the advice Lily had given him and was happy speaking with Joanie about the ordeal. Joanie's eyes flashed with admiration and Lily hoped that something nice was going to result from this at least.

Then Chloe gave her a nod. It was time to question the suspect.

~

Lily walked into the room, trying to push the fear as deep down as it would go. The man looked desolate as he sat slumped in his chair, his wrists cuffed together. His hair was tousled and there was a rough shadow of stubble peppering his face. Shadows ran under his eyes and his lips were chapped, while his skin was pale.

"His name is Colin Stewart, ring any bells?" Chloe asked. Lily shook her head.

"Why did you do it? Why did you want Lily dead?" Chloe asked.

The man looked up and had a stupefied look on his face. He furrowed his brow and his gaze passed between the two women. "Lily? I… I didn't want to kill you," he said. His voice was softer now, a far cry from the angry growl that had rumbled against her ear.

"Then why did you attack me? Why did you try and drag me away?" Lily asked. It took all her willpower to keep her voice steady rather than tremble with emotion.

"I thought… I thought you were *her*," he said.

"Her? Who do you mean?" Chloe asked.

The man raised his hands to rub his eyes. The handcuffs clinked as he moved them. "My wife," he groaned. Chloe once again asked him to elaborate on the details. The man was distraught and they had to wait for him to compose himself enough to continue speaking. There were many long pauses between the words, and he seemed so uncertain of himself. It wasn't the kind of profile Lily thought the killer would have. She imagined him to be ruthless and cunning, but this man was so… so lost. Perhaps it was just because the light was shining on him now. It was funny how so many fears seemed smaller and less threatening in the light.

"I found out recently she was having an affair. It almost destroyed me. She left me with nothing, nothing but a note. She said I wasn't good enough, that I wasn't... well, it doesn't matter now. I managed to track down the man she was having an affair with. He lived here. I came here in the hope of finding her, and then the fog set in. It made it harder to see, but I was sure I had found her when I was out walking. She seemed so happy. I assumed she was going to see him and I just... I lost control. There was a time when she was that happy about seeing me. I'm her husband. When did she lose that sense of love? I just... I don't understand. All I wanted was to talk to her about it, to ask her why and to see if there was anything that could be done to save our marriage. But then I saw her and I was just filled with so much rage. I called out to her and she pretended not to know me, at least I thought that's what happened. In the fog I couldn't see her clearly. I dragged her away. I just wanted to get her to talk to me, but I guess I had a hold on her too much. She was dead, but I thought I could send a message to anyone else who was thinking about cheating on their husbands. I left a message for her and ran back and that was the end of it.

But then the next day when the fog began to lift it was clear that it wasn't her at all. I hadn't intended to kill anyone, I promise. I was just... I thought it was her. This other woman... I feel so bad."

"If you felt so bad then why didn't you come forward?" Chloe asked in a flat, blunt voice. Lily thought about Hayley, about how she had just been walking through the fog to visit her boyfriend, how she had been planning to marry him and begin their lives together. There had been so much to look forward to and it had all been taken away because of something as basic as a mistaken identity. It sickened her to think that a life could have been ended all because someone had been in the wrong place at the wrong time.

Colin gnawed at his lower lip and didn't seem to be able to answer that question. "I wanted to I just... I was confused. I didn't know what to do. But I thought I had figured out where the man she was cheating with lived. I watched their house and I saw her leave. I wanted to talk to her and tell her what was happening. She deserved to know. I followed her and tried to get her attention, but then she ran from me. I thought if we could speak with each other then we could figure out what was going on. I thought she might be able to help me find my wife as well. But she was gone, and then I saw him arrive and I just... I can't believe he can get away with being married, with living one life where everyone thinks he's perfect."

Lily remembered the lecherous glances Will had given her and thought to herself that not everyone thinks he's perfect. She filed that away for later though, because it wasn't Will who was on trial here.

"So you killed one woman, stalked another, and then when Lily arrived to find you, you tried to attack her as well. You keep saying that this was all about your wife, but it seems to me as though plenty of other women are getting hurt because of this," Chloe said.

"I know, but I didn't mean it to be that way. I promise. I'm not a bad man I'm just... I'm just hurt and upset and angry. I just want to find her, to speak to her. I want my old life back," he groaned.

"You're never going to have that life back again," Chloe said. "It doesn't matter that the fog clouded your judgment. The light is shining on your now Colin, and there's no way for you to hide the crimes. You're going to jail for murder. Your wife might come out of the woodwork and speak to you, but I wouldn't count on it. If I was her then I'd be getting as far away from you as possible."

"No, please, it was a mistake..."

"A woman is dead because of you," Lily added. "She had a life too. You have to pay the price."

He looked at Lily, and perhaps he saw enough of his wife in her to feel ashamed. He hung his head and then tearfully nodded, accepting his fate and his role in these despicable actions. Lily was glad that they had caught the killer, but there was still something nagging at the back of her mind.

"Wait a minute Colin, if these attacks weren't directed at me, then why did you leave parcels and treats outside my door?" Lily asked.

Colin blinked away the tears and looked at her blankly. "I have no idea what you're talking about," he said. Lily had no reason not to believe him at this point as he had confessed to everything else, but it left her wondering who had left the parcels.

"It seems as though there's still one more mystery to solve," Chloe murmured under her breath as she took Colin away. Lily left the room in a bit of a stupor, returning to Bob and Joanie, who were deep in conversation. Lily updated them on what happened, and they were both shocked to hear the truth.

"Do you want to come back home or are you going to stay with Bob for the night?" Joanie asked.

"I think I'd like to stay with Bob, if that's alright," she said, turning to the burly man. "I just don't feel safe in the cottage. There's still someone out there who is looking for me and I can't shake the feeling that it's someone bad. I want to find them. I need to understand why they're doing this, because until then I'm not going to be able to sleep at night," she said, her voice cracking with emotion. The killer had been found, but that wasn't the end of things, at least not for Lily. She stepped outside and looked out toward the fog that continued to thicken in Didlington St. Wilfried. Within the wisps there was someone who knew who she was, someone who was stalking her, but who?

9

A few days had passed and the fog had begun to lift. Lily had spent most of the time in Bob's house, aside from a brief journey to Pat's when she took Zeus back and told him about what had happened. He said he wished he could help more with her predicament, but she said she was just waiting for the day when the fog was gone completely so the security cameras would work again.

Lily had another matter to attend to as well. She had left it a few days just in case any other information about the case came to light, but Colin had admitted everything in that initial questioning session. Lily had seen David walking through the village on a few occasions. He looked like a ghost. She wanted to go up to him and tell him that everything was going to be okay, but she knew it was going to be a lie and that her words were going to be hollow, so she left him for now. Nothing was going to be able to bring back his dead girlfriend, and he would likely always torture himself about why he had invited her over on that foggy night, and why he had not gone out himself to walk her safely to his door.

But there was another heart she might save.

She walked up to a red door and gave the handle of the knocker a good firm rap. There was a sharp rap that echoed through the house. The door opened and Jo was standing there. She greeted Lily with a smile, and a surprised look in her eyes.

"Hello Lily, is there anything I can help you with? I thought they caught the killer," Jo said. As Lily gazed into the woman's eyes she was once again taken aback by the resemblance between the two of them. She also remembered the way Will had looked at her, and if Colin's accusations about Will and his wife were true, then Will certainly had a type.

"They did, but I wanted to talk to you about something else. Do you mind if I come in?" Lily asked. Jo opened the door widely and ushered her in.

"If you want to speak to Will as well you're going to have to come back another time. He's gone to work," Jo said. Lily smiled, as though the information was new to her. She had already observed Will and knew what time he was likely to be away.

"I just need a few moments with you, if that's alright," she said. Jo showed Lily into the living room and poured two cups of tea. Lily took the opportunity to look at the happy family photos that were peppered around the room, and she hated herself for the bombshell she was about to drop on this picture of a happy life. She told herself that it wasn't her who was ruining things though, it was Will. She was only showing the truth, a truth that already existed.

She thanked Jo for the tea as it was handed to her and took a sip.

"What can I help you with?" Jo asked.

"Actually it's something that I can help you with, at least I hope it's a help. During Colin's confession he told us something that I think might be of interest to you. Now, I must say that it is a delicate matter and it's not necessarily something that you might want to hear. I think in your position I would want to know the truth though, so if you're okay with it then, shall I continue?"

"I think so," Jo said, nerves flickering across her face.

Lily nodded. "The reason why Colin was here in the first place was because he was trying to track down his wife, who cheated on him. He didn't manage to find her, but he did manage to find the man she was cheating with. He also found that man's wife, and tried to speak to her, to tell her what was going on," Lily said. Jo remained puzzled. "That man is Will, Jo," Lily revealed softly. "I think that Colin's wife was having an affair with your husband."

Lily's voice was gentle, but there was no way those words could ever be anything less than sharp and scathing. For a moment Jo was still, silent, and remarkably composed in the face of seeing her world crumble around her.

"Did you hear me?" Lily said.

"I did," Jo replied. Her lips became a thin line as she stared into her cup of tea.

"I'm sorry if I shouldn't have told you. I wasn't sure whether I should or not, but I thought if I was in your position I would like to know."

"No, you did the right thing," Jo said, and then offered a weak smile as she lifted her gaze to meet Lily's. Lily saw the sadness swimming in Jo's eyes. "The truth is I've suspected something for a long time. Will has always been the kind of man to want whatever he wants and not care who he hurts when taking it. As far as he's concerned life is meant to be lived. I used to think I was enough for him, but I suppose not," she sighed. "I think the warning signs were there, but I tried to ignore them. I told myself that I was being paranoid, that he wouldn't do this again."

"He did this to you more than once?" Lily asked, unable to keep the surprise from her voice.

Jo shrugged. "I know it might not make much sense, but he is my husband and we are married. I just can't bring myself to bring it to an end. But now I suppose this might be it. I need to talk to him about it. I always used to think that when I fell in love and got married my life was going to be easy. I thought getting a husband was the complicated part, but it's not. It's always quite unfair how nobody tells you how difficult life is going to be, isn't it?" she said.

"It is," Lily said, but in the back of her mind she knew that she had been warned. From a young age her mother had always told her that she was good for nothing and never going to accomplish anything in life, and while it had shocked Lily that Jo had stayed with Will despite knowing he had cheated on her before, she was only shocked because she assumed English women were a little more sensible than the scatter brained ones she had known in America, such as her mother. Lily's mother would always say that trouble followed her around, but often Lily wondered if the opposite was true and it was her mother who sought out trouble. She also wondered if she was going to repeat the same mistakes as her mother as well, or if she could escape the curse that her mother had placed on her. Was she ever going to be able to accept that someone could love her, that she might be good enough for someone like Blake?

She left Jo's house not long after this, unsure if Jo was actually going to speak to Will about the revelation or not. It made her think about marriage. It was still something Lily wanted one day, even though her romantic life had never been anything to write home about. This whole ordeal had taught her that it was better to live life with someone to count on though, and she thought to herself that if Bob could be brave enough to declare his interest in Joanie then she could be brave enough to declare her interest in Blake. After all, he was never going to be able to share her feelings for him if he didn't know what those feelings were, and she was never going to be happy if she kept everything locked inside.

Lily was deep in thought as she returned to Starling Cottage, for she had been through quite a lot and had even questioned her place in the village. It still wasn't over yet as she didn't know who was leaving the parcels, but she had once again brought a killer to justice. Although she hadn't been able to prevent Hayley's murder in the first place, she was at least glad that she had been able to play some role in apprehending the suspect. It made her realize that she wasn't going to be happy living a normal life. If she could not help solve murders then she would be leaving the world in a worse place, and if she ran away then she would be just like her mother. Leaving would be accepting defeat, and Lily had spent too long becoming the kind of person who would fight against the inevitable.

She would focus. She would continue to write her stories, and if there were other killers who cropped up in Didlington St. Wilfred she would help the police bring them to justice, and she would also try her best to believe that she was worthy of romance.

The fog was beginning to clear. The wisps were receding and sunlight was beginning to spread across the land, as though God was blessing it with his caress. As this happened Lily's mood began to lift and she was eager to review the footage from Pat's camera, sitting and waiting for the mysterious stalker to reveal themselves.

Lily was looking intently at the cameras when she saw movement. She stiffened suddenly, tension running through her body. She was about to call Joanie as a pair of legs walked toward the door, but a different kind of excitement thrummed through her when she saw that it was Blake. She left the cameras behind and opened the door before he had even got to knock. He tilted his head toward her and looked a little confused.

"That was quick, how did you know I was here?" he asked.

"Let me show you," Lily said, an excited gleam in her eyes as she took Blake through to the camera room. "Since Colin wasn't the same person who was leaving behind the parcels I have to try and figure out who it was, and this is the best way."

"Any idea yet?"

"No, unfortunately," Lily sighed.

"Well, I'm sure that the truth will reveal itself soon. Until then I thought that you might like to accompany me for dinner. I thought you might use the distraction considering what you've been through. I didn't like how troubled you seemed when you came to me. You've always been so unflappable Lily, and to see you like that was… well, it wasn't something I've been used to at all."

"I'm sorry for disappointing you," Lily said in a sharp tone.

Blake held out his hands and shook his head. "No, I didn't mean it like that. I just meant that, well, I didn't like knowing there wasn't anything I could do to make things better for you. So I hope that by taking you out to dinner I can make up for that, if that's something you'd like?" he asked.

A wide smile spread upon Lily's face at the suggestion, and at the fact that he had been concerned for her. She assumed Blake had wanted someone strong and invulnerable, but perhaps that wasn't the case at all. Maybe he liked having someone to protect.

"I'd love to Blake, are you going to do the cooking, or are you going to get your staff to cook us a meal?" Lily asked with a teasing gleam in her eyes.

Blake laughed. "Neither. We're going to go out. Only the finest restaurant will do for Lily McGee! So I hope you have a dinner dress that's suitable. We'll go into the city, I know a few good places."

"That sounds wonderful," Lily said, her eyes sparkling with the idea of going on a real, genuine dinner date with a man she thought the world of. There were a number of dresses hanging in her closet that she hadn't had a chance to wear during her time in England, and there was no better occasion than this. Her stomach swam with excitement and her heart leaped with anticipation at having an evening with Blake all to herself. She was eager to spend a night speaking with him about his past and his future, really learning about the man rather than their usual conversations, which revolved around the murder cases she was working on. Other than those she had only been able to see him when they were at church, and with so many people around it was hardly an atmosphere to create intimacy.

So finally after Lily's struggles with her place in the world she was beginning to feel a little better about things. The past was ready to be left in the past. Lily could only move forward in life and get the things she wanted if she actually took them when the opportunity presented itself to her, and she thought she was doing herself a disservice if she missed out on them simply because she was afraid. Her thoughts turned to Hayley, as well as the other people who had been murdered. Their lives had been cut short. They did not have the opportunity to fall in love or even regret not falling in love, so Lily felt as though she owed them. She wanted to make sure she lived because they could not.

Her eyes sparkled. "Just let me know when," she said. "But I want us to agree to something first; promise me that there won't be any talk about murder."

"I promise," Blake said. He said something else after this, but Lily didn't hear what he said because she had become distracted by something else, and only something dramatic would have been able to tear her attention away from Blake. Beyond his shoulder she saw a figure approaching the cottage. A shawl was pulled around the shoulders, obscuring the face. The body was hunched over. Lily's throat tightened and fear curdled in her heart. How she wished that Zeus was by her side. She turned away from Blake, who appeared confused. He followed her outside, but she was racing away, each step propelling her with nerves. She flung open the door to Starling Cottage. A parcel had been left on the front doorstep and the figure was turning away.

"Hey! Hey wait! Who are you? What do you want?" Lily cried out, moving so frantically that she almost tripped over the package that had been left. The shrouded head looked back, but the face was cast in shadow. The figure seemed panicked at being seen and began to shuffle away hurriedly, breaking into a run down the sloping road. Lily ran through the garden path. By this point Joanie had sensed the commotion and had joined Blake outside. Lily's feet pounded against the ground, the impact reverberating through her bones. The shawl of the mysterious person fluttered as they ran away, veering toward a cluster of buildings that led to a forest. Lily wasn't going to let them get away, not when she had come this close to catching whoever it was. She needed to see them. She needed to understand who they were and why they were doing this, why they had come to her cottage. The anger burned away her fear. Adrenaline seared her soul and she wasn't about to back down now that she had the mysterious figure in her sights. So much of the terror had come from the unknown, but now that she could see the person she felt stronger again, she felt capable, and she would not continue to be afraid.

Lily closed the distance to the figure, bearing down on them, sensing that soon she would have an answer. The figure darted through a narrow opening between two houses. Lily squeezed in too, the brickwork scraping her arm. The figure darted around a corner and Lily was afraid that she would lose them. Every step brought her closer though. She reached out, her fingers stretching as far as they could reach in the hope that she could haul back the shawl, but it remained tantalizingly out of her reach.

"Stop! Stop! I just want to know who you are," Lily said, the words riding on panting breaths. The fear had given way to desperation now, a deep seated need to understand the truth, to cast the veil of mystery away and see exactly who was plaguing her. There were trees ahead, but the figure wasn't going to make it to the trees. Lily summoned her strength and ignored the sweat that was stinging her eyes. She groaned and muttered and prayed to God for the strength to catch this person, and catch them she did. When Lily was just about in reach she launched herself at the figure and knocked them to the ground, tackling them like a rugby player. Lily was caught in a tangle of messy hair and limbs. She exhaled deeply, blowing the hair out of her face so that she could see. For all she knew this person could be dangerous. They might have a knife. She twisted to protect her abdomen, but still fought to pull away the shawl. The figure moaned and writhed, trying to protect themselves, trying to squirm away, but Lily had waited too long for this. She had been through too much. The fog had been lifted from the village, and soon it would be lifted from her soul.

It was time to unmask the truth.

She finally gripped the shawl, pinching it in between her fingers and she yanked the hood back. What greeted her was a mass of long, light brown hair. The woman shook her head and kept screaming 'no!'. It took a few moments for Lily to get a good look at her face, to finally understand who this woman was and why she had tried to hide herself from Lily's sight. The shock was so much that Lily had to back away, leaving the woman free to leave.

"Mom?" Lily asked, the word a stone in her throat, "What are you doing here?"

A SNEAKY SIBLING MYSTERY IN A QUIET ENGLISH VILLAGE

1

---

ily stood in the alley where she had caught her mom, with Blake just behind her. There was a dumbfounded look on her face as she gazed at the culprit who had been leaving mysterious packages on her doorstep. They were filled with baked treats, and on the surface there had been nothing suspicious about them, aside from the fact that they were left with no indication who they were from, and nothing suggesting that this person wanted to get in touch with Lily. Given her role in life, both as an amateur sleuth and a writer of cozy mysteries, her mind often turned to the grisly realities of crime and it was easy to become paranoid with the thought that someone was targeting her with these parcels. Her mind had gone in many directions, trying to figure out if it was perhaps a disgruntled person who had been placed under suspicion during one of her investigations, or a crazed fan who wanted to appreciate her, or perhaps even an admirer who wanted to make his intentions known.

However, it happened to be someone she had never thought of, someone who she assumed had been out of her life completely and she would never see again.

It was her mother.

Lily was rigid with a myriad of emotions; fear, anger, suspicion, rage all burned inside her to such an extent that she did not know how to react at all, so she seemed frozen, as though she had been turned to ice. Her mother turned around. She had been in the process of running away, returning to whatever rock she had crawled out from under and whichever shadow had been hiding her. It was the same woman as Lily had seen leave her all those years ago, although the once bright face that shone so brilliantly with makeup was now creased with lines. The light in her eyes had faded, but she was not as gaunt as she was before. Her hair was the color of straw, her clothes not as outlandish as they used to be, although still far more glamorous than was usual in the quaint English village of Didlington St. Wilfred. In fact it was remarkable that nobody had mentioned her to Lily, or that Lily had not seen her mom herself. It just went to show that sometimes you didn't see something unless you were looking for it.

"Hey Lily," her mother said.

Lily was still silent, and seething with rage. Blake, ever the gentleman, stepped forward and held out his hand.

"Ms. McGee, it's a pleasure to meet you, and quite a surprise. Welcome to Didlington St. Wilfred," he said in his clipped, erudite accent, introducing himself by name.

"Why, it's a pleasure to meet you Blake, you can call me Kelly, or, gee, a man as handsome as you can call me just about anything you like," Kelly said, fluttering her eyelashes. It seemed as though she had not changed as much as Lily hoped she would have. Of course she was charmed by Blake, because who wouldn't have been charmed by him? But not her mother, anyone but her mother.

"Why Lily, you've really struck the jackpot here," Kelly said, again making eyes toward Blake. Blake, to his credit, smiled wryly and withdrew to stand beside Lily, still waiting for her to respond.

Lily's fists turned into tight balls. Now more than ever she wished that Zeus was by her side so she could have commanded the dog to scare her mother away. Instead she had to subdue her own temper.

"Mom," she said tersely, trying her best to quell the rage that was welling inside her, but it was like a boiling volcano that was ready to erupt. Her cheeks flushed with heat and she trembled so fiercely that she thought the ground was shaking beneath her feet, only disabused of the notion as she realized that nobody else around her was reacting to the quaking earth. "What are you doing here? How did you ever know I was here?"

The easy smile fell from Kelly's face. She clasped her hands together and looked a little bashful. "Well, uh, I knew from your books, of course. Oh Lily, I'm so proud of you, I truly am. And I just, well, I wanted to say… that is, I hoped that after all this time we might—"

Lily did not give her a chance to finish. She mustered all the composure she could to control the timbre of her voice, but there was no mistaken the deep ire that coated her words.

"Mom. Leave. I don't want to see you. I don't want to hear what you have to say. I don't want anything from you. You shouldn't be here. Just go back where you came from."

"But Lily—"

"I mean it Mom. I've made a good life for myself here and the last thing I need is for you to come here and screw it up," her voice threatened to rise on the last words, and her eyes glared brightly as she stared at her mother. She thought back to the last time she had seen her mom and how different her reaction had been. Then she had pleaded and her eyes had been filled with warm tears, tears that spilled all over her pillow as she had been left alone. She had taken that sorrow and turned it into strength, but when she saw Kelly again she felt just as she had all those years ago, so weak and fragile and lonely.

Kelly closed her mouth, forming a thin line. She had a resigned look on her face and spoke in a low, gentle voice.

"I'm sorry for disturbing you Lily. I hope that we can find some time to talk while I'm in the area," she said, and then walked away from Starling Cottage. Lily watched her with a hawklike gaze to make sure that Kelly did indeed leave, and the tension only began to loosen itself from her body when Kelly disappeared from view.

When she did, Lily turned as well and walked back to the cottage. As she did she heard Blake and Joanie whisper to each other, wondering aloud why Lily had acted like that.

Lily was in the kitchen. She opened cupboards with so much force that the hinges snapped and threatened to break. The kettle bubbled as the water boiled, but it didn't seem to boil quickly enough for Lily's liking. She slammed a mug down on the counter. It almost shattered. She stood with her palms resting against the counter, grinding her teeth. She thought she had worked through all of her issues about this, but the past had come back and hit her like a lightning bolt. Didlington St. Wilfred had been a refuge for her, a sanctuary. It was a place where she had fled her old life and begun anew, and now a dark chapter had reared its grim head and Lily wondered if people were ever able to escape their past.

She had no doubt that Joanie and Blake were discussing her reaction. Her behavior was out of character for what they knew of her, and she hated that Blake had to see her like this, especially when they had just arranged to go out to dinner. She noticed movement out of the corner of her eye. It seemed as though Blake had drawn the short straw when it came to who was going to check on Lily. He had a meek look on his face, as though he was trying to gauge whether it was safe to come in.

"Are you okay?" he asked.

The kettle boiled, whistling fiercely. A jet of steam burst into the air.

"Oh I'm just fine, just fine indeed," Lily said in a tone that suggested the complete opposite of what she declared. Blake moved a little deeper into the kitchen, although he still kept his distance. His tone was reserved and he picked his words carefully.

"I'm guessing you had no idea that your mother was here?"

"No," Lily spat. She poured the boiling water into the mug and breathed in the scent of tea, even though she wasn't particularly thirsty.

"Is it something you want to talk about?" he asked. Lily held the mug to her face and pressed her lips to the rim. The hot liquid slipped against her lips, threatening to burn her insides.

"I'm sure she'd love that. She always likes being the center of attention," Lily said sarcastically. Blake arched his eyebrows. This wasn't going to do anything to endear her to him, but Kelly had a way of raising the most violent of emotions within her. "I'm sorry," Lily sighed. "I know this probably comes as a shock to you. I just... I didn't expect it to be *her*."

"Would you rather it had been someone else?" Blake asked.

"Yes, no, I don't know," Lily shook her head and waved a helpless hand in the air. "It's just that I thought the business I had with her was done a long time ago. I don't have a place for her in my life now. I have everything just as I want it," as she said this she noticed Blake's brow furrow. She supposed she didn't quite have *everything* as she wanted it, and she wished that she had chosen another phrase to convey what she meant. It was too late now though, as the words had already been spoken.

"I don't mean to pry in your business because it's not my place, but would you mind if I offered you a piece of wisdom as a friend?" Blake asked.

Lily winced at the word 'friend'. "Sure," she said.

He tapped his fingers against the kitchen counter as he approached her. "I don't know what happened between you and your mother, but it's clear she came here hoping for a second chance. I know that not every situation is the same, but if I were you I would give anything to have another conversation with my mother. It's not every day that you get these opportunities, and perhaps it's not the worst thing in the world to speak with her for a while. She came all this way to see you after all, the least you could do is hear her out."

"Believe me, there's a lot less that I could do," Lily snapped, although Blake's words did tug on her heartstrings a little. It was perhaps a little spoiled of her to turn away her mother when Blake would have cherished some time with his own. She took another sip of tea and placed the mug on the counter. Her rage had been quelled somewhat, although her blood still ran hot. She couldn't be mad in the presence of Blake though. He had such a reassuring air and a kindly look on his face that put her at ease. If only there were more people like him in the world, she thought, then perhaps there wouldn't be as many crimes.

"I appreciate what you're saying Blake, and I don't mean to sound insensitive, but not all mothers were created equal. I'm quite sure that your mom was a wonderful, charming person who doted on you and would have done anything for you. I imagine she lit up when you entered the room and always made you feel as though you were the most important person in the world," she spoke with an inquiring tone. Blake nodded sheepishly. Lily continued, "Well, for me it was different. Mom always treated me like an afterthought. I don't even think she wanted to have a kid, at least not for the reality of it. She just wanted a kid so she could show off that she was a mom. There were always things that were more important than me, and then before I was old enough to look after myself she left me."

Blake's countenance changed to one of concern. "She what?" he asked in a voice that was half of a gasp.

422

Lily nodded. "She had found a new man. Fallen in love. She said he was everything she had ever wanted in a man and she wanted a new life with him, a new life without me."

"She didn't take you with her?"

"He didn't want me. So she made the choice to go ahead and go with him. She told me that I was old enough to take care of myself. She said that the world was cruel and that I had to take any opportunity that came my way. And she said that she knew I would hate her for this, but that she was teaching me a valuable lesson. She said that nobody could trust anyone in this world. I wanted to ask her how she could trust him then, but I was too busy crying. She said that men would be kind to me if I learned how to smile, and then she hugged me, and she was gone. She never wanted to be a mother to me and I don't care what she wants to say to me now because it's too late. It's been too late ever since she left me. I was better off without her then, and I'm better off without her now," Lily said. Her words were not imbued with rage any longer, but rather they were heavy with sorrow. It could have been so much different if Kelly had made another choice, but she had chosen someone else over Lily when it had been important, and that choice had defined the course of their lives and their relationships. It was too late to change it now.

2

S taying in Starling Cottage was beginning to drive Lily crazy, and it only forced her to dwell on everything that was happening. She was also half afraid that Kelly would make another appearance. To combat this, Lily decided to go to the police station to update Chloe on the situation.

Lily used to feel at peace whenever she walked through Didlington St. Wilfred, except when the fog had lingered over the land. Ever since the parcels had been arriving at her doorstep she had been filled with the eerie sense that someone had been watching her. Her mind had conjured a sense of danger and foreboding about this. Blake had asked her if she would have preferred it to be anyone else, but she honestly didn't know. At least if it had been some criminal she could have put her talents to the test and figured out a way to solve the puzzle. There was no solving her mother though. Now when she walked through the village she peered into every window, wondering if her mother would look back. Had her mother been in the crowd when they had discovered the hanged body? Had Kelly been lurking in the backdrop as the crypts had been explored? Had she been a face in the crowd at Reverend Brown's church service?

She flushed crimson as she thought about the Reverend, knowing that she was deliberately not seeing him. No doubt he would say that it was very un-Christian of her to rebuke her mother like this, and he was right. But in her mind Kelly had rebuked her first and forgiveness was not as easy to come to as Reverend Brown might suggest. Coming to Didlington St. Wilfred had brought Lily closer to her faith, but there were still some rough edges that needed to be smoothed away and this was one of them.

When she reached the police station she walked in without any trouble at all, and greeted Chloe with a polite smile. The first time these two had met they had not gotten on well at all, but over time they had developed a mutual respect for each other, although it wasn't quite close to being a friendship. Lily's opinion of Chloe had changed after finding out that there was no romantic connection between Chloe and Blake, although Lily wondered if she had just sabotaged her own chances by telling him that she didn't want to spend time with her mother.

"Are you okay Lily?" Chloe asked. "Is another case developing?" She looked concerned, and it must have been from the look on Lily's face.

"No, actually I've solved one."

"Really? You have been busy. Should I be offended that you didn't invite me along?"

"Actually it's one that you're familiar with. I discovered who has been leaving the mysterious parcels on my doorstep. I just wanted to come and tell you that the matter has been resolved so you don't need to keep it in mind any longer."

Chloe put down the stack of papers she was holding and flicked a strand of dark hair behind her head. "Who was it?"

"My mother," Lily replied.

Chloe arched her eyebrows and nodded in understanding. "I see."

Lily was a little surprised that Chloe could deal with things with such a lack of curiosity. "Are you not going to ask about her?"

"Do I need to?"

"I just assumed you would want to know why she was doing it."

"I think I can tell just by the look on your face. It's quite clear that you don't have the best relationship with her."

"No, I don't."

"I don't either, with my mother. That's how I can tell. I'm sure you'd see the same look on my face if my mother came into town," Chloe said with a slight laugh. Lily felt relaxed and was glad to be in the company of someone else who had a difficult mother.

"What was yours like, if you don't mind me asking?"

"Oh, she was opinionated and never seemed to want me to live my own life, but rather the life she had picked out for me. When I told her I wanted to be a police officer..." Chloe gave a low whistle.

"She wasn't a fan?"

"Definitely not," Chloe shook her head and sighed. "She still thinks it's a job that only men can do. Even though she spent her life saying that men aren't good for anything and that she wouldn't trust them as far as she could throw them."

"What did she want you to be?"

"Anything other than a police officer. She would have been happy if I worked in some small office somewhere, steadily collecting my paycheck. What she really wanted from me was grandchildren, but to be honest she wasn't the best mother so I don't know what kind of grandmother she's going to be. I'm almost a little envious of the children I'm going to have because what if she's kinder to them than she ever was to me? I think that's a little unfair, but I try not to think about those things until they happen. There's still a long time until then, I think. I add an extra year on whenever she nags me about it," Chloe said, laughing.

"You're still in touch then?" Lily said.

"I am. I take it you weren't, up until these pastries?"

"I hadn't seen her for a long time. She left me when I was younger. I had to fend for myself. She was never a very nice person to me, and definitely not a good mom. I just don't understand why she's come here now, and I don't particularly want to know. I wish she had never turned up."

"At least you solved the mystery, although I assume you won't be using this one as an inspiration for one of your books?" Chloe asked.

Lily smiled. "I'm not sure I want to give my mom the attention. It was bad enough when I was younger."

"Do you have an example you might be willing to share?" Chloe asked, getting comfortable in her chair. It was the most at ease Lily had ever felt with the interim police constable, and she wondered if this was perhaps the first steps toward a friendship between the two of them. However, before she could answer, the door to the police station was flung open. A woman was standing there, panting and red faced. Her strawberry blonde hair was matted to her forehead, and her freckled cheeks burned crimson. Her dress was skewed and she trembled terribly.

"Please, please help me," she gasped in a rush of breath, "there's been a murder. Please… it's my mother. Please help…"

Lily and Chloe both rushed to the door and helped the woman to a seat. She was weak and as soon as she sat down the color paled from her cheeks. She was caught in between weeping and gasping in deep breaths of air. She leaned forward and groaned, clutching her hair. She was a tall woman, and slender. Her fingers were long. They looked like spider's legs as she clutched her skull.

"It's okay. You're safe here, but you need to calm down if we're going to get to the bottom of this. Are you saying that your mother has been killed?" Chloe asked.

The woman nodded and glanced toward the door, her eyes wide with fear. She brushed strands of hair away from her pink lips.

"What's your name and what happened?" Chloe asked.

"I'm Harriet, and my mother... oh... she's dead!" Harriet cried again. Lily gazed upon her with sympathy and shared a look with Chloe. It could often be so difficult to question someone who was in this emotional state. It was natural to want to comfort them when they were in shock, but it was also so difficult to pry for worthy information. Harriet glanced toward the door again.

"Harriet, is the murderer chasing you?" Lily asked.

Harriet nodded again.

Lily and Chloe suddenly went rigid with tension. If there was a killer brazenly running around Didlington St. Wilfred then they were going to have to do something about it. Lily doubted that anyone would be so brash as to charge into the police station just after killing someone, but perhaps anyone who committed such a vile act was not in their right mind anyway. Chloe ran to a wall and picked up her baton, a long, thick, black rod that could crack skulls and make any criminal think twice about continuing with their crimes. But if this murderer had succumbed to the mists of rage then there was no telling what they might do. Lily was glad of the distraction though. It was better to think of anything than her mother. She began putting together a profile of the killer. She assumed it was a man given the way Harriet was scared. Perhaps it was a robbery gone wrong and the thief was chasing Harriet to prevent her from identifying him, or maybe it was something more sinister. She quickly glanced at Harriet and noticed that she did not have any marks on her wrists or face, suggesting that she had been clever enough to flee from the murderer without confronting them.

Some people would not have been so rational, especially if they had just seen their mother being killed.

Lily helped Harriet out of the chair and went to escort her to the far end of the room. Chloe was taking up a defensive position to the side of the door, waiting for the murderer to burst in. If they did then Chloe would be ready to strike them with the baton. If not, then Chloe would have to call their forces and begin a manhunt. Even though Didlington St. Wilfred was a small village it was surprisingly easy for someone to lose themselves, either in the nooks and crannies of the village, or in the thick woods that ran around the village. But Lily promised that justice would find them eventually.

"It's going to be okay Harriet. Nobody is going to attack you here," Lily said, glancing uneasily at the door. At least she wasn't in America, she thought. If so then she would have been afraid of someone bursting in with a gun. As it was she could relax, for such a crime was far rarer over here, especially in this setting rather than a city. But still, she worried. The unpredictable nature of criminals was such that she could not truly guarantee any of their safety. She was about to escort Harriet to the back of the police station, but she did not want to leave Chloe alone in case she became overpowered.

"I hear movement. They're coming," Chloe said in a low voice.

Lily braced herself. There was a thunder of footsteps outside, but something was off to Lily's ear. It soon became clear what. The door flung open. Chloe raised her baton, holding her strike for any sign of an attack, but none was forthcoming. Lily's assumption had been wrong; it wasn't a man at all, but rather a woman standing in the doorway. Her face was a picture of rage. Just like Harriet she was red faced, although she was more angry than scared.

She raised a hand and pointed a finger toward Harriet, as though she was cursing the poor woman.

"You," she hissed. "You murderer!" She ran into the station, the movement taking Chloe by such surprise that the police constable was a heartbeat slower to react than normal. Lily put herself in between Harriet and this newcomer as Chloe gathered herself and wrestled the woman back, interlocking their arms and dragging her aside.

"Let me go! Let me go!" the woman cried out. "I'm not the enemy here! She's the murderer! She killed my mother!"

3

_____

L ily gasped.

They were *sisters*. Lily would not have guessed from their appearance. Where Harriet was tall and slender, this other woman was shorter and had a rounder face. Her lips were full and ruby, while her eyes were a bright blue. Her hair was shorter, cut to reach her chin, and it was as dark as the night. Her hands were small and she had a button nose. Lily was caught in between the two of them, shifting her gaze between them. They were both sisters who were accusing each other of the same murder.

"I didn't do it! How can you say that!" Harriet cried.

"Because I saw you! Oh don't try this woe is me act. Sure, act scared, it's not going to help you because I know the truth, and soon enough these people will too," the other sister spoke harshly, her words laced with venom. Chloe was struggling with her, still trying to keep her from flinging herself at Harriet.

"We're not going to learn anything if you don't calm down and stop struggling," Chloe said in exasperation as she used her strength to haul Harriet's sister away, joining Lily in standing between the two. Chloe adjusted her collar and pressed her lips together. "Now then, let's talk about this calmly so we can figure out what's going on. What's your name?" she asked the new arrival.

"I'm Steph," she said, but she only had eyes for Harriet.

"Okay then. Steph and Harriet. So let me get this straight, you are sisters and your mother has been killed, and you're both accusing each other of doing it?" Chloe asked.

"Yes," the sisters replied in unison. Lily shared an uneasy glance with Chloe. This was not going to be an easy case at all.

Chloe and Lily were in a room with Harriet. They had decided that the only way to proceed was to question both women individually. Steph had not been a fan of this, accusing Chloe and Lily of giving Harriet a chance to sway them with her lies first. Chloe assured Steph that both accounts would be scrutinized and that the truth would be reached. In the meantime Steph was taken to a cell and locked there to prevent her from going anywhere she should not go. She did not take kindly to that either. Lily wanted to have sympathy for her because her mother had just died, but it was difficult when she might well have been the killer. Then again, so could Harriet. Lily could not come to a conclusion simply because she had received some information first, or because Harriet seemed the most inconsolable. Grief affected people in different ways, as did guilt.

Chloe leaned forward and clasped her hands on the table, while Lily folded her arms.

"I know this has been a traumatic day for you Harriet, but we really need to get this straightened out," Chloe began.

"She's lying! She killed Mom!" Harriet cried, gesturing toward the door.

Chloe's tone was measured. "I understand that, but right now it's your word against hers. We're going to need to go through the evening so that we can discover what has happened. Why don't you tell me a little bit about Steph and the relationship between the two of you and your mother. Why would Steph have wanted her dead?" Chloe asked.

This seemed to calm Harriet down. She snorted and rolled her eyes. "The same reason she's hated me all my life, because Mom loved me more than her," Harriet said.

"Would sibling rivalry really drive her to murdering your mother?" Lily asked.

Harriet's eyes were bloodshot. "You don't understand. It's different between us. I'm adopted."

Lily and Chloe glanced at each other. "I'm sorry," Chloe said, "but I don't quite see how that makes it different to any other sibling rivalry?"

Harriet sighed and ran her hand through her hair. "It's different because Steph was never planned. She came into Mom's life by chance, because one night she and Dad got a little too drunk and lost their sense of responsibility. But with me it was different. They had to fight for me. They rescued me. They actively went out and picked me from all the other kids they could have had. Steph has always seen herself as a random quirk of fate, while I was the one that our parents chose to have. She has always seen herself as second best and recently things have just been getting worse." Harriet sniffed and attempted to collect herself. "People always used to say that when we got older it would be easier. Even Mom used to tell me that when we got older we would end up being best friends. I wish that had been the case, but it wasn't true. Steph and I never got on. She never let me in. Even at her wedding she had to be ordered by Mom to have me as a bridesmaid, but she wouldn't have me as her maid of honor."

"She's married? Where's her husband?" Chloe asked.

"Well that's the thing; recently things have been strained. She got divorced and had to move back home. I thought maybe it would be another chance for us to find some common ground, that maybe we could be proper sisters and I could help her through it all, but she just shut herself away and wanted to ignore me. The house became crowded. Mom wanted to help her, of course, but she became caught in the middle. I tried to tell Steph that this enmity between us wasn't any good for Mom, but she wouldn't listen."

Lily nodded. It certainly sounded like a strained situation. She had always wondered what it would be like to have a sister. For much of her life she had been so lonely, but to hear the tragedy of this relationship was awful and she considered herself fortunate to be as she was. Then again, she realized that Harriet probably thought Lily was fortunate to have a mother still as well, and Lily felt uneasy for keeping Kelly at arm's length, just as she had when Blake had mentioned how he would have loved to have another chance to see his own mother. She tried to push these thoughts out of her mind though and focus on the case. There was a murderer who needed to be caught.

"How did your mother die?" Chloe asked softly.

Harriet's lower lip quivered and she seemed to lose a part of herself. Her voice wavered and her eyes glistened. "I was in the other room. I heard Steph and Mom arguing. I had had enough of it. I stayed out of it, but I should have gone in there. Maybe if I had gone in there…" she trailed away.

"It's no good thinking about things like that," Lily said gently. "There are always things we think we could have done differently, but there's no guarantee that would have changed the outcome. And even if it would have, it's not going to bring her back now. All we can do is try to get justice for her, so let's try that, okay?"

Harriet sniffed again and nodded, exhaling deeply. "They were in the kitchen. I heard a great clatter and then a scream. I ran out of the room and found Steph standing there with a pan in her hand

and Mom was on the floor. She looked so... so helpless. I had never seen her like that before. She had always been the strong one. She never got ill. She never needed help. She was always the one who gave the help. She was the one who rescued me, but I couldn't do the same for her."

"And that's when you decided to come here?" Chloe asked, trying to keep Harriet talking because otherwise the poor girl would have descended into a flow of tears again.

"Yeah. I looked at Steph and there was something different about her, something so bad inside. I ran away and I came here and then... I just wish things had been different. I wish that I had done something differently," she tilted her head down in an attempt to hide the warm flow of tears that trickled down her cheeks. It didn't quite work and Lily saw them glisten upon her freckled skin. Lily was moved by the sight and felt a well of emotion build as well.

"There is something else we need to talk about," Chloe said as she was making notes. "Obviously this is a unique case. What you have told me so far sounds entirely reasonable, but Steph has made the allegation against you as well. For the sake of thoroughness I need to ask you about it."

Harriet's head shot up and she looked aghast. "You think I could have done this? I never would have done this!" she cried.

"I understand that. I just need to talk to you about the accusation and about why Steph might be trying to blame you," Chloe asked.

"Because it's all she's done all her life. She's just trying to get away with it. She wants me to suffer and wants me to take the blame because it's the only chance that she has to be free. It's my word against hers. We were the only two in the house, aside from Mom," as soon as she mentioned that last word the emotion overwhelmed her again and she doubled over, clutching her sides.

"I think that's all for now," Chloe said, glancing across at Lily. "If you wait outside we'll talk to Steph next."

"What's going to happen to Mom?"

"Speak to the officer outside. He'll collect your mother and take her to the morgue," Chloe said. Harriet nodded and left the room. Chloe exhaled and tapped her fingers against the surface of the table. "I've never experienced a case like this," she said.

"I haven't either," Lily replied. "Did you believe her?"

"I have no reason not to yet, but we'll see where things stand after we've spoken to Steph," Chloe said. Lily nodded and sipped some water. Harriet's testimony had been filled with emotion, emotion that was difficult to fake. But guilt and sorrow were so often intertwined, and Lily could not allow herself to blindly believe someone's words without knowing all the information.

Steph was brought in shortly after this. She scowled and dragged her feet, before slumping into the chair. Her arms were folded across her chest and she did her best to avoid making eye contact with Lily and Chloe. It was certainly a contrast to the way Harriet had been sitting so rigidly, almost on the verge of shattering completely.

"Steph, could you tell us in your own words what happened?" Chloe began.

Steph grunted. "What's the point? It's not like you're going to believe me. She got here first," Steph muttered.

Lilly glanced at Chloe. "You're younger than Harriet, yes?" Lily asked.

Steph nodded.

"Harriet did mention that there was some rivalry between you two," Lily continued.

Steph burst into life at this. Her face became animated and her arms flailed out beside her. "I bet she did. And I bet she blamed it all on me as well." Her words dripped with bitterness.

"How would you characterize your relationship?" Chloe asked.

"I would characterize it as awful," Steph replied.

"Did you never get on, even as children?"

Steph barked a dry laugh. "How could we? Harriet always resented that I came along. For a few years she was the only one and had all the attention, then I came along and ruined it. It didn't help that Mom played up to that. I tried to break out on my own and have my own life, but Mom always brought it back to Harriet. She always used to tell me that Harriet was so lonely and that she just needed to belong and that we were family so I had to make allowances for her, but I never signed up for that. Mom and Dad adopted her. I didn't. But I was still forced to have her as my sister."

"Why couldn't you get on well?" Lily asked.

"Harriet and I are just too different. Mom thinks she's an angel, but she's not, not really. She's as bad as the rest of us, just in different ways," Steph said.

"How do you mean?" Chloe asked.

"She's just always been a little off. You know? Like at school she never had any friends. Mom kept forcing me to bring her to hang out with mine, but even then she just stood there awkwardly like she didn't know how to talk to people. Mom said I had to be patient because of what she had been through as a child, but I don't think that can be an excuse for all the behavior. Mom was always too soft with Harriet. It hasn't done Harriet any good either. She still lives at home. She's so dependent on Mom," Steph said.

"You do understand what that says to me is that she loved your mother a great deal and would be unlikely to kill her," Chloe said.

Steph rolled her eyes and nodded. She brought herself forward and leaned on the table. "Of course it does. That's the way it works with Harriet. She's very good at playing the victim. And I'm not saying she hasn't been through some stuff, but we all have and we don't all go around acting like this and thinking that we should be able to get away with everything because of it."

"When you say 'acting like this' what exactly are you referring to?" Lily asked.

"Oh, everything really. She's always tried to drive a wedge between Mom and me. Whenever something broke as a kid I was always the one blamed. If I ever stayed out longer than I should have or went somewhere I shouldn't have Harriet was always the one to tell Mom. Mom always used to say nobody should be ashamed of telling the truth, but it wasn't the truth, it was just Harriet's truth. I always tried to protest, but Mom would never listen to what actually happened, and the same thing is happening now. It's like I'm talking to her…" she trailed off at this point and something changed within her. The sorrow took hold now, spreading through the dark grief. She moaned and then put a hand to her mouth, attempting to suffocate a sob. She wasn't entirely successful. She closed her eyes and her dark hair fell across her face like a veil. Her shoulders shuddered. The anger she felt toward Harriet had finally given way to sorrow.

"Tell us what actually happened then Steph. We're listening," Chloe said.

Steph wiped her eyes and swallowed her emotions deep down. Her voice had a different tone when she spoke now. It was clearer and harsher, but there was a sense of distance as well. Lily had to remind herself that just because Steph's manner was blunt that didn't mean she was guilty.

"I was talking to Mom, trying to tell her what Harriet was doing," Steph began.

"What was Harriet doing?" Chloe asked.

"She was trying to steal my inheritance away from me. Even though she was adopted she's always seen herself as the true daughter, the first daughter, and she thinks she should be given the larger share. She thinks because she's lived in that house non stop she should inherit it. She thinks that I should receive less because I'm more capable of looking after myself. She's been working behind Mom's back to try and see if there are any loopholes she can use to get more. Mom wouldn't hear any of it though. She thought it was a lie. I was yelling at her, trying to make her see sense. Sometimes I don't know if Mom really was blind to all this, or if she was just trying to deny the truth because then she could pretend that everything was perfect," Steph said.

"Was your Mom ill?" Chloe asked.

"No," Steph replied.

"If she wasn't in danger of dying any time soon then why was the inheritance such a big deal?" Lily asked.

"It had come up in conversation recently. When I returned home we talked about things. Harriet hated my being back. She finally had things the way she wanted them all these years, just her and Mom, and then I had to come back and ruin it. Well... that's what she deserved. Anyway, Mom was saying that she was planning for the future after Dad had died because she wanted to make sure we were both taken care of. I don't think Harriet had ever really thought of it like that before. I think she always just assumed that she would live in that house for the rest of her life, but of course if it became ours then that meant we would either have to live there together or sell it, and split the proceeds. I told her that I was in favor of selling it because then we could finally go our own way in life, but to Harriet that was the worst thing I could have said. The next day I heard her speaking on the phone to a solicitor asking if there was anything she could do to protect herself. I told Mom, but Harriet denied it. Mom believed her, and then when we were having the argument Harriet came storming in filled with rage. I yelled at her and she yelled back and Mom tried to get in between us. Harriet grabbed a pot and she... she swung at me, but Mom was in the way. Harriet dropped the pot and ran away. I was too stunned to move. Then I followed her and came here. That was it," Steph said. Her words were more measured now than they had been before, as though the flow of sorrow within her was slowing them down.

Chloe continued taking notes. Lily thought about the woman's testimony. There were so many similarities between what she said and what Harriet said, and to her dismay there was equal reason to believe her. Money was one of the most common driving forces between crimes like this, and when passionate emotions were involved it created a deadly cocktail. But there were still more questions to ask.

"You said that you were forced to take Harriet out with you when you went to see friends. When you did so, did you make her feel part of the group?" Chloe asked.

"I tried to, but like I said it was always impossible with her. She never put herself forward. She never tried to get anyone to like her. She had these walls up that would never come down, but when they did... well... then you got to know that she wasn't the person she seemed," Steph said. "I really dreaded going back home after what happened. It was the only place I had though. Mom kept telling me that I had to forgive her. I don't think she could ever understand what happened."

"What do you mean? Harriet did mention that you had recently gotten divorced," Lily said.

A strange smile came upon Steph's face. "I don't suppose she told you why I got divorced, did she?" Both Chloe and Lily shook their heads. "Because he cheated on me. With Harriet. She never could cope with the fact that I knew how to have a life while she didn't. So she tried to steal my husband from me."

Lily was stunned by this revelation. If this was true it meant that Harriet had kept things from them in her statement, and if she had kept this a secret then what else was she hiding?

"I'm going to need the contact details of your ex-husband so that we can speak to him about this," Chloe said. "Do you happen to know which solicitor firm Harriet was speaking to?"

"I don't," Chloe said. "Look, I know Harriet got here first and I'm sure that what she told you was very convincing, but she was the one who killed Mom and she has to pay. I've had to live with her getting the benefit of the doubt all my life and it's not going to happen again here."

"I think that's all we need to ask you for now," Chloe said. "If you go outside there will be an officer waiting for you. I'm afraid that until this situation has been resolved we're going to have to hold you, both of you."

"I figured you were going to say something like that. I'll do whatever you like, just as long as I don't have to share a cell with her," Steph said bitterly. She rose from the seat and walked toward the door, pausing for a moment as her strength left her. She leaned against the door frame, but then composed herself again and left the room.

"I wasn't expecting that," Chloe said now that she and Lily were alone.

"They both made compelling arguments. There's no love lost between them at all, and it's quite curious how Harriet did not mention that she was the one who broke up Steph's relationship," Lily said.

"That's if Steph is telling the truth about it all."

"One of them is guilty. We know that for sure. We just have to figure out which one."

"We'll have a conversation with the husband first. I'll call the phone company to see if we can get a record from the house and find out which solicitor was contacted. While I set all that up would you mind going to their house and examining the scene? The evidence might show us something that those two are hiding," Chloe said. Lily nodded and walked out of the police station while Chloe took care of the paperwork. It was harrowing to think that a family could be torn apart from the inside like this. Families were such precious things, and once again Lily felt a pang of guilt. She pushed it aside though. Kelly had abandoned her a long time ago. Lily wasn't the one breaking apart her own family.

L ily walked through Didlington St. Wilfred. The house was located at the far end of the village. It was a low roofed cottage with a well maintained garden, although that was not out of the ordinary as most of the gardens in Didlington St. Wilfred were well looked after. There was a police ambulance on site as the dead body was being taken away. Lily walked up to the house and stepped inside. There was always an eerie sense of foreboding in a house where death had occurred, as though some shadow of horror lingered in the place. Lily ignored the chill crawling upon her skin and stepped through, beginning in the kitchen. A chalk outline had been drawn on the floor, showing where their mother had died. A heavy pot was beside the chalk outline. The pot was one of the old ones, a thick, cast iron thing that could bludgeon anyone to death. Lily looked up and stepped through to the nearest room, which was the living room. The doors were both open and it wouldn't have taken much for the sound to reach that room.

It was a tidy house, and homey. There were plenty of pictures up on the walls, and photo albums stacked on shelves. Lily noticed that in practically all the pictures of the girls together there was a clear sense of enmity between them. The smiles were forced, and in their eyes there was nothing but frustration. Lily could almost imagine their mother acting like a mother hen, forcing them to stand together, telling them how to pose and that they had to smile brightly. Kelly had said the same thing to her, because she always said that a smile could make people trust you, that a smile was the best thing in the world. Most of Lily's memories of her mother were troubled ones, but there had been some good times. One such memory had been in the mall. Kelly had taken Lily into a photo booth and they had made funny faces as pictures were taken of them. It was something that Lily hadn't thought about for a number of years, and she found herself smiling.

She then put the photo albums away and continued through the house. She came into Harriet's room, which was organized and tidy. There were few pictures of her on the walls, at least not with other people. There were plenty of books though, including one that Lily had written. She thought it a little strange that Harriet hadn't mentioned it. Perhaps Harriet hadn't gotten around to reading it yet, or that she was not a fan. On the bedside cabinet there was an address book. Lily opened it, and she felt forlorn when she saw all the blank pages. It appeared that Steph had not been lying when she said that Harriet had enjoyed her solitude. In fact she wondered why Harriet even had an address book at all. She thought that perhaps the number of a solicitor would be in there, but there wasn't. It was likely inconsequential. Perhaps Harriet's mother had just wanted her to have it in the hope it would encourage her to have more of a social life.

Lily looked around a little more, but did not find anything of interest. She ventured into Steph's room next, which was far more haphazard than Harriet's. There were still unpacked boxes in the room. Shelves were filled with various knick knacks that looked as though they had been placed without any thought to style or fashion. There were bright and colorful dresses hanging in the closet, while the photos that were placed on the bedside table were ones filled with people, and in all of them Harriet was absent.

Lily realized that growing up like this must have been hard for both girls. Harriet was ostracized, perhaps in some ways by choice, but still she would have grown up seeing her sister being popular, while she was always pushed into spending time with them by a mother who only wanted the best for her. Lily knew that forcing a child to do something wasn't the best way to get them to enjoy something. From Steph's perspective she would have wanted to enjoy her own life without having to be beholden to what Harriet wanted. She must have been vulnerable after the divorce as well. It couldn't have been easy to come back into the home where the woman who had had an affair with her husband, lived.

Lily then moved into the mother's bedroom. The walls were adorned with pictures, memories crystallized in these images that would never cease to exist. There were plenty of pictures of her and the girls together, as well as ones with their father. It was saddening to see these pictures. The people in them had no idea that tragedy was going to strike like this. Lily wondered if there was anything regarding the will in this room. Perhaps there would be some clue to show who was getting what. It was grim to rummage through a dead woman's personal belongings, but Lily had not taken up this endeavor because it was easy.

Eventually she found a folder that looked important. When she opened it she realized that it was Harriet's adoption folder, with a report of the child. Lily read with interest. It described how Harriet had come from a home with a violent father and a neglectful mother, and that although she was young the trauma may have lasting repercussions. They couldn't say how these things were going to manifest themselves later in life, but the general likelihood was that she was going to have intimacy issues and emotional issues. Was it possible that her childhood trauma had ended up being the driving force for murder? It was tragic to think that even after giving her a stable home and a better life than she could ever have received with her biological parents, the initial trauma was so powerful it had reverberated through the years and still had this awful effect.

445

Lily closed the folder and sat on the bed for a while, thinking about how this home had once been filled with so much hope and so much promise. Children had been brought into the world, and it must have been the happiest of occasions. Instead it had ended up in tragedy, and only one of the daughters was innocent, but which one?

～

Lily had been in the home for a while, although nothing had jumped out at her as being notable evidence. She walked through Didlington St. Wilfred at a slow pace, with much on her mind. As she passed Marjorie's bakery she caught the sweet scent of pastries being baked and her stomach rumbled. It was then that she realized she hadn't eaten for hours. She walked inside and Marjorie greeted her with a warm smile.

"Lily! It's good to see you. Have you had any luck in finding out who has been sending you these mysterious parcels?" Marjorie asked. Lily updated her on the latest events. Marjorie listened, intrigued, as she wrapped up a creamy éclair, Lily's favorite. Lily did describe how she wasn't exactly enthused about Kelly's reappearance, but as she did so she was acutely aware that Marjorie reacted much like Blake did. Lily glanced toward the wall, on which hung a picture of Marjorie's dear husband, who had departed the world before Lily had arrived. Lily felt it had always been a shame that she had never had a chance to know the man.

"I'm not sure what to do now," Lily confessed. "She's come all this way, but I don't want to let her think that I can just forgive her easily. I'm worried she might only be here because I've finally made a success of myself."

"But you don't know that for sure. I know in your line of work it's easy to assume the worst of people, but that's not something the Lord preaches."

"I know," Lily sighed. "I want to be a good Christian and open my heart to forgiveness, but it's always harder when it's someone who has hurt you so badly."

"That's why it's a struggle, and it's why it's so important. But I think you're going about this all wrong Lily. There's nothing wrong with having a conversation with her, and that doesn't mean you're forgiving her. She's come all this way and maybe it's a blessing that you can have the opportunity to talk to her again. But talking is just talking. If you don't like what she has to say then you can tell her that you'd rather she leave. You're not obligated to give her anything, but sometimes when someone makes the effort like this it's a good thing to give them a chance to speak. You never know, it might be good for you too. I'm sure there are things that you've wanted to get off your chest?"

Lily nodded as she thought of all the words that had careened through her mind, the words she had been unable to speak when Kelly had left.

"I suppose you're right. I just think it's a tragedy when a family has to be ripped apart," she said, and then went on to speak about the current case. Marjorie listened intently and nodded.

"I know the girls," she said. "They were always so different. I never thought either of them would be capable of this though. It makes you wonder just what people are truly capable of."

"What would you do in my situation?" Lily asked abruptly.

Marjorie considered the matter for a few moments. "It's difficult for me to put myself in your shoes because I was blessed to have a lovely relationship with my mother, but I would definitely hear what she has to say. I think when people make the effort the least we can do is allow them the chance to speak. You never know, she might surprise you. When you're open to things like this it can lead to good things. Shutting them away only means you close off the possibilities to happiness. And I'm not saying that this is going to be happy because sometimes it's sad, but the great beauty of life is that there's a fine line between happiness and sorrow, but it's all real, and it's all we have at the end of the day, at least until we pass on," Marjorie spoke in kindly words, although there was a wistful sadness behind her smile. Lily nodded. Eventually there would be a finality to everything and then she would never have the chance to see her mother again, but that didn't necessarily mean she was ready to speak to Kelly yet. It just meant that she was more open to the possibility.

She still had a case to solve after all, but she told herself that once this was resolved she would deal with her mother properly.

5

A couple of days had passed before they could speak to the husband. Chloe was also still waiting on the phone records. Apparently the two women had been quiet and obedient in their cells. They had shifted from wailing in sorrow to being morose. Neither of them had come forth with any more information, and the autopsy on their mother had shown that they were both telling the truth about the way she had died, although it was still yet to be determined which one of them had been holding the pot. Lily was frustrated and flustered. Her dreams had been troubled and she had not been able to prevent her mind from returning to her childhood, which annoyed her because she was supposed to be thinking of other things. Joanie had tried to talk to her about it, but Lily had made a tactical retreat as she was afraid that talking about it too much would mean that she would never be able to forget about it.

She did, however, consider speaking to Blake. Things with him had not gone as she had hoped, and now she wondered if their dinner together was in dispute after the way she had behaved. There was no difficulty in seeing why he had been offended by her treatment of Kelly, but she didn't want this to be the only perception he had of her. She wondered if she had kept him too distant recently, and if this was going to cost her a chance to be happy.

She sighed as she wondered what was going to happen next with him, and she did not have the courage to go and see him immediately as she was afraid that he was going to cancel their plans. Instead she returned to the police station to see if there had been any updates on the case. When she arrived she found Chloe sitting at her desk, reading a letter.

"Do we have a time when we're going to meet the husband?" Lily asked as she walked in. Chloe didn't reply. Lily approached her and cleared her throat, and then uttered Chloe's name. It was only then that Chloe glanced up at her. The look on her face wasn't one that Lily could easily comprehend.

"What's wrong?" Lily asked.

"I'm not sure anything is wrong. At least I don't think it will be from your point of view," she replied, and then held out the letter for Lily to take. Lily took it. Her brow was furrowed, for she didn't understand how a letter for Chloe could have any bearing on her. However, as she read she felt her heart soar with hope and a smile appeared on her face. The letter was giving Chloe new orders. She was being called back to the city and Constable Dudley was being reinstated, as the inquiry into police negligence had returned with a verdict that he was innocent of all charges.

"This is wonderful!" Lily said, although she quickly caught herself as she realized this meant that Chloe was going to have to leave. It surprised her that she felt this way. When they had first met they had clashed, and Lily hadn't thought she would ever be friendly with Chloe. Lily's gaze rose and she saw that Chloe looked somewhat pensive.

"I mean, it's wonderful for Bob," Lily added.

"You don't have to do that Lily. I knew my time here was going to be temporary. The truth is that when I first arrived here I thought of this as a punishment, but I've learned a lot and I think I'm a better police officer because of it. I'll actually be disappointed to leave here," Chloe said.

"Well, you know that you're always welcome to visit whenever you're free. You'll always have a special place in Didlington St. Wilfred," Lily said earnestly.

Chloe smiled. "I appreciate that, and one day you're going to have to visit me in the city. Maybe you can help me on a case there."

"If you think you need help," Lily said with a grateful smile. Chloe placed the letter on her desk and sighed.

"I suppose that we should get on with this case then, since it will be my last. I'd like to go out on a high note. I have managed to arrange a meeting with the husband. His name is Harry Crump," Chloe said. Lily nodded and they went outside, getting into Chloe's vehicle. He lived in a town near Didlington St. Wilfred and had arranged to meet them at his office. He worked in a small, but successful printing company.

He greeted them with a handshake, although there was a solemn look on his face. He closed the door behind them and they sat down. Behind his seat was a wide window that let in a great deal of light.

"So, you want to talk about Steph and Harriet?" he asked, leaning forward and clasping his hands together.

"We do indeed," Chloe said after introducing herself and Lily, as well as the nature of the crime. "At the moment we're trying to ascertain which of the sisters is telling the truth and it seems as though you were intimately familiar with both of them."

Lily wondered if Harry would take offence at that, but he didn't seem too. In fact he smirked and held his head. "I guess the truth came out then. So you want to know what happened between us three?"

"It would be helpful, I think, to shed some light on the dynamics between the two women. Is it true that you and Harriet had an affair?" Chloe asked.

"I'd hardly call it an affair," Harry began, although he stopped mid-sentence as he realized that he was just equivocating. "We did," he said. "We always used to have a private joke between us because I'm Harry and she's Harriet. We got on well together. And then things between Steph and I became strained. I'm not sure that married life was what either of us thought it would be. She was always so outgoing and I guess I thought she might calm down a bit after we'd gotten married and enjoy domestic life. It didn't help that planning the wedding was full of strife. I know we're supposed to take our wife's side, but I never got the anger she had for Harriet. I always just thought they should get on with it and forget about whatever happened in the past, but that didn't happen. I know Vicky (that was their mother) really wanted Harriet to be Steph's maid of honor. I tried to talk her into it as well, but she wasn't having it at all. Steph wanted to begin a new life away from the village."

"So you didn't understand why Steph felt the way she did?" Lily asked.

"I understood it alright. I've known them for some years of course, and you get to know people a lot in that time. If it was the same as in their childhood then, well, it must have been tough for all of them," Harry said.

"What do you mean?" Chloe asked.

Harry shifted his gaze from side to side before he answered. "Harriet has always needed a lot of attention. She's terribly insecure and it takes a lot to get her out of her shell. I suppose it's to do with what happened when she was a child. Steph has never had any of those problems. She vivacious and confident. It was what attracted me to her in the first place, but her problem has always been that she's never been able to understand why people don't act the same way as her. To her it comes naturally, so she thinks that other people should be able to turn it on and off just like a switch. Anyway, growing up Harriet needed more attention so Steph was left to her own devices, and I guess she thought this meant that Vicky loved Harriet more than she loved Steph. I kept telling her that it was ridiculous, and that Vicky probably just loved them differently rather than more or less, but Steph would never hear it. She just wanted to leave it all behind. I felt pretty sorry for Harriet really, I always figured that the two of them should have been best friends. Steph could have brought Harriet out of her shell, and Harriet could have reined Steph in when she got a bit wild. As it was they just brought out the worst in each other."

"So how did this affair begin then?" Chloe said. "I'm sorry about the personal nature of the question, but we really do need to try and understand their relationship as much as possible."

"Its okay," Harry said with a wave of the hand. "I've had to speak about it enough during the divorce that it hasn't been a sore subject for a while. The truth is I didn't cover myself in glory. I regret the part I played in it. Like I said, Harriet and I have always had a good rapport and I know she doesn't have that with many people. When I first met her I wanted her to know that it was possible to have a friend. It took me by surprise when Steph told me that she didn't want me speaking to Harriet outside of family occasions. But I tried to make her feel as included as possible. The wedding was difficult for her. She was a bridesmaid, but she didn't feel as included as the others. Sometimes, even in the pictures, it was as though she was an afterthought. Anyway, I was walking through the hotel and I saw her sitting on the stairs, crying. I couldn't ignore her. I went to sit on the stairs with her and I put my arm around her. She cried on my shoulder and told me that she was afraid she was never going to get married, that nobody would ever fall in love with her. I told her that she was being ridiculous, that she was beautiful, funny, and kind, and that she just needed to meet the right man who would be able to see those qualities. I told her that if she ever needed to talk about anything then she could always speak to me. She was worried that Steph wouldn't like it, but I told her not to worry about Steph.

So over the following few months we were messaging each other and I could tell that she hadn't had a good friend for a while. She spoke to me about a lot of personal things, things that she said she had never told anyone else. I think even then I knew I was crossing a line, but I couldn't stop it. I couldn't abandon her. Little did I know that I had been texting her a little *too* much, so much that Steph had noticed. She became suspicious and accused me of having an affair."

"So you weren't actually having an affair, you were just talking?" Lily asked.

Harry's cheeks flushed red and he looked a bit haggard. "I think you'd better let me get to the end of my story," he said. "So yeah… Steph told me that I was acting suspiciously and that she wanted to know who I was talking to. I told her it was nothing to worry about and that it was just a friend, but that I was sorry if I had neglected her and I would try to make amends. She wasn't having it at all. Now, I know I hadn't covered myself in glory at this point, but I thought I deserved the benefit of the doubt. It's not like I had done anything to hurt Steph before. But she flew off the handle and told me that I was an awful husband and that marriage wasn't what she thought it was going to be like. Then I told her that maybe if she didn't go out with her friends every weekend and actually spent some time together it might start feeling like a marriage. The argument escalated and she told me to leave because she needed some time alone to figure things out. I got in the car and drove, and ended up in Didlington St. Wilfred. I asked Harriet if she wanted to go for a drive. It was stupid of me. I don't know if I wanted it because I wanted to spend time with someone who was happy to see me, or because I wanted to spite Steph. Maybe it was a little bit of both. Anyway, we went for a long walk and we talked about everything under the sun. Up until that point it had always been her confiding in me, but this time I let everything spill out. It was the kind of conversation I had never been able to have with Steph, and I began to wonder if I had picked the wrong sister. I felt awkward at first because I was talking about Steph, but Harriet told me not to worry. I talked to her about all the difficulties we were having and how I was feeling lost in my own marriage. She listened well, and although she couldn't offer much advice because she had never been in a relationship before, yet she made me feel safe and calm. Then we went back to the car. Before I drove her home we hugged and then... then I guess the emotion got the better of us and we kissed. That should have been the end of it. Perhaps it could have been if I had been stronger. But after that we kept meeting up and it became a full blown thing. Steph was already suspicious so it didn't take her long to figure out, and when she found out who it was... well... that was all she wrote. We were divorced within a few months," Harry said.

It must have been such a traumatic experience, and yet he spoke about it in a matter of fact tone, as though it was a simple mundane matter.

"So can I ask you a question then; if you and Harriet got on so well has the relationship continued?" Chloe asked.

"No, it didn't. When Steph started the divorce proceedings she told me that she was going to move back home. I thought it was going to be too difficult for me to continue seeing Harriet. I thought it would be better for her if I took myself out of the situation," Harry said.

"And how did she take the news?" Lily asked.

"She was devastated. She was basically as upset as Steph had been. She begged me to make it work between us, but I knew it was just going to be too hard. She needed someone who was going to be unencumbered by all of this, someone who could offer her a simple life. That wasn't me."

"And have you heard from her recently?" Chloe asked.

Harry shook her head. "No, neither of them have tried to contact me, and I've just been minding my own business."

"Harry, now that you know what's happened, what is your assessment of the situation? As someone who knows both of them can you help shed any light on this? Which sister do you think is more capable of murder?" Lily asked.

Harry leaned back in his chair and sucked in a breath. "That's some question. It's not one I ever thought I would have to answer." He tapped his fingers against his stomach as he considered the question. "The thing about them is that they're more alike than they think. They both have a bad temper, and they're both sensitive, just in different ways. If they thought they were being wronged then maybe they would lash out, although they both loved Vicky so much. She was the only one who could ever calm them down, although I suppose she wasn't able to calm them down quickly enough this time," he said, shaking his head as the reality of the situation sunk in.

There was just one question that Chloe had left for him.

"Harry, do you know if Harriet was in touch with a solicitor about potentially challenging Steph's claim on the will?" she asked.

Harry shook his head. "I honestly couldn't tell you. If she did she never mentioned it to me, but if it was a recent development then I wouldn't have any knowledge of it anyway," he said. With that Lily and Chloe returned to the car and drove back to Didlington St. Wilfred.

"Well, it seems as though Steph is the one with more anger, and with perhaps more cause to be angry. After all she's been through it wouldn't surprise me if she lost her temper. It must have been humiliating for her to not only get a divorce, but have to return home to live in the same house as the woman with whom her husband was cheating," Lily said.

"And it would have been doubly hard if Vicky wasn't willing to listen to her side of the story," Chloe said.

"In their house there were pictures of them together, but you can tell they were forced to pose. I don't think there's one natural picture of them together where they're both happy. I can imagine that Vicky might have wanted them to be friends so much that she kept trying to push them together, even after this happened. But something Steph said has been playing on my mind."

"What's that?" Chloe asked.

"It's about Harriet playing the victim. I was just thinking that in this situation it's quite easy to sympathize with Harriet, but she still began an affair with her sister's husband. She could have stopped it too, and maybe if she had then this whole mess wouldn't have occurred," Lily said.

"She might have wanted to finally steal something away from Steph, to get something that she never would have otherwise. But that doesn't mean that she was the one who killed her mother," Chloe said.

"Then it seems as though we still have some work to do," Lily replied.

6

Speaking with Harry had shed more light on the situation, although it didn't help Lily determine which sister was the guilty party. Steph would have been filled with anger about the situation, but Harriet would have been angry too as she might well have blamed Steph for preventing her having a relationship with Harry. Given what Lily had read in her assessment, she wasn't ready to ignore the idea of Harriet being the killer, despite her being the most distraught. She began to think that there was some truth to Steph's words. Harriet had already crossed one social taboo by falling in love with her sister's husband. Was murder another line she would have crossed? Lily decided she needed to have another conversation with the sisters in the hope that she might be able to pluck some vital knowledge out of the air.

Harriet was sitting in a cell. It was a small room with a padded bunk, a toilet, and a small desk. There were heavy bags under her eyes, but she brightened when she saw Lily.

"Is it over? Have you arrested her?" Harriet asked.

"I'm afraid not. We're still trying to collect more evidence," Lily watched the disappointment fall across Harriet's face. "I'm actually here to ask you a few more questions."

Harriet swallowed a lump in her throat. Perhaps she knew what this was going to be about.

"Harriet, why didn't you tell us that you were the reason why Steph and Harry got divorced?" Lily asked.

Harriet pressed her lips together and when she spoke her voice was a raspy whisper. "I wasn't the reason. They were having trouble before Harry and I started speaking to each other. It was never going to work between them. I could have told them that from the beginning. She didn't understand him like I—" she caught herself before she finished the sentence. Lily didn't need to be an expert sleuth to tell that Harriet still had deep feelings for the man, and deep feelings could often lead to unpredictable results.

"Harriet, if we're going to get through this then I need you to be honest with me, completely honest, okay?" Lily said. Harriet nodded in response. "I need to ask you about what happened. How did you feel when you and Harry began to get close?"

Harriet shifted her weight uncomfortably. "I felt awful, but at the same time I felt happy. Harry is... he understood me like nobody else had before. He could tell exactly what was happening in my mind and he always knew the right things to say. I never had to worry about being afraid when I was with him. I had never been that close with anyone before and I knew it was a sham because there was no way he spoke to Steph the same way he spoke to me. He couldn't have been that close with her. I felt guilty though, because she was still my sister, but she always had everything. Things came so easily to her and I thought for just once it was nice that something was coming my way for a change."

"Did you not think that something bad was going to happen?" Lily asked, trying to hide the incredulous tone in her voice.

Harriet merely shrugged. "We were in love. I thought that when two people love each other things are going to be simple. I figured that Steph couldn't have been happy with him so she wouldn't have minded. She needed to find someone who could make her as happy as Harry made me. But things didn't work out like that," she said despondently.

"No. She was angry," Lily said.

"She was furious," Harriet said. "I wasn't prepared for how angry she was going to be. She ranted and raved and she told me that I was just a... well... I can't repeat the word," Harriet blushed at this.

"It must have been difficult when she moved back home," Lily said.

"I thought she was going to find somewhere else to live. Mom told me that it was going to be difficult and that she was disappointed in what I had done, but that she understood how lonely I was. I think she knew that I must have loved Harry dearly because otherwise I never would have done that to Steph. Despite what she thinks of me I don't hate her. I just... I just get sad sometimes because I know that she can handle life better than I can. It would be easier for her to find another husband than for me to find someone who can love me."

"And how did you feel when Harry told you that he wasn't going to see you anymore?" Lily asked. Harriet's face darkened and her eyes became haunted with pain.

"I couldn't understand it. I thought it was all some horrible joke. I assumed that we were going to be together and that we were the ones who were going to be married, but he said that it was going to be too difficult. I told him it was going to be okay, but he just kept saying that he wasn't right for me and I didn't understand because I thought if he wasn't right for me then who was? And then it hurt even more, and Steph was smug because even though she was the one who got divorced she knew that I hadn't ended up with him either. It ended up being me who suffered again, because that's the way my life has always been. All he had to do was say he loved me, but he couldn't do that," Harriet said, breaking down into sobs.

Lily listened to her weeping words and felt pity for the woman. She had been shattered as a child and she hadn't quite been able to be put it back together in the same way. There was a lack of understanding about the world and social etiquette and Lily wasn't sure how she was going to cope by herself. It was clear that she had leaned on Vicky for the majority of her life, and now she was looking for someone else to lean on.

But it apparently wasn't going to be Harry. It certainly wasn't going to be Steph.

"I miss her so much," Harriet lamented. Lily wished she had something else to say, but she was rendered silent. The pitiful sight of this grieving woman reminded Lily of how she had reacted after Kelly had left. At first she had been shocked, then she had been angry, until finally the emotions had given way to sadness and she had wept fretfully. She had always hated this, as she felt that her mother wasn't worth the tears after the way Kelly had left. But the sorrow had been uncontrollable.

She moved away from Harriet and went to speak with Steph. Steph laid on the bunk with her hands cradling the back of her head. She stared at the ceiling and had a vacant look on her face.

"Steph, can I talk to you?" Lily asked.

"Sure, it's not as though there's anything else to do around here," Steph said.

"I wanted to ask you how much you know about Harriet's past," Lily asked.

Steph rolled her eyes. "Why, because it will give you some great insight into her behavior and you will see how much she has gone through and how brave she's been to make it this far?"

"I just thought it might help contextualize the situation."

"Of course I know what she's been through. You don't think I heard it every time I got annoyed with her or every time I tried to get back at her for something she did? Mom was always quick to remind me of it. I heard the story over and over and over again, so I don't need to hear it from you as well."

Lily furrowed her brow. "How can you have such a lack of sympathy for your sister?" she asked.

"I am sympathetic to her. I hate what happened to her. Nobody should have to go through that, but that doesn't mean she gets to get away with everything. The problem I have with Harriet is that she's always so quick to blame everyone else for her problems. She's never taken a chance on herself to actually make something of her life, and Mom didn't help. I love her, but she indulged Harriet too much. There's a reason why Harriet has never moved out of the house. She's terrified of going out into the real world. She was able to live a little bit of a fantasy with Harry, but that ended swiftly enough as well. If she had just listened to me all those years ago then maybe she would be better prepared now, but she didn't, and this is all her fault. If she had just controlled her aggression then this would all have been different. And maybe if Mom hadn't gotten in the way..." she choked up at this and tilted her head to the side, hiding her face from Lily's gaze. Lily assumed that she was crying. There were two stories. In one of them Steph had gotten so angry that she had killed Vicky. In the other Harriet had swung for Steph, but Vicky had gotten in the way. It was so difficult to find the truth that hid behind the blurred lines.

Steph swung her legs over and leaned forward. Her eyes were dark and ringed red. "I know that I can be difficult and you probably think that I'm an awful sister, but the only thing I've ever wanted to do is love my parents and have them love me. Harriet needs help, a lot of help. She lives in her own world sometimes. I'm not even sure if she realizes she did anything wrong with Harry. She's capable of more than you think, and she's talented at making you think that she's been wronged. You need to look past her lies and look at her for what she is; a damaged person who is capable of throwing back the trauma she's faced throughout her life. I know this might sound harsh, but Mom's life would have been better off if she had never adopted Harriet. Actually I know it would have been, because she still would have been alive," Steph's voice became hoarse then. "I think I'm done talking now," she said, and turned to lay flat on her stomach. Lily quietly stepped away. She wasn't sure that speaking to the two women again had done much to help her with the case at all.

7

L ily was sitting with Chloe. The two women were going over everything they had learned about the case so far, but they were still far from a solution. Both sisters had a motive. Both had history of aggressive behavior and a long history of being angry with each other. It was so difficult to separate them, especially when they were so adamant that the other had done it.

"Is there a possibility we're missing something here?" Chloe asked suddenly. "Do you think they might have done this together?"

"Why?" Lily asked.

"I don't know, to get this inheritance they were talking about. Maybe Steph wanted it so that she could begin a new life. The sisters could finally go their separate way," Chloe said.

"But why would they now be trying to pin the crime on each other?" Lily asked.

Chloe sighed helplessly. "I don't know. Maybe one of them had a change of heart. Even if one of them confesses it's not going to be evidence unless they both confess. There are no witness, there's nothing apart from the fact these two women hate each other."

"And yet they both loved their mom," Lily said in a quiet voice.

"It's a tragedy that she was the one who died. It makes me wonder if there's something else at play here. They were angry at each other, not at their mom. It makes me think that Steph's story is more plausible because in her version it was an accident."

"I don't know, if Vicky was still trying to get Steph and Harriet to be friends then Steph might well have had enough. She's been through a lot of emotional turmoil recently. I don't think Harriet is the only one who has been traumatized, and Steph's trauma is more recent," Lily said.

Chloe nodded along. It still wasn't getting them any closer to the solution.

"When are the phone records coming in?" Lily asked.

Chloe checked the time. "Hopefully soon. I've been meaning to ask you by the way, have you decided what to do about your mother yet?"

"I think I'm coming closer to making a decision. It's just so sad seeing how this family has devolved so quickly. I wonder what they would say if their mom could come back," Lily said.

Chloe nodded. Before she could reply the phone rang. She looked at Lily with expectant eyes and beckoned her over. Lily sat beside Chloe as they scanned through the numbers, looking for anything that seemed unusual. Lily was searching the numbers as they appeared while Chloe read them out. Most of them were local numbers. It seemed as though Vicky was a person who enjoyed keeping up to date with her friends using the telephone. Eventually they came to one number that was linked to a solicitor's office. Lily and Chloe's hearts were in their mouth as they knew it was the key to this whole case. But Lily's heart did sink. She knew that one of the sisters had to be guilty, and perhaps it was bias on her part because Harriet had been the first one to arrive at the station, but her whole demeanor had been of someone who was innocent.

However, it seemed as though she was guilty. This backed up Steph's case.

"Here we go," Chloe said as she dialed the number and put it on speaker phone.

A receptionist answered. "Hello, this is Constable Chloe Hargreaves from Didlington St. Wilfred. I was hoping to get some information about a woman who might have called you recently regarding her mother's will. It's for a murder case, so I would appreciate anything you could tell us."

"Of course," The receptionist stuttered on the other end of the phone as soon as Chloe mentioned the word murder. "Is this person a client of ours?"

"I'm not entirely sure. They may just have been making polite enquiries. The last name is Harrison," Chloe said.

The receptionist asked them to hold as she looked through the records. "I guess this means we have our murderer," Lily said reluctantly. Her words were confirmed when the receptionist came back.

"Yes, we do have records of a Miss Harrison calling us," the receptionist said. Lily and Chloe nodded. It seemed as though the trauma of the past had been inescapable for Harriet and her life had never truly been her own. "A Miss Stephanie Harrison."

Lily wasn't sure if she had heard the receptionist correctly. "I'm sorry, could you repeat that please?" Lily asked. Chloe's mouth had dropped open as well. The receptionist did indeed confirm the name.

"May I ask what she was enquiring about?" Chloe said.

"I'm sorry, but that information isn't available on here, and the solicitor she spoke to is out for lunch at the moment. He's due to return in an hour. I could get him to call you if you would like?" the receptionist offered.

"That would be wonderful," Chloe said, although they weren't going to wait an hour. "I think we need to speak to Steph again."

This time Chloe and Lily went to speak to Steph together. Lily's mind was still reeling from the revelation that it was Steph who had been in contact with the solicitors rather than Harriet, as Steph had claimed. For a moment Lily had been convinced that Harriet was the murderer, but now the pendulum of justice was beginning to swing in another direction.

"What do you want now?" Steph groaned.

"We just had some phone records come through. We know that a solicitor was called," Chloe said.

Steph shifted into a sitting position, almost leaping off the bed. "So you know what I said was true!" she exclaimed, but the excitement faded from her face as she gazed into Chloe and Lily's stony expressions.

"We know that a solicitor was called, and who called them. It was you," Chloe said.

Steph's eyes were filled with panic, but she quickly recovered her composure. "Well I was calling the solicitors office a lot when I was trying to finalize my divorce. You must have gotten the records a bit confused."

"There's no confusion," Lily said. "You even used your maiden name. If you were arranging your divorce then you would have used your married name. Besides, the date on the call is a long time after your divorce. It was you who called the solicitors, not Harriet, but why would you need to speak to them? What could you gain from it?"

"I don't need to tell you that," Steph said bluntly.

"Steph, you've lied to us. We can drag you in front of a judge if you want, or you can just tell us the truth now. Whatever scheme you

467

were trying to hatch didn't work. Just tell us the truth so we can end this," Lily said.

Steph folded her arm across her chest and kept her lips tightly shut. Lily stared at her though, and when she did she put herself in Steph's position. She thought about how all her life Steph had felt second best, and how even when she had gotten married that had been ruined by Harriet.

"Do you have any thoughts?" Chloe asked.

"Actually I do," Lily said, her voice razor sharp. "I think that you have a deep seated inferiority complex Steph. All your life you've been trying to have your own life without having to worry about Harriet, and when you finally escaped, marriage wasn't all it was cracked up to be. But that's because you didn't love Harry, isn't it? It's because you just wanted to marry the first man who could offer you a way out. So you took it and hoped for the best, only it ended up worse than you could imagine. Harriet stole your mother's attention, and now she stole your husband as well. You had everything taken from you and were forced to return home, never allowed to grow up, never allowed to forget that you're the younger victim. And you knew it was never going to change because Harriet certainly wasn't going to move out. She was too dependent on your Mom. So you wanted to try and force the issue. There was the conversation about the inheritance and you didn't like the idea of sharing with Harriet. She may have been angry when you suggested selling the house, but I think you had another idea," Lily watched Steph's body language change as she spoke, and Lily knew she was getting closer and closer to the mark. "I think you finally saw an opportunity to take everything from Harriet, just as she had taken everything from you. You wanted to test the waters and see if you could challenge her claim to the inheritance. You thought because she was adopted that might not make her as legitimate a daughter as you are."

As soon as Lily spoke these words Steph's fury lashed out into the air as well. "Yes, fine! I admit it. I called the solicitors. Harriet has always had everything she's wanted. I finally tried to break away from that but she still ruined everything for me. She broke apart my marriage. She was supposed to be my sister, but what good has she ever done for me? It was always me having to help her, Mom was always trying to make me feel bad for her and I'm just tired of it. Didn't she realize that I was suffering as well?" there was a choking cry, almost a plea for salvation.

"And that's why you killed her, isn't it?" Lily said in a cold voice. "You wanted to get revenge, you wanted to just make her understand that she couldn't force you to be sisters. All this anger that had been boiling up inside you for all these years finally exploded out and you hit her. You wanted to make her see, but the only way you could get her attention was by doing something drastic. Maybe you didn't really mean it, but maybe deep down a part of you did. Maybe you wanted to hurt her so that she could understand how much she had hurt you by always treating you as the second best daughter."

"No… no… I love Mom! I never wanted to hurt her!" Steph cried.

"But you did. You felt the anger inside you because she had always been there. She had always acknowledged Harriet's pain, but what about yours? Didn't she know how you had always hated being forced to take photos with Harriet? Didn't she know that your complaints were valid? You loved her, but sometimes it just felt as though she didn't understand you, and you couldn't let go of this anger. You couldn't forgive her," Lily said, her words becoming more frayed with emotion as she spoke, for she couldn't tell if she was speaking about Vicky or her own mother.

Perhaps both.

"No… no… it was an accident!" Steph eventually cried, and then she realized that she had just made a confession. For a moment she went rigid and tense, but then the strength left her and she sank down, all the breath leaving her in one rushing torrent. "Fine," she croaked, staring at the floor with her hands hanging between her legs. "Fine... I'll admit it. I did it. But it didn't happen the way you thought. I did call the solicitors. I just wanted to see how iron clad it was. Harriet always got everything and she was Mom's favorite. I'd lost a lot in the divorce and I wanted to see if there was any chance that I could get a bigger share. Mom had overheard me. We had a row in the kitchen. I tried to tell her again how I felt and what Harriet had done, but Mom said that Harriet wasn't to blame. She blamed it all on Harry, thinking that he had somehow tricked Harriet into sleeping with him. She could never just see the truth. And then Harriet came in and asked us what we were arguing about. I told her. It was stupid, but I was angry and I wanted to see her face. She started to cry and yell and say that she was just as much a part of the family as anyone else and that I couldn't wipe her away simply because she was adopted. We were ready to fight and Mom had to get in the way. I grabbed the nearest thing to me and swung it in the air. For once in her life I just wanted Harriet to shut up about her problems and how nothing was ever her fault. I just wanted to make her feel guilty that she had stolen my husband. Do you know she had never once said that she was sorry?

Anyway, Mom threw herself in the way. I hit her and as soon as I felt the impact I knew what had happened. I stood there, absolutely stunned by what I had just done. Harriet ran. I wanted to try and catch her, to tell her that we could figure something out. Then I saw that she had arrived here and I knew that she had already told you. I had to think of something quickly."

"So you blurred the lie the with the truth, hoping that we wouldn't figure out which sister had called the solicitors," Lily concluded. Steph nodded sadly.

"I never wanted Mom to die. I never wanted anyone to die. It was just so... I was so angry," Steph said.

Lily and Chloe left her to mull on her sorrow. They returned to the office. It wasn't long before the solicitor called them and confirmed what Lily had suspected. They then returned to Harriet's cell. Chloe opened it.

470

"We know everything now. Your sister confessed," Chloe said. Harriet nodded. Tears ran down her cheeks. She looked at the open door of the cell and staggered through. Her voice was reedy and weak.

"What am I supposed to do now?" she asked. "What am I supposed to do without Mom?"

She sounded forlorn, like a lost child. Lily could hear the words reverberating through time. The same thought had passed through her mind when Kelly had first left her.

"You stay strong. You keep to the lessons she taught you and you learn more on your own. You keep yourself safe and you try to make a good life for yourself, because it's what she would have wanted," Lily spoke with forceful emotions. They coursed through her body and made her blood sing. Harriet looked at her, but she didn't seem convinced. Lily wasn't sure she had the strength to make it through the world alone. Harriet walked out of the police station and stood outside, turning to the right and the left, as though she was trying to figure out which way to go. It took her a long time to decide.

Chloe and Lily returned to Steph's cell and officially arrested her. Now Steph finally broke down into sorrowful tears and was inconsolable. She kept wailing for her mother, craving one final moment to speak with her. It was then that Lily made her final decision.

∼

"So that's that," Chloe said. "My final case in Didlington St. Wilfred."

"I'll be sorry to see you go," Lily said.

"Me too. I know that we didn't get off on the best foot and I'm sorry for that. When I head back to the city I'm going to be a lot more open minded about how I solve cases. And like I said, if you're ever

471

in the area come look me up. I'll be glad to see you again," Chloe said.

Lily smiled. "I hope that we can see each other again someday. I'm sure you're going to rise up the ranks and become a big deal in the police force."

"We'll see," Chloe said. "There is one thing I want you to promise me before I go though."

"What's that?"

"Make a move on Blake. You're crazy about the guy, and if I've learned anything on this job it's that things can end far sooner than we'd like them to, so it's important to make the most out of every opportunity."

Lily recoiled in shock and her cheeks were flushed with a deep crimson shade. "How did you know?"

Chloe chuckled and tossed Lily a look. "You know, you're not the only talented detective around here," she said. Lily broke out into a smile and laughed as well. Perhaps it was time to make some changes in her life. She came to Didlington St. Wilfred for a fresh start, and while she had enjoyed her thriving career as a writer and her exploits in helping investigate crime, there was still something missing, and that something was Blake. She didn't want to be like Harriet and Steph, who had let their issues fester over a lifetime and eventually rule over them. She wanted to confront her trauma head on and then overcome it, with the help of the most wonderful man in the world.

But before she went to speak to him there was another person she needed to speak to.

8

L ily's heart was in her mouth as she met her mother in a small café. Kelly wore an awkward smile as she entered the café. Lily did not get up to meet her. Lily wrapped her hands around the mug of cocoa, while Kelly just ordered a sparkling water. Lily was about to comment that Kelly usually ordered something alcoholic, but she held her tongue. She didn't want to get this meeting off to a bad start.

"I'm so glad that you decided to see me," Kelly said. Her eyes were glistening and her lips trembled. Lily had never seen her this emotional. She was the kind of person who was always in control, and liked being that way.

"It took me a while to get my head around this whole thing, but I want to make sure there are some ground rules, okay? I'm willing to listen to what you have to say, but that doesn't mean I'm just going to accept you back into my life. You really hurt me when you left Mom, and I didn't expect to ever see you again. If this is some way to get back into my good books because I'm successful then I'm just going to walk right out of here because I don't need some leech."

"It's nothing like that at all!" Kelly said, looking horrified. "I really just... I missed you so much Lily. I wish I could explain everything. I wish my explanations made sense. When I was younger I thought I had to live life a certain way and I thought you were getting in the way of that. I wanted to be free and easy and live this fantasy, but the fantasy was never as good as the real thing. It didn't take long for me to realize I had made a mistake. I felt horrible. I punished myself. I drank to try and push the demons away, but they never left. They kept stabbing at my mind and I was low, real low."

"Did the guy you left with help you?"

Kelly wore a sardonic smile and shook her head slowly. "He kicked me to the sidewalk quickly enough. I begged and borrowed what I could. I scrounged every penny, but it all went to drugs. I kept wanting to make my way back to you, to apologize and tell you how wrong I had been, but I knew you would never forgive me. And you were right not to. I was a terrible mother and for years I kept hurting myself because it was the only way I could make things up to you. I told myself that I didn't deserve anything else because I had abandoned my daughter and I just..." she choked up with tears. Lily tried to keep the wall of emotion up, but it crumbled at the sight of her mother crying. She remembered Kelly being proud and angry and upbeat, but had rarely seen her cry like this. It was harrowing to see this woman be so broken by the world, and Lily realized that this trauma wasn't one that she had endured by herself. Being apart from Lily had affected Kelly deeply as well. It was something they shared, and Lily couldn't find herself still being completely angry with what had happened.

"Mom, it's okay, you're here now," Lily said.

Kelly pinched the bridge of her nose and breathed in deeply, trying to compose herself. She sipped her drink.

"So what happened after that? How did you get out of it?" Lily asked.

"I was walking along the street one night and I came to a church. I wasn't even intending to go there, but I guess you always end up where you're meant to be. There was a man inside and he gave me a warm drink and showed me kindness. We spoke all night and I told him about all the troubles I had caused in my life, and how I had treated you. He told me that there was still hope, that as long as I was alive and breathing there was always hope, as long as I was strong enough. I told him that I wasn't strong, but he said that it was okay, because when we're not strong Jesus Christ gives us strength and we can learn from Him.

He invited me back the next night, and the night after that. I started becoming a regular member of the congregation. I learned everything he had to teach me and I managed to kick the habit. It was the hardest thing I had ever done, but I started to realize that it was penance for the way I had treated you. I had to go through that, to be broken down to realize what was important. The best, most incredible thing I have ever done is bringing you into the world Lily, and I'm so sorry that I didn't realize how precious that duty was. I keep going back to that moment in my mind and I keep telling myself to stay, to just stay with my daughter. But I can't change that. I can only change the future, and I want to change Lily. I'm not expecting you to forgive me or anything, but I just wanted you to see me like this, to know that I'm not the same failure as I always was, and I wanted you to know that I'm proud of the person you've become. I know that probably doesn't mean much coming from me, and it's not as though you need me to be proud of you anyway, but I am all the same, and I'm glad that I'm still here to be able to read your books and see you make a name for yourself."

"You've read my books?" Lily asked.

"Each and every one of them," Kelly said with a wide grin. Lily cringed a little because she had put a lot of herself into those books, especially the main character's troubled relationship with her mother, but she hoped that had gone over Kelly's head.

"Well, thank you, and it does mean more to me than you might realize Mom," Lily said, feeling humbled by Kelly's story. It was a reminder that the spirit of God and the love of humanity could bring people together and save people from the darkest places. She took a sip of her drink before she spoke, and reminded herself that it was important to be honest with Kelly. "You know Mom, I was really angry with you for a long time. That's why I reacted the way I did when you showed up. I had spent so much of my life knowing that I was never going to see you again and the last place I expected you to turn up was here. You really hurt me Mom. That moment when you walked away from me was the worst moment of my life. I had to figure out how to grow up all by myself, which is probably why I've done such an awful job of it."

"You seem to be doing well from what I can see," Kelly said.

Lily offered a grateful smile. "It's just hard for me to be around you again because the last image I have of you is you abandoning me to leave with some guy."

"I know," Kelly's head hung in shame. "I wish I could go back and make everything right. I'm not expecting anything from you Lily. If this is the only time we see each other then so be it. But I wanted to at least talk to you once, you know, in case we never got the chance again. If I hadn't found Christ then I probably would have ended up in some ditch somewhere, but He gave me another chance, and I wanted to use that chance for the best."

Lily nodded. It was certainly a noble effort and she was glad to know that her mom was taking these things to heart. So many people wasted second chances, but she had definitely turned her life around. There was a certain hollowness within Lily's heart though, for there would always be the pain at knowing those years of being her mother's daughter had been taken away from her. Nothing could give them back, and nothing could take away that pain. But there were always other things that could overwhelm the pain.

476

"Mom, there is one question I've been wondering about. Why did you leave me all these parcels? Why didn't you just come to my door directly?"

"I thought about it," Kelly said after a moment's pause. "The first time I knocked on your door I wanted to stay there, but then I got scared. I wasn't sure how you were going to react. I had to do a lot of praying to summon the courage. I wanted to leave you these parcels because I knew that I had failed as a mother before, and I wanted to make it up to you. Mom's bake their daughters sweet things, right? It sounds stupid, but in my head it was a way for me to make up for all the time I had missed with you. Cooking really helped me when I was struggling with my addictions."

Lily nodded slowly, but found it difficult to stop a smile from breaking out upon her face. "Mom, I really wish you had just knocked on my door. For so long I was convinced that someone was stalking me!"

Kelly looked horrified, but then she burst out laughing. The tension had been broken and Lily was glad that she had come to meet Kelly today. They shared stories about their lives. Kelly was extremely interested in learning about how Lily had begun writing cozy mystery stories. Lily also told her about the real life crimes she had helped investigate, and about the friends she had made along the way. Kelly was enchanted by the stories and all that had happened to Lily during her time in Didlington St. Wilfred. It was amazing how much had happened, and as Lily recounted all the stories she realized how far she had come in life, and as a person. She had gone from an annoyed Hollywood assistant to a successful writer and a part of a community. She felt guilty that she had ever entertained the notion of leaving this village and the home she had made here.

"How long are you staying around for?" Lily asked.

"Not long. I don't have much money left. It took a lot of saving to get over here," Kelly said. When Lily asked her what she did for a living, Kelly revealed that she worked in a diner and lived in a small apartment with one of the other members of her congregation. In Didlington St. Wilfred she was staying in a small bed and breakfast, and mostly kept to herself, explaining that she had been so afraid of Lily seeing her before she was ready.

"I kept telling myself that I had to see you, but I could never quite make that final step. I'm glad you caught me though," Kelly said.

"I am too," Lily replied, and she meant it. There were still many feelings for Lily and her mother to work through, but they were on the right track now and Lily felt better for having taken the steps to see Kelly. It helped put the trauma in the past.

"My life is still in America. This place is wonderful though. Have you ever thought of coming back?" Kelly asked.

"The thought has crossed my mind, but in all honesty I've felt more at home here than I ever have before. I get to be myself here, and I get to make a difference in the world. It's more than I've ever been able to do before and I think if I left, then I would be less than I am," Lily said.

"Well, it certainly helps when you have such gorgeous men around. The English are always so refined and charming, aren't they?" Kelly said with a wide smile. There was a glint in her eyes that reminded Lily of the way she used to be. It was comforting to see that her mother hadn't entirely lost the essence of who she once was. "So when are you going to tell me who that charming man at your cottage was?"

"Blake?" Lily said.

"Yes, Blake," Kelly said.

"Well, he's a friend. His father owns a lot of the land in the village. He's been involved in some of the cases that I've worked on. One of them had to do with his father, and we just struck up a good relationship. We see each other at church every Sunday as well."

"He sounds like the total package. How long have you been seeing each other for?"

"Oh, we're not seeing each other," Lily said, getting a little embarrassed.

"Whyever not?" Kelly asked.

"Well, it's just, you know, these things take time and I'm not really sure that I'm the type of woman he wants. He does have some pedigree when it comes to the English noble class and I'm just... well... I'm just me."

"Lily McGee," Kelly said, taking on a maternal tone. "I know I did you wrong in so many ways when you were growing up, but I never taught you to be down on yourself. You're a beautiful, talented, successful young woman who could have any man she wants. If you want Blake then you go and get him. Life can be incredibly lonely, but it doesn't have to be. I was never lucky enough to meet the right person, but I think you may have, and so you need to go to him and tell him whatever you need to tell him. Don't waste this opportunity," Kelly said.

Lily squirmed in her seat. "But—" she began, although Kelly cut her off.

"There are no 'buts' here Lily. I'm not going to have my daughter ever doubt herself when it comes to a man. Go and get him, and have a better life than I've had. That's the only thing I want for you," Kelly said.

It meant a lot to Lily to hear Kelly say this. She felt emotion welling inside her. This was always the woman that she wanted her mother to be. It had taken more years than Lily would have liked to get to this point, but at least she had reached it. She nodded and decided that she would indeed go and talk to Blake. It was about time she laid all her cards on the table.

A COOKED UP CRIME IN A
QUIET ENGLISH VILLAGE

1

The mood in Didlington St. Wilfrid was joyous, for once. Instead of being shrouded in a cloud of gloom because of a murder, people were celebrating the resolution of the case against Constable Dudley. It had been a long time to wait for the result of the investigation into police neglect, and throughout Constable Dudley had been worried that he would be made the scapegoat for the failings of his superiors. In truth, Lily McGee had been worried about the same thing. Once the results of the investigation had been revealed, the mayor had announced that there was to be a celebration in honor of Constable Dudley's resilience and innocence. The city hall was decorated and now filled with the residents of the small country village, and Constable Dudley was bashful at receiving all the attention. He glowed when Lily and Joanie showed up.

"This is marvelous, just marvelous," Bob Dudley said, rocking back and forth on his heels. His time away from his job had not always been kind to him, but recently Lily had noticed him once more making an effort in his appearance.

"It really is Bob. I'm so glad that they finally got back to you about this. I had no doubt that it was going to come in your favor," Lily said.

"Thank you Lily. That does mean a lot coming from you, and I'm looking forward to working with you again. I just hope that I haven't lost my instincts," Bob said.

"I'm sure you haven't," Lily said. Joanie then moved forward and kissed Bob on the cheek, and Lily thought that Bob was going to faint.

"Would you allow me to be on your arm for the evening?" Joanie asked. Bob blinked and almost couldn't believe his ears.

"Of course," he said, sounding a little flustered. Joanie linked her arm in with his and they strolled around to talk to the other people who were there. Lily smiled. They made an odd couple; the stocky, blustering English policeman and the glamorous, flighty American, but perhaps sometimes it was the oddness that made relationships work. Lily wasn't sure how serious Joanie was about this, but it certainly brought Bob a great deal of happiness and pride. Lily spent some time speaking with Marjorie, who had produced a great number of sweet treats for the occasion, and Pat, who was smiling and seemed to be enjoying the atmosphere of being with a crowd. She never thought she would see the day when this happened, but a near death experience could change the nature of a man. God had the grace to give Pat another chance, and Pat was taking it with both hands. As always, Zeus was by his side. As soon as the dog saw Lily, he came bounding up to her and began licking her cheeks. Lily ruffled his cheeks and kissed him on the nose.

"I miss you Zeus," she said, and she meant it with all of her heart. She and Zeus had formed a strong bond while Pat had been recovering from his wound, but it was only right that Zeus should be with his owner. Still, it had left a hole in Lily's heart and she wasn't quite sure how to replace it. Reverend Brown was there as well, and was happily speaking with everyone in turn, seeming to enjoy the fact that he could socialize without having to give a sermon.

It was the kind of ceremony that everyone could enjoy and that the village needed after what had been a difficult time. Lily watched as Lord Huntingdon and his son, Blake arrived. Their appearance was always an occasion that people took notice of. Lord Huntingdon walked with two canes, and his gait was slow. He was a man who sought to battle the effects of aging, and was not going to let the years have mastery over him. Lily's gaze turned to Blake. To see him standing beside his father was almost to see an echo of the future come to her vision. She wondered if Blake would be the same when he was older; crotchety and charming in equal measure, and then she wondered how she would be.

It left a wistful feeling behind.

She wondered if her mother was going to show up. Lily had spent a little more time with her, and although she wasn't ready to forgive her mother for all the crimes of the past, she could at least sit with her and accept the possibility that they might have a new relationship. Lily had seen enough people lose those closest to them. She would have been a hypocrite if she decided to turn her back on someone who was begging to be forgiven. Lily smiled at all the familiar faces she saw, some of whom she had helped find justice for. They looked healthy and hale, and it was a sign that time could heal all wounds, time, and the love of the Lord.

The mayor took to the stage and cleared his throat. The microphone amplified his voice, although he had such a booming tone that he didn't really need it.

"Now, tonight we are here to honor the good and upstanding and loyal resident of this village, Constable Bob Dudley," the mayor said. There was a round of applause and buoyant cheers from those in attendance. The mayor waited patiently for them to die down. "Constable, you have had to endure hard times recently, and I am glad to see that your patience has been rewarded. Your superiors may have had to investigate your honor, but to us here in the village there was no doubt that you were innocent, and we are all glad that you have returned to your rightful place as protector of Didlington St. Wilfrid. However, before we continue with the honors of tonight, there is someone to whom we owe a great debt, and someone who I would like to make a personal thank you to. Officer Chloe Hargreaves, who took over this constabulary and served with distinction, please come here so I could have a few words."

The mayor's words lingered in the silence of city hall. People began to mutter as he called Chloe to the stage again, but no Chloe appeared. Lily had had a personal goodbye with Chloe, but she was surprised that Chloe had not attended this function. She had been invited after all. The mayor waited for a few moments, but when Chloe did not present herself he carried on with his speech. Lily frowned though, and crept away. She darted through the houses and made her way toward the house where Chloe had been staying. There she found Chloe packing up the last things into her car.

"Are you not coming to the party? The mayor was just asking for you," Lily said.

Chloe shook her head and laughed a little. "Those things really aren't for me. I don't do this for public recognition you know."

"I know, but it's still nice to be told that you're appreciated. I'm sure Constable Dudley would appreciate you coming for one drink, just to say goodbye."

"Are you sure? After all, I was the one who took his job."

"That wasn't your fault. That was just his superiors. He doesn't bode you any ill will. None of us do. You're a part of this village Chloe."

Chloe smiled and nodded. "Thank you Lily, that does mean a lot, but I'm not sure how true it is. The truth is that you can all give a part of yourselves to this place and to each other, and you're all better because of it. I'm just not that type of person. I do appreciate you coming to find me though, and maybe one day I'll come back here, perhaps to enjoy my retirement."

Lily arched her eyebrows in surprise. "I didn't realize you were so forward thinking," she said.

Chloe laughed. "Speaking of forward thinking, have you thought anymore about what I said?"

"I've thought about it," Lily said, wishing she could hide the crimson blush on her cheeks. Chloe gave her a knowing smile and laughed, then got in her car. Lily stepped back as Chloe drove away. She had left Didlington St. Wilfrid as quickly as she had arrived, and Lily wondered if she would ever see the woman again. There were some people in life who were friends forever, and some who you only knew for a small glimpse of time. Both could be deep friendships though, and Lily wished Chloe well.

She returned to city hall to hear the tail end of the mayor's speech. She smiled to herself, for the mayor did enjoy the sound of his own voice. But at the end he welcomed Bob Dudley up onto the stage. It would be one of the few moments that night when Bob would not be found with Joanie on his arm.

Bob's eyes glistened with tears and the smile on his face was a wide one. He was not a man accustomed to public speaking, so he faltered over some words and mumbled a few others. There was nothing that could be confused about the earnest emotion in his voice though.

"I'd just like to thank you all for coming today and for standing beside me. When this first happened I was afraid because I thought I would have to go through this alone, but there have been so many of you who have stood beside me and it has truly touched my heart. I have always worn this badge with pride, and I am glad that I have officially been exonerated of any suspicion of negligence. I hope to continue serving this village with distinction for many years to come. Being a police officer is the only thing I've ever wanted to be, and these past few months have been so challenging. I've had to think about what is truly important in my life, and it's this," he said, pointing to his badge. "I will make this badge proud, and I will make you all proud too. Thank you," he said, presumably stopping because he was on the verge of tears.

He descended from the stage and Joanie linked arms with him again. Lily clapped hard along with the crowd, and there were a few whistles thrown the constable's way as well. She turned and intended to go and treat herself to one of Marjorie's delicacies, but then she smelled the scent of a familiar aftershave, one that often left her weak at the knees. It was Blake.

"Good evening Miss McGee," he said with a twinkle in his eye. Lily returned the smile, although she felt a little awkward toward him considering what had happened recently.

"Good evening Blake. How are you doing?"

"I'm very well thank you. I was waiting for you to come up and speak to me, as is so often the case, but I was afraid I might have to wait for the entire evening."

"I wasn't sure that you wanted to speak to me again after what happened with my Mom," Lily said sheepishly.

Blake nodded and his expression turned serious. "I have to admit that I was a little taken aback by your reaction to seeing her."

Lily hadn't had the chance to talk to Blake about her relationship with her mother, Kelly. But Kelly had given her some advice; to go after what she wanted because life was far too short to do anything else. Lily had resolved to do just that. It was time to be honest with Blake.

"I know, and I know how it must have seemed to you," Lily said. Blake had lost his mother, so he must have thought her unusually cruel.

"I'm not one to judge. I know that people can have different relationships with their mothers, but as someone who lost theirs I would advise you that even if you are angry with them it's better to deal with that anger, because you don't want that anger to be the only thing that's left."

"I know, that's what I was thinking about too, and I did speak with her in the end. We actually sorted a lot of things out, and while we're not quite perfect yet I think we might be getting to the point where we can be civil with each other, perhaps even friendly. But I did want to apologize to you as well. I didn't mean for me to sound so heartless, it's just that Mom and I had such a hard relationship. It wasn't easy for me to see her again, especially when I had left her, and my entire American life, behind."

Blake nodded. "I understand."

Lily swallowed a lump in her throat. The other thing she wanted to bring up was the date that Blake had suggested they go on. It was something Lily had yearned for, yet now she didn't know if Blake was seeing her in a different light. "I just don't want you to think I'm a monster or anything," she blurted out.

Blake burst out laughing. "I don't think anything could make me think that of you," he said. Lord Huntingdon was calling Blake from across the other side of the room. Blake gently touched Lily's arm and leaned in, lowering his voice to an intimate whisper. "It seems as though I'm being summoned. However, I believe I still owe you a dinner Lily, perhaps we could go somewhere tomorrow evening?"

"Of course, where did you have in mind?" Lily asked, although in truth there were not too many choices in the small village.

"I was thinking somewhere in the city. It might be nice to take a break from Didlington St. Wilfrid for a while. I'll come and pick you up around six," he said. Lily glowed and watched him leave. A smile spread across her face and she felt as pleased as punch.

2

Lily smoothed down the pale blue dress that flowed along her body. It was the most glamorous thing she had worn since arriving in Didlington St. Wilfrid, and it reminded her of the days when she would be attending high powered parties with the Hollywood elite, although she would be there to run errands rather than actually able to enjoy the festivities. Her hair was curled into tresses that framed her elegant face, and she wore a touch of makeup to accentuate her natural beauty rather than as an act of vanity. She tilted her body as she looked in the mirror, hoping that Blake would like what she wore. This hadn't been her first choice, but Joanie had decided to take matters into her own hands, suggesting that Lily did not have the requisite skills to pick out a perfect outfit. Lily found it difficult to argue with her. Lily knew plenty about crime and murder, but romance was where she fell down. She had rarely been close to anyone and had always been guarded, but it was time to change all that. Being in Didlington St. Wilfrid had shown her that she was capable of belonging to a community, but there was certainly something missing, and that something was love.

Of course, she was not sure that she loved Blake yet. However, if there was a man she could picture herself falling in love with then it was certainly him. He was tall and handsome with the typical English gentleness that she found so charming. He was intelligent and witty, and he saw her as an equal rather than looking down upon her as some from his position in life would have done. He was the kind of man that would have girls swooning if he appeared in a film, and she felt honored that she had the chance to go on a date with him.

He came to pick her up at six on the dot and arrived with a small bouquet of flowers, which she thanked him for and placed in the kitchen. He also told her she was beautiful and kissed her on the cheek. He was wearing a pale blue shirt and a navy blue blazer, although he was not wearing a tie. Lily thought the outfit framed his physique well, and she cast an admiring glance toward him. He led her to the sleek car and they drove away from the village, heading toward the city.

It was akin to traveling between worlds, Lily thought. The city reminded her of her life in L.A. There were tall buildings and cars rushing by, as well as people who were too busy with their own lives to stop and ask the time of day. It took a little longer than expected to reach the restaurant as the roads were packed with cars. Apparently there was some sporting event this evening that had attracted a huge gathering of people to the city. It brought into stark contrast the way her life had changed, and she couldn't believe she had actually contemplated leaving Didlington St. Wilfrid.

They parked outside a restaurant which was wide and had dark windows.

"I've heard good things about this place," Blake said. They walked in and were greeted with the sound of meat sizzling on a grill. The smell of spice and herbs drifted toward them, and it made Lily's stomach rumble. There was a server who came to greet them and asked if they had a reservation. She was a petite girl who looked younger than she probably was, and introduced herself as Betsy. When Blake gave his name she pulled a couple of menus and took them toward the rear of the restaurant, to a small square table. A candle had already been lit and the flame flickered. The table was covered with a thick cream colored tablecloth. Blake pulled out Lily's chair for her and then took the liberty of ordering some wine while they perused the menu. Lily's eyes nearly popped out of her head when she saw the prices on the menu. While she was making a good living from her books, she still had the mindset of someone who struggled with finances.

"Are you sure you want to eat here?" she said, gesturing toward the price.

Blake laughed. "Don't worry about it, remember, I'm taking *you* out. You can choose the place next time. It has been a while since I've had some fish and chips."

Lily laughed and began to relax. If Blake was already thinking about a second date then that was a good sign indeed.

The wine was poured and they placed their food orders. Blake opted for a sirloin steak, cooked medium rare. When Lily chose the same, Blake seemed surprised.

"I am American, it's kind of in our blood. You said you wanted fish and chips," she said. Blake nodded.

"I know but the steak was calling my name.

Sometimes it is easy to forget that you come from another country, aside from your accent of course. But you have acclimated well to England. Is there anything you miss from your homeland?"

Lily pondered the question for a moment. "To be honest there's not much I miss. Some of the snacks, of course, but other than that I like Didlington St. Wilfrid better. It's far quieter here. I'm sure there are places in America that are relaxing, but not in L.A."

"So you would never go back?"

"Never say never, but I don't have any plans to. I like my life as it is right now. How about yourself? Do you have any plans to take up traveling again?" Lily asked, recalling that Blake had been jet setting around the world before his father had been taken ill and involved in a murder plot. Since then Blake had remained in the village, abandoning his previous life.

"To be honest I don't think so, at least not for a while. I couldn't do that to Father. I hate to say it, but I'm not sure how many years he has left in him, and I wouldn't want to miss out on any time with him. He would be so lonely without me."

Lily nodded, feeling foolish that she had brought it up. "I should think he has quite a bit of time left. He certainly has the spirit of a fighter."

Blake smiled sadly. "He does, and he puts on a good show for everyone, but behind closed doors…" Blake trailed away for a moment, "… he doesn't quite seem the same as he used to. He's starting to notice it himself as well, which leads to him getting frustrated. I suppose it's time that I started to settle down as well. I'm a little ashamed that I haven't given him a grandchild yet."

"I don't think you should be ashamed of that. It's the kind of thing you should only do when you are ready, and with the right person," Lily said.

"I suppose."

"Have you never come close to settling down yourself?" she asked.

"That's Bernard Ferrabeau," Blake spoke in a low voice. Lily frowned.

"You were going to settle down with Bernard Ferrabeau?" she asked.

Blake shook his head. "No, over there, he's a food critic. This place must be good if he's coming to review it. He's a tough one though. They call him Ferocious Ferribeau." Lily craned her neck to see as a man walked into the restaurant. He held his head high and had an air of superiority and power. She supposed that being a critic would indeed have that affect in a place like this. It seemed as though Blake was not the only one who knew who this man was either, for whispers danced around the restaurant. Ferrabeau was a stocky, short man with a bald scalp in the back and wiry black hair that was swept back in a friar's cut. He had a thin mustache that looked gelled, a squat nose, and two beady eyes. He sat at a table in the middle of the restaurant and snapped open his napkin.

Lily tried to get Blake's attention again. "I was asking you if you had ever come close to settling down?"

Blake sipped his wine, and did not seem distracted by the critic now. "My father actually arranged a marriage for me in my youth. I was only twenty," he laughed, as though he couldn't quite believe it had happened to him. "I was quite mad really. Father had it in his head that he still needed to arrange a marriage for me and so he took it upon himself to arrange a deal with a friend of his."

"What happened?"

"Well, when I met Fiona it was a disaster! We were completely mismatched and there was no hope of it ever working. Father kept telling me that sometimes marriage takes work, but I think there has to be at least a little something there. I don't believe that all relationships are easy and that once you find someone you like it's all going to magically work out, but there has to be some kind of connection, something that sets them apart when you look at them, don't you agree?"

"I do," Lily said, and gazed into his eyes. She wondered if she was set apart from the rest when he looked at her.

"Have you ever been close to settling down?"

Lily blushed and shook her head. "No... no I haven't. I had a boyfriend back in America, Tommy, but it wasn't- he wasn't... you know. It was just something to pass the time more than anything else."

"I see. Well I certainly don't want anything to simply pass the time."

"Neither do I."

"And what is it you want Miss McGee?" he asked.

Lily's smile faltered and she was almost afraid of saying the truth because she did not want to scare him off, and yet if she tried to make a joke of it and downplay things then he might think she wasn't as serious as she was. It frustrated her because if she had been in front of a murderer she would have no qualms in asking the hard questions and knowing what to say, yet when romance entered the picture she became addled. It was at this point that the waitress came over and served their steaks. The meat was sizzling and juicy, the fries were chunky and crispy, and the salad that was served with it was colorful and fresh. However, the waitress seemed to be distracted by the food critic, and kept glancing away from the table.

"I suppose when someone like Ferrabeau comes to visit the staff always gets distracted," Blake commented. They tucked into their steaks, which were a perfect, juicy pink in the middle. Flavor burst in their mouths and they both looked at each other with complete satisfaction.

"I think it's safe to say that this place lives up to its reputation," Lily said. Her gaze flicked toward Blake. She wondered if she should answer his question now, or if he would ask it again. But what did she want? Back in America she had never seen herself settling down, but here it was different. In her mind she had a vision of living with Blake in Didlington St. Wilfrid, going to church on Sundays, and perhaps even having children one day. It was a gentle dream, and one that made her heart happy, but she was almost afraid to admit to herself that this was what she wanted, and more than that; she was afraid that she would become like her mother. She was about to tell Blake all this, but there was a sudden commotion that startled everyone in the restaurant.

3

Lily turned around sharply, resting on the back of the chair. The sight that greeted her disturbed her greatly. Bernard Ferrabeau did not look so ferocious as he lay on the floor of the restaurant, staring up at the ceiling. He had collapsed suddenly, taking the whole tablecloth with him, including the glass and the cutlery. Water had spilled onto the floor, and his plate of food had taken a tumble as well, cracking upon impact. The salad he was eating was a mess now. The cream tablecloth covered him like a blanket, resting over his bulbous stomach, while his brow was drenched with sweat and his cheeks were flushed.

He was not breathing.

The waitress, Betsy, had screamed and dropped the tray of glasses she had been holding. They shattered around her feet. The other guests all gasped in shock, unsure what to do or say.

"Betsy! What happened-" the manager said as she came rushing out of a back office only to freeze as she saw Bernard Ferrabeau laying on the floor. Lily watched the panic sweep across her face. She was young, but there was a determined look on her face and she wasted no time in leaping into action. The manager commanded Betsy to clean the glass up, and called to another waiter to help her. Then she sank to the floor and began giving Ferrabeau CPR, although after a few moments it was clear that it was not going to help. The man was gone. Even so, she called out for an ambulance and had her head in her hands. She pulled the tablecloth over the food critic and then rose to her feet.

"Everyone, please, be patient with us. I'm not sure exactly what happened. It's going to take us some time to figure out what's going on. I understand that this is quite distressing and it's not what any of us planned this evening, but please bear with us and if there's anything we can do, do not hesitate to ask," she said, and looked in the direction of the kitchen. Unlike most restaurants Lily had been to, this one had an open kitchen so the chef and his staff were all visible to the guests. The manager looked directly at the head chef, while Lily turned to Blake.

"You're going to get involved in this, aren't you?" he asked. Lily took a bite of her steak and rose. It wasn't in her character to simply sit by and let things unfold. The manager was at the bar, so Lily approached her.

"Excuse me," Lily began.

"I know that you probably want a refund or something, but please just bear with us," the manager said brusquely. Lily forgave her tone as she was evidently under a lot of stress.

"Actually I was hoping that I might help," Lily said.

"Help? Are you a cop?"

"I have done some work with the police before. I'm more of a concerned citizen with a specialist interest in crime," Lily said.

"She's being modest," Blake interjected, "this is Lily McGee, author of mystery books. I've been a witness to a number of crimes she has solved and if she's offering you her help then I suggest you take it."

The manager looked toward one of her employees who was on the phone. "They're saying that they won't be able to get anyone here for a few hours. It's the match, it has everything locked down and they have everyone available working to prevent a riot," he said. "It's the same for the ambulance."

"I appreciate the offer, but I'm really not sure that there's anything you can do. You may be a crime author, but I'm not so sure there has been a crime here," the manager said.

Lily offered a polite smile and turned back to look at Ferrabeau. The other guests were unable to take their eyes off him either, but none of them dared approach him. Lily did. She walked up to the table and looked down at him. It would never fail to strike her as tragic how quickly something as precious as life could be lost. It was chilling to think that moments ago he had been sitting in his chair eating his meal, only for his life to end without a chance for him to even say goodbye. Lily bent down and closed her eyes for a moment to offer him some respect. She almost wished she could have apologized to him for what she was about to do, for his body held valuable information that she needed. The faint smell of citrus was in the air.

She pulled the tablecloth back, exposing his face, and looked at his pupils. They were dilated. She pressed the back of her hand to his forehead and noted how warm it was, and then she picked up a fork and gently opened his mouth. She did not see any food stuck in his gullet, and she had not heard any choking sounds either. She closed his eyes and recovered him, standing again, placing the fork back on the table.

"Did anyone see what happened?" Lily asked, her words ringing around the entire restaurant.

It took a few moments, but an older gentleman who was sitting directly opposite Ferrabeau spoke up. "He took a sip of water and then he started to eat that salad. He took a few bites and I suppose he was enjoying it, but then a strange look came over his face. He looked down at the plate and his eyes widened. I think he wanted to call for help, but I don't suppose he could muster any words. Then he grabbed the tablecloth and dropped to the floor. I was about to help him, but it all happened so quickly," he said, his words faltering a little as he spoke. Lily knew the memory would haunt him for a long time.

She turned to the manager.

"It seems to me that this is the result of an allergic reaction," Lily said.

"No! No it can't be. We were careful. We knew all about his dietary requirements, right?" the manager said, twisting her head around toward the head chef. He wore a scowl.

"Of course. I wouldn't let anything like this happen in my kitchen," the chef said.

"Well it has, hasn't it!" the manager cried. She pressed her hand to her head and rubbed her temples. "Are you sure about this?" she asked Lily.

"I think under the circumstances it's reasonable to believe he died from an allergic reaction than from suddenly dying. I think you should find out what's in that salad. You need to deconstruct this and figure out what exactly it was that killed him," Lily said. "In the meantime I think that you and I need to have a word in private."

The manager nodded her head. She had light red hair that looked like strands of fire, especially when it caught the light. Her face was colored in freckles and her fingers were long and slender. She barked an order to anyone who was listening to pick up the salad and figure out what went wrong. Then she and Lily went to the back, where her office was located.

~

Lily passed the staff room that was filled with lockers, and then went into the small office. There was a small desk and a narrow filing cabinet, with a rota pinned up on the wall behind the manager. Now that she was sitting down, she allowed her head to drop and looked extremely haggard. When Lily walked in, she grabbed a sheet of paper and crammed it in a drawer.

"What's your name?" Lily asked gently.

"Melinda," she replied.

"Okay Melinda. I know that this is difficult for you and it's probably overwhelming, but we need to figure out what happened here," Lily said. Since she had spent a while working with the police she sometimes forgot how distressing it could be for a civilian to be confronted with a crime scene, and to see someone die before their eyes. There was never any easy way to deal with it and there was never anything helpful to say. The only way Lily could cope with it was by trying to find the truth behind what had happened.

"We know what happened. There was a mistake with the order. That's all."

"A mistake that cost a man his life. We need to find out what happened. And are you sure it was a mistake?" Lily said.

Melinda's gaze shot up and she paled. "Do you think someone could have meant this?"

"I often find it helpful to keep my mind open to all possibilities. You mentioned that you had a list of his dietary requirements, which means that your staff knew what would have harmed him. Do you think any of them might have had a reason to use it?"

Melinda was silent for a moment, and then she snorted. "How much do you know about the hospitality scene here?"

"I'm afraid not much. My... companion knew of the critic, but it's not something I tend to keep up with myself," Lily admitted.

Melinda nodded and pressed her lips together before she answered. "Bernard Ferrabeau is one of the harshest critics in the industry. Every person at every restaurant he visits sweats because a bad review from him can tank a place. He never holds back and he's never shy if anything is less than perfect. But if you get a good review from him... that can shoot you to stardom. Suddenly everyone wants to be there. It's why a lot of places think it's worth the risk. I guess dying is the worst review he could possibly give," she said, her voice trailing to a whisper.

"So from what I gather nobody would want to kill Mr. Ferrabeau because it would reflect poorly on the restaurant?" Lily said.

"Poorly? Someone dying from the food is a death knell for the restaurant. This is the worst thing that could happen. If I don't lose my job from this it'll be a miracle," Melinda said.

"You don't own this place?" Lily asked.

Melinda stared at her again with incredulity. "Do you know how much it costs to own and operate a restaurant? I'd have to work for a lifetime," she said, shaking her head, "Brian Samson owns this place. He's a hands off owner, at least when he wants to be. But when he wants his say, well, he makes sure he's heard."

"What do you mean by that?"

"I mean that it was his idea to have Ferrabeau come in to review the restaurant. I told him it was a bad idea, but he just wouldn't listen."

"Why would it be a bad idea, just because the risk is too high? Did you not have faith in your chef to cook a perfect meal?" Lily asked, asking probing questions to see if there was any possible rift between Melinda and her staff.

"Of course I have faith in Stephen," Melinda said indignantly. "He's the best chef I've ever met, and we're lucky to have him. I just..." a look of realization struck her face and she nodded. "Yes, you wouldn't know. I'm sorry, it's just that this kind of thing is pretty common knowledge around here. Bernard Ferrabeau used to work here, before Brian bought the place. Brian wanted to revamp things and get a younger, more modern chef in. Ferrabeau never even got a chance to prove himself. He was just ousted and then he decided to take a break from being a chef and became a critic instead. I can't blame him really, he probably got paid the same amount for a lot less work."

"Except that it ended up killing him," Lily reminded her.

Melinda's tone softened.

"Why would your owner invite Mr. Ferrabeau back to review this restaurant if there was such bad blood between the two of them?" Lily asked.

"That's exactly what I said!" Melinda clapped her hands together and leaned back in her chair. "It's risky enough for a normal restaurant, let alone for one that Bernard already hates... hated," Melinda corrected herself and shook her head. "I begged Brian not to do this because I knew it would only end badly for us. I'm sure Bernard had his professional integrity and stuff, but how could he possibly give us a good review when so much has happened? Anyway, Brian told me not to worry and that he believed in Bernard's professionalism. He thought that if the meal was good enough then Bernard would not in good conscience be able to lie about it, and if we of all places got a good review from Bernard then it would be great publicity and nobody would ever be able to deny the quality of our service. I don't know how I'm going to tell him about this."

Lily thought on the matter for a moment. It certainly seemed as though nobody would have wanted this place to fail, and yet a man was dead nonetheless. Even if it was a mistake somebody was responsible for it, the only question was who? She filed Brian Samson away for later. Since he wasn't present it was unlikely that he had anything to do with the killing, although she could not rule out his involvement from afar as he had arranged for Ferrabeau to come to the restaurant in the first place. If he had not been asked here, then he would not have died.

"I will let you get on with that, and give you a moment to collect yourself. I know how distressing this can be. And I hope you understand that if I ask you difficult questions it's only because I have to. I have another one that you may not like answering, but who could have been responsible for placing an allergen in his food?" Lily asked.

"You mean if it wasn't an honest mistake?" Melinda said with a hint of defensiveness in her voice. Lily found it admirable just how far Melinda was willing to go to defend her staff. But how far would she go if one of them was indeed guilty? Lily inclined her head anyway, accepting Melinda's point.

"Well there's Stephen of course, and Amelie, but really the kitchen is so busy and chaotic that anyone in there might have had the opportunity," Melinda said.

"Thank you," Lily said, rising from the chair. It wasn't entirely helpful, but she would get to work narrowing down the possibilities.

"There is one other person who might have done it," Melinda said as though she had been struck by a sudden flash of inspiration. Lily sat back down to listen. "There was a man I recently had to fire; Ash Bailey. He was our sous-chef, but I discovered he was secretly working for the competition. He was sharing our secrets, and once I found out I sacked him on the spot. He denied it all of course, but he had no leg to stand on. He might have found some way to sabotage the review," Melinda said.

"I'll certainly look into it," Lily replied, although she thought this was unlikely. Although the kitchen was chaotic she imagined it would have been difficult for anyone to sneak in, especially someone who had been fired and clearly wasn't welcome. Lily left the office and returned to the main restaurant. Guests had risen from their seats and were mingling together, murmuring their theories and discomfort. People were giving the dead body a wide berth. Betsy was sitting on a chair, trembling. She looked ghostly pale. A glass of water sat beside her. Perhaps she really was as young as she looked, Lily thought to herself.

"Any luck in there?" Blake asked.

"It seems as though Ferrabeau had a grudge against this restaurant, so I find it odd that he would come back here, but perhaps he just wanted to put the final nail in the coffin."

"Little did he know that the coffin would be his own," Blake remarked wryly. "Do you really think we're talking about murder here, or could it just be an accident?"

Lily sighed and placed her hands on her hips. "I'm not quite sure yet. I'm going to have to ask some more questions, but if this was a mistake then it's the biggest mistake of someone's life. It has already cost a man his life, and it might well cost these people their jobs."

"Is there anything I can do to help?"

"I would say just sit back and enjoy your steak, but I suppose you've lost your appetite," Lily said. Blake nodded and looked sad. "The best thing you can do is try and keep people calm. If they begin to dwell on what's happened they're going to start coming up with all kinds of wild theories and they might start to panic. Until the police get here we're the closest thing to them, and we have to try and keep things under control," Lily said.

"I understand. You go do your thing," he said, and squeezed her arm reassuringly. Lily smiled. It was nice to know that she had Blake to count on, and she wondered what it would be like to know that he was there every day of her life. There was time to think about that later though. Right now she had a job to do.

Lily began by approaching the chef.

"Stephen? May I have a word with you?" she asked. Stephen was a man of below average height. He had taken off his chef's hat, revealing a layer of fuzzy black hair that was cut as short as it could be without disappearing entirely from his head. His face was round and lined with deep creases, while his eyes were so dark they were black.

"I don't really have the time right now. A man has died and I'm trying to keep this kitchen together," he snapped.

"I understand that, and I understand it must be distressing for you, but I think it would help everyone if we tried to figure out what happened. May I speak with you in private?" Lily spoke in her quietest, gentlest voice, hoping to disarm Stephen while also showing him that she was not going to be deterred. He seemed to sense this and muttered under his breath. He walked out from behind the kitchen counter and led Lily out the back door of the restaurant. There was a small alley area. A bucket filled with cigarette butts was by the door, and beyond them were the trash bins. The stale smell of food waste lingered in the air.

"Alright, what do you want to know?" Stephen asked, standing with his legs wide and his arms folded across his chest.

"I just want to talk about what happened."

"What do you mean? You know what happened, you were there with me!" he said, spreading his arm out as he gestured back toward the restaurant.

"I understand you're distressed at the moment, but getting agitated is not going to help," Lily said.

Stephen rolled his eyes and gnawed his lips. He began pacing back and forth. "Well nothing is going to help. Getting agitated is about the only thing I can do! You don't get it, do you? A man died and I'm head chef. I'm going to get crucified for this. I'm never going to work again! That's it, my career is over. All my training, all my experience… it's all for nothing. They wouldn't even hire me as a chef in prison after this. Do you get what this means? I'm going to have nothing! It's not like I can do anything else. I can't just march back into college and start taking another course. I put everything to this."

"Mr. Ferrabeau has less than nothing, and the least I can do for him is to try and find out what happened," Lily said.

Stephen stopped pacing and took a deep breath. He reached into his pocket and pulled out a cigarette. He lit his own and then, as an afterthought, he offered Lily one. She declined. The end of the cigarette was a red glowing ember as he inhaled, then he tilted his head back and exhaled a plume of smoke that rose into the air as a wispy cloud, and kept climbing higher and higher until it dissipated entirely. He seemed calmer after this.

"I know that. I get that, and I'm sympathetic about what happened, it's just a shock, you know? I have to think about what's happening next. Of course I'm devastated that it happened. I didn't want this. Nobody did."

"Because it's going to tank the restaurant?" Lily asked.

"Exactly. This whole thing was supposed to put us on the map. It was all Brian's idea. imagine the headlines!" he said, Stephen shook his head and took another drag of the cigarette.

"Melinda didn't seem to think that it was the best idea. Do you agree?"

Stephen shrugged. "It doesn't matter what I think. It's Brian's place, and it's his money. He can do what he wants with it."

"Did you have any kind of rivalry with Bernard Ferrabeau?"

"Rivalry?" Stephen furrowed his brow for a moment before he realized what Lily was getting at. "Right, yeah, in case I poisoned him. Look, of course we were rivals. I was the one who got his job after all. I never had anything but respect for him. I used to study his recipes all the time. In fact I was a little shocked when I was told

506

that I would be replacing him instead of working with him. I was looking forward to learning from him. But hey, that's the business, right? All I wanted tonight was for him to like my food. That's all any chef cares about; respect."

Lily nodded. What he said seemed to make sense. "Melinda suggested to me that it might have been an ex-employee here, your old sous-chef Ash Bailey? She said there was some kind of dispute after he was sacked and that he might have wanted to take revenge."

"She said that did she?" Stephen took another drag of the cigarette, this time flicking ash down into the bucket. "I guess it's possible. He was angry when he left."

"Have you seen him tonight? It must have been difficult to come back in here without being noticed."

"Things were so busy I wouldn't have noticed if he slipped in. I'm usually concentrating on the food. And I can tell you exactly what I was making if anyone asks, our salmon with our famed lemon drizzle. Anyone who moves around me is just a body," he said.

Lily nodded. "Now, I'm going to ask you a question that's going to make you angry. This isn't intended as anything personal, it's just a question that I have to ask in order to reach the truth, okay?" Lily asked, preparing him as best she could, although given the way he was acting she doubted it would do much to dilute the fire of his anger. He nodded anyway, and looked at her with narrow eyes. "Do you think you were under so much pressure to cook the perfect meal for Mr. Ferrabeau, a man you respected greatly, that you might have made a mistake?" she asked.

Stephen dropped the final, dying piece of the cigarette into the bucket and stared at her. "I'm going to pretend you didn't ask me that," he said. "Look, if you want me to be candid with you then there is one person you want to speak to."

"Who is that?" Lily asked.

"I hate doing this, right? We're all supposed to be a team, a family, and I'm not one to throw my family under the bus. But this is, well, this is serious business. He was eating the salad, right? Well Amelie was the one preparing the salad, and if there's anyone here who might have had a motive to hurt this place then it's her."

"Why is that?" Lily asked.

"Because she thinks she's better than she is. It's all because of some article that named her as one of the hottest chefs in the country. She's getting too big for her boots and she's upset that she hasn't taken over from me yet. She's young and she doesn't understand patience. But if she was so inclined then she had the opportunity to send this place to the ground." He then walked back into the restaurant. Lily hadn't really expected him to answer any differently.

4

Lily returned to the restaurant and as she did she heard the sound of a woman moaning. It was a guest. She was doubled over, holding her stomach.

"I feel so ill. I think I might have been poisoned too!" she cried. She was helped to her seat and given a glass of water. Blake was beside her and gave Lily an anguished look. The staff all looked worried that a second person might lose their life. Lily walked up to the woman and pressed her hand against her forehead. There was no noticeable difference in her temperature.

"I think you might be overwhelmed by the occasion. It's natural for people's minds to work in overdrive right now. I know it's hard for everyone, but I'm sure that if we just try and remain calm the police will be here soon. At the moment there's no reason for anyone to believe they have been poisoned, other than Mr. Ferrabeau. And even then it may have just been a freak accident," she said this not only to set people at ease, but also to give the perpetrator a sense of arrogance, in the hope that they might make a mistake.

The woman nodded and took a sip of water. She pulled her hand away from her stomach and seemed a little more comfortable. Lily turned to the staff, who were mostly huddled behind the bar with stricken looks on their faces. Drinks had been poured, but nobody seemed in the mood to taste any of them. She asked after Amelie, who was apparently in the restroom. While Lily waited for Amelie, Lily asked them if any of them had seen Ash.

"I don't think so, but if anyone knows what's going on with him it'll be Paul," the barman said, nodding to the rear of the kitchen that was hidden from view. Lily walked back there and saw a sink that was piled high with dishes. A tall, young, dark skinned man leaned against a wall, his eyes glued to his phone.

"Excuse me, are you Paul?" she asked.

He looked up and gave her a cheery grin. "I am!"

"Could you tell me if you've seen Ash Bailey around here recently?" Lily asked.

"Ash?" Paul laughed. "Ash wouldn't come around here after what happened. Besides, he's busy tonight," he said.

"And how would you know that?" Lily asked.

Paul turned his gaze back to his phone and scrolled through, before turning it to Lily. It was a picture of a man at the soccer event that was taking place elsewhere in the city, the event that was drawing attention away from everything, including a murder. "That's him. And look, you can see the time on the scoreboard," he said, drawing his fingers across the screen to zoom in on the photograph. Indeed, there was no doubt that the picture had been taken tonight, at around the time when Bernard Ferrabeau had been killed. It did not preclude the possibility that Ash could have snuck in earlier, but she doubted any food would have been left out long enough to be served to the food critic, and surely Ash would have had no guarantees that it would have been used anyway. No, whoever had added the allergen must have been in the restaurant, and Lily wondered if it was the person she had not spoken with yet.

Lily thanked Paul for his time. He smiled again and went back to amusing himself with his phone. He didn't seem disturbed by what had happened, but people dealt with this kind of thing differently, especially young men. She noticed that they tended to hold the opinion that nothing like this was ever going to happen to them, only to other people, so they had an air of invulnerability. Lily used to feel the same thing until she came to Didlington St. Wilfrid. Now she knew that it could happen to anyone, anywhere, at any time.

When Lily returned inside, Amelie still hadn't come back from the restroom. This made Lily wonder if anything suspicious was going on. Amelie had not used the restrooms meant for the guests, but instead had gone back to the locker room that Lily had passed earlier. As Lily made her way to the locker room she heard Melinda's voice behind the closed door of the manager's office. Lily pressed her ear against the door momentarily, but the words were muffled. The one thing Lily did notice was that Melinda's voice was raised. It must have been a terse conversation. She could not imagine how difficult it would be for Melinda to tell the owner that the famed food critic had died. Although Lily did not know Brian Samson, the impression she got from Melinda and Stephen is that he wanted things done a certain way, his way, and men like that usually did not react at all well when things spun out of their control.

There was nothing that Lily could do about that though. The only thing she could do was try and solve this murder, although as she entered the locker room she thought about Blake and felt a twinge of guilt. This was supposed to be an evening where they would get to know each other and learn about each other, but a murder had gotten in the way. Was this going to be Lily's life from now on? Was she going to drop everything when someone died? It wasn't fair to Blake, and she began to wonder if this was a mistake. Should she pursue a romance with him if she could not promise herself to him fully? Was she instead married to murder? It was a chilling thought and she wasn't exactly sure how to process it.

The locker room was stuffy and hot. Jackets were hung up on hooks and the lockers were grim, the paintwork having been reduced to flecks. Amelie was sitting on a bench in the middle of the room. She was on the phone, speaking in frantic French, which Lily did not understand. As soon as Lily entered, Amelie finished her conversation and put the phone down.

"Can I help you?" she asked. Her English was fluent, but it was decorated with a French accent that was pleasing to the ear. Amelie was a slim woman with arching eyebrows and short brown hair. Lily imagined she would have been gorgeous with longer hair, but probably kept it short on account of her job.

"Actually you can. I'm just speaking with everyone about the events of the night."

"Ah yes, the author, perhaps you are researching for a new book, yes?" Amelie said, a flutter of laughter followed her words. "Although if the man you are with is any indication then your next book must be a romance." She wore a suggestive look on her face and smiled wickedly, evidently approving of Lily's choice in men. Lily cleared her throat, unsure how to respond to that.

"I'm just trying to find out what happened before the police arrive. It'll make their job easier, and it will mean that everyone can go home a little earlier."

"That will make a change," Amelie said, stretching her arms out and yawning.

"You often work late?"

Amelie chuckled. "It is the nature of the beast. We rise, we work, we sleep, we rise again. Anyone who works in a kitchen knows that they do not get a social life."

"That sounds depressing."

"We are all slaves to our passions, it just happens that mine is cooking."

"Indeed. And it seems as though you're rather gifted when it comes to cooking as well. I heard that you were mentioned in some kind of article recently," Lily said.

Amelie's face lit up and she beamed widely, but this radiant smile only lasted for a moment before it faded again. "Yes, yes. I am one of the hottest chefs to watch out for in the country, apparently, but what am I to do? Stephen is here and he is not going anywhere. I have to wait. Wait and wait that is all we do in life."

"And do you have a problem with waiting?"

"I have a problem with waiting when I am ready for things. I do not think it's fair that I should still be making these salads. But the industry is, you know, men still do not like taking orders from women. I have tried for other jobs, but they say I need experience of running my own kitchen, so I tell them how can I get experience without having my own kitchen, and they shrug and say there is nothing they can do. It is filled with old men who cling to their jobs, but one day I will be able to shine," Amelie said.

Lily wondered if she was trying to make that day come quicker than it perhaps might. "I suppose if there was a scandal here the head chef would be the one responsible, and you would be first in line for the position," Lily said, trying to mask it as an offhand comment, but Amelie was not so naïve. The warm look fell from her face and her gaze narrowed.

"Are you implying that I would sabotage my head chef?"

"I'm implying that you might be willing to do whatever it takes to get ahead in your career, to get what you've earned and you've deserved."

"You think I would betray my professionalism? You think I would throw away all I have worked for? I take pride in my work. Yes, I do not like making the salads, but it is what I must do, so I make the salads the best I can make them. When Bernard Ferrabeau ordered the salad I put my all into it because if this restaurant becomes better then I become better and we all win. It does me no good to kill a man. You will find nothing like that here," she said, and rose, indicating that the conversation was over.

5

L ily left the locker room and heard that Melinda was still on the phone. Lily knocked on the door and to her surprise it swung open. Melinda was sitting at her desk and looked strained. Her voice was terse.

"Come on Ryan, just hold the story for a little while. Even the cops haven't been here yet! You can't print anything until you've had an official statement. How did you even hear about this?" she asked. Lily couldn't hear what the other person was saying on the other end of the phone, although she gathered that it was a reporter. Melinda then noticed that Lily was standing there and ended the call. She glared at Lily without saying anything and then pushed past her, returning to the restaurant. Before she did so, however, she tossed her gaze back toward the desk, and Lily wondered if there was any reason why she was looking there in particular.

"Right, which one of you has been talking about this?" Melinda yelled, glaring at her staff and the guests in turn. Lily had noticed that plenty of them had been on their cell phones this whole time. She would have thought that they would have the discretion to not talk about this, but apparently it was too much for people to keep to themselves. Melinda placed her hands on her hips and continued glaring at them. Eventually a few hands rose, and then a few more. Gossip was rife, and word had already made its way to a reporter.

Melinda threw her hands up, rolled her eyes, and shook her head. She let out a nervous, humorless laugh and then stared at the dead body that was still lying in the middle of the restaurant.

"Great, this is just great. Thank you all for making this wonderful night even better," she said, clearly cracking under the pressure. Lily approached her, ready to try and calm her down, but Melinda was already back in the office. Lily walked up to Blake and gave him a sympathetic smile. She leaned into him and his hand slipped around her back. He radiated warmth.

"I'm really sorry that this has happened," she said. "I know this night was supposed to be special."

"It still is. I get to see the great Lily McGee in action once again," Blake smiled broadly, although she wondered how genuine he was being. Surely he couldn't be happy with this arrangement?

"I know, but I don't *have* to do this. If we wanted to we could just walk outside and return to the village and forget this ever happened. This doesn't have to be my problem," she said.

"But would you be able to live with yourself if you didn't? I've come to know you quite well since you arrived here Lily, and I know that you're compelled to help people. It's one of the things that I find most interesting about you. Ordinary people would walk away and forget about it, but not you. It's not who you are, and I for one like that about you."

But will you still like it in a year's time? In ten years' time? Will you still like it when we're old and grey and we're having a cup of tea, when suddenly someone rings at the door and says there's been a murder? Lily's thoughts ran through her mind, although she did not give voice to them yet. Instead she simply smiled at him and was grateful for the support she was getting from him. She supposed that just as she could have walked away, so too Blake could have walked away from her, but he wasn't. There was something that was keeping him around, and she would be a fool if she denied that that wasn't special.

It was at this point that Amelia burst out of the locker room in a rage. Where she had been relaxed and almost playful when speaking with Lily, she was now tense and furious. Her face was flushed red and she held a hand out in front of her, pointing directly at Stephen.

"You... you accused me! You really think I could have done this? You really think I could have killed him? Have you no respect?!" Amelie shrieked.

Stephen glared at Lily. "I never said that. I just said that you were preparing the salad he ate. That's all."

"Ah yes, the salad. I know what I am doing with the salad. Do you really think I would make a mistake? I am not like you. I am still trying to make something of myself."

"What's that supposed to mean?" Stephen replied, frowning deeply. His voice raised too.

"You know what it means. I don't see you in any lists of the hottest chefs. You are going to be forgotten before too long."

"And you're flame is going to cool, believe me. You think you're so hot? They make those lists every year, and you wouldn't be anything without me. I'm the one who has mentored you. I'm the one who has given you this opportunity and the best thing you can do is listen to me."

"Listen? Listen to you moan and complain and keep me on salads? When are you going to give me a chance to show what I can do? You never let me cook anything complicated. You keep it all to yourself because you are afraid that I will do better than you. You are all afraid of the future, of revolution!" Amelie cried. Melinda heard the commotion and came out again. Lily was stuck in the middle of it all, forced to watch the scene unfold. Tensions were high and it was hardly professional to display these raw emotions in front of the guests, but everyone was so frayed they didn't seem to care.

"What's going on out here? What is this about?" Melinda asked.

"Oh Amelie is just losing her rag again," he spat and rolled his eyes. "It's the same old story. You know what Amelie? You just need to learn to pay your dues like the rest of us."

"You mean like you did when Bernard was fired? You could not jump into his whites quick enough. As soon as the possibility of the job came up you were in there, snipping and snapping and you never gave him a chance to fight you off. You told Brian that the only way was the future, and that Bernard was the past. Maybe I will tell him the same thing. If he listened once he can listen again," Amelie said.

Stephen's nostrils flared and his eyes bulged. He did not get a chance to say what was on his mind though, as Melinda stepped in between the two of them.

"You two need to cut this out now!" she yelled. "I know it's a stressful time, but we have to remember that there are still guests in the building. Let's keep this all in house. Go and cool off, both of you. Stephen, I want to see you in my office," Melinda said. Stephen wore a hangdog expression and followed her into the back, while Amelie pulled out a packet of cigarettes, and Lily watched her go outside for a smoke. There were murmurs and mutterings between the people left in the restaurant. It was unusual for the guests to get such an insight into the goings on of a business, but death often brought out the most intense emotions. Blake moved around the guests with his usual diplomatic charm and tried to keep them occupied, asking them about their days and their lives, trying to keep their minds on anything other than what had occurred. Things were so bleak and everyone had been so distracted that none of the tables had been cleaned, and the half eaten meals were just sitting there, waiting to be thrown away.

Lily took the opportunity to ask some questions of the other staff, although they didn't have much information to offer. They were as shocked as anyone else. They seemed divided about whether it could have been an accident or not. They confirmed that Amelie had indeed been working on the salad, so it seemed as though she was the prime suspect. Lily just didn't think she had a motive though, unless she was putting on a convincing act. She certainly had an axe to grind against Stephen, but then what if Stephen was trying to implicate her?

But then something curious came her way. The barman, Connor, was a striking man with a flaming beard and a puff of ginger hair. He told Lily that there was someone else who had a problem with Bernard Ferrabeau, and it wasn't the person who Lily would have expected. It was the person who served his food; Betsy.

6

Lily approached Betsy quietly. She was sitting where she had been after the death had occurred; at a table in the corner, where the shadows shrouded her. It was in stark contrast to how Betsy had greeted her and Blake when they had entered. Then Betsy had had a winning smile, but now mere hours later she was distraught. Her eyes were puffy, her cheeks flushed, and her skin was wan. Her hands trembled. Lily noticed that there was less water in the glass beside Betsy than there had been previously, which showed she was drinking at least. Tears glistened in her eyes. Her white blouse clung tightly to her figure, while her straight black hair framed her round face.

"May I talk to you?" Lily asked.

Betsy nodded and sniffed. She pushed a few errant strands of hair away from her face, although they swung back like a loose curtain almost immediately.

"Are you okay?" Lily asked.

Betsy shook her head. "I'm worried."

"What are you worried about?"

"My job. This place. Everything."

"It's okay," Lily said, smiling gently. "It's all going to be okay. It's natural to be worried. Everyone else here is worried too. We're all in the same boat. You're not alone." Lily gestured to the other people in the restaurant as she said this, although Betsy looked as though she was lost.

"It's just that Brian is going to be so upset," she said.

Lily's ears pricked up at this. It was one thing for the manager and the head chef to refer to the owner by their first name, but was it natural for Betsy to do so?

"I'm sure whatever happens he will be okay too. Would you like to tell me what you know?" Lily asked.

"I don't know anything. Just that one minute I was taking his order and the next he was dead and I just... I... I don't know what could have happened. He was there and then he wasn't and how can it just happen like that? How can it all just end?" she asked, her words on the precipice of becoming sobs.

"I know it's difficult, but we have to try and focus on the things we can do to make a difference. Unfortunately for us life goes on, and right now we have to find out what happened to Mr. Ferrabeau. I know it's difficult, but is there anything else you can tell me?"

"I don't... I don't think so," Betsy said, shaking her head.

"Okay," Lily replied gently. She would come back to that later. "As you know I've been asking people questions about their relationship with Mr. Ferrabeau. I was just informed that you might have something against him?"

"Me?!" Betsy asked, her voice coming out in a squeak. "No... I mean... that was so long ago!"

"What was long ago?" Lily asked.

Betsy reached out for her glass and held it with trembling hands. She brought it to her lips and took a sip. A pink tongue nipped out to lick the remnants of the liquid away. She placed the glass back down on the table, taking a deep breath as she did so.

"It happened just after I first started here, before Brian bought the place. I was only fifteen. I had a Saturday job waitressing. I thought I was doing well, but the chef... Bernard... he got angry at me whenever I made a mistake. He would berate me in front of everyone and he said that I should stick to a paper route or something. He wanted to fire me, but the other owner took pity on me. Then a couple of years later Brian bought it and Bernard thought this was his chance to finally get rid of me. He told Brian that I should go, and then Brian talked to me and he... he showed me kindness," Betsy's eyes lit up when she spoke about Brian, and Lily suspected that something else was going on there. "He said that I would never have to worry about losing my job while he was in charge, and he said that I wouldn't have to worry about Bernard either."

"So Bernard was out and Stephen was in," Lily said.

Betsy nodded.

"But I never hated Bernard for it," she added quickly, as though she knew that this line of questioning had put her under suspicion. "I know he was just doing his job and it was because his standards were perfect. To be honest," she added with a wry smile, "I wasn't really very good at my job then. I was so nervous and I did make mistakes, and I wasn't strong enough to carry many plates. But I got better. I just needed a little bit of time. Brian gave me that time," she said, again with adoration in her voice.

"So when you knew that Bernard was going to come here you didn't want to get revenge for the way he made you feel a few years ago?"

Betsy shook her head and looked adamant. Now that she had been talking a little more, the tears had subsided. "I'm not like that. I would never be like that. I have to admit I was a little nervous when

I found out that he was going to be here. It wasn't what I expected when I turned up for work today."

"You mean you didn't know?"

"No, and I hated Brian for keeping it a secret from me. Only he, Melinda, and Stephen knew. The rest of us were told when we came in. It's something to do with the integrity of the review. Anyway, I was a little nervous and I really hoped that I didn't do anything to jeopardize things. I wanted this to work out for Brian. He's been working so hard to make this place a success. He knew that if Bernard gave us a favorable review then nothing could stop us, and he could finally do all the things he wanted to do."

"What kind of things?" Lily asked.

"Well, there was a refurbishment he was talking about doing, and he definitely wanted to do more publicity and more events here to raise the profile, and then there are the staff costs."

"What staff costs?"

"Oh, just pay raises, you know. Apparently Stephen has been angling for one for a long time. Brian says there's no room in the budget though, and he wouldn't lie about that. He wants the best for the restaurant. We just need to make more money before he can spend more money," Betsy said. Lily tried to hide her surprise, but inside her heart flared with suspicion. Greed was one of the oldest motivations there was, and if Stephen had been angry that he didn't get a pay rise then that certainly gave him reason to want to send a message to Brian Samson.

"I see, well, that is all very helpful, and interesting. I always like learning about how things work. Do you think you might be able to tell me now what happened when you took Mr. Ferrabeau's order? Anything would be useful, no matter how insignificant it may seem," Lily said.

Betsy nodded. She had calmed down a lot so her mind must have been clearer. She tilted her head back and raised her finger to her chin, before glancing back toward the counter. She retraced her steps.

"I greeted him like anyone else. I thought about mentioning my name, but he didn't react when he saw me so I don't even know if he remembered me. I showed him to his table and I gave him the special menu we had prepared for him. I remember asking him what wine he wanted. Melinda had gone through what to say beforehand. It was like being an actress. Anyway, after that I went back to the bar and I made sure to make another note about the allergens he could not have."

"Did you need to do that even though you had a special menu?" Lily asked.

"We always do it. It's good practice, so I saw no reason to break the habit. And Melinda said that Bernard would not only be reviewing the food, but also all of us, so we had to make sure that we took care of the tiniest detail. I wasn't about to let that ruin our chances," Betsy said.

Hmm, Lily thought, so that meant the chefs were doubly aware of Bernard's allergies. It seemed unlikely that they would make a mistake if they were told to be this careful. There was no doubt in her mind now that this was an intentional murder.

"And what happened then?"

"The order came to be fetched. I don't know if you heard or not, but we have a bell the chefs ring when an order is ready to be served."

"Who handed it to you?" Lily asked. Betsy did not pick up on the urgency of her voice. This was a key question, because the chef who touched the dish last was most likely the one who had tainted it.

"It was just sitting at the counter, waiting for me. Nobody handed it to me," Betsy said. A ball of frustration pulsed inside Lily's chest. There was going to have to be another way to the truth.

"After that I took the order to Bernard and wished him a good meal, and then I walked away. The next thing I knew…" her voice slowed and she was facing the horror of death again. Before she could dwell on the death Lily interrupted her thoughts.

"Thank you so much Betsy. You've been incredibly helpful. I know that this is going to be difficult for you, but why don't you go and stand with your friends? Everyone else here has been shaken by what's been going on and I'm sure it will help to be around people," Lily added a smile to her words. Betsy returned the gesture and nodded, walking around toward the bar. Lily scanned the restaurant and had a pensive look on her face.

Blake noticed and came toward her. "Is everything okay?" he asked.

"It was definitely murder," she spoke in a low whisper. "And I'm pretty sure it was one of the two chefs. I just have to figure out which one it is."

7

As Lily stood in the middle of the restaurant, Melinda and Stephen returned from their private chat. Lily watched the head chef closely, noticing that he seemed more relaxed. Was he the true culprit; killing Bernard Ferrabeau to send a message to Brian Samson about pay, or was he the target, the one being framed by an ambitious rival who was getting tired of waiting her turn?

"Lily," Melinda said in a soft voice, clasping her hands together. "Could I have a word with you?" she pulled Lily aside. "I know that things have been tense around here and I'm so sorry that you've been involved, but I really can't ask you to stay any longer. You shouldn't have your evening interrupted like this. I really think it's best if we just wait for the police. I'm sure it won't be long before they're on their way."

"I appreciate that, but I am making headway here. I think it would be for the best if I continued."

"But you're not even a police officer. I can't ask you to give over more of your time to this. It's not fair to you and your friend. I'm sure you have better things to do than wait around here and sort out this morbid affair," Melinda said. Lily noticed a tone of anxiety in her voice. Perhaps it was just borne from the strain that she was under because of the events here tonight, but there might have been something else to it. Either way, Lily McGee never reacted well to being told that she could not do something.

"Actually I don't mind at all. I believe if you can lend your skills to something then you are obligated to, and I really don't think it would be best for me to leave now."

Melinda wore a forced smile and continued speaking in a low voice. "This is private property. I would be within my rights to ask you to leave, if it came to that. I really only have your best interests in mind. It can't be healthy to spend so much of your time investigating murders like this."

"What's going on here?" Blake asked, striding to Lily's side before she could answer. She looked up at Blake with a wry smile.

"Well Blake, it appears as though they would like me to leave the premises," Lily said.

"And why would that be? It seems to me as though Lily is actually managing to get some answers here," Blake said. Lily suspected that was the exact reason why Melinda was asking her to leave. She remembered what Stephen had said about the kitchen being like a team, a family, and they would protect each other. Maybe the whole restaurant was like that and Melinda was just trying to protect them. They might have thought the police would be less tenacious.

"Of course not, it's just that we know that you've done a lot here and we really can't ask for more. You're clearly here for a special occasion and I would never forgive myself if this got in the way of that," Melinda said, pressing a hand to her heart in a most insincere gesture.

"Nothing can stand in the way when a murder has been committed," Blake said. "But if you are insistent then perhaps I could speak to the owner of this place and we can see if he appreciates Lily's help. I might even be able to talk to him about some investment opportunities myself. I've always wanted to own some shares of a restaurant. Perhaps if I provided a cash injection Lily might be permitted to stay," Blake said as he pulled out his phone and held it open expectantly, waiting for Melinda to provide him with Brian Samson's number. Melinda hesitated though, clearly weighing whether it was worth the risk. It did not seem to be.

She forced a smile. "I'm so sorry, of course you can stay for however long you like. We really do appreciate the help you've been giving us," Melinda said.

Lily gave a haughty smile and returned to the bar. She stood there for a moment, trying to put herself in the moment when it all happened. She cursed herself a little for having been engrossed in conversation with Blake, for usually she was observant of everything around her. With Blake it was difficult though. Even now she found herself wanting to look toward him, to gaze at his smile and lose herself in his eyes. She took a deep breath and focused herself. She imagined how the process went. Betsy would have taken the order. Amelie would have prepared the order. She worked with her back to the restaurant so nobody would have been able to see what she was doing. People would have passed her by, already too focused on what they were doing. As Stephen said, other people were just bodies amid the chaos of the kitchen. The bar staff was busy pouring drinks. Melinda was moving around the place. Amelie would have placed the salad on the counter and rung the bell, ready for Betsy to collect it. Betsy was a little delayed, for she did not take the plate right away. That left a brief moment where the dish was open for anyone to taint the dish, but it couldn't have just been anyone. It must have been someone who knew what they were doing, who knew how to hide their tracks.

Her gaze rose to Stephen. There was still an order left that was ready to serve when Bernard Ferrabeau had died. It was positioned just to Stephen's left. He would have had the opportunity, but he said he was already making another dish. Lily furrowed her brow, thinking hard about what might have happened. She thought about all she had learned, discarding the useless information as it flashed in her mind, and only keeping that which was pertinent. She opened her eyes and they darted toward the dead body again. Was there a clue that she had missed? It required further examination.

She returned to the middle of the room and bent down. She pulled away the tablecloth and looked at the body, but the expression was unchanged. If there was an answer here then it would be around the body. The salad had been tidied up, but there was still some residue on the floor that would be left there until the carpet was cleaned properly. Lily bent down and breathed in hard, trying to pick up every scent she possibly could. The faint smell of lemon drifted into her nostrils. Something gnawed at the back of her mind. Beside the body was the short, specialized menu that Bernard Ferrabeau had been served. Lily glanced at it. He had the salad to start, and then he had ticked off the steak. For dessert there was a chocolate tarte. Nowhere on the menu was anything with lemon.

Lily rose and walked to the kitchen. Stephen eyed her suspiciously, while Amelie stood with her hands folded and leaned against her workstation.

"I don't know what you are hoping to find here," Amelie muttered. Lily didn't respond. They may have been chefs and this kitchen may well have been their domain, but it was also a murder scene, and that meant she would be the one to rule here. She went to Stephen's station and looked out at the restaurant. She had a direct view of Bernard's table. The dish would have been placed to the left of her.

"What was it you said you were doing while Bernard was eating his salad?" Lily asked in a quiet, innocuous voice.

"I was cooking," Stephen muttered.

"What were you cooking?" she asked.

"The salmon."

"The salmon with lemon drizzle," Lily said. "Amelie, did the salad have any citrus ingredients at all?"

"No," Amelie said, standing up straight. "That is our secret sauce. We were not going to use that with him. It had one of the ingredients he was allergic too."

"Interesting," Lily said, "because when I go to the body I get the distinct hint of lemon. I posit that you," she turned to Stephen, "doused his salad while it was waiting to be collected. You were already working with it. Nobody would have noticed what you were doing because it would have only taken an instant. The only question is why. Were you trying to frame Amelie, or were you trying to get back at Brian Samson for not giving you a pay raise?" Lily said.

Stephen's face twitched with disdain. He muttered and denied it, but his grumblings were only a delaying tactic and it was clear that he had been caught. His gaze darted everywhere, as most peoples' did when they were caught dead to rights. As it happened though, he had one last trick up his sleeve.

"Right, this has gone on long enough," Stephen said. "You're not with the police and I don't have to answer anything that you ask me. I think it's time that you leave. Melinda was too polite before. You're not welcome here," he said, and clamped his strong hands around Lily's bare arms. She cried out in pain as she tried to wrest herself free, but Stephen was stocky and he was already dragging her away. Amelie was shocked and cried out, while Melinda barked at Stephen. He was determined though, and took her toward the back alley where the smoker's were, hauling her through the door as though she were a bag of trash.

"I've had enough of your questions. This is my kitchen and I say what goes," he said as he threw her out. Lily staggered and spread out her palms, hitting them against the wall. Pain stung them and it reverberated up her arms. She turned to see Stephen about to slam the door, when suddenly a powerful hand grabbed the chef's arm and yanked him back inside. Panic flooded Stephen's face as he flailed around and turned, trying to attack his assailant. Blake stood there unfazed though. He blocked Stephen's blow and hit back with one of his own, jabbing Stephen in the stomach. Stephen folded like a sack of potatoes, groaning.

"Is this true? You killed a man for *money*?!" Amelie exclaimed, shaking her head profusely. The disdain on her face showed that she was a true chef, an artiste, and what little respect she had for Stephen had plummeted. Stephen scowled, shrugging off Blake's grip on his shoulder. Blake's face was like thunder and he loomed over the chef like a specter of death. Lily balanced herself and rubbed her palms together.

"It's not about money, it's about respect. I wanted to get paid what I was worth, and I'm not the only one," he said. As soon as he said this Melinda stepped forward.

"I can't believe this Stephen. I will not have this conduct in my restaurant. Let's get you to the back. I don't want to hear any more of your lies. You'll say anything to bring this restaurant down, and after we worked so hard to build it up! This place will go on without you, mark my words. We're going to survive this," she said, and tried to drag him away. Stephen scowled again, but before he could speak Melinda grabbed a bread roll and stuffed it in his mouth. "I don't want to hear anything from you! Let's get some tape and gag him before he can poison the air with his vicious lies," Melinda yelled.

Lily frowned. She walked back into the kitchen and went past everyone who was standing there. The guests all looked shocked at what was happening. Lily strode toward the office and opened the door. Melinda had been so busy with Stephen that she hadn't realized what Lily was doing at first. She soon did though, and was at the door looking like a wraith. Lily was already in the desk though, searching through the drawer. She had a sneaking suspicion that Melinda had hidden something important there, perhaps something incriminating. The desk was empty though.

"You don't need to be here Lily. We got the culprit!" she said.

Lily frowned. Everything pointed to the fact that something was hidden here; Melinda's general demeanor, her panic when Lily had first gone into her office, the glance toward the drawer when Melinda had left. Instead of looking for whatever it was that Melinda wanted to hide, Lily instead looked at the manager. She followed Melinda's gaze. Often people revealed things that they would rather have kept hidden without realizing it. The look went to the trash. Lily dived down and saw a piece of paper that looked freshly placed.

"No!" Melinda yelled, diving into the room. Lily turned her back and managed to defend the paper from Melinda. Melinda's arm were like a spider's legs, long and reaching, her fingers inching toward the paper. Lily tried to protect it as best she could, but Melinda managed to pinch the edges. Lily wrenched the paper away. It tore, but not enough to prevent Lily from reading what it said. She spun free of Melinda's grip and darted around the other side of the desk, rushing out of the door and back into the restaurant.

"Give me that!" Melinda screamed. Her face glowed red as she came toward them. Her hair danced about her face as though she was lost in a fire. She was panting and looked manic, and utterly vulnerable. When she realized everyone was looking at her she stopped moving. Stephen wore a smirk.

Lily looked at the letter and read it out loud, the bit that had caught her eye. "If you don't listen to my demands then I'll make this restaurant's name a mockery, and you'll never make it as big as you think it's going to go. Stop thinking about your little pet Betsy and pay the rest of us some attention. We're tired of being unappreciated. Pay us what we're worth or I'll make sure that this restaurant will never make money again," Lily finished reading, and then looked up from the letter.

All eyes turned to Betsy, who went as red as a beetroot and squirmed under the gaze of everyone. Lily turned to Melinda, who was now smoothing down her hair and her dress.

"I just got a call. The police are on their way," Connor, the barman, yelled out in a quiet voice. Melinda held her head high and laughed dryly.

"Oh fine... okay. I guess this is how it really ends then," she said, as though nothing really mattered at all. Lily turned her gaze between her and Stephen, who was scratching his chin and glancing toward the window, presumably waiting for the cops.

"So this is the part where you tell us what happened. I assume the two of you were working together?" Lily said.

"We both wanted the same thing. More money. Being an author, I don't think you know what it's like to work for someone else. We put in all the hard work and he reaps the rewards. I've seen the accounts. I know how much this place makes. He says that he can't afford to pay us more," she let out a derisive laugh. "And don't think I'm in this alone. I've been trying to get raises for all of you," Melinda said, sweeping a finger across the staff, who looked nonplussed. It was almost as though she was trying to include them in her crime, but that wasn't going to work.

"We never asked you to do that Melinda. Some of us are happy with our lot," Amelie said.

"Well you shouldn't be. Money is the only thing that matters and we needed more. It's not our fault Brian is tight fisted," Stephen moaned.

"But it is your fault that a man is dead. An innocent man, who did nothing wrong," Lily said, looking sadly toward the fallen figure of Bernard Ferrabeau. Melinda wore a stern expression, as though she was trying to remain dignified when really she felt nauseous.

"It wasn't supposed to go this way," she muttered.

"Then what way was it supposed to go?" Lily asked with little patience. Criminals always had their plans and always seemed to complain when they went awry, but they were never the victims. They never lost their lives.

Melinda didn't seem struck with guilt so much as she was struck with a sense that things could have been different. "We were just supposed to scare him that was all. Ferrabeau wasn't supposed to die. He was just supposed to get a little ill. *Someone* put too much in. We just thought if we would get a bad review then Brian would be forced to pay us more, or at least know that we weren't messing around. I never meant for any of this to happen."

"Well it did happen. A man died because of you, both of you," Lily said in turn, looking at both Melinda and Stephen. They began to look somber now, and somehow the fact that it was a mistake made it worse. There had been no reason for Bernard Ferrabeau to lose his life. It had all been because these two people were greedy. Lily could almost have cried.

8

The excitement hadn't ended there. The burly barman Connor made sure that Melinda and Stephen weren't going to do anything to escape. It wasn't long before the police finally arrived, the sirens blaring through the air as they approached the restaurant. Uniformed officers placed yellow tape around the building, while a beleaguered detective came in.

"Right, what's going on here then?" he asked, looking at the covered body. He leaned down and pulled back the tablecloth, and a sad expression came upon his face. Lily went to speak to him.

"Has it been a long night?" she asked.

"You can say that again. It always is whenever a game like this is played. Are you the manager here?" he asked.

"Actually no, my name is Lily McGee, and there are a few things we ought to talk about," she said. The detective looked at her with interest and introduced himself as Don Jones. At first he seemed amused when she started speaking to him about the murder investigation, probably assuming that it had been treated as some sort of parlor game to keep everyone entertained. However, as Lily explained her process and the way the evidence had been discovered, Don realized that she was being serious. He arched his eyebrows and nodded along with everything she was saying. Lily described it all in minute detail, and the expressions of Stephen and Melinda were shrouded in guilt as well.

"I see, well, it seems as though you've done my job for me Miss McGee, I owe you a debt. I'm sure you've earned a relaxing night somewhere along the way," he turned to one of the officers beside him. "I hope that you got that down as a statement," he said. The officer nodded. "Well then, I suppose all that's left for me to do is clean up here," he said. The other officers were already moving around the restaurant, taking statements from the other people who had been present, while Don signaled for Stephen and Melinda to be placed in handcuffs. They looked sullen as they were taken away and Lily shook her head. It seemed ridiculous to her that they would risk a man's life for money. Now they had destroyed their own careers and been left with nothing. Had there really been any point to it at all?

Lily and Blake waited around for a little while, just to make sure that nothing more was needed from them.

"Another case solved. Lily McGee has done it again," Blake said with a warm smile.

"Yes, but in the process I completely ruined our evening," she said.

"The evening isn't over yet," Blake replied with a wry smile. He was about to say something else, but before Lily could discover what he meant by that, there was a commotion at the door as someone tried to burst into the restaurant, but was stopped by the police barrier outside.

"I'm the owner!" he cried, and once the police had verified his identity, Brian Samson was let in. He was shorter than Lily had imagined, but he moved with confidence. His clothes were crisp and smart. Betsy ran to him and flung her arms around him, and then seemed to think better of her actions and stepped back. Brian squeezed her arm.

"Right, what on earth has been happening in my restaurant?" he asked, and moved toward the detective. Don Jones explained everything to him, and then pointed to Lily. Brian gave her an inquiring look and then approached her.

"Are you the one who sorted all this out?" he asked, his accent came from a rough part of London.

"I am, yes."

"Well, thank you very much," he said, holding out his hand. Lily took it and he shook it once, sharply. "Those two are crafty and I can imagine they would have tried to pin it on anyone else."

"They did," Lily said, and nodded toward Amelie.

Brian sighed. "I can't believe they would go so far as to do this. I know I was a bit slow in giving them the raises they asked for, but they were coming! It's not easy running a business you know," he shook his head and then glanced toward the body. "It's such a shame as well. Bernard and I had our differences, but I never wanted this to happen to him." He had a thoughtful look on his face and then beckoned Amelie over. "Right, Amelie, you've been angling for a promotion, you've got it."

"I have?" she asked, her face lighting up.

"Too right. If we're going to get through this debacle then we're going to need to keep going. If you begin to show weakness then it's all going to crumble. We can't let that happen. The public needs to know that there has been a change around here and that this is never going to happen again. The first thing I want you to do is create a new dish, and call it the Ferrabeau. I want something classy, something traditional, something that he would be proud of," he said, his voice heavy with emotion as the weight of what had happened sunk in. Amelie nodded sharply and stood to attention as though she was a soldier. She rushed off, as though she was going to start her new assignment immediately. Brian then turned back to Lily.

"I mean it when I say I owe you a debt. Let me fetch you something," he said, and walked over to the counter where the cash register was. Lily was about to refuse, saying she didn't do this for a reward, when Brian returned. He handed her a gold card. "It's not much, but look, whenever you want to come in for a meal you just use this and I'll make sure that you get whatever you like. You're always going to be welcome here Lily McGee," Brian said.

"Thank you Mr. Samson, it is appreciated," Lily said. She took it gratefully, believing that it was the polite thing to do. After all, it had no monetary value so she wouldn't have to use it if she did not want to. "If I could suggest one thing too? The value of a place is not in the money it makes, but in the joy it brings people. Perhaps it would be prudent to review the pay structure and ensure that everyone is being given what they are worth."

"Believe me, there are going to be a lot of changes around here," Brian said, nodding in agreement with her. He then moved away and went to talk to his staff. Police still swarmed the restaurant, while the other guests were beginning to leave, the excitement of the night finally dying down. While Brian was optimistic about the restaurant recovering its reputation Lily knew it wouldn't be that easy. Perhaps people might come here out of morbid curiosity. Maybe they would even want to sit at the table where a man had died. People often expressed strange attitudes when it came to death. It was usually something people ran from, hid from, pushed away from, in the hope that it would never darken their doorstep. And yet there was something that fascinated people as well. It was quite the paradox, and Lily supposed that she was just as vulnerable to it as anyone else, for she had made it her living.

She turned to Blake and arched her eyebrows. "I suppose we should head back to the village," she said.

"Actually, I think there's one more stop we should make first, before we head home," he said.

Lily was intrigued by what he meant. They walked outside into the cool air of the night. Goosebumps prickled on her skin as the temperature had dropped outside since they had entered the restaurant. The traffic had thinned. An ambulance had pulled up and the medics were striding into the restaurant with a gurney, ready to finally take Bernard Ferrabeau away so that he could rest eternally. She thought it sad. Bernard had not realized that coming here would be his last meal.

9

Blake's car slalomed through the city, swerving around the tight corners of the narrow roads. The buildings gleamed in the bright lights of the night. Even though it was dark the city still glowed. Bars were open with raucous crowds. Lily assumed that the local team had won the match, for revelry was everywhere. Blake mentioned that he was still hungry, so he ordered a pizza to an address that Lily was unfamiliar with. It turned out that they liked the same toppings; a spicy pizza with ham and pepperoni.

"You know, if my Mom had done a better job of raising me then she would probably have warned me about getting into a strange man's car and being taken to a strange man's place," she teased.

Blake smiled. "My mother always taught me to put a woman at ease, so I can assure you we are not going anywhere untoward. You are completely safe with me," he said. Lily relaxed in the passenger seat, she believed him. They pulled up in the parking lot of a huge office building. It towered above them, rising high into the sky. The windows were dark as they reflected the night. Blake pulled out a set of keys and opened the door. They took the elevator up to the top floor and emerged into a deserted office where the desks were silent and empty. Computers had been turned off, and the only source of light was coming from outside. They moved to the window where there were a couple of chairs. Blake did not sit down though. The pizza arrived shortly after they did. Lily could tell the delivery person found the situation strange, as he scurried away as soon as the pizza was delivered. Blake placed the pizza box on the table and opened it, offering Lily the first slice. When she smelled the food she finally realized how hungry she was. When she worked a case she always had to remind herself to eat because she was so focused on other things.

"This isn't exactly a steak, but it's still delicious," Lily said through a mouthful of pizza. Blake smiled as he began eating to.

"So what is this place?" Lily asked after she had sated the initial, most nagging pang of hunger.

"This is the place where I spent most of my life Lily. This was my world," he said, opening his hands up to the window, seeming to encapsulate the entire city. From this vantage point all the buildings below were tiny, and the cars were like pinpricks. The night sky glittered with stars and the moon was full, hanging in the sky like a silver coin. It was breathtaking in its beauty, and it was so high that it was closer to God, although Heaven was still an eternity away.

"It's not a bad world to live in," Lily said wryly as she stood beside Blake and gazed out of the window.

"It's not. I used to think myself fortunate for being able to work in a place like this and have this view every day. I thought to myself that I was one of the most privileged men in the world to be able to look upon this whole city and, if I wanted, charter a plane to take me to anywhere in the world. It was all at my fingertips, but then I realized that something was wrong."

"What was that?"

"Being up here is all well and good, but it does tend to leave you removed from the world, distant... alone. I think sometimes when you take too broad a view of the world it can leave you stranded. You end up seeing everything, but feeling nothing. I had gotten used to it, until I returned to the village. Everything is so small there, but it's so connected. There's something wonderful about that, something precious. I don't know if it's anything that can be replicated. I certainly haven't found that sense of community anywhere else. When I was younger I was foolish. I used to want to do anything to escape. Mother told me to appreciate what I had because one day I would be sorry that it was gone. I didn't think she understood me at all. I was a child of ambition, a man ready to conquer the world, but deep down she knew me better than I knew myself. She knew what I needed, and I was never going to find it out here."

"What is it you needed?" Lily asked, surprised at the intimate tone that Blake was using. She got the sense that he was telling her something he had not told anyone else, perhaps ever. His voice was low and quiet, and in the beautiful darkness it felt as though they were the only two people in the world. She had never felt this close to him before, but she knew she could be even closer.

"A home," he said. It was a simple word, and yet the depth of emotion contained within his voice was profound, and in her heart Lily knew that was what she needed as well. Blake tore his gaze away from the vision of the city stretching before him and looked at Lily. She felt the weight and the heat of his gaze upon her. Goosebumps prickled on her flesh again, but this time it was not due to the chill of night, but rather a deep heat that burned inside.

"I have to admit something to you Lily," he said, his voice soft and rolling, like velvet. "I lied to you before."

"You lied to me? About what?" she asked, wondering what he had to confess.

"When I told you that I decided to stay in Didlington St. Wilfred because I needed to look after my father. That wasn't the entire truth. He had staff in the house. I could have gotten away with paying him a weekly visit, and I could have employed a nurse to take care of him. It would have perhaps been cold hearted of me, but the option was there. However, I met a remarkable person who made me look at the world in a different way. She was a curious American woman who seemed so ill fitting to the world of the quaint village, and yet at the same time seemed to be exactly what it needed. She was like a breath of fresh air, and even after I spent time with her I found myself thinking about her all the time."

"She sounds like a remarkable woman," Lily said, a smile twitching on her face. Blake returned the gesture.

"I think that goes without saying," he said. "And the more I've gotten to know this woman the more I've come to realize that perhaps she is the kind of person I need in my life as well. Ever since my mother died I have felt guilty that I haven't continued the legacy of our family. All she wanted for me was to settle down. I told myself that it was only because I was keeping myself too busy with work, and that if I simply put the effort in I would be able to find a wife with ease. But I wasn't sure anyone would be quite suited to me. I always found it difficult to picture myself with a woman, but you are spirited and fearless and the way you always put other people before yourself is admirable. I look at the way you have inspired loyalty and friendship within the people of the village, even Pat, and I can tell that the world is a brighter place for having you in it Lily. My life is brighter. I don't know whether it was fate or chance or God's grace that led you here, but I would be remiss if I did not tell you how I feel."

Lily's heart fluttered and her stomach swam. Tears glistened in her eyes and her lips trembled. She almost wanted to pinch herself in case this was a dream, but then she would have risked waking up. However, she knew this was not a dream. There was no mistaking the spice of Blake's aftershave in the air, or the warmth radiating from his body, or the sheer exhilaration that sang within her veins.

"But tonight was a disaster," Lily said weakly. "Our date was ruined."

"I knew from the moment we met that you weren't going to let injustice stand Lily. You're not the type of person to step aside and let other people make a difference in the world. You are the person who makes the difference, and I'm not sure you realize how rare that is. You should have heard all the people speaking about you in glowing terms. Everyone was impressed by you, and I think you definitely won a few fans over. Your books are inventive and interesting, and on top of this you help people even though you put yourself in danger by doing so. That is the mark of a great person, and one that I feel honored to have in my life."

His words were sweet, but Lily was still hesitant about giving herself over to these soaring feelings within her heart.

Stop being so foolish! A voice cried in her mind. It was the voice of Joanie, of her mother, of Majorie, of all the sensible people telling her that she deserved to be loved and deserved to be happy and should not by any stretch of the imagination turn away this proclamation of affection.

"I still owe you an answer from earlier," Lily spoke in a trembling voice. She was far from the assured, confident investigator that she had been at the restaurant. Now she was just a young woman whose fragile heart was on the line. "You asked me what I wanted from life and I didn't give you answer. The truth is I've never really known what I wanted, at least not until recently. I never had a proper family growing up and a part of me thought I didn't really deserve one. I suppose a part of me still thinks that. But recently I've been feeling that something is wrong, something is missing. I was so busy in America that I never gave much thought to what was going to happen tomorrow. In the village it's different. I've encountered so many different people and so much sorrow. If there's one thing I've learned it's that this could all end at any time, and the only way we can properly respect the dead is to make sure that we live as much as we can and cherish every moment. And it's important to be honest with each other as well. I want to be honest with you Blake, and I want to be honest with myself." Lily took a breath to compose herself. Her head was spinning with dizzying emotions and it felt as though her tongue was swelling. She had been doing a lot of thinking, and she knew exactly what she wanted now.

"I want to stay in Didlington St. Wilfrid for the rest of my life. I want to fall in love and have a family, with the right man of course."

Blake arched his eyebrow and his eyes sparkled. "And who would that man be?" he asked.

"Well, I do think the butcher is quite cute, and I'd always be sure of getting the best cuts of meat," she said, chuckling. Blake laughed too, but then the laughter faded and the air was laced with tension. "A man who is distinguished and handsome, a man who makes me feel unpredictable, a man who from the very first moment we met captured my attention in a way that no other person has before. I've always been afraid of these big feelings Blake. You say that I'm brave, but not when it comes to emotion. I've been hurt before. I've been betrayed by my own mother and left to fend for myself. Trust doesn't come easily to me, and it scares me how much I want to trust you. But I realize that it's not worth living in fear. Nothing good comes of it. We have to be brave enough to go after what we want and that's what I'm doing now. I want to have a relationship Blake. I want something lasting and faithful that leads to having a family. I want to be a part of something bigger than myself, and I want to prove to myself that I'm not going to have the same failings as my Mom. I want to know that I'm not going to end up hurting my kids the same way as she hurt me."

"Is that why you've been so afraid to surrender yourself to love?" Blake asked. Lily nodded mutely. It was something that she had never admitted to anyone else before. Blake reached out and caressed her cheek lightly. Tingles spread all over her body in one sweet burst of exquisite delight. "I think that when good people get hurt they do all they can to make sure nobody else suffers the same way. I don't believe you could ever abandon a child Lily, or make anyone feel anything less than special. But there's no need to become overwhelmed with the future and with talk of children.

Let's simply take it moment by moment. All we have to focus on now is a kiss..." his words drifted into the air and they were lost as he leaned into her. His hand dropped from her cheek and rested against her neck, tilting her head back a little. Her heart thrummed so fiercely she thought it might burst out of her chest, and then his warm breath swam over her like a tsunami. She flung her arms around him as they embraced in front of the stars and the moon, and she felt truly blessed by God. The path He had given her had not always been an easy one to walk, but it had led to Blake, and she knew now that she would not have wanted anything more.

She surrendered to the kiss and the world become more vivid and vibrant than it had ever been before. Lily's heart burst with happiness and she became so overwhelmed with emotion that she thought she might cry. There were so many times in her life when she thought she was meant to be alone, but she realized that wasn't the case any longer. She loved Blake and he loved her, and nothing was going to stop them being together.

WHO KILLED THE COLONEL IN A QUIET ENGLISH VILLAGE?

1

"Well Mom, I'm not really sure where to begin. I've been making good progress on a couple of novels. The newest one should be out soon. I've told the publisher to send you an advance copy. I'd check the dedication if I were you, I've slipped a little mention in there that you'll appreciate. They've also spoken to me about potentially coming over to America for a book tour. I think these English cozy mysteries have charmed the American audience and they are like to want to see me, so if that happens I'll be sure to let you know and maybe we can meet up. I might even bring Blake along as well.

It might sound strange saying this, but I hope my books don't become too popular. This might come across as elitist, but I know what's going to happen. They're going to try and turn it into a TV show or a movie, and my heroine is going to end up being some dolled up girl who swears and wields a gun. Some things change, but Hollywood always stays the same. I suppose I'll just have to try and protect my characters as much as I can. You'd probably tell me to not *get stressed about the problem until it actually happens. Maybe I'll start listening to you.*

And yes, I know that I mentioned Blake earlier and you're probably reading this thinking to yourself 'when is she going to tell me what's happening with Blake?' Don't worry. Everything is fine. Everything is great, in fact. We haven't stumbled into any other crimes. We've just been, well, we've been enjoying being together. We went on a trip to Yorkshire the other week. Blake has told me that he wants to show me all the nicest parts of Britain, so we're getting them done piece by piece. Even though many of these places are beautiful there's none that are quite like Didlington St. Wilfrid though. I don't know whether it's the place as such, or the people in the village that make it so special, but I truly have fallen in love with the place. I can without any hesitation say that it is my home, in every way that a home can be. I'm sure you must understand this after your visit as well.

You know, I find myself being more open in writing to you than I would be if we were sitting together. I know in many ways it's difficult for us to be so separated, especially after we have only just reconnected, but I hope that this is still a fulfilling relationship for you. I'm sure that we will see each other again in person, but for now I am enjoying the routine of writing to you every so often. Letter writing is a lost art form I think, and it feels more special than simply receiving an email. On that note, I must admit that I am finding it a rewarding challenge to be a part of a relationship. Love was always something I thought about in the abstract, and as an outside observer it always seemed chaotic and unpredictable. There are elements of that with myself and Blake, but in the fundamental aspects he is quite reliable. He is always there when I need him, and I have no cause to doubt his sincerity or his loyalty. He surprises me with gifts and small romantic gestures, and every moment I spend with him feels special, even if we are simply spending time together. It almost makes me wish that I had taken the risk and plunged myself into love before this, although I daresay there is not another man like Blake in the world. I am blessed that he was patient enough to wait for me to be ready, and I certainly hope he is as pleased to be with me as I am with him.

We still continue to go to church every Sunday and I find myself taking great solace in the teachings that Father Brown gives us. There is much time to reflect here, and I find myself thinking about the deeper questions in life. Back in America I was always pulled this way and that, but here I can ponder more spiritual matters. Blake and I spend many evenings discussing the teachings of the Lord and various moral quandaries. I can picture you smirking now, likely thinking to yourself that I am making a fool of myself for not going out to bars and parties and enjoying my time when I am young.

To be honest with you Mom, I may be young, but I do not feel young. Spending time investigating the criminal element and trying to solve murders does indeed take its toll. I wish that it was easier to give up, but whenever I think about it I come to the conclusion that I have a moral imperative to use my skills for good. That being said, the good Constable Dudley has not come to my cottage for aid since he was reinstated. It is my hope that the only crimes I am involved in for the foreseeable future are the fictitious ones I invent for my novels. Didlington St. Wilfrid is a far better place when it is at peace.

I hope that this letter finds you well, and I look forward to hearing about your tour of the national parks. You have quite a task ahead of you, but I'm sure you will find joy and peace in cataloguing the natural beauty of the country.

I shall see you soon, I am sure, and I will write to you even sooner than that. And please let me know what you think of my new book.

With love,

Lily

~

L ily set her pen down and waited for the ink to bleed into the page. She passed her eyes over what she had written and furrowed her brow at a few things she would have liked to rephrase, before reminding herself that this wasn't one of her novels and she did not have to worry about an editor returning the manuscript with a lot of red ink spreading over the pages. She folded the page in half and then slid it into an envelope, before addressing it to her mother and placing a stamp upon it. When Kelly had left Didlington St. Wilfrid it hadn't seemed right for it to be a goodbye, so Lily had suggested they write to each other. It was more personal than exchanging emails, and more special than calling each other, plus they didn't have to worry about the difference in hours. She leaned back in her chair and smiled, gazing out at the leaves falling outside. She gathered her coat and wrapped a scarf around her neck in an attempt to shield herself from the chill of the wintry months. It was only going to get worse, Blake had warned her, for England was always guaranteed to have a bitter Christmas.

She called out to Joanie, but there was no response. Lily assumed that she was probably out capturing some photographs of the local area for a new collection. Joanie had been speaking about showcasing the changing of seasons with photographs. It struck Lily as ironic that someone as gregarious and personable as Joanie would take up the vocation of a wildlife photographer, a job that required her to spend most of her time alone. In truth Lily had not expected Joanie to stay in Didlington St. Wilfrid this long. Lily had thought that Joanie would soon tire of the quiet life that this small corner of England offered, but it seemed to have charmed Joanie as much as it had Lily. She smiled as she thought about how happy she was here, and knew that she was going to stay here for the remainder of her life.

Even though this thought had been in her subconscious for a while, it was the first time she had actually thought about it directly and it made her consider her future. Before it was vague and unclear, and in a way she liked that because it allowed her some freedom and flexibility, but now here she was, in love with the place around her and the man she was spending most of her time with. Her cheeks had a rosy glow, and it wasn't because of the warmth in the house.

She picked up the letter and opened the door, but as she did so she saw Constable Bob Dudley standing there with a raised fist, about to knock. His cheeks were ruddy and his nose was red. When he breathed his breath rose like wisps of smoke through the air.

"Ah, Lily, I'm sorry. I didn't mean to catch you when you were running an errand. Are you off to see Blake?" he asked.

"Not at the moment, I was just about to post a letter, although I was planning to see him later. Are you here for me or Joanie?" Lily asked.

"You, I'm afraid. I know it's been a while, but there's been a murder, and I could use your help," he said.

Lily's heart sank. She had been dreading this day, yet somehow she knew it would always come. Murder seemed to be stitched into the fabric of life, and she had been placed here to help solve them. She looked crestfallen.

"Of course Constable. I'll be with you in a moment. Just let me make a quick phone call," she said, for she knew that these things were never dealt with quickly. She stepped back into Starling Cottage without closing the door. Constable Dudley turned his back and rubbed his hands together, blowing into them to fight off the cold. Lily pulled out her phone and found Blake's name in her contacts.

She braced herself for the response because she always hated doing this to him, despite him having reassured her that he didn't mind her solving mysteries. However, it had been a while and she and Blake had had so many uninterrupted months of pure romantic bliss. The last thing she wanted to do was rock the boat when they were in such a good place.

"Morning Lily, I wasn't expecting to hear from you until later, are you okay?" he asked. His voice was always enough to make Lily's heart melt.

"I'm fine Blake, but Constable Dudley just popped by…" she began.

"Ah," he said. It was only one syllable, but it contained a lot of emotion.

"I know. You know how long these things can take. I'm not sure when I'll be done. I might have to take a rain check tonight," she said.

"Are you sure? I don't mind calling the restaurant and getting them to hold the reservation for a little later."

"I wouldn't want to let you down if I can't make it. After all, it's been a while. I might have to take a little bit of time to get used to the procedure."

"I'm sure you'll get back in the swing of things without any trouble at all," he said. Lily worried that she could hear a certain edge to his voice.

"Blake, are you okay?" she asked.

"I'm fine," he said tersely. "Just do what you do and we'll get together again when you have a chance. I hope that it goes well. Keep me updated. I love you."

"I love you too," Lily said, although he hung up almost before she finished saying those three simple words that had come to mean so much to her. She stared at the phone and frowned before she slipped it back into her pocket. It wasn't like Blake to be so annoyed with her. Perhaps she was making things worse in her mind, or perhaps he really did have a problem. She tried to put that out of her mind as it wasn't going to do her any good until she could speak with him again. If he did have a problem with her solving cases then they would have to talk about it like reasonable adults, as that was how they handled their problems.

Constable Dudley turned to greet her as she closed the door to Starling Cottage behind her. She smiled. "Lead the way then Constable," she said, and followed him down the garden path, looking with regret at the flowers that were dying, as they were deprived of the summer heat.

2

Lily McGee sat in the passenger seat of Constable Dudley's car as it snaked its way out of the narrow roads of Didlington St. Wilfrid and turned through the outskirts. Great autumn fields stretched out as far as the eyes could see, framed with lush evergreen trees. The sky was a chill blue, almost as though cold water was dripping over the world.

"Where are we going?" Lily asked.

"To Colonel Rawlings' house," Constable Dudley asked.

"The Colonel is dead?" Lily raised her eyebrows in surprise.

"That's what I've been told. I can hardly believe it myself. I thought he was going to live forever."

"He has lived an awfully long time. Didn't he serve in World War II?"

"He served in just about every conflict he could," Constable Dudley said. "He is a great man. *Was* a great man," the Constable seemed pained that he had to correct himself. Lily remained quiet for a moment and thought about what she knew about Colonel Rawlings.

The Colonel was well known in the local area as he was a part of the natural landscape. He was a decorated military officer who had served for many years in the army, and retired gracefully. Since retirement he spent his years on his estate, as well as making many charitable donations to the local area. Lily had never met the man, although she had seen him in passing during certain special functions that the mayor organized. He had a good reputation among the people of the village though, and she was certain that some of them would have been bewildered to learn that he was still alive. Some people just knew how to hang on though. If anyone was a survivor then the Colonel was. However, although he had been able to survive so many years of military service, something had claimed his life, and murder was murder, no matter how old the victim was.

The manor house was stately, although it was not quite as big as Lord Huntingdon's. The iron gate was open, and the stone pillars that held it in place were weathered. The house itself looked in a state of disrepair. It seemed as though it needed to be invigorated. If there was one word that was apt for this house, it was *old*. Lily could almost hear it creaking as she got out of the car. The house was three stories tall and the windows were imposing. She walked up to the front door and tested the handle. It opened without any difficulty to a lobby and a wide wooden staircase that led up to the next floor of the house. Faded pictures hung on the walls, and as Lily turned her gaze around she noticed that there were strange pockmarks on the lower half of the wall.

"It's amazing to think what this place must have been like in its heyday," Constable Dudley said, taking a moment to place his hands on his hips and look around the house, which seemed to go on and on forever. Houses in America were big, not that Lily had ever lived in one, but this one dwarfed even them. However, there was a chill that crept in from the outside and seemed to linger. It made it all the more eerie that it was a murder scene. "I can't imagine what the parties must have been like. You should asked Lord Huntingdon about them the next time you see him. He would have attended a few."

"Lord Huntingdon and the Colonel knew each other?" Lily asked.

"Oh yes. I mean, people like that go around in similar circles, don't they? It was only natural. They're probably the two most important men to ever have graced our village, at least in recent memory. I don't want to speak ill of the men who founded it, of course," Constable Dudley said.

It was at this point that they were interrupted by a soft voice. "May I help you sir and madam?" he asked. The man was old with a stooped back and looked as though he should have embraced retirement years ago. His hair was as white as snow and his mustache was the same. His face was lined with wrinkles. He wore white gloves and a suit that seemed far too rigid for modern times, but Lily supposed old habits die hard.

"I'm Constable Dudley. I'm here about the Colonel. This is my consultant Lily McGee," the Constable said.

"My name is Charles," the man said, bowing his head in the direction of his guests. "Please come this way," he turned on his heels and then led them to another room. He moved slowly and stiffly. Lily was tempted just to burst out and walk in front of him, but she resisted the temptation. She did not want to embarrass the man. He led them through to a small sitting room that had two other people within it. One was a young girl, as thin as a waif with golden hair and a pinched expression. Her eyes were bleary and the skin on her lower lip was very red. She wrung her hands together and looked up at the newcomers, startled when she saw the Constable. The other woman was sitting perched forward in a chair, her dark hair pulled back in a tight ponytail. She had her hands clasped together and her face was set into a scowl. While the other girl was clearly young, this woman looked to be middle aged.

She did not seem intimidated by the Constable's presence, or Lily's.

"This is Ava and Violet," Charles introduced the two women in turn, explaining that Ava was the housekeeper and Violet was the chef. Charles himself was the butler. His voice wavered as he spoke, although Lily could not tell if this was due to age or the emotion of losing Colonel Edward Rawlings.

"Perhaps you could start by describing to us what happened?" Lily asked. The Constable pulled out a notepad and a pen. She looked at each of them in turn. None of them seemed inclined to speak. Her mind began to turn already, and wondered if any of them would have dared to murder their employer.

"Which one of you found the body?" Lily asked, since nobody was inclined to say anything. Charles remained silent, but Violet looked toward Ava, who raised her hand as though she was seeking to ask a question at school.

"I did," she said. Her voice was touched by an Eastern European accent. "It was a terrible sight." Ava choked back a sob.

"I'm sure it was. I know this has probably been quite sudden for you all. Perhaps you could take me to the scene of the crime. It might help us if we can see where it happened," Lily said. Investigating murders was always a patient game. It was important to never rush into anything .These people were still shocked. They would talk to her, in time. She just needed to let them feel more comfortable with her.

Ava nodded and rose. She kept her hands clasped together, aside from when she wiped her eyes. Her movements had a kind of frantic energy. They walked through to the Colonel's office, which was located toward the rear of the building and looked out upon an expansive field that stretched out for as far as the eye could see, although Lily did not have time to admire the view. The most striking thing about the room was the dead body, of course. Colonel Rawlings was sitting in his hair, cradling a rifle in his arms. A desk was before him, filled with various papers and files. A cream cloth had been placed upon his face. Lily noticed how Ava tensed as soon as she entered the room. It was likely that it was brought on by the trauma of it all, but it may have been guilt as well. Portraits hung on the walls and there were other ornaments dotted around the room, giving it a cluttered feel. She wasn't quite sure how a house of this size could have felt cluttered, but it did.

"Do you know what he was doing in here?" Lily asked.

"He was writing his memoirs. He used to stay up late writing them, and he got up early every morning as well. He liked to say he did his best writing when the rest of the world was sleeping," Ava said.

"I see, and when do you think he died?" Lily asked.

"It was through the night. We weren't sure what to do. I wanted to make him comfortable. We all did. We called the police first thing," Ava said. Lily glanced toward the Constable. It wasn't very reassuring to know that they had waited to call the police for hours after the crime had been committed. However, given the state of them she might have assumed that they were simply overwhelmed. It seemed to be just the three of them managing this entire house, and she could not have imagined them coming together to be clear headed about everything. She decided not to comment on the matter directly, but filed it away for later, in case further information suggested it was important.

Lily walked around the desk, angling her body to move through the narrow path, and stood next to the Colonel. An acrid stench filled the air. His fingers were laced around the handle of the rifle. The metal barrel was pointed at his head. It was an ominous sight. Lily reached out and pulled the cloth away slightly. In the background she could hear Ava whimper. Lily peeked into the shadows and her stomach turned at the grisly sight. It looked unholy, and she put the cloth back. Part of her wished she hadn't put herself through that, but she thought she ought to see what the victim looked like. She lifted her gaze and noticed the blood spatter on the ceiling.

"I don't mean to cause offence Ava, but I thought you said this was murder? From what I can see here it looks like suicide," Lily said.

Ava's eyes went wide and her skin turned an even lighter shade of pale. She shook her head slowly. "Oh no, no that's not possible at all. The Colonel would never do such a thing. Why would he kill himself when he was in the middle of writing his memoir? He never liked leaving a project half finished," she said. Lily thought there were many reasons why a man might kill himself, even a man like Colonel Rawlings. But since she was here she decided she might as well investigate anyway. Sometimes crimes weren't about who did it, but why it had been done in the first place.

"I should call for some medical assistance. We need to get this body moved," Constable Dudley said, nodding toward Lily.

Although she and Constable Dudley hadn't worked closely together since he had been reinstated, they had fallen back quite naturally into their rapport. She sometimes missed the brash, more straightforward Chloe Hargraves, but after what Constable Dudley had been through Lily was glad he was back where he belonged.

While he stepped outside to make a call, Lily took a closer look at the office.

"Can you tell me a little more about these ornaments?" Lily asked.

Ava stiffened at the question. "They are things the Colonel was rewarded with over the years, from all over the country. He liked those the most. They were from Africa," she pointed to the far wall, toward some tribal carvings that stood on wooden stands. Lily went over to inspect them. A lifetime's worth of accomplishments were contained within this office, and likely throughout the mansion as well. She noticed that a lot of the photos were of the Colonel in various stages of his career.

"And these?" she asked.

"Pictures of when he was younger. He called them the glory days. He liked them because he said it reminded him of when he could still look forward to the future," Ava said. It was a curious phrasing, and Lily noticed that aside from his ancestors there were no pictures of anyone but the Colonel. He had always come by himself to the events in Didlington St. Wilfrid, but she assumed a man like him must have had a family somewhere.

"Does the Colonel have any children?" she asked.

"No, he does not," Ava said. Lily pressed her finger against her lip and tapped it thoughtfully. At first glance it did seem as though it was a suicide, but why would such a proud man as the Colonel kill himself now? It didn't make much sense to her when he had lived such a long life, or when he was in the middle of writing his memoirs. There was definitely a mystery here, Lily simply wasn't sure what kind of mystery it was.

3

L ily had returned to the sitting room with Ava while Constable Dudley waited for the medical personnel to arrive when the body would be handled. She informed them that she was going to speak to each of them in turn, and she decided to begin with Violet. Ava seemed as though she deserved a break.

They walked through to yet another room that was small and darker. It served Lily's purpose, even if the chairs were not comfortable.

"Right, let's get this over with then so I can get out of here," Violet said in a harsh tone. Lily arched an eyebrow.

"Am I keeping you from something?" she asked, surprised that Violet would be so uncaring to one who had died mere hours ago.

"Just the rest of my life. Look, I didn't kill him and I don't know who killed him. Frankly he might well have offed himself. I don't particularly care. I just want to get home so I can take care of everything that I need to take care of."

"Such as?" Lily asked.

"Well, finding a new job for one thing. I'm just glad I have a couple of interviews lined up already," Violet said.

"So you were planning on leaving?"

Violet barked a laugh. "You could say that."

"Did you have some sort of conflict with Colonel Rawlings?"

"You don't have a conflict with Colonel Rawlings. You have a war. He still thinks he's in the bleeding army. Anyway, yes, I did."

"Would you mind telling me what this conflict was about?"

"I don't mind at all. It's because he's a stingy bugger. I got tired of working for a man who couldn't understand the concept that the cost of things rises over the years, and our pay needs to rise to match it. I'm amazed he didn't try and pay us in shillings and crowns," Violet said, rolling her eyes. "It's been a squeeze, you know, with the price of everything rising. I kept asking him and telling him that we need more money, but he wouldn't listen. He's lived here for too long. He doesn't understand what normal people need, to survive."

"You do realize that by admitting this you're putting yourself in line as a suspect?" Lily asked.

Violet merely barked a laugh again and slapped her palm on the table. "Why would I want him dead? That's not going to get me a pay rise, is it? I already put in the paperwork to leave. I was just waiting for him to hire a new chef. I didn't want him to starve. I'm not that heartless, and what would Ava and Charles do without me? They'd be ghosts without me here. That poor girl doesn't know how to boil an egg, and Charles... well... I don't know how he's still managing to stand upright at his age, but he couldn't cook a meal. I'm the one who held this place together. But you don't see me crying about it, do you?" Violent then dipped her head and genuine emotion seemed to break through the anger. "Look, aside from the money I never had any problem working for the Colonel. I don't know why anyone would want to kill him either. I just want to get on with my life so that I can have a decent job where I don't need to worry so much about my bank account," Violet said.

"I just have one more question for you Violet. Where were you through the night?"

"I was sleeping, wrapped up nice and toasty with my hot water bottle. I'm all the way up there," she pointed upwards, and Lily assumed she meant she lived on the top floor of the house. "If you ask me he might well have killed himself. If I made it to be that old, I'd probably kill myself too."

She got up without Lily announcing that the interview was over, but Lily wasn't about to argue with her. Violet's aggressive nature perhaps made her more suspicious than she actually was. There was merit to her argument that she gained little by the Colonel's death. It wasn't as though he was keeping her constrained to a contract. Even so, Lily knew that sometimes money could lead to deeper feelings of resentment, and Violet was eager to paint herself as the most competent member of the trio. It was possible that she might have convinced them that the Colonel needed to die, or that she committed the crime without them knowing.

Lily then invited the butler into the small room. She waited patiently for him to shuffle in. His feet barely left the floor as he walked, for he lacked the strength to properly propel himself. His shoulders were rounded, but he had not lost his sense of duty. He held the door open for Lily and pulled out a chair for her before he sat down himself. In fact he was going to remain standing until Lily told him that he must sit down. He sighed with relief as he took the weight off his feet, and she wondered how much pain he put himself through for the sake of his household duties.

"Charles, could you tell me a little bit of what happened?" Lily asked.

"Of course Madam," Charles said, inclining his head. He spoke softly and slowly, each word enunciated perfectly, with no syllable missed. "It was a night like any other, up until the terrible events. There was nothing amiss at all, and believe me I would have known if there was anything troubling Sir."

"Have you worked for him for long?"

Charles straightened at this and wore a look of pride on his face. "For all my adult life. My father served the Rawlings family before me, and his father before him. It is a solemn duty and now... now I suppose I must retire," he trailed off and his brow furrowed, as though he had only just realized that he was going to have to give up work.

"Do you mind if I ask you a personal question Charles?"

"You may ask me anything you wish Madam," Charles replied.

"How old are you?"

"I am seventy five," he replied.

"Why have you not retired before this?" Lily asked.

"Because Sir needed me. It was my duty to serve him. I could not let him down, and unfortunately I never had any heirs of my own to take over for me. It has always been a sense of shame," he hung his head and looked embarrassed. Lily thought to herself that he should have retired a long time ago. Working so hard like this had evidently taken its toll on him. But there was something else that occurred to her.

"Did Colonel Rawlings feel shame that he did not have an heir?" she asked.

Charles folded his hands in his lap. "I do not think it's appropriate to speak about his personal feelings like that."

"Charles, the man is dead. He's not going to be offended, and it might help me figure out what has happened here," she replied.

When she said the word 'dead' Charles flinched and he paled, as though it hadn't truly sunk in that the Colonel was gone.

"Yes I... I suppose you're quite right," he said, averting his gaze. Lily noticed his eyes glistening with liquid. She felt sorry for the butler in that moment. He had given his whole life in servitude to Colonel Rawlings, and had taken great pride in that fact. Now he was stripped of his duties and likely his entire identity too.

"At times he did seem to be struck by melancholy when he thought about the legacy he was leaving behind. I think he would have liked children, but he gave so much of himself to the army that there wasn't much left, and by the time he retired... well, he was far too old to be a father then," Charles continued.

"Do you think that would have made him depressed enough to kill himself?"

"Oh no madam! Not at all!" Charles' mouth fell open and he looked offended. "Colonel Rawlings would never have done such a thing. Suicide is a cowardly act, and the Colonel was the farthest thing from a coward there ever could be. No, someone definitely killed him, and whoever did it was a coward themselves. Colonel Rawlings was not the same man he used to be. If this had happened in his prime you and I would not be having this conversation because the Colonel would have dealt with them with a swift hand, you mark my words. But I suppose time does come for us all, and it will come for me soon enough as well. I suppose I shall have to retire. I can't fathom working for anyone else," he said. Lily got the feeling he was going to continue mumbling if she didn't stop him.

"So you didn't have any problems with your pay or anything like that?"

"Of course not," Charles said, and again seemed flustered. He must have thought that speaking about money was inappropriate. "I was paid a fair wage for a fair job. I had lodgings here, and I have never been short of anything to eat." He took a moment to look around the room, although Lily had a feeling that he was thinking about the entire mansion. "This has been the only home I have ever known," he said. Lily pitied him. His entire life had been linked to this one man, and now that man was dead. All the foundations that Charles had built his life upon had now crumbled, and he was so old it would be difficult to begin again. Given what Violet had said, Lily didn't think Charles had been given a fair wage, but he didn't seem upset about it and he might never have known. Living here would have sheltered him from the outside world, and since he didn't need to take care of any of the bills or buy any of the food what did money really matter to him?

"And where were you when this happened? Were you sleeping?" Lily asked.

"Oh no madam. I never retired until the Colonel did. I was taking care of some chores. There is always something that needs doing around here," he said. Lily again felt sorry for the man. She had learned that Colonel Rawlings stayed up late and got up early, which meant Charles had to do the same. No wonder he looked so haggard and worn. Perhaps retirement would bring him the respite he deserved, although she worried that he was one of those people who required perpetual motion to keep living, and when all their routine was taken from them they began to slow down like a clock that was ticking away its last moments.

"What happened when you realized something was wrong?"

"I moved to the study as soon as I heard the gunshot," he paused for a moment and looked sheepish. "Unfortunately I am not as spry as I once was and it took me a few minutes to reach the study. Ava was already there... the poor girl. I'll never forget her scream. It was almost as loud as the gunshot itself," he said.

Lily then asked him where exactly in the house he had been. He said he had been in the lounge, which was on the same floor as the

Colonel's office, but on the other side of the house. At the slow speed he walked it would have taken Charles a while to reach the office, and if Ava had been in the office before Charles she couldn't imagine Charles going to the office, killing the Colonel, then leaving in a rush before Ava reached him after hearing the gunshot, and then coming back. His frailty was certainly counting against him as a suspect.

"I just have one question to ask you Charles, who do you think killed Colonel Rawlings?" Lily asked, wondering if his duty to the Colonel would trump the camaraderie he had with his colleagues.

The answer he gave surprised her though. "Well, I suppose it could have been the writer."

Lily's ears pricked up. "Excuse me, the writer?"

Charles nodded.

"What writer? This is the first I'm hearing of any writer," Lily said, a little annoyed that this information had not been forthcoming.

"I'm sorry madam, but things have been rather hectic here tonight. Sir, was working with a writer on his memoirs. He wanted to write them himself, but he thought he should get a writer to add a bit of dramatic flair to the events of his life. He was here that day. He came by every so often, whenever Sir had a new section of the book completed. His name is Wallace Handel," Charles said. Lily made a note of the name and thanked Charles, telling him that that was all she needed from him. She would have to find this writer and see if he could give her any insight into the Colonel's mindset, and perhaps he had noticed anything amiss with the staff as well.

But before she did that there was one final member of the household staff to whom Lily needed to speak.

Ava walked in and was still trembling. She looked washed out and pale, and her eyes were bloodshot. Her body closed in on itself, so she had the appearance of a scared mouse as she sat across from Lily.

"What is going to happen now?" Ava asked.

"Well, we're going to try to find out what happened to the Colonel. After that it really depends.

"I mean with me. What is going to happen to me, to this house?"

"Well I really don't know. I assume that depends on the Colonel's will."

"Then I shall have to find somewhere else to live, and another job. It was already difficult to get this job, and where am I supposed to work? Will anyone want to employ me when they know the last man I worked for died? It is not as though I can ask him for a reference," she said, her words falling into sobs. She bowed her head and ran her hand along her forehead. Lily gazed at her, wondering if she was being selfish in thinking of herself at this moment in time, or if she was being wise. It was natural for Ava to worry about the future given that she did not have many prospects, and if Violet was telling the truth then she may not have been able to save up much money either.

"I just don't know what I'm going to do," Ava said, and began weeping more openly. Her entire body began to shake and her cheeks were flushed red. Streaks of tears cascaded down her face. Lily spoke comforting words and held out a tissue. Ava took it and mopped her eyes, looking sheepish.

"I am sorry for this. It has been a long day and I know there are going to be longer ones ahead," she said.

"It's perfectly fine," Lily replied. "You just take your time. You've been through a lot. In my experience these things are very hard on the people who find the body. But I'm sure there are other people who are looking for a housekeeper. I happen to know Lord

Huntingdon, perhaps I could ask him if he has any vacancies. I'm sure he would be happy to hire you, as long as you don't turn out to be the murderer," Lily said. It was her attempt at a joke, but it was clearly the wrong time, and the wrong audience. Ava paled and looked horrified. She shook her head vehemently.

"Oh no, please no, I would never..." she said, placing her hands on the table and looking toward Lily with an earnest plea. Lily noticed that her nails had been bitten and the skin around them was all white and torn.

"I'm sorry, that was my poor attempt at a joke," Lily said. "Let's just continue with the questions. Where were you when you heard the gunshot?"

"I was in my room," Ava said, and then told Lily that it was on the first floor, just by the stairs. "I raced down the stairs as soon as I heard it and went into the office."

Lily tilted her head because something about this didn't quite add up. Ava seemed on edge and so scared, Lily couldn't imagine her running into danger. "That was very brave of you Ava. Weren't you concerned that you might have been shot yourself?" Lily asked.

Ava licked her lips and wore a nervous smile. "I just thought the Colonel was hunting again," she said.

Lily arched an eyebrow. "He used to hunt in the house, at night?" Lily asked.

"There has been a bit of a problem with mice recently. The Colonel hated them. He called them rodents and said that they had no place in his house. Whenever he saw one he grabbed the nearest gun and shot at it. He was beginning to keep a gun on him at all times just in case. Whenever he saw one that was it, and he didn't seem to care that he was doing more damage to the house than he was to the mice," Ava said. Lily thought back to the pockmarks she had noticed when she entered the house. She wondered if the Colonel might have been trying to shoot a mouse, but it was unlikely a mouse had been on the ceiling.

"What happened when you entered the office?" Lily asked.

Ava's eyes went wide and her voice dropped, taking on a haunting tone. It had been a terrible night for all three members of the Colonel's household staff, but it was going to have a devastating effect on Ava. Lily knew that this was going to stay with Ava her entire life.

"When I saw him I screamed. I… I wasn't quite sure what to do at first. I couldn't believe what I was seeing. Then I… I took a cloth and placed it over his head. I thought it was the least I could do, for his dignity," she said.

"I understand it's very difficult for you to talk about this. You're doing really well, and I'm sure the Colonel would be proud of how brave you're being," Lily said. Ava brightened at this and beamed with pride.

"But I am going to need to ask you some questions that you may find difficult, but I want you to be completely honest with me. Do you think that the Colonel could have killed himself, perhaps because he realized he was getting old and was depressed that he was losing his vigor?" Lily asked.

Just like Charles though, Ava shook her head. "I don't think that could be the case at all. The Colonel did not think like that, I am sure of it. He had been getting more tired recently though. I think it drained him to be writing his memoirs. He spent so long writing, and then he had to speak about it all again when Wallace came."

"The writer?" Lily asked, her ears pricking up as the writer was mentioned again. "Could you tell me a little bit about him?"

"Oh yes, he's very smart and kind," Ava said, her smile widening. "He knows a lot about everything and he has no end of stories. He's met some very famous people as well. He even signed one of his books for me! I have never had anyone do that before," she spoke with a tone of awe, and Lily suspected that she might be smitten with this man. She wondered if the feelings were returned, and she didn't bother asking Ava if she thought Wallace Handel could have had anything to do with the Colonel's death, as she doubted she would get an unbiased answer. Instead she investigated a different thought.

"The Colonel had a lot of pictures of himself," Lily said. It had struck her as odd because there were just so many, and she wondered how the Colonel felt about having so many depictions of himself staring back at him.

"Yes, he did. He liked to document things that happened to him, although he had turned to using Polaroid's and other cameras in recent years rather than getting photographers to come in and take proper pictures of him. It was always fun to take pictures for him though, especially on his birthday when he got dressed up. He used to take some pictures every day. I don't like to think about the money he used to spend on film."

"Do you think I could take a look at the latest photo album that he made up?" Lily asked, thinking that there might be some clue there. Photographs had the magical ability to capture a moment in time, freezing it and enabling it to be revisited again and again. There might have been something, no matter how small, in one of the photographs.

Ava said, of course, and then went to fetch the album for Lily.

While she was doing this, Lily returned to the office just as Constable Dudley called out to her that the medical personnel were going to be arriving soon. She went into the office for another look around. The pictures that hung on the walls showed the Colonel when he was most proud and in his prime. She wondered how it must have felt for him to be decrepit and yet to stare at the picture of him in his youth, and then to speak about his life and relive all the days that would never come again. She wondered if all the household staff were wrong, and that the Colonel was indeed the type to kill himself if he allowed himself to wallow in despair. Perhaps this writer would be able to give her more insight into the Colonel's state of mind.

She then looked at the Colonel himself, glad that Ava had the quick thinking to cover his face. As she looked more closely at him she noticed that there was a residue on his hand from where the gun had fired. On the one hand the evidence suggested that it was a suicide, but on the other the Colonel just didn't seem the type at all.

It wasn't long before the medical personnel arrived and gathered the body. Charles stood by the door, as if to give a guard of honor. Ava was sobbing again, and stayed near Lily. Violet was nowhere to be seen.

"What are you thinking after you've spoken to them?" Constable Dudley asked in a low voice. Lily was careful about what she said as Ava was within earshot.

"I'm not quite sure. To be honest I'm more worried about the three of them, well, perhaps apart from Violet. But Ava and Charles' lives were so wrapped up with the Colonel's. Without him it's as though they're lost. I'm not sure what either of them are going to do next. My first instinct is that I don't believe they killed him though."

"Then it might turn out that this was suicide after all," Constable Dudley said.

"Perhaps," Lily replied.

Now that the Colonel's body had been taken away, Lily could move around to the other side of the desk. She asked Ava if she could see a copy of the Colonel's memoir, but Ava replied that Wallace Handel had the only copy. Then, Lily turned and looked outside the window. An endless field stretched into the horizon, a smooth emerald plain that was gentle and calming, with trees toward the limits of her vision. She looked down and noticed that the latch was damaged.

"Has this always been like this?" Lily asked.

"Yes. Charles was meant to fix it, but he hadn't gotten around to it yet. It's been like that for a long time," she said. Lily continued staring out at the horizon, furrowing her brow. Sometimes the simplest explanations were the best ones, but she could not just assume that this was the case. A man had died, and it was up to her to find out exactly how.

4

Constable Dudley had taken Lily back to Didlington St. Wilfrid. He had official business to take care of; a benefit of being a consultant to the police rather than actually being employed by them was that Lily did not have to suffer the burden of paperwork. While she waited for Constable Dudley to do this, and for him to learn where Wallace Handel lived, she went to see Lord Huntingdon and Blake.

She arrived at their manor house and greeted the serving staff with a smile, wondering if any of them would have had the same problem as Ava if Lord Huntingdon ended up dead. She supposed they wouldn't as Lord Huntingdon had an heir, unlike Colonel Rawlings.

"Good day Lily! Are you well?" Lord Huntingdon asked, kissing Lily on both cheeks. Blake was more reserved, which was strange for him. She thought back to their earlier conversation and wondered if there was anything wrong.

"I'm okay, although I'm working a case again. It's something I wanted to speak to you about actually. I don't know if you heard or not, but Colonel Edward Rawlings is dead," she said.

Lord Huntingdon clapped his hands and rocked back, wearing a delighted smile. Lily furrowed her brow.

"I'm sorry, did you hear me properly? I just told you that Colonel Rawlings has died. I thought he was your friend?" Lily asked. She glanced toward Blake, who merely shrugged. She could feel the tension radiating off him, and it gave her cause for concern. As far as she was concerned everything had been fine between them. Was there some problem that she was missing?

"A lot of people think that because they don't bother to learn the truth about anything," Lord Huntingdon said.

"But he's a beloved member of the community," Lily said.

"Only to those who never knew him properly," Lord Huntingdon replied, this time his voice was dry and his eyes narrowed. He beckoned for Lily to sit down and join him. "I was in the army when I was younger, you know. I never particularly wanted to fight, but it was expected of men like me at that age, and my father certainly wanted me to join. I think he wanted to straighten me out, I had a bit of a reputation back then," Lord Huntingdon had a twinkle in his eye and he winked at Lily. She looked toward Blake, who rolled his eyes. "Anyway, who should have been my commanding officer? But Rawlings," he spoke the name in a heavy breath, evidently showing his dislike for the man. "That man's focus was discipline. It was as though he knew nothing else. There was no give with him. He was as rigid as a brittle tree, and he often snapped in anger as well. I remember pitying his children, until I learned that he had none. It was a blessing too. I can't imagine what they would have gone through to have him as a father," Lord Huntingdon shook his head and shuddered as he thought about the memory.

"What a bully he was... the army is the only place a man like that can be celebrated, well, that or as a prison warden," he continued, cackling at his own joke. "I think it says a lot about the institution that he was able to rise so high in rank even though none of the men who served under him would have ever wanted to reward him. I'm glad times have changed and that we aren't forced to serve anymore. It's something I'm glad my own son was spared," he said, looking at Blake.

Lily wondered how Lord Huntingdon could be so cavalier about another man's death, but she supposed that when people reached a certain age they had a different outlook on death.

"What was it he did exactly?" Lily asked.

Lord Huntingdon sighed as the laughter faded. "He just liked bullying us. He locked onto any weakness and focused on it in the name of discipline, but really he was just cruel. There was a lad with us, his name was Ronald. His father had put him into the military in the hope that it would make him fitter and more robust. He should never have been there. He couldn't finish the drills, so Rawlings had him running laps out in the wind and the rain. The boy barely got any sleep. In the end he was turned into a wreck. I used to hear him sobbing at night, although he always denied it. I tried to stand up to Rawlings, but he took this as an insult and punished me. I had to peel potatoes, paint walls the same color over and over again, and it was never ever good enough. It was the worst time of my life. I was glad when I came out and I vowed that no son of mine would ever serve, and that's a promise I kept," Lord Huntingdon said, looking proud of himself. "I know that Rawlings has a good reputation around here because of some of the things he's done, but I know what kind of man he was deep down, and I think he knew it too. I don't know why it's always the cruel ones who live the longest. It never seems very fair to me," he said.

This was revealing though, for if it was true it meant that perhaps the Colonel had more enemies who might wish him ill. She was going to have to go through the military record to see if there was anyone who might have wanted to take revenge for some treatment at the hands of the Colonel, although she thought it odd that a grudge would have taken this long to be confronted. She thanked

Lord Huntingdon for his time and then rose, gesturing for Blake to leave the room with her as she wanted a moment in private.

"How is it going?" he asked.

"It appears that it's suicide, but there are still a few things I need to explore before I make that determination myself. But never mind all that, how are you?" Lily asked, reaching out to touch his arm.

"I'm fine," he said, although his body was rigid with tension and he didn't seem like the warm, comforting Blake that she was used to.

"Are you sure? Because I'm getting the feeling that there's something wrong here. Is it about the mystery? I thought we spoke about this before Blake. If anything has changed then please let me know because I don't want to run around solving murders if you're going to be unhappy with it."

"I'm not unhappy with it Lily. I know that this is what you do."

"That's what you said before, but it has been a long time since I've been involved in solving a crime."

"Lily, it's fine," he said tersely.

Lily sighed and turned her head to the side. "You know Blake, I have enough mysteries in my life. I don't need you to be one too. If there's something going on that I need to know about then I'd prefer that you just tell me."

"There's nothing going on Lily, and there's nothing you have to be concerned with. I'm fine, I promise," he said, and punctuated his words with a kiss. He pinched her chin in his hand and she looked into his eyes, but she wasn't quite satisfied with his answer. There was something in the way he looked at her that was a little off, but she supposed if he wasn't going to tell her what it was then she was going to have to wait to find out. The dead needed her full attention right now. He told her that he would see her later, and that he had cancelled the reservation at the restaurant. It was in moments like these when Lily wished she had more experience in relationships because she might know what to do or say to make things better, but somehow things were more complicated than the murder cases she worked on. At least with them she could look at the suspects and their motivations. Here there were only two people involved, and yet things were murky. She just wished Blake would tell her what was wrong. It wasn't like him to keep things from her.

5

W ord was abuzz with the news of Colonel Rawlings' death. He was an eccentric man, but well known throughout the village. In fact he was one of the most notable men to come from the local area, and he had made it his life's mission to ensure that the heritage of the area was maintained. Although he did not have a family, he made sure that a lot of his wealth went into local heritage sites in order to keep old buildings repaired and in as good a condition as they could be in, which Lily found ironic considering the state of his own house. She heard nothing but good things about the Colonel, which contrasted with Lord Huntingdon's reaction, but perhaps it was because none of these people knew the Colonel as Lord Huntingdon had.

Constable Dudley had been busy and had learned that the Colonel's will noted that all his wealth was to be donated to the community of Didlington St. Wilfrid, ensuring the maintenance of historic buildings and the grounds of the village. This was in addition to the money he had donated to local causes while he had been alive, and to many people the Colonel was a man that would be missed. Lily wondered if these donations had been a way for him to make up for the fact that he did not have a family, or if it was just a matter of pride for him. Perhaps he saw it as his responsibility, since his family had been in the area for a long time.

She would never be able to learn the truth though, because Rawlings was dead.

The mayor had sent word around the village that there would be a memorial service for the Colonel to celebrate his life and his achievements, as well as his philanthropic work for the community of Didlington St. Wilfrid. Lily was certain that there would be a grand turn out, although she wished she had some better information regarding the cause of the Colonel's demise so that people could have proper closure. However, she reflected, perhaps in a memorial service it was not his death that mattered, but his life.

While this was being arranged, Lily decided to take a trip to see the writer, Wallace Handel, in the hope that he might be able to shed some light on the Colonel's mindset on the day that the man died.

Lily enjoyed the drive through the countryside. Wallace Handel lived in a cottage a few hours south of Didlington St. Wilfrid. When she and Constable Dudley had learned of this, he had been surprised when she said she was going to drive down. In England any drive over an hour was seen as a marathon, whereas in America it sometimes takes an hour just to get to the nearest store. It had been a while since Lily had gone on a long drive, and she enjoyed the scenery around her. It reminded her of times in her youth when she would take road trips; loading up on snacks and embarking on a journey into the unknown. She would leave the city behind and go into the desert, heading toward the Grand Canyon. It was dry and brittle, with copper sand and purple mountains stretching as far as the eye could see. England was different. It was lush and vibrant with life. The scenery was dotted with sheep and cows and horses, all grazing on the wide rolling fields. Farmhouses and small villages dotted the landscape, small pockets of civilization housed within this flourishing environment. Her heart swelled with a profound sense of awe. England may have been a small country, especially compared to

America, but it had a lot of heart and she loved being nestled in the bosom of nature.

However, while she appreciated the natural beauty, she was glad that she lived in modern times for she could use the benefits of technology. The roads were winding and narrow, twisting through hedges and forests with barely any signs along the way, so she relied on her Satnav. If this hadn't been possible she didn't know what she would have done, for she was hopeless without maps. She drove slowly, worried that she might run into a car coming the other way, or a tractor that would plow into her. The roads were only wide enough for one vehicle at a time, and occasionally there were small bays cut into the bushes to allow cars to stop and let another vehicle pass it by. Aside from this she also passed people riding horses too, and they always waved to her. These sights were never visible in L.A., and she felt blessed that she was able to see them now.

Eventually she reached Handel's house. The cottage was tucked away in a copse of trees, far from the nearest city and a good walk away from the nearby village. She actually almost missed the entrance as a weeping tree stretched its branches over the sign, obscuring it from view. She braked sharply and turned the car into the driveway, hearing the sound of bracken cracking under the weight of the tires. The cottage was picturesque, larger and broader than Starling Cottage. Its thatched roof was wide and smoke rose from a chimney. Outside there was a large, sleek car, its gleaming shape at odds with the rustic scene in which it was found. It was the perfect place to write, Lily thought, and she thought that she might need to get a place similar to this one day.

After knocking at the door it opened, and a man stood there. He was tall, and was likely in his mid-thirties. He had blonde hair that was cut fairly short, and he wore jeans and a plain white t shirt. He was in good shape, and had a strong jawline. His chin showed a dimple. He did not greet Lily with a smile.

"If you're looking for the village, it's that way," he said, gesturing with his thumb.

"Actually I came here to find you. You are Mr. Wallace Handel, are you not?" Lily asked.

His eyes narrowed. "I am of no concern to anyone. I'm sorry, but I do not appreciate any fans who come here. I don't know how you found where I live, but please respect my privacy." He spoke with a faint European accent, although Lily couldn't quite place it. It wasn't the same as Ava's.

"No, you misunderstand. My name is Lily McGee. I'm not here because I'm a fan. As it happens I'm a writer too," she said.

"And what do you write?" he asked.

"I write crimes, you know, mystery fiction," Lily said, hoping the fact they shared a profession would lead to a common understanding. Those hopes were quickly dashed by the look of derision in his face.

"I have no interest in fiction. I write the truth," he said.

"Well, I'm sure you could argue that every story has an element of truth in it," Lily said, although Wallace made sure she knew he had no intention of discussing the philosophy of writing with her.

"I am very busy. I don't know why you thought you could come here unannounced. I would like you to leave," he said, and went to walk back inside his cottage.

"I'm here about Colonel Rawlings. He's dead," she blurted out. Wallace hesitated and did not close the door.

"I see," he said calmly. "I suppose you had better come in then."

Lily walked through the doorway into a vast lounge. Awards hung on the wall. On the far side of the room was a bookshelf that was crammed with books. A large coffee table was in the middle of the room, with two couches on opposite sides. She could not see a television. There was a fireplace in the room, although it was not lit at the moment, and a thick rug laying on the floor before it. She imagined it would be incredibly cozy during the winter, and quite romantic. She thought the place was wasted on him, for it didn't seem as though there was a female presence in the cottage.

He sat down and looked at her expectantly. He did not offer her anything to drink.

"What is it you would like to know?" he asked when she did not say anything. She perched on the couch and folded her hands into her lap.

"You were working with the Colonel on his memoirs, yes?" she asked, and Wallace nodded. "How did this arrangement come about?"

"He knew of my work with some other retired military officers and thought that I could breathe life into his words. He said he was impressed with the way I managed to capture the honor and nobility of these men, and he wanted me to do the same for him. He was willing to pay well, which I was glad about as I did not realize the scope of the project until I met him for the first time."

"What do you mean?"

"Usually when I write a biography of someone they like to focus on a particular part of their life so that I can drill down deeply into it and find the truth. The Colonel wanted to capture everything, from birth to the present moment."

"To his death now," Lily said. Wallace paused for a moment, but otherwise he did not react. Lily found this to be strange, but he seemed to be a reserved man and she wasn't sure he would readily show his emotions about anything, unless that emotion was contempt.

"Indeed."

"How long have you been working for him?"

"Only a few weeks. We have had long sessions together, but the project is still in its beginnings. There is a long way to go, although now I suppose I shall have to wait and see if there are any instructions concerning them. I don't know if he would have wanted them to be published if he could not see the final draft."

"Could I see them?" Lily asked.

This time Wallace did flinch. "I am not sure that would be wise. I have not organized them well. As a fellow writer you must know how delicate a story can be, if the words are wrong, well, it can ruin everything."

"I think the fact that a man has died has ruined a great deal," Lily said. "I wouldn't be reading it for artistic merit anyway. I just wish to know if there was anything in there that might show something about his state of mind, especially anything from the day he died. At the moment I'm still trying to figure out what kind of man the Colonel was."

"He was a proud man," Wallace said with a curt smile. "He was a man who took great pride in what he did and what he achieved in life, and he had a surprisingly good memory as well. He followed his duty and he found his meaning in his loyalty to the crown. I think he wished that he could have served right up until his death. The army was everything to him."

"Well, perhaps I could read that for myself?" Lily asked. She got a sense of respect from Wallace. He must have had an affinity for soldiers as well if he had written biographies about other military personnel.

"I'm afraid you can't. I have not transcribed our conversations yet. You see I made tape recordings of all our meetings and then wrote them up afterwards, weaving my own words in with the Colonel's."

"Then perhaps I could listen to the tapes," Lily said evenly, meeting Wallace's gaze. His lips parted, as though he was going to say something, but then he thought better of it. Lily had another card to play anyway. "Since this is a murder investigation I could return here with a warrant, or even with a search party to find all the tapes, but I'm sure you would hate having police officers tramp through your home."

"I would indeed," Wallace said after a moment. He rose and went to a cabinet. It was filled with cassette tapes, all of which were neatly labeled. He pulled one out and handed it to her. The date written on the label was the day Colonel Rawlings died. Lily tucked it in her pocket, while Wallace remained standing.

"Is there anything else?" he asked, annoyance in his voice when he realized that Lily wasn't leaving.

"Actually there is. Did you notice anything amiss about the Colonel that day? Was he on edge? Was he nervous?"

"He seemed normal to me," Wallace replied. "And I know what you are going to ask next. Perhaps you would not ask it directly, but I can tell the question is on your mind. You crime writers are always the same. Everything is a puzzle to be solved, and everyone is under suspicion. It can't be pleasant to wonder what secrets people are hiding whenever you look at them."

"What am I going to ask?" Lily said in a challenging tone. He was beginning to get under her skin.

"You are going to ask me where I was when the Colonel was killed, and by that you are really asking if I killed him," he said, scratching the back of his hand as he did so. Then he let out a small chuckle. "Believe me, it would do me no good to kill him. I have only been paid my advance. I would not want to miss out on the rest of the generous fee he was paying me. If you have any doubts then you can contact my other clients, who are all still alive. As far as I'm concerned I was supposed to meet him to discuss the next part of the book. I shall have to think about what I am going to do next," he said.

"Is that all you're really concerned about Wallace? Don't you care that he died?"

"Of course I do, but I have to maintain a certain level of detachment with my subjects. It is a fine balancing act, but one that I have mastered through the years. If you wrote about truth rather than these novels then perhaps you would understand," he said dismissively, and Lily decided that she had had enough of his company for now.

6

When Lily returned to Starling Cottage she shouted out to Joanie, who was enjoying a bath. They had a brief chat about what they had both been up to, and Lily confessed to Joanie that she was a little worried about the way Blake was behaving.

"I wouldn't worry about it too much. You know that Blake is crazy about you. Maybe he just has something that he needs to deal with. I'm sure he'd tell you if it was important," Joanie said. Lily wished she could believe her without experiencing any doubt at all. She left Joanie to her bath and shut herself in her study after pouring herself a mug of tea and making herself a sandwich. She drew the curtains, plunging herself into darkness, and then switched on a lamp. She opened the photo album and plugged the cassette into a tape recorder, which she had borrowed from the police station.

The photo album was filled with pictures of the Colonel and his household staff. It seemed that usually Ava was the one taking pictures. The Colonel looked happy in the pictures, and he seemed to enjoy the grounds of his home as many pictures were taken outside. There was one picture that struck her the most though, and this was one taken alongside an old portrait of him. It showed just how much he had been diminished by time. The broad shouldered soldier with a stern face had withered into an old, frail man. Time could be the cruelest thing of all sometimes, she thought.

It was eerie to hear his voice coming through the tape recorder as well, for she knew that she was listening to the words of a dead man. His voice was thin and reedy, but it still retained some of the bombastic qualities he must have had during his time in the military. He spoke well, with a clipped upper class accent that was being lost to time. Lily found her attention drifting as much of the talk was on his reverence for the military and how it had inspired him as a boy. He listed off names of soldiers who had been his heroes and battles that he had recreated with his toys, although Lily had little to no interest in military history so the list of names and dates didn't do anything to help her focus. Instead she wondered about the man speaking, and if he had realized how empty his life had become. He had devoted himself to the military, but had it really been that beneficial for him? He had no family to take care of him, no loved ones to continue his legacy, nobody to teach his wisdom too. She thought that perhaps that was why he wanted to write his memoir, so that a piece of him would be left behind for someone to latch onto. But perhaps going over his life had made him realize how lonely he was, and he had decided to end it all before the misery could take hold.

In the recordings the Colonel went on to show his love of the military by saying that it should have been kept as a larger part of society, and that if it was it would have solved plenty of the problems that plagued modern society. He said if it was a major force then there wouldn't be as many criminals, and more people would be able to take care of themselves. It was clear he was disappointed in the way the world had turned out after all the measures he had taken to defend people's freedom, and he wished there were stronger men who could ensure that the morality of the world was kept intact.

Lily winced at some of the things he was saying. She supposed she shouldn't expect such an old man to have progressive views, especially since there was nobody in his life to challenge them, but even so some of the things he was saying were quite controversial. She imagined Wallace scribbling furiously, having to cut certain things out. But as she was leafing through the photo album her eyes caught something that was amiss. She looked more carefully at the photo of the Colonel in his office and frowned. She closed her eyes and thought back to the crime scene, and then looked at the photo again. Then she spotted what was different.

Near the desk there was a case. When she had been there in person the case had been empty, but in the picture there was a necklace. She wondered if robbery could be a motive.

It was a new lead, but as she peered out of the curtains she realized that time had moved swiftly and it was too late to return to the Colonel's house now. Instead, she decided to listen to more of the recordings. When the Colonel spoke about something he was passionate about his voice grew richer and she imagined he sounded more like he did when he was younger. Sadly, the things he was passionate about were controversial. He went on a tangent about modern society and had unkind things to say about minorities and people he considered to be weak. He said that sterner discipline was needed and that he was in favor of bringing back the death penalty. Lily found these things abhorrent, and she was beginning to agree with Lord Huntingdon's assessment of the man.

The thing that chilled her most though was when the Colonel shared some of his experiences serving in the war. He spoke about men dying in a matter of fact manner, as though he was describing a trip to the store. Her blood ran cold as he relived killing men. Did he feel guilt? Did he feel any remorse at all? Had remembering this made him want to punish himself?

If so, she wasn't hearing it in his voice. He wasn't boasting about it either though. It was just simply something that had happened in his life, and he was recounting it like any other memory. Something told Lily that it shouldn't be like this though. It shouldn't be just any other memory. It all seemed so senseless as well, and Lily couldn't help but feel her dislike for the man growing. Yes it was war, and yes he was a soldier, but was that really any excuse? There had to be a line drawn somewhere, and after investigating so many murders Lily wasn't sure she could accept any reason for killing to be moral.

By the time she went to bed she was left chilled by all she had heard, and she was grateful that the world had changed to such an extent that the vast majority of people did not have to endure what Colonel Rawlings had been through.

The memorial service had been arranged swiftly by the mayor, who had used all his contacts to ensure that the Colonel was given a good send off. The whole village had pitched in to cater for the event as they all wanted to repay the Colonel for the help he had given to the community, and the help that he would continue to give through the trust that was going to be set up after his death. Lily attended with Blake. She did not ask him if anything was wrong, but she could still sense some tension between them.

Reverend Brown gave a lovely service and many tears were shed over the Colonel. A large portrait of him had been brought from his home and stood before everyone. The Reverend mentioned that he was survived by no family, but that in a way the entire community was his family, and that he would always be remembered. There were those who had been more affected by him than others. These people stood up and talked about how donations to their businesses had helped them get through lean times, or to expand. They spoke of the Colonel's generosity and his willingness to help those who showed a spirit of endeavor. To Lily's surprise, Marjorie rose as well, and spoke of how the Colonel had donated some funds to her bakery after her husband had died to help give her the chance to grieve him properly, without having to keep working through this difficult time.

It was an emotional service, but Lily couldn't help wondering if people would have still felt the same if they had heard what she heard. There were many facets to people though. The Colonel was a complicated man. On the one hand he was bigoted, authoritarian, and a bully, and on the other hand he was generous to the community and philanthropic with his money. It was strange to think that this singular man had all these conflicting characteristics, but that was part of the beauty of life she supposed, and she would have to keep the truth to herself. It wouldn't serve any purpose for her to ruin the memory of the man for all these people.

When the service was over, Reverend Brown came over to speak to Lily and Blake.

"How was your dinner out the other night?" he asked with a sense of anticipation.

"We had to cancel. That's the day that the Colonel was discovered," Blake said sternly. The Reverend looked a little disappointed. Lily wasn't sure why. She did not fail to detect the tension within Blake's voice though, but he kept saying it was fine. The sooner the matter with the Colonel was over the better, she thought, because then she could confront Blake and perhaps he would finally share what was on his mind.

But before that she had to return to the Colonel's house and see if she could find the truth behind this necklace.

Lily waited for a good while until the memorial service was over to allow for Violet, Charles, and Ava to return to the house too. When Lily arrived Violet scowled.

"This really isn't the right time. Haven't you asked us everything you need?" Violet said. Lily narrowed her eyes at her. Violet had already admitted that she wasn't happy with her pay. Had she tried to steal a valuable item of jewelry to make up for all the money she thought she was owed? Lily wasn't going to come out with direct accusations yet, but they were swimming in her mind.

"The case is ongoing and new information is being discovered all the time. I'm afraid sometimes things come out that I wasn't aware of before," Lily replied. Violet merely rolled her eyes. Lily turned to Ava, who was trembling with nerves and sadness again. The memorial service must have brought all these feelings to the surface. Lily hoped that now that it was over Ava would be able to find some closure. However, there was still this matter to attend to.

"Ava, I noticed in one of the photos that there was a necklace beside the Colonel's desk. However, when I was in the office the other day it was gone. Do you know anything about this? I thought it might have something to do with the way he died," she said.

The words were barely out of Lily's mouth when Ava broke down completely. Tears flowed from her eyes and her words were cracked and brittle things. Her breath heaved and when she spoke her cadence faltered and stumbled. "I did it," she cried, and let her face drop into her hands. Lily was confused, for she didn't think that Ava was capable of murder.

7

L ily stood in front of Ava. Charles pulled out a handkerchief and handed it to Ava, while Violet stood with her hands folded across her chest.

"I think you might need to elaborate on this Ava. What happened?" Lily said gently.

Ava dried her eyes, but she could not stem the flow of tears. "I went into his office and he was just sitting there and I... I couldn't help myself," she said.

"But why? Why would you do this? I thought you were afraid of having to find a new job?"

"I am! That's why I did it!" Ava said. She blew her nose and breathed in deeply. Her face was swollen and puffy, but that did not help clear things up for Lily.

"I'm sorry Ava, but I still don't understand. Killing Colonel Rawlings wouldn't have saved your job. Why did you kill him?" Lily asked.

It was at this point that Ava stopped crying. She blinked and looked at Lily, and that same look of fear that had been on her face when Lily had first met her was back again.

"I... I didn't kill him," Ava said.

"But you just said you did," Lily replied, wondering if she had missed something.

"No, no I didn't kill the Colonel! I could never do something like that. No I... I took the necklace," she said, bowing her head in shame. "I was the only one in the office. When I saw him there I knew that I was going to have to find another job, and I was so worried. I don't have that much in savings and then I saw all these beautiful things. I knew it was going to take a while for Violet and Charles to reach me, and that necklace was just staring at me. The Colonel had so many beautiful things I didn't think he would mind if I took one, not when it was going to help me so much. I could almost hear him telling me that he wanted me to have it. He always said that he would take care of me, and it seemed like the right thing to do in the moment, but I haven't been able to do anything with it. Then you came and I was so worried that you would find out. I've been distraught. It was stupid. I panicked and I wanted to put it back, but then I worried that someone would see me and I just... I didn't know what to do. I'm so sorry. I don't want to go to jail. I didn't mean to do anything wrong, I was just so scared. Are you going to arrest me?" she asked, looking up at Lily as though Lily had Ava's entire life in her hands.

Lily exhaled deeply. It was a crime, yes, but she wasn't sure that Ava deserved to be punished for this.

"You know Ava, I'm only a consultant with the police. I can't actually arrest anyone, and Constable Dudley isn't here. I think that if you put the necklace back then there's no harm done. I'm not certain you should be punished for simply moving a necklace from one room to another. I'm sure you just meant to take it away for safekeeping," Lily said with a reassuring smile. Ava sighed with relief and Lily almost thought that she was going to faint. She nodded and quickly scuttled away, presumably to put the necklace back where it belonged. That was that mystery solved, Lily thought, although it didn't bring her any closer to the real culprit. There had been no robbery. The only thing stolen was Colonel Rawling's life, but why would anyone have done this? The only person she had met with a real, solid motive was actually Lord Huntingdon, but she wasn't about to believe he had killed Colonel Rawlings. Even though Violet was a callous woman she had already admitted that she was looking for other work. It wasn't as though Colonel Rawlings was keeping her captive. Charles was too old, and far too devoted to the man, and Ava was too terrified. Wallace Handel had only begun the project and hadn't been paid his full amount yet, so it didn't do him any good to have the Colonel dead either.

Who was she missing? What was she missing?

By the time Lily returned to Starling Cottage and had something to eat it was late. She went into her bedroom, but she couldn't sleep immediately. Her mind was perplexed by this case. It had seemed so simple at first, and she wondered if this all really was a suicide. Then she thought she should listen to the rest of the tape just in case there was something she missed. Even if she didn't then she would have exhausted all the clues and she could tell Constable Dudley that there was nothing to learn, and then she could face Blake to try and figure out what was happening with him.

She lay in bed and played the tape. The Colonel's voice filled the room, occasionally interspersed with Wallace asking a question. It was more of the same to begin with, with the Colonel recounting some aspects of his life. He was a young soldier in World War II, having lied about his age to get a commission, but rising to a position of responsibility quickly after taking control of his squad when his commanding officer was killed. Back then, he said, positions were given on merit. If you were good enough then you were old enough. Lily wondered if she was going to reach the point where he trained Lord Huntingdon. She was curious to learn what Colonel Rawlings was going to say about him.

She could feel herself drifting off to sleep when suddenly she was jerked awake. The rhythm of the conversation up until this point had been lulling and constant, but suddenly Wallace's tone had changed. She wound the tape back to find the point at which it happened. The Colonel was talking about a group of enemy soldiers he had been escorting back to a prisoner of war camp.

Where was this? Wallace asked.

The Colonel searched his mind. Lily heard Wallace mutter something under his breath. He asked for a date, and again the Colonel was able to supply it. Lily listened as Wallace's tone became terse and agitated. He pressed for more details, gunning for specific information. He had not been this focused during the other anecdotes, preferring to allow the Colonel to speak at length about anything he wanted. This was different though. Colonel Rawlings was barely able to get a word in before Wallace was asking him another question, and his accent became more pronounced. Lily had no idea what had gotten under his skin to make him act like this. When she had met him he had been unflappable and she couldn't imagine him being flustered, but there didn't seem to be anything different about this anecdote than some of the others. The Colonel didn't seem to show any remorse about this and even said that the prisoners got what they deserved. Wallace then suddenly berated the Colonel with a flurry of vicious words, and then the tape shut off.

Silence filled the room. Lily's mouth hung open in shock. Whatever tiredness had made her soul weary was shaken from her body. She rushed to her computer, switching it on and typing quickly into the search engine to see if she could find any further information about this incident. After putting in the dates and the location she found reports of what had happened. God bless the Internet, she thought, it helped unearth so many secrets that would otherwise be lost to time. Through this she discovered that one of the prisoners was named Wilhelm Handel, and although the details were sparse, after reading what information was available and having listened to the

Colonel's recollections it was clear that they had been escorted under harsh conditions. She dreaded to think what had actually occurred there, especially after what she had learned from Lord Huntingdon. If Colonel Rawlings was that strict with his own men she couldn't imagine what he was capable of with the enemy.

But now she knew there was another motive for Colonel Rawling's death; revenge.

8

———

ily rushed to Constable Dudley's house and roused him, waking him from his slumber. His eyes were filled with sleep, but once he realized what was happening he gained awareness and rushed to his car, while Lily jumped in the passenger seat. This time she wasn't going to go without Constable Dudley as backup. They raced through the bleak night, and it didn't take as long to get to Wallace's cottage as it had before, as there was little to no traffic on the roads. The headlights of the car blasted through the darkness and caught small animals in the headlights. They scurried away, running free from the danger of impact.

Lily wondered if Wallace was going to be there when they arrived. She thought herself foolish for telling him everything that was happening, for it might have given the game away if he had indeed killed the Colonel. In her mind she was beginning to piece things together. During the drive Lily spoke about some of the things she had learned about the Colonel.

"Perhaps he did all his community work to make up for the things he did during the war," the Constable said.

"I'm not sure anything can make up for that. He was such a young man too… he had his whole life ahead of him. He can't have been able to see a future for himself as anything other than a soldier," Lily said. War had forged him into the kind of man he had become, that young man who had wanted to represent his country had been put into a position of power and he was transformed, changed into the cruel commander who saw discipline as the only way to move forward in life.

They arrived at the cottage and hammered on the door. Wallace answered. He had shadows under his eyes, and he didn't appear to have been sleeping at all. He sneered when he saw Lily.

"I know you did it," Lily said.

Wallace sighed and curled his lip. His gaze moved from Lily to Constable Dudley, perhaps considering his options. They were blocking his way to his car though. He didn't make any motion to move, at least not yet.

"I'm surprised it took you so long to come and talk to me again. I assume you've listened to the tape?" he said.

"Yes, I have," she said, and she noticed that Wallace was scratching his hand again. A rash had appeared in the skin between his thumb and forefinger, and now she knew what it was from. "I take it you're allergic to the residue from the gun?" she asked, framing it as a question even though she already knew the answer.

"I would not act so smug. You only have the answer because I gave it to you," he said, and sighed again, shaking his head slightly. "This is why writing the truth is much better than fiction. You have to cope with what is presented to you, unaltered, rather than twisting things to fit your agenda."

"The only thing that you've twisted is a man's life," Lily said.

"I did no such thing," Wallace said.

"Then what happened after the tape shut off?"

601

Wallace's voice dropped and he spoke more slowly. "He realized then that I was more invested than I should have been. It took him a couple of moments longer than it should have, but he put the pieces together. He realized I had the same last name. The moment he spoke my great grandfather's name I roared and then he knew, he knew that I shared the same blood he had spilled all those years ago. I remember my family talking about the way he had been treated. All these years the Allies have spoken about how cruel the Germans were, and we have had to carry that burden with us. But war is cruel on both sides. There are no angels on the battlefield, and Rawlings certainly was not one. It was not fair that he survived this long without suffering any consequence. The man had no remorse either. He sat there and said he was proud of what he did. I gave him a chance to apologize, but he didn't think he did anything wrong. I left him then, but I knew I had to make it right. I thought about it long and hard for the rest of the day. I returned at night, sneaking across the field in the darkness. I opened his window and came to confront him. I had to make it right because I am the only one left. I didn't want him dead, I only wanted to wipe that smug look off his face.

He saw that I was about to attack him and he had a rifle at the ready, but he was old. I pushed it back, aiming it away from my face. I closed my eyes and heard the shot, and when I looked up I saw what had happened. I knew nobody would believe me, so I escaped out of the window again and ran across the field. I assumed people would think it was a suicide, but then you came to speak to me."

Lily knew that when Ava entered the room she would have been so concerned with the body of the Colonel that she wouldn't have thought to look out of the window, and by the time anyone did Wallace would have been long gone. But there was something that didn't sit right with her.

"But you had the tape. You could have destroyed it. You and the Colonel were the only ones that knew it existed. You could have told me that it was lost and you only had tapes up until that day. I hate to admit this, but you could have gotten away with it. Why did you give that tape to me?" she asked.

Wallace gave her a sad smile.

"Because I shouldn't get away with it. If I did then I would be just as bad as the Colonel. I have spoken with other soldiers Lily and some of them have shown remorse about their actions in the past. I have seen how war haunts them. I was hoping that Rawlings would be the same, but the man was impossible. Victory is celebrated when a war is over, but nobody truly wins. I have seen the shame carried by my family over the actions of my great grandfather, even though at the time he only thought he was defending his nation. Our pride has been broken, but nobody is there to ask him about his story. I just wanted a chance for the truth to be heard, to put an end to this legacy of shame."

These things were all old, Lily thought, and in the past, but some things should never be forgotten.

"I did this for my ancestors, and for all the victims who could not confront those who victimized them. Even if I brought the truth to light the Colonel was so old he never would have been arrested or even been put on trial. He was celebrated even though he had hurt people. I only wanted to scare him, but things went tragically wrong. But I am not going to hide from my crimes like him, nor am I going to carry guilt and shame like those who walked before me. I want the truth to be known. I want people to understand why I did this, and that there are still victims out there who deserve justice, who deserve an explanation and an apology."

"But you're throwing away your career," Lily said before she left Constable Dudley to arrest him. In a strange way Wallace seemed relieved.

"I'm sure my publisher will pay a lot of money for a book written about life in a prison, and I think they will appreciate a book written that tells both sides of a story as well. I will still write about the Colonel, but I will also write about my great grandfather as well.

Then people can make up their own minds about the Colonel," he said.

Lily left Constable Dudley to arrest him and walked back to the car. He called for another car to arrive at the cottage, which would take Wallace Handel to the nearest police station. Lily leaned against a tree and thought about everything that had happened. The Colonel had not been a good man at least not in the war, but he had still accomplished good deeds and inspired people to mourn him after he died. Was that enough to wipe away his crimes in the past? No, it was not, she thought, but it was a shame that he had died rather than been able to understand why his actions had hurt people, and why he needed to show a little recrimination. It was a shame about the mice, she thought as well, because if they hadn't been scurrying around the Colonel's house he might never have had that gun on hand, and this wouldn't have ended in his death.

In a way though a war had been responsible for the Colonel's death, his war on mice.

9

Lily eventually returned to Didlington St. Wilfrid and was about to be able to go to sleep, but just as she was standing outside Starling Cottage she exhaled sharply and shook her head. She had promised Blake that she was going to speak to him when the murder had been solved, and she did not want to wait until the morning. She got into her car and drove the short distance through the village toward Lord Huntingdon's manor. She parked outside the gate and walked to the side of the house, where Blake's bedroom was located. There was a small balcony jutting out from the house, and a large window beyond. Lily picked up a handful of small stones from the ground and heaved them high. They rattled against the window and she hissed Blake's name in a sharp whisper. It took a couple of handfuls of stones, but eventually a light was switched on and flooded out into the darkness of the night as Blake opened his balcony doors. He was tying a robe around him and rubbing his eyes.

"Lily? What are you doing here?" he asked. Even though his voice was low it came through loudly and clearly in the still night. He chattered his teeth and rubbed his hand together. "It's freezing.

Lily had a scarf wrapped around her neck and a woolen hat that covered her honey brown hair.

"I need to talk to you," she said in an urgent tone.

"Now? It's the middle of the night. Can't this wait until morning?"

"No, it can't wait until morning. Something is the matter and I need to know what it is. You've been acting strange around me and I don't like it. I've finished working on this case now, so I'm here to listen to what you have to say."

"Well I'm not going to do it out here. Are you mad? Come inside," he said, closing the balcony doors behind him. Lily waited by a side entrance. It opened shortly after she and Blake had finished speaking outside, and she was secretly glad that he had insisted on her coming inside. The chill touching her cheeks was now touched with warmth. She unwrapped her scarf, took off her woolen hat, and her coat. Blake was still in his robe. They moved to the kitchen where he made some cocoa.

"So what happened with the Colonel?" he asked. She was a little frustrated because she was there to talk about Blake's feelings, not the case, but she entertained him for a few moments and told him everything that had happened.

"Wow, I can't believe it was an old thing like this that cost him his life," Blake said.

"I know, and I'm not sure how I feel about it. I'm not looking forward to the moment when Wallace's book comes out because people here are going to have to face the truth about the man the Colonel really was."

"I think men can be more than one thing to different people. I'm sure he'll still be remembered fondly by the people of this village, even if he may have done unsavory things to other people," Blake said.

"I'm just not sure if that's right though," Lily said.

"I suppose that's not for us to decide," Blake said. "All we can do is pray on it if we're struggling that much."

"And I will do just that," Lily said, enjoying the rich, thick taste of chocolate that swam across her tongue. "And have you been praying a lot recently?"

"I pray every day. I'm a good Christian," Blake said, acting as though he was taking offence.

"I'm sure your father will be so proud," Lily said, mirroring his tone, but then her gaze dropped. She spoke more seriously. "Blake, I know that something has been different recently. I don't know what exactly, but I meant what I said before. I don't want your feelings to be a mystery to me. Frankly I don't understand what's happened. I thought we were getting on really well. I thought we were solid. We were in a really good place and getting closer and we've been taking all these trips... I don't know, am I wrong here? Have I misinterpreted things? Is this not as good as I think it is?" she asked, and as she spoke fear rose within her. Now she resembled Ava, who had been wide eyed with terror, and Lily was worried her entire world was going to fall apart. She had invested so much of herself in Blake and had given so much of herself to the relationship. It had been filled with risk to put her heart on the line, and was it all really going to end now?

She looked into Blake's eyes and wasn't sure what she could see.

"Lily," he began, but she cut him off before he could say anything else.

"I've been thinking about it Blake and I'm ready to give up mysteries. I won't solve the murders if you don't want me to. I'm sure that Constable Dudley can handle things himself. It was months in between cases and the only thing I can think is that you were unhappy that I took on another case. I know at the beginning you said you didn't mind, but maybe after spending so much time together you changed your mind. I know that it's dangerous and that I'm always putting myself into risky situations, and you might not want that if we're going to take the next step. That's if you want to take the next step of course. Maybe I'm all wrong and you don't want to do that at all. I don't know. I just know that I love you Blake and I don't want things to end now. Please... please just tell me what's going through your head," she said, the words tumbling out of her in a thick torrent, the anxiety growing within her mind. It felt as though she was on uneven ground, as though it was shaking underneath her and she was losing her footing. As she looked at Blake she regretted coming here. Now it was all hanging in the balance and she worried she was forcing the issue.

Was it all really going to come to an end here?

Blake grabbed her hands and tried to steady her.

"Lily," he repeated in his calm, steady tone. He kissed her hands and smiled. "Lily, I don't have a problem with you solving these mysteries. You're helping people, you're bringing justice to criminals. How could I possibly be selfish enough to mind you doing that?" he asked.

Lily wasn't sure if it was a rhetorical question or if he wanted an answer. "But then what is going on here? Why have you been so distant with me? And don't say that you haven't. I know there's been something you've been holding back. There's something going on in your head right now and I don't know what it is. Have you fallen out of love with me?" she asked, her voice trembling with emotion as she spoke.

"No Lily, of course I haven't fallen out of love with you. I don't believe I could ever fall out of love with you," he said.

"Then what is it? What's wrong?" Lily asked.

Blake smirked. "The truth is while I don't mind you solving crimes there was something about this one that had some unfortunate timing."

"I know I missed our date but—" Lily said. Blake silenced her by placing a finger upon her lips.

"Let me speak Lily," Blake said. "That night was a special night. I had been looking forward to it for a while and when you got the call that Colonel Rawlings died it got in the way of my plans. I'm sorry that I was off with you, I was just trying to figure out what I was going to do instead."

"Instead of what? It's not as though you haven't seen me Blake."

"I know, but it wasn't the right time. I was hardly going to use the Colonel's memorial service."

"Use it for what?" Lily asked, scrunching up her face.

Blake sighed again. He set his cup of cocoa aside and put hers on the counter as well. "Lily, you are the most chaotic person I think I have ever known. When I returned to Didlington St. Wilfrid my father told me that he had a new tenant who I simply had to meet. I wasn't sure what to expect, and when I first met you I was just astounded by your beauty and your spirit and everything that God has blessed you with. Over time I've gotten to know you better, and my appreciation for you has only grown, as well as my love. You know, over the years I felt guilty that I hadn't gotten married and started a family. After my mother died it hit me hard, and I always regretted that I had never been able to make her a grandmother. I didn't want the same thing to happen to my father either, but I also knew that I didn't want to be with anyone unless I truly wanted to be with them, and I didn't want to bring a child into this world unless I was with a woman I could trust with all my heart.

And then I met you, and I realized that there was another reason why I haven't been married until now; because I've been waiting for you. There is no other woman who can hold a candle to you, Lily McGee, at least not in my eyes. You always know how to make me laugh, and since our romance has blossomed I've been happier than ever. You light my days with pure joy and I know that I never want to spend another day without you in my life. In fact I think I've known it for a while, I just wanted to wait until the proper moment to do this. But again you ruined my plans because I had a whole thing planned. I wasn't expecting you to come to my house tonight, and I can't delay this any longer," Blake said.

Lily had some inkling of where he was going with this, but she didn't want to hope in case she was off base. But a smiled widened on her face and light danced in her eyes.

Then Blake dropped to his knee. He clasped her hand with one of his, and then reached into his pocket to produce a ring box. He held it up and she gasped. He opened it and the ring sparkled.

"Lily McGee, will you marry me?" he asked. Lily's gaze shifted between the sparkling ring and his gorgeous face. Tears welled in her eyes and she sank to her knees as well, before she flung her arms around the man she loved and kissed him with all the strength she could muster. She peppered him with kisses.

"Yes, yes, a thousand times yes!" she cried out over and over again, relishing in the joy of the moment. Blake smiled as he took her slender hand and slipped the ring onto her finger. She admired it for a moment, and then they kissed again. Lily then nestled against his chest, feeling herself melt against the warmth that emanated from his body. She sighed happily and he kissed her head. His fingers entwined with hers and they both gave contented sighs.

"This is perfect," she said. Blake looked down at her.

"I can't believe you thought there was something wrong," he said.

"Well, you were acting off… and I suppose my mind does tend to gravitate toward the more negative result," she said.

"You're going to have to do something about that. Good things can happen in life Lily," he replied.

"I'm beginning to see that," she said, smiling widely. She closed her eyes as she held him tightly.

"I love you so much Blake," she said.

"I love you too Lily. I can't wait to begin the next phase of our life together," he said. "I suppose we should go and have a conversation with the Reverend."

"We're going to have conversations with lots of people," Lily said, smiling joyously. "I can't believe I'm actually going to be able to plan my own wedding. I never thought I'd ever be doing something like this!" she cried. Her heart swelled with the things Blake had said, and the moment had been so overwhelming that she felt tears trickling down her cheeks.

"Everything you said was so special and perfect Blake, and I can't believe you feel that way about me. I feel the same way about you as well. All my life I've felt as though I've been drifting. Back in America I had the feeling that I was waiting for my real life to begin, and although I loved everything I accomplished when I arrived here I still felt as though something was missing. You've been that missing piece Blake, and I think I knew that from the moment I first met you. I started loving you then, and I haven't stopped. I don't think I'm ever going to stop."

"I really hope you don't," he said with a wide grin.

"You just make me feel so special Blake, and I hope that I make you feel special too. I want to promise you that I'm always going to be by your side, and I'm always going to stand by you and support you, no matter what. I want to be the best wife and the best partner that I could ever be," she said.

"And I want to be the best husband for you Lily," Blake said, his deep voice making Lily's heart rumble. "And maybe in time we can think about being the best mother and father as well," he added. Lily smiled widely, and then she nodded too. Yes, she was ready for a family, and yes, she was ready to be his wife. She was ready for so many wonderful things, and she was so happy that God had chosen to bless her in this way.

EPILOGUE

After Blake had proposed to her and Lily had gathered her emotions, she had asked him what he originally had planned. He said he had intended to take her to a high point on the rolling hills and have a picnic under the stars, where they could gaze upon God's glory and make the solemn promise with the natural world all around them. She felt a little guilty for ruining that surprise, but they had the picnic there anyway.

Once the news was out and Lily told people, she was surprised to learn that many people had already known. Joanie's response was to roll her eyes and say 'finally', since apparently she had gone ring shopping with Blake to help him choose an appropriate engagement ring. Lily couldn't believe Joanie had been so good at keeping this secret from her, and Joanie admitted as much, saying that it had almost killed her. She was just glad that Blake had finally done the deed, and of course she accepted when Lily asked her to be her maid of honor.

Marjorie also knew, because Blake had commissioned a celebration cake to be made, which had to be cancelled when the Colonel died. Marjorie admitted that she had scraped off the icing and served it as a plain cake at the community hall on one of the tea mornings.

Nobody had seemed to mind. She was pleased as punch for Lily's news and sighed as she looked up at the picture of her husband.

"I hope you are as happy as we were," Marjorie said. It was the sweetest thing she could have said, and Lily's heart melted.

It turned out that the Reverend had also known as well, which is why he had come up to them during the memorial service and asked them how their dinner was. It also accounted for Blake's curt reply, as he hadn't wanted the surprise to be given away. Father Brown was excited to be performing the ceremony, but Lily was astounded that so many of her friends had been involved in this mystery, and that it was one she had been unable to solve. Even Pat had been involved, because Blake's original plan was to have Zeus carry the ring up to her during their picnic, but that had not come to fruition either.

However, there was one person who Lily had the privilege of telling first, and that was her mother.

 Mom,

I know this letter is going to reach you before you've had a chance to reply to my last letter, but I have urgent news. I'm probably going to call you to tell you anyway, but I wanted to chronicle this because it's never going to happen again. Blake proposed to me! I'm going to be his wife! I'm going to have a little family of my own, Mom...

She continued writing and as she did, so many emotions flowed through her. She knew she had never been this happy before and it was intoxicating. Before when she thought of the life she was going to have with Blake it had mostly been a fantasy, but now it was real. He was going to be her husband and their relationship was going to be sealed in a hallowed way under the eyes of the Lord. And while she was writing she looked down at her hand and saw the ring sparkling, almost as though it was winking at her. The smile on her face was wide and the only mystery that awaited her now was how her life was going to turn out, but that was one she had no interest in solving before it was all revealed to her.

All was right in Lily McGee's world, and now she had a wedding to plan. She set her pen down and picked up a bridal magazine, feeling an excited tingle in her heart.

A HAIR RAISING HAPPY EVER AFTER IN A QUIET ENGLISH VILLAGE

1

S pring had swept across England, but in Didlington St.
Wilfrid the nights were still cold with echoes of the winter
frost. Most people remained in their homes, cozying up by a
hearth to warm themselves and fight back against the chill. Pat was
not such a man. He was hard and brittle, and had suffered much
sorrow in his life. He had sacrificed a lot for the opportunity to
serve Queen and country, he had protected people without their
even knowing it, and his name would not be remembered in the list
of heroes who had given so much for the sake of the world.

But that was just the nature of the job. Pat had always been the type
of man to grit his teeth and get on with things, no matter how bad
they were, but when his sister died it had been one injustice too
many. Pat had turned bitter, his heart charred and black, containing
little more than ash. He became prickly and was always ready for an
argument. His only friend had been Zeus, the faithful dog who was
always by his side.

And then Pat had almost died. The Lord had saved him for some
reason, although he wasn't quite sure what. He didn't think of
himself as special or worthy enough for such a cause. However, the
fact remained that he had been spared when so many other people

had not. He had a second chance at life, and although it took a lot to pull himself out of bed each morning and to bear the pain that ached in his head he still did it. He faithfully went to church each Sunday, and instead of arguing with people he spoke with them, thinking that this was the way his sister would have wanted him to be.

He still missed her terribly though.

Since he had returned to Didlington St. Wilfrid his recovery had been slow. At first he had barely been able to walk, and Lily McGee had taken care of Zeus. Ah, Lily. She had had as much effect on Pat's life as much as anyone else. He wasn't sure he could ever show her how grateful he was. He loved her as much as a man was capable of loving a woman, purely platonically of course. She had saved his life and she had taken care of his dog. When she had first arrived in the village he had scorned her and rolled his eyes, wondering why on earth some American had shown up in this quaint part of England that time had forgotten, but she had proven that she belonged here.

Now time had rolled around and Pat was able to walk again. His movements were labored and he struggled against the pain, but he wasn't going to let it defeat him, and he wasn't going to make Zeus suffer because of his own shortcomings. He liked walking at night because it was cold and quiet. It could sometimes feel as though he was the only man in the world, and as the wind whirled around him he thought he could hear a whisper, and he liked to pretend that it was his sister. He would tilt his head back to look up at the twinkling stars and the crescent moon, and he would smile, knowing that she was looking down upon him.

On this night, as he looked toward the horizon and the world that stretched beyond Didlington St. Wilfrid, he felt the presence of the Lord around him and wondered how he had been so blind to it before. Rage and bitterness had clouded his vision, preventing him from feeling the cleansing, pure love of God that now radiated within him. He and Zeus walked along the edge of the cliff. The sea ebbed and flowed below him. The glassy surface of the water was almost as inky black as the night sky itself, although of course the sky was dappled with twinkling stars. It was incredible that this world had been given to him, had been given to all of them by God, and he was annoyed at himself for how he had allowed himself to be so angry for so long.

"I have been a fool," he muttered. Zeus looked at him with dark, soulful eyes, tilting his head in a way that suggested he understood. Pat reached down and patted Zeus, while he looked up at the stars. He wondered if God had forgiven him for the way he had acted.

Pat had traveled around the world during his life, but there was nowhere quite like Didlington St. Wilfrid. It was charming and quaint and it had the most kind hearted people that Pat had ever known. They had all rallied around him during his illness and he wasn't sure he could ever repay them. He wished he knew how to make it up to them, but he only knew one way, and that was to be the protector of the village. There were dangers lurking in the world that most people were unaware of, but Pat knew the truth. He had seen the places of the world and the dark hearts of men who did not feel the touch of God's love. Lily McGee saw it too, and perhaps it took people who were outsiders to defend the sanctity of this place. It was paradise, but in every paradise there were serpents that tried to slither in and take advantage of the innocent.

That moment was when Pat noticed something strange below in a cove. It was a blinking light, the pattern too orderly to be natural. Pat's old instincts kicked in. He had lost a lot during his time as an invalid, but he would never lose his instincts. Hairs prickled on the back of his neck, and Zeus growled as well. Animal and man were in sync. Pat's hands curled into fists and he grit his teeth. The sensible thing would have been to return to the village and fetch Constable Dudley, but by the time he did that whoever was in the cove might well be gone.

Pat was on his own, and he had a job to do.

"Come on Zeus. Let's see what's going on down there," he said, and crept toward the cove. He and Zeus walked sure-footedly down the mountain. There was a small winding path that had been made over time, weathered into the rock, although it was still difficult to see in the shrouded darkness and Pat had to tread carefully. He allowed Zeus to go ahead, for the dog found it easier to navigate and did not show any sign that he was in danger of slipping down.

The light had stopped blinking now. A lesser man would have turned back and convinced himself that he had not seen anything, but Pat's heart was filled with certainty. Then, in the far reaches of the sea, he saw another light signaling back. Something was out there, and it was coming straight for the cove. As he descended the path he started to hear voices.

"They're taking their sweet time tonight," one of them said in a rough, coarse accent. It didn't sound as though he was local.

"They'll be here. We've got plenty of time to wait anyway. It's not like anyone is going to disturb us. These people don't know what's happening even though it's right under their noses. Give it half an hour and we'll be done. Another night of good work," another voice said, this one sounding satisfied with himself.

Pat scowled, wishing that he had given into the technological revolution and had a phone so that he could contact someone, anyone. By the time he went back to the village and returned with help these men would be gone, if their estimates were correct. It was too dark and the men were hidden by the shadows, so Pat didn't even have any description to go on. Whatever they were doing though, it was illegal, and Pat wasn't about to stand by and not stop it. God had given him another chance at life, another chance to do the right thing, and Pat was not going to waste that chance.

He crept down, keeping low, trying to remain out sight. The darkness could be his ally just as much as it was for these men, men who did not belong there. Pat overheard their voices and wished that they would say something incriminating or something that would tell him what they were doing here, but mostly they just talked about soccer. Pat intended to wait until whatever they were expecting came toward the shore. But then, by sheer misfortune, a loose cluster of rocks cascaded from the cliff edge and rattled near Pat. One of the men swung around and the beam from a flashlight scarred the air.

"Put that thing out! You're going to confuse them, or attract attention," one of them yelled in a terse voice, but it was already too late. Pat had tried to duck down, but the top if his head was visible.

"There's someone there!" a voice cried.

Pat's heart sank. Before he knew it a group of men were surrounding him, each one looking meaner than the last.

"Please don't hurt me, I was just out walking my dog. I don't want to cause any trouble," Pat said, holding his hands up, trying to make himself look weaker than he actually was. He kept his breathing steady though. He remembered the training that had forged his muscles many decades ago, the training that said to never panic, to always remain in control. Zeus bared his teeth and narrowed his eyes. Pat was more scared for his companion than he was for himself.

"Well it seems as though you have caused a trouble, and unfortunately for you, you've wandered into something that you shouldn't have. Nobody was supposed to be hurt by this, but too many people are depending on this and I can't have anyone going about telling people they saw us here."

"I won't tell anyone, I promise!" Pat said.

"In my experience a man's promise isn't worth anything," the other man spat, and with a nod he signaled to the rest of the men to circle in on Pat. Pat kept his head bowed, but when they were close enough he leapt into action. His fist flew and caught one of them in the throat, sending them staggering back. He winded another, and then on the back swing jabbed his elbow straight into the nose of another, cracking bone. There was a yelp of pain and the man was thrust back, holding his nose, dark blood streaming from the wound. Zeus was nipping through the crowd of men, a storm of violence, while Pat was fighting valiantly. The surprise had given him an advantage, as had his years of special training, but he was not as strong as he used to be, and time was the enemy of every man.

He suffered a punch in his gut that drove the air out of his lungs, and then a hard fist cracked the side of his face. He felt the warm, metallic taste of blood swimming over his tongue. Another blow made him sink to his knees. The world was becoming even darker than it had been before as the ground rushed up to meet him. His heart sank because he knew he had failed, and he wished he had been stronger. Through blurred vision he could see Zeus struggling with two men as they tried to drag him away, and all Pat could think of was Lily. She was their only hope now. She had to figure out what was going on. He closed his eyes and darkness overwhelmed him, but before he lost consciousness he felt a tear trickle down his cheek, hating that he had let down his faithful friend. Zeus was in God's hands now. God's and Lily McGee's.

2

There was a knock at the door of Starling Cottage. The place had become something of a mess in preparation for the wedding. There were stacks of bridal magazines everywhere, as well as bits and pieces of table decorations and invitations. After going through this process Lily began to understand why people made a career of being a wedding planner, because there were so many little things to keep track of, and she kept having the niggling feeling that she was missing something. The cottage had also had an extra visitor as well. Now that the wedding was impending, Kelly had come across from America to stay with Lily, and mother and daughter had enjoyed spending a bit more time together. There had been a time when Lily never thought she would get married at all, and where Kelly thought that she would not be welcome at a wedding if it did take place, so both women were pleased at the way things had turned out for them.

"Who could it be at this time of the morning?" Kelly asked. Joanie was still living in Starling Cottage too, although she was not awake yet.

"It's probably one of the lovely people of the village who want to come and offer me something else for the wedding," Lily said in a slightly strained voice.

"I think it's nice that people want to be a part of your special day," Kelly replied.

"I know, but there does come a point when it stops being *my* special day and it becomes theirs," she said, although she wasn't really that upset. She felt blessed that so many people wanted to celebrate with her. Originally the plan had been for Blake and her to be married in a small, intimate ceremony with only their closest friends and family in attendance, but once word spread through Didlington St. Wilfrid that the wedding was happening, keeping it small was an impossibility. Everyone was looking forward to it, and everyone wanted to be involved. It was a sign of how much Lily, and Blake, were respected and cared for, and how tightly the community was knitted together. Didlington St. Wilfrid had suffered more than its fair share of sorrow as well during Lily's stay, and they needed something to celebrate to help counter these grim events.

"Well it certainly best not be that cop of yours. The last thing you need is to be involved in some grisly crime," Kelly said.

Lily flashed her a smile. "You don't have to worry about that Mom. I've already warned Constable Dudley that I'm not on call for the time being." It had been a hard decision for Lily to make, but ultimately she decided that a wedding only happened once in a lifetime and she shouldn't let anything overshadow it, not even murder, so she had made Constable Dudley promise that he wasn't going to ask for her help even if a murder did occur. It might have been selfish of her, but she was only going to get married once, and she intended to enjoy every moment of it.

Lily reached the door and opened it to find a woman standing there, one who could have been a reflection of herself. The woman was slightly taller, a little more slender, with fair skin and lustrous honey brown hair. She had a bewitching smile and the kind of aura that Lily had only experienced in one place; Hollywood.

"It's you, it's really you!" the stranger said in an America accent. "I'm so happy to meet you. I'm playing Lilah McDrew."

Lily's heart sank, and she quickly regretted basing the character in her novels so heavily on herself. "You're Lilah McDrew?" Lily asked skeptically.

"I am," the woman beamed, "well, my actual name is Charlotte Robins, but you can call me Lottie."

"What are you doing here?" Lily asked.

"Oh, didn't you get my email? I checked with the studio if it was okay to come and shadow you for a while. I thought it would make my performance more authentic if I spent some time with you and got used to your surroundings. I can see where you draw your inspiration from. Being here is like being in one of your books! I'm a big fan as well. I couldn't believe it when they announced they were making a movie. As soon as I heard it I just knew I had to audition for it, and I feel so blessed to have gotten the part. I hope it's not too much of an imposition, but I was hoping that I might be able to stay here as well so that we can spend as much time together as possible," she said, and it was then that Lily noticed the suitcases standing behind Lottie.

"I see. Well, you really should have checked with me rather than the studio. I'm actually a little busy at the moment. You see, I'm getting married."

"You are?" Lottie's face lit up in a beaming smile. "Oh that's wonderful! Nobody at the studio mentioned it, and I am here now... I do think it would make my performance in the film better," she said. Lily sighed. She was beginning to regret signing away permission to have a film made of her books. Blake had talked her into it, so he was going to have much to answer for.

"Okay," Lily sighed, feeling as though it would have been rude to turn the woman away since she had come so far. "But don't expect me to be working a mystery while you're here. At the moment all I'm focused on is getting married," Lily said.

"Just being here will help my performance tremendously, I'm sure," Lottie said, and then settled in. Lily shot a weary glance at her mother, who could only shrug in reply.

Lily decided that she was too happy to let Lottie's presence disturb her, so she wasn't about to let it dominate her mood. However, almost as soon as Lily had a free moment Lottie was there, hoping to catch a few words with her. Since Lily didn't have a good reason to deny Lottie the opportunity, she found herself sitting down with Lottie in the garden. It was a calm day, with a gentle breeze drifting through the air. Colorful flowers stood all around them, vivid bursts of blues and purples and reds, each one prettier than the last.

"What is it you would like to know?" Lily asked.

"Well, firstly, how is it that you come up with all these stories? You've written so many books now I find it amazing that you haven't run out of ideas!"

"Some of it is drawn from real life while others I just … it's not really easy to describe. It comes to me like breathing. The most important thing is that I stay true to the characters though. I know people like the mysteries, but to me the characters are the most important thing. If I take care of them then they're going to take care of me, at least that's what I've always believed."

"And what's it like to work a mystery? It must be so exciting!" Lottie said.

Lily pursed her lips and exhaled slowly. "I suppose reading about it can be, but living it isn't. It's hard. You see so many people whose lives have been destroyed, you're touched by their grief and you know there's nothing you can really do."

"You can solve the crime."

"Yes, there is that, and I know that closure can help, but nothing I do can bring the dead back. There are days when I feel proud that I'm making a big difference, and some days when I wonder if I'm ever making a difference at all. And then there are the murderers. Some of them let their emotions get the better of them. Some of them are unlucky. And some of them... some of them have the devil inside them," Lily felt her throat tighten and a cold chill run down her spine.

"I don't do this for fun," she continued. "I don't do this for a hobby. I do this because I feel as though it's my moral duty to help people. God gave me this gift, and I have to try doing my best while I'm here. I have to keep telling myself that even though I haven't been able to bring back the dead I've at least been able to prevent more people from being killed."

"That's a really great line, do you mind if I make a note of that so I can use it in the film?"

"Go ahead," Lily said, shaking away a broad smile.

It was around midday when Blake arrived at the cottage, bringing a wide smile to Lily's face and a jolt of happiness to her heart.

"She seems like a firecracker," Blake remarked after being introduced to Lottie.

"Yes, it seems as though even though I'm not going to be involved in a mystery Didlington St. Wilfrid is still going to throw up a surprise or two for me."

"Well, what would life be without surprises?" Blake said.

"I don't know what you're smirking about, this is all your fault."

"My fault?"

"She wouldn't be here if you hadn't convinced me to let them make a film."

"Well, how often does someone get the opportunity to see a film made of their work? We're actually going to be able to go to the premiere and see your work on the big screen. It's going to be incredible!"

Lily's lips formed a thin smile. "Hmm," she said, still unconvinced. She then wore a teasing smile as she walked up to him and draped her arms around his neck, taking a moment to enjoy the warmth of him, this man who was filled with love for her, as though their two hearts had been incomplete before they had met each other. "I'm starting to wonder if you're more excited about this film than you are about the wedding," she murmured.

Blake smirked. "Well, you know, the movie premiere is only going to be for one night while I get to spend the rest of my life with you..." he said, his eyes gleaming with joy. Lily playfully smacked his arm and pursed her lips, shaking her head. Blake merely chuckled.

"Speaking of the wedding, how are things going?" he asked.

"I think I'm on top of things. Joanie has been going around town to check what people are doing. Mom is rehearsing her reading. Marjorie is excited that she's baking the wedding cake. It seems to all be falling into place. I keep having this feeling that I'm forgetting something, but I don't know what it is. I'm probably just trying to convince myself that something is wrong."

"Well, we still need to have that sit down with the Reverend," Blake reminded her, "but other than that there's nothing I can think of that you've forgotten. However, aside from the wedding we do need to think about the honeymoon as well. Have you made a decision about where you'd like to go?"

"Yes, I have. I think it would be wonderful to cruise the Norwegian fjords and see the Northern Lights. I was looking at some pictures on the Internet last night and it just seems so wonderful," she said.

Blake nodded, "then the fjords it is."

Lily looked at him and smiled. "You know, I'm really, really happy that we're getting married," she said.

Blake laughed. "I'm happy too. There's nothing I want more in the world. You're the only one for me, Lily McGee," he said, and took her into his arms for a tight embrace. Lily allowed herself to sink into it. For the longest time she hadn't thought she deserved love or was worthy of this grace of God, but with Blake she knew that she had found her soulmate, her one true love, and here in Didlington St. Wilfrid she had everything she could ever possibly need. Her life was perfect, and at the moment she knew that there was nothing that could stop her feeling this way.

But then there was a knock on the door. It was a doleful, dolorous sound, as heavy as the toll of funeral bells.

When Lily opened the door she saw Constable Dudley standing there with a haggard look on his face.

"Oh no Bob, what are you doing here? You know we agreed to keep Lily out of this," Blake said. He stood steadfastly by Lily's side.

"I'm sorry Constable, but I must insist we keep to the promise. I feel awful, but you have to solve this one yourself," Lily said. She hated acting like this, but she couldn't allow her wedding to be overshadowed by a murder.

"I wouldn't have come to you if it wasn't important," Constable Dudley said. He wrung his hands together and his voice trembled, threatening to crack apart under the weight of emotion. Lily had rarely seen him like this before. The last time he had been this distraught was when he had been suspended from duty. Something serious must have happened to make him feel this way.

"Bob," Blake said in a warning tone, as though he knew that if Lily heard what the crime was about then she would be tempted to try and solve it. Lily placed a hand on his arm to silence him though. Constable Dudley was clearly troubled.

"Carry on Constable," she said gently.

"It's just... well... I'm not quite sure how to say this. I promise you Lily, I know how important your wedding is and the last thing I want to do is to put this on you now, but I thought you would want to know. It's about Pat and Zeus..."

"What about Pat and Zeus?"

"They... Pat is in hospital in critical condition, while Zeus has... Zeus has been taken."

Lily's heart shattered into pieces. "What happened?" she gasped.

Constable Dudley shook his head. "Details are still sketchy. Pat was barely conscious when we found him. He was laid in the cove, left for dead as far as we can tell. The life was almost beaten from him. Someone found him this morning. I spoke to him, but all he could say was something about a light. Maybe it was just the light that he was seeing at the end, the light that we're all going to see. I'm sorry for telling you about this..."

"No, no Constable it's fine. You're right. I do want to know," Lily said, her shattered heart already being bound together again by the rage that built inside her, the righteous fury that was borne from the love for two dear friends. Knowing that Pat was in hospital was one thing, but hearing that Zeus had disappeared was quite another. The faithful companion had become a loyal friend to Lily during the time that she had looked after him, and she would fight to the ends of the earth to protect him, because she knew that Zeus would do the same for her.

"Do we know anything else? Anything that's happened?" Lily asked.

Constable Dudley shook his head. "Only that there was something about a light, that Pat has been hurt, and that Zeus is gone," he said.

Lily looked up at Blake.

"I already know what you're going to say Lily, and I'm not going to try and argue with you."

"I can't stand by when Zeus is missing, and whoever did this to Pat needs to pay," Lily said. She had not intended anything to interrupt the planning of her wedding, but some things were more important. If anyone had hurt Zeus... she trembled with a mixture of rage and fear, and she wasted no time in beginning the process of investigation.

3

"I can't believe I'm actually getting to see you at work!" Lottie said. She was sitting in the backseat of the car, leaning forward excitedly. It was only a short drive to the coast, but Lily was already irked by her mood. Constable Dudley was busy writing up a report and creating a flyer for Zeus, in case anyone saw him, so Lily, Blake, and Lottie were out investigating the crime scene. Lily wished that Lottie hadn't come with her, but Lottie had a way of charming people into getting what she wanted. Kelly and Joanie were back at the house, taking care of anything to do with the wedding.

"You shouldn't sound so excited Lottie," Lily snapped. "Two of my friends have been hurt. One of them is in the hospital and the other is… I have no idea where," a lump formed in her throat as she thought about Zeus. "This isn't some adventure. This is real life."

She glared at Lottie in the rear view mirror and watched the actress sink down into her seat, looking as chastised as a naughty child.

They arrived at the cove. The parking area was high up on the cliff. The wind breezed along, caressing the long grass at the top of the cove, while the narrow beach below shone bright and golden. The rocks were white and light grey, jagged in places and smooth in others, and the sea was a clear blue, stretching out to meet the sky far on the horizon, and if Lily stared long enough she could not tell where the sea ended and the sky began. The air was salty and briny. She pointed down to the cove, which was where Constable Dudley said that Pat had been found. Blake led them down a narrow trail, holding his hand out to Lily whenever the path became rocky. Lily offered her hand to Lottie, although she chose to speak few words to the actress.

"What was Pat doing out here? Why would he be so stupid as to come down here by himself if he thought there was danger?" Lily said, her frustration borne of worry. The stones and pebbles crunched under her shoes, and in the daylight the cove looked peaceful and idyllic. It was almost impossible to think that a crime had been committed here. The sea lapped against the shore, the shallow, foamy cresting waves rippling gently, the cool sun beating down upon the weathered mountain.

"Maybe he didn't know. He might have just stumbled onto something without knowing about it," Blake said.

"Without knowing about it? Blake, we're talking about Pat *and* Zeus here. I don't think there's anything that could have escaped their attention. No, they must have seen something. I just can't understand why they wouldn't have come back to the village to get some help."

"It was late, perhaps he didn't want to disturb anyone," Blake said.

Lily sighed and placed her hand against her mouth, tapping her lips. She walked along the bay to where the sea formed a shadow upon the beach and cast her gaze around. As she stood there she began to piece together what might have happened. In the back of her mind she could hear the grunts of pain as Pat was beaten up, and the whining yelps of Zeus as he was taken away. Whoever had done this was cruel, and from what she was seeing they left no trace of their presence here. If anything happened to Zeus, Lily wasn't sure what she was going to do, and even though she had no idea how anything could have been different, she still felt guilty for not being here with Pat and Zeus.

"I just don't know," she said, feeling forlorn. "What if this is it Blake? What if we can't find anything? What if nobody saw anything?" she asked.

Blake walked up to her and placed his hands around her shoulders. "Hey, come on Lily. I know you're worried, but we've only just begun. These things take time. I'm sure something will turn up. It always does."

"But what if it's too late? We don't know where they've taken Zeus. What if they starve him? What if they sell him? What if they…"

Blake squeezed her tightly. "Lily, worrying about what happened to Zeus isn't going to help him right now. You know this. You're Lily McGee. Solving mysteries is what you do. There is a way to understand what happened here, and if anyone can do that, you can."

Lily nodded and composed herself. Emotions swirled within her in a twisting maelstrom, but she knew that Blake was correct. The only way to find Zeus was to figure out what happened here, and she would comb every grain of sand if she needed to.

"Maybe they sailed away," Lottie said. She was standing a few feet away from Blake and Lily, standing on an outcropping of rocks, gazing out to the glistening sea that was dappled with sunlight.

"What did you say?" Lily asked.

"They might have sailed away," Lottie pointed out to sea.

"Pat did mention a light. Maybe it was something to do with a boat," Blake said.

"If that's true then they could be anywhere. And who would want to sail up to this cove? It's not as though we're a major port. There's nothing to steal here. Why would anyone want to bother with Didlington St. Wilfrid? It just doesn't make sense," Lily said.

"Sounds like smugglers to me," Blake said in an offhand comment, chuckling to himself as he wandered along the beach, looking for anything that was out of the ordinary.

"Smugglers?" Lily asked.

"Oh, yeah, don't worry. It's just stories I was told as a boy. Apparently smugglers used to come here and used tunnels in the mountains. I suppose they liked it precisely because it's out of the way. I don't know if it was ever true or not, but I enjoyed the stories. There's something romantic about pirates and smugglers, isn't there?"

"Only the fictional ones," Lily said, but her brow was already furrowed as the cogs of her mind were whirring. She paced up and down the beach, thinking and thinking. "Pat said there was a light, what if someone was signaling down here? It was enough to get his attention. I know Pat. He wouldn't walk Zeus down here normally, he'd have kept up to the ridge," she glanced up. "He would have been walking up there, and he'd have looked down to see a beam of light. Perhaps it was a signal."

"And if it was, then the question remains who or what were they signaling to? And did it signal back?" Blake finished her thought for her.

"There's certainly something going on here, something that people were willing to hurt Pat to keep it a secret," Lily said. "I think you're right Blake. I think there are smugglers using Didlington St. Wilfrid. Pat found them and now they've taken Zeus. If they were here maybe there is truth to the stories that you were told. Quick, spread out and look for tunnels," Lily said.

"How thrilling. You never get anything like this in America! Perhaps this could be the sequel," Lottie said.

Lily grit her teeth and bit back her reply. The one thing she lamented about the movie was that it might glamorize what she did. There was nothing thrilling or exciting about her life. People were in danger, and every time she set out to solve a mystery she put herself in danger too.

Blake could sense that Lily was tense, so he came over and placed a hand on her shoulder.

"Are you doing okay?" he asked.

"No, I'm not," Lily said. "And she's not helping either," she tossed a glance toward Lottie, who was poking around a cluster of rocks.

"She's harmless," Blake said. "She just doesn't understand, that's all."

"Blake, people are in danger here. There's a real crime being committed. The last thing I ever want is for people to believe that I do this for fun. If she doesn't understand that this thing is real then she might end up getting hurt, and I'm the one that's going to be responsible. I should tell her to go back to the cottage. She can shadow me later."

"Maybe the one thing she needs is to see the reality of the situation Lily. If she doesn't begin to understand how harrowing this process can be then how is she going to express that in the film? If you want it to be as true to life as possible, if you want people to truly understand what you go through when you investigate these crimes then she's going to have to go through it as well."

Lily's heart sank because she knew Blake was right.

"Well I hope she learns quickly," Lily said. Blake caressed her shoulder and then walked back to the part of the cove he was looking at.

They spent a good while investigating the cove, searching the rocky surfaces for any tunnel entrances, but if there was one it eluded them. It shouldn't have been so difficult as there weren't even any plants covering anything, so she begun to think that something fanciful had happened. Perhaps the smugglers had camouflaged it somehow, or found some way to paint over the entrance. She traced her fingers along the grainy rock, trying to sense if there was any difference in texture, but as far as she could tell there wasn't. She prayed to find some door or another entrance, and then wondered if there was something buried under the sand. She groaned at the prospect of having to brush away all the sand in search for a trap door.

"There has to be something more we can do before we go to those extremes?" Blake said, wiping away a glistening trail of sweat that had formed upon his forehead.

Lily placed her hands on her hips as she traced her gaze along the cove, following the path of the rocks as it curled around the beach, allowing it to nestle in its bosom. The water lapped tightly against the rocks, creating a barrier that could not be crossed, except, she thought, when the water wasn't there.

"Blake, do you know when the tide goes out?" she asked.

"I don't I'm afraid," he replied. Lottie shook her head and looked at Lily blankly. Lily walked to the water's edge and peered as far as she could around the cove. The cliff edge wrapped around and part of it was covered by the water, so there was only so far she could see. If the tide went out though it would be entirely different, and perhaps an entrance would be revealed. Lily pulled off her shoes and then rolled up her jeans. Grains of sand trickled in between her toes, tickling her, and then she stepped into the water. It only took a few steps before the water reached her ankles.

"Lily, what are you doing?" Blake called out.

"I'm going to see if there's an entrance around here that's been hidden by the tide," she replied.

"We should wait until the tide goes out again then, it's too dangerous otherwise."

"Zeus might not be able to wait that long. I'm not going to sit here and watch the water. I'll be fine," Lily said, and was already feeling the water around her thighs. It ebbed and flowed and the pull of it threatened to take her out to sea. Lily kept her hands on the rock edge. She was on her tip toes by now, only just managing to feel the surface below the water. Beyond her the sea called, the currents trailing away. If she lost control she knew she was in danger of being carried away to sea and before too long she would be beyond help. She wondered if that was what had happened to Zeus, and she couldn't bear to think of him drowning.

As far as she was concerned he was alive until proven otherwise.

She pressed her hands against the cliff edge and heard Blake and Lottie muttering behind her. Then there was a yelp as Lottie slipped and crashed into the water, drenching herself completely. Blake dove in after her as she flailed her arms around. She clung to him as he carried her toward Lily.

"What are you doing Lottie? Can't you swim?" Lily asked.

Lottie wiped water from her eyes and shook her head. "No, I can't, but I always said I would do anything for a role."

Lily rolled her eyes. People in Hollywood were mad, but she wasn't going to waste her time on making sure that Lottie was safe. She was a grown woman, and if she wanted to put herself in danger then so be it.

"I think I've found something," Lily said after stretching around the cove. She saw a narrow black opening that looked as though it had been cut into the rocks. Water lapped into it, but by this point Lily had just about lost the seabed, so going forward was going to require her to swim. It had been a long time since she had taken to the water, and as she pushed off she heard Blake call out to her, but she wasn't going to stop. If the smugglers had left by land they might well have taken Zeus into this tunnel. He could be in there now, whimpering and scared.

Lily gave herself to the water, and then fought against it. She kicked out and then slammed her hands down, crawling through the sea. She could feel the current around her, trying to coax her away. It was stronger than she had anticipated, so she grit her teeth and fought against it, gasping as she pushed herself toward the tunnel. It looked like a black abyss that was always just a little bit out of reach, before she felt the sloping rocks under her feet once again and was free of the current. She gasped as the spray of water hit her face and seeped underneath her clothes. The chill of it was shocking and she shuddered, taking more than a few moments to get used to it. As soon as she stepped into the dark tunnel she took a moment to tug at her clothes, trying to wring the water out of them and warm herself up. It was hard to do in such a dark and dingy tunnel, but she wore a triumphant smile as she knew that she was closer to finding Zeus.

4

―――――――――

"**L**ily? What are you thinking?" Blake said when he emerged spluttering into the tunnel. Lottie had been clinging to his back like a limpet, and now coughed out some water as she leaned against the inside of the tunnel.

"I'm thinking that Zeus needs my help! We could have been waiting all day for the tide to go out. He's already been held for hours. I can't imagine how scared he is!" Lily said. Blake frowned and pursed his lips, but he didn't say anything else. He merely looked around at the narrow tunnel. The sea flowed outside, creating a cascading echo that rang around the damp cave. The rocky surface was wet, but mesh matting had been placed down, which helped them to keep their balance

"I don't think the smugglers of old would have put this stuff down," Lily said, her eyes gleaming with excitement as they always did when she was on the trail of a clue. When she sniffed the air she could smell nothing but the brine of the ocean, and the darkness of the tunnel beckoned ahead of her.

"Are you really sure you want to do this?" Blake asked in a beleaguered tone. Lily looked at him as though he shouldn't have even bothered to ask the question.

She turned to Lottie. "Are you okay to continue?"

Lottie nodded, putting on a brave smile. Lily arched her eyebrows at Blake, who merely shook his head. "Right, let's just be cautious. We have no idea where this tunnel leads, or who might be waiting for us at the end. For all we know the smugglers are still here. Let's be quiet and listen for any noise. I don't want to suffer the same fate as Pat," Blake said.

"Given that the tide is in they won't be expecting us, if they're waiting up there," Lily said. She hoped that they might at least be able to get some information about who these men were and what they wanted. Blake pulled out his phone and accessed the torch. A bright dot of light sparkled like a moving star, and it illuminated the narrow tunnel. It sloped up at a gentle incline and Lily found herself moving closer to him.

"I'm sorry that we're embroiled in a mystery again, especially so close to our wedding," she whispered.

"Somehow I had a feeling something like this was going to happen. You just can't live a simple life can you," he said, flashing her a soft smile to make sure she knew he was joking.

"And you're the one who is going to have to put up with this for the rest of your life."

"It's almost enough to make me have second thoughts," he teased, and then he grew serious. "I know why you have to do this Lily, and I want to do it too. Pat is a good man and Zeus is, well, Zeus is Zeus. I just get worried, that's all. These men are clearly dangerous and if anything happens to you…" he trailed away, unable to continue the thought.

"I feel the same way about you Blake. We just have to take care of each other, okay?"

"Okay," Blake replied, and then they fell into silence.

⁓

Lily wasn't sure how long they were walking for. Blake took the lead, while Lottie followed behind Lily. Their footsteps were bunched together and slow, for nobody wanted to take the risk of slipping. Lottie panted behind Lily, and occasionally there was a whispered moan from the actress. Lily began to feel a bit guilty about leading Lottie here. Perhaps she should have been firmer and insisted that Lottie stay behind in the cove. She didn't think there would have been any danger in the daylight.

"Are you okay?" Lily whispered.

"I'll be fine," Lottie replied, although Lily wasn't sure if she believed her.

The tunnel was narrow and uncomfortable. There were parts where Blake had to stoop low to avoid hitting his head upon the ceiling. The jagged walls caught Lily's clothes. The air in the tunnel was cold and dank. It smelled stale, and it made her shiver. She was already thinking about what it would be like to return to Starling Cottage and slip into a warm relaxing bath and feel the chill seeping away. That wouldn't be for a while yet though, so she would have to put up with the uncomfortable trickle of water dripping down to her feet, and the chill that made her teeth chatter. Blake's light pierced the darkness. On occasion it felt as though the tunnel was never going to end. There were points where it veered sharply at an angle, and Lily became quite disoriented. She had been trying to follow the twists and turns, but to no avail.

"Do you have any idea where we are?" she asked, still whispering.

"It could lead all the way down to London for all I know," Blake said.

Lily had been expecting the tunnel to open up on some smuggler's hideaway, but it just seemed to reach deep into the land. There were no other sounds other than the ones this group was making, and

Lily was afraid that she might not find Zeus at the end of it after all. The more they walked the more she grew afraid that this had been a dead end.

"Perhaps we were wrong about the smugglers after all," Lily whispered, and began to think about what else might have flashed that light at night. Nothing came to mind though, and she feared that this might be one mystery she would never be able to solve. If Lottie was right and someone had sailed away from the cove with Zeus she might never see the dog again, and whoever had sailed near Didlington St. Wilfrid might feel compromised, and never come to this place again.

But they had come too far into the tunnel to turn back now. All that was left was to move forward and see where it ended. If nothing else they would at least be able to say they had explored a tunnel. However, it was clear to her that it was used for *something* because of the mesh underneath their feet. It hadn't gone all the way up the tunnel, for now there was smooth rock below them, but clearly someone had put it there for a reason. She wondered if they had missed another passageway or something that would indeed lead to a smuggler's den, and if so they would have to take a closer look on the way back.

The tunnel reminded her of the crypts, although this tunnel did not have the same eerie, haunted feeling as the crypts had had. Even so, she remained close to Blake and her breath was taut and shallow.

Eventually they saw another source of light. Blake immediately shielded his phone and turned the light off so that nobody would be able to see them. The tunnel widened slightly and as far as they could tell they had reached its end.

"I have to admit that I was hoping for something more," Lily said.

"We might have found something though," Blake said in a soft whisper, and indicated for her to keep her voice down, before he pointed to a square grate that was beside them. They pressed their faces to the grate, the old metal was cast iron and as cold as the stone around them. The shadows hid their faces, and Lily gasped when she realized what she was looking at. She did not have the greatest vantage point of the room, but she could see a map of Didlington St. Wilfrid hanging on the wall above a desk, and a man was sitting at the desk. He was no ordinary man though, he was the mayor. The tunnel led right to the mayor's office in the town hall, which was one of the oldest buildings in the village, along with the church.

The mayor's wide black hat was beside him on the desk. He had his head bowed, revealing a pale bald scalp, flecked with a few bright red spots. The mayor scratched his head and leaned back, yawning. There was something forbidden and voyeuristic about watching a man like this. Lily cast a sharp glance toward Blake, although she did not dare say a word. She knew Blake was wondering the same thing; why did the tunnel lead to the mayor's office, and did the mayor know anything about it?

Lily pressed her hand against the grate and then pointed to the mayor with one finger, indicating that she was going to enter the room. Blake shook his head, pointing at the mayor and then at the door, meaning that he thought they should wait for the mayor to leave. Lottie turned her gaze between the two of them, not able to understand the meaning behind this sign language.

While Lily and Blake were having this silent discussion there was activity in the room. The door opened and heavy footsteps entered. Lily could feel the vibrations through the grate. She saw a fearsome man looming above the mayor's desk, and the mayor was frozen. He leaned back and placed his palms on his desk. His lips trembled and eyes went wide with fright. Lily saw that the man had stringy long black hair and a mean curled lip. A tattoo snaked down his right arm, the vivid colors depicting thorned roses and fanged snakes.

The eyes of one of those snakes seemed to stare at Lily directly, and she felt a shiver of fear pass through her, as though this serpent was going to alert the man of their presence.

For the moment though it seemed as though the man was only concerned with the mayor.

"W-what are you doing here? I thought we had an agreement. You were never supposed to come here during the day," the mayor stammered out, attempting to be firm, but it was clear he had no authority with this stranger.

The man licked his lips and leaned over the desk, placing his balled fists onto the desk. He spoke in a low, gravelly voice, and each word was audible to Lily. She could feel the venom dripping from him.

"I wasn't intending to, but we have a bit of a problem, don't we? And you know that problems aren't good for business."

"I don't know anything about any problems. I promise!"

"Oh I'm sure you don't, which is why I'm telling you, because you're here to make sure that we don't have any problems, unless you've forgotten."

"I haven't forgotten at all!" the mayor said, continuing to cower behind his desk. The stranger reached over and picked up the mayor's hat, turning it over in his hands as if inspecting it.

"Good, good, because if you had then we'd have had an even bigger problem. The thing is that last night someone saw us. One of your little people here, and it makes me wonder if anyone else has seen us. I need you to find out. The reason we chose to make this little place our haven is because the people here mind their own business. If they start poking their nose into ours then we're going to have some trouble, and we don't want trouble, do we?"

"N-no," the mayor stammered. "But I shouldn't think anyone would have seen anything. The people here aren't like that. I promise. They don't like meddling. They're just simple folk."

"That may be, but I need you to find out for sure."

"And how am I supposed to do that? If I ask questions people might start becoming suspicious."

"You're just going to have to be careful, aren't you?"

"And what… what would you like me to do if someone did see anything?" the mayor asked.

The man put the hat back onto the desk and leaned over, fully obscuring the mayor from Lily's gaze. There was nothing but hatred in the man's voice. Lily's blood curdled and she found herself clutching Blake's hand.

"Then me and the boys will take care of it. Remember, the business comes before anything else. If you want to keep the gravy train running then you're going to have to do your part. There is a reason why we pay you as much as we do. It's time for you to earn your keep. Find a way to get it done and do it," the stranger said. It was clear that it was an order, and the mayor did not look like a mayor at all. He looked like a trembling, frightened man who had no authority whatsoever, and Lily couldn't understand what had driven him to this state.

As the man left, Lily knew that he was the key to rescuing Zeus.

It had taken all of Lily's willpower to not jump out of the grate immediately, but there was more information that needed to be gathered. She needed to find out the mayor's involvement, and there were clearly more people working with this stranger. She glanced at Blake, who held up his finger, indicating for them to wait until the coast was clear. If the cruel stranger should burst in again then he would know that Lily was onto them.

The mayor was leaning forward, head in his hands, when Blake and Lily pushed the grate out. Blake held out his hand for Lily to climb in. Lottie followed, and then Blake pulled himself in, before pulling the grate back into place. The mayor had a second shock and it was fortunate that his heart was as strong as it was, otherwise he might well have suffered an attack.

"What's going on here?" Lily asked.

The mayor was so stunned he almost fell from his chair and stared at Lily, wide eyed. She must have looked like something from the depths of the sea, for her clothes were still dripping wet and her locks of hair were clumped together, while her skin had the sickly pale shade that came with being soaked and cold. The mayor let out a whimper and breathed deeply.

"What are you doing? Where did you come from?" he cried.

Lily marched toward him. "None of that matters now. Who was that man? What's going on here?" she asked sharply.

"I... I don't know what you mean? What man?" the mayor said. He was flustered and his eyes darted around, searching for a way out. Blake moved toward the door, ready to guard it in case someone should come in, or if the mayor decided he wanted to make a break for it. Lily swept her hand across the desk, knocking his hat off onto the floor.

"You know what man," Lily snarled, raw fury pouring out of her eyes. "I know you're involved in something *Mayor*," she spat the title as though it was an insult, "and whatever it is, I'm going to find out. You can't protest your innocence here because I don't have the time to believe you. I need to know what's happening and what that man wants. Remember, I've gotten to the core of every case that I've worked on in this village, and this is going to be no exception. I'm going to expose you, it's only a matter of time, so you can either do the honorable thing now or you can wait and everyone will see what kind of a snake you are."

"I... I'm not a snake! I promise! I haven't done anything," the mayor cried.

"The people you work for certainly have though. Haven't you heard Mayor? Pat is in the hospital and Zeus is missing. Do you really think I'm going to stand by and let whoever is responsible get away with this? I've a good mind to call Constable Dudley right now," she said.

The mayor's eyes widened in fear and he shook his head, but it wasn't as a result of anything Lily had said. All color had drained from the mayor's face and he looked weaker than she had ever seen him before.

"You've bitten off more than you can chew this time Lily. This isn't like anything you've faced. I'm sorry for what happened to your friends, but this... you're going to have to give up on any chance of seeing Zeus again. I'm sorry, but it's just... you need to give up on this," he said, bowing his head again.

Lily moved around the desk and leaned in so that the mayor could feel the heat of her breath. "I don't care how scared you are Mayor. I don't care what these men are capable of. I'm not giving up on the people I care about. I'm not just going to forget about finding Zeus. You're going to talk, and you're going to talk *now*."

"I can't," the mayor croaked weakly. "They'll hurt me if I do."

"Is that really the most important thing right now?" Blake said, "I remember the day you became mayor. Everyone was happy for you. You swore that you would do everything in your power to maintain the dignity and the traditions of the village. You said that you were going to be a man of the people. I haven't always agreed with the decisions you've made, but I could at least always respect you. That's gone out of the window though. What use is respecting a man who has no shred of decency?"

Lily had never heard Blake speak with such disdain to anyone, and there was nothing she could say to add, so she simply continued glaring at the mayor.

"There are people who have cheered you on, who have considered you a friend, and you've betrayed them all. This office is supposed to mean something. You took an oath to God, and instead you're making deals with criminals," Blake continued. "What happened to you? Maybe nothing happened and you've always been like this."

"No, no... I wasn't always like this," the mayor said. "I'm a good man, I am!" His pleas fell on deaf ears. "I am a good man," he repeated weakly, as though he was trying to convince even himself.

"Good men can still do evil things. The test of a good man is if he takes responsibility for them. At the moment you're not doing that," Lily said.

The mayor sighed. "It wasn't my fault. I... they blackmailed me a long time ago. This office is all I ever wanted. I couldn't lose it. I couldn't take the chance. I thought it was just one job, to look the other way. Nobody was even getting hurt by it. It just happened though and then it became another job and then another, and each time I got sucked deeper into their world. I knew I should try to escape, I knew I should try to tell someone, but their threats are not to be taken lightly. I tried to be good. I promise you. Blake, Lily... you know the kind of man I am. Please have mercy on me. Please try and understand."

Lily looked at him coldly. "If you are the man we think you are then you'll tell us what's going on here," she said.

The mayor sighed. Drops of sweat glistened on his forehead and he wiped them off with a handkerchief, although it did not stop the slick sheen from making his head shine.

"I don't know who they work for, you know, up high," he said, pointing a finger into the air. "I just know that they have a lot of resources and they're very powerful. They came to me one day and they told me that all they wanted was to bring a ship into the cove from time to time. I suppose the bigger ports are too tightly monitored and there are more restrictions and security than there used to be. I got the sense that they were going to do this anyway, whether I liked it or not, and since they had this thing over me... well..."

"You thought you could profit from it as well," Blake said sharply. The remark was cutting. The mayor mumbled out an errant few words to try and maintain his innocence, but the façade of that had long since been dispelled.

"What are they doing?" Lily asked.

"They're smuggling things, drugs I suppose. I haven't dared to ask. They give me a cut off the top. I think they deliver it to the cove and then distribute it throughout the country. There's no coastguard out here. There is no water police. I thought it was a victimless crime."

"No crime is victimless," Lily said sharply and glanced toward Blake. This was bigger than she had first suspected, and if what the mayor was saying was true then they were up against criminals with a long reach.

"Look, if they think that anyone is onto them they're going to silence them. It's best if we all just leave well enough alone. I'm sure at some point they'll stop whatever they're doing. It can't go on forever after all," the mayor said.

"I'm not about to leave a friend to suffer alone," Lily said. "We need a plan. When is the next shipment going to arrive?"

The mayor looked so tired he might have passed out. "I don't know. I'm not privy to that information."

"Then tell me your best guess," Lily barked, rolling her eyes.

"Oh, er, well, I suppose if they were interrupted last night then it might have been delayed. They might try again tonight. It's so hard to tell with these people though..." he trailed away.

"We'll assume it is tonight, and if not tonight then we'll try again tomorrow. We'll watch that coast every night until they arrive," Lily said.

"What do you have in mind?" Blake asked. She looked evenly at the mayor.

"I'm going to get him to take me to the smuggler's den," she said. The mayor looked even more frightened and shook his head, protesting vehemently.

"I don't want any part of this!" he said.

"Too late Mayor. You've been a part of this for a long time now. It's time that you did something good for the village," Lily replied, determined that she was going to get to the bottom of this. She didn't know who these men were, but they had Zeus, so she was going to get them back. They had also been using Didlington St. Wilfrid for nefarious purposes, and she did not like the thought of this. It was a slight against the village that she had come to call her home, and it made her feel tense inside.

5

They waited until nightfall before they were ready to leave. They gathered at Starling Cottage. The mayor was a nervous wreck. He kept mumbling to himself under his breath. Lily wondered if he was praying for salvation, and not for the first time she found herself thinking that the mayor was lucky, for the Lord was far more forgiving than she was. The more she thought about Zeus the more the tension twisted in her gut like a knife. All she wanted was to get him back and rid Didlington St. Wilfrid of these criminals.

It was then that Lottie came down the stairs, dressed in more suitable clothing.

"Are we ready to go?" Lottie asked.

Lily shared an uncomfortable glance with Blake and pulled Lottie aside. "I appreciate your enthusiasm Lottie," Lily said in an attempt to be as diplomatic as possible, "but I really don't think this is the place for you. I'm sure you must have gained enough information from shadowing me today?" she asked.

"It was very informative, but this is going to be even more so. I mean, I get to actually watch you hot on the trail of criminals. I'll get to see the moment where you apprehend them," Lottie said, unable to hide the excitement from her voice.

"But this is more dangerous than earlier Lottie. I can't promise that you'll be safe. I can't promise that I'll be able to watch over you. I shouldn't have let you even get into the water today. You should stay here. I promise that I'll tell you all about it once it's over and I'll make sure to give you all the details."

"And I really appreciate that Lily, but it's not quite good enough. I think in order to make my performance as authentic as it can be I need to experience these feelings for myself. I want to know what it's like so that I can show the audience every single emotion that you go through."

"I know you want to make the film as good as it can be and I can certainly respect this level of professional pride, but I really must insist—"

Lily was interrupted as Lottie reached out her hands and clasped Lily's tightly. "I know how much this means to you Lily. I understand that this is more than just a hobby. It's a calling. When you were in the mayor's office I could really feel your emotion, and you've made me believe in how important this is. I don't know this Pat or Zeus, but you've made me care about them. I want to help them. These people are horrible and you have no idea what's waiting for you. All we know is that there are plenty of them, so if there's anything I can do then I want to make sure that I'm there to do it. Please, just let me help. That's all I want," Lottie said.

Lily was struck by her earnestness and so, despite her better judgment, she ended up acquiescing to Lottie's demands. Lily also had the thought that it was more useful to spend time thinking about the plan rather than trying to convince Lottie that coming with them was a bad idea.

Once Constable Dudley had arrived they were ready to move.

"Right," Lily said. "The mayor is going to lead us to where the smugglers are. I want to make sure that Zeus is safe. I'm hoping that we can take them by surprise and arrest them before they have a chance to run. Constable Dudley, this is going to be a big night for you so I hope that you're ready," she said.

"We'll get them," Constable Dudley said, nodding firmly in Lily's direction. She knew that he was as offended as she was by the thought that crimes had been permitted to happen in Didlington St. Wilfrid right under his nose. She had caught him staring daggers at the mayor. More people were going to share the sentiment. People placed a lot of trust in their public officials, and yet so many of them seemed to want to betray that trust. Whoever became mayor next was going to have to make every effort in getting and regaining the trust of the public.

They made their way to the cove, hoping to catch the smugglers in the act. The mayor was fidgeting and kept wringing his hands together. They parked some ways away, before slipping through the darkness. Lily kept praying that Zeus was still safe. Even though he was her friend she had to remind herself that he was still an animal and he didn't understand fully what was happening. He might think that nobody was going to come for him at all, and she didn't have faith that these smugglers were going to take care of him.

They walked down the cliff, keeping quiet so as not to alert anyone who might have been in the cove. Lily's heart hammered in her chest and she thought herself reckless for taking this chance so close to her wedding, but again she knew she had no choice. Whenever it came to a dispute between her own safety and the safety of others there had never been any real choice. She had always sought to save those who needed help before herself.

When they arrived in the cove they crouched down, but it quickly became clear that nobody was present. Lily furrowed her brow and walked to a standing position. She headed toward the water, watching the tide flow out in its endless rhythm. She gazed into the horizon for a boat. The others followed her, and they all looked puzzled. The only one who hadn't moved with them was the mayor. He stood in the middle of the beach and had his head downcast.

"I'm sorry Lily," he said, his hushed voice barely audible over the whisper of the sea. "But I don't want to die."

He then let out a high whistle and Lily's heart sank. She should have seen it coming. She should have been more careful. She had tried to believe that the mayor had some honor left in him, but she had been wrong. He had betrayed them because of cowardice, just as he had betrayed all of Didlington St. Wilfrid before them. Bright lights beamed down into the cove, as glaring as a spotlight. Lily held her hand over her eyes while they adjusted to this brightness that was shining through the night. She saw the silhouettes of men walking down into the cove, and heard their rough voices laughing.

"Good job mayor," she heard the cruel stranger say. The mayor stepped back into the shadows and turned his back, for he did not want to witness what was going to happen next.

"Well well," the smuggler said, turning his attention on Lily. "It looks as though we have some meddling villagers after all. You should have known better to keep to yourselves. I'm afraid you've just made a costly mistake." Then he looked at Lily directly and took a step closer toward her. The water curled around Lily's ankles. "The mayor warned me about you. I guess around here they don't know the simplest way to deal with a problem," he said, and lifted his arm. Extending from his hand was a pistol, as black as death itself. The barrel pointed directly at Lily and her heart clenched inside.

"No!" Blake cried out.

"Yes," the smuggler said, but just before he was about to pull the trigger Lottie leaped in front of Lily.

"Wait! She's not really Lily McGee. She's just an actress. I'm the real Lily!" Lottie cried.

Lily's mouth hung open in shock. "No!" Lily said, and tried to push Lottie out of the way, but Lottie was surprisingly strong and Lily found it difficult to wrench her out of the way. They struggled, but the smuggler was just laughing his brittle, haunting laugh.

"It doesn't matter to me which one is which. I'll just kill you both," he said, and then the gentle beauty of the cove was disturbed by the thunder of a gunshot that cracked the air. It all happened so fast. Lily screamed and Lottie fell onto her. There was a feeling of warmth that crept through Lily's clothes, and then Lottie went limp in her arms. Lily looked at the sadness in Lottie's eyes as she slumped toward the sand, Lily cradling her wounded body.

The smuggler was taking aim again.

Lily felt the world tumble and rush around her as she was pulled back. There was another thunder of gunfire, but this time the bullet passed by her harmlessly. But her world now was the swirling water, foamy and salty, stinging her eyes and filling her mouth. She barely had a chance to breath before she was plunged into the depths, pulled down by what felt like a sea creature, but what she would soon realize was Blake and Constable Dudley. They pulled her back and swam under the surface of the water. Lily could hear the muffled cries of the smugglers as they shone their lights onto the sea, but the surface of the water offered no hint to the mysteries that lay beneath. The smuggler fired wildly. Bullets surged through the water and left trailing bubbles in their wake, embedding harmlessly into the rocks and sand below the water.

They swam away as quickly as they could, although Lily felt as though the current had more power than they did. Her lungs were burning and she thought they were going to explode before suddenly she burst into the open air and wiped away the water. She turned her head around and then clung to the rocks nearby, pressing herself against them so tightly that she might as well have melted into them. Blake then shielded her with his body. They were away from the cove, near another, smaller inlet. The shadows cloaked them. The night was their friend, but Lily's heart had broken because Lottie was gone.

"It's all my fault," Lily moaned.

"She saved you," Blake said. "She did exactly what you would have done under the same circumstances. Now let's find a way to make these people pay."

He and Constable Dudley helped her swim back to the shore, for she was in a state of shock. By the time they had climbed back up the mountains (the path leading up from this inlet was not as true as the one near the cove) the smugglers had gone, as though they were ghosts in the night. Blake, Lily and Constable Dudley rushed back to the cove. Lottie had been left there. Lily sank to her knees and wept, but as she pulled Lottie's body over to put her on her back, she noticed that Lottie was still breathing.

"We need to get her to the hospital," she said, and with Blake and Constable Dudley's help they rushed Lottie back up to where the car was parked. Constable Dudley had already called ahead for an ambulance to be ready.

They were all drenched and were trembling. Lily was still shaking. One of the other police officers made them some tea.

"It's all my fault," Lily said, her teeth chattering together.

"She's alive Lily. She's not dead," Blake said.

"But she might not last. And she's still been shot. I should never have trusted the mayor. I should never have let her come with us. I'm supposed to protect people, not put them in harm's way!" she cried.

Blake squeezed her shoulders. "This isn't your fault Lily. It's the mayor's fault and these smugglers. They're the ones who did this, and they're the ones who are going to pay. We have to find another way to catch them, that's all. There's always a way to catch the criminals. You just have to find it. Zeus still needs you Lily. He's out there and we're going to find him. We just need to figure out how."

Lily nodded and sipped her tea, beginning to calm down. Her heart was still thumping in her chest though, and she couldn't escape the fact that now two people she knew were in the hospital and one was missing, while another had betrayed them all. This was certainly not how she expected the build up to her wedding to go.

6

"I don't understand how the mayor could have done this. To betray the village for this long is one thing, but to go behind our backs like this..." Constable Dudley said, shaking his head. He, Blake, and Lily had all changed their clothes. Blake and Lily were wearing spare police uniforms. She remarked that she had done so much work with the police that it had only been a matter of time until she wore a uniform. It was not a funny joke, but Blake and Constable Dudley expressed bleak laughs nonetheless.

"It's clear that he only cares about himself, nothing else," Blake said. "You know, I always thought that I didn't love this place enough to stick around. I always thought that I didn't belong here, but to know that the mayor has been corrupt for so long..."

"This town is going to need a lot of healing," Constable Dudley said.

"There will be time to heal when we've finished this," Lily said. The fresh clothes and the time to compose herself had reinvigorated her. "The smugglers have to be somewhere nearby. They must be storing their goods somewhere. The cove isn't large enough. If this is worth a lot of money then we're talking about trucks and big shipments.

Something like that would have been noticeable. They must be storing everything near Didlington St. Wilfrid, but far enough away that nobody is going to notice a huge convoy. If the mayor has been working with them for a long time then maybe we'll find something in his office," she said.

～

They made their way to the town hall. Constable Dudley forced his way into the mayor's office.

"Does this count as treason?" he asked.

Neither Lily nor Blake had an answer for him. Blake switched the lights on. Lily immediately looked toward the grate, wondering if anyone would be lurking in wait for them. She could hear nothing though. They quickly began rummaging through the closets and drawers, pulling out files and sheafs of paper in the hope of looking for anything that seemed out of the ordinary. As Lily put it when Constable Dudley asked them what they were looking for; they would know it when they saw it.

They were there for a good amount of time though and they turned nothing up. They stood in the middle of the office and stared at each other.

"I'm not sure we're going to find anything here," Blake said.

"But if we don't then how are we going to find them? Perhaps we should just make a list of all the likely places in the area that they could use as a base and investigate them one by one," Constable Dudley said.

Lily shook her head sharply. "That would take too much time. They're spooked now. They know we're onto them so they're going to pack up and move on, and they'll probably find some other place to use. Didlington St. Wilfrid might be rid of them, but I'm not about to let them go free so they can haunt another village. And I still want to know what they've done with Zeus…" she trailed away and Lily thought for a moment. Then she snapped her fingers. "What about his house?"

Constable Dudley and Blake looked at each other and shrugged. "It's worth a try," they said,

The mayor's house was only a short walk from city hall. It was as pretty and picturesque as all the other houses in the village, and was always used by the incumbent mayor. Constable Dudley once again forced his way into the house and they crept through the rooms, examining them one by one. It was a difficult balance to strike between moving swiftly and searching thoroughly. It might be something written on a scrap of paper, or a note that had been tossed into the waste paper basket.

"I've found something!" Blake said. He was in the mayor's office and had managed to wrench open a locked cabinet. Files had spilled out and he spread them over the desk. There were shipping manifests detailing dates when the cargo could be expected, while there was also an address for a quarry. A quarry would be the perfect place for clandestine deals to be agreed on and for illegal shipments to be stored, and the mayor would have had the information needed and the sway to make sure that the quarry was clear whenever he needed to use it.

"This has to be where they are. And it must be where Zeus is as well. I'd wager that the mayor is still with them. We can get them all!" Lily said.

"Then we have to move fast," Blake added. "If they leave now then we're never going to be able to track them down."

And if that happened then all chances of finding Zeus again were gone, Lily thought to herself, and that was even if he was still alive.

She swallowed the bilious dread and then excused herself for a moment, saying that she had to make a quick phone call before they left for the quarry.

7

They drove toward the quarry, but stopped short of driving right up to it in case any of the smugglers were watching, although Lily hoped they were being overconfident and too preoccupied with moving all their supplies away. Night was still in full strength, with the moon hanging like a silver coin in the sky. The stars twinkled and on any other night Lily would have been able to admire the beauty of it, but on this night she had too many things to worry about. The quarry was a huge bowl in the land, carved out of rock and silt and dirt. There were abandoned cranes standing in the quarry, as well as trucks and trailers. Huge bags of building materials were present, as well as massive crates. At the far end of the quarry Lily could see lights and flickering shadows, which she took to be the smugglers. She nudged Blake in the ribs and nodded. He returned the gesture, for he had seen it as well.

"I suppose this is it then," she said in a quiet whisper. She slipped her hand into his and felt their fingers entwine. "I just want you to know before any of this happens that I'm sorry. I know you would have preferred a normal build up to our wedding."

Blake smirked. "Nothing is normal when it comes to you Lily. It's always something better. But I know you have to do this. *We* have to do this. Zeus needs our help. We wouldn't be able to get married in earnest if we let him down, if we let these people get away with this. Just look out for yourself, okay? Don't make me a widower before we're even married," he said, trying to make a joke out of it, but Lily could see the pain behind his eyes. She nodded and felt herself choking up as well. In the blitz of this investigation they hadn't taken a moment to stop and take stock of what they were doing. These men were dangerous, and both Lily and Blake were risking their lives, and as such their marriage. They had not officially pledged themselves to each other in front of God yet, and Lily just hoped that this happy ending would not be taken away from her at the final moment. She had promised Blake so much and she hoped that she would be able to live up to this promise. She had dreamed of the life they were going to have together, and she squeezed his hand tightly, knowing that he would be able to feel her love.

Constable Dudley cleared his throat. "What are we doing Lily? What's the plan?" he asked. They crept a little closer to the cluster of men. Thankfully Lily's hopes had come to fruition and they were all preoccupied, so none of them were watching for any intruders. The mayor was sitting on an upturned barrel, looking disheveled and sorry for himself. Lily might well have pitied him had he not tried to save his own skin at the expense of theirs. She took stock of everything around her, but still she could not see Zeus, and she began to fear the worst. She closed her eyes and made a silent prayer to God, asking him for courage and bravery, and for as much protection as he could give to herself, to Blake, and to the good Constable.

Once she had finished praying she heard a small whimper, and her eyes were drawn to the right, into the shadows. As she strained her eyes and peered into the darkness she could see a small square cage, and the form of a dog within it. Her heart lifted as she realized that Zeus was alive.

"Okay, I'm going to free Zeus. You two make a distraction. Once I've got Zeus I'll try and get the mayor as well. We might not be able to get all of these men, but if we can run back to the car we can get back to the village and the mayor will be able to incriminate all of them," Lily said.

It wasn't the best plan she had ever come up with, but right now it was the only one she could think of. These men outnumbered them and they were ruthless. They would shoot them on sight, so the only chance of living was to flee the quarry and head back to the safety of Didlington St. Wilfrid. Then she looked to the Heavens, and hoped that God had indeed heard her prayer.

Blake and Constable Dudley crawled away, while Lily moved through the shadows, keeping herself low. It seemed as though nobody was paying attention to Zeus. The closer she got the more she could hear his sorry whimpering and it tugged at her heartstrings.

"Zeus, Zeus you good boy. It's Lily. I'm here to rescue you," she whispered. Zeus' ears pricked up as soon as he heard her voice. He rose from his slumped position where it looked as though he had melted into the floor and pushed himself up onto his paws. It took her a few moments to see him in the darkness, but his nose twitched and he caught Lily's scent in the air. She reached in between the narrow bars of the cage to pet him. He licked her hand profusely, although she was struck with hurt when she felt some of his matted fur, for he had been wounded in the struggle.

"You good boy Zeus. We're going to get you out of here, okay? You're such a good boy and you're so brave. We'll get you away from these nasty men," she said, having to stop herself from crying as she was so overwhelmed with relief at finding Zeus alive. She quickly struggled with the lock. It was tight, but she managed to twist it away. The smugglers must have thought that a complicated lock wasn't needed considering he was just an animal. It was little more than a bolt along the door. The cage door swung open and Zeus bounded out, almost bowling Lily over. Lily wrapped her arms around him and hugged him tightly, while he pawed her and licked her face. This was one victory at least. A small one certainly, but still a victory.

"Let's go and get the mayor, and then we can get out of here Zeus," she said, continuing to pet him. The matted blood ran down his flank, but it didn't hinder his movements any. As Lily rose to a standing position she felt more confident at having Zeus by her side again. She just hoped that Blake and Constable Dudley had done their job.

Zeus was about ready to bound out and attack the smugglers, so Lily had to command him to heel. He obeyed, but only barely. His sharp teeth were bared and his lip snarled, while an angry growl rumbled out of him. It was just about all Lily could do to stop herself from rushing out as well. With her hand on Zeus' neck, she crept back out to the main area of the quarry, eager to see what was happening.

Then her heart dropped.

Blake and Constable Dudley's distraction hadn't worked. They were standing before the smugglers, and now Lily was their only chance at escaping.

She glanced around quickly, annoyed that she did not have any weapons, before she realized that she was surrounded by them. She picked up large rocks and skulked within throwing distance, before hurling them at the smugglers. The rocks rained down and Zeus was finally allowed to rush forward. He darted in amongst the smugglers, snapping and biting at their heels. He was like a storm unleashed. Lily continued pelting the smugglers with rocks. They cowered and protected their heads, but just as she thought she was getting the upper hand they drew guns and stopped her in her tracks. Lily found herself standing near Blake and Constable Dudley, while she called Zeus to heel. She pressed herself against Blake and had a look of defiance upon her face as she stared into the cruel eyes of the smuggler. This man had beaten Pat, captured Zeus, and shot Lottie. Now he had his attention firmly upon her.

"From what the mayor said I really thought you'd have been smarter than this. I'd love to chat with you, but I have a schedule to keep to," he said, and raised his gun. Lily grit her teeth, unable to believe that she was going to die so close to her wedding. Had this really been one crime too many?

Then Blake stepped forward, trying to put himself between her and a bullet. It wouldn't have made a difference in the grand scheme of things, but she appreciated the gesture.

"Wait. I don't know if you know who I am, but your friend there," Blake looked at the mayor as he said this, "will tell you that I'm important and, more importantly, I'm wealthy. I'll make this worth your while. Call it a ransom or whatever you want, but I'll pay it. We'll go our separate ways and you'll never have to see us again. We don't even know your name."

"But he does," the smuggler said, jerking his head toward the mayor. "And given the lengths you've gone to find me so far I can't be confident that you'll ever stop. It's too late for money. I can't be guaranteed of your silence," he said.

"But you're killing innocent people. Doesn't that mean anything to you?" Lily asked in one last, desperate appeal to his humanity, if he had any.

The smuggler wore a wry grin. "This is going to be the least of my crimes. Three more bodies aren't going to make a difference," and then he looked back at the mayor and recalculated. "Four," he added. The mayor whimpered, his eyes darting back and forth. Even Zeus showed less fear.

The man raised his gun, but before he could shoot Lily raised her eyes to the Heaven and began speaking.

"Our Father who art in Heaven..." she began, reciting the Lord's prayer. Her voice was deep and mellifluous, and it seemed to take on a holy power as it echoed through the quarry. The smuggler smirked and mocked her with laughter, apparently allowing her these final words before he was going to shoot her. Blake and Constable Dudley stood beside her and joined in, and she noticed that even the mayor was praying along with her. She never took her gaze off the night sky as she reached the crescendo of the prayer. Her voice trembled with emotion and tears glistened in her eyes. "… deliver us from evil. **FOR THINE IS THE KINGDOM, THE POWER, AND THE GLORY! AMEN!**" she shrieked, and just at that moment a bright, beaming light shone upon them. The smugglers looked shocked and cowered. They had been mocking Lily, but now they feared her, for they believed that an angel was descending from the heavens to smite them for their wicked ways.

In a way, that was exactly what was happening. Lily smirked as she ducked, taking the opportunity to flee from the smuggler's line of fire. The churning thunder of a helicopter's rotor reverberated through the quarry and the light illuminated the smugglers. The criminals all sought to flee, but the helicopter descended into the quarry and in the distance sirens wailed as police cars joined in. Police officers jumped out of the helicopter and yelled to the smugglers, who realized they had been outmatched and outthought. They sank to their knees and placed their weapons on the ground. Police swarmed through the quarry, and one in particular came up to Lily and the others.

"Hello stranger," Blake said.

"Hi, looks like congratulations are in order," Chloe Hargreaves said.

Lily smirked. "If you hadn't arrived when you did there might not have been a wedding at all," she leaned her head against Blake's chest and let out a long sigh of relief.

"Well, when I received your tip I mobilized as quickly as I could. It took a while to put a taskforce like this together though. We've been trying to nail this smuggling ring for a long time. Looks like you've done it again Lily," Chloe said with a respectful smile.

"We all did it," Lily said, reaching down to pet Zeus.

"I'd make sure that one doesn't get away if I were you," Blake said, nodding toward the mayor.

"Oh, don't worry. I'll make sure that anyone who was involved gets what they deserve," Chloe said.

"You know, since you're in the area you should stay for a while and come to the wedding," Lily said.

Chloe smiled. "Thanks, but I think I have my hands full," she said, glancing toward the prisoners. She nodded with respect and then walked back toward the cluster of criminals they had caught. Lily could imagine it was going to take them a long time to sort through all the evidence and question everyone involved, but at least the smuggling ring was going to be unraveled.

"How about we get out of here," Blake said. Lily thought it was the best idea she had ever heard. She took his hand and held it tightly, and as they moved away from the quarry she prayed a silent thank you to God for keeping her and Blake safe. She had been a hair's breadth from death, but Lily McGee still had plenty of life to live yet.

8

Lily was still tired from the exploits of the previous night, and now that the adrenaline had worn off shock was beginning to set in. She was shaking a little from being held at gunpoint, and fearful because she knew how easily things could have gone differently. But Didlington St. Wilfrid had been rid of this cesspool of crime and she was still intact, and there were people she needed to see.

Blake drove her to the hospital where she walked in with Zeus. She popped in to see Lottie first, who was wan, but conscious. She smiled when she saw Lily and tried to prop herself up, but lacked the strength to do so.

"Lottie, I'm so sorry," Lily began. "I shouldn't have ever let you come with us. I should have been more insistent."

Lottie shook her head. "Don't be silly Lily. I was the one who insisted on coming with you, and it was my choice. In that moment I just… I knew that it was imperative you survived. I wouldn't have been able to stop them. Is everything sorted now?"

"It is," Lily said. "How are you feeling?"

Lottie exhaled slowly. "You know, I finally understand what you go through when you investigate these crimes. I didn't have a true sense of the danger until that man was standing before us. I have a lot of respect for you Lily, and you can be sure that I'm going to make this film as authentic as possible. I'm even thinking about writing a book about it myself, although I'm not sure I could do it as much justice as you could. The director is going to have his hands full with me though because I'm going to make sure that everything is accurate as possible."

"Just be careful, you don't want to make directors angry," Lily said, thinking back to the very first case she had been involved with, and she smiled at the thought of how far she had come since then. She bid her farewell to Lottie for the moment and then went to another part of the hospital. Pat was in bed and he looked worse for wear. He was hooked up to machinery. His eyes were bleary, his skin blotchy, and he looked to be half the man he used to be. He wheezed when he breathed, and he looked dazed. He was aware of Lily and Blake when they walked in though, and when he realized that Zeus was there as well he smiled. Zeus leaped onto the bed. Lily was about to pull him off as she didn't think the nurses were going to like it, but when she saw tears appear in Pat's eyes she didn't have the heart to pull Zeus away. The dog was licking Pat and Pat was making a fuss over Zeus, and it was a heartwarming sight, at least until she realized what Pat was saying.

"... you have to be a good boy, okay? I know it's not going to be easy and I'm going to miss you so much," he said.

"Pat... is there something you need to tell me?" she asked. When he looked at her she could see the pain in his eyes. "Lily... they got me. My body has been through too much. I need to ask you—"

"No... no Pat. You can still recover. You're in the hospital," she said, her voice cracking on her raw emotion.

Pat shook his head sadly. "It's okay Lily. It's just my time. I was given a second chance, a second chance to find myself again. You helped with that. You helped me become a better person Lily and now I can go and see my sister again. And I'll see you again too, eventually, in a long time though," he said, and whatever strength had returned to him by seeing Zeus left him almost immediately. "I needed to hold on to see you again, to ask you..." he coughed, choking on his rasping voice. "Take care of Zeus for me. He's yours now," Pat said.

"No... no Pat he's yours. Please don't do this. Fight for a little longer. Come on Pat. I know you can do it. It doesn't have to be this way."

"It does Lily, it does," he said, the words slipping out of him in a final breath. He had hold of Lily's hand, and his other hand rested on Zeus' head. Then Lily felt his hand slip away. A flood of sorrow poured from her eyes and Zeus lifted his head, nuzzling Pat with his nose as though such a simple gesture would have been enough to rouse the man again. Zeus then looked at Lily, confused. He jumped off the bed and nestled into her. She wrapped her arms around Zeus, and Blake wrapped his arms around her. There was much to be thankful for, but it seemed there was always a price.

They wept in the hospital room, mourning a man who had been indignant and bitter when Lily had first met him, but who had turned out to be one of the finest men Lily had ever had the pleasure of knowing. Indeed, this was borne out by the fact that plenty of people turned out to his funeral to pay their respects. Although Pat had kept to himself he was a fundamental part of the community and would always be remembered. Everyone seemed to have a fond memory of a time that Pat berated them or argued with them about something, and they joked that he would even be complaining about things in Heaven.

Lily stayed with Zeus for a while as they stood by Pat's grave. The dog was reluctant to leave, so Lily stayed with him until Zeus was ready to move. Then they went back to Starling Cottage, and thankfully it was familiar to Zeus so he didn't have to take too long to settle into his new home.

9

About a month had passed after the funeral. Lily had to take some time to process all that had happened, and she did not want to get married when she was still in the throes of grief. The village had been shocked by the revelation that the mayor had been involved in such a villainous act, so when it finally came time for the wedding everyone was happy to rally around a celebratory event. It seemed as though everyone needed it to cleanse their palate and remind themselves that life could still be beautiful.

And, speaking of beauty, Lily made the perfect bride. Her hair was curled slightly, cascading down in soft tresses. She wore a slim, elegant gown that had patterns of thin flowers stitched into the fabric. She wore a thin veil that covered her head, and a wide smile as she walked down the aisle. The church was filled with people from the village and they gave her a rousing ovation as she walked in, making her heart swell. Blake was waiting for her at the end of the aisle, dressing in his finest suit, and looking as handsome as any man ever had. When she locked eyes with him she felt love touch her heart, and she knew that she would never have to go without love, not for all her life.

Kelly walked her down the aisle, mother and daughter linked arm in arm. When they reached the end of the aisle Kelly turned to her and hugged her, whispering in her ear. "I wasn't sure this day would ever arrive. I'm so happy that you allowed me to be a part of your life Lily. I want you to know that I'm never going to leave you again," she said. Lily squeezed her tightly. Then her hand was placed into Blake's. He smiled his charming smile, and she noticed how relaxed he seemed. Despite the church being crammed to the rafters with people she did not feel self-conscious. As she stood there with Blake everything else seemed to melt away, and only Reverend Brown's voice filtered through, guiding her and Blake through the vows that were going to bind them together for the rest of their lives.

As Lily spoke she felt this strange energy crackle and tingle within her, and it truly felt as though she was blessed by God. She felt weightless, as though she was soaring up to dance among the stars, and yet at the same time she was anchored to the man standing beside her. Lily's heart was not her own any longer. She was more than she had ever been before, because she was in love and she was married to Blake Huntingdon, because she had promised her life and her love and her faith to him, and he had done the same in return. In truth she hadn't been sure that anything would have been different after she became married because she and Blake were still the same people, but it was transformative. The spirit of God washed over her, and this love that was forged from the finest things He had ever created now stitched her and Blake's hearts together.

The final vows were said with the rings. Lily looked down the aisle as the ring bearer bounded up toward them. Zeus's coat was glossy and his tongue lolled out. The rings were tied to his collar, and he stayed by Lily's side as the rings were removed, almost as though he was reminding Blake that if he ever dared to cross Lily, Blake would have Zeus to answer to. The rings were then exchanged, and then there was a kiss to seal the marriage. Lily felt as though she was floating through the air as the congregation erupted in a huge, uproarious cheer as she and Blake were declared husband and wife. The music of the organ filled the church as she and Blake walked outside. The ground was strewn with confetti and the people of Didlington St. Wilfrid were ready with their good cheer. Her friends were there, including Marjorie, Joanie, and Constable Dudley. Her mother was there too, as was Blake's father. There were other people she recognized, people she had helped during her time in the village, and it was a reminder of how many lives she had touched.

But the most important one was, of course, Blake. Her heart was filled with love and this happy, joyous feeling poured out of her from every pore. It was the finest feeling she had ever experienced and she knew that there was no other man she would ever feel this way about.

There were plenty of photos taken, some of which would later be displayed in the town hall as this day would go down as one of the most significant in the village's history. When Lily had arrived nobody would ever have assumed that this American stranger would become such a vital part of the community, but she had earned her place in history, and for years to come there would always be stories of Lily McGee, later Lily Huntingdon, who helped out anyone who was in trouble, and never swayed from her sense of justice. Didlington St. Wilfrid would become something of a tourist spot because of her, as the motion picture would bring new eyes to Lily's lives, and people would seek out her home to try and understand where she found inspiration from.

But that was all to come. On the wedding day there was still plenty of excitement to be had. Photos were taken and by the end of that Lily was exhausted, but she still had to dance and eat and enjoy the rest of the day. Marjorie's cake was delicious, while Joanie caught the bouquet and ran up to Constable Dudley, making him blush. The children of the village amused themselves with games while the adults danced in the town hall. Lily made sure to spend time with each and every person she could, because she wanted all of them to have a fond memory of this day that was so important to her, and she wanted to remember them all as well. Time moved in an ever turning wheel and there was so much that could be lost, but life was never about the loss. It was all about the memories she captured with people. These were the things that lasted, and although some people like Pat could not be there in person, she knew they were watching over her in spirit, so she took a moment to look up to the Heavens and thank them.

For many people, especially outsides, Lily would be most known for her work in helping to capture criminals, but to her that was never what life in Didlington St. Wilfrid was about. It was all about the people she met, the friends she had made, and the life she had fashioned for herself. When she had arrived she had been confused and lost, looking for her place in the world. She had well and truly found it now, and she was never going to leave.

Toward the end of the night Blake beckoned for her to join him. She took his hand and they sneaked out of the hall, walking away to a rising hill that overlooked the village. Zeus walked beside them. He would be their constant companion now. They sat down on the soft grass and Blake wrapped his arm around her. The church steeple rose high into the sky, while the rest of the land looked lush and verdant. It was the most beautiful sight she had ever seen.

"What a day," Blake said, chuckling.

"What a day indeed. It's gone by so quickly. I wish we could relive it over and over again."

"But that would mean that we never get to live any other days, and I know there are going to be so many happy days left for us to live," Blake said, kissing her softly on the head.

"True, but there are going to be sad days as well."

"Life wouldn't be life without a mixture of the two. But today has been perfect. I can honestly say that it's been the happiest day of my life. There's nothing else I would have wanted to happen today to make it any better. It's been incredible, and you've been absolutely beautiful. I think I've fallen in love with you again Lily," he said. She tilted her head up to meet his, and their lips met in a tender sharing of love.

"I'm looking forward to falling in love with you a thousand more times over the rest of our lives," she replied.

The kiss deepened, but was interrupted as Zeus was fussing around their laps. Lily felt his tongue lapping against her hand and she broke off the kiss, laughing. "Yes Zeus, and we love you too," she said. Zeus seemed to settle after this. Her gaze drifted down toward the ring upon her finger. It sparkled in the soft light of the evening. The sun was setting now, and the sky was streaked red. The ring symbolized the unending love she had for Blake.

"It makes you wonder what's going to happen next, doesn't it?" Blake said.

"Are you really ready to move on from today already?" she teased.

Blake smirked. "No, no of course not. I'm just thinking, you know, about the village and things. So much has happened here recently and we both know that more is going to happen in the future. There might even be another wedding soon."

"There might?"

"You saw the way Joanie and Bob were."

"Oh, yes," Lily said, and let out a light laugh. "I'm still not sure I quite understand them, but as long as they're happy that's all that matters I suppose."

"Yes, I imagine there are some people out there who wouldn't understand us, but we worked out alright."

"You mean people would have been surprised that a distinguished Englishmen ended up marrying to some American girl who turned up here from Hollywood out of the blue?"

"Something like that," Blake said. "But I wouldn't have had it any other way. I think I knew from that first moment that you were going to have a huge impact on my life. Can I make a confession to you?"

Lily looked down at her ring. "If you had any secrets you really should have told me before you put this ring on my finger. Is this something that's going to change the way I feel about you?"

Blake laughed. "No, at least I hope not. No Lily, all I wanted to say was that you know at the beginning, when I said that I was going to stick around to help Dad out with managing the land and everything? Well… that wasn't entirely the case. I could have done that from afar you see, but I just wanted to be around you and see what happened. You managed to bring me back home, and that was something I wasn't sure would ever happen."

"I'm glad. You know, there's going to come a time when we're sitting here old and grey, thinking about this moment. It's going to make me smile."

"It's going to make me smile too."

They spent a few moments cuddled together like this in this cozy, intimate moment. Lily closed her eyes and focused on the rhythm of his breathing and the warmth of his body, falling deeper and deeper into this wonderful obsession with him. Then she opened her eyes a crack and let the beauty of the world pour into her vision.

"Speaking of what's next," she said in a quiet voice, her hand idly stroking Zeus and playing with his ear, "have you thought anymore about it?"

Blake pursed his lips. She felt a sliver of tension run through his body. "I've been kind of preoccupied with the wedding," he said.

"I know, but it must have been in the back of your mind at least," she asked, and when she looked up at him she knew that it had been.

"I just wouldn't want to take on too much too soon. We have just been married I want to make sure that I spend a lot of time with my wife."

"We'll have plenty of time Blake. This is important. The village can't go on for too much longer without a mayor, and you know that you'd be perfect. You'd bring respect, honor, and dignity back to the title, and I think in a way it's your birthright. Your family has so much history in this area it only seems fitting that you should become the mayor."

"*Our* family," Blake corrected, and this made Lily smile. "The more I think about it the more I like the idea. I just don't want people to think that I'm using my family name to get the position."

"I don't think anyone is thinking that Blake. I'm sure that if you went back there and asked everyone they would all be happy for you to be mayor. In fact I'm sure that most of them would actually nominate you themselves."

"In that case it's settled then. I'll run for mayor."

"You certainly have my vote," Lily replied, offering him a kiss to show how proud of him she was. "You'd have his vote too, if he was able to vote," Lily said, looking down at Zeus.

"Maybe that will be my first act, to make it possible for dogs to vote."

"Oh dear, now I'm starting to reconsider!" Lily said. She then glanced back. "I suppose we should think about getting back to the party."

"I'm sure they won't miss us for another moment or two. It's nice to have this time to ourselves."

"I know, but they have made the effort to turn out for us. It feels rude to sneak out here."

"They're fine. They've got plenty to occupy themselves with. We'll get back there soon enough. I don't think anyone will begrudge me spending a few moments with my wife."

Lily smiled, a warm feeling entering her heart at being called his wife. "I'm going to enjoy getting used to being called that."

"And I'm going to enjoy getting used to you being my wife," Blake said, adding an extra intonation to the word. He leaned in to kiss her again and then pressed his forehead against hers. She could feel the emotion within him, so strong and pure, so radiant, and she knew the same thing was reflected in her. "There's something else I wanted to talk to you about as well."

"Is this about me moving out of Starling Cottage? Because I will get around to it. I just wanted to wait until we were married first. As soon as we're back from our honeymoon I'll make sure I'm out of there, although I will miss the place," Lily said. Joanie planned to stay there for the time being, and Kelly was also going to stick around for a little while.

Blake laughed. "No, it's not about that. It's about... well, you know how Zeus has been the most recent addition to our family... I was thinking that maybe we should start thinking about another one."

"Another dog?" Lily asked.

Blake prodded her playfully in a spot that he knew was ticklish. "Now you're just being obtuse," he said.

Lily leaned into him and then tilted her head. Another strong wave of emotion swam through her. "I suppose it is the thing to do after all. And you are going to need someone to carry on the Huntingdon name."

"I am indeed. Didlington St. Wilfrid could use a new arrival as well," Blake said.

Lily's smile turned a little wistful. "Do you think we could name him or her after Pat?" she asked.

"I think that's a wonderful idea. Patrick for a boy, Patricia for a girl," he said, and Lily thought it was perfect. They kissed again and then agreed that it was about time they returned to their party. Blake rose, and then helped his wife to her feet. She slipped her hand in his and they walked back to the hall, turning their backs on the beautiful sunset evening. It looked as though an artist had painted a watercolor sky. They walked slowly, making sure to enjoy every moment, but Lily also had one eye on the future. She could picture a child walking with them, and who knew what else the future held? It was so uncertain, but that was the beauty of it. There were things in life she could be certain of though, such as the love that Blake had for her, and the love of God that they were blessed with.

They returned to the cacophony of the party and moved back to the dance floor. Marjorie's cake was almost all gone, while Joanie was making her best effort to get Constable Dudley to release his inhibitions and dance as though nobody was watching. Lily smiled widely as people gathered around her, and once again she was filled with the feeling that she was home.

THANK YOU FOR CHOOSING A PUREREAD BOOK!

We hope you enjoyed the story, and as a way to thank you for choosing PureRead we'd like to send you this free Special Edition Cozy, and other fun reader rewards...

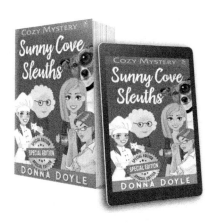

Click Here to download your free Cozy Mystery PureRead.
com/cozy

Thanks again for reading.

See you soon!

A TASTY BONUS BOOK...

You are clearly a dedicated sleuth!

We'd love to bless you with this delicious bonus book from Donna Doyle's bestselling series, Baker's Dozen Cozy Mysteries

Here's the first book in the series for your enjoyment. **Sammy Baker's first complete mystery, Dying For Cupcakes...**

DYING FOR CUPCAKES

CHAPTER 1

S ammy Baker glanced at her phone, hoping the GPS was right this time. It had already been a very long drive from New York to Illinois, and the little electronic voice had forced her to make several wrong turns. Road construction season seemed to be in full swing, and neither the new roads nor the ones that had been blocked off seemed to have been updated in the system yet. The rainy September weather hadn't helped at all, and her wiper blades scraped across the windshield with a grating sound that made her cringe.

But at least her Toyota RAV4 seemed to be holding up. Greg had said she didn't need an SUV, that if she insisted on getting a car she should get something smaller and easier to maneuver. But Sammy was grateful for the extra cargo room now that she had packed up her entire life into the car. There were a few things she hadn't been able to fit inside, but she wasn't about to go back for them. Greg could have them. He could have their apartment and all the furniture inside it except for her grandmother's old jewelry chest. He could have his big screen television that he seemed to love so much. He could also have Sammy's best friend, apparently, since he wanted the skinny little brunette more than he had wanted his own wife.

Sammy sighed and focused on the road, trying not to think about everything she had left behind. It was behind her for a reason, after all. She had been thinking about returning home to Sunny Cove ever since her father had passed away from a heart attack the previous year, and now she had the perfect excuse.

If only the road would get her there a little faster. Her lids were heavy as she watched the numerous dollar stores and fast food chains slide by. Things were definitely different here than they were in the big city. Sammy had forgotten just how much more space there was. The buildings were never more than a story or two high, huddling modestly against the ground, and she could actually see the sky all around her. There were actual parking lots instead of massive garages. Trees and strips of green grass interrupted the pavement and asphalt at regular intervals, reminders that the country wasn't very far away.

Sipping the last of a Coke that had gone flat and watery in the bottom of a Styrofoam cup, Sammy noticed that there were more and more churches the farther she got into the Midwest. The small towns she passed through still had their truck stops and bars, but even the smallest villages had at least one steeple rising up into the cloudy sky.

She absently touched the cross that hung around her rearview mirror. It had been a long time since she had been to church. At one time, she had been just as devout as anyone else in her hometown, but that had changed a long time ago, and in less than a month. Every now and then, she would turn to Jesus when her heart was heavy, but she usually ended her prayers with a sense of skepticism. Where had her Savior been when she had needed him so much back in high school? Why hadn't he saved her from a horrible marriage, wasting almost ten years of her life on someone who had ended up throwing her away like yesterday's garbage?

"No. I'm not supposed to think like that. I'm not supposed to feel sorry for myself." She'd read numerous self-help books, sneaking them home from the library once her marriage had started falling apart. Greg hadn't been interested in any type of counseling, saying that was for losers, and so Sammy had decided she could do the work herself.

While she had found a lot of affirmations, Bible quotes, and reminders of just how wonderful a person she was, none of that had helped her when Greg finally found one of the volumes under her pillow. "What is this all about?" he had demanded as he slammed the book down on the kitchen table. "Are you cheating on me or something?"

Sammy had instantly started shaking, feeling the adrenaline surging through her system. She knew she shouldn't have felt as though she was "in trouble" for wanting to improve their marriage, but Greg never did anything unless he could be convinced it was his idea first. "I was just doing a little reading. It says that every marriage can use a little help, even if it seems like a good one."

"Yeah, right. You're just trying to brainwash me." Greg had picked up the book again, this time chucking it into the kitchen trash.

"Hey, that belongs to the library!" She had immediately fished it out, wiping it off with a paper towel and hoping they wouldn't end up charging her the twenty-five-dollar cover price for a few new stains.

"And it's not about brainwashing. It's about communicating." She had started to get angry then. This wasn't the first fight they'd had, not by a long shot.

"Does it say anything about all the time you spend at your charities and volunteer functions instead of at home with me?" he had demanded.

Sammy's chest had tightened then. She had carefully scheduled anything she wanted to do around Greg's schedule, and even when she had wanted to spend time with him during the evening or on a weekend he always had something else to do. She had considered telling him that the book did say something about what a controlling husband he was, but she had just scowled at him instead. It was only a few weeks later that Sammy had discovered just what Greg was doing while Sammy was volunteering at the library or at the local animal shelter.

Just the memories of those times were making her hands shake again. She swung into the next parking lot, needing a moment to pull herself back together. The lot was nearly empty, with only a few cars near the front. Sammy looked up at the sign over the entryway: First Church of Calmhaven. She checked the GPS, since she didn't even remember a town like that on the map. It showed up, but only as a small blip. She must have entered the city limits while she was reminiscing about her past.

Her stomach rumbled. She could see a café just down the street, but something kept her where she was for the moment. The clouds in the sky were dark and ominous, but the light that came through the front doors of the church gave off a warm, yellow glow. But Sammy didn't belong to this church, and it would be rude to just go wandering in, wouldn't it?

That was when she saw the sign just to the left of the door. It showed the times for Sunday and Wednesday services as well as open hours, which were from four to eight. Sammy glanced at the clock on her dash board. It was just a little past five. She could go on down to the café and get back on the road, knowing she still had a ways to go before she reached the motel where she planned to stay the night, but instead she put her car in park and got out.

She stepped tentatively into the lobby, half-expecting someone to appear and demand to know just what she thought she was doing there. But she only found a bulletin board with upcoming events, a table with stacks of flyers, and a donation bin for canned food. A large set of double doors in front of her could only be the sanctuary, and Sammy slowly and quietly opened one of them.

As she walked up the aisle, she noticed that she wasn't the only one there. The handful of worshipers were scattered among the pews, each having their own moment with God, and they didn't turn to look at the newcomer. Sammy sat near the back and tried to slow her breathing. The air smelled faintly of coffee; there was probably a carafe of the cheap stuff somewhere around. The pew was cushioned, and soft organ music played throughout the sanctuary. Sammy pulled in a deep breath and let it out slowly, practicing the techniques she had read about.

"Hello, there."

The voice next to her was a friendly one, but it startled her nevertheless. Sammy jumped in her seat, turning to find that an elderly woman had joined her in her pew. She had friendly blue eyes that matched her voice, and her hair was a cloud of white around her head. "I don't think I've seen you around here before."

"Oh, no. I mean, no, you haven't." Sammy suddenly felt that shaking feeling return. "I was just passing through town, and I thought it was okay, I can leave, I'm sorry." She started to rise.

A cool hand on her arm stopped her. "Don't be silly, dear. Sit. I didn't come over here to scare you off. It's just that Calmhaven is a small town, and we tend to notice when someone new comes by. What's your name?"

Sammy blinked, trying to quiet herself once again. She had overreacted. Everything seemed to make her a little nervous. She hadn't been like that when she was younger. While Sammy didn't want to give Greg any credit for impacting her life, she knew that he had. She cleared her throat. "Sammy Baker."

"Is that short for Samantha?" the other woman asked.

"It is."

"Ah. You never know. I knew a little girl once named Maddie, and she spent a lot of time telling everyone that it was absolutely *not* short for Madeleine. My name is Molly Gertrude. Molly Gertrude Grey." She held out her hand.

Sammy shook it, somewhat calmed but still feeling uncertain. Maybe she should have just gone on instead of stopping. She would have to learn to get past her nerves eventually. "It's nice to meet you."

Molly Gertrude tipped her head slightly to the side. "What brings you here tonight? You look like something is weighing heavily on your heart."

She gave a derisive snort, knowing that her heart might as well be made of lead. "You could say that."

Miss Grey nodded and pressed her lips together slightly. "The pastor is with someone at the moment, but I'd be happy to talk to you about it if you'd like."

Sammy had no business unloading her sob story onto some stranger. This woman didn't know her, her ex, or what it had been like to be stuck in that dead-end marriage for almost a decade. But she suddenly found the urge to talk about it, more so than she had in a long time. "My husband left me for another woman, who just happened to be my best friend. I decided to go back home to start my life over, but I'm not really sure what to do with myself anymore." She choked up a little on the words. When things had started getting bad with Greg, Sammy had realized that her entire life was about to change. The places and people she loved were all different somehow, and she didn't quite know how to handle it.

"And you stopped in to talk to God for a little while? That makes perfect sense to me." Molly Gertrude nodded toward the front of the sanctuary, where a massive wooden cross was softly backlit. "I take great solace in telling Him my burdens, no matter how big or how small."

A wave of guilt came over Sammy. She had been selfish in coming here, not taking into account just how important the church was to those who came here regularly. "I'm not so sure He's listening."

Molly Gertrude's forehead creased with concern. "Why not? God loves all of his children."

"Even those who haven't talked to him for a long time?" She felt tears stinging the backs of her eyes. "I used to believe, you know. I sang in the church choir when I was little, and I went to every single event the youth group put on. But then my life changed. I realized that there were some things even God couldn't fix for me." She swiped angrily at her cheek with the back of her hand, removing the hot tear that had spilled over her lashes. So many times, Sammy had thought she had cried until she was out of them, but there always seemed to be more waiting for her.

"You know, dear, it's often most difficult to believe in God when we need him most. But I know he heals the brokenhearted. I've seen it happen many times. It's not instant, and Lord knows that's what we all want these days, but it does happen. And my good friend Jesus has never let me down, not once!"

"I wish I could have that kind of assurance." The other woman's words had been nice ones, but they weren't enough to convince Sammy that everything would be okay again.

Molly patted her on the knee. "You will. You just have to believe that God has a plan for you, and remember to embrace forgiveness. You'll never be able to move on until you can forgive those who have wronged you. Let them and all that negative energy out of your heart so you can make room for love again."

Sammy looked down at her lap and smiled. "I'll try. I guess we'll see what happens when I get to my old hometown again. I'm a little nervous about it, to be honest. I haven't been back much."

"You'll do just fine, dear. I know you will. Here." She produced a business card out of her purse and handed it over. "I don't know where you're headed, but if you come this way again be sure to look me up. I'd like to know how things have worked out for you."

"Thanks." She took the business card, noting that Miss Grey was the owner of a bridal agency, of all things. Sammy wouldn't need her services anytime soon, that was for sure. "I'll do that. Right now, I'd better get a sandwich or something and hit the road."

"Henry and Harriet's is just down the street. They've got the best food in town, and they're cheap, too. Tell them I sent you. They're good friends of mine." She smiled again, her eyes twinkling.

"I'll do that."

1

The next morning, Sammy walked tentatively into Just Like Grandma's. A little café situated in the middle of town, it was bustling with customers. Young families sat in the booths, and men on their way to work lined the stools in front of the counter. The scent of frying bacon and coffee filled the air, and there was barely enough floor space for Sammy to wind her way to the counter.

"Excuse me," she said to the woman frantically running the cash register. "I'm looking for Helen Honeycutt."

"I'll be with you in just a minute, honey." She turned away, calling down the counter. "Frank! Come down here and get your change! I know you've got two feet, and you can use 'em!"

A man at the end of the counter quickly hustled to the cash register, but he folded the woman's hand around the dollar bills she tried to hand him. "Just keep it, Helen. I think you deserve it, especially today."

"I don't care what anybody says, Frank. You're a good guy." She pocketed the money, waved to the man as he headed out the door, and turned to Sammy. "Sorry about that. Things are a little crazy around here. If you want to place an order, you might want to make it to go. I don't think there are any seats free."

Sammy could see just how busy the café was, and it made her feel bad for picking this particular hour to come in. "No, I'm Sammy Baker. I'm here about the job and the apartment?" One night as she had taken a break from packing up her entire life, Sammy had gotten online and managed to find the classified ad from Just Like Grandma's. She hadn't heard of the place when she had still lived there, but these small restaurants had a habit of changing hands often. The job was one she knew she could do, and it was a big bonus that a small apartment above the café would be included in her pay.

Helen threw her hands in the air. "Oh, thank goodness you're here! I was hoping you wouldn't back out. People who say they're coming in from out of state usually change their minds before they even cross the city limits, you know. But Jill just quit, and I need you like yesterday. Here." The owner bustled Sammy around the counter and into the back. "Grab an apron and get started. Table 5 has been waiting on some ketchup for about fifteen minutes."

Startled, Sammy plucked a frilly apron off the wall that looked like it belonged to the 1950's and quickly tied it on. "Is there any training I'm supposed to do? Or paperwork to fill out?"

"Honey, the best training you'll ever get is to dive in and get started. As for the paperwork, we can wait until I don't have people lined up out the door. Just make yourself at home, and you can ask myself or Johnny if you don't know where something is." She waved to a man in a greasy apron who was cooking an entire griddle full of food.

Sammy nodded. She grabbed a bottle of ketchup off a nearby shelf and headed out into the dining area. She'd had her share of food service jobs, but she noticed right away that the tables weren't tagged with their numbers. Taking a guess, she delivered the condiment and got started. Since there wasn't any other waitstaff visible, she started in one corner and worked her way around the room.

"Hey, you're pretty good," Helen said with a crooked smile several hours later. It was hard to believe that Just Like Grandma's was the same restaurant it had been earlier in the day. Only a few customers were scattered around, and the black-and-white tile floor was actually visible again.

Sammy wiped down a chrome-trimmed table and smiled. "Thanks. It was a lot of fun."

"Fun?" Helen's eyes widened, and she swung her steely braid over her shoulder. "You must like to torture yourself, honey. That rush we just went through is exactly why Jill quit. I wouldn't let her have breaks to go play games on her cell phone."

"I like to be busy," Sammy said with a shrug. "And the tips weren't too bad, either." She knew from experience that everything in Sunny Cove was going to be cheaper than in New York, so she hadn't been too concerned about making a living here. Still, the cash in the pocket of her apron let her know that she had been right.

"Well, if you want to stay busy, then how about you get in the back and show me some of those baking skills you bragged about over the phone. Or was that just a play on words with your last name?" Helen smiled, letting Sammy know that she was just messing with her.

It had always seemed a little ironic that she was not only a talented baker but had married a man with the last name of Baker. She had considered changing back to her maiden name, but it almost seemed easier to keep her married name. It was already on everything she owned, and it felt like more of a clean start to have an unfamiliar name in her old hometown. "What would you like me to bake?"

The older woman waved toward the kitchen. "Look at what we have in there and make anything your heart desires. I've been buying it all commercially, but I don't like to do it that way. There's nothing like a homemade muffin or piece of pie."

"You really don't have a preference?" The bakeries where she'd worked in the city always had certain flavor combinations they were going for, always trying to stay on the very tip of the trends.

Helen pushed her gently toward the swinging door. "As long as it tastes good and the customers want to buy it, then I don't care. I'll be cleaning up out here while we wait for the next rush."

Smiling to herself, Sammy stepped into the kitchen. Johnny was scraping the griddle, and she smiled shyly at him while she began looking through the fridge and the pantry. It took a few minutes to familiarize herself with everything in the kitchen, but she soon lost herself in what she did best. Sammy wasn't worried about the fact that she still had all her earthly possession in her car and would need to haul them upstairs to her new apartment at the end of the day. She wasn't thinking about all the wrongs she had suffered with Greg. Instead, she was mixing and stirring, crafting something that she truly cared about.

An hour later, Helen came into the kitchen. "What smells so good?"

Johnny shrugged and gestured toward Sammy with one hand.

"Nothing too fancy," Sammy said as her new boss approached. "I made lemon cupcakes with homemade whipped cream frosting. I thought something light and fresh would be a nice contrast to the gloomy weather." She picked one up off the tray and handed it over.

Helen immediately peeled back the wrapper and took a big bite. Her brow creased and then raised, her eyes sharp as she looked at Sammy. "This is excellent!" She took another bite before picking up a second cupcake and handing it to Johnny. "Try this."

He did as he was told, smiling and giving Sammy the thumbs up.

She blushed, flattered that her first creation had gone over so well. "What would you like me to do next?"

The older woman laughed as she picked up a platter and loaded half the cupcakes onto it. "I think you'd better keep baking. As soon as word gets around that we actually have something besides supermarket cookies, we'll be running out in no time. I'll holler for you when we get busy up front again."

"You really like it?" Sammy asked the cook once their boss had disappeared. He didn't seem to talk much, if at all. She would have to ask Helen about that some other time.

He nodded enthusiastically and smiled, even going as far as to rub his belly with one hand.

Feeling better than she had in a long time, Sammy went back to work. This time, she started in on some homemade dinner rolls. She wanted to show Helen that she wasn't all about the sweets, and she knew her own creations had to be better than the brown-and-serves that the restaurant was currently using.

Just as she was checking on her dough to see how well it was rising, Sammy heard a crash from the front of the restaurant. She and Johnny peered through the serving window to see a man running through the dining room. He was a blur of mismatched clothes as he dodged around stools and customers. Helen was calling out to him, telling him to calm down, and the other customers stared with wide eyes.

"What's going on?" Sammy mused, but of course Johnny didn't answer.

To her surprise, the kitchen door swung open. The man burst through and into the back. He looked frantically around him, no doubt searching for a way out. Sammy froze, unsure of what to do. The intruder's eyes met hers for a moment, wild and fearful, before her turned and crawled under a prep table. The remaining lemon cupcakes that sat on top of the table shimmied a little at the movement.

Johnny picked up a pot and a spoon, banging the two together at the man, but he didn't seem to get the hint.

More commotion from the front had them looking through the serving window again. "He's back there," Helen said to a man in uniform. "You'd better find a way to keep that young man off the streets, Alfie. It looks like he's getting himself into trouble again."

"I'm doing my best, Helen." The officer that came through the door was tall and broad-shouldered. His hair, dark brown but peppered with just a little bit of dark silver, was cut close underneath his policeman's cap. He glanced around the kitchen, quickly finding the perpetrator under the table. A garbled message came over his radio, one Sammy couldn't understand, but the officer seemed to. "Yeah, I got him," he responded before bending down. "Austin, you know you can't be back here."

The man under the table watched him warily. "Alfred Jones, 465 Watson Street. Gray with white shutters."

The officer's shoulders sagged. "I wish you wouldn't do that."

"Christmas in 1994 was on a Sunday."

"Is that so?" the uniformed man responded kindly. "Why don't you come down to the station and we can talk about it a little more?"

Austin started to come out from under the table, but he paused. "Coffee with two sugars?"

"Yep, we can have a cup of coffee. As many sugars as you want." He helped Austin off the floor and guided him toward the kitchen door.

"Two," Austin insisted.

"Just two, then." The officer escorted him out of the café and to his squad car that waited by the curb. He stood near the door, talking to him for a moment before coming back inside. "I'm sorry about that, Helen. You know how it is."

Sammy stood in the kitchen doorway, holding the swinging door at bay and dying to know what was happening. The policeman noticed her. "Are you all right, ma'am?"

"I am," she said, nodding and feeling a little dumb. She tried to attribute it to the odd situation that had just happened, but the fact that this man was so good-looking didn't help. She dusted her hands on her apron and held one of them out. "I'm Sammy Baker."

"Sheriff Jones," he replied, enclosing her fingers in his warm ones. "You look familiar."

"Oh, um, I grew up around here. I've just moved back into town."

The sheriff's dark blue eyes sparkled. "I'll see you around, then."

"Wait!" She didn't know exactly what prompted her to do it, but Sammy dashed back into the kitchen and grabbed a cupcake off the table. She handed it to Sheriff Jones, feeling that she somehow owed him for what he had done. "On the house."

He raised it in the air in thanks. "I'd better get him to the department and get his cup of coffee before he gets too antsy." Jones disappeared through the front door, chomping into the cupcake as he went.

Sammy watched him go. After a moment, she realized Helen was watching her from her position by the cash register. "What?" she asked, but she turned and dove back into the kitchen before her boss could say anything.

2

"Sammy girl, you saved my life today," Helen said when the dinner crowd had eaten all the stew they wanted and left their tips on the counter. "And so did those rolls. They went perfectly with today's special. Did you plan that?"

"Not really." Sammy dumped a load of dishes into the sink and began washing up. Nobody had told her that she was a dishwasher, but the job needed to be done. "I just felt like doing it."

Helen nodded as she began rinsing plates. "I'll give you a copy of the menu for the week, so you can plan all your deliciousness around it. And you give me a list of any ingredients you need if we don't already have them."

"Thank you. I'll do that." It was flattering to know that she had done such a good job on her first day that Helen was willing to give her a little bit of control over what Just Like Grandma's offered.

"What are you thanking me for? There's not a cupcake or a roll left in the house. I'll be lucky not to have people breaking down the doors to get more," she said with a smile.

Right on cue, a knocking sounded on the front door.

Helen rolled her eyes. "I'll go tell them we're closed. Doesn't seem to matter that you post your hours and lock the door." She returned a moment later. "Sammy, there's someone here to see you."

Sammy looked past Helen's grim face to that of the man who accompanied her. Like Alfred Jones, he wore a uniform, but he didn't wear it quite as well. An older man with a paunch and a dour look on his face, he frowned at her. "Samantha Baker?"

"Yes?"

"I'm Sheriff J.J. Barnes from the Calmhaven Sheriff's Department. I'm here to arrest you for the poisoning of Alfred Jones."

The edges of her vision instantly grew dark. Sammy's hands shook, and she tucked them into the pocket of her apron. "What are you talking about?"

He glanced at Helen and Johnny, who stood staring at the scene unfolding in the kitchen. "Maybe we ought to head down to the station so we can talk in private."

"Um, okay." Slowly, Sammy untied her apron and hung it on the wall where she had gotten it, hoping she would have a chance to wear it again. This couldn't be happening. She hadn't poisoned anybody. She'd barely been in town for a full day. In shock, she followed Sheriff Barnes to his squad car.

Fortunately, Barnes hadn't elected to use the pair of handcuffs that dangled from his belt. Sammy hadn't put up a fight, though, knowing it was much smarter to cooperate while they figured all this out. The sheriff had shown her to a private office with Sheriff Jones's name on the door and told her to have a seat. He lowered his heavy body into the chair behind the desk.

"Can you please tell me what's happening?" she asked.

He pursed his lips and leaned forward over the desk. "I think you already know, young lady. Sheriff Alfred Jones is currently in the Sunny Cove Hospital under treatment for poisoning. The last thing he ate was a cupcake, a cupcake that *you* baked and that *you* handed him personally."

"I gave him a cupcake," she stammered, "but I didn't poison it. I was just trying to thank him for doing his job."

"The only thanks you gave him was sending him to the hospital," Barnes grumped. "And you happened to leave Sunny Cove without a sheriff while you were at it. That's why I'm here, that and the fact that this case involves one of the Sunny Cove Sheriff's Department's own. Now, why don't you tell me why you did it?"

"Aren't I supposed to have a lawyer present or something?"

"All right, fine." Barnes picked up the phone and held a chubby finger over the buttons. "Who's your lawyer?"

He had her there. "I just got back into town. I don't have one right now." She was sure her divorce attorney back in New York didn't count.

Barnes set down the receiver. "Then let's talk. And I'll tell you right now, you might as well confess. It'll go a lot easier for you when you go in front of the judge. Besides, I already know why you did it."

"Then maybe you should tell me, because I certainly don't know." She didn't mean to sound snarky, but this was getting ridiculous.

Barnes picked up a thick file folder that lay on the corner of the desk and let it fall to the surface again, making a soft thump. "Almost twenty years ago, your father was sent to prison."

Sammy's jaw dropped. This had to be some sort of nightmare. "What does that have to do with this?"

"You see," Barnes continued as though she hadn't said anything, "you thought you were fooling someone by coming into town with a different last name. But people around here talk. It's a small town, and you should know that. It didn't take much digging to find the connection between yourself and Sheriff Jones."

"I wasn't aware that there was any connection." Sammy wondered if Barnes was deaf or just that stubborn.

"Jones's father was a county judge, and he just so happened to be the same judge who sent your father to jail for five years. He was later acquitted, but you have a grudge against the man. After all, a teenage girl who sees her daddy in an orange jumpsuit is going to be a little affected by the experience. You wanted to get revenge on Judge Jones, but since he's dead you poisoned his son."

Sammy shook her head, refusing to let all of this information into her brain. "I never did any such thing! I didn't even know who Sheriff Jones's father was. And even if I did, I'm not vindictive enough to hurt an innocent person over it."

"But you did bake those cupcakes?" Barnes fired back.

"Well, yes." She shrank back a little in her seat.

"And you personally gave one to Sheriff Jones?" He didn't look like the sort of man who could be taken seriously, but his eyes were hard as he questioned her.

"Yes."

He tipped his hand over, spreading his fingers. "Then it seems pretty obvious to me."

Sammy bit her tongue to hold back her tears. This was so unfair. She had come to Sunny Cove to start her life over after her divorce, maybe to rediscover her roots and find her way again. For a few hours, she thought that was happening at Helen's café, but now her past was being thrown right in her face. She clearly remembered those long weeks back in high school when her father had been arrested for fraud and money laundering. It had been traumatizing to see him convicted and hauled away, and that was the exact moment that Sammy had stopped believing in the religion that had been a part of her entire life up until that point. "You know my father died last year, right?"

"I do." Barnes pointed at the folder he had indicated before. "It's all in here."

"Then why would I take revenge now? Wouldn't it have made more sense to do it a long time ago? Like when my dad was in jail? Or maybe right after he'd been acquitted and released?" Anger stirred inside her at the Sheriff's insinuation.

"People do crazy things when they're grieving," was his explanation. "You came back home for your father's funeral, it reminded you of how unfair the entire thing was, and you started plotting your revenge. These sorts of things take a little time to plan, I suppose."

"I wouldn't know, since I didn't plan anything."

A gentle tap sounded on the door. Barnes rose to his feet and waddled around the desk. "Excuse me a minute."

Sammy could hear his voice and another one just on the other side of the door, speaking quietly but urgently. She wasn't normally in the habit of eavesdropping, but this seemed like a good time to start.

Barnes returned, looking frustrated. "We don't have enough evidence to hold you right now, but you should consider yourself detained to the city limits," he said as he sat back down with a sigh. "You're not to leave Sunny Cove under any circumstances while we investigate this matter, do you understand?"

Sammy couldn't quite let herself feel relieved. She might not have to spend the night in a jail cell, but she was still a suspect in a crime she didn't commit. For a brief moment, she let herself wonder if the sense of disbelief that overwhelmed her was the same thing her father had felt when he was in a similar situation. "I understand."

"I'll have one of the deputies give you a ride back to Helen's. She tells me you're staying there?"

In all the commotion, Sammy hadn't even had a moment to think about that fact. "Yes." At least, she hoped so. This wasn't a very good impression to make on her first day.

3

The next day, Sammy walked into the Hamburger Hideaway and immediately wished she hadn't. The fast food joint was just as greasy and ugly as she remembered it. The hard benches in the booths were all bright red, a stark contrast to the white walls and bright yellow napkin dispensers. The floor was greasy enough that it was dangerous to walk across.

"Well, well. You made it. I honestly figured you wouldn't show up," said the woman across the booth from her as Sammy sat down. "I mean, why would someone who had actually gotten out of this crappy place ever come back?" Heather Girtman had been one of Sammy's good friends back in high school, and she was the first person Sammy had thought to call to get together with now that she was back.

"Lots of reasons." They stepped up to the counter and ordered their food before sitting again, greasy brown sacks in hand. The Hamburger Hideaway didn't have any dishes. "Mostly, though, I got divorced. I didn't want to stick around for the fallout."

Heather raised one dark eyebrow as she picked at her fries, which were limp with grease. "Someone divorced *you*? Why would they do that? Doesn't he like a goody-two-shoes?"

Sammy wondered how she had never before realized that Heather spoke in nothing but questions. "Well, things happen, you know." She thought she would spend an evening dishing out all the emotional trauma of what had happened between herself and Greg, but it suddenly didn't seem like something she wanted to discuss. "What about you? What have you been up to?"

"Me? You don't think I would actually get out of this town, do you? I mean, I've tried, but there's just nothing else out there, you know? Sometimes you have to pick the evil you know instead of the one you do know."

"I suppose that can be true sometimes." Sammy grabbed a napkin to catch the grease that dripped from her burger onto her hand. "I guess I'm doing a little of the same, coming back to my hometown like this. Where are you working?"

Heather sighed as she picked at the lid of her cup. "I'm still at the bowling alley. Some old place I've always been. I was engaged to Billy York for a little while, but we broke it off."

"Oh? How come?" But Sammy already knew. Billy York was bad news, and even though Heather seemed like she maybe wasn't the most upstanding citizen in town these days, she was probably better off without him.

Her old friend waved a fry through the air, brushing off her former relationship. "We didn't do anything but fight. I didn't mind for a while. It was kind fun, you know? But who wants to be in a relationship like that? It might be good enough for some, but not for me."

"That much I can understand." As much as Sammy had been looking forward to coming out and having dinner with an old friend, she was starting to regret it. Heather had changed a lot. Maybe they both had, but either way it made it feel like she had joined a stranger at their table. They could get along, but they weren't necessarily friends any longer. She wondered if she would find anyone in town that was still the same person she had known back when she lived here. "Nobody needs that kind of drama."

"Speaking of drama," Heather said with a mischievous glint in her eye, "I heard you got taken down to the police station the other night." She grinned as though this was a good thing while she reached for the bottle of ketchup.

Sammy pulled in a deep breath. She knew the subject would be unavoidable. A small town like Sunny Cove was full of grapevines, and people kept their ears constantly pressed to them. She remembered just how bad it had gotten when her dad had been arrested. For an entire week, she had stayed home from school just so she wouldn't hear all the kids gossiping and spreading lies. "That's true. But I was only questioned, not arrested." It seemed important to make that point clear.

"So did you do it?" Heather leaned across the table, awaiting Sammy's confession.

"Of course not!" It shocked Sammy that her old friend could even think such a thing. "Why would I poison someone?"

"Well, the rumor is that you did it out of revenge for your dad." Somewhat disappointed, Heather sank back into her side of the booth and poked a long-nailed finger at her fries.

Adrenaline and anger mixed in Sammy's veins. "People are saying that?" Somehow, she had imagined that her conversation with Barnes had been a private one. She had been a fool.

"It's just a rumor," Heather replied dismissively. "I wouldn't worry about it. And old Judge Jones is dead, so you don't have to worry about him locking you away just for fun."

Sammy swiped her hands down her face. This was awful. "I'm not worried about that, because I didn't do anything wrong. I baked the cupcakes, I gave one to Sheriff Jones, but that was it. I didn't poison it."

"Hey, you don't have to justify it to me. I believe you." She took a long pull of her soda and swished it around her mouth. "I think they need to add more syrup. This tastes awful."

But Sammy already knew that nothing at the Hamburger Hideaway was going to taste as good as what they served at Just Like Grandma's, and she didn't really care if Heather believed her or not. "It'll all blow over. I've got a job and a place to stay, so I'm already doing much better than most people who move halfway across the country without much notice."

"Yeah, I heard you were at Helen's. Every time someone mentions that poisoned cupcake, they turn right around and mention how good your baking is." Heather sounded almost sarcastic, but that seemed to just be the normal tone of her voice.

"I'm lucky that she's been so supportive of me. She waited at the restaurant until Barnes brought me back and made sure I got into my apartment okay. Things have been a little bit awkward, but I think they'll be okay." Poor Helen had to endure the crime scene investigation team swarming her restaurant, swabbing and sampling everything in sight, and yet she had still made sure that Sammy was taken care of when she got back. It gave Sammy a lot of hope to know she wasn't completely alone. Her apartment wasn't huge, but it was enough for just her. She still had a lot of work before she was completely settled in, but she would get there.

"So that's it? You're not going to do anything?"

Sammy looked across the table at Heather. "Do anything about what?"

"About the poisoning!" Heather looked indignant at having to explain it. "You're the prime suspect, Sam. I don't know this Barnes guy, but you can't just rely on him to clear your name. He wants to put someone behind bars just so he can say he did it. You need to get an attorney or find the real criminal or something."

The thought hadn't occurred to her, and Sammy felt the weight of it sink onto her shoulders. Maybe Heather was right. If it was up to Barnes, she would have stayed in jail the night Alfred had eaten that cupcake. "I'm not sure what to do."

"Me, neither, but I wouldn't just sit around. You'll think of something. In the meantime, you want to go have some fun?" Heather waggled her eyebrows.

"I wasn't aware there was any fun to be had in Sunny Cove." As a kid, she'd spent her time riding her bike around the neighborhood or playing at the park. Once she was in high school, she and the other kids drove laps around town and went down to the lake on hot days. The options for fun and entertainment had been few and far between.

"Oh, trust me. There's plenty of fun to be had in this town if you know where to look for it." A slow smile spread across Heather's face.

Sammy wasn't sure what her old friend had in mind, but she didn't make it sound particularly appealing. It was early in the evening yet, and she had the rest of the night off from Just Like Grandma's, but she didn't want to go running off to some house party or get drunk in a barn. "I have to work in the morning, and I still have a lot of boxes to unpack. I think I'll head home. Thanks anyway."

"Okay." Heather stood and put her trash in a nearby can. "Suit yourself. Don't unpack too much in case you end up in the pokey." She grinned and waggled her fingers as she walked out the door.

Sammy sighed. Maybe she should have just gone to a completely different town where nobody knew her at all.

4

It was easy to throw herself into her work the next day. Sammy went downstairs early to make glazed donuts, something that she knew would go over well with the breakfast customers. Then she helped Johnny with some dishes before stepping out onto the dining floor to bus tables and bring out plates of bacon and eggs.

In a way, it was nice to have so many jobs to do. She never had to slow down or stop simply because the lunch rush was over or because she had baked enough for the day. There was always something to do, and she liked it that way.

"Got any of those donuts left?" asked a man at the counter in a business suit. "I already had one, and I'm on a diet, but I don't think I would turn down another one."

Flattered, Sammy stepped into the back to get a new tray of them. "Coming right up. I'm glad you're enjoying them. I made them myself." She wasn't trying to brag, but she knew that customers appreciated knowing where the food came from.

He had immediately snatched the treat from his plate, but he looked up at her with surprise before he took a bite. "Really? You know how to make donuts? I thought they all came in from factories."

She giggled. "I know. We're so used to getting them in boxes, but they're really not too hard to make, especially when they're just plain glazed donuts."

"Well, those are my favorite. You should start selling them by the dozen. I would buy them to take to the office."

Sammy refilled his coffee. "That's a good idea. I'll talk to Helen about that."

"You don't need to talk to her about anything. You could start your own bakery. If the other stuff you make is half as good as this, then you'd have it made." He bit into his donut, immediately following it with a sip of coffee, sat back, and closed his eyes with satisfaction.

"I appreciate that, but this is her place, after all." Maybe this guy was right, and she could make it on her own. But not only would that involve a lot of work and overhead, she felt she owed Helen a little something for being there for her. The older woman had allowed her to live and work in her building even though she was suspected of attempted murder. Most people wouldn't be so understanding.

"Suit yourself. I'll be back later in the week for more." He put his payment on the table and left.

"Waitress!" called a shrill voice from across the room.

Sammy turned to see a woman in her forties holding up an empty coffee mug. "Are you going to get me a refill, or would you rather stand there and flirt all day?"

She most certainly hadn't been flirting with the man at the counter, but her denial wouldn't make the customer any happier about it. "Here you are. Thanks for your patience. Is there anything else I can get you?"

"No, I…" the woman broke off as she studied Sammy's face. "You're Sammy Beaumont, aren't you?"

It was no surprise that someone would recognize her. Sammy had actually been more surprised to find out just how many people in Sunny Cove she didn't know anymore. Some of the older generation had passed away, those her age had brought in people from other towns to be their husbands and wives, and there was the fact that she had been gone for so long. The backdrop was the same, but so many of the characters were different. "It's Baker now, but yes."

"Well then, get back! Helen! Helen! I want you to wait on me!" The customer shooed Sammy away with her hands, her diamond rings glittering.

Helen came bustling over. "What's the matter, Julia?"

The customer flicked her fingers toward Sammy as though she was a filthy stray dog on the street. "How can you let this woman touch the food? We all know what she did!"

The owner heaved a sigh, and from the tired look in her eyes, Sammy could tell this wasn't the first time she'd had this conversation. "Sammy didn't do anything wrong. I ate one of those cupcakes myself, and I'm standing right here in front of you."

But Julia didn't seem convinced. "Still, I'd rather know exactly who's been near my food, if you know what I mean."

"Sammy, could you go into the kitchen and see how Johnny's doing with the soup of the day, please?" It was Helen's nice way of dismissing her, since they both knew Johnny didn't need any help with the soup. He might not talk, but he was a very competent cook.

"Of course." Sammy went through the swinging door and turned to lean against the wall. "I swear, if one more person mentions poison cupcakes, I'm going to scream."

Johnny gave her a sympathetic look and returned to his grill. He had several hamburgers going, all of which looked much better than the grease patties the Hamburger Hideaway served up.

Helen joined them a few minutes later, a crease of concern on her brow. "Sammy, honey, I think we need to keep you in the back for a while," she said quietly.

"She doesn't bother me," Sammy assured her. "I can handle it when customers don't think I'm good enough to wait on them. I won't get much in the way of tips, but they'll figure it all out once this mess is over with."

Helen didn't look so convinced. "I hate to put it this way, but I think it might be bad for business to keep you out on the floor. It's funny, because people are certainly going crazy over your food, but I don't know that they've all put two and two together."

"But—"

"No, now don't make this any harder on me than it already is. I never could have been more blessed than to have you come work for me at the time you did. I know my restaurant and that apartment up there are in good hands, and I know you didn't do anything to Alfie besides give him a little treat. I ate a cupcake myself, so I know they weren't poisoned."

Johnny rubbed his stomach in agreement.

"Why do you call him Alfie?" That was at least the second time Sammy had heard Helen refer to the sheriff that way, and it struck her as funny despite the dire state of affairs.

Ms. Honeycutt smiled fondly. "I knew him when he was just a little boy. His parents lived across the street from me, and little Alfie Jones was always chasing his ball into my front yard. One time, I found him climbing the apple tree in my front yard. As he got older, he insisted that everyone start calling him Al. Old habits die hard, though, and I was always a little amused at how angry he would get when I still called him Alfie."

Sammy couldn't help but smile back. It was a cute story, even if the man they spoke of was in tentative condition at the hospital. "Speaking of, I wonder if you might help me. Someone suggested to me that I should look into this whole case and find out who *really* poisoned Sheriff Jones. That way I can clear my name."

Helen pressed her lips together and tipped her head to the side, her long braid swinging across her back. "That sounds like a dangerous thing to do."

"It might be," Sammy shrugged. "But I don't know if I can count on Sheriff Barnes to find the real culprit. He seemed more interested in throwing me in jail than in hearing the truth."

"And you want to avoid the same thing happening to you that happened to your father. I can understand. But you're already in some trouble just by being in the wrong place at the wrong time, and I would hate to see that get any worse for you."

She made a good point, but Samantha wasn't quite ready to let this go. "I'll be in real trouble if Barnes finds a way to put the blame on me. I don't want to say he's a bad guy; I'm sure he's just doing his job. But I'm scared, Helen. You say you've known Alfred Jones since he was a little kid. Is there anyone else who might have had it in for him? Maybe that guy he chased in here that day?" It seemed as though Sheriff Jones had been very kind to the man who tried to hide under the prep table, but that was the only possibility she knew of.

Tapping one finger against her lips, Helen rolled her eyes up to the ceiling. "That's an interesting thought. I can't say I've ever thought of Alfie as having very many enemies. He's a wonderful man. I'm surprised he's not married, to be honest. Any young woman in their right mind would be lucky to have him. I mean, he dated Sonya McTavish for a little while, but it was a good thing they broke it off, if you ask me."

"What happened?" Sammy pounced on the idea. A broken heart could easily be the motive behind a contaminated cupcake.

But Helen brushed off the idea. "Sonya isn't my favorite person, but she couldn't have done anything like that. She and the sheriff only went on a few dates, I heard they had an argument of some sort, and then she up and moved to California or Florida or somewhere. But she hasn't been in town for months."

"Okay. So could she have someone here, a friend who could have done this for her?" Sammy knew she was grasping at straws, but she had to find some lead to follow.

"It's possible, but I don't know who it would be."

Sammy leaned against the fridge. "There's got to be someone else. I can't imagine a sheriff would be friends with everyone."

"I think Alfie would have been a few years ahead of you in school, so you might not have known him very well when you still lived here. But he's always been a good man, and that's why people voted him in as sheriff. Sure, he's made some arrests and pulled people over, but as far as I've heard he's always been nice about it."

Glancing at the prep table across the room and reliving her first evening in Sunny Cove, Sammy asked, "And you're sure that Austin guy wouldn't have done it? He was right near the cupcakes. I don't know if he touched anything or not, because I was just trying to keep up with everything."

Helen shook her head. "Austin Absher is a soul to be pitied. He does get himself in trouble quite a bit for stealing, but I have my suspicions that he only does it because he's not taken care of very well at home. He's got the mind of a child, and he doesn't really know any better. He steals when he's hungry, Alfie picks him up and calms him down, and then he takes him back home."

"He seemed to know an awful lot of other facts, even if he's not sure about right and wrong," Sammy pointed out. "He recited the sheriff's address, and he knew what day of the week Christmas was on over two decades ago."

"A rare talent, but not one that he's able to use very well, I'm afraid. The boy has a mind like a steel trap, and he remembers every single thing he's ever seen or heard. It's a wonder, and I suppose that could mean he knows how to poison someone, but I don't think he ever would. You saw how willingly he went with the sheriff once he knew he had been caught."

Sammy felt a wave of sadness wash over her. "Isn't there anyone who can help him?"

"His uncle is supposed to be his guardian, but like I said, I don't think he's doing a very good job of it." Helen wrung her hands and shook her head.

That was a horrible situation, but it also wasn't getting her anywhere in her investigation. It was something Sammy knew she would have to come back to, though. "What about you, Johnny? Do you know anybody who might be so mad at Alfred Jones that they would do this to them?"

The cook shook his head.

"What if it wasn't the cupcake?" Sammy asked out loud. "That was the last thing that anyone *saw* him eat, but that doesn't mean there wasn't something else he had in the meantime."

"What are you saying?" Helen asked, and Johnny raised an eyebrow.

"Just that there could be more to the story than what we're being told. I wonder if anyone else saw Jones around town later that afternoon, because it sounds like there might have been a little bit of time between when he left here and when he actually went to the hospital. It's at least worth checking into."

"Sammy." Helen's voice had a warning tone to it. "I don't want to hear that you've dressed in scrubs and snuck into the hospital to check their records, or that you're some computer genius and you've hacked into the computer system at the sheriff's department."

717

"Well, I don't think I'm that talented," she replied, her face flushing. "But the scrubs...and a surgical mask...that might work."

Now even Johnny was shaking his head emphatically.

"Okay, no, I won't do that. But I'm not going to leave this alone until my name is cleared." She knew she couldn't. She owed it to herself and to her father.

"I know you won't, dear. And I'll do everything I can to help you. But you have to be sure you don't incriminate yourself for something else in the process. God willing, it will all work out."

Sammy took a deep breath. "I hope you're right."

5

It was difficult to stay in the back when Sammy knew how busy Helen was out front. She had stayed up half the night trying to come up with a plan for this new "investigation" she had unofficially launched, but she hadn't been able to come up with much. A computer search of recent newspaper articles showed that Helen had been right about Jones. He had done his job as sheriff, but he had never made any big drug busts or hunted down a psychopathic genius. In a place like Sunny Cove, crimes were usually restricted to fender benders or the occasional break-in.

After putting in another batch of dinner rolls—which had become such a hit that they were practically a staple at Just Like Grandma's —and pouring herself another cup of coffee, Sammy grabbed the trash and took it outside.

The sun was too bright, even in the alley behind the building. It shone down on Sammy's Toyota, reminding her of just how far she had driven only to land in this mess. She wasn't sure how much of a friend Heather really was these days, but she had been right about getting her name cleared. Sammy couldn't hire a private investigator, but she could become an amateur one if need be. Unfortunately, she didn't seem to be a very good one yet.

The trash bag was heavy, and she shuffled across the gravel to the dumpster. She stopped when she heard something inside it, her heart thumping in her chest. When she had lived in the city, there had been rats galore in places like this. But this was a small town in the middle of nowhere. It was just as likely that she would find a raccoon digging through yesterday's scraps. When Sammy tentatively peered into the receptacle, two wild green eyes stared back at her.

She leaped back, letting out a little yelp of fear, but she decided she didn't feel nearly as scared as Austin looked as he peeked up over the edge of the dumpster. His dark hair stuck up in all directions and he wore a bright blue scarf around his neck.

"What are you doing here?"

"Americans throw away over $165 billion dollars worth of food annually," he replied quietly.

"Are you hungry?" She took a step forward, horrified that this poor man might actually be digging around in the trash for his next meal.

This time, Austin didn't respond with a fact but simply nodded.

"Why don't you come inside? I'll buy you a meal. Even a coffee, if you'd like," she added, remembering how excited Austin had been to have coffee with Sheriff Jones.

But the man shook his head emphatically. "People don't like me in there."

It was the first thing she had heard him say that wasn't a random fact, and it caught her attention. "What do you mean?"

"People don't like me in anywhere."

"Austin, I'm sure that can't be true. Maybe they just don't understand you." She paused for a moment, her heart aching for this poor man. "And I heard that you're usually stealing food. If you're hungry, and I'm buying you a meal, then you won't need to steal anything. It'll be okay."

But Austin shook his head once again.

"Okay." She wasn't going to push him. Sammy had done her share of volunteer work, but she was no professional when it came to someone like Austin. Still, she wanted to help. "What if I brought some food out here? You could eat it here, where there aren't any people, and maybe you and I could talk."

He stared at her for a moment before he finally nodded.

She couldn't stop the smile that spread over her face. "You come on out of there, sit here on the step, and I'll be right back with a big plate of food for you, okay?"

"Promise?" He stood up a little inside the dumpster, looking hopeful.

"Cross my heart." She slipped inside, eager to help the poor soul in the alley. The timer went off on her dinner rolls, so she pulled them out of the oven. Grabbing a large plate, Sammy loaded it down with chicken casserole, green beans, and two of the rolls slathered in butter. As she poured a to-go cup of coffee, she noticed Johnny watching her curiously. "Don't worry. I'm paying for it. This is for a friend of mine."

He nodded and went back to his work, looking unconcerned. It didn't seem that there was much that rattled Johnny.

Returning to the alley, Sammy found that Austin had climbed out of the trash bin as she requested. But instead of sitting on the back step, he was standing in front of her car. His eyes were intense as he looked over the blue SUV.

"I've got your food."

Austin turned, looking startled again until he realized it was her. He immediately trotted over, yanked the plate out of her hand, and immediately began eating.

"You really were hungry," Sammy mumbled to herself. She sat down next to him, setting the coffee on the concrete step for whenever he was ready.

In between bites of casserole, Austin suddenly pointed at the nametag on her apron. "Sammy."

"That's right."

His finger swiveled to point at her car. "Sammy from New York, SRB 811."

"I see you've memorized my license plate." The idea might have unsettled her if she hadn't already come to know a little bit about him. "Is it fun for you to do that?"

Austin didn't answer her question, instead asking, "Sammy where?"

"I'm right here."

He shook his head so hard she thought he might make himself dizzy. "Where?"

Sammy suddenly remembered what Austin had said to Sheriff Jones on that fateful evening. "You want to know where I live? I live here. There's an apartment upstairs, over the café." She pointed up and behind her, toward the second-story windows of the brick building.

He stared up for a moment. "Sammy from New York, SRB 811, 1214 Main Street."

"That's correct." When she had lived in the city, she never would have told a stranger her address. But Austin seemed harmless, and this definitely wasn't the city. "You know just about everyone around here, don't you?"

His mouth full of green beans while he scooped up more with his fork, Austin nodded.

"I wonder if you can help me with something. You know Sheriff Jones?"

He gave that insistent nod again.

"Well, someone tried to hurt him. No, it's okay. He's in the hospital, and he's being taken care of by some very good doctors. But I want to find out who did that so they can't do it again." Sammy hoped she was putting this in simple enough terms for him. She wasn't sure just what Austin was capable of understanding. "Do you know of anybody who might want to harm Sheriff Jones? Or make him go away?"

Austin swallowed and set down his fork, staring off into the distance for a moment before he turned those viridescent eyes on Sammy. "The bad man with the big wheel."

Considering the way Austin usually spoke, something this vague was unusual. "Do you know the bad man's name?"

"No. Big wheel." He spun his finger in the air and made a ticking noise.

"Like on a game show? On TV?"

Austin shook his head.

Sammy was just about to ask him if he knew an address when an old truck came down the alley. Austin instantly froze, watching as the tan vehicle crept down the narrow gravel lane. He set down the plate, grabbing the two remaining rolls and stuffing them in his pocket. Every muscle in his body was tense, like he was ready to run.

The truck came to a stop right behind Just Like Grandma's. The grizzled man behind the wheel looked at them with tired eyes. "There you are, Austin! Get in this truck right now. I'm sorry if he was bothering you, ma'am. Please don't call the police on him this time. I'll find a way to pay you for whatever he stole."

"Oh, no. He didn't steal a thing." Sammy rose to her feet, hoping she could talk to the driver about Austin's situation, but Austin was instantly bolting toward the truck. "Austin, wait! At least take your coffee with you." She picked up the mug and handed it to him.

The driver, a man whom Sammy had to assume was his caregiver, watched the exchange with surprise. "How much do I owe you?"

"Nothing," Sammy assured him. "It's on me. Austin was helping me out with a problem."

The man looked skeptical. "Well, I'm sorry he bothered you." The truck sped off down the alley, kicking up a small cloud of dust in its wake. Austin waved at her from the rear windshield.

Sammy waved back, feeling sorry for the man. He was so misunderstood; that much she knew for sure. When she turned back to the restaurant, she saw Helen waiting for her on the back step.

"Feeding the strays, are you?" she asked.

"I'm sorry. I'll pay for his meal myself. But he was so hungry he was looking for food in the dumpster." Even though she knew he had a full belly now, she felt tears at the backs of her eyes to think that anybody had to do such a thing just for food.

"You'll do no such thing," Helen assured her, a warm smile creasing her cheeks. "I knew you were a good girl as soon as I met you. But I'll give you a word of warning. If you feed him now, he's just going to keep coming back. I used to let him come in, and I'd give him a peanut butter sandwich and a cup of milk every day, no charge. But he was scaring all the customers away, and I have to make a living."

"I'll make sure he stays out of sight," Sammy promised. It didn't seem fair, but she understood that this was Helen's business. And she also hoped Ms. Honeycutt was right and that Austin *would* come back. She wanted to know more about the bad man with the big wheel.

6

Later that evening, Sammy sat upstairs in her apartment. She had been so busy with the restaurant that she hadn't had much time or energy for getting settled in, but she couldn't live out of cardboard boxes and garbage bags forever. She had always sworn that the next time she moved—which was supposed to be when she and Greg got a house in the suburbs—she would have everything organized and labeled in plastic totes, so the unpacking would be nice and smooth.

Instead, she had driven down to the local supermarket to load her car up with any cardboard boxes they had available. As soon as she had gotten them back home, Sammy had stormed through the house, chucking her things into any box she had on hand. There was no organization, no one certain box for clothes and another for books. She had been so desperate to get out and get away from the man who had hurt her so much that she had only grabbed the most important things and left everything else behind.

Fortunately, the apartment over Grandma's came fully furnished, and Helen had even left a bucket of cleaning supplies under the sink. Though it wasn't very big, it was a nice little place that she was quickly coming to enjoy. The living room and kitchen were one large space, with a counter and the cabinets on the left and a couch and chair on the right. There was room on the short wall to hang a television once she got one. Just past the big set of built-in bookshelves was the door to the bathroom on the right. A small space with tiny tiles on the floor and a cast iron tub, it was nothing like the modern baths she had seen on television. Still, Sammy thought it had an old-fashioned charm and great lighting for the days when she actually felt like putting on makeup. Past the bathroom was the bedroom, where a big quilt covered the queen-sized bed.

Except for the wooden floors, everything in the apartment was light and airy in color. The tall, flat ceilings were painted white, as was all the trim work. The walls were an incredibly pale shade of green, like the eggshell of a wild songbird, and it went nicely with the dark wood of the floors. Someone had put down a shabby rug in the middle of the living space, and the bright blue of it gave just the right pop of color to the apartment.

Sammy knew she could be happy here, but she still didn't enjoy the idea of unpacking. She hadn't even taken the time to label her boxes, and there hadn't been much point since their contents were in such disarray. With a sigh, she hoisted a big box with pictures of bananas on the outside onto her kitchen table and opened the lid, finding a few books, a yoga DVD, and a New York Yankees sweatshirt. Underneath that was a notebook that she must have thought was important as she was packing, the extra charging cable to her phone, and a framed picture of her parents.

"This is going to take forever," she said to them as she set the frame on the bookshelf. "I'll be running all over the apartment with every box."

As she picked up a carton that had formerly held eggs, she remembered Heather's words as she suggested Sammy not unpack too much in case she ended up in jail. She shook her head, trying to knock the thoughts away. She hadn't done anything wrong, and somehow she would find a way to prove it. She just wasn't sure how yet.

The next box was one she wished she hadn't opened. Right on top were her old scrapbooks. Sammy had kept them because it seemed appropriate, even though she rarely ever looked through them. They simply moved around with her from place to place, often stuffed under her bed or in a cabinet where they wouldn't be noticed and nobody would ask her about them.

Now, she lifted the old blue binder out of the box and carried it to the couch. Even though it had been years since she had flipped through the pages, she knew exactly what would be on each one. Lifting the cover, she saw the newspaper clipping that showed her father's arrest. Sammy couldn't even say now why she had saved it, considering that the night the police had come to their house had been such a horrible one. But she had continued to do so, keeping anything that was even slightly related to the case. If she had been older at the time, if she had been an adult, maybe she would have tried to discover the truth behind what happened.

"Bernie Philpot Testifies against Rick Beaumont," said one headline. It was accompanied by a picture of a man as he walked down the steps of the courthouse, covering his face with his hand against the flash of the cameras. A young girl walked next to him, but she wasn't as reserved at the attention. Her big eyes, tipped up slightly at the outer corners, stared boldly into the camera, and her face was carefully framed with thick, dark hair. "Beaumont Takes the Stand," said another. Sammy didn't need the big bold letters to remind her of exactly what happened, but "Beaumont Sentenced" still put a knife in her heart.

Rick Beaumont had taken a job in the next town at a bank. It was a good job with good hours and even better benefits. Every day, Sammy had gotten up at the same time as her father. He would get ready for work while she got ready for school. It had been just the two of them since her mother had passed away when she was three, and they had a companionable relationship that Sammy had never thought would change. She got home from school about an hour before he got home from work, and they would make dinner together every evening, usually eating in front of the television. Samantha had often wished she still had her mother with her, but she was always grateful for her father.

But then one day, he didn't come home from work. She waited and waited, calling his office several times to see what was keeping him. Sammy made dinner by herself, wondering if her father had gotten into a car accident or if he had just been stuck in a late meeting. A police officer had finally showed up, explaining that her father had been arrested and that Sammy was to stay with her aunt across town.

He'd been charged with fraud and money laundering. The bank accused him of loaning money to those who didn't qualify in order to fund illegal business that he had a share in. The money that those businesses made from selling illegal goods was accepted back into the bank under Rick's approval. Sammy hadn't understood all of it as a kid, and she had never been interested in knowing the details as an adult. Even now, she preferred to do all of her banking online just to keep as much distance from it all as possible.

Though things had eventually worked out five years later, when the judge who had sentenced Rick to jail confessed on his death bed that he had been forced to do so in fear of his son being harmed, the damage had already been done.

A soft knock on the door brought Sammy back to the present. She set the scrapbook aside and crossed the room, looking through the peephole. She opened the door. "Hi, Helen. What brings you by? I thought you went home hours ago."

The older woman was breathless from climbing the stairs. "I did, but I've been thinking about you."

Sammy held the door open and stood aside to let her in. "How come?"

"I don't really know," she puffed. "God just put you in my heart, and I thought at least I would come by and see how you were coming along with unpacking."

"It's a slow process," Sammy confessed, choosing not to explain that she had just been reminiscing over one of the worst times in her life instead of actively doing anything. "You'd think I would have better things to do with my free time, considering I'm not allowed to put a toe over the city limits. But somehow that has only made things harder. Have a seat. I'll make you a cup of tea. It's one of the few things I've been able to find among my stuff."

"I won't turn that down." Helen pulled out a chair at the little kitchen table while Sammy removed the big box and started the kettle. "Are you making any progress on your little investigation?"

Sammy dug through the cabinet to look for the bottle of honey she had purchased down the street at the grocery store. "Not really. I haven't broken into the hospital or hacked any computers, if it makes you feel any better."

"In a way, yes. In another way, no."

This answer surprised Sammy enough to make her turn around. "Really? I thought you wanted me to be careful."

"I do. But I've been thinking about it a lot, and it probably would be a very good thing if you managed to clear your name." She accepted the mug of hot tea and stirred it absently.

Sammy sat down. "I'm just not sure who to talk to. I mean, I'm from this town, but I don't really know anybody anymore. Not like I used to. I spent some time talking to Austin Absher today, though. I hoped he would know something, since he seems to have the entire population of Sunny Cove memorized."

Helen raised an eyebrow. "I see. And did he have any hints for you?"

"None that I could understand." Sammy noticed that her companion's lips were tight, and she didn't have the homey attitude she usually did while working. "What's wrong?"

"I want to tell you something, but I'm concerned that it might be more than you can handle. Not that you aren't grown-up enough to hear it, but that you might try to pursue it. I'm not sure it would be safe." Her hand shook a little as she picked up her mug and took a sip.

This only intrigued Sammy more. "Just tell me. I promise I won't do anything stupid."

Helen hesitated before leaning forward and saying in a loud whisper, "I've heard rumors that there's some sort of illegal gambling den around here. I don't know where, but I assume it must be hidden. It's not like you can just go driving down Main Street until you see the big sign telling you to come waste your money on a game of Black Jack."

"And you think Jones was looking into it?"

"Well, he never told me anything about it directly," she said dismissively. "He was always dedicated to his job, and he wouldn't gossip about things he wasn't supposed to. But you hear things when you work in a place like this, you know? Little snippets here and there about folks placing bets, how much they lost the night before, or what their weekend plans are. I can't exactly ask them outright.".

"Do you think they would tell you if you did?" Sammy's mind was churning, trying to find a way to locate this underground casino herself. She had never been much of a gambler, not even buying lottery tickets. But a club like the one Helen referred to couldn't stay open if there weren't any patrons.

"I don't know. Maybe not. Even so, I don't want to get involved. No offense to you, dear, but I've got a business to run. I can't scare off my customers when they think I'm accusing them of going against the law. They come to Grandma's because it makes them feel like they're at home, where they're safe." She wrapped her hands around her cup and stared down into the light brown liquid. "I'm sorry."

"Don't apologize," Sammy insisted. "It's still more of a lead than I've had with anything else. I know I didn't put poison in that cupcake, so that means someone must have. Either that, or the doctors are wrong about what happened. Maybe the sheriff had some sort of reaction to something. I am due for a physical, so if I find out what doctor Jones sees—"

"Oh, here we go again!" Helen said with a smile. "You keep talking like that, and one of these days I'm going to believe you're really going to do it."

"And maybe one of these days I will," Sammy said with a mischievous smile. "I'll have to work on my spy skills. I'll have to get a television so I can rent some secret agent movies."

"Just don't practice with your grappling hook while you're in the apartment," Helen laughed. "That could make for some big holes in the wall."

"Not a problem." She shook her head as she studied her current home. It really was nice to have someone to talk to, and even if she didn't feel like Sunny Cove had exactly welcomed her with open arms, she knew there was potential for her here. "I really appreciate everything you've done for me, Helen. You've got to be the nicest person in the whole town."

The older woman reached across the table and patted her hand. "I'd like to see things get a little better for you. I remember all of that business with your father, and from what you told me before, things weren't any better in New York. Sometimes, you have to come back home and get a good look at your roots before you can figure out where to grow your new branches."

"Speaking of new branches, I had a customer suggest to me the other day that I start selling donuts by the dozen. He talked about it like I should just leave here and open my own business, which is definitely not something I want to do at this time, but I thought it wasn't a bad idea. It seems like we do well with serving meals, but it wouldn't take much to expand. All it would take is to order some boxes to put the donuts in, and people could purchase them to go." She bit her lip, wondering if her new boss would think she had crossed the line with the suggestion. Just because Helen had asked for her help with the menu didn't mean she owned the place.

But Helen's sapphire eyes were sparkling. "That's a wonderful idea! There's space in the corner of the dining area for a shelf or a table where we could set out the baked goods. And you wouldn't have to limit it to donuts. I imagine plenty of busy wives would like to buy a bag of your rolls to serve with dinners at home." She folded her arms and sat back, clucking her tongue. "If I'm completely honest, I still just can't believe how crazy people are for your confections. Not that they aren't good, but we all know what the rumor is."

"Fortunately, at least as far as I know, nobody else has gotten sick. You and Johnny each ate a cupcake. Heck, we served almost all the rest of them to customers that day. People have eaten rolls and donuts. Maybe I should point that out to Barnes. If I was so interested in poisoning people, then I could have easily done it to far more folks than just Jones." She crossed the room to the bookshelf, where she had set the notebook she'd found earlier. "I need to start keeping track of all this."

"That's fine, dear. You play private eye in the evenings, but don't stay up too late. I'll need you to start doing a lot more baking during the day, and I don't want you to fall asleep in the dough." She chuckled as she stood from the table. "Thanks for the tea. I'll see you in the morning."

"Goodnight, Helen." Sammy let her out, trying to be a good hostess, but then she immediately sat back down to write down everything she knew about the case. It was disappointingly little, but it was a start.

Then she flipped the page and started a new list, this one consisting of supplies she would need to start her mini baking business.

7

I t turned out that Helen was right. Her boss had let her come out to waitress during the busy hours. As she delivered plates of grilled cheese and bowls of tomato soup, her customers made idle small talk. She responded in kind, occasionally asking them what they would be doing over the weekend or if there was anything exciting to do in town. Nobody came right out and told her about the alleged casino, and she never heard any mention of bets, chips, or poker tables. Sammy knew she had to be discreet, as well. She didn't want anyone realizing just what sort of information she was after.

"All right. I've got one last customer at the counter and then that's a wrap for the night," Helen announced when she bustled into the kitchen at seven.

Johnny gave a thumbs up and began cleaning his griddle.

"Have any big plans?" Helen teased as she picked up a broom and began sweeping the floor. "Maybe reading an old Agatha Christie novel?"

"Actually, I'm thinking about going to the movies." Sammy had driven past the old Stargazer Theater several times as she ran her errands in town, and it brought back fond memories of hanging out with her friends back in high school. The show only cost a dollar fifty back in those days, and the concession wasn't much more expensive, which made it perfect for young girls living it up on their allowance.

"Really? No special missions tonight? I'm surprised, given how determined you sounded last night." Ms. Honeycutt reached for the dustpan and collected the debris from the floor.

Sammy shrugged. "I know. And I'm definitely not giving up on this whole idea. But when I sat down and wrote out all my leads, I realized that I'm not going to get anywhere without people to talk to. Right now, the only thing people around here know about me is that my dad had been in jail for a while and that I might or might not have tried to kill the sheriff. That's not going to encourage anyone to give me any information."

"A wise observation," Helen said with a nod of approval. "You can get out and mingle a little, remind people that you're just a regular person."

"Exactly." Sammy placed a plate in the drying rack. "It's not the kind of thing that will get me any immediate results, I know. But I'll feel like I'm at least making a little progress and having fun in the meantime."

"Why don't you go ahead and go?" Helen washed her hands and picked up a nearby towel to dry them. "There's not much left to do, and Johnny and I don't mind cleaning up."

The cook shook his head in agreement.

"I don't want to skip out on you."

"Nonsense." Helen flicked the towel at Sammy. "Get on out of here. Go to the movies and enjoy yourself. Lord knows you deserve it."

"Well, okay. If you're sure." Sammy had never had a boss who would let her leave early just for fun, and it was almost hard to believe. "Any good movies you can recommend?"

"Honey, I don't remember the last time I went to the theater. Sitting still in those little seats makes my knees hurt, so it's easier for me to just put my feet up at home. But you go find something good, and then you can come to work tomorrow and tell me all about it."

"I'll do that." Sammy jogged up the stairs to grab her purse and straighten her hair before she headed out. The Stargazer, just like most of the other downtown businesses in Sunny Cove, was within easy walking distance, but she decided to drive since it would be late when the movie let out. She soon stood under the dazzling yellow lights of the marquee amongst couples and teenagers, all keen for a little bit of Friday night entertainment.

A movie was just letting out, as evidenced by the flood of people coming out the double doors. Sammy waited patiently for the path to clear before she dove in and threaded her way to the ticket counter. She kept her eyes open for anyone she might know, perhaps an old friend from her childhood whom she had lost track of or someone she knew from Just Like Grandma's. But the sea of faces all seemed to be strangers.

The Stargazer had kept the old-fashioned theater look, unlike the larger venues she had been to in the city. The patterned red carpet and the red curtains that hung from the walls were a tribute to old Hollywood. The air smelled of popcorn and candy, which reminded her of days long ago.

"One for Changing Seasons, please." It looked like a romantic comedy, which was probably just what she needed to get her mind off of real life.

The young clerk punched in her order, fiddling with the ticket printer. "I'm sorry. It's acting funny tonight. Give me just a sec."

"No problem." Sammy patiently occupied herself by people-watching. There were young couples with their arms around each other and parents trying to get their children to decide if they wanted buttered popcorn or nachos. Over by the restrooms, an old man leaned against the wall as he waited for his wife, looking like he was about to fall asleep.

To the left of the concession stand, near a door that must have led to an office, a tall brunette wrapped her arms around an older man. "Daddy! I'm so glad to see you!"

The man looked tentatively around the lobby. "What are you doing home, Sonya? You're supposed to be in California."

"Oh, I know, but I wanted to come home for your birthday."

"Okay, but let's not celebrate here." He put his arm around her waist and escorted her off to the side, disappearing through an unmarked door. Sammy noticed how beautiful the woman was, with big, catlike eyes and waves of dark hair. She was a little familiar, but Sammy couldn't figure out where she knew her from.

"Here we go." The clerk finally handed her a ticket. "Sorry about that. You're in theater two on the right."

"Thanks." Sammy skipped the concession counter and headed straight into the big room. The rows of seats were already filling up quickly, and it was difficult to find a seat where she wouldn't feel like she was in the way. Settling down in the middle with a couple of seats between her and the next viewer, she pulled her phone out of her purse to turn off her ringer.

"Hey!" came a loud whisper to her right. "Have you seen this movie before?"

Sammy turned to see a man in his forties looking at her and pointing at the screen. He wore a white polo shirt with a Cricker Dairy Farms logo on the breast.

"Um, no. I can't say I have." She refrained from explaining that most people didn't go to the theater to watch a movie they had already seen.

"It's a good one," he enthused, his words a little slurred. "I won't spoil it for you, but you're going to cry at the ending." A wave of his pungent breath slammed into her face, letting Sammy know just why he was acting this way. The man had been drinking. Heavily.

"Okay, well, I'll try not to cry too loudly then."

"Nah, you're okay. Just let it out. That's what you've got to do when you're feeling it. Excuse me, I gotta get more popcorn." He got up and stumbled past her, nearly falling as he stepped over her feet. As he made his way down the aisle, he bumped into a woman coming the opposite way. "Sorry, ma'am. They made this floor all slanted."

"You're drunk again, Bob," the woman retorted, rolling her eyes.

Sammy settled in to watch the movie, glad that Bob, whoever he was, must have decided to sit somewhere else. The ending didn't make her cry, and instead she felt quite happy and relaxed as she made her way out of the theater with the rest of the crowd.

But then she noticed a man break off from the surge toward the front door and head off to the right. Everyone else was either going towards the bathrooms on the left or leaving, and the man looked furtively behind him before disappearing through an unmarked door. It was the same door that the man from the office and his daughter had gone through.

Sammy couldn't say just what had made her so curious, but she turned off toward the left as well. Her heart thumped in her chest, even though she knew she was jumping to conclusions. She would probably find a hallway that led to a back exit or a break room just for employees. Still, she hoped nobody was watching her as she opened the door and slipped inside.

It took her eyes a moment to adjust from the brightness of the lobby to the dark interior of the room she now stood in. It was so dark that she didn't think anyone noticed her as she stepped to the side and stayed near the wall. Long tables were arranged around the room, single lights hanging over them to illuminate the cards and chips on the table but keeping the players who stood around them in the shadows. Girls in short skirts carried trays of drinks to the people who tossed dice and pulled at slot levers. Across the room, a man threw a ball into a roulette wheel, chanting at the machine to win his bet for him.

Sammy didn't need to ask around to know exactly where she was. This was the illegal gambling den Helen had speculated about. The roulette table was the "big wheel" Austin had referred to, although how he had come to know about it was a new mystery. Her throat closed as she looked around. What was she going to do about this?

She turned to leave, but a man stepped into her path before she could reach the door she had entered through. It was the same man who had come out of the office. His daughter was nowhere in sight. "Hi, there. Can I get you some chips?" He waved her toward a table.

"Oh, no thanks. I'm not very good at gambling. I'm just here because…a friend recommended this place to me." She hoped he couldn't see just how much she was shaking now. The dim light would hopefully help.

"Wonderful!" He held out his arms as though he was truly excited to hear such a thing. "I'm Henry McTavish, the owner of this place, but you probably already know that if you're here. Can I ask the name of your friend?"

"My friend?"

"The one who recommended my little establishment." His smile faded slightly. "We don't let just anyone come in off the streets. As a matter of fact, I'll have to speak to my man about why he wasn't at the door."

"Oh, right. Well, I don't know if I should say." She didn't want to get anyone in trouble. Heather was the only person she could think of who might know about this place, but even so Sammy wasn't going to throw her under the bus.

"An understandable concern. Tell me, what's your game of choice, Miss...?"

"Baker," she replied automatically. At least most of the people who had known her from before would think of her as Beaumont. "And I'm not sure."

"Come on over here, then." He showed her to a bank of slot machines and practically forced her onto a stool. McTavish pulled a token out of his pocket and pressed it into her palm. "The first one is on the house. After that, you can use your debit or credit card right here in this slot, or you can use the machine over there to get chips. I'll come check on you in a little bit, but remember: the higher the bid, the bigger the win!" He chuckled to himself as he strolled off across the room.

Sammy watched him go, wondering just what kind of security they had in this place. Was there really supposed to be a bouncer at the door? Or had they gotten rid of him once they knew she was on her way in? Was McTavish aware that she was onto him, just acting friendly enough to keep her there while he decided what to do with her? Sammy wasn't going to stick around and find out.

When she saw the owner head behind the bar on the opposite side of the room, she stood up as calmly as she could and turned toward the door she had entered through. A big man dressed in black stood there now, and he watched her closely as she approached. "Have a nice night," she choked out. The hairs on the back of her neck stood up as she opened the door, sure that he would stop her, but she made her way out into the lobby and to her car without incident.

8

She wanted to get away from the theater as fast as she could, but the bright red-and-blue lights that showed up in Sammy's rearview mirror a few minutes later showed her just how fast she was going. But at least if she was being pulled over, then anybody from the casino who might be in pursuit of her wouldn't dare to stop.

And of course, it was Sheriff J.J. Barnes who showed up at her driver's side window. "I should have known it was you, with that out-of-state license plate."

"Sheriff, I was just coming to the department to talk to you. I've discovered something really important."

He raised a thick gray eyebrow. "So important that you were speeding on your way to the sheriff's department? That doesn't seem very likely to me. Step out of the vehicle."

"What? Why?"

"You were going at least fifteen miles per hour over the speed limit. You're coming with me." He beckoned at her with his finger.

Panic swept over her. "I can't just leave my car here on the side of the road. Can't I follow you to the station or something?"

"I'll call someone in to tow it, and you can pick it up at the impound lot when you get released. Besides, I'm already on my way to a call, and I don't have time to sit down and chit chat. Let's go."

With little choice but to comply, Sammy turned off the ignition, grabbed her purse, and got out. Barnes escorted her once again into the back set of his squad car. She listened impatiently as he made the call for the tow, and then they were on the road again.

"Listen, I really need to talk to you about something," she started, figuring that if Barnes was going to make her sit in the back of his car, she would at least use that time wisely. "I think I might have just found an underground gambling den in the Stargazer Theater. No, I mean, I *know* I did."

"Right," Barnes replied from behind the wheel. "And I suppose the people who run this place are also the ones who poisoned Sheriff Jones?"

"Well, maybe. I don't quite know that part yet, but it seems possible. You see, Jones was dating Henry McTavish's daughter, but I heard they broke up. Now she's back in town, and her dad didn't seem all that happy to see her. Maybe he doesn't want a cop coming around when he's hiding a casino in the back of the theater. He could have poisoned him to get rid of him." The timing didn't quite seem to work out, but it was the best theory she had so far.

"Save it. I've got more immediate concerns right now. Jones is on the mend, and I've got a call about an intruder at Cricker Dairy Farms. The deputies tell me that Bob's probably just drunk and imagining things again, but I've still got to check." Barnes swung into a long driveway, pulling up in front of a cluster of farm buildings. He turned to glare at her over his shoulder. "I'm not sure I trust you to stay in here by yourself. You're coming with me."

"If you insist." Sammy wasn't particularly looking forward to seeing Bob Cricker again, but she climbed out when Barnes opened the door for her.

The two of them approached a big red barn. It looked like it was used for storage, since the buildings full of cattle were a little further back on the property. Barnes had a flashlight in one hand and held the other over his gun, just in case.

"Come on in," said an inebriated voice from just inside the door. Bob Cricker stumbled into the light from the officer's flashlight and reached inside the door, flooding the interior with light.

Inside, Sammy noticed a lot of dairy equipment. She didn't know exactly what all the big steel vats were for, but she knew they must have had something to do with processing the milk. The walls were lined with big shelves. These were weighed down with various bottles, jugs, and crates.

Barnes shut off his flashlight. "What seems to be the problem, sir?"

"That dang kid has gotten in here again," Bob said, shaking his head.

"What kid is that?"

"That Absher boy!" Cricker shouted. "I found him in my cattle barn last winter, curled up in the corner. My guess is he tried this barn so he could steal some milk or cream." He stumbled between the rows of equipment, looking behind the big vats as if he expected to see Austin hiding behind one of them.

Sammy's stomach curled in on itself. If Austin really had been here —something she was a little doubtful of considering Mr. Cricker's drunken state—then Barnes probably wouldn't handle him as nicely as Jones had. He didn't know Austin or his situation. "Are you sure?"

Barnes turned to look at her. "I can handle this, thank you very much. Mr. Cricker, are you sure it was him? I don't see anyone here."

"Oh, trust me. It was him. I'd recognize him anywhere. I feel sorry for the kid, but what am I supposed to do when he's coming onto my property without permission?" Bob stopped at an old fridge in the corner, extracted a beer, and popped the top.

"I'll take a look around," Barnes promised. He glanced at Sammy. "You stay right here. Don't move."

"I won't," Sammy promised. This whole thing was ridiculous, but she wasn't going to go against the law. She stood in the middle of the barn, studying the big room and looking for any sign of Austin. Somehow, she was going to find a way to help him.

After Barnes had thoroughly looked over the premises without finding anyone, he came back to report it to Cricker. The dairy farmer had finished his beer and was reaching for another one. "Doesn't that just figure. It takes the law so long to get out here that the criminal is already gone. I've got a lot invested in this place, you know. I can't just let people come in and out willy-nilly." He put out a hand to lean against one of the shelves on the wall, but he missed. Instead, Bob knocked over a big green jug. It fell to floor with a thump. The black cap popped off, sending a greenish liquid sloshing over the floor.

"Oops!" Bob laughed. "Guess I'd better watch where I'm going. That happened a few weeks ago, too. But it fell straight into a batch of cream, and I had to throw the whole thing out. What a waste!"

Sammy watched as the last of the liquid glugged out onto the floor. "What is that stuff?"

"Just some pesticide," Bob replied with a shrug. "It's not even the commercial-grade stuff, 'cause I can't afford that. But it works around the sides of the barns and on the fence lines."

"I see." And she was wondering just how much she was seeing. The gears in her mind were clicking and whirring. All she had to do was get someone to listen to her, and she knew she would already have the rest of the evening to chat with Sheriff Barnes.

9

A week later, Sammy bustled from one table to another. The lunch rush was just about over, but she had already spent all morning in the kitchen baking. The shelf in the back, with a big sign advertising their new bulk bakery items, had been thoroughly stocked with donuts, cupcakes, and rolls. Helen no longer restricted her to the back, and the customers were tipping her generously.

"Samantha," Helen said from behind her, "there's someone here to see you."

"Okay, just a sec. I've got to take these orders." Sammy had just seated a table of four.

"Let me do that, dear. I think this is important." The older woman took the pad and pen out of Sammy's hand and gestured with her head toward the door.

She hadn't seen him since that first encounter, but Sammy recognized Alfred Jones immediately. She instantly smiled at seeing him standing in the doorway, looking perfectly healthy. "Sheriff Jones," she said as she approached him. "What brings you by today?"

"I have a few things I'd like to speak with you about," he replied. "Is there someplace private we can talk?"

She bit her lip, thinking. "Yeah. Follow me." Sammy led him through the restaurant, to the kitchen, and out the back door. She wasn't about to bring him up to her apartment, and the alley was the quietest place she could think of. "Is everything all right?"

"Well, you're not being brought back in for questioning again," he replied, one dark eyebrow raised.

Sammy laughed. "That's a good start. I take it Barnes finally listened to what I had to say." It had been a very long night when they had gotten to the sheriff's department after leaving the Cricker Dairy Farm. Barnes had brought her in not only because of her speeding but because he had a few more questions to ask her. Regrettably for him, she had done all the talking.

"He did, and as soon as I was cleared by the doctors, he passed all the information along to me before he headed back to Calmhaven. The department has been very busy in the meantime tracking everything down. We had to call in to the state boys when we raided McTavish's place. It sounded like a pretty big job compared to the size of our force."

"And?" Sammy had been anxiously listening to the news on the radio, waiting to hear something. She knew the police had gone in; that much was obvious from what she'd overheard from her customers. But the details were still fuzzy.

"And you were right," Jones confirmed. "There was quite the operation going on in the back. I guess McTavish must have been very select about who he let come in, otherwise he would have been busted a long time ago. I didn't tell him you were the one to come to us with the information, but he guessed. Of course, he blamed it all on his bouncer for taking a break when he was supposed to be working."

"I see. Does that mean I'm in any danger?" She'd never done anything like this before.

Jones scratched at his square jaw. "No, you don't have to worry about that. McTavish is behind bars, and he's not going anywhere anytime soon. There was a lot of paperwork to be gone through at the Stargazer, some of which was very interesting to the state police. They called in some special investigators, and they have evidence that the casino was just a small part of a much bigger criminal organization."

"Oh." She didn't know what else to say. "I'm glad I didn't stick around there any longer, then."

"Still, it makes you quite the hero for discovering it all. But there was some more information they uncovered, and that's really what I'm here for today." He shifted his weight from one foot to the other.

Sammy swallowed. "That doesn't sound good." She suddenly wished she'd asked Jones to stay somewhere in the restaurant with her, so at least she could be washing dishes or wiping down tables while they talked.

"In that paperwork, the detectives discovered that Henry McTavish isn't Henry McTavish at all. His real name is Bernie Philpot."

She whipped up her head to look at him. "What did you say?"

"I see you recognize the name. I didn't really know you when we were younger, but I knew of you. Everyone did, with all that in the papers. I recognized the name right away, too, and I knew he was the man who had testified against your father and set him up. He had disappeared after the trial, but apparently he just changed his name and went out of the country for some plastic surgery. He's still the same man who framed your father. Do you want to press any charges against him?"

Never in her life had Sammy imagined that she would have a chance like this, but she still didn't want to take it. She'd been through enough drama, and she was ready for things to be normal again. She knew the right thing to do was to forgive him, even if his actions had been very wrong. "No. What's done is done. My father isn't even alive anymore, so there's no point in dragging it all back out into the light. Besides, it sounds like he's going to spend plenty of time behind bars with everything else he's got going on."

"That's true. And so is Bob Cricker. It turns out your guess about the pesticide in the cream was correct. All the cupcakes you made were contaminated, and so was all the rest of the cream from that batch. The amounts were small enough that it didn't affect most people beyond a little bit of a stomach ache, but apparently I'm allergic to one of the chemicals used in it. Lucky me."

"Does that mean you forgive me for giving you a cupcake?" she asked, hoping that she wouldn't forever be known as the one who gave a tainted treat to the sheriff.

"I admit that laying in a hospital bed for several days isn't exactly what I would call a vacation, but yes, I forgive you. And I'm glad that nobody else ended up as bad off as I did. The health department did a thorough inspection of Cricker Dairy Farms, and everything should be all right from now on. Bob signed the place over to his nephew, who doesn't like to drink as much as his uncle does."

"Sounds like everything is all wrapped up, then." It felt odd to know that it was all over. Sammy was glad she wouldn't be going to jail, but she wished she'd had more time to conduct a proper investigation. If she'd had time to do some of the things she'd talked to Helen about, like getting information from the local doctor or finding a way to break into the sheriff's computer, then it could have been fun. It had mostly been an accident that she had helped solve this case, but maybe it didn't have to be in the future.

"It is," Sheriff Jones agreed, "but I want you to know that cupcake was the best cupcake I had ever had, poisoned or not."

She blushed, her cheeks hot despite the cool September air. "I'll have to make you another one sometime. Just let me know your favorite flavor."

Later that night, when Sammy had worked a full day and she hauled herself up the stairs to her apartment, she knew there was one final task she had yet to take care of. It took a while to dig the business card for the Cozy Bridal Agency out of her purse, but she called Molly Gertrude Grey as soon as she found it tucked inside her wallet.

"Hello?"

"Um, hi. This is Sammy Baker. We met a couple of weeks ago in church, when you came and sat next to me." Now that she was actually on the phone with her, she felt incredibly silly. Why would some stranger care what had happened to her?

"Oh, yes! I remember you, dear! How are you?"

"I'm all right. And that's just the thing, Miss Grey. When you and I met, I didn't think I would be all right at all. I had been through so much, and it just seemed like I wouldn't get over it. Things have still been very interesting here in Sunny Cove, but they're working out. I've got a good job, a nice apartment, and my boss is wonderful." A job and a home weren't everything, but they provided a certain sense of security that Sammy really needed. "I even helped solved a crime, and it led to finding the man who framed my father. It's such a weight off my chest." She went over the events of the past couple of weeks, hoping she wasn't talking too much.

"I'm delighted to hear it, dear. I think about you often, you know. I even put you on the prayer list at church. I hope you don't mind."

"I don't," Sammy replied honestly. "And I'll be praying for you, too, even though I'm sure I don't need to."

"And why is that, dear?"

"Because I'm quite certain that God himself sent you to me."

Molly Gertrude laughed. "That's a wonderful thought, dear, and it's that sort of thinking that's going to keep you going in life. I wish you the very best. Just be careful when you track down your next criminal."

"I think for now I'll stick with baking. It's a lot less dangerous. I'll save the high-speed chases for the professionals." They said their goodbyes and promised to talk again soon before they hung up. Sammy settled down on the sofa with a Mary Higgins Clark novel she had picked up at the library. Yes, life was much easier when she wasn't trying to track down a wrongdoer. For now.

CONTINUE READING MORE OF SAMMY'S ADVENTURES THE COMPLETE BOXSET...

A Full shelf of Culinary Cozy Mystery with amateur sleuth and master baker, Sammy. Enjoy all 13 Sammy Baker Cozy Mysteries in one massive box set, plus one extra bonus mystery!

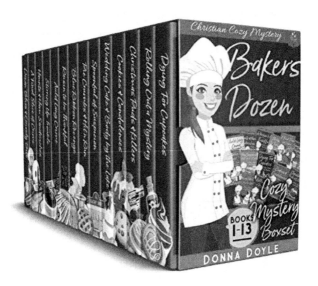

Start Reading The Baker's Dozen Boxset Now

OTHER BOOKS IN THIS SERIES

If you loved this story why not continue straight away with other books in the series?

OR READ THE COMPLETE BOXSET!

Start Reading On Amazon Now